Synthetic Biology Conference 2015 (SEED 2015)

Engineering, Evolution, and Design

Boston, Massachusetts, USA
10-13 June 2015

Volume 1 of 2

ISBN: 978-1-5108-1577-3

Printed from e-media with permission by:

Curran Associates, Inc.
57 Morehouse Lane
Red Hook, NY 12571

Some format issues inherent in the e-media version may also appear in this print version.

Printed by Curran Associates, Inc. (2015)

AIChE
120 Wall Street, FL 23
New York, NY 10005-4020

Phone: (800) 242-4363
Fax: (203) 775-5177

www.aiche.org

Additional copies of this publication are available from:

Curran Associates, Inc.
57 Morehouse Lane
Red Hook, NY 12571 USA
Phone: 845-758-0400
Fax: 845-758-2634
Email: curran@proceedings.com
Web: www.proceedings.com

TABLE OF CONTENTS

VOLUME 1

GENERAL SUBMISSIONS

VOLUME 2

POSTER SESSION

Learning the Sequence Determinants of Alternative Splicing from Millions of Random Synthetic Sequences

Alex Rosenberg, Rupali Pathwardan, Jay Shendure, Georg Seelig

Electrical Engineering and Computer Science & Engineering,
University of Washington

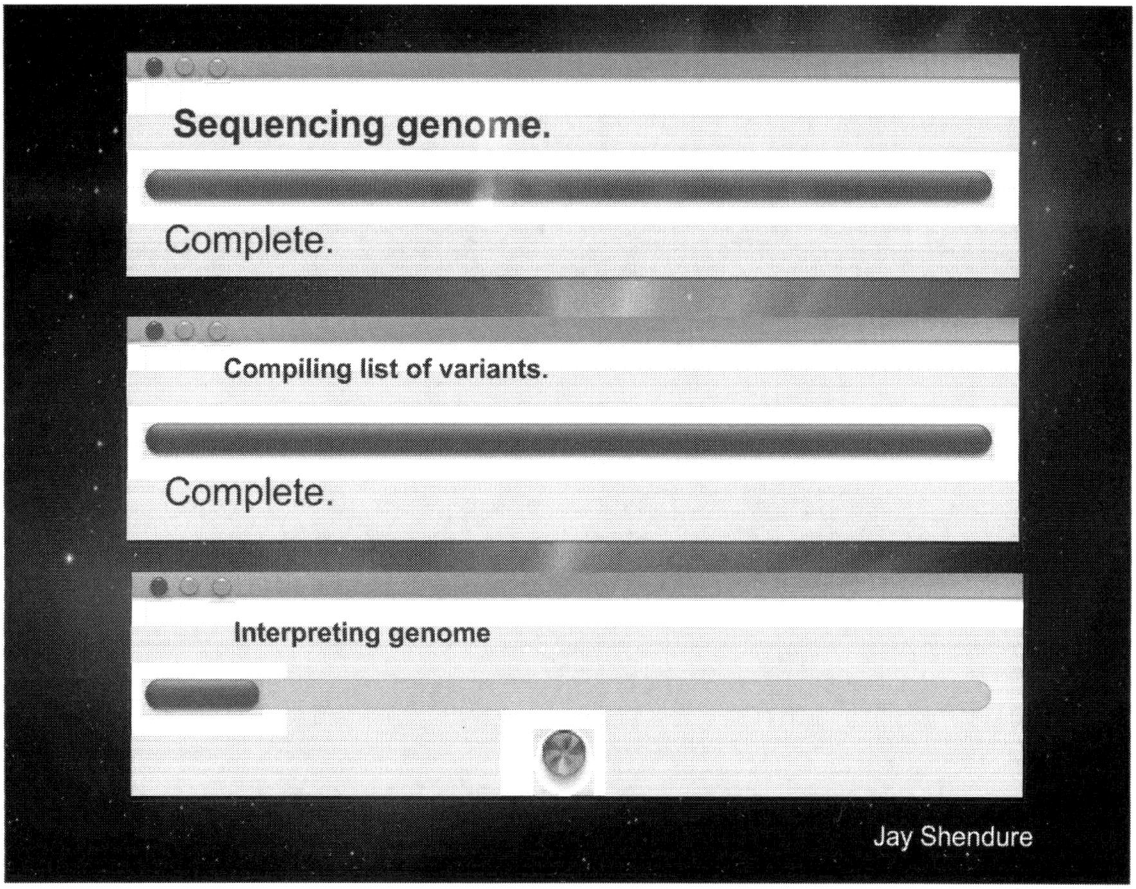

Understanding the impact of variants with machine learning

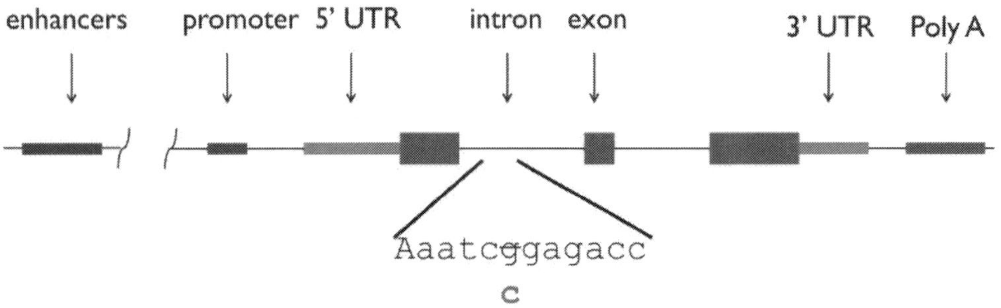

- Build a sequence-function model using machine learning
- Model are limited by data (e.g. "only" 50K splice events)

More data is better

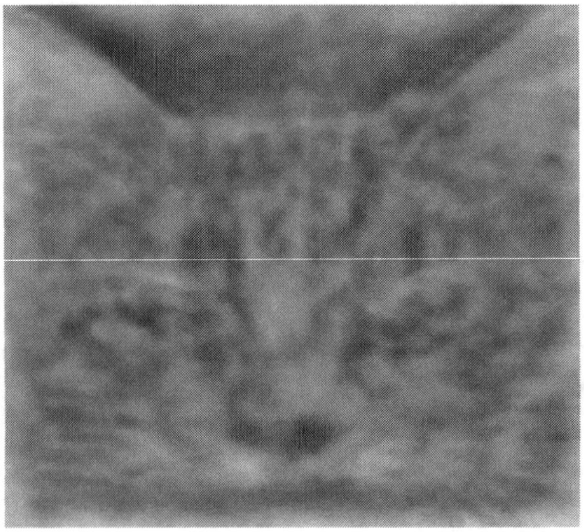

A massively parallel approach to understanding the genome

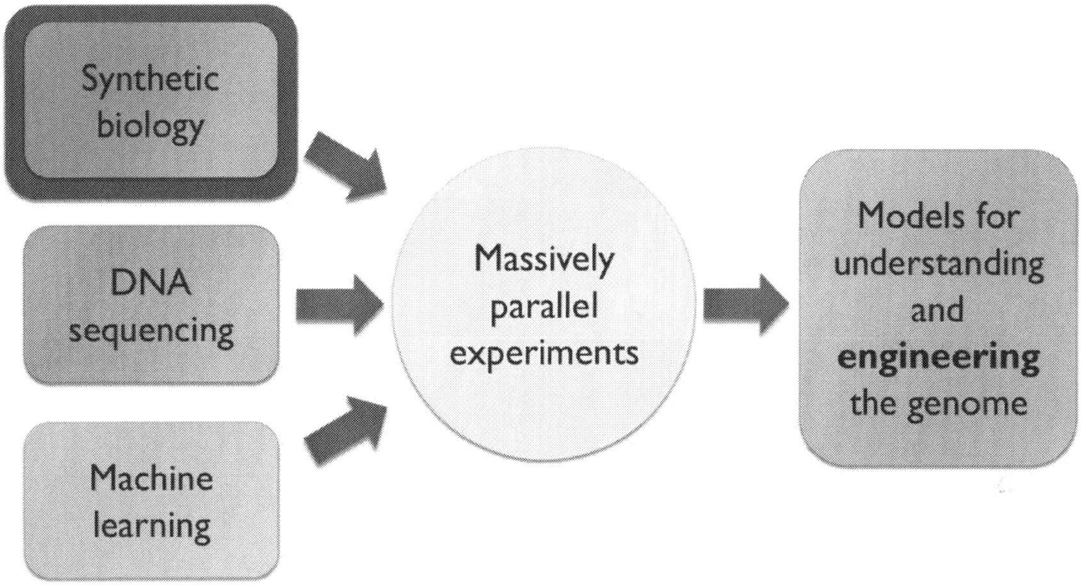

RNA-Splicing

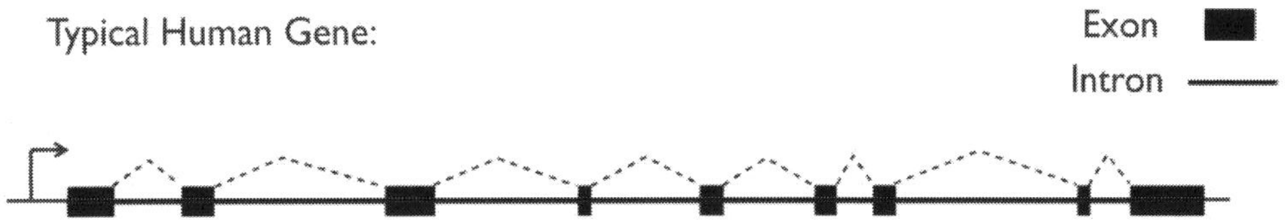

Alternative Splicing

- Different isoforms can have distinct protein functions
- 95% of coding genes are alternatively spliced
- Misregulation of splicing can lead to disease and cancer

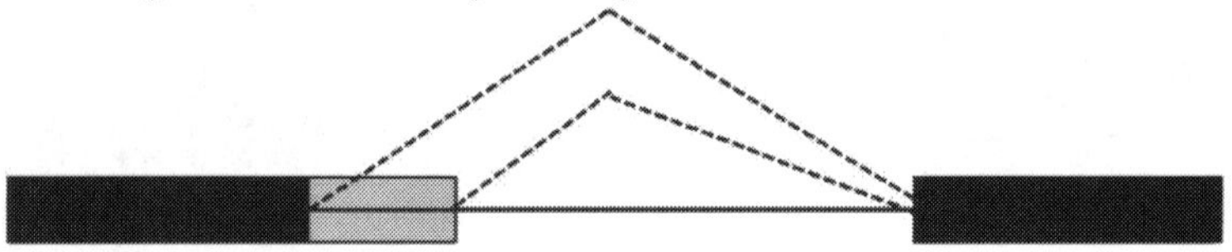

Isoform A

Isoform B

Regulation of Alternative Splicing

What are the sequence determinants of alternative splicing?

- The splice site sequences (splice donors)
- Sequences around the splice sites

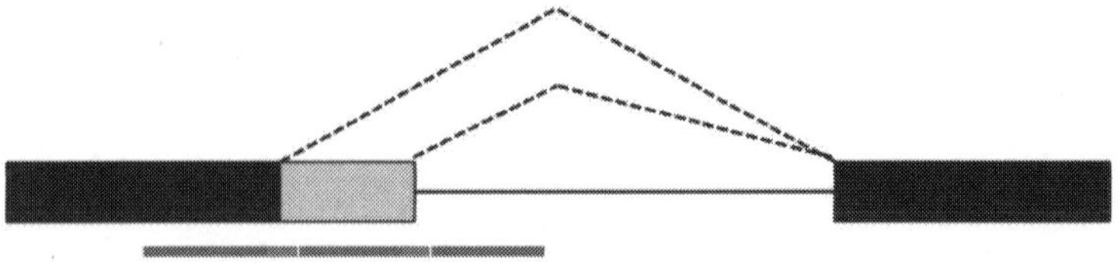

Massively Parallel Splicing Assay

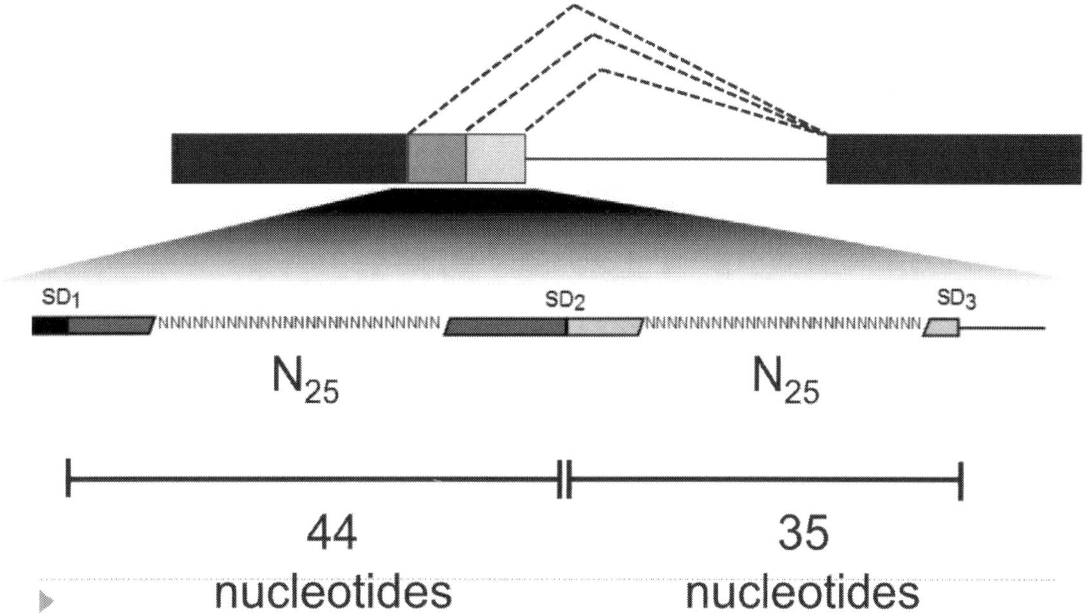

Massively Parallel Splicing Assay

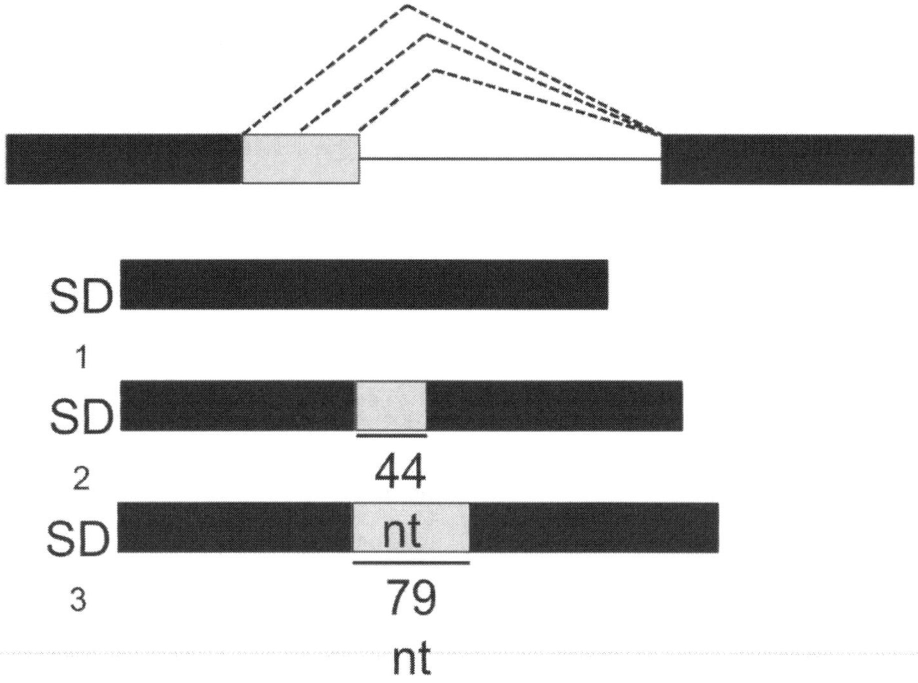

Massively Parallel Splicing Assay

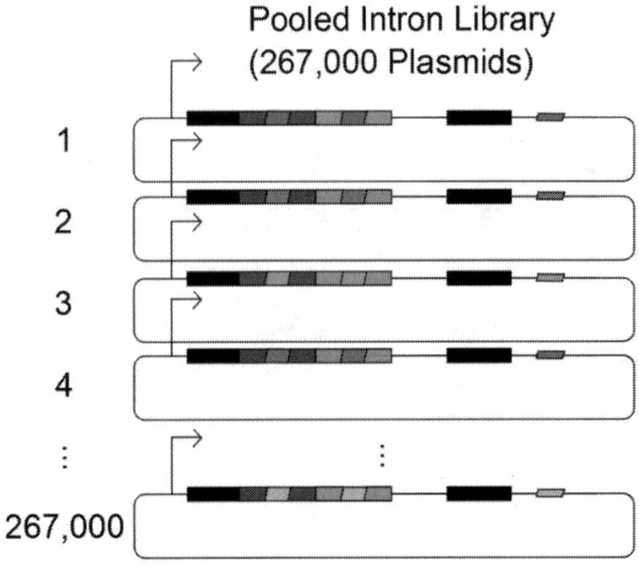

Let's give a cell lots of DNA sequences and record what happens

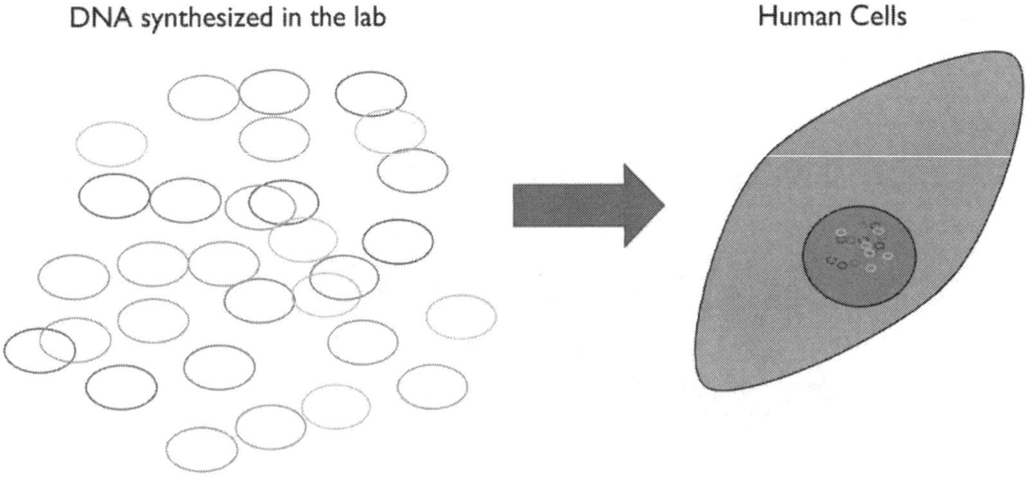

Massively Parallel Splicing Assay

- Used RNA-seq to quantify isoform levels
- For every mRNA molecule that we sequenced we determined:
  - how it spliced
  - which plasmid variant it was transcribed from (barcode in 3'UTR)

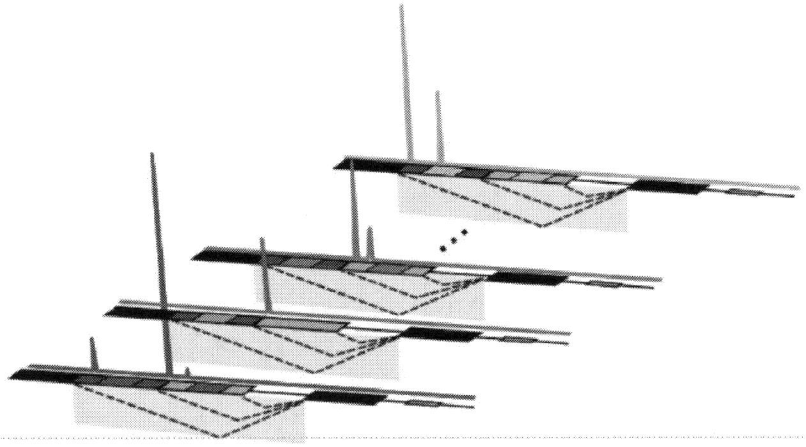

Resulting Data

	SD_1	SD_2	SD_3	SD_{NEW}
⬭	0	26	0	0
⬭	0	2	0	27
⬭	113	4	1	0

⋮ ⋮

267,000 Different Sequences

Resulting Data - Summary

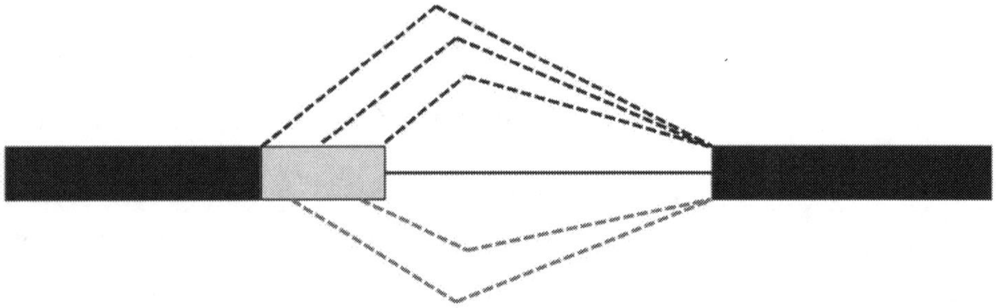

SD_1	SD_2	SD_3	SD_{NEW}
28%	47%	6%	15%

Short Sequence Motif Effect Sizes

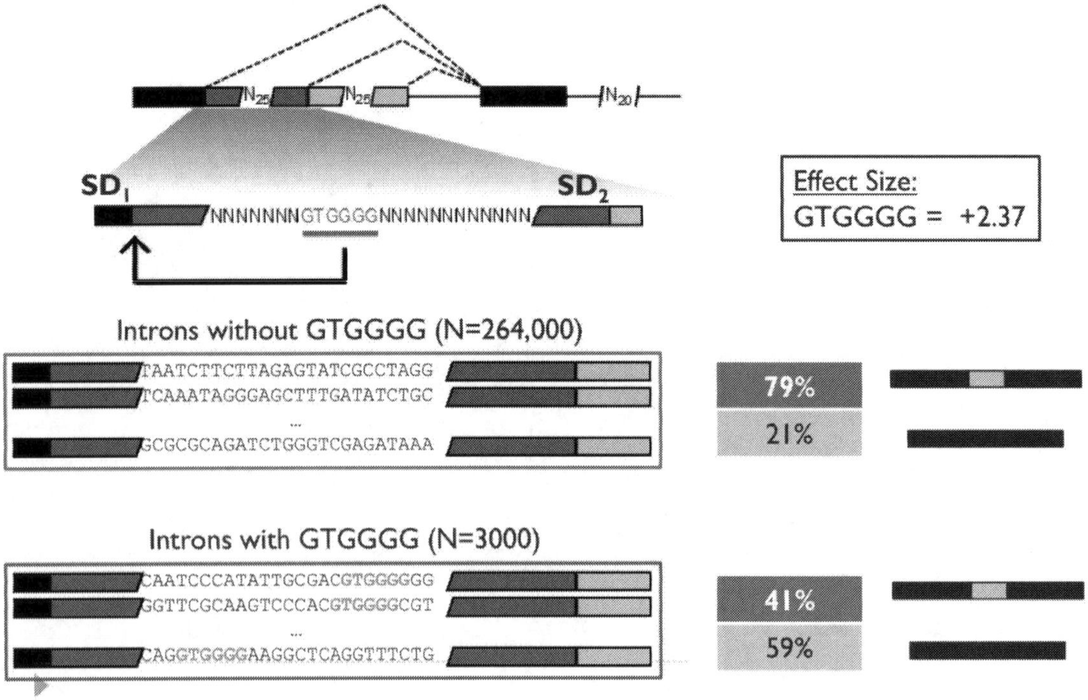

All 6-mer Effect Sizes

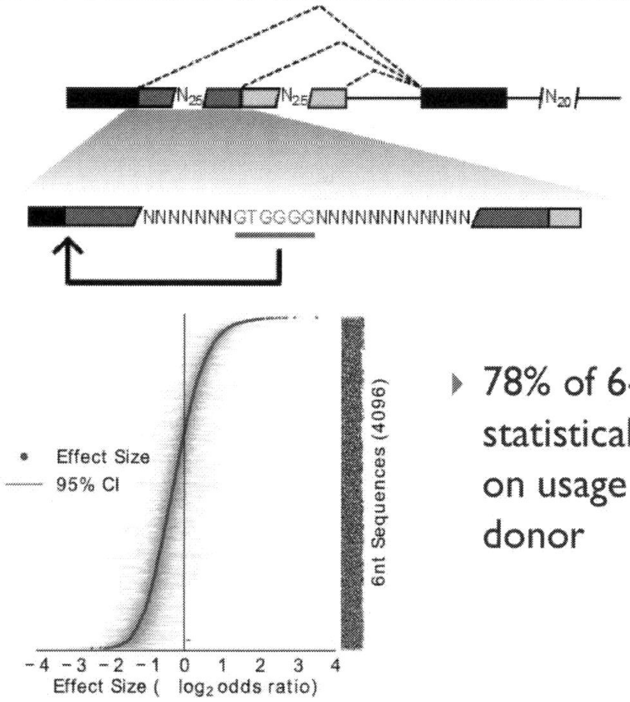

- 78% of 6-mers have statistically significant effect on usage of the first splice donor

RBFOX1/2 Binding Site Differences in HEK293 and MCF7 Cells

Rank	Motif
1	TGCATG
2	GCATGC
3	CGCATG
4	TCGCCT
5	ATGCAT
6	ACGACA
7	ACGACG
8	AGCCCC
9	CTCGGC
10	CATGCA
11	CCCCAC
12	AGCATG
13	AACGAC

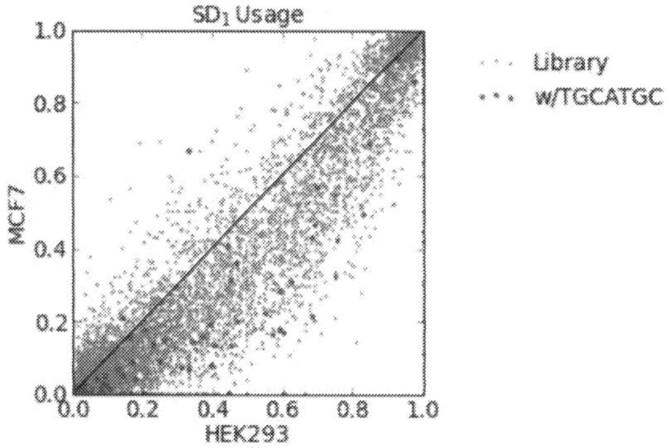

RBFOX2 Expression in HEK293 vs MCF7

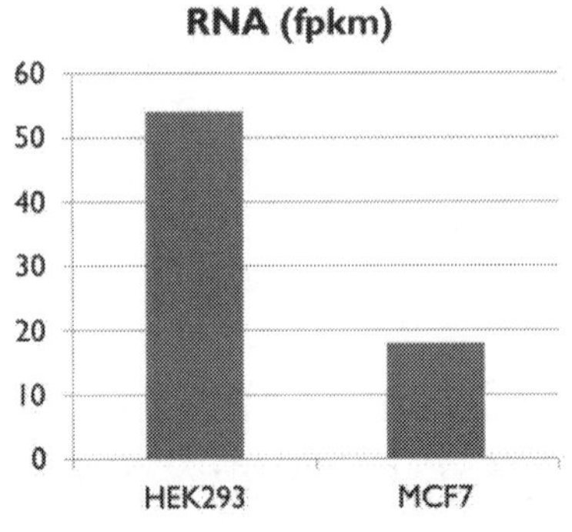

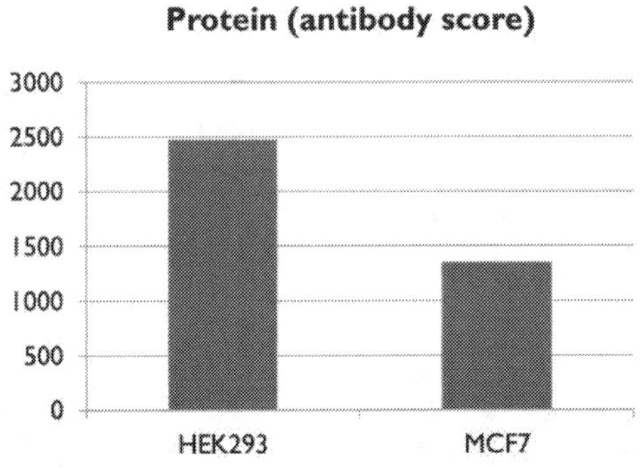

The Human Protein Atlas

RBFOX1/2 Binding Site Differences in HEK293 and MCF7 Cells

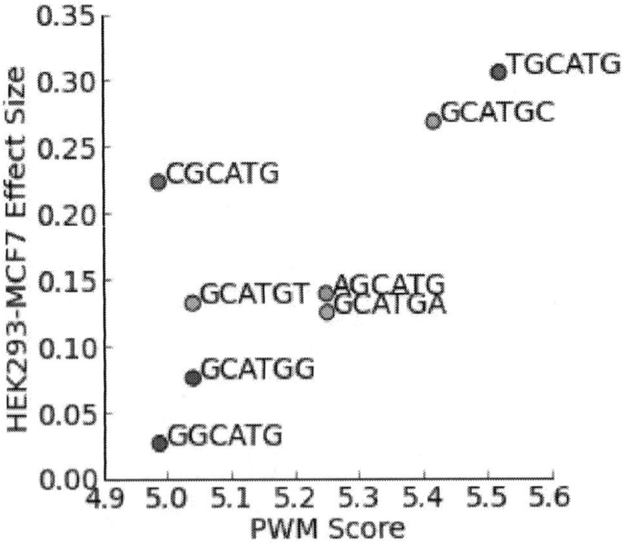

Ray, Debashish, et al. "A compendium of RNA-binding motifs for decoding gene regulation." *Nature* 499.7457 (2013): 172-177.

Combinatorial Regulation of Alternative Splicing

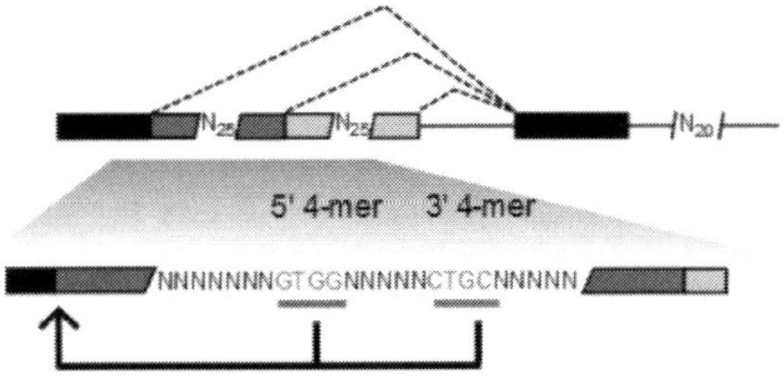

Two Possible Models of Combinatorial Sequence Regulation:

- Additive: Sequence motifs act independently of each other
  - Effect Size(GTGG & CTGC) = Effect Size(GTGG) + Effect Size(CTGC)
- Cooperative: Sequence motifs interact with other motifs

Combinatorial Regulation of Alternative Splicing

- Short motifs act additively and independently of each other

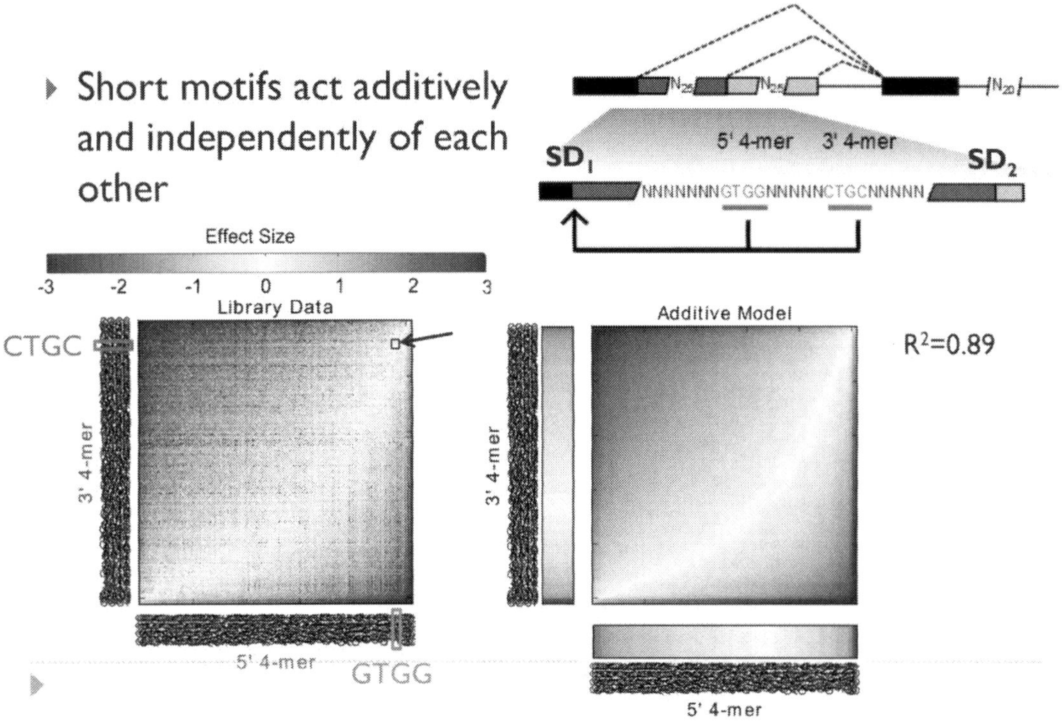

Building an Additive Model of Splicing

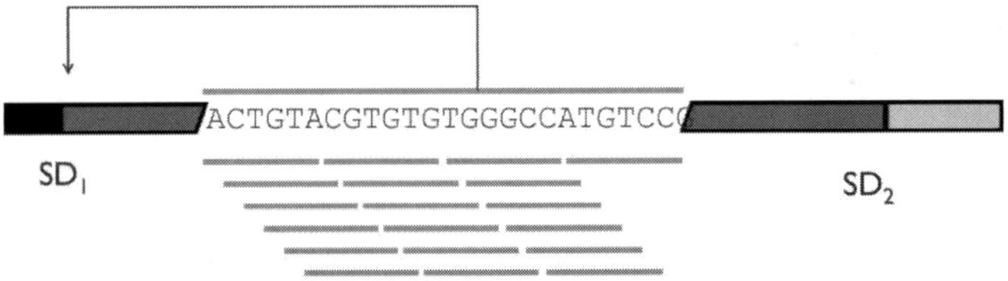

- Effect Size(ACTGTACGTGTGTGGCCATGTCCG) = Effect Size (ACTGTA)
  + Effect Size (CTGTAC)
  + Effect Size (TGTACG)
  ...
  + Effect Size (TGTCCG)

Effects of Single Nucleotide Polymorphisms (SNPs) on Alternative Splicing in Humans

- Can our model predict the effects of nucleotide changes on alternative splicing?

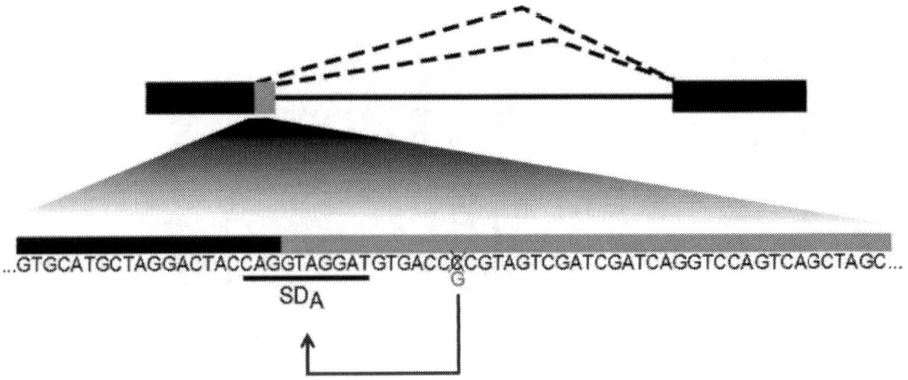

Individual Contribution of a Nucleotide to Splicing

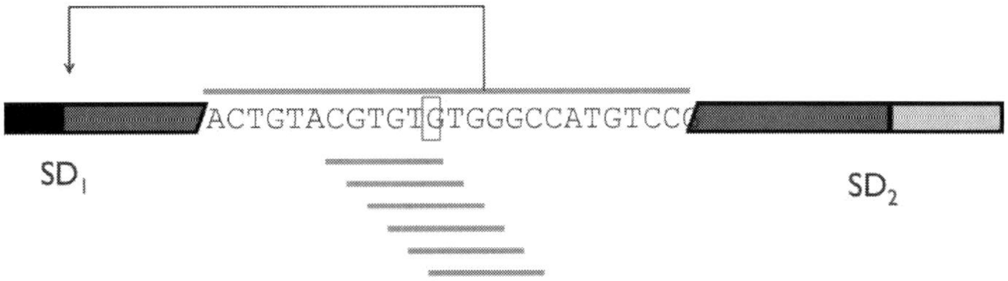

- Effect Size(G at position 12) = (Effect Size (CGTGTG)
  + Effect Size (GTGTGT)
  + Effect Size (TGTGTG)
  + Effect Size (GTGTGG)
  + Effect Size (TGTGGG)
  + Effect Size (GTGGGC)) / 6

Alternative 3' Splicing Libary

- N = 1,670,000

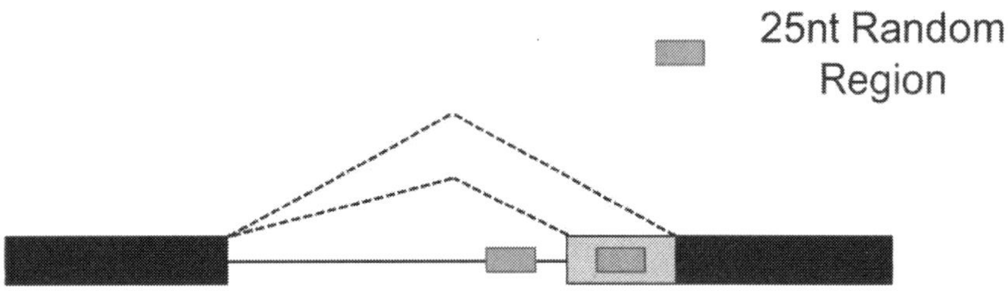

6-mer Effects in the Alternative Exon

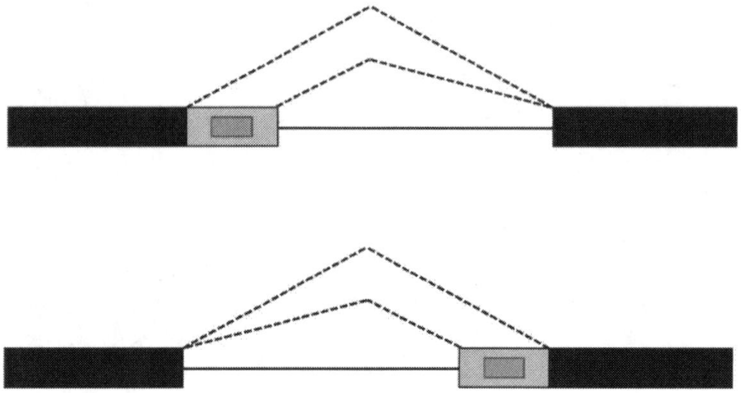

6-mer Effects in the Alternative Exon

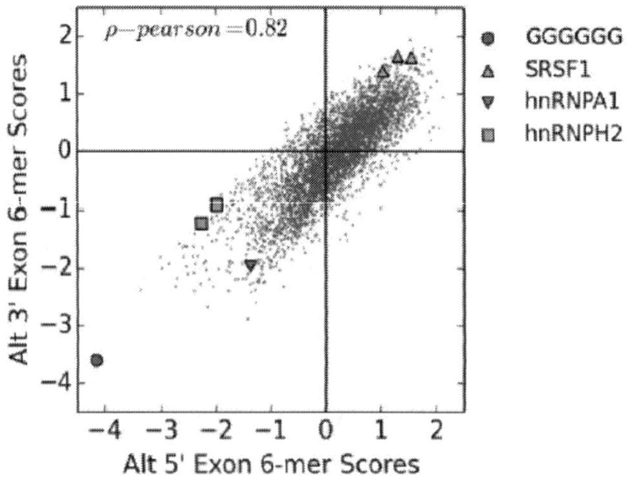

Common Role For Motifs in Alternative Exons?

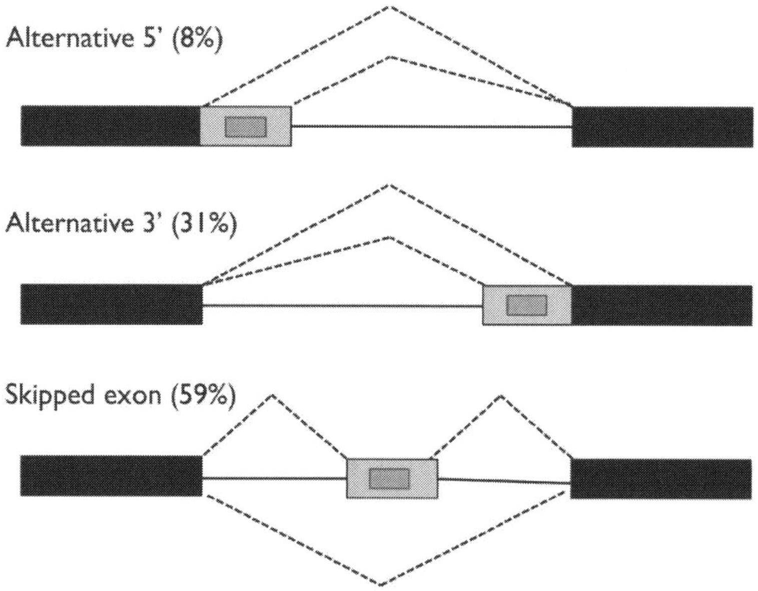

- Bradley, R., et al. "Alternative Splicing of RNA Triplets Is Often Regulated and Accelerates Proteome Evolution." *Plos Biol* 10 (2013): e1001229

Skipped exons

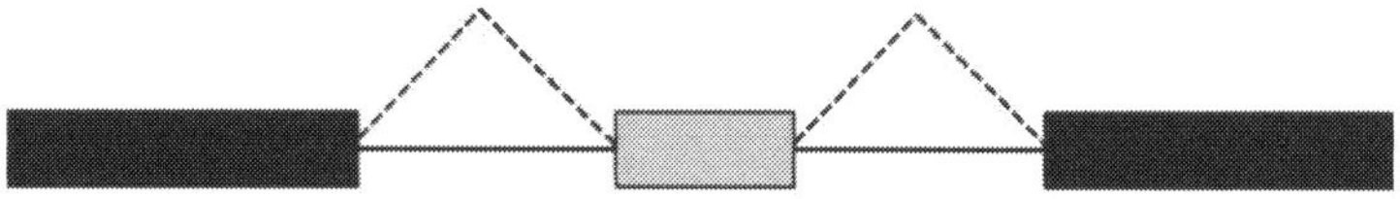

Skipped exons

- Exon skipping

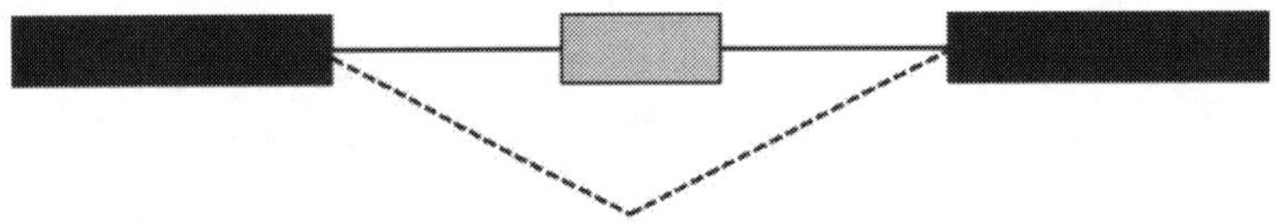

Skipped exons

Predicting the Effects of Mutations in Skipped Exons

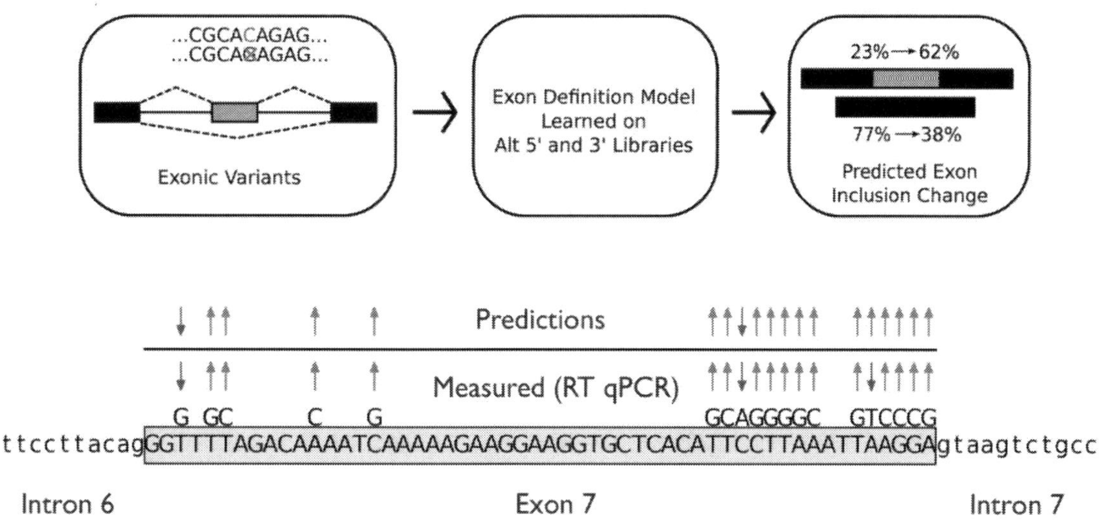

Predicting the Effects of Mutations in SMN and CFTR proteins

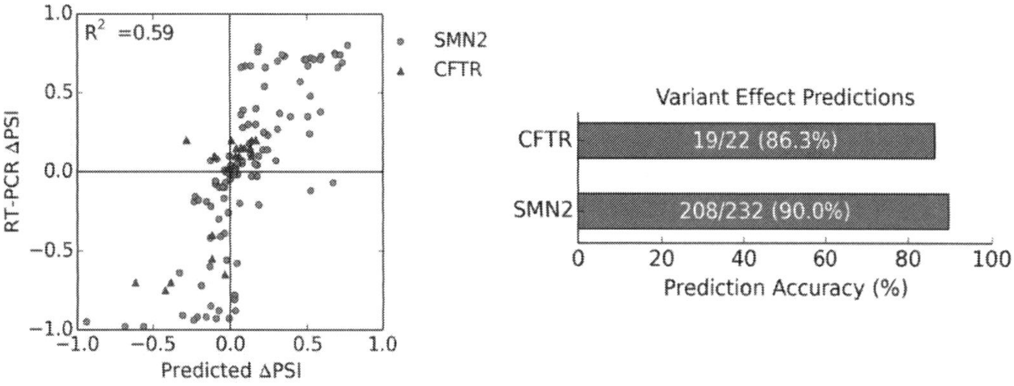

Nearly identical exon definition in 3' and 5' alternative splicing

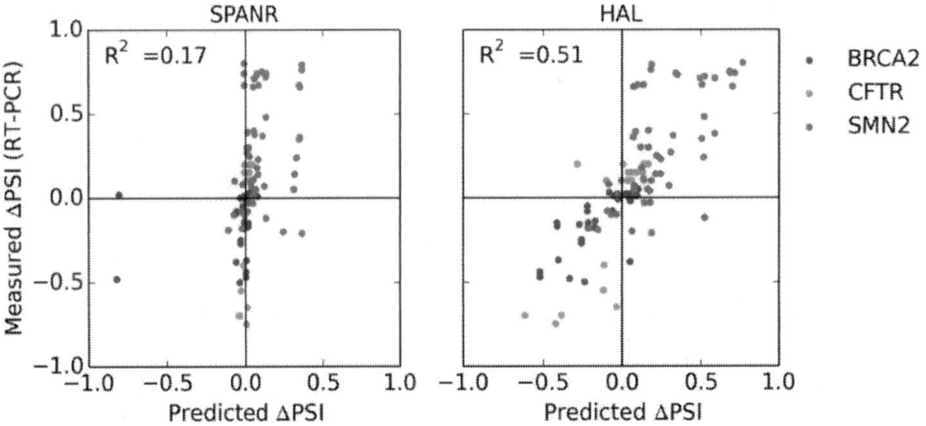

SPANR: Ailpanahi et al., Science (2015)

Summary

- We presented a new approach to learn the regulatory rules governing alternative splice site selection
- A model that was trained only on synthetic data predicts splice site selection better than any previous model directly trained on the genome
- A model that was not trained on skipped exon can predict the effect of mutations in skipped exons
- Outlook: use model for forward engineering

A broadly applicable method for understanding gene regulation

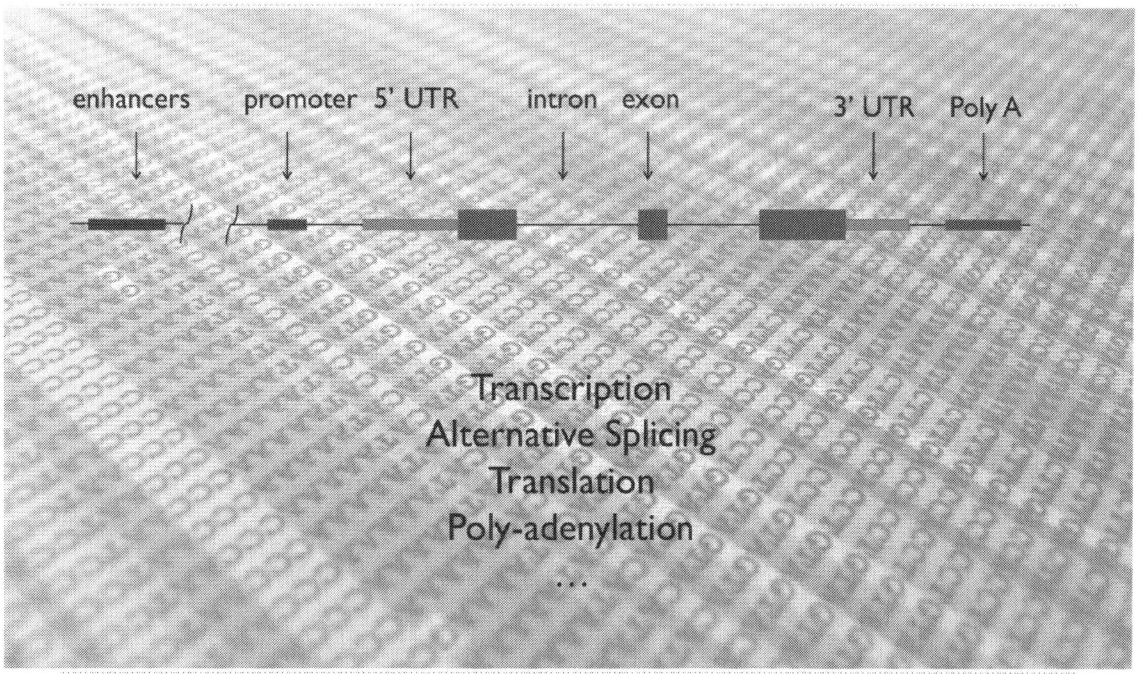

Acknowledgements

Yuan-Jyue Chen, Sergii Pochekailov, Ben Groves, Gourab Chatterjee, Rebecca Black, Alex Rosenberg, Paul Sample, Sundipta Rao, Alberto Carignano

Alex Baryshev, Sumit Mukherjee, Sifang Chen, Nick Bogard, Arjun Khakhar, Randolph Lopez, Ban Wang, Anna Kuchina

BURROUGHS WELLCOME FUND · NSF · Alfred P. Sloan Foundation · Department of the Navy ONR Science & Technology · DARPA

Scalable System-wide Design of TF- and dCas9-Based Genetic Circuits

Howard Salis

Penn State University

http://salislab.net

PENNSTATE

SEED 2015. Boston, MA.

The Salis Lab

We develop predictive biophysical models & design algorithms to engineer cellular sensors, circuits, and pathways

Genetic System Sequence → Cellular Function

biophysical modeling (thermodynamics & kinetics)

RBS Calculator
RBS Library Calculator
Riboswitch Calculator
Operon Calculator

Targeted Behavior → a Specific DNA Sequence

computational optimization

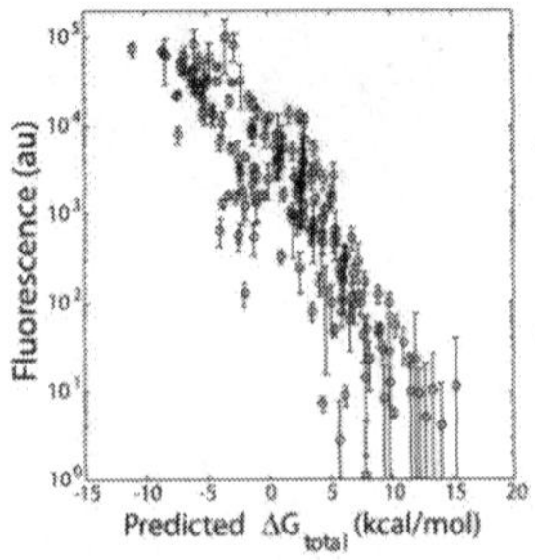

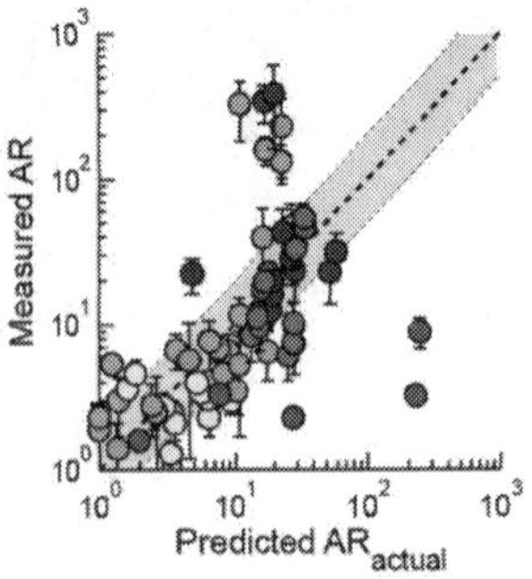

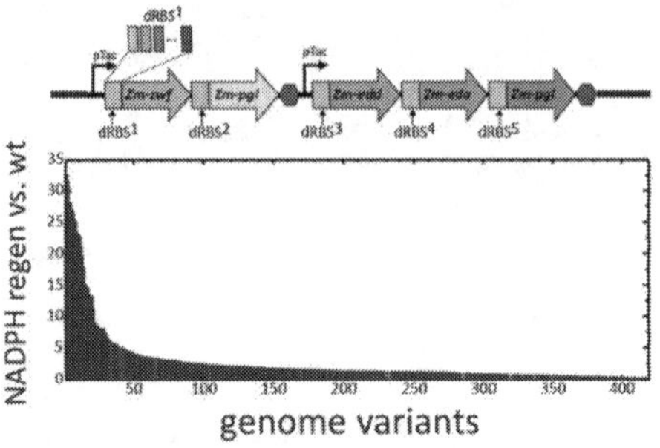

Translation & Translational Coupling **Riboswitch Regulation** **Pathway Engineering**

The Salis Lab

We develop predictive biophysical models & design algorithms
to engineer cellular sensors, circuits, and pathways

Kushwaha M. and H.M. Salis. A Portable Power Supply for Cross-Species Engineering of Genetic Circuits and Metabolic Pathways, *Nature Communications, accepted*

Tian T. and H.M. Salis. A Predictive Biophysical Model of Translational Coupling to Coordinate and Control Protein Expression in Bacterial Operons, *Nucleic Acid Research, accepted*

Ng C.Y., I. Farasat, C. Maranas, and H.M. Salis. Rational Design of a Synthetic Entner-Doudoroff pathway for Improved and Controllable NADPH Regeneration, *Metabolic Engineering*, v29

Farasat, I., Kushwaha M., Collens J., Easterbrook M., Guido M., and Salis H.M. Efficient Search, Mapping, and Optimization of Multi-protein Genetic systems in Diverse Bacteria, *Molecular Systems Biology*, v10 (6)

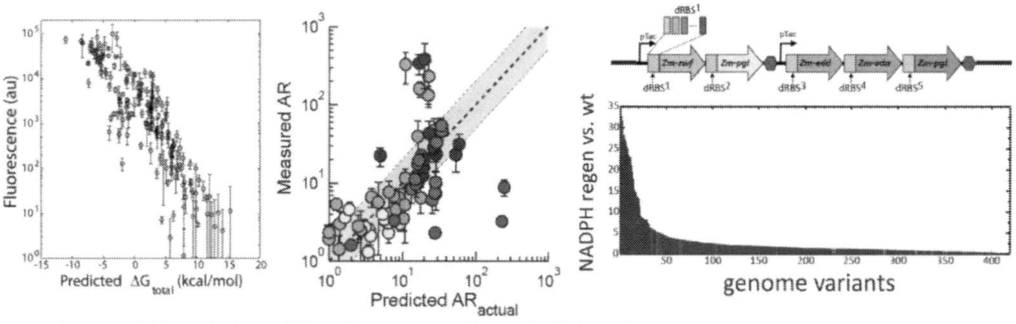

Translation & Translational Coupling **Riboswitch Regulation** **Pathway Engineering**

Genetic Circuit Engineering

Genetic circuits create and manipulate information with actionable outcomes

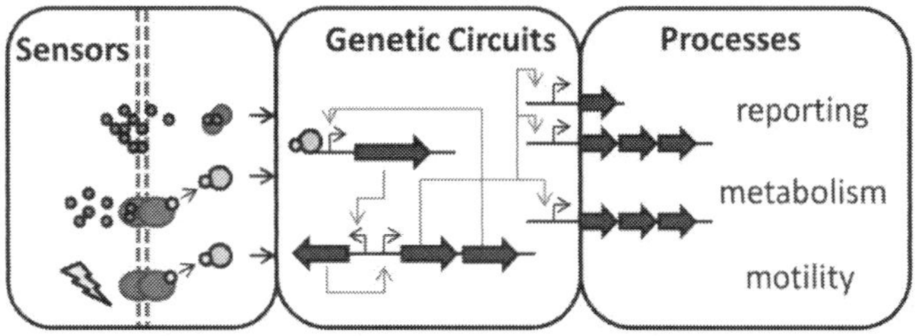

Applications

Decision Making

Signal Processing

Feedback Control

Example Objective: design a circuit to amplify the dynamic range of a sensor

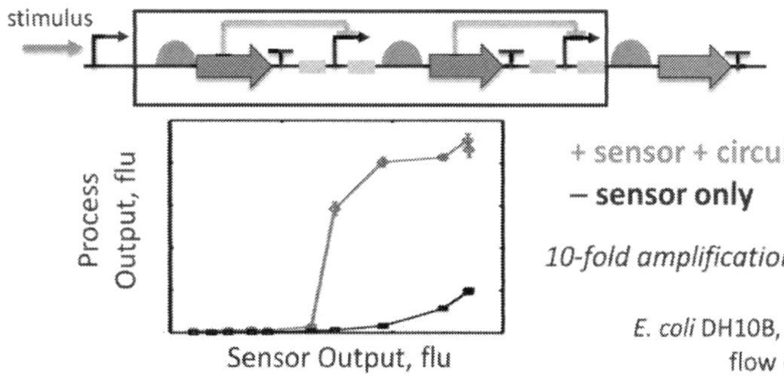

+ sensor + circuit

– sensor only

10-fold amplification

E. coli DH10B, LB media, flow cytometry

Genetic Circuit Engineering

Genetic circuits create and manipulate information with actionable outcomes

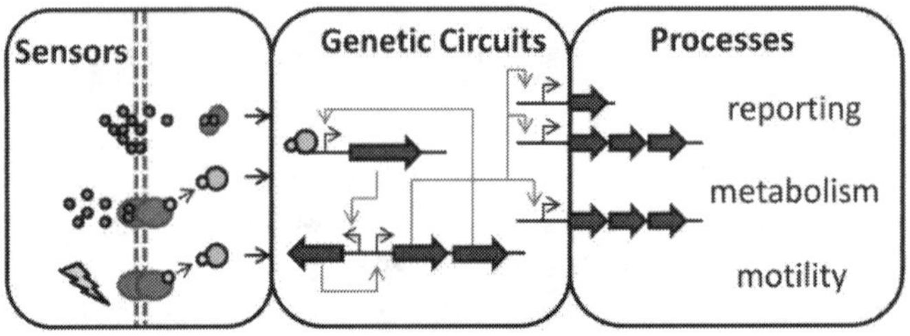

Applications — ***an Ecosystem of Engineered Genetic Circuits***

Decision Making	digital logic: NOT, OR, NOR, AND, NAND, XOR gates
	analog: addition, subtraction, D2A/A2D, feedback control
Signal Processing	analog & digital memory: toggle switches & DNA flip switches
	oscillators: repressilator, mixed feedback loops, multiplexing
Feedback Control	algorithms: traveling salesmen, edge detection, weighted sort

a growing toolbox of parts: over 30 transcription factors, 11 recombinases, 500 terminators, and 1000s of promoters & RBSs ... and CRISPR/Cas9

The Challenges of Genetic Circuit Design

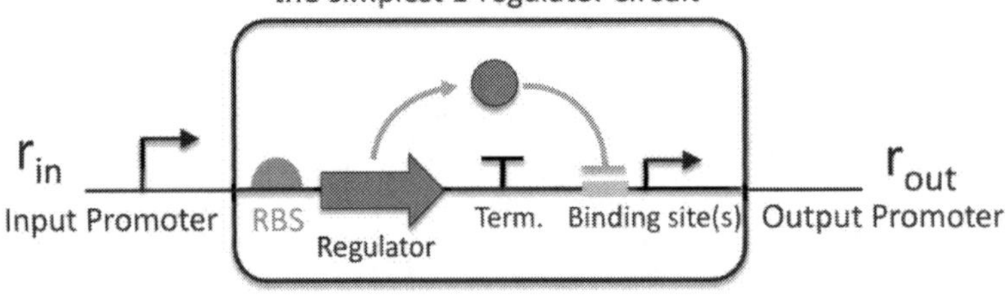

Big Q: How do we predict the **function** of a genetic circuit according to its **parts**?

Genetic Circuit Parts/Sequences ➞ Static/Dynamic Response

The Challenges of Genetic Circuit Design

the simplest 1-regulator circuit

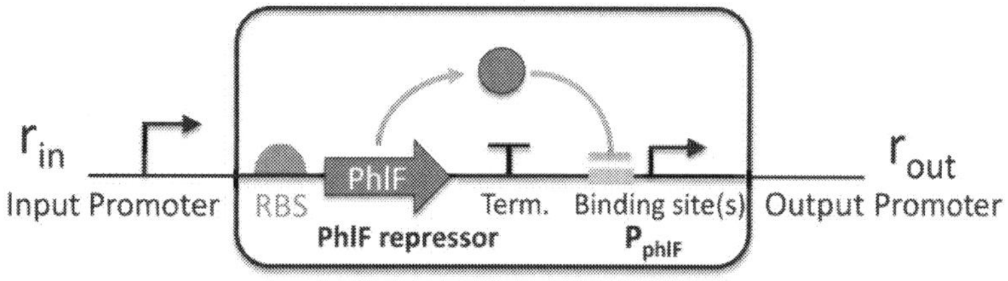

One A: Characterize a Circuit's Input-Output Behavior & Re-use It Everywhere

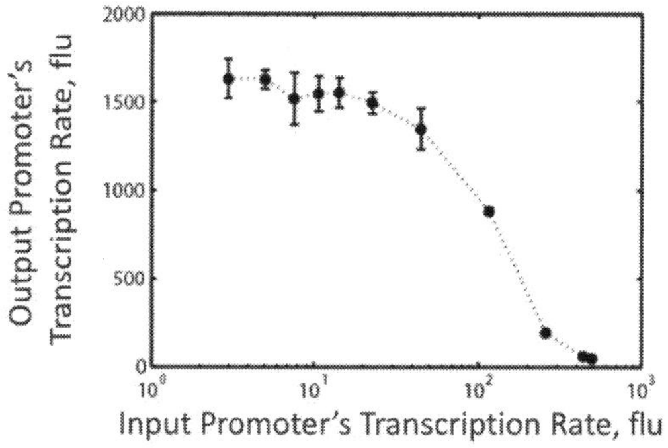

RBS sequence = ...ATTAGTAACATAAGCAAGAGCCTC

regulator = PhlF repressor
(a TetR homolog from *P. fluorescens*)

Terminator = rrnB T1

Output promoter = P_{phlF}

Characterized in *E. coli* DH10B,
LB media, steady-state, flow cytometry

The Challenges of Genetic Circuit Design

the simplest 1-regulator circuit

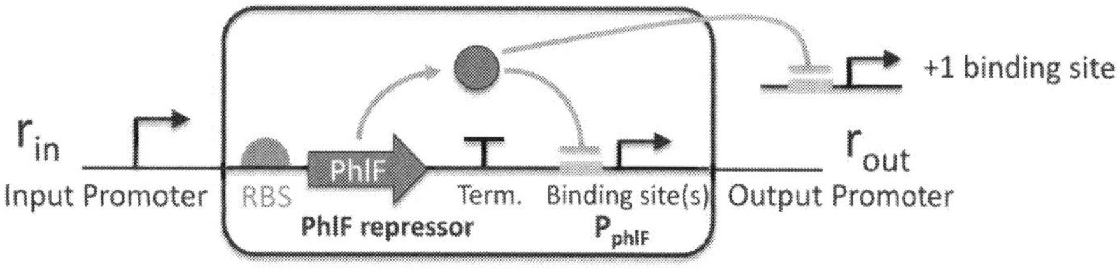

One A: Characterize a Circuit's Input-Output Behavior & Re-use It Everywhere

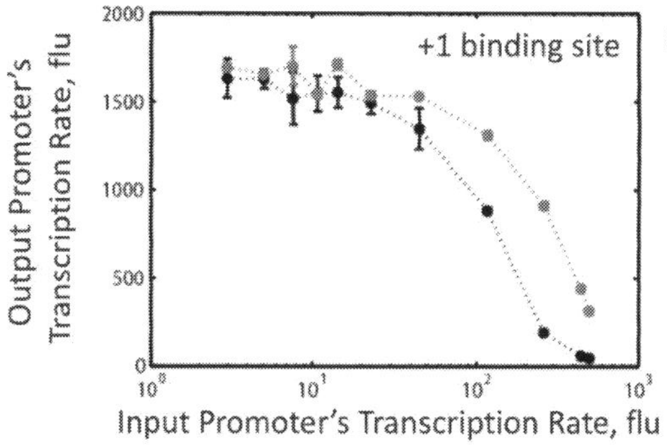

RBS sequence = ...ATTAGTAACATAAGCAAGAGCCTC

regulator = PhlF repressor
(a TetR homolog from *P. fluorescens*)

Terminator = rrnB T1

Output promoter = P_{phlF}

Characterized in *E. coli* DH10B,
LB media, 37°C steady-state, flow cytometry

The Challenges of Genetic Circuit Design

the simplest 1-regulator circuit

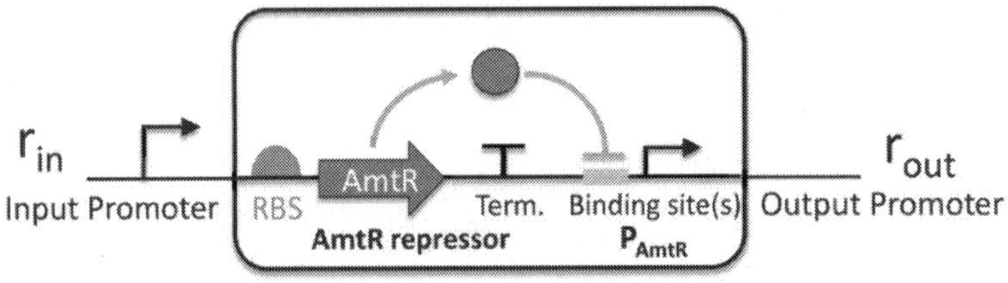

A Circuit's Input-Output Behavior is Highly Sensitive to Many Factors

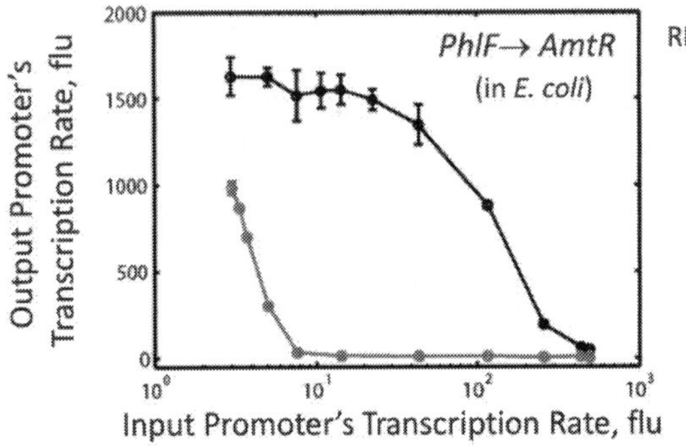

RBS sequence = ... AAGTTTACGCTAGGAGCTCAGA

regulator = AmtR repressor
(a TetR homolog from *C. glutamicum*)

Terminator = rrnB T1

Output promoter = P_{AmtR}

Characterized in *E. coli* DH10B,
LB media, 37°C, flow cytometry

The Challenges of Genetic Circuit Design

the simplest 1-regulator circuit

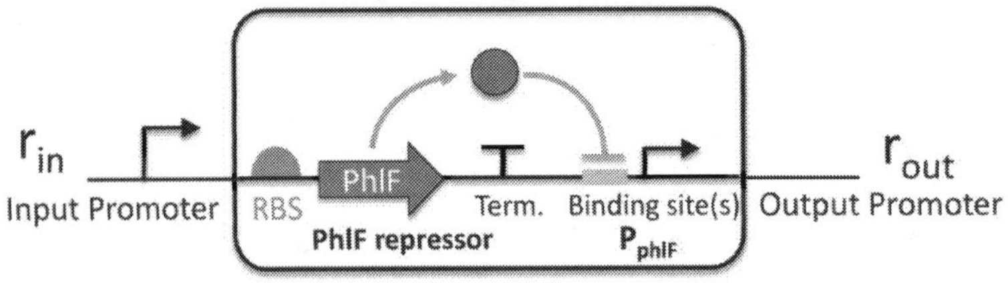

A Circuit's Input-Output Behavior is Highly Sensitive to Many Factors

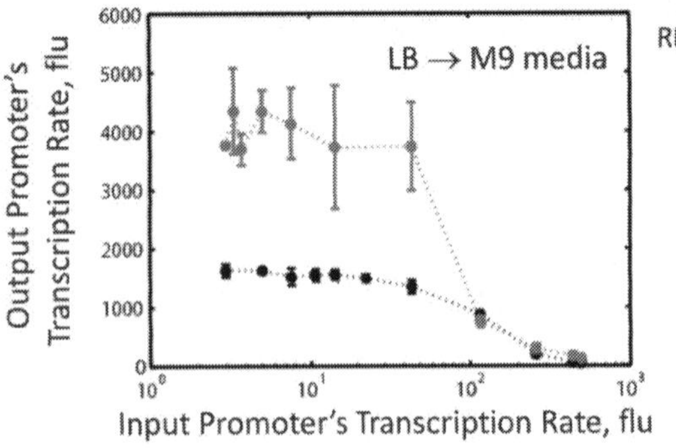

RBS sequence = ...ATTAGTAACATAAGCAAGAGCCTC

regulator = PhlF repressor
(a TetR homolog from *P. fluorescens*)

Terminator = rrnB T1

Output promoter = P_{phlF}

Characterized in *E. coli* DH10B,
M9 + 0.4% glucose, 37°C, flow cytometry

The Challenges of Genetic Circuit Design

the simplest 1-regulator circuit

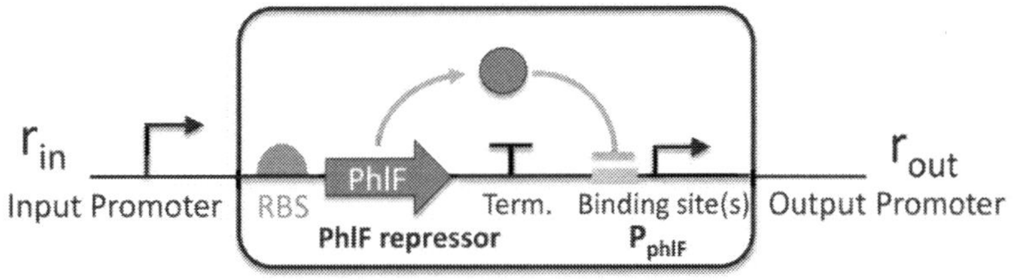

A Circuit's Input-Output Behavior is Highly Sensitive to Many Factors

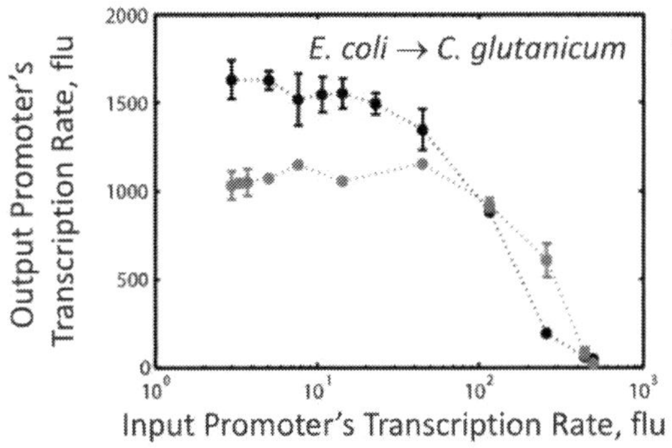

RBS sequence = ...ATTAGTAACATAAGCAAGAGCCTC

regulator = PhlF repressor
(a TetR homolog from *P. fluorescens*)

Terminator = rrnB T1

Output promoter = P_{phlF}

Characterized in Corynebacterium glutamicum,
LB media, 30°C, flow cytometry

The Challenges of Genetic Circuit Design

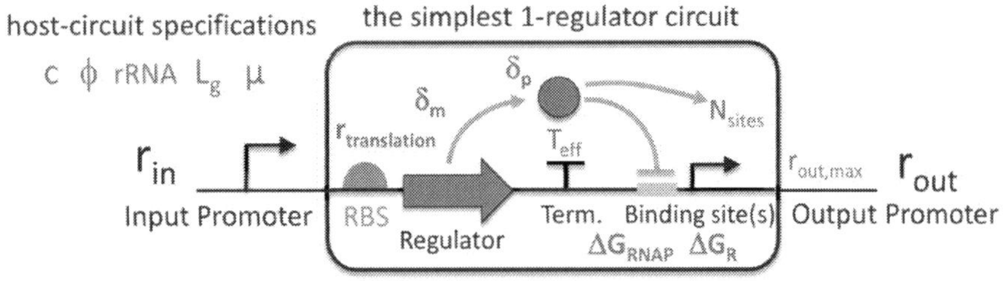

Big Q: So how do we predict the **function** of a genetic circuit according to its **parts**?

circuit DNA copy number, c   mRNA degradation rate, δ_m regulator degradation rate, δ_p
mRNA translation initiation rate, $r_{translation}$ terminator efficiency, T_{eff} host growth rate, μ

Genetic Circuit Parts/Sequences ➡ Static/Dynamic Response
biophysical modeling (thermodynamics & kinetics)

repressor binding energy, ΔG_R number of repressor binding sites, N_{sites} host genome size, L_g
RNA polymerase binding energy, ΔG_{RNAP} promoter's maximum transcription rate, $r_{out,max}$
host ribosomal RNA   circuit orientation (RNAP vs. DNA replication), ϕ

Modeling Circuit Function

host-circuit specifications

c ϕ rRNA L_g μ

the simplest 1-regulator circuit

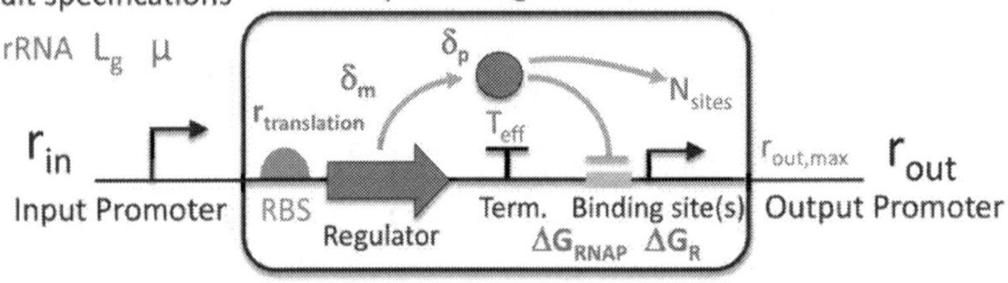

Simplest Steady-State Model Equations

$$R_{total} = c\,\phi\, r_{in}\, r_{translation}\,\mu^{0.7}\left(\delta_m \delta_p\right)^{-1}$$

How much repressor inside each cell?

$$p_{R,1} = \exp\left(-\beta \Delta G_R\right)/L_g$$

$$p_{RNAP,1} = \exp\left(-\beta \Delta G_{RNAP}\right)/L_g$$

Probability of one repressor or one RNAP binding?

How much unbound repressor?
(a partition function calculation)

$$R_{free} = \frac{R_{total}}{1 + c(N_{sites}+1)P_{R,1}}$$

$$\frac{r_{out}}{r_{out,\max}} = \left(1 + P_{RNAP,1}P\right)^c \;\Big/\; \left(1 + P_{RNAP,1}P\right)^c + \left(1 + R_{free}P_{R,1}\right)^c - 1$$

What is the transcription rate of the output promoter?

Modeling Circuit Function

host-circuit specifications

c ϕ rRNA L_g μ

the simplest 1-regulator circuit

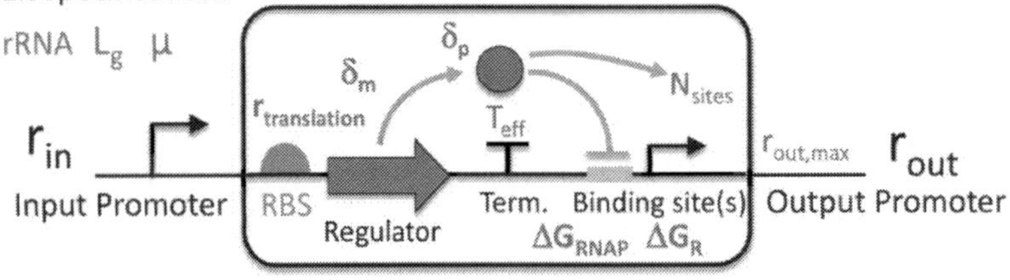

Q: Where do we obtain these numbers?

circuit DNA copy number, c

mRNA translation initiation rate, $r_{translation}$

number of repressor binding sites, N_{sites}

Known / Measured / Predictable

(with reasonable precision)

terminator efficiency, T_{eff}

host growth rate, μ

host ribosomal RNA   host genome size, L_g

sequenced host genomes

biophysical models

RBS Calculator v2.0

well-characterized toolboxes

Characterization of 582 natural and synthetic terminators and quantification of their design constraints

Ying-Ja Chen[1], Peng Liu[1], Alec A K Nielsen[1], Jennifer A N Brophy[1], Kevin Clancy[2], Todd Peterson[2] & Christopher A Voigt[1]

and several simple measurements

(e.g. growth rates of organisms in various media)

Modeling Circuit Function

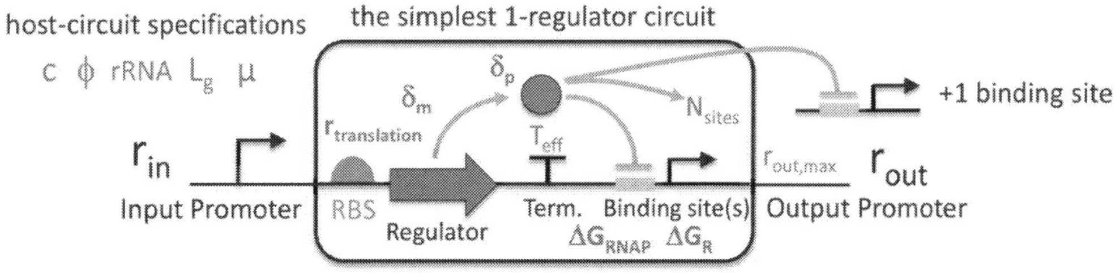

A: We used +binding site measurements to measure *in vivo* circuit biophysics

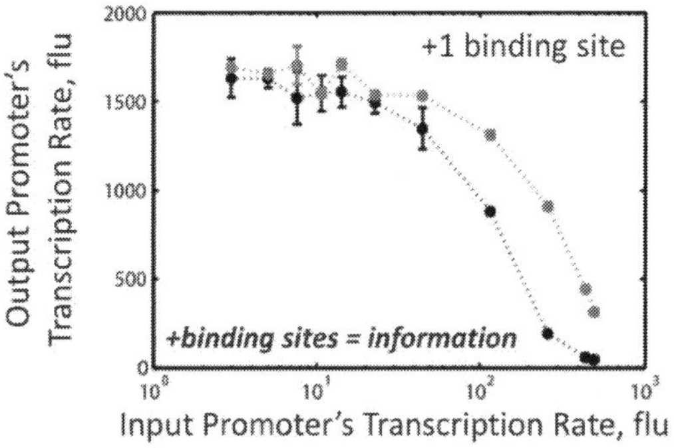

repressor binding energy, ΔG_R

RNA polymerase binding energy, ΔG_{RNAP}

Not Known

mRNA degradation rate, δ_m

regulator degradation rate, δ_p

promoter's maximum transcription rate, $r_{out,max}$

Measuring Circuit Biophysics

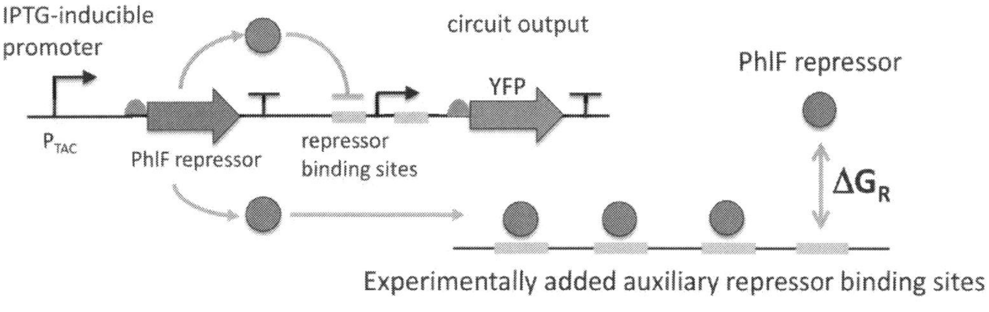

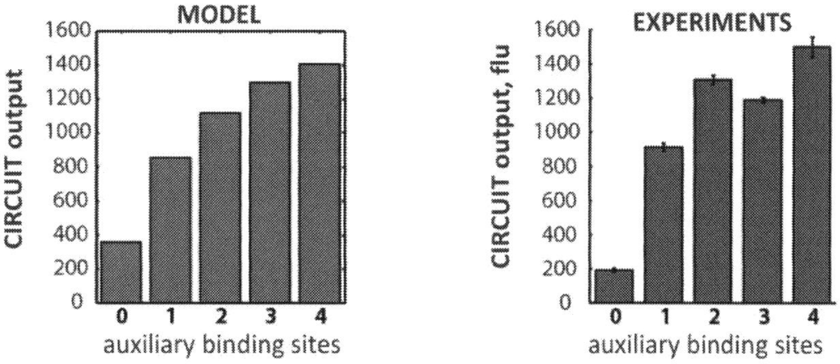

0.2 mM IPTG
triplicate
E. coli MG1655
flow cytometry

Measuring Circuit Biophysics

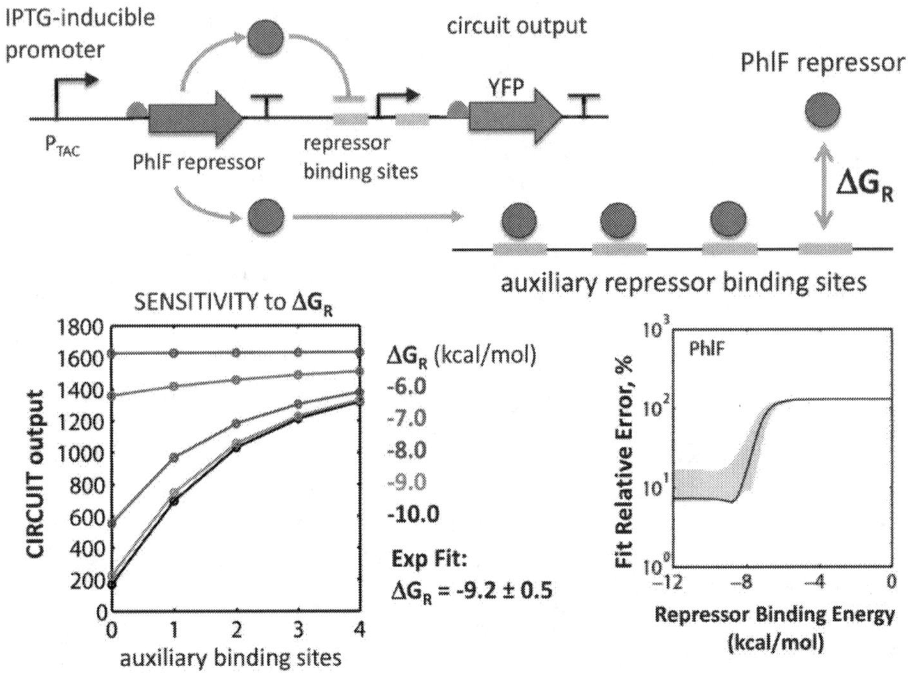

Measuring Circuit Biophysics

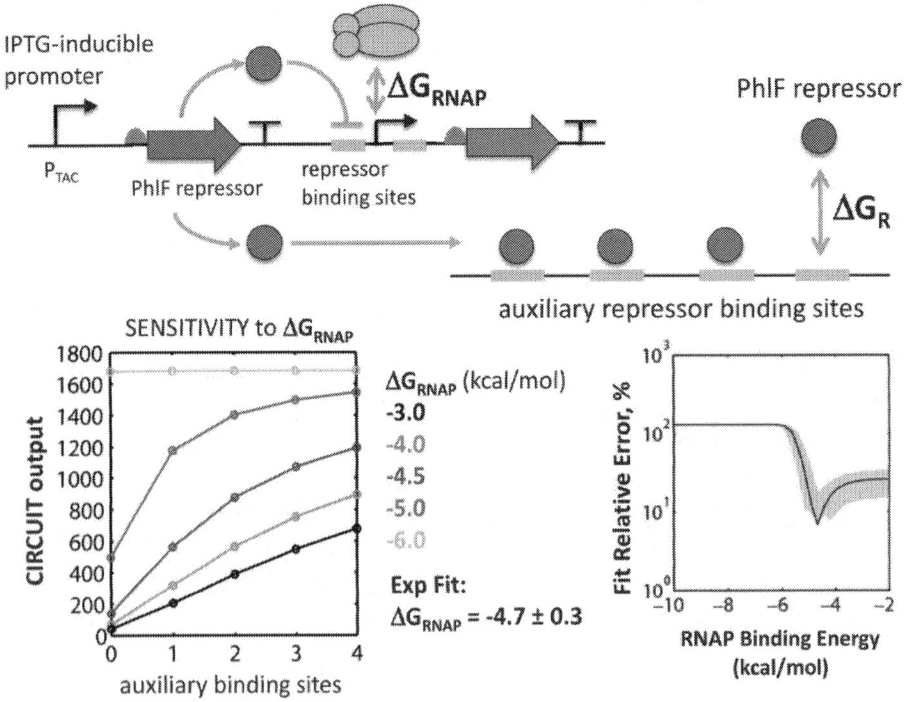

Measuring Circuit Biophysics

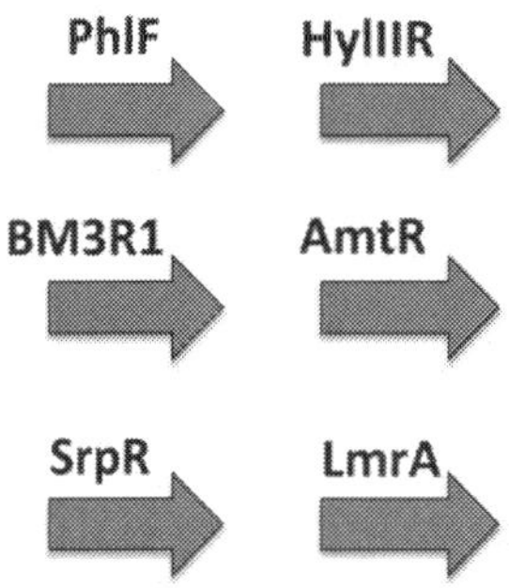

From the Voigt lab repressor toolbox
Stanton et. al.,
Nat. Chem. Bio., 2014

Experimental Measurements of Repressor and RNAP Binding Free Energies

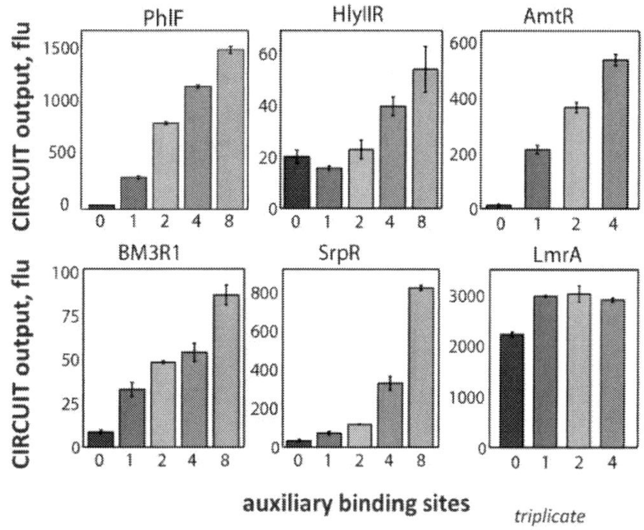

triplicate
E. coli MG1655
flow cytometry

Measuring Circuit Biophysics

Experimental Measurements of Repressor and RNAP Binding Free Energies

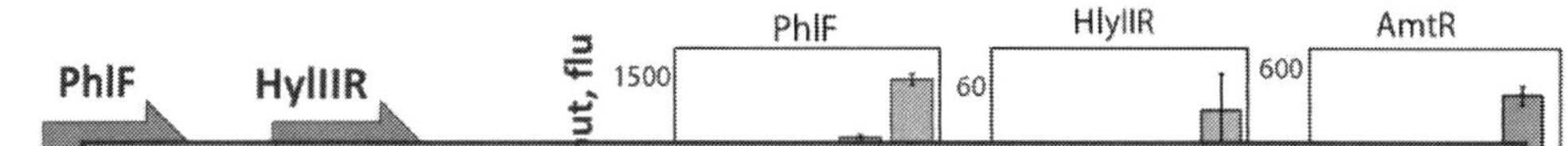

Repressor	$\Delta G_{Repressor}$ (kcal/mol)	ΔG_{RNAP} (kcal/mol)	$r_{out,max}$ (mRNA/min/copy)	$(\delta_m \delta_p)^{-1}$ (min^2)
PhlF	-9.2 ± 0.5	-4.7 ± 0.3	170.4	3.2
BM3R1	-7.6 ± 0.2	-4.7 ± 0.3	7.8	75.3
AmtR	-8.9 ± 1.0	-4.8 ± 0.4	85.2	12.2
LmrA	-8.1 ± 0.4	-4.6 ± 0.4	388.2	38.3
SrpR	-7.4 ± 0.3	-4.4 ± 0.3	175.2	2.2
HlyIIR	-6.4 ± 0.3	-4.7 ± 0.4	144.6	14.0

P_{lacZ} (wt, ΔLacI) = 18 mRNA/min; LacI $(\delta_m \delta_p)^{-1}$ = 14.4 min^2
(Kennell & Riezman, *JMB*, 1977)

E. coli MG1655
flow cytometry

Predicting 1-Regulator Circuit Function

RBS Library Calculator 1.0

efficient expression optimization for multi-protein genetic systems

A: The **model** predicts the circuit output when changing the repressor mRNA's translation rate
Experiments confirm these predictions.

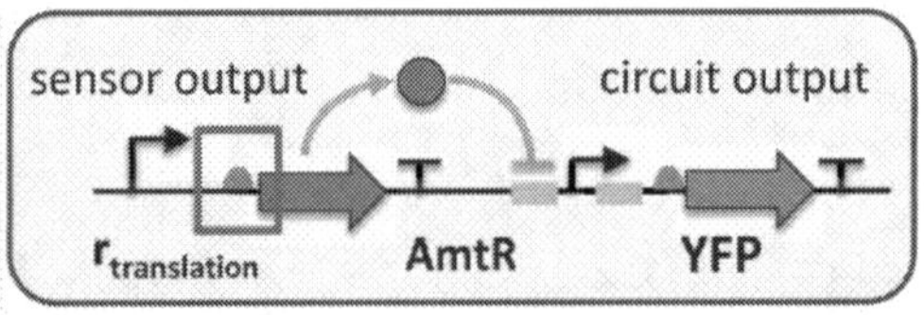

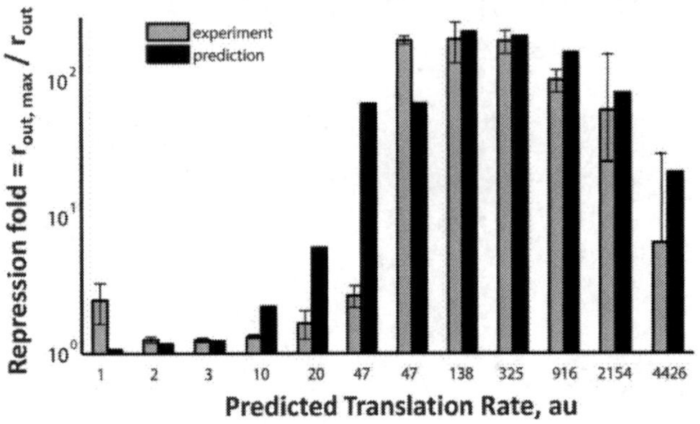

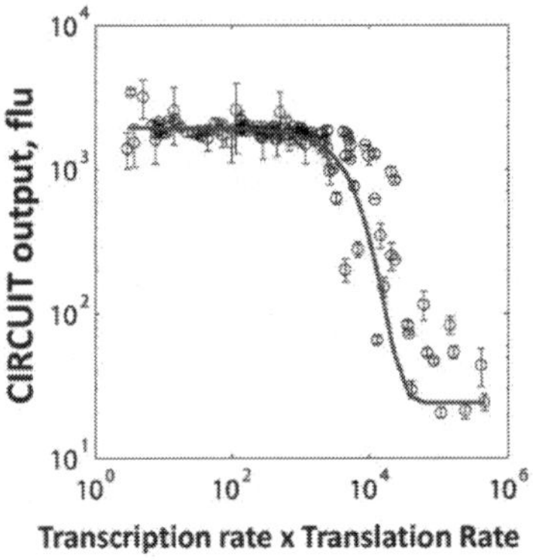

Predicting 1-Regulator Circuit Function

A: The **model** predicts the circuit output
when experimentally changing circuit, system, and host parameters

3 vectors, 5 hosts, 6 regulators
12 regulator expression levels
847 circuit-system-host combinations

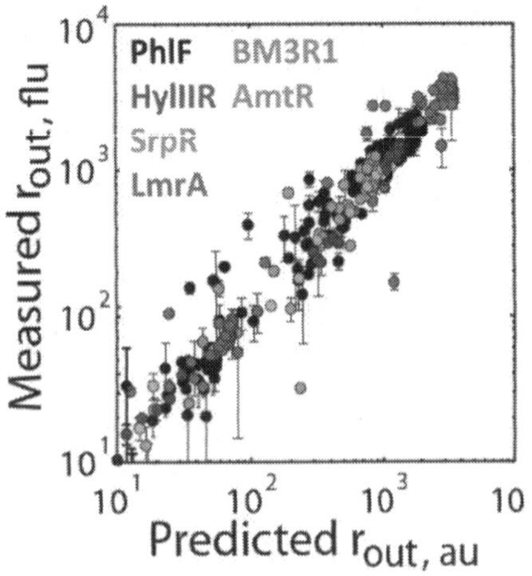

E. coli DH10B, E. coli BL21, P. fluorescens, C. glutamicum, S. Typhimurium LT2

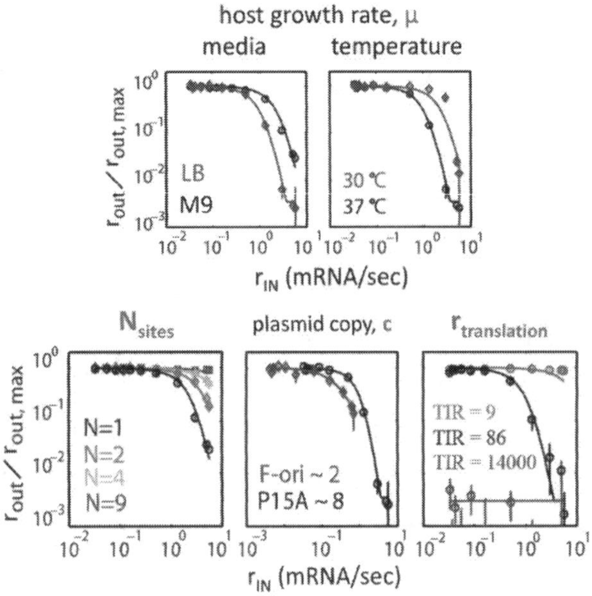

Predicting 2-Regulator Circuit Function

Signal Amplifying Circuit

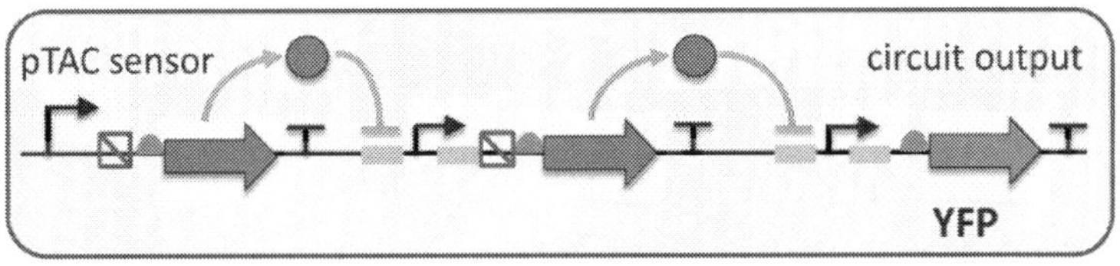

Design Objective

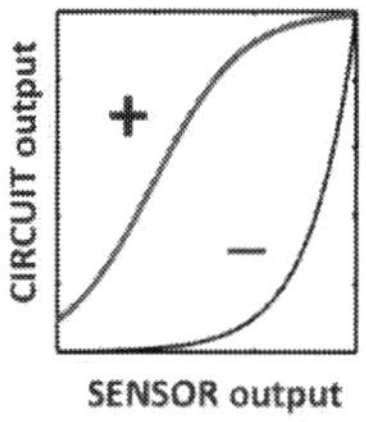

We use our model to select repressors and mRNA translation rates to construct & characterize the best signal amplifying circuits

Repressor Toolbox with Measured Biophysical Parameters

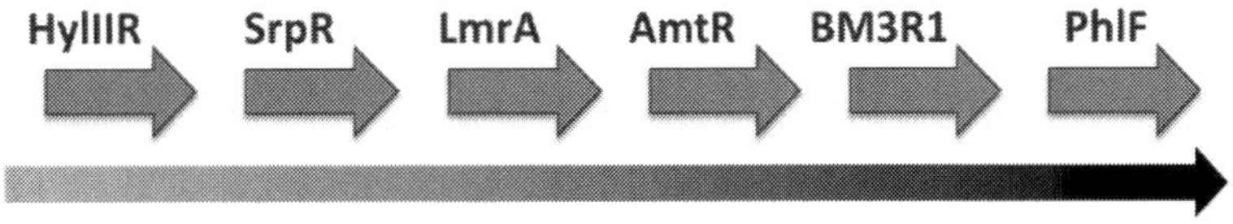

Weak Repressors Increasing repressor binding free energies **Strong Repressors**

ribozyme

We also use ribozymes from the ribozyme toolbox for insulation (Voigt Lab)

Predicting 2-Regulator Circuit Function

We use our model to select repressors to construct & characterize the best signal amplifying circuits

Design Objective

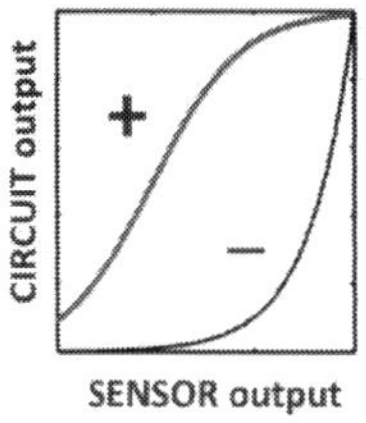

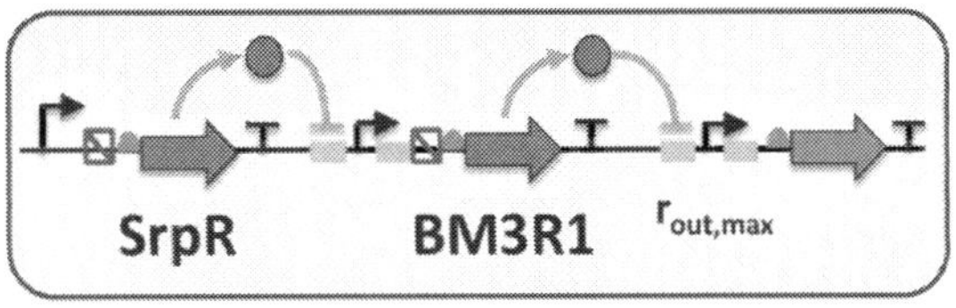

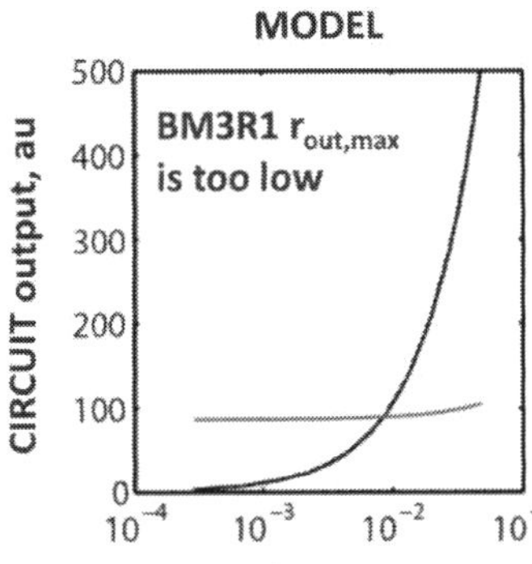

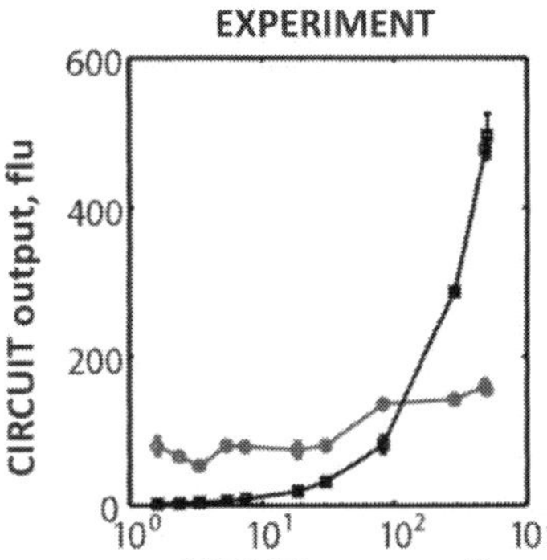

Signal Amplification SA = 0.77

BAD

Predicting 2-Regulator Circuit Function

We use our model to select repressors to construct & characterize the best signal amplifying circuits

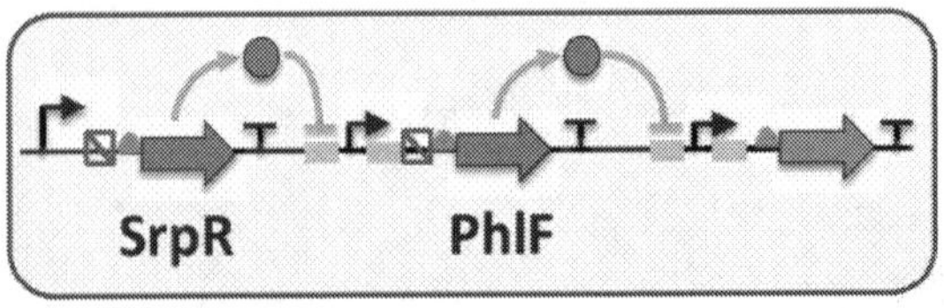

Design Objective

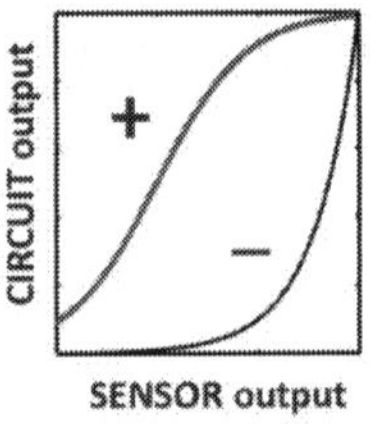

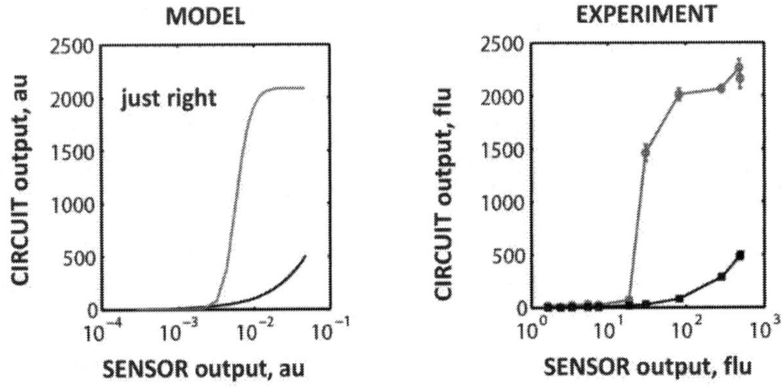

Signal Amplification
SA = 26

GOOD

Predicting 2-Regulator Circuit Function

We use our model to select repressors <u>and mRNA translation rates</u> to construct & characterize the best signal amplifying circuits

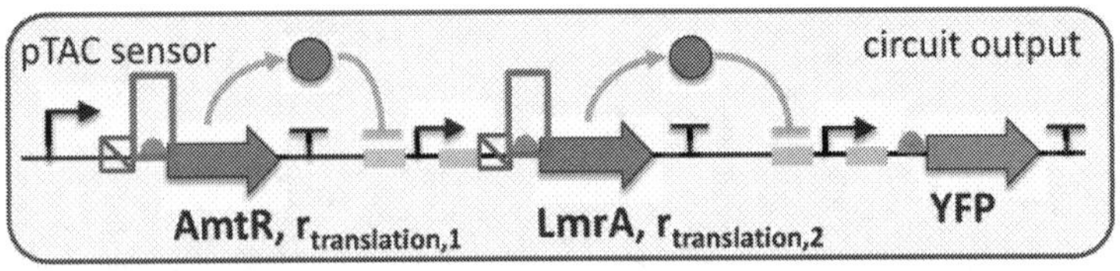

Design Objective

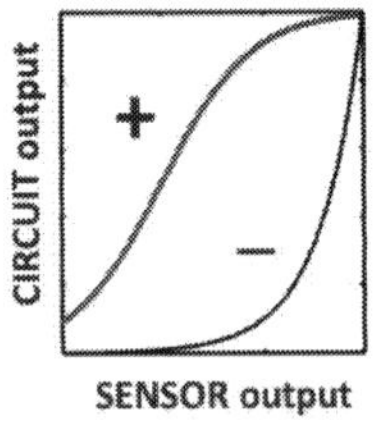

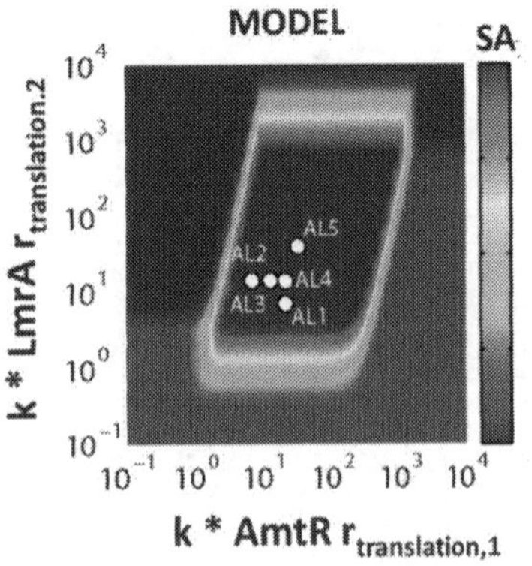

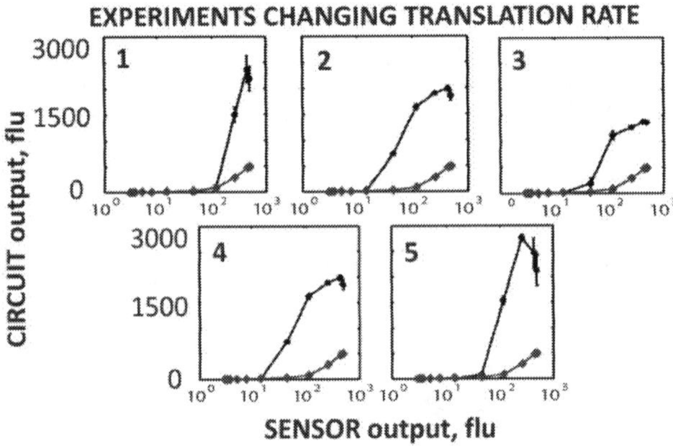

Designing Many-Regulator Circuit Function

Q: How do we scale the design of genetic circuits to <u>many regulators</u>?

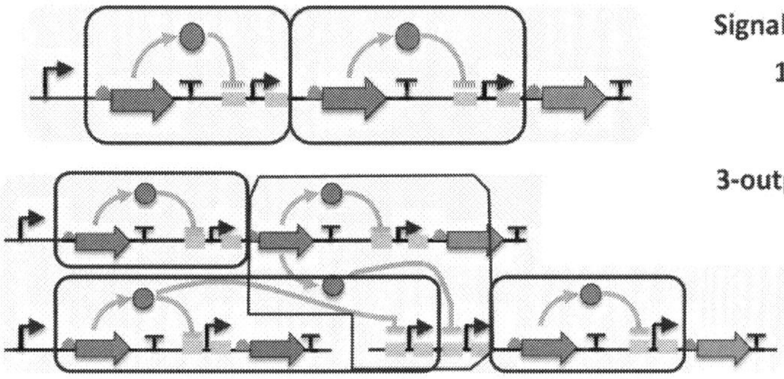

Signal Amplifier, 2 regulators

10+ model parameters

3-output Analog to Digital Converter

4 regulators

20+ model parameters

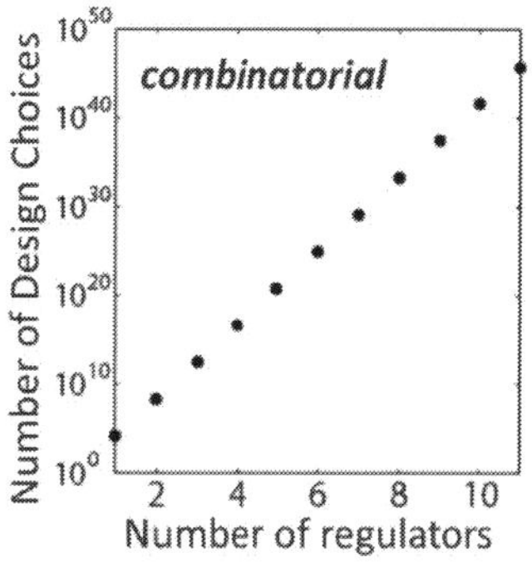

at least 5 model parameters per regulator!

5D space

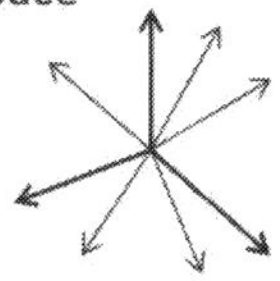

TF binding energy, ΔG_R

TF binding sites, N_{sites}

RNAP binding energy, ΔG_{RNAP}

mRNA translation rate, $r_{translation}$

mRNA and protein degradation, $\delta_m \delta_p$

+host-dependent variables

Designing Many-Regulator Circuit Function

Q: How do we scale the design of genetic circuits to <u>many regulators</u>?

A: We introduce a new dimensionless unit that <u>greatly reduces the design space</u>

We would like to name the first dimensionless unit for genetic circuit engineering after **Prof. Mark Ptashne** whose contributions to understanding gene regulation have become the foundational basis for Synthetic Biology

The Pt number

5 x N Model Parameters

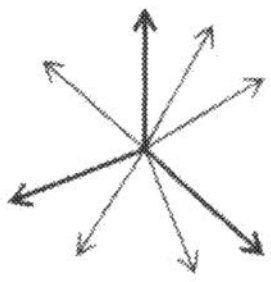

TF binding energy, ΔG_R

TF binding sites, N_{sites}

RNAP binding energy, ΔG_{RNAP}

mRNA translation rate, $r_{translation}$

mRNA and protein degradation, $\delta_m \delta_p$

+host-dependent variables

1 x N Model Parameters

1 Pt number per regulator

$$\mathrm{Pt} = \frac{\exp(-\beta \Delta G_R)\dfrac{R_{total}}{L_G}}{1 + \dfrac{cN_{sites}}{L_G}\exp(-\beta \Delta G_R)}$$

$$R_{total} = c\,\phi\, r_{in}\, r_{translation}\, \mu^{0.7} (\delta_m \delta_p)^{-1}$$

Designing Many-Regulator Circuit Function

Q: How do we scale the design of genetic circuits to <u>many regulators</u>?
A: First, determine your optimal Pt numbers. Then select your parts / sequences.

There is a simple relationship between regulator binding occupancy and Pt number
***that stays the same across 847* circuit-system-host combinations**

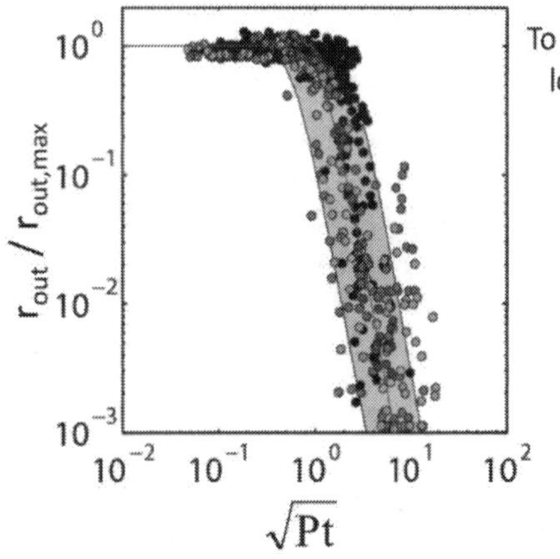

To lower binding occupancy by **10-fold**,
lower the regulator's Pt number by **4-fold**

Your design choices (in "linear" regime) are:
(i) lower translation rate by 4-fold
(ii) select a regulator with higher ΔG (2.25 kcal/mol)
(iii) increase N_{sites} by 4-fold

All three design choices are equivalent

Designing Many-Regulator Circuit Function

Q: How do we scale the design of genetic circuits to <u>many regulators</u>?
A: First, determine your optimal Pt numbers. Then select your parts / sequences.

What is the design space for the best-possible 2-regulator signal amplifiers?
The model predicted the best combinations of regulators (with different expression levels)

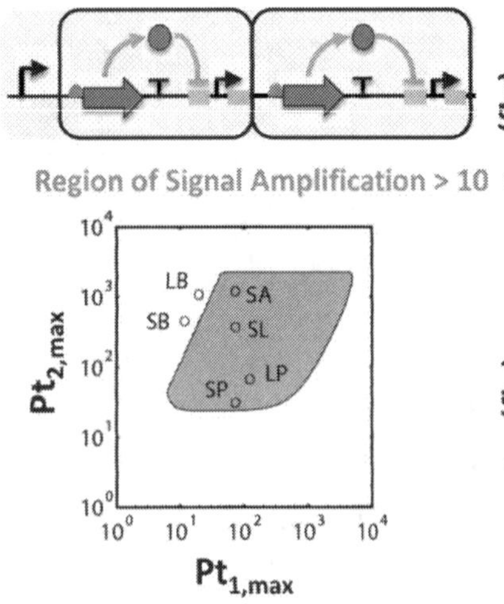

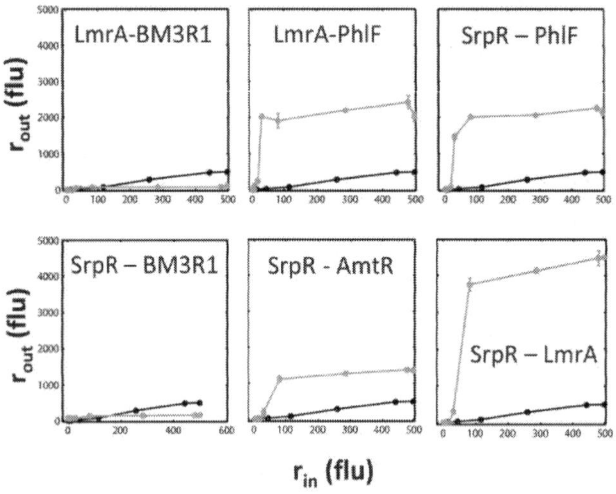

Designing dCas9-based Circuits

Q: Can we apply the same concept to design dCas9-based analog genetic circuits?

Objective: Construct 2-sgRNA genetic circuits to amplify the dynamic range of a sensor

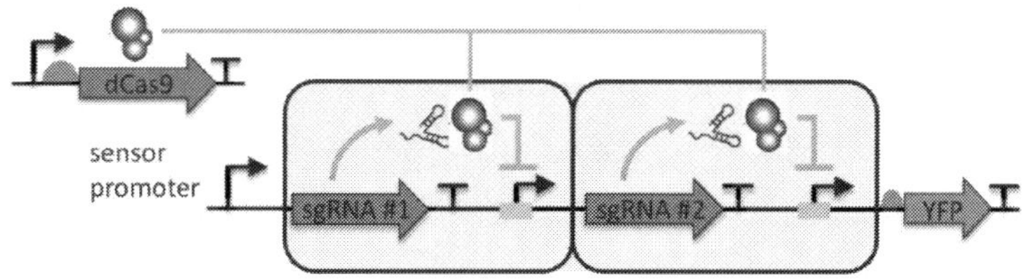

Design Objective

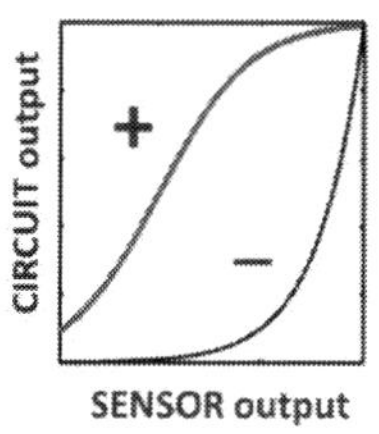

We introduced combinatorial mutations into the sgRNA guide sequences to vary sgRNA : DNA site binding affinities

TGGACTGAGC**KNK**CCGTC**NN** — 256 guide sequences

CTCTA**KK**AACC**KK**AACGA**NN** — 256 guide sequences

65536 possible 2-sgRNA circuits

Designing dCas9-based Circuits

Q: Can we apply the same concept to design dCas9-based analog genetic circuits?

Ad-hoc design of multi-sgRNA analog circuits does not work

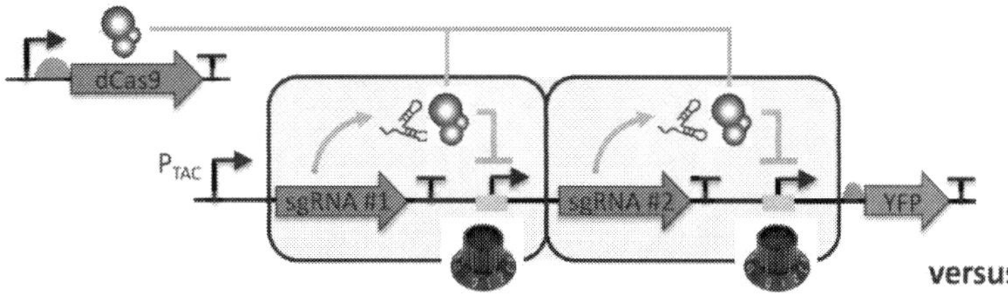

96 2-sgRNA circuits with varying guide RNA sequences

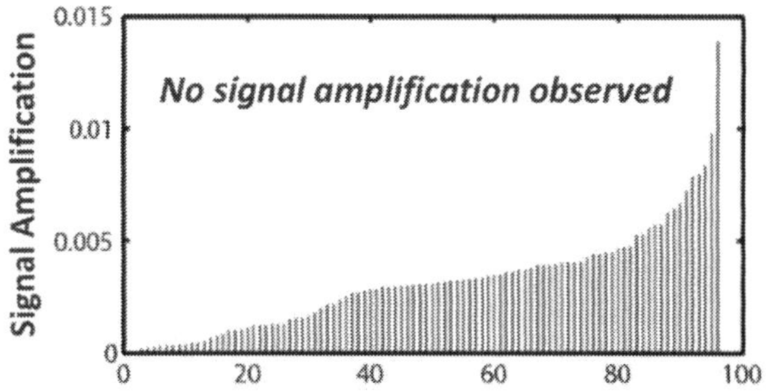

versus SrpR – LmrA circuit

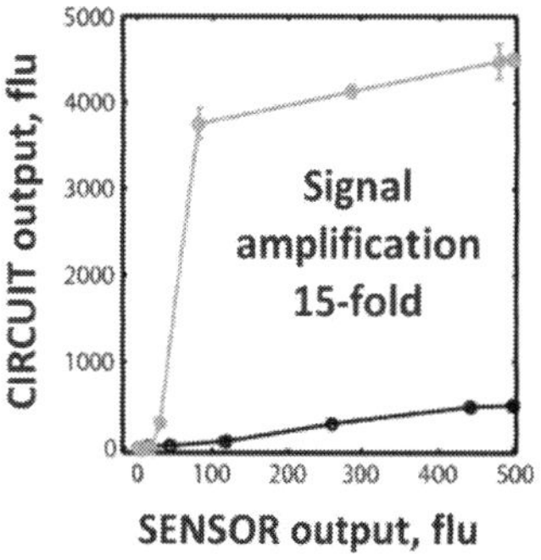

Designing dCas9-based Circuits

Q: Can we apply the same concept to design dCas9-based analog genetic circuits?

We developed a biophysical model of (d)Cas9-based gene editing/regulation

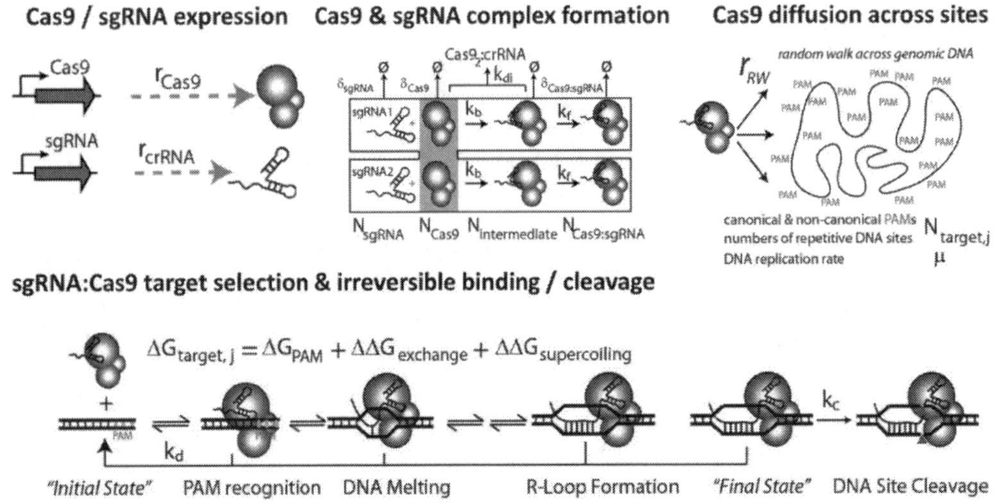

Designing dCas9-based Circuits

Q: Can we apply the same concept to design dCas9-based analog genetic circuits?

We parameterized the model using over 4000 measurements

Parameterization using several data-sets:
Sternberg et. al. *Nature* 2014
Jiang et. al., *Nat. Biotech.* 2013
Szczelkun et. al., *PNAS*, 2013
Pattanayak et. al., *Nat. Biotech.*, 2013
Wang et. al., *Cold Spring Harbor Press*, 1983
Depew et. al., *PNAS*, 1975

cleavage activity vs. time (model & *in vitro* data)

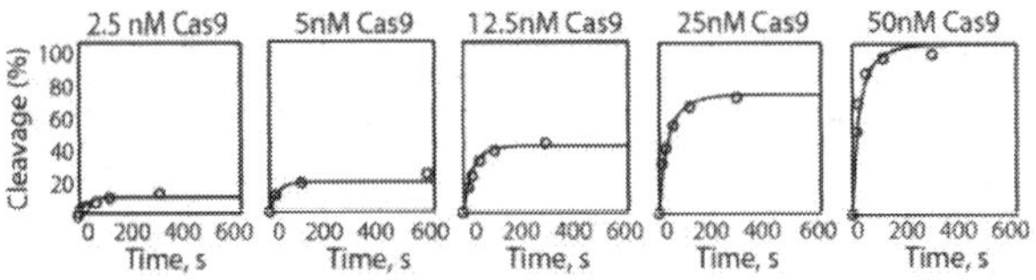

Canonical PAMs	
PAM	ΔG_{PAM}
NGGA	-9.4
NGGT	-9.6
NGGG	-9.5
NGGC	-9.7
	kcal/mol

Non-canonical PAMs	
PAM	ΔG_{PAM}
NCGG	-7.7
NAGG	-7.7
NAGT	-7.7
NAGA	-7.5
NAGC	-7.7
NTTT	-7.2
NGAG	-6.9
NGAT	-6.9
NGAA	-6.8
NGAC	-6.8
NGCT	-6.7

Model calculates binding at non-canonical PAMs

Model calculates binding at mismatched DNA sites

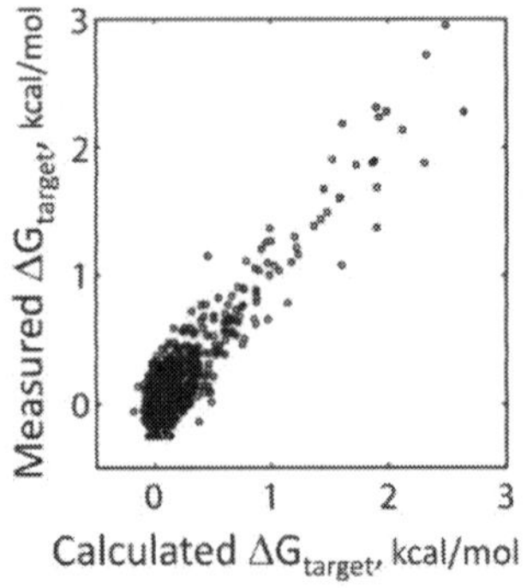

Designing dCas9-based Circuits

Q: Can we apply the same concept to design dCas9-based analog genetic circuits?

The model calculates sgRNA binding occupancies for a many-regulator circuit

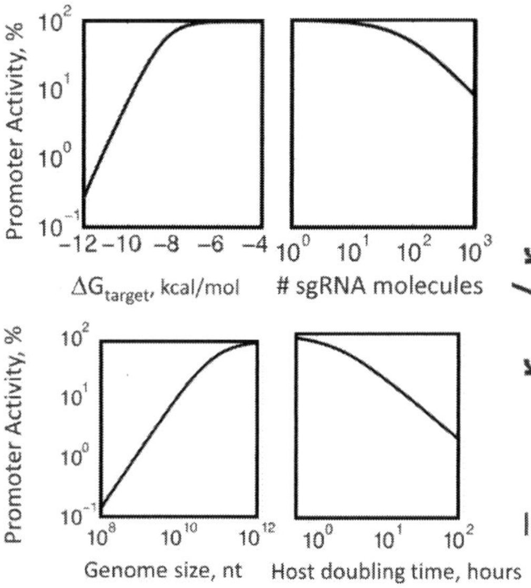

Experimental validation
sgRNA repression vs. auxiliary binding sites

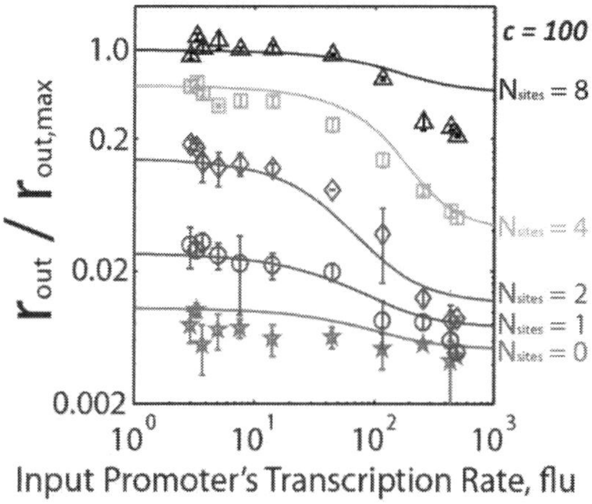

E. coli DH10B, flow cytometry

Take Home Messages

Big Q: How do we predict the **function** of a genetic circuit according to its **parts**?

1. Circuit-host modeling is needed to predict circuit function
   at least 5 model parameters per regulator
   + host specifications (e.g. growth rate, genome size)

2. *in vivo* circuit biophysical quantities can be measured by adding
   auxiliary regulator binding sites, measuring, and analyzing

3. The Ptashne number (Pt) greatly reduces the circuit design space
   from ***five*** degrees of freedom per regulator to ***one***

Ptashne number for repressor TFs

$$Pt = \frac{\exp(-\beta\Delta G_R)\frac{R_{total}}{L_G}}{1+\frac{cN_{sites}}{L_G}\exp(-\beta\Delta G_R)}$$

Ptashne number for sgRNAs

$$Pt_{[i,j]} = \frac{[\text{sgRNA}_i\text{:dCas9}]_{total}\frac{\exp(\square\square G_{target,[i,j]})}{L_G}}{1+\sum_{n=1}^{M_{sites}} cN_{sites,n}\frac{\exp(\square\square G_{target,n})}{L_G}}$$

Acknowledgements

Happy to answer questions!

Graduate students
Iman Farasat
Amin Espah Borujeni
Tian Tian
Long Chen
Chiam Yu Ng
Grace Vezeau
Sean Halper
Alex Reis

Undergraduate students
Ashlee Smith, Walker Huso,
Sam Klug, Justin Miller
Clay Swackhammer, Emily Sileo
Selina Roman-White

Funding
Office of Naval Research
Air Force Office of Scientific Research
National Science Foundation
Penn State Institute for the Energy & Environment

Modeling Circuit Function

host-circuit specifications

the simplest 1-regulator circuit

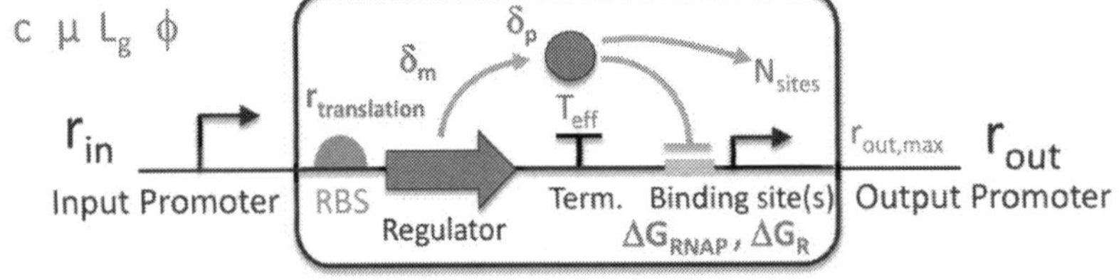

Q: Where do we obtain these numbers?

circuit DNA copy number, c mRNA degradation rate, δ_m regulator degradation rate, δ_p
mRNA translation initiation rate, $r_{translation}$ terminator efficiency, T_{eff} host growth rate, μ

Known / Measured / Predictable (with reasonable precision)	**Not Known**

repressor binding energy, ΔG_R number of repressor binding sites, N_{sites} host genome size, L_g
RNA polymerase binding energy, ΔG_{RNAP} promoter's maximum transcription rate, $r_{out,max}$
host ribosomal RNA

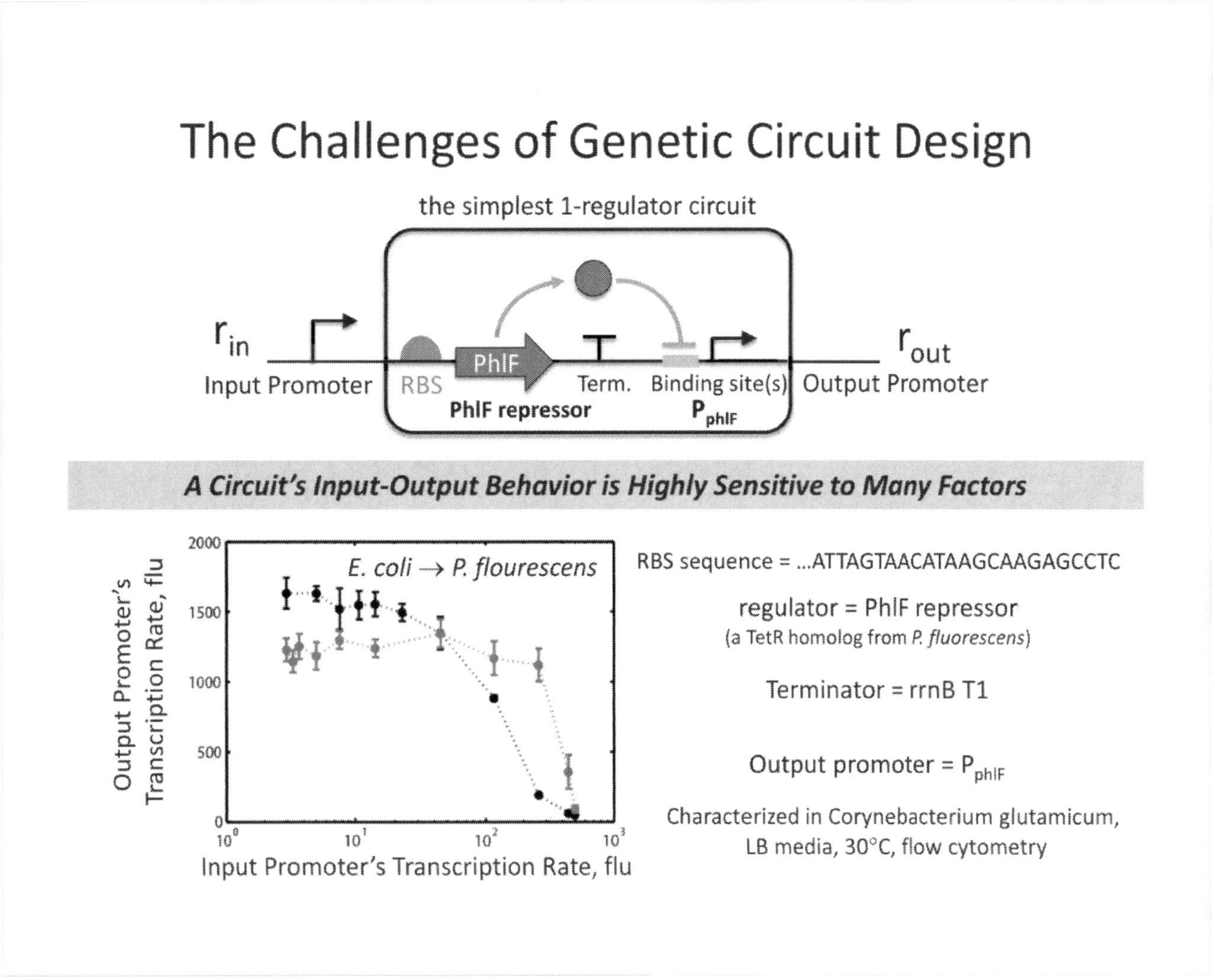
The Challenges of Genetic Circuit Design
the simplest 1-regulator circuit
r in
Input Promoter
RBS
PhlF
PhlF repressor
Term.
Binding site(s)
P phlF
r out
Output Promoter
A Circuit's Input-Output Behavior is Highly Sensitive to Many Factors
E. coli → P. flourescens
Output Promoter's Transcription Rate, flu
Input Promoter's Transcription Rate, flu
2000
1500
1000
500
0
10^0
10^1
10^2
10^3
RBS sequence = ...ATTAGTAACATAAGCAAGAGCCTC
regulator = PhlF repressor
(a TetR homolog from P. fluorescens)
Terminator = rrnB T1
Output promoter = P phlF
Characterized in Corynebacterium glutamicum,
LB media, 30°C, flow cytometry

Quantifying host-circuit interactions with a mechanistic chassis model

Andrea Weiße

SynthSys - Synthetic & Systems Biology

University of Edinburgh

andrea.weisse@ed.ac.uk

THE UNIVERSITY *of* EDINBURGH   SynthSys   DFG

Trade-offs in cellular physiology globally affect gene expression.

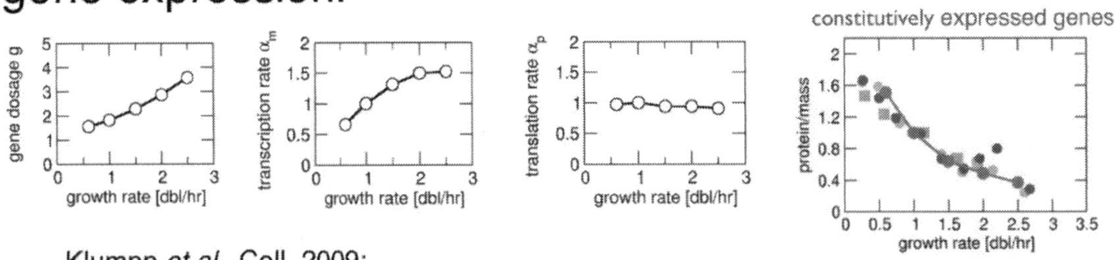

Klumpp *et al.*, Cell, 2009:
Growth rate-dependent global effects on gene expression in bacteria

Keren *et al.*, Mol Sys Biol, 2013:
Promoters maintain their relative activity levels under different growth conditions

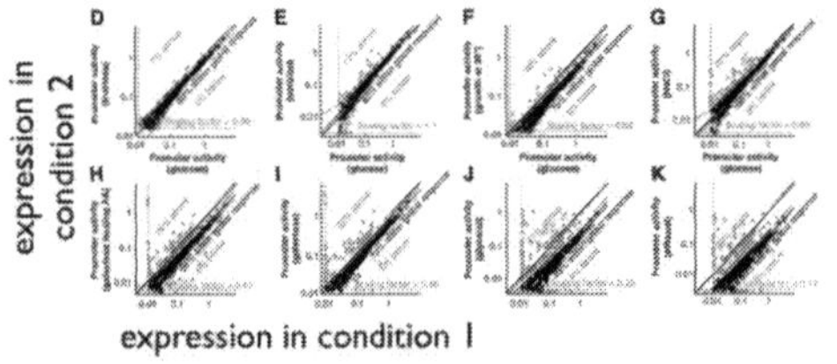

Berthoumieux *et al.*, Mol Sys Biol, 2012:
Shared control of gene expression in bacteria by transcription factors and global physiology of the cell

Gerosa *et al.*, Mol Sys Biol, 2013:
Dissecting specific and global transcriptional regulation of bacterial gene expression

Cross-talk between the cell and synthetic circuits can alter function and host fitness.

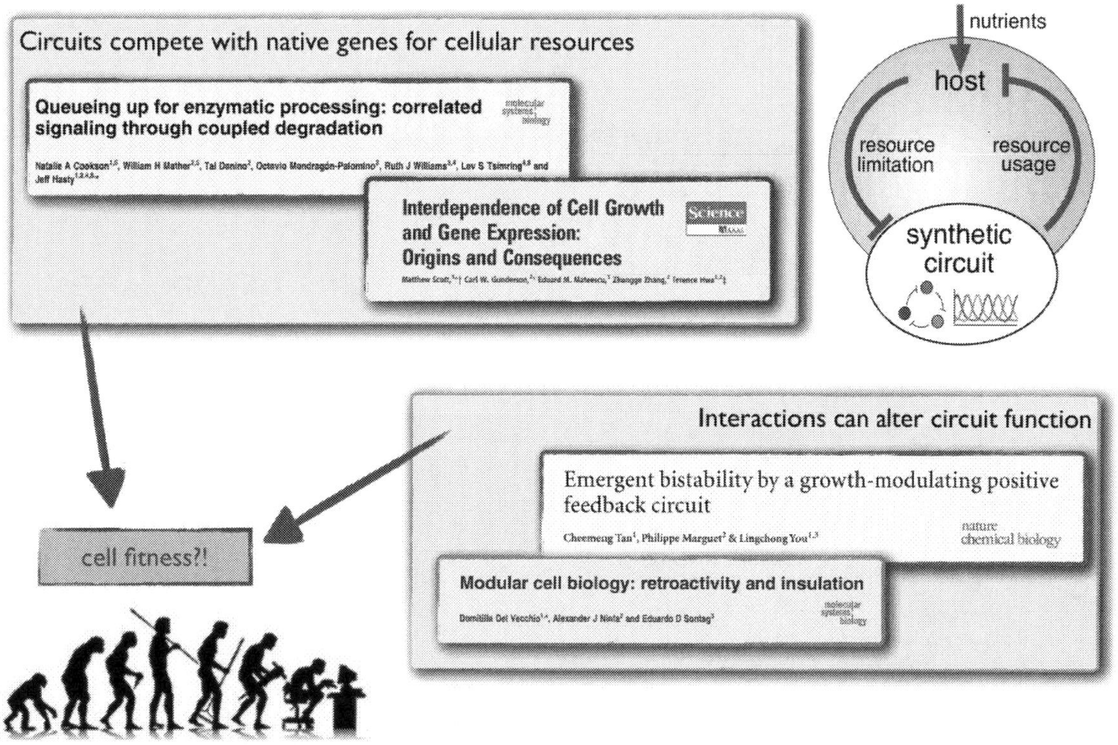

A model of the cell can help us understand global effects between gene expression and growth.

$$\phi_R + \phi_P = 1 - \phi_Q \equiv \phi_R^{max}$$

Scott *et al.*, Science 2010:
Minimal model to describe phenomenology

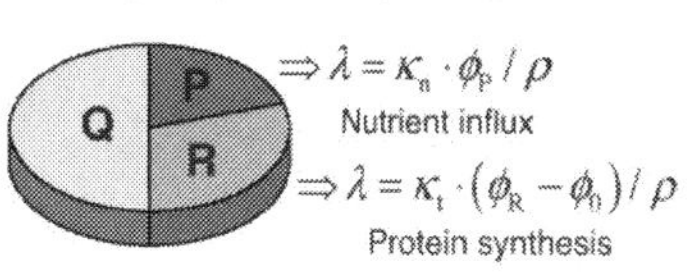

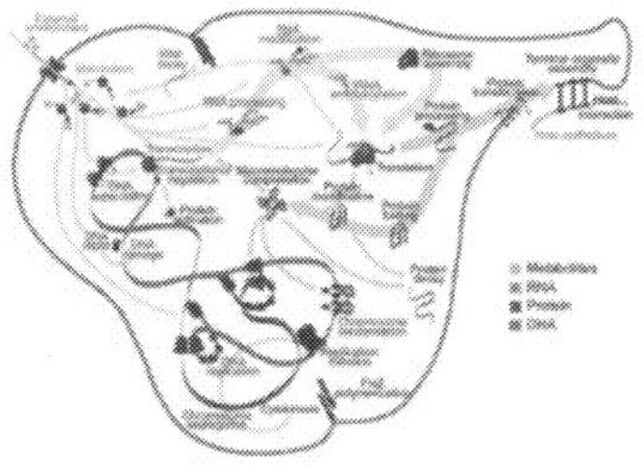

Karr *et al.*, Cell 2012:
Whole-cell model

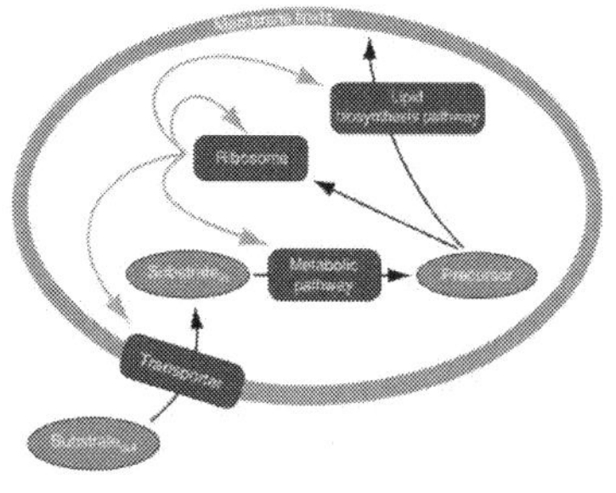

Molenaar *et al.*, Mol Sys Biol, 2009:
Resource allocation based on growth rate maximisation

We built a mechanistic model around 3 cellular trade-offs.

All living cells face trade-offs in:

1. Finite energy
2. Finite number of ribosomes
3. Finite mass/proteome

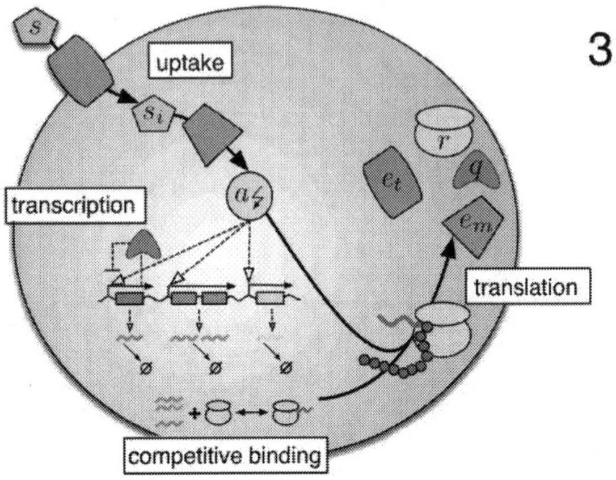

Cells have a finite pool of energy to fuel biosynthetic processes.

$$\text{energy} \quad \frac{da}{dt} = n_s \nu_{\text{cat}}(e_m, s_i) - \sum_x n_x \nu_x - \lambda a$$

nutrient quality/ energetic yield

consumption by protein synthesis

rate of catabolism

energy trade-off

dilution

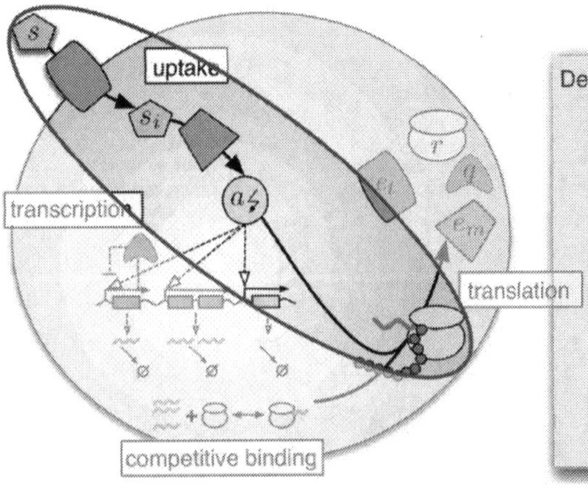

Derive rate of translation/energy consumption mechanistically:

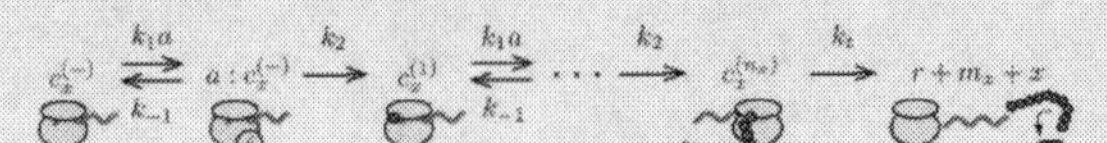

bound ribosomes

$$\nu_x(c_x, a) = \frac{c_x}{n_x} \cdot \frac{\gamma_{\max} a}{K_\gamma + a}$$

translation elongation

length of protein

We built a mechanistic model around 3 cellular trade-offs.

All living cells face trade-offs in:

1. Finite energy
2. Finite number of ribosomes
3. Finite mass/proteome

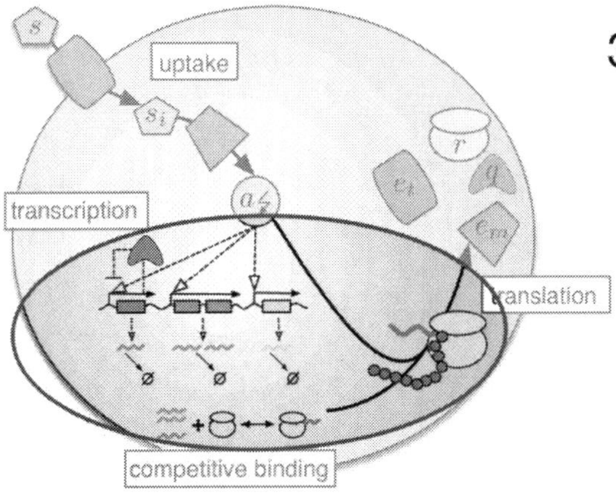

Different types of mRNA compete for a finite pool of free ribosomes.

free mRNA (transcription, ribosome un-/binding, degradation, free-up from translation, dilution)

$$\frac{dm_x}{dt} = \omega_x(a) - k_b m_x r + k_u c_x + \nu_x - d_m m_x - \lambda m_x$$

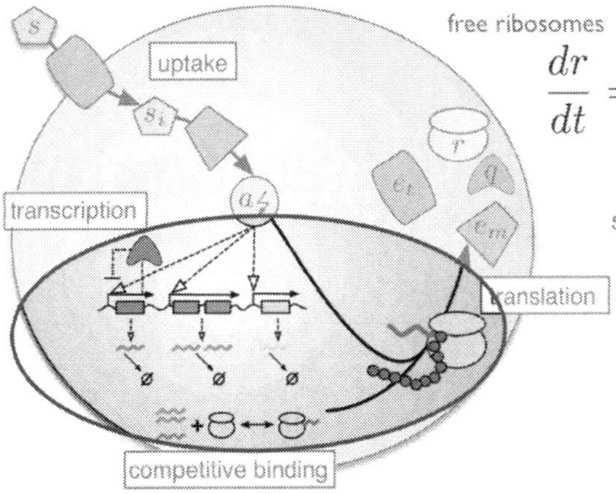

free ribosomes (synthesis, dilution, free-up, ribosome trade-off, competition)

$$\frac{dr}{dt} = \nu_r - \lambda r + \sum_x \left[\nu_x - k_b m_x r + k_u c_x\right]$$

transcription rate (q auto-inhibition)

$$\omega_x(a) = \frac{w_x a}{\theta_x + a} \times \frac{1}{1 + (q/K_q)^{n_q}}$$

We built a mechanistic model around 3 cellular trade-offs.

All living cells face trade-offs in:

1. Finite energy
2. Finite number of ribosomes
3. Finite mass/proteome

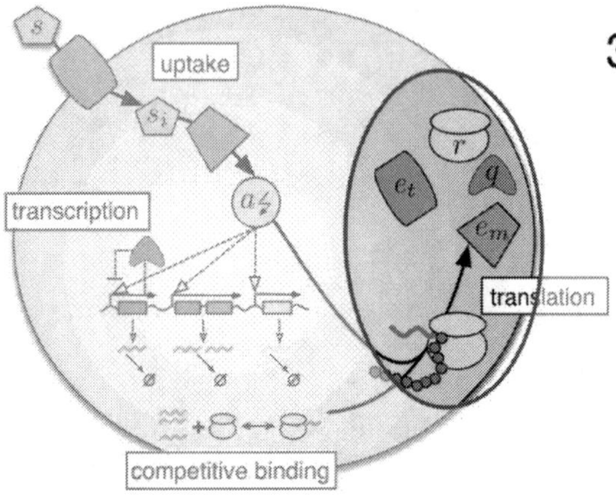

Cells have a finite proteome.

proteins # translating ribosomes

$$\sum_x n_x x + n_r \sum_x c_x = M^*$$

protein length ribosome length

mass trade-off

then at exponential growth

translating ribosomes

$$\dot{M} = \gamma(a) \sum_x c_x - \lambda M$$

translation elongation

We define the growth rate as

rate of mass replication

$$\lambda = \frac{\gamma(a)}{M^*} \sum_x c_x$$

typical mass of exponentially growing cell

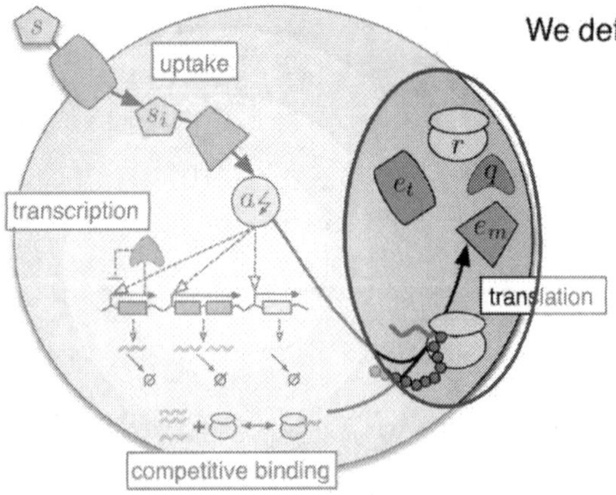

internal nutrient $\dot{s}_i = \nu_{\text{imp}}(e_t, s) - \nu_{\text{cat}}(e_m, s_i) - \lambda \cdot s_i$

energy $\dot{a} = n_s \cdot \nu_{\text{cat}}(e_m, s_i) - \sum_{x \in \{r,t,m,q\}} n_x \cdot \nu_x(c_x, a) - \lambda \cdot a$

free ribosomes $\dot{r} = \nu_r(c_r, a) - \lambda \cdot r + \sum_{x \in \{r,t,m,q\}} (\nu_x(c_x, a) - k_b \cdot r \cdot m_x + k_u \cdot c_x)$

transporter enzyme $\dot{e}_t = \nu_t(c_t, a) - \lambda \cdot e_t$

metabolic enzyme $\dot{e}_m = \nu_m(c_m, a) - \lambda \cdot e_m$

house-keeping $\dot{q} = \nu_q(c_q, a) - \lambda \cdot q$

free mRNA $\dot{m}_x = \omega_x(a) - (\lambda + d_m) \cdot m_x + \nu_x(c_x, a) - k_b \cdot r \cdot m_x + k_u \cdot c_x,$

bound mRNA $\dot{c}_x = -\lambda \cdot c_x + k_b \cdot r \cdot m_x - k_u \cdot c_x - \nu_x(c_x, a), \quad x \in \{r, t, m, q\}$

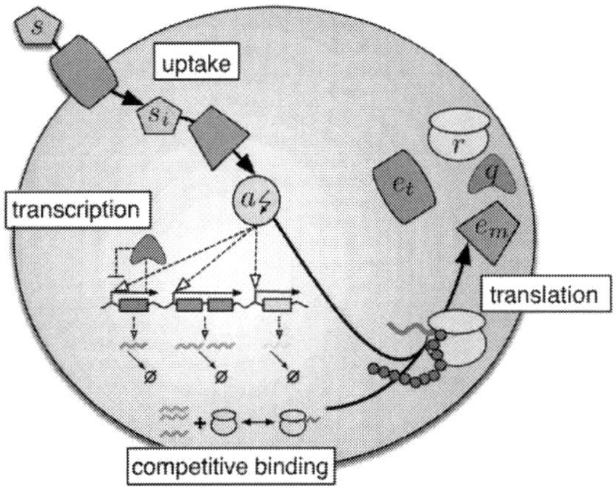

14 ODEs, 21 parameters (many from literature)

The model robustly recovers empirical growth relations.

We fitted the model to the data by Scott et al. (Science 2010) on linear growth relations.

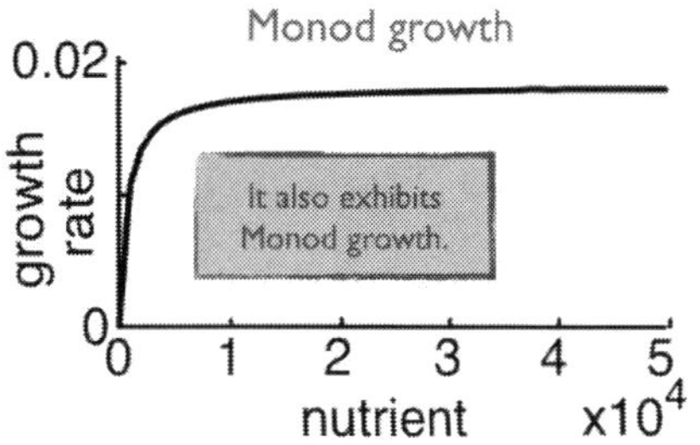

2 linear growth relations

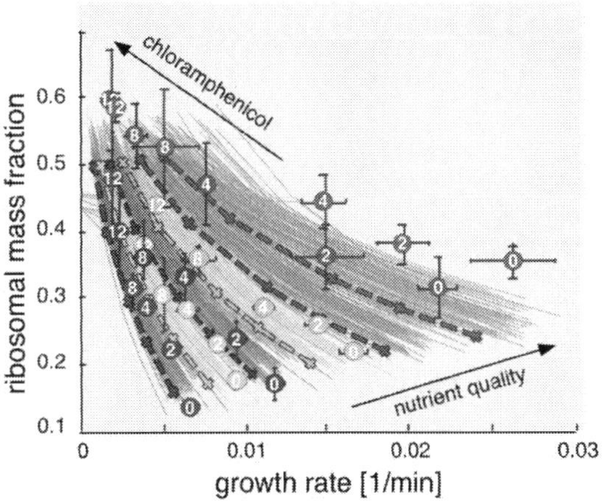

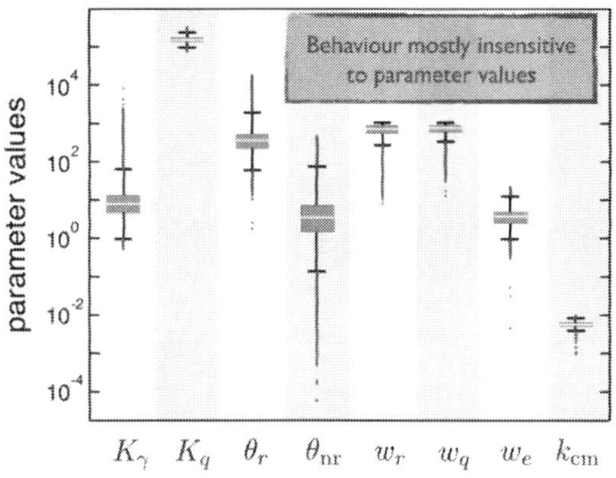

We can derive the empirical growth relations analytically.

1. When varying nutrient conditions

$$\lambda = \frac{1}{\tau_\gamma}(\phi_R - \phi_r)$$

(ϕ_R: mass fraction translating ribosomes; τ_γ: time to translate a ribosome)

2. When inhibiting translation

$$\lambda \simeq \frac{1}{\tau_e}(1 - \phi_q - \phi_R) \cdot \frac{s}{K_t + s}$$

(ϕ_q: housekeeping load; ϕ_R: ribosomes; τ_e: time to import energy needed to make enzymes; $\frac{s}{K_t + s}$: constant environment)

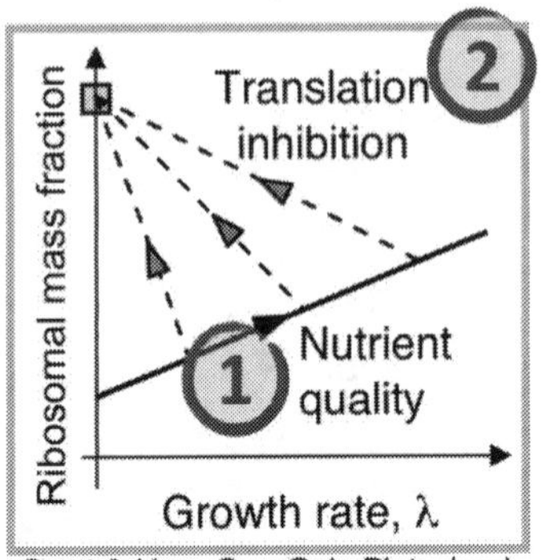

Scott & Hwa, Curr Opin Biotechnol, 2011

3. When changing amounts of external nutrient

$$\lambda \simeq \frac{(1 - \phi_q)s}{K_t\tau_e + (\tau_e + \tau_\gamma)s}$$

(ϕ_q: housekeeping load; K_t: import threshold; $(\tau_e + \tau_\gamma)$: enzyme & ribosome efficiency)

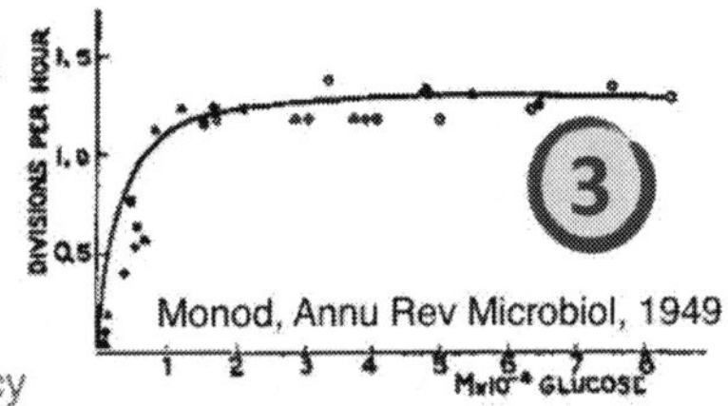

Monod, Annu Rev Microbiol, 1949

We can use the model for **host-aware** design in synthetic biology.

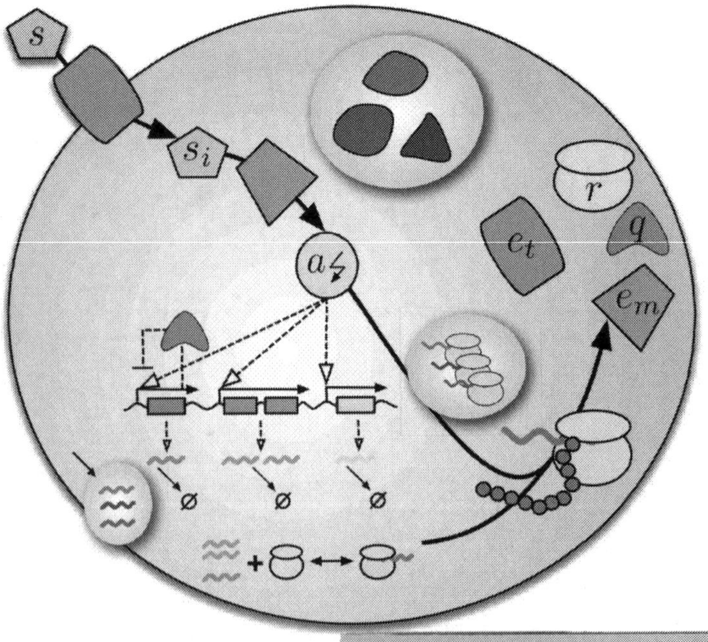

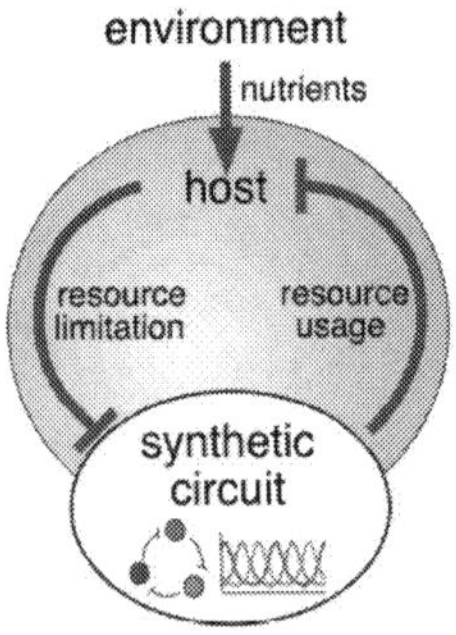

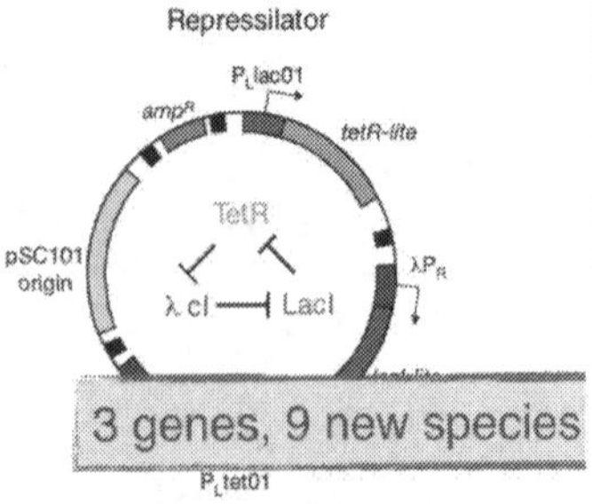

3 genes, 9 new species

Other trade-offs (e.g. degradation, TF-competition ... Elowitz & Leibler, Nature 2000

Levels of synthetic proteins oscillate while cellular components are stable.

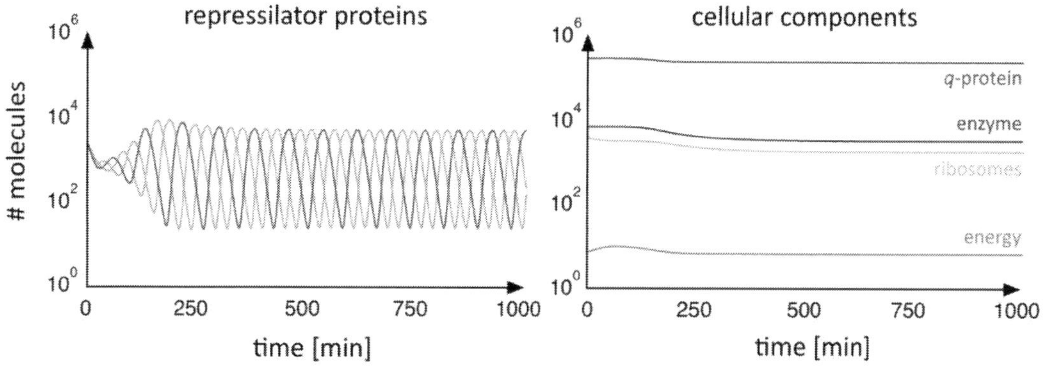

We can quantify the three-way interaction of circuit, host and environment.

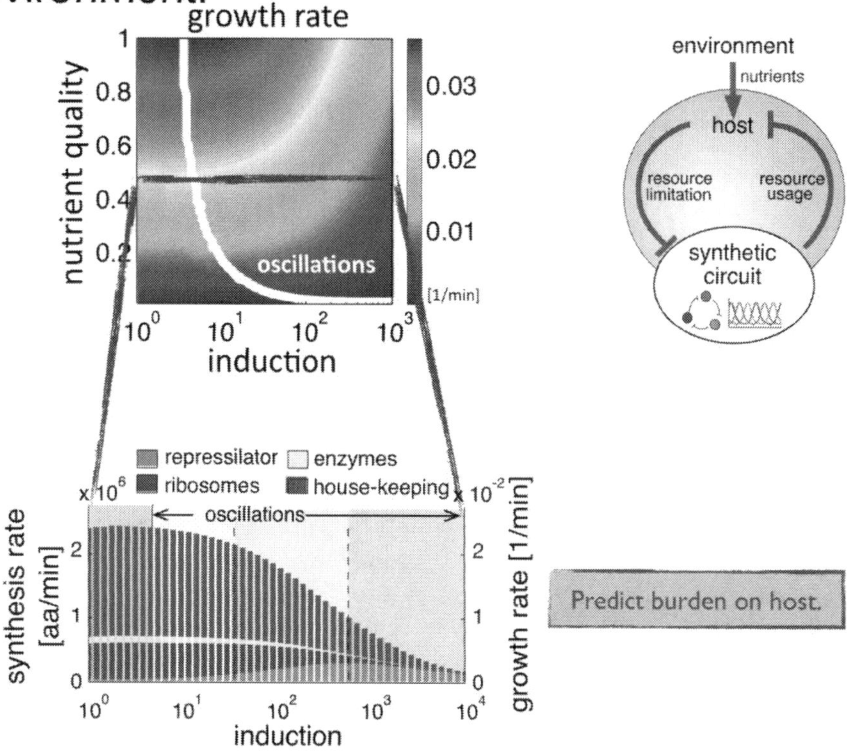

Predict burden on host.

We can quantify the effect of host-physiology on the circuit function.

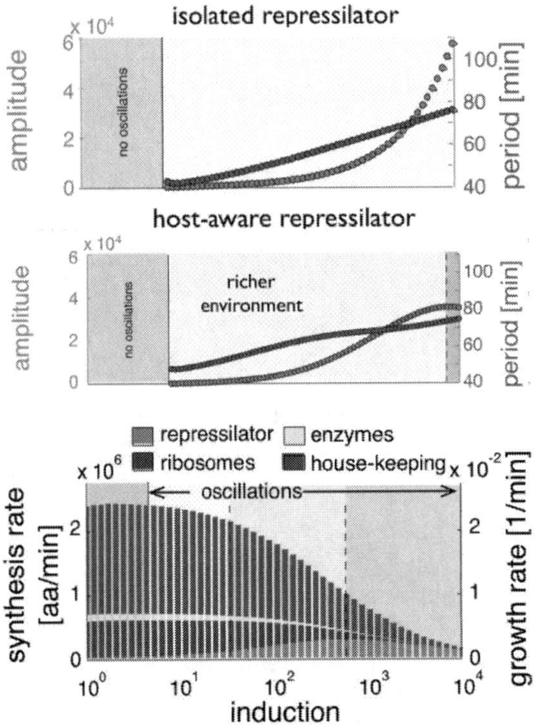

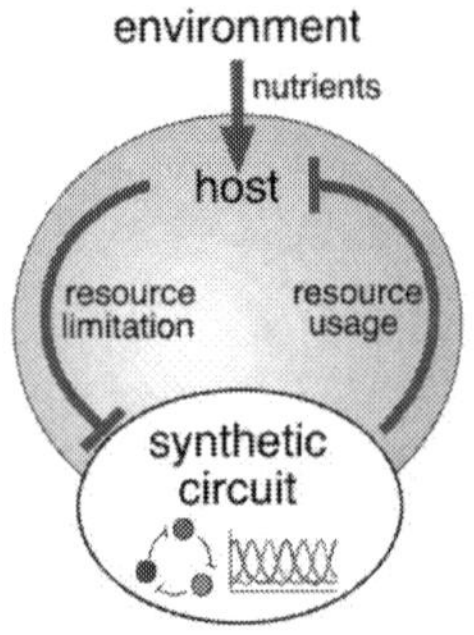

In summary

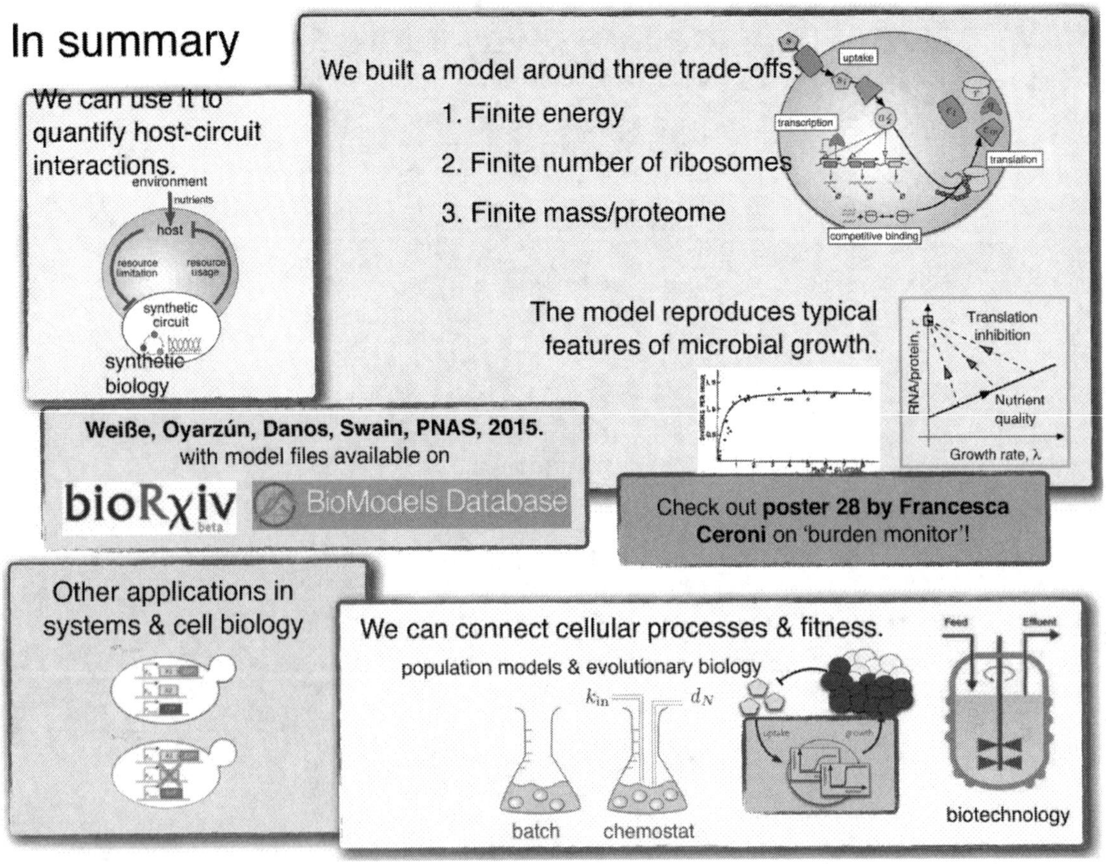

Many thanks to

SynthSys - University of Edinburgh

Peter Swain

Vincent Danos

Imperial College London

Diego Oyarzún

Biomathematics group

and also thanks to

Chris Josephides (SynthSys)

Guillaume Terradot (CRI, Paris)

Philipp Thomas (SynthSys)

'eiße, Oyarzún, Danos, Swain, PNAS, 2015.

andrea.weisse@ed.ac.uk

THE UNIVERSITY of EDINBURGH SynthSys SYNTHETIC & SYSTEMS BIOLOGY DFG

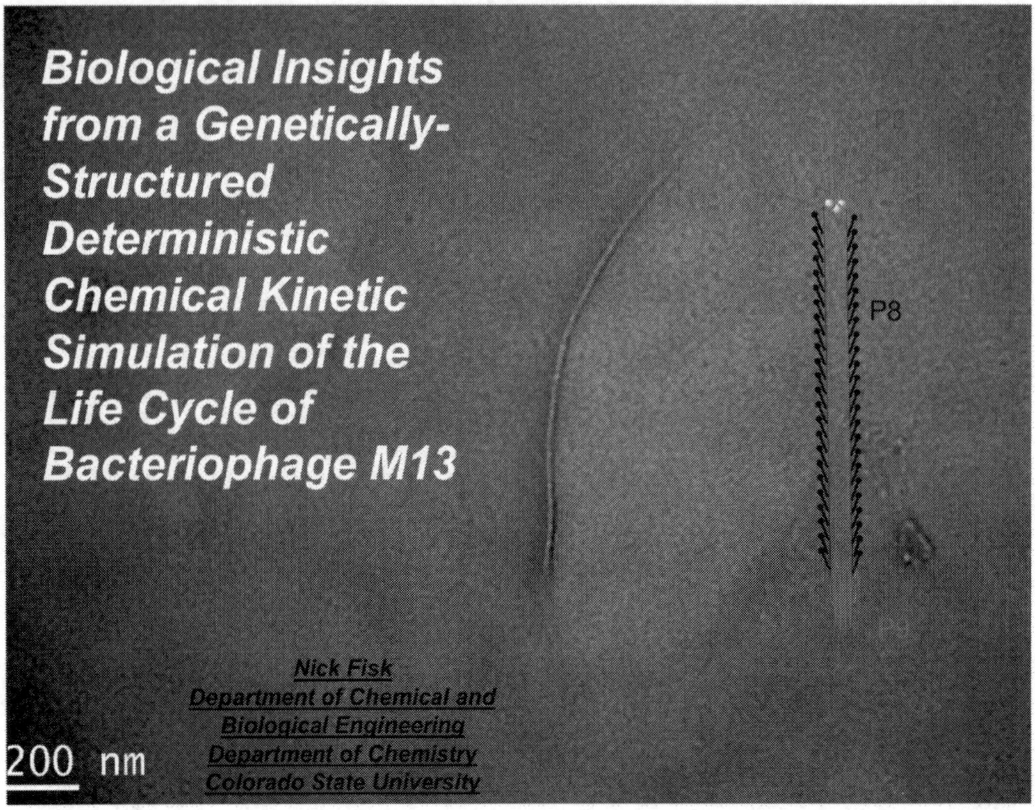

Filamentous Phage Sensors, Diagnostics, and Materials

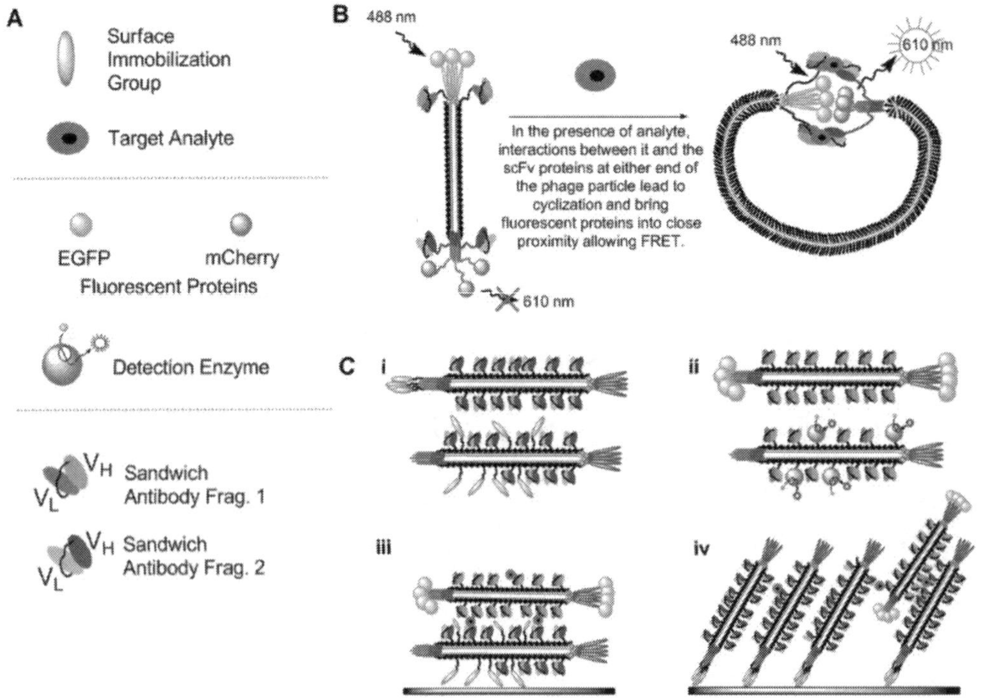

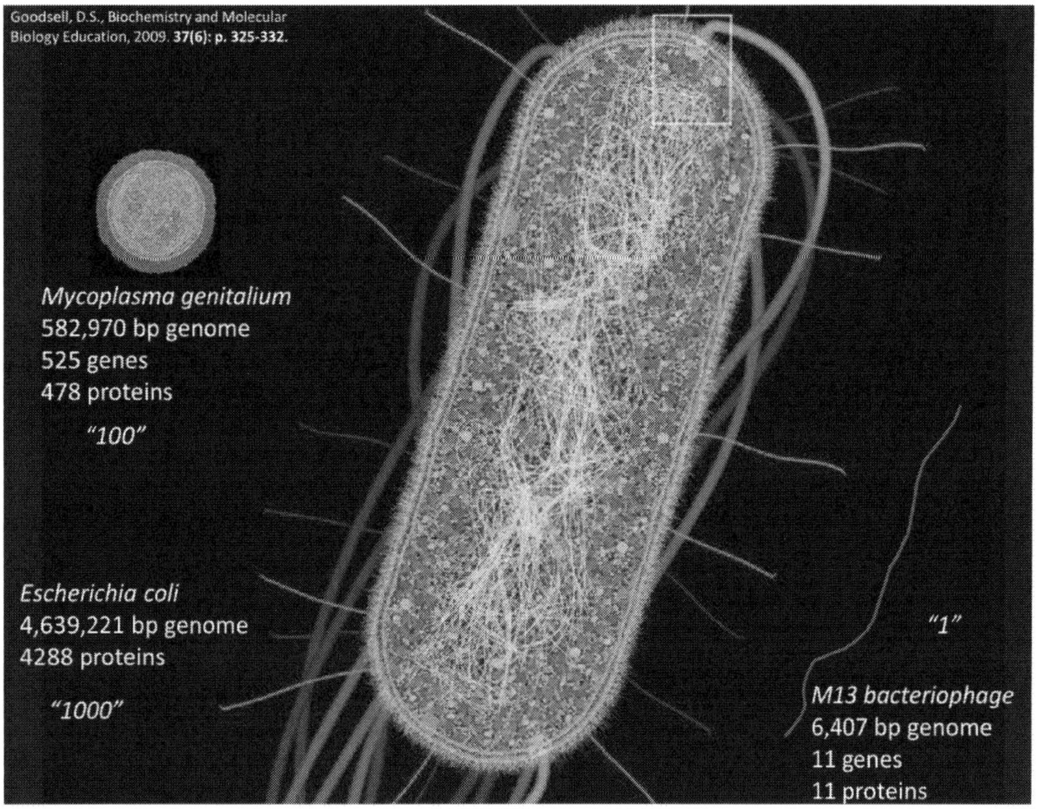

The M13 Life Cycle

1) Describe the M13 bacteriophage life cycle
2) Genetically structured chemical kinetic model
3) Insights into control of the phage life cycle

REFERENCES

M13 Biology Reviews:
Model, P.; Russel, M., in *The Bacteriophages*, R. Calendar, Ed. 1988, Plenum. p. 375-456.

Rakonjac, J.; et.al. Curr. Issues Mol. Biol., 2011. **13**(2): p. 51-75.

Phage DNA synthesis:
Meyer, T. F.; Geider, K., Nature, 1982. **296**(5860): p. 828-832.

Phage mRNA synthesis:
Cashman, J. S.; Webster, R. E.; Steege, D. A., JBC, 1980. **255**(6): p. 2554-2562.

Phage Protein Levels:
Salis, H. M.; Mirsky, E. A.; Voigt, C. A., Nature Biotechnology, 2009. **27**(10): p. 946.

Alberts, B.; Frey, L.; Delius, H., JMB, 1972. **68**(1): p. 139.

P5 Attenuation:
Model, P.; McGill, C.; Mazur, B.; Fulford, W. D., Cell, 1982. **29**(2): p. 329-335.

Yen, T. S. B.; Webster, R. E., Cell, 1982. **29**(2): p. 337-346.

Zaman, G. J. R.; et. Al.., Biochimica et Biophysica Acta, 1991. **1089**: p. 183-192.

Carrier State
Merriam, V., Journal of Virology, 1977. **21**(3): p. 880-888.

Lerner, T. J.; Model, P., Virology, 1981. **115**(2): p. 282-294

Ploss, M.; Kuhn, A., Physical Biology, 2010. 7(4).

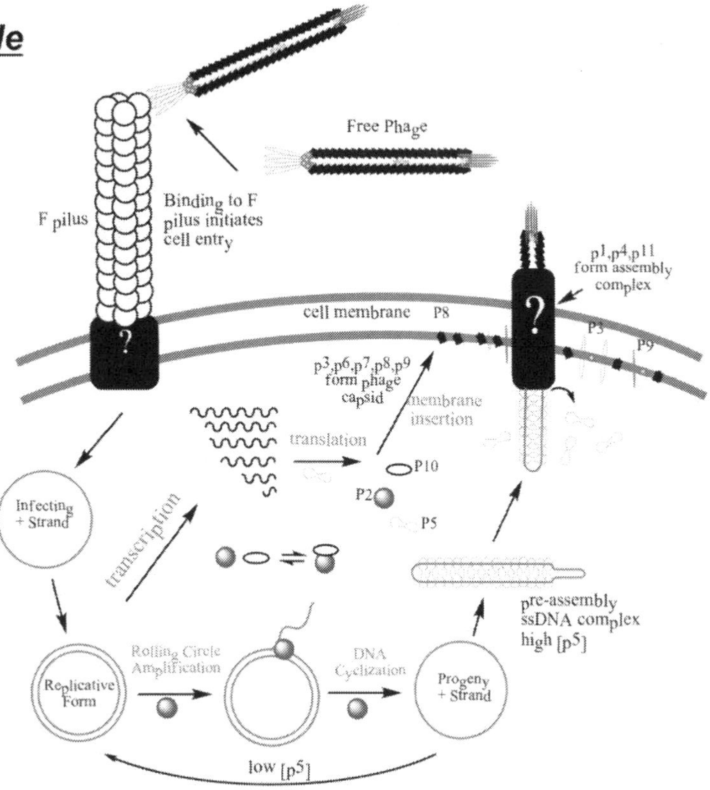

The M13 Life Cycle: Cartoon to Quantitative Simulation

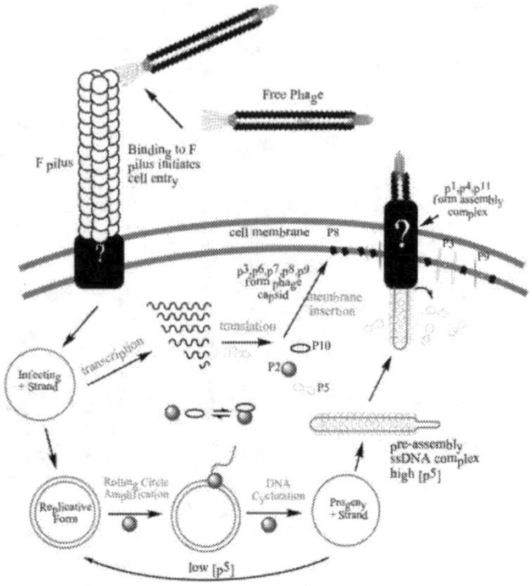

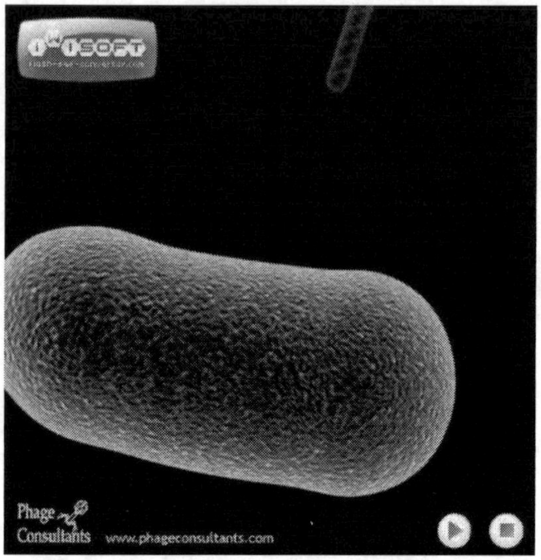

Genetically Structured Chemical Kinetic Simulation

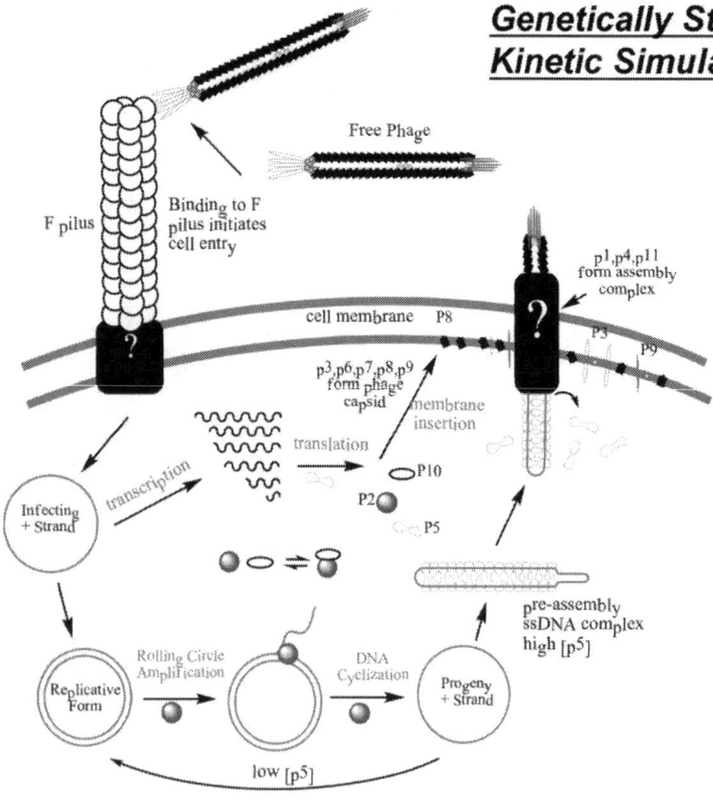

Experimental Biochemistry & Biology

Organized Structure of Biology

Differential Equations

Quantitative Predictive Model of Biology

Synthetic Biological Systems with Designed Functions

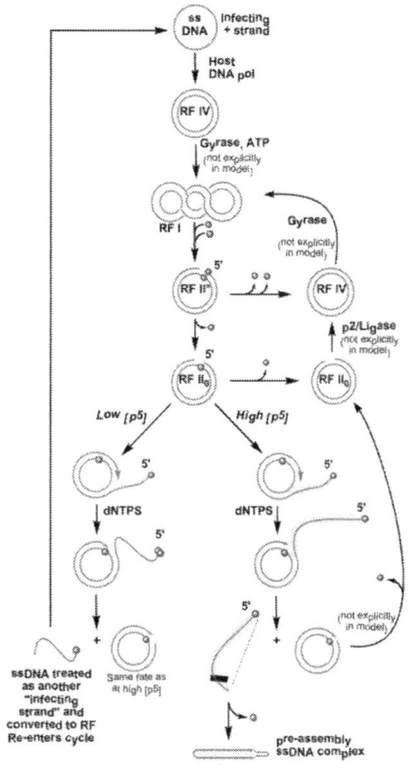

Reaction and Rate Equations

ssDNA to RF I

$$\text{ssDNA} + \text{DP3} \rightarrow \text{ssPDNA} \qquad [\text{R1}]$$

$$\text{Rate} = C_1 * \text{sDNA} * \text{DP3} \qquad [\text{E1}]$$

$$\text{ssPDNA} \rightarrow \text{RF1} + \text{DP3} + \text{DA} + \text{DB} + \text{DH} + \text{DZ} + \text{DW} \qquad [\text{R2}]$$

$$\text{Rate} = C_2 * \text{ssPDNA} \qquad [\text{E2}]$$

$$\text{RF1} + \text{P2} \rightleftarrows \text{RF2} \qquad [\text{R3}]$$

$$\text{Rate} = C_3 * \text{RF1} * \text{P2} - C_4 * \text{RF2} \qquad [\text{E3}]$$

Rolling Circle Replication

$$\text{RF2} + \text{DP3} + \text{DA} + \text{DB} + \text{DH} + \text{DZ} + \text{DW} \rightarrow \text{RF2DP3}$$

$$\text{Rate} = C_1 * \text{RF2} * \text{DP3}$$

$$\text{RF2DP3} \rightarrow \text{P2} + \text{DP3} + \text{RF1} + \text{ssDNA} + \text{DA} + \text{DB} + \text{DH} + \text{DZ} + \text{DW}$$

$$\text{Rate} = C_2 * \text{RF2DP3}$$

Utility of Life Cycle Models of Phage Systems

► Evaluate the Consistency and Completeness of the Current Understanding of the Phage

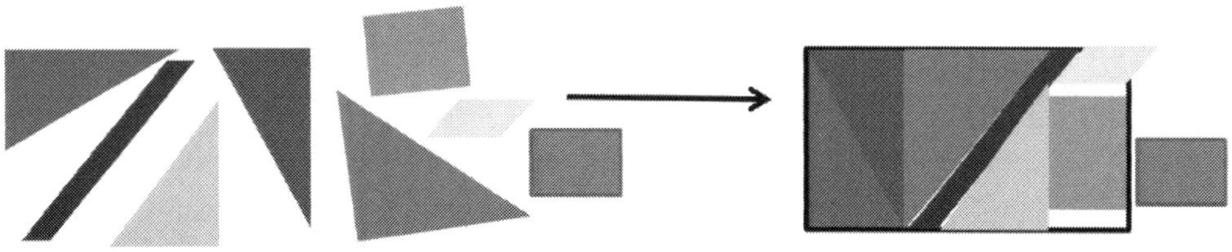

► Characterize the Emergent Properties of the Integrated System

► Test Complex Hypothesis (where experiments are impractical)

► Guiding Experiments and Providing a Framework for Understanding Results

Rich history of Bacteriophage Simulations

λ

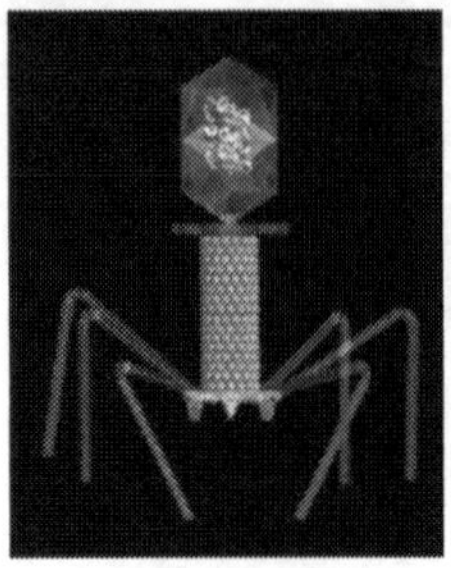

McAdams, H. H.; Shapiro, L., *Circuit Simulation of Genetic Networks.* Science, 1995. **269(5224): p. 650-656.**

Heidtke, K. R.; Schulze-Kremer, S., *Design and Implementation of a Qualitative Simulation Model of Lambda Phage Infection.* Bioinformatics, 1998. **14(1): p. 81-91.**

Arkin, A.; Ross, J.; McAdams, H. H., *Stochastic Kinetic Analysis of Developmental Pathway Bifurcation in Phage Lambda-Infected Escherichia Coli Cells.* Genetics, 1998. **149(4): p. 1633-1648.**

Tian, T. H.; Burrage, K., *Bistability and Switching in the Lysis/Lysogeny Genetic Regulatory Network of Bacteriophage Lambda.* Journal of Theoretical Biology, 2004. **227(2): p. 229-237.**

T7

Endy, D.; Kong, D.; Yin, J., *Intracellular Kinetics of a Growing Virus: A Genetically Structured Simulation for Bacteriophage T7.* Biotechnology and Bioengineering, 1997. **55**(2): p. 375-389.

Endy, D.; You, L. C.; Yin, J.; Molineux, I. J., *Computation, Prediction, and Experimental Tests of Fitness for Bacteriophage T7 Mutants with Permuted Genomes.* PNAS, 2000. **97**(10): p. 5375-5380.

You, L. C.; Suthers, P. F.; Yin, J., *Effects of Escherichia Coli Physiology on Growth of Phage T7 In Vivo and in Silico.* Journal of Bacteriology, 2002. **184(7): p. 1888-1894.**

Chan, L. Y.; Kosuri, S.; Endy, D., *Refactoring Bacteriophage T7.* Molecular Systems Biology, 2005. **1.**

And other virus and phage systems:

HIV, Vesicular Stomatitis Virus, Influenza, phage MS2, phage Q-beta

The Simulation by the Numbers

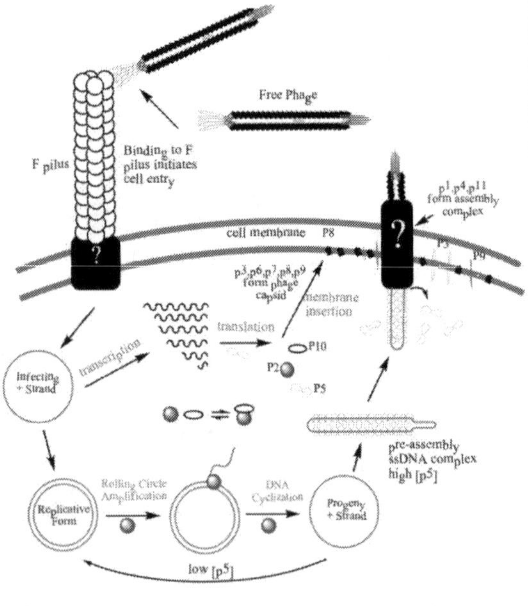

Process	No. of Eqns	No. of Kinetic Parameters	No. of Unknown Kinetic Parameters
DNA Replication	7	7	4
Transcription	15	11	1
mRNA Degradation	15	10	3
Translation	29	25	1
Assembly	5	5	4
Total	**71**	**58**	**13**

10

Control Mechanisms in the Persistently Infected State

- Filamentous phage are unique among bacteriophages in that their life cycle does not at any stage involve lysis of host cell.
- Phage are continuously produced as the host cell grows and divides
- How does the phage control the levels of genomes and proteins as its host grows and divides?

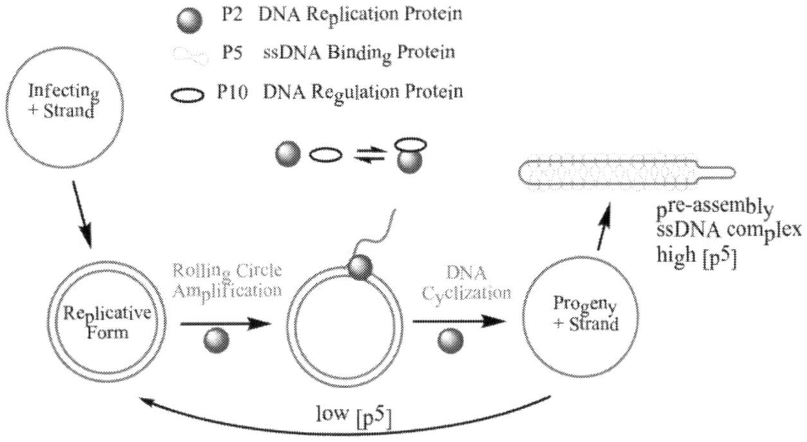

Control Phage Protein Levels:

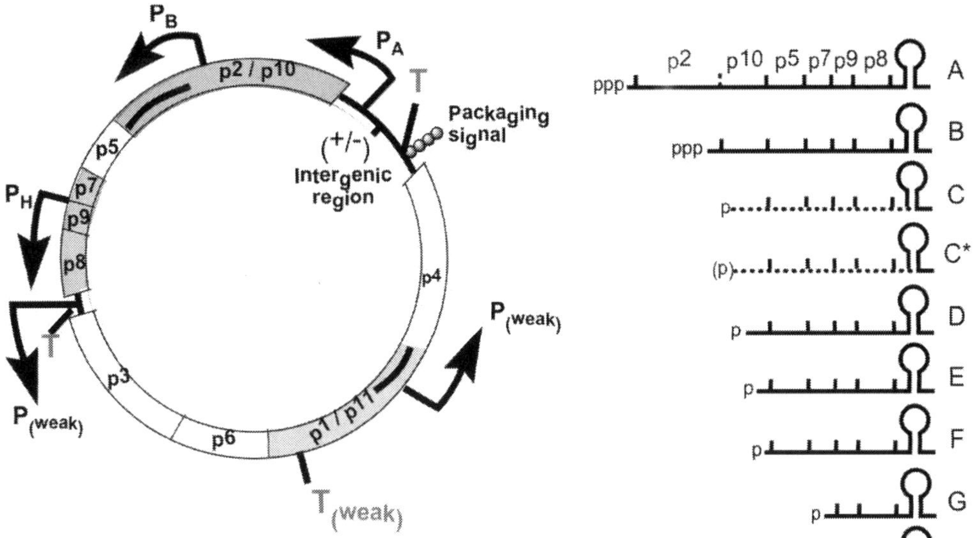

- ►Transcription
- ►mRNA processing and degradation
- ►Translational Repression

Cashman, J. S.; Webster, R. E.; Steege, D. A. Journal of Biological Chemistry, 1980. **255**(6): p. 2554-2562.

The Persistently Infected State is not a Steady State

- Careful measurements show that the percentage of infected cells decreases over time
- After a initial period of stability the number of infected cells decreases with increasing cell growth
- How does the phage control the levels of genomes and proteins as its host grows and divides?

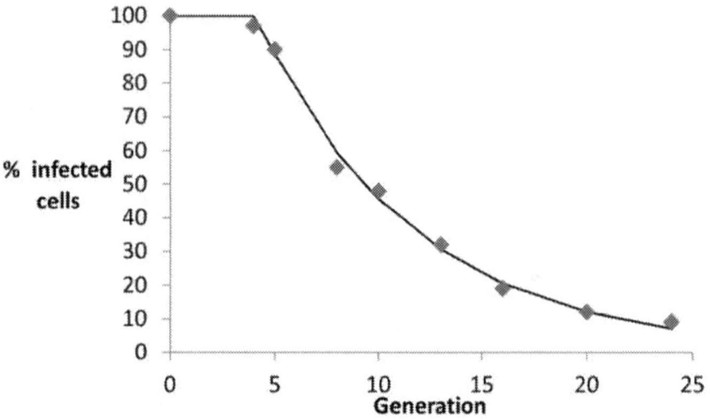

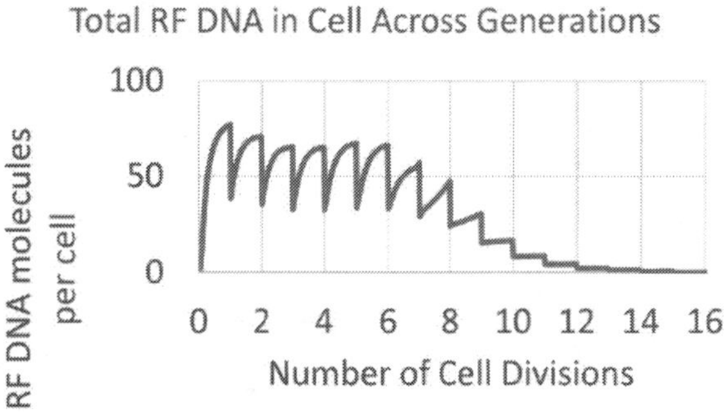

Merriam, V., *Stability of Carrier State in Bacteriophage-M13-Infected Cells*. Journal of Virology, 1977. **21**(3): p. 880-888.

Proposed Role for Translational Repression: Control of Replicative Protein Levels

Translational repression and p10 regulation control the levels of phage DNA replication

Experiments suggest that translational repression control of P2 is unimportant in phage replication and absent in related bacteriophage

Our model agrees with the conclusion that P2 translational repression is unimportant to the phage but finds a role for translational repression in controlling phage resource allocation. In particular production of p5 versus p8 late in infection.

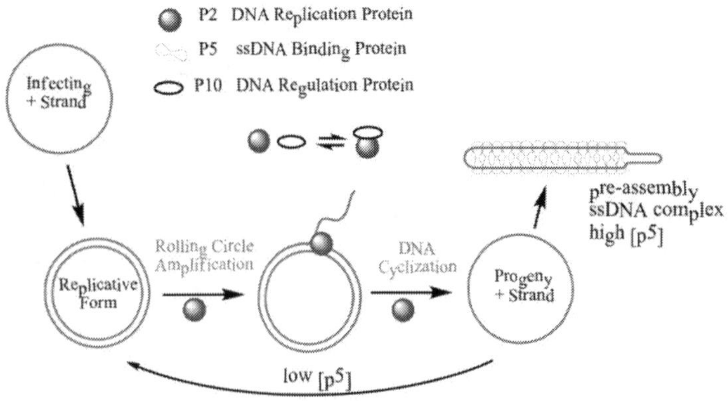

The Simulated Course of the Persistently Infected State Under Different Control Regimes

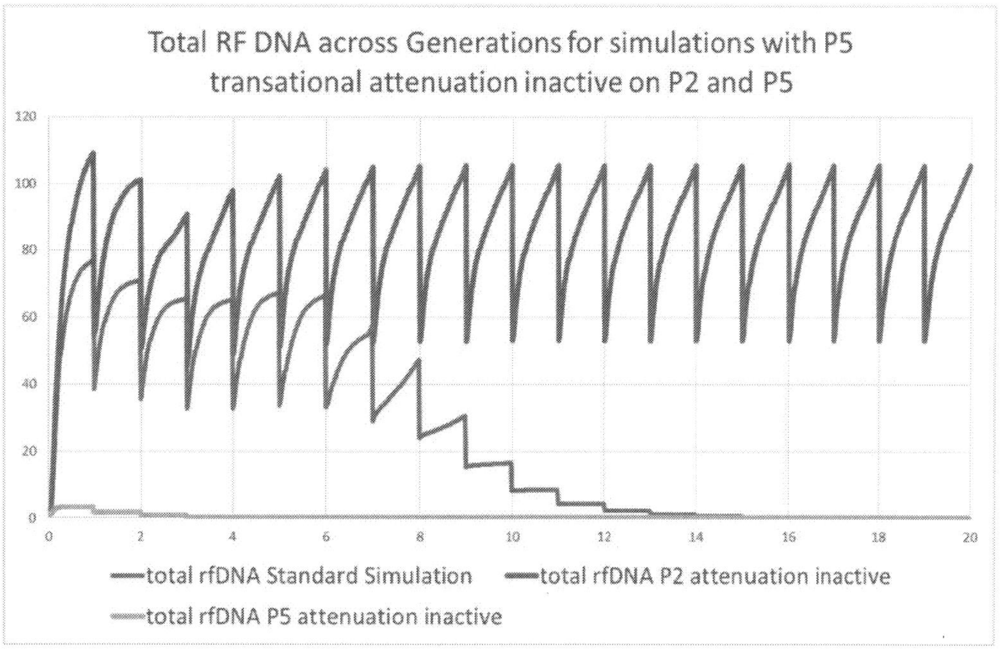

Conclusions

► Developed a chemical kinetic simulation of M13 life cycle, parameterized as much as possible from experimental measurements of phage biology

► The Simulation reproduces many aspects of quantitative phage biology and provides a framework for exploring the connections between disparate components of phage biology

► Simulation provides insights into functions of control mechanisms that suggest specific experiments

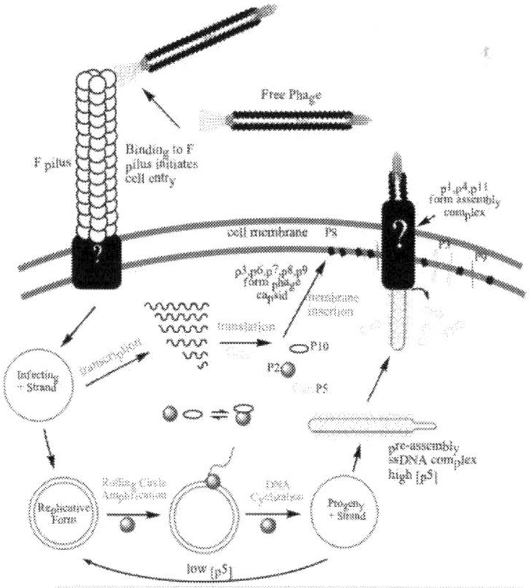

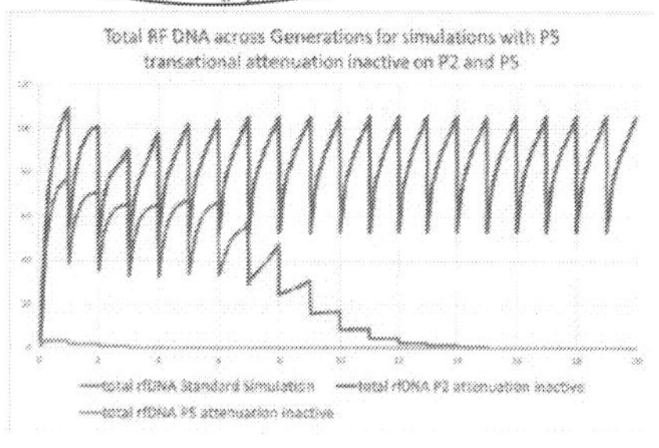

Acknowledgements:

Steven Smeal

Dr. Meg Schmitt

Ashok Prasad

Ning Zhao   Wil Biddle
Dave Schwark   Jen Bjerke
Zach Butz

Sam Saccamano   Gilman Plitt
Colton Feller   Jeff Eichman
Ryan Schwark   Bret Beaversted

BILL & MELINDA GATES foundation

BOETTCHER FOUNDATION

Colorado Office of Economic Development and International Trade

Colorado State University

NSF

Conclusions

► Developed a chemical kinetic simulation of M13 life cycle, parameterized as much as possible from experimental measurements of phage biology

► The Simulation reproduces many aspects of quantitative phage biology and provides a framework for exploring the connections between disparate components of phage biology

► Simulation provides insights into functions of control mechanisms that suggest specific experiments

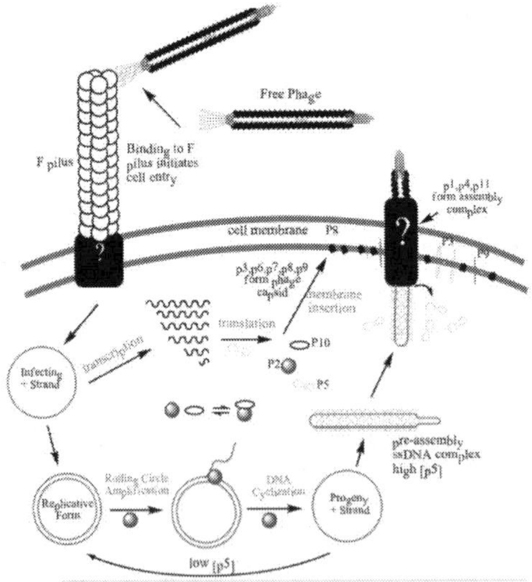

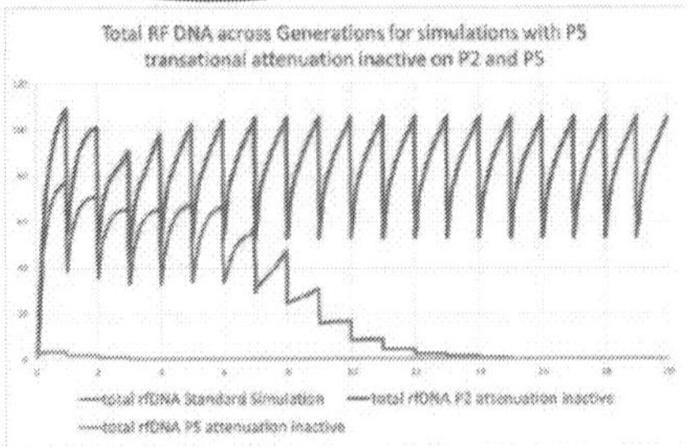

REFERENCES

M13 Biology Reviews:
Model, P.; Russel, M., *Filamenteous Bacteriophages*, in *The Bacteriophages*, R. Calendar, Ed. 1988, Plenem. p. 375-456.

Rakonjac, J.; Bennett, N. J.; Spagnuolo, J.; Gagic, D.; Russel, M., *Filamentous Bacteriophage: Biology, Phage Display and Nanotechnology Applications*. Curr. Issues Mol. Biol., 2011. **13**(2): p. 51-75.

Kay, B. K.; Winter, J.; McCafferty, J. D., *Phage Display of Peptides and Proteins a Laboratory Manual*. 1996, Academic Press.

Phage DNA synthesis:
Meyer, T. F.; Geider, K., *Enzymatic-Synthesis of Bacteriophage-Fd Viral-DNA*. Nature, 1982. **296**(5860): p. 828-832.

Phage mRNA synthesis:
Cashman, J. S.; Webster, R. E.; Steege, D. A., *Transcription of Bacteriophage-Fl - the Major In vivo RNAs*. Journal of Biological Chemistry, 1980. **255**(6): p. 2554-2562.

Phage Protein Levels:
Salis, H. M.; Mirsky, E. A.; Voigt, C. A., *Automated Design of Synthetic Ribosome Binding Sites to Control Protein Expression*. Nature Biotechnology, 2009. **27**(10): p. 946.

Alberts, B.; Frey, L.; Delius, H., *Isolation and Characterization of Gene 5 Protein of Filamentous Bacterial Viruses*. Journal of Molecular Biology, 1972. **68**(1): p. 139.

P5 Attenuation:
Model, P.; McGill, C.; Mazur, B.; Fulford, W. D., *The Replication of Bacteriophage-F1 - Gene-V Protein Regulates the Synthesis of Gene-II Protein*. Cell, 1982. **29**(2): p. 329-335.

Yen, T. S. B.; Webster, R. E., *Translational Control of Bacteriophage-F1 Gene-II and Gene-X Proteins by Gene-V Protein*. Cell, 1982. **29**(2): p. 337-345.

Zaman, G. J. R.; Smetsers, A.; Kaan, A.; Schoemakers, J. G. G.; Konings, R. N. H., *Regulation of Expression of the Genome of Bacteriophage M13. Gene V Protein Regulated Translation of The mRNAs Encoded by Genes I, III, V and X*. Biochimica et Biophysica Acta, 1991. **1089**: p. 183-192.

Zaman, G. J. R.; Kaan, A. M.; Schoenmakers, J. G. G.; Konings, R. N. H., *Gene-V Protein-Mediated Translational Regulation of the Synthesis of Gene-II Protein of the Filamentous Bacteriophage-M13 - a Dispensable Function of the Filamentous-Phage Genome*. Journal of Bacteriology, 1992. **174**(2): p. 595-600.

Carrier State
Merriam, V., *Stability of Carrier State in Bacteriophage-M13-Infected Cells*. Journal of Virology, 1977. **21**(3): p. 880-888.

Lerner, T. J.; Model, P., *The Steady-State of Coliphage-F1 - DNA-Synthesis Late in Infection*. Virology, 1981. **115**(2): p. 282-294.

Ploss, M.; Kuhn, A., *Kinetics of Filamentous Phage Assembly*. Physical Biology, 2010. **7**(4).

Rates of Phage DNA Synthesis

The model produces concentration vs time profiles for the phage species and intermediates involved in the M13 life cycle.

• The model's predictions of intracellular levels of phage DNA, mRNA and proteins
match the order of magnitude and are generally within 5-fold of the experimentally
reported values.

• The experimentally reported values typically have large ranges, most likely
because of the dynamic nature of phage infection.

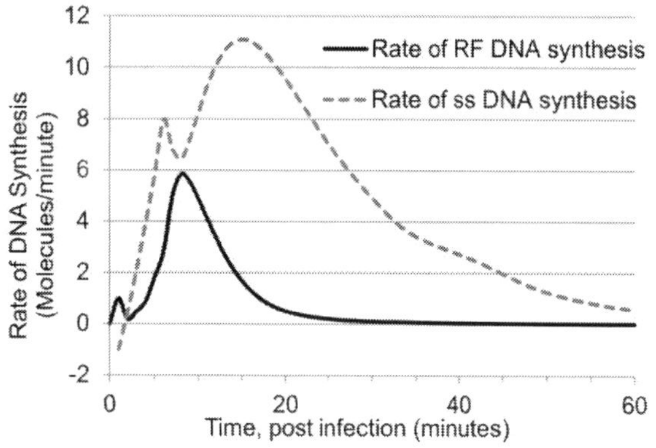

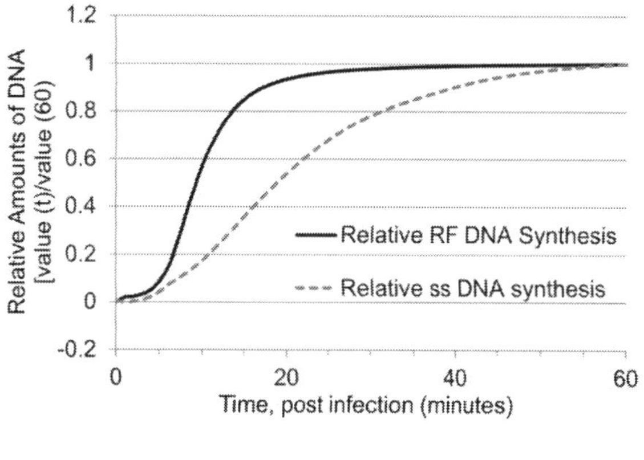

Time Course of Phage Protein Synthesis

• The model's predictions of intracellular levels of phage DNA, mRNA and proteins match the order of magnitude and are generally within 5-fold of the experimentally reported values.

• The experimentally reported values typically have large ranges, most likely because of the dynamic nature of phage infection.

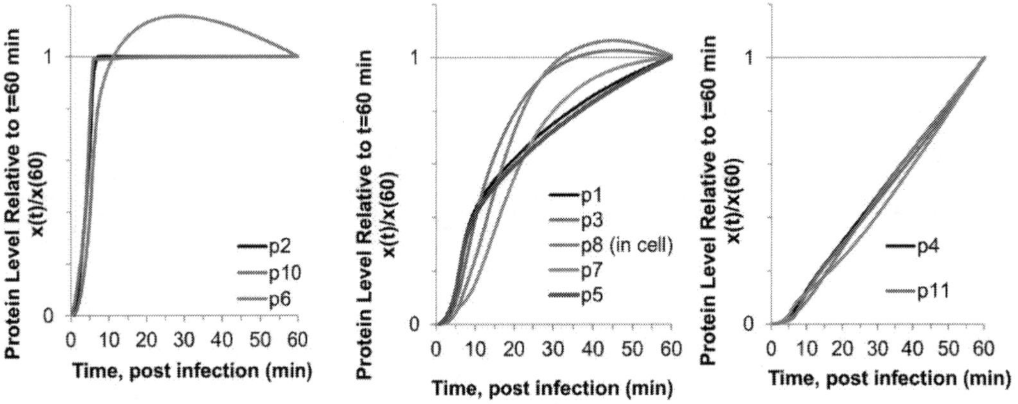

Rates of Phage DNA Synthesis

• The model's predictions of intracellular levels of phage DNA, mRNA and proteins
match the order of magnitude and are generally within 5-fold of the experimentally
reported values.

• The experimentally reported values typically have large ranges, most likely
because of the dynamic nature of phage infection.

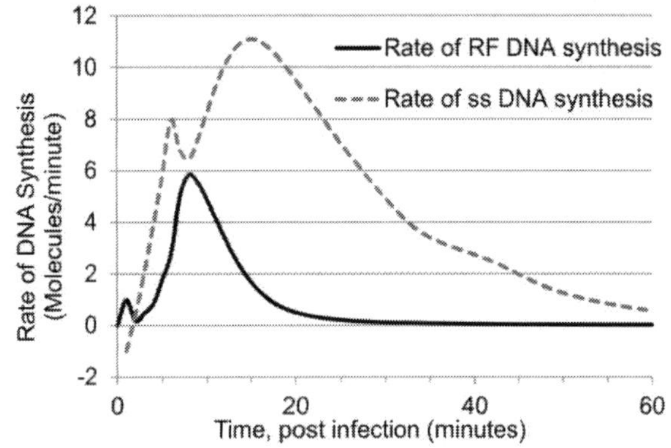

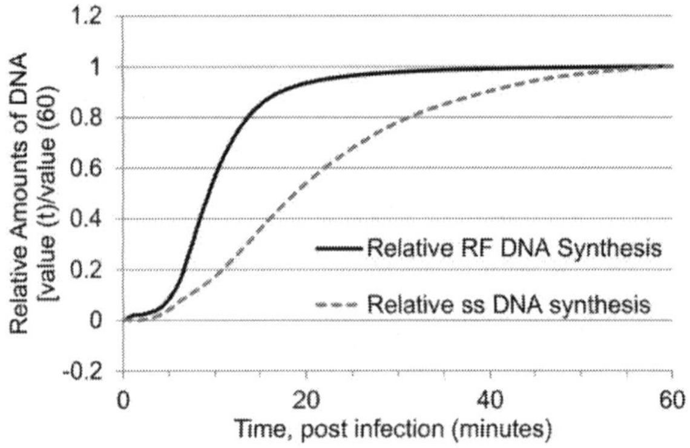

Effect of Translational Attenuation on Phage

• P5 attenuation is a feedback mechanism that involves binding to mRNA to repress translation

• P5, the ssDNA binding protein, represses the translation of p2, p10, p5, p3, and p1.

p5 Attenuation	Phage	dsDNA	p5DNA
Std Simulation (all proteins ON)	1.00	1.00	1.00
No Attenuation	0.00	0.02	0.00
p2 Off	0.77	1.20	1.80
p10 Off	0.00	0.02	0.00
p5 Off	0.51	0.19	1.85
p3 Off	1.00	1.01	1.01
p1 Off	1.00	1.00	1.00
Reported Values *	**N/A**	**N/A**	**N/A**

p5 Attenuation	p1	p2	p3	p4	p5	p6	p7	p8	p9	p10	p11
Std Simulation (all proteins ON)	1.00	1.00	1.00	1.00	1.00	1.00	1.00	1.00	1.00	1.00	1.00
No Attenuation	2.80	21.3	3.52	0.90	2.11	5.11	2.29	11.0	0.85	123	1.77
p2 Off	0.78	55.1	1.10	0.77	1.72	0.55	1.84	1.63	0.80	1.45	0.73
p10 Off	1.24	0.42	0.70	1.24	0.17	10.4	0.15	15.5	1.54	219	2.52
p5 Off	0.41	0.95	0.13	0.78	4.26	1.78	4.75	0.03	0.52	0.89	1.21
p3 Off	0.98	1.00	2.38	0.97	1.01	0.96	1.01	0.47	0.97	1.00	0.97
p1 Off	1.21	1.00	1.00	1.00	1.00	1.00	0.99	0.97	1.00	1.00	1.00
Reported Values*	**6.5**	**20.2**	**18.7**	**N/A**	**3.6**	**N/A**	**N/A**	**N/A**	**N/A**	**9.0**	**N/A**

Phage Life Cycle Across Multiple Cell Generations

Unlike the majority of other bacterial viruses, infection with M13 does not kill the cell

• Phage particles are continuously produced and extruded through the cell membrane as the infected cell continues to grow and divide

• Persistently infected state evolves through cell generations

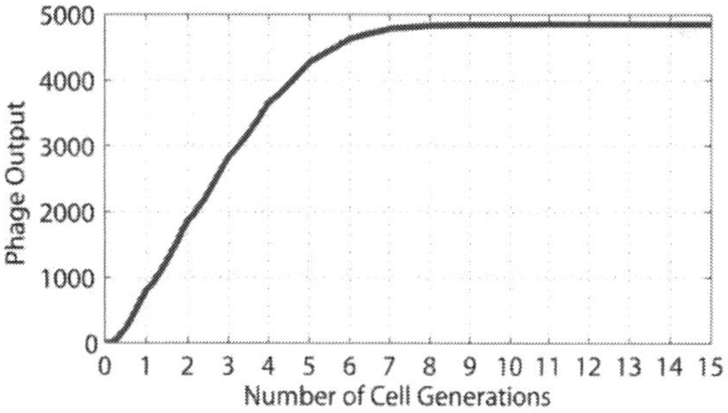

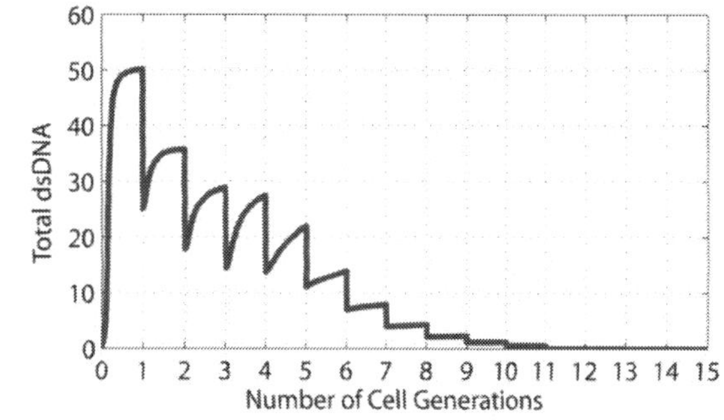

Spatial and Quantitative Optimization of Engineered Multi-Element Therapeutic Proteins

Way, J. C., Harvard University
Burrill, D. R., Harvard University
Robinson-Mosher, A., Harvard Medical School
Collins, J. J., Massachusetts Institute of Technology
Silver, P. A., Harvard Medical School
Originally presented on: 6/10/2015 14:00:00 - 14:30:00

The current paradigm of drug development is to create single-function molecules that bind tightly to their targets and inhibit their function. In contrast, natural processes, including defense against pathogens and disease, involve many elements – multiple protein domains, weak interactions that function cooperatively, geometrically optimized elements, and so on. The spatial and quantitative aspects of such systems are tweaked in the course of evolution to create defenses that look quite different from the simple molecules designed by humans. Our goal is to create complex therapeutic proteins and cells that mimic natural systems. We have engineered proteins that deliver an activating hormone or cytokine, either erythropoietin or interferon alpha, to specific target cells. The quantitative aspect is to mutate the cytokine and reduce its activity, so that cell binding is driven by an antibody element to which the cytokine is fused. The spatial aspect is primarily to use a linker that allows simultaneous binding of the cytokine and antibody elements to their receptors, and also maximizes the rate of the cytokine binding after the antibody element has already bound. We also developed course-grained molecular dynamics software to simulate the movement and binding of such proteins, with the goal of optimizing the engineering choices before making the proteins.

Erythropoietin (Epo) is a hormone, produced by the kidneys in response to low blood oxygen, that stimulates red blood cell production. Commercial Epo (e.g. Procrit, Epogen) is used to treat anemia in kidney failure and cancer patients. In 2004, sales were about $10B, but since then the use of Epo has dropped significantly due to clinical trials showing that treated patients suffered from clotting events leading to heart attacks, strokes, and deep vein thrombosis. These effects are due to Epo action on cells other than red blood cell precursors, such as platelet precursors and blood vessel endothelial cells. To avoid these side effects, we constructed a form of Epo that is fused to an antibody and targeted to late red blood cell precursors. Upon treatment of mice, the engineered protein stimulates production of only red blood cells, while controls and commercial forms of Epo stimulate red blood cells and platelets to a similar extent.

Interferon alpha (IFNalpha) is a cytokine that represents the primary response to virus infection. IFNalpha is used in treatment of hepatitis and certain cancers, but a major side effect is 'flu-like symptoms', so that long-term high-dose treatments are burdensome. We generated a fusion protein consisting of a weakened IFNalpha fused to antibody V regions that bind to a tumor-specific surface marker. Animal testing is in progress.

These experiments illustrate how protein engineering can be added to the repertoire of synthetic-biological tools to create novel functions that go beyond what is possible by manipulating transcriptional circuits.

SYNTHETIC GENOMICS®

Engineering Bacteriophage to Address Multidrug-Resistant Pathogens

Synthetic Genomics Seminar Series
Wednesday June 10th 2015

Kyle C. Cady, Ph.D.
kcady@syntheticgenomics.com

SGI Engineering

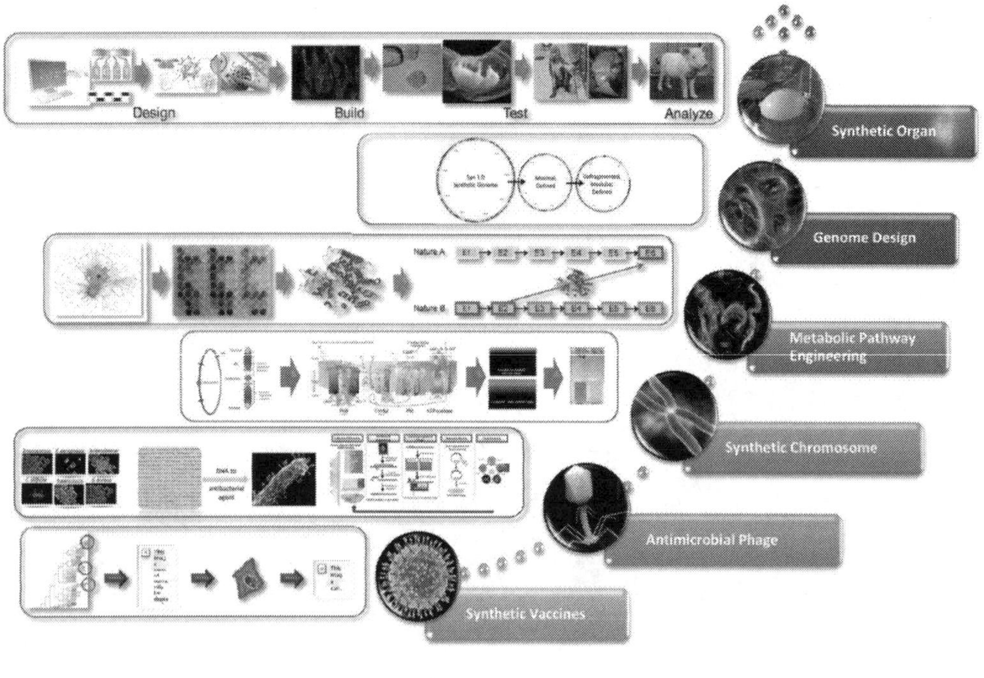

SGI Engineering

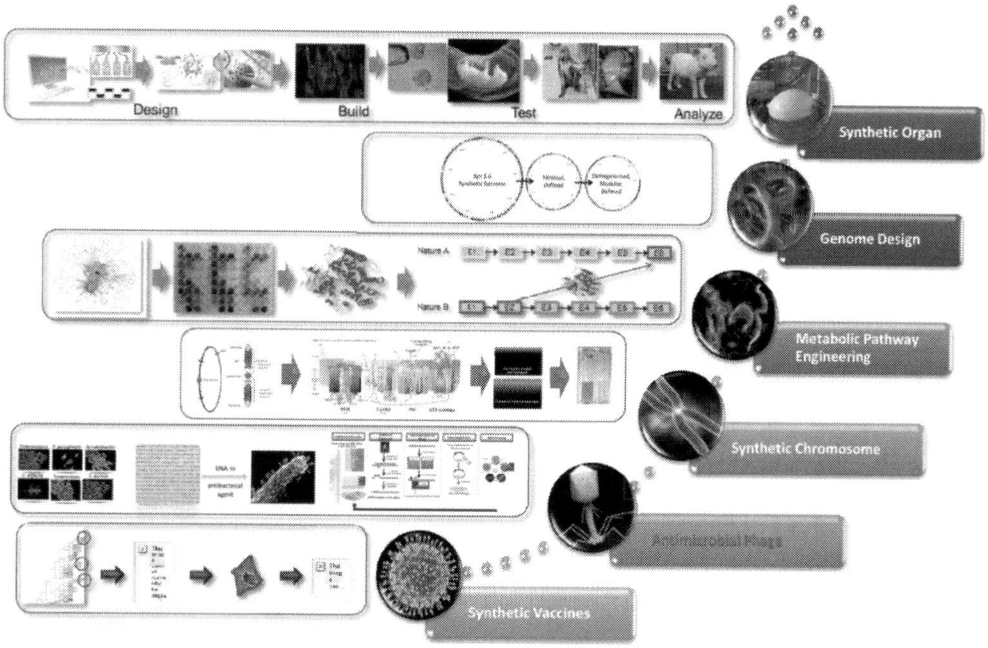

Antimicrobial Resistance – A Global Problem

- Antibiotic resistant pathogens cause >2M illnesses and 23,000 death in US annually
- Economic impact in US estimated at $20-35B/year in excess direct health care costs and additional $35B/year in lost productivity
- Resistance to almost all available antibiotics has been developed by multiple pathogens; some have mortality rates of 30-40%
- Global issue as resistant strains for multiple organisms are spreading around the world with clear evidence that resistance genes are being shared between pathogens
- New antibiotic pipeline is very limited

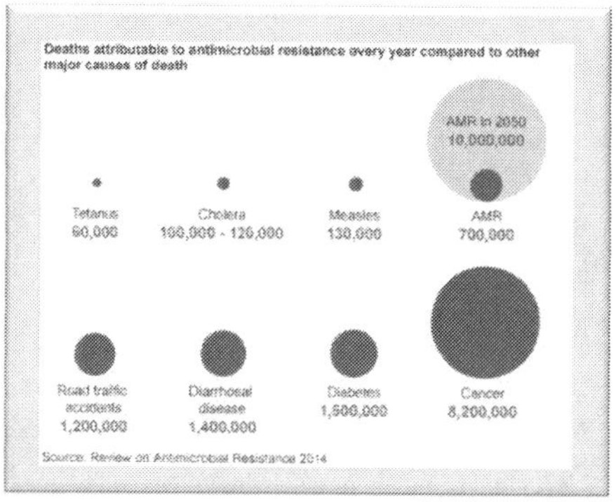

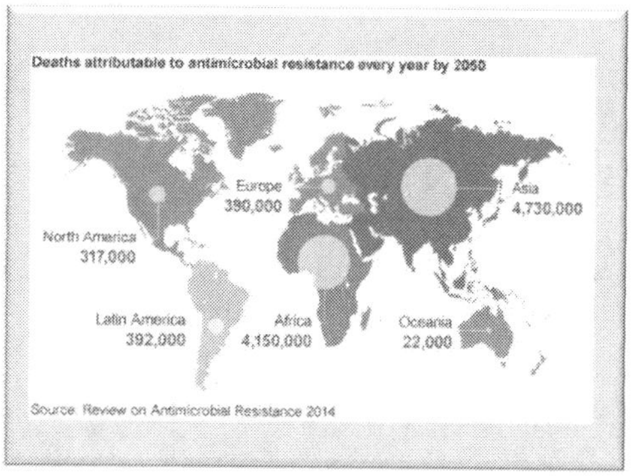

5

Antimicrobial Resistance – R&D Road Block

The discovery dates of distinct classes of antibiotics. No new classes have been discovered since 1987.

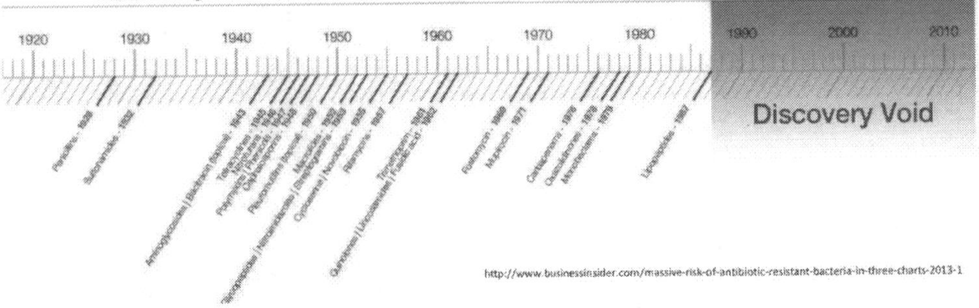

http://www.businessinsider.com/massive-risk-of-antibiotic-resistant-bacteria-in-three-charts-2013-1

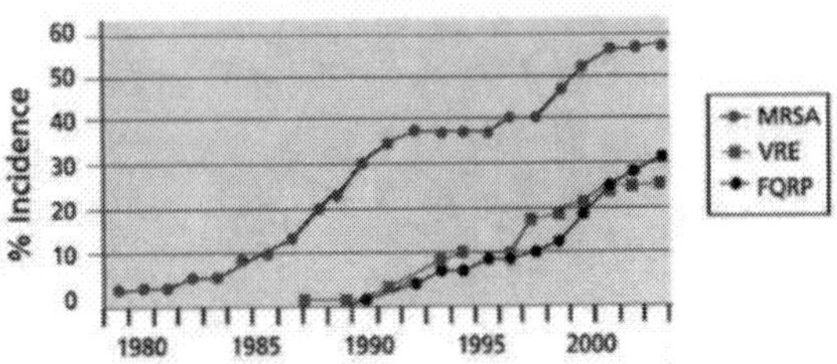

MRSA=methicillin-resistant *S. aureus*
VRE=vancomycin-resistant *enteroccoci*
FQRP=fluoroquinolone-resistant *P. aeruginosa*

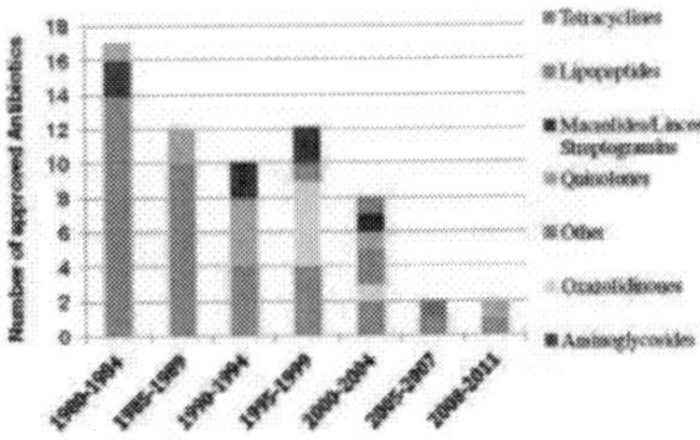

Number of approved antibiotics during the last 30 years. Taken from *Bassetti et al.*, 2013

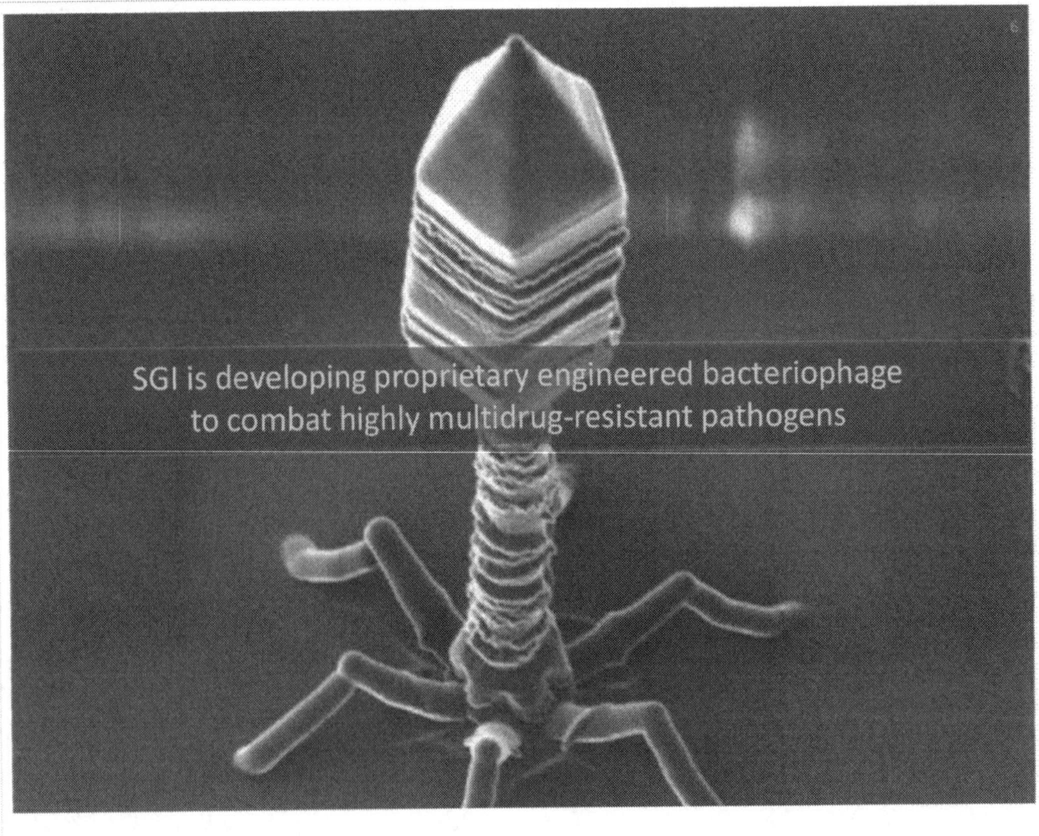

Bacteriophage Therapy

The use of bacterial specific viruses to treat pathogenic bacterial infections

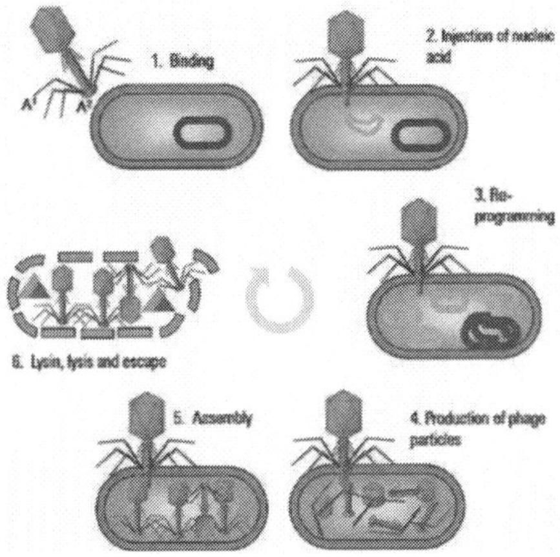

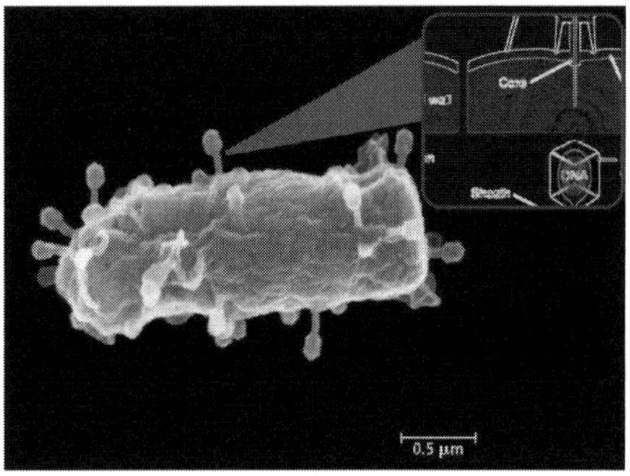

Phage Therapy Background

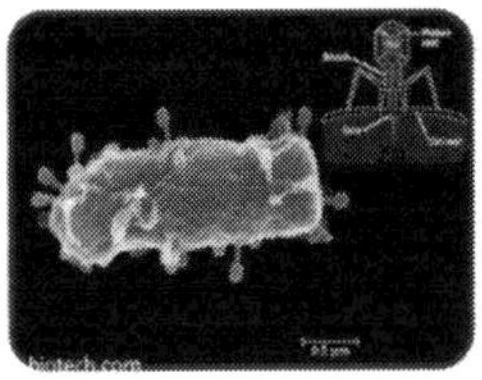

- Phages are the most abundant biological entities on earth
- Target and kill bacterial hosts with high specificity
- Potential advantages over antibiotics include targeting specificity, self-producing and self-limiting

Brief History

- Standard part of healthcare system in parts of former Soviet Bloc
- Staphylococcal bacteriophage lysate approved in US for human use in 1959 (later limited to veterinary use in 1990)
- Approved products for food safety in US (ListShield, EcoShield and SalmoFresh)
- Human therapeutics with natural phages or phage-based products in development for *S. aureus, P. aeruginosa* and *E. coli*

Next-generation Phage Therapeutics

SGI is building a platform for engineering phage as a new class of antimicrobials to combat multidrug-resistant pathogens

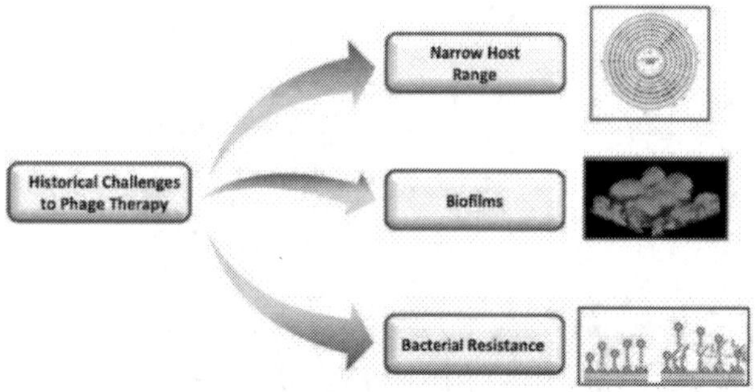

Addressing challenges through engineering; resulting in new products protected with new intellectual property

Next-generation Phage Therapeutics

SGI is building a platform for engineering phage as a new class of antimicrobials to combat multidrug-resistant pathogens

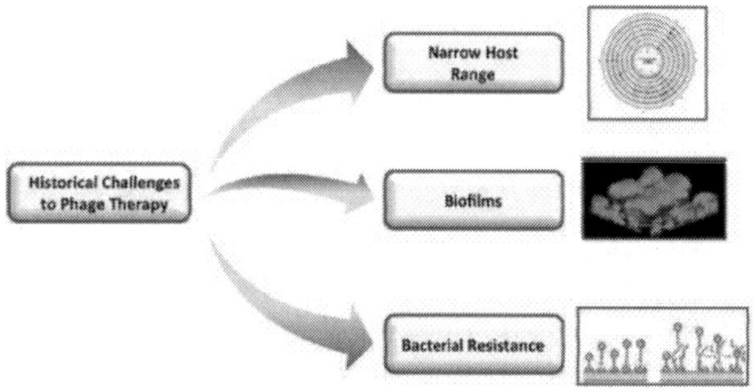

Addressing challenges through engineering; resulting in new products protected with new intellectual property

Initial target: ***Pseudomonas aeruginosa***

Significant unmet medical need:
- Life-limiting infection in Cystic Fibrosis
- Leading cause of hospital acquired pneumonia
- Significant pathogen in UTIs, wounds, diabetic ulcers
- High mortality in systemic infections

Antibiotic resistance:
- Multi-drug resistance common
- Significant evidence for increasing incidence

Phage starting points:
- Many lytic phages isolated and characterized

Key models and existing data:
- Pre-clinical respiratory, wound/burn and bacteremia models and protection data
- Limited but encouraging clinical data

Phage Engineering Overview

Previous engineering breakthroughs largely in *E. coli*

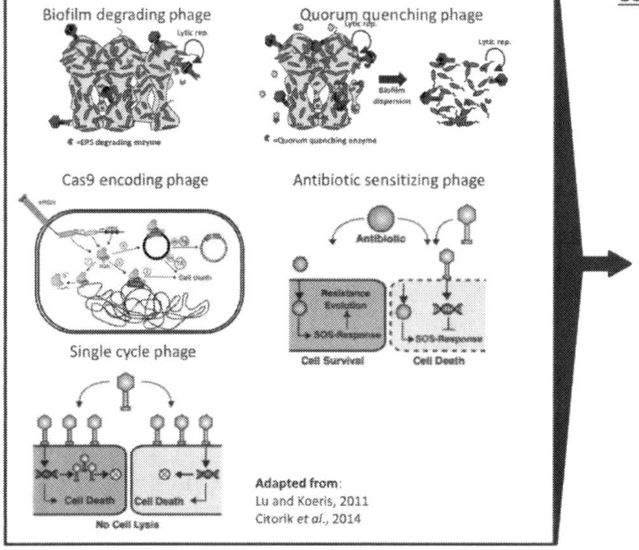

Advance and adapt platforms to economically important pathogens

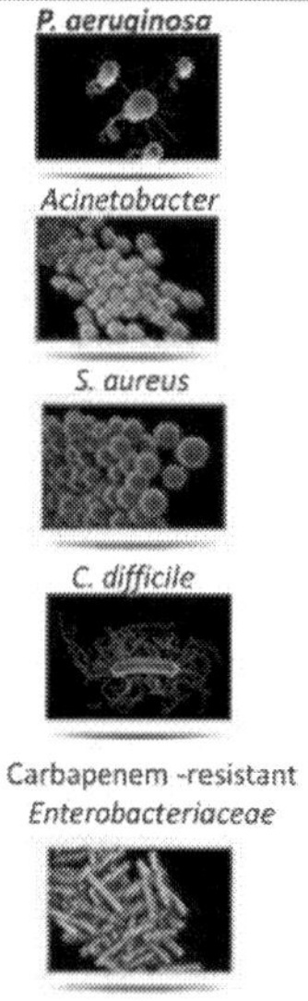

Bioinformatics – Understanding the Bacterial Host Genome

MDR *PA* diversity

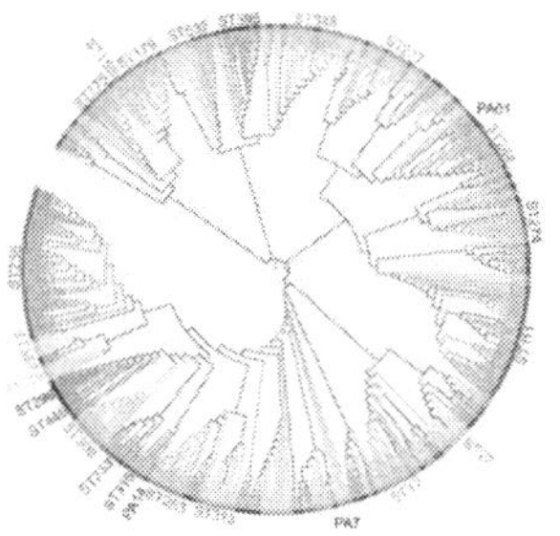

Phylogenic tree of clinical strains

Phage resistance systems

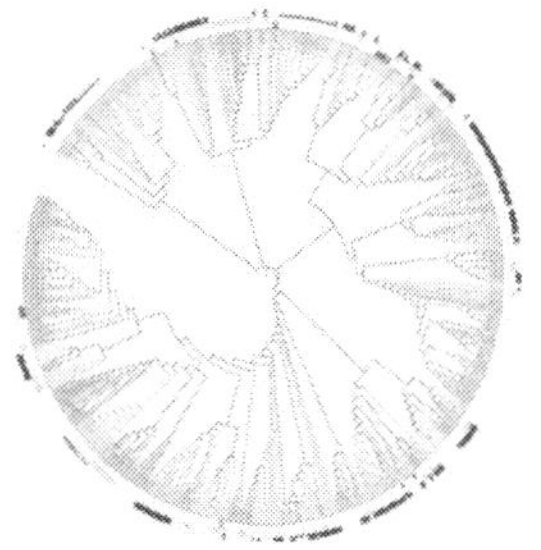

Phylogenic distribution of phage resistance systems

Common accessory genome sequenced

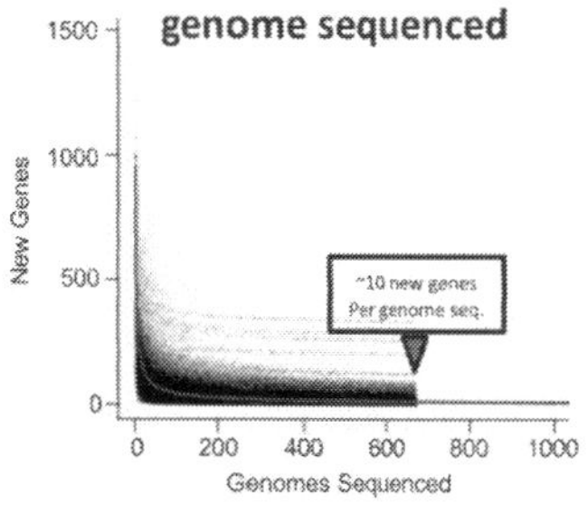

Few new accessory genes identified with additional sequencing

Bioinformatics – Understanding the Phage/Host Interaction

Identifying phage factors responsible for efficient killing of MDR *P. aeruginosa*

- Screened natural phage against broad MDR *P. aeruginosa* collection
- Developed key assays for functional assessment of phage
- Mapped sensitivity and resistant phenotypes against genetic sequencing information
- Findings are driving phage engineering and resulting in desired phenotypes

Functional Screening Assays

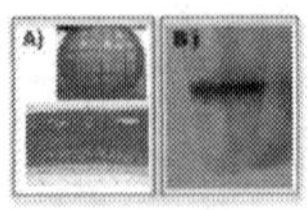

1. Multiple screening assays are used to assess efficacy of phage for bacterial killing or biofilm disruption. These included plaque assay and *in vitro* human keratinocyte infection model (A) and static biofilm assays (B).

Host Range Analysis

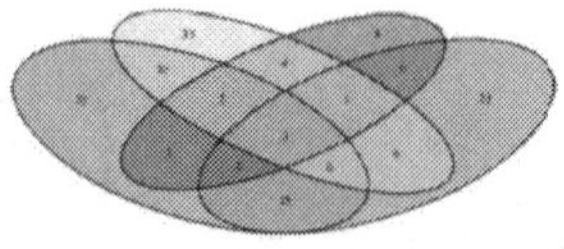

2. Screening data is mapped across phage to identify shared and independent host range patterns. Venn diagram (above) and phage host interaction maps (below) are two of the visualization tools used.

Bioinformatics Assessment

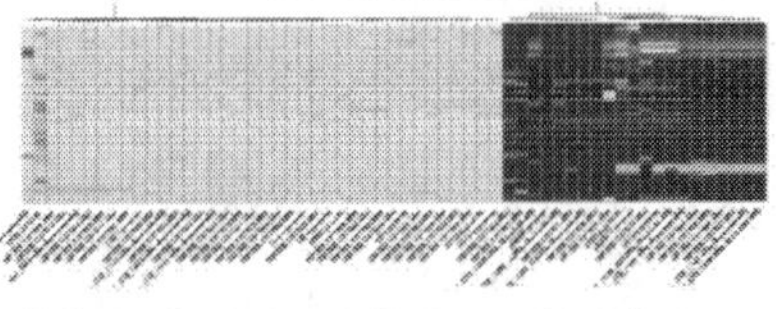

3. Bioinformatics software integrates screening data with sequence analysis to identify unique genetic elements that correlate with desired phenotypes. This information feeds into engineering process. As one example, automated full-genome homology analysis (gene by gene) of phage shown above.

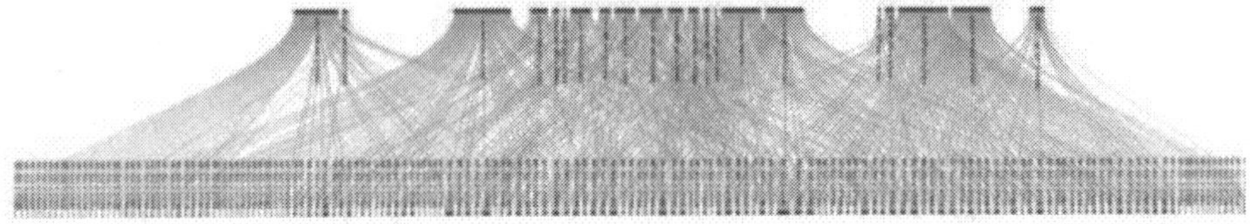

Engineering Target Identification

Utilize phage/host screening and co-infection/natural evolution, with sequencing and comparative genomics, to identify genes that are driving desired phenotypes

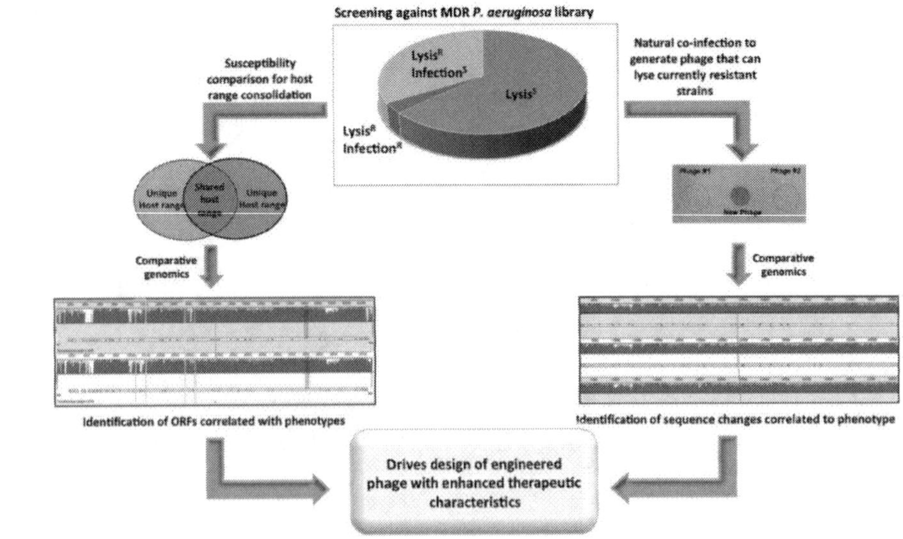

15

Molecular Approaches to Phage Engineering

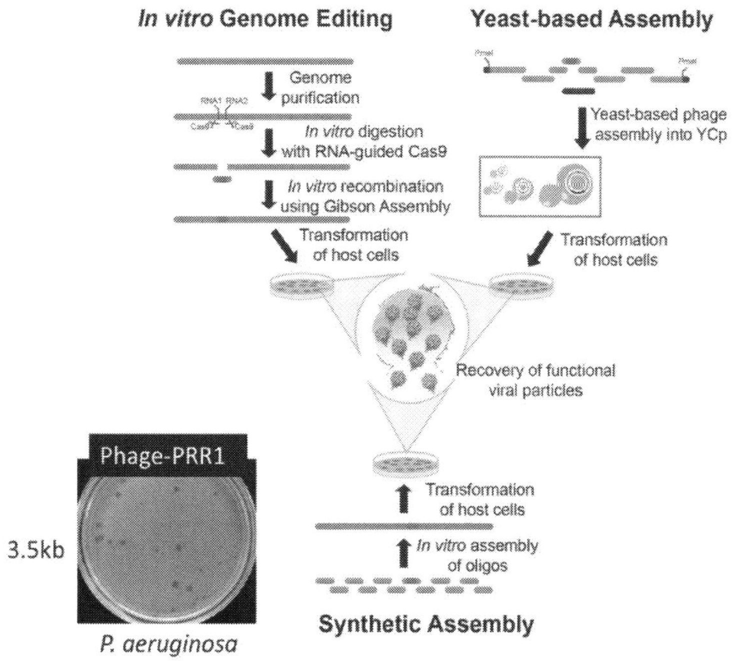

16

Molecular Approaches to Phage Engineering

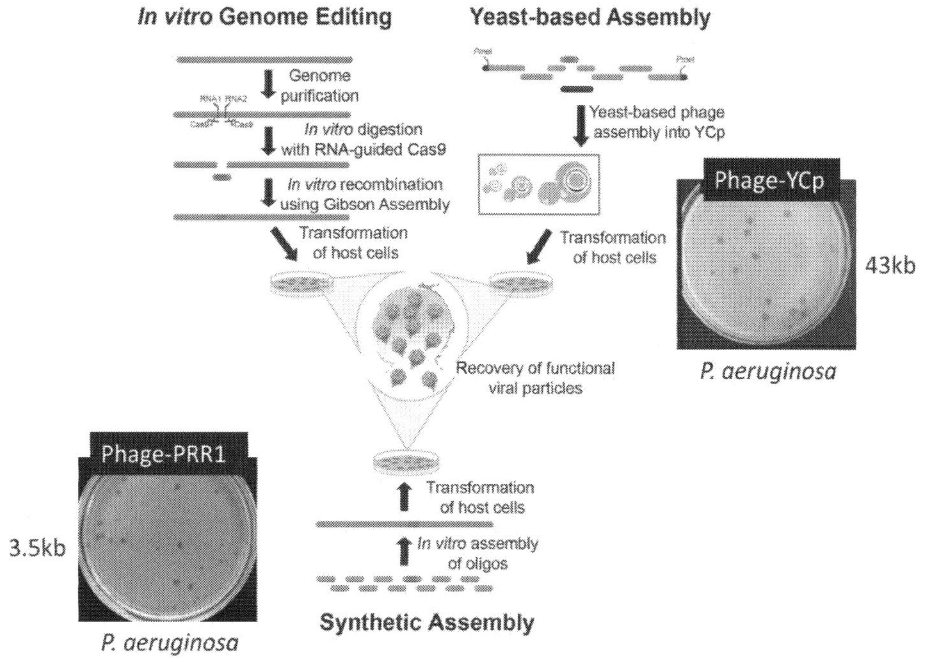

Large Phage Cas9-based Genome Editing

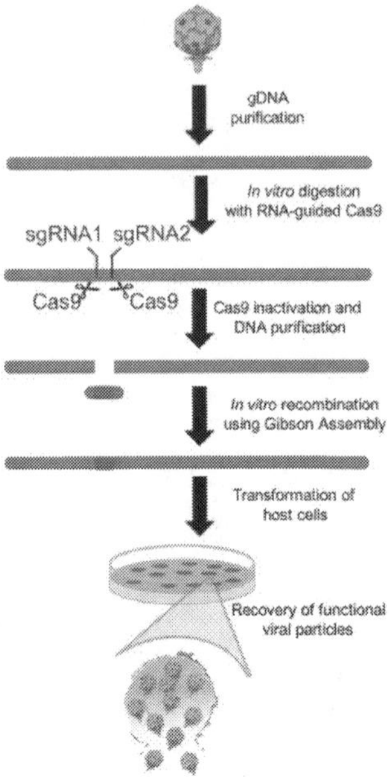

RNA-Guided *In vitro* Engineering (RNA-give)

- New and rapid engineering with Cas9
- Complete system enables manipulation of previously genetically intractable lytic phage genomes
- Broadly applicable for high- throughput viral engineering

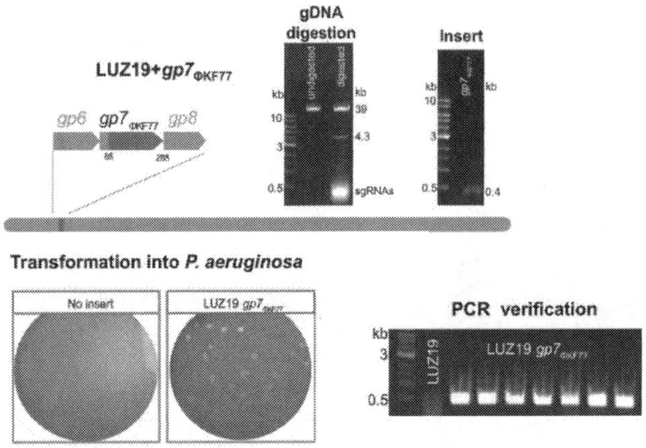

Host Range Expansion

Overcoming need for large phage cocktails through host range expansion

- Evaluated genomes of related phage with different host range
- Identified region potentially responsible for differences in specificity
- Expanded host range of phage by transitioning a single gene

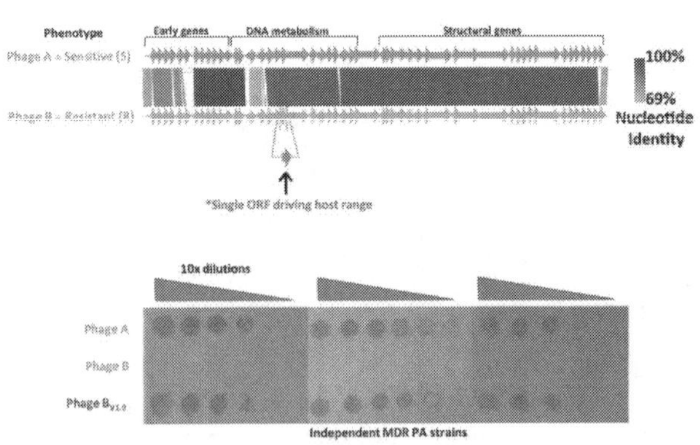

Eng. Phage $B_{V1.0}$ has single ORF from phage A

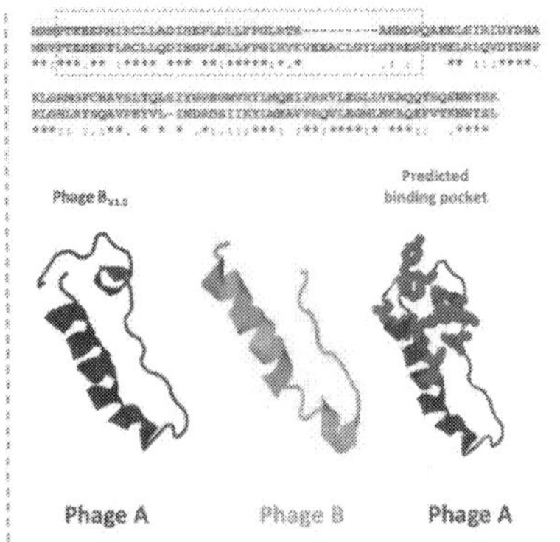

Structural modeling of ORF indicated in host range specificity. Phage B contains a deletion that is predicted to map to the binding pocket of the full-length protein.

Early Biofilm Disruption

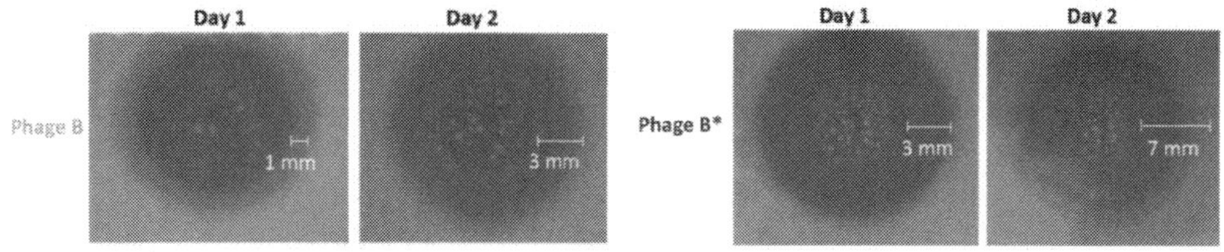

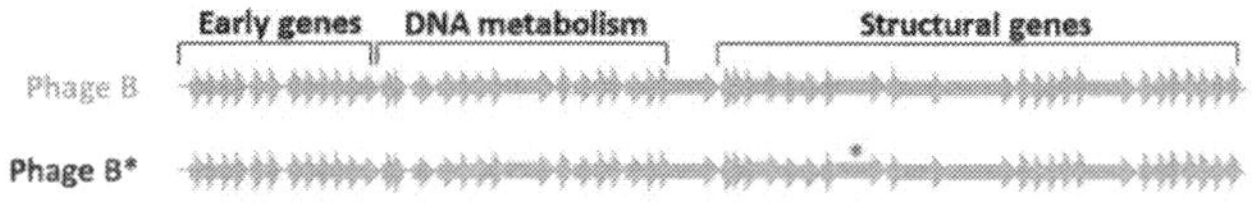

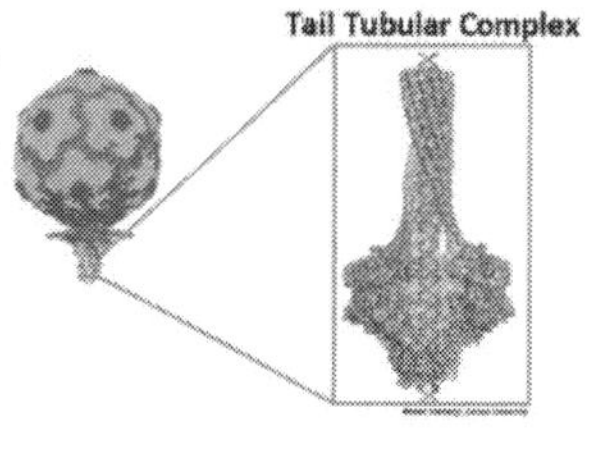

Phage	Adsorption	Latent period	Burst size
Phage B	~17	30 min	70
Phage B*	~12	30 min	122

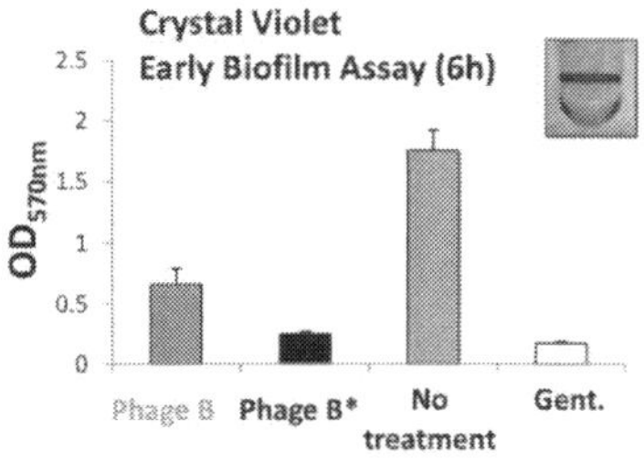

Enhancing phage activity

through improved replication

Iterative Engineering

Consolidated initial host range expansion and early biofilm disruption phenotypes into a single phage and demonstrated in two *in vitro* models

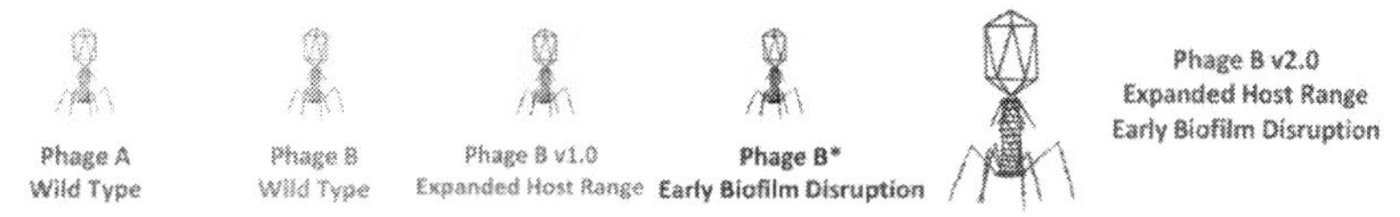

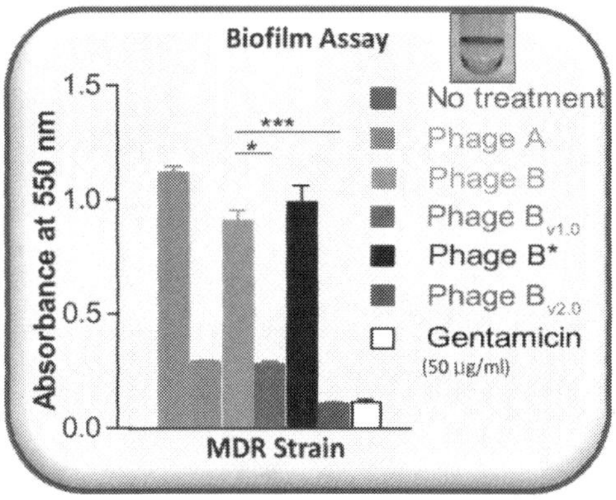

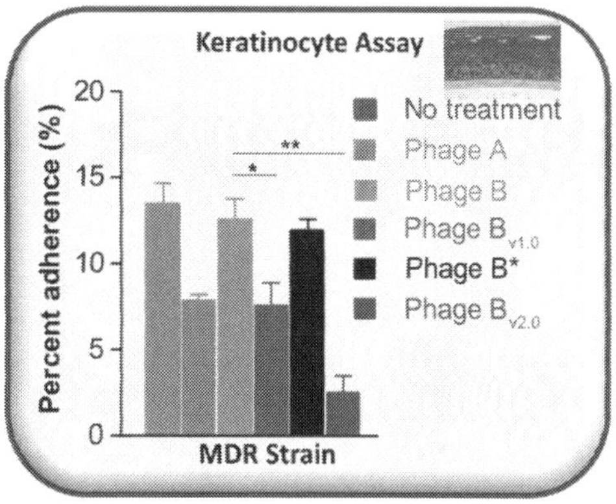

Engineered Phage Platform

Leveraging bioinformatics, comparative genomics, viral engineering and DNA synthesis technologies to rationally develop an engineered phage platform

Key Phage Platform Features

Synthetic Production	Modular and Multigenic Design	Addresses Genetic Diversity	Bacterial specificity	Bypasses evolved resistance

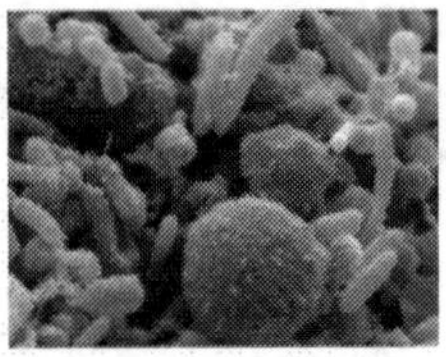

Microbiome Engineering

Drug-Resistant Pathogens

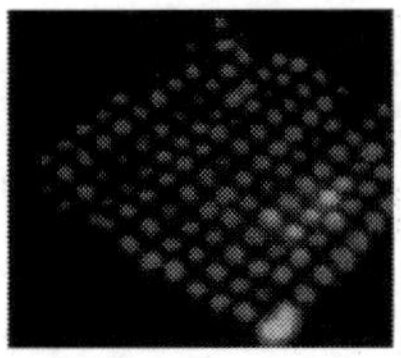

Medical Coatings

Acknowledgements

Key Contributors (SGI and SGVI)

- Srividya Akella
- **Magda Barbu**
- Rob Brown
- Yvette Del Rosario
- Sammy Farah
- Martina Felderman
- Dan Gibson
- Heather Gouvis
- Bolyn Hubby
- Kurt Kamrud
- Kristofer Miller
- Todd Peterson
- Toby Richardson
- Ariel Schwartz
- Hamilton Smith
- Leah Soriaga
- Billyana Tsvetanova
- J. Craig Venter

Collaborators

- Rob Lavigne
- Pieter-Jan Ceyssens
- Mark Enright
- Alex van Belkum

BIOMÉRIEUX

The Economist

Synthetic biology

Strange medicine

A way to treat bacterial infections with artificial viruses

Feb 28th 2015 | La Jolla | From the print edition

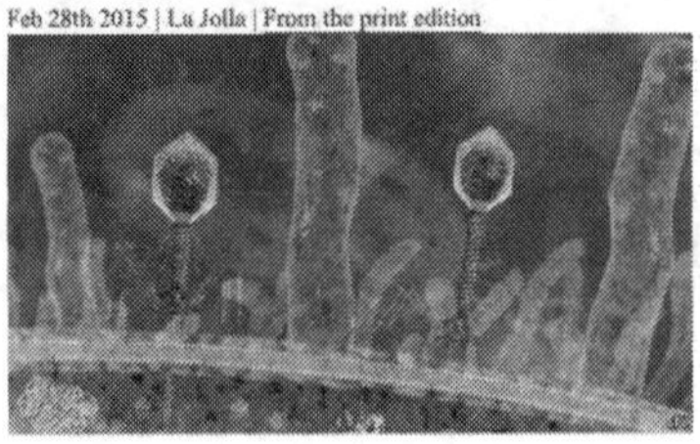

SYNTHETIC GENOMICS®

Engineering Smarter and Stronger T Cells for Cancer Immunotherapy

Chen, Y., University of California, Los Angeles

Originally presented on: 6/10/2015 15:00:00 - 15:30:00

Adoptive T cell therapy for cancer has demonstrated exciting potential in treating relapsing cancers. In particular, T cells that express synthetic chimeric antigen receptors (CARs) specific for the B-cell marker CD19 have shown impressive results in clinical trials for various B-cell malignancies, prompting avid interest from both scientific and entrepreneurial communities in recent years. However, CD19 CAR-T cell therapy remains the only robustly effective T-cell immunotherapy to date, and several obstacles remain to be overcome before the full potential of adoptive T-cell therapy can be realized. My laboratory is pursuing several strategies for the engineering of T cells with stronger anti-tumor functions and greater robustness against evasive mechanisms employed by cancer cells. I will discuss the design, construction, and implementation of multi-input CARs to increase tumor specificity and decrease the probability of mutational escape by tumor cells. I will present the design of synthetic circuits to reroute signaling pathways triggered by tumor-secreted cytokines, thus negating the immunosuppressive effects of the tumor microenvironment. Finally, I will discuss efforts to engineer a cytotoxic protein that triggers target-cell death upon recognition of intracellular oncoproteins, thus expanding the repertoire of detectable tumor markers beyond surface-bound antigens. These strategies combine to address critical limitations facing adoptive T-cell therapy, providing potential treatment options for diseases that are otherwise incurable with current technology.

jbei
Joint BioEnergy Institute

U.S. DEPARTMENT OF ENERGY

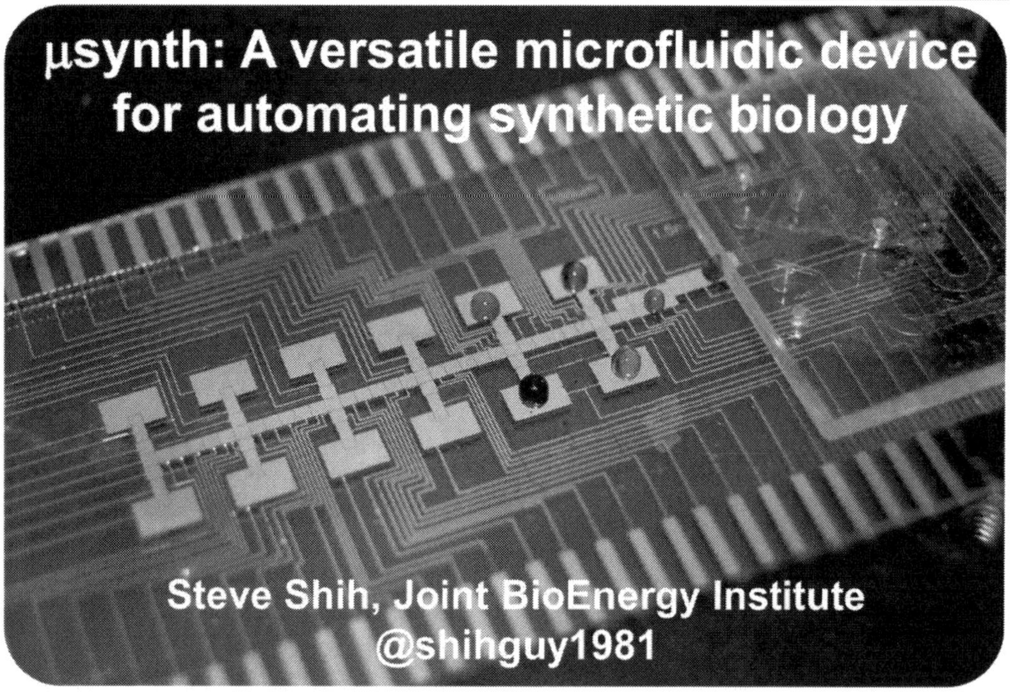

jbei ENERGY

Outline

- **Synthetic Biology**
  - Design-build-test-learn cycle
- **Microfluidics: Introduction and Theory**
  - Microfluidic paradigms, General mechanism, Electromechanical model
- **Microfluidics for Synthetic Biology**
  - Device Design
  - Automation system
  - DNA assembly and Transformation
- **Conclusions**
  - Final notes

2

Outline

- **Synthetic Biology**
  - Design-build-test-learn cycle
- **Microfluidics: Introduction and Theory**
  - Microfluidic paradigms, General mechanism, Electromechanical model
- **Microfluidics for Synthetic Biology**
  - Device Design
  - Automation system
  - DNA assembly and Transformation
- **Conclusions**
  - Final notes

3

Synthetic biology

❑ Synthetic biology is to design and to build genetic circuits in living cells

Biofuel | Medicines | Chemicals

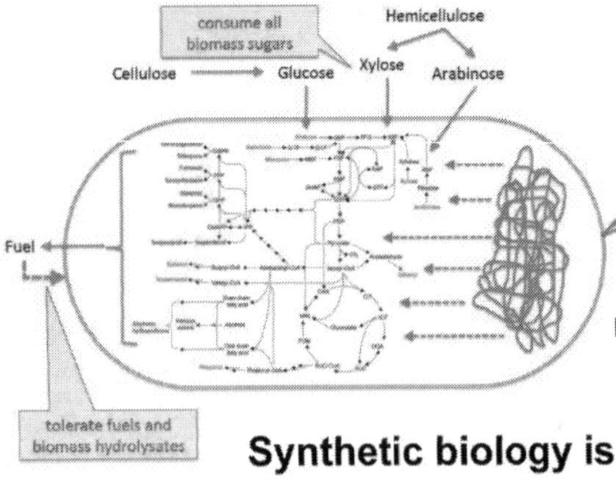

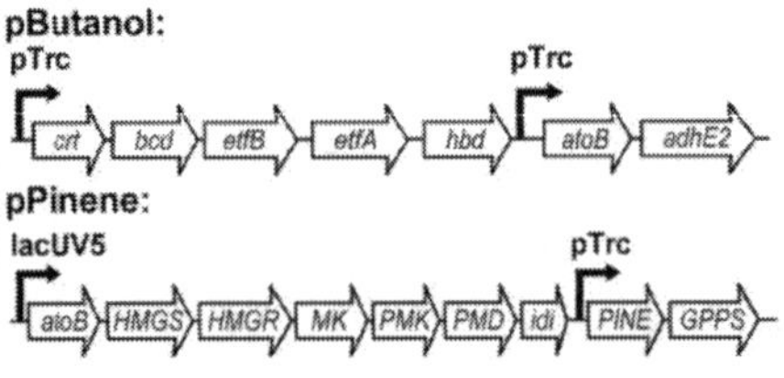

Bokinsky et al. *PNAS*. 2011, *108*, 19949-19954

Synthetic biology is becoming a necessity!
We need to explore synthetic alternatives…

4

jbei

Design-build-test-learn cycle

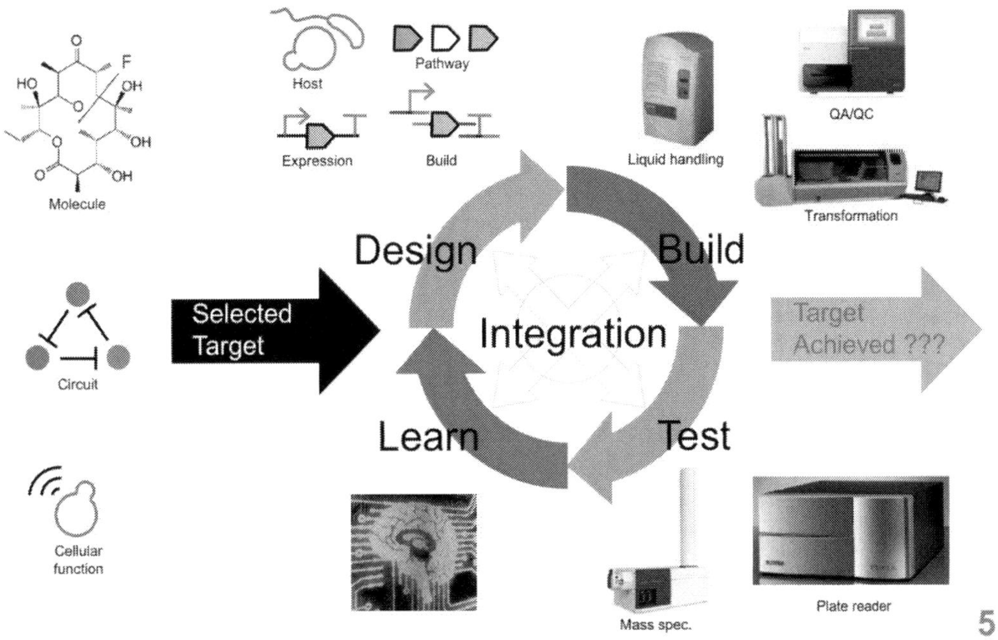

5

jbei

Microfluidics – Automating Synthetic Biology

- Microfluidics has the potential to automate liquid handling with low volumes (pL-nL) for various biological applications

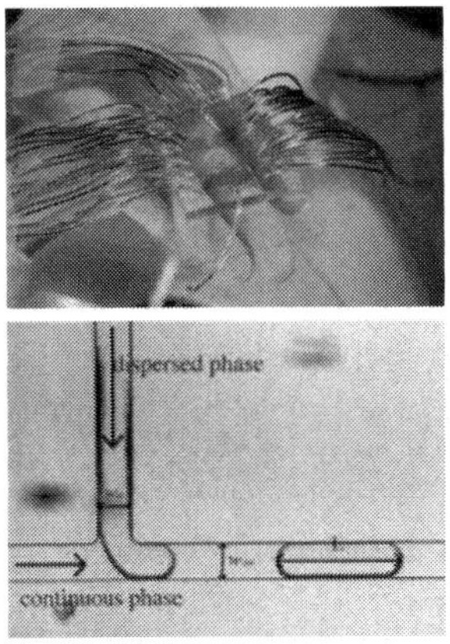

"Lab-on-chip"

- Smaller samples – *analyze precious specimens*
- Reduced reagent use – *"green" and less expensive*
- Fast analysis – *higher throughput*

6

jbei ENERGY

Microfluidics – Automating Synthetic Biology

- Microfluidics has the potential to automate liquid handling with low volumes (pL-nL) for various biological applications

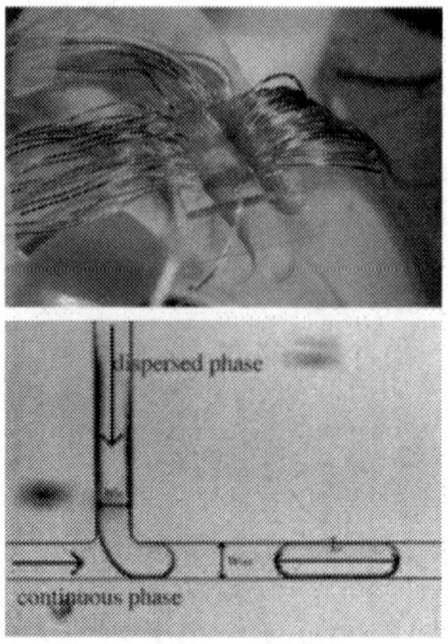

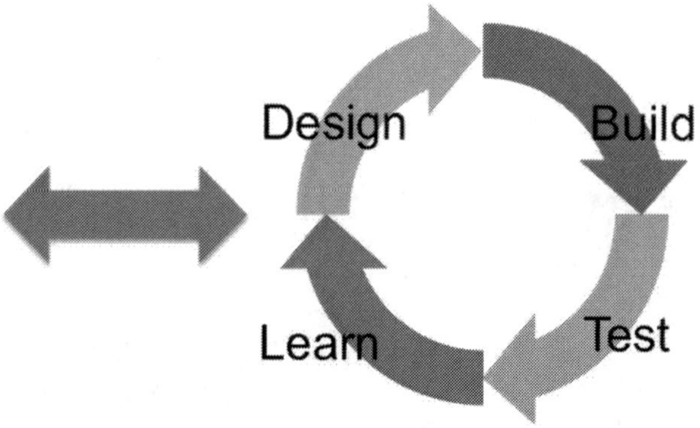

Use microfluidics to automate the synthetic biology cycle – to learn more efficiently!

7

jbei ENERGY

Outline

- **Synthetic Biology**
  - Design-build-test-learn cycle
- **Microfluidics: Introduction and Theory**
  - Microfluidic paradigms, General mechanism, Electromechanical model
- **Microfluidics for Synthetic Biology**
  - Device Design
  - Automation system
  - DNA assembly and Transformation
- **Conclusions**
  - Final notes

8

Labs on a chip platforms

- There are many different microfluidic platform technologies being used to form Labs on a Chip

Microchannels (continuous)

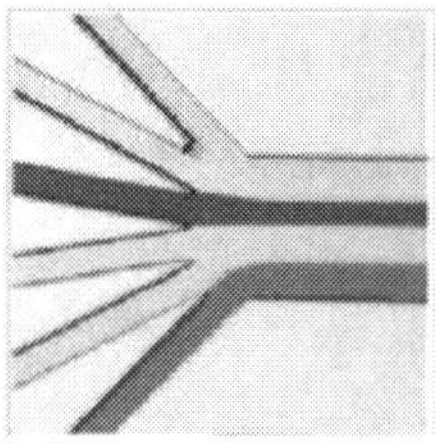

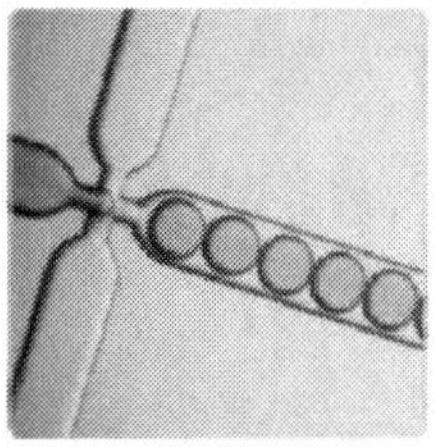

Microchannels (droplets)

Paper Microfluidics

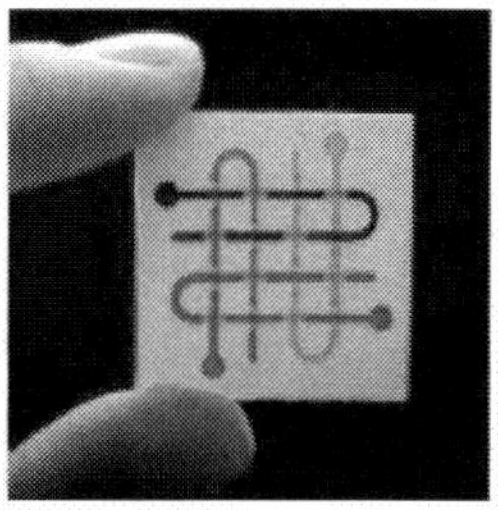

Slip Chips

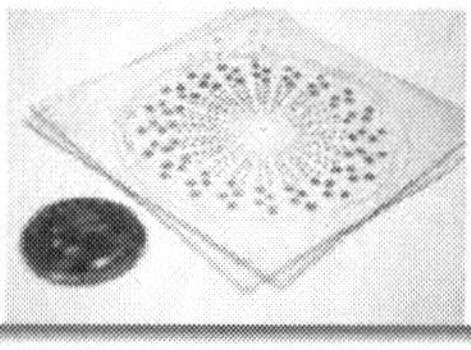

Digital Microfluidics

9

Digital microfluidics

- Digital microfluidics (DMF) is one of a group of techniques used to miniaturize fluid-handling for "lab on a chip" applications
- A unique feature of DMF: samples are manipulated as droplets on an open or closed surface

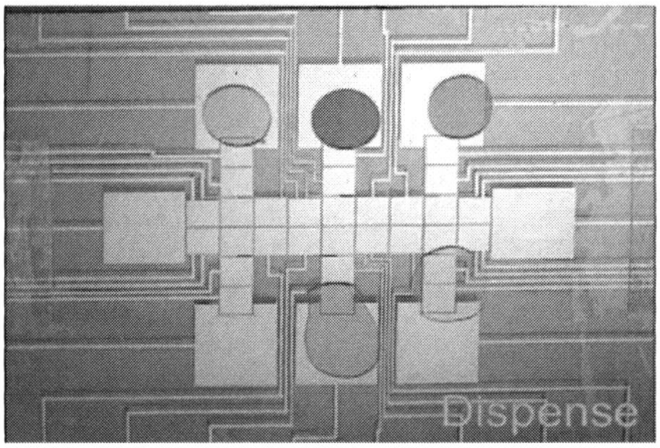

Typical operations include dispense (precision of ~ 2-3% CV), move, merge, mix, and split

10

jbei | U.S. DEPARTMENT OF ENERGY

DMF: Mixing

- DMF and microchannels share many characteristics, including laminar flow (i.e., no turbulence)
- Thankfully, the act of DMF droplet translation brings droplet constituents together faster **10-100x** faster than diffusion

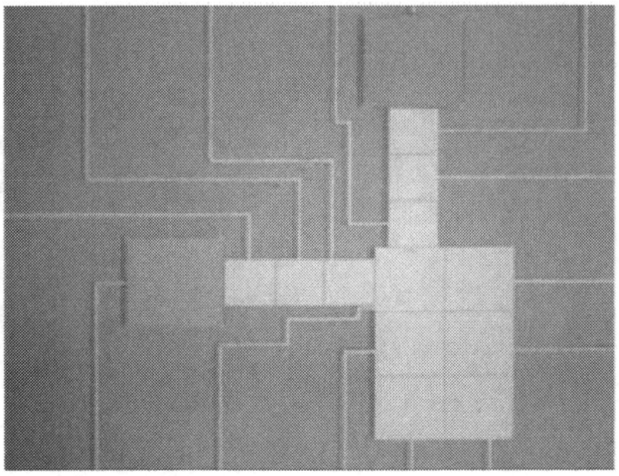

for more, see: Paik et al. *Lab Chip* 2003, 3, 28-33, Paik et al. *Lab Chip* 2003, 3, 253-259

11

jbei | U.S. DEPARTMENT OF ENERGY

DMF: Automation

- DMF enables a "programming" approach to sample processing, allowing the user to build, combine, and execute "functions" in parallel

Programmed function for "Move"

Droplets executing "Move"

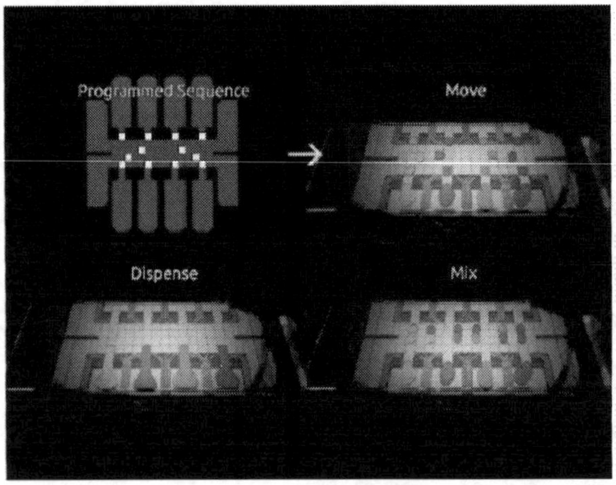

Droplets excecuting "Dispense"

Droplets executing "Mix"

12

jbei ENERGY

How does DMF work?

- In the most common implementation of DMF, the device includes an array of electrode coated by an insulator and hydrophobic layer

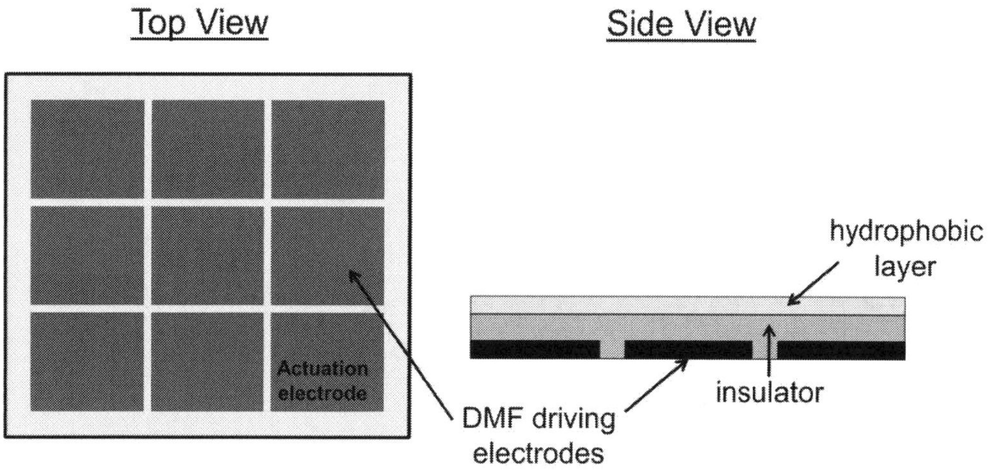

13

jbei ENERGY

How does DMF work?

- In the most common implementation of DMF, the device includes an array of electrode coated by an insulator and hydrophobic layer

- Driving electrodes are typically referenced to a "top-plate" counter electrode (often transparent)
- Droplet are surrounded by a filler fluid (often air or oil)

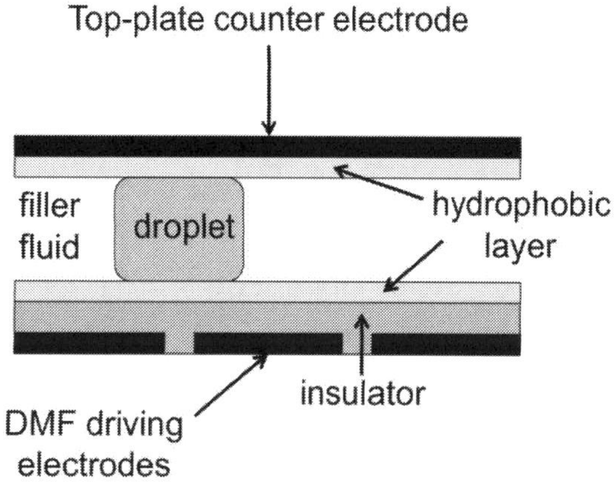

14

How does DMF work?

- In the most common implementation of DMF, the device includes an array of electrode coated by an insulator and hydrophobic layer
- An electrostatic force *pulls* the droplet toward the electrode

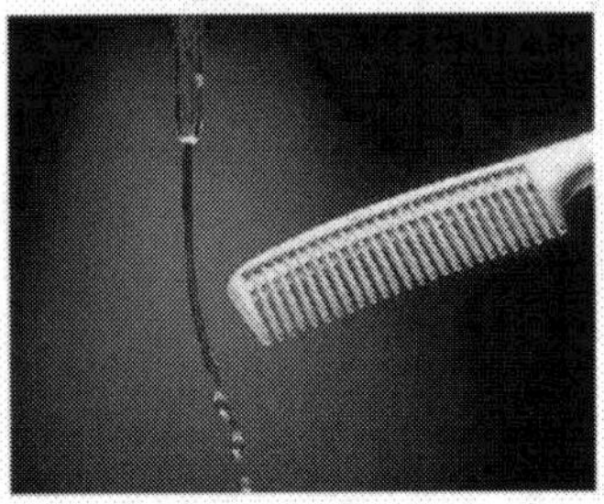

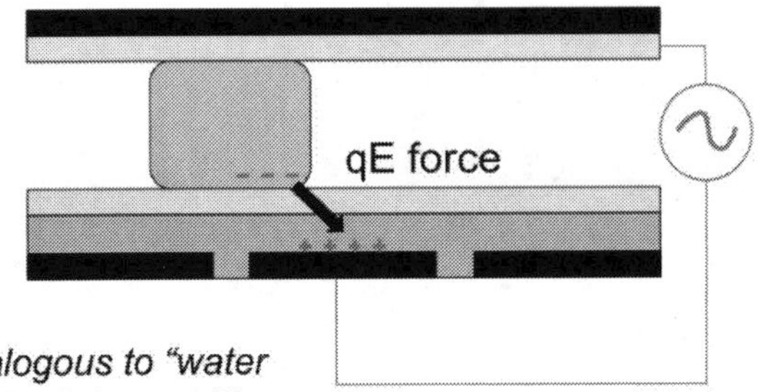

Analogous to "water bending" demo with a charged comb

15

How does DMF work?

- In the most common implementation of DMF, the device includes an array of electrode coated by an insulator and hydrophobic layer
- An electrostatic force *pulls* the droplet toward the electrode

- We can switch the field to successive electrodes – droplets can be positioned anywhere on the array

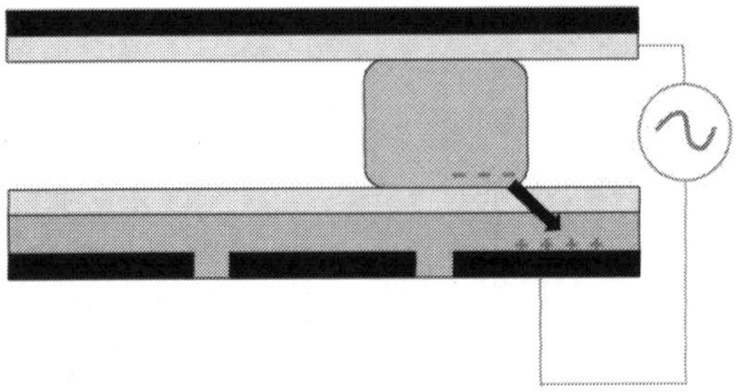

16

Electromechanical model

- A more general description of DMF driving force can be made using electromechanics and lumped-sum model

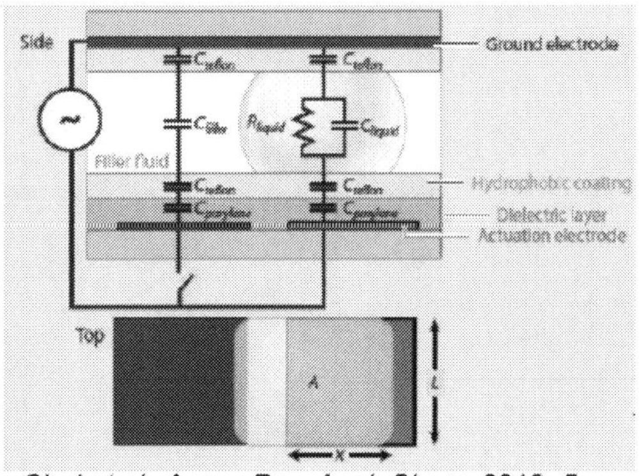

Choi et al. *Annu. Rev. Anal. Chem.* 2012, *5*, 413-440

- Electromechanical energy in the system is:

$$E(f,x)=\frac{L}{2}\left(x\sum_i\frac{\varepsilon_0\varepsilon_{\mathrm{r,i,liquid}}V^2{}_{\mathrm{i,liquid}}(j2\pi f)}{d_i}+(L-x)\sum_i\frac{\varepsilon_0\varepsilon_{\mathrm{r,i,filler}}V^2{}_{\mathrm{i,filler}}(j2\pi f)}{d_i}\right)$$

$\varepsilon_{r,i,liquid}$, $V_{i,liquid}$, and $\varepsilon_{r,i,filler}$, $V_{i,filler}$ are the relative permittivity and voltage drop for the liquid and filler fluid portion of the electrode respectively, and d_i is the thickness of layer i

17

Electromechanical model

- A more general description of DMF driving force can be made using electromechanics and lumped-sum model

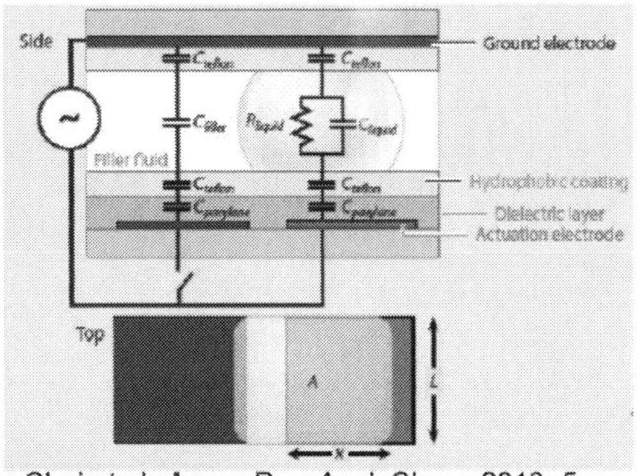

Choi et al. *Annu. Rev. Anal. Chem.* 2012, *5*, 413-440

- Electromechanical energy in the system is:

$$E(f,x)=\frac{L}{2}\left(x\sum_i\frac{\varepsilon_0\varepsilon_{\mathrm{r,i,liquid}}V^2{}_{\mathrm{i,liquid}}(j2\pi f)}{d_i}+(L-x)\sum_i\frac{\varepsilon_0\varepsilon_{\mathrm{r,i,filler}}V^2{}_{\mathrm{i,filler}}(j2\pi f)}{d_i}\right)$$

$$F(f)=\frac{\partial E(f,x)}{\partial x}=\frac{L}{2}\left(\sum_i\frac{\varepsilon_0\varepsilon_{\mathrm{r,i,liquid}}V^2{}_{\mathrm{i,liquid}}(j2\pi f)}{d_i}-\sum_i\frac{\varepsilon_0\varepsilon_{\mathrm{r,i,filler}}V^2{}_{\mathrm{i,filler}}(j2\pi f)}{d_i}\right)$$

18

jbei ENERGY

Electromechanical model

- We can estimate the forces generated on different reagents at different frequencies

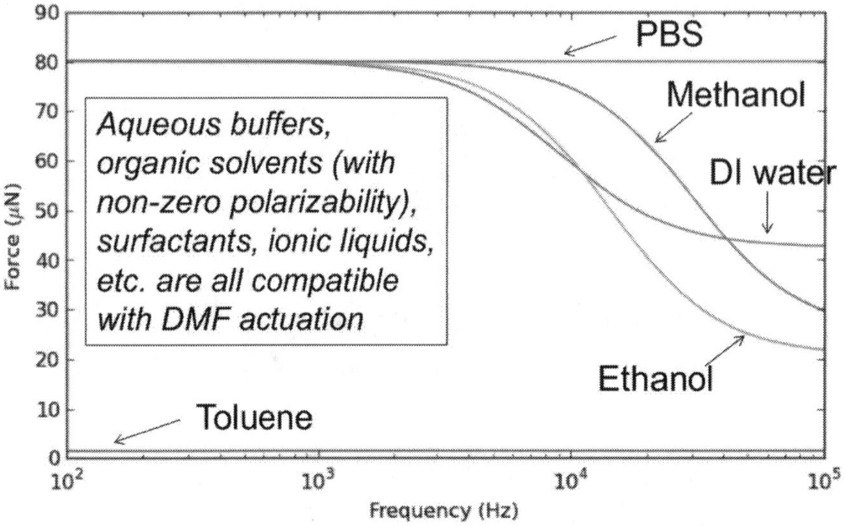

19

jbei ENERGY

Droplets-in-channel microfluidics

- Many researchers in the droplet microfluidic area that are using this technology for high-throughput applications

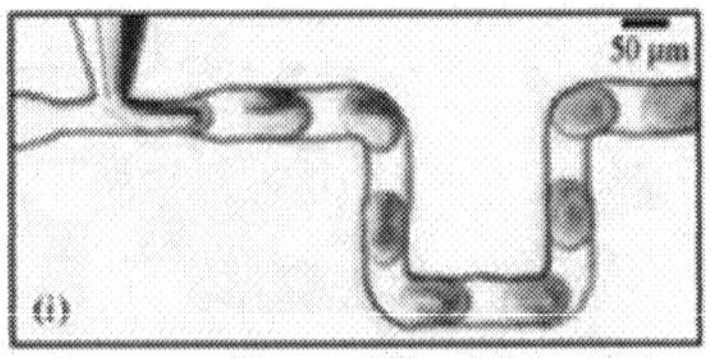

Ismagilov (Cal Tech)

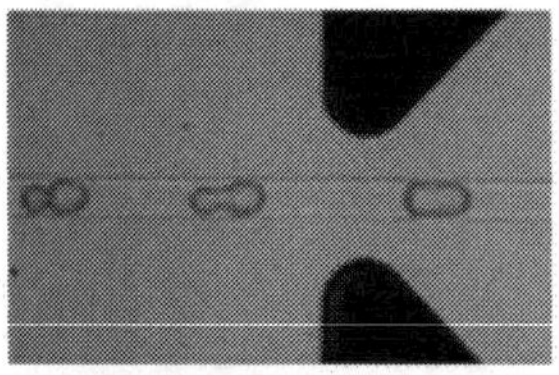
Weitz (Harvard)

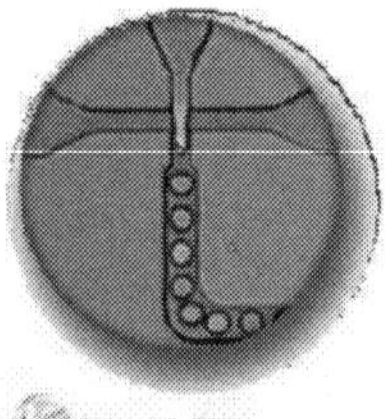
RainDance Technologies

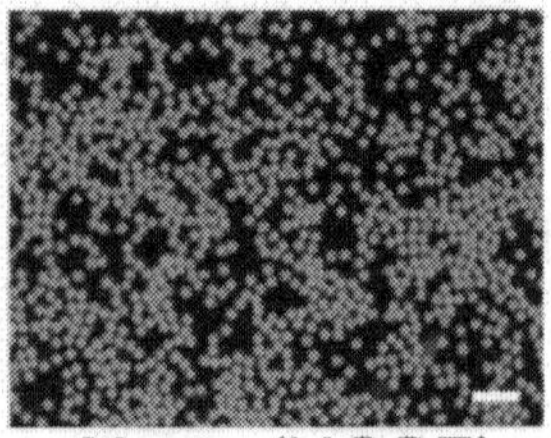
Abate (UCSF)

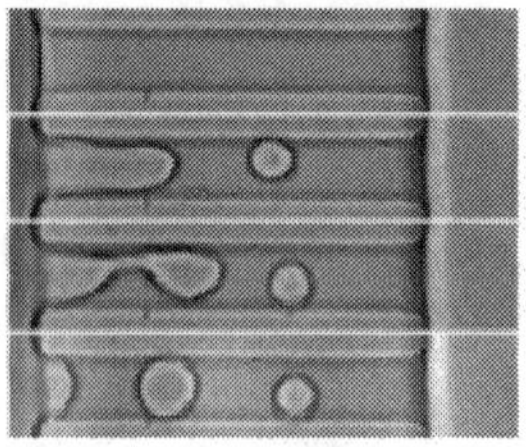
Chiu (Washington)

20

DMF vs. Droplets-in-channels

- Note that DMF (as defined here) is distinct from alternative methods in which droplets are manipulated in channels…

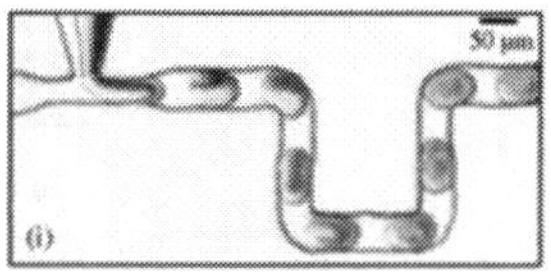

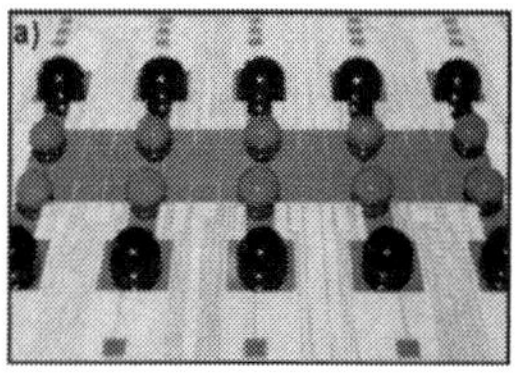

- The key distinction:
  - In DMF, each droplet is independently addressable (in parallel), but often with modest throughput
  - In droplet-in-channels, droplet are manipulated in series, but often with very high throughput

21

Outline

- **Synthetic Biology**
  - Design-build-test-learn cycle
- **Microfluidics: Introduction and Theory**
  - Microfluidic paradigms, General mechanism, Electromechanical model
- **Microfluidics for Synthetic Biology**
  - Device Design
  - Automation system
  - DNA assembly and Transformation
- **Conclusions**
  - Final notes

22

Microfluidics and synthetic biology

- **Build:** To assemble DNA parts using a variety of assembly methods and transform into microbes
- DNA assembly methods: Golden Gate, Gibson, TAR and others…

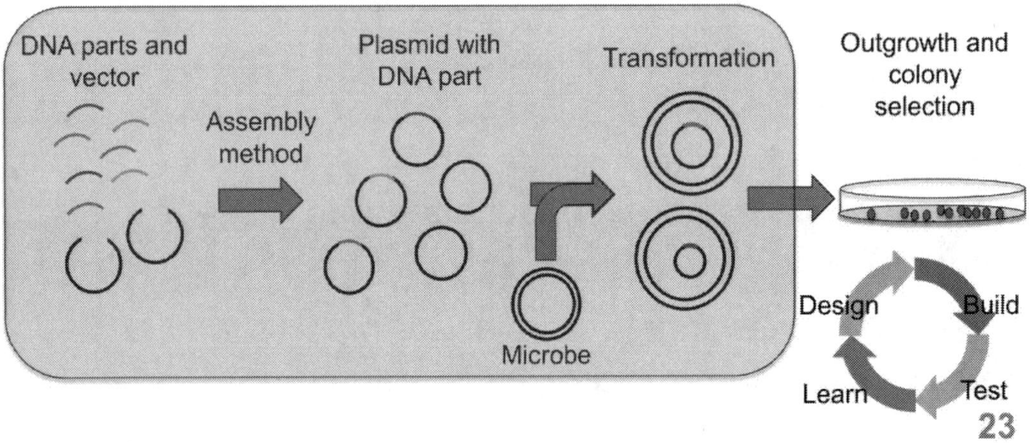

23

μsynth: Microfluidics and synthetic biology

- Using a digital-to-droplet microfluidic device that will assemble different DNA fragments and transform them into microbes

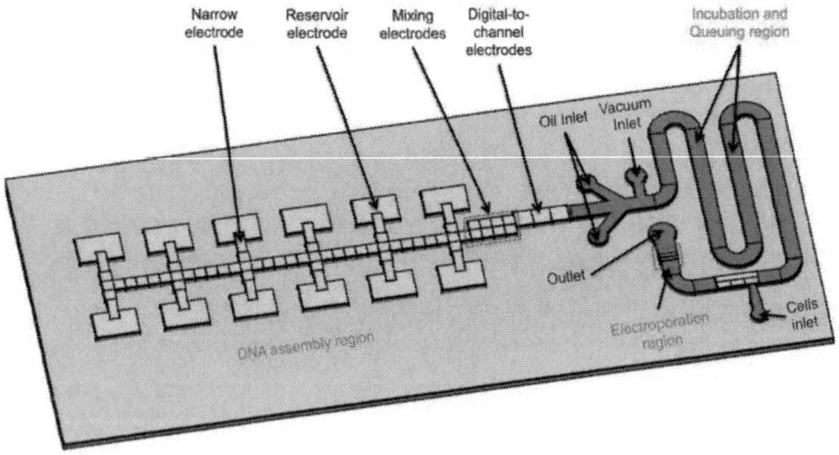

Shih et al. *ACS Syn. Biol.* 2015, In press 24

jbei ENERGY

Model system: Bicistronic design

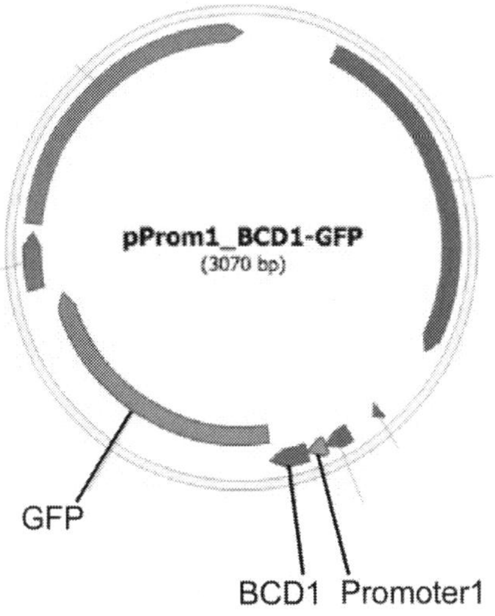

- 3-part assembly (gfp-BCD, Promoter, vector backbone)
- BCD: modulate gene expression

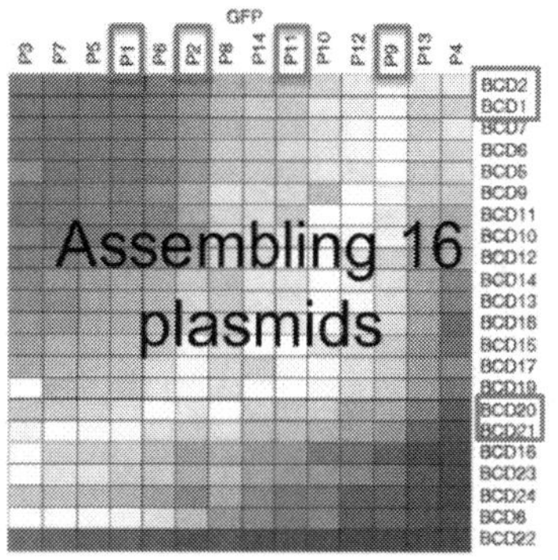

Mutalik et al. *Nat. Methods* 2013, 10 354-360

25

jbei ENERGY

DNA assembly and transformation

- We tested out three DNA assembly methods: Gibson (50°C), Golden Gate (room), and TAR (yeast) cloning (room)

Need a surfactant to prevent coalescence between droplets containing DNA fragments

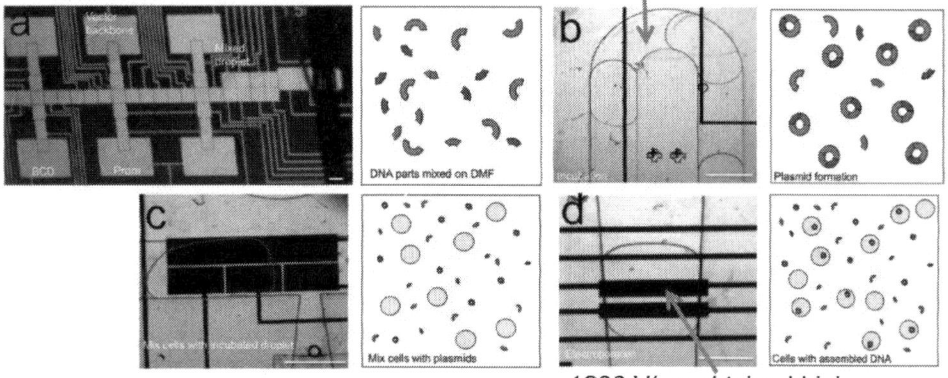

~1600 V/cm obtained high transformation efficiency

Shih et al. *ACS Syn. Biol.* 2015, In press.

26

jbei ENERGY

Automation system

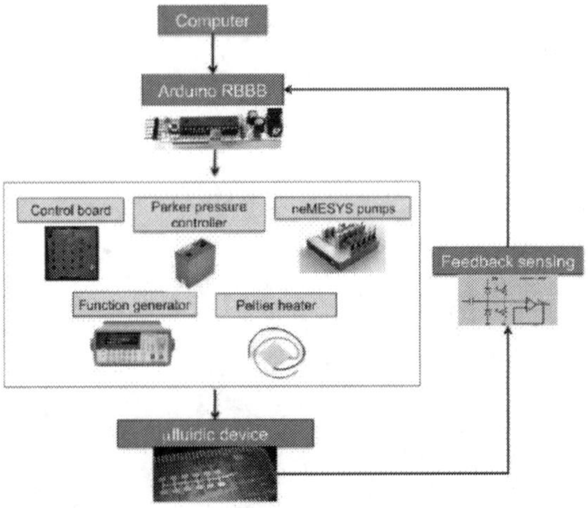

- Control system to automate droplet movement
- Software in C++ and hardware using I^2C communication

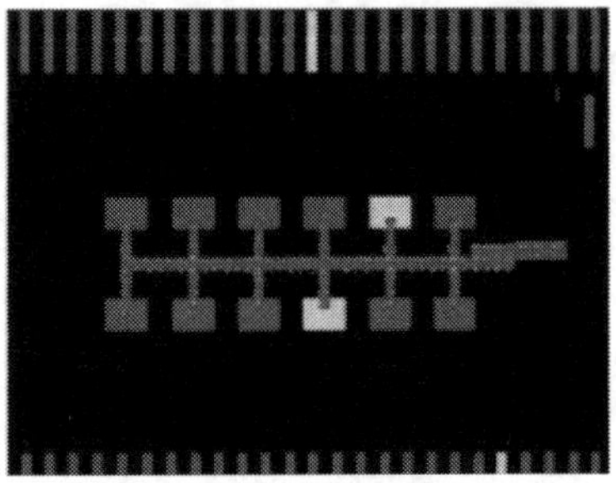

Shih et al. *Biosens Bioelec.* 2013, 42, 314-320
Shih et al. *Lab on Chip.* 2015, *15*, 225-236

27

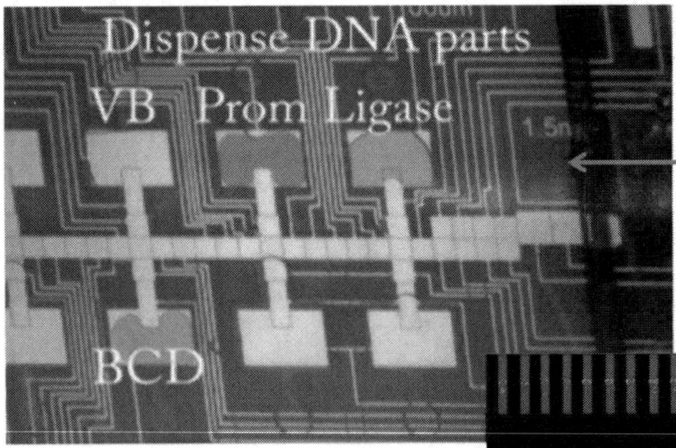

Dispensing and mixing of DNA parts

User-specified sequence for droplet movement

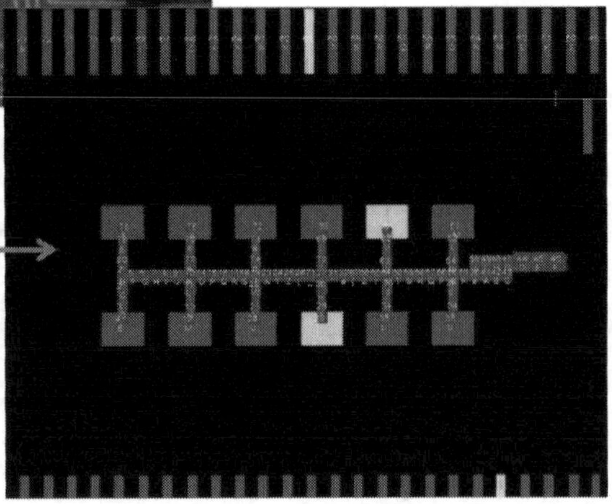

Valve and temperature control

- A valve system to control pressure inside microfluidic channel
- A Peltier heater to ensure regio-specific temperature control on microfluidics

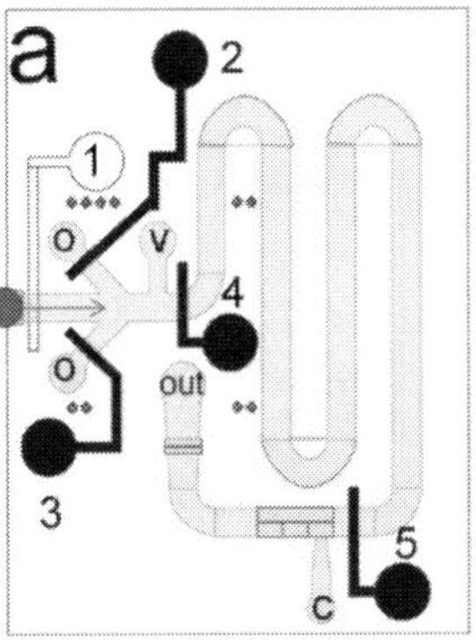

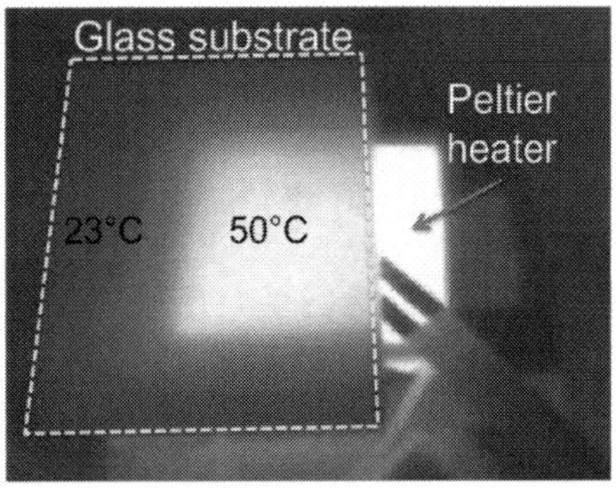

Shih et al. *ACS Syn. Biol.* 2015, In press.

DMF feedback system

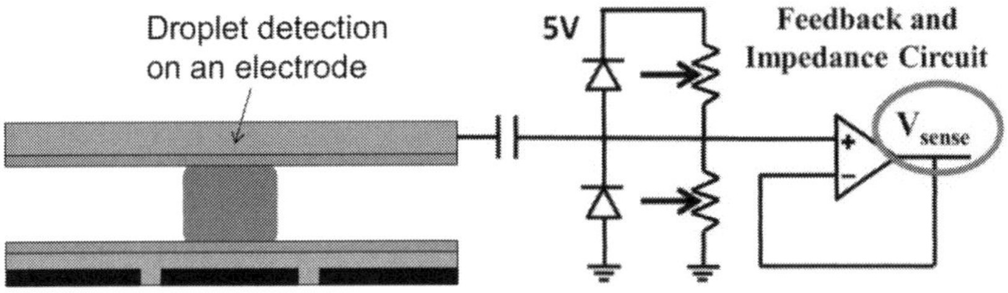

- Measure the impedance to monitor droplet movement.
- Feedback - continuously applies the voltage until droplet is on the desired electrode

By measuring the potential drop, V_{sense}, we can determine the dispensed volume

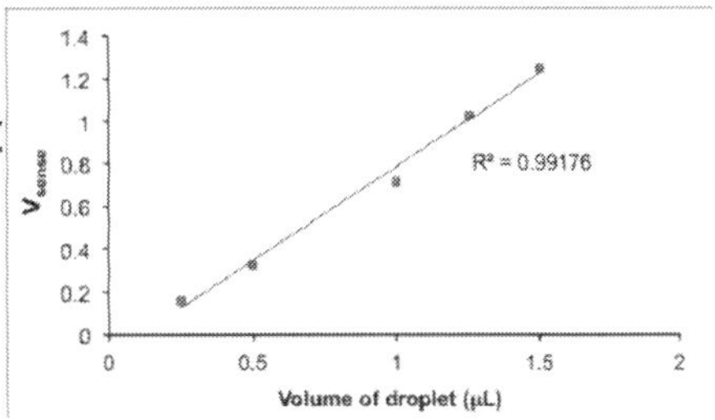

Shih et al. *Lab on Chip.* 2015, *15*, 225-236
Shih et al. *ACS Syn. Biol.* 2015, In press.

Advantage: Reproducible droplet dispensing

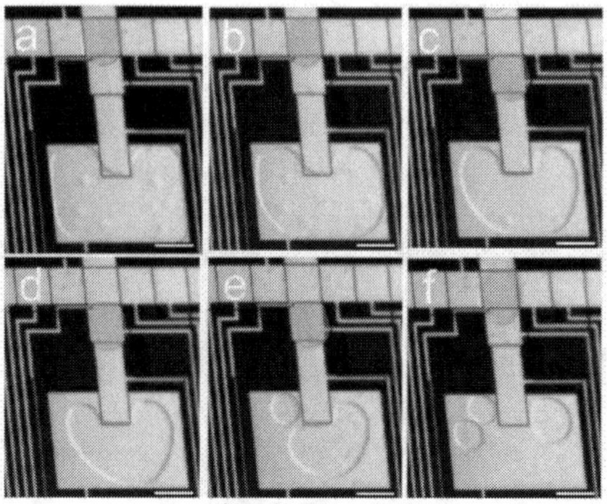

We can use feedback to dispense droplets with a target volume

- Droplets are successive dispensed to measure reproducibility
- Use Image J to calculate droplet area
- For six droplets, we obtain **~1.5%** droplet volume deviation

Shih et al. *ACS Syn. Biol.* 2015, In press.

31

Cross-contamination study

- We assemble 2 plasmids: 15 times for one plasmid and one time for the other plasmid (representing 16th assembled plasmid)

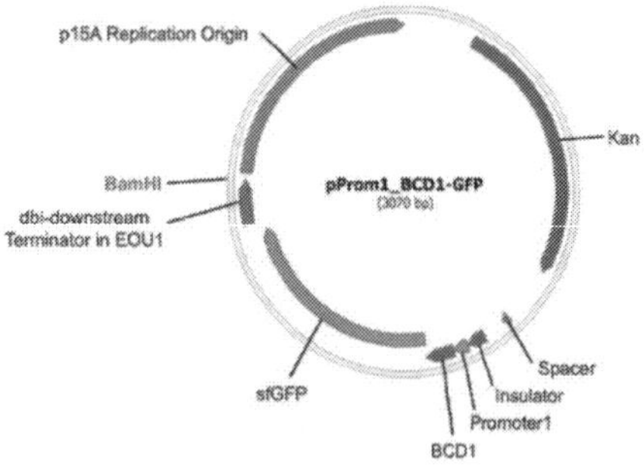

Assembled 15 times

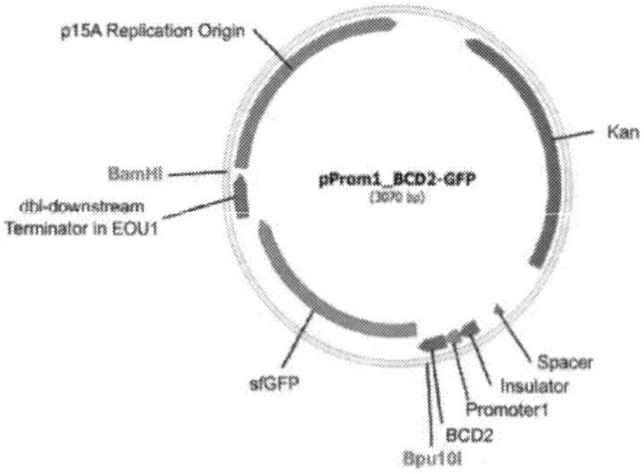

Assembled one time

Three part assembly: BCD-gfp, promoter, and vector backbone (kanamycin selection)

32

Gel profiles

- We picked 30 colonies from pProm1_BCD2 and performed a restriction digest
- With 95% confidence, we showed cross-contamination is < 9.5%.

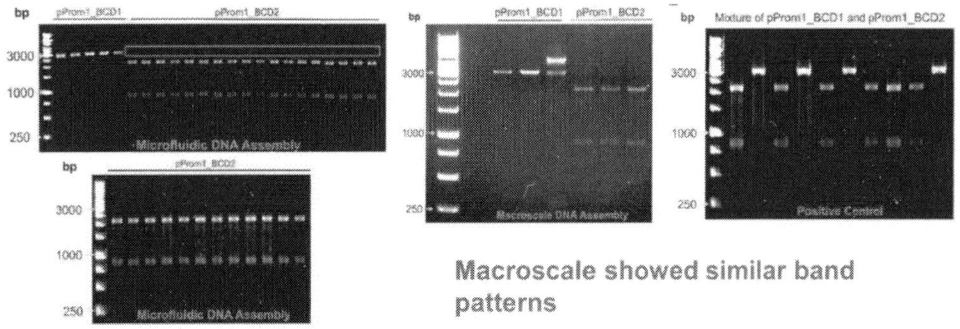

No colonies picked from pProm1_BCD2 showed pProm1_BCD1

Macroscale showed similar band patterns

Positive control show equal mixtures

Shih et al. *ACS Syn. Biol.* 2015, In press.

33

Sequencing results

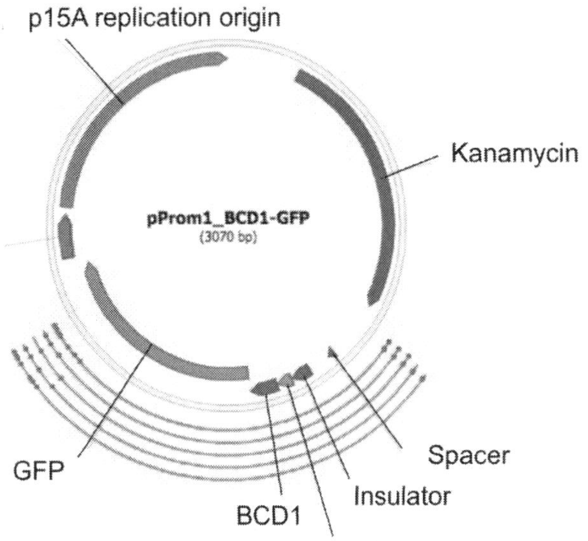

- Golden-Gate, Gibson, and yeast assembly
- Obtained 5 colonies for each plasmid
- Green arcs – matching sequence

34

Sequencing results

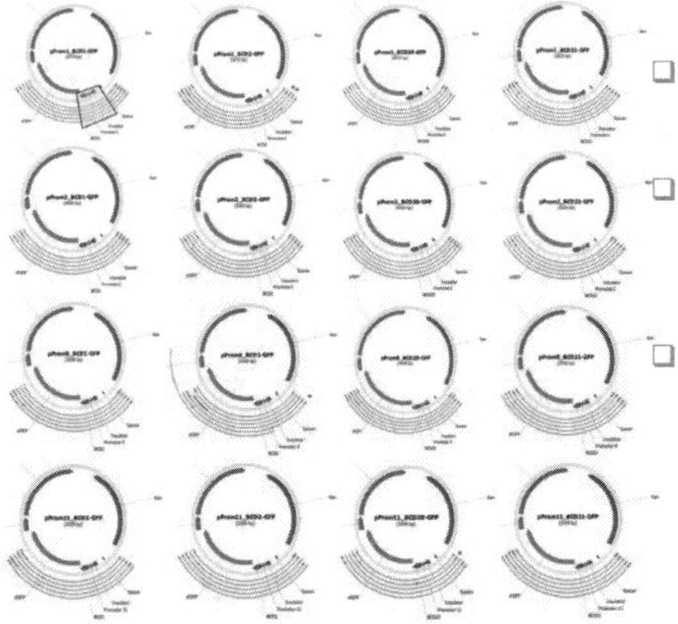

- Golden-Gate assembly
- Obtained 5 colonies for each plasmid
- High sequencing match for the ligation region (~95%)

Shih et al. *ACS Syn. Biol.* 2015, In press. 35

Outline

- **Synthetic Biology**
  - Design-build-test-learn cycle
- **Microfluidics: Introduction and Theory**
  - Microfluidic paradigms, General mechanism, Electromechanical model
- **Microfluidics for Synthetic Biology**
  - Device Design
  - Automation system
  - DNA assembly and Transformation
- **Conclusions**
  - Final notes

36

Summary

- We designed a microfluidic device for automating DNA assembly and transformation
- We assembled 16 plasmids using three DNA assembly methods: Golden Gate, Gibson, and Yeast assembly
- With 95% confidence level, cross-contamination on our devices is < 9.5%
- Sequencing results showed high sequence clones (> 95%)

37

Poster Advertisement

- **Philip Gach (Postdoctoral Scientist – JBEI)**
- Synthetic Biology in a Chip: A Digital Microfluidic Platform for Cell Transformation, Culture and Expression
- Poster 58 – Thursday June 11

38

Acknowledgements

Nathan Hillson

Anup Singh

Garima Goyal

Peter Kim

Nicolas Koutsoubelis

- DOE for funding

U.S. DEPARTMENT OF ENERGY

Charles Denby
Philip Gach
Hector Martin Garcia
Anna Lechner
Christopher Petzold
Jess Sustarich

39

Questions?

Throughput challenges

- We start to "run out" of space on the device if we are to increase the number of electrodes.
- m x n electrodes require **m x n connections**

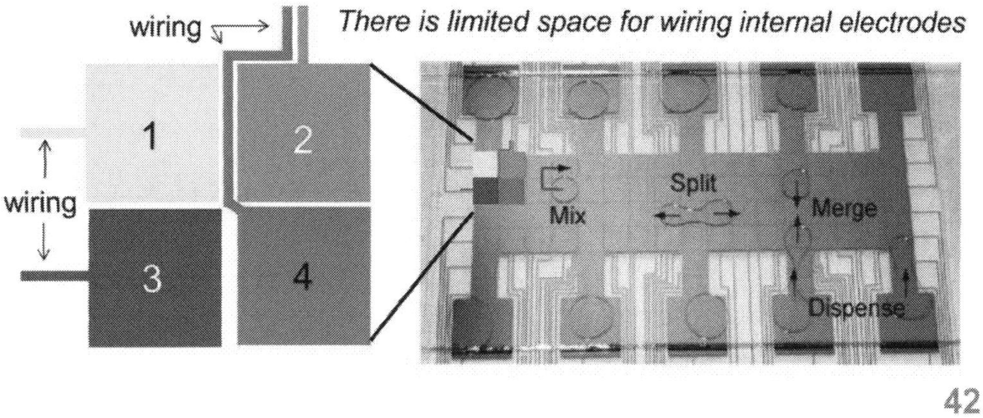

42

Electrowetting

- For high surface-tension liquids, two effects are observed: shape change and position change

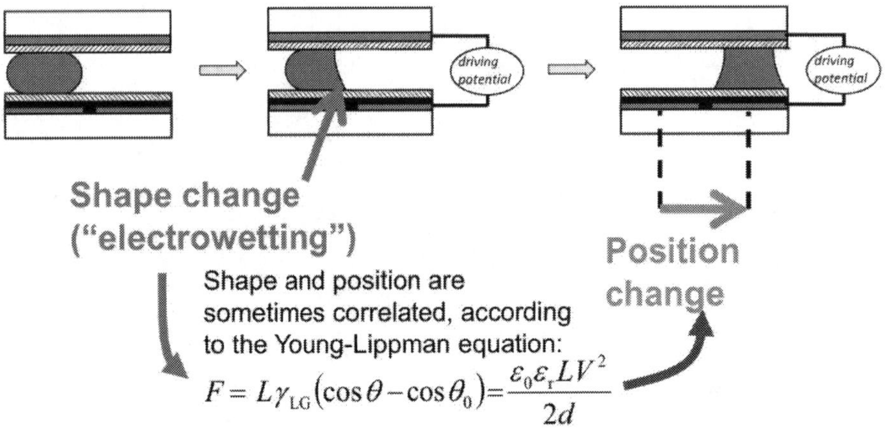

Electrowetting

- For low surface-tension liquids, little or no shape change is observed; only position change occurs

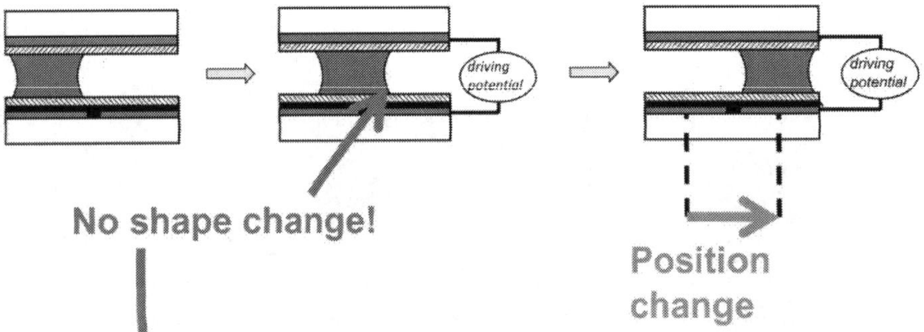

$$F = L\gamma_{LG}(\cos\theta - \cos\theta_0) = 0 \quad ???$$

Electrowetting

- In fact, simple modeling of *qE* driving forces for droplet position (with no estimation of shape) generates useful predictions of droplet position change, regardless of surface tension

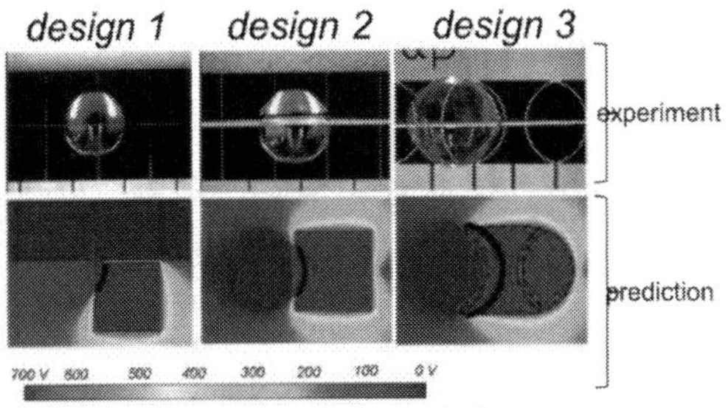

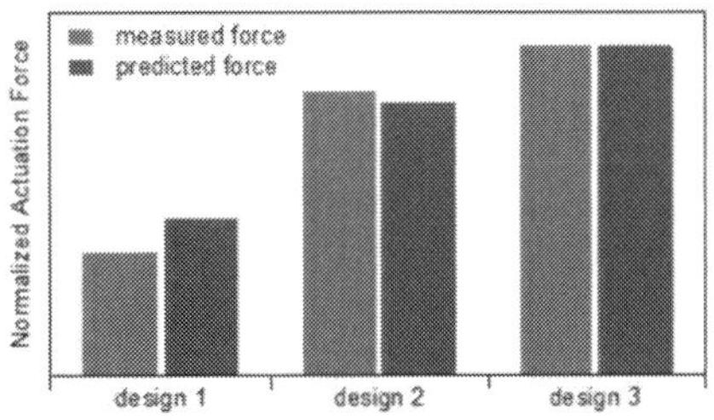

-from Abdelgawad et al. *Journal of Applied Physics*, 2009, 105, 094506.

Table 1 Organic solvent and solution droplet movement feasibility

Liquid	μ/D[a]	k_d[a]	σ/S m^{-1}[a]	γ/dynes cm^{-1}[a]	$\Delta\theta$/°[a]	Movable[b]
Formamide	3.7	111	3.5×10^{-3} [52]	57	16.1 ± 0.4*	Y
Water	1.9	80.1	8.7×10^{-4}*	72	30.0 ± 0*	Y
Formic acid	1.4	51.1	7×10^{-3} [53]	37	26.3 ± 2*	Y
DMSO	4	47.2	3×10^{-5} [52]	43	15.3 ± 0.3*	Y
DMF	3.8	38.3	3.2×10^{-5} [52]	37	6.9 ± 2*	Y
Acetonitrile	3.9	36.6	1.9×10^{-5} [52]	29	9.8 ± 0.6*	Y
Methanol	1.7	33	1.7×10^{-4}*	22	9.8 ± 1*	Y
Ethanol	1.7	25.3	7.4×10^{-5}*	22	10.5 ± 0.7*	Y
Acetone	2.9	21	5×10^{-7} [54]	23	6.4 ± 0.2*	Y
Piperidine	1.2	4.3	1×10^{-5} [55]	29	8.9 ± 0.6*	Y
1-Pentanol	1.7	15.1	8×10^{-7} [56]	25	12.8 ± 0.8*	Y◇
1-Hexanol	1.8	13	1.6×10^{5}*	26	14.6 ± 0.7*	Y◇
Dichloromethane	1.6	8.9	1×10^{-7} [57]	27	3.7 ± 0.6*	Y◇
Dibromomethane	1.4	7.8	2.6×10^{-6}*	39	7.3 ± 1*	Y◇
THF	1.6	7.5	5×10^{-8}*	26[58]	4.9 ± 0.9*	Y◇
m-Dichlorobenzene	1.7	5		35	0.1 ± 0.1*	Y◇
Chloroform	1	4.8	7×10^{-8} [57]	27	0.5 ± 0.4*	Y◇
65% Toluene, 35% 1-hexanol	1.1§	3.7*	3×10^{-8}*	28*	6.1 ± 3*	Y◇
70% Toluene, 30% 1-hexanol	1§	3.4*		28*	0.3 ± 0.3*	Y◇
4-Methyl-3-heptanol	1.23	3.3		25[59]	0.1 ± 0.3*	Y◇
75% Toluene, 25% 1-hexanol	0.9§	3.1*		28*	0.0 ± 0.2*	Y◇
4.7 mM TBATFB in toluene		2.3*	1.8×10^{-7}*	28*	5.6 ± 0.6*	Y◇
80% Toluene, 20% 1-hexanol	0.8§	2.9*		28*	0.1 ± 0.3*	N
4-Methyl-4-heptanol		2.9		25[60]	0.2 ± 0.2*	N
Toluene	0.4	2.4*	8×10^{-14} [58]	28	0.2 ± 0.4*	N
Carbon tetrachloride	0	2.2	4×10^{-16} [58]	26	1.0 ± 0.0*	N
Cyclohexane	0	2	7×10^{-16} [58]	25	0.2 ± 0.5*	N
Decane	0	1.8*		23	0.5 ± 0.6*	N
p-Dichlorobenzene dissolved in toluene		1.5*				N

46

Pluronics

- Pluronics: Triblock Co-polymers

$$[PEO]_x - [PPO]_y - [PEO]_x$$

Hydro: Philic - Phobic - Philic

- May coat surfaces or accumulate at interfaces to prevent biomolecule adsorption

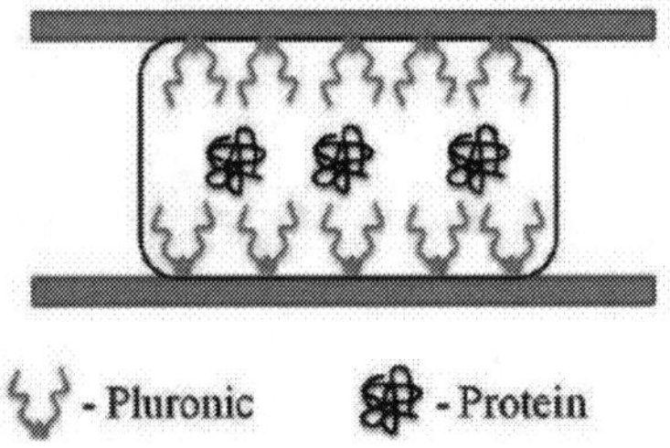

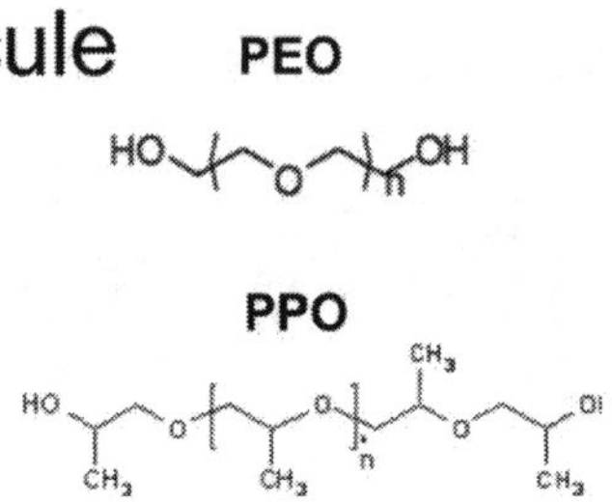

47

Preventing surface fouling

- The use of amphiphilic solution additives at low concentrations nearly eliminates protein adsorption of proteins from droplets onto surfaces

Fluorescent images of surfaces after exposure to droplet containing:

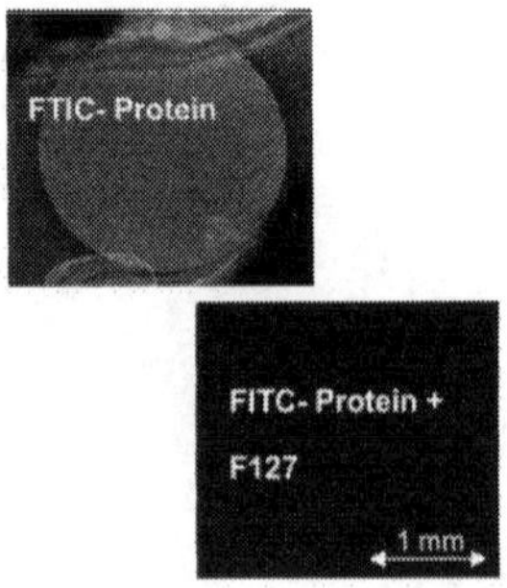

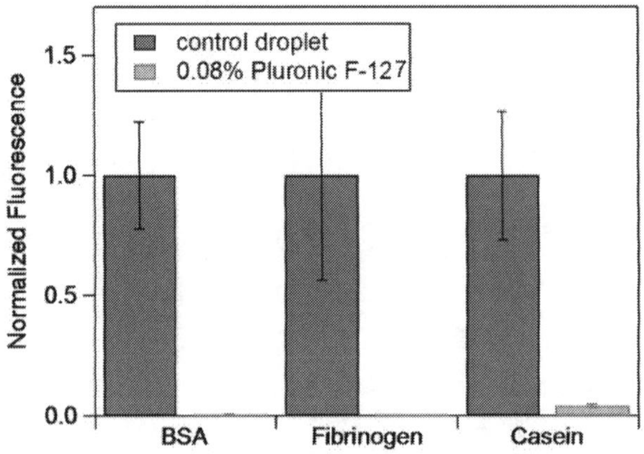

Luk, et al., *Langmuir*, **2008** 24, 6382-6389

48

jbei

U.S. DEPARTMENT OF ENERGY

Pluronics – Physical Properties

Pluronic	average molecular weight[a]	ave. PPO chain length (n)	ave. PEO chain length (m)
L35	1900	16	11
F38	4700	16	46
L44	2200	21	11
L62	2500	30	8
L64	2900	30	13
F68	8400	30	75
L92	3650	47	10
P105	6500	54	38

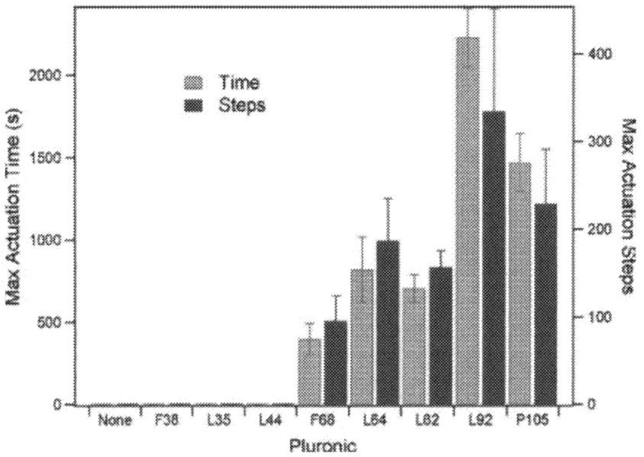

- PPO chains must be above ~30 molecular units
- Increase in PPO ~ increase in steps and actuation time
- PPO is the greater effect than PEO

Au, et al., *Langmuir,* **2011** 27, 8586-8594

49

Device Fouling

Non-specific adsorption of biomolecules on surfaces –leading to cross-contamination or droplet sticking

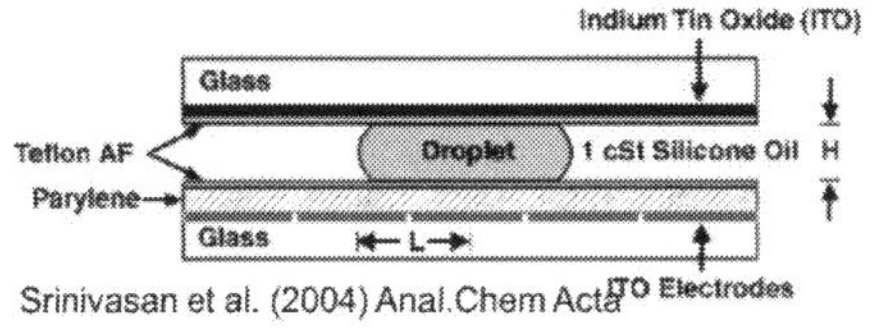

Srinivasan et al. (2004) Anal.Chem Acta

Oil as a filler fluid

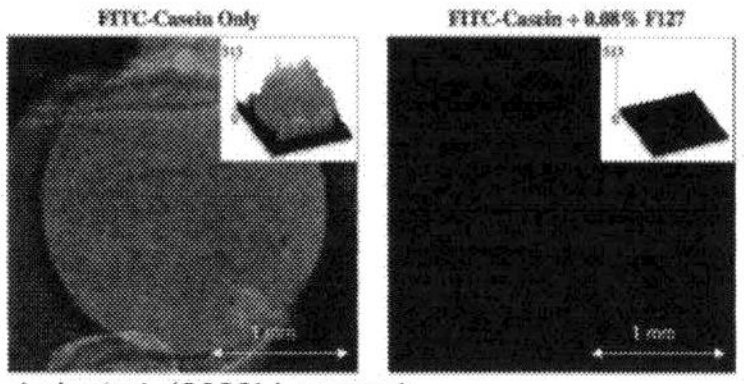

Luk et al. (2008) Langmuir

Amphiphilic polymer additives

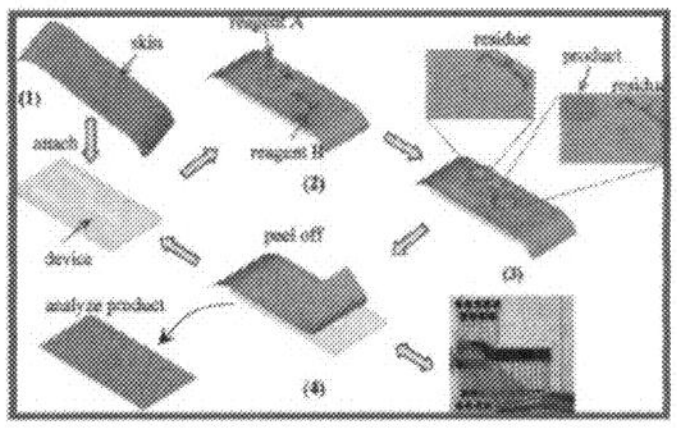

Removable hydrophobic insulator

Yang et al. (2009) Anal. Chem.

50

Droplet evaporation

Raises concentration of salts and other buffer constituents (hypertonic solution)

SOLUTION:

Positioning devices in a humidified atmosphere (stacked petri dishes)

-No negative effects on cell vitality

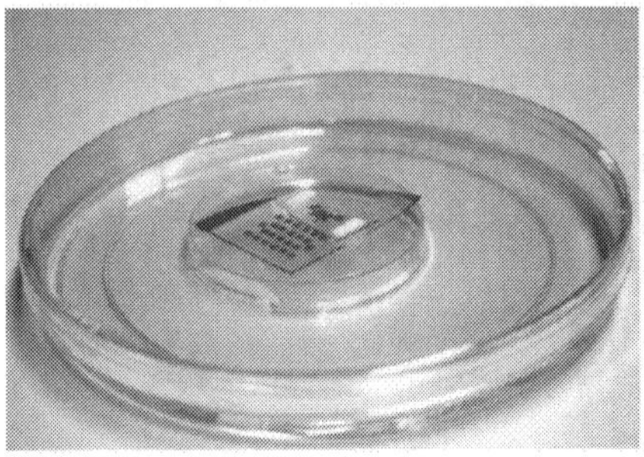

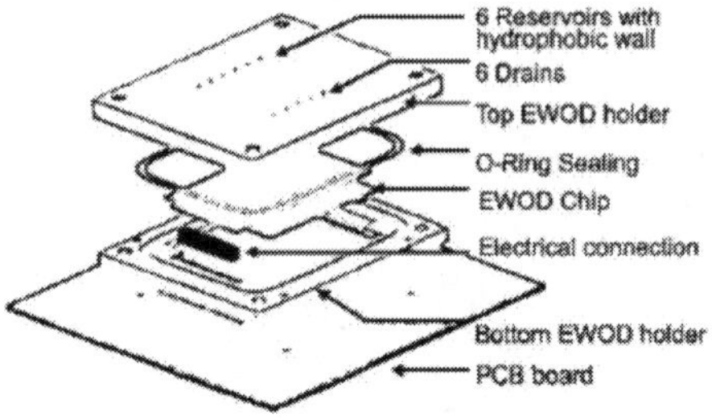

Other solutions: in-house humidifier chamber; sealing device with O-ring

51

Non-uniform dispensed liquid

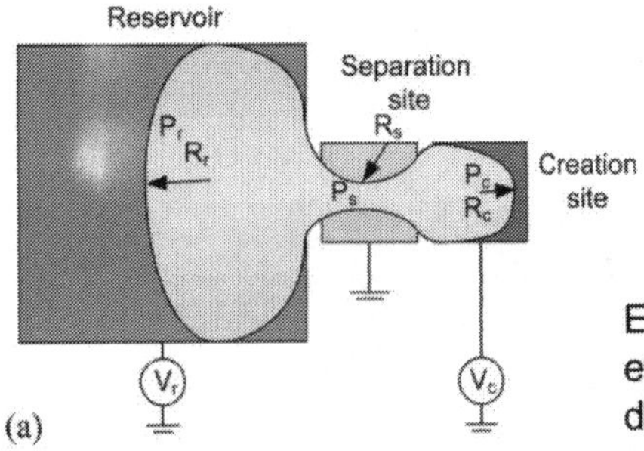

(a)

$$\Delta P_{sr} = P_s - P_r = \gamma_{LG}\left[\frac{1}{R_s} - \frac{1}{R_r} + \frac{\varepsilon_0 \varepsilon V_r^2}{2\gamma_{LG} t}\frac{1}{d}\right] \quad (1)$$

$$\Delta P_{sc} = P_s - P_c = \gamma_{LG}\left[\frac{1}{R_s} - \frac{1}{R_c} + \frac{\varepsilon_0 \varepsilon V_c^2}{2\gamma_{LG} t}\frac{1}{d}\right] \quad (2)$$

EWOD term must be greater than Capillary effects – we may have accumulated error due to capillary forces

Errors include: surface defects, dielectric charging, fabrication variations, operation environment - these can offset the contact angle change or droplet shape causing an offset in droplet volume.

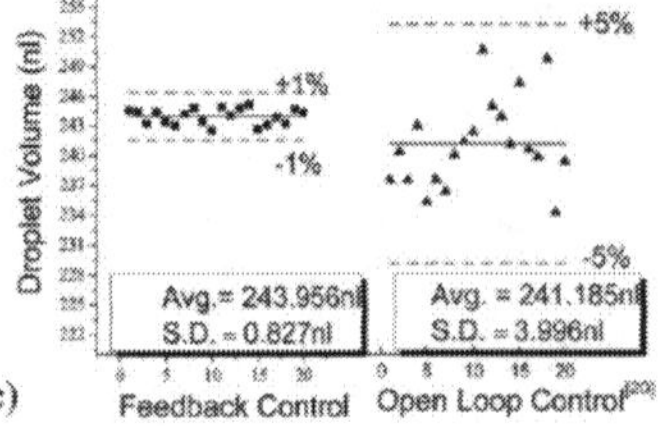

c)

52

DMF for Suspension Cells

- Are cells manipulated by DMF harmed? We hypothesize "no"

The E-field drops across the insulator!

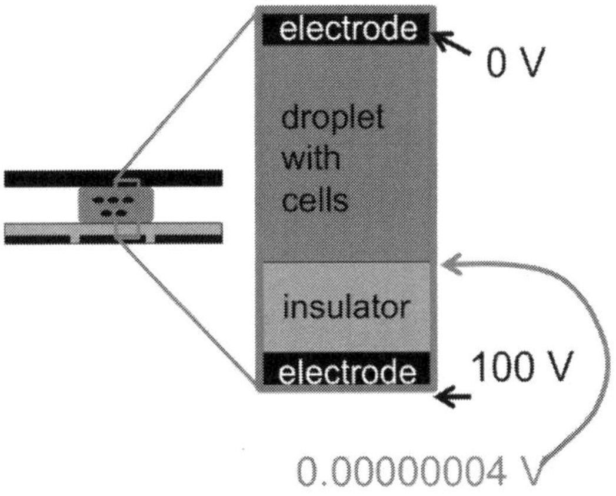

Viability and proliferation tests support this hypothesis

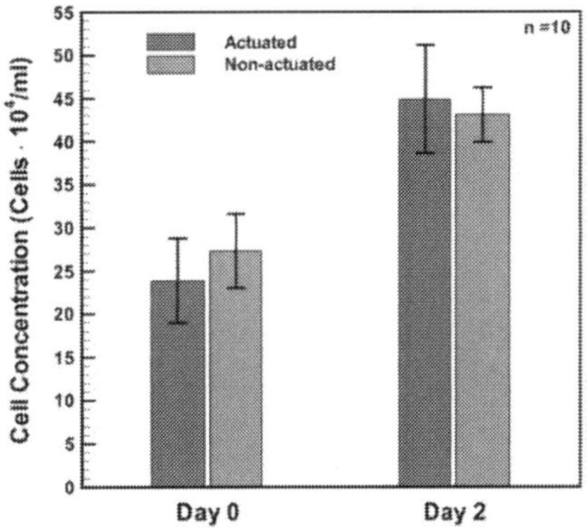

Barbulovic-Nad et al, *Lab Chip* **2008**, *8*, 519-526

DMF for Suspension Cells

- Are cells manipulated by DMF harmed? We hypothesize "no"

<u>Ba/F3 cell Microarray Heat Map (~22,000 genes)</u>

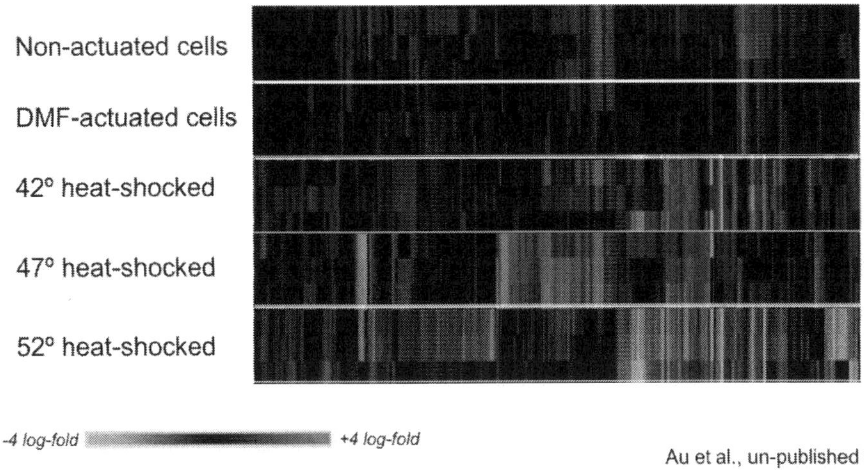

Au et al., un-published work

DMF for Suspension Cells

- Are cells manipulated by DMF harmed? We hypothesize "no"

Genes Expressed Diff. Relative to Control:

Non-actuated cells	-	
DMF-actuated cells	3	**Ube2c** (cell division, ubiquitination), **Dstn** (cell motility, actin polymerization), **EG633570** (?)
42° heat-shocked	65	some common proteins: stress response (e.g., **Osm**, **Ler3**, **Plk3**), transcription regulators (e.g., **Fos**, **Fosb**, **Axud1**), MAPK cascade (e.g., **Osm**, **Ccl3**)
47° heat-shocked	48	
52° heat-shocked	82	

Au et al., un-published work

DMF for Suspension Cells

- Are cells manipulated by DMF harmed? We hypothesize "no"
- The bottom line: the effects of DMF actuation appear to be modest (if any), and are likely no worse than those caused by other common cell handling techniques (e.g., FACS)

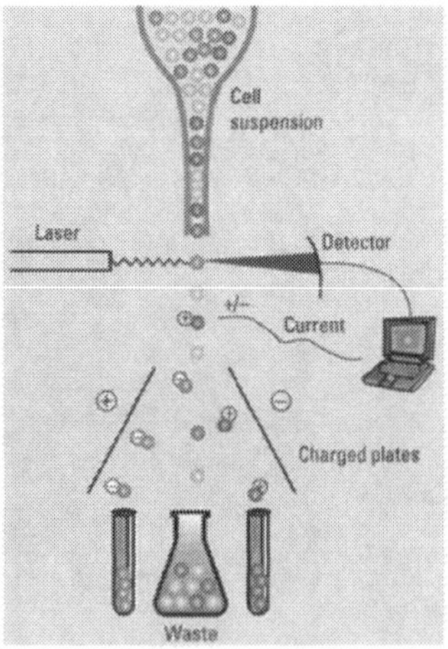

Fluorescence Activated Cell Sorter (FACS)

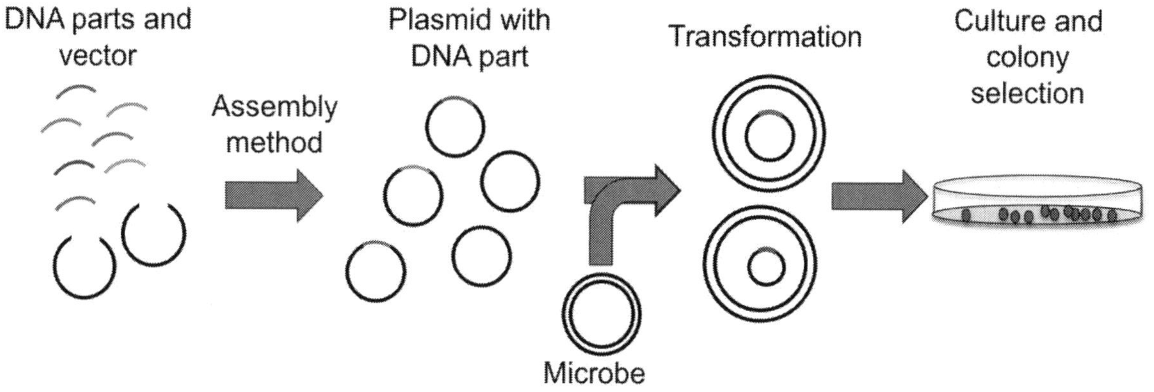
DNA parts and
vector
Assembly
method
Plasmid with
DNA part
Microbe
Transformation
Culture and
colony
selection

Use of an Enzyme-coupled Biosensor to Engineer a BIA Fermentation Pathway in Yeast

John Dueber, U.C. Berkeley

Natural Family of Benzylisoquinoline Alkaloids (BIAs) Produced in Variety of Plants

- *~2,500 Natural BIAs – analgesics as well as anti-cancer and antibiotic candidates*
- *Genetics is difficult in these plants*
- *Microbial production promising (much faster doubling time)*

Adapted from Hagel and Facchini. Plant and Cell Physiology. 2013

BIA Pathways Contain Multiple P450s

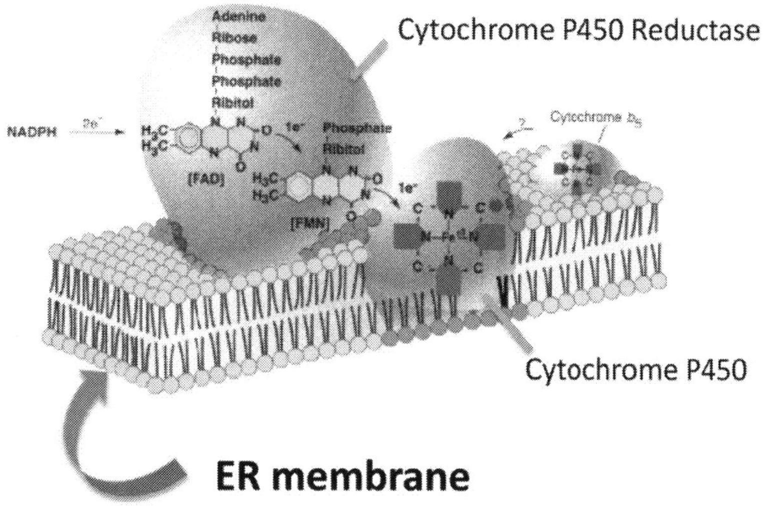

P450s usually require ER membrane for high activity

Yeast is the Most Promising BIA Fermentation Host

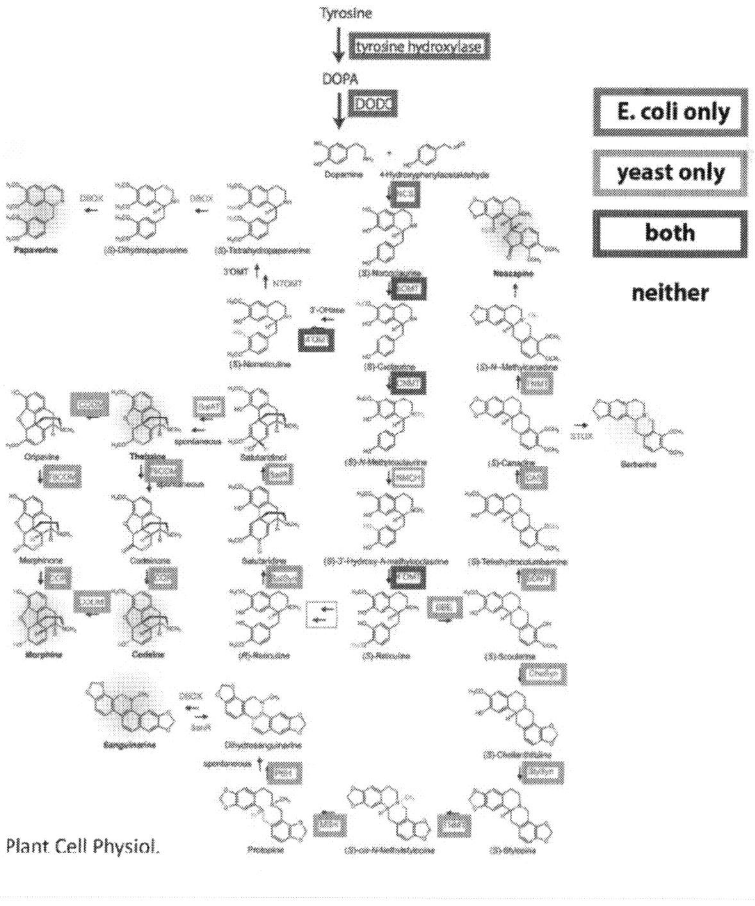

Adapted from Hagel and Facchini. 2013. Plant Cell Physiol.

BIA Biosynthesis Requires a Tyrosine Hydroxylase

Design an Enzyme-coupled Biosensor for Tyrosine Hydroxylase Activity

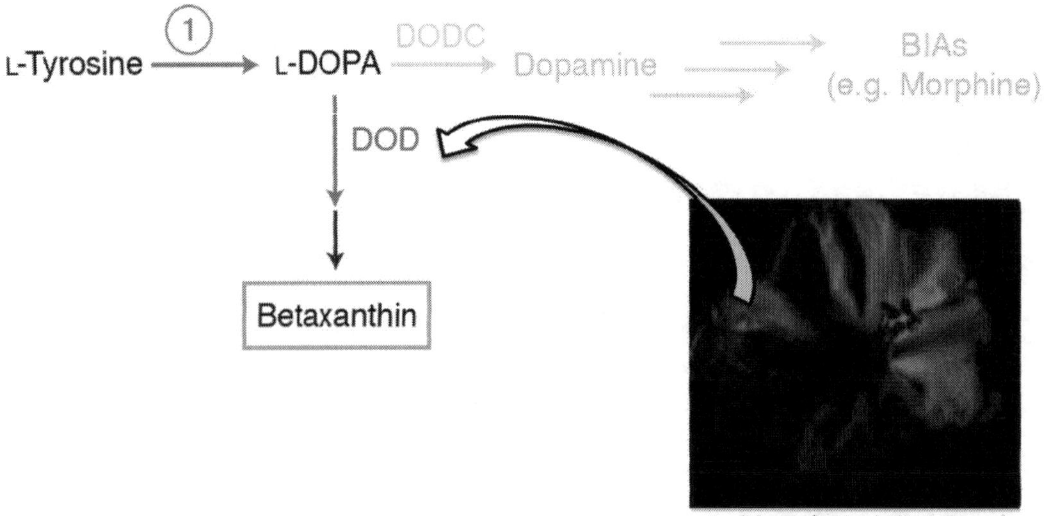

4 o'clock (Mirabilis jalapa)

Betaxanthin as a Biosensor

Microscopy of fluorescent product compartmentalized in vacuole

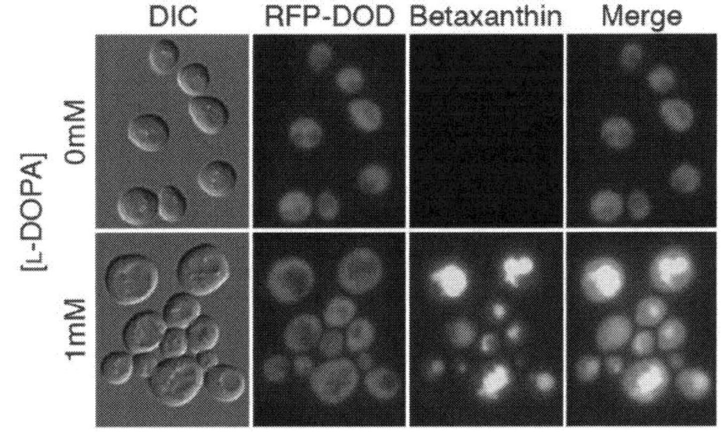

Betaxanthin as a Biosensor

Fluorescent plate reader measurements of cultures

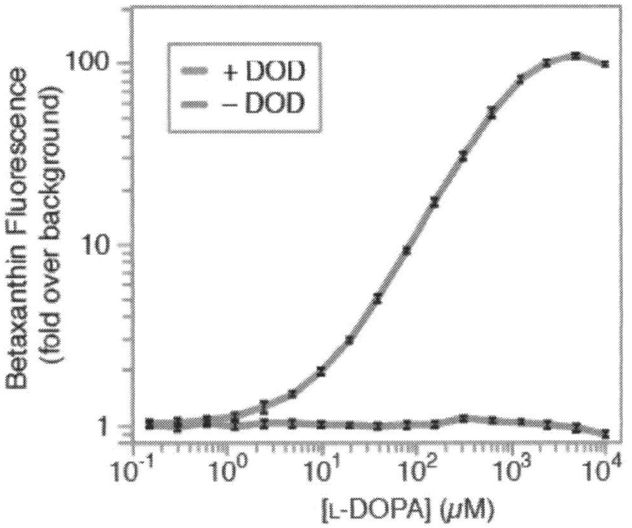

Betaxanthin as a Biosensor

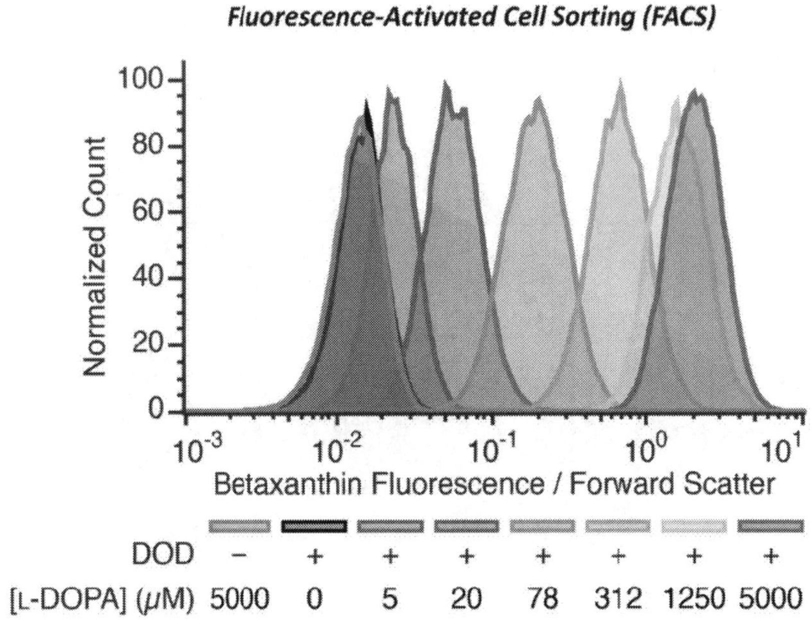

Where to Look for a Tyrosine Hydroxylase?

Beets.

Tyrosine Hydroxylation Occurs in Beets

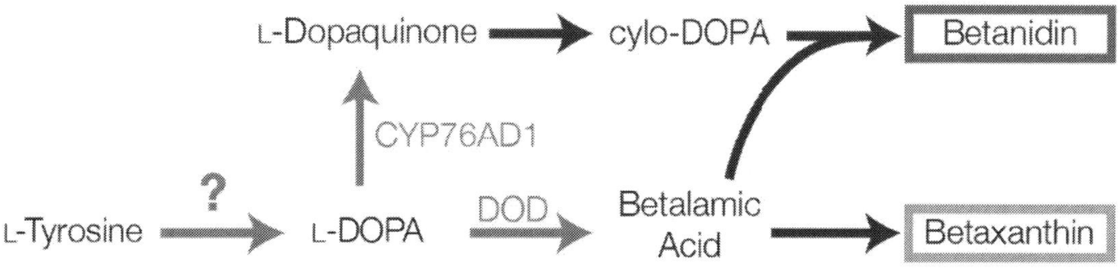

CYP76AD1 is Also a Tyrosine Hydroxylase

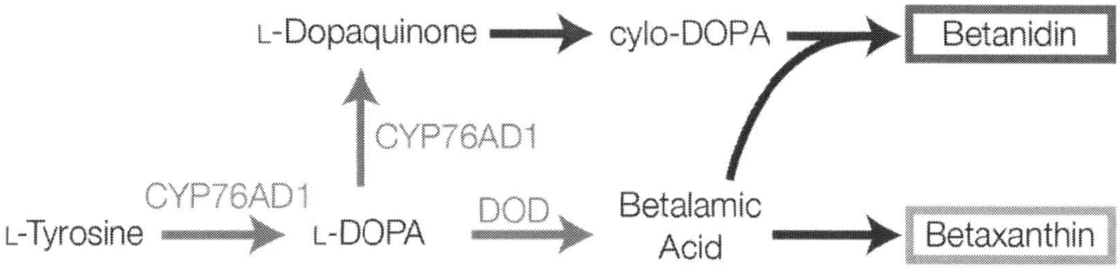

DOD

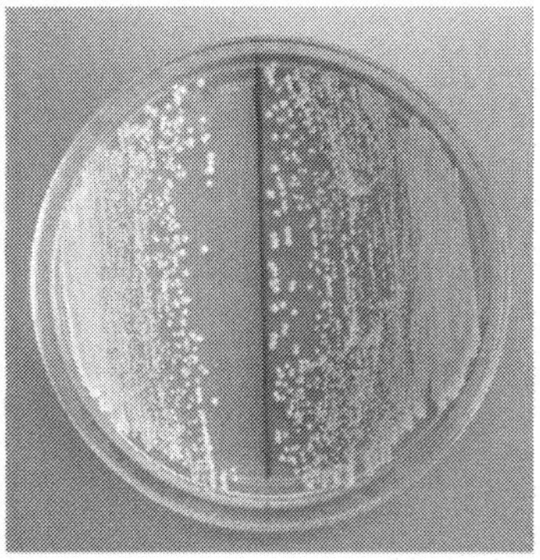

CYP76AD1 + DOD

DeLoache et al. 2015. NCB.

Improving CYP76AD1 Activity via Mutagenesis

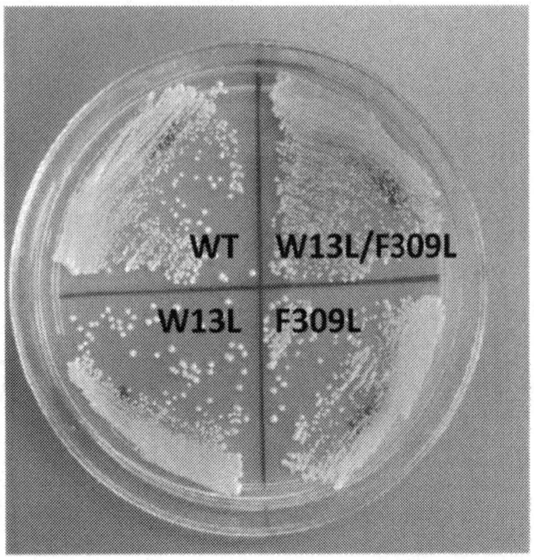

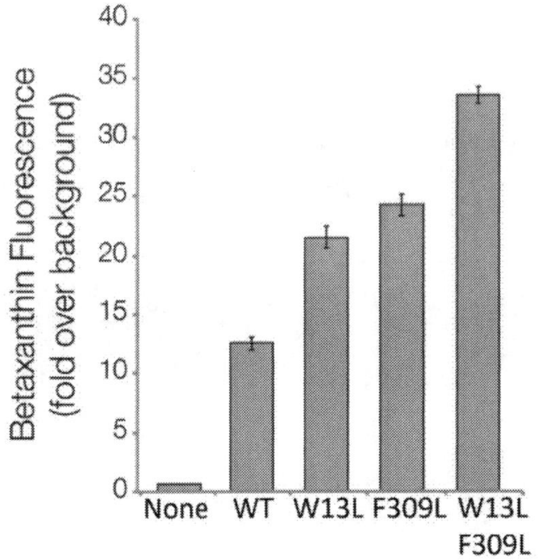

L-DOPA Production Improvements Translate to Dopamine

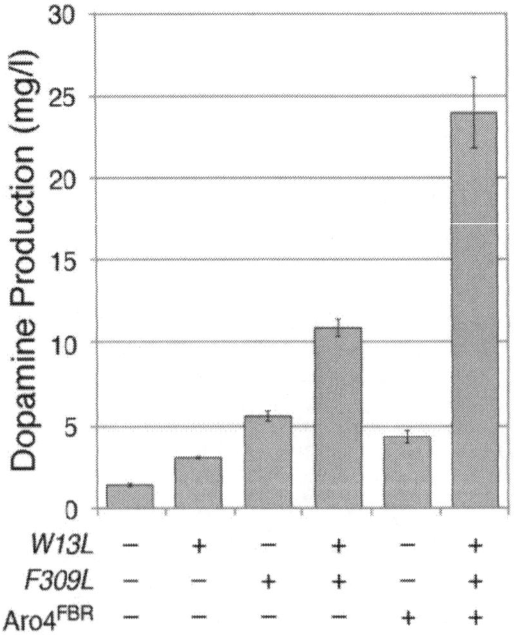

Dopamine Levels Correlate Very Well with Betaxanthin

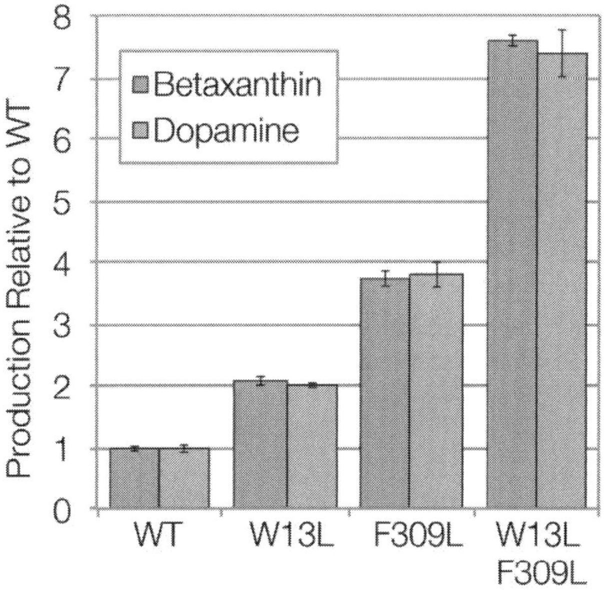

F309L Mutation Reduces Off-Pathway Flux

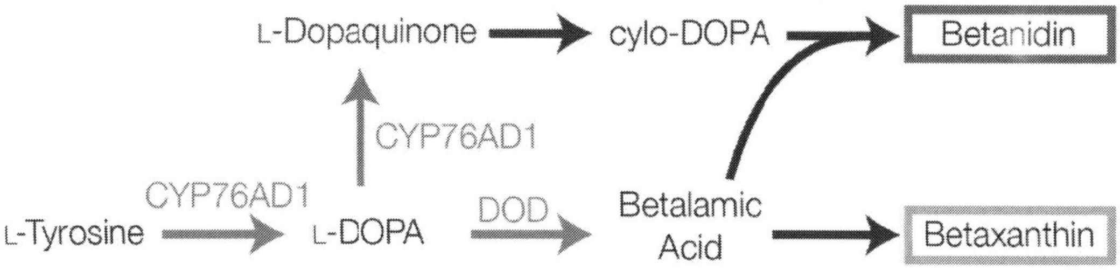

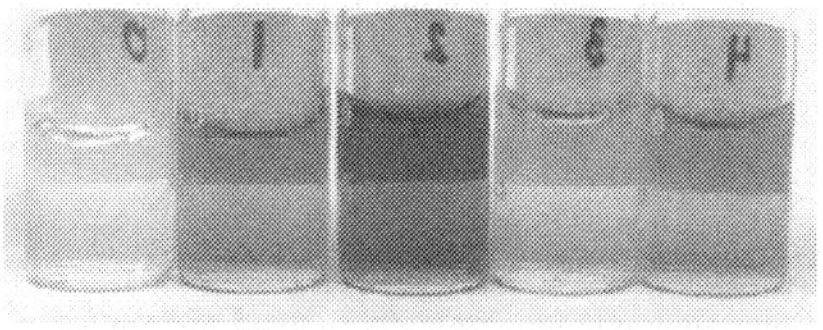

Neg. Control | WT | *W13L* | *F309L* | *W13L F309L*

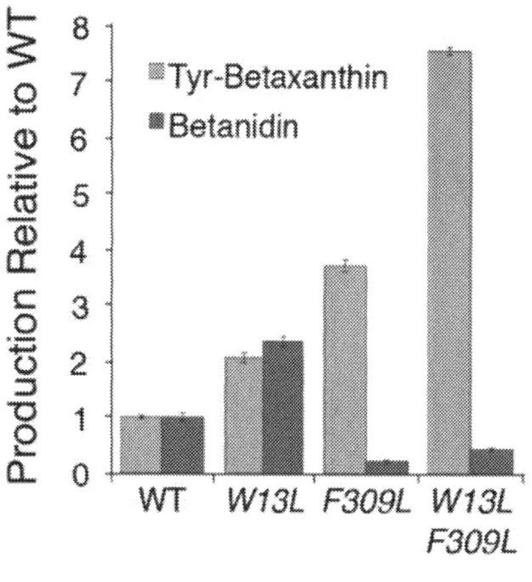

F309L is Found in Multiple Paralogs

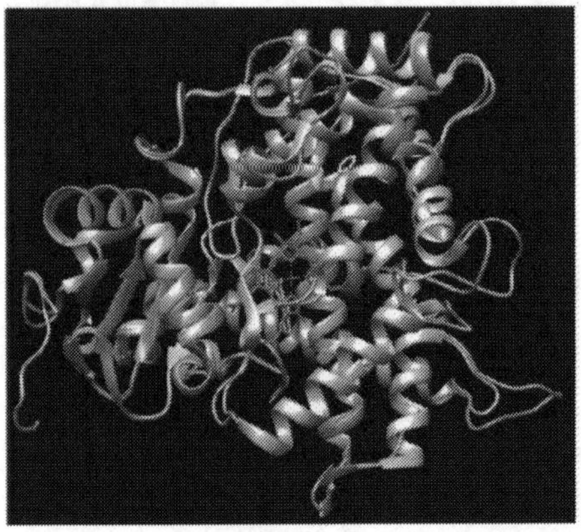

```
Bvlu_022460_qtnn    HMLMDLFLAGTDTTSITLEWAMTELLRNPEKMEKVQIELDKVLGKD-SSLQESMISKLPY 343
Bvlu_022470_qtfp    HLLADLFIAGTDTTSNTLEWAMAELLRNPEKMEKAQAEINKVLGKD-SSMQENDISKLPY 342
Bvlu_022400_apow    HMLFDLPLAGTDTTSSTLEWAMTELLRNPKVMEKAQIEIDQVLGKD-GSNQELDIAKLPY 462
Bvl_001290_xkhk     HLLMDFFTAGTDTTSSTLEWAMTELLRNPEKMAKAQVELEQVLGKN-KVVGEFDISKLPY 355
Bvl_001300_sidn     HLLMDFFTAGTDTTSSTLEWAMTELLRNPEKLAKAQAELEQVVGKN-KVVKEADISKLPY 355
Bv9_228610_yqeq     HLLVDIFDAGTDTTASTLEWAMAELVKNPEMMTKVQIEIEQALGKDCLDIQESDISKLPY 352
Bv9_228860_ickx     HLLVDIFDAGTDTTASTLEWAMAELVKNPYMMVKVQDEIEKAIGKGCSMVQESDISKLPY 314
CYP76AD1            HLLVDIFDAGTDTTSSTFEWVMTELIRNPEMMEKAQEEIKQVLGKD-KQIQESDIINLPY 350
                    *:* *:* ******: *:**.*:**::**  : *.* *:.:.:**.   : *  * :***
```

F309L Mutation Causes Same Substrate Shift in Orthologs

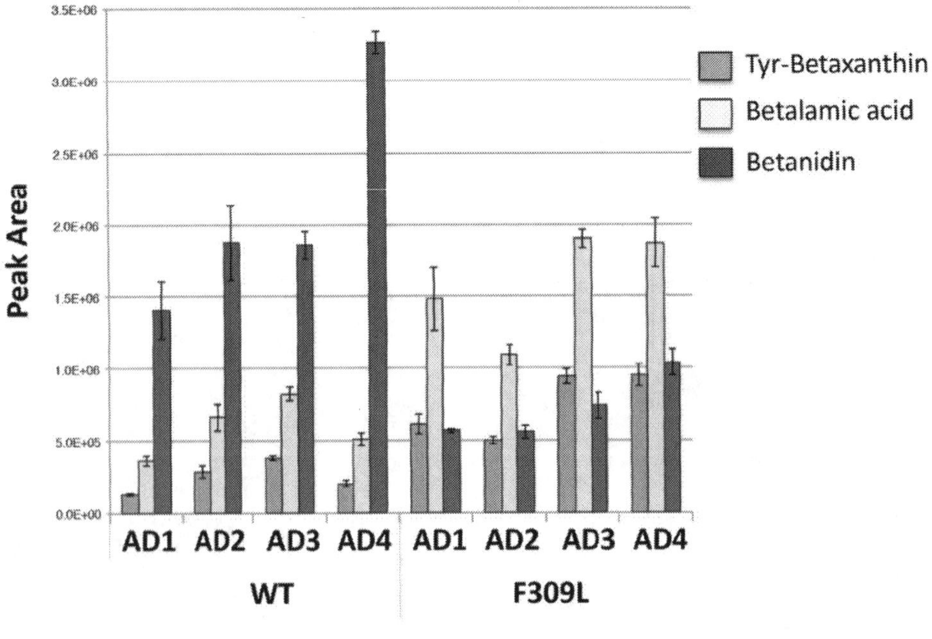

Reticuline Biosynthesis from Glucose

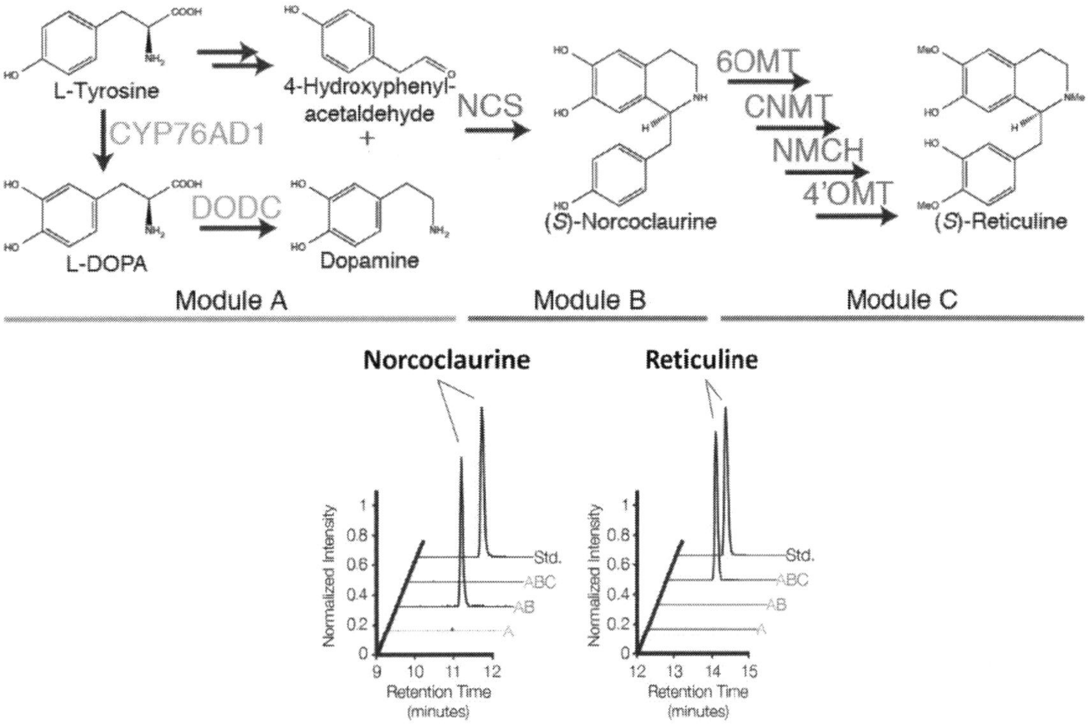

Reticuline is a BIA Hub

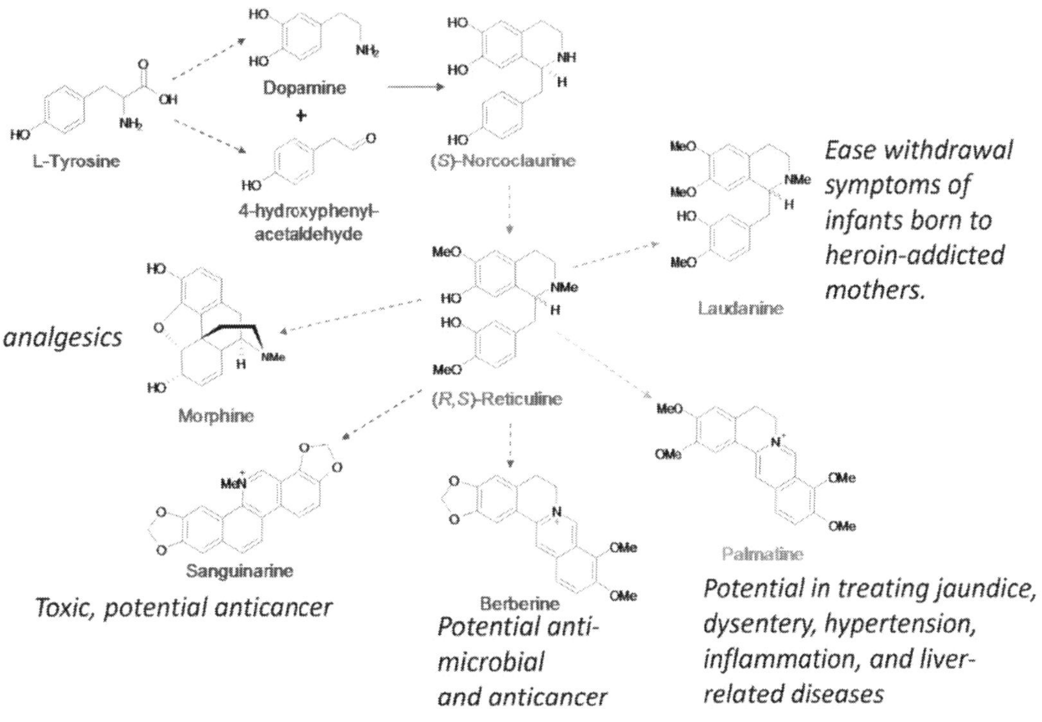

Could this Strain be used to Home-brew Morphine?

Oye, Bubela, Lawson. Nature. 2015

How Far Away are We?

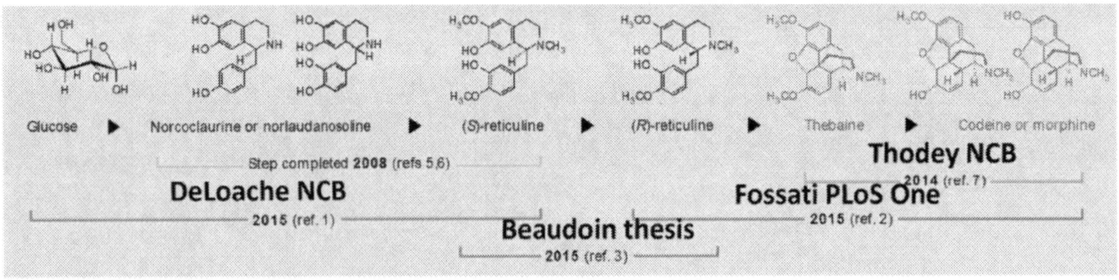

- *Pathway is ~15 enzymes, all activities have been identified*
- *A few very low efficiency steps*
- *Probably unexpected challenges of putting all enzymatic reactions into a single strain.*
- *Detectable amounts of morphine in < 3yrs?*

Summary

Coupling a biosensor to a intermediate of interest is a powerful for isolating/improving an enzymatic activity.

S-reticuline production should enable:

- *~2500 natural products*
- *"unnatural" natural products*
- *But also potentially illicit production*

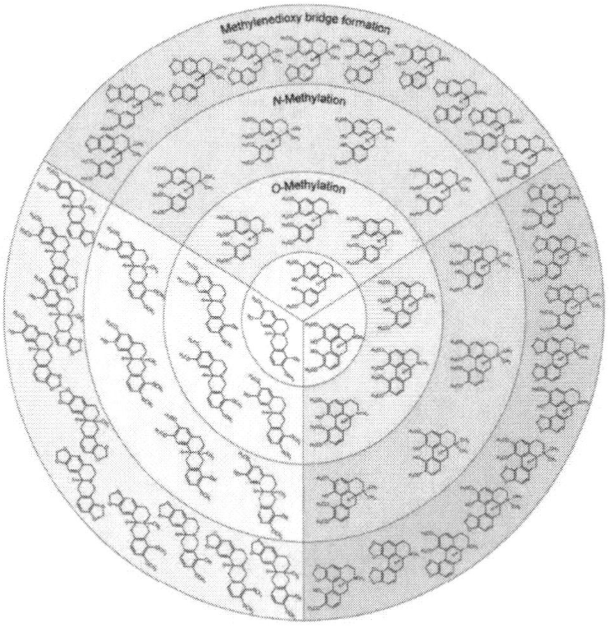

Hagel and Facchini. 2013. Plant Cell Physiol.

Collaborators:
Claire Tomlin, Kristala Prather, Dan Nomura, Amy Keating, Lauren Narcross, Vincent Martin

Energy Biosciences Institute

Bakar Fellows Program

NSF

DEPARTMENT OF ENERGY UNITED STATES OF AMERICA

~30 mg/L Morphine Homebrew?

- *Start at 25mg/L dopamine (although this titer could likely be further increased).*
- *Rest of the pathway is currently ~10^6 off from a strain that could produce this from a pint of home-brew*
- *Commercial fermentations often difficult – but homebrew would require only mg/L scale.*

FACS of Mutants

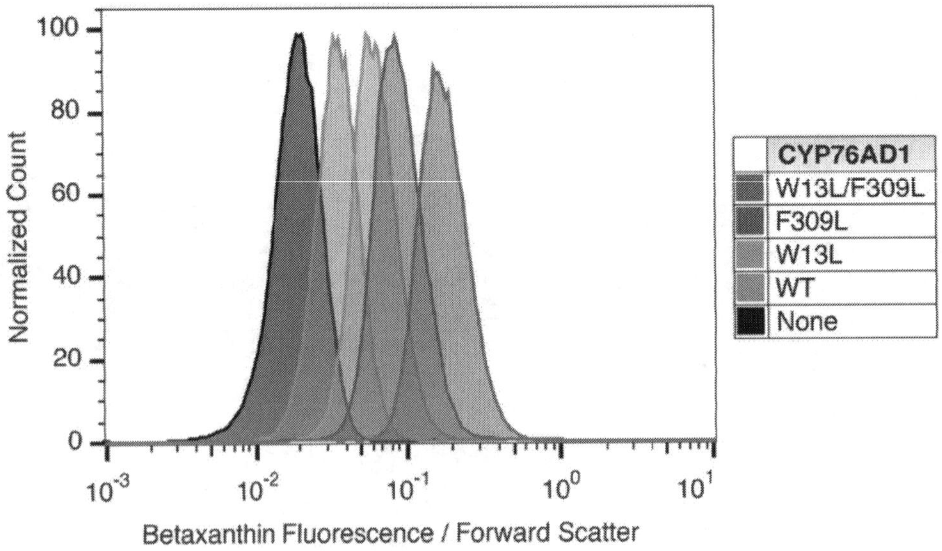

F309L Mutation Reduces Off-Pathway Flux

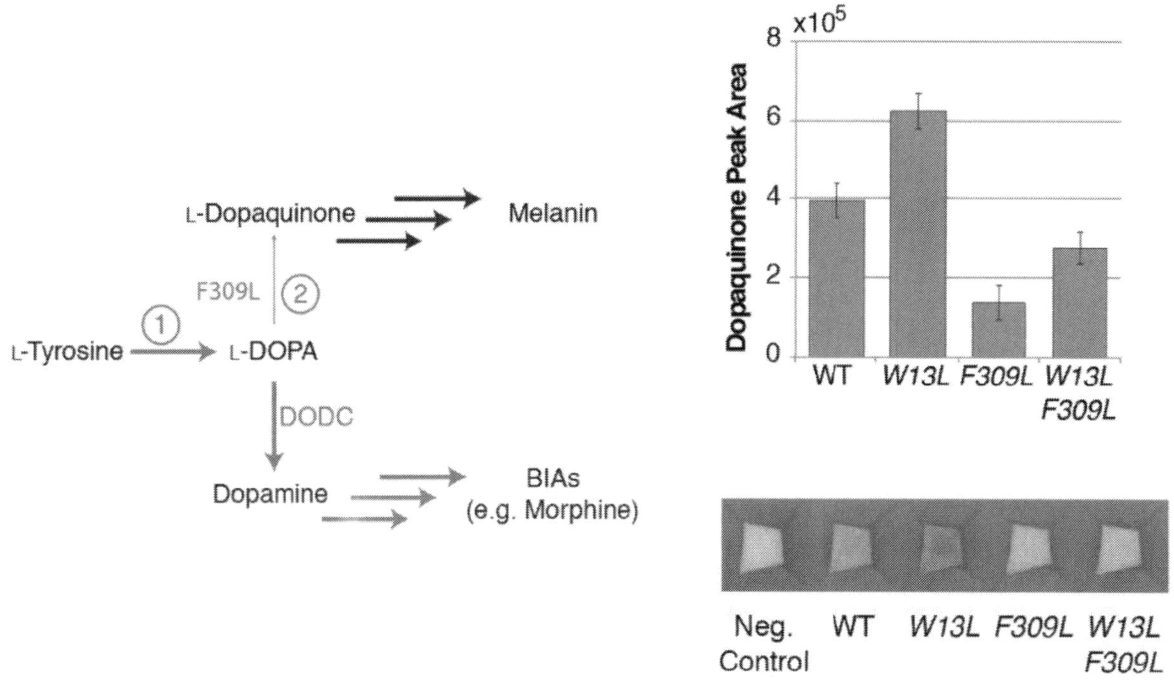

Synthetic Biology with CRISPR: Controlling and Reprogramming the Genome

Stanley Qi
Assistant Professor
Department of Bioengineering
Department of Chemical and Systems Biology
Stanford University

Stanford MEDICINE
Stanford | Bioengineering
Schools of Engineering & Medicine

One challenge of synthetic biology: complexity

Need robust tools to control complex biological systems

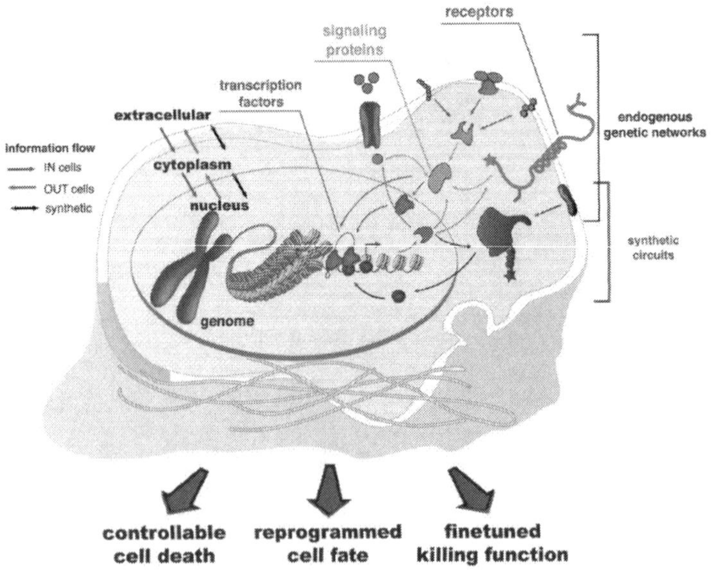

How to reprogram cells with designed useful functions?

How to use synthetic systems to understand biology?

Need useful tools to control cells on different levels:

- **genetic/epigenetic**
- **signaling**
- **cell-cell interaction**

How to robustly control the genome function?

Manipulate complexity of the genome

Engineering the genome on different scales?

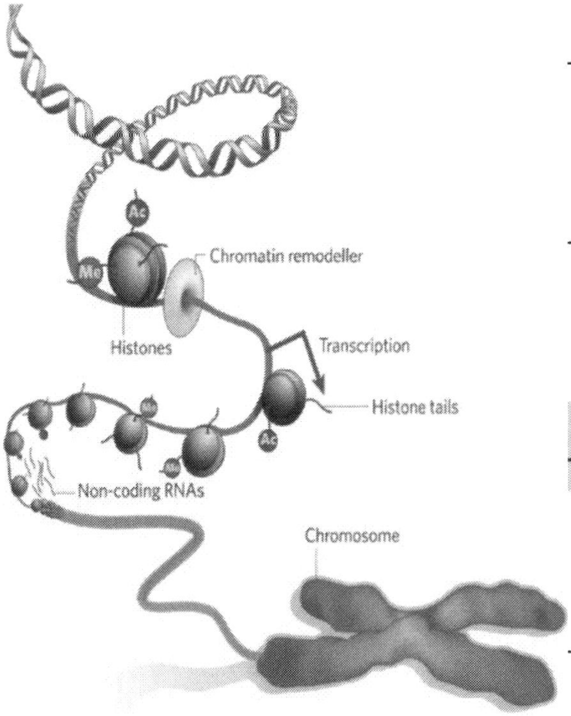

Genome editing
engineering base codes of DNA

Epigenome editing
changing modifications of DNA

Transcription regulation
controlling transcription of DNA

Chromosome reorganization
Patterning spatial organization of DNA

The CRISPR pathway for sequence-specific genetic immunity

CRISPR = Clustered Regularly Interspaced Short Palindromic Repeats

- **RNA-guided DNA recognition**
- **Highly specific & efficient**

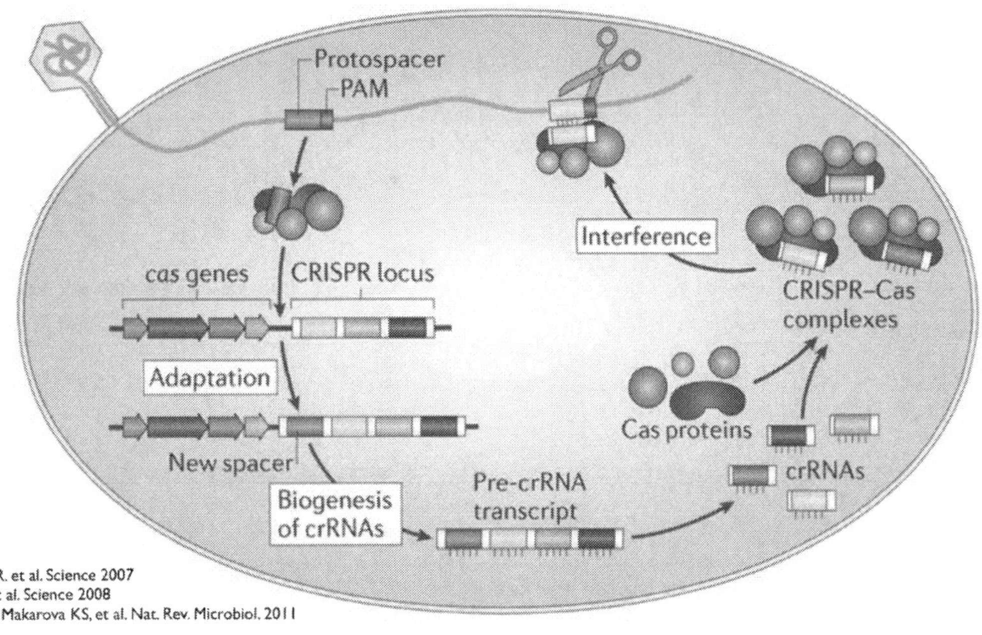

Barrangou R. et al. Science 2007
Brouns JJ, et al. Science 2008
Image from Makarova KS, et al. Nat. Rev. Microbiol. 2011

Repurposing Cas9 for genome regulation

Cas9 is a programmable RNA-guide DNA binding protein

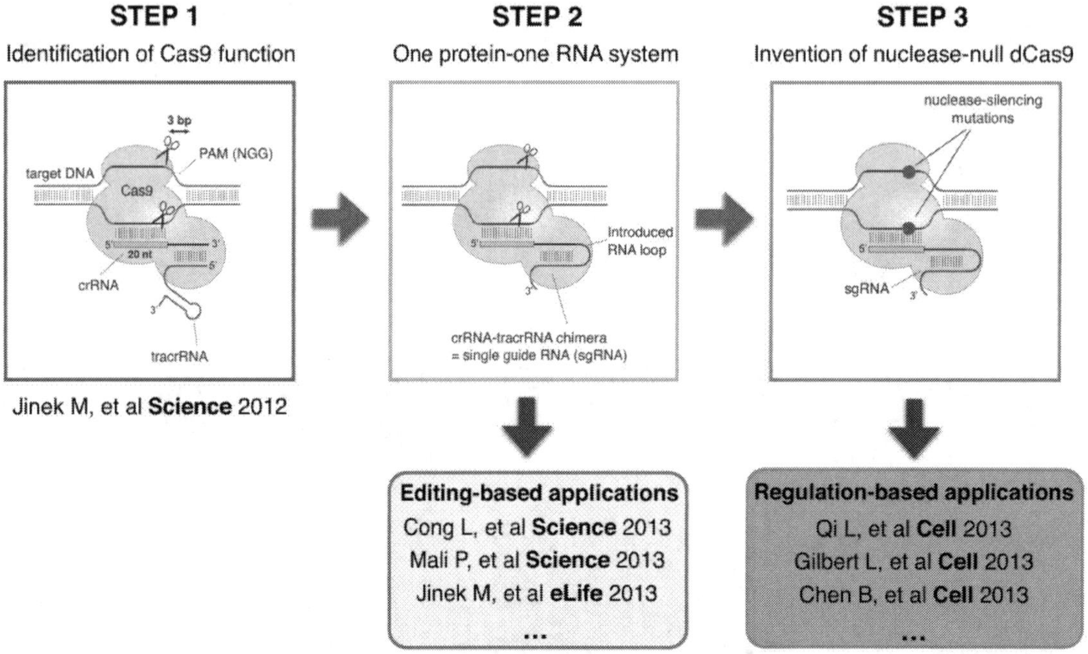

CRISPR is a genome engineering platform

RNA-based technology for editing, regulation & other purposes

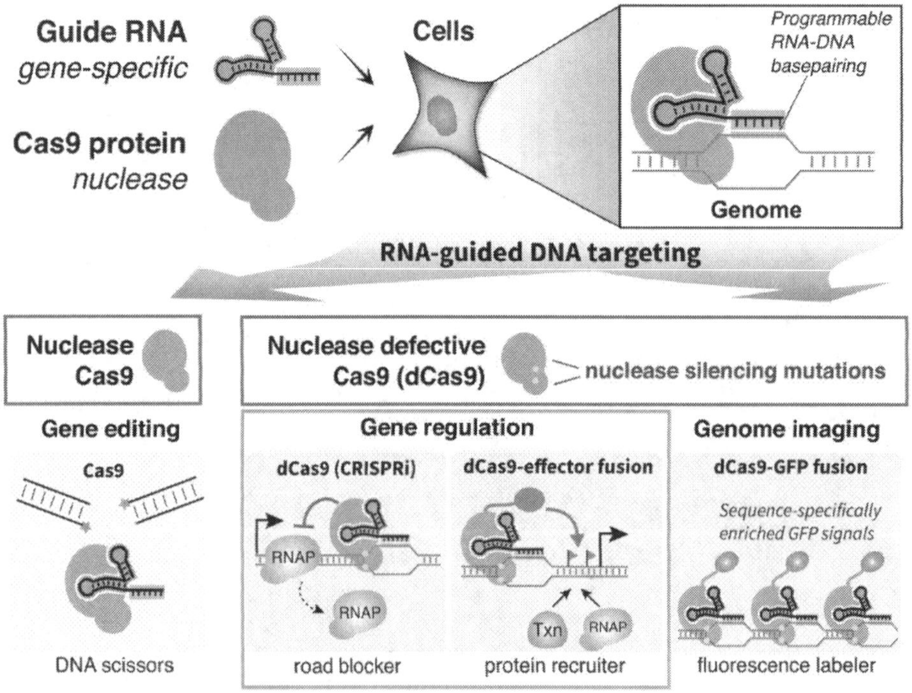

Topics to be covered

1. Development of CRISPR tools for gene regulation

2. Applications to engineer the genome function

Stanford | Bioengineering
Schools of Engineering & Medicine

Topics to be covered

1. Development of CRISPR tools for gene regulation

2. Applications to engineer the genome function

Stanford | Bioengineering
Schools of Engineering & Medicine

Three generations of CRISPRi/a tools

Constantly improving the dCas9 complex for better efficiency

1st generation
CRISPR interference (CRISPRi) based on transcription blocking

2nd generation
CRISPRi/a based on effector recruitment

3rd generation (several forms)
CRISPRa based on multi-effector recruitment

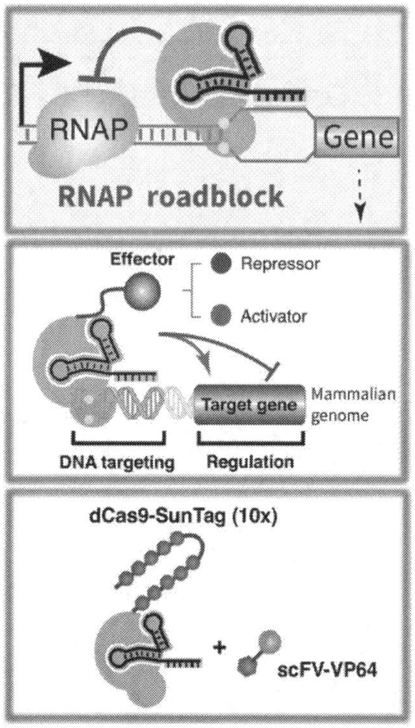

CRISPR tools for gene regulation

1st generation: CRISPRi tools based on transcription blocking

dCas9 (CRISPRi)

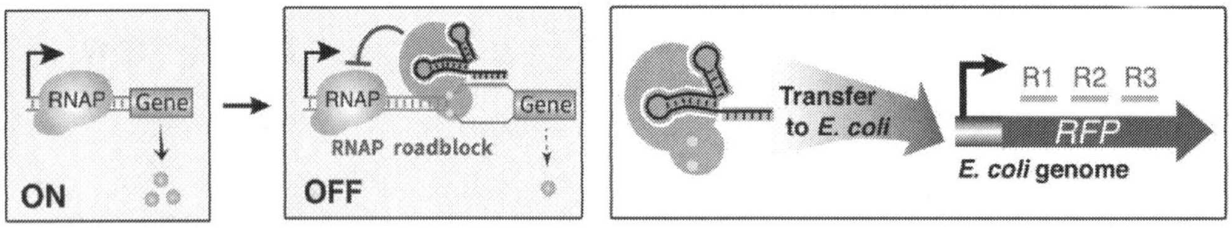

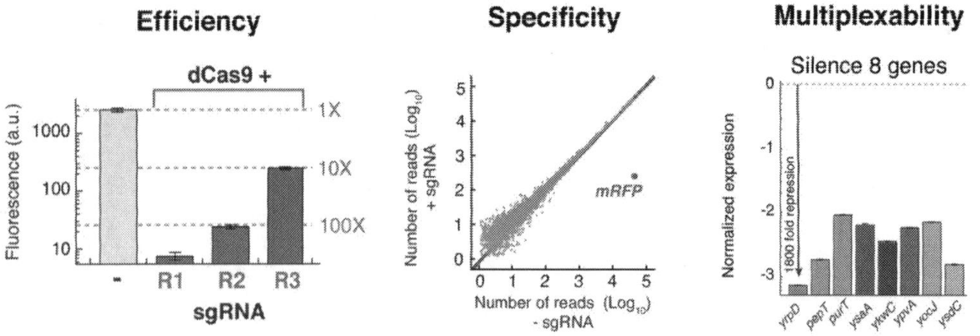

- CRISPRi is very efficient & specific in bacteria
- Not very efficient in mammalian cells

Qi L et al. ***Cell*** *2013 Feb*

CRISPR tools for gene regulation

2nd generation: CRISPR tools for regulator recruitment

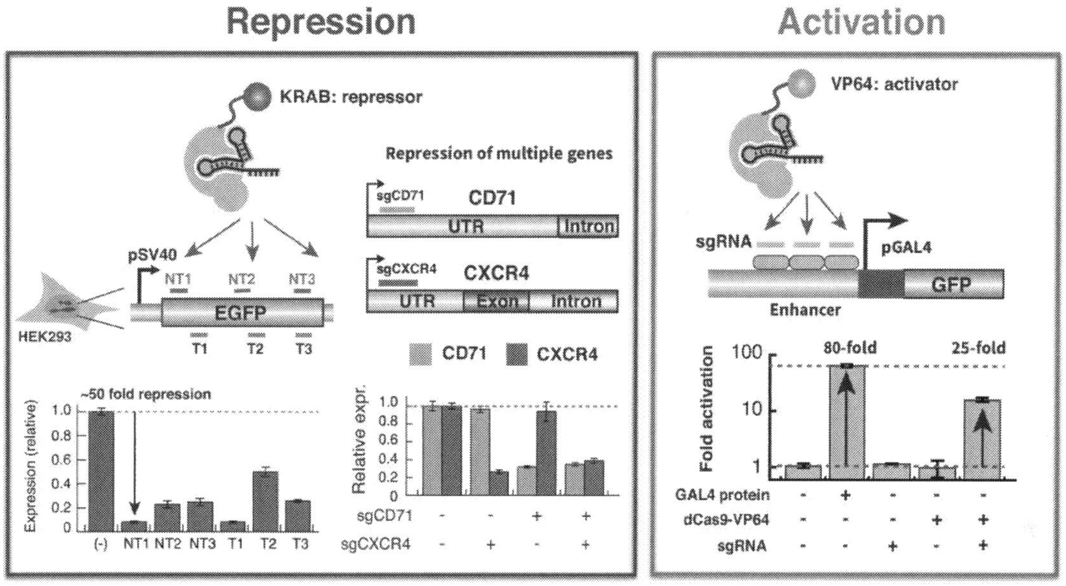

Gilbert LA, et al. **Cell** 2013 July

CRISPR tools for gene regulation

2nd generation: dCas9 gene regulation is specific in human cells

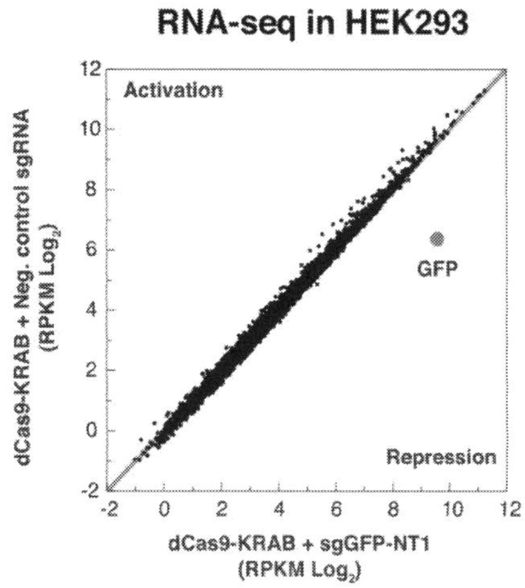

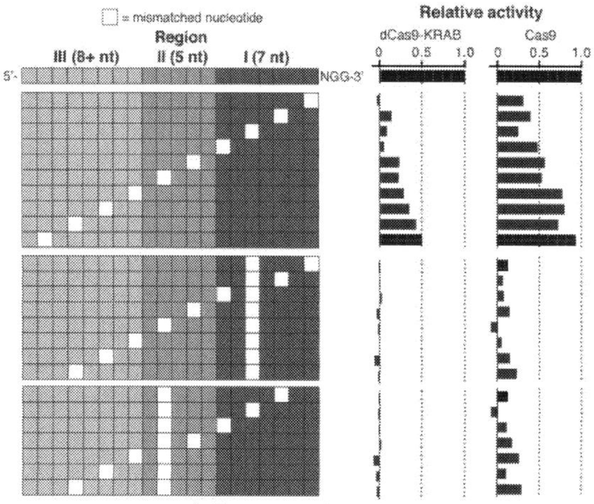

assay of 49 genes that affect growth:
dCas9 is more sensitive to mismatches

Gilbert LA, et al. **Cell** 2013 July

Gilbert LA, et al. **Cell** 2014 Oct

dCas9 for gene regulation is specific!

CRISPR tools for gene regulation

2nd generation: activation is not very efficient

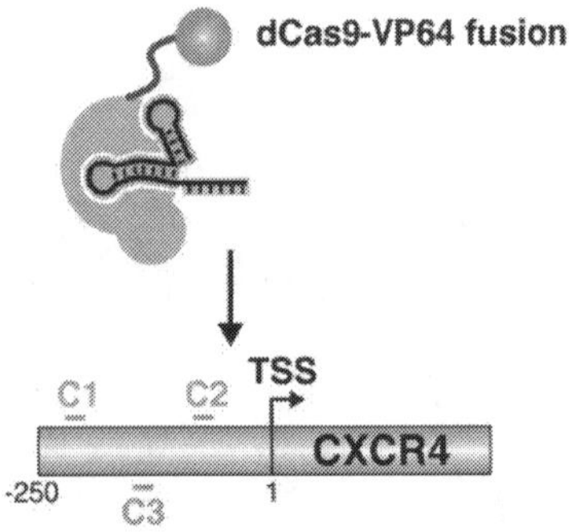

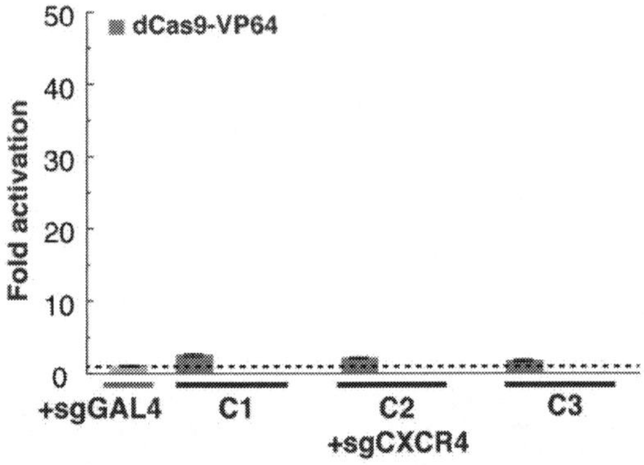

- Average 2-3 fold activation of endogenous genes
- Co-delivery of multiple sgRNAs enhances activation

Recruitment of multiple or synergistic activators?

CRISPR tools for gene regulation

3rd generation: multiplexed regulatory recruitment

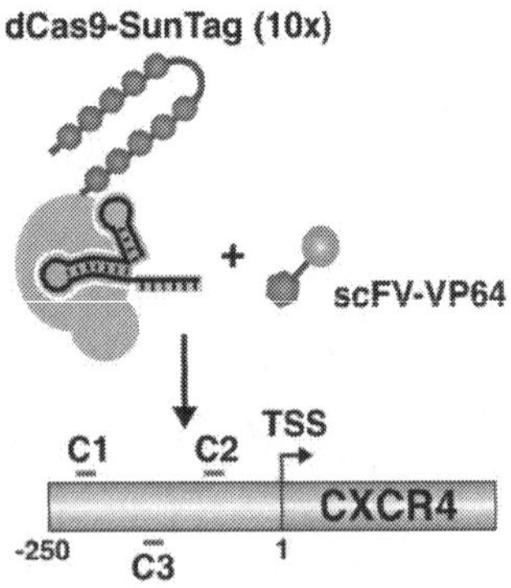

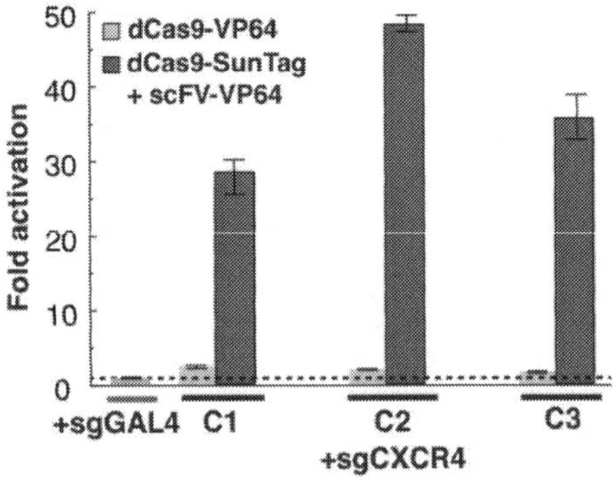

Tanenbaum M, et al. **Cell** 2014 Oct

K562 cells **Multiple activators greatly amplify regulatory effects!**

CRISPR tools for gene regulation

CRISPR is efficient in regulating noncoding RNAs (lncRNAs)

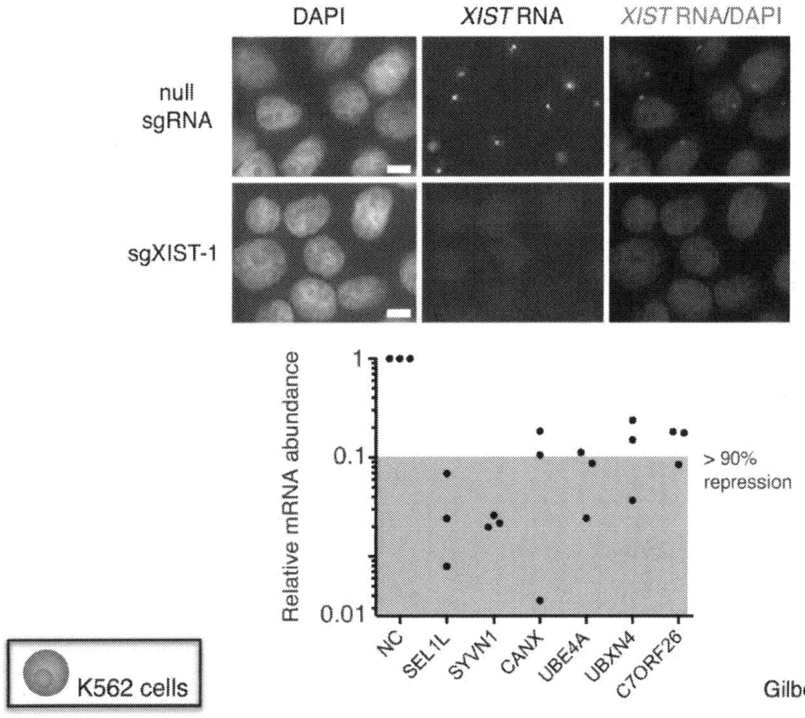

Gilbert L, et al. **Cell** 2014 Oct

Topics to be covered

1. Development of CRISPR tools for gene regulation

2. Applications to engineer the genome function

1) Quantitative study of gene functions
2) Regulating metabolic pathways
3) Inducible and reversible transcriptional logics
4) Genetic screening

Applications to engineer the genome function

Application I: Quantitative study of gene function: drug resistance

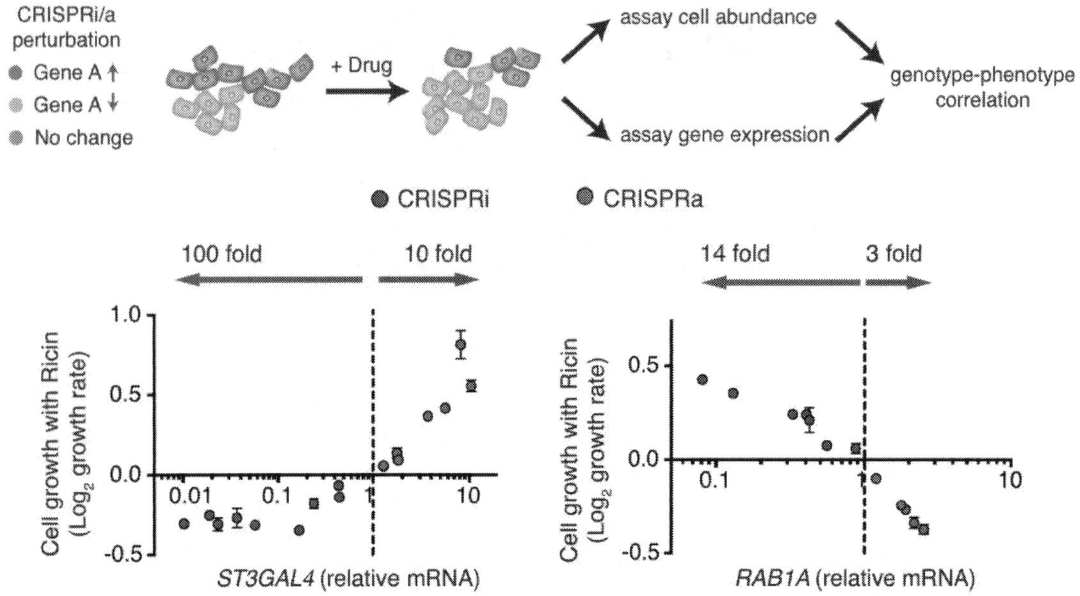

K562 cells

In the presence of Ricin

Gilbert LA, et al. **Cell** 2014 Oct

Applications to engineer the genome function

Application I: Quantitative study of gene function: cell migration

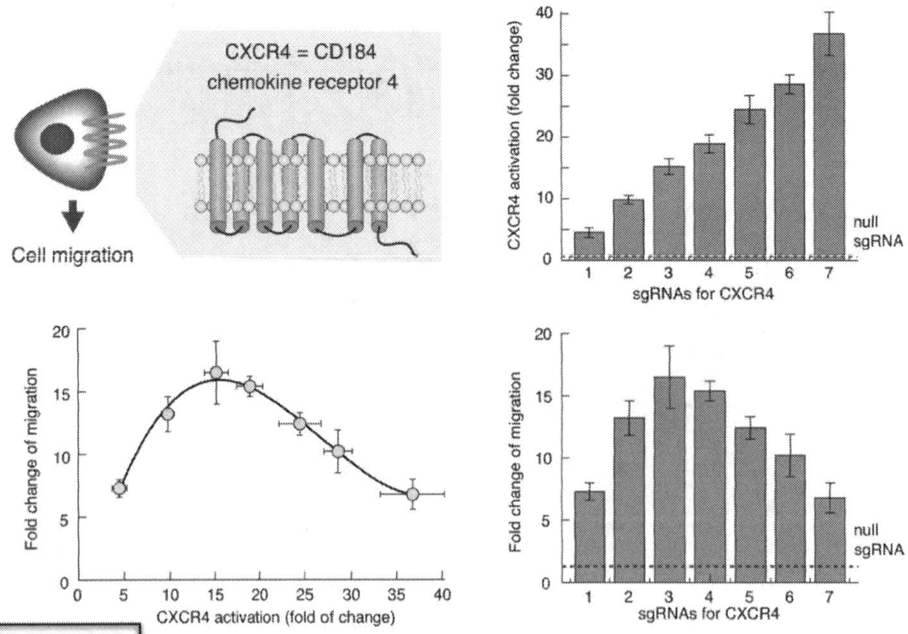

K562 cells

Applications to engineer the genome function

Application II: RNA circuits to modulate metabolic pathways

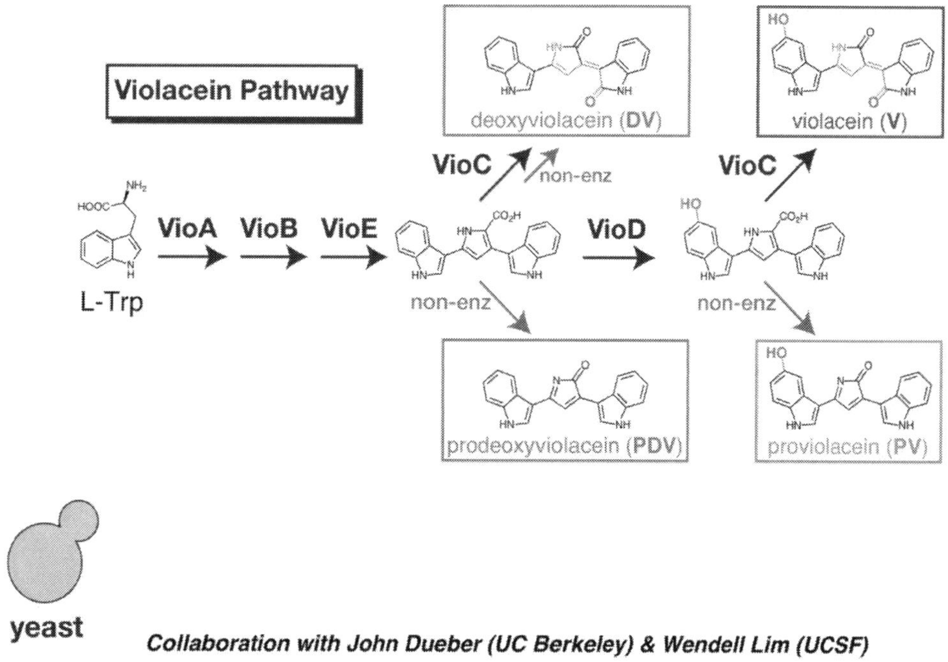

Collaboration with John Dueber (UC Berkeley) & Wendell Lim (UCSF)

Applications to engineer the genome function

Application II: scaffold RNAs for multiplexed gene control

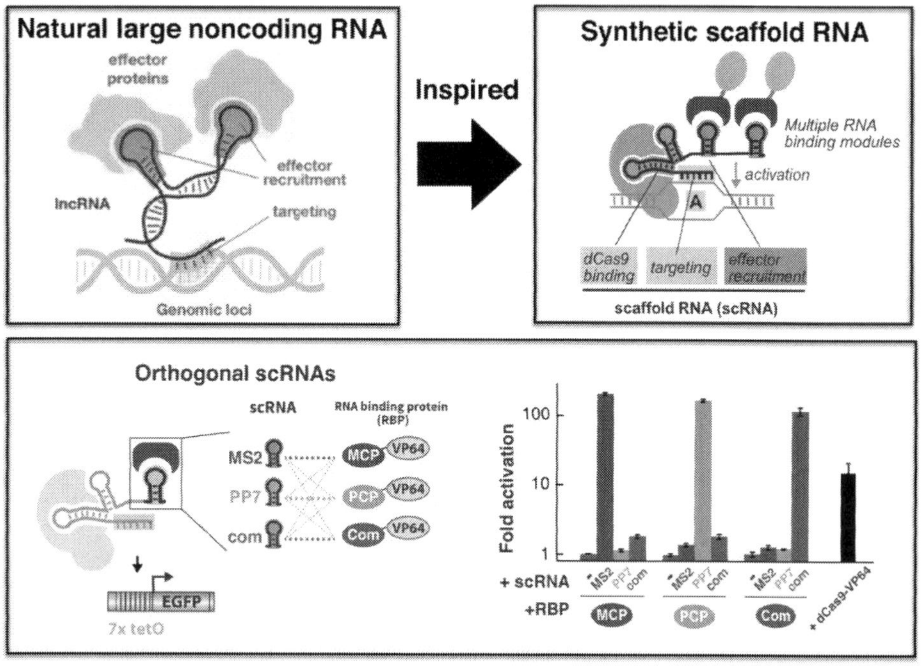

Zalatan J, et al. ***Cell*** 2015 Jan

Applications to engineer the genome function

Application II: Control multi-gene metabolic pathways

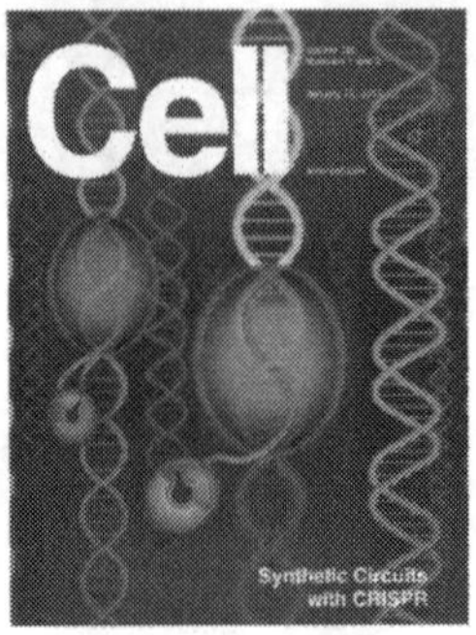

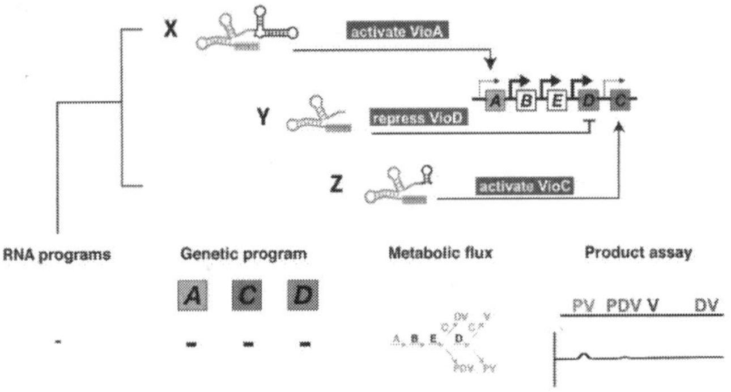

Collaboration with Michael Lee & John Dueber (UC Berkeley), Jesse Zalatan (currently UW), Wendell Lim (UCSF)

Applications to engineer the genome function

Application II: Multiplexing gene regulation with different patterns

Parallelled activation & repression

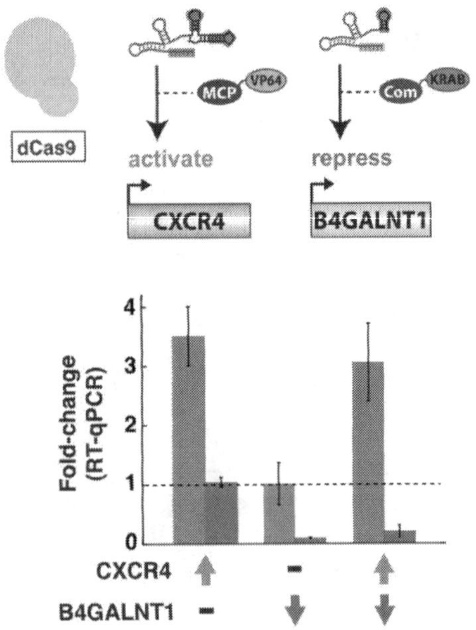

Local combinatorial control

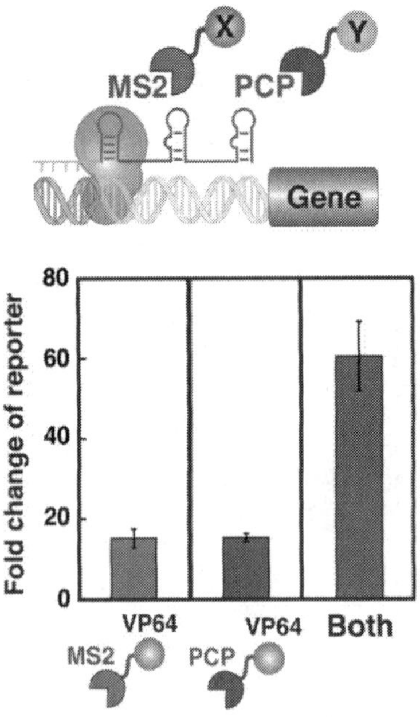

Marie La Russa

Applications to engineer the genome function

Application III: inducible and reversible transcription control

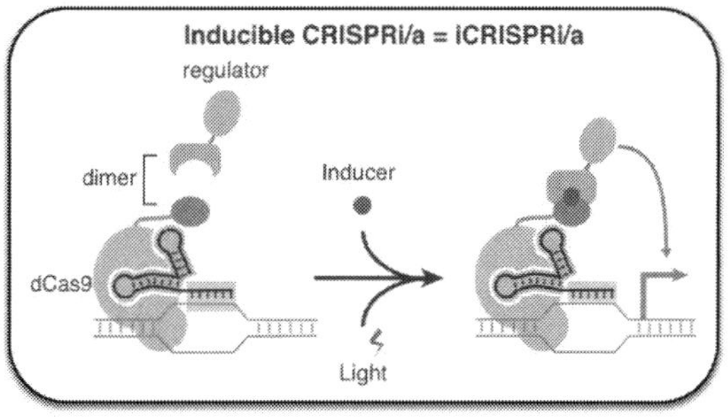

1. Particularly useful for *in vivo* studies
2. The dynamic regulation allows interrogation of timing and duration of gene expression

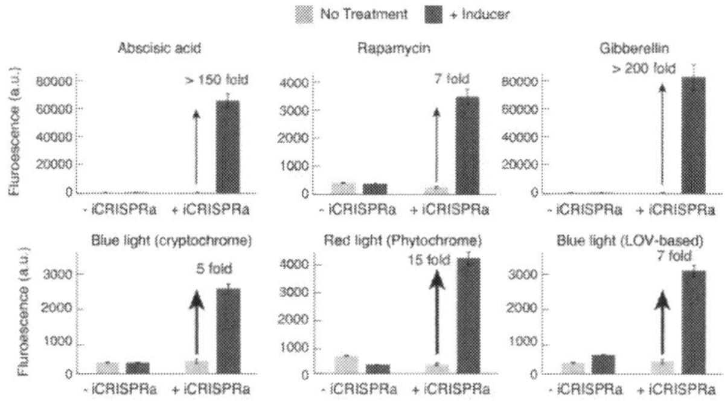

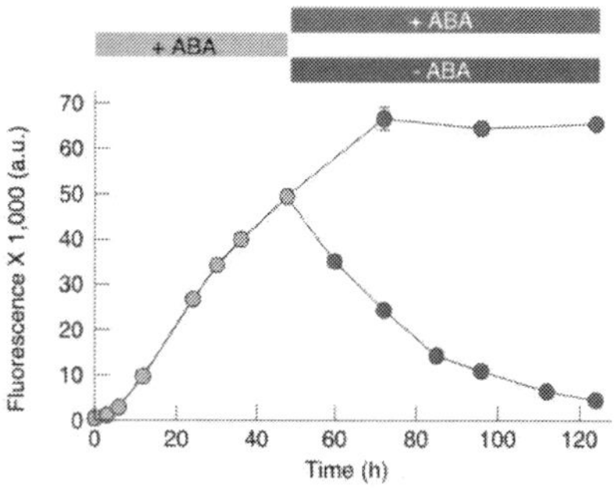

Tony Gao & Xin Xiong

Manuscript in preparation

Applications to engineer the genome function

Application III: inducible and reversible transcription logics

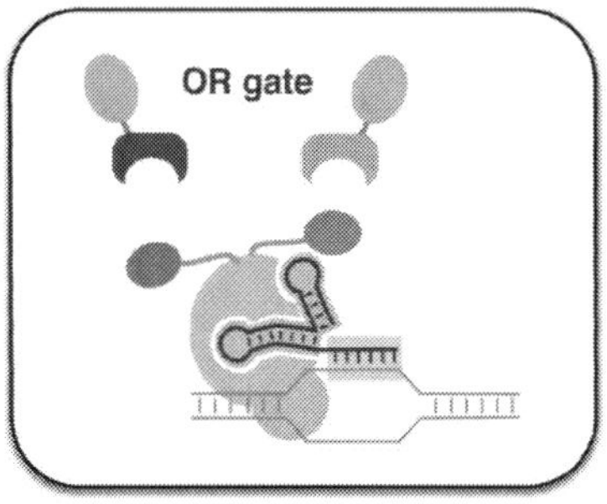

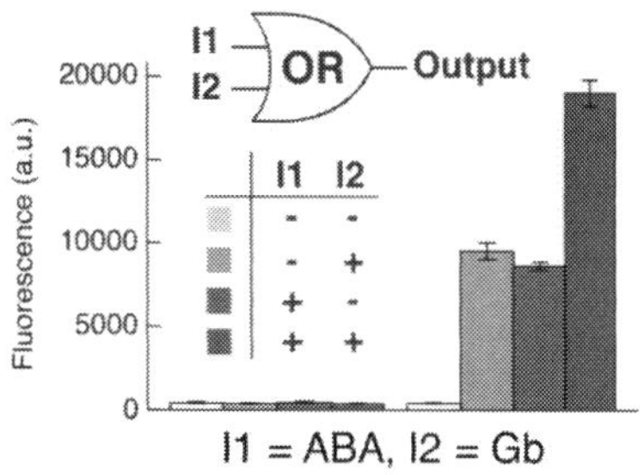

I1 = ABA, I2 = Gb

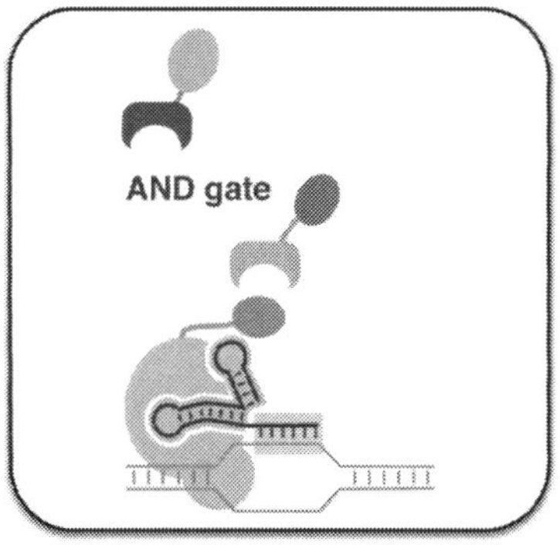

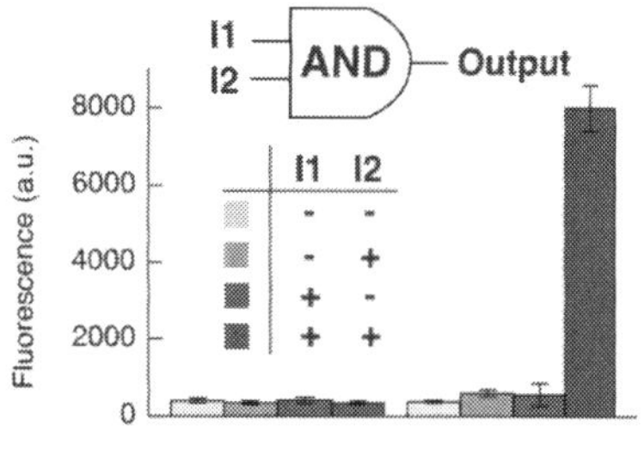

Applications to engineer the genome function

Application IV: functional screening using multi-well plates

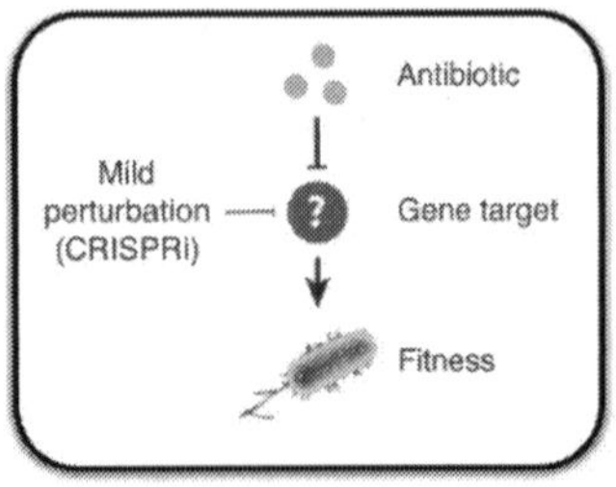

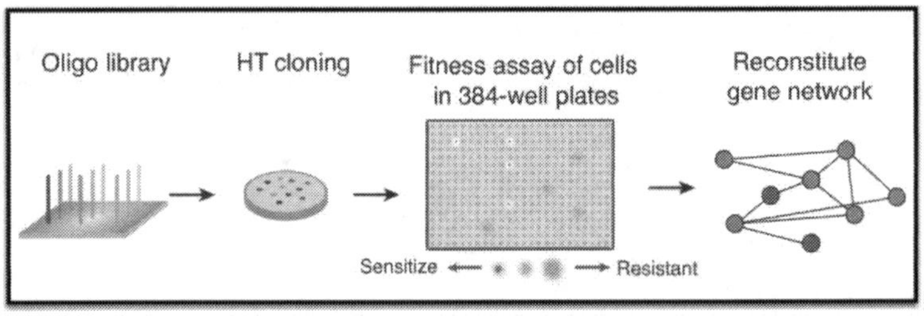

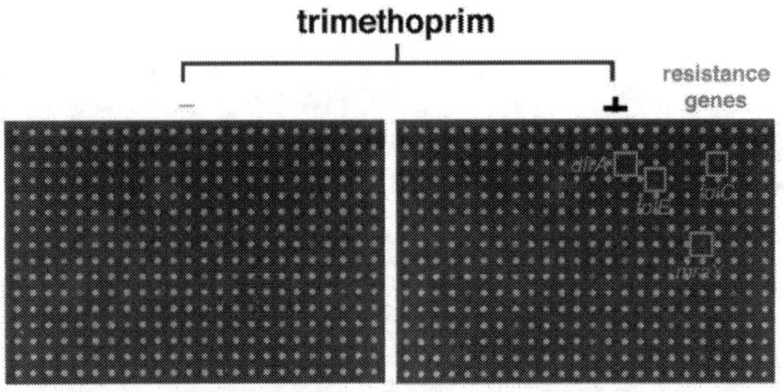

Partial knockdown of 288 essential genes in *B. subtilis* (2~3 fold)

Screened for 35 antibiotics in 2 weeks
Each antibiotic has 3 concentrations

Antibiotic	Gene Target
Aztrenonam	pbpB
Cefoxitin	pbpB
Cerulenin	fabF
Cycloserine D	ddl
Fosfomycin	murAA
Mupirocin	ileS
Trimethoprim	dfrA
Sulfamonomethoxine	sul
...	...

Collaboration with Carol Gross (UCSF)
Manuscript in preparation

Applications to engineer the genome function

Application IV: genome-wide functional screening using cell pools

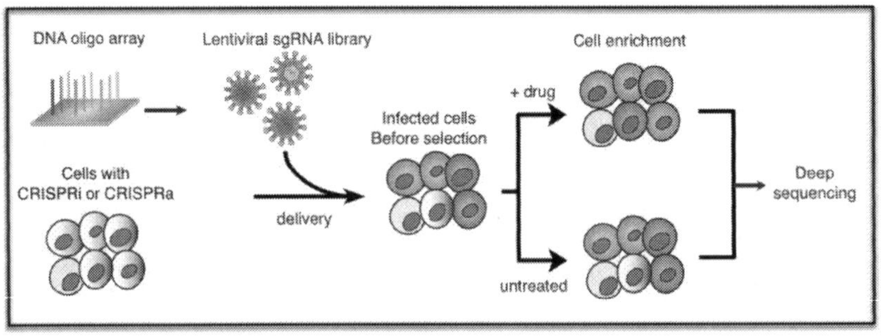

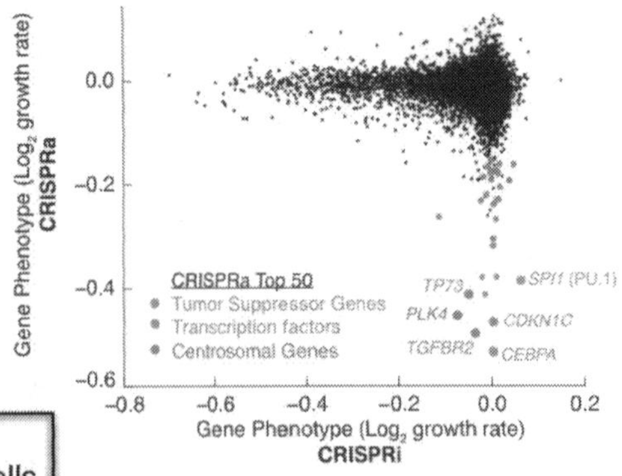

1. CRISPRi and CRISPRa allows both loss-of-function and gain-of-function screenings
2. Rapid profiling of genes and their combinations

Gilbert LA, et al. **Cell** 2014 Oct

Applications to engineer the genome function

CRISPR-era: an online algorithm facilitates sgRNA design

era = editing, repression, activation

VISIT: http://CRISPR-era.stanford.edu

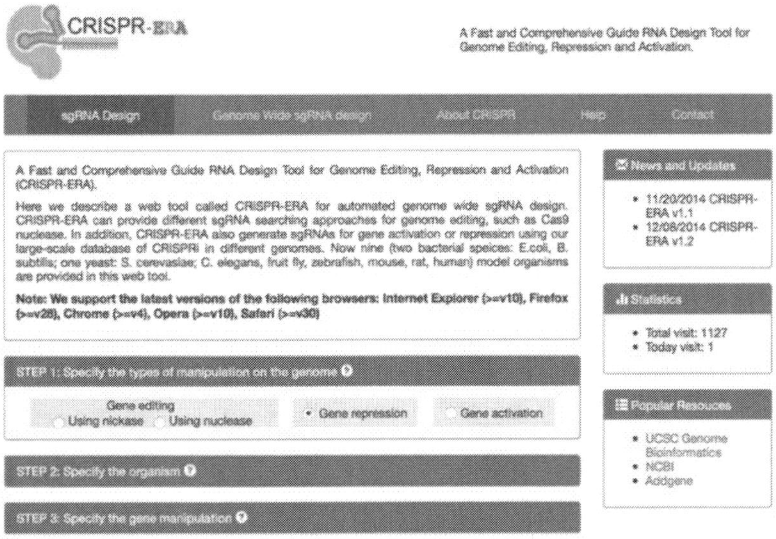

Liu H et al. Manuscript in review, with Xiaowo Wang (Tsinghua)

Summary

RNA-based technologies for genome editing, regulation & imaging

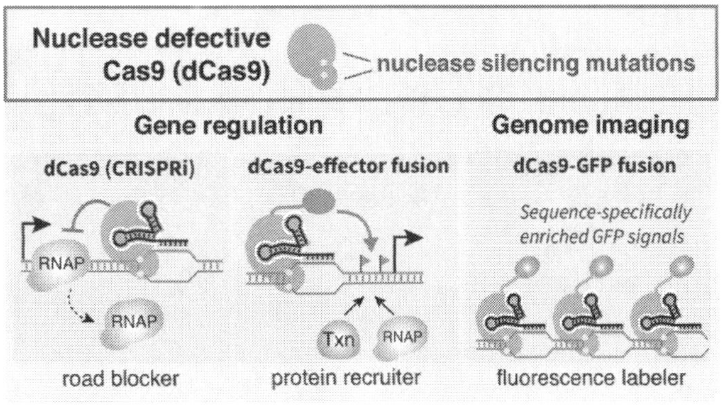

1. Highly designable (basepairing) and reprogrammable (up or down regulation)
2. Nicely interfaces with the genomic programs or any locus in the genome
3. Easy to generate library of “orthogonal” wires with unique binding sites
4. Likely predictable in terms of numbers
5. Work across organisms

Acknowledgments

Antonia Dominguez | Haifeng Wang | Dehua Zhao | Marie La Russa | Nathan Kipniss | Yanxia Liu | Tony Gao | Chris Still

Qi Lab: http://qilab.stanford.edu

Collaborators

Wendell Lim (UCSF)
Jonathan Weissman (UCSF)
Bo Huang (UCSF)
Carol Gross (UCSF)
Nevan Krogan (UCSF)
Jennifer Doudna (UC Berkeley)
Adam Arkin (UC Berkeley)
Sheng Ding (Gladstone)
Deepak Srivastava (Gladstone)
Xiaowo Wang (Tsinghua)

Fundings

NIH Director's Early Independence Award (DP5)
NIH: Roadmaps of the Epigenome (R01)
UCSF Catalyst Award
Gladstone Institutes BioFulcrum
QB3: California Institute for Quantitative Biosciences

Stanford MEDICINE Stanford | Bioengineering Schools of Engineering & Medicine

Toward Computation with Multicellular Systems

Eric Klavins

Posters: Miles Gander, Arjun Khakhar, Chris Takahashi, William Voje, and Yaoyu Yang

University of Washington
Seattle, WA

Synthetic Development

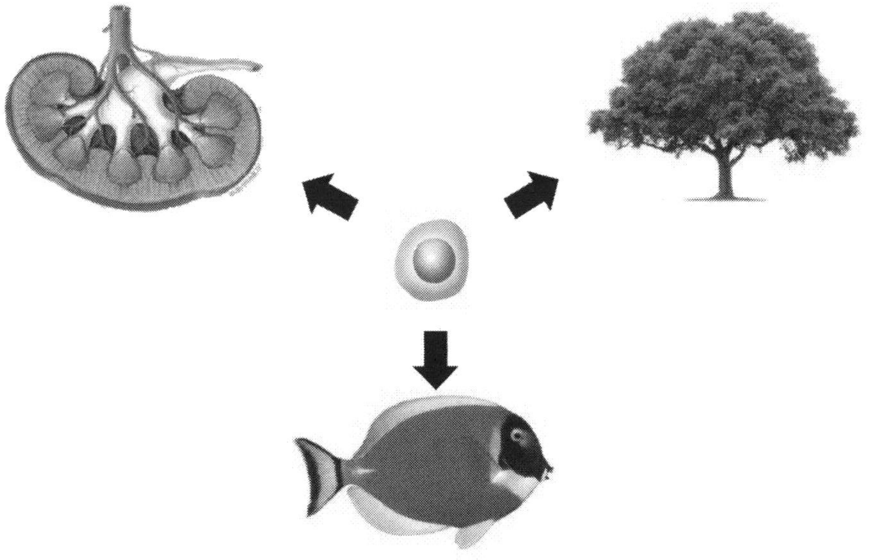

Developmental processes are the computations performed by distributed multicellular programs written in DNA. Think Turing and Von Neumann.

Toward Synthetic Development:

A CompE Approach

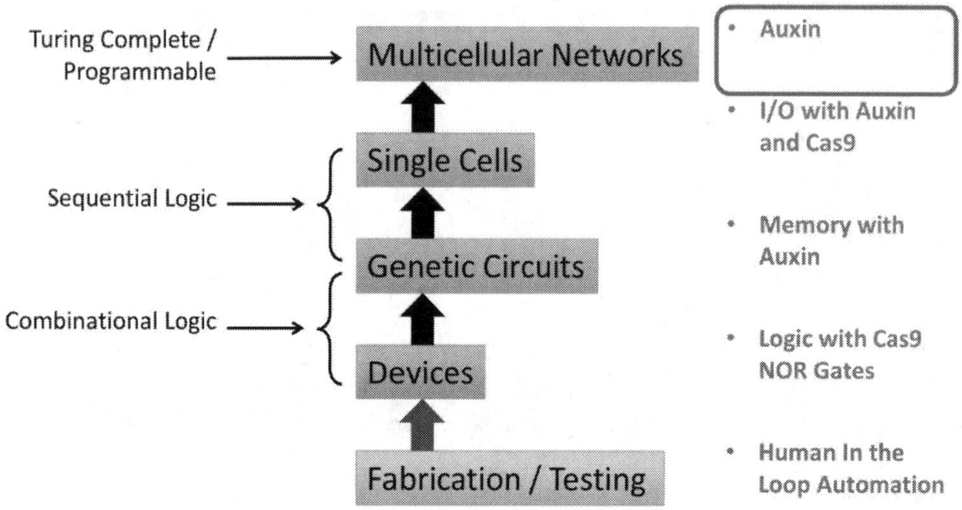

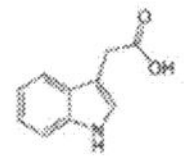

Auxin Developmental Control

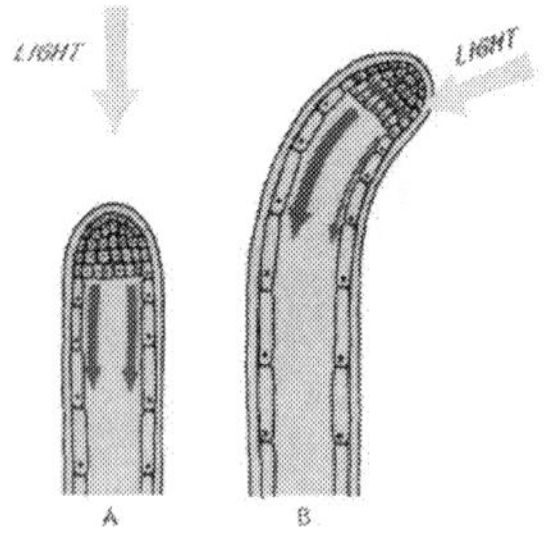

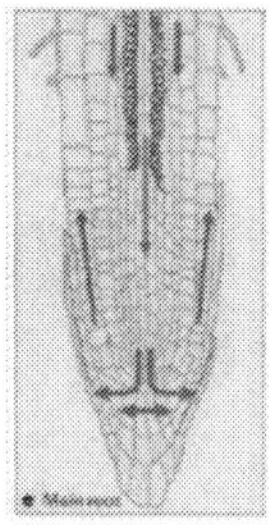

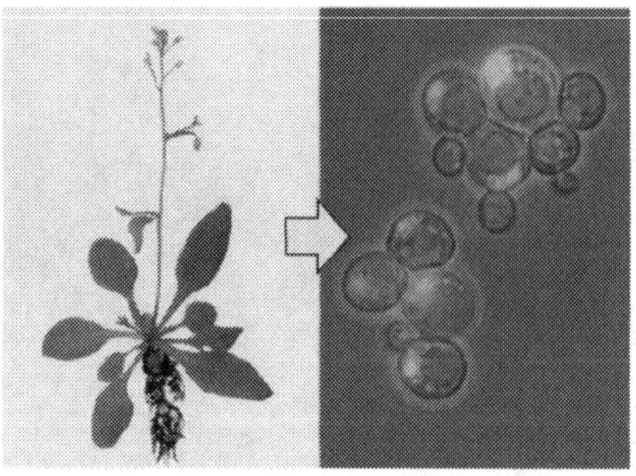

AUXSYNBI/O

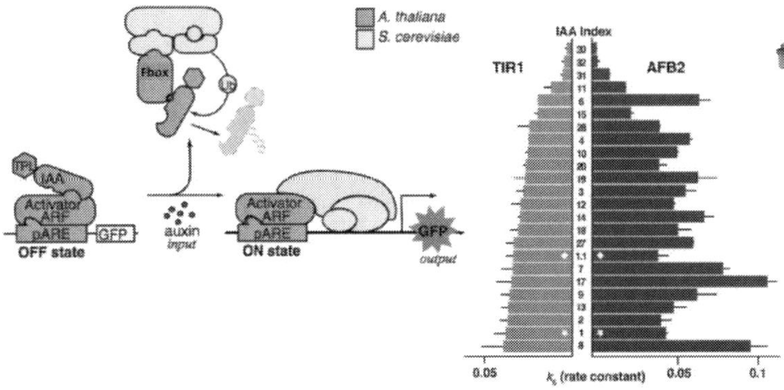

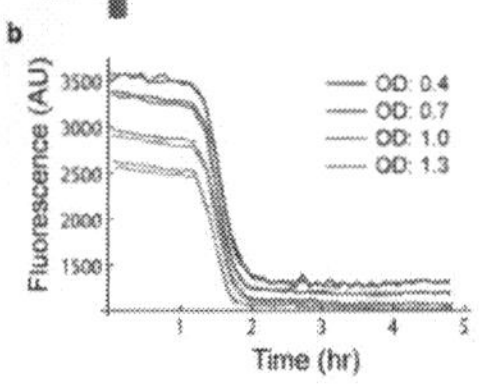

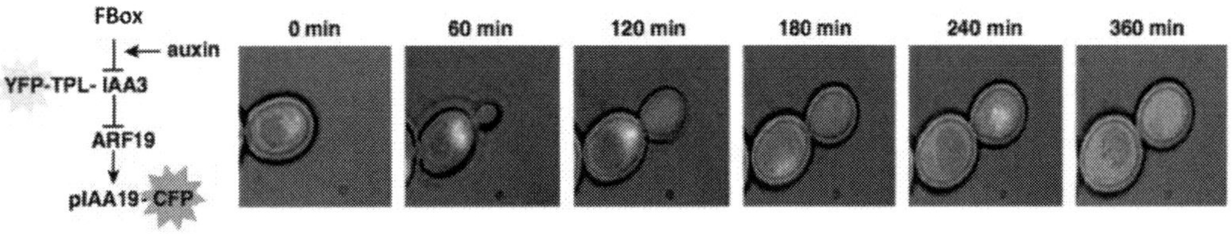

C. Takahashi, A. Miller, F. Ekness, M. Dunham and E. Klavins, ACS Syn Bio. 2014.
E. Pierre-Jerome* S. S. Jang*, K. A. Havens, J. L. Nemhauser and E. Klavins, PNAS, 2014.
H. Yu, B. Moss, S. Jang, M. Prigge, E. Klavins, J. Nemhauser, and M. Estelle, Plant Phys. 2013.
K. Havens*, E. Pierre-Jerome*, S. Jang*, J. Guseman*, N. Bolten, E. Klavins, and J. Nemhauser, Plant Phys, July 2012.

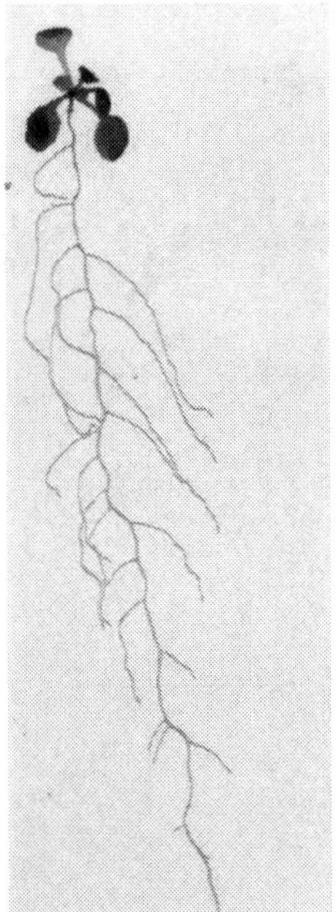

Tuning Development of Lateral Roots

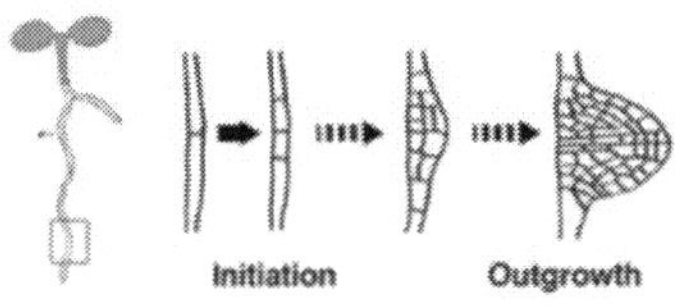

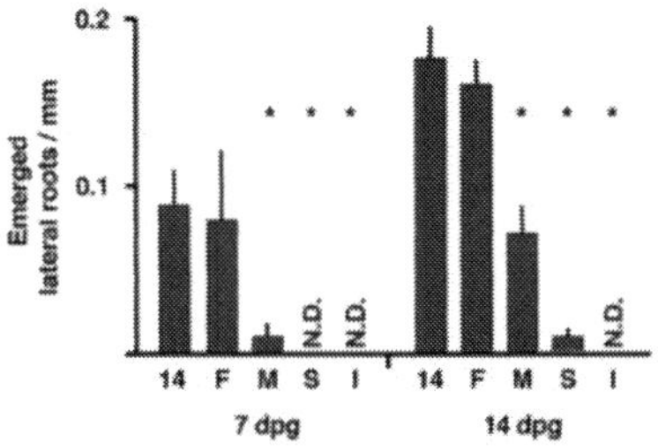

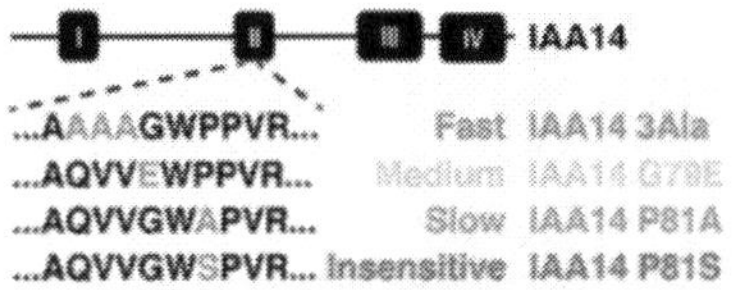

J.M. Guseman, A. Hellmuth, A. Lanctot, T.P. Feldman, B.L. Moss, E. Klavins, L.I.A. Calderón, Villalobos, J.L. Nemhauser. 2015.

Outline

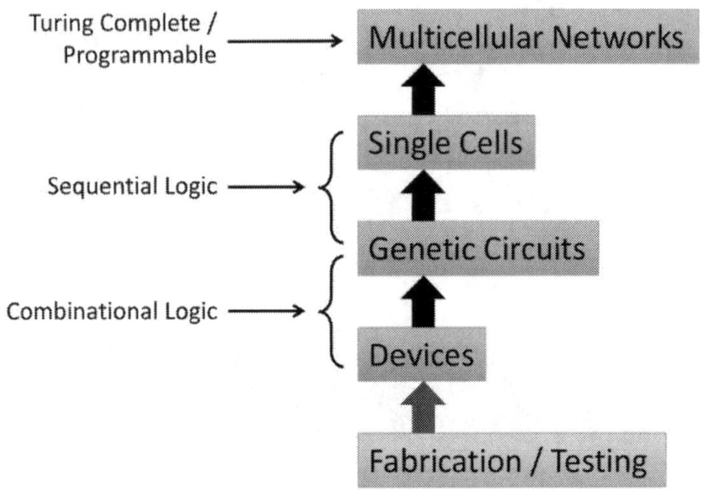

- Auxin
- I/O with Auxin and Cas9
- Memory with Auxin
- Logic with Cas9 NOR Gates
- Human In the Loop Automation

Receving Auxin Signals with CRISPR/Cas9

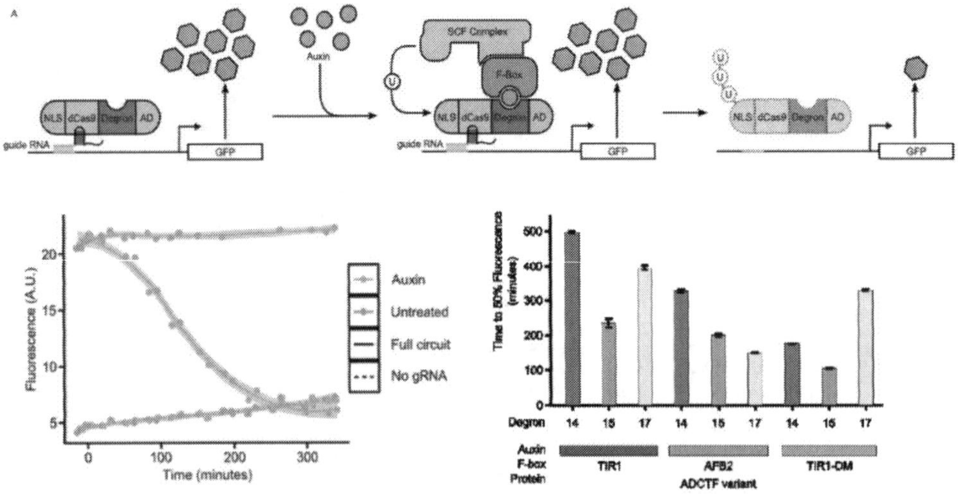

A. Khakhar, **N. Bolten**, J. Nemhauser, and E. Klavins. ACS Syn Bio. bioRxiv. pp. http://dx.doi.org/10.1101/020487. 2015.

Auxin Biosynthesis

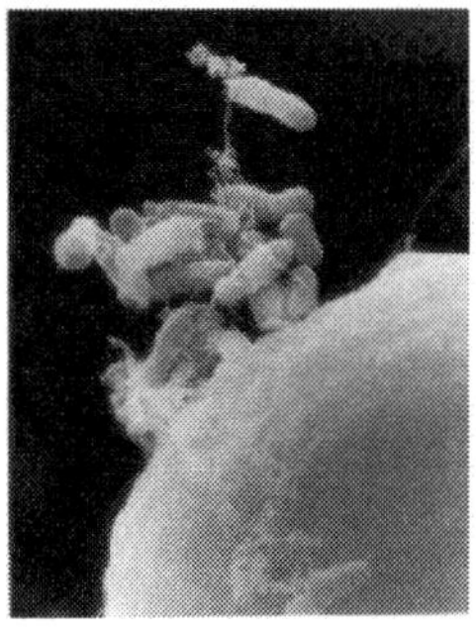

A. tumefaciens attached to a carrot cell (Wikipedia)

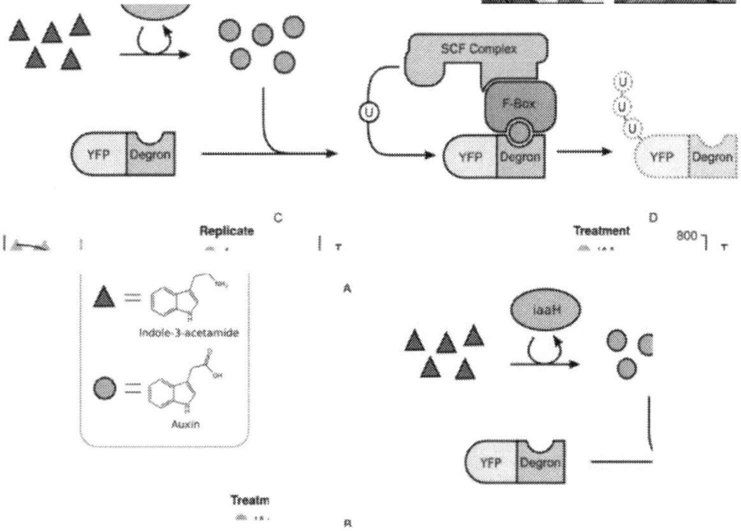

A. Khakhar, N. Bolten, J. Nemhauser, and E. Klavins. ACS Syn Bio. bioRxiv. pp. http://dx.doi.org/10.1101/020487. 2015.

Synthetic Cell to Cell Communication

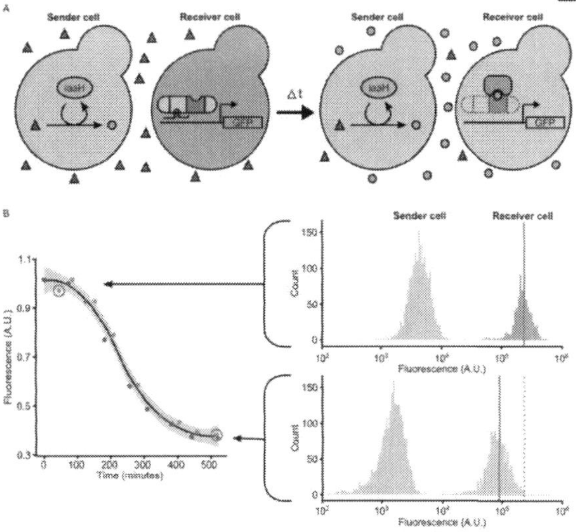

A. Khakhar, N. Bolten, J. Nemhauser, and E. Klavins. ACS Syn Bio. bioRxiv. pp. http://dx.doi.org/10.1101/020487. 2015.

Outline

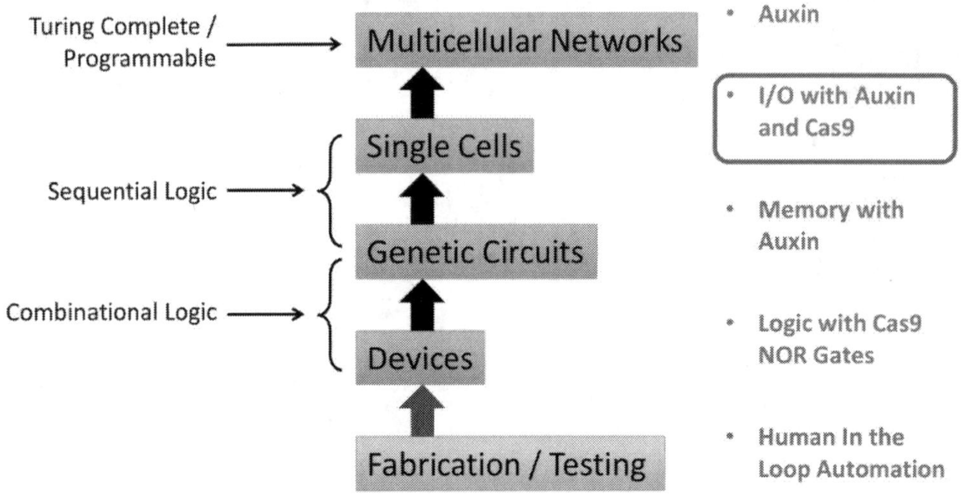

One Bit Memory with Auxin

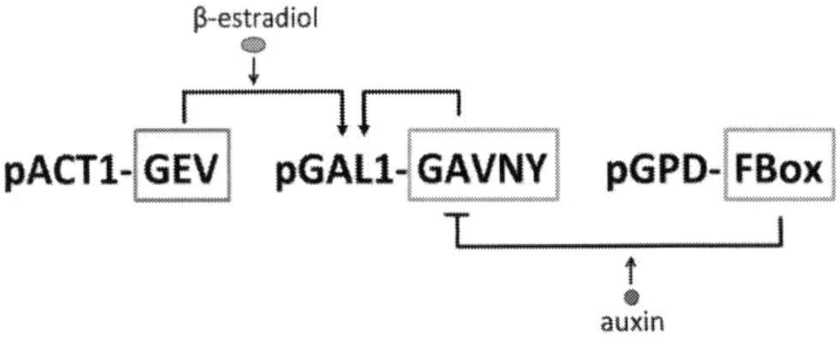

G. GAL4 DNA binding domain
E. Estrogen Receptor
V. VP16

G. GAL4 DNA binding domain
A. Auxin degron IAA17.T2
V. VP16
N. Nuclear localization signal (2x SV40)
Y. EYFP

Y. Yang, J. Nemhauser, and E. Klavins. SEED 2015 Poster.

One Bit Memory with Auxin

$$\frac{dA}{dt} = \frac{\lambda_a A^n + \lambda_b \pi_b v}{1 + \rho_a A^n + \rho_b \pi_b v} - \frac{\lambda_f \gamma_{af} A u}{\delta_f + \gamma_{af} A + \omega_f u} - \delta_a A$$

A: **GAVNY** u: auxin v: β-estradiol

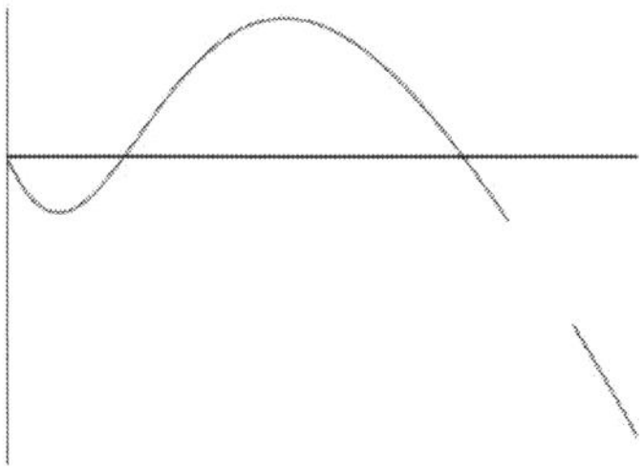

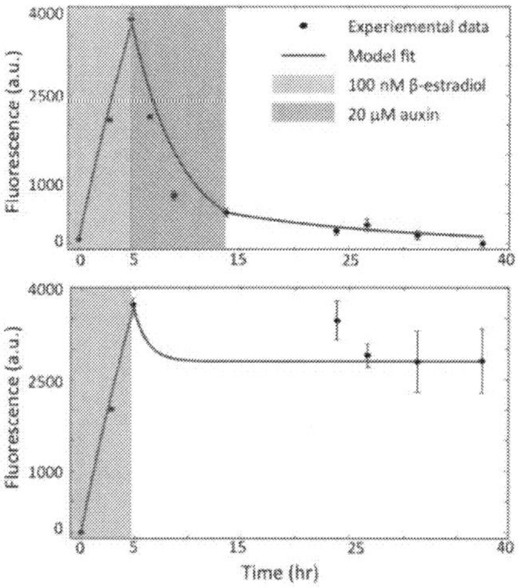

Y. Yang, J. Nemhauser, and E. Klavins. SEED 2015 Poster.

Outline

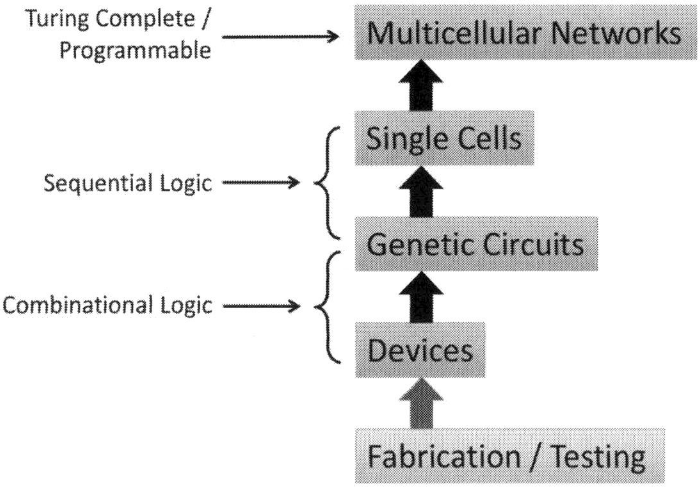

- Auxin
- I/O with Auxin and Cas9
- Memory with Auxin
- Logic with Cas9 NOR Gates
- Human In the Loop Automation

State More Generally

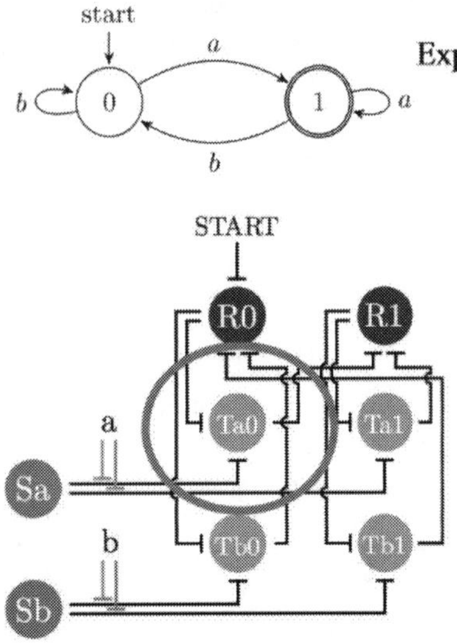

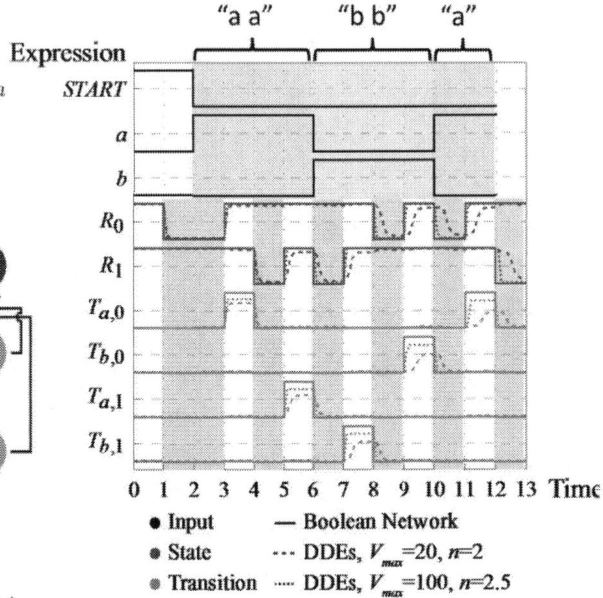

K. Oishi and E. Klavins, ACS Synthetic Biology, 2014

Sequential Logic = Finite State Machines = Combinational Logic with Feedback

Cas9 NOR Gates in Yeast

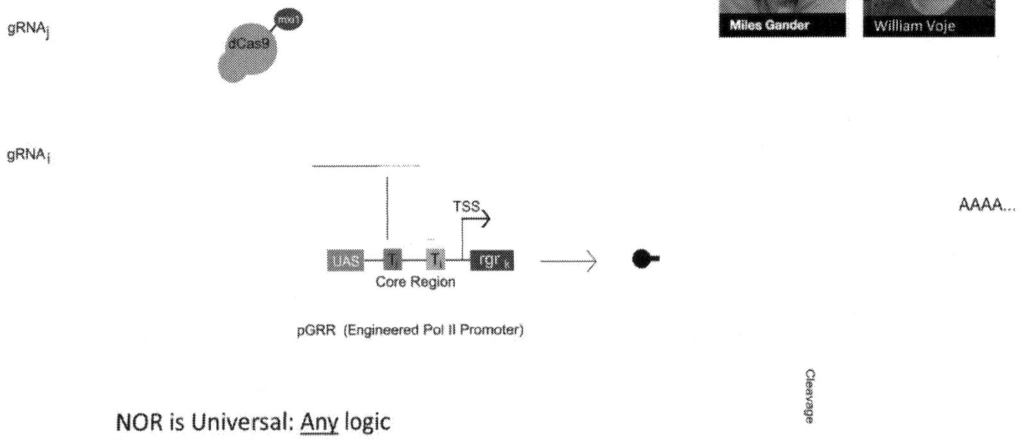

NOR is Universal: <u>Any</u> logic function can be constructed from a network of NOR gates.

M. Gander, William Voje, J. Vrana, James Carothers, and E. Klavins. SEED 2015 Poster.
William Voje, M. Gander, James Carothers, and E. Klavins. SEED 2015 Poster.

20 Wires

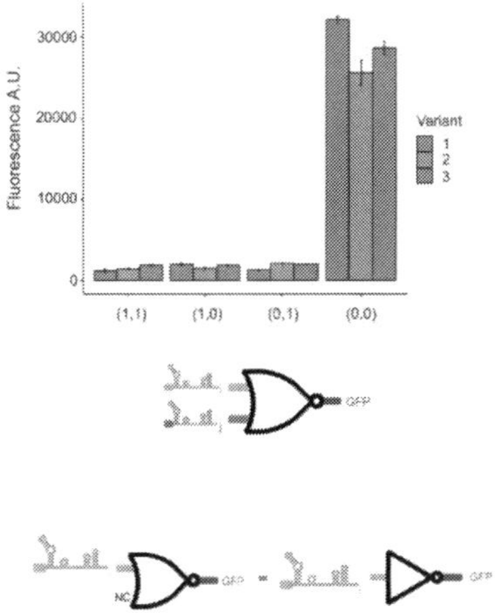

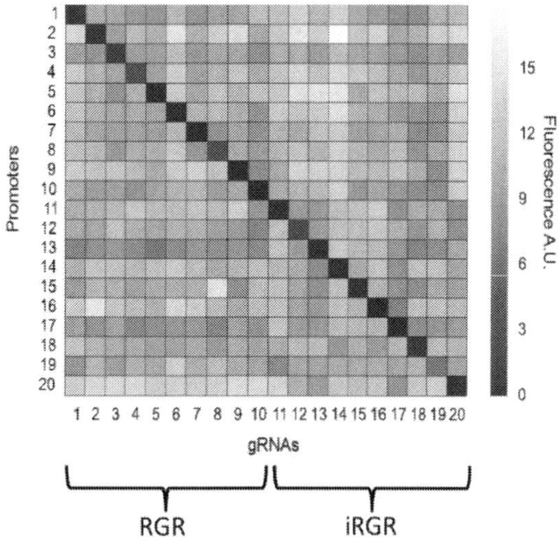

M. Gander, William Voje, J. Vrana, James Carothers, and E. Klavins. SEED 2015 Poster.
William Voje, M. Gander, James Carothers, and E. Klavins. SEED 2015 Poster.

Combinational Logic Circuits

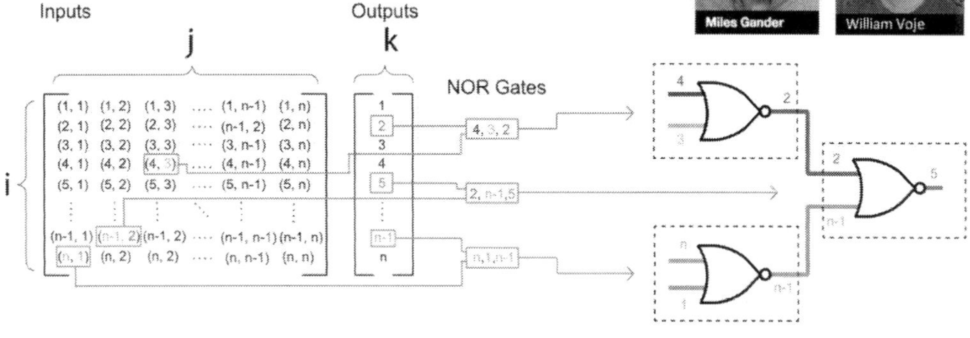

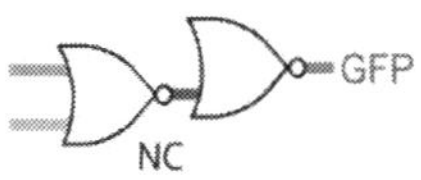

M. Gander, William Voje, J. Vrana, James Carothers, and E. Klavins. SEED 2015 Poster.
William Voje, M. Gander, James Carothers, and E. Klavins. SEED 2015 Poster.

Deep Cascades

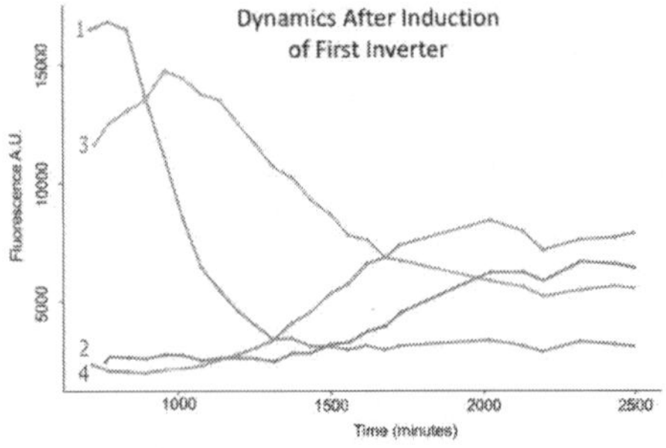

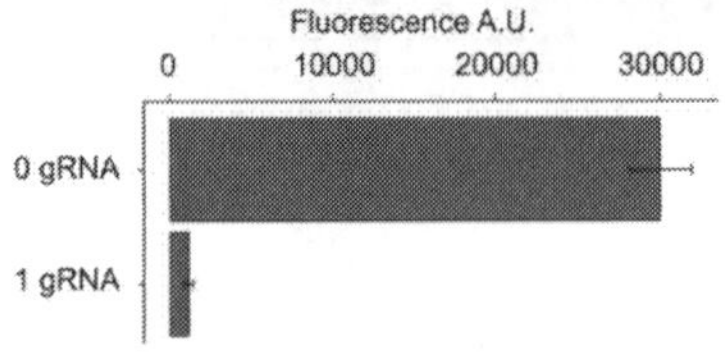

M. Gander, **William Voje**, J. Vrana, James Carothers, and E. Klavins. SEED 2015 Poster.
William Voje, **M. Gander**, James Carothers, and E. Klavins. SEED 2015 Poster.

Outline

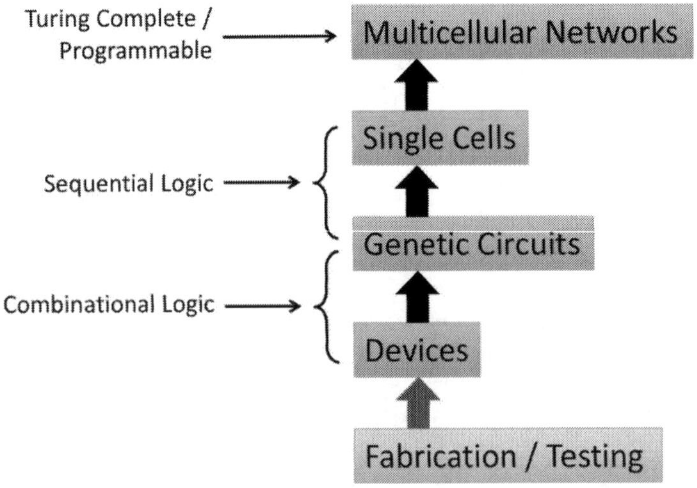

- Auxin
- I/O with Auxin and Cas9
- Memory with Auxin
- Logic with Cas9 NOR Gates
- Human In the Loop Automation

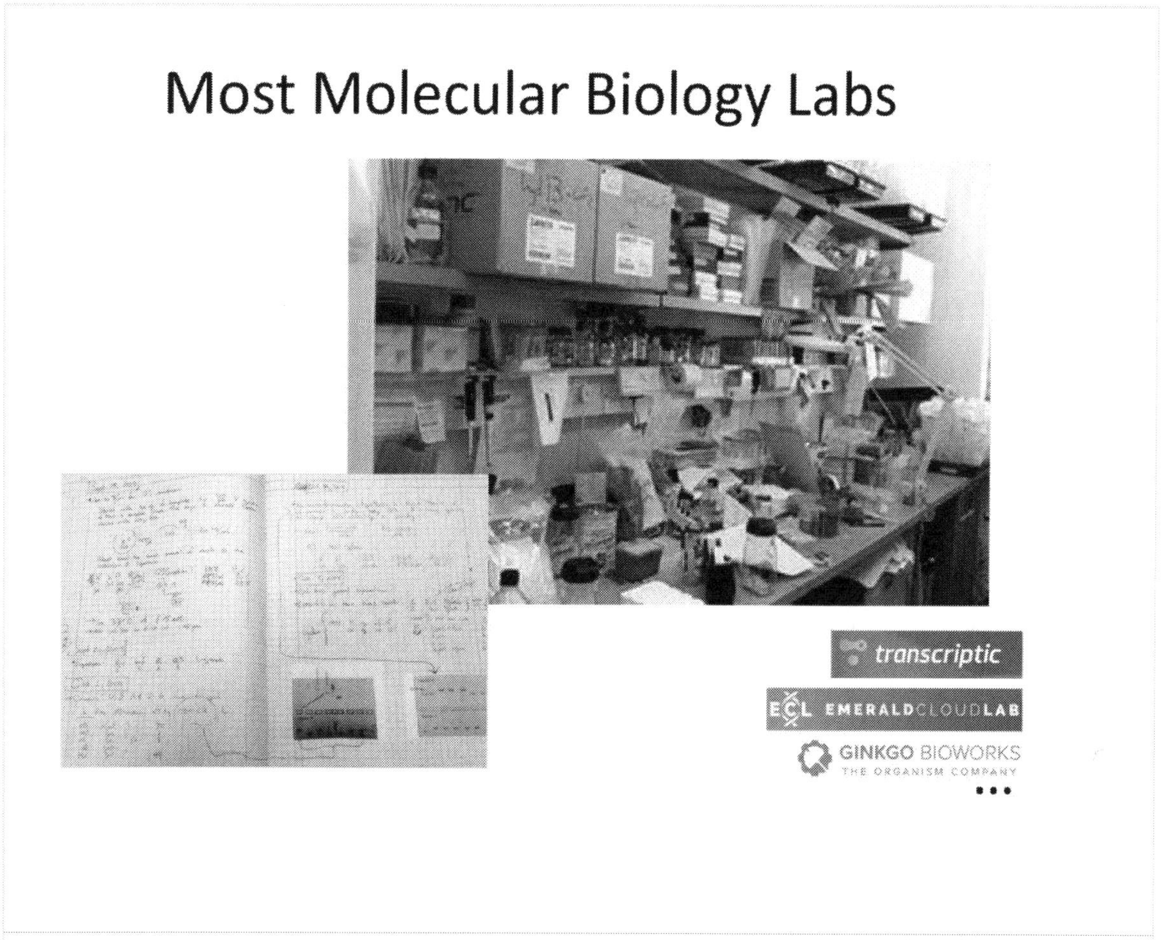
Most Molecular Biology Labs
transcriptic
ECL EMERALDCLOUDLAB
GINKGO BIOWORKS
THE ORGANISM COMPANY
...

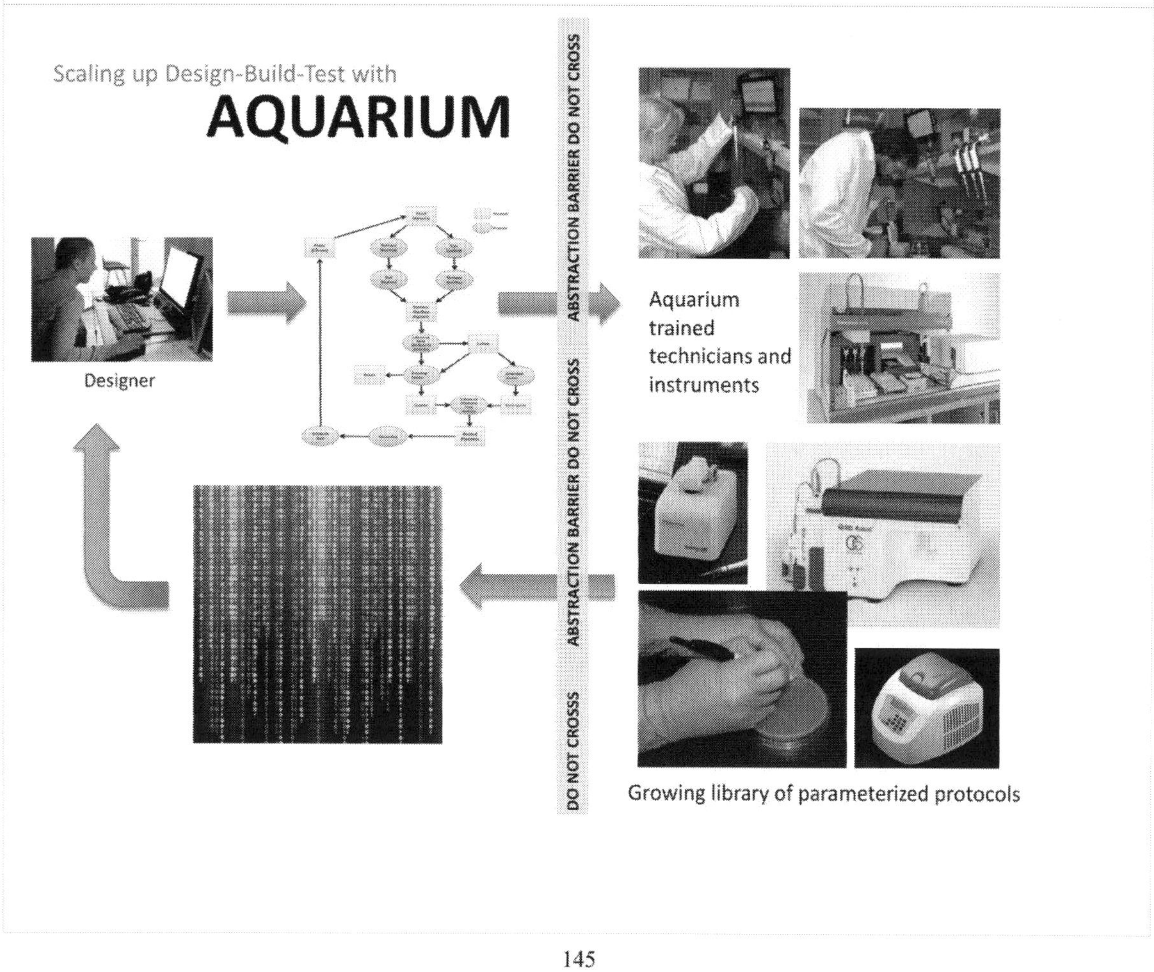
Scaling up Design-Build-Test with
AQUARIUM
Designer
ABSTRACTION BARRIER DO NOT CROSS ABSTRACTION BARRIER DO NOT CROSS ABSTRACTION BARRIER DO NOT CROSSS
Aquarium trained technicians and instruments
Growing library of parameterized protocols

The **AQUARIUM** Prototype

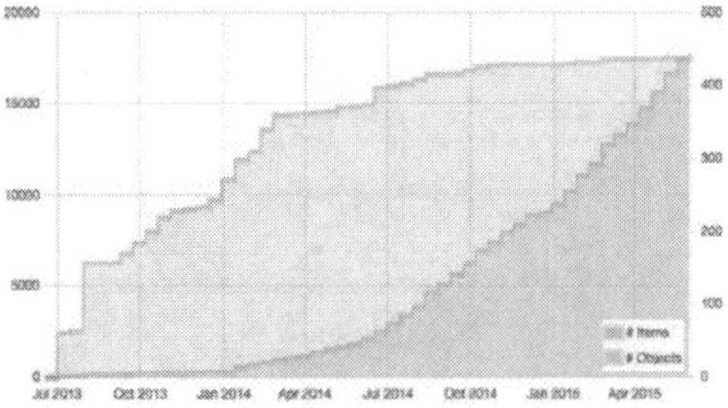

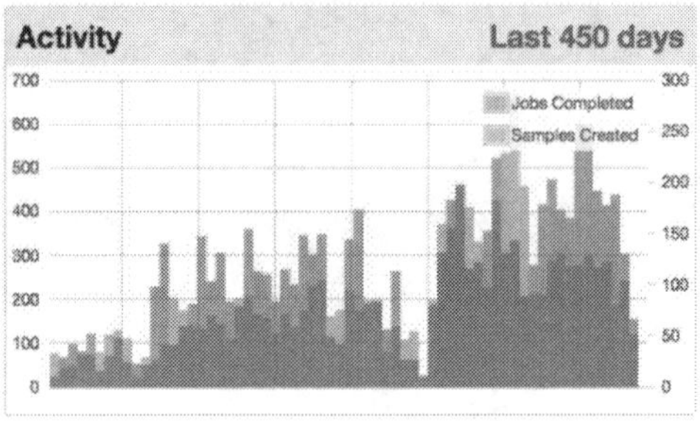

Work study students do 100% of Klavins lab cloning, 50% of yeast work.

Protocols as Code

A language for semi-formal protocol descriptions

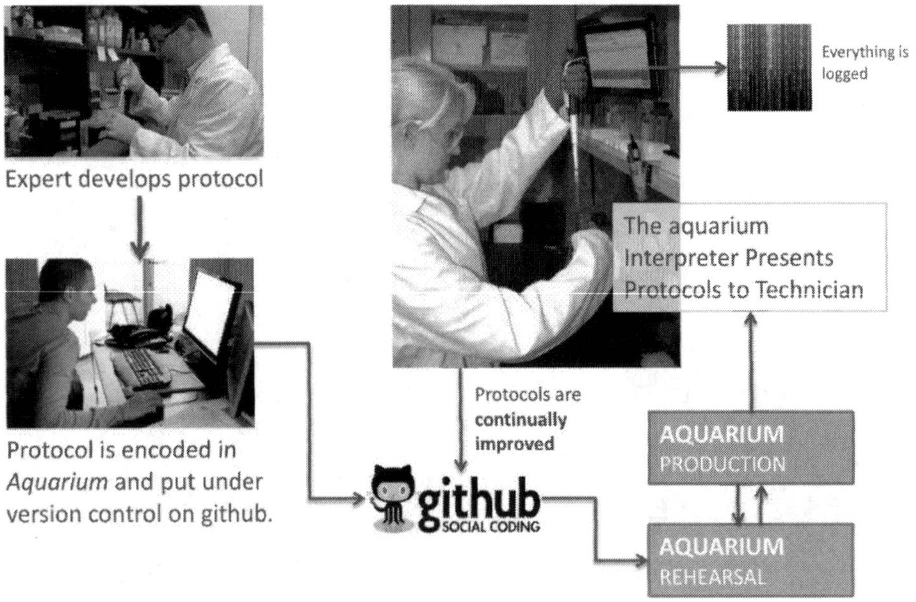

Krill Interpreted by Aquarium presents instructions to technicians

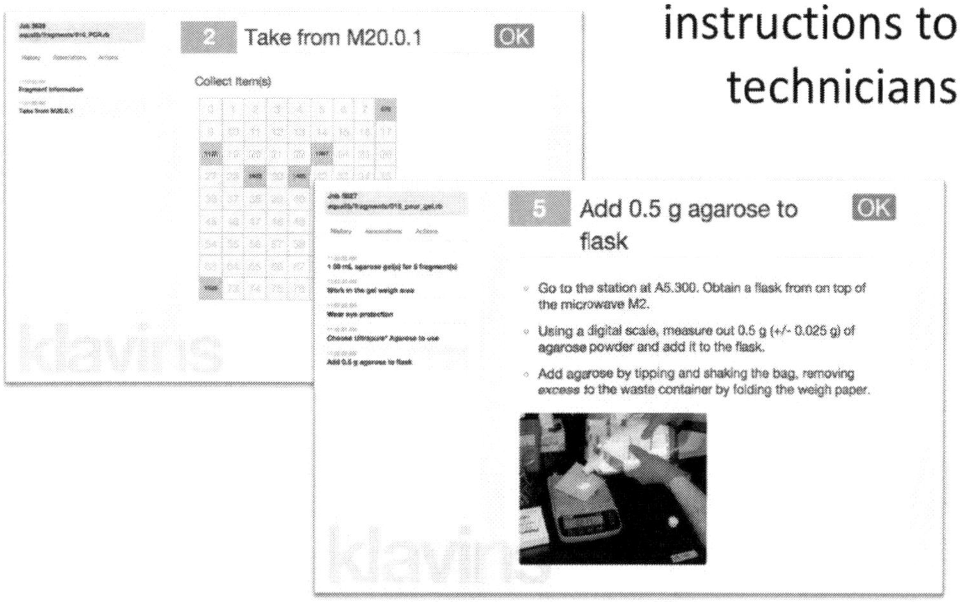

Programmable Workflows

Process Control

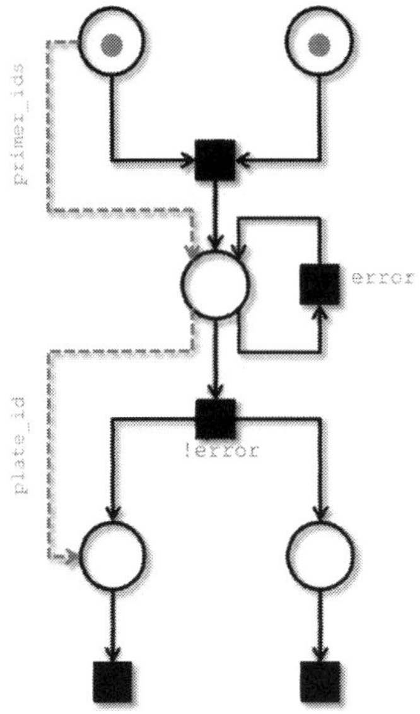

- Based on Petri Nets
- Places are jobs associated with
  - a protocol
  - parameters
  - a user group
  - a desired start / stop time
- Transitions have conditions
  - errors/cancelations
  - inventory status
  - user availability
- Inputs and outputs can be wired

Workflows Currently in Aquarium

Training	Plasmid Verification
Media, supplies, reagents	Sequencing Prep
Daily, Weekly Tasks	Mutagenesis Libraries
Primer Order / Stocking	Yeast Integration
Fragment Construction	Yeast Mating
Gibson Assembly	Yeast Sporulation
E coli methods	Cytometry
Recombineering	Plate Reader

A Programmable Lab is an Enabler

Training

Newbies can be trained on our entire workflow in 5 weeks.

Projects involving multiple plasmid constructions, yeast integrations, matings, and characterization in another five weeks.

Large datasets!

Miles Gander

A molecular 'wire' that can be composed into very *deep* synthetic genetic circuits.

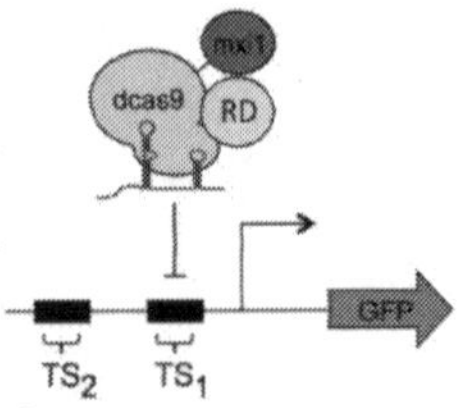

Arjun Khakar and Nick Bolten

Novel cell-cell communication in yeast using plant hormones and a new inducible CRISPR.

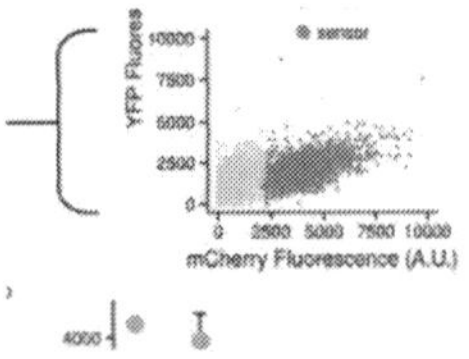

Yaoyu Yang

A synthetic genetic switch in yeast using parts from plants.

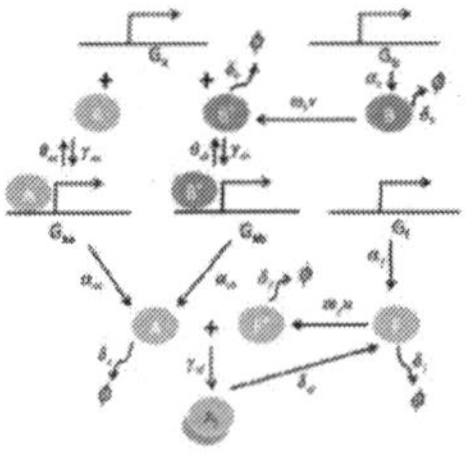

1,000s of yeast strains built and screened in last six months.

Summary

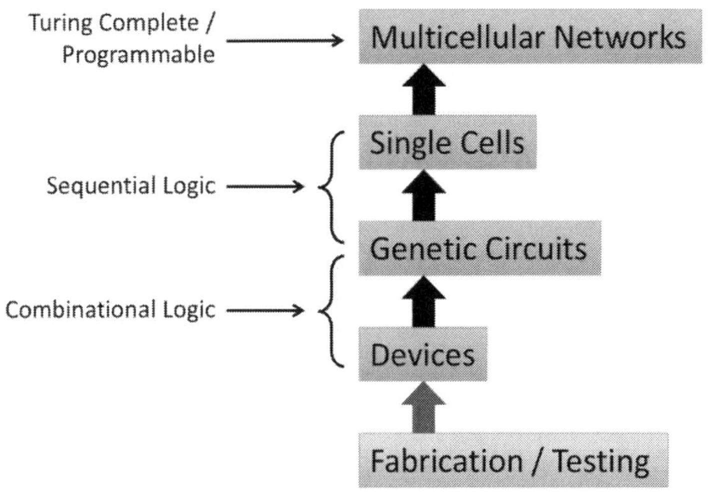

- Auxin
- I/O with Auxin and Cas9
- Memory with Auxin
- Logic with Cas9 NOR Gates
- Human In the Loop Automation

Klavins Lab

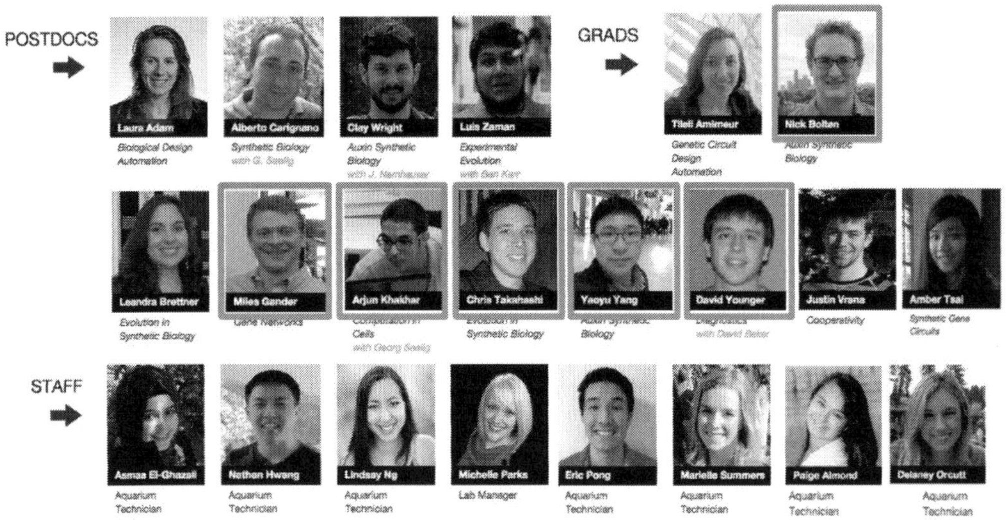

James Carothers
Jennifer Nemhauser
Georg Seelig
Jeff Tabor
Dieter Fox

Ben Kerr
David Baker
Oleg Igoshin
Erik Winfree
James Rehg

NSF: EFRI: Multicellular Signaling and Pattern Formation
NSF: Expeditions: The Molecular Programming Project II
NSF: AUXSYNBI/O
NIH: Control of Protein Degradation Dynamics
SRC: Finite State Machines in Yeast
INTEL: ISTC for Pervasive Computing
UW Center for Innovation

Synthetic Approaches To Understanding Cis-Regulation

Sri Kosuri
UCLA Chemistry & Biochemistry
http://kosurilab.org sri@ucla.edu @srikosuri

"What I cannot create, I do not understand"

– Richard Feynman

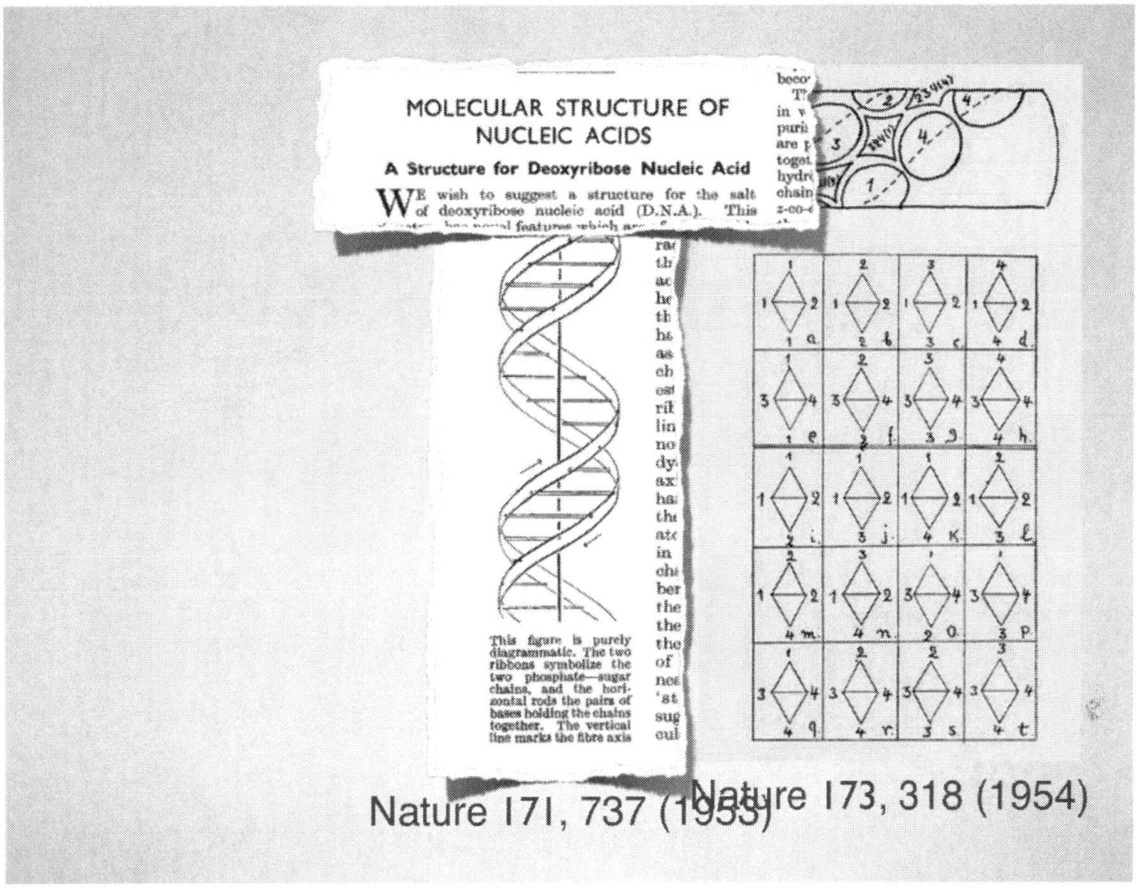
MOLECULAR STRUCTURE OF NUCLEIC ACIDS
A Structure for Deoxyribose Nucleic Acid
WE wish to suggest a structure for the salt of deoxyribose nucleic acid (D.N.A.). This
This figure is purely diagrammatic. The two ribbons symbolize the two phosphate—sugar chains, and the horizontal rods the pairs of bases holding the chains together. The vertical line marks the fibre axis
Nature 171, 737 (1953)
Nature 173, 318 (1954)

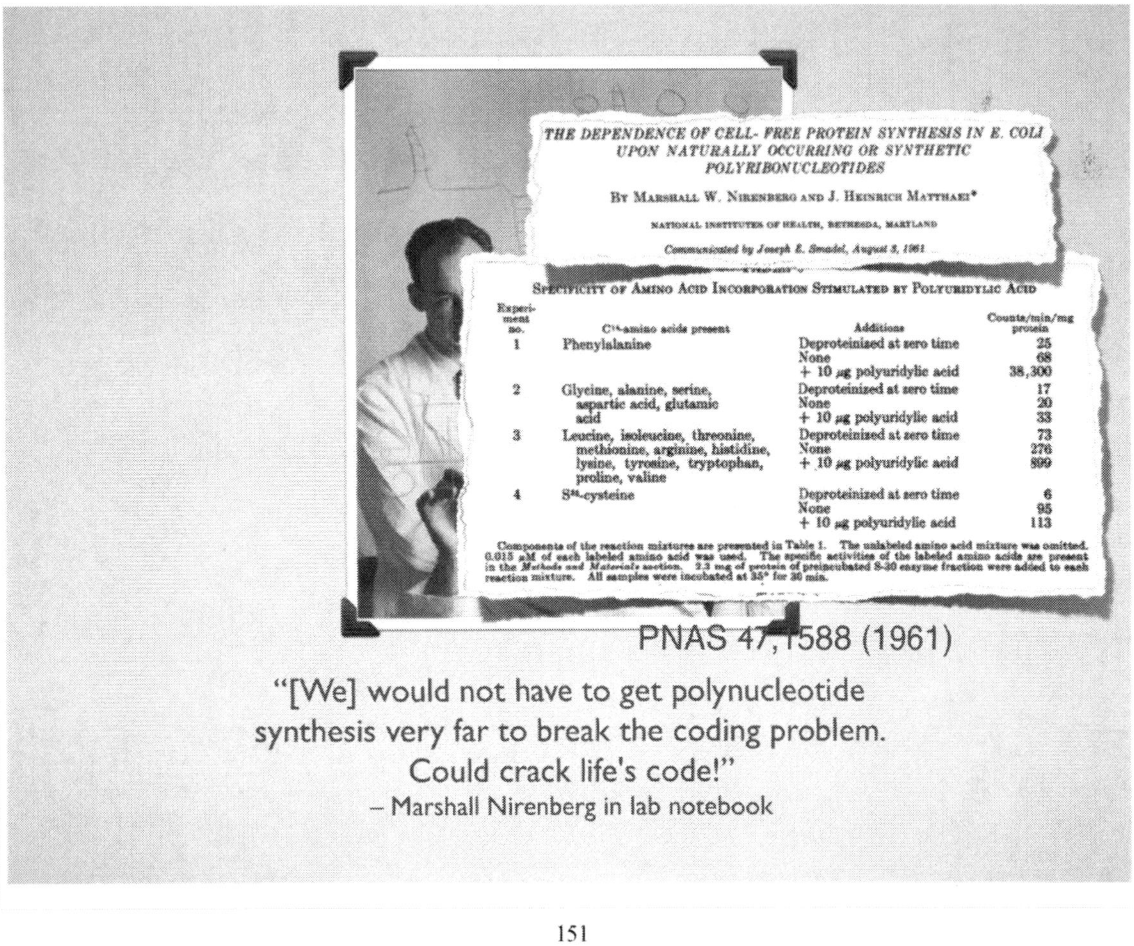
THE DEPENDENCE OF CELL-FREE PROTEIN SYNTHESIS IN E. COLI UPON NATURALLY OCCURRING OR SYNTHETIC POLYRIBONUCLEOTIDES
BY MARSHALL W. NIRENBERG AND J. HEINRICH MATTHAEI*
NATIONAL INSTITUTES OF HEALTH, BETHESDA, MARYLAND
Communicated by Joseph E. Smadel, August 3, 1961
SPECIFICITY OF AMINO ACID INCORPORATION STIMULATED BY POLYURIDYLIC ACID
Experiment no. | C14-amino acids present | Additions | Counts/min/mg protein
1 | Phenylalanine | Deproteinized at zero time | 25
None | 68
+ 10 μg polyuridylic acid | 38,300
2 | Glycine, alanine, serine, aspartic acid, glutamic acid | Deproteinized at zero time | 17
None | 20
+ 10 μg polyuridylic acid | 33
3 | Leucine, isoleucine, threonine, methionine, arginine, histidine, lysine, tyrosine, tryptophan, proline, valine | Deproteinized at zero time | 73
None | 276
+ 10 μg polyuridylic acid | 899
4 | S35-cysteine | Deproteinized at zero time | 6
None | 95
+ 10 μg polyuridylic acid | 113
Components of the reaction mixtures are presented in Table 1. The unlabeled amino acid mixture was omitted. 0.015 μM of each labeled amino acid was used. The specific activities of the labeled amino acids are present in the Methods and Materials section. 2.3 mg of protein of preincubated S-30 enzyme fraction were added to each reaction mixture. All samples were incubated at 35° for 30 min.
PNAS 47, 1588 (1961)
"[We] would not have to get polynucleotide synthesis very far to break the coding problem. Could crack life's code!"
– Marshall Nirenberg in lab notebook

The Genetic Code

Second letter

First letter	U		C		A		G		Third letter
U	UUU	Phenylalanine (Phe)	UCU	Serine (Ser)	UAU	Tyrosine (Tyr)	UGU	Cysteine (Cys)	U
	UUC		UCC		UAC		UGC		C
	UUA	Leucine (Leu)	UCA		UAA	Stop	UGA	Stop	A
	UUG		UCG		UAG	Stop	UGG	Tryptophan (Trp)	G
C	CUU	Leucine (Leu)	CCU	Proline (Pro)	CAU	Histidine (His)	CGU	Arginine (Arg)	U
	CUC		CCC		CAC		CGC		C
	CUA		CCA		CAA	Glutamine (Gln)	CGA		A
	CUG		CCG		CAG		CGG		G
A	AUU	Isoleucine (Ile)	ACU	Threonine (Thr)	AAU	Asparagine (Asn)	AGU	Serine (Ser)	U
	AUC		ACC		AAC		AGC		C
	AUA		ACA		AAA	Lysine (Lys)	AGA	Arginine (Arg)	A
	AUG	Methionine (Met)	ACG		AAG		AGG		G
G	GUU	Valine (Val)	GCU	Alanine (Ala)	GAU	Aspartic acid (Asp)	GGU	Glycine (Gly)	U
	GUC		GCC		GAC		GGC		C
	GUA		GCA		GAA	Glutamic acid (Glu)	GGA		A
	GUG		GCG		GAG		GGG		G

Gene Expression in *E. coli*

Coding Sequence:

Synonymous codon changes can result in 250-fold changes in expression levels

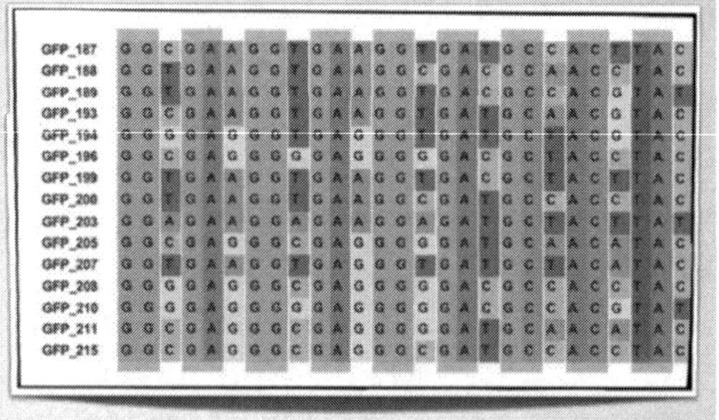

Most variance can be traced to 5' end

Kudla et al., *Science* 324, 255 (2009)

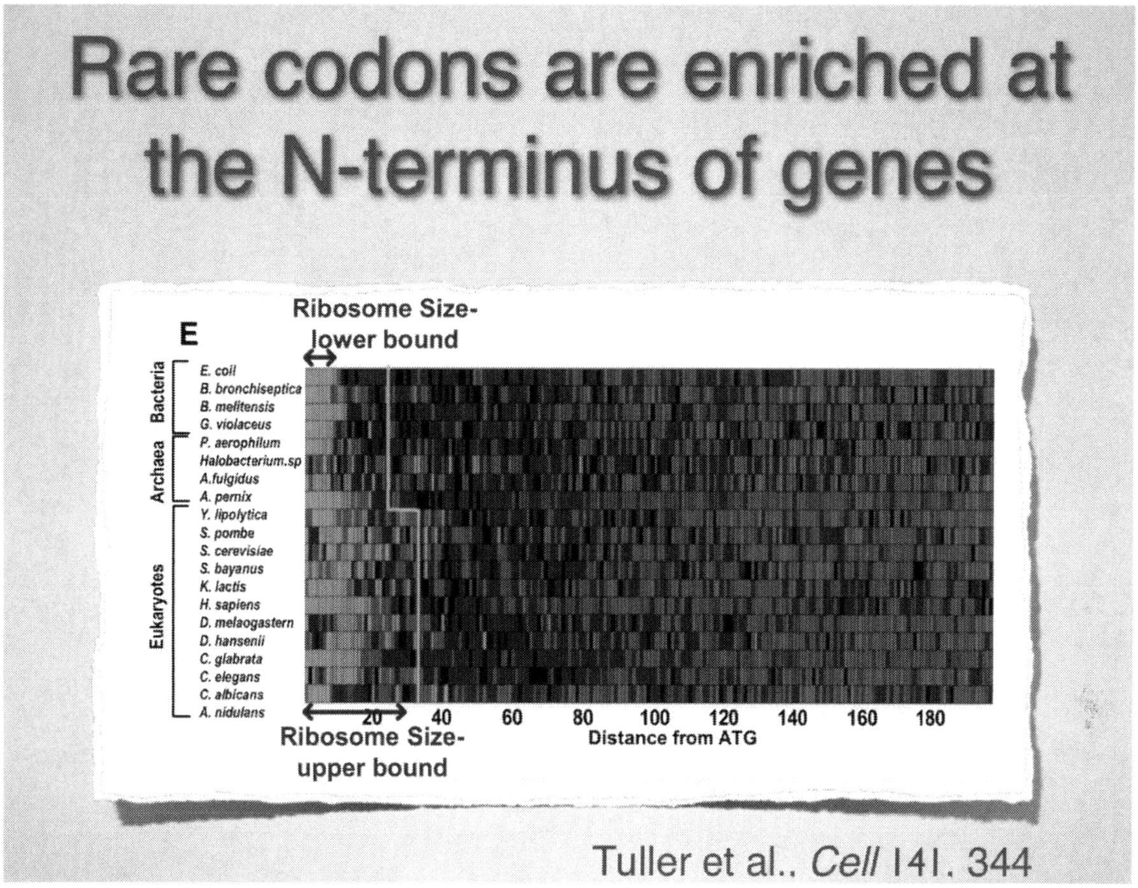
Rare codons are enriched at the N-terminus of genes
E
Ribosome Size-lower bound
Bacteria
Archaea
Eukaryotes
E. coli
B. bronchiseptica
B. melitensis
G. violaceus
P. aerophilum
Halobacterium.sp
A.fulgidus
A. pernix
Y. lipolytica
S. pombe
S. cerevisiae
S. bayanus
K. lactis
H. sapiens
D. melaogastern
D. hansenii
C. glabrata
C. elegans
C. albicans
A. nidulans
20 40 60 80 100 120 140 160 180
Distance from ATG
Ribosome Size-upper bound
Tuller et al., Cell 141, 344

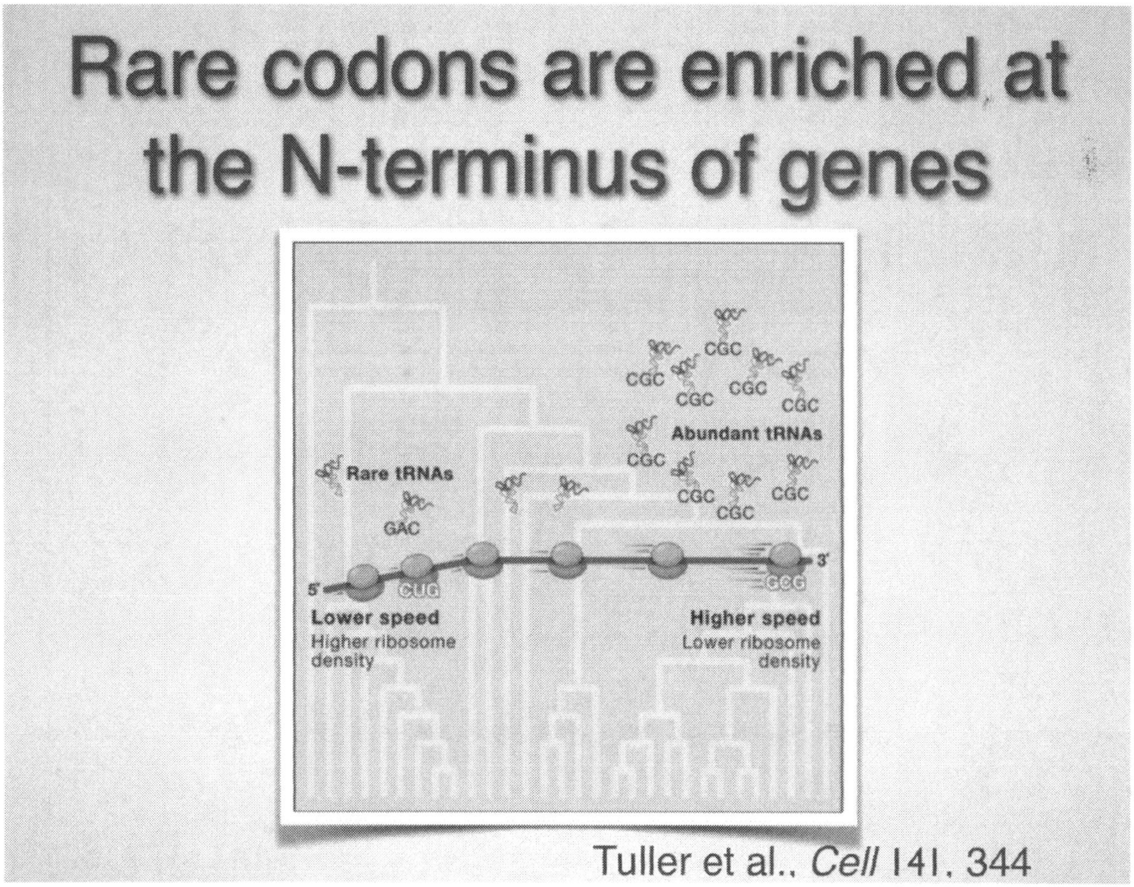
Rare codons are enriched at the N-terminus of genes
CGC
Abundant tRNAs
Rare tRNAs
GAC
CUG
CGG
5'
3'
Lower speed
Higher ribosome density
Higher speed
Lower ribosome density
Tuller et al., Cell 141, 344

Rare codons are enriched at the N-terminus of genes

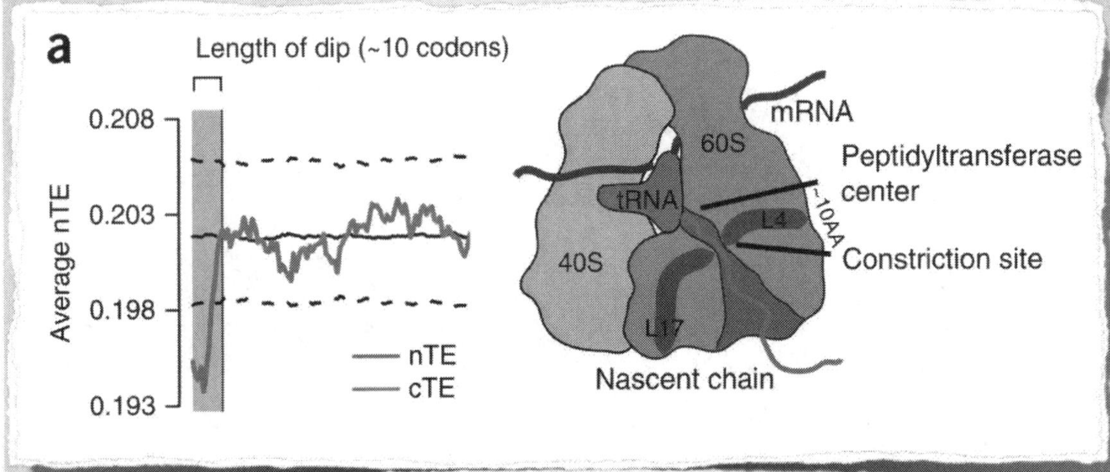

Pechmann & Frydman *NSMB* 20, 237

Codon Usage

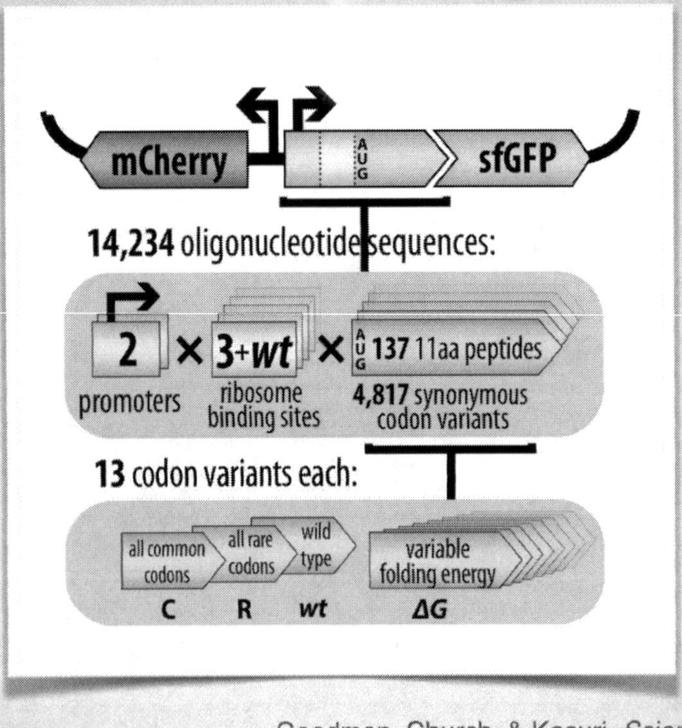

Goodman, Church, & Kosuri, *Science* 342, 475 (2013)

DNA from Microarrays

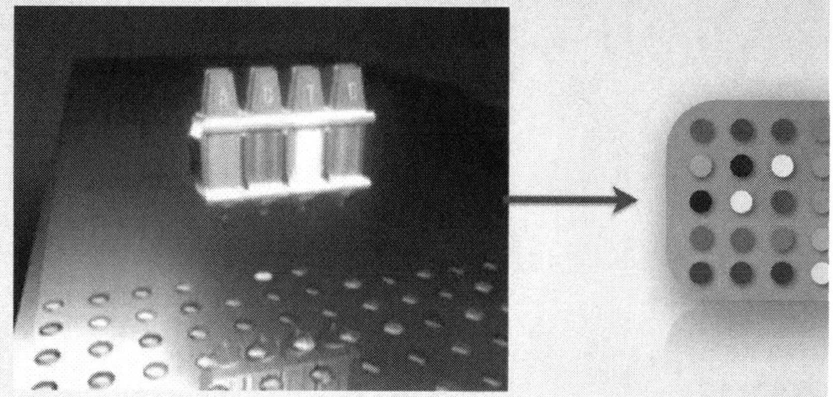

Library shows large dynamic range

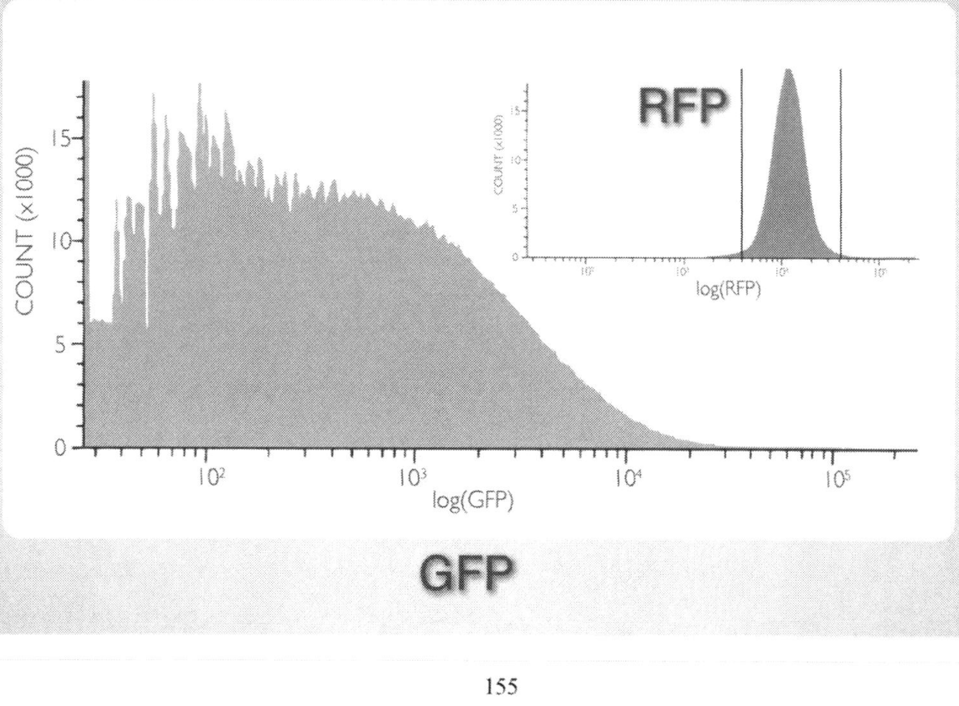

GFP

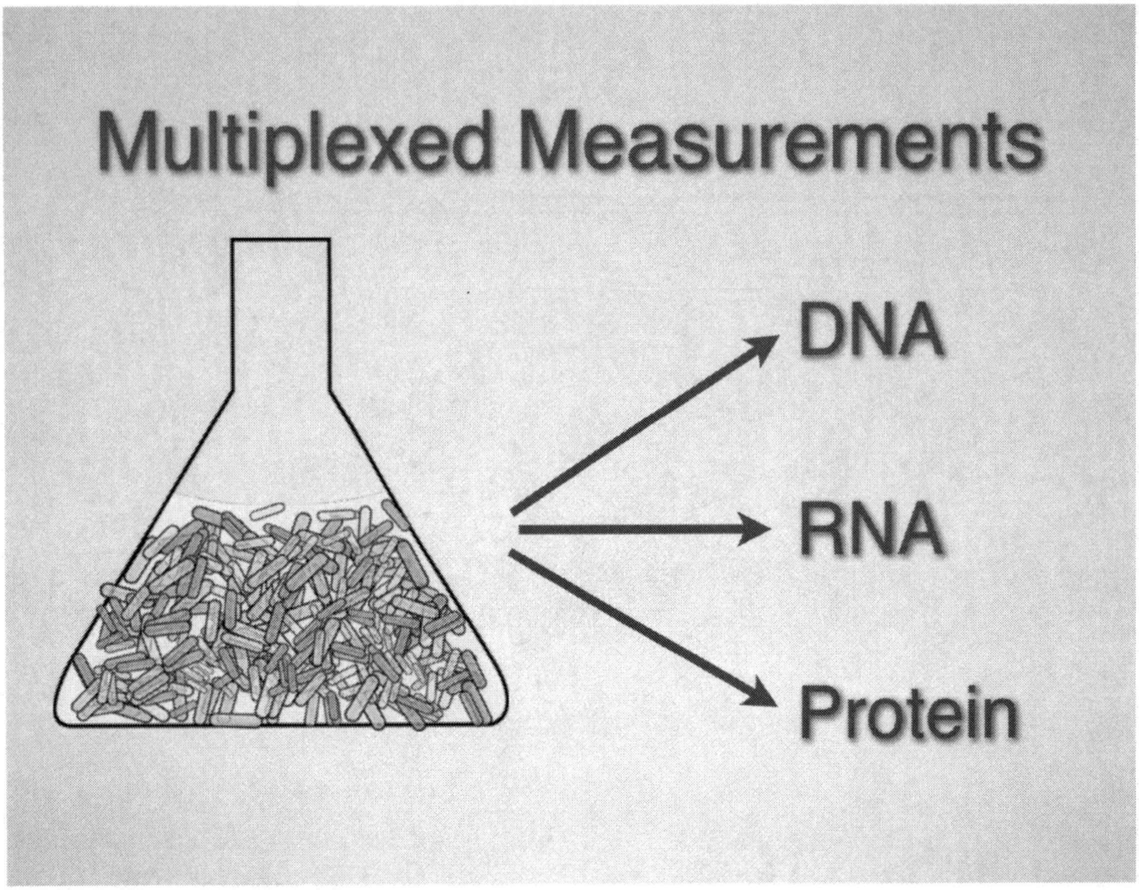
Multiplexed Measurements
DNA
RNA
Protein

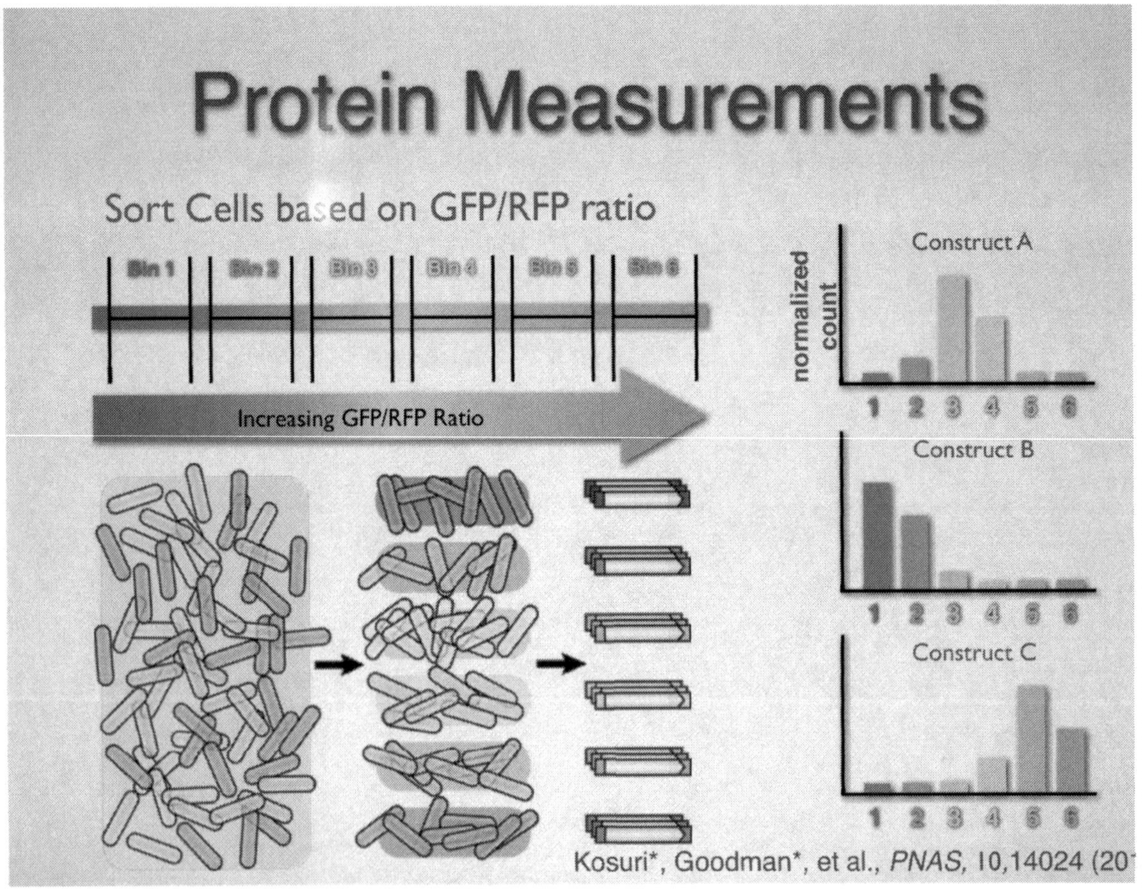
Protein Measurements
Sort Cells based on GFP/RFP ratio
Bin 1
Bin 2
Bin 3
Bin 4
Bin 5
Bin 6
Increasing GFP/RFP Ratio
normalized count
Construct A
1 2 3 4 5 6
Construct B
1 2 3 4 5 6
Construct C
1 2 3 4 5 6
Kosuri*, Goodman*, et al., PNAS, 10,14024 (20

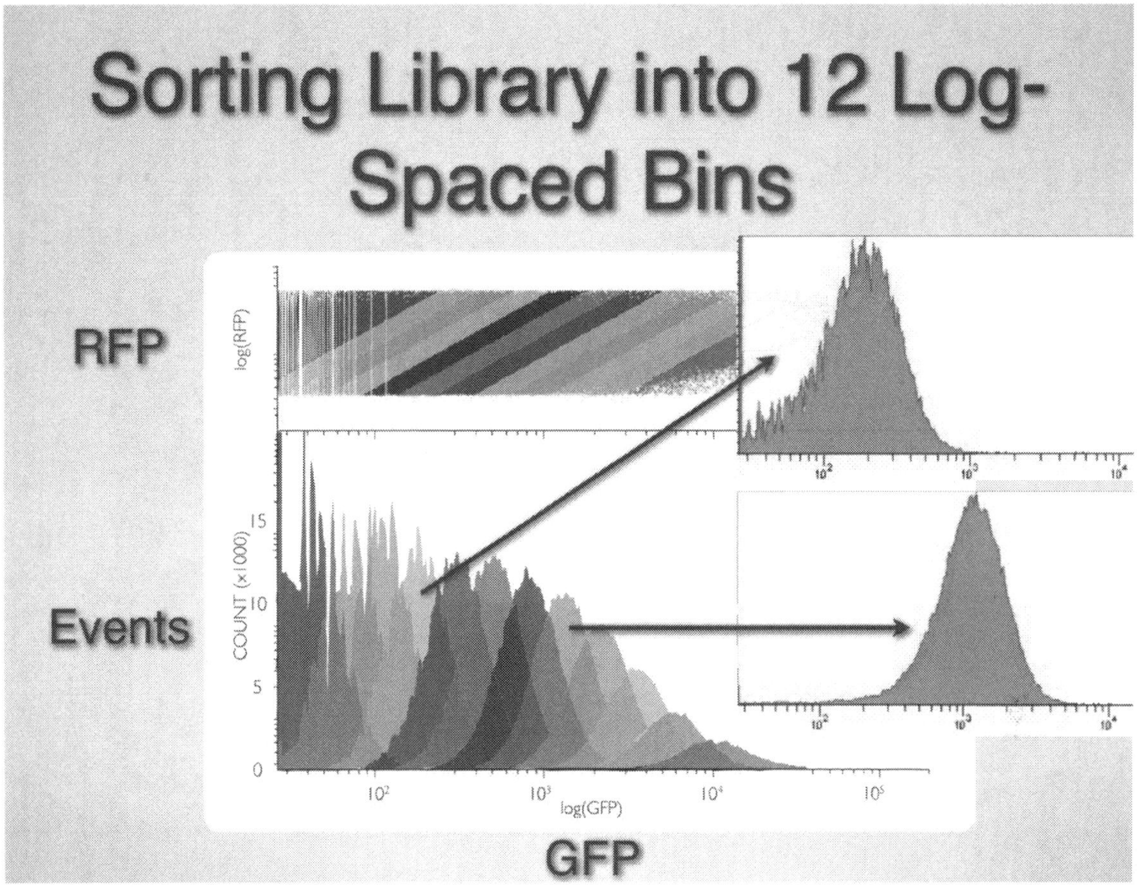
Sorting Library into 12 Log-Spaced Bins
RFP
log(RFP)
Events
COUNT (×1 000)
15
10
5
0
10²
10³
log(GFP)
10⁴
10⁵
GFP

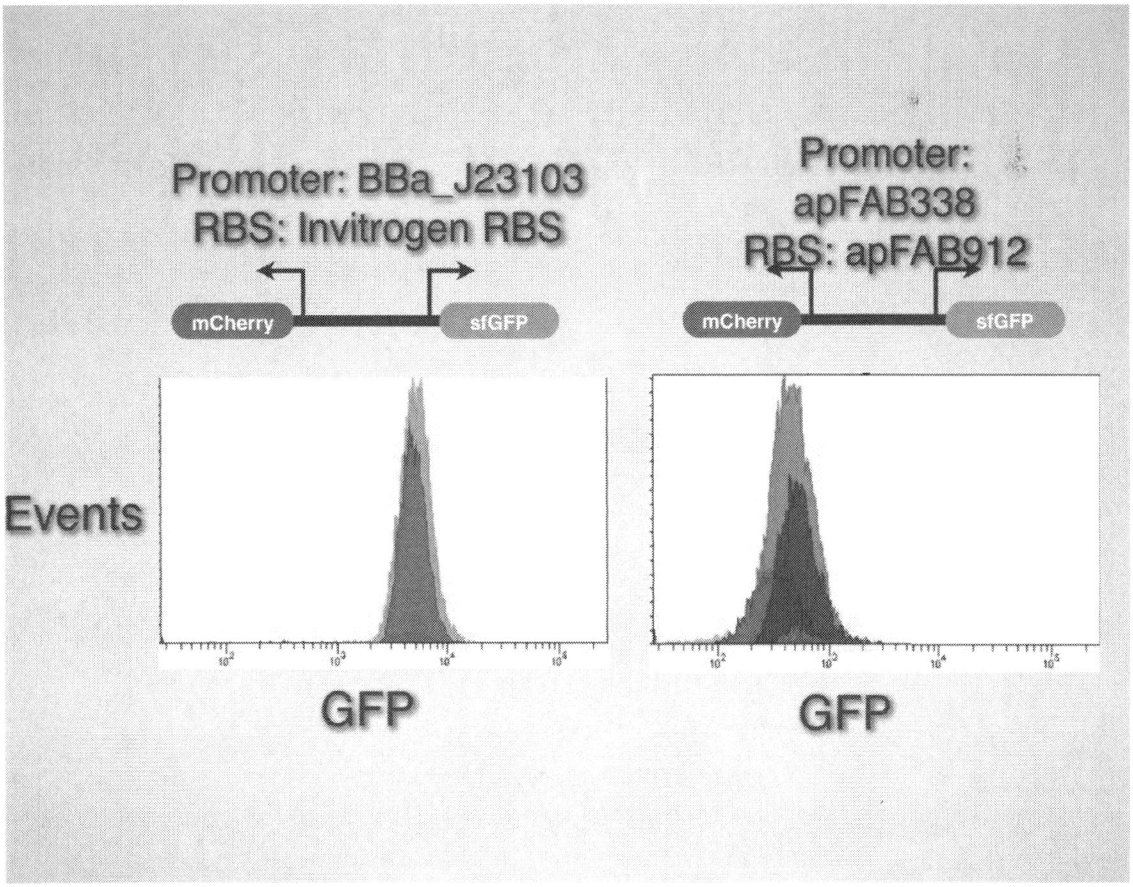
Promoter: BBa_J23103
RBS: Invitrogen RBS
Promoter:
apFAB338
RBS: apFAB912
mCherry
sfGFP
mCherry
sfGFP
Events
GFP
GFP

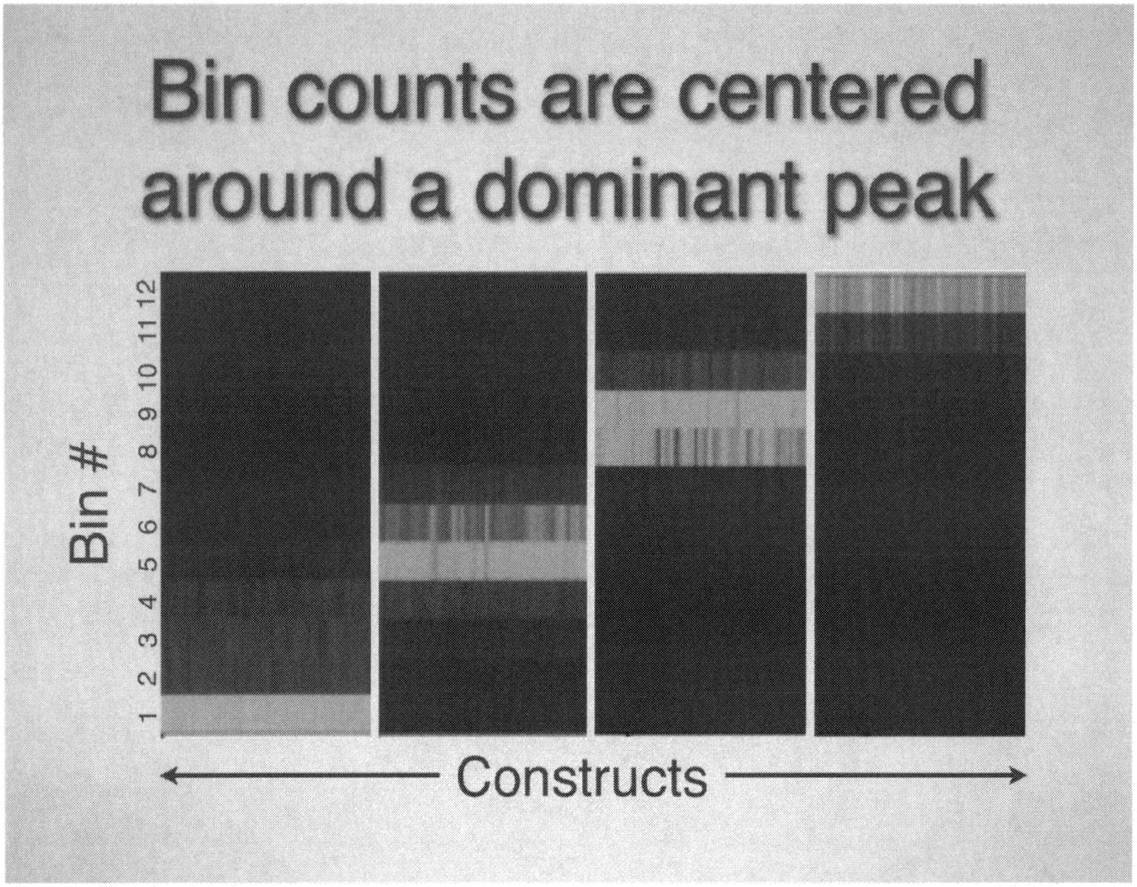
Bin counts are centered around a dominant peak
Bin #
1 2 3 4 5 6 7 8 9 10 11 12
Constructs

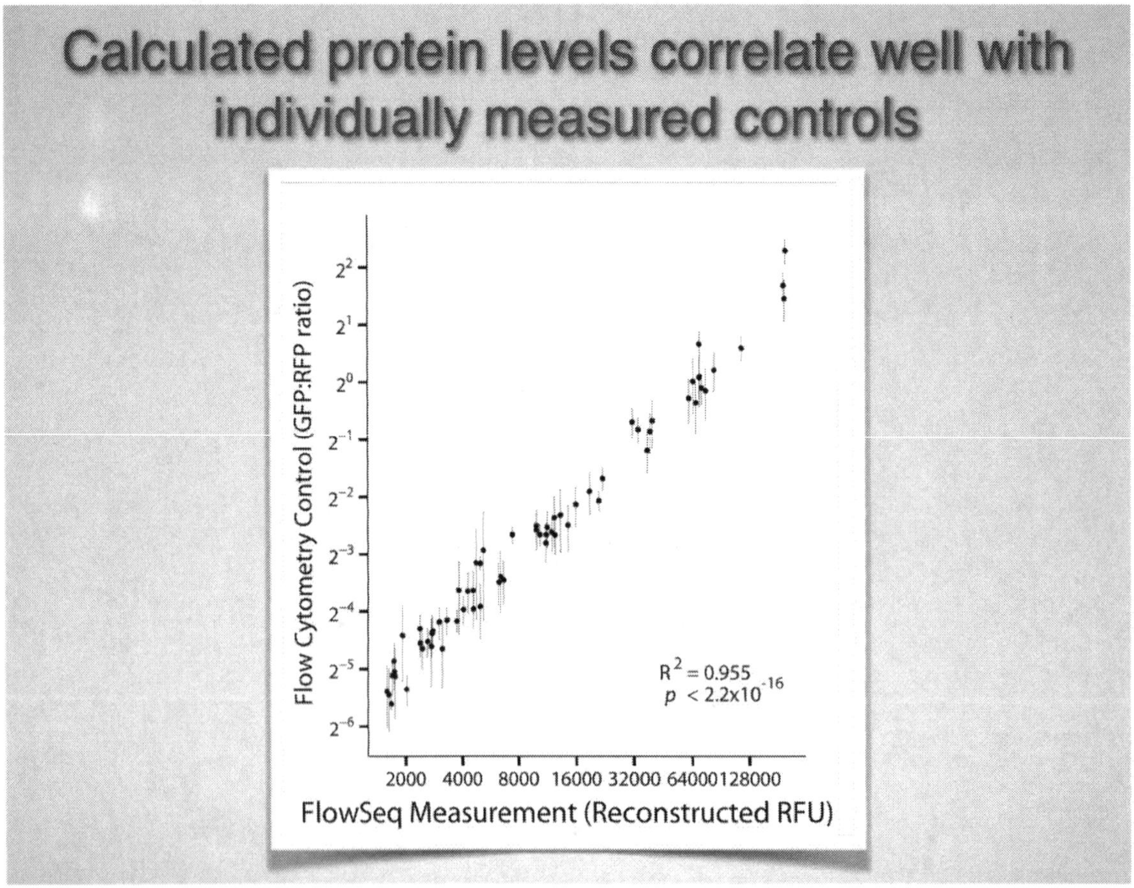
Calculated protein levels correlate well with individually measured controls
Flow Cytometry Control (GFP:RFP ratio)
2^2 2^1 2^0 2^-1 2^-2 2^-3 2^-4 2^-5 2^-6
R^2 = 0.955
p < 2.2x10^-16
2000 4000 8000 16000 32000 64000128000
FlowSeq Measurement (Reconstructed RFU)

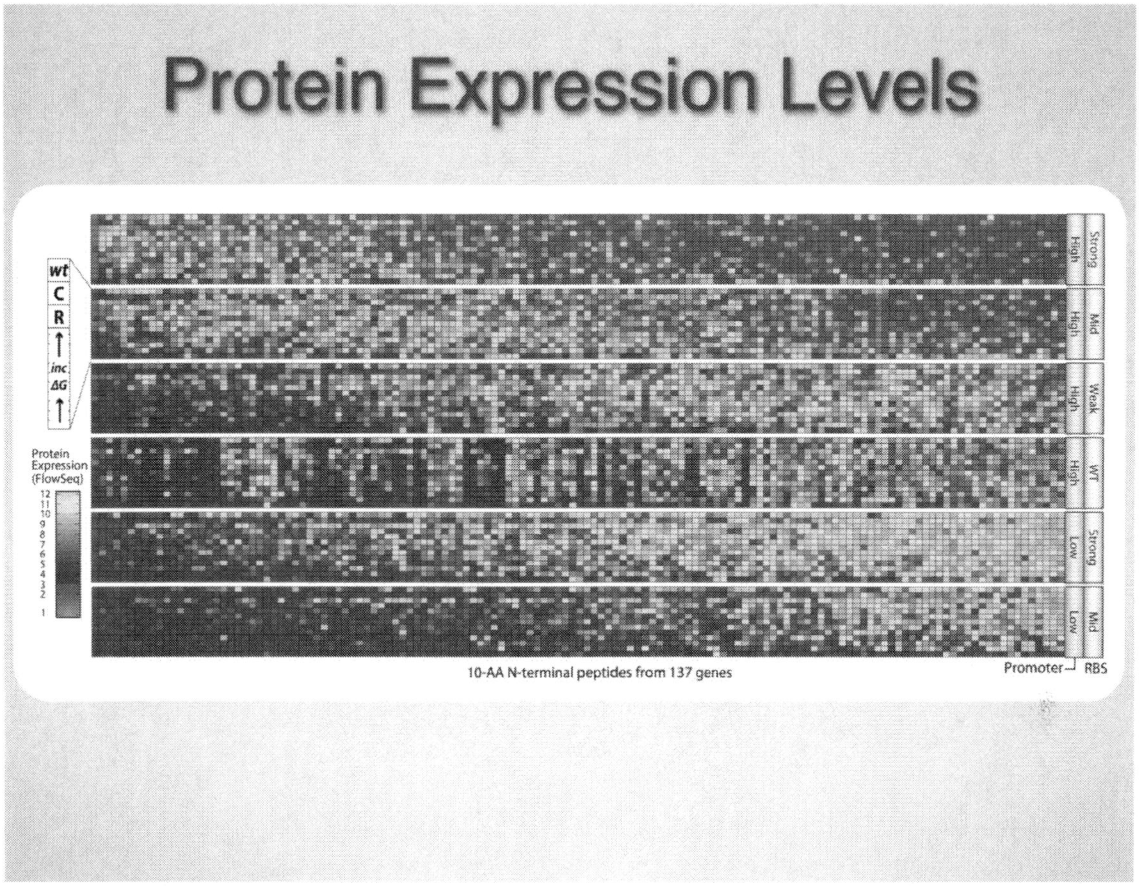

Protein Expression Levels
wt
C
R
inc ΔG
Protein Expression (FlowSeq)
Strong High
Mid High
Weak High
WT High
Strong Low
Mid Low
10-AA N-terminal peptides from 137 genes
Promoter
RBS

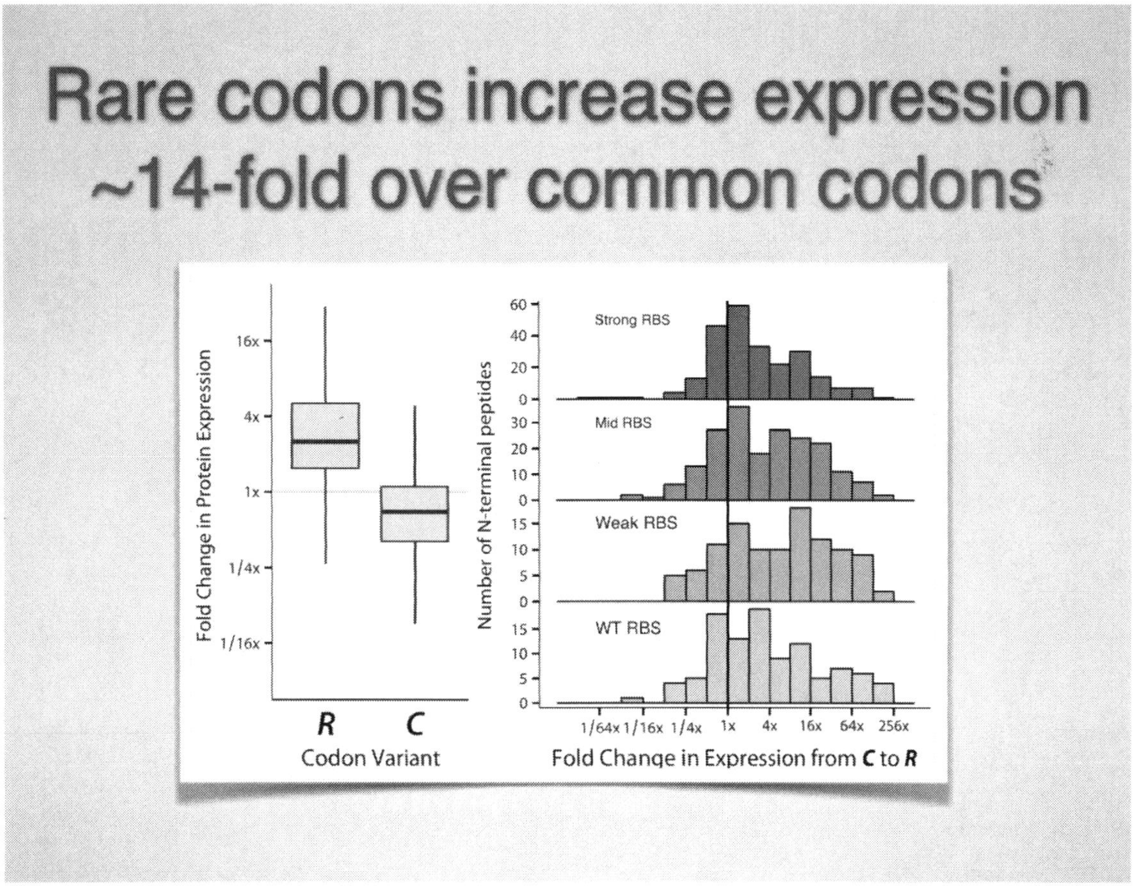

Rare codons increase expression ~14-fold over common codons
Fold Change in Protein Expression
16x
4x
1x
1/4x
1/16x
R
C
Codon Variant
Number of N-terminal peptides
Strong RBS
Mid RBS
Weak RBS
WT RBS
1/64x 1/16x 1/4x 1x 4x 16x 64x 256x
Fold Change in Expression from C to R

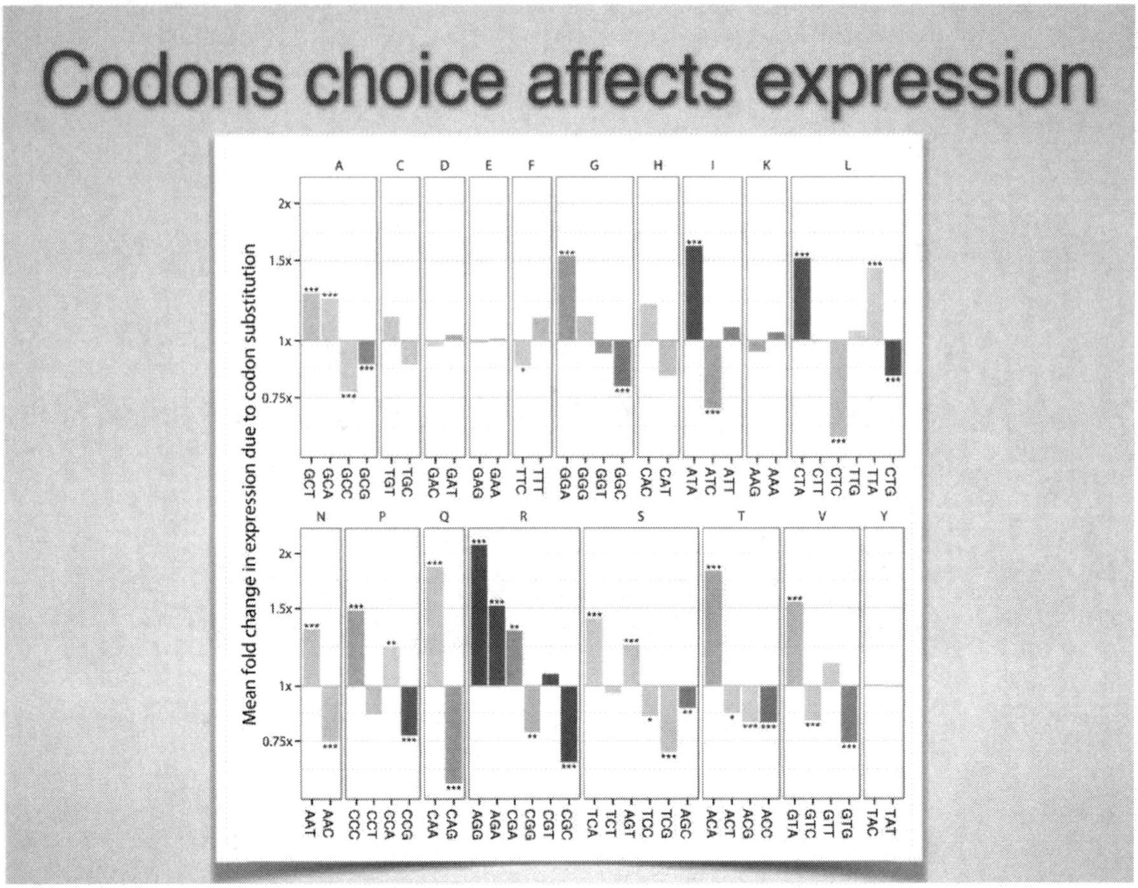
Codons choice affects expression
Mean fold change in expression due to codon substitution
A C D E F G H I K L
N P Q R S T V Y
2x
1.5x
1x
0.75x

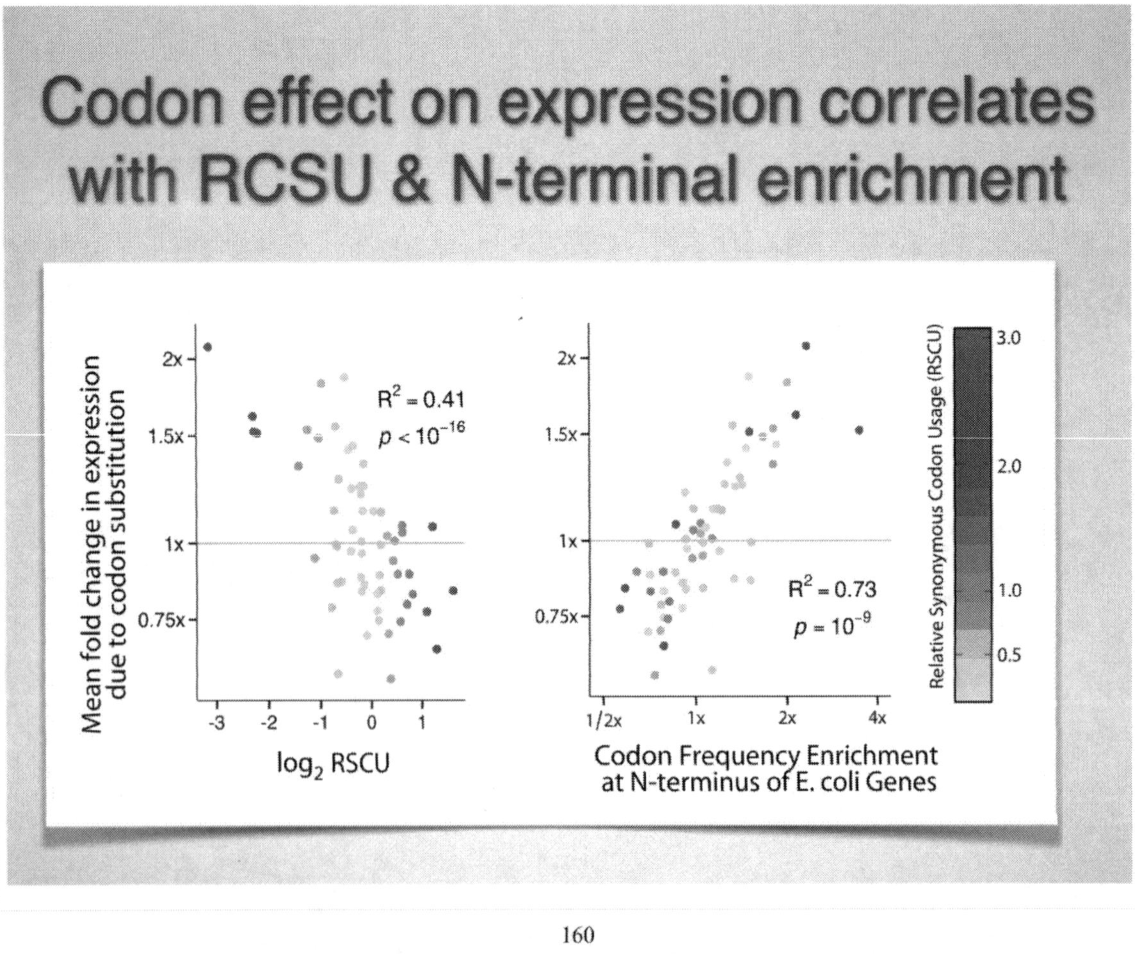
Codon effect on expression correlates with RCSU & N-terminal enrichment
Mean fold change in expression due to codon substitution
2x
1.5x
1x
0.75x
$R^2 = 0.41$
$p < 10^{-16}$
log$_2$ RSCU
$R^2 = 0.73$
$p = 10^{-9}$
Codon Frequency Enrichment at N-terminus of E. coli Genes
Relative Synonymous Codon Usage (RSCU)
3.0
2.0
1.0
0.5

The expression effect is mostly due to secondary structure

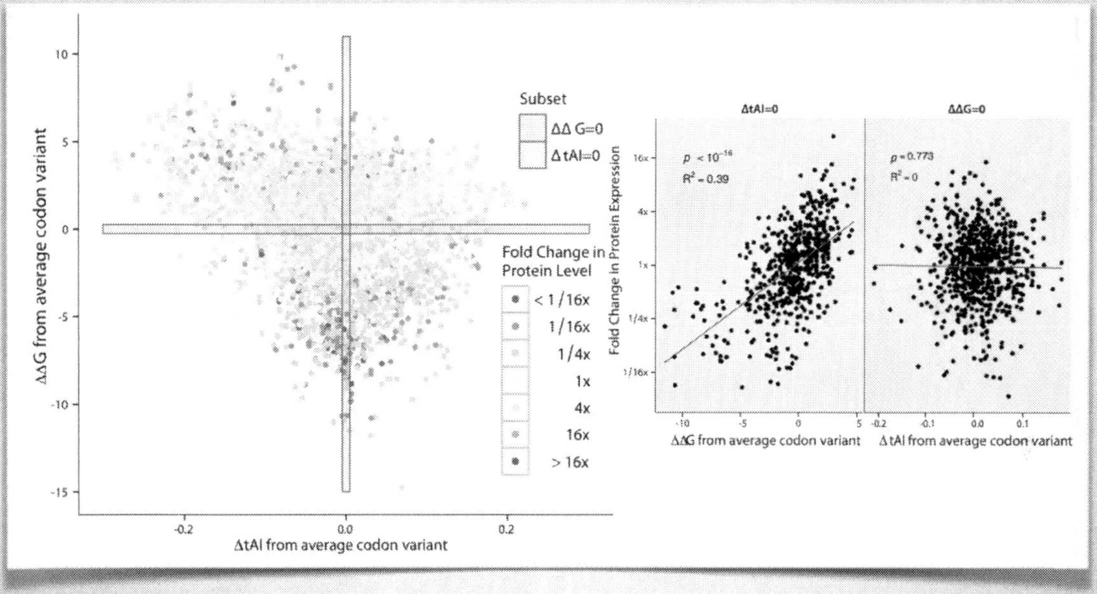

The expression effect is mostly due to secondary structure

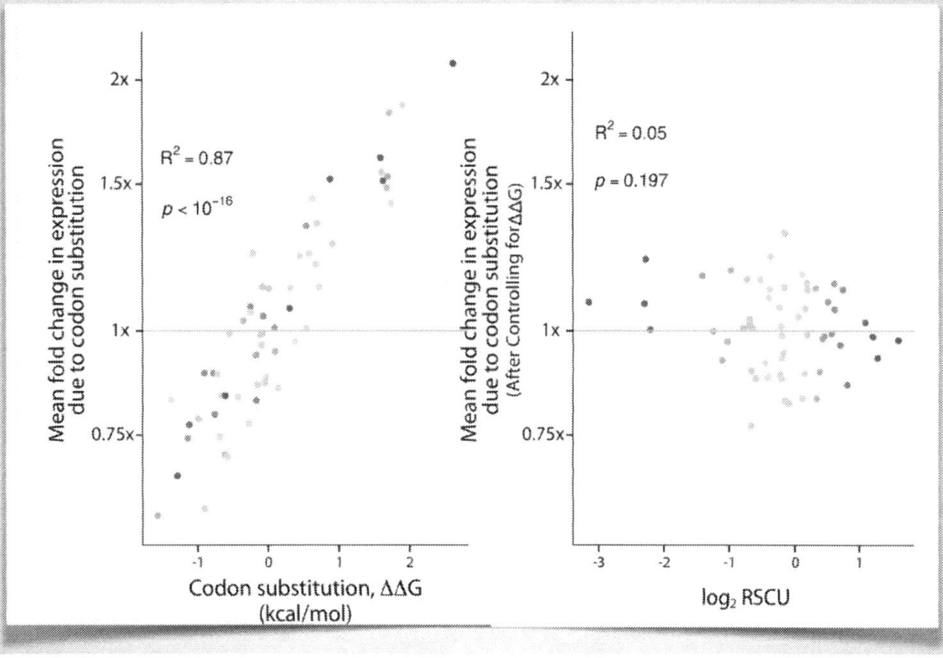

Secondary structure 10bp downstream of start codon is most correlated with expression increases

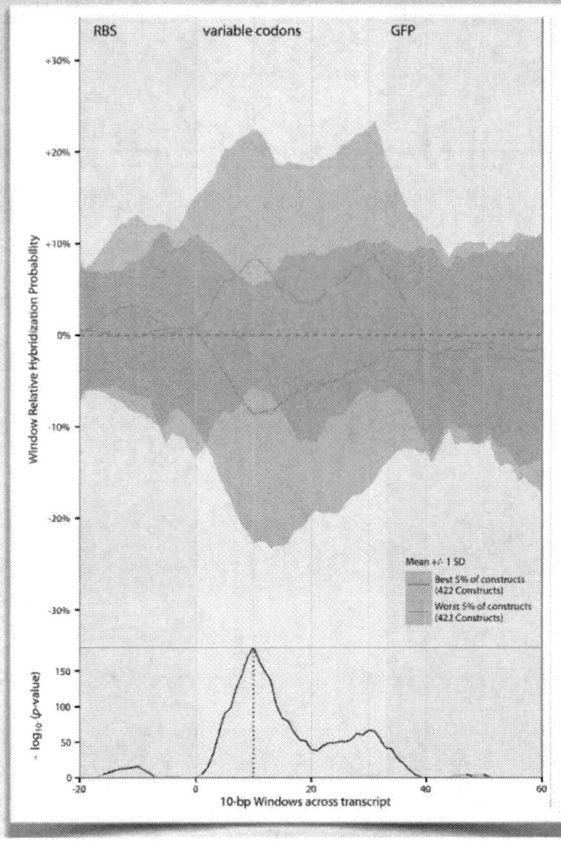

Codon Usage Conclusions

Changing N-terminal 11-aa codon usage strongly affects expression (Avg. of 60-fold in our library)

Using subset of codons, usually rare ones, result in large increases in expression (~14-fold on average)

These same codons are enriched at the N-terminus of natural genes.

This effect is mediated by reduction in secondary structure, and not inherent to codon usage itself.

Kosuri Lab @ UCLA

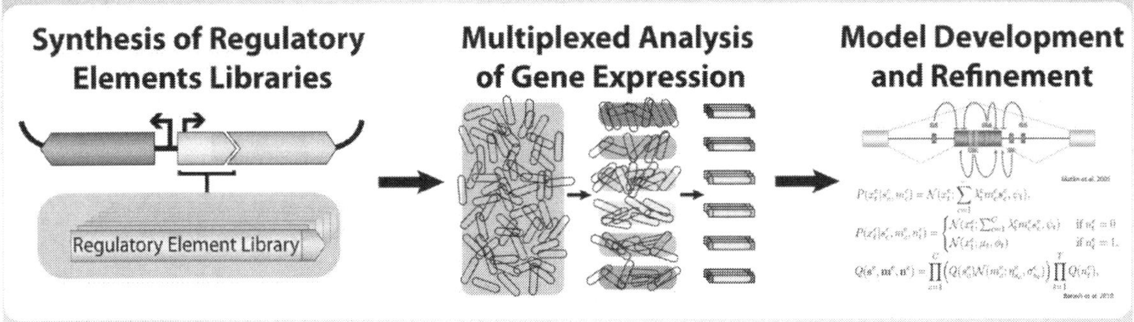

Bacterial

Promoters, RBSs, CDSs, Codon Usage, sRNA, Regulation, Terminators, ...

Human

Promoters, Enhancers, RNA splicing, ncRNA, Histone Remodeling, Methylation, Editing

3679 promoters (3103 unique positions)

5658 TF binding sites

AcrR, Ada, AgaR, AidB, AlaS, AllR, AllS, AraC, ArcA, ArgP, ArgR, ArsR, AscG, AsnC, AtoC, BaeR, BasR, BetI, BirA, CRP, CaiF, CbI, ChbR, CpxR, Cra, CreB, CsgD, CspA, CueR, CusR, CynR, CysB,, CytR, Dan, DcuR, DeoR, DicA, DinJ-YafQ, DnaA, DpiA, EnvR, EvgA, ExuR, FNR, FabR, FadR, FeaR, FhlA, Fis, FlhDC, Fur, GadE, GadE-RcsB, GadW, GadX, GalR, GalS, GcvA, GlcC, GlpR, GlrR, GntR, H-NS, HU, HcaR, HipAB, HipB, HyfR, IHF, IclR, IdnR, IlvY, IscR, KdgR, KdpE, LacI, LeuO, LexA, LldR, LrhA, Lrp, LsrR, MalI, MalT, MarA, MarR, MazE, MazE-MazF, McbR, MelR, MetJ, MetR, MhpR, Mlc, MlrA, MngR, MntR, ModE, MprA, MqsA, MraZ, MurR, Nac, NadR, NagC, NanR, NarL, NarP, NhaR, NikR, NorR, NrdR, NsrR, NtrC, OmpR, OxyR, PaaX, PdhR, PepA, PhoB, PhoP, PrpR, PspF, PurR, PutA, PuuR, QseB, RbsR, RcdA, RcnR, RcsAB, RcsB, RcsB-BglJ, RelB, RelB-RelE, RhaR, RhaS, Rob, RstA, RutR, SdiA, SgrR, SlyA, SoxR, SoxS, TdcA, TdcR, TorR, TreR, TrpR, TyrR, UhpA, UidR, UlaR, UxuR, XapR, XylR, YefM, YefM-YoeB, YehT, YiaJ, YpdB, YqhC, YqjI, ZntR, ZraR, Zur

Codon Usage
Wyss Institute

Dan Goodman George Church

Funding: Department of Energy (DE-FG02-02ER63445), NSF SynBERC (SA5283-11210), Office of Naval Research (N000141010144), Agilent Technologies, Wyss Institute

Rocky Cheung
Eric Jones
Nathan Lubock
Hwangbeom Kim
David Yao
Mike Corrado
Jessica Davis
Kim Insigne
Taylor Ward
Guillaume Urtecho

http://kosurilab.org

Funding: NIH (DP2GM114829, U01HG007912), NSF (MCB-1445112), Department of Energy, BSCRC Innovation Award, Searle Scholars Program, Allen Foundation, Fred and Linda Wudl

Affiliations: Dept. of Chemistry & Biochemistry, Institute of Quantitative and Computational Biosciences, UCLA-DOE Institute for Genomics and Proteomics, Broad Stem Cell Research Center, Jonsson Comprehensive Cancer Center, Molecular Biology Institute, UCLA iGEM

Applying synthetic biology to real world problems

Using interactions between microbes, humans and machines

- Health: Reprogramming the microbiome

- Sustainability: Sunlight to stuff

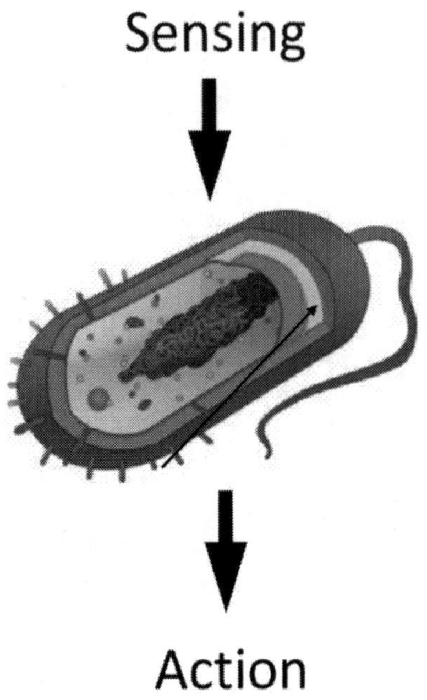
Sensing
Action

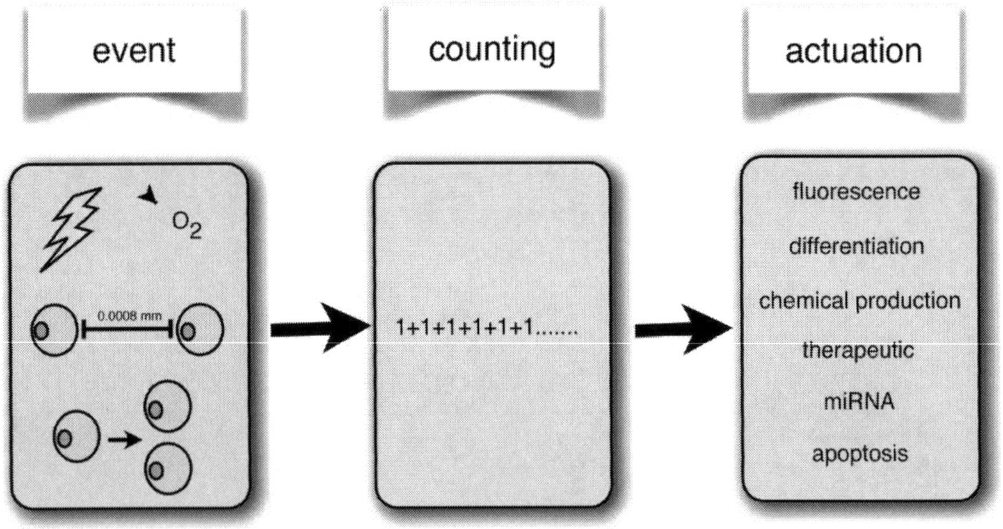
event
counting
actuation
O_2
0.0008 mm
1+1+1+1+1+1.......
fluorescence
differentiation
chemical production
therapeutic
miRNA
apoptosis

Interface with the environment

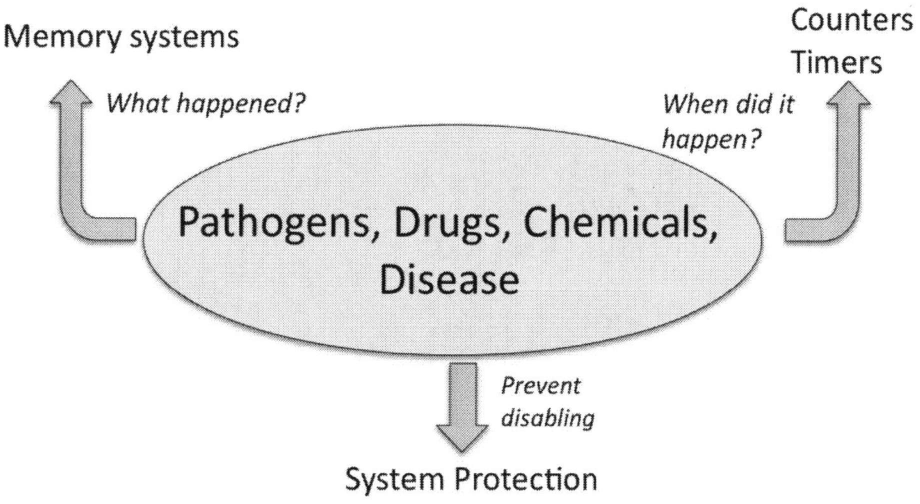

Detect and Remember Exposure

- Construct a 'memory loop' that is activated when bacteria enters mammals in response to gut environment

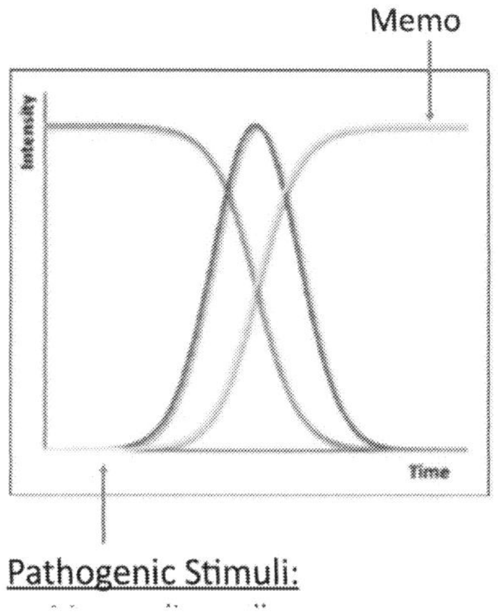

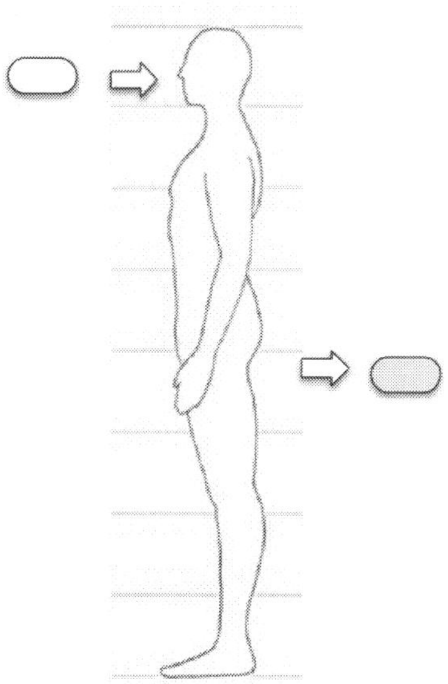

Dual use –
Biology for Diagnostics and Therapeutics

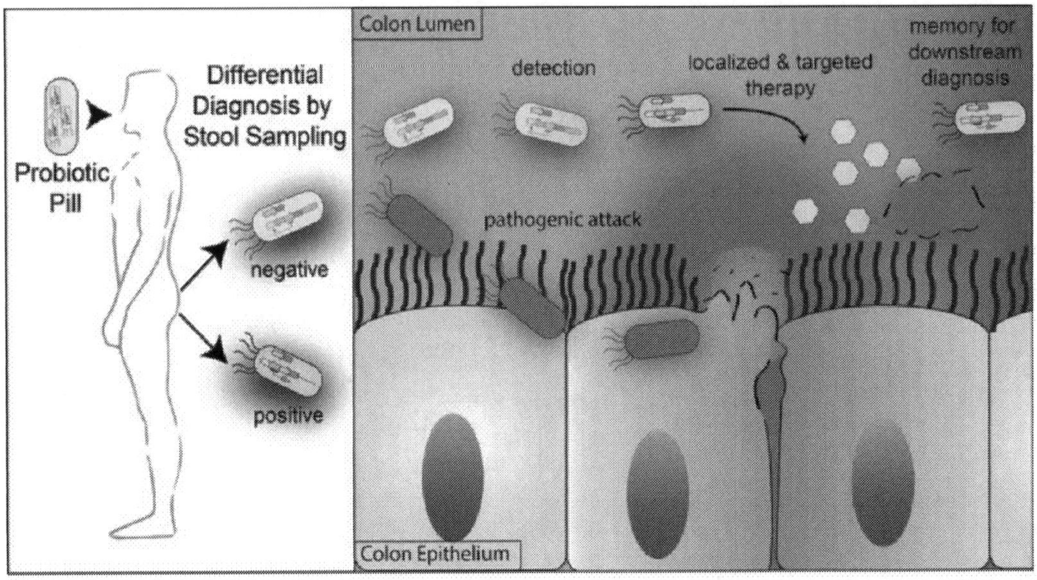

Natural vs. human approaches

Human approach

- Signal → Effector gene(s)
- Effector gene expression level varies with signal concentration & duration

Nature's approach

- Signal → Switch control
- Switch → Effector genes
- Effector genes are expressed at correct level & duration

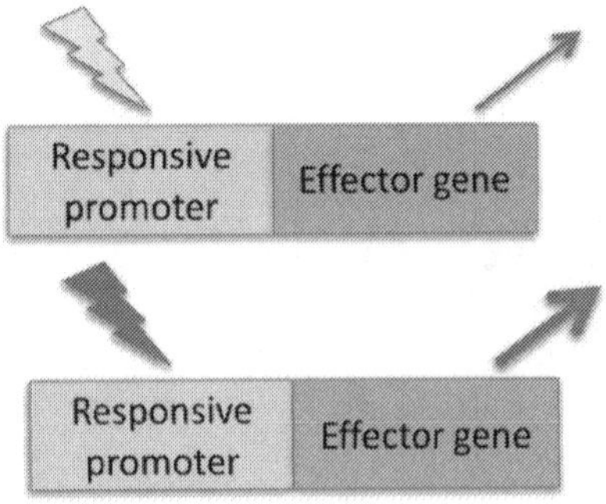

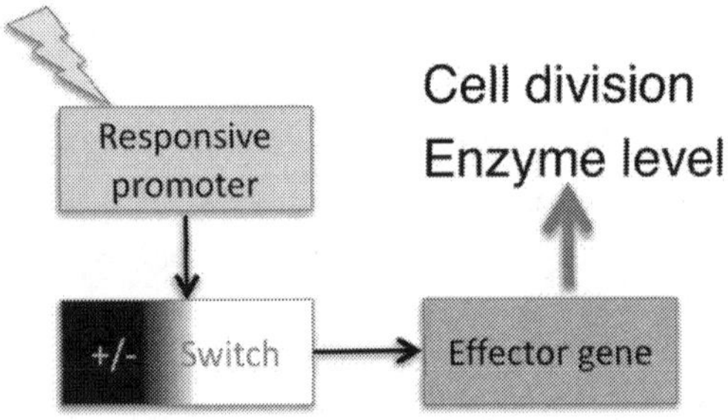

Memory Element Design

Criteria:

- Highly stable states
- Low-host Burden
- Antibiotic selection not required
- Integration in the bacterial host genome

Constraint:

- Extreme state stability with low protein copy numbers requires highly cooperative switch mechanics = ***lambda lytic/lysogenic switch***

Design of the Memory System

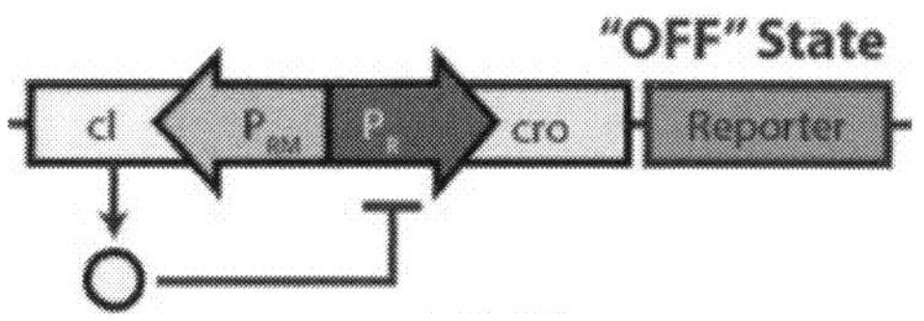

CI state

Design of the Memory System

'Trigger' –
drives either Cro

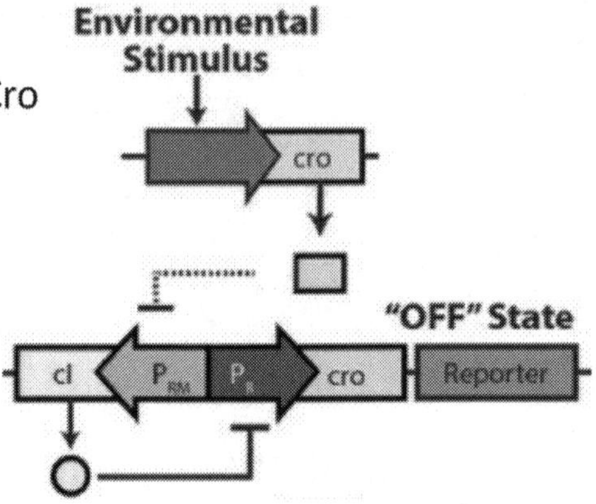

CI state

Design of the Memory System

'Trigger' –
drives either
Cro or CI-d/n

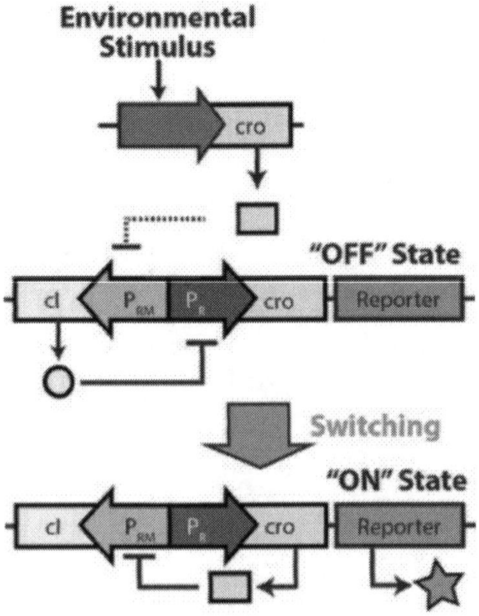

CI state

cro state

Memory of antibiotic exposure

in vitro

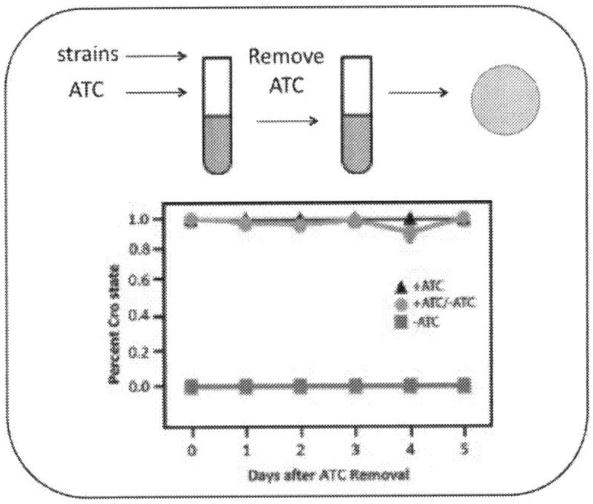

in vivo

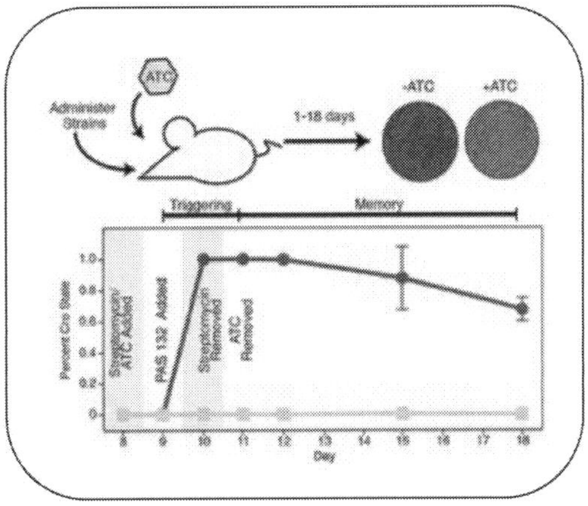

Memory system is stable after antibiotic exposure in a test tube and a mouse.

Kotula and Kerns, PNAS, 2014

Natural Gut Bacteria with Synthetic Circuit do better

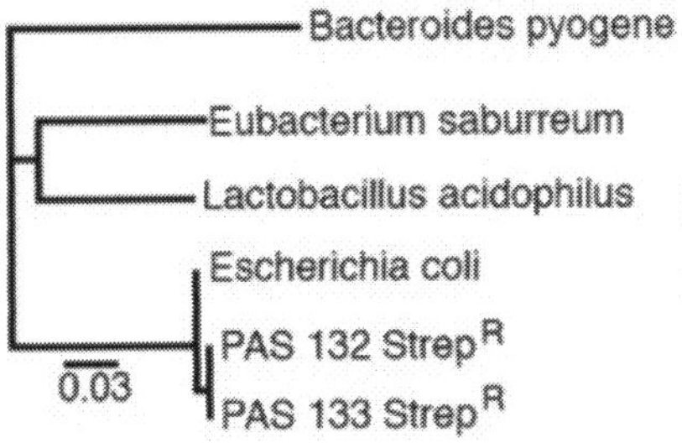

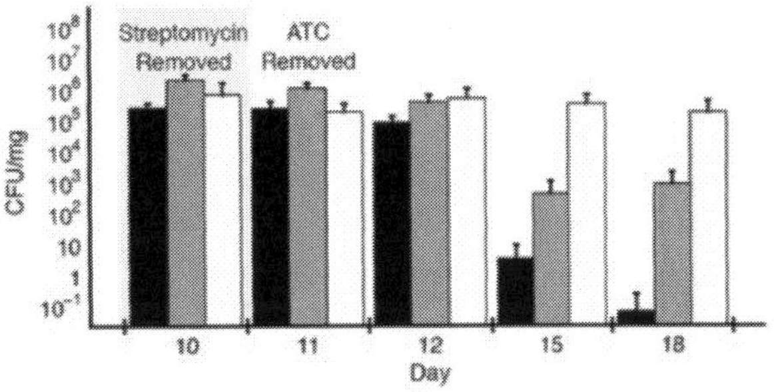

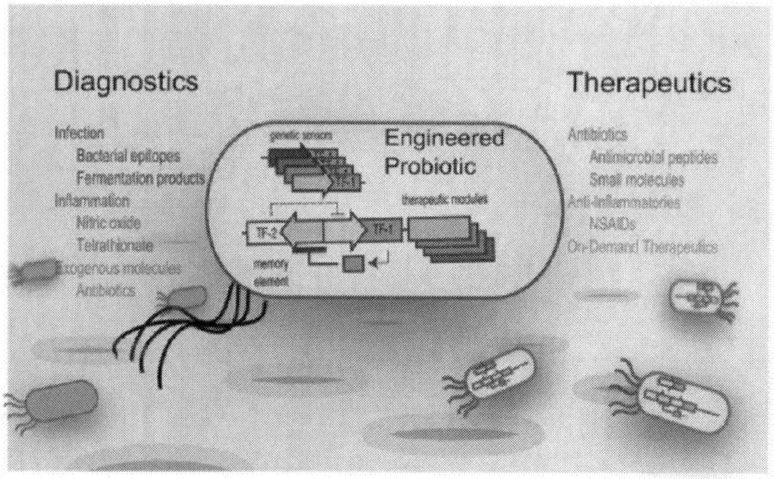

Mechanism for tetrathionate production

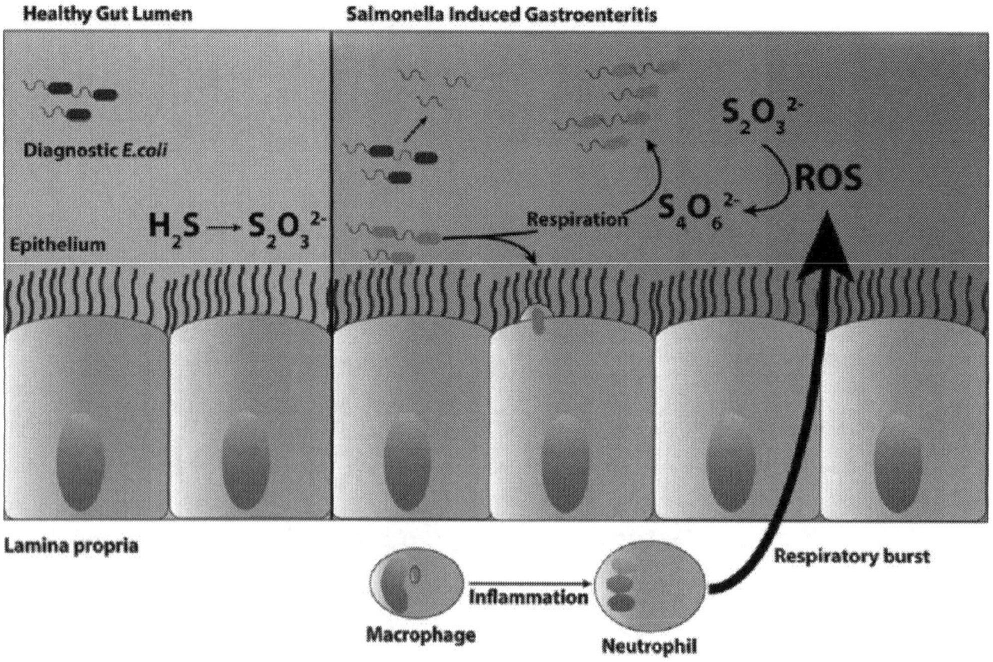

Engineering E.coli to sense and remember tetrathionate exposure

ttrRS and P_{ttrB} cloned from *Salmonella*

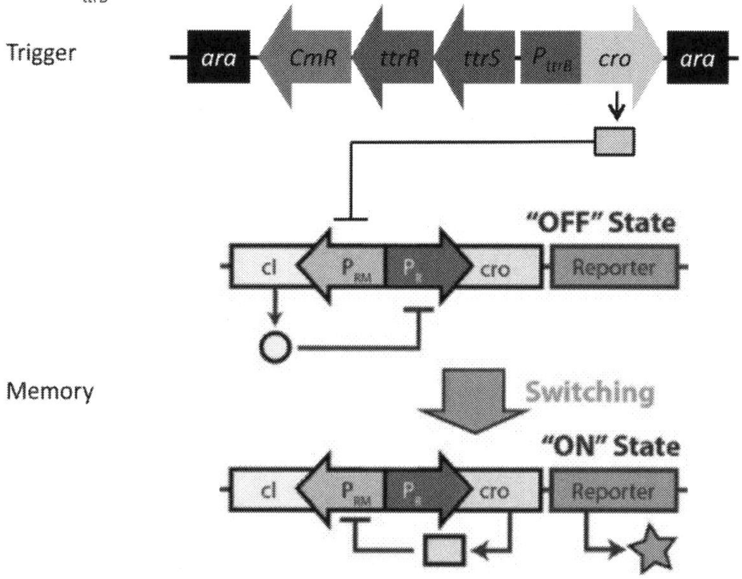

Tetrathionate Sense and Respond

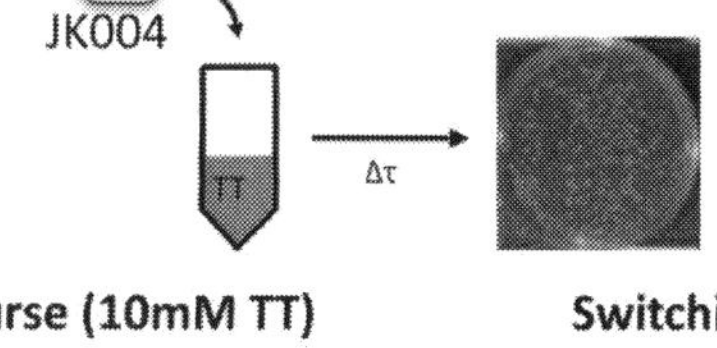

Switching Time course (10mM TT)

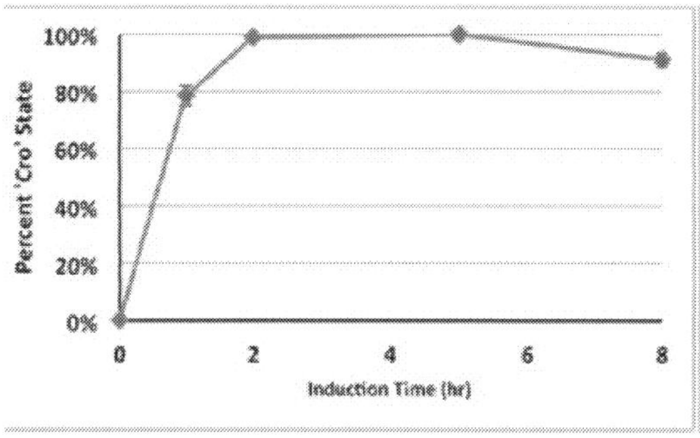

Switching Dose-Response (5hr)

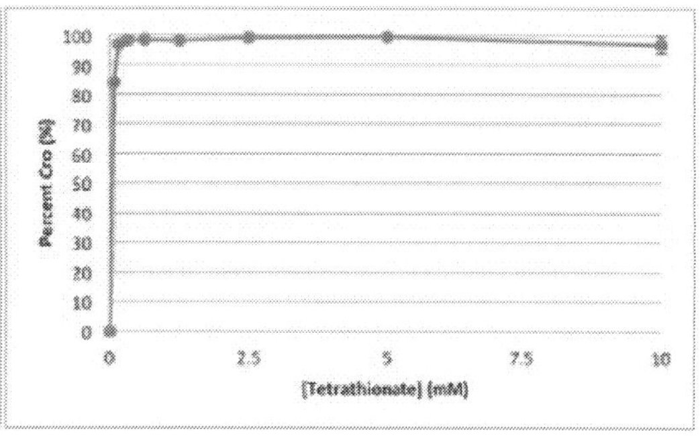

Jordan Kerns

Engineered tetrathionate memory system retains memory

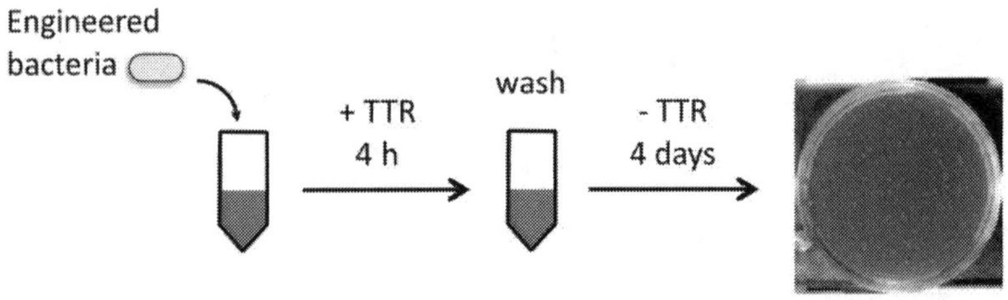

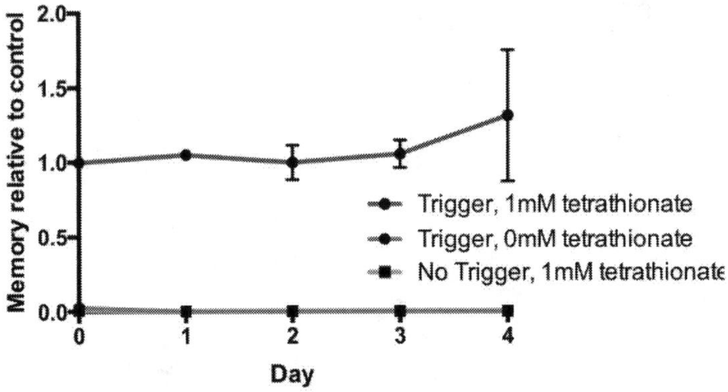

David Riglar

Tetrathionate triggered memory in mice exposed to Salmonella infection

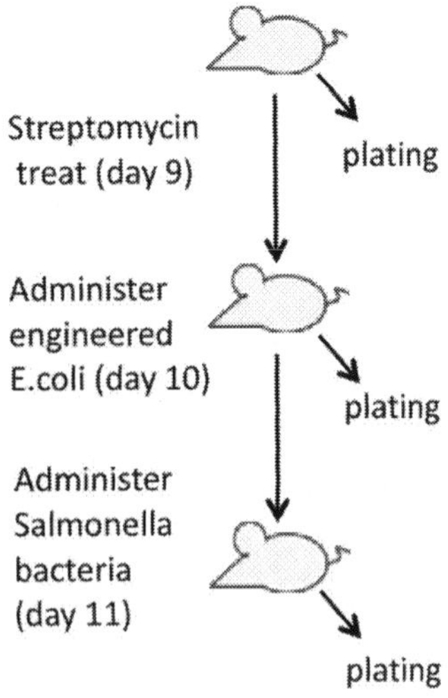

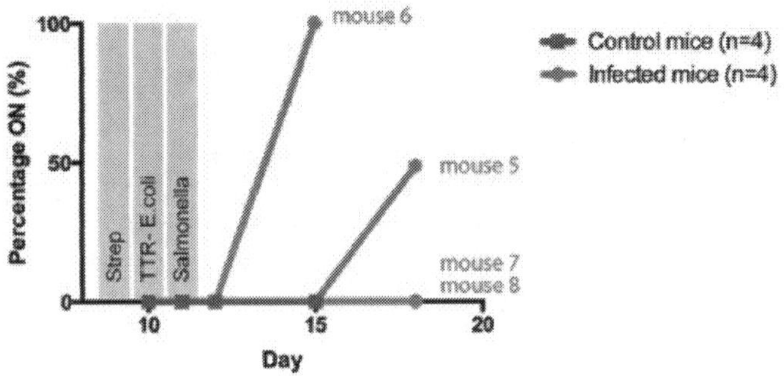

Natural Gut Bacteria Programmed to Diagnose and Respond

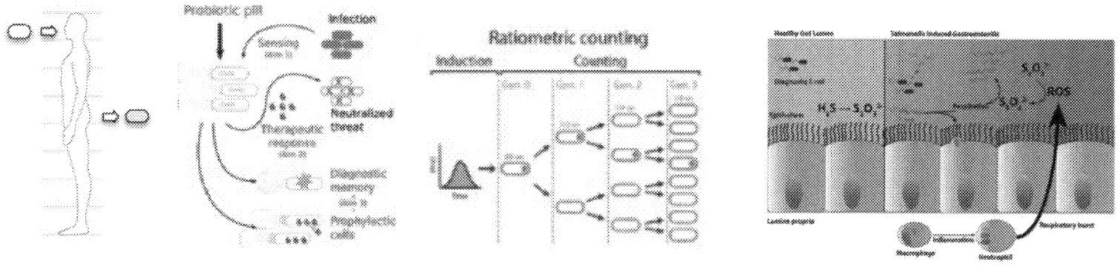

Re-booting the Gut Microbiota

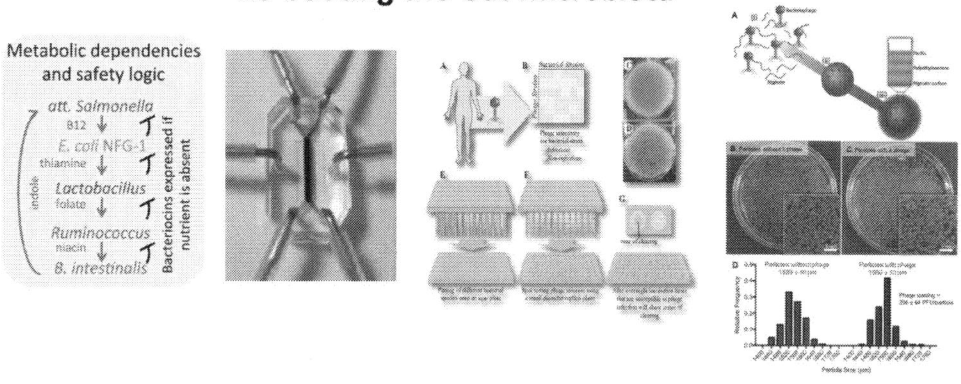

Synthetic Probiotics for global health

Circadian Clock

Oscillators that enable anticipation of diurnal variations in environmental conditions and thereby couple physiological processes to geophysical time.

24-hour periodicity

Free-running

Entrainable

Applications: Chronotherapeutics

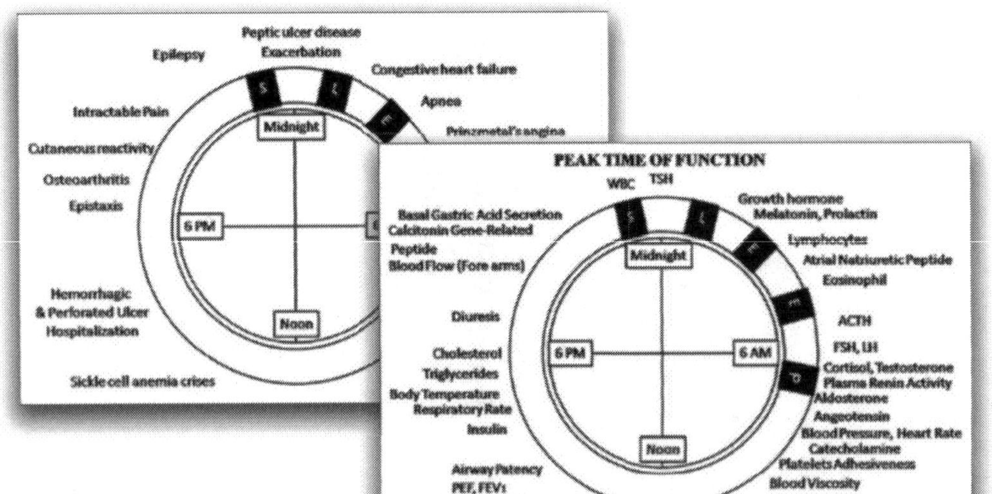

Applications: Chronotherapeutics

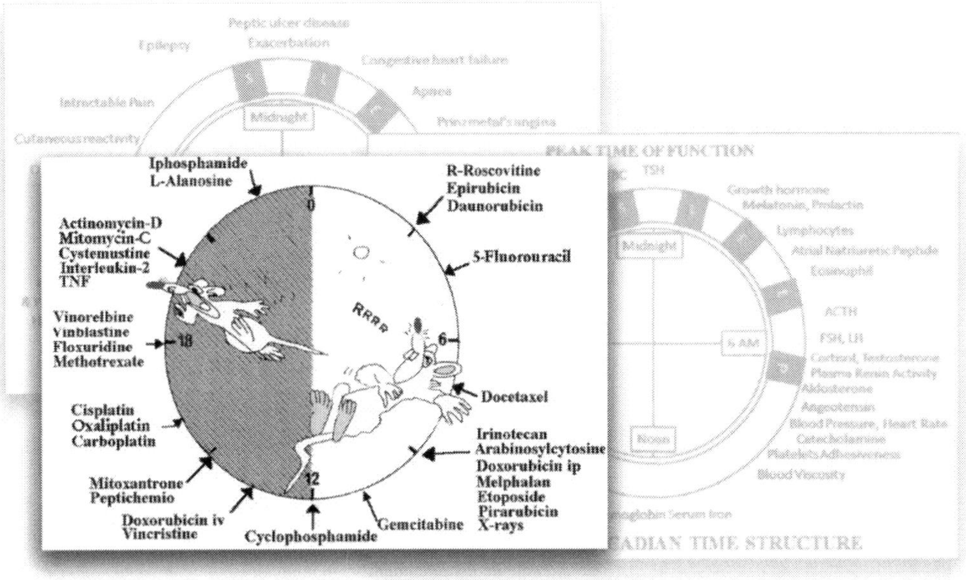

Circadian interaction between the microbiota and host

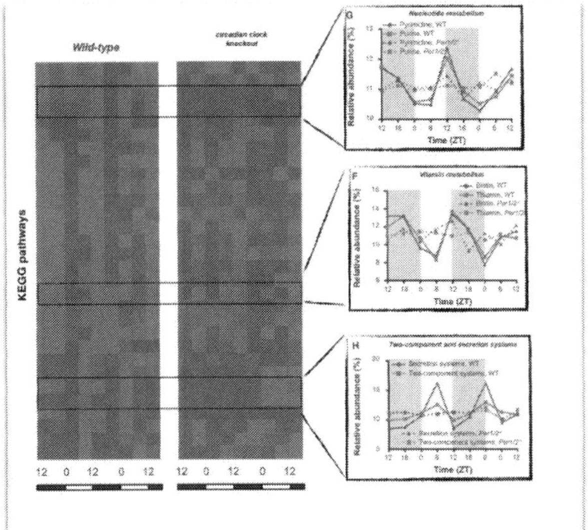

Functional capacities of the intestinal microbiome change over circadian rhythm.

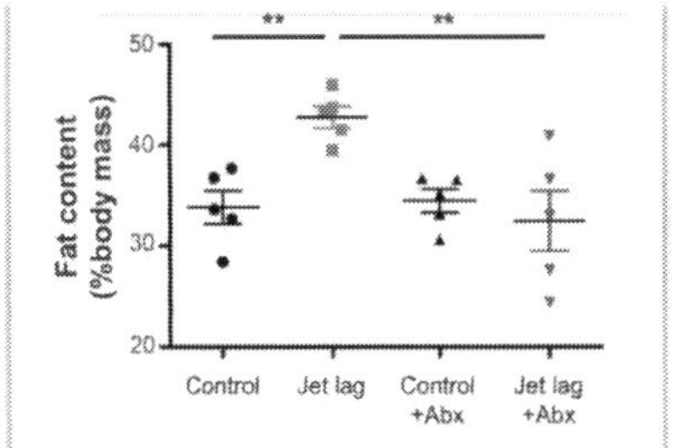

Microbial circadian rhythm changes have metabolic consequences

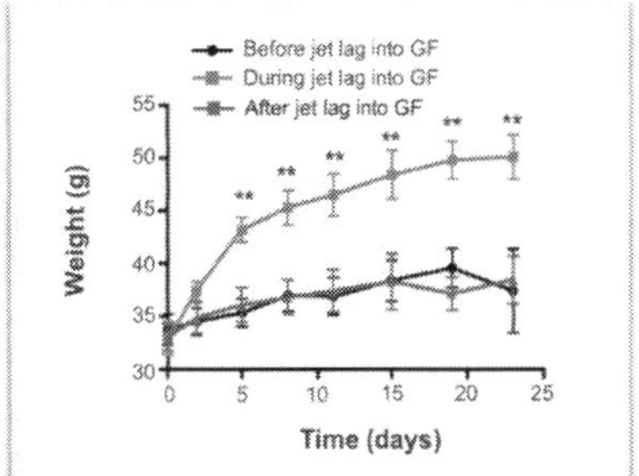

Effects are transferrable between organisms via the microbiome

Circadian clocks occur throughout Nature

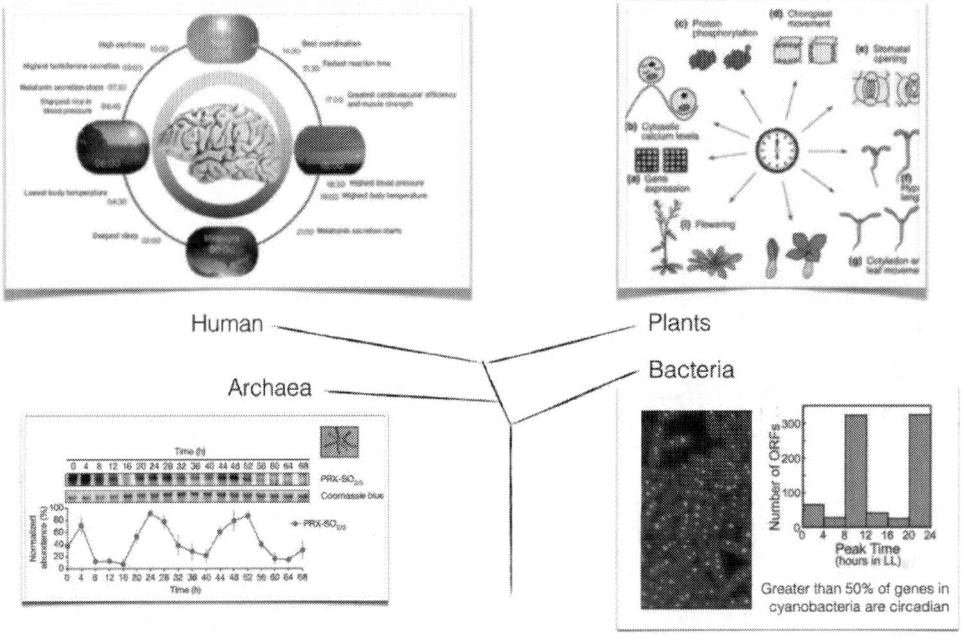

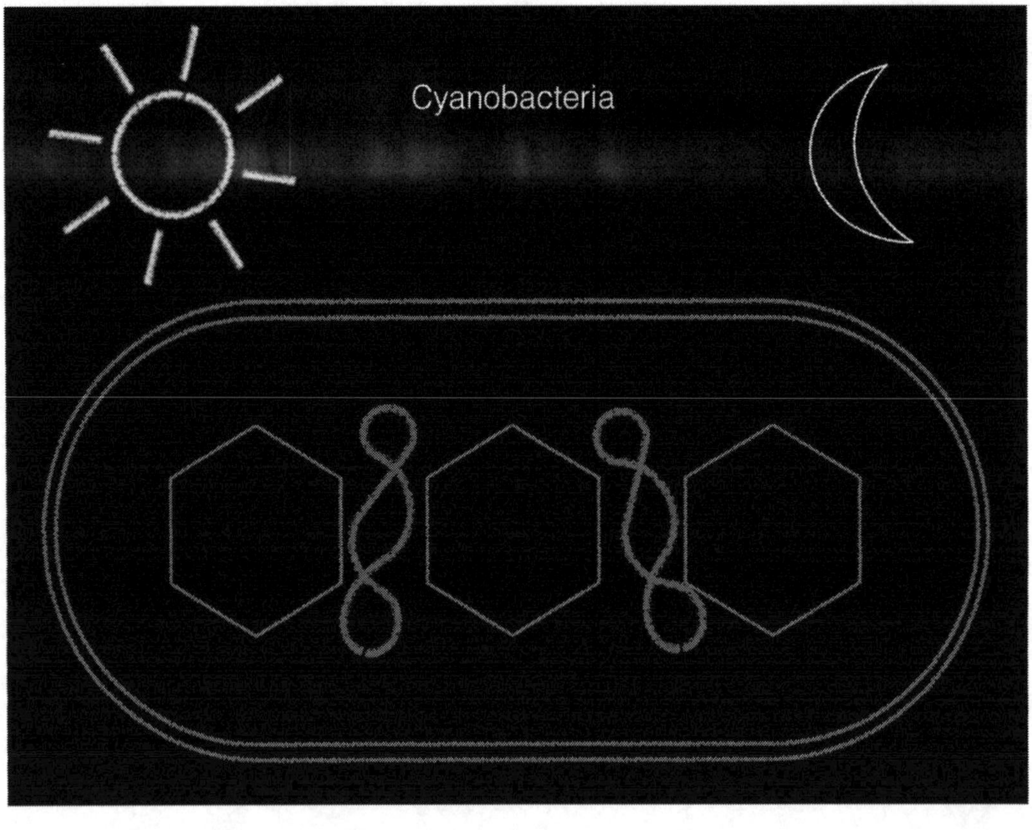

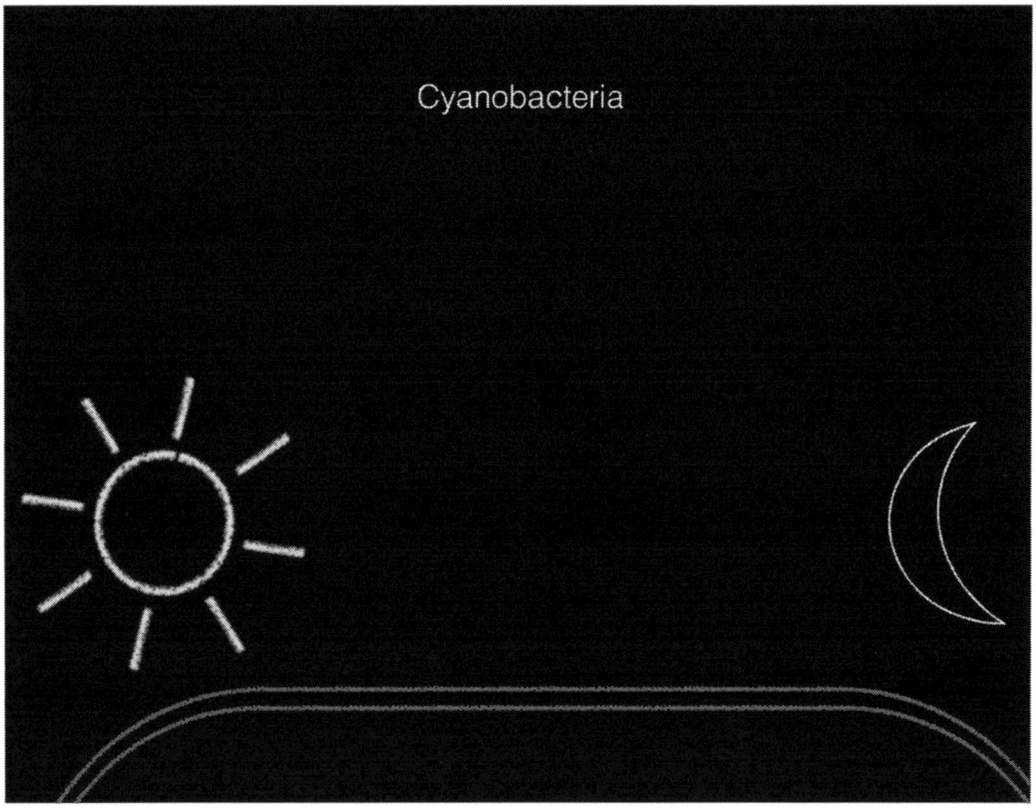
Cyanobacteria

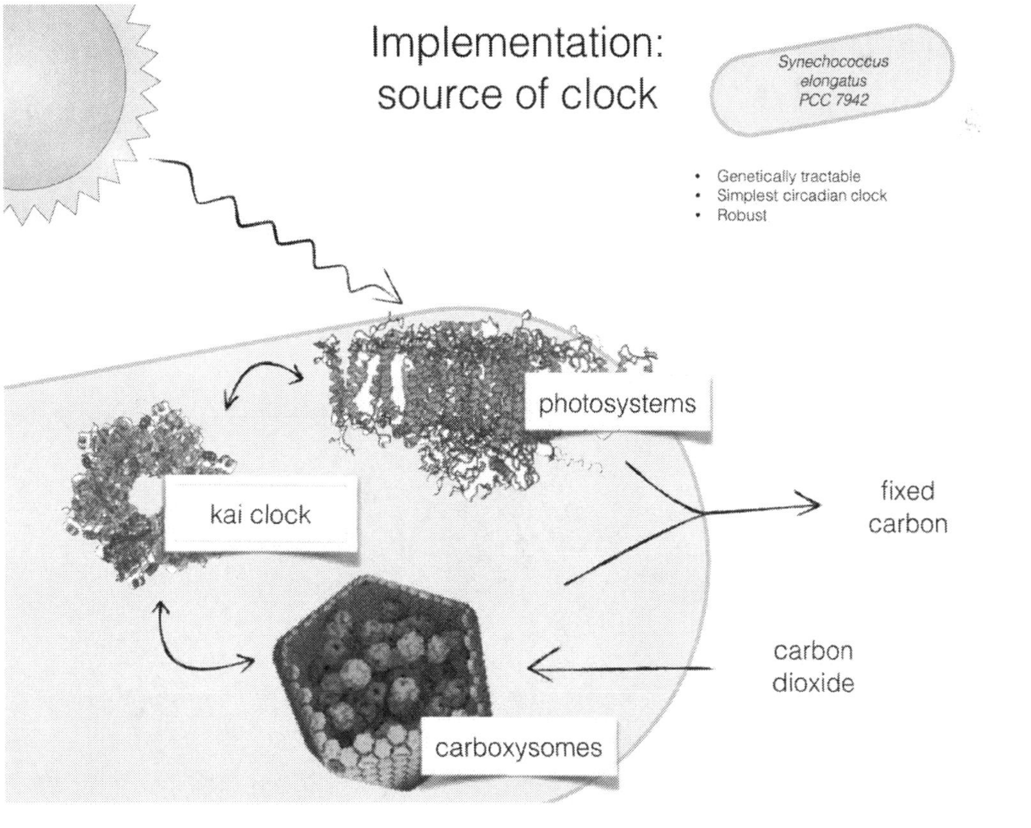
Implementation:
source of clock
Synechococcus
elongatus
PCC 7942
• Genetically tractable
• Simplest circadian clock
• Robust
photosystems
kai clock
fixed
carbon
carbon
dioxide
carboxysomes

KaiC

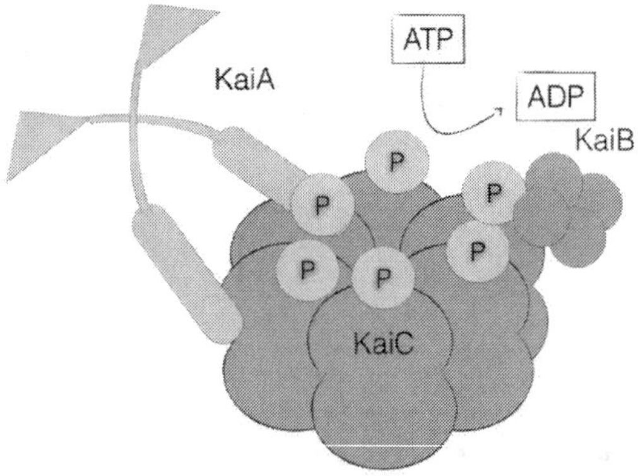

KaiA
ATP
ADP
KaiB
P
KaiC

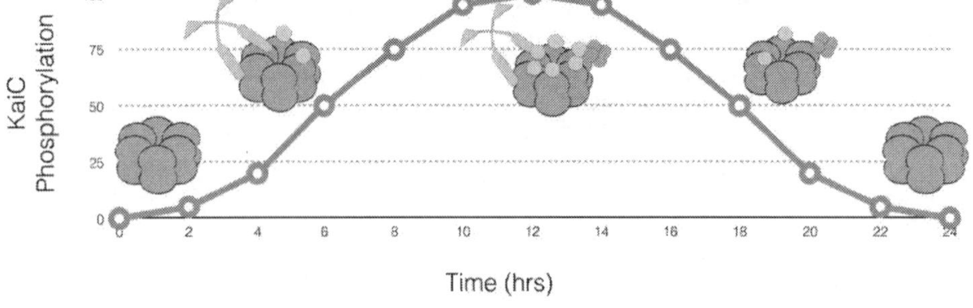

KaiC Phosphorylation
100
75
50
25
0
0 2 4 6 8 10 12 14 16 18 20 22 24
Time (hrs)

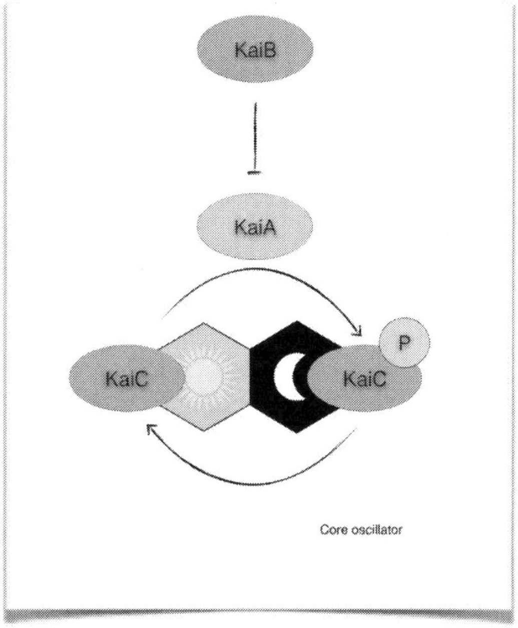

Kai oscillator functions *in vitro*

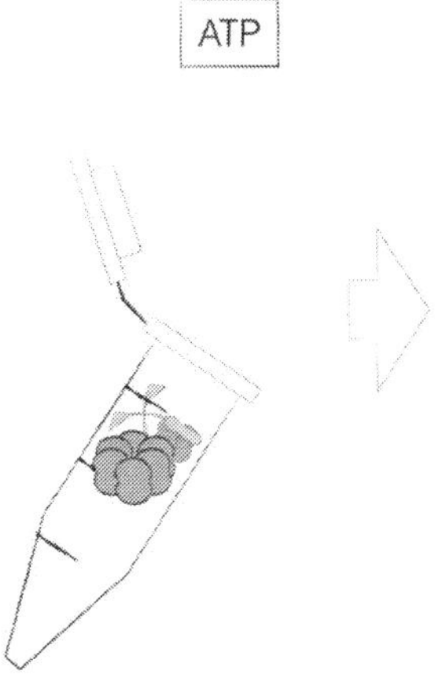

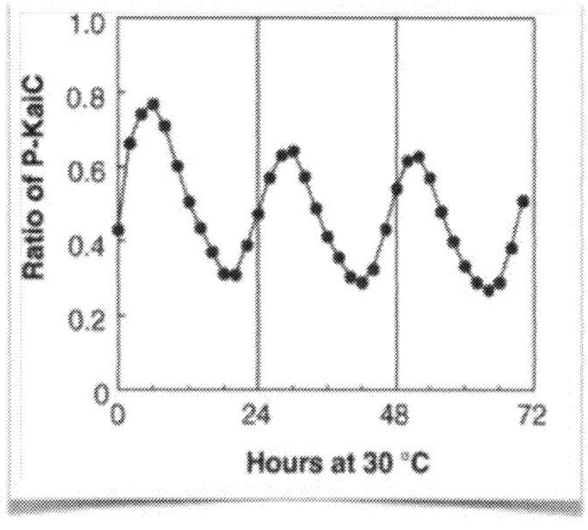

Can we do this in a heterologous system *in vivo*?

Experimental Method

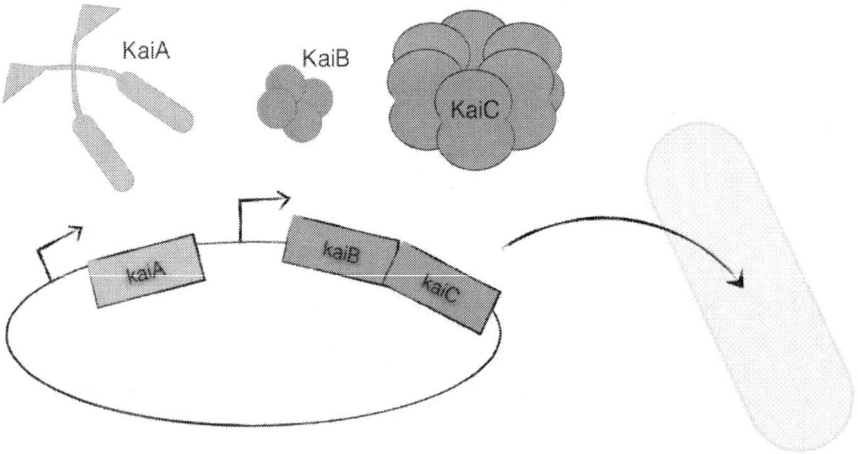

Experimental Method

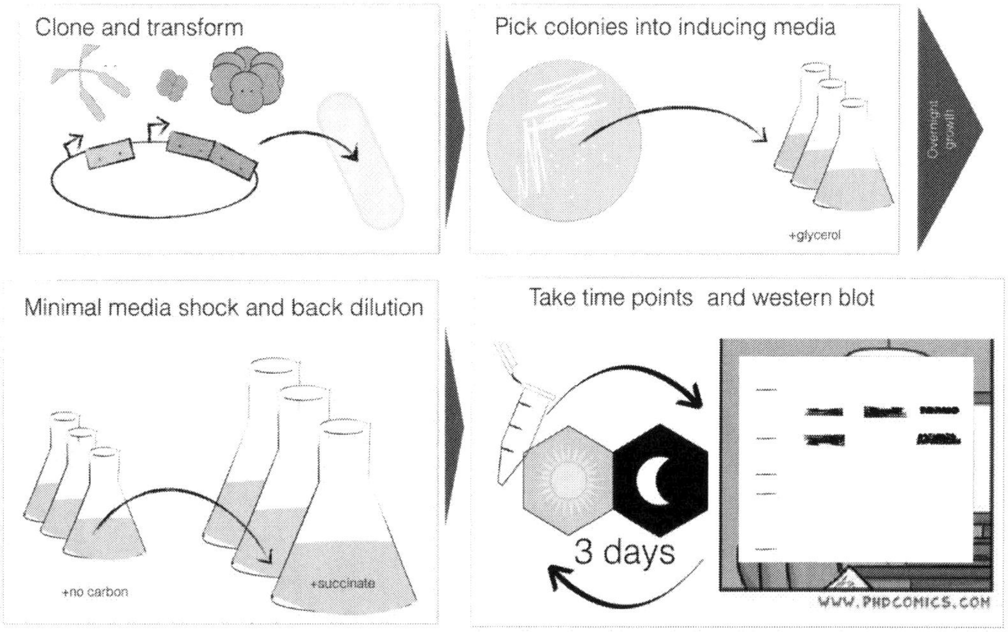

Native Circadian Pathway

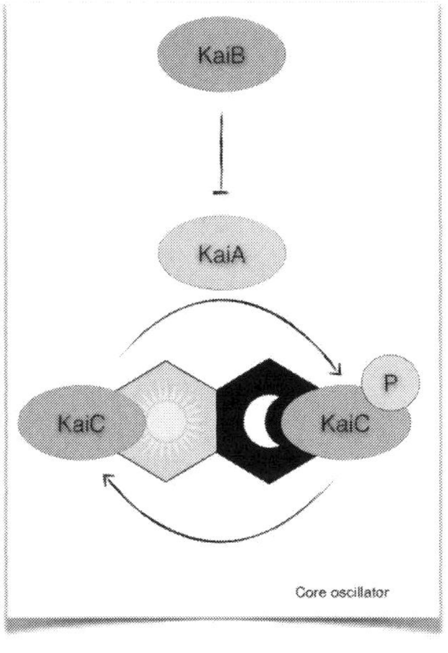

Circadian Pathway Functions in E. Coli

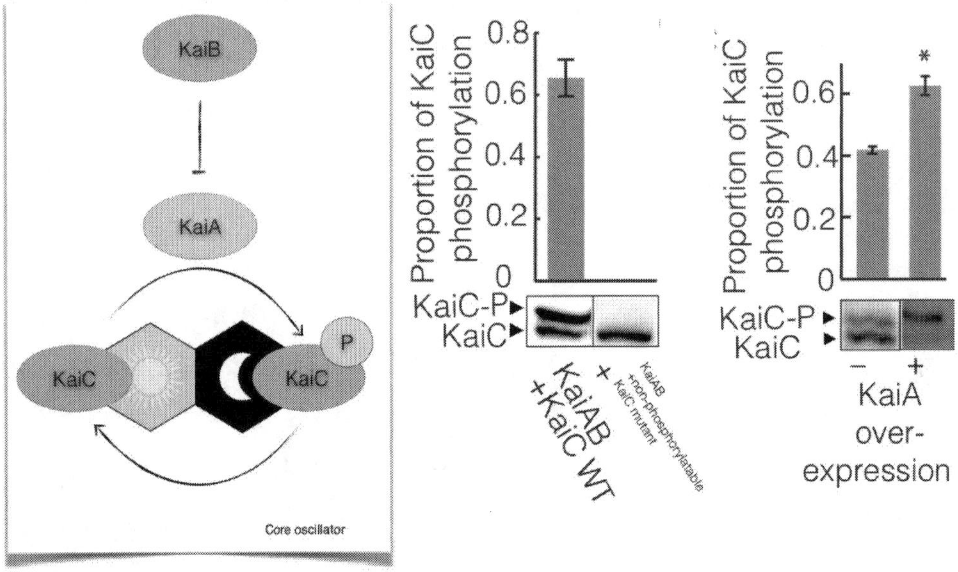

Chen *et al.* 2015 (accepted)

Oscillations in KaiC Phosphorylation *in vivo*

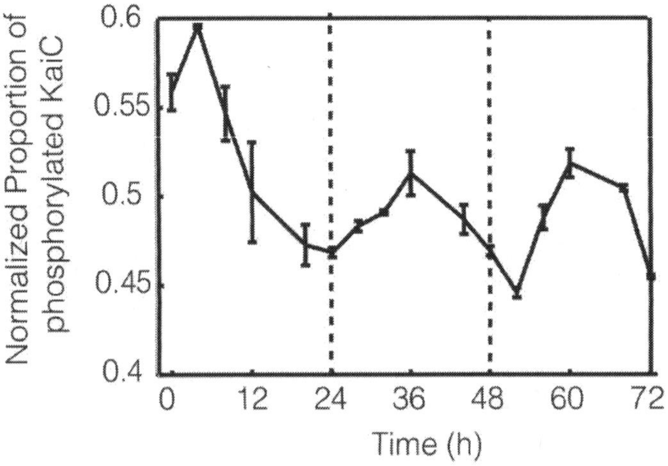

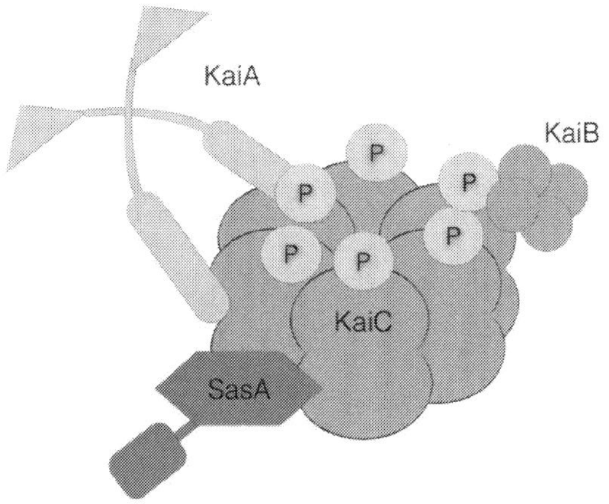

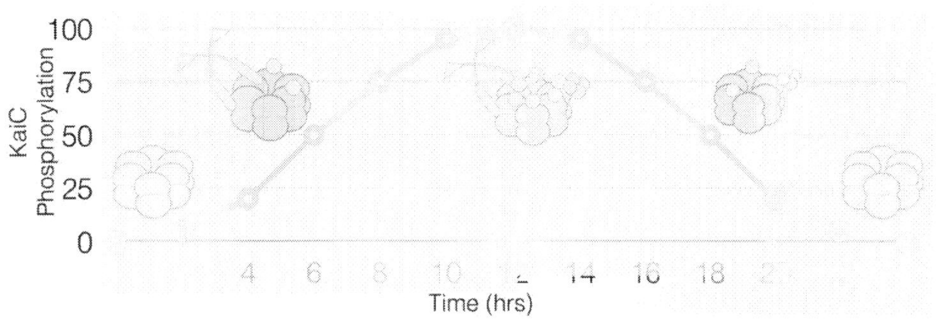

Transcriptional Readout for the Kai Clock

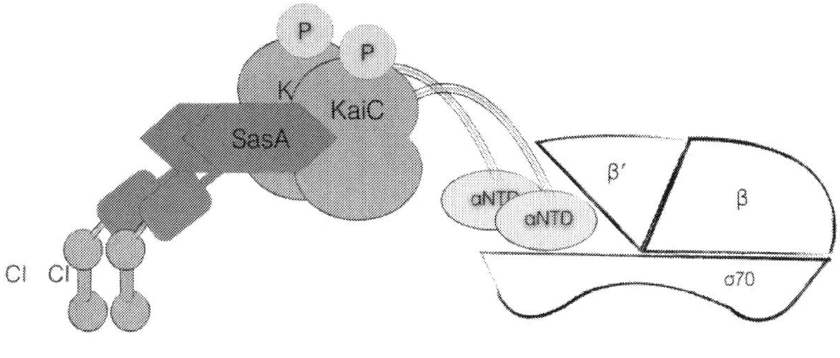

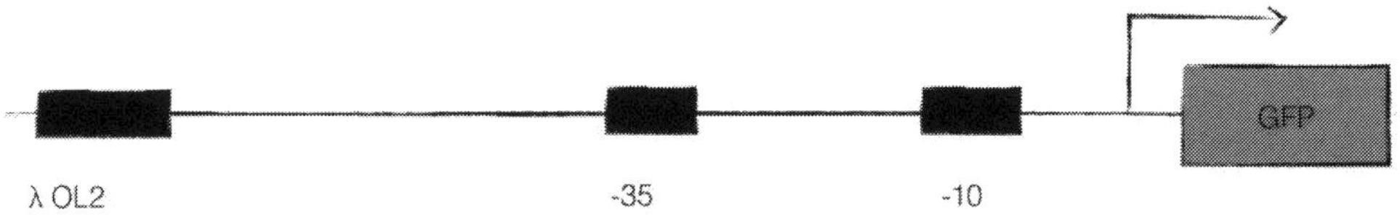

Transcriptional Readout for the Kai Clock

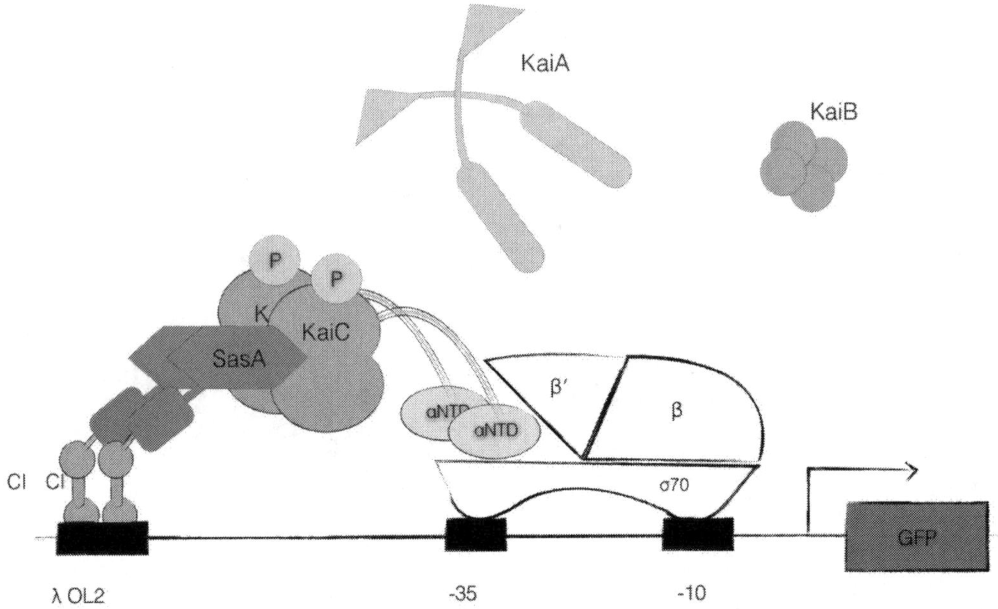

Experimental Method

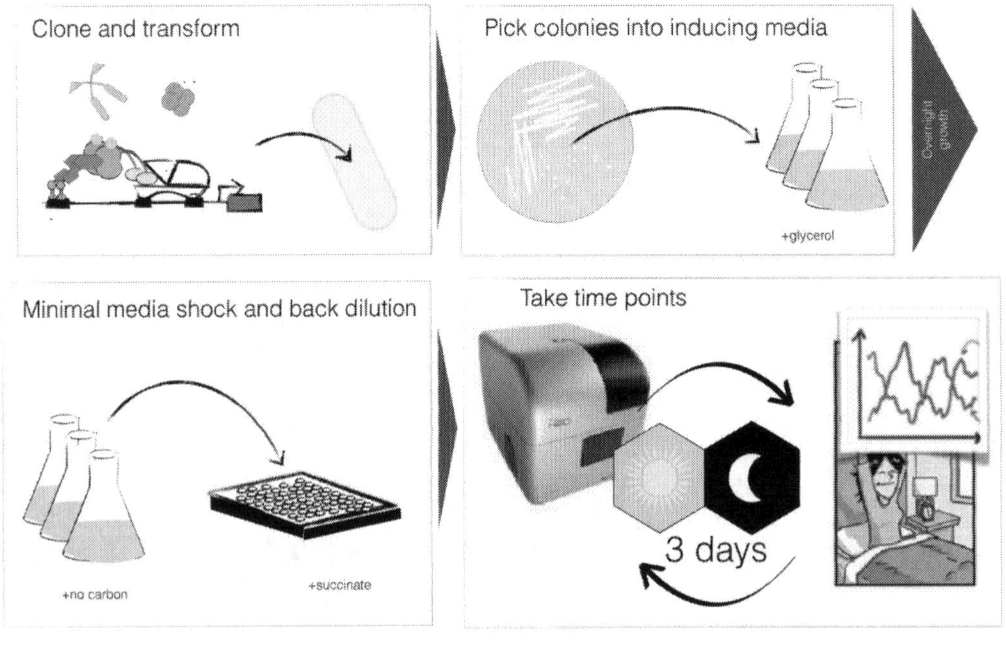

A Synthetic Circadian Oscillator

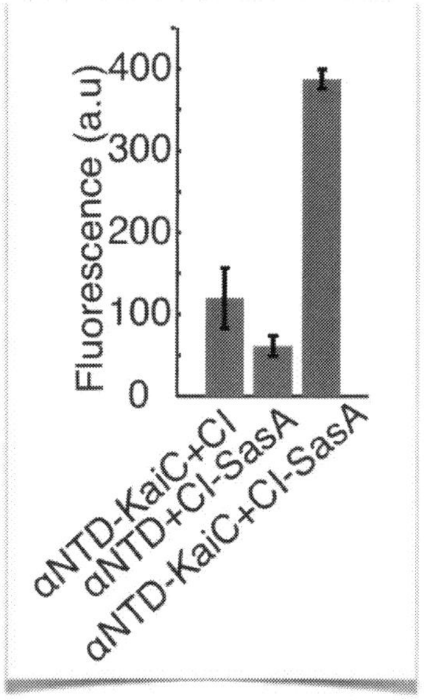

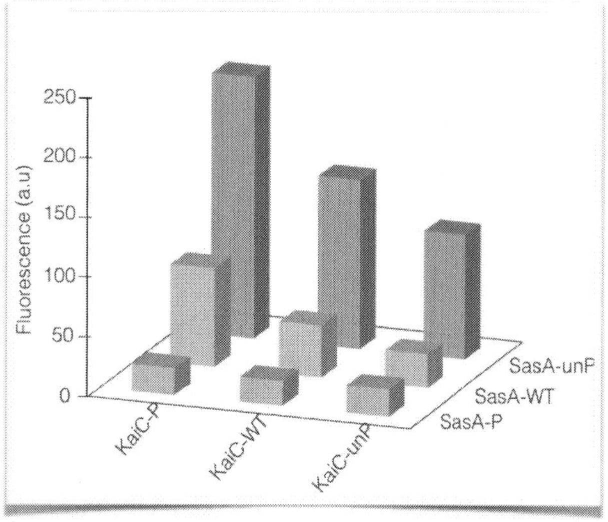

A Synthetic Circadian Oscillator

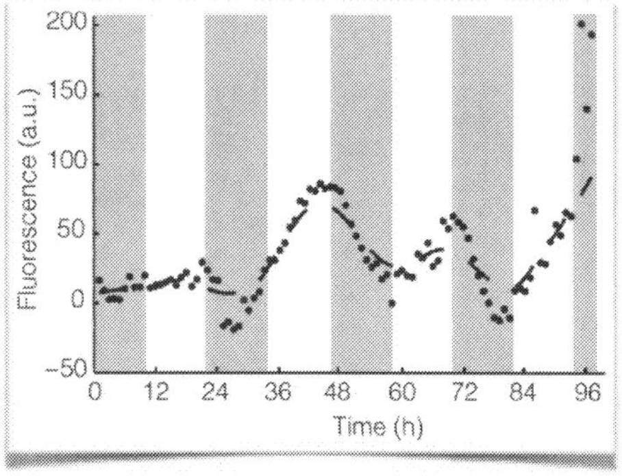

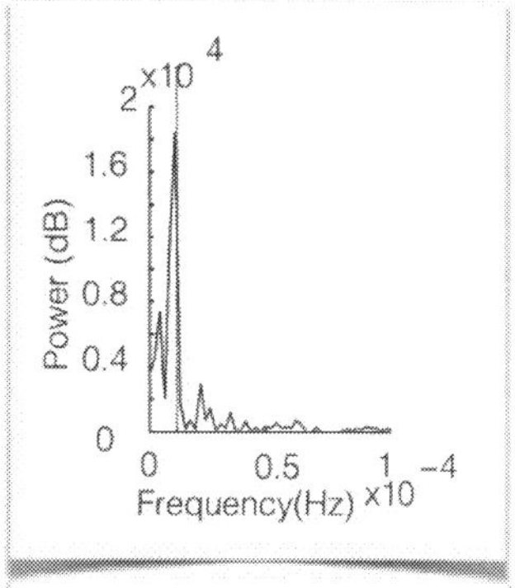

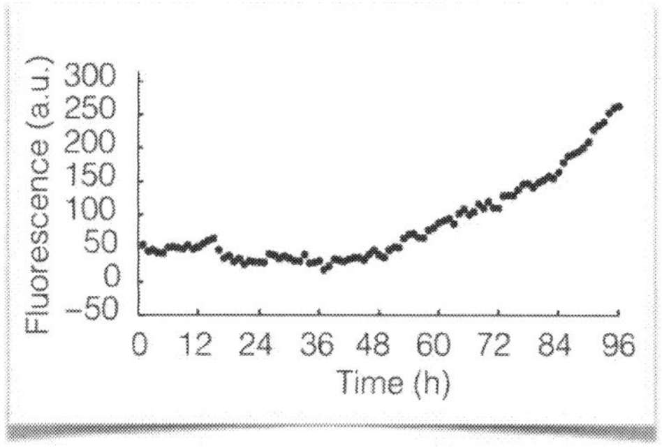

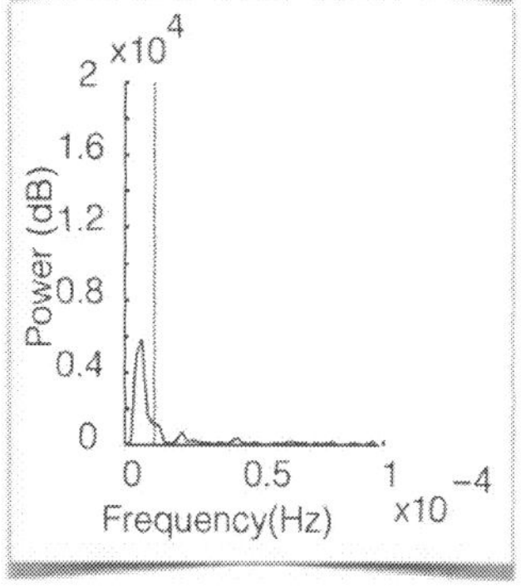

Oscillations in single cells

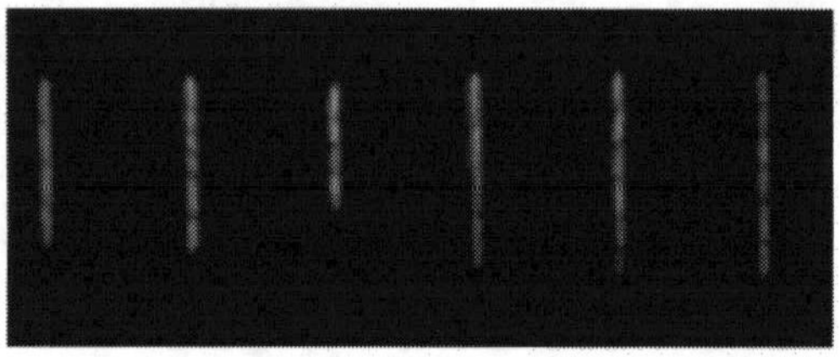

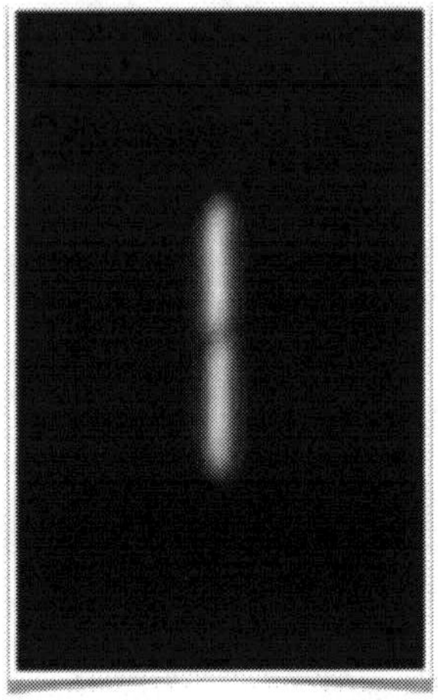

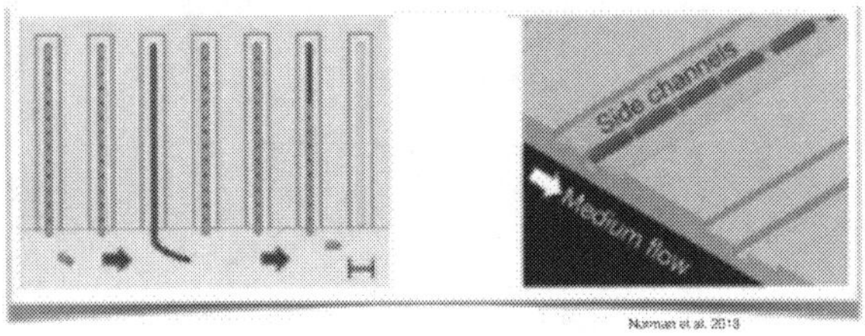

Norman et al. 2013

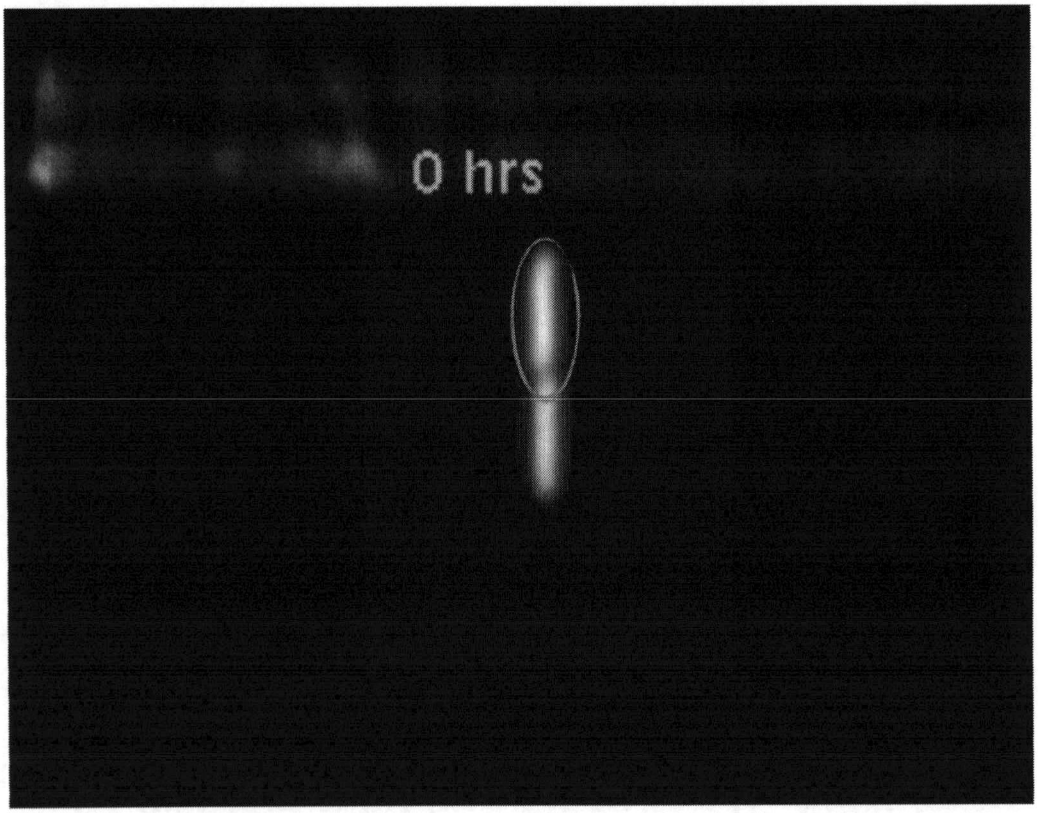

Oscillations in single cells

Time

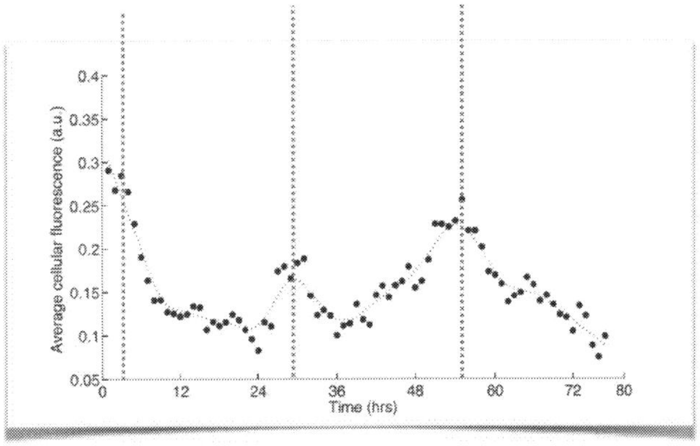

A circadian clock is modular and transferable

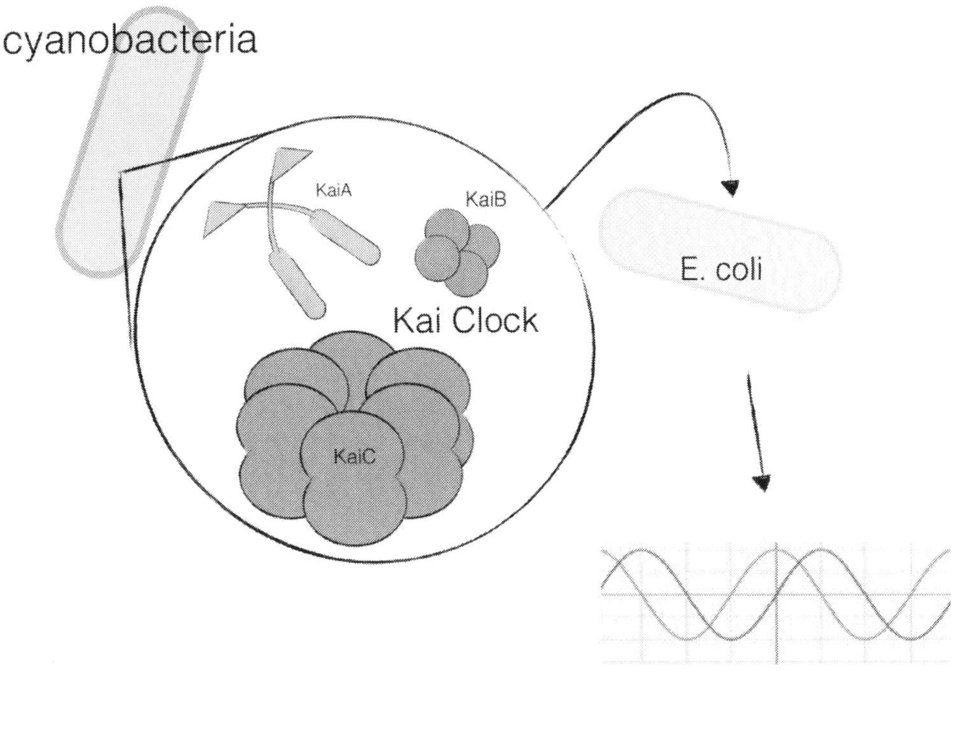

Solar power is sustainable

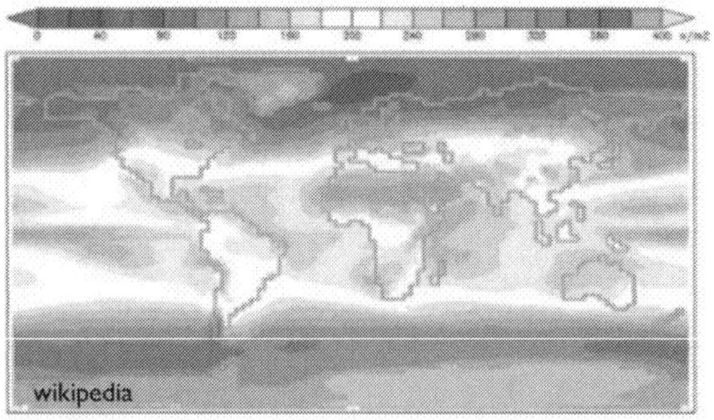

Incident solar radiation averages 200 W / m²

5×10^{14} (Earth) m² ⟶ Solar power total is: ~100,000TW

Worldwide power usage is 15 TW

Biology meets chemistry - the Bionic Leaf

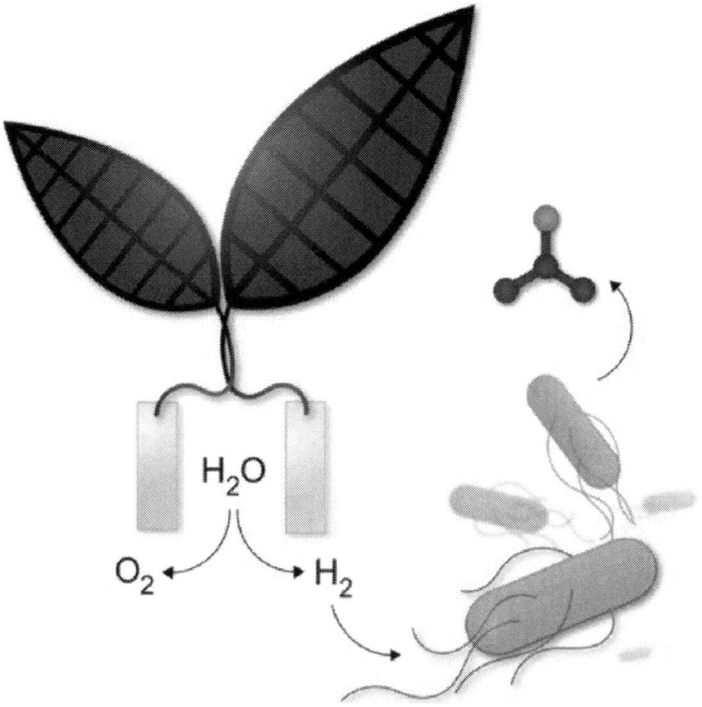

How can photosynthesis be inefficient?

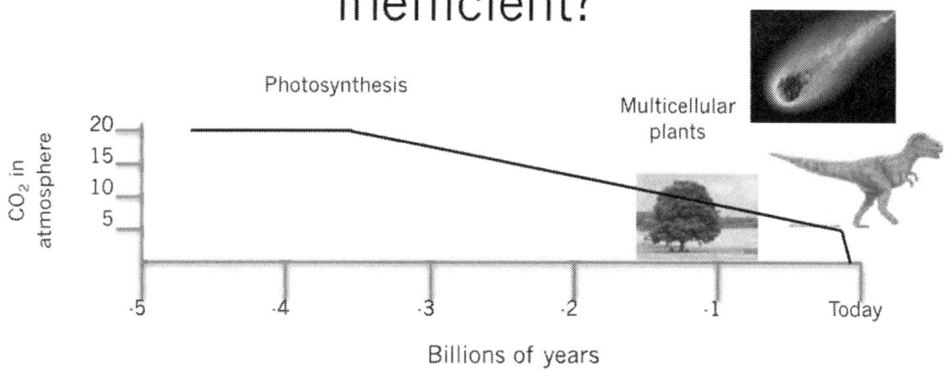

- Photosynthesis evolved ~ 3.5 billion years ago
  - Atmosphere: No oxygen, ~20% CO_2
  - RubisCo evolves in high CO_2
- Asteroid hits earth, killing dinosaurs
  - CO_2 levels drop rapidly
  - RubisCo is oxygen sensitive
  - Photosynthetic microbes deal more efficiently with high O_2/CO_2

We think we can do better than plants.

✓ Solar capture >18%	✗ Solar capture 1-8%
✗ Very limited chemical production	✓ Produce carbohydrates, lipids, small molecules

Water splitting is the heart of photosynthesis.

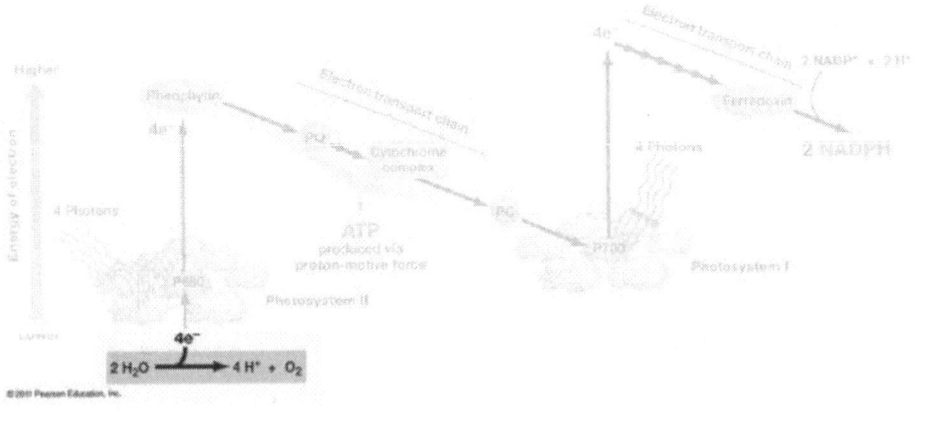

The Artificial Leaf – Dan Nocera

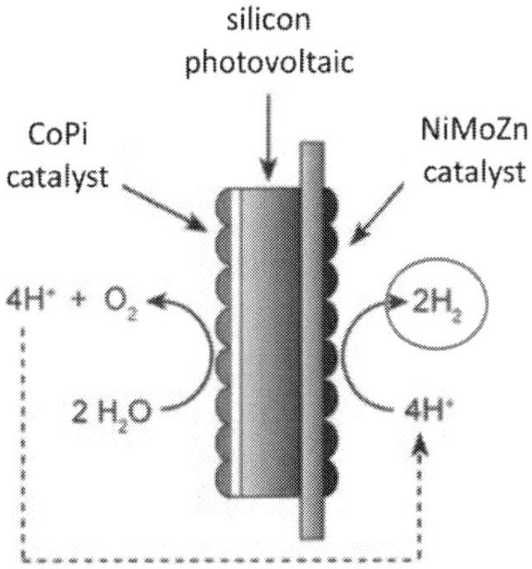

Building a "Bionic Leaf"

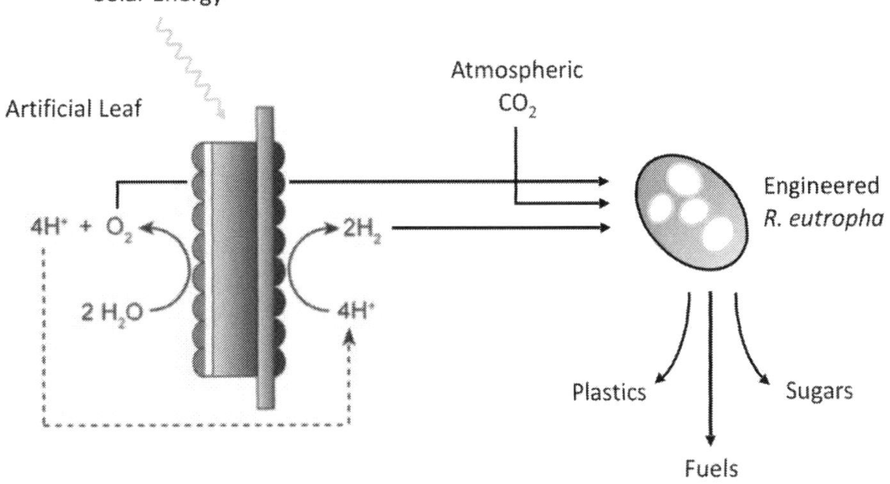

The Bionic Leaf in Action

59

Ralstonia grows when E_{cell} > 2.7 V

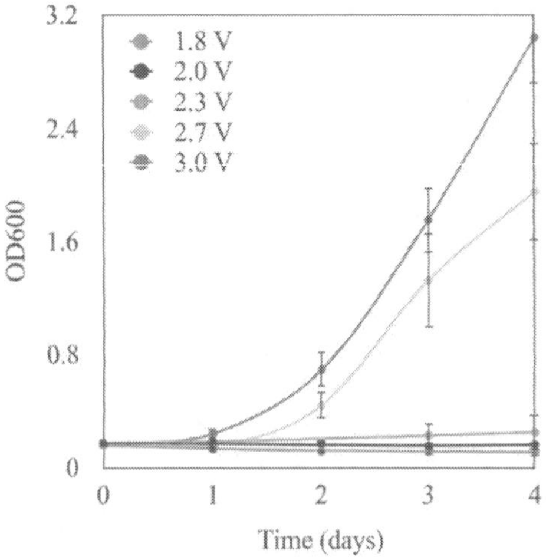

- **8% maximum theoretical yield of biomass from electricity (best achieved at the time)**
- **Due to efficient, low-potential electrolysis using inexpensive earth-abundant electrodes**
- **For a 18% efficient PV, this gives ~ 1.4% solar-biomass efficiency.**
- ***Plants: < 1% solar biomass efficiency***

Bionic Leaf – Interface Solar with Biology

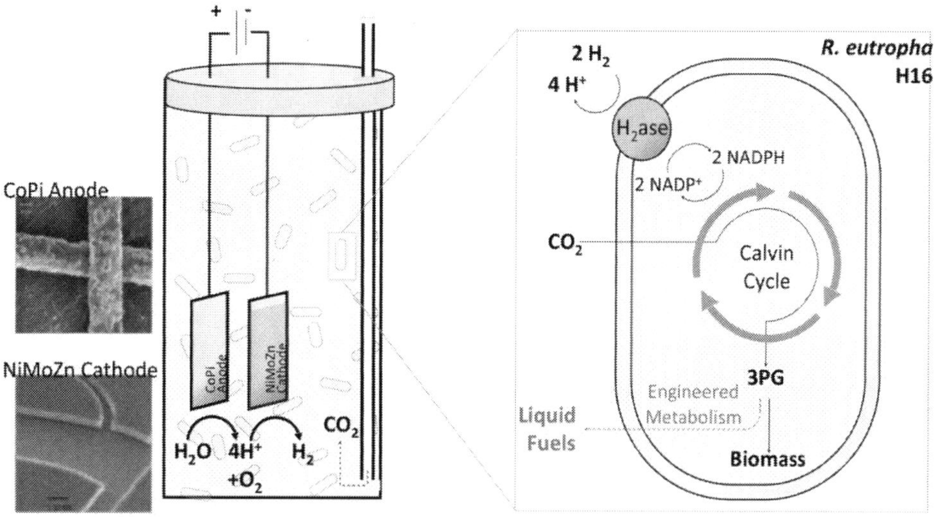

Isopropanol Production

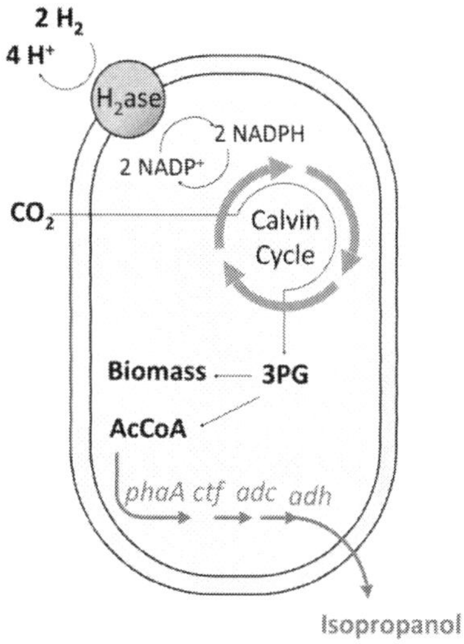

Isopropanol Production

	Previous Studies	Our Work	
Fuel Produced (mg/L)	*150*	*220*	*~48% Improvement*
Biomass Produced (OD)	*2.2*	*2.5*	*~14% Improvement*
Current Density (mA/L)	*720*	*340*	*~47% Improvement*
Overpotential (V)	*2.77*	*1.47*	*~50% Improvement*
Electrode Material Cost ($/g)	*$39.00*	*$0.05*	*~99.9% Improvement*

500% Total Improvement in Electricity→Fuel Efficiency

From electricity to biomass, making it better by improving:

– The biology: Evolving better strains

– The chemistry: Creating better catlysts

– The hardware: Redesigning

A cheap, high yield drop-in system

Long Term s omass:

Close to 5% or better
(comparable to algae)

For comparis
Cc -1.8%

Scaleable!

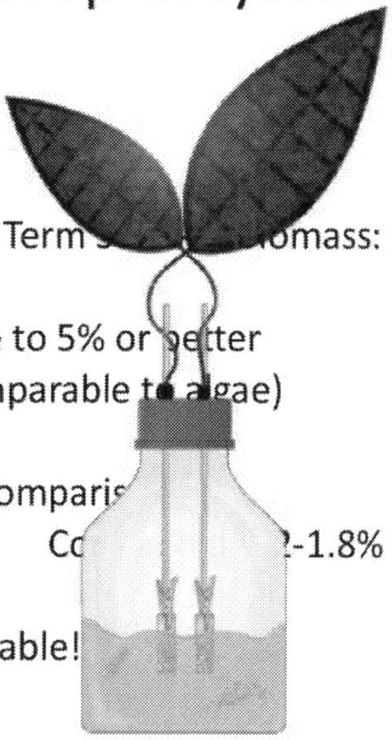

Local and personalized production

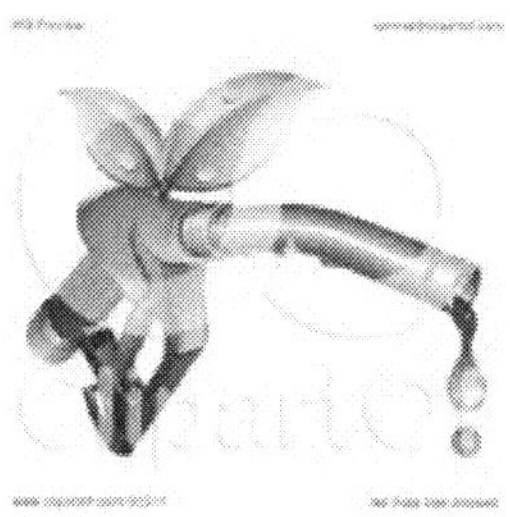

Living Materials

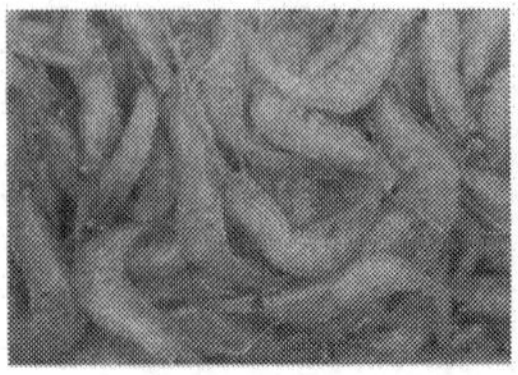
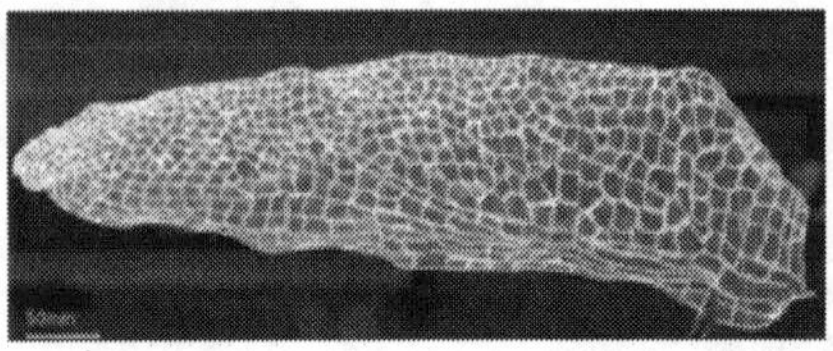
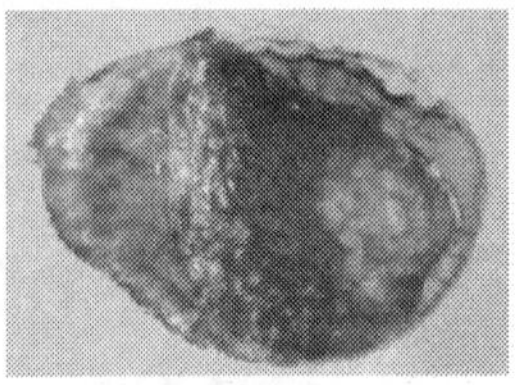

3D Printing of Chitosan from shrimp

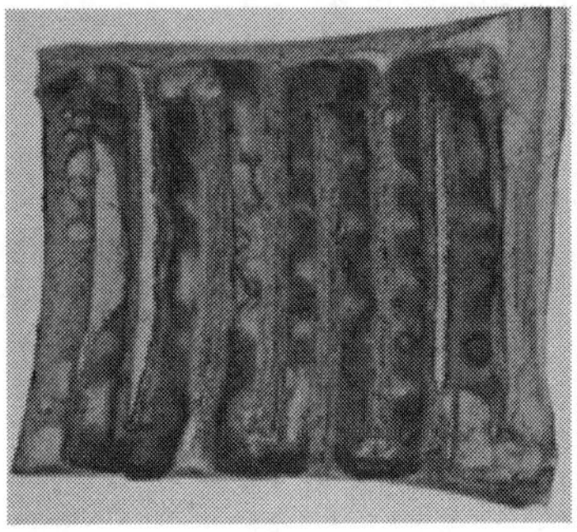

Photosynthetic wearables

with Neri Oxman MIT

Mushtari explores using synthetic biology in design.

How can we design relationships between synthetic organisms and humans?

We designed a wearable to contain photosynthetic cyanobacteria cultures. The form was grown algorithmically and then 3D printed. The piece contains 60 meters of channels, which were filled with luminescent fluid and cyanobacteria.

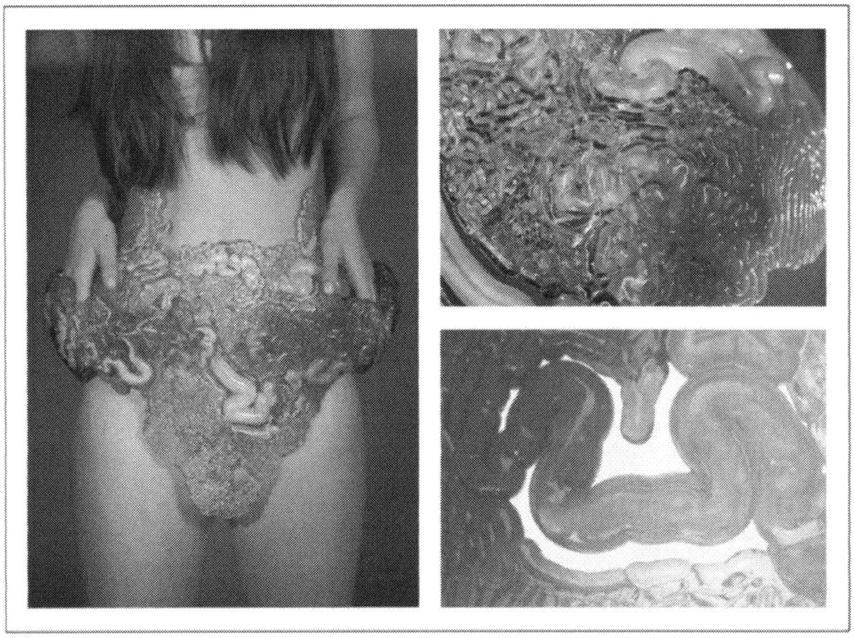

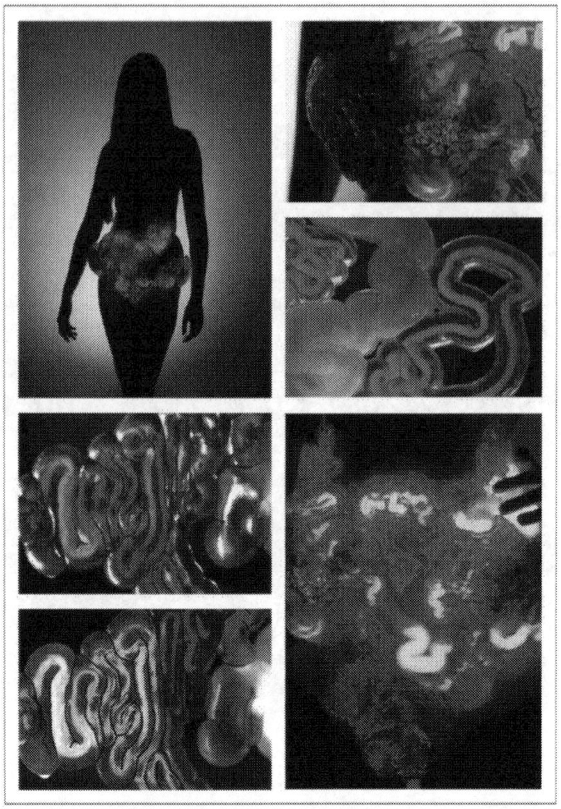

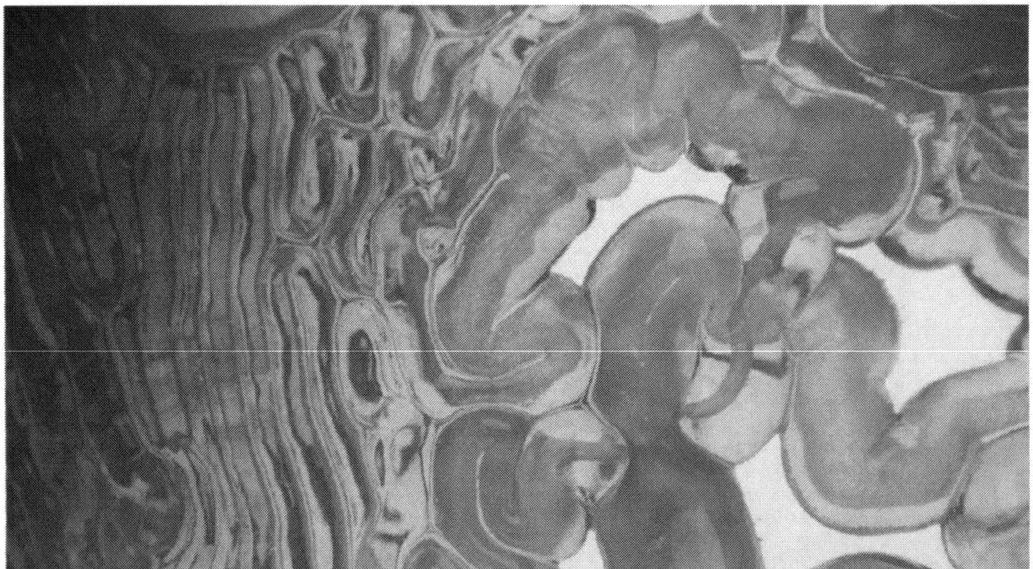

Thanks to many people both past and present – Here are just a few

Manipulating the Gut:
David Riglar, Andy Schumaker, Bryan Hsu, John Kotula, Jordan Kerns, Jeff Way

Synthetic circadian clock:
Anna Chen, Roger Chang

Bio-ionic Leaf:
Brendan Colon, Chong Liu, Joe Torella, Chris Gagliardi, Marika Ziesak, **Dan Nocera** (HU Chemistry)

Synthetic Wearables:
Steph Hayes, Will Patrick, **Neri Oxman** (MIT)

Funding:

The development of platform-based technologies for the optimization of sustainably produced chemicals

Culler, S., Genomatica

Originally presented on: 6/11/2015 10:00:00 - 10:25:00

Genomatica has established an integrated computational/experimental metabolic engineering platform to design, create, and optimize novel high-producing organisms and bioprocesses. Here we present the use of our platform to develop *E. coli* strains for the production of the industrial chemical 1,4-butanediol (BDO) from carbohydrate feedstocks. These efforts have culminated in commercial scale production of over 5M lbs of BDO from dextrose. BDO is a chemical intermediate (>3B lbs/yr) that goes into a variety of products including automotive, electronics and apparel (such as spandex), and is currently produced commercially through energy-intensive petrochemical processes using hydrocarbon feedstocks. Therefore, this product represents an opportunity to make a significant impact on the replacement of traditional petrochemical processes with benign bioprocesses using renewable feedstocks.

We are continuing to enhance our technology platform, enabling faster development of new processes with increased chance of success. Here we describe the application and development of one such new technology, cell-free transcription-translation (TX-TL), towards improving the design-build-test-learn cycle for the construction of a high-performing microorganism capable of producing BDO and other industrial chemicals from carbohydrates. We used systems biology approaches including ^{13}C-flux analysis, proteomics, and metabolomics to characterize the production of BDO in TX-TL as compared to traditional fermentation processes. The presentation will highlight the successful integration of the TX-TL technology into our rational and high-throughput strain engineering platform, and will show the potential advantages for using cell-free systems towards the optimization of sustainably produced chemicals.

Metabolite Valves: Dynamic Control of Metabolic Flux for Pathway Engineering

Prather, K. L. J., Massachusetts Institute of Technology

Originally presented on: 6/11/2015 10:25:00 - 10:50:00

Microbial strains have been successfully engineered to produce a wide variety of chemical compounds, several of which have been commercialized. As new products are targeted for biological synthesis, yield is frequently considered a primary driver towards determining feasibility. Theoretical yields can be calculated, establishing an upper limit on the potential conversion of starting substrates to target compounds. Such yields typically ignore loss of substrate to byproducts, with the assumption that competing reactions can be eliminated, usually by deleting the genes encoding the corresponding enzymes. However, when an enzyme encodes an essential gene, especially one involved in primary metabolism, deletion is not a viable option. Reducing gene expression in a static fashion is possible, but this solution ignores the metabolic demand needed for synthesis of the enzymes required for the desired pathway.

We have developed "metabolite valves" to address this challenge. The valves are designed to allow high flux through the essential enzyme during an initial period where growth is favored. Following an external perturbation, enzyme activity is then reduced, enabling a higher precursor pool to be diverted towards the pathway of interest. We have designed valves with control at both the transcriptional and post-translational levels. In both cases, key enzymes in glucose metabolism are regulated, and two different compounds are targeted for heterologous production. We have measured increased concentrations of intracellular metabolites once the valve is closed, and have demonstrated that these increased pools lead to increased product yields. These metabolite valves should prove broadly useful for dynamic control of metabolic flux, resulting in improvements in product yields.

*** Formatting Issue - Best Available Paper/Graphic ***

Imperial College
London

Gene circuits for self-tuning metabolic pathways

Diego Oyarzún
Biomathematics
Department of Mathematics
Imperial College London

d.oyarzun@imperial.ac.uk

SEED Synthetic Biology, Engineering, Evolution & Design

> Traditional metabolic engineering is "static":

- > Add new enzymatic genes to host for chemical production.
- > Diverts part of the native flux and causes growth defects.
- > Accumulation of toxic intermediates.

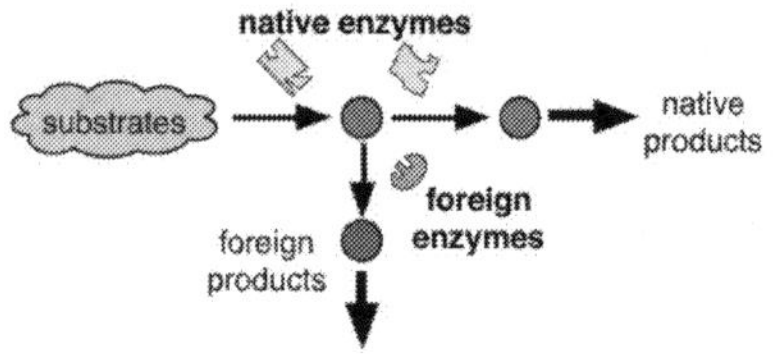

> Self-tuning pathways:

- > Principle: sense an intermediate and control expression through feedback

Some basic schemes

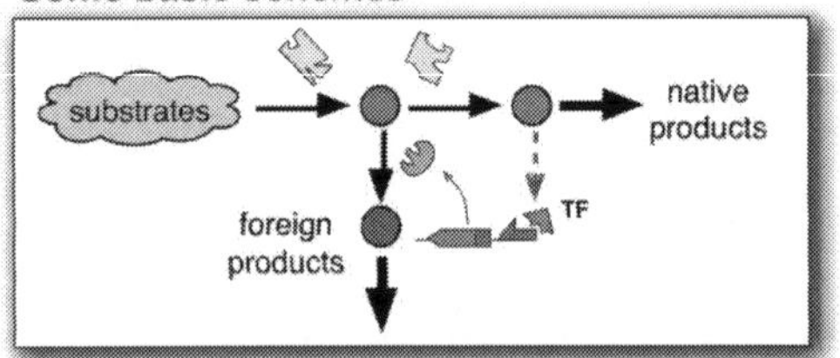

IF excess of native product
THEN upregulate foreign enzymes

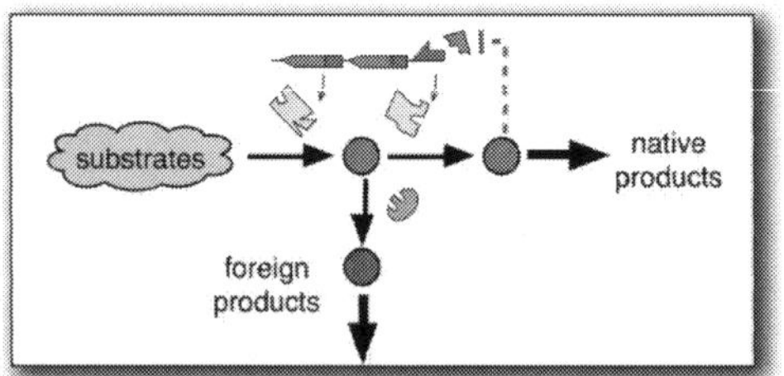

IF native product drops
THEN upregulate native enzymes

- > Expression levels self-tune to growth conditions.
- > Minimizes impact on native fluxes.
- > Alleviates growth/yield trade offs.
- > Prevents accumulation of toxic intermediates.

*** Formatting Issue - Best Available Paper/Graphic ***

> Success stories:

Improving lycopene production in *Escherichia coli by* engineering metabolic control

William R. Farmer and James C. Liao

Main challenges

> Very few molecular mechanisms to sense intracellular metabolites.
> (metabolite-responsive TFs, riboswitches, etc)
> Lack of model-based design methods & theory.

PNAS | August 5, 2014 | vol. 111 | no. 31 | 11299–11304

Improving fatty acids production by engineering dynamic pathway regulation and metabolic control

Peng Xu, Lingyun Li, Fuming Zhang, Gregory Stephanopoulos, and Mattheos Koffas

Dynamic phenotype switching

Soma & Hanai, *Metabolic Engineering*, 2015

Williams *et al*, *Microbial Cell Factories*, 2015

Recent reviews

Brockman & Prather, *Biotechnology Journal*, 2015

Jones, Toparlak & Koffas, *Current Opinion in Biotechnology*, 2015

Venayak *et al*, *Current Opinion in Biotechnology*, 2015

Nature has evolved complex, non-obvious, regulatory systems

Transcriptional regulation in E. coli's carbon uptake

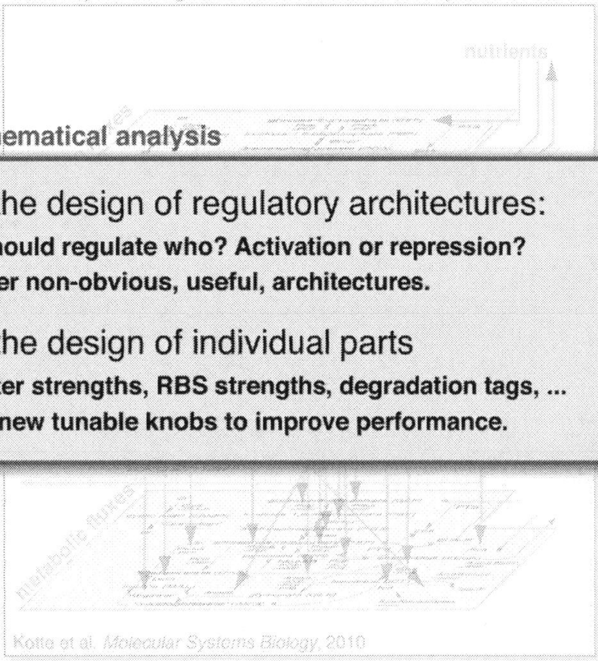

Role of mathematical analysis

> Guide the design of regulatory architectures:
> - **Who should regulate who? Activation or repression?**
> - **Discover non-obvious, useful, architectures.**

> Guide the design of individual parts
> - **Promoter strengths, RBS strengths, degradation tags, ...**
> - **Detect new tunable knobs to improve performance.**

Kotte et al, *Molecular Systems Biology*, 2010

Chubukov et al, *Nature Reviews Microbiology*, 2014

*** Formatting Issue - Best Available Paper/Graphic ***

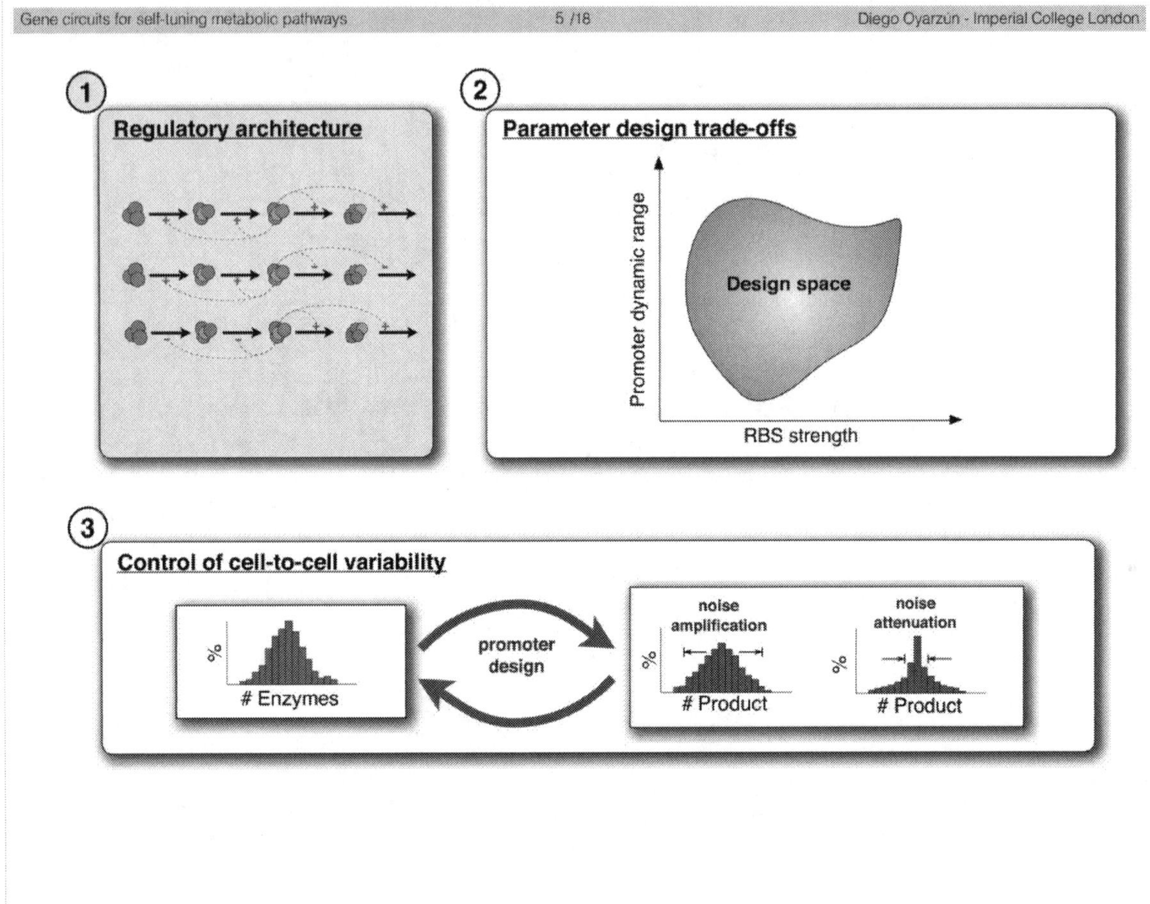

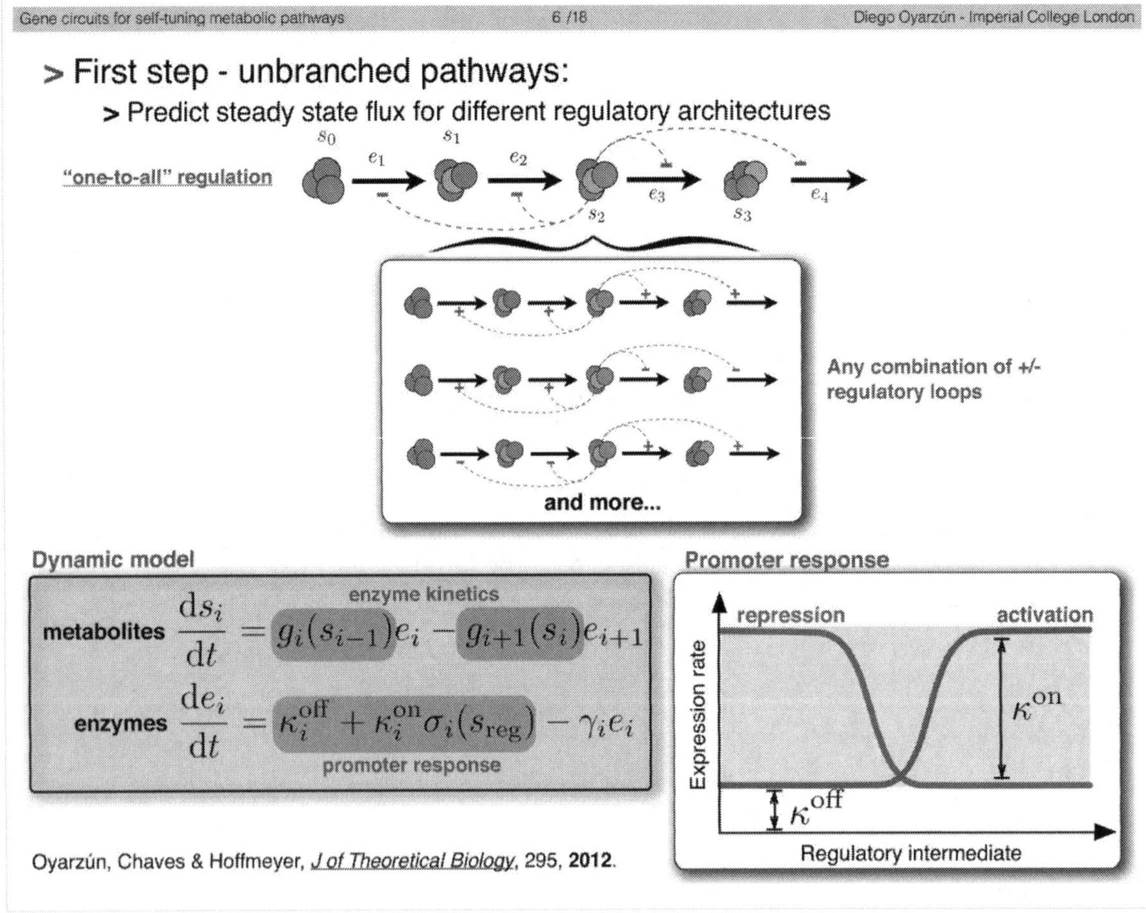

*** Formatting Issue - Best Available Paper/Graphic ***

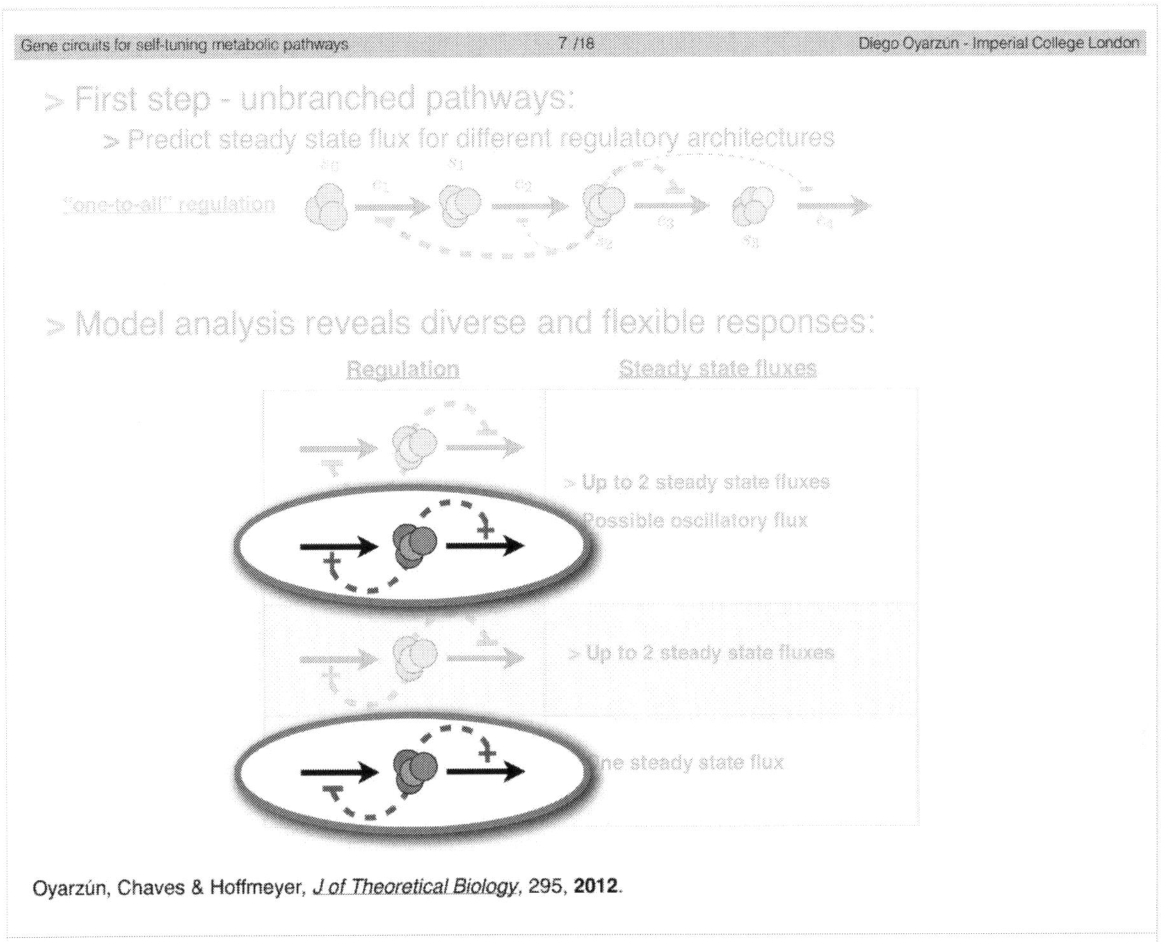

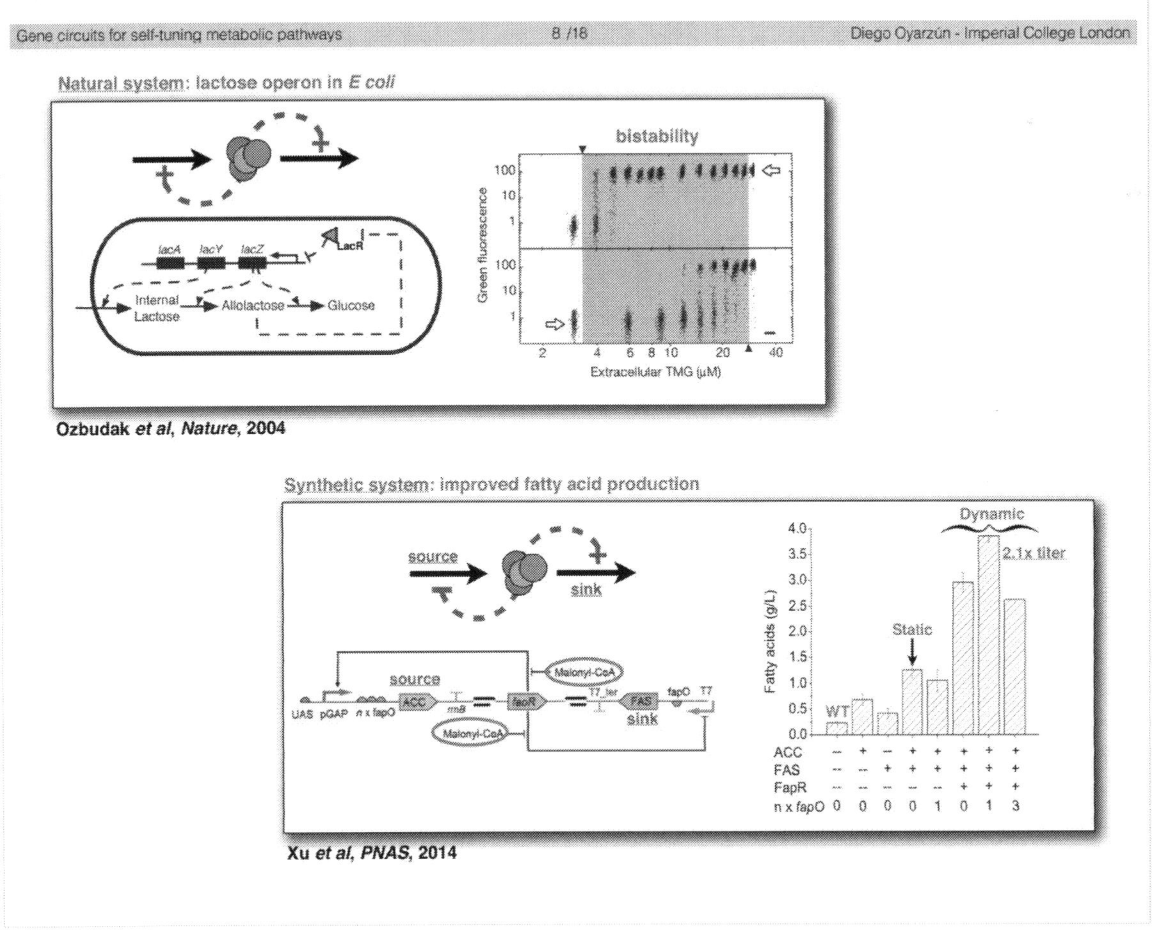

*** Formatting Issue - Best Available Paper/Graphic ***

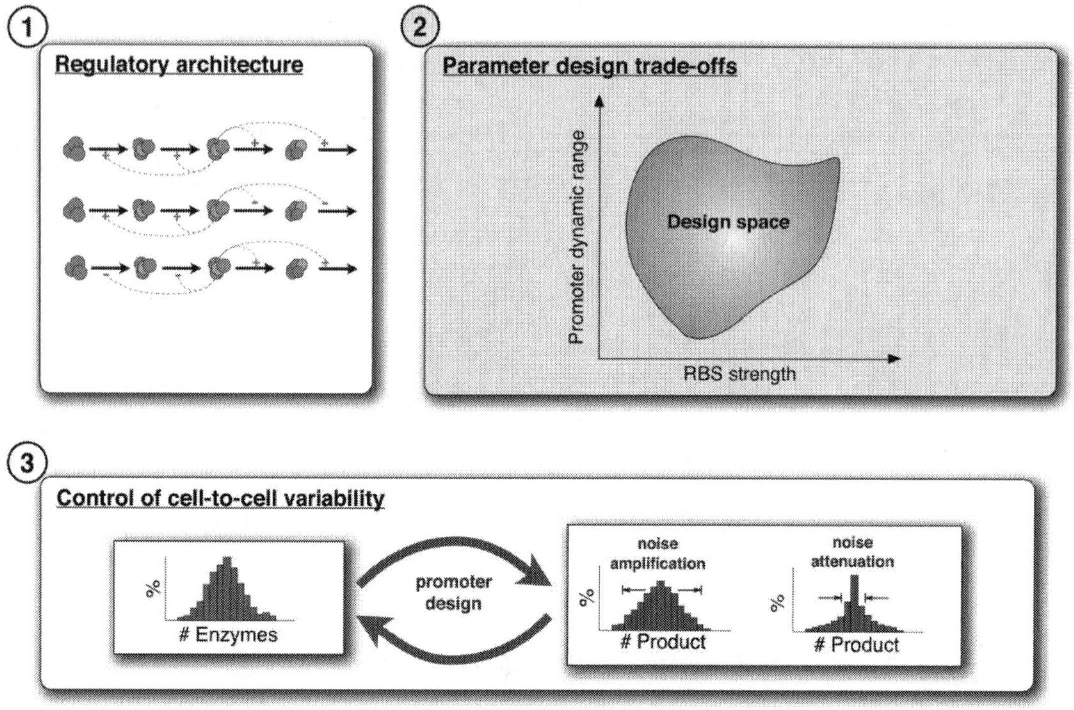

> Main tunable knobs in Synthetic Biology (as of now):

> Promoter response.

> Ribosomal binding site strength.

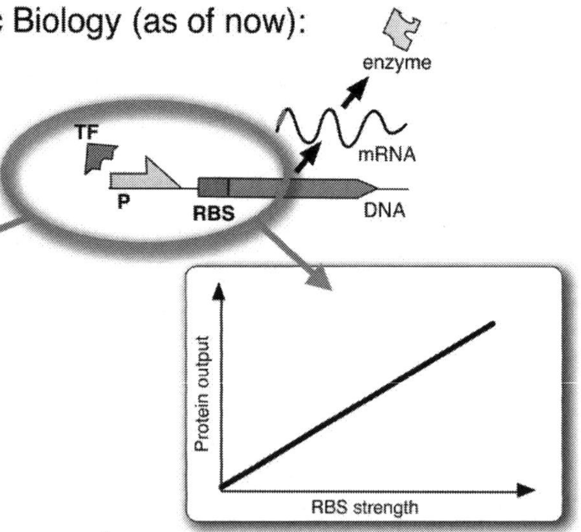

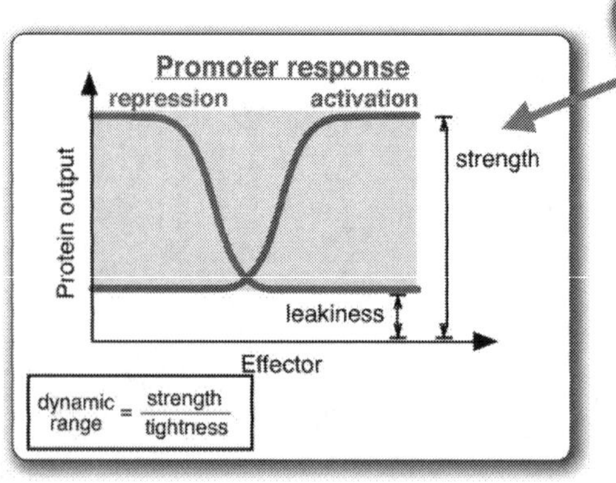

> How do parameters shape pathway performance?

> How do they constrain each other?

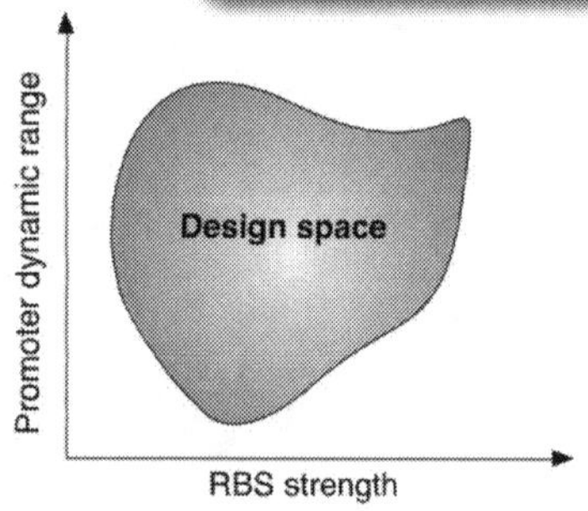

*** Formatting Issue - Best Available Paper/Graphic ***

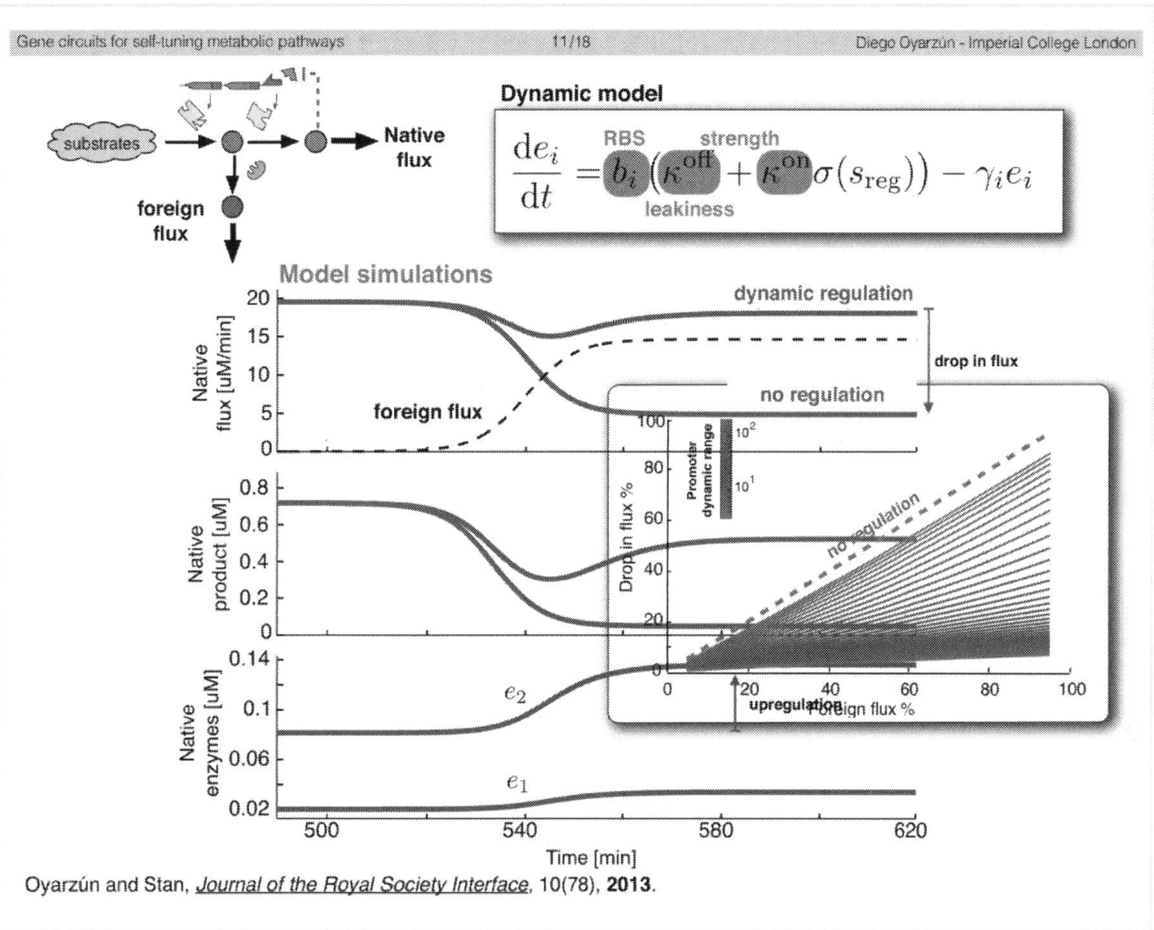

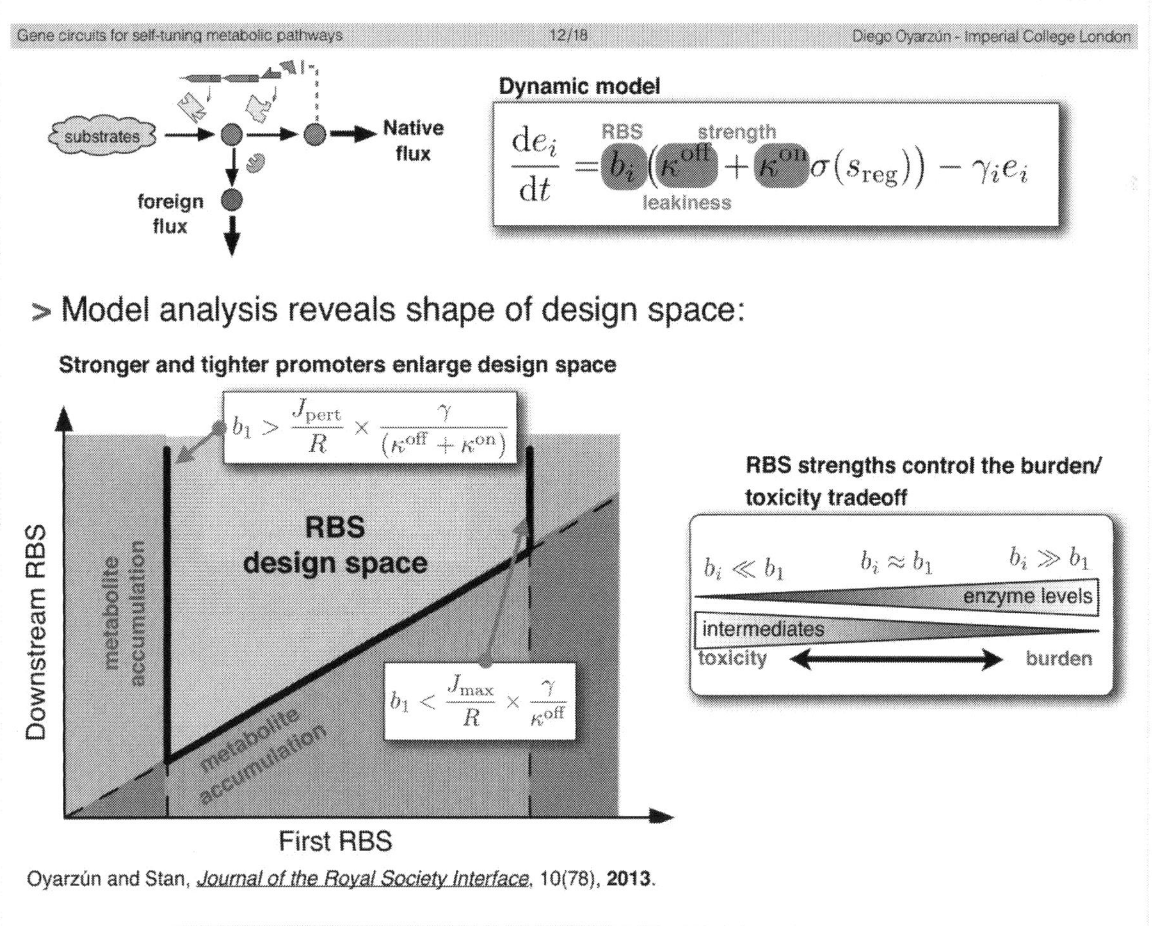

*** Formatting Issue - Best Available Paper/Graphic ***

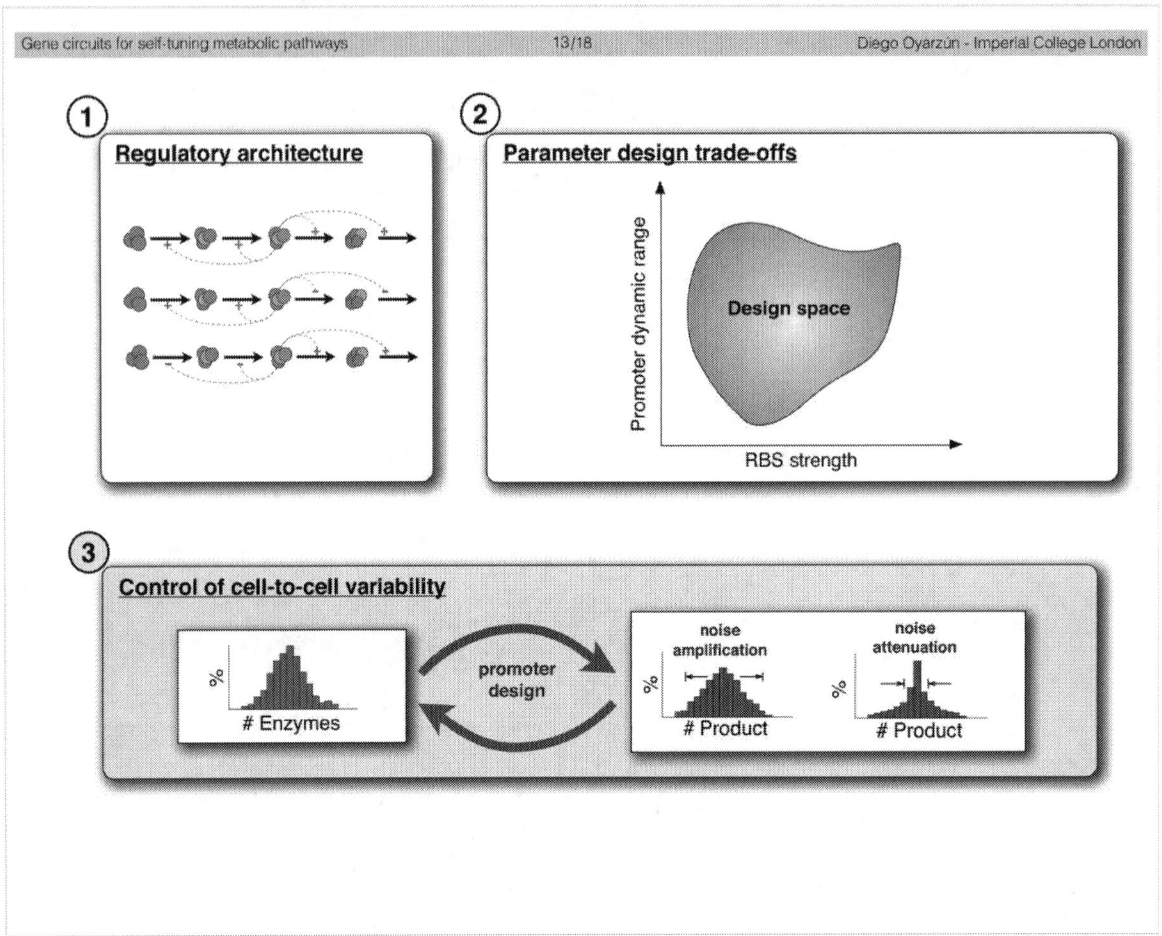

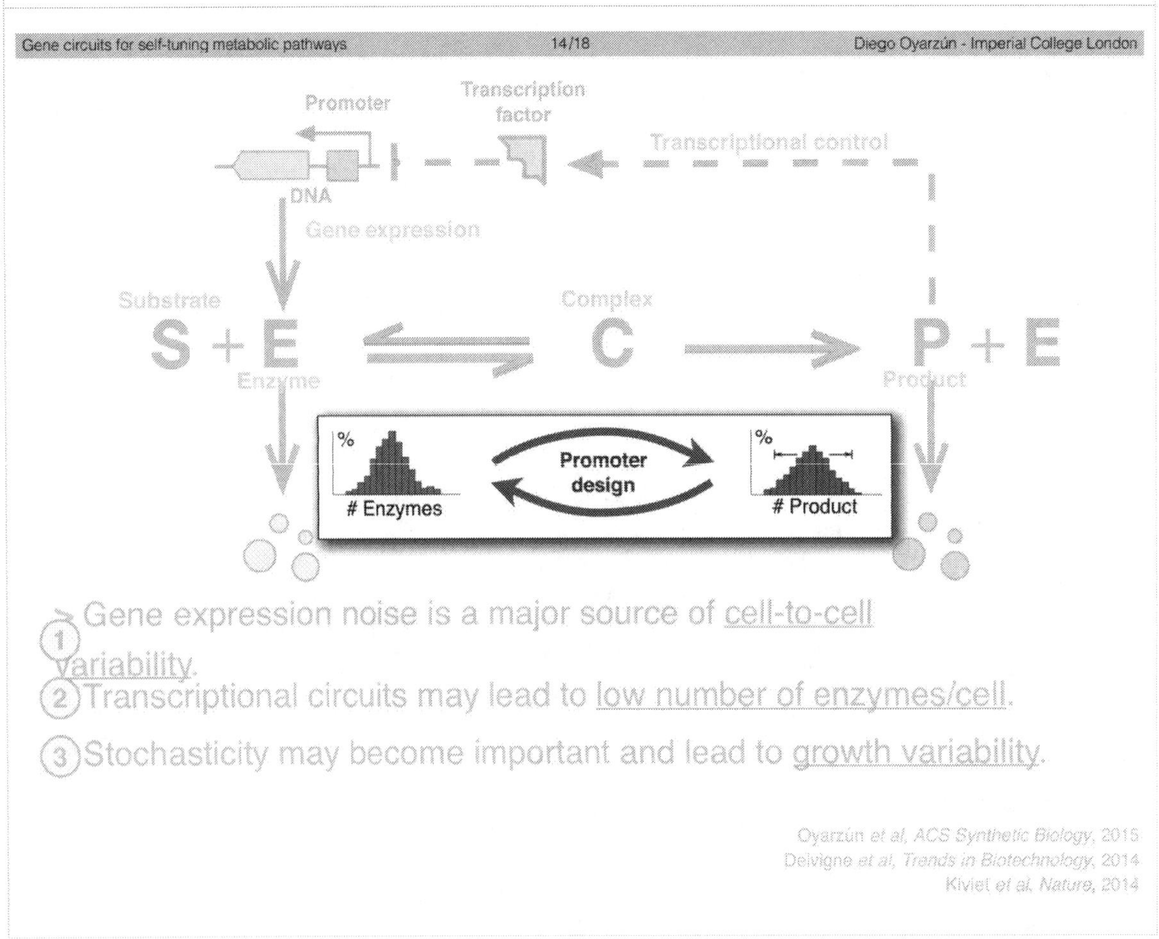

*** Formatting Issue - Best Available Paper/Graphic ***

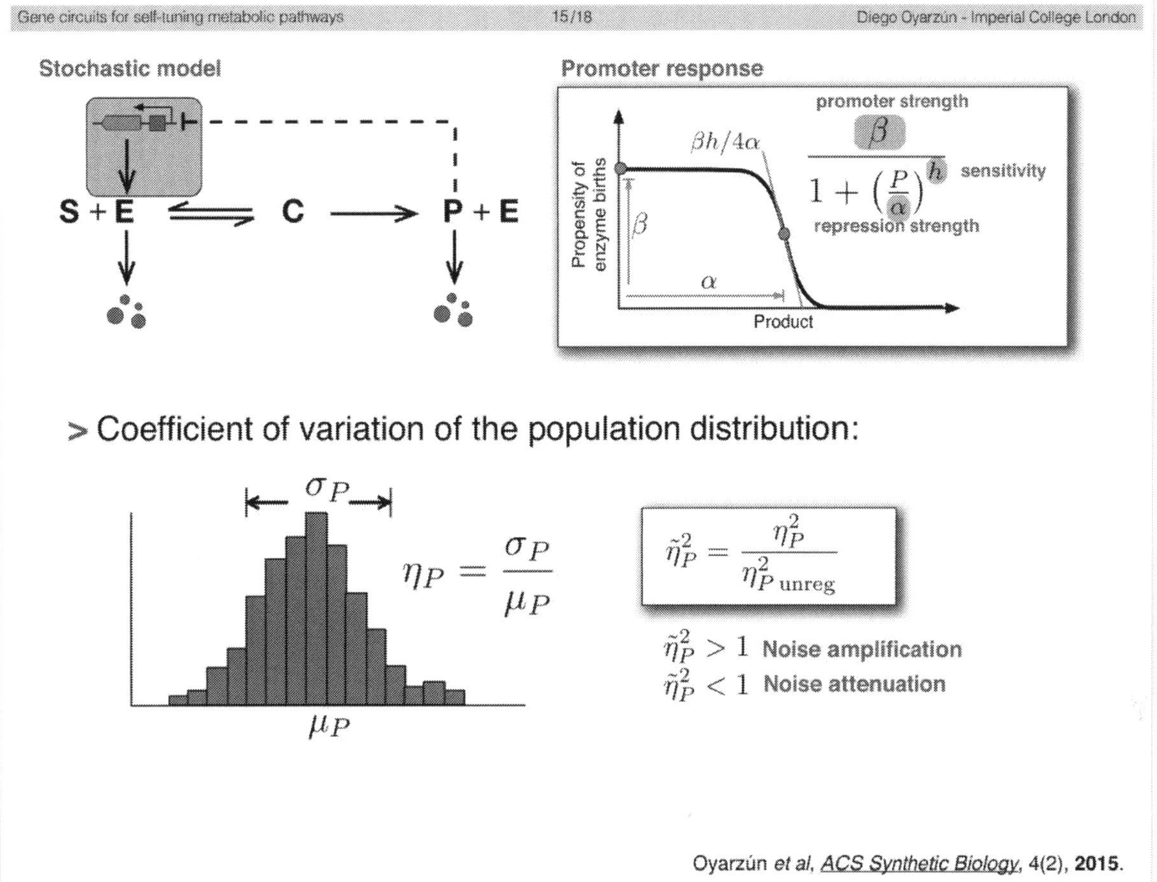

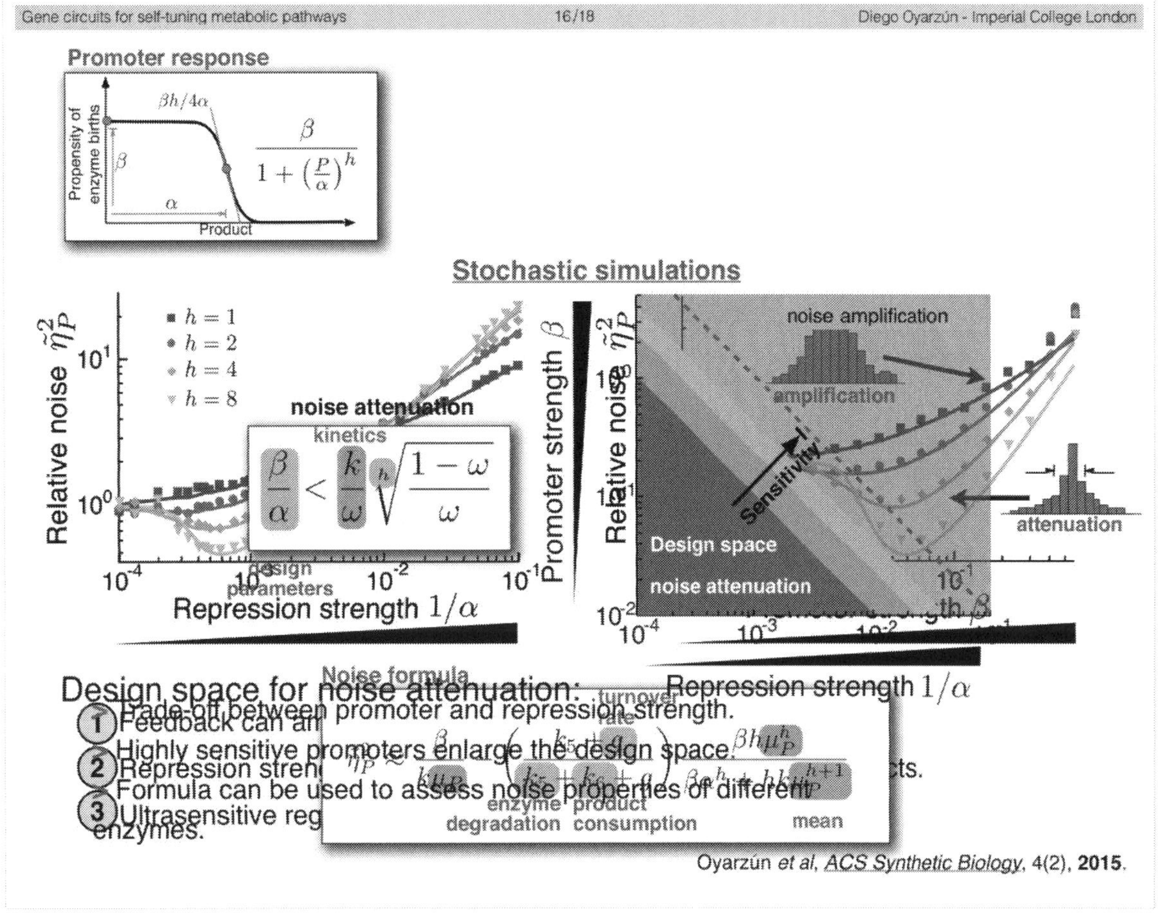

*** Formatting Issue - Best Available Paper/Graphic ***

Summary

① **Regulatory architecture**

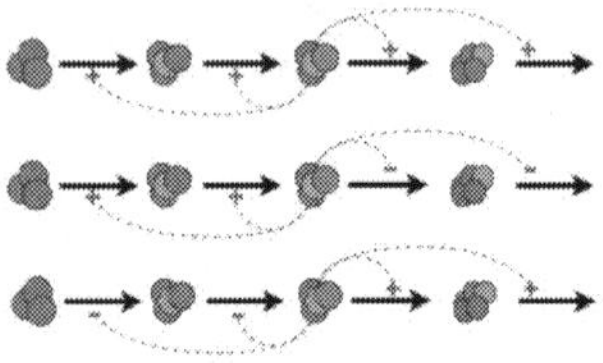

② **Parameter design trade-offs**

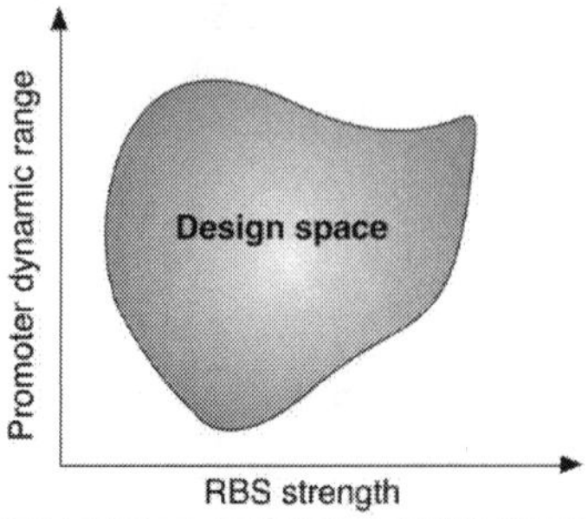

③ **Control of cell-to-cell variability**

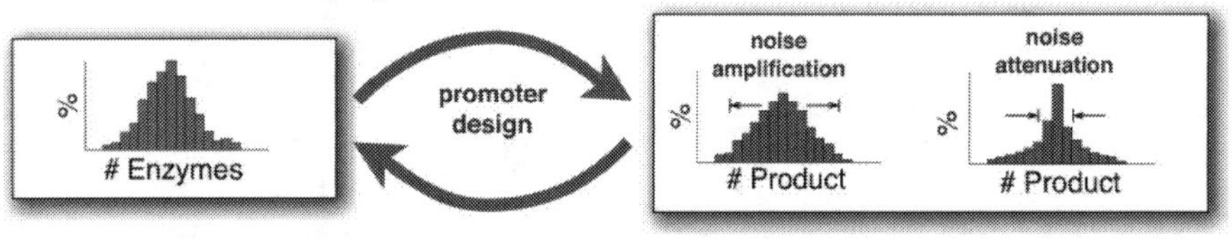

Acknowledgements

Students

> Matt Haines (**experimental work at Baldwin lab**)
> Jean-Baptiste Lugagne, now at INRIA Paris.
> Marit Hoffmeyer, now at U Bremen.
> **? (looking for 1 PhD student now)**

Collaborators

Experiments

> John Heap, Imperial College London
> Geoff Baldwin, Imperial College London
> **Fuzhong Zhang, U Washington St Louis**

Theory

> Andrea Weiße, U Edinburgh
> Madalena Chaves, INRIA, France
> Peter Swain, U Edinburgh
> Guy-Bart Stan, Imperial College London

HFSP

Theory-experiment partnership
Young Investigator Grant, Human Frontier Science Program

Imperial College London

d.oyarzun@imperial.ac.uk
www.imperial.ac.uk/people/d.oyarzun

Protecting Groups for Improved Control of Indigo Biosynthesis

Tammy Hsu

Dueber Lab

June 11, 2015

40,000 tons of indigo per year

Steingruber E, *Indigo and Indigo Colorants*, 2004

Indigo synthesis is environmentally unfriendly

- Chemical synthesis of indigo
  - Aniline
  - Formaldehyde
  - Sodium amide
  - Hydrogen cyanide
  - KOH
  - NaOH

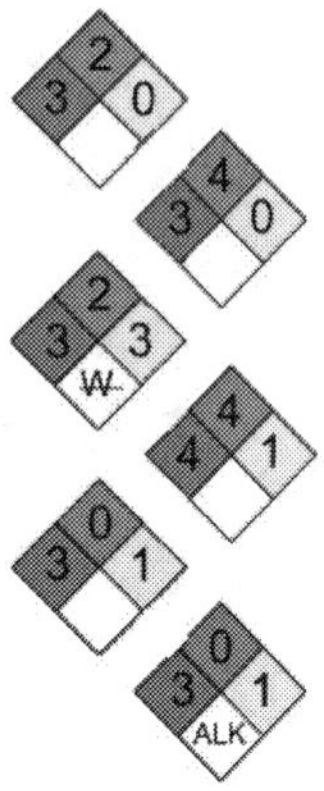

Steingruber E, *Indigo and Indigo Colorants*, 2004

Indigo dyeing is environmentally unfriendly

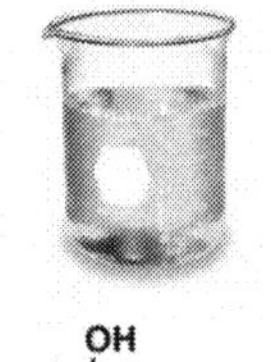

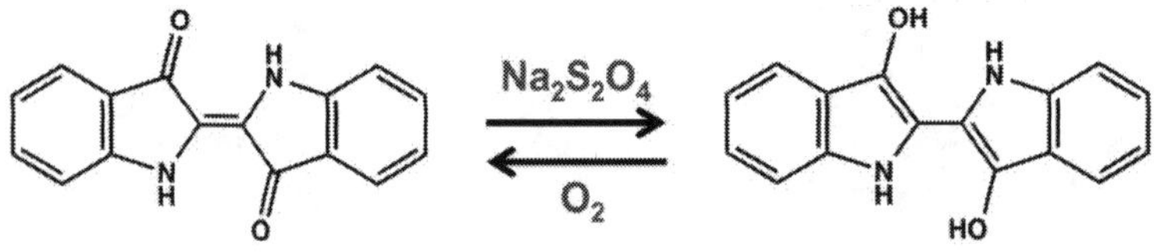

indigo

leucoindigo

Natural indigo from plants

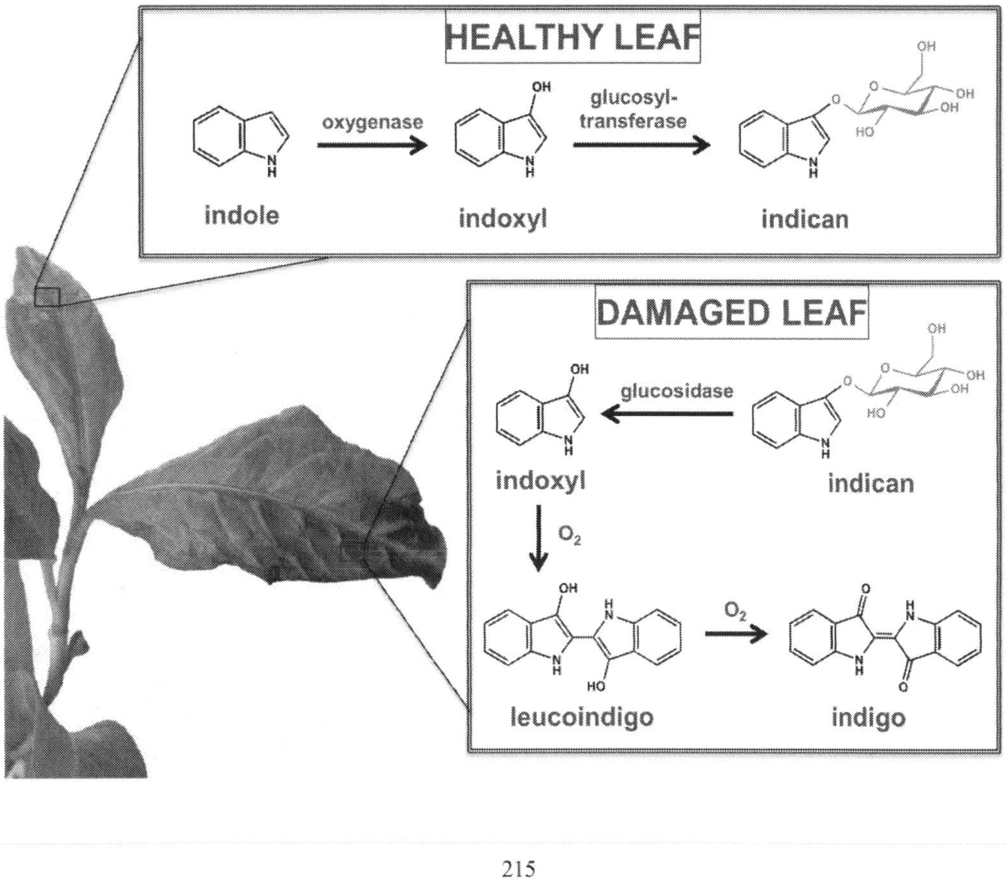

Indican is a potential dye

- Soluble
- Stable
- One enzyme away from indigo
- No reducing agent required

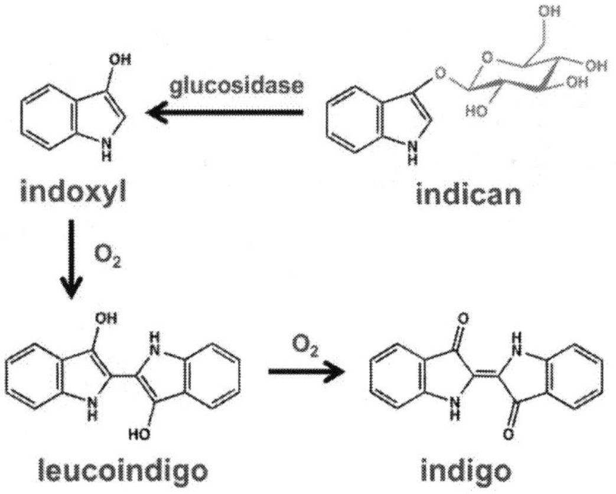

Berkeley iGEM 2013

Indican can dye cotton

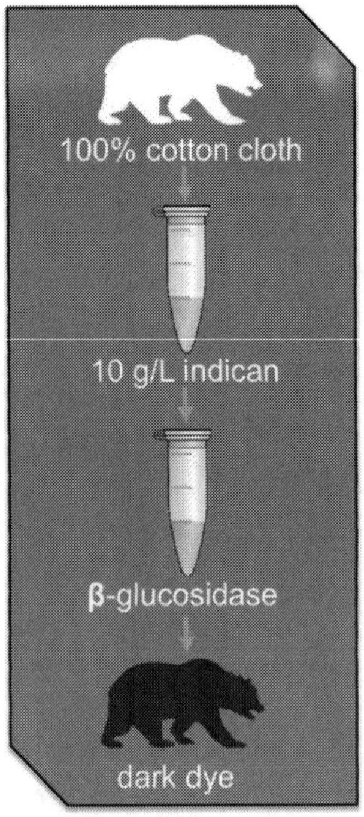

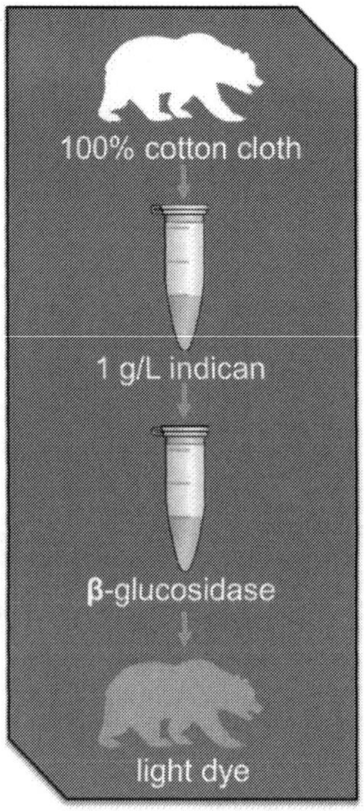

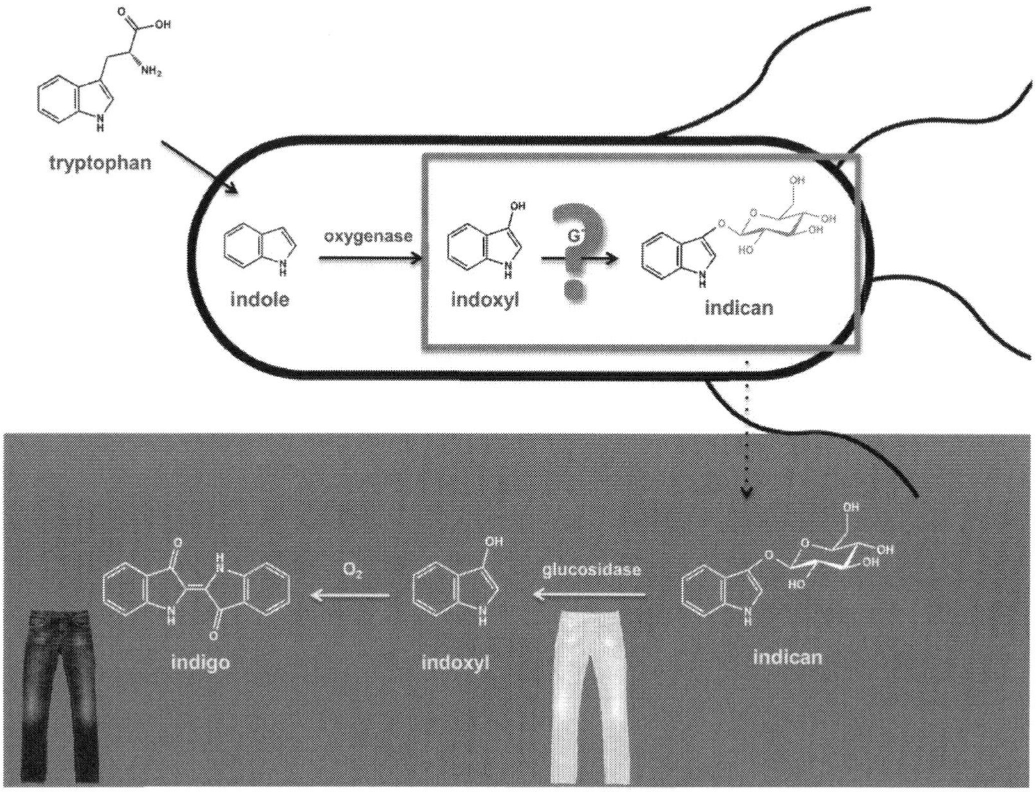

How to find the GT

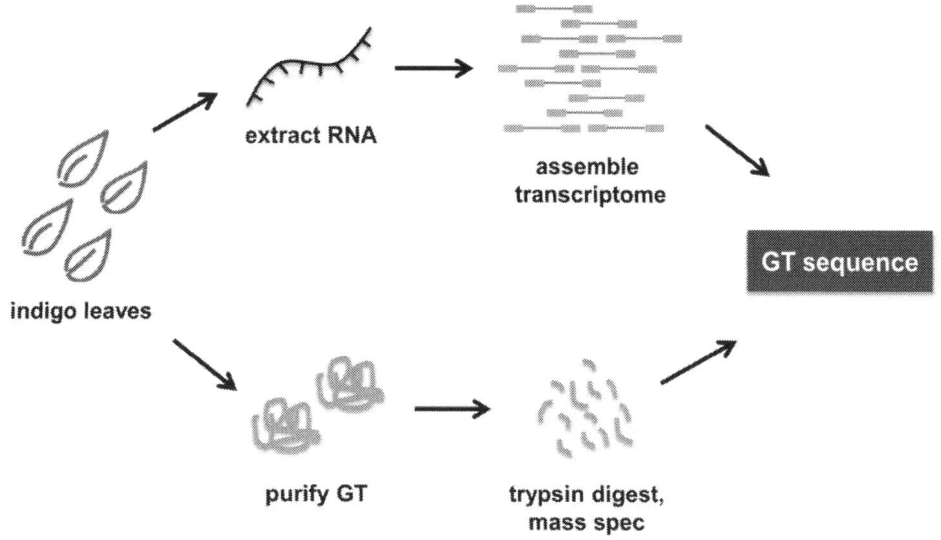

Harvest leaves **Flash freeze**

Grind to powder **Extract proteins**

Minami Y et al, *Plant Cell Physiol.*, 2000

Identified the GT

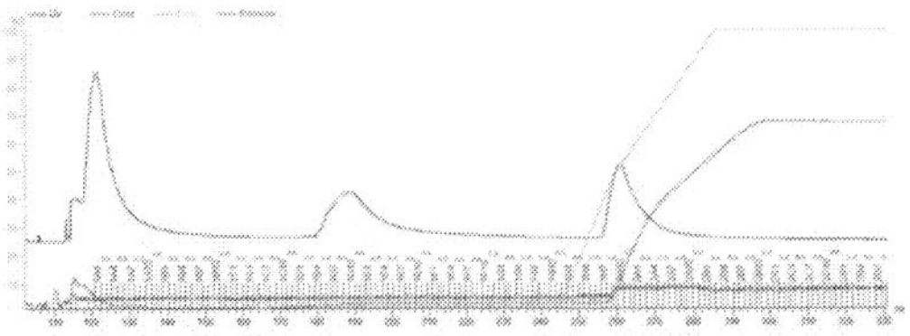

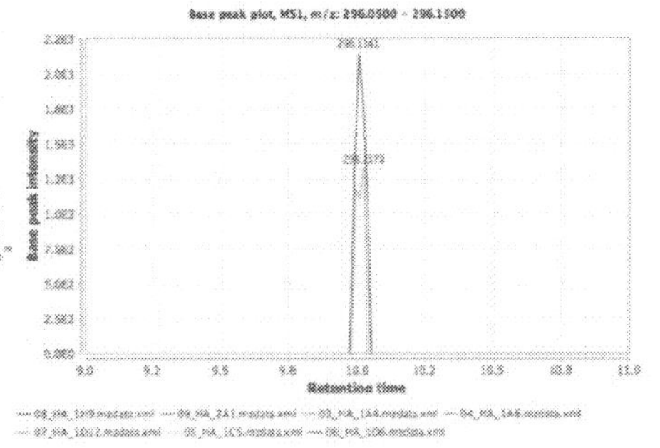

ladder

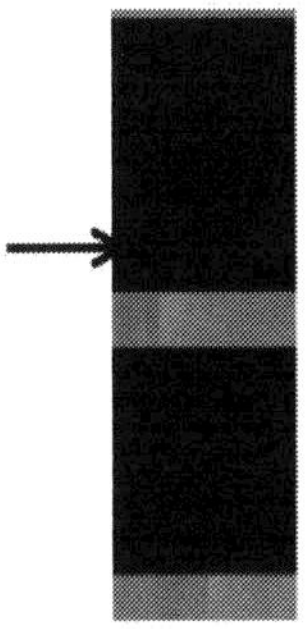

- LC-MS/MS, align to transcriptome
- Found most likely sequence

Indican production in *E. coli*

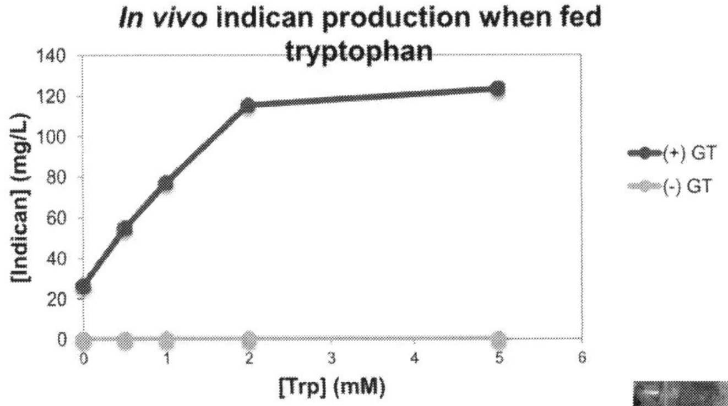

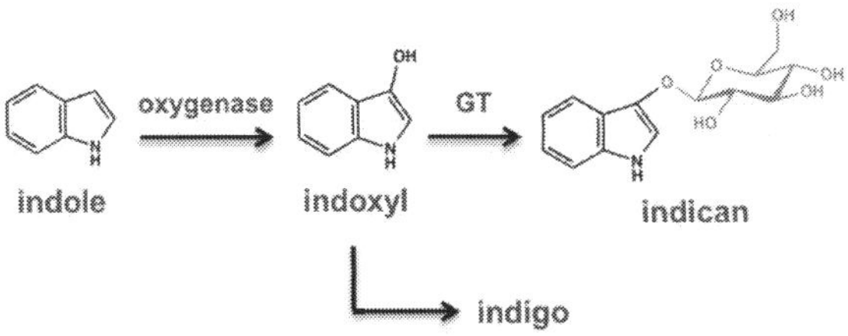

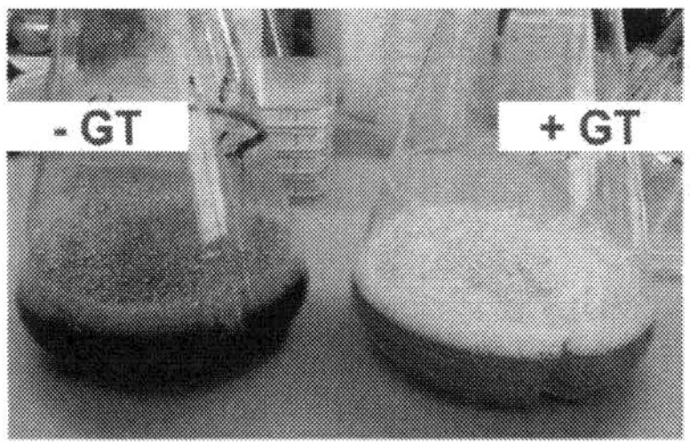

Increasing titer with glucosidase knockout

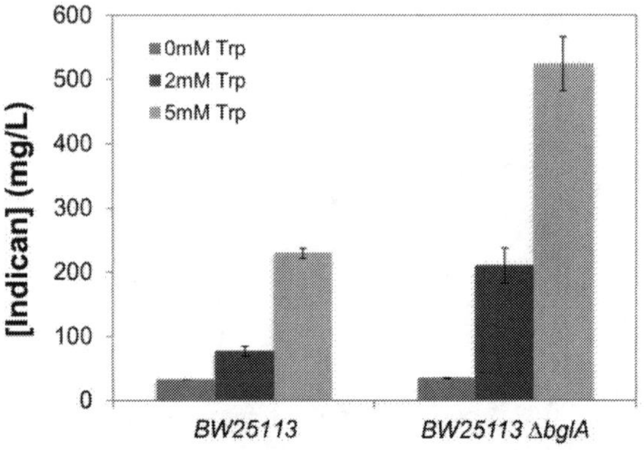

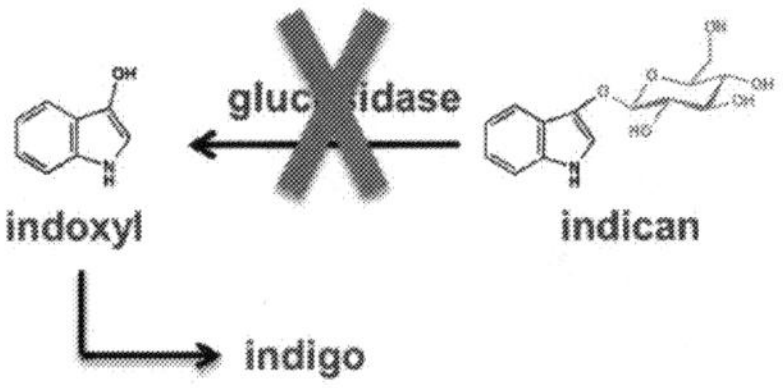

Currently produce 1 g/L indican

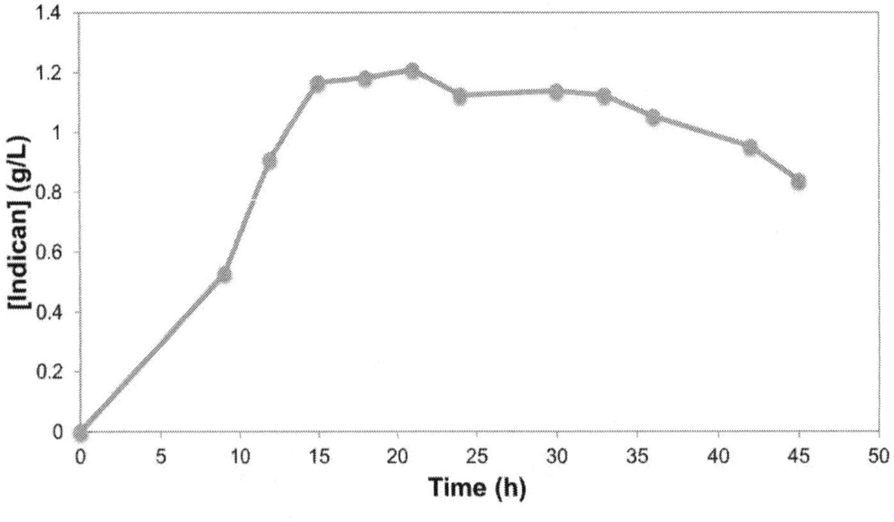

Parry Grewal

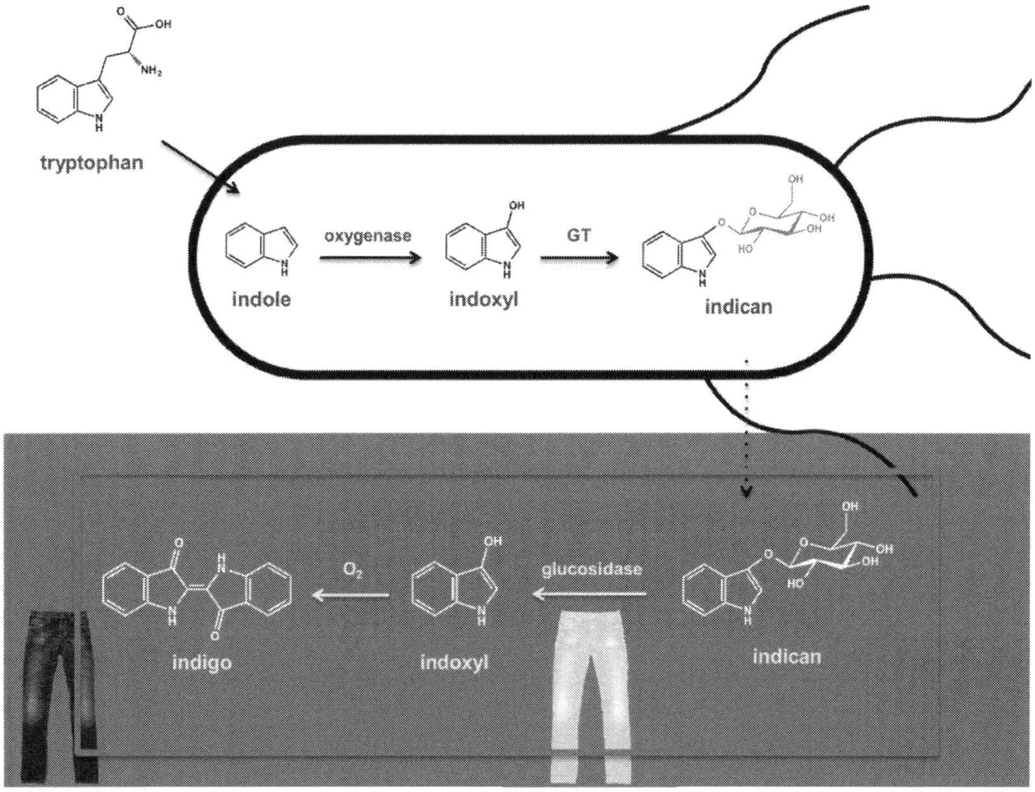

Dyeing with pure indican

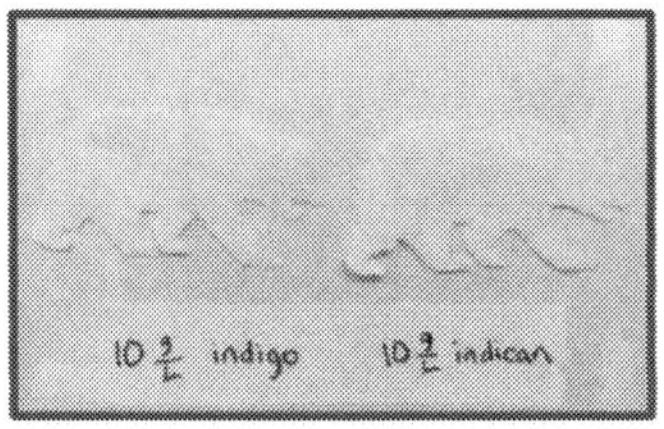

before dyeing

after indigo/indican after glucosidase after wash

Concentrating culture supernatant

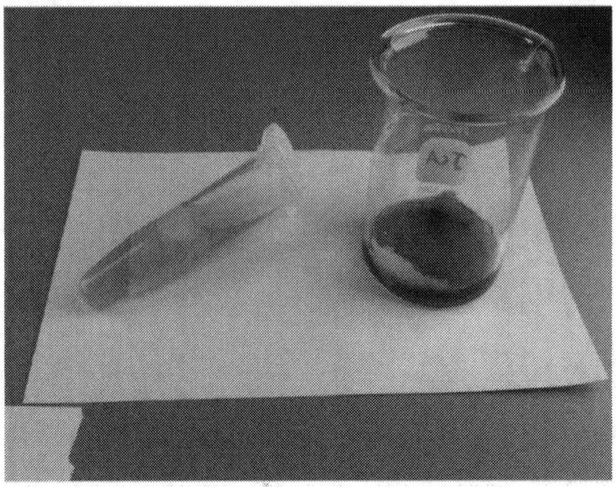

culture supernatant	boiled culture supernatant
1x	8.3x
0.6 g/L	3.4 g/L

Dyeing with *E. coli* culture supernatant

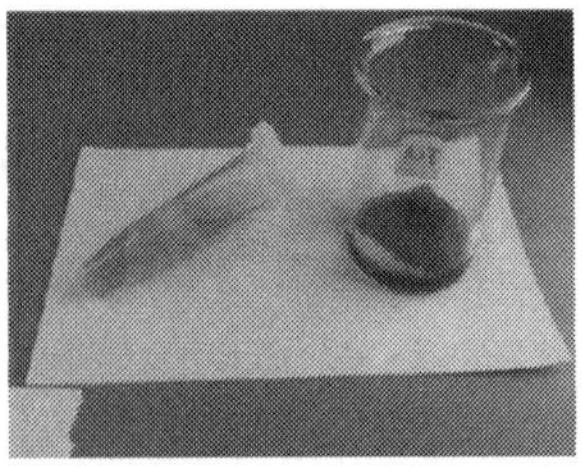

0.6 g/L 3.4 g/L

after indican after glucosidase after wash

Summary

- Identified glucosyltransferase
- Showed *in vivo* production of indican

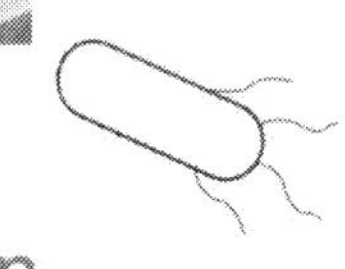

- Dyed cloth with biosynthesized indican

Next Steps

- Tryptophan production in *E. coli*
- Scaling up

Acknowledgements

John Dueber

Zach Russ

Dueber Lab

Berkeley iGEM 2013

Parry Grewal
David Stanley
Nadav Sorek, C. Somerville Lab
QB3 Proteomics/Mass Spec Lab

NDSEG Fellowship
Bakar Fellows Program

THE DUEBER LAB

SPSB

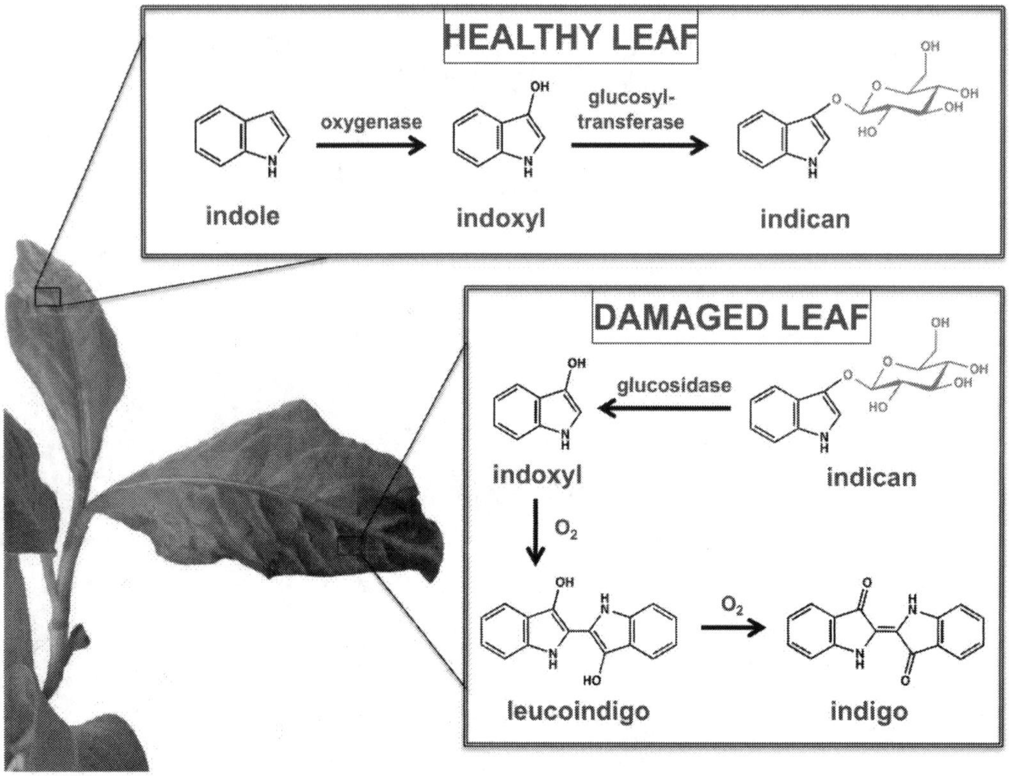
HEALTHY LEAF
oxygenase
glucosyl-
transferase
indole
indoxyl
indican
DAMAGED LEAF
glucosidase
indoxyl
indican
O_2
O_2
leucoindigo
indigo

Indigo

iGEM 2013

Berkeley iGEM 2013

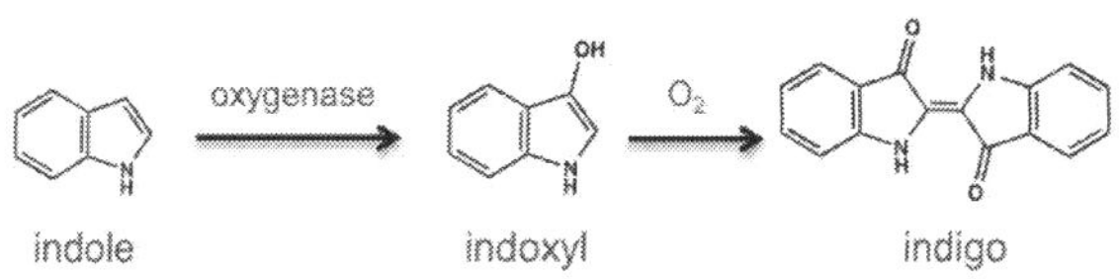

Carbon source

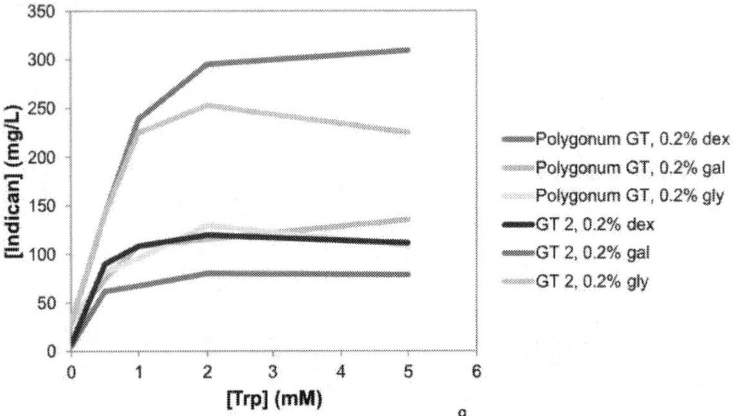

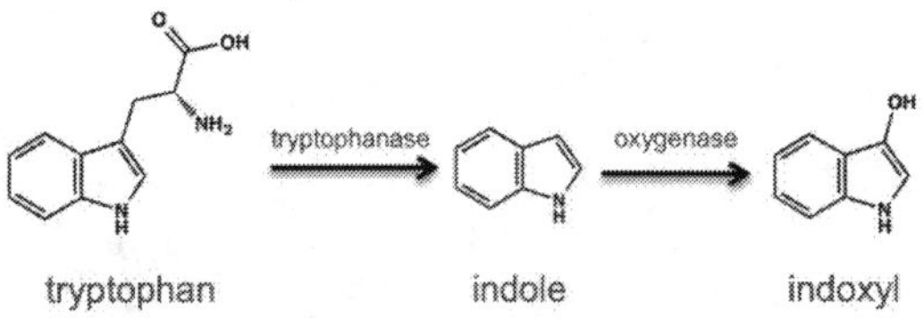

Oxygenation can matter

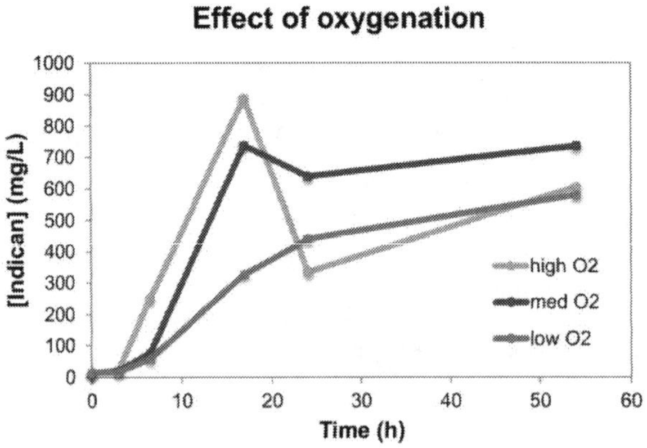

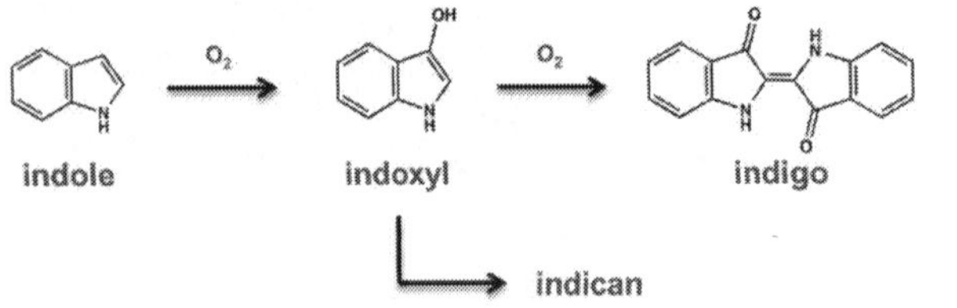

Parry Grewal

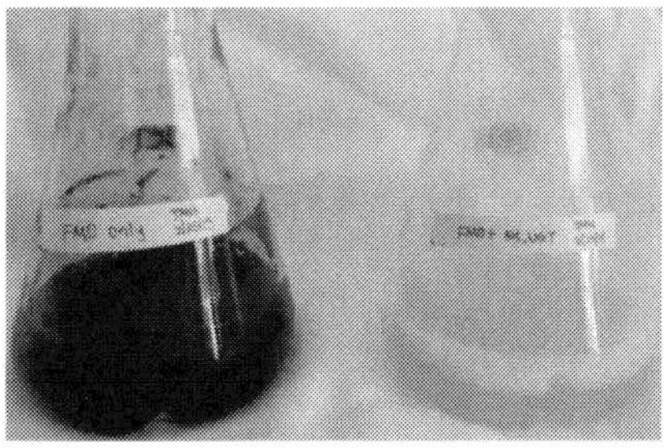

Other protecting groups

- Sulfonation

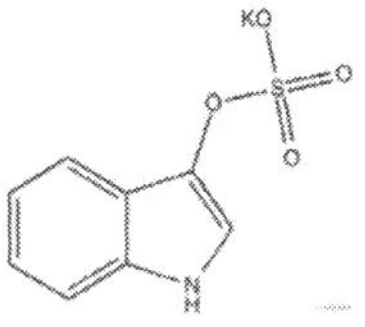

SPSB

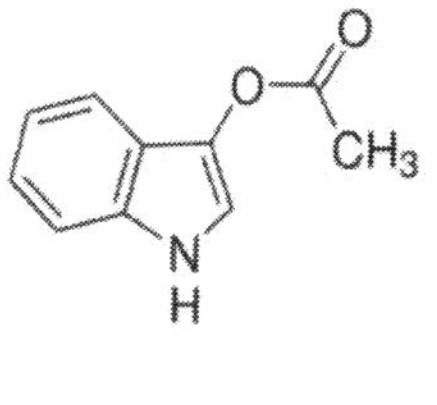

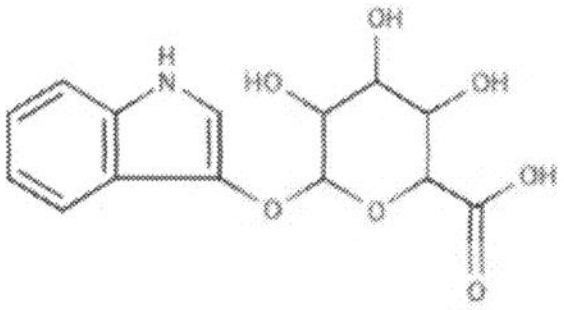

SPSB

What do Soybeans Have to do with Sustainable use of our Oceans?

Ursin, V., Monsanto
Hass, J.,
Froman, B.,
Valentin, H.,
Originally presented on: 6/11/2015 13:30:00 - 13:55:00

Essential fatty acids, or EFAs, are fatty acids that humans and other animals must ingest because the body requires them but cannot synthesize them. Two fatty acids are known to be essential for humans, the omega-3 fatty acid, α-linolenic acid (ALA) and an omega-6 fatty acid, linoleic acid (LA). Other fatty acids are "conditionally essential," meaning that humans can synthesize them from EFAs, but they can become limited under some conditions. These include two long-chain derivatives of ALA, eicosapentaenoic acid (EPA), and docosahexaenoic acid (DHA). Multiple health organizations recommend specific intake levels of ALA, EPA and DHA to promote optimal health. One such recommendation is the consumption of two fish meals per week, preferably fatty fish. This recommendation is problematic for a number of reasons, including sustainability and relative safety of the marine fish which are highest in EPA and DHA. Through the metabolic engineering of the fatty acid biosynthesis pathway in soybean seeds, we have created a land-based, sustainable source of biologically active omega-3 fatty acid, well suited for human and animal diets. The resulting product is soybean oil rich in the omega-3 fatty acid, stearidonic acid (SDA), a naturally occurring precursor to EPA and DHA. We have shown that when ingested, SDA increases beneficial omega-3 levels, specifically EPA, in humans, poultry and some fish species. Compared to EPA and DHA, SDA combines superior stability and sensory characteristics for use in a variety of applications, making this oil suitable for incorporation into a wide variety of foods. Through incorporation into commonly consumed foods, SDA Soy oil can provide health benefits of fish consumption, while decreasing dependence on marine sources of omega-3 fatty acids.

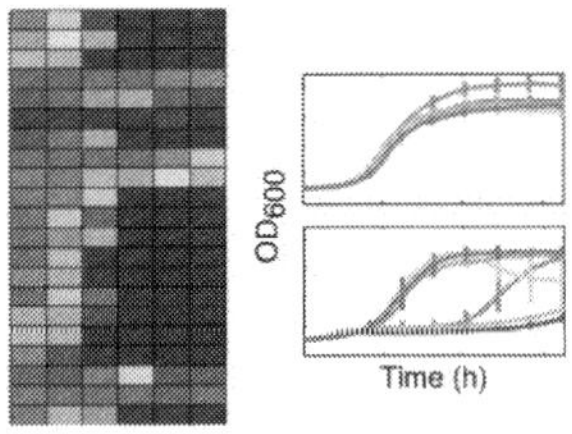

Improving Microbial Biofuel Production using Efflux Pumps

Mary Dunlop
School of Engineering
University of Vermont

Biofuels are Often Toxic to Microbes

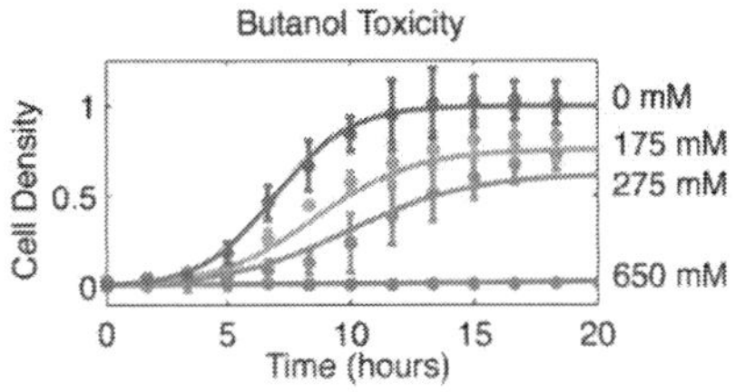

Efflux Pumps Can Export Biofuel-like Compunds

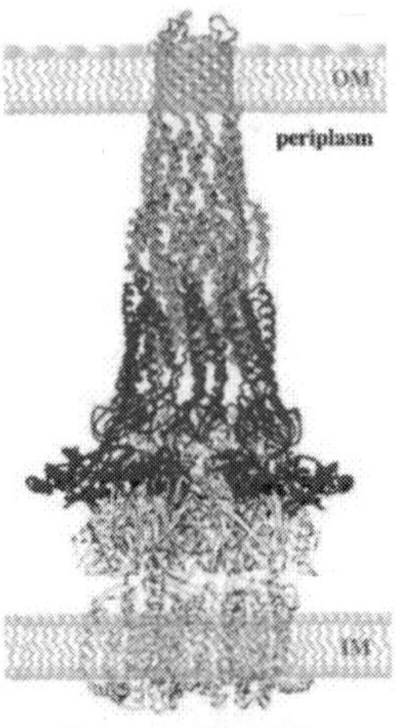

Higgins et al, PNAS 2004

Pumps export toxins using the electrochemical gradient across the cell membrane.

AcrAB-TolC (*E. coli*)

Tetracycline
Chloramphenicol
Beta-lactams
Novobiocin
Hexane
Pentane
Xylene
Pinesol
Pine Oil
Fusidic acid
Nalidixic acid
Fluoroquinolones
SDS
Triton X-100
Bile salts
Ethidium Bromide

SrpABC (*Pseudomonas putida* S12)

Pentane
Hexane
Heptane
Octane
Nonane
Propanol
Butanol
Pentanol
Hexanol
Heptanol
Octanol
Toluene
Benzene
Styrene
P-Xylene
Ethylbenzene
Propylbenzene

Bioinformatics Search for Biofuel Pumps

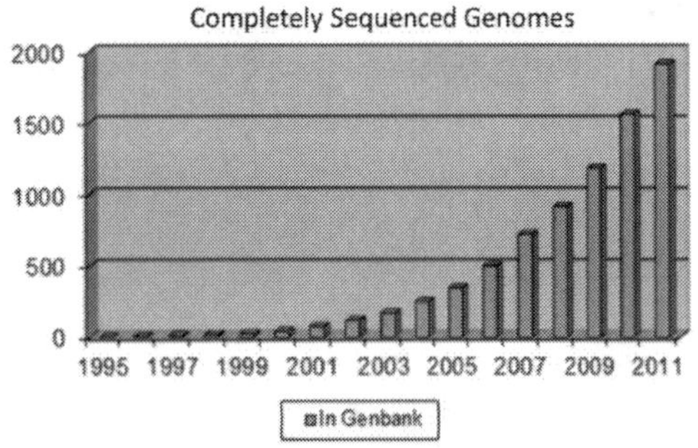

Constructed library of 43 efflux pumps, transformed into *E. coli*

Marine microbes from oil-rich environments

Dunlop, Dossani, Szmidt, Chu, Lee, Hadi, Keasling, Mukhopadhyay – Molecular Systems Biology 2011

Competition Experiment

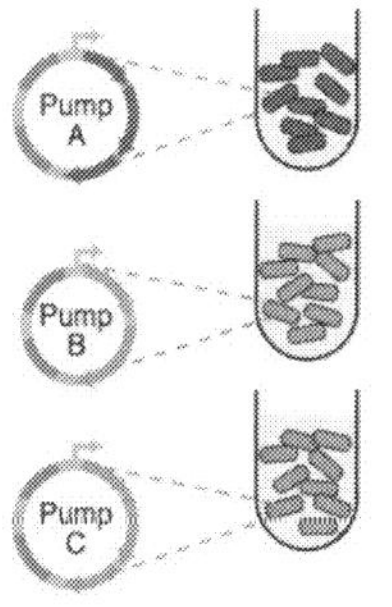

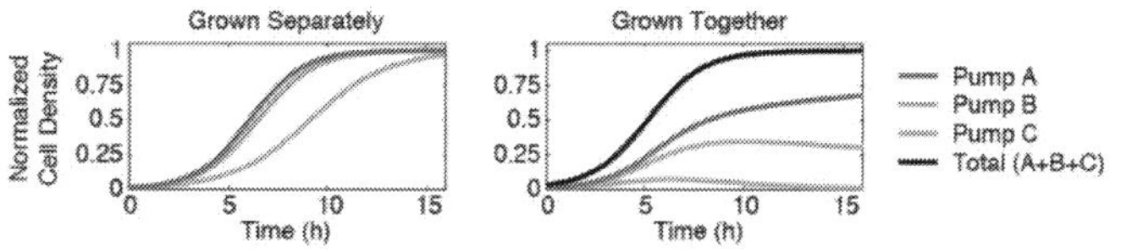

Competition Selects for Biofuel Pumps

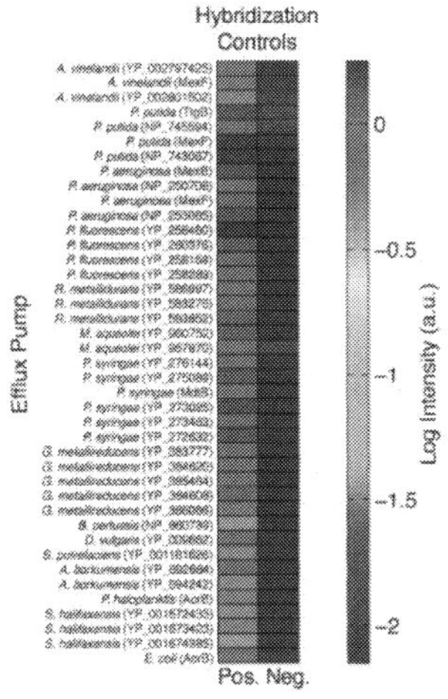

Pumps Can Export Biofuels

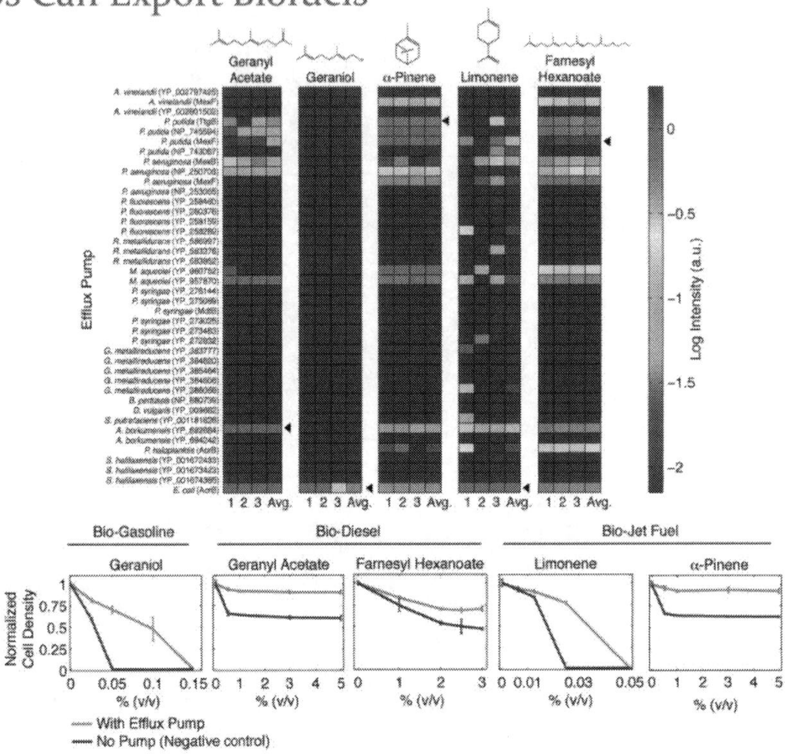

Sources of Biofuel Export Pumps

Host	Pump
Azotobacter vinelandii	MexEF-OprN
Pseudomonas putida KT2440	TtgABC
Pseudomonas putida KT2440	Putative pump (NP_745594) Only inner membrane & periplasmic proteins
Pseudomonas putida KT2440	MexEF-OprN
Pseudomonas aeruginosa PA01	MexAB-OprM
Pseudomonas aeruginosa PA01	Putative pump (NP_250708) Only inner membrane & periplasmic proteins
Pseudomonas aeruginosa PA01	MexEF-OprN
Marinobacter aqueolei	Putative pump (YP_960752) Only inner membrane & periplasmic proteins
Marinobacter aqueolei	Putative pump (YP_957870)
Alcanivorax borkumensis SK2	Putative pump (YP_692684)
Pseudoalteromonas haloplanktis TAC125	Putative pump (YP_341810)
Escherichia coli K12	AcrAB-TolC

Soil bacterium

Soil bacterium, has been used for bioremediation

Pathogen, known for antibiotic resistance

Marine microbe, isolated from offshore oil rig

Marine microbe, oil-degrader, prevalent after oil spills

Marine microbe

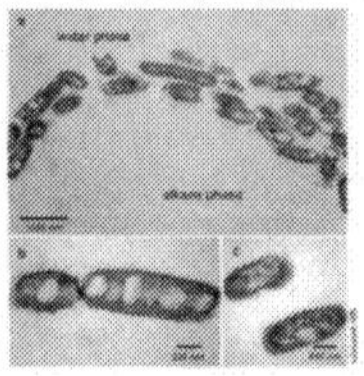

Competition Sorts Pumps Correctly

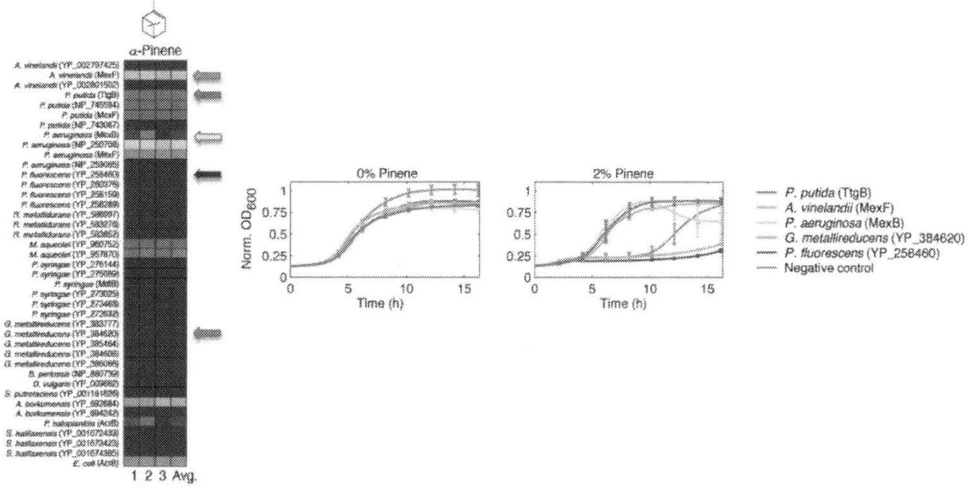

In a Limonene Production Strain

With Taek Soon Lee, Zain Dossani (JBEI)

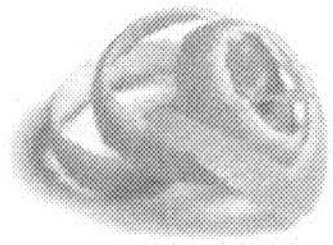

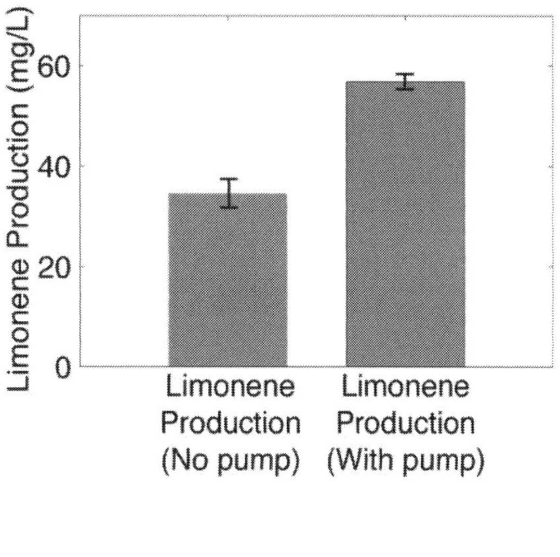

Tradeoffs in Pump Expression: Toxicity Due to Pump Overexpression

Will Turner

Biofuel Toxicity

No Pump | With Pump

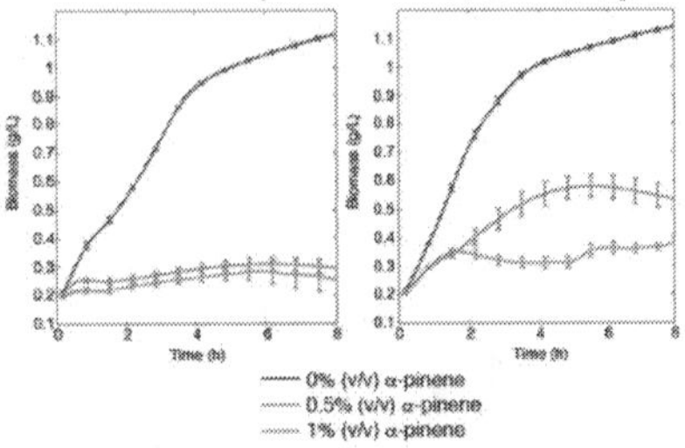

Pump Toxicity

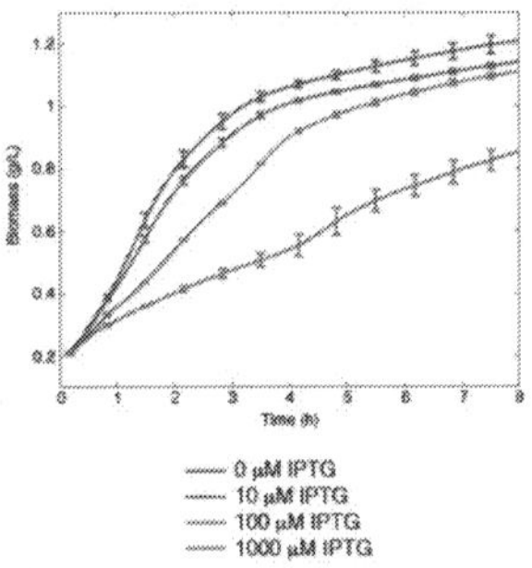

Model

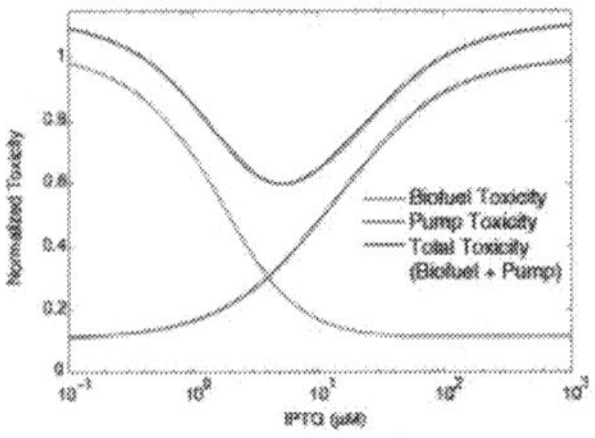

Experimental Data

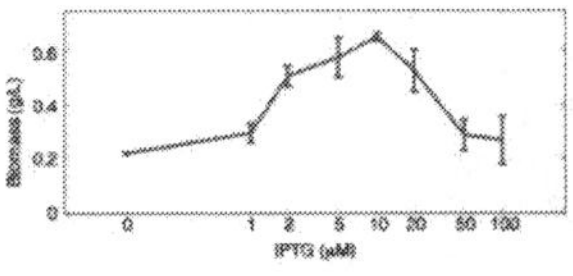

Turner, Dunlop – ACS Synthetic Biology 2014
Dunlop, Keasling, Mukhopadhyay – Systems & Synthetic Biology 2010

Characterizing Tradeoffs

Maqu_3494

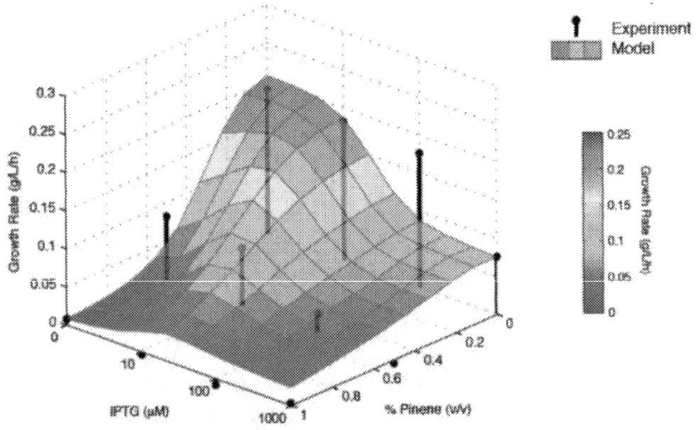

No single IPTG value works well for all pinene levels
Dynamic control would allow pump expression to be adjusted automatically

Dynamic Control of Pumps

Jessica Lindle

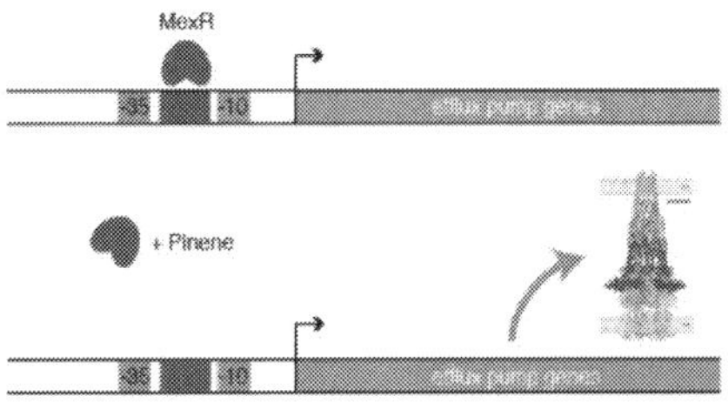

Synthetic Promoter Library

MexR Binding Site	-35	MexR Binding Site	-10	MexR Binding Site
MexR Binding Site	-35	MexR Binding Site	-10	no Binding Site
no Binding Site	-35	MexR Binding Site	-10	MexR Binding Site
no Binding Site	-35	MexR Binding Site	-10	no Binding Site
MexR Binding Site	-35	no Binding Site	-10	MexR Binding Site
MexR Binding Site	-35	no Binding Site	-10	no Binding Site
no Binding Site	-35	no Binding Site	-10	MexR Binding Site
no Binding Site	-35	no Binding Site	-10	no Binding Site

Vary number and location of binding sites
Also included degenerate bases in the promoter sequence

Initial inspiration: Zhang, Carothers, Keasling – Nature Biotech 2012

Convergence to Top Performers is Fast

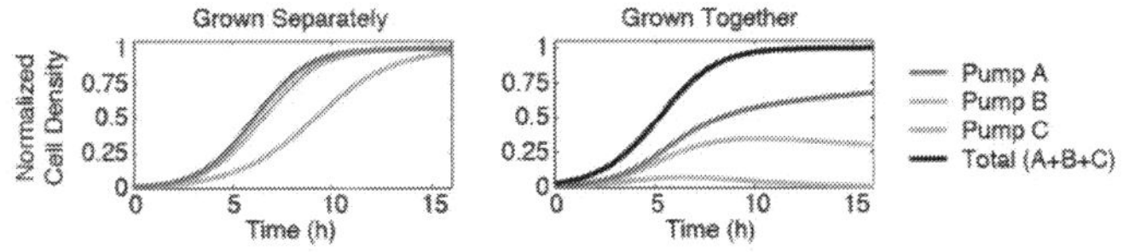

1000 library members

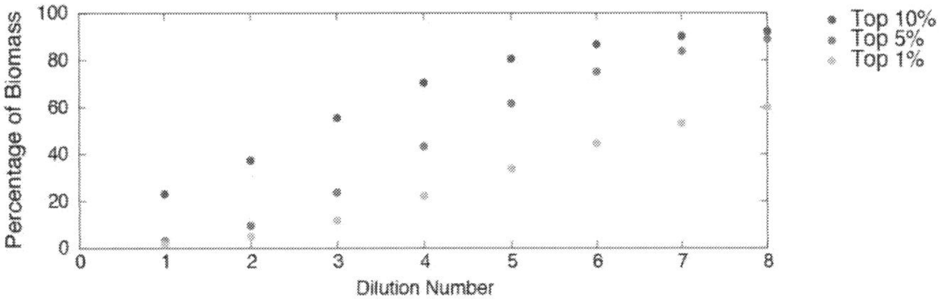

Library Members Each Have Biofuel/Pump Toxicity

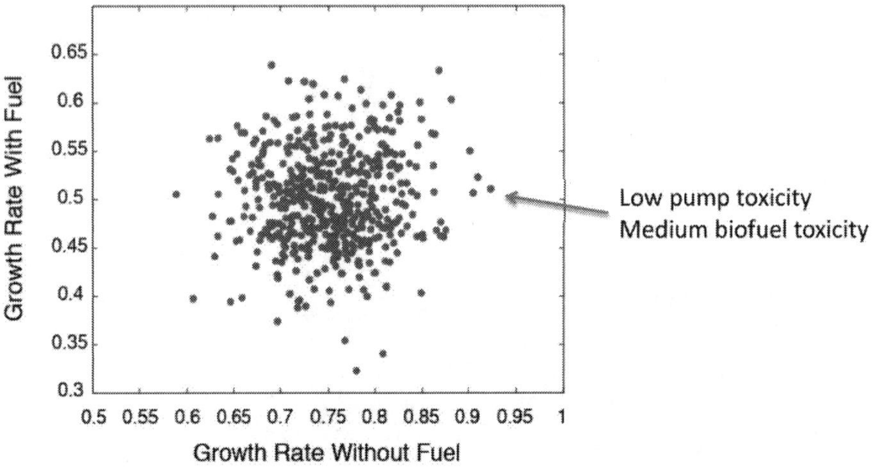

Model: Selecting for a Biosensor

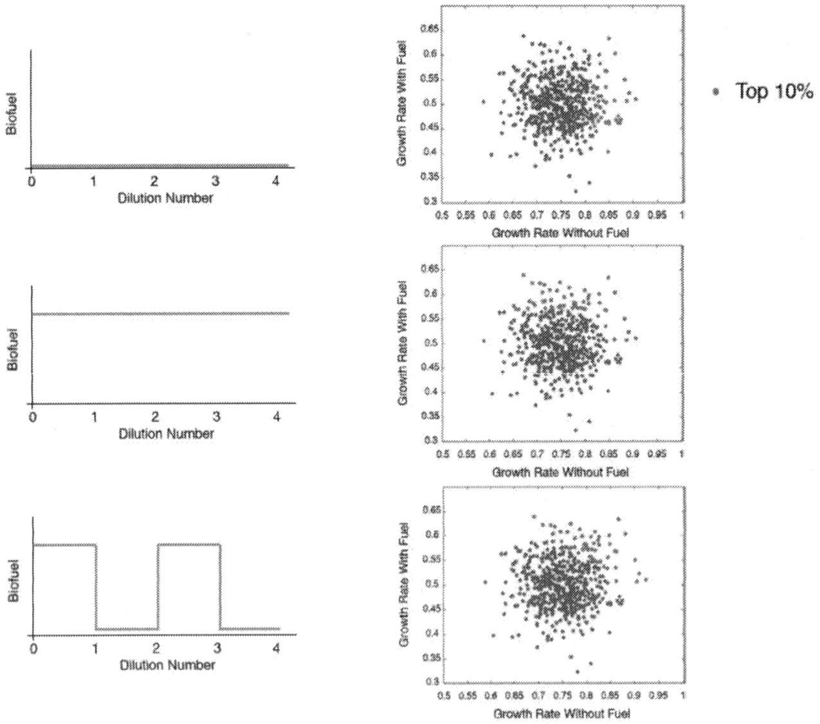

Experiments: Selecting for a Biosensor

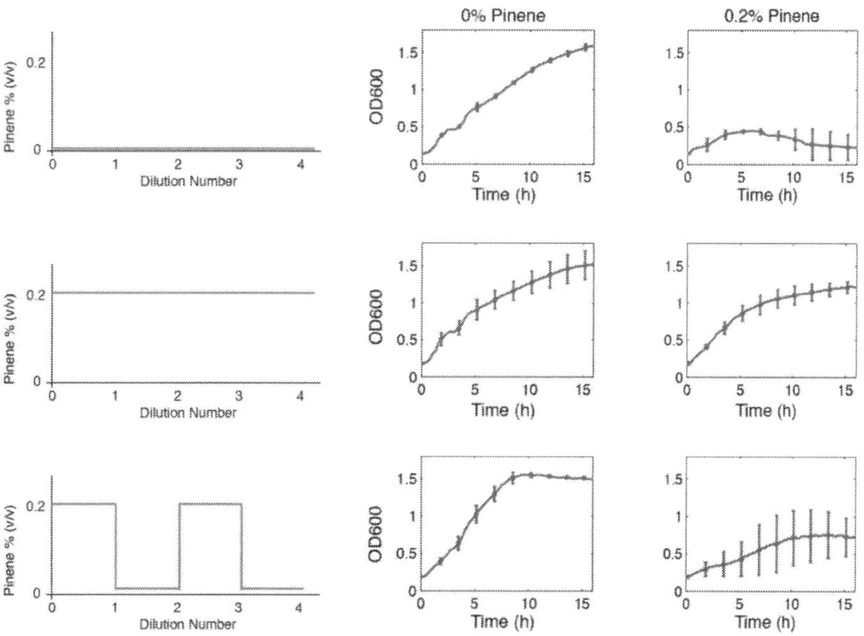

Acknowledgements

<u>University of Vermont</u>

Current Lab Members
Imane El Meouche
Jessica Lindle
Nick Rossi
Tim Tomko
Tiebin Wang
Alexa Bartel
Yik Siu

Former Lab Members
Ana Camposeco-Jacobs
Javier Garcia-Bernardo
Mary Harrison
Will Turner

<u>Joint BioEnergy Institute</u>
Zain Dossani
Heather Szmidt
H. Cheng Chu
Taek Soon Lee
Masood Hadi
Jay Keasling
Aindrila Mukhopadhyay

Posters (Non-Genetic Antibiotic Resistance)
Yik Siu - #158, Thursday
Tiebin Wang - #173, Friday

<u>Funding</u>

R01-AI102922 NSF CAREER Award DOE Early Career Award VT Space Grant

Food security challenges
- Growing global population
- Rising middle class
- Limited agricultural lands
- Limited water

✓ Increasing crop productivity
  - High yield potential
  - Resistant to pests & diseases
  - Resistant to herbicides
  - Tolerate abiotic stresses

✓ Enhancing the nutritional value of staple foods

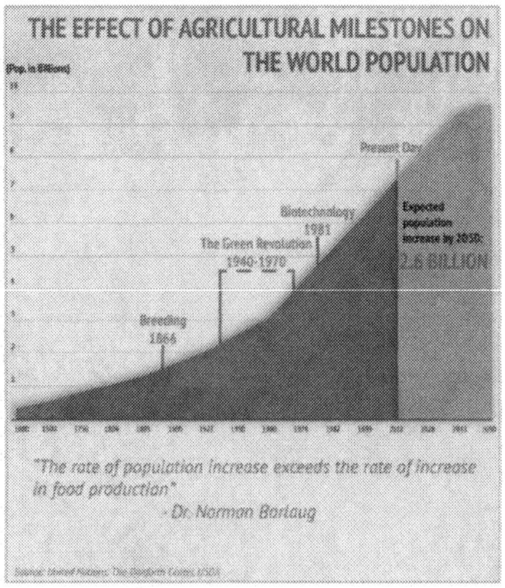

DUPONT
PIONEER.

2

Approaches to increase crop yield

- **Breeding to produce high yielding varieties**
  - Seed production technology (SPT)
  - Marker assistant selection (MAS)
  - ..
- **Conventional transgenic products**
- **Genome modification with double strand break (DSB) technology (Cas9-gRNA)**

DUPONT
PIONEER.

3

Historical perspective of plant transformation

- **Inability to modify the genome to suit product needs**
- **Random transgene insertions**
- **Extremely difficult and expensive to bring multiple transgenic traits (stacking) to the market by breeding**

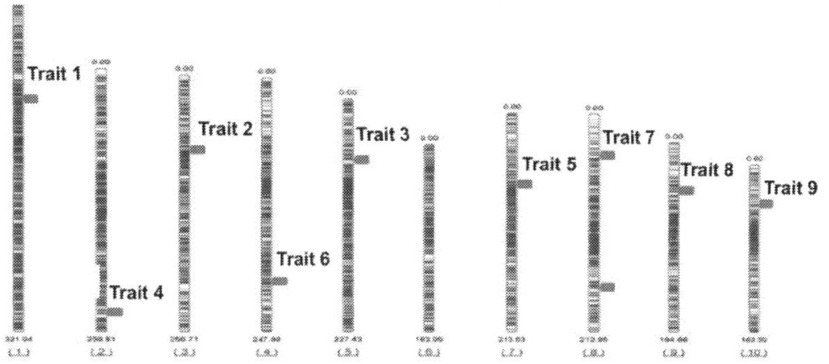

DUPONT
PIONEER.

4

Genome modification with double strand-break (DSB) technologies

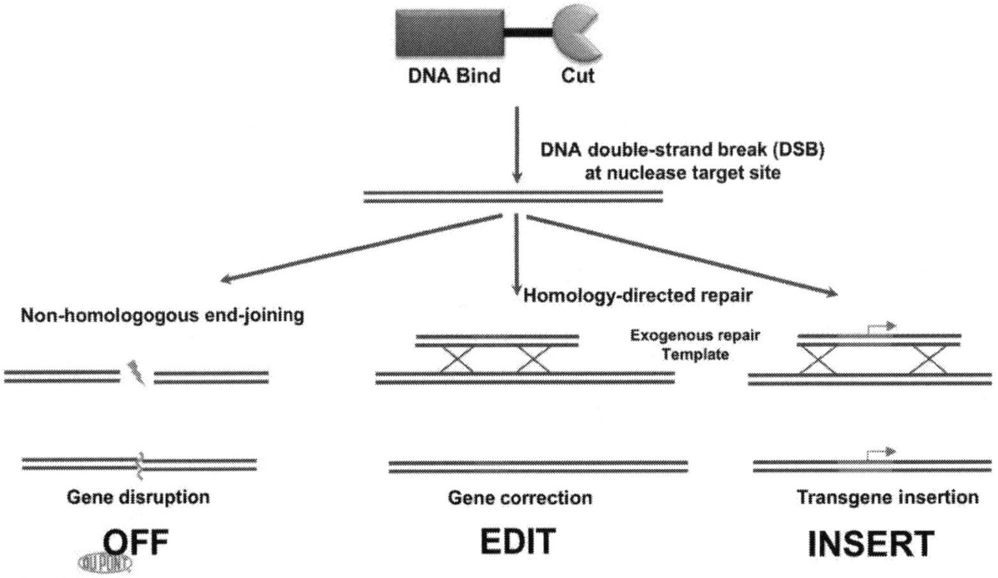

DUPONT PIONEER

5

Double strand break platforms

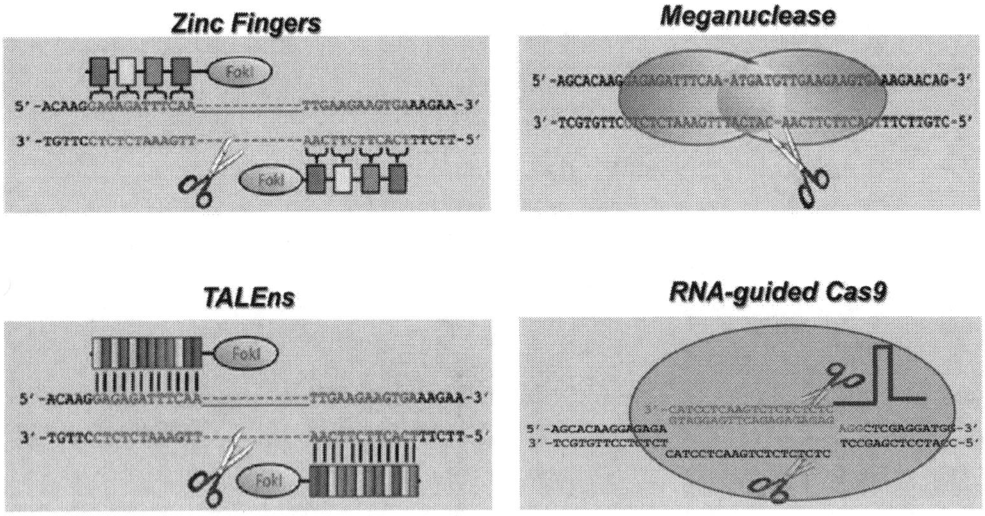

DUPONT PIONEER

6

I-CreI Meganuclease

Homing endonuclease, *Chlamydomonas reinhardtii*

recognize and cleave 22 nt sequences

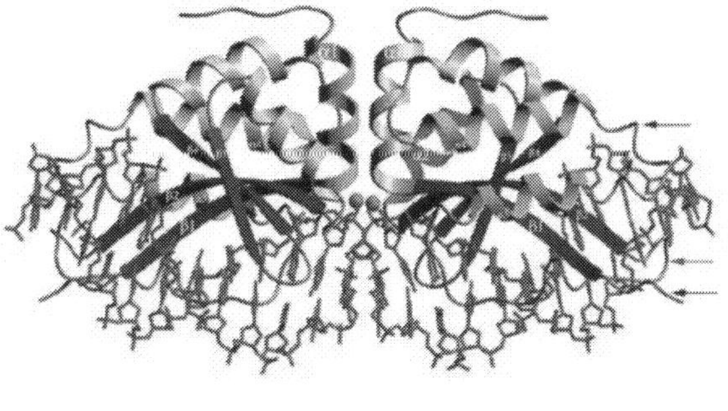

Precision Biosciences Inc.

Novel heterodimeric single chain I-*Cre*I variants

DUPONT
PIONEER

CRISPR-Cas

Yogurt cultures
Natural immunity

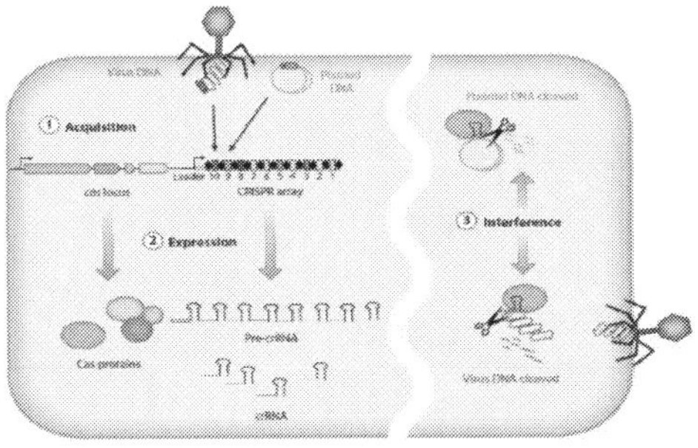

Cas9-gRNA

Mammals	Mouse
Silkworm	Rabbit
C. elegans	Frog
Zebrafish	
Drosophila	Plants

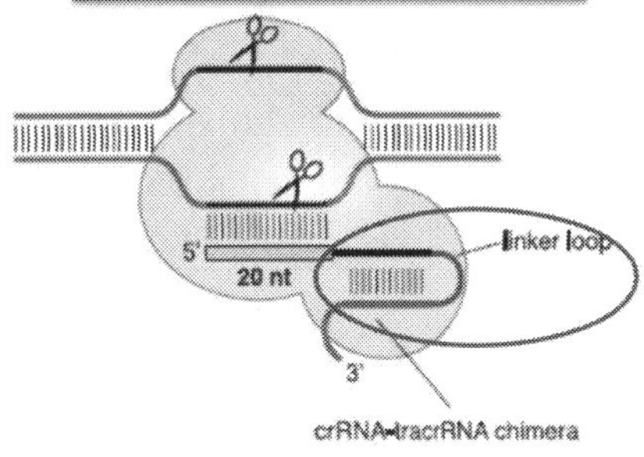

RNA-DNA base-pairing

DUPONT
PIONEER

Components for maize gene modification with CAS9-gRNA

CAS9-PROTEIN

ZM-UBI Pro **Cas9**

S. pyogenes, maize codon-optimized

GUIDE-single strand of RNA

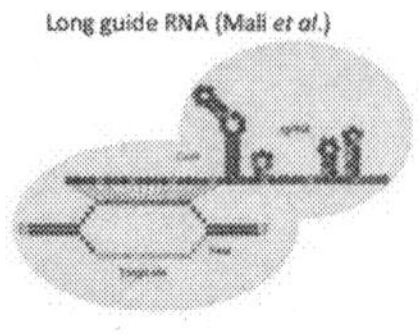

U6 PolIII snRNA promoter

PolIII Pro

Single guide RNA (20 nt + 77nt)

+/- Repair template

homology homology

DUPONT PIONEER

Plant genomes edited by Cas9-gRNA

DUPONT PIONEER

Maize Transformation

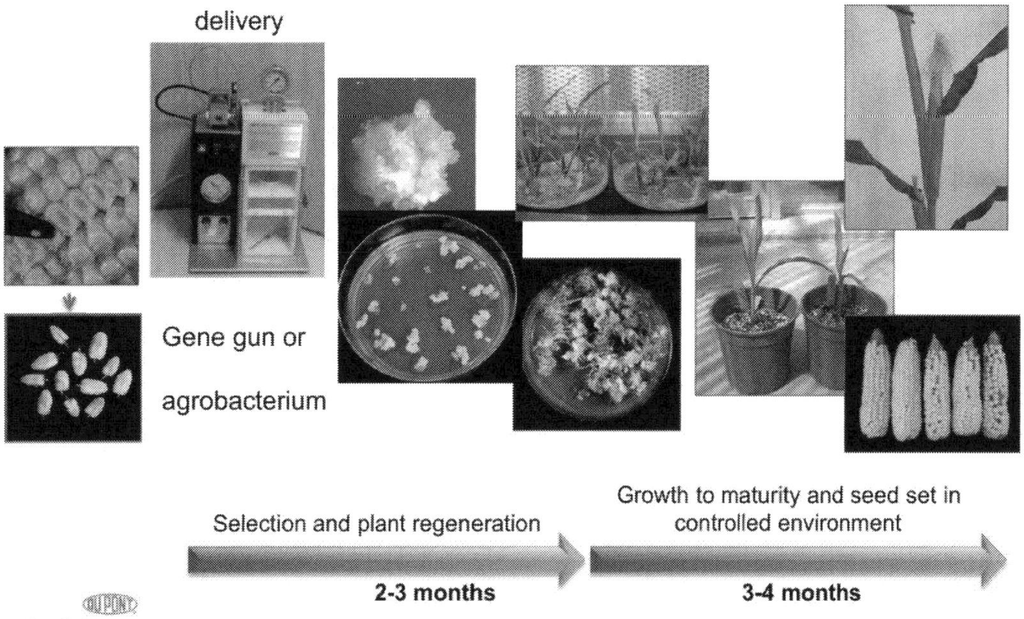

11

Fast assessing genome modification

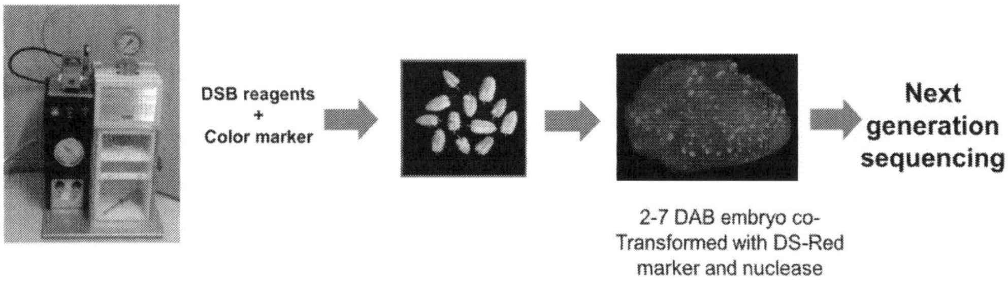

12

Genome engineering applications in maize

- Gene knockout
  - Discovery tools for determining gene function
  - Create new plant varieties
- Gene editing
  - Precise modification of endogenous genes (amino acid changes, regulatory element replacements...)
- Gene insertion
  - Integrate transgene into pre-defined genome locations
    - Complex Trait Loci (CTL) enable efficient trait stacking

DUPONT
PIONEER

13

Targeted insertion at two maize loci using Meganuclease

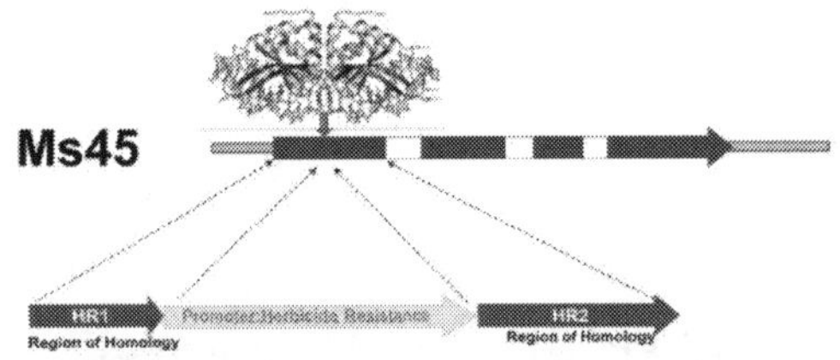

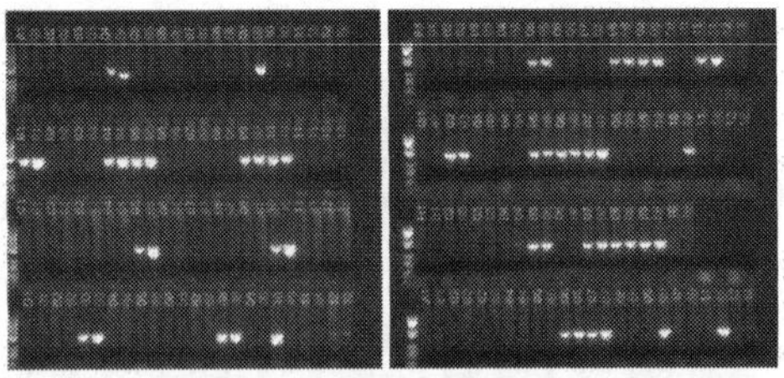

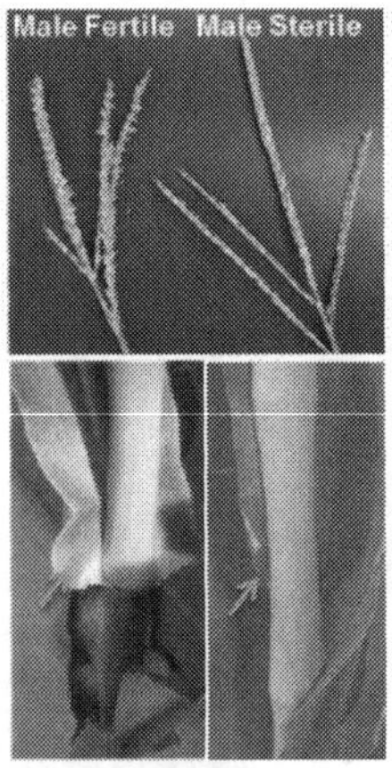

MGN	Events Recovery	TS Mut Frequency	Insert Frequency (HR1+HR2)
MS45(1-2.36)	8-16%	87-92%	13-14%
LIG34 Meganuclease*	15-17%	4-24%	0.7-2%

*Near liguleless locus
Gao et al P Journal 2010

DUPONT
PIONEER

14

Cas9-gRNA vs meganuclease

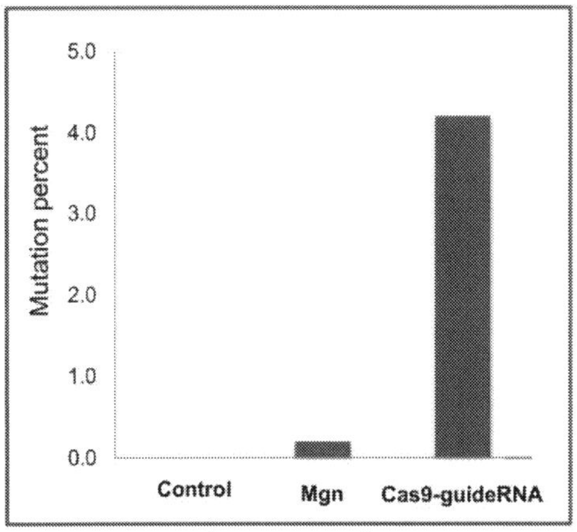

Cas9-gRNA

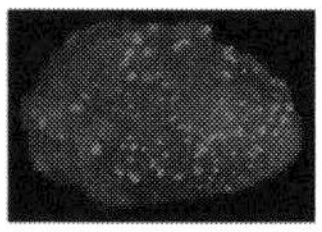

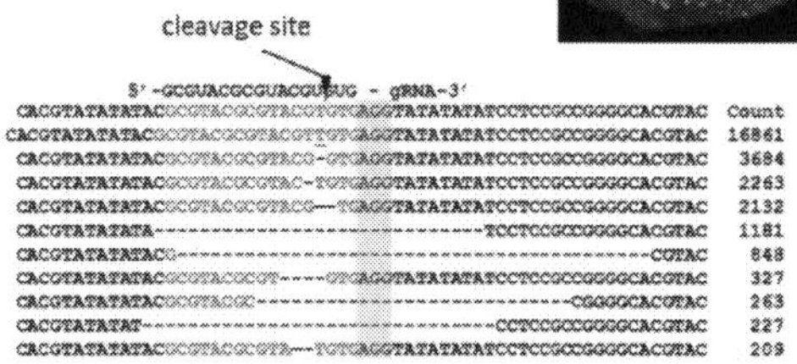

Mgn

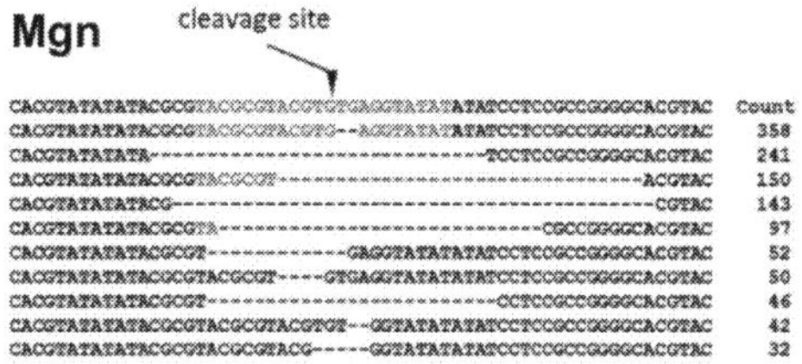

Cas9-gRNA yields high mutation frequency

DUPONT PIONEER

Josh young

Simultaneous targeting of 3 plant genes

- ***Liguleless* gene region (LIG)**
- **Male Fertility Genes; *Ms26* and *Ms45***

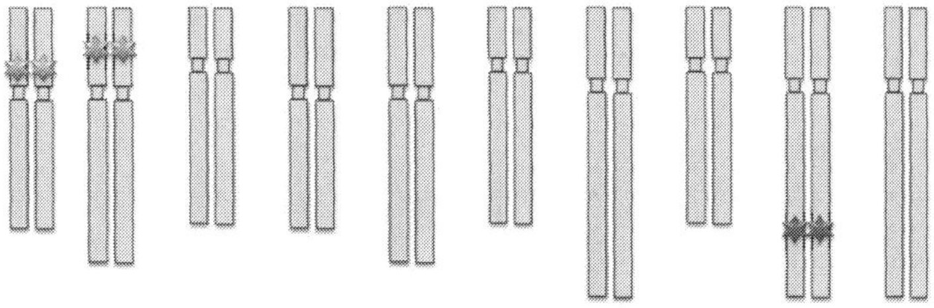

	Chrom1 *MS26*	Chrom2 LIG	Chrom9 *MS45*
LIG + MS26+ MS45	Mutant	Mutant	Mutant

DUPONT PIONEER

Sergei Svitashev

Precise sequence deletion

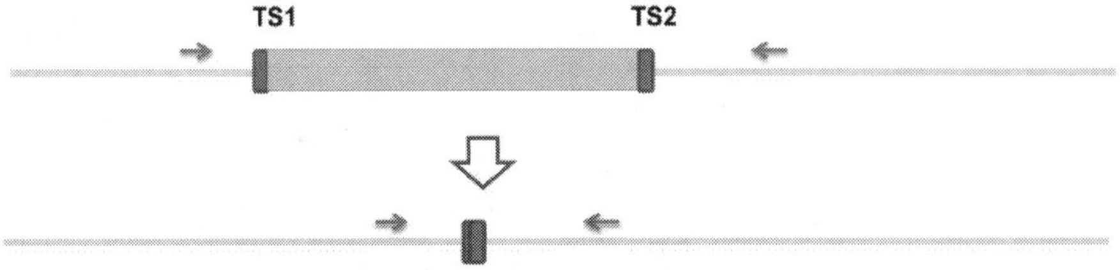

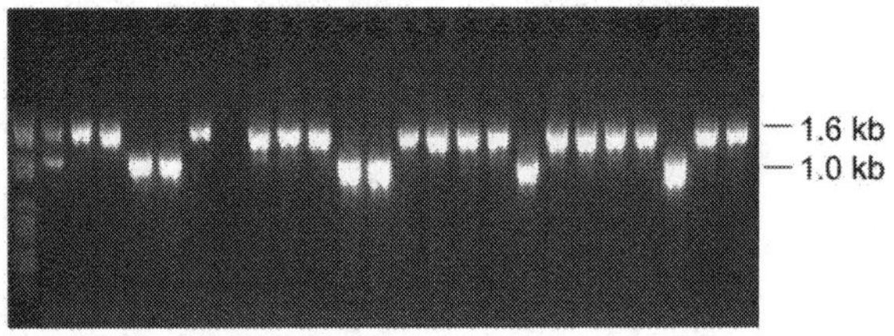

Range from 50bp to 20kb deletion

DUPONT
PIONEER

17

Replacing promoters

Editing native genes with other plant promoters

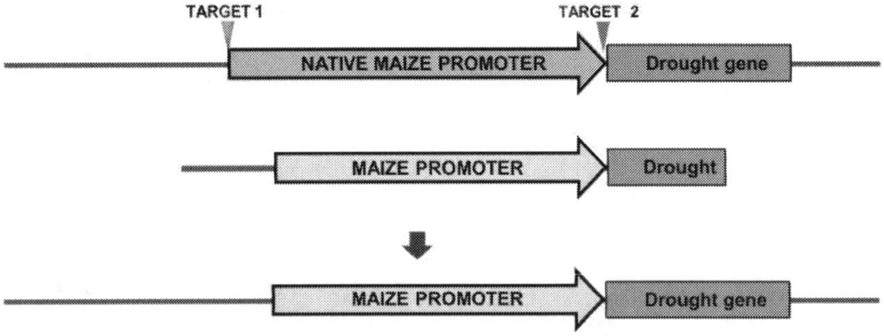

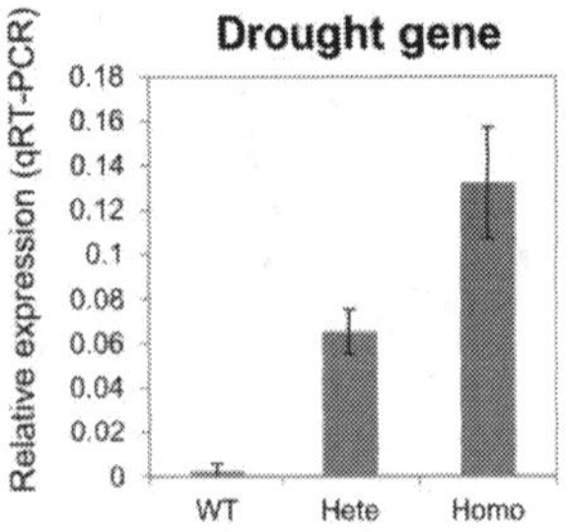

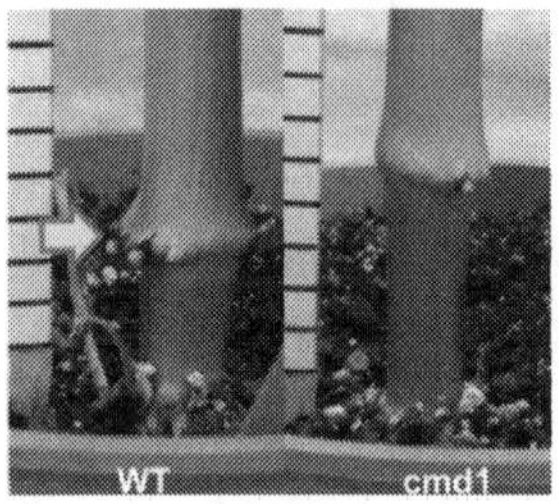

DUPONT
PIONEER

18

Complex Trait Loci (CTL)

- Pre-defined genomic region
- No endogenous gene disruption
- Introduce important genes into one locus
- Localize traits and simplify breeding

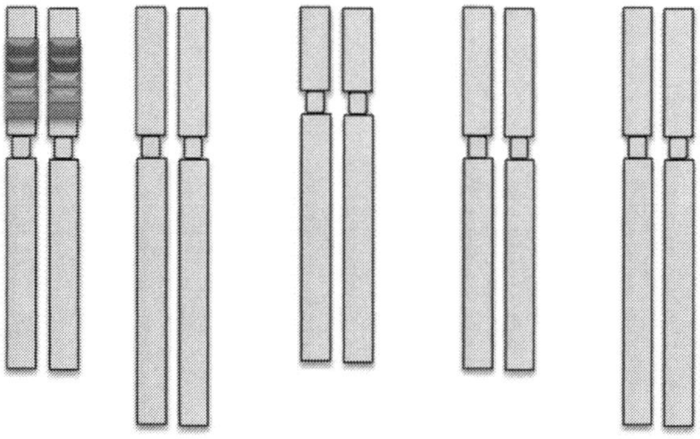

PIONEER.

19

Insertion frequency varies among sites

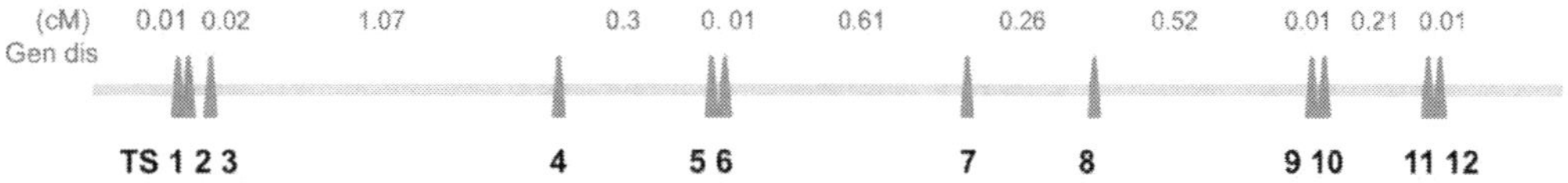

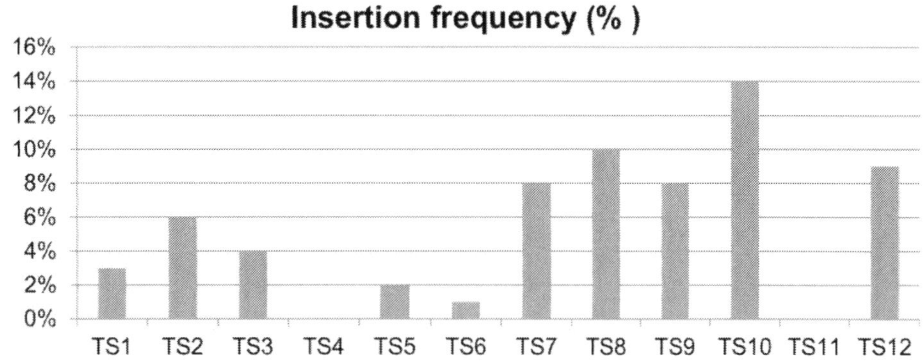

NHEJ (%):	97	94	90	91	84	91	94	95	81	70	5	88

PIONEER.

20

Directed plant genome modification at DuPont Pioneer

- ✓ Generated mutations at endogenous sequences in maize, rice, sorghum and soybean
  - Multiplex in maize (3), soybean (9)
- ✓ Removal of unwanted genomic sequences
- ✓ Modification of native gene and its regulatory elements
  - Generated new alleles
- ✓ Site specific gene insertion at selected genomic regions
- ✓ Complex Trait Loci enable efficient trait stacking

DUPONT
PIONEER

21

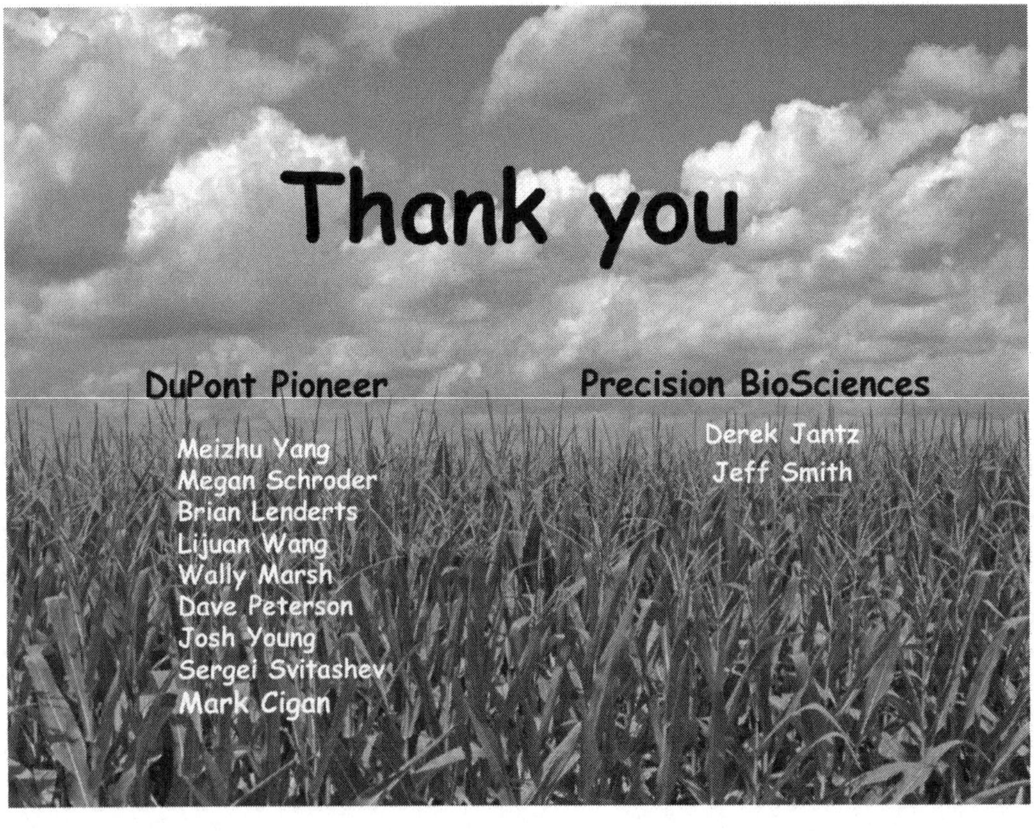

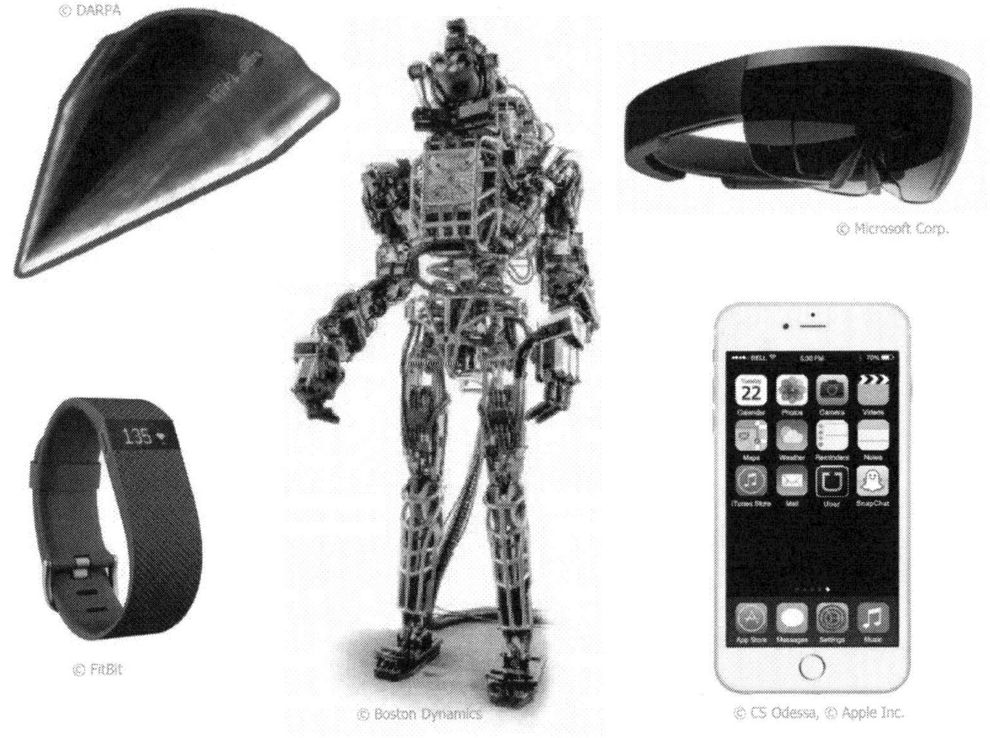

© DARPA © Microsoft Corp. © FitBit © Boston Dynamics © CS Odessa, © Apple Inc.

Approved for Public Release, Distribution Unlimited

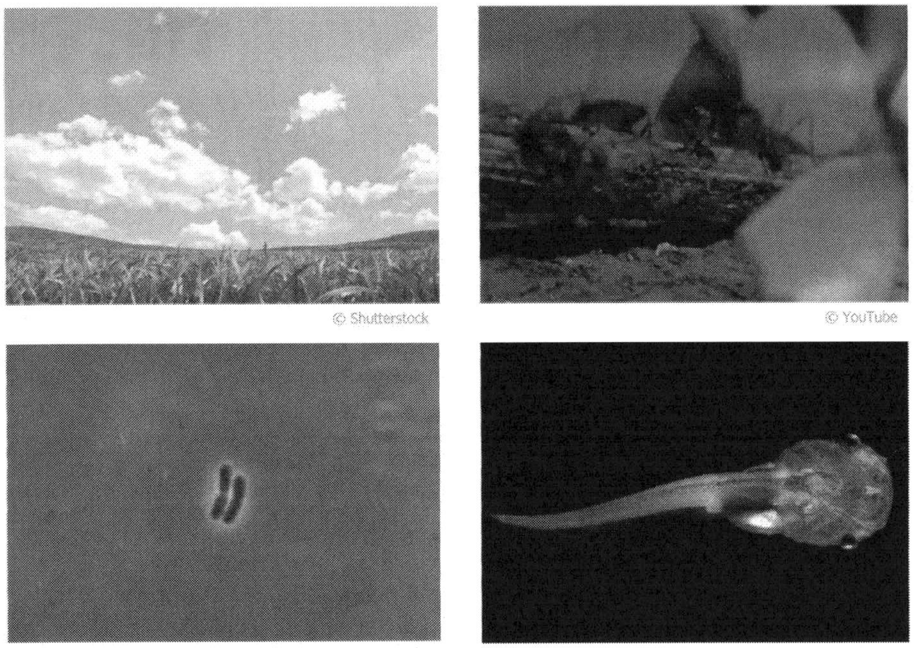

© Shutterstock © YouTube © YouTube © YouTube

Approved for Public Release, Distribution Unlimited

BTO

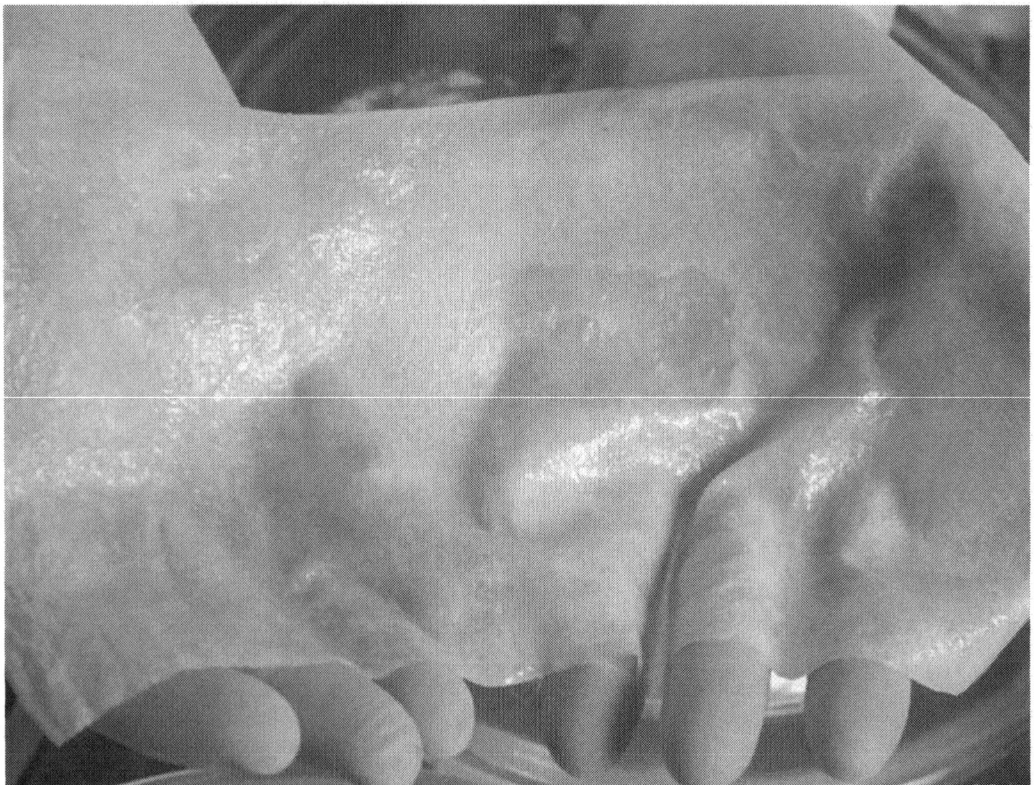
Approved for Public Release, Distribution Unlimited
© Corbis

Living Foundries: Differentiated U.S. technology base

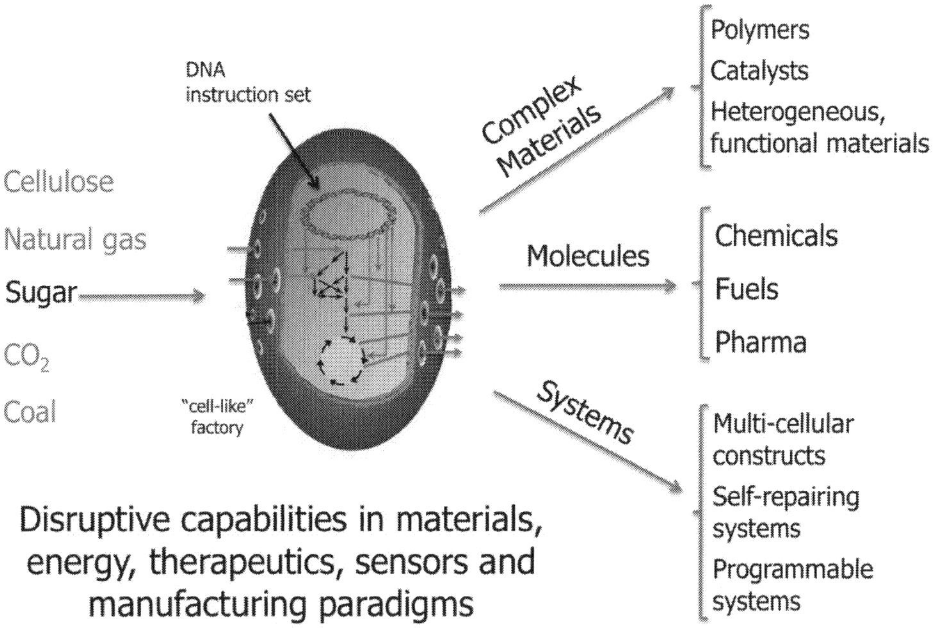

Disruptive capabilities in materials, energy, therapeutics, sensors and manufacturing paradigms

Approved for Public Release, Distribution Unlimited

DNA information -> Message -> Material Things

Apply an Engineering Approach

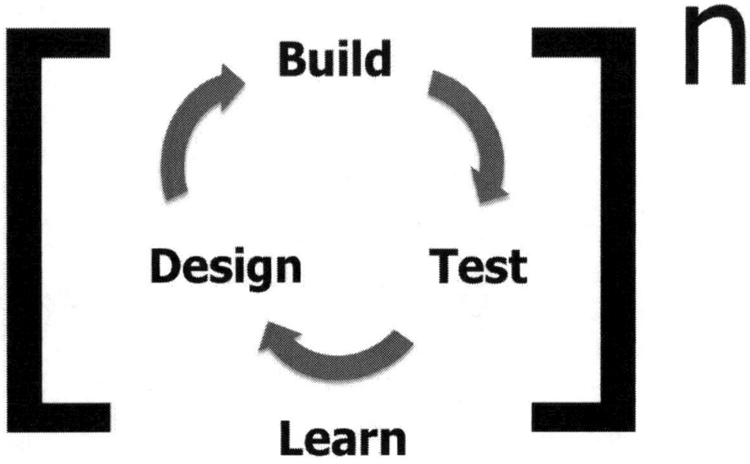

Approved for Public Release, Distribution Unlimited

New Tools to Forward Engineer

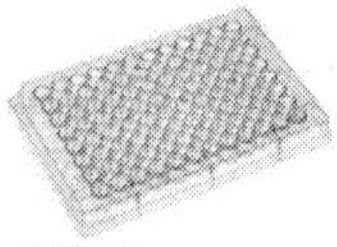

© EdgeBio

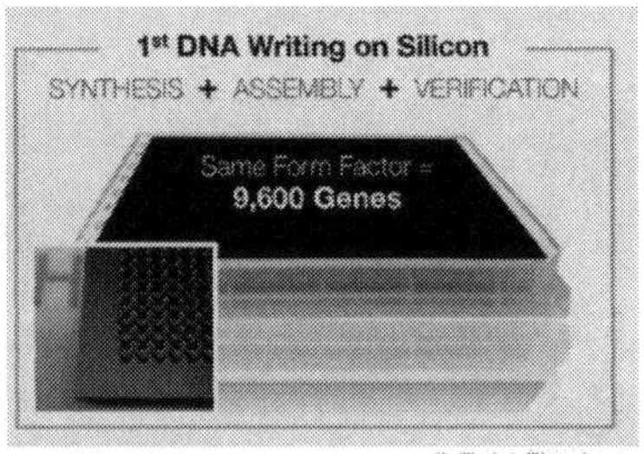

© Twist Bioscience

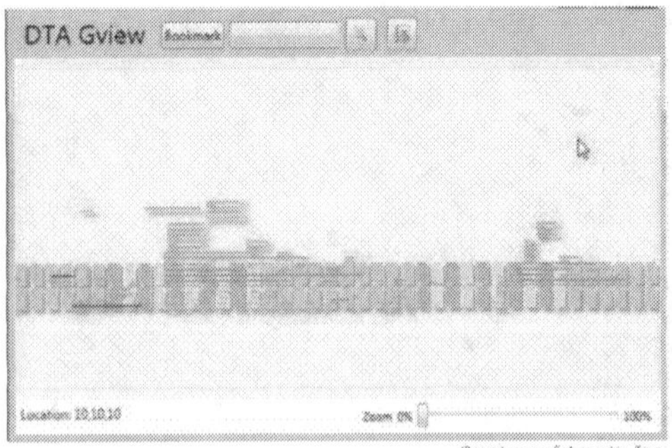

Courtesy of Amyris, Inc.

© Illumina

Approved for Public Release, Distribution Unlimited

Impact of Tools

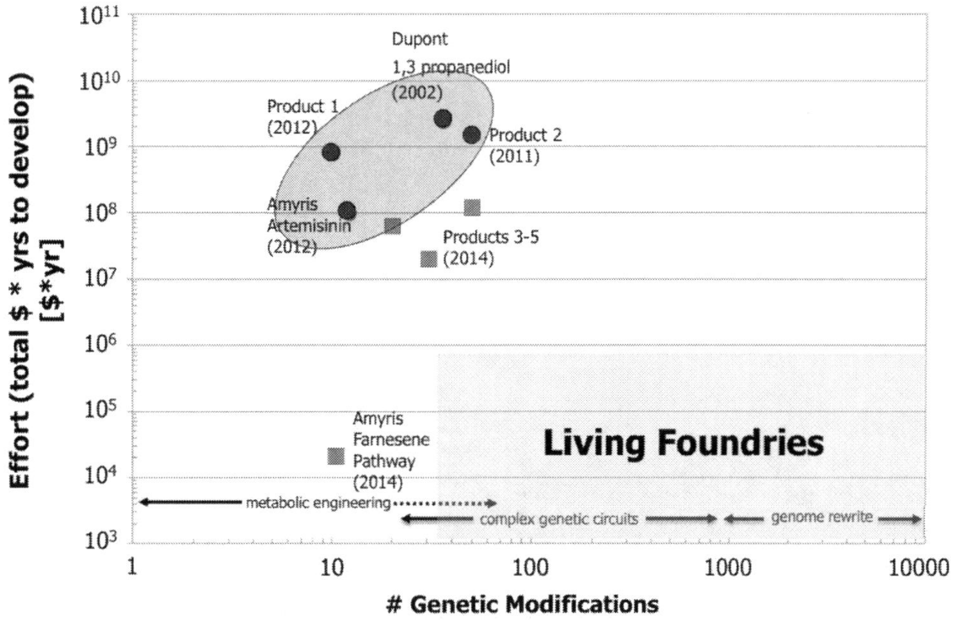

Approved for Public Release, Distribution Unlimited

Infrastructure: Scaled, rapid prototyping of genetic designs

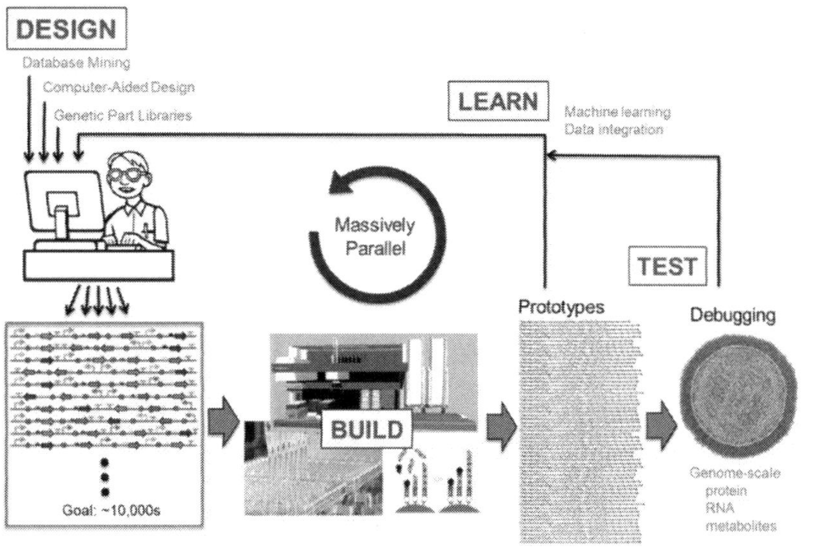

Courtesy of MIT-Broad

1000 new chemical building blocks

Approved for Public Release, Distribution Unlimited

10

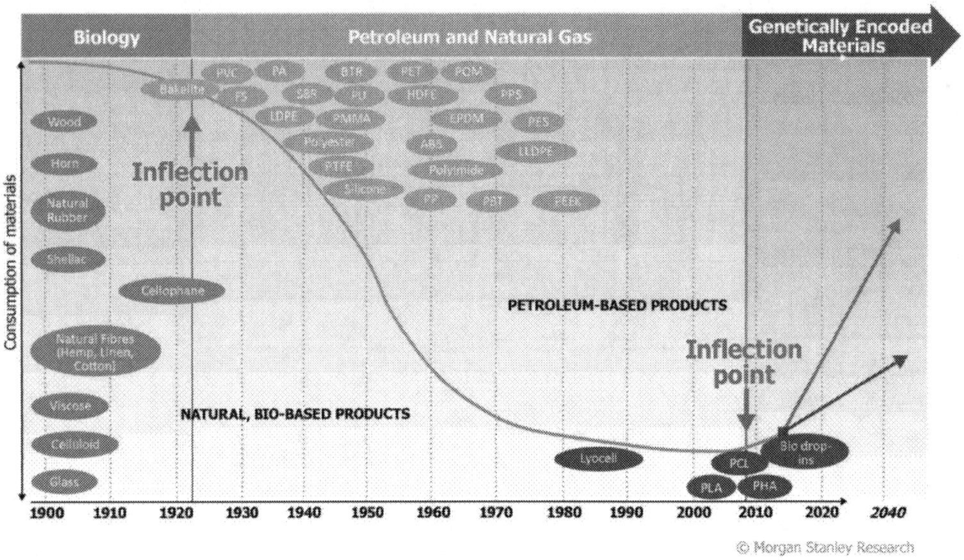

DARPA

www.darpa.mil

BROADBAND DNA SYNTHESIS

June 2015

@TwistBioscience

AGENDA

- Twist Bioscience overview
- Broadband gene synthesis
- Assembly design optimization
- Library synthesis
- Alpha Access Program

AGENDA

TWIST BIOSCIENCE

- Twist Bioscience overview
- Broadband gene synthesis
- Assembly design optimization
- Library synthesis
- Alpha Access Program

JULY 9, 2015 3

TWIST BIOSCIENCE

TWIST BIOSCIENCE

Synthetic DNA Company

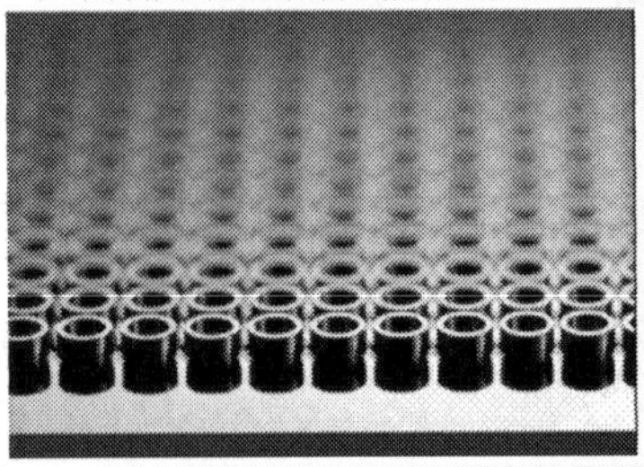

Transformative Platform
for DNA Production

DNA Writing on Silicon

Delivering Game-Changing
Throughput, Speed, Cost and Quality

Investors: $72.1M Funding

ARCH Venture Partners PALADIN CAPITAL GROUP illumina

$5.1M Contract

DARPA

JULY 9, 2015 4

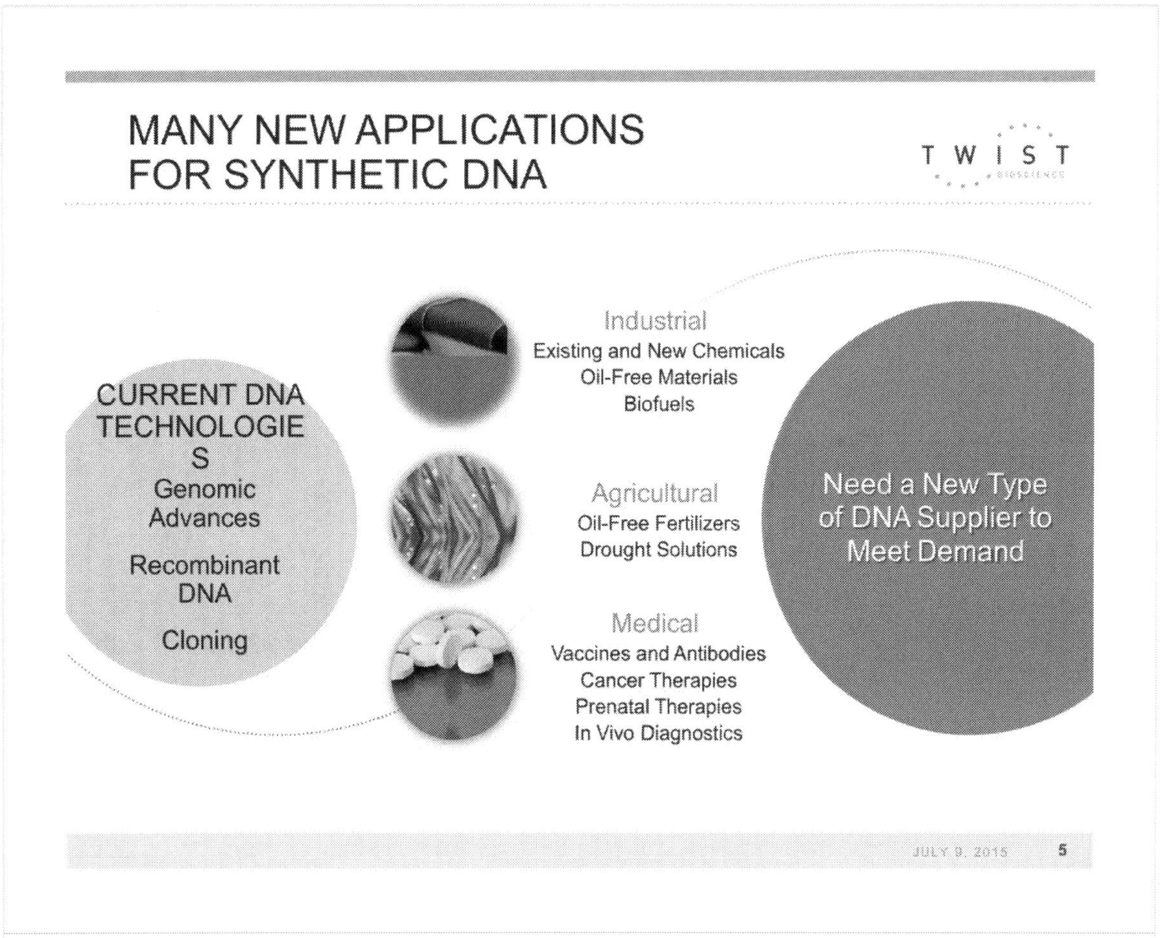
MANY NEW APPLICATIONS
FOR SYNTHETIC DNA
TWIST BIOSCIENCE
CURRENT DNA TECHNOLOGIE S
Genomic Advances
Recombinant DNA
Cloning
Industrial
Existing and New Chemicals
Oil-Free Materials
Biofuels
Agricultural
Oil-Free Fertilizers
Drought Solutions
Medical
Vaccines and Antibodies
Cancer Therapies
Prenatal Therapies
In Vivo Diagnostics
Need a New Type of DNA Supplier to Meet Demand
JULY 9, 2015
5

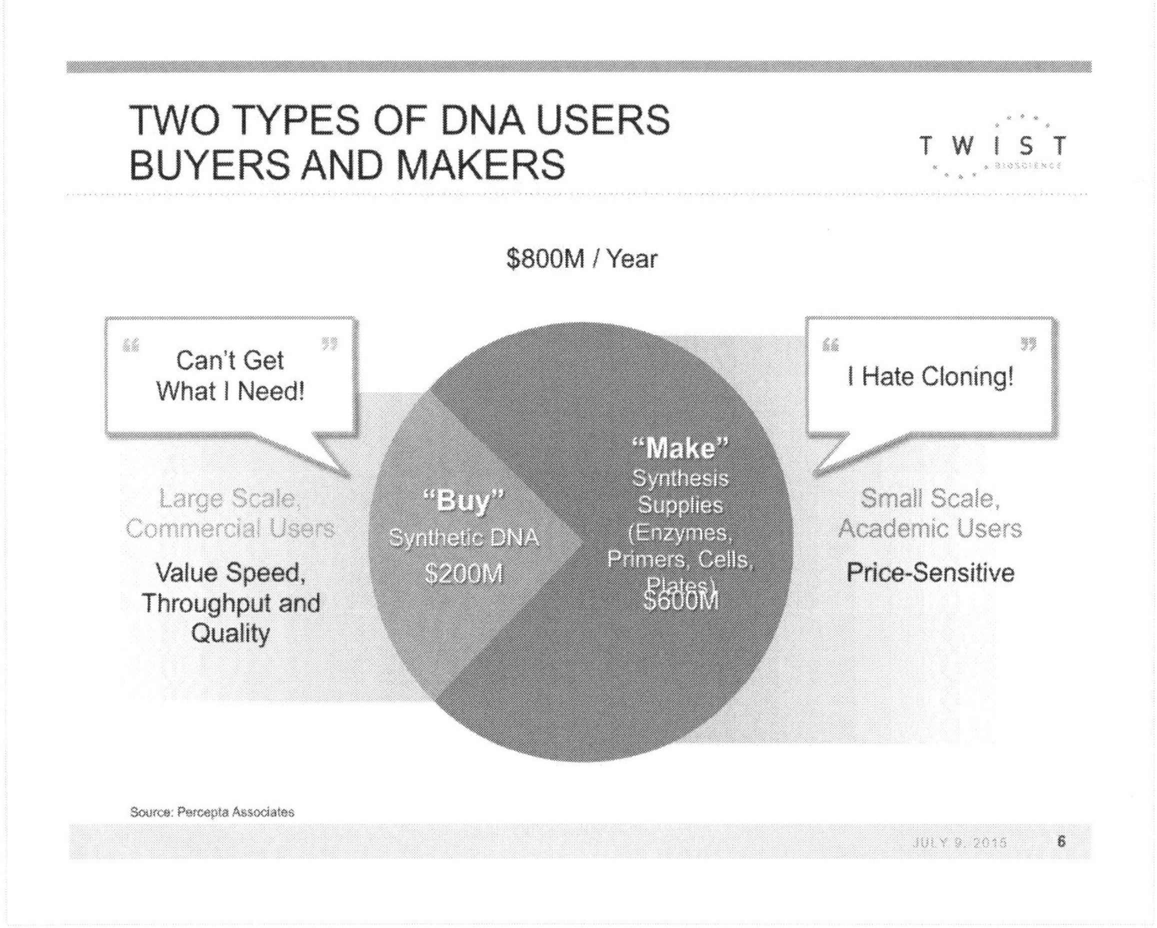
TWO TYPES OF DNA USERS
BUYERS AND MAKERS
TWIST BIOSCIENCE
$800M / Year
"Can't Get What I Need!"
Large Scale, Commercial Users
Value Speed, Throughput and Quality
"Buy"
Synthetic DNA
$200M
"Make"
Synthesis Supplies (Enzymes, Primers, Cells, Plates)
$600M
"I Hate Cloning!"
Small Scale, Academic Users
Price-Sensitive
Source: Percepta Associates
JULY 9, 2015
6

DNA APPLICATION DEVELOPMENT PROCESS

TWIST BIOSCIENCE

Design Gene Sequences

- 1 to 10,000+ Variants Possible per Cycle

AGCTGCTCGAATACGATAG

Build

Manufacture Synthetic Gene Constructs

- Genes (1k Basepairs)
- Parts (3 - 10 Genes)
- Pathways (10 - 100 Genes)
- Chassis (100 - 1000 Genes)

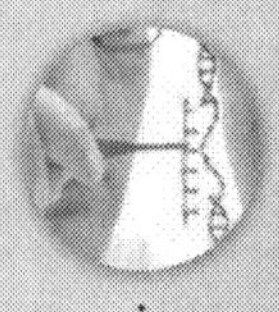

Cycles 5-10x

Test

Select/Screen for Desired Function

- E.g., Bacteria, Yeast, Algae Bio-Factories, Plants, Mammalian Cells

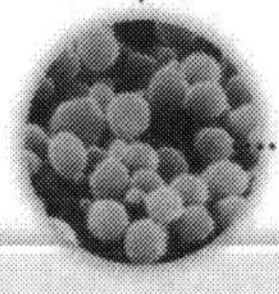

Process Bottleneck

- Too Long
- Can't Scale Up or Down
- Expensive

Want 1 Gene in 2 Weeks → $0.2-$1/bp or Non-Clonal + Work

Want 10,000 Genes in 2 Weeks → Use 3 Vendors and/or Wait 3 Months

AGENDA

TWIST BIOSCIENCE

- Twist Bioscience overview
- **Broadband gene synthesis**
- Assembly design optimization
- Library synthesis
- Alpha Access Program

CLASSICAL GENE SYNTHESIS

Stemmer (1995) *Gene* 164:49

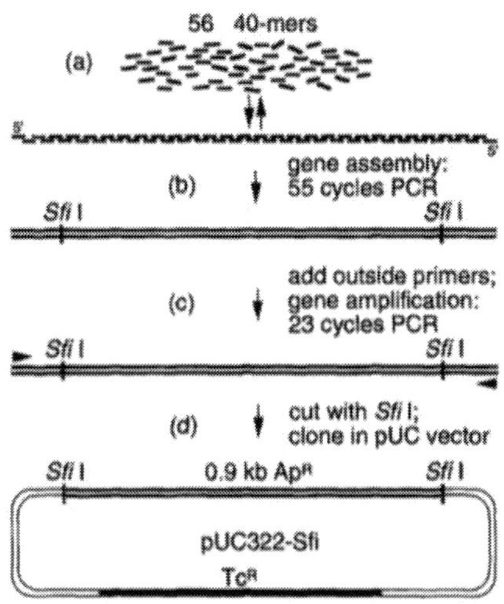

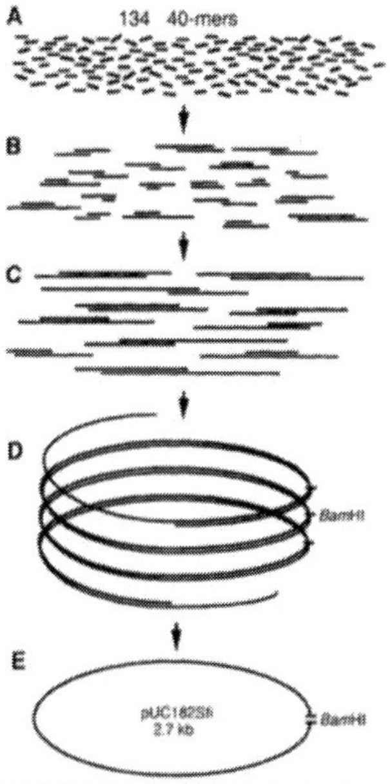

OLIGONUCLEOTIDE SYNTHESIS

TWIST BIOSCIENCE

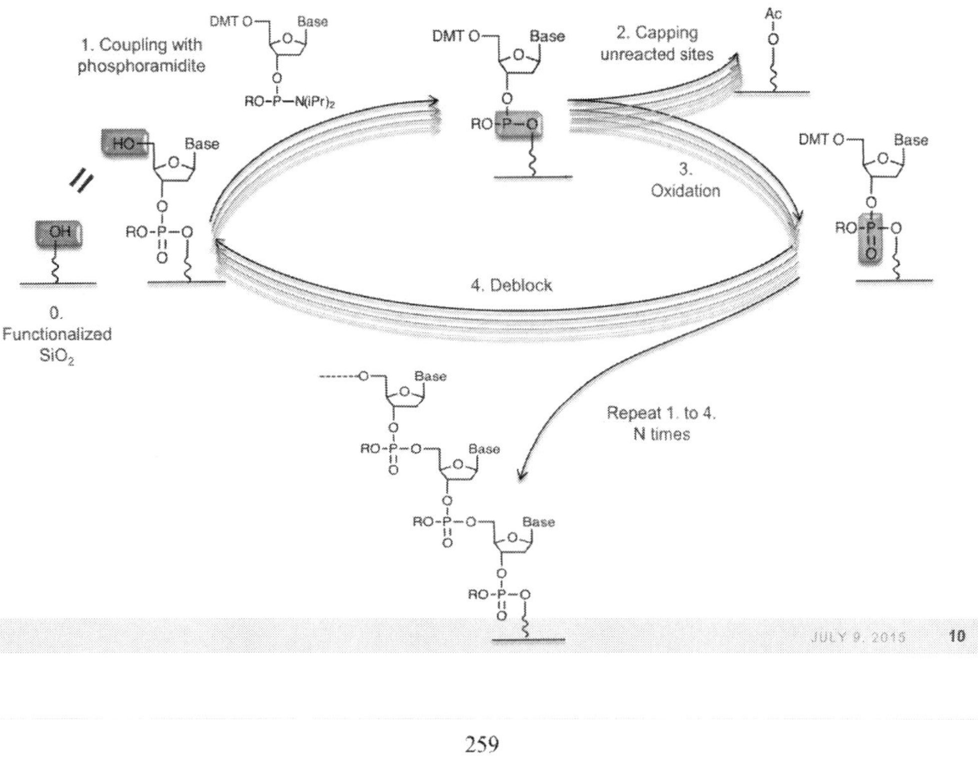

CLASSICAL OLIGO SYNTHESIS

MERMADE 96E

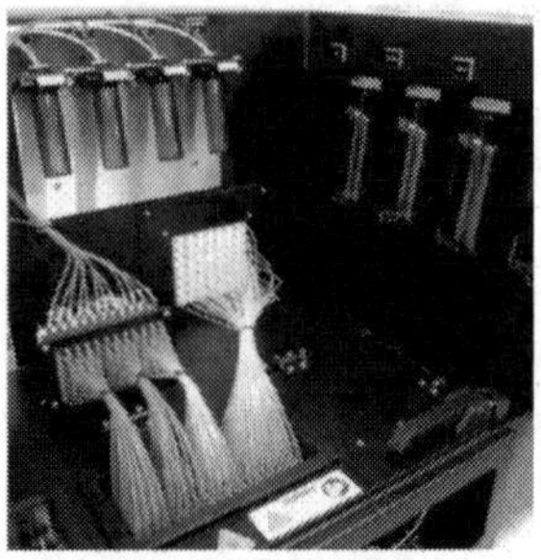

Dr. Oligo 192

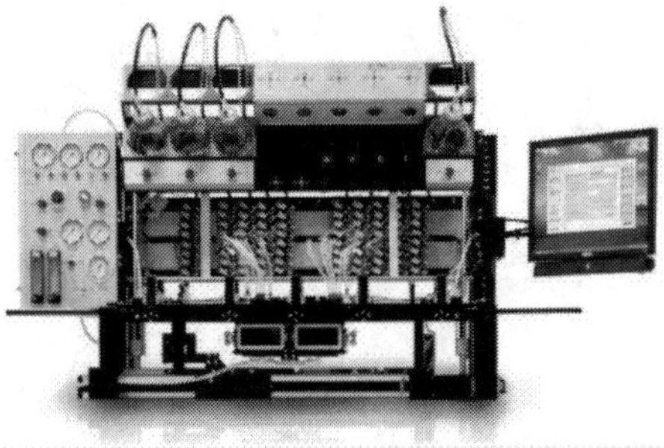

SYNTHESIS AND ASSEMBLY COMPARISON

TWIST BIOSCIENCE

	Classical
Amount of DNA	Waste (Nano-mol)
DNA processing	Pooling stoichiometrically
Gene(s) per 96-well	~1

MICROARRAY FABRICATION

TWIST BIOSCIENCE

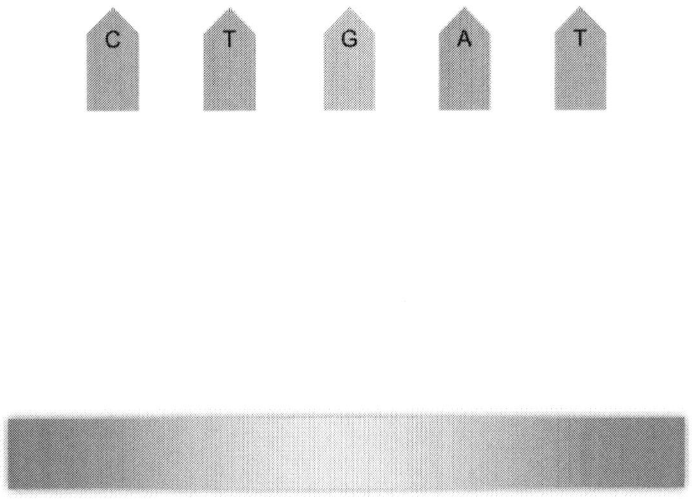

MICROARRAY FABRICATION

TWIST BIOSCIENCE

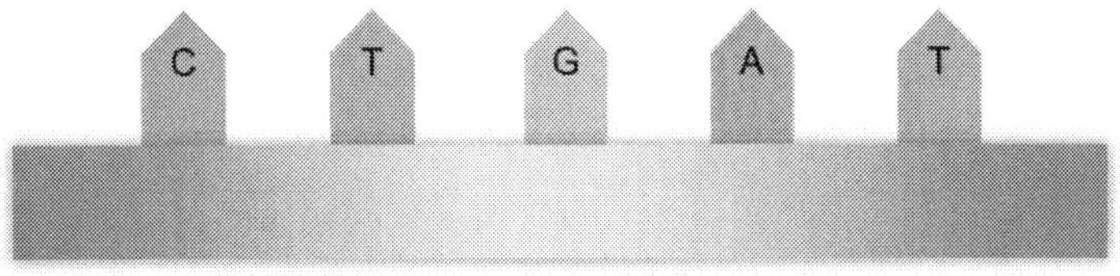

MICROARRAY FABRICATION

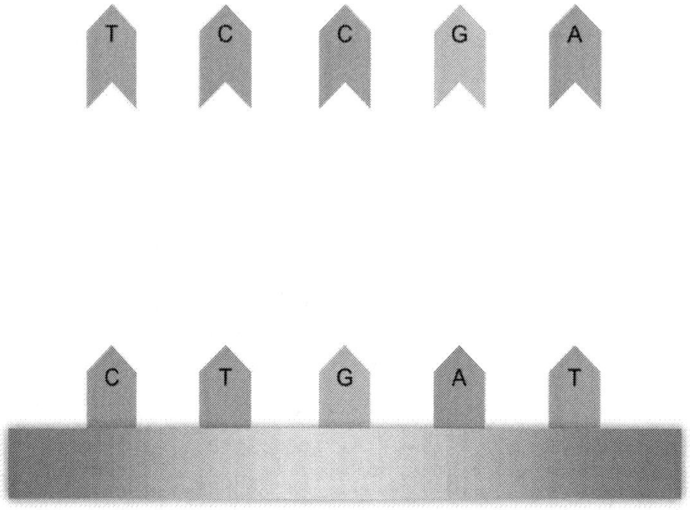

MICROARRAY FABRICATION

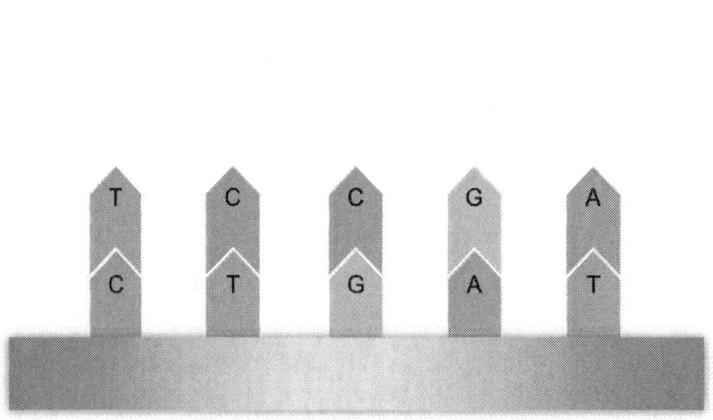

MICROARRAY FABRICATION

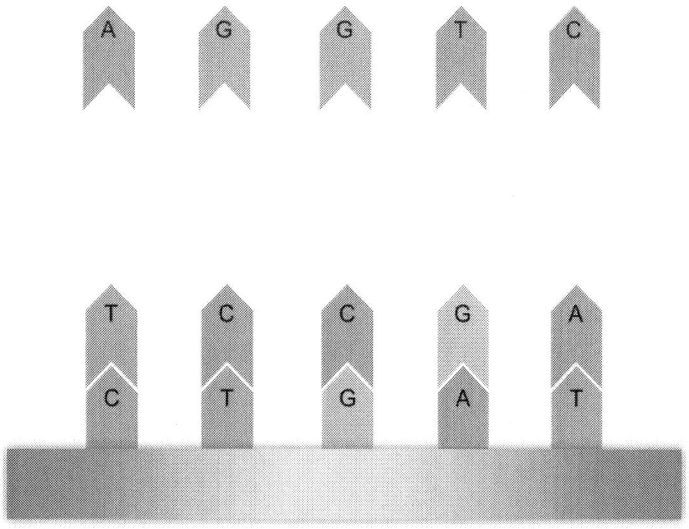

JULY 9, 2015 17

MICROARRAY FABRICATION

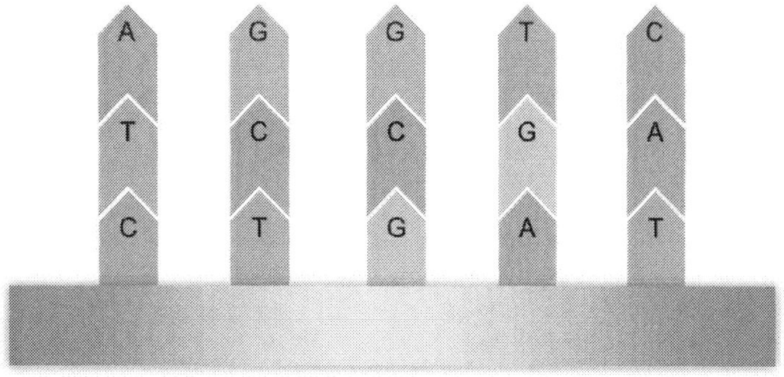

JULY 9, 2015 18

MICROARRAY FABRICATION

TWIST BIOSCIENCE

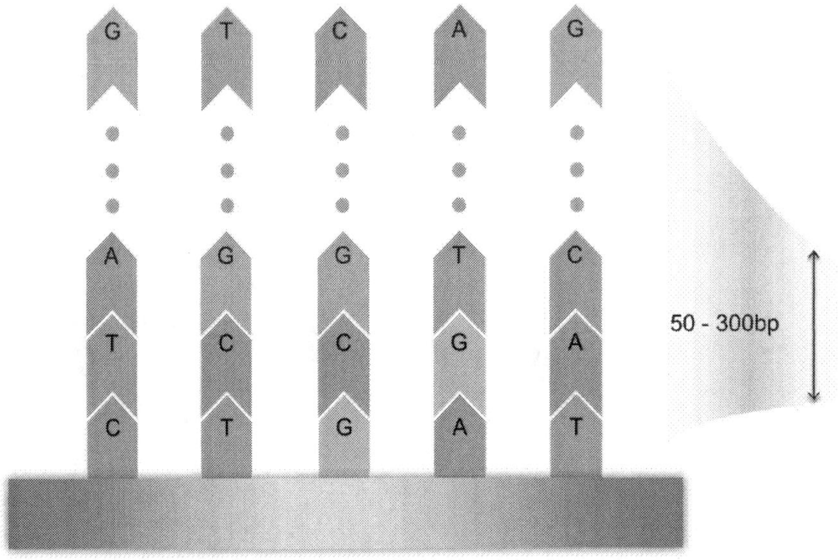

OLIGO LIBRARY

TWIST BIOSCIENCE

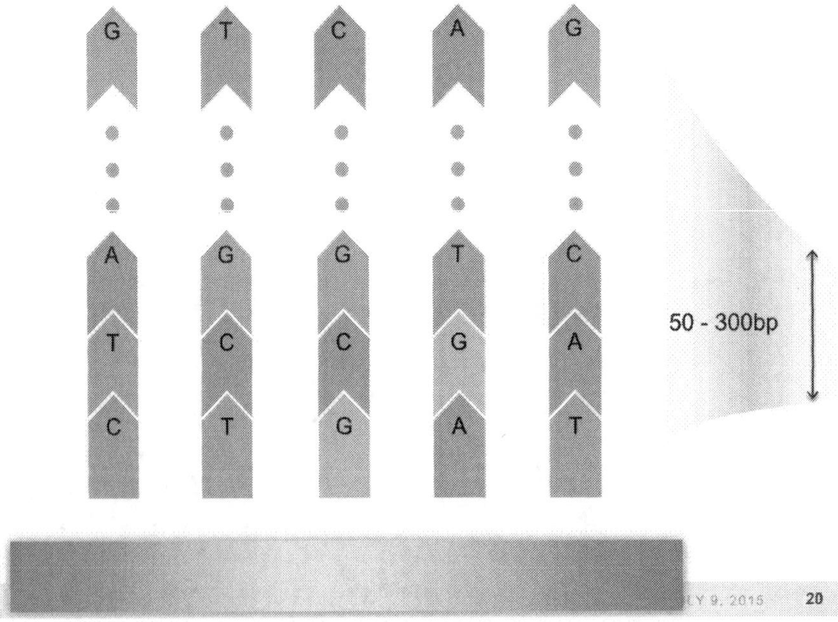

MICROARRAY-BASED OLIGONUCLEOTIDE ASSEMBLY PROCESS

TWIST BIOSCIENCE

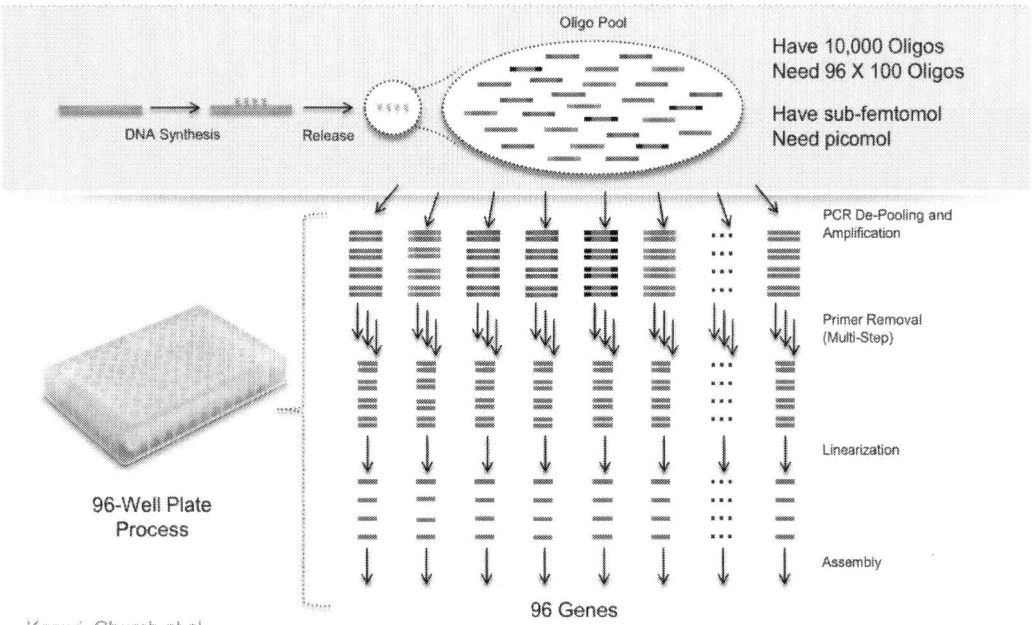

Kosuri, Church et al.
Nature Biotechnology (2010) 1295–1299

SYNTHESIS AND ASSEMBLY COMPARISON

TWIST BIOSCIENCE

	Classical	Microarray
Amount of DNA	Waste (Nano-mol)	Amplification (< Femto-mol)
DNA processing	Pooling stoichiometrically	De-pooling
Gene(s) per 96-well	~1	96

LIMITATIONS OF 20-YEAR OLD TECHNOLOGY

TWIST BIOSCIENCE

96 Well Plastic Plate

96 Oligos = 1 Gene

Chemistry Works
but Does Not Scale to Meet
Market Need for Throughput
and Cost Efficiency

OUR APPROACH

TWIST BIOSCIENCE

NO AMPLICATION	3D Silicon surface for mass of DNA "just right"
NO DEPOOLING	Subpool of ~100 oligos, pre-mixed at perfect ratio
NO PLASTIC	All steps in high-density Silicon nano-wells Miniaturize volume and increase throughput

Vertical integration Silicon → Oligos → Genes → (NGS)

TWIST TRANSFORMATIVE PLATFORM

TWIST BIOSCIENCE

96 Well Plastic Plate

96 Oligos = 1 Gene

1st DNA Writing on Silicon

SYNTHESIS **+** ASSEMBLY **+** VERIFICATION

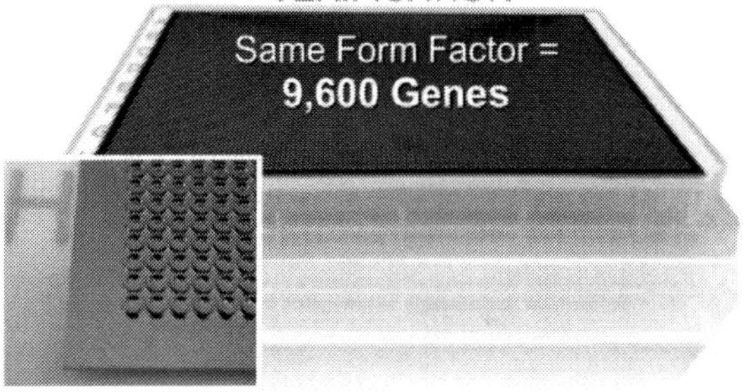

NO DEPOOLING	Vertically integrated end-to-end process
NO AMPLIFICATION	3D Silicon for higher surface / volume ratio vs. 2D
NO PLASTIC	Silicon nanowells to miniaturize volume and increase throughput

JULY 9, 2015 25

TWIST'S 3D SILICON TECHNOLOGY

10,000 GENE SYNTHESIS PLATFORM

TWIST BIOSCIENCE

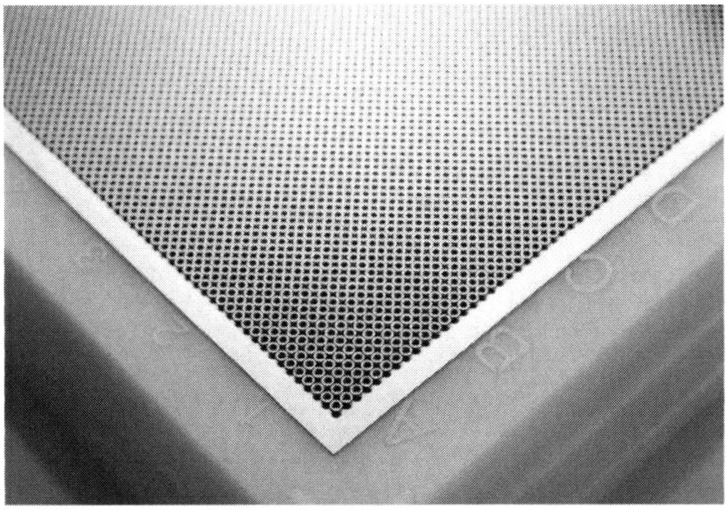

- High throughput
- High quality
- Ultra long
- High mass
- Fast
- Affordable

JULY 9, 2015 26

SYNTHESIS AND ASSEMBLY COMPARISON

TWIST BIOSCIENCE

	Classical	Microarray	TWIST BIOSCIENCE
Amount of DNA	Waste (Nano-mol)	Amplification (< Femto-mol)	No Amp, no waste (Pico-mol)
DNA processing	Pooling stoichiometrically	De-pooling	No pooling No de-pooling
Gene(s) per 96-well	~1	96	10,000

Delivering Game-Changing
Throughput, Speed, Cost and Quality

TWIST'S 3D SILICON TECHNOLOGY

ALPHA PROCESS : 256 GENES

TWIST BIOSCIENCE

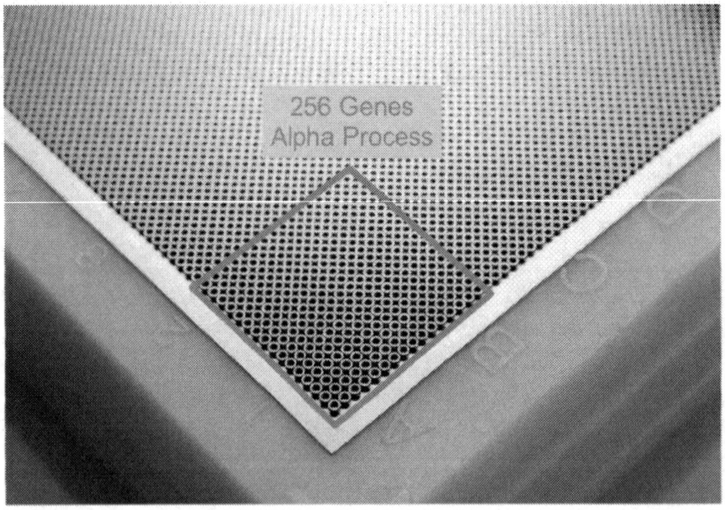

- High throughput
- High quality
- Ultra long
- High mass
- Fast
- Affordable

EXAMPLE OF SYNTHESIS

- 240 genes ranging from 838bp to 1470bp
- 16 controls
- 29,040 unique oligos

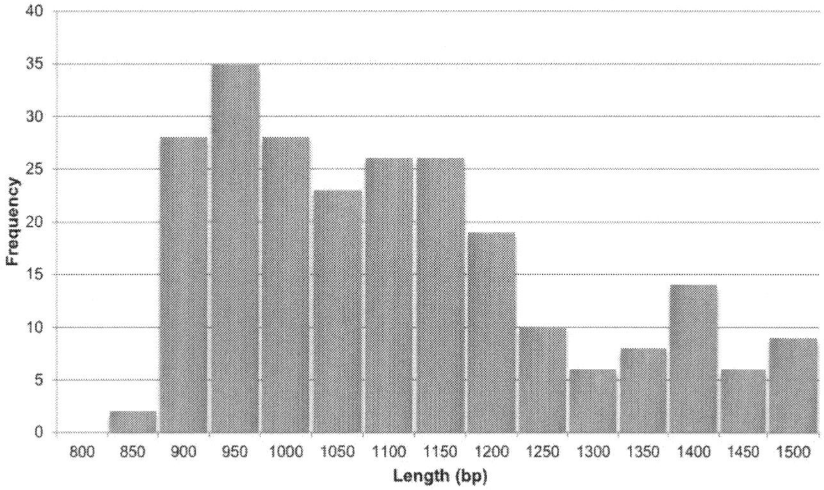

OLIGO REPRESENTATION

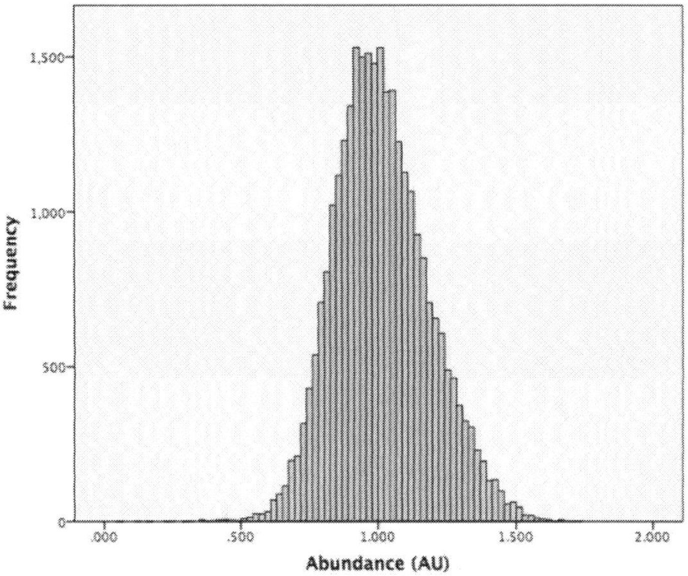

29,040 unique oligos from 240 clusters

- 99% are 2x from the mean
- 100% oligos present

OLIGO REPRESENTATION

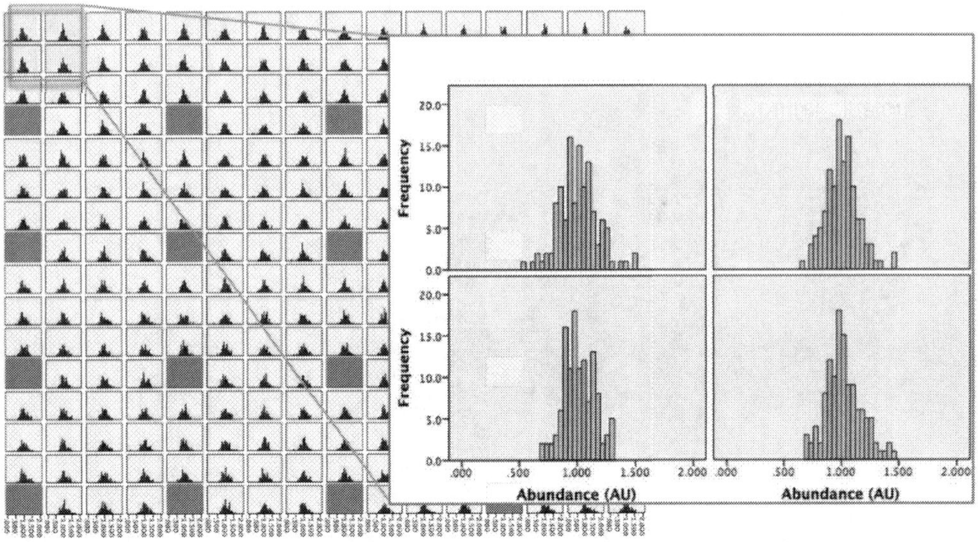

OLIGO ERROR RATE

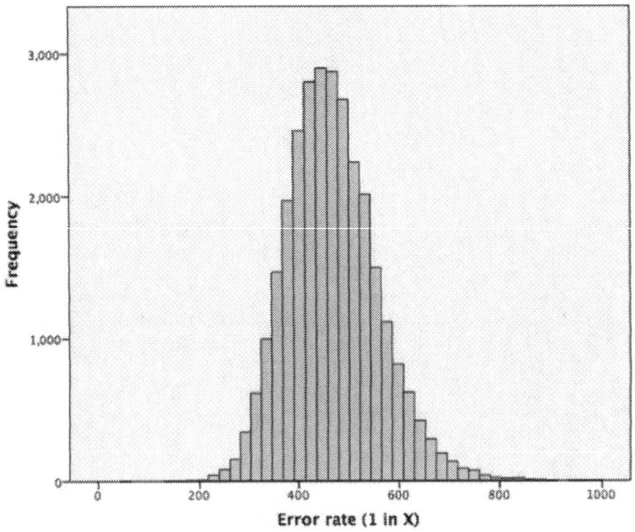

29,040 unique oligos from 240 clusters

- 1 in 500bp error rate
- Off-shelf chemistry
- Not optimized

OLIGO ERROR RATE

T W I S T BIOSCIENCE

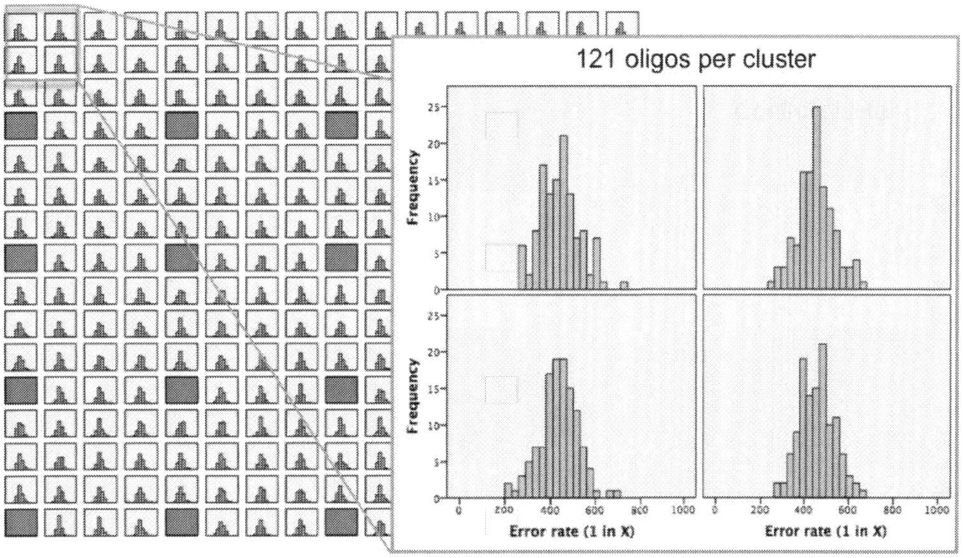

FRAGMENT SIZE ANALYSIS

T W I S T BIOSCIENCE

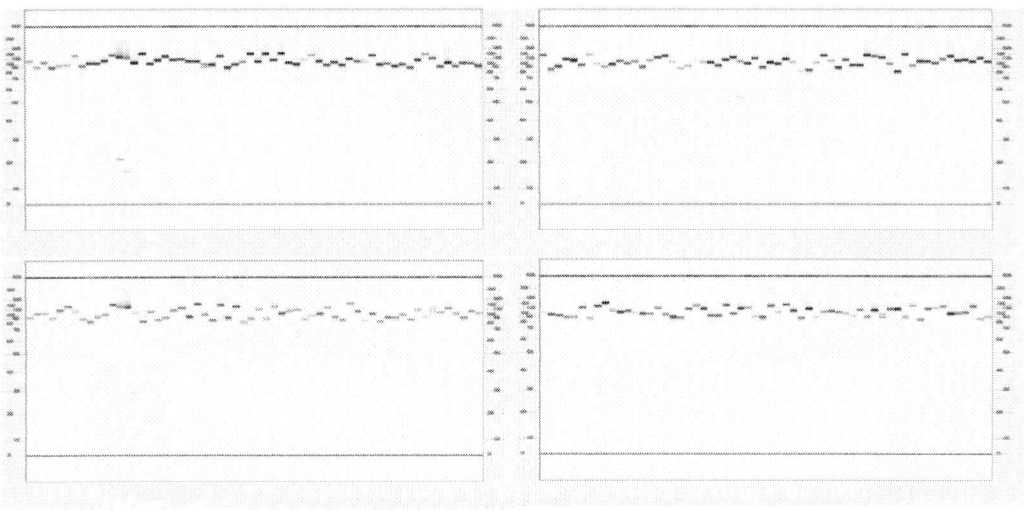

240 successfully assembled genes out of 240 total

NGS VERIFICATION

TWIST BIOSCIENCE

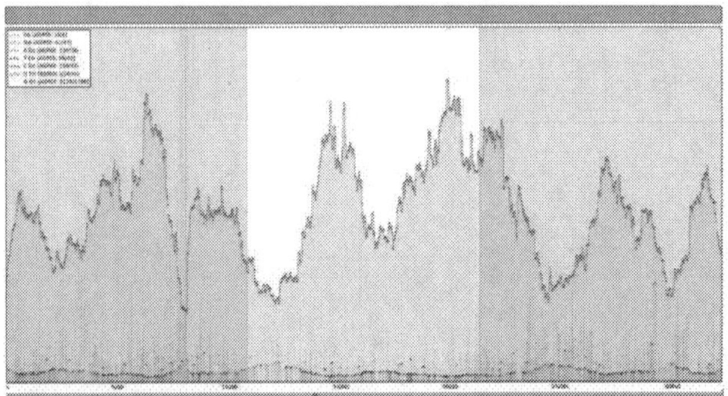

Perfect Clone

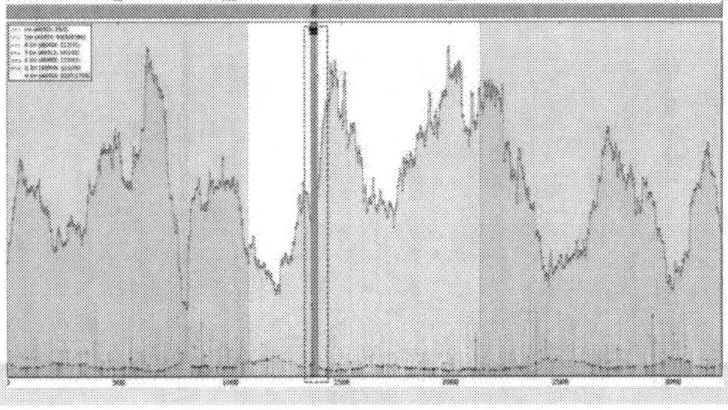

Clone with Multi-base deletion

NGS VERIFICATION

TWIST BIOSCIENCE

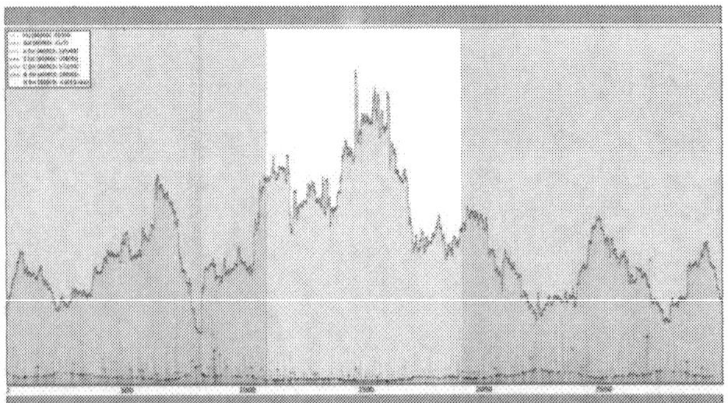

Perfect Clone

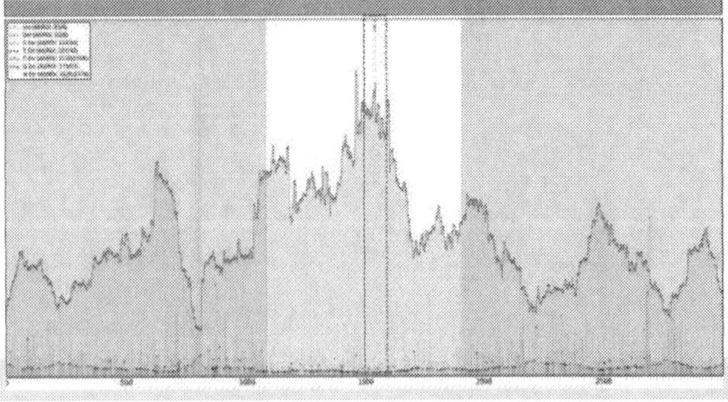

Clone with a Mutation

NGS DATA SUMMARY

TWIST BIOSCIENCE

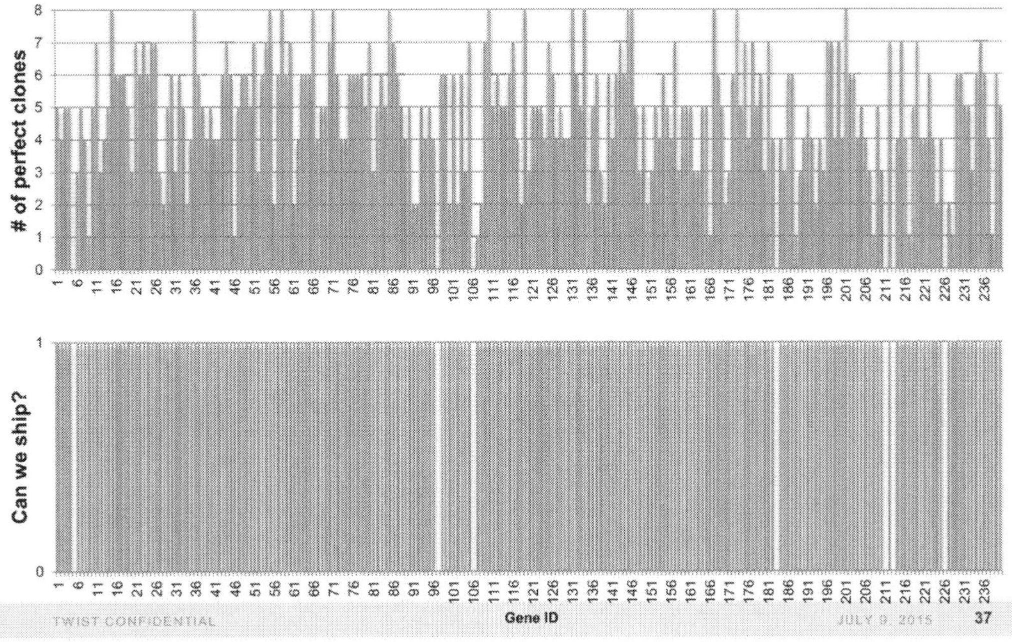

TWIST CONFIDENTIAL JULY 9, 2015 37

FREQUENCY OF PERFECT CLONE

TWIST BIOSCIENCE

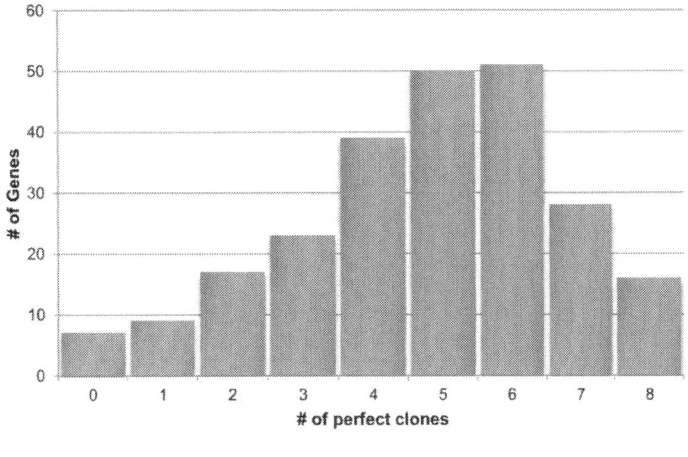

TWIST CONFIDENTIAL JULY 9, 2015 38

GENE ASSEMBLY SUMMARY

TWIST BIOSCIENCE

Step	Success Rate	Stepwise Success	Cumulative Success
Oligo Synthesis	240 / 240	100%	100%
Gene Assembly	240 / 240	100%	100%
Cloning	238 / 240	99%	99%
NGS Verification	233 / 238	98%	97%

233 genes (260,646bp) delivered out of 240 total requested

AGENDA

TWIST BIOSCIENCE

- Twist Bioscience overview
- Broadband gene synthesis
- **Assembly design optimization**
- Library synthesis
- Alpha Access Program

GENE ASSEMBLY PROCESS

TWIST BIOSCIENCE

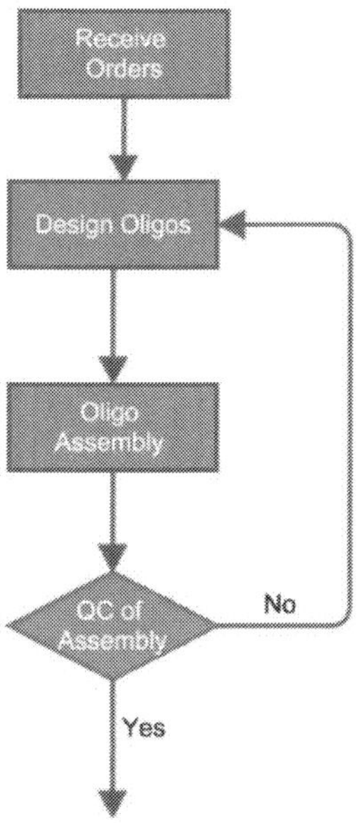

GENE ASSEMBLY PROCESS WITH MACHINE LEARNING

TWIST BIOSCIENCE

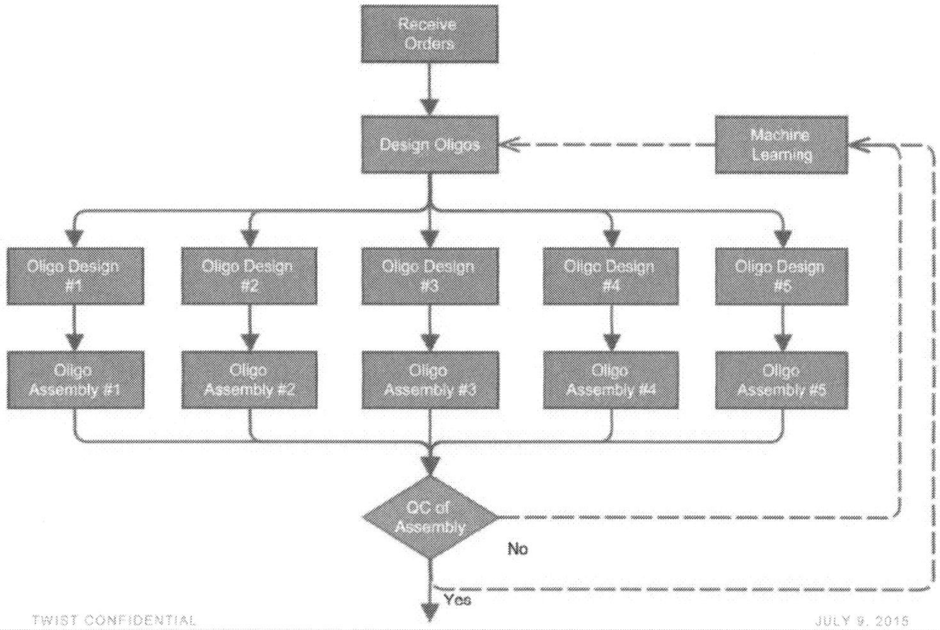

INFORMATICS PIPELINE

TWIST BIOSCIENCE

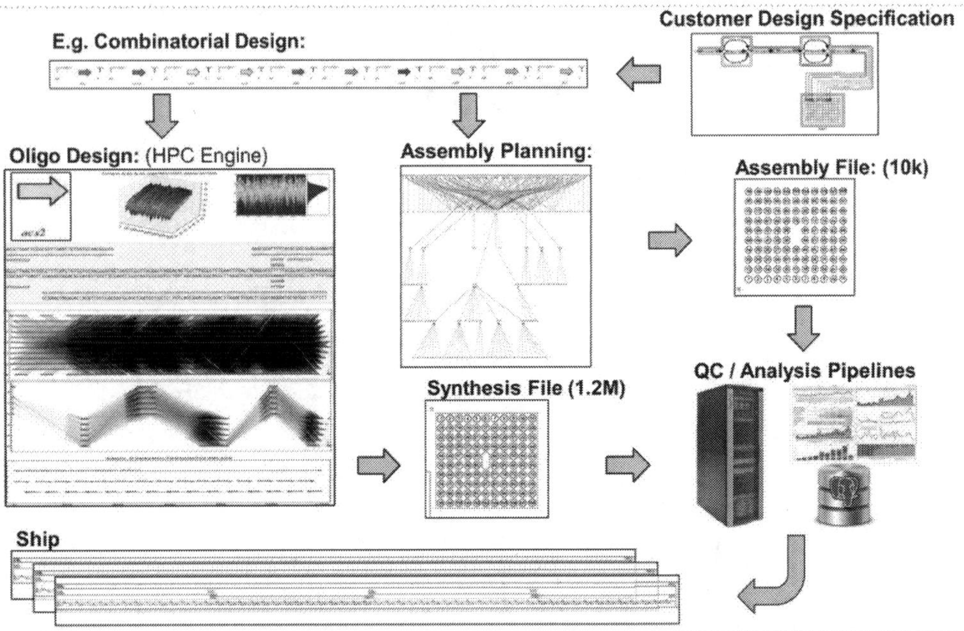

GENE ASSEMBLY PROCESS

TWIST BIOSCIENCE

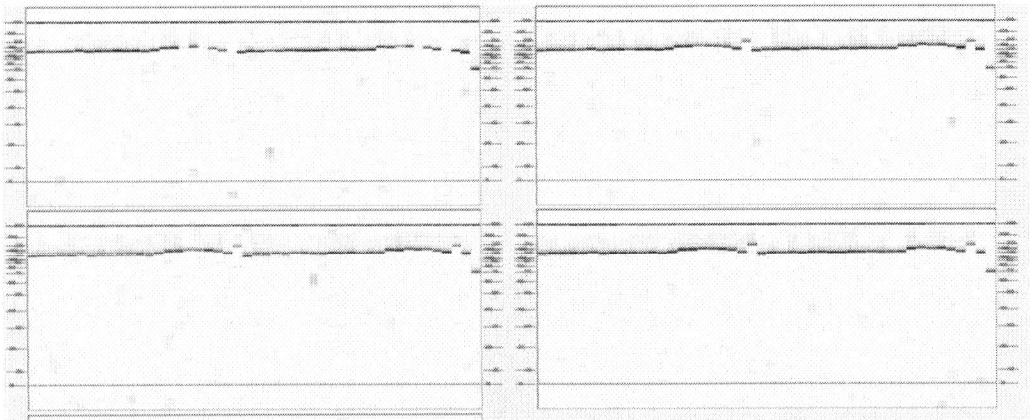

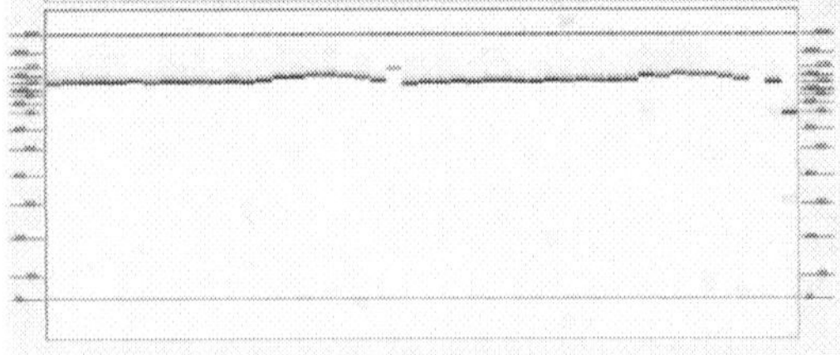

- 48 difficult with 30~35% GC content.
- Each designed and assembled in 5 different ways (oligo design and assembly conditions).

GENE ASSEMBLY PROCESS

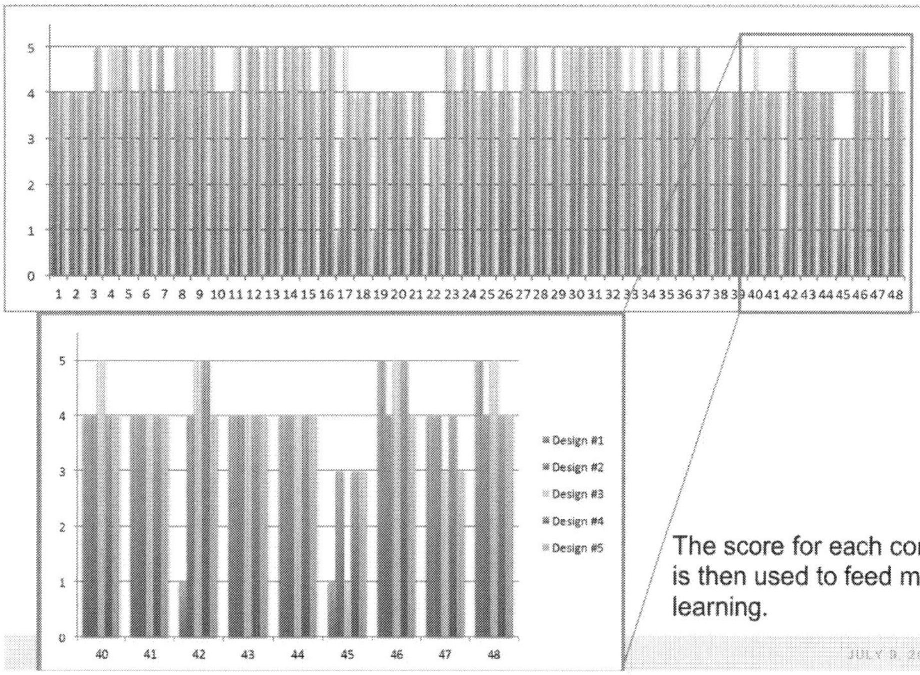

The score for each condition is then used to feed machine learning.

MACHINE LEARNING PIPELINE

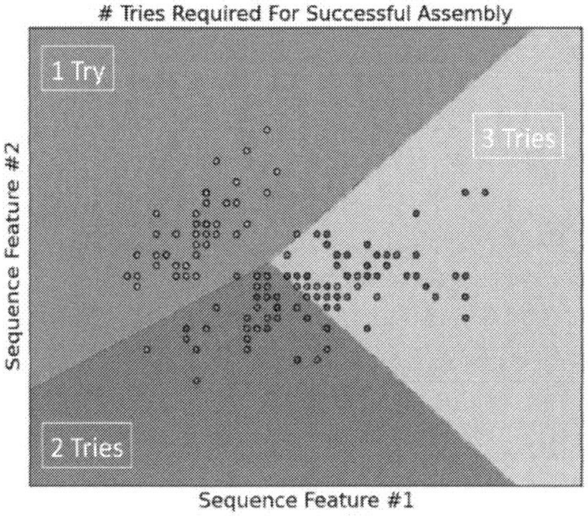

- Predict assembly success based on primary sequence, given a design strategy and backend process
- Classification improves with more sequence features, and over time with more data
- Better design choices, more accurate pricing and scheduling

AGENDA

- Twist Bioscience overview
- Broadband gene synthesis
- Assembly design optimization
- Library synthesis
- Alpha Access Program

SITE SATURATION MUTAGENESIS

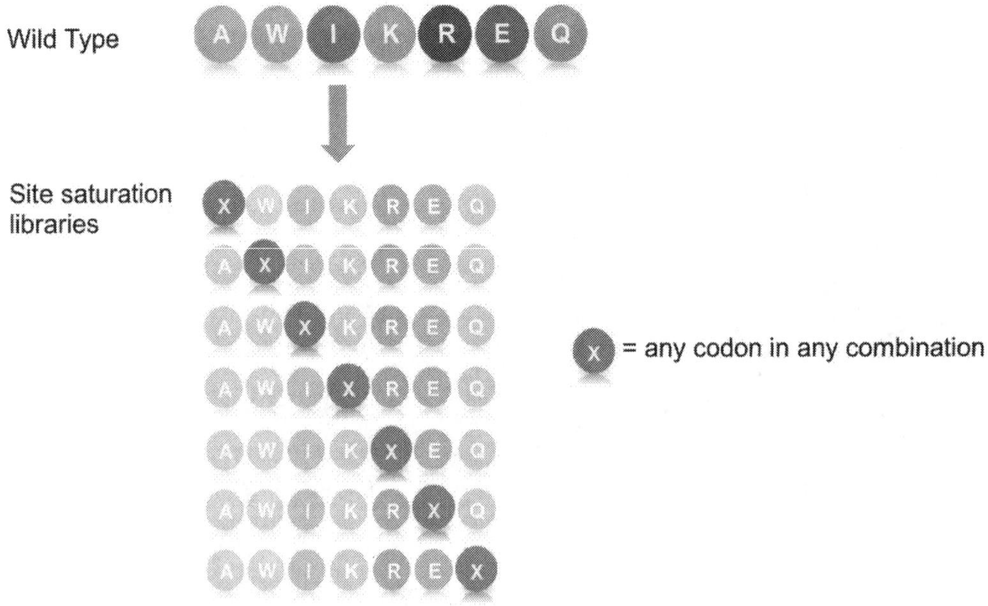

SITE SATURATION MUTAGENESIS

T W I S T BIOSCIENCE

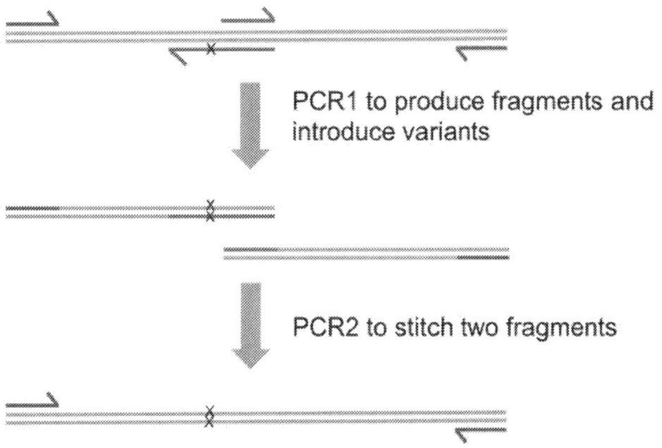

- High-quality and high-throughput Twist platform produces mutagenic primers.
- Optimized primer design and PCR protocol to ensure high-quality amplification.

SITE SATURATION MUTAGENESIS

T W I S T BIOSCIENCE

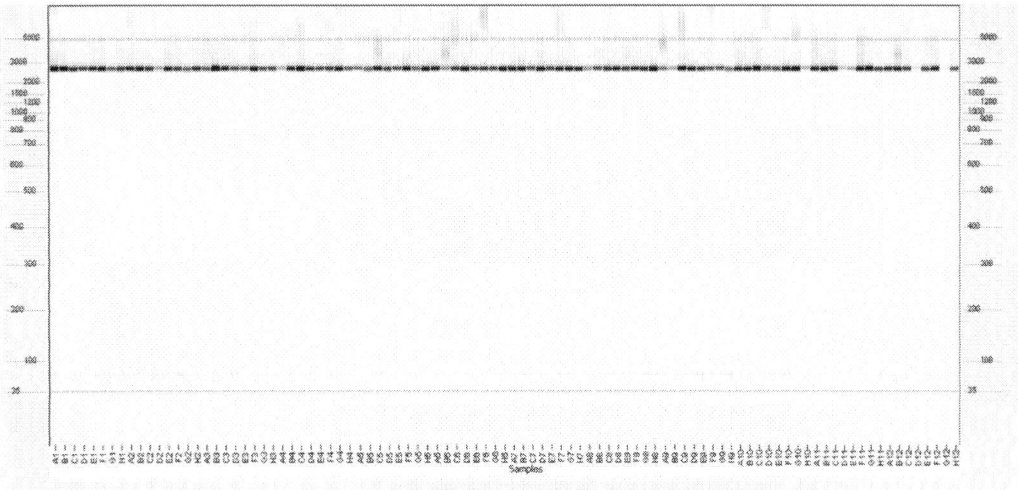

- Representative electropherogram for saturation mutagenesis on 96 positions, 19 variants per position
- 100% success rate based on gel electrophoresis

SITE SATURATION MUTAGENESIS

TWIST BIOSCIENCE

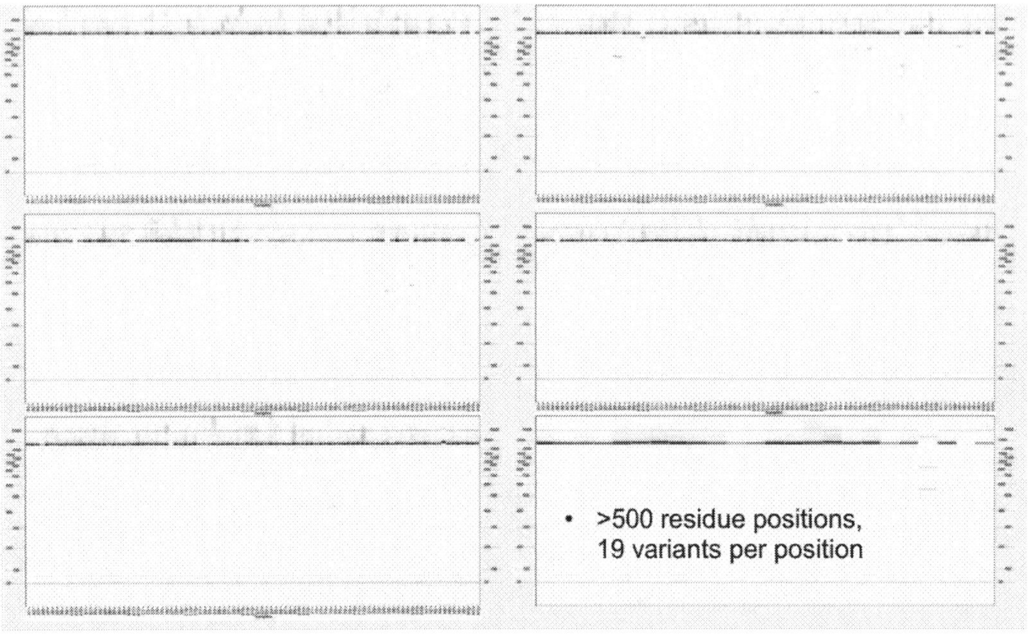

- >500 residue positions, 19 variants per position

SITE SATURATION MUTAGENESIS

TWIST BIOSCIENCE

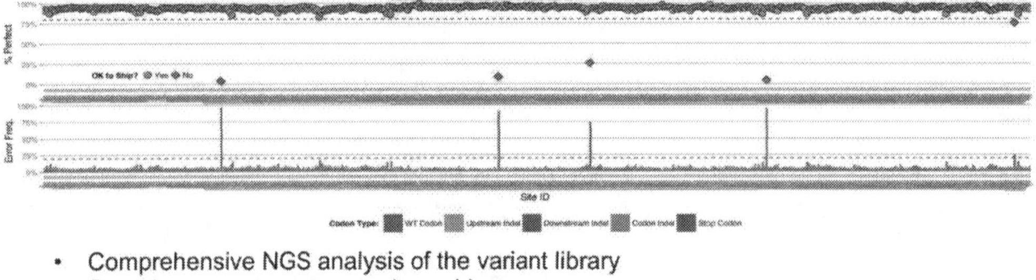

- Comprehensive NGS analysis of the variant library
- Superb evenness among amino acids
- Rational design. Any codons in any combination
- >99% success rate, >95% perfect, <2% WT

TWIST CONFIDENTIAL JULY 9, 2015 52

SITE SATURATION MUTAGENESIS

T W I S T BIOSCIENCE

Per-site frequency by amino acid

AGENDA

T W I S T BIOSCIENCE

- Twist Bioscience overview
- Broadband gene synthesis
- Assembly design optimization
- Library synthesis
- Alpha Access Program

ALPHA ACCESS PROGRAM

TWIST BIOSCIENCE

Limited Openings

- Genes up to 1.8 kb (longer soon!)
- Perfect sequence, 100% NGS verified
- Linear or cloned (catalog or custom)
- No flanking sequences, no Type IIs sequence limitations
- Will attempt all genes, charge only successful ones, feed machine learning
- 1 to 10,000 sequences per order
- Typically 20 business days or less (faster soon!)
- Industry leading pricing

MAKING CLONING OBSOLETE

TWIST BIOSCIENCE

Phase I
DISRUPT "BUY"

Phase II
CONVERT MAKERS to BUYERS and GROW MARKET

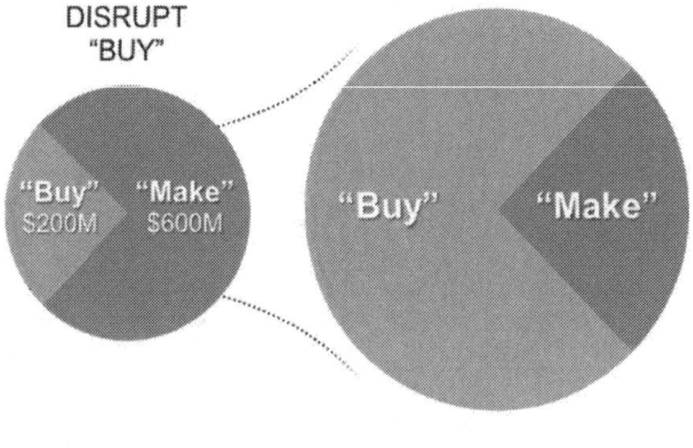

Phase III
OPEN NEW MARKETS

Other DNA Synthesis

RNA Synthesis

Protein Synthesis

DNA as Long-Term Data Storage

"DNA on Silicon" Tools – Lab Automation

SUMMARY – THANK YOU!

TWIST BIOSCIENCE

Large Growing
Synthetic DNA Market

Proven Technology
Alpha Access Available

Game Changing
Momentum
Speed, Throughput, Cost, Quality

1st DNA Writing
on Silicon Platform

Experienced Team
with Strong Backing

TWIST BIOSCIENCE

JULY 9, 2015 57

TWIST BIOSCIENCE

POWERING THE SYNTHETIC BIOLOGY REVOLUTION

Appendix

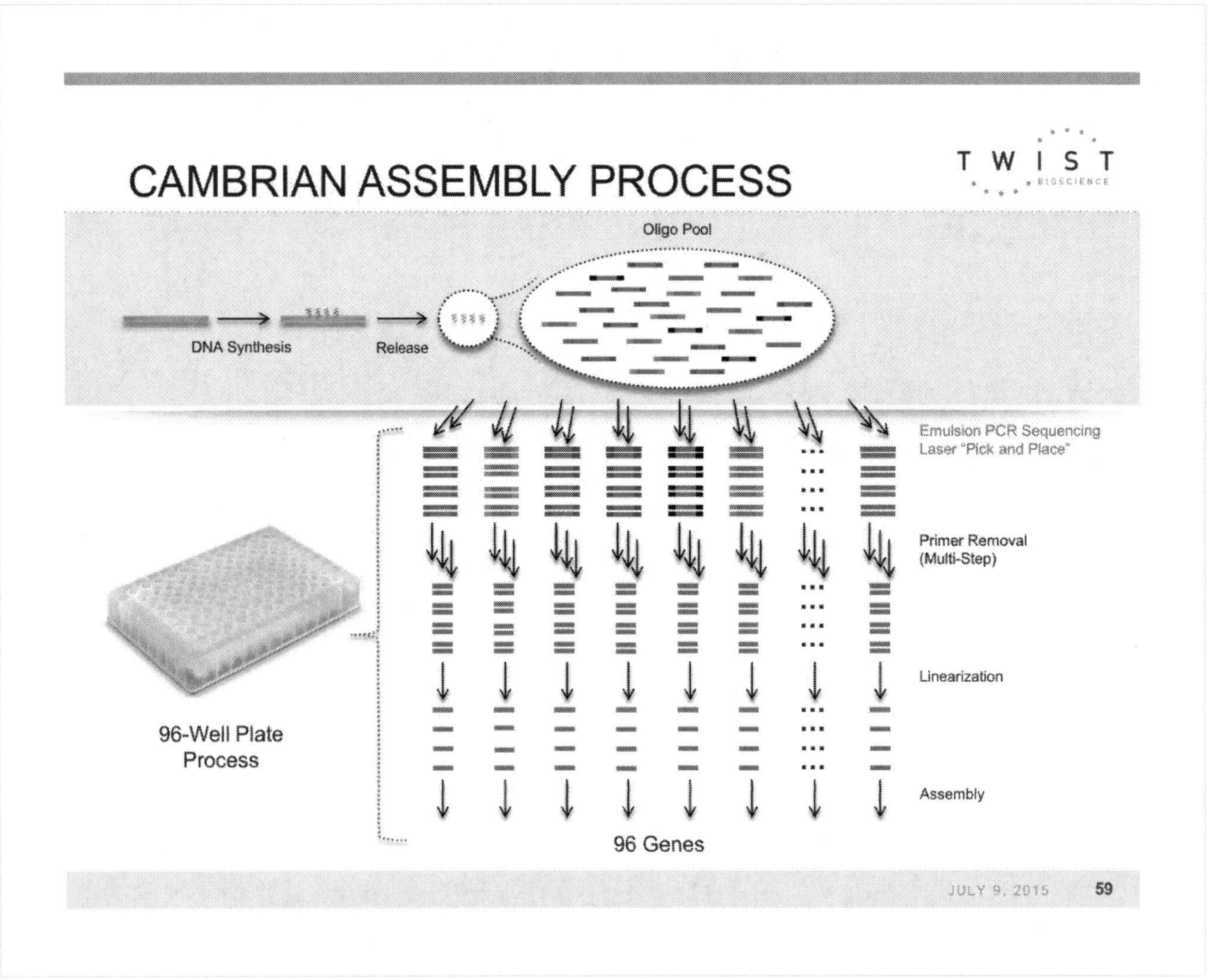
CAMBRIAN ASSEMBLY PROCESS
TWIST BIOSCIENCE
Oligo Pool
DNA Synthesis
Release
Emulsion PCR Sequencing
Laser "Pick and Place"
Primer Removal
(Multi-Step)
Linearization
Assembly
96-Well Plate
Process
96 Genes
JULY 9, 2015
59

Rapid Prototyping of Genetic Programs

Ben Gordon

MIT-Broad Foundry

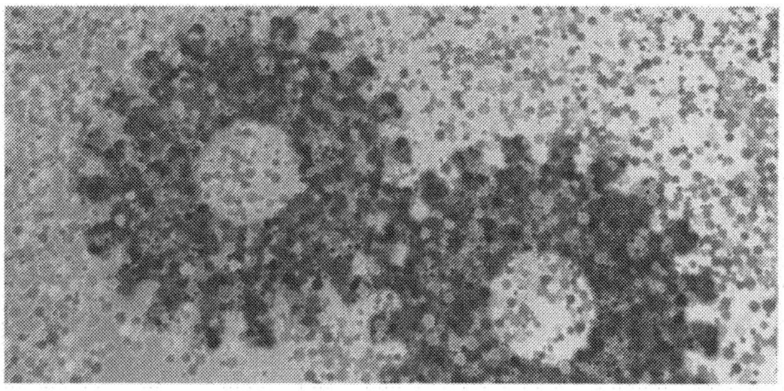

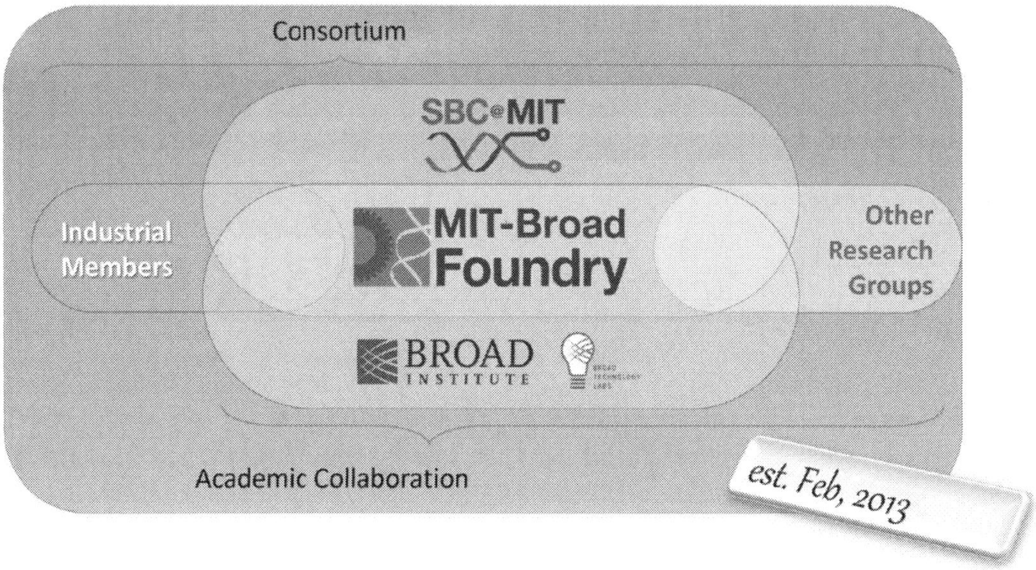

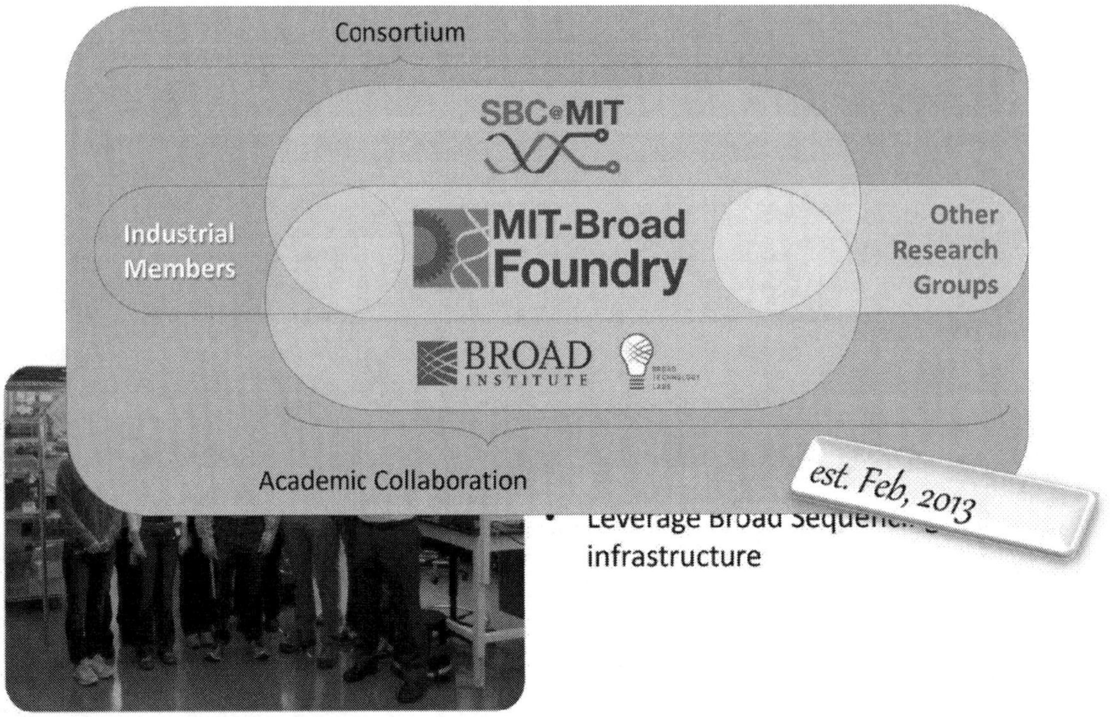

Accessing complex biological functions

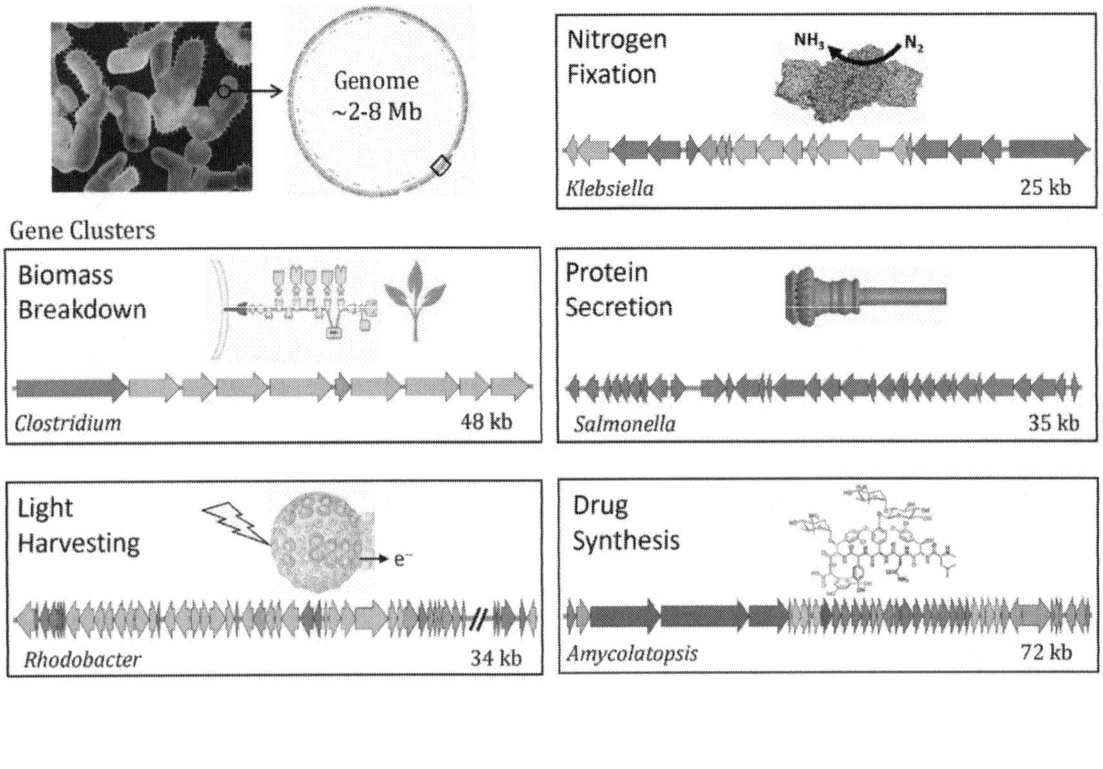

Gene Expression

Same genetic program. Different expression levels

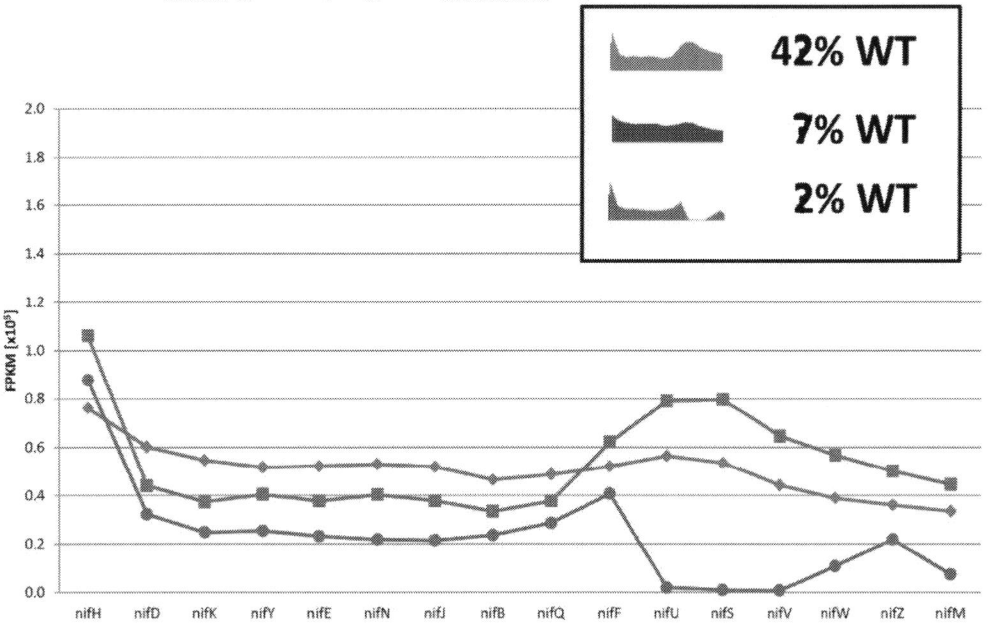

Smanski et al. Nat Biotech, 2014

Surface Response / Empirical Modeling

Identifies:

- Key Factors (e.g. key genes)
- Interactions (e.g. Gene-Gene)
- Gradient (toward optimum)

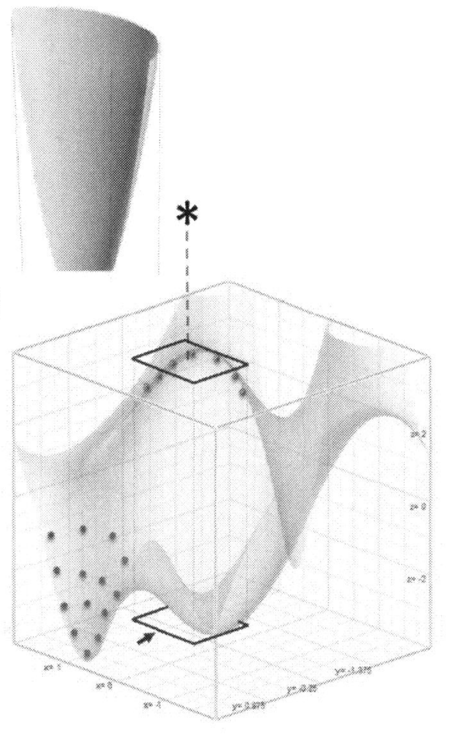

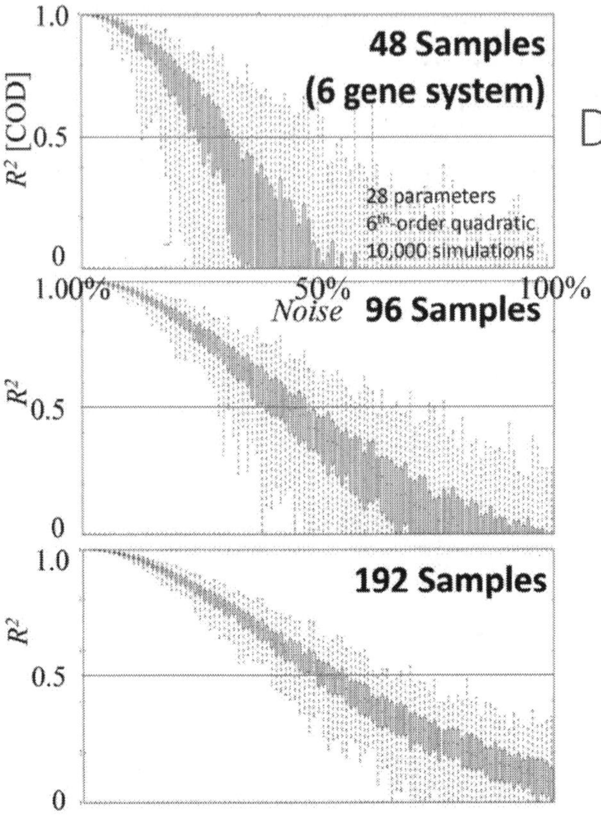

Simulation:
Determine Requirements

Noise / *# Genes* / *# Samples* — *Model Complexity* / *Functional form* / *Domain/Range*

Critical Factors

- Precision expression control
- Wide dynamic range
- High-throughput Assembly
- Efficient Iteration

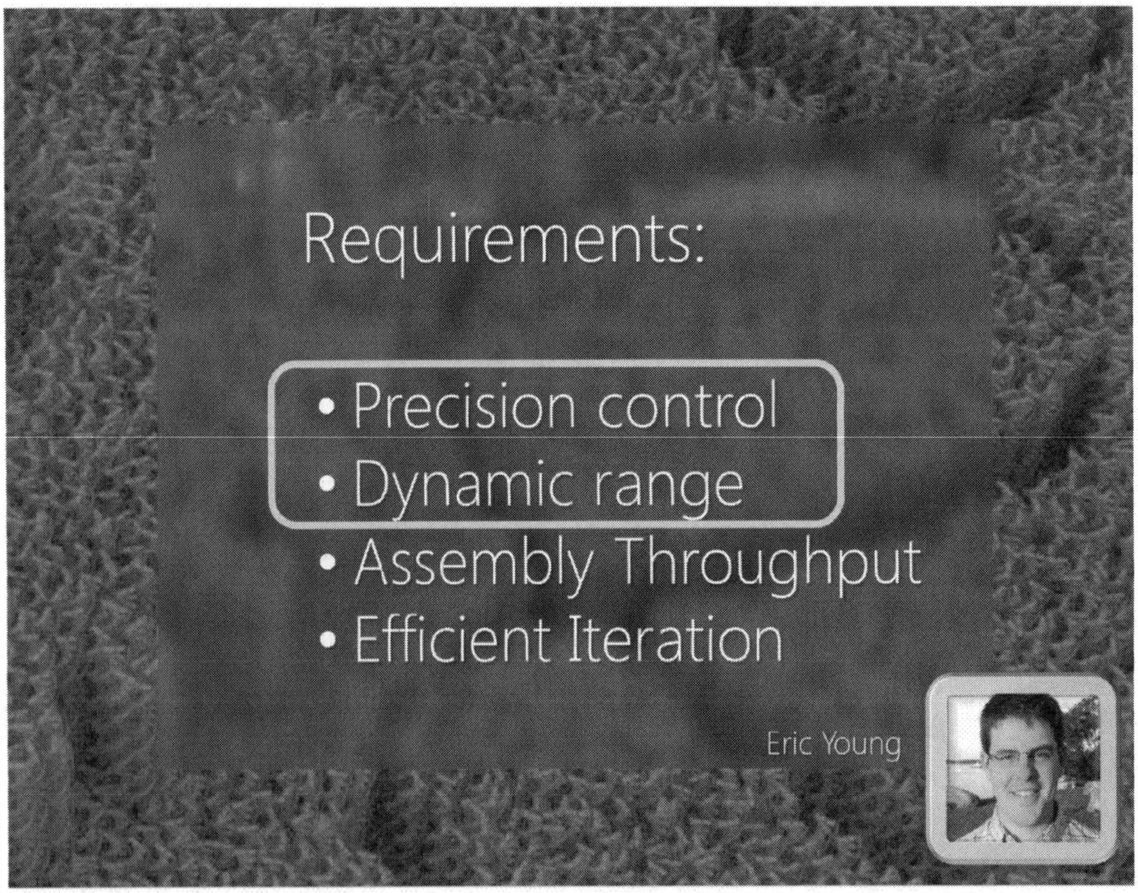

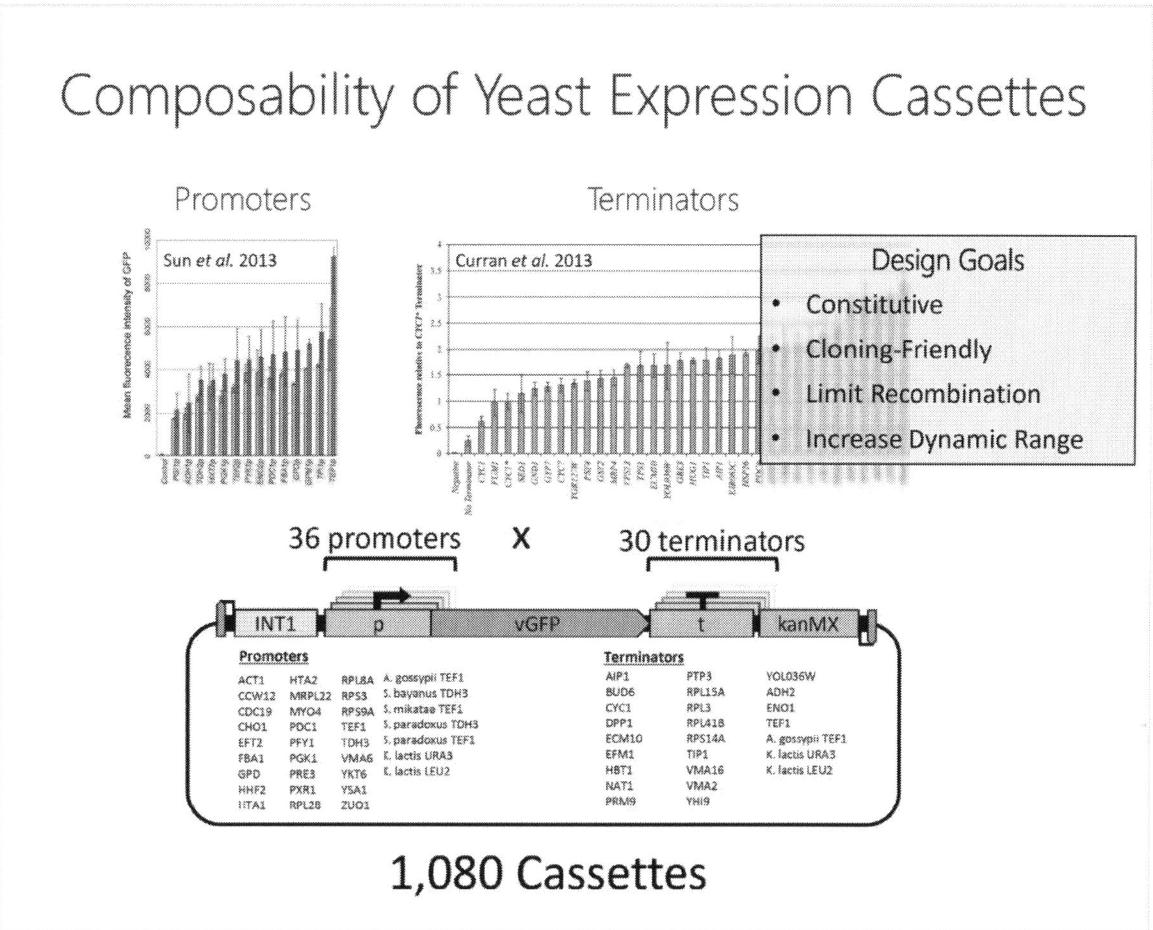

Composability of Yeast Expression Cassettes
Promoters
Terminators
Sun et al. 2013
Curran et al. 2013
Design Goals
Constitutive
Cloning-Friendly
Limit Recombination
Increase Dynamic Range
36 promoters
X
30 terminators
INT1
p
vGFP
t
kanMX
Promoters
ACT1 HTA2 RPL8A A. gossypii TEF1
CCW12 MRPL22 RPS3 S. bayanus TDH3
CDC19 MYO4 RPS9A S. mikatae TEF1
CHO1 PDC1 TEF1 S. paradoxus TDH3
EFT2 PFY1 TDH3 S. paradoxus TEF1
FBA1 PGK1 VMA6 K. lactis URA3
GPD PRE3 YKT6 K. lactis LEU2
HHF2 PXR1 YSA1
HTA1 RPL28 ZUO1
Terminators
AIP1 PTP3 YOL036W
BUD6 RPL15A ADH2
CYC1 RPL3 ENO1
DPP1 RPL41B TEF1
ECM10 RPS14A A. gossypii TEF1
EFM1 TIP1 K. lactis URA3
HBT1 VMA16 K. lactis LEU2
NAT1 VMA2
PRM9 YHI9
1,080 Cassettes

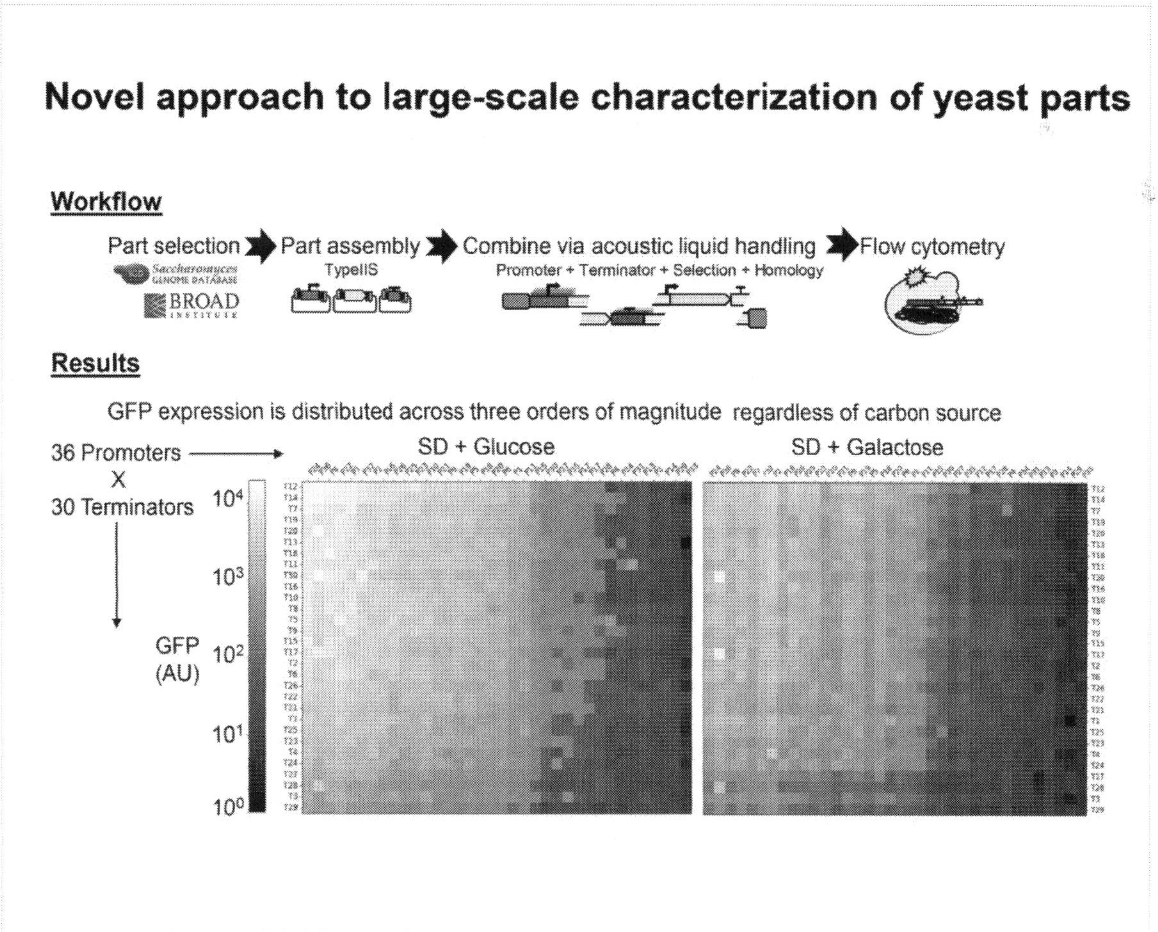

Novel approach to large-scale characterization of yeast parts
Workflow
Part selection
Part assembly
Combine via acoustic liquid handling
Flow cytometry
Saccharomyces GENOME DATABASE
BROAD INSTITUTE
TypeIIS
Promoter + Terminator + Selection + Homology
Results
GFP expression is distributed across three orders of magnitude regardless of carbon source
36 Promoters
X
30 Terminators
SD + Glucose
SD + Galactose
GFP (AU)

Yeast part behavior fits a simple relationship

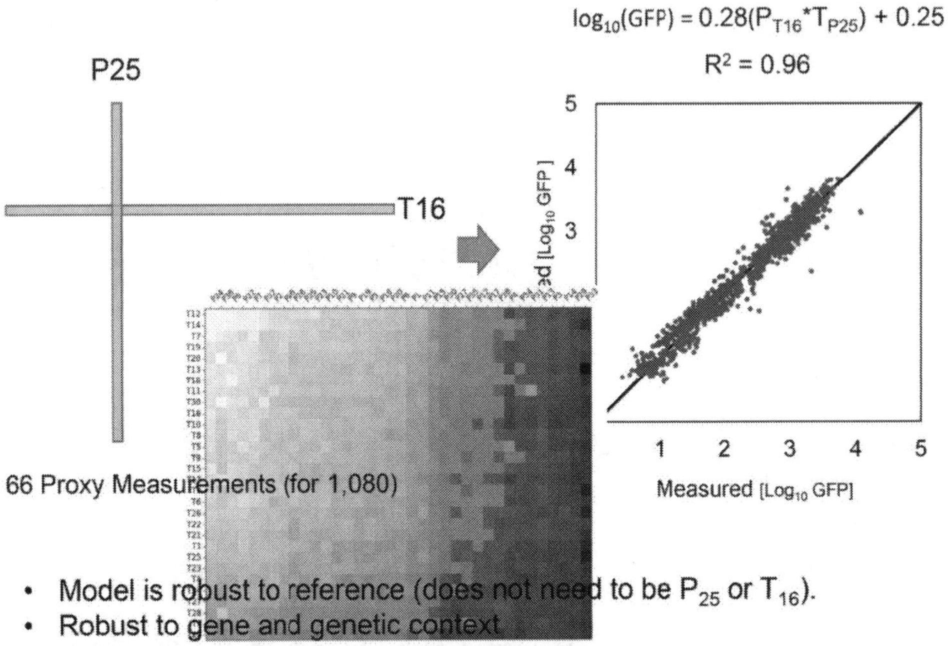

- Model is robust to reference (does not need to be P_{25} or T_{16}).
- Robust to gene and genetic context
- New parts require only one measurement
- Hints at mechanistic independence

Requirements:

- Precision control
- Dynamic range
- Assembly Throughput
- Iteration Efficiency

Eric Young

Application: learning itaconic acid pathway dynamics

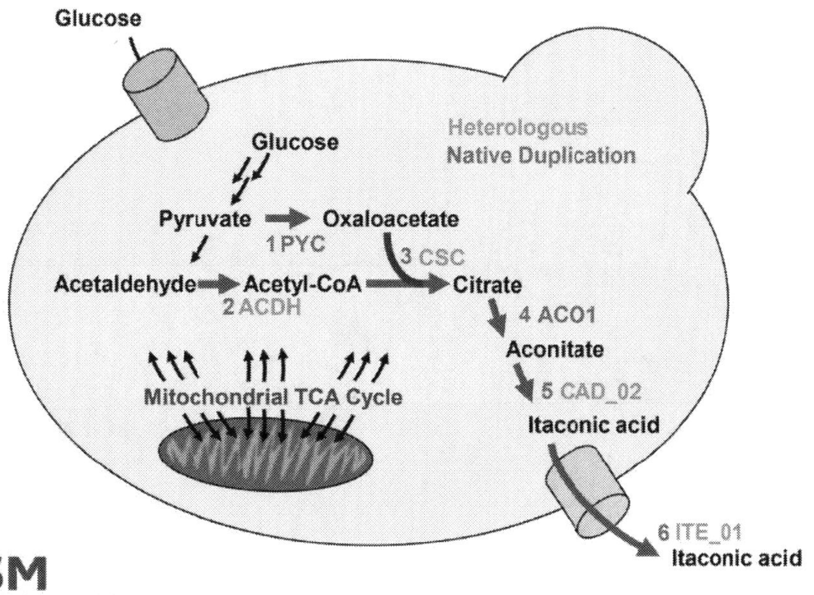

DSM
BRIGHT SCIENCE. BRIGHTER LIVING.

Design and build pathway variants

Design

I. Design DOE variant library

Genetic Library: 3-Level; 2nd Order Model
with Low, Medium, & High expression

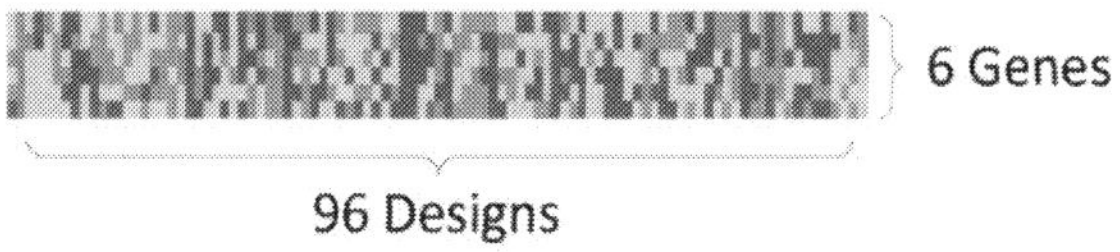

II. Assign parts to genes

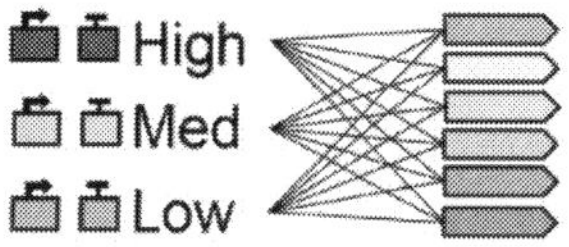

Unique transcription units:
No risk of recombination

Build

I. Construct Transcription units via Type IIS assembly

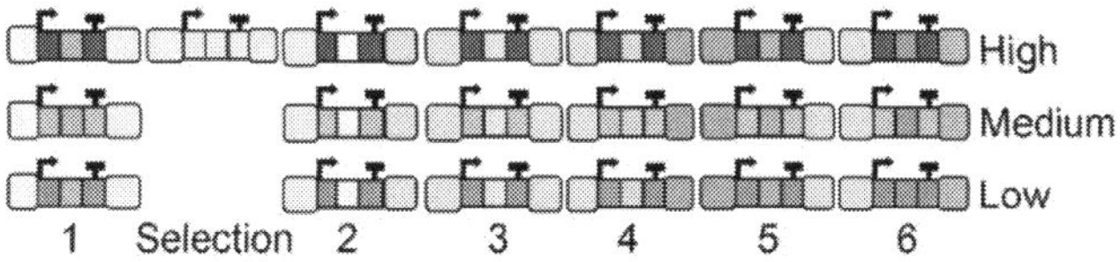

II. Assemble designs via acoustic liquid handling & recombination

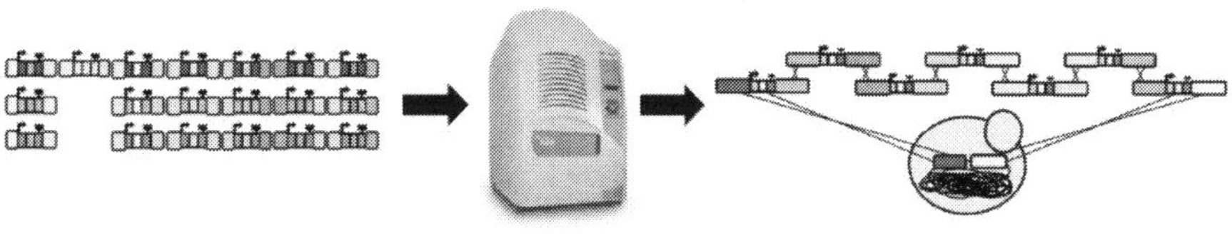

Assembly Tree: Response Surface

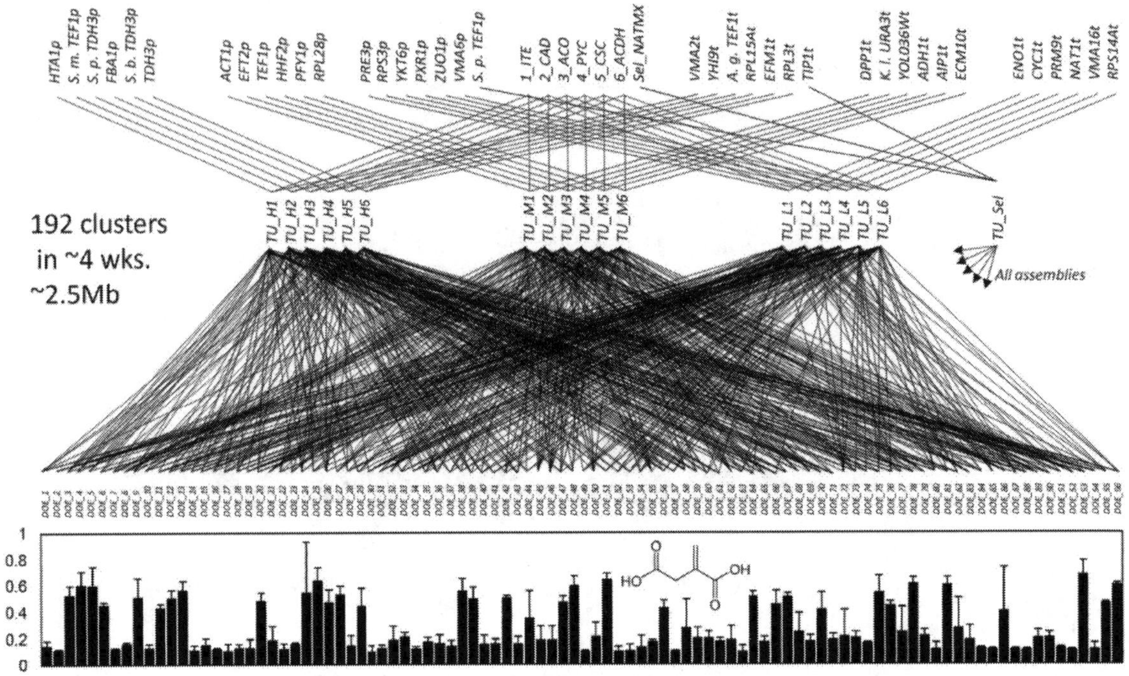

Analyzing total itaconic acid titer with 2nd order model

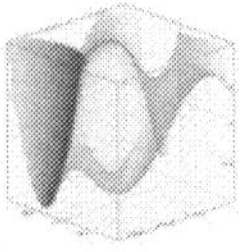

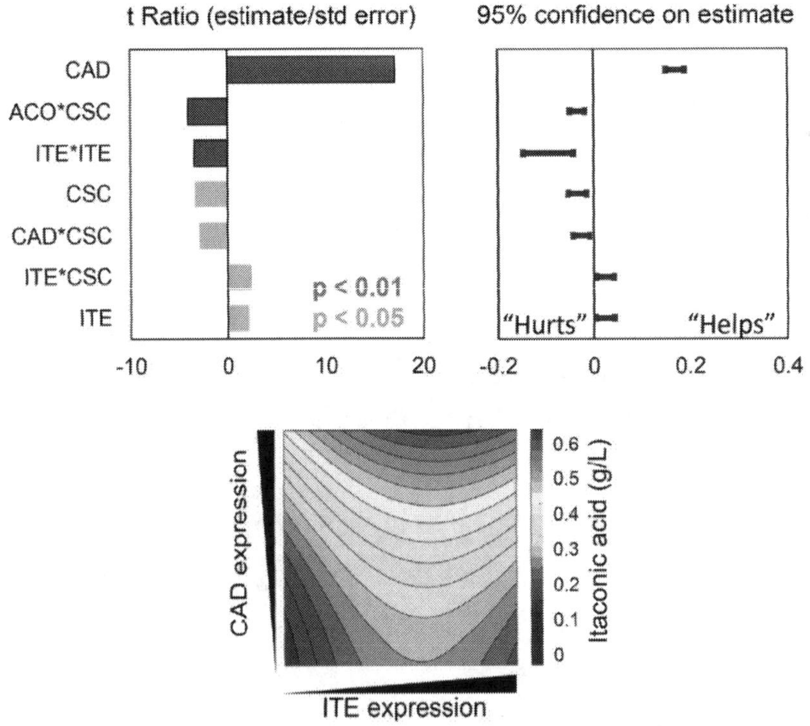

Application: learning itaconic acid pathway dynamics

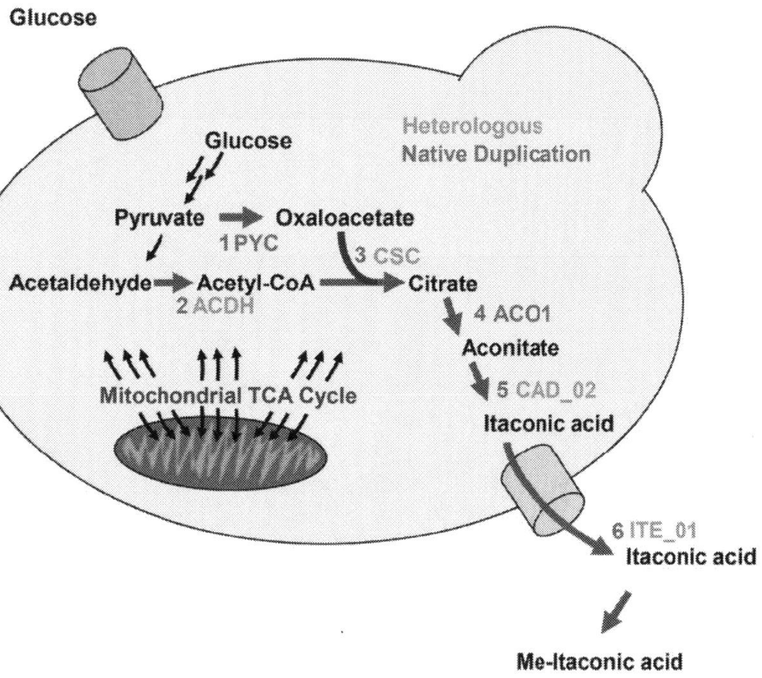

Assembly Tree: Response Surface

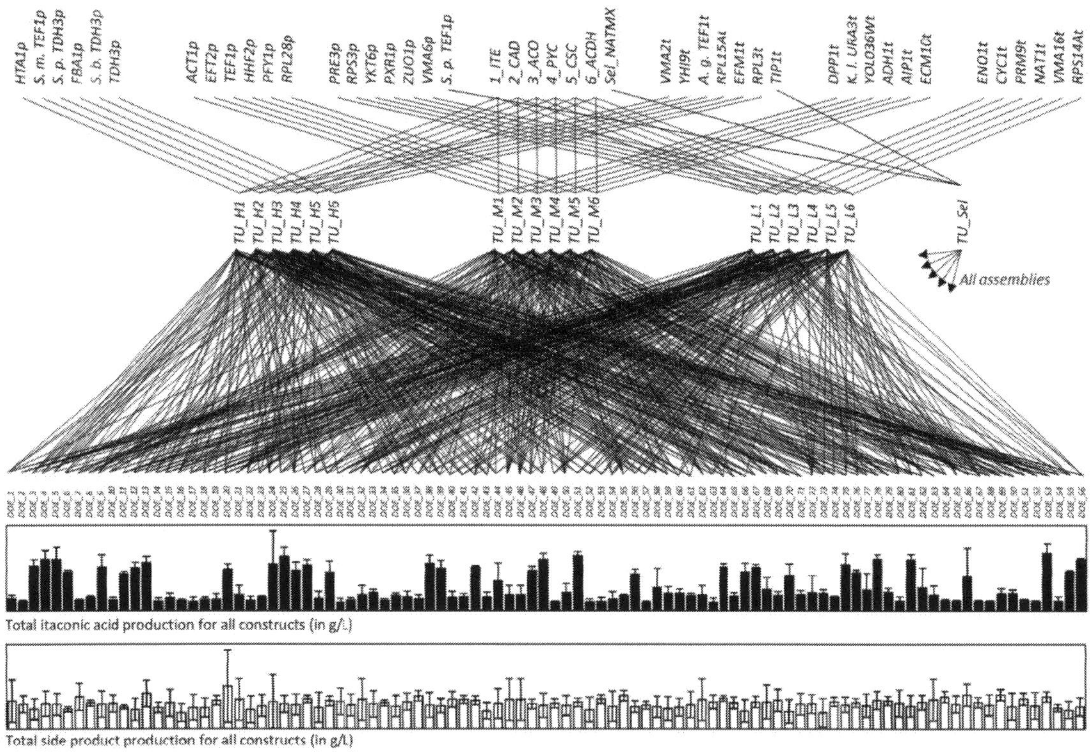

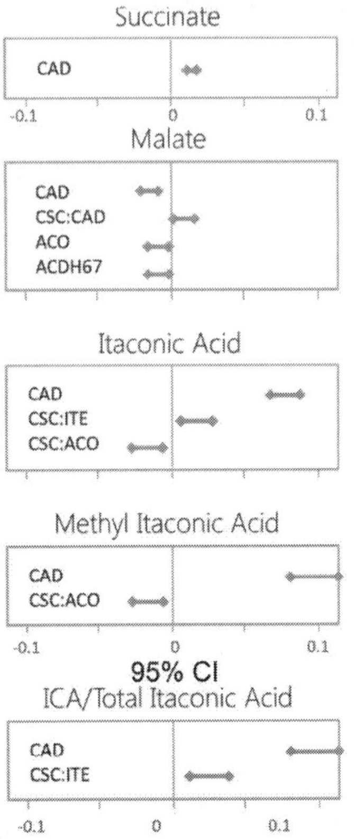

Multiresponse Tuning Analysis

- **Metabolites**
- **Side-products (purity/efficiency)**

	Succinate	Malate	Iconitate	me-Iconiate	Ico/Total Ico
CAD	↑	↓	↑↑↑	↑↑↑	↑↑↑
ACO		↓			
ACDH67		↓			
CSC:CAD		↑			
CSC:ACO			↓	↓	
CSC:ITE			↑		↑
			Maximize		Maximize

To increase titer & purity

- [Major Factor] **Increase ACO**
- [Minor Factor] **Increase CSC:ITE**
- [Minor Factor] **Do not increase CSC:ACO**

Requirements:

- Precision control
- Dynamic range
- Assembly Throughput
- Iteration Efficiency

Tarjei Mikkelsen

Lauren Woodruff

Combinatorially Optimizing Gene Expression

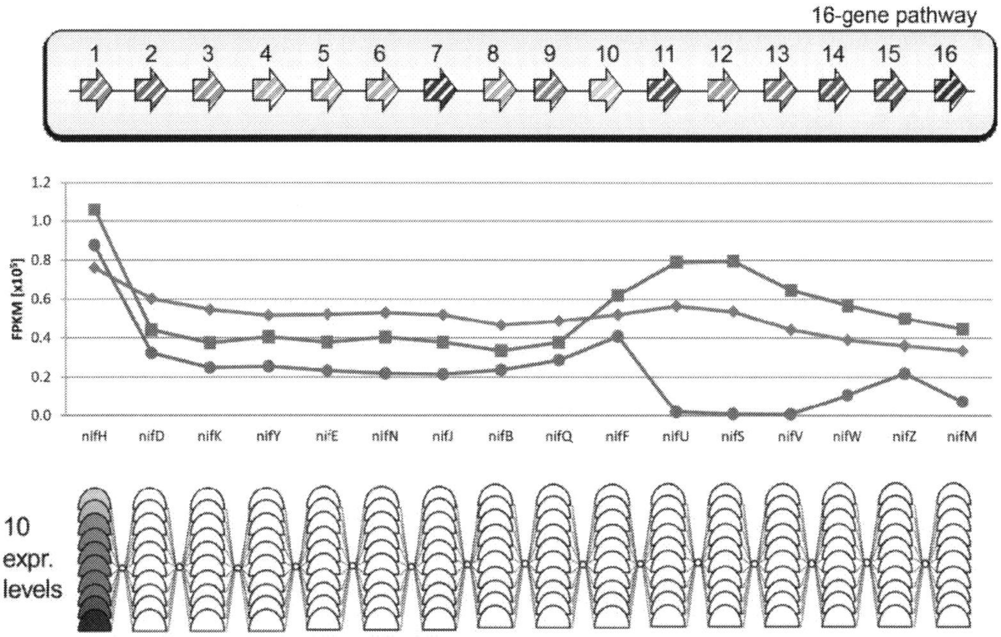

10^{16} Combinations

Building permuted clusters hierarchically from parts

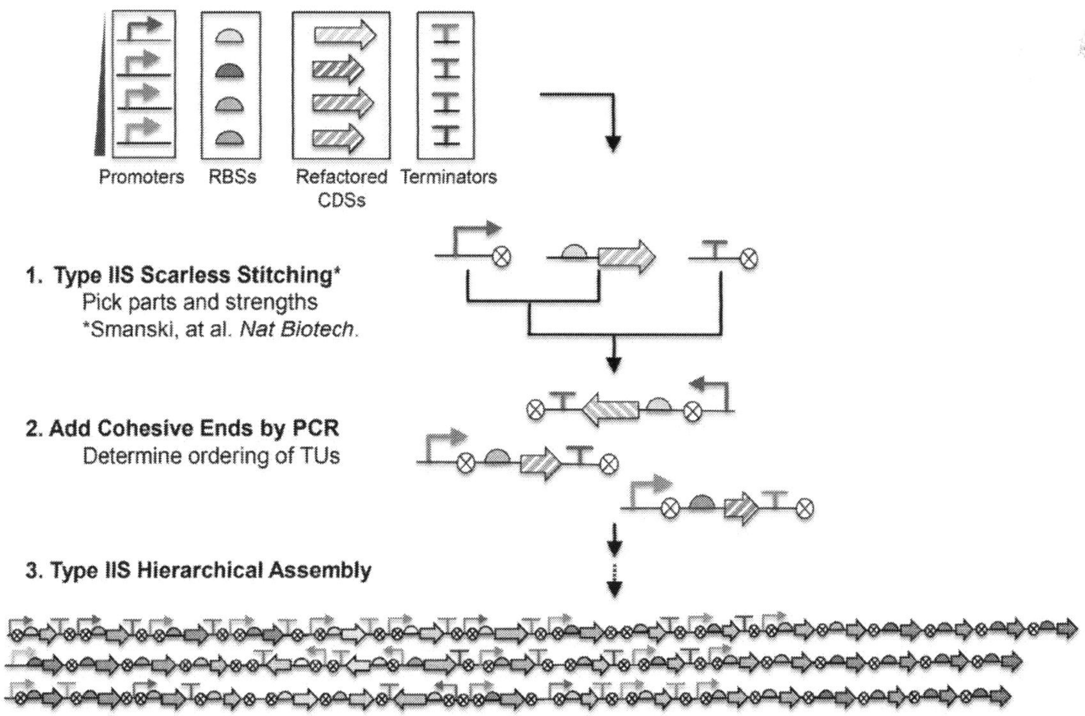

Smanski et al. Nat Biotechnology, 2014

Streamlined construction enables larger, more complex libraries

Multiplex Retrievable Assembly (1,000s retrievable variants assembled in 1 tube)

>100-fold increase in library size

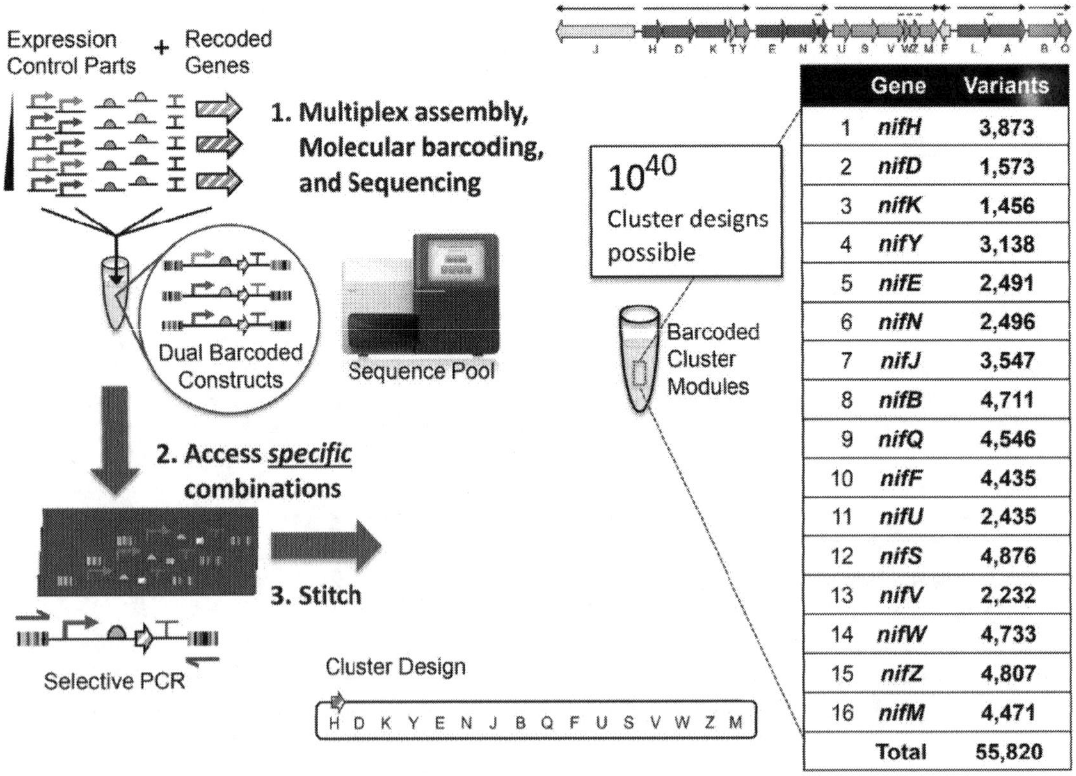

	Gene	Variants
1	*nifH*	3,873
2	*nifD*	1,573
3	*nifK*	1,456
4	*nifY*	3,138
5	*nifE*	2,491
6	*nifN*	2,496
7	*nifJ*	3,547
8	*nifB*	4,711
9	*nifQ*	4,546
10	*nifF*	4,435
11	*nifU*	2,435
12	*nifS*	4,876
13	*nifV*	2,232
14	*nifW*	4,733
15	*nifZ*	4,807
16	*nifM*	4,471
	Total	55,820

Rapid iterative permuting of multi-gene pathway

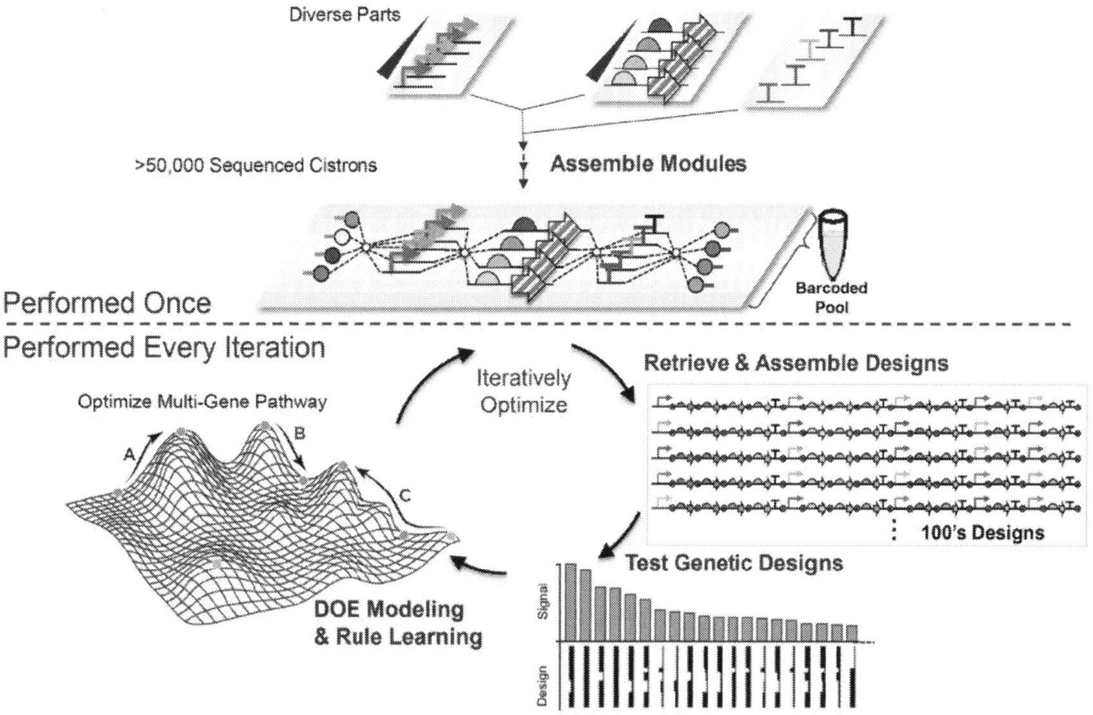

Multilevel DOE Library to Combinatorially Tune Transcription

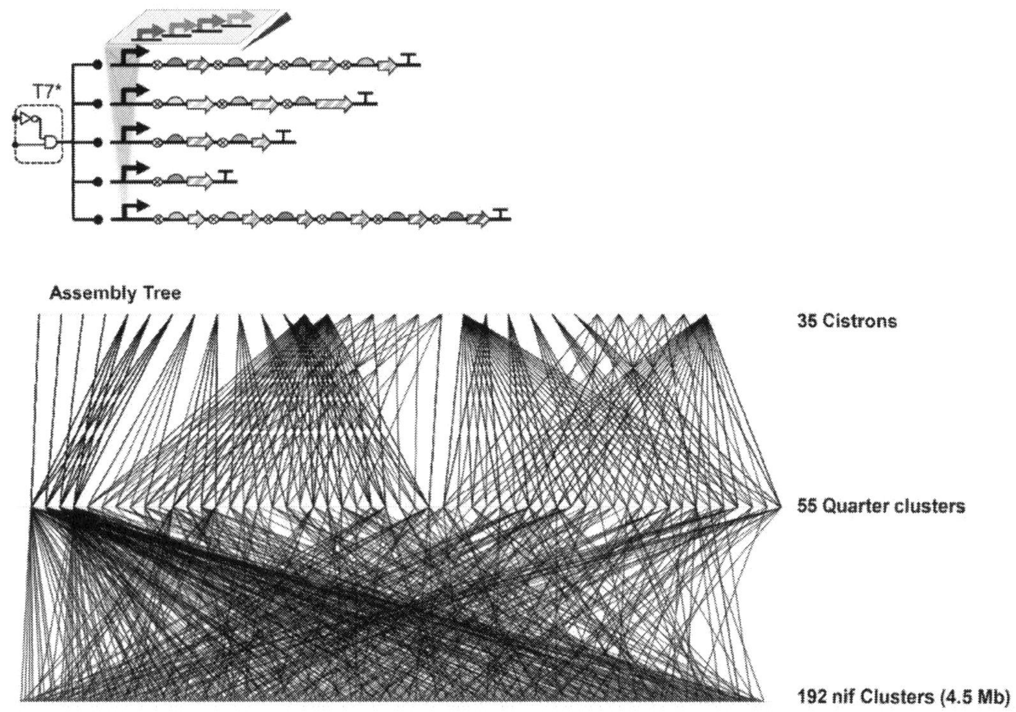

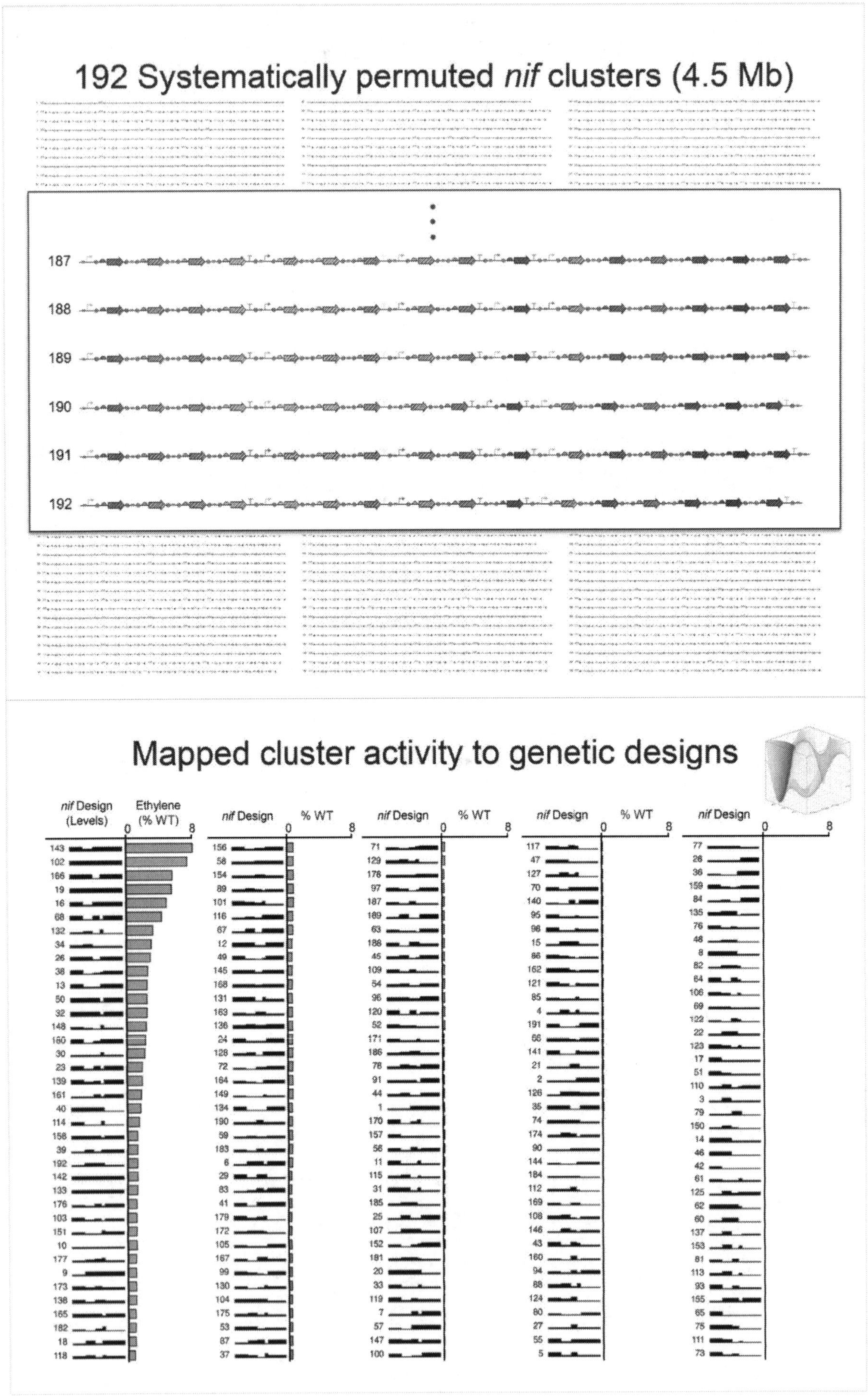
192 Systematically permuted nif clusters (4.5 Mb)
187
188
189
190
191
192
Mapped cluster activity to genetic designs
nif Design (Levels)
Ethylene (% WT)
nif Design
% WT
nif Design
% WT
nif Design
% WT
nif Design
0
8

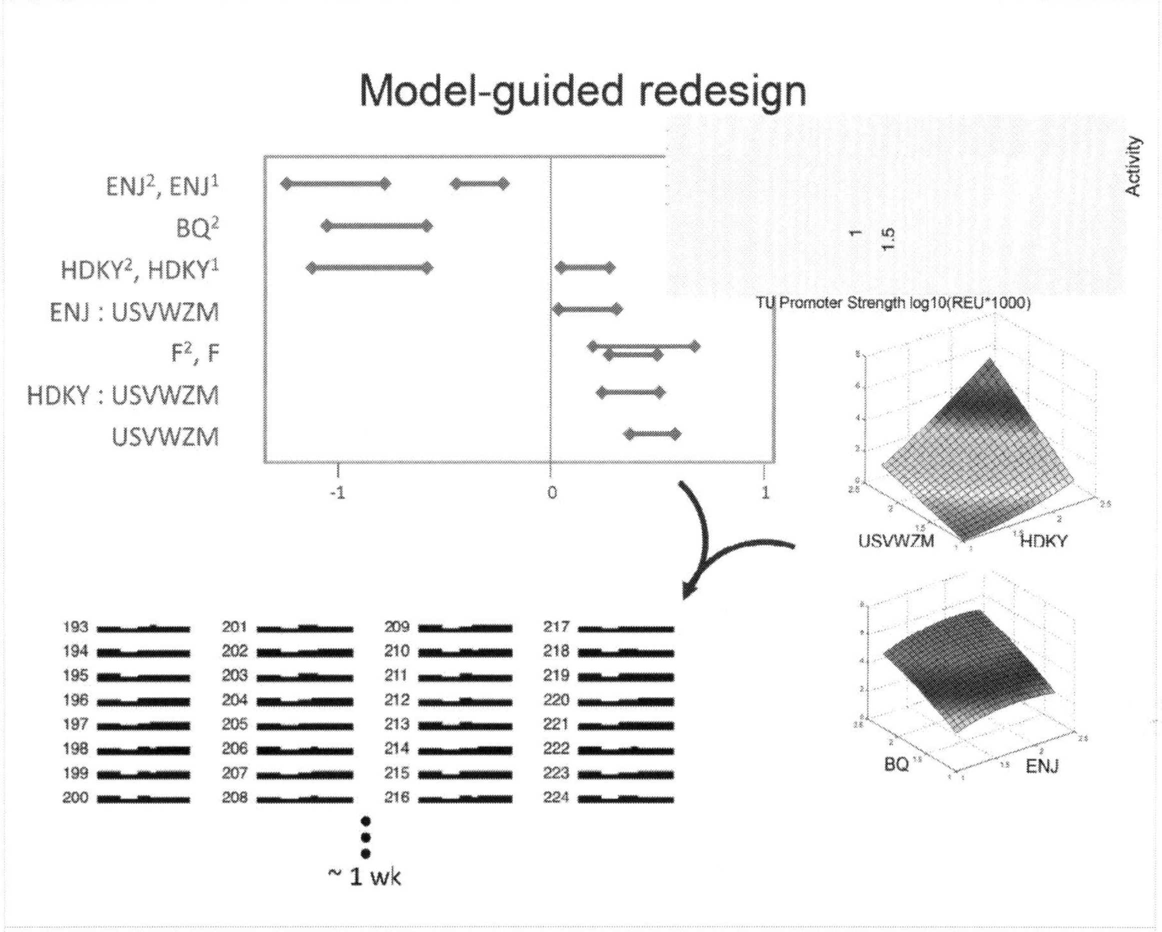
Model-guided redesign
ENJ², ENJ¹
BQ²
HDKY², HDKY¹
ENJ : USVWZM
F², F
HDKY : USVWZM
USVWZM
-1
0
1
Activity
TU Promoter Strength log10(REU*1000)
USVWZM
HDKY
BQ
ENJ
193 194 195 196 197 198 199 200
201 202 203 204 205 206 207 208
209 210 211 212 213 214 215 216
217 218 219 220 221 222 223 224
~ 1 wk

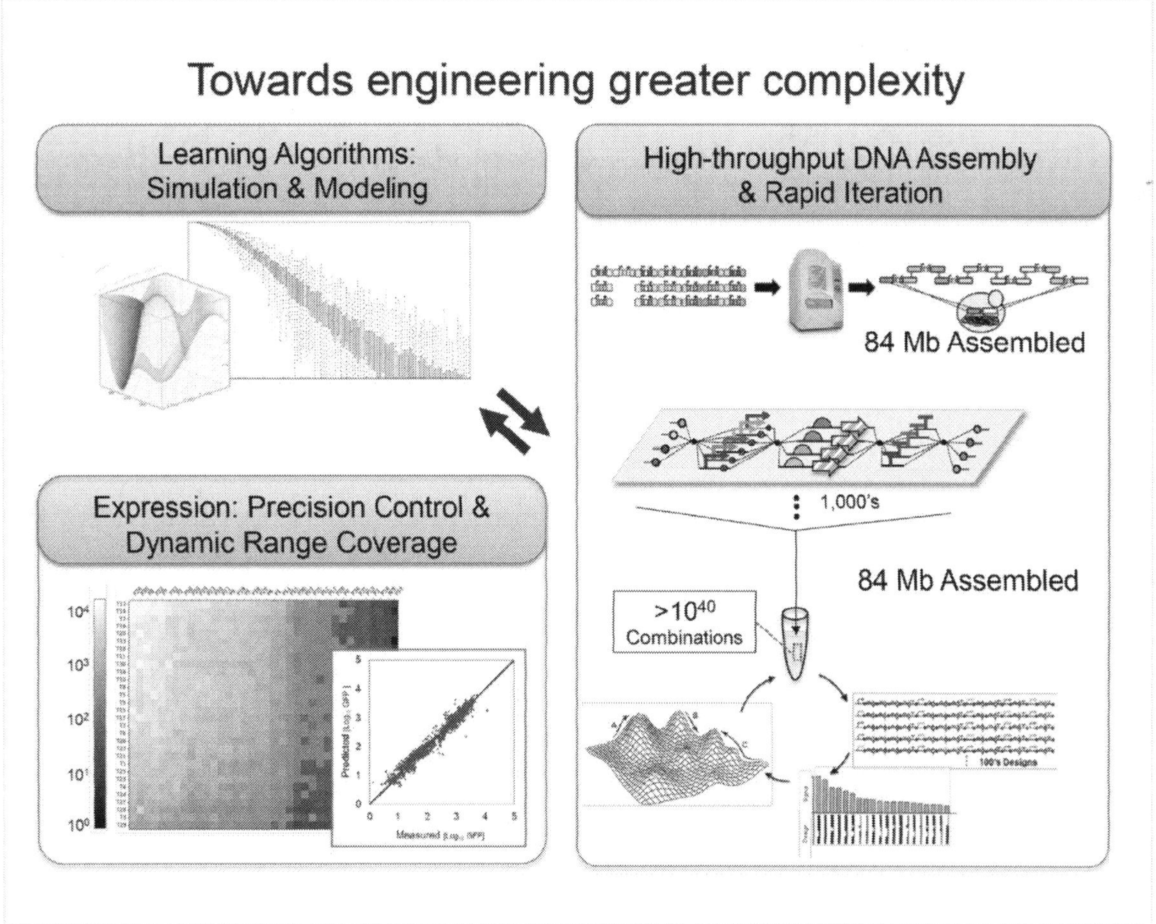
Towards engineering greater complexity
Learning Algorithms: Simulation & Modeling
High-throughput DNA Assembly & Rapid Iteration
84 Mb Assembled
1,000's
84 Mb Assembled
>10⁴⁰ Combinations
Expression: Precision Control & Dynamic Range Coverage
10⁴
10³
10²
10¹
10⁰
Predicted
Measured
100's Designs

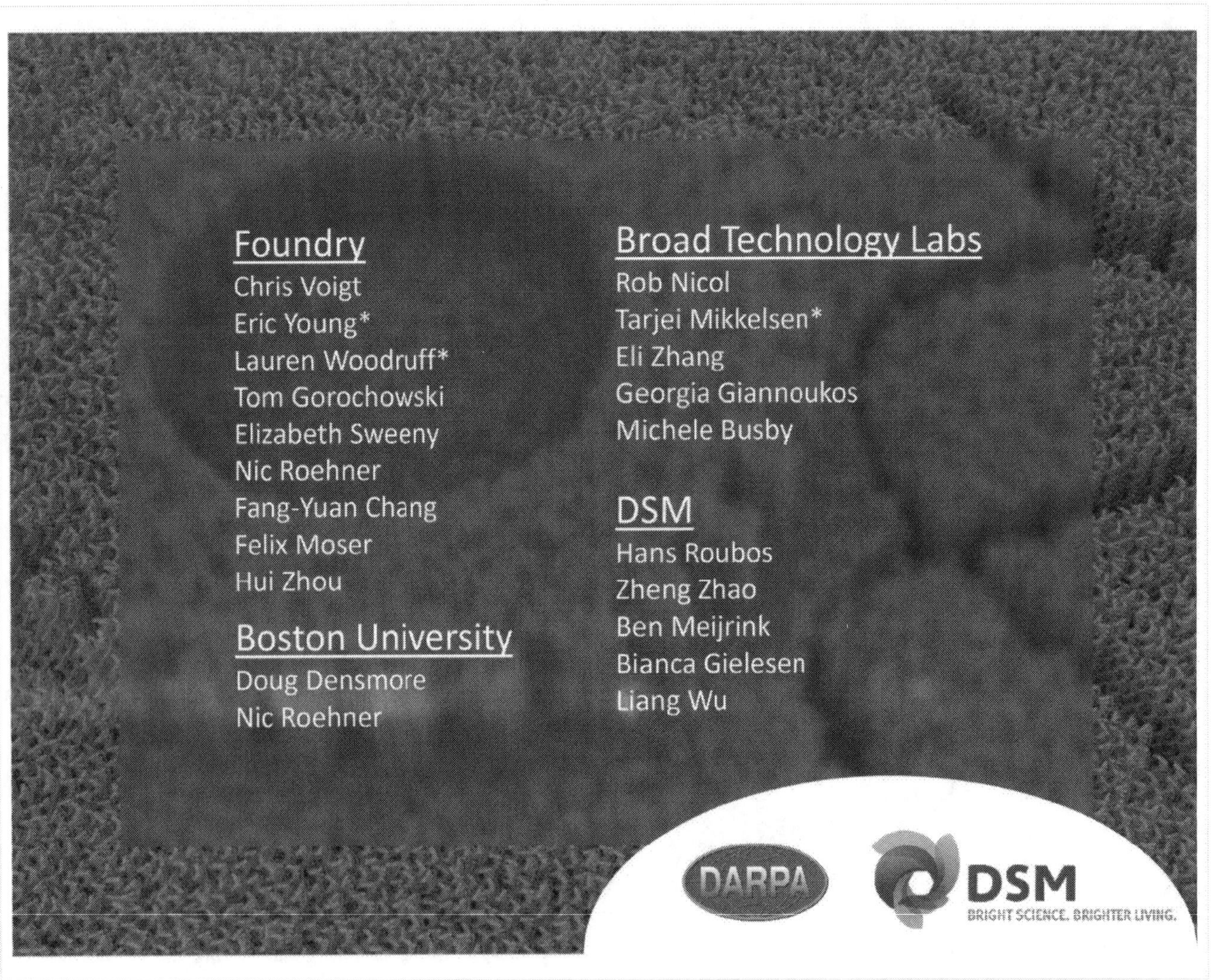
Foundry
Chris Voigt
Eric Young*
Lauren Woodruff*
Tom Gorochowski
Elizabeth Sweeny
Nic Roehner
Fang-Yuan Chang
Felix Moser
Hui Zhou
Boston University
Doug Densmore
Nic Roehner
Broad Technology Labs
Rob Nicol
Tarjei Mikkelsen*
Eli Zhang
Georgia Giannoukos
Michele Busby
DSM
Hans Roubos
Zheng Zhao
Ben Meijrink
Bianca Gielesen
Liang Wu
DARPA
DSM
BRIGHT SCIENCE. BRIGHTER LIVING.

Synthetic Biology Open Language (SBOL): Community-Driven Standard for Communication of Synthetic Biology Designs

Jacob Beal, *Bryan Bartley, Kevin Clancy, Goksel Misirli, Nicholas Roehner (Editors), Herbert Sauro (Chair), & SBOL community*

SEED Conference
Boston, MA
June, 2015

SBOL
Raytheon
BBN Technologies

SBOL **The SBOL Community**
Raytheon
BBN Technologies

Major funders: NSF EPSRC Engineering and Physical Sciences Research Council

- 100+ people from all around the world
- 30 universities, 14 companies, 8 others
- Ongoing development starting 2008

SBOL Synthetic Biology Open Language

Raytheon BBN Technologies

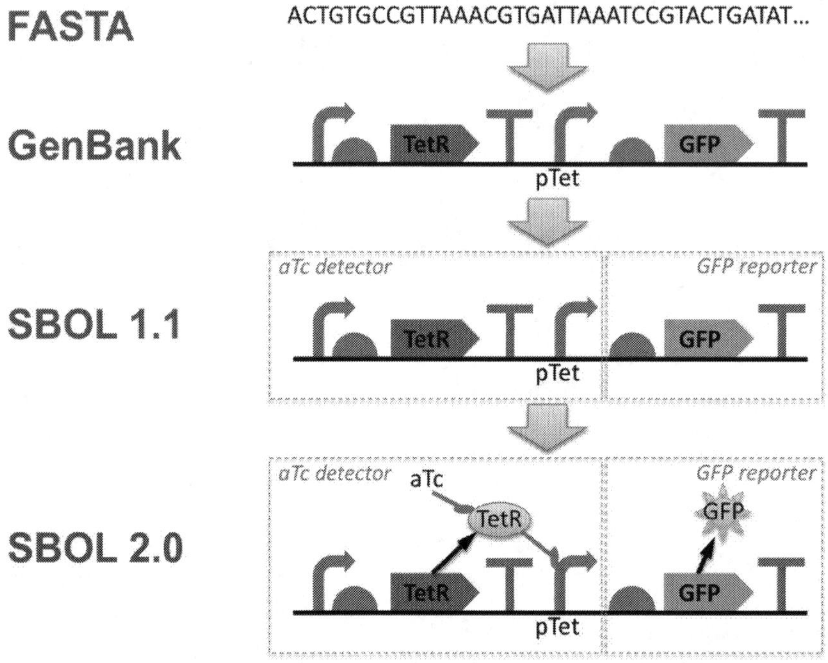

SBOL Purpose of SBOL 2.0

Raytheon BBN Technologies

Lots of different synthetic biology resources...

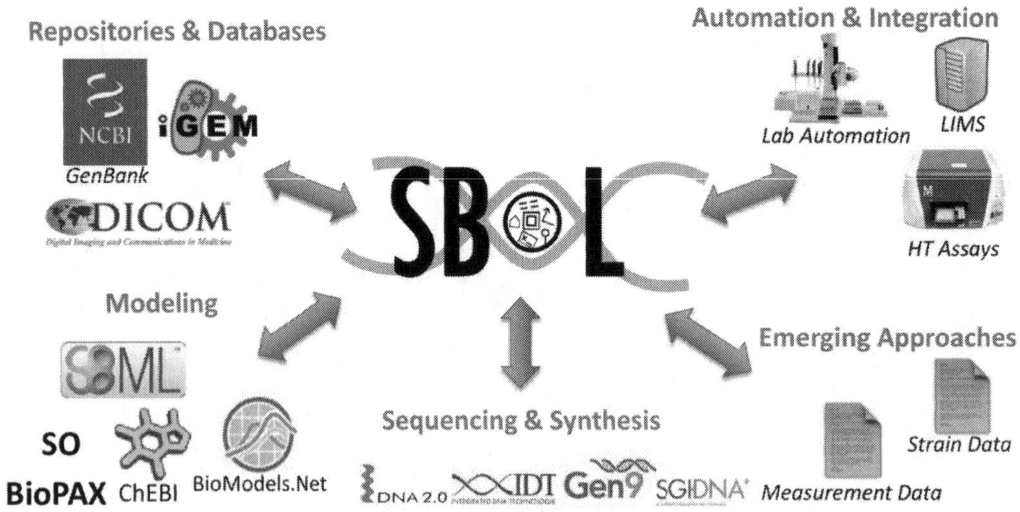

... SBOL is a "hub" for linking them together

Components of SBOL

Raytheon BBN Technologies

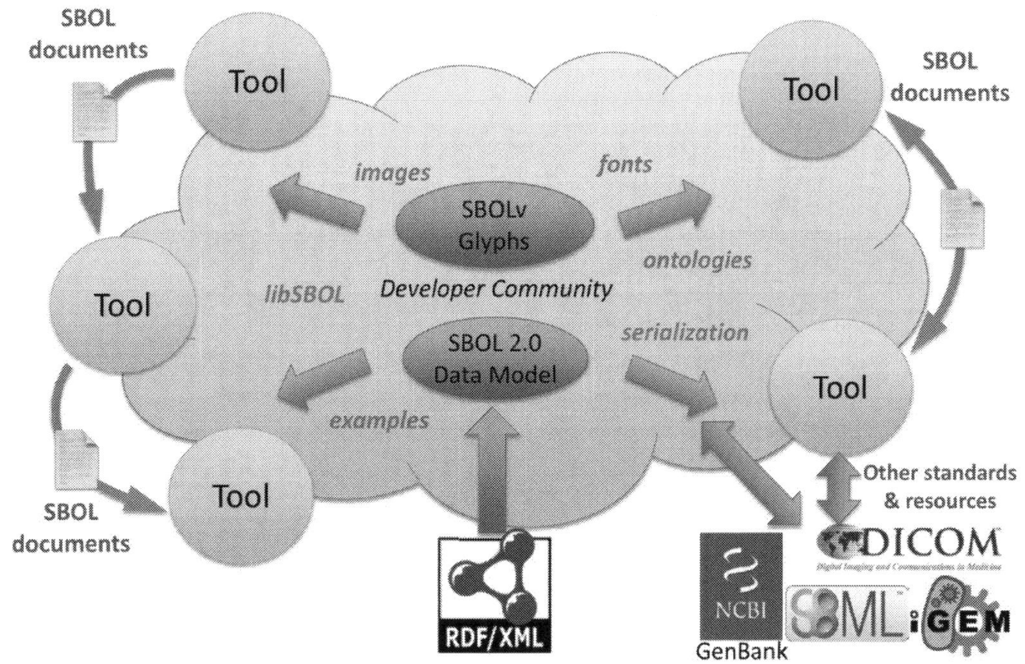

What SBOL 2.0 Represents

Raytheon BBN Technologies

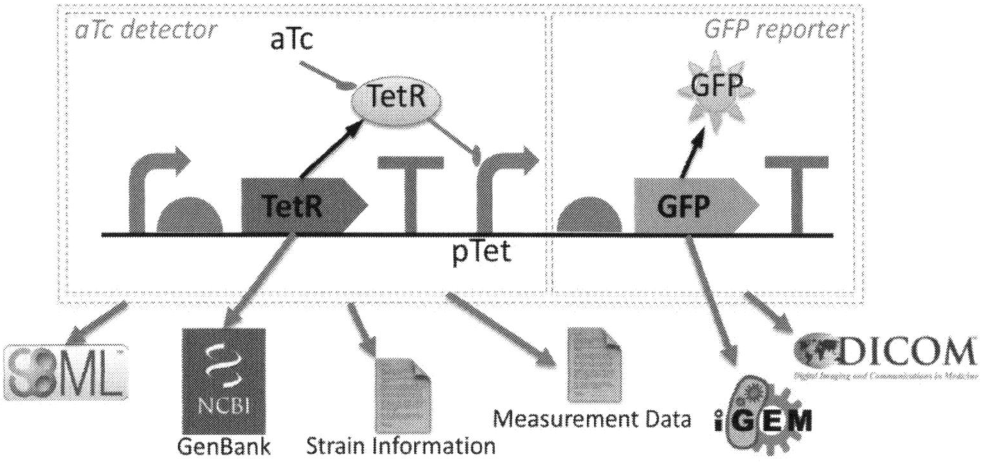

- Sequence structure
- Incomplete designs
- Qualitative function
- Heterarchical composition
- Links to other information

SBOL Integration with non-SBOL Data

Raytheon BBN Technologies

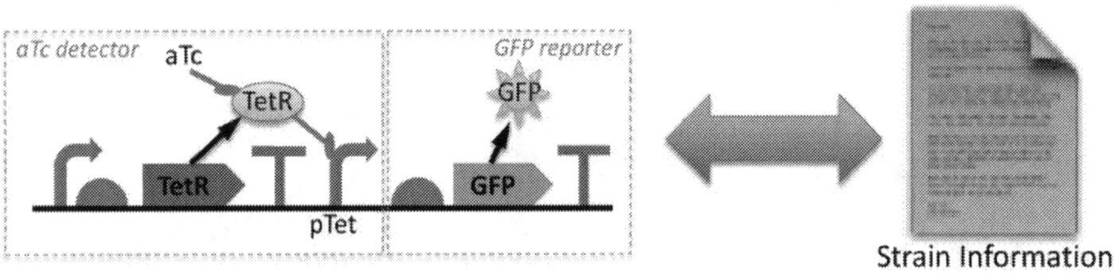

- Embed SBOL in data
- Embed data in SBOL
- Link SBOL to data
- Links data to SBOL

Whatever model works best for the tools...

SBOL SBOL supports many workflows

Raytheon BBN Technologies

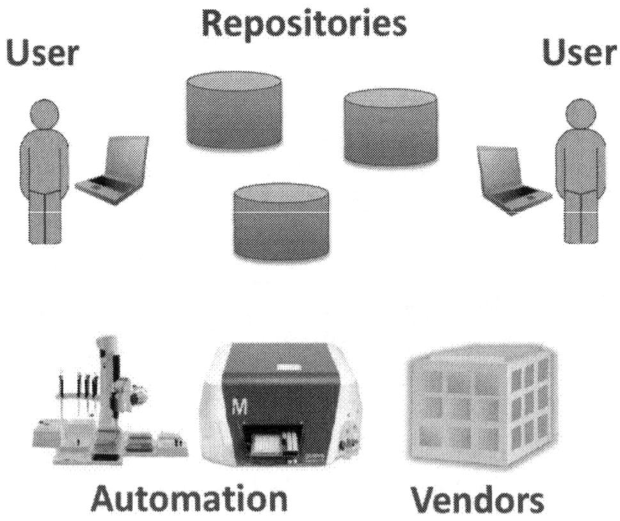

SBOL supports many workflows

Raytheon BBN Technologies

- Reuse/Share

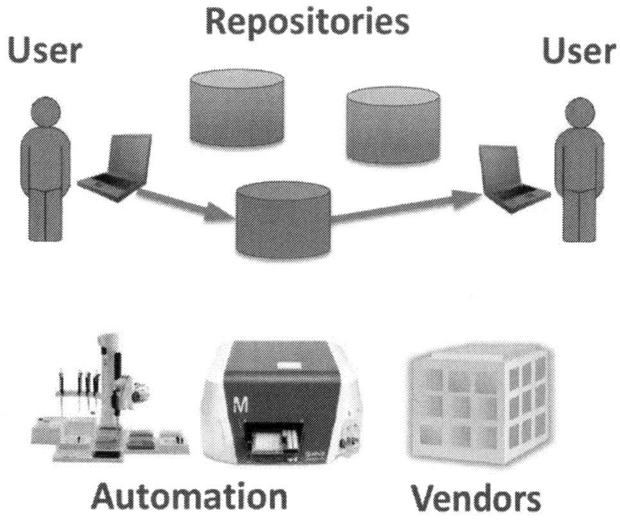

SBOL supports many workflows

Raytheon BBN Technologies

- **Reuse/Share**
- Visualize

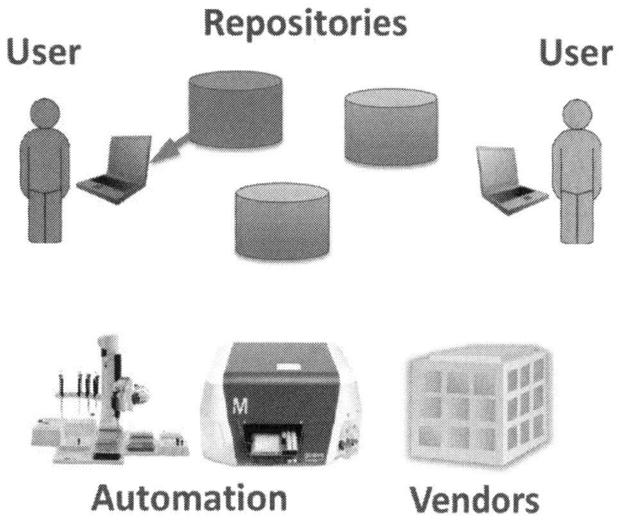

SBOL supports many workflows

Raytheon BBN Technologies

- **Reuse/Share**
- **Visualize**
- Design/Model

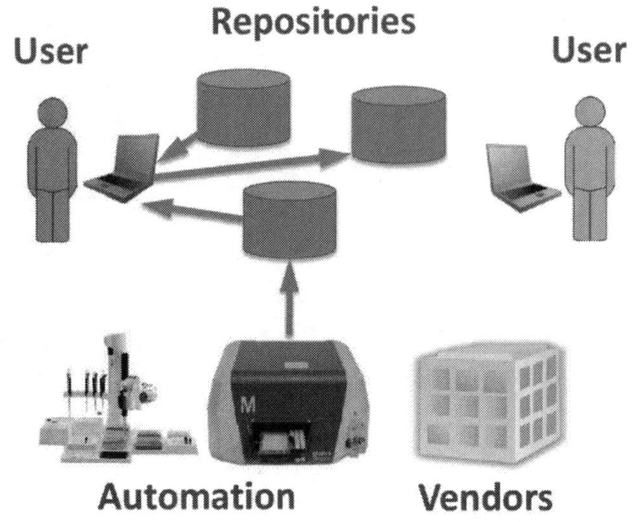

SBOL supports many workflows

Raytheon BBN Technologies

- **Reuse/Share**
- **Visualize**
- **Design/Model**
- Synthesize/Build

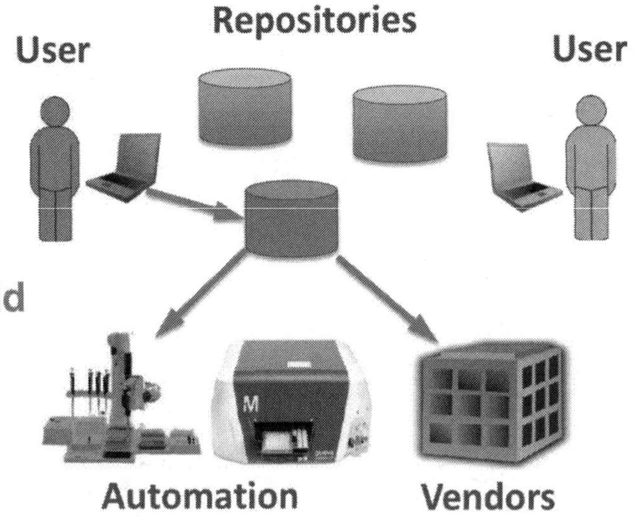

SBOL supports many workflows

Raytheon BBN Technologies

- **Reuse/Share**
- **Visualize**
- **Design/Model**
- **Synthesize/Build**
- Automate

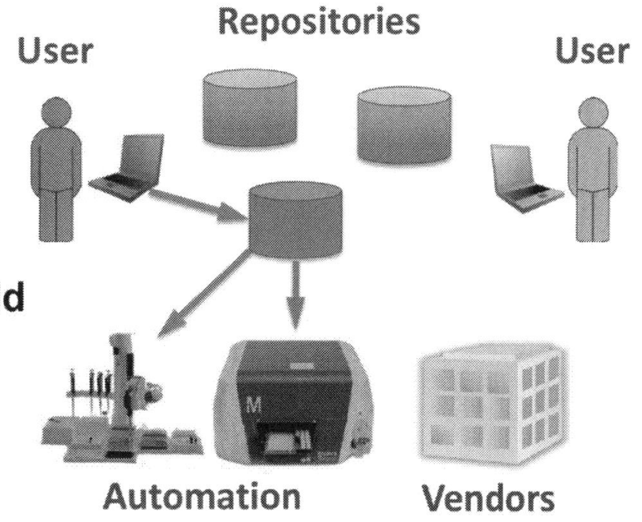

Workflow: Reuse & Share

Raytheon BBN Technologies

SBOLhub

- Searchable repository for public, private designs
- Journal integration for "one-click" private review
- Import SBOL/FASTA/GenBank, Exp

Newcastle University

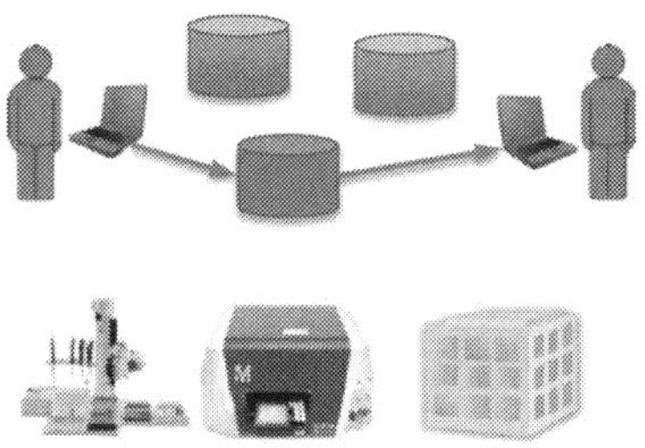

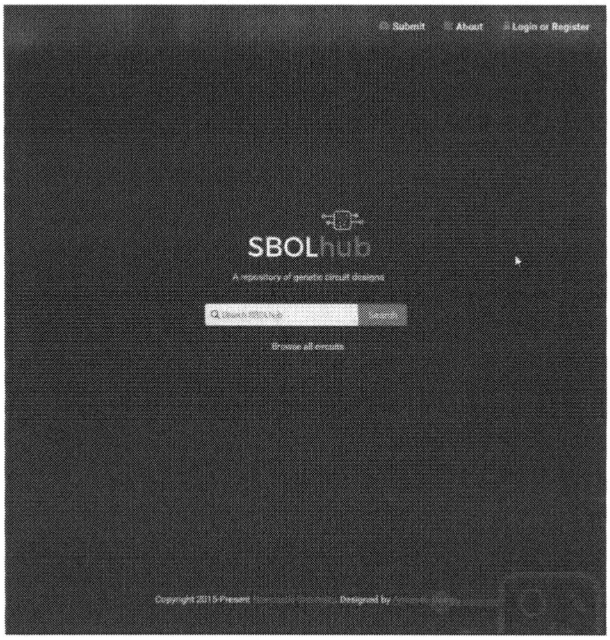

SBOL Workflow: Reuse & Share

Raytheon BBN Technologies

ICE

- Web of integrated public and private design registries
- Sequences, plasmids, strains, visualization
- SBOL, Genbank, FASTA

http://github.com/JBEI/ice

jbei Joint BioEnergy Institute

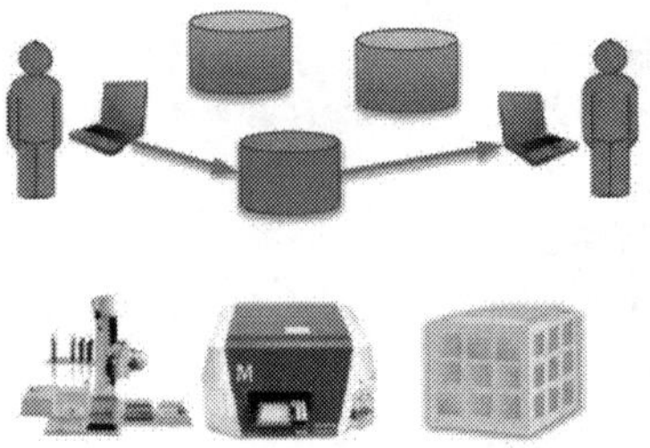

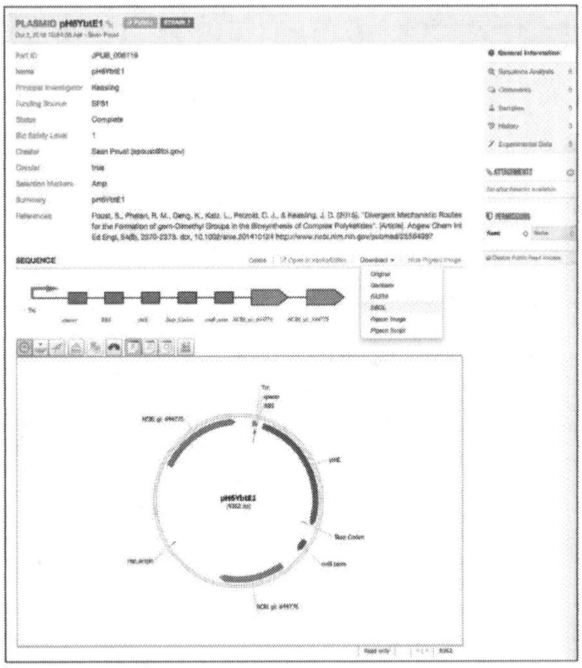

SBOL Workflow: Visualize

Raytheon BBN Technologies

Pigeon

- Input simple network description language
- Design visualization using SBOL glyphs

http://pigeon.synbiotools.org/

BOSTON UNIVERSITY

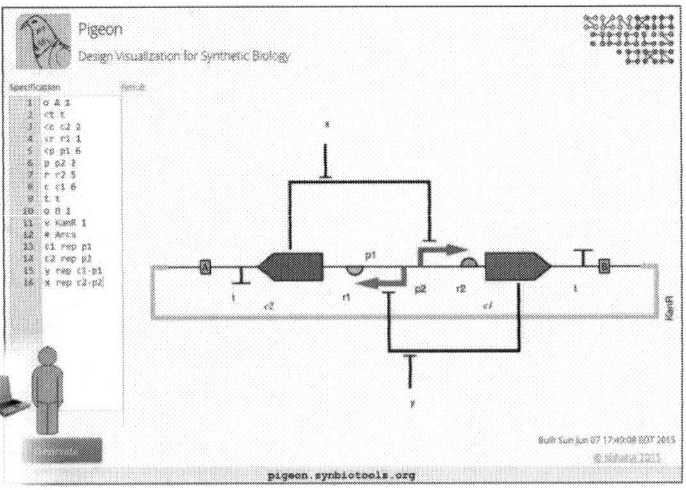

SBOL Workflow: Design & Model

Raytheon
BBN Technologies

iBioSim

- Graphical design of reaction networks
- Import or export both SBML and SBOL 2.0
- Simulate ODE, SSA

www.async.ece.utah.edu/iBioSim/

THE UNIVERSITY OF UTAH

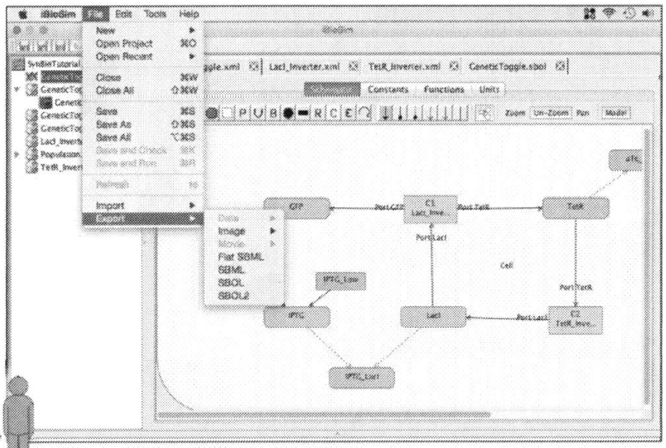

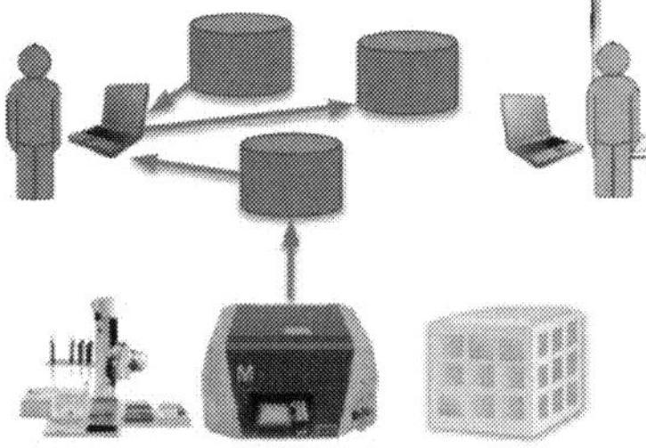

SBOL Workflow: Design & Model

Raytheon
BBN Technologies

TinkerCell

- Modular graphical design of GRNs
- SBOL import/export
- Simulate ODE, SSA

http://www.tinkercell.com/

W
UNIVERSITY of WASHINGTON

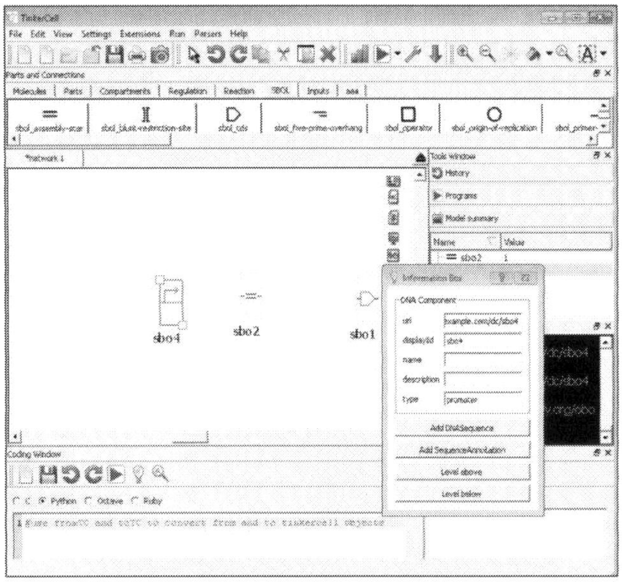

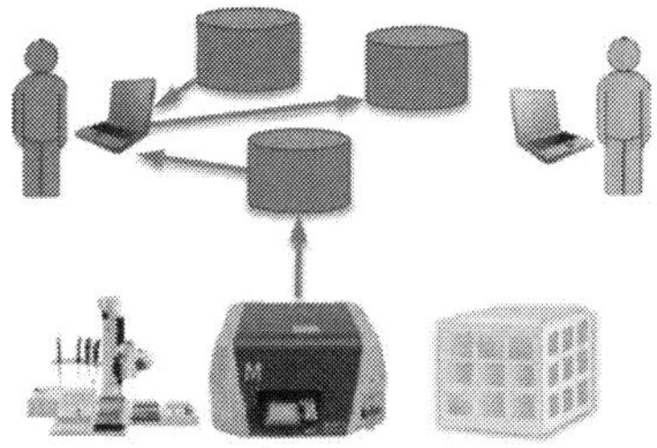

SBOL Workflow: Design & Model

Raytheon BBN Technologies

BioCompiler

- Generate GRN design from functional program
- SBOL visualization and design output
- ODE simulations

https://synbiotools.bbn.com/

Raytheon BBN Technologies

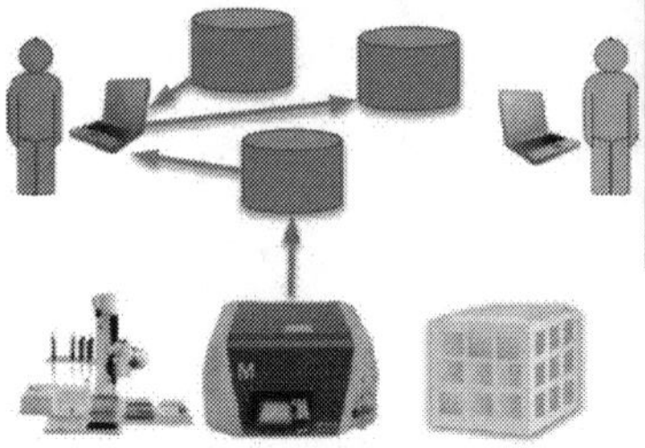

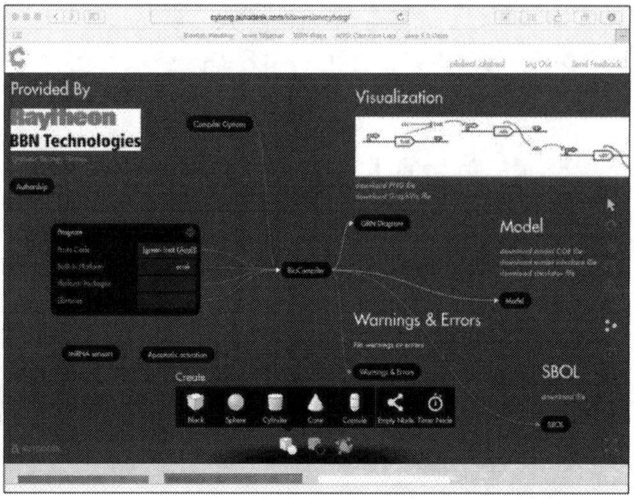

SBOL Workflow: Design & Model

Raytheon BBN Technologies

Cello

- Generate repressor design from logic spec
- SBOL encodes gate library and outputs
- Used w. TetR homologs

MIT BOSTON UNIVERSITY

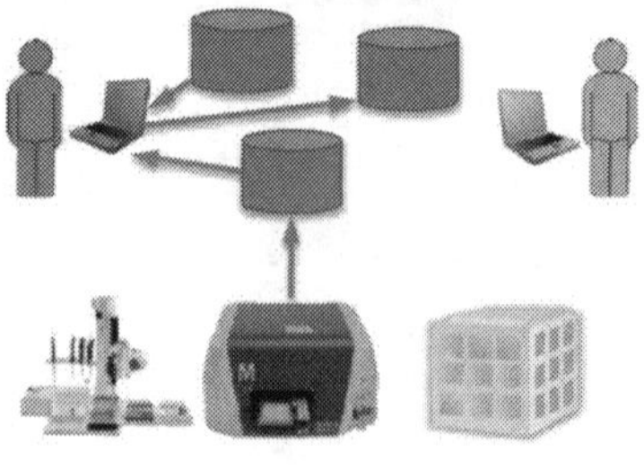

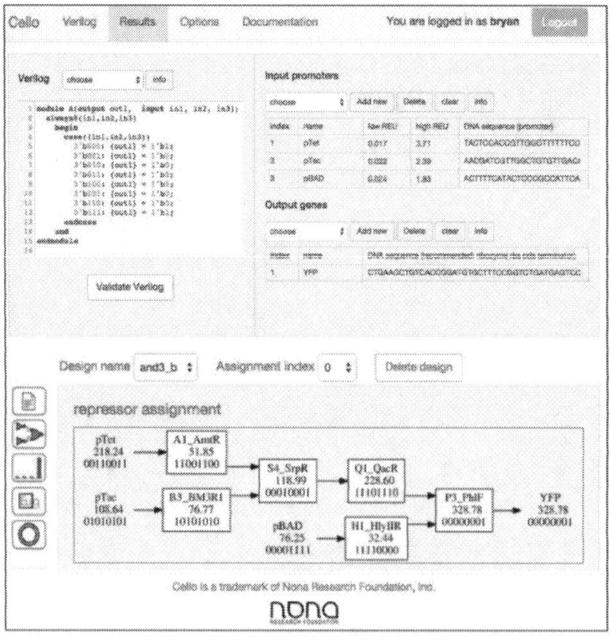

SBOL Workflow: Synthesize/Build

Raytheon BBN Technologies

VectorNTI

- SBOL import/export
- Sequence analysis, editing, and design
- Primer design for cloning, assembly

http://www.thermofisher.com/

ThermoFisher SCIENTIFIC

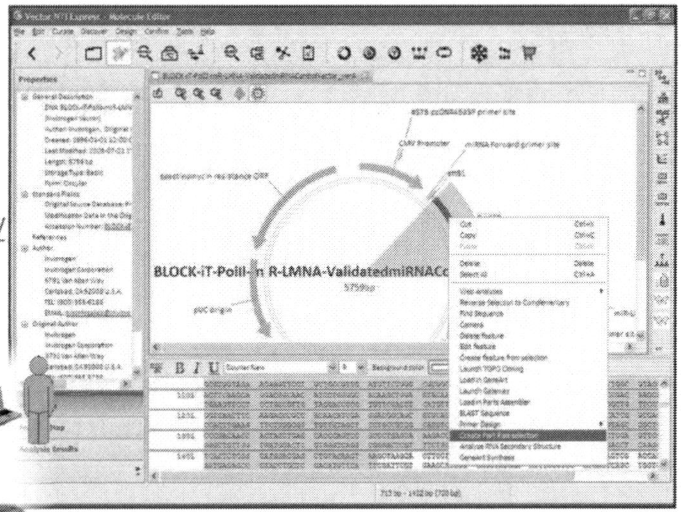

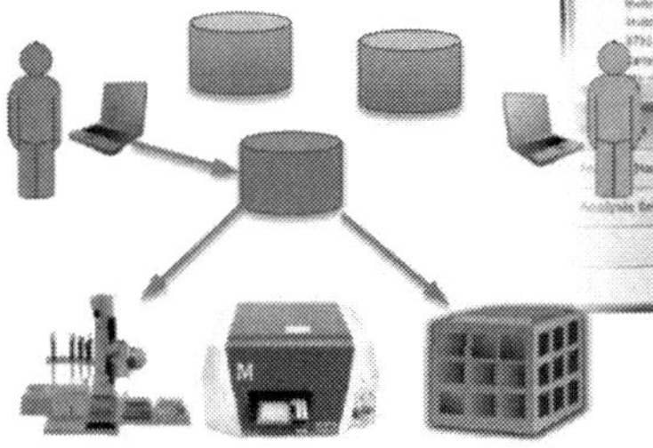

SBOL Workflow: Automate

Raytheon BBN Technologies

j5

- Import SBOL designs
- Generate assembly plans, oligos, protocols
- Support for onward laboratory automation

http://www.teselagen.com

teselagen BIOTECHNOLOGY

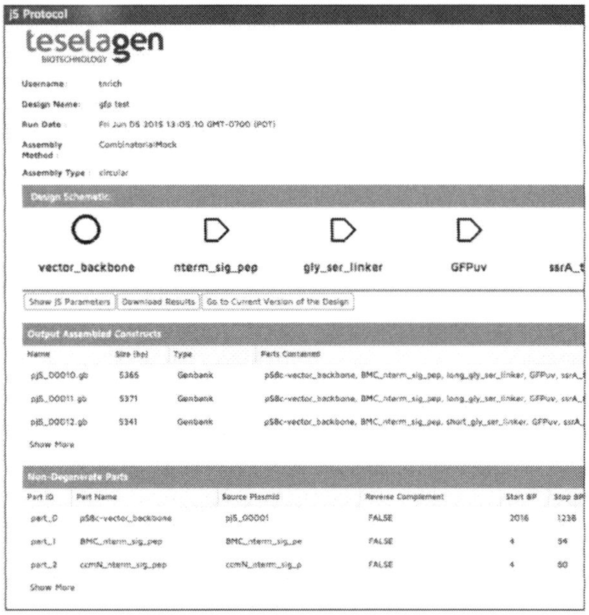

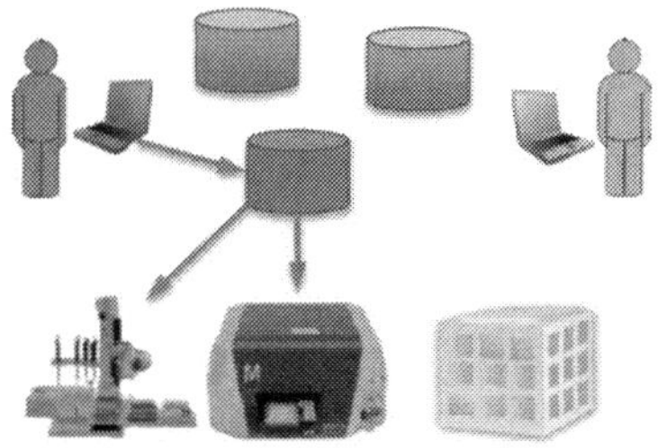

SBOL Resources for Users/Developers

Raytheon BBN Technologies

http://sbolstandard.org/

SBOL Call to Action!

Raytheon BBN Technologies

- Tool builders, use SBOL
- Laboratory practitioners, send your designs and tell us about what tools you need
- NIST S.B. Standards Consortium
  - Digital Biological Information WG
- Community meeting at IWBDA
  - August 19-21, Seattle

NIST

ACS SyntheticBiology

IWBDA 2015 August 20-21 Seattle, Washington

http://sbolstandard.org/

Mammalian Synthetic biology:
foundations and applications

Ron Weiss

Director, Synthetic Biology Center
Department of Biological Engineering
Department of Electrical Engineering and Computer Science
MIT

SEED, June 2015

MIT BE BIOLOGICAL ENGINEERING · SBC@MIT Synthetic Biology Center @ MIT · MIT CISB | Center for Integrative Synthetic Biology — An NIGMS-funded National Center for Systems Biology · EECS

Our synthetic biology workflow

Today:
- RNA as a platform for therapeutic applications
- Programmable organoids

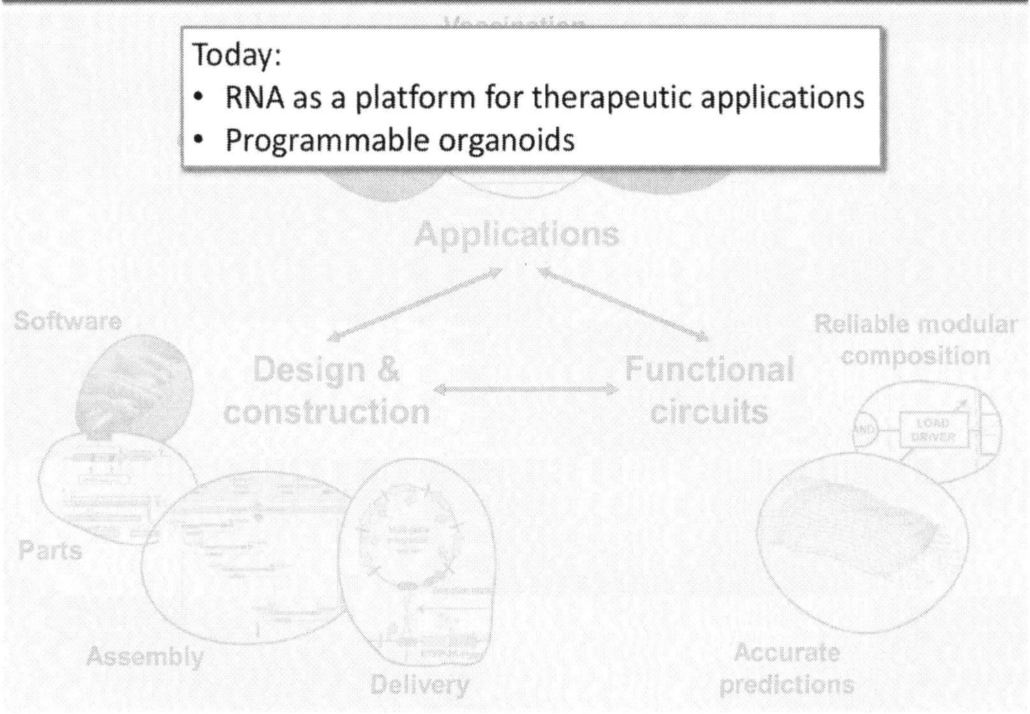

RNA therapy advantages

- No genomic integration
- No nuclear barrier
- Immediate expression
- Expression in non-dividing cells
- No epigenetic silencing
- Non-permanent expression

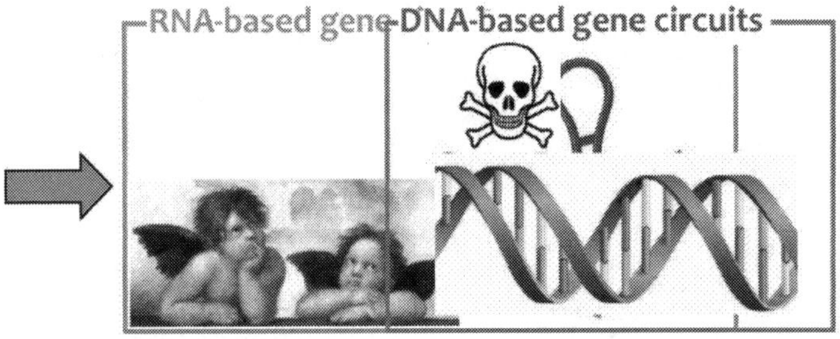

mRNA-based therapy: applications

mRNA used in several applications
- Cancer immunotherapy
- Infectious disease vaccination
- Genome engineering
- Genetic reprogramming

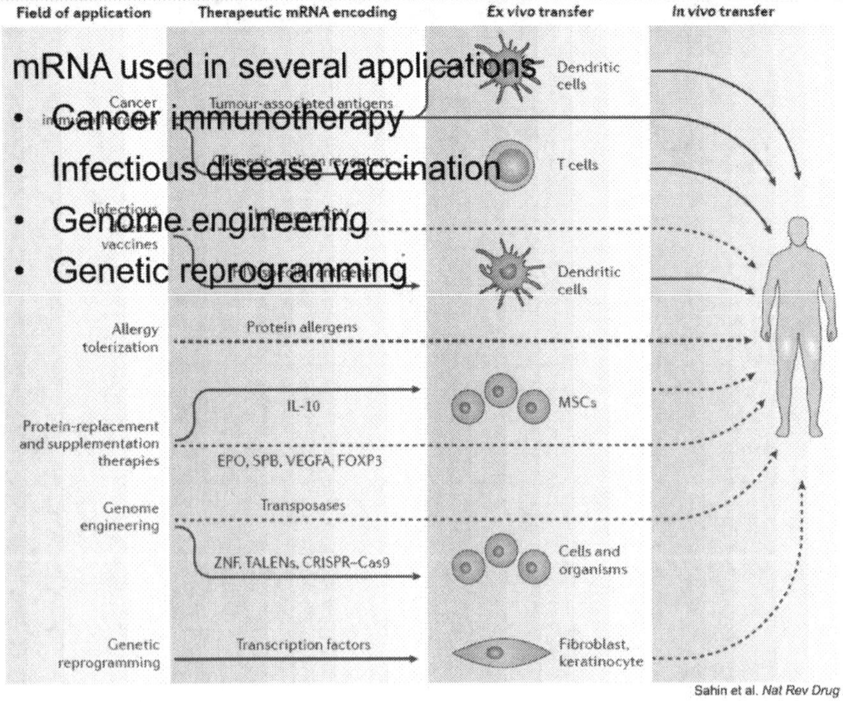

Sahin et al. *Nat Rev Drug Discov.* (2014)

RNA expression platforms

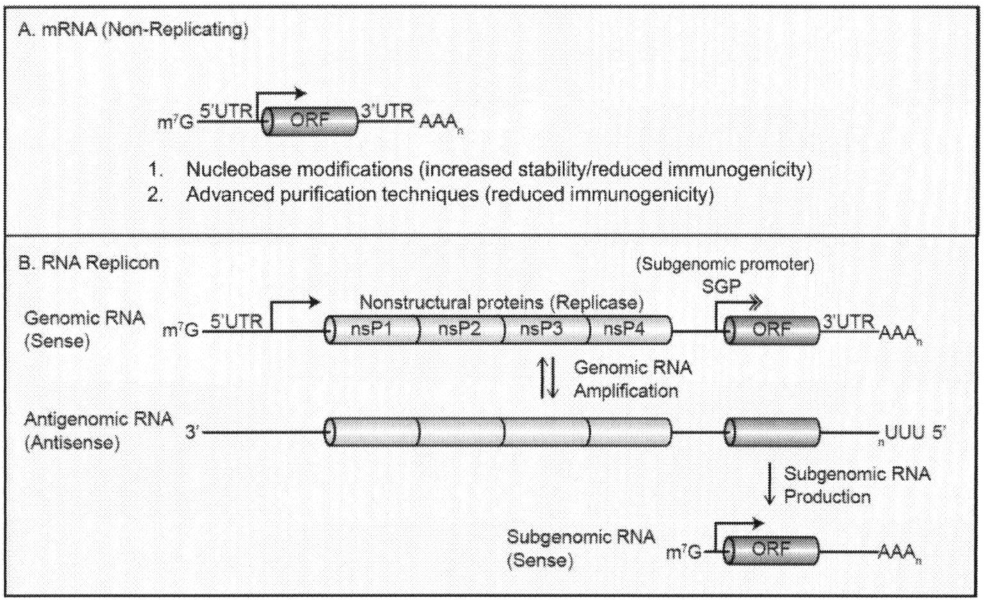

Andries*, Kitada* et al. *Expert Rev Vaccines* (in press)

Lipid nanoparticle delivery gives sustained replicon expression *in vivo*

1 µg replicon RNA encoding luciferase:
L: Naked RNA; R: LNP+RNA

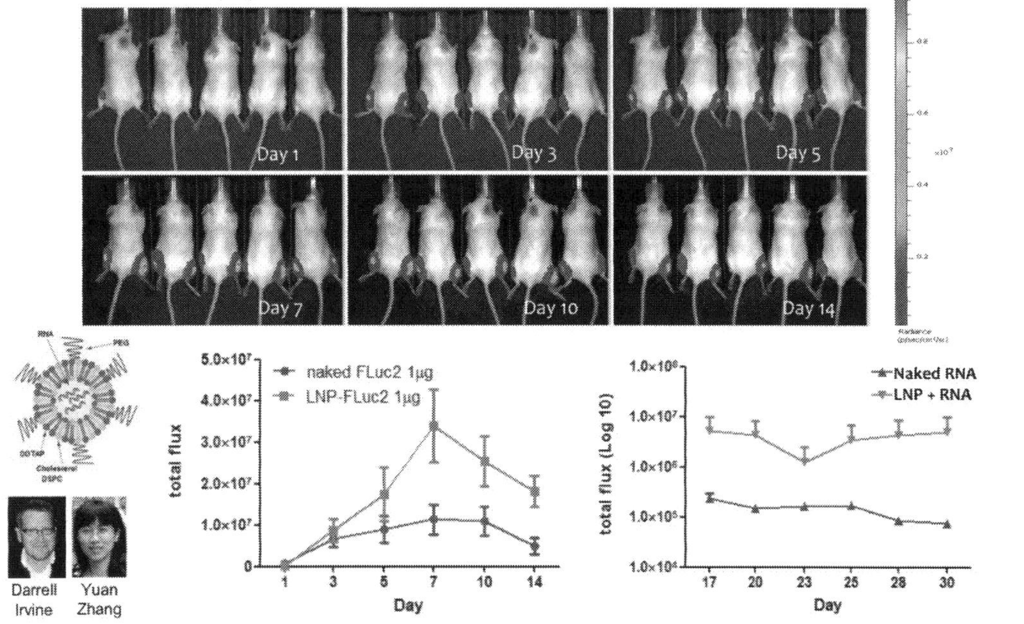

Replicons expressing SIV gag induce potent immune responses

Mice were immunized with 6ug RNA replicons encapsulated in LNP via i.m. injection

* miniSIVgag = HIV antigen

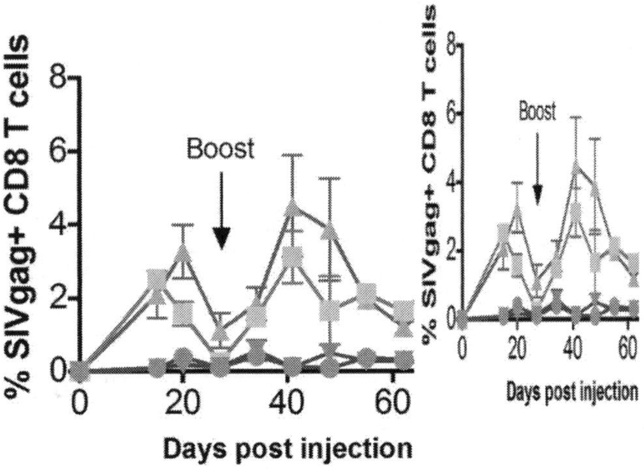

Darrell Irvine   Mariane Bandeira Melo

In vivo mouse model of breast cancer

Quantitative imaging of tumor cells *in vivo*

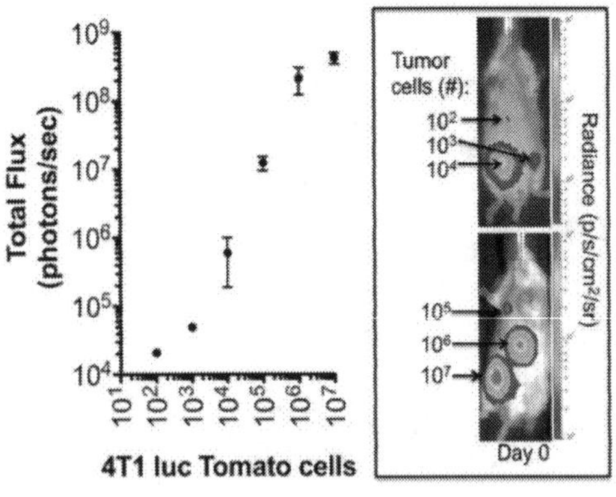

Tumor growth kinetics *in vivo*

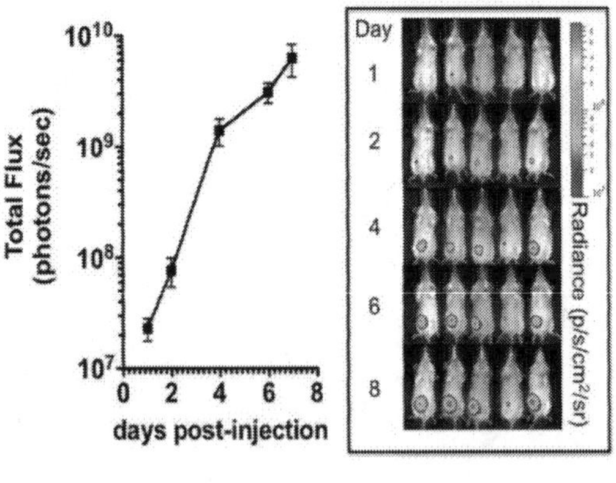

Initial experimental results for our cancer immunotherapy with constitutive expression of various immunomodulators (IL-12, IL-15, TGFβ, MLKL):

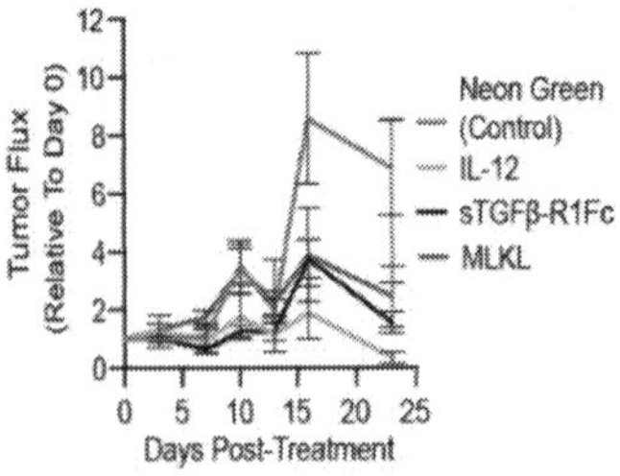

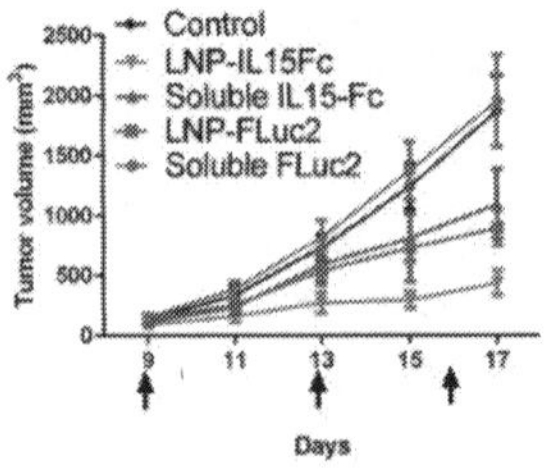

mRNA-based therapy: challenges

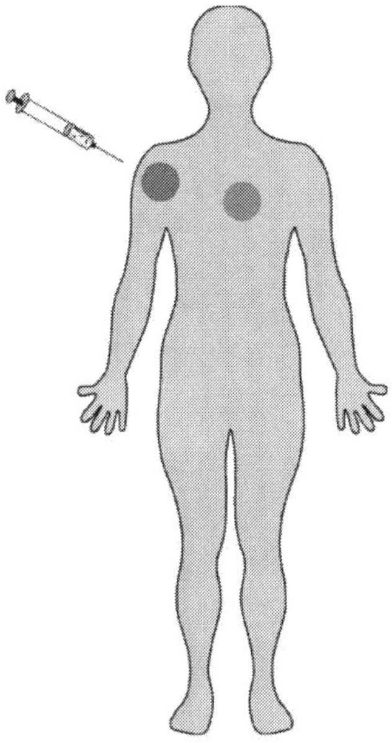

1. Ensure tissue specific expression of RNA
   - To prevent expression in non-target tissues
2. Control the quantity of protein expression
   - For toxic proteins
   - For a protein with narrow therapeutic window
   - To deal with inter/intra-individual variation

Devices for post-transcriptional gene regulation

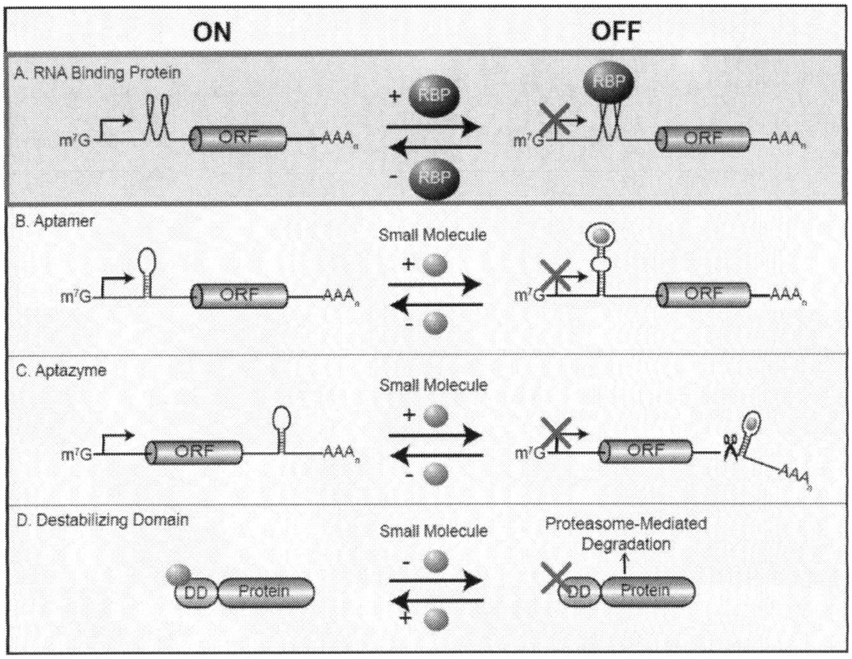

Andries*, Kitada* et al. *Expert Rev Vaccines* (2015)

ynthetic mpli i d NA (SafeR) circuits

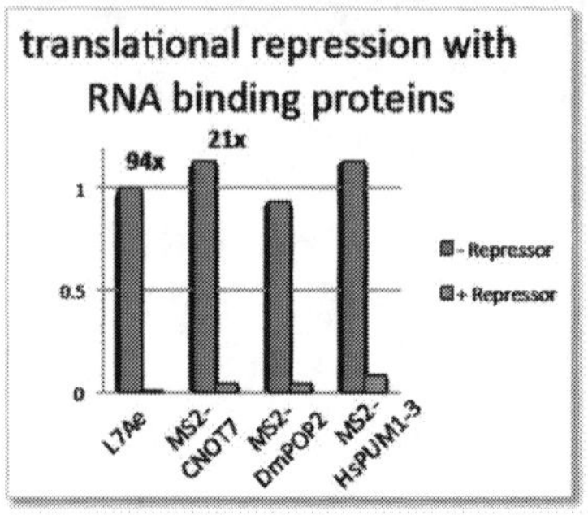

Feedback regulation: two-state switch

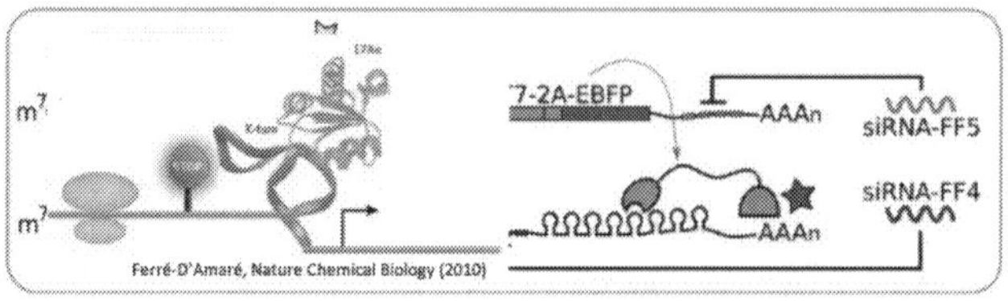

Multi-input microRNA sensor

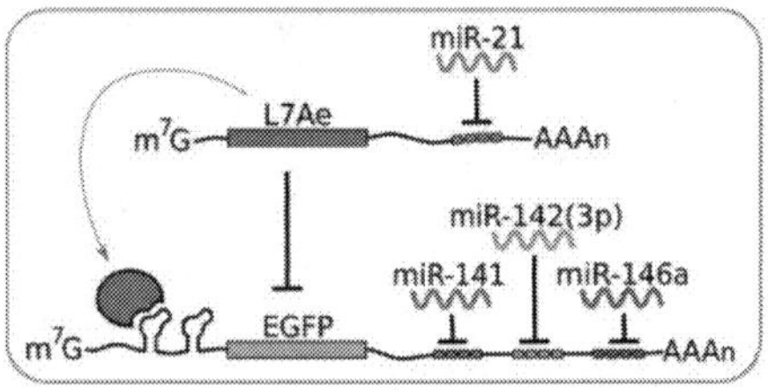

Information transmitter: cascade

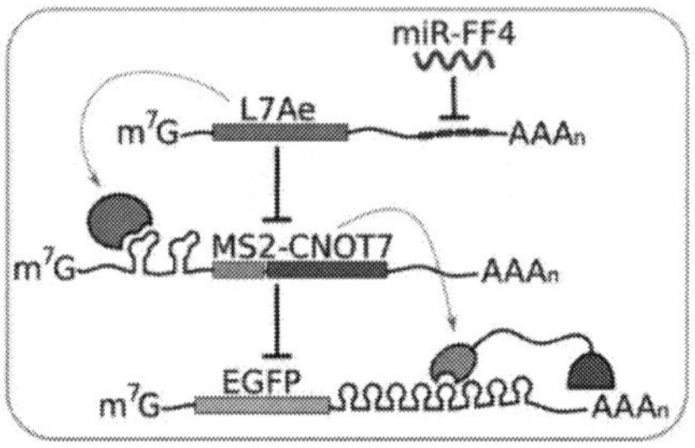

Establish an **RNA based framework** for circuits that sense, process, and actuate in mammalian cells: RNA replicon encoded SafeR (synthetic amplified RNA) circuits

Multi-input microRNA/RBP classifier for HeLa cells

HeLa cancer cells

- ↑ miR-21
- ↓ miR-141
- ↓ miR-142
- ↓ miR-146

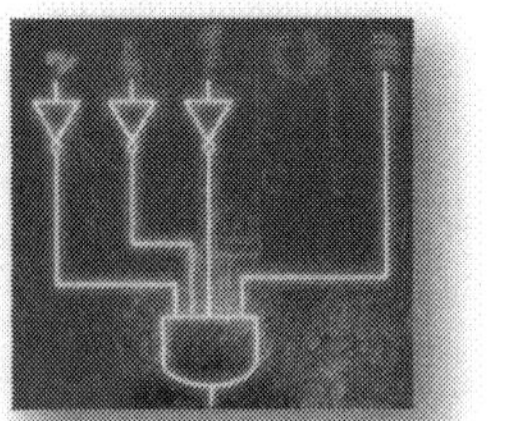

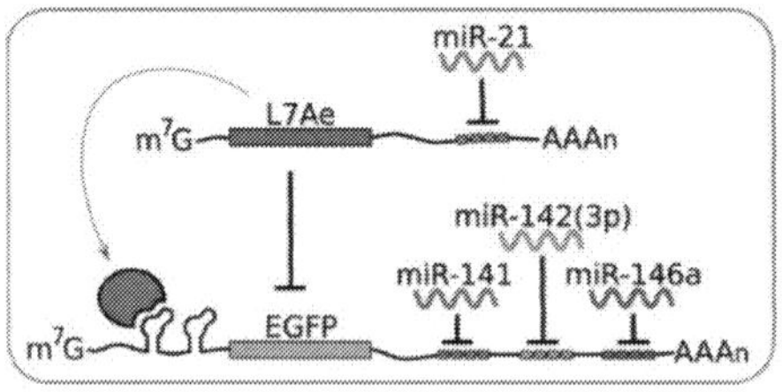

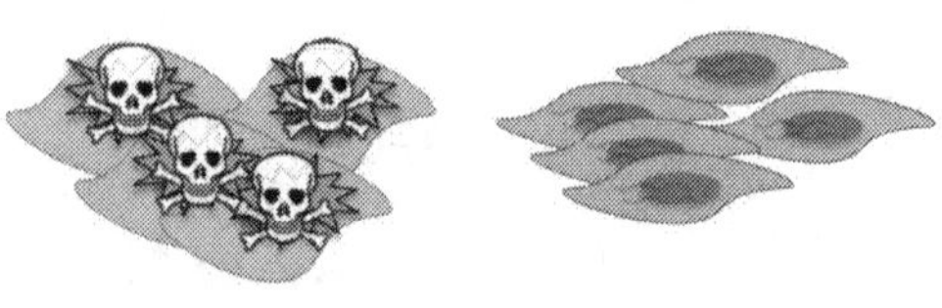

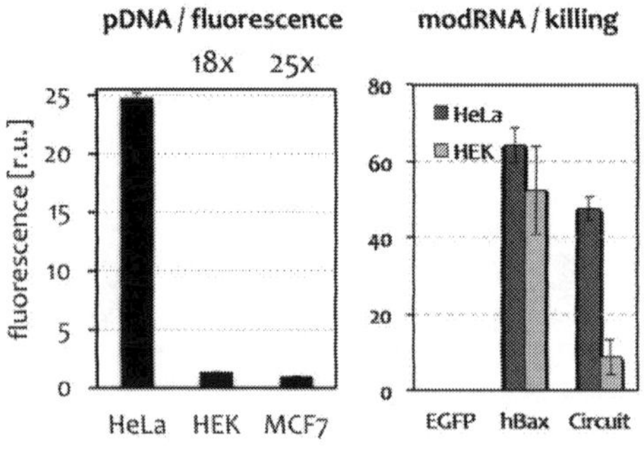

Liliana Wroblewska, Tasuku Kitada, Velia Siciliano, Bre Stillo, Hiro Saito

Wroblewska, *et al.*, Nature Biotech (*accepted*)

Cascade: information transmitter

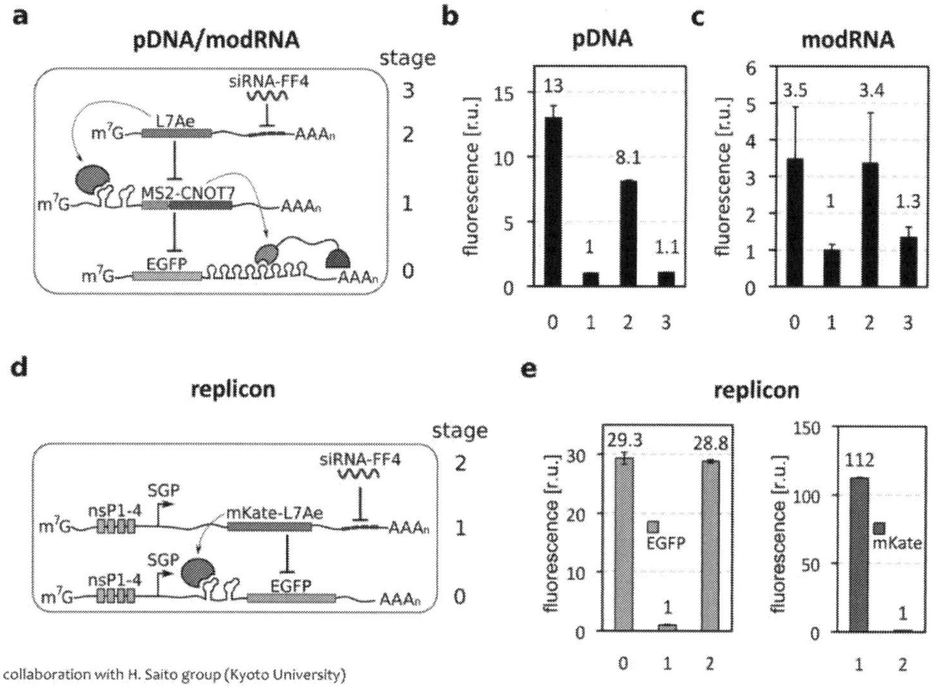

* in collaboration with H. Saito group (Kyoto University)

Two-state RBP switch: feedback regulation

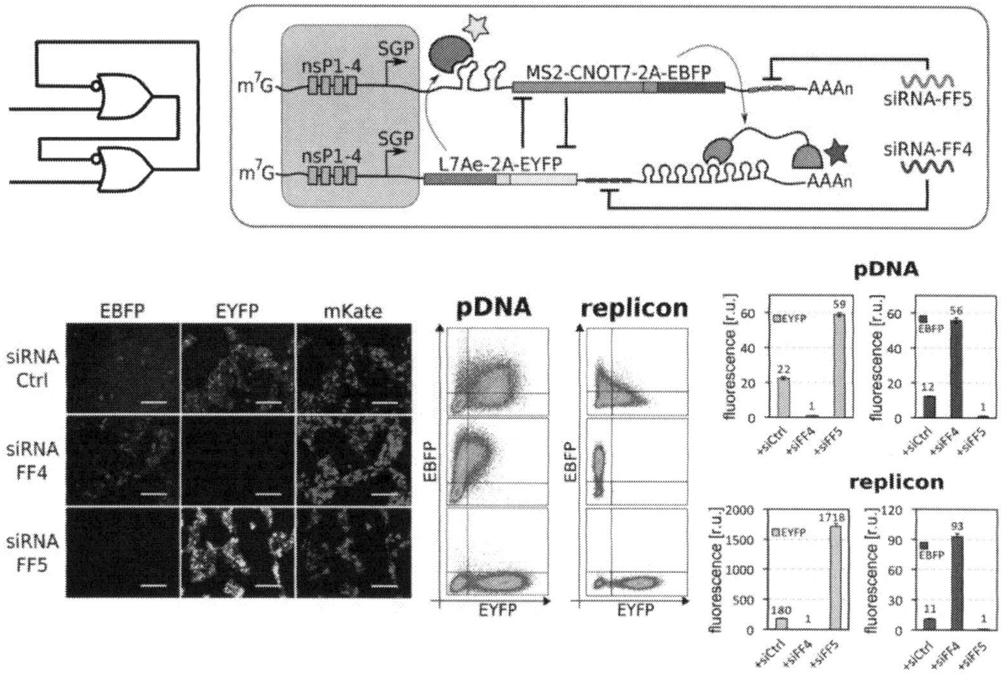

Computational model for toggle switch

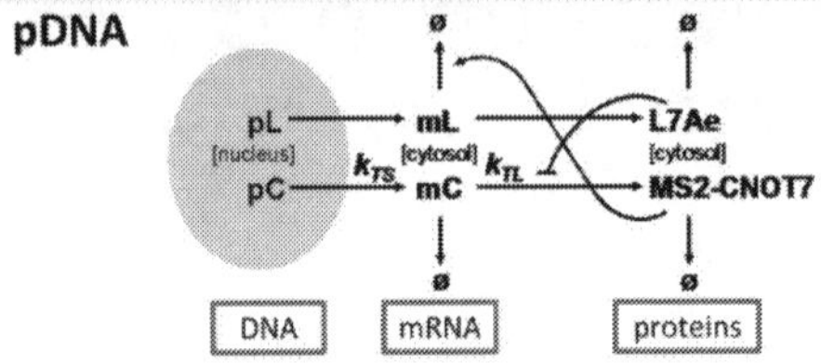

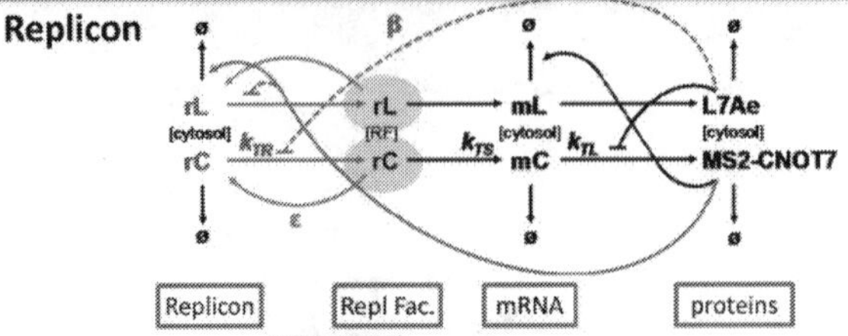

pDNA Parameter Sensitivity Analysis
Copy Number
Transcription Rate
Copy Number Variance
Protein Degradation Rate

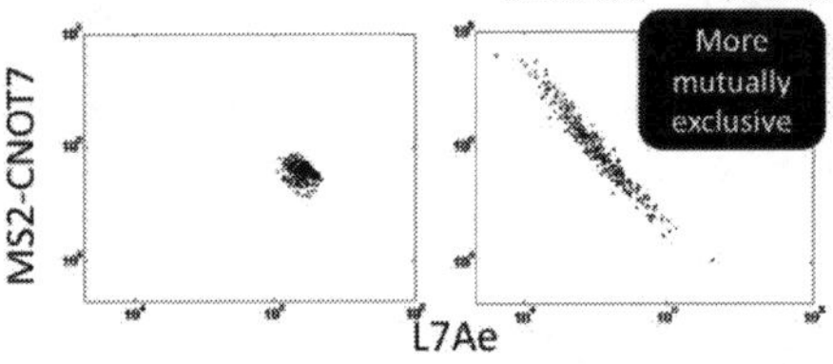

Replicon Parameter Sensitivity Analysis
Copy Number
Transcription Rate
Positive feedback
Negative Feedback
Replication Factory Formation

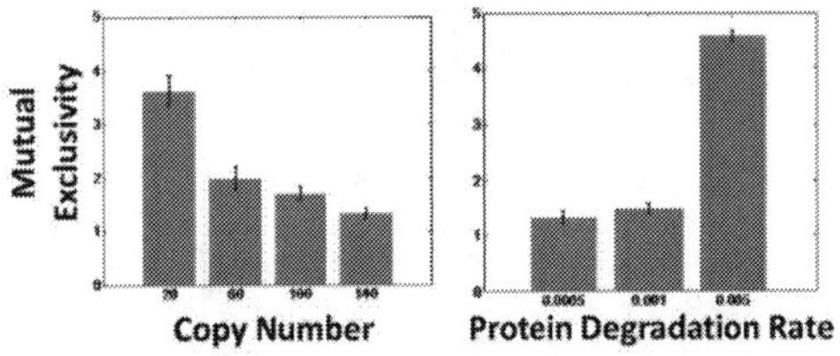

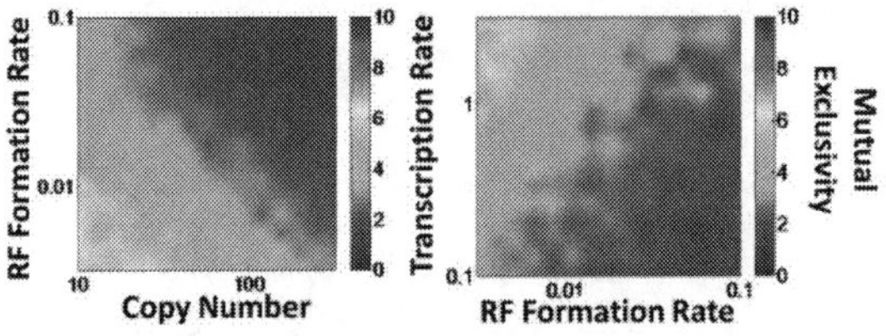

pDNA shows low initial mutual exclusivity at **high copy number** and **slow protein decay**

Replicons show high mutual exclusivity for **stochastic RF formation** and **fast transcription**

Post-transcriptional regulation

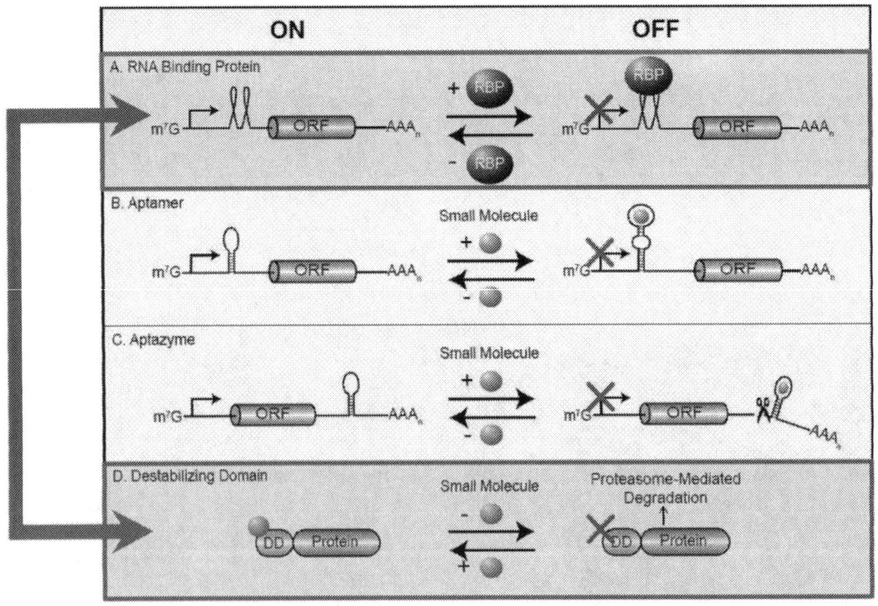

- DD + RBP = small molecule dependent RBP repression

In vivo small molecule induction from replicon circuits

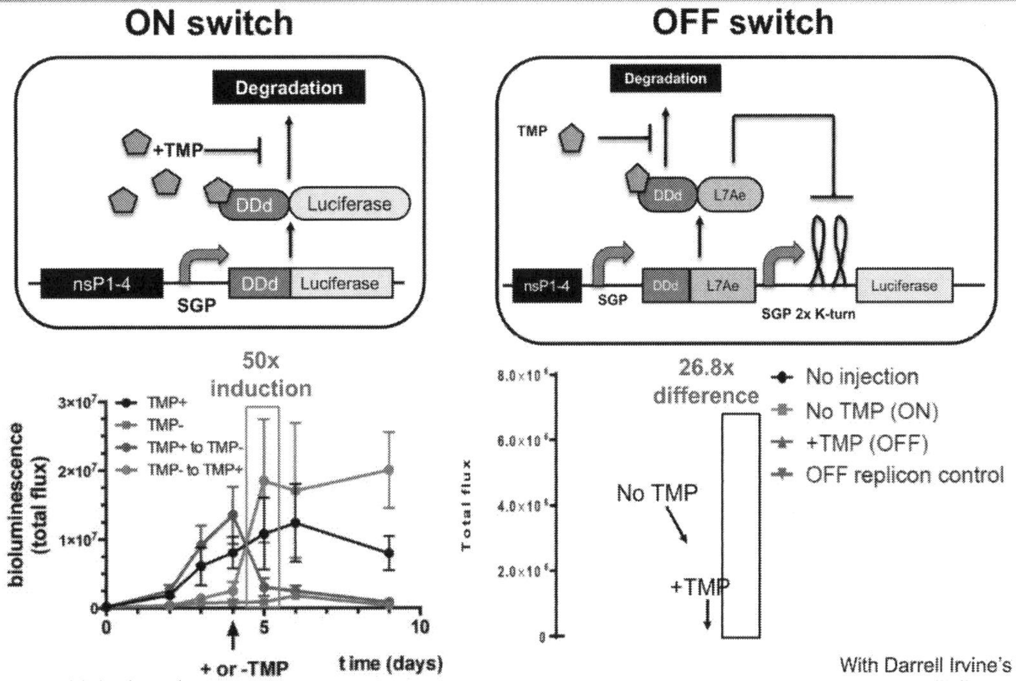

Unpublished results

With Darrell Irvine's group (MIT)

Summary of RBP-based circuits

- RNA-binding proteins exhibit good properties for circuit implementation
- RBP devices can be composed into more sophisticated circuits
- Long-term expression with replicons *in vivo*
- Effective (FDA-approved) small molecule / RBP regulation *in vivo*
- Opportunity for improved safety over DNA delivery for a variety of therapeutics applications

Tissues-by-design / Organoids

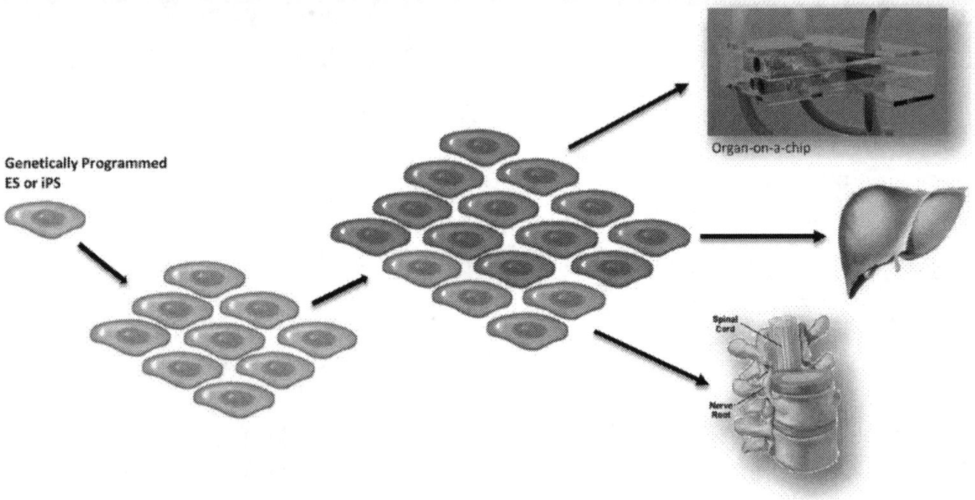

- Can we create and maintain large scale spatially defined functional tissues?
- Possible applications:
  - ✧ Drug development
  - ✧ Organ transplantation

Stem Cell-derived Organoids

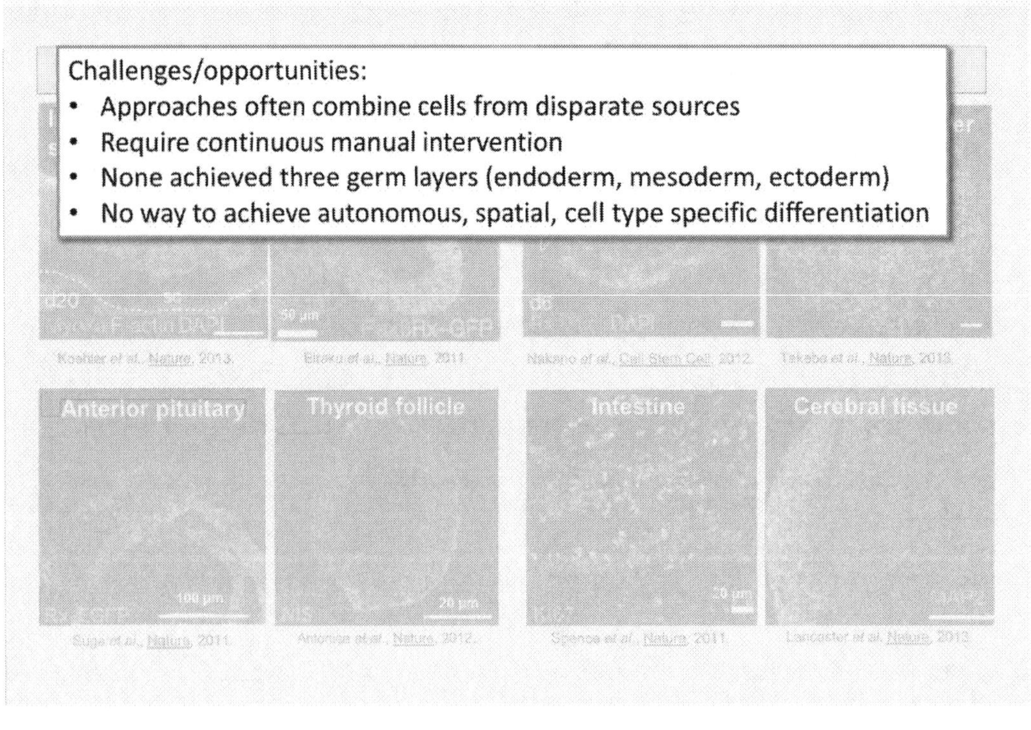

Challenges/opportunities:
- Approaches often combine cells from disparate sources
- Require continuous manual intervention
- None achieved three germ layers (endoderm, mesoderm, ectoderm)
- No way to achieve autonomous, spatial, cell type specific differentiation

A self-timed genetic program for β cell differentiation

High level behavior:

```
"Self-timed, multi-step
β cell differentiation"
```

Bio-program for iPS:

```
input: DOX
if (DOX)
  express(Endoderm-CFR)
if (cell-is-Endoderm?)
  express(Islet-CFR)
if (cell-is-Islet?)
  express(GFP)
```

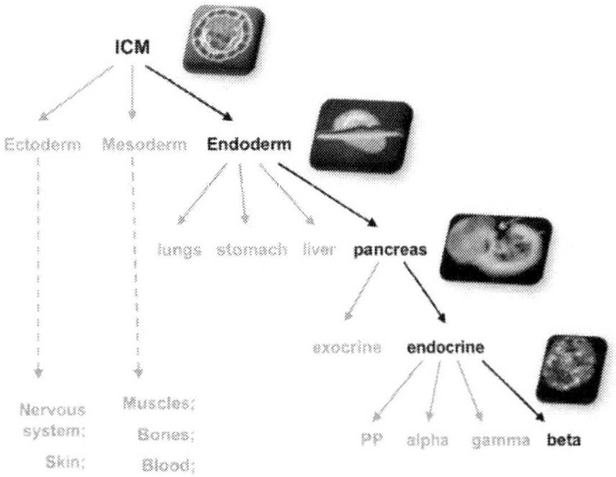


```
CFR = Cell fate regulator
```

Step 1: Genetically Engineer hiPS to Endoderm fate with Gata6

Dev Cell. 2014 May 27;29(4):454-67, PMID: 2483 5466

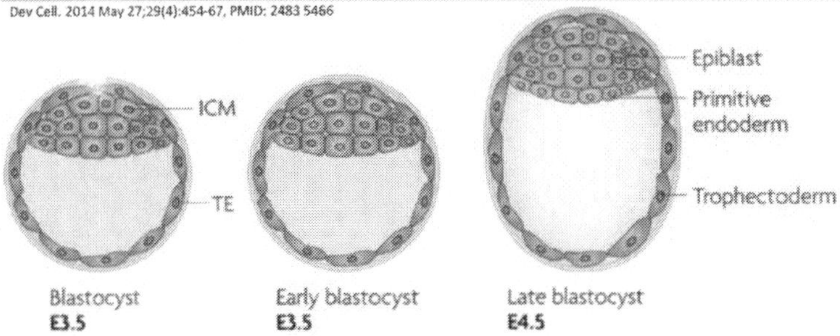

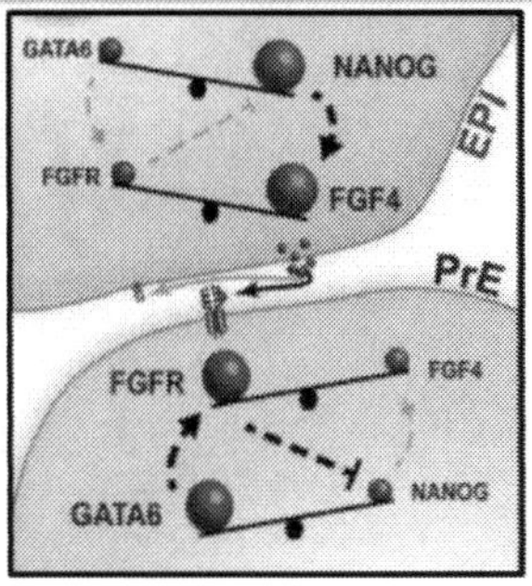


```
If (Dox)
  [→ become Endoderm]
  express(Gata6)
Else
  [→ remain pluripotent (later ectoderm)]
  [Nanog remains high]
```

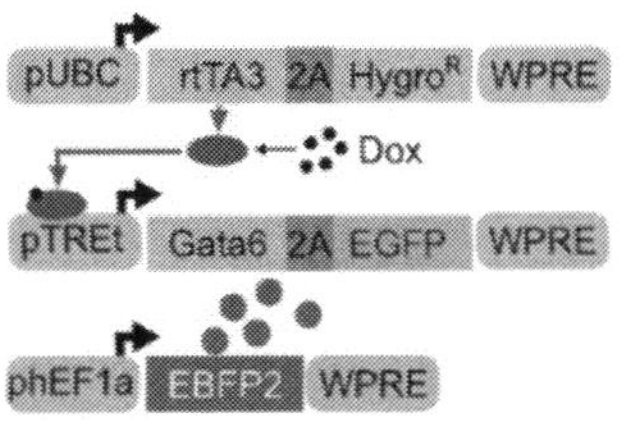

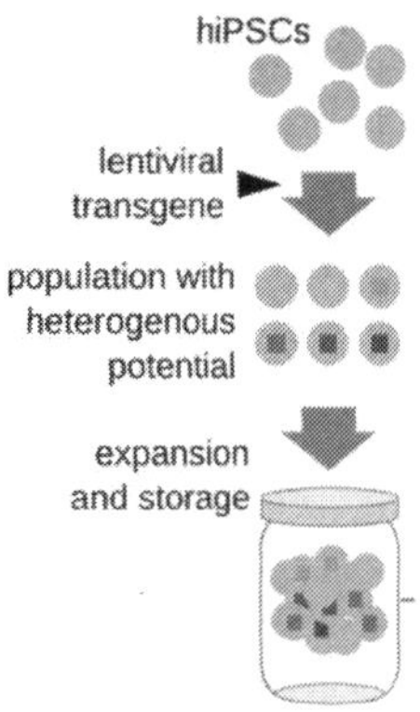

Patrick Guye

Generation of endoderm sub-populations thru symmetry breaking

hiPS Cell line A: destined to be endoderm

hiPS Cell line B: remain pluripotent initially (later ectoderm)

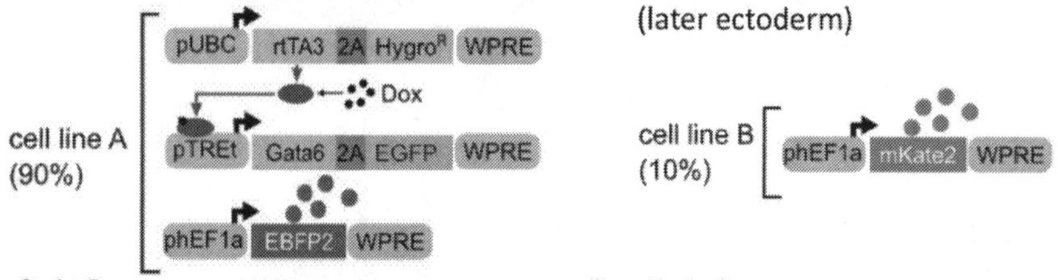

Day 0: A+B

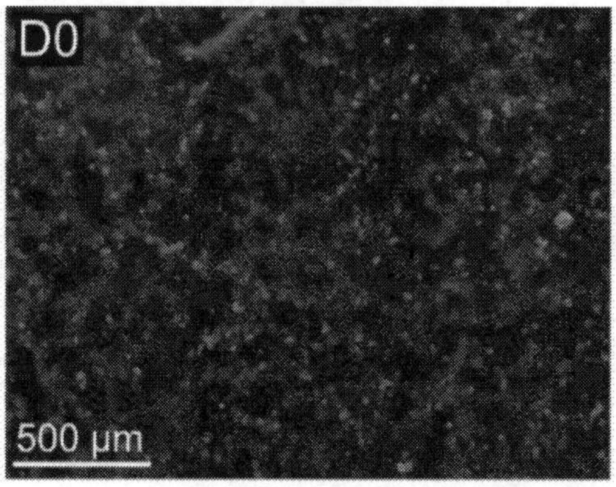

Day 5: A+B

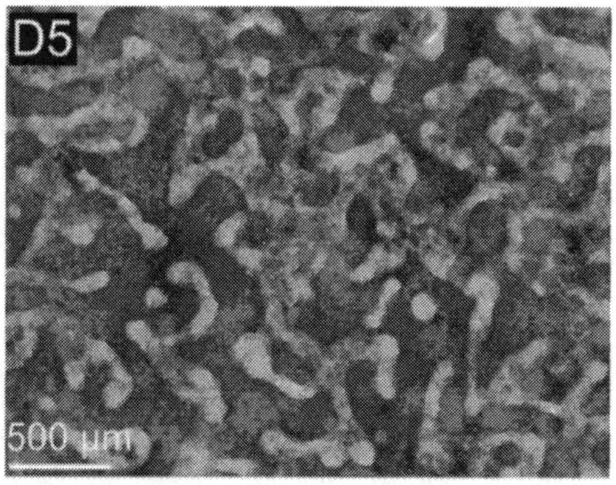

Even in cell line A: critical symmetry breaking!

Cell fate decision switch: Gata6 levels + nearby cells

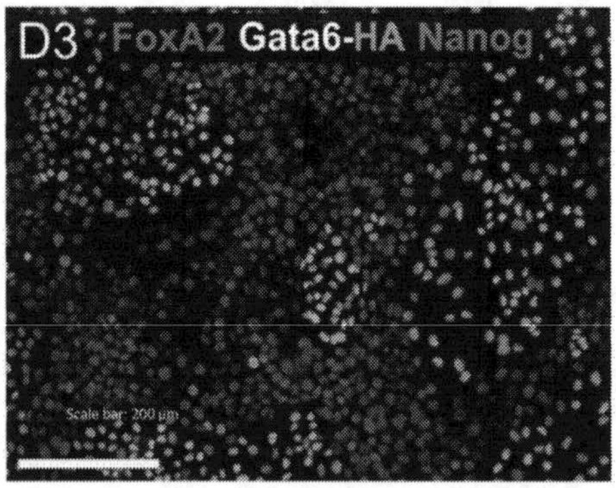

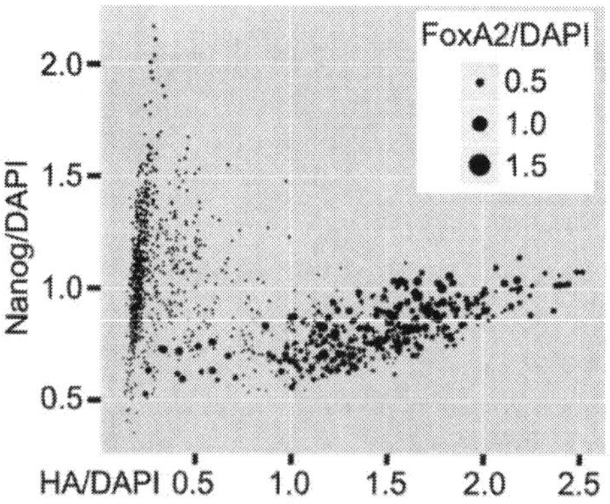

Cell fate switch	
High Gata6:	Endoderm
Medium Gata6:	Correlates with neighbors
Low Gata6:	Pluripotent

FoxA2 = Endoderm / hepatic marker

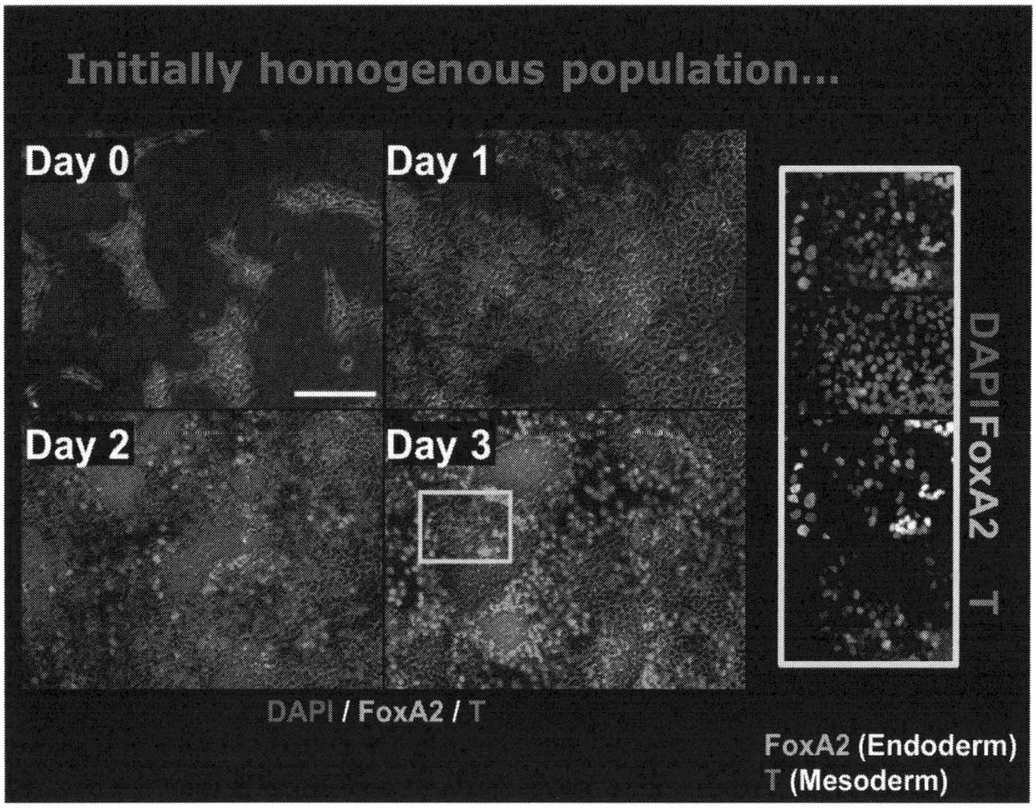

Further development of cell fates

Mesoderm → Vasculature (CD31, CD34)

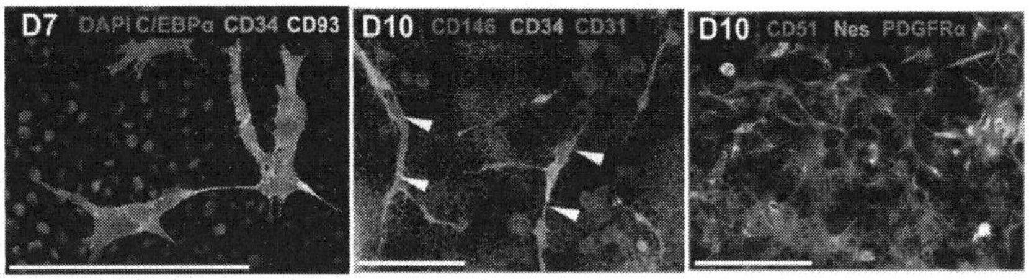

Endoderm → Hepatocytes (AAT, C/EBPα)

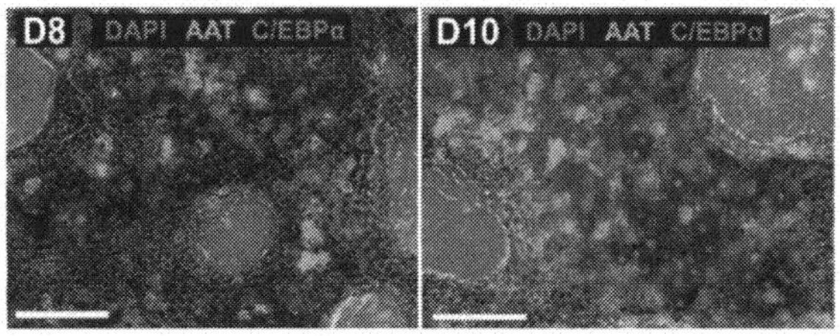

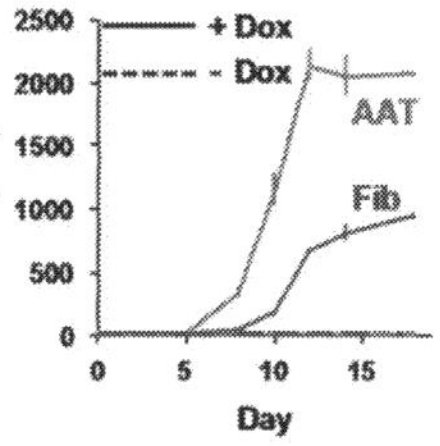

AAT and Fibrinogen indicate hepatocyte activity

Liver-like tissue

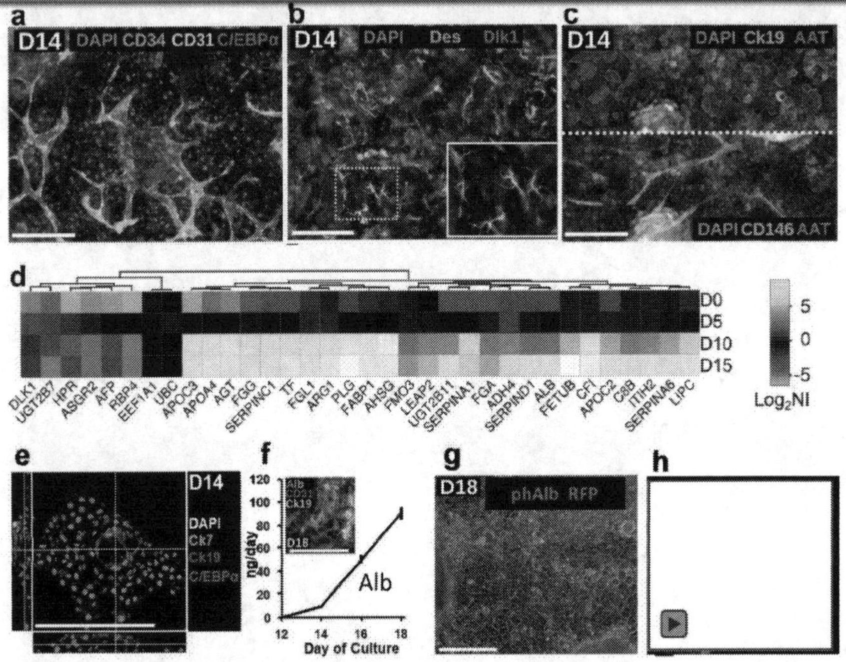

Alb = Albumin (indicates hepatocyte activity)

Initial Vasculature and Hematopoiesis

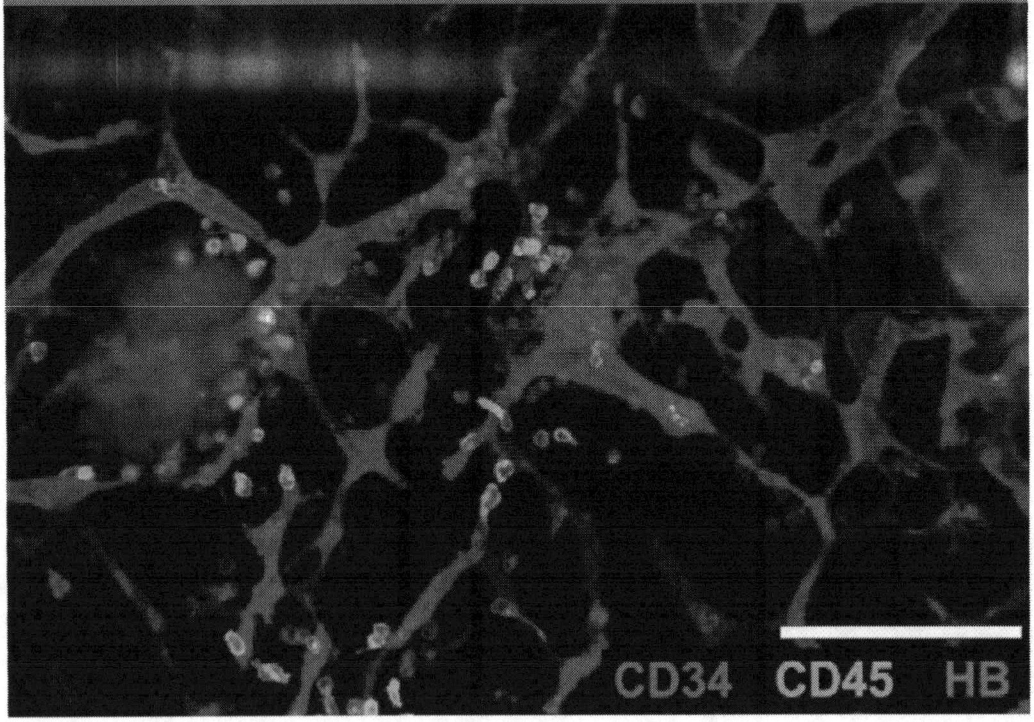

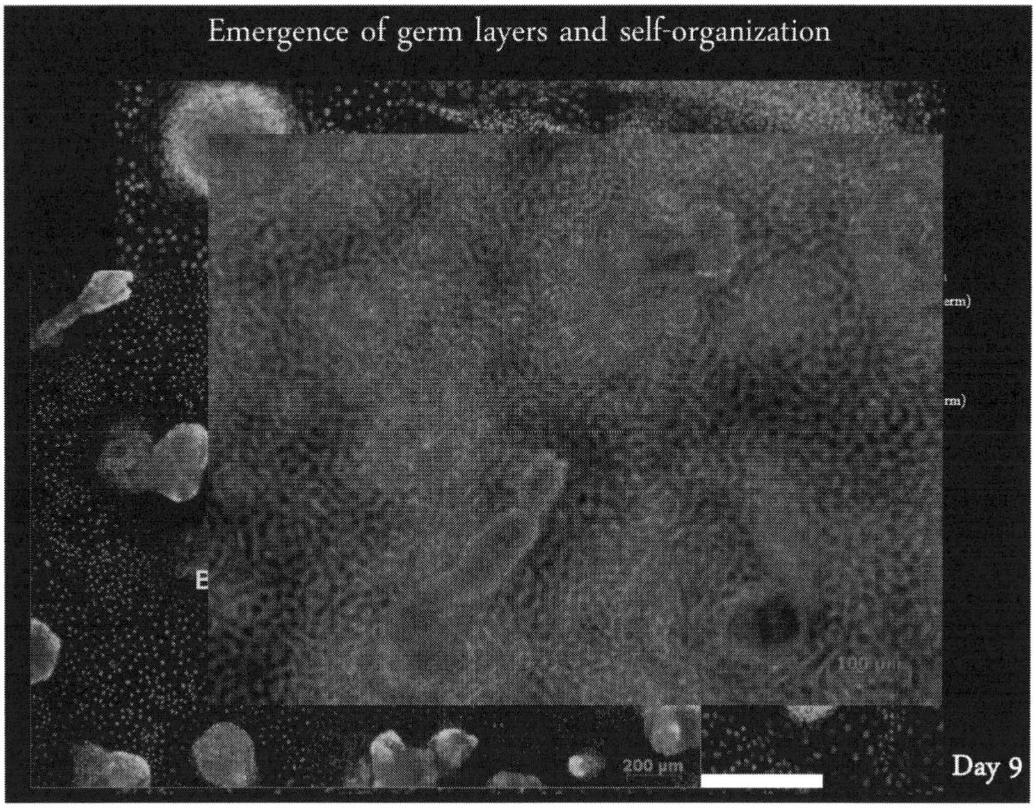

Neuronal progenitors

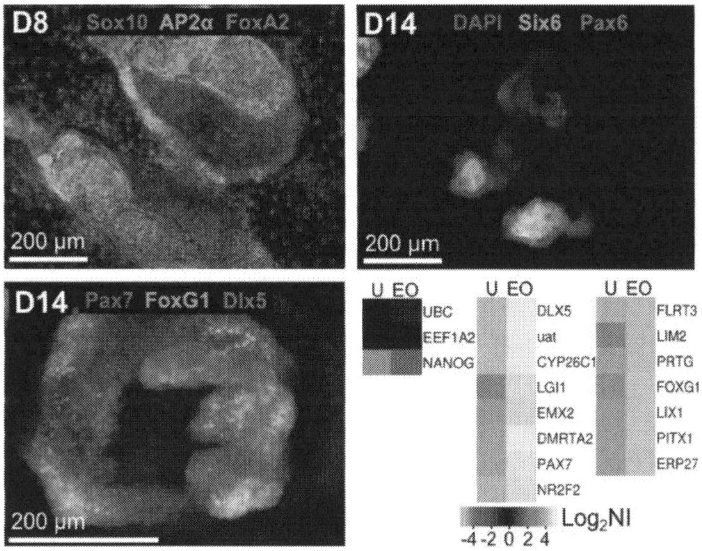

Generated all cell types known to exist in liver bud next to neuronal progenitors from Human Pluripotent Stem Cells

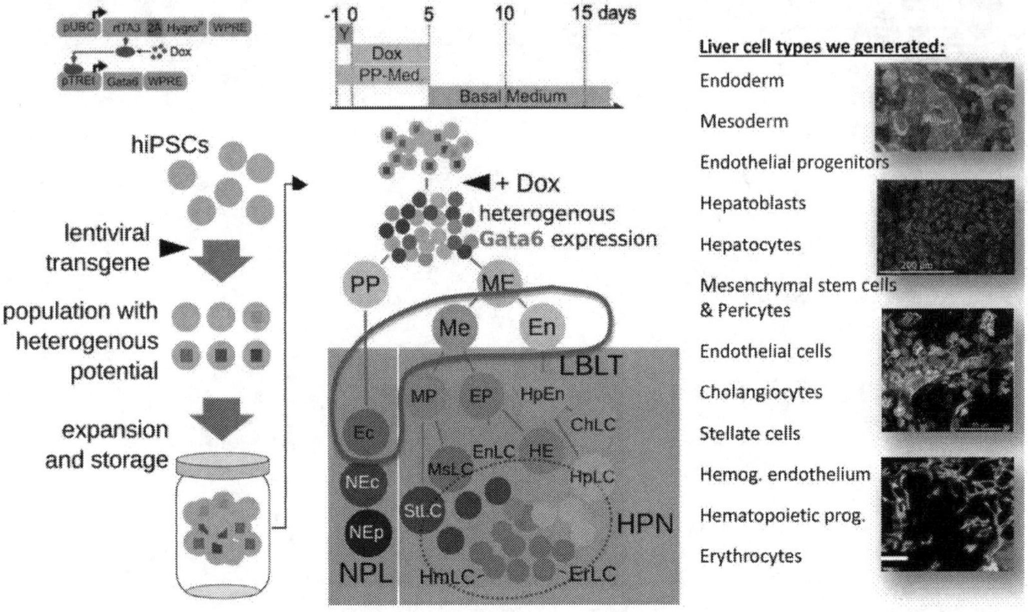

Liver organoid as drug toxicity testing platform

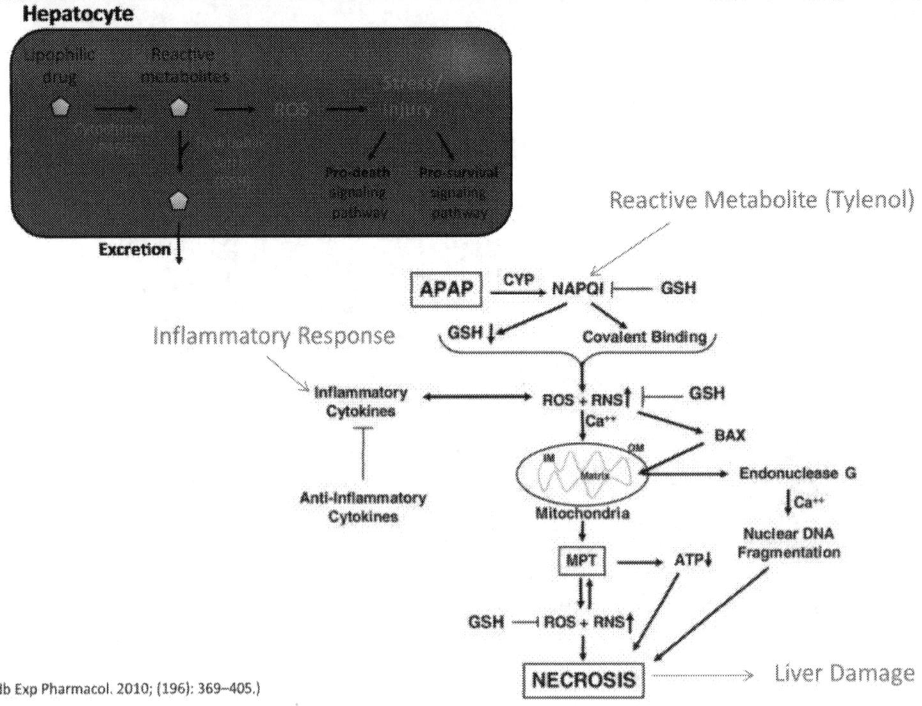

(Handb Exp Pharmacol. 2010; (196): 369–405.)

Acetaminophen (Tylenol) Assay in Organoid

Low Sensor

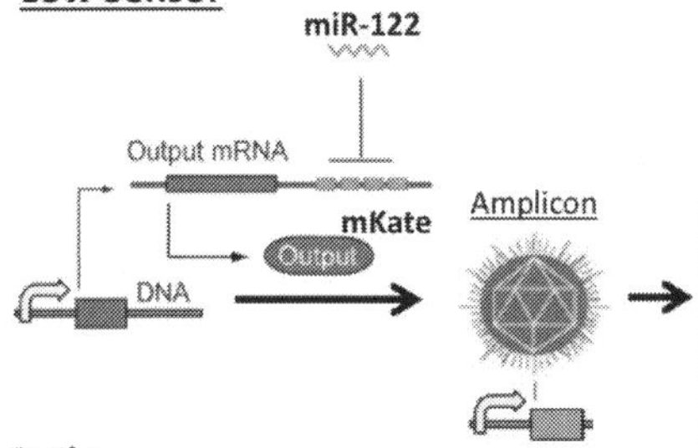

Pathway

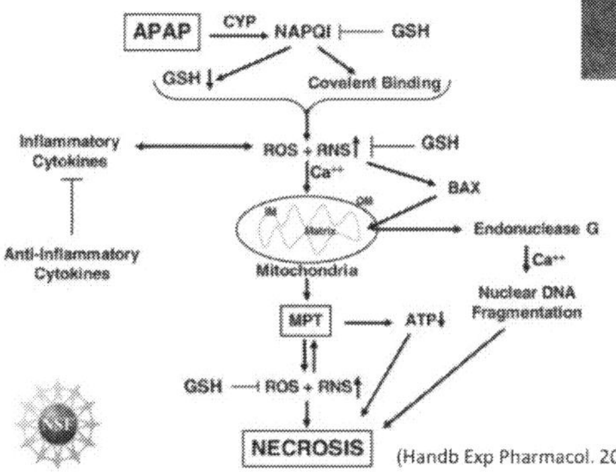

(Handb Exp Pharmacol. 2010; (196): 369–405.)

*Note LD50 in primary hepatocytes is ~12.5mM

Screening
@15 hour:

Control (no dose)	Low dose: 0.5mM APAP	Medium dose: 2mM APAP

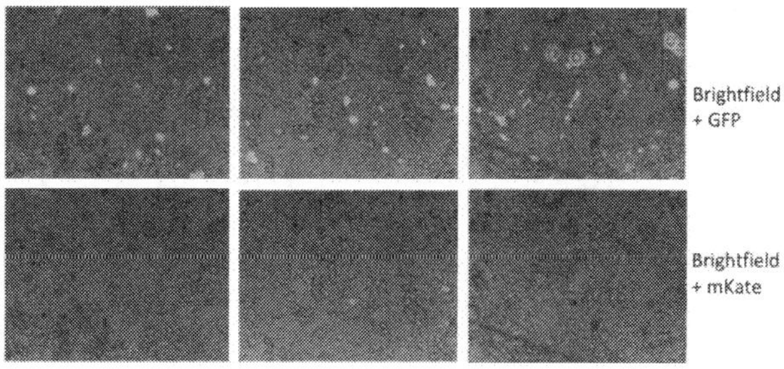

Result Initial miR-122 low sensor observation suggestive of a pathway effect that may be related to liver damage

Elvira Vitu   Sebastian Palacios

Acetaminophen (Tylenol) Assay in Organoid

Control (no dose)	Low dose: 0.5mM APAP	Medium dose: 2mM APAP

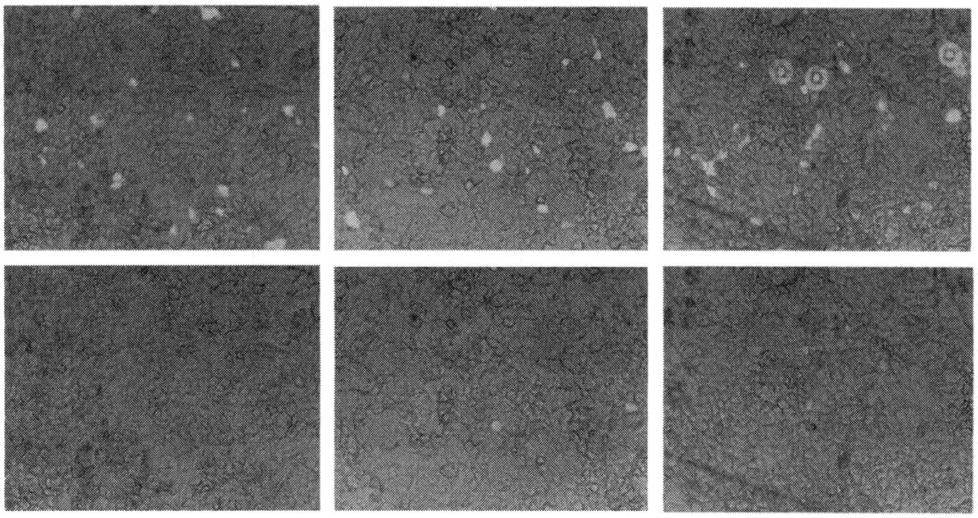

Summary of programmable organoid

- From a single hiPS cell population, generated
  - Endoderm, mesoderm, ectoderm
  - All cell types known to exist in liver bud
  - Neuronal progenitors
- Engineered symmetry breaking
- Working on programmed multi-step differentiation
- Initial results with genetically encoded sensors for real-time, single-single, patient-specific drug toxicity assays

Acknowledgments

- MIT:
  - Jon Babb
  - Noah Davidsohn
  - Xavier Duportet
  - Steve Firsing
  - Cristian Grecu
  - Saurabh Gupta
  - Patrick Guye
  - Tasuku Kitada
  - Yinqing Li
  - Ting Lu
  - Deepak Mishra
  - Adrian Randall
  - Velia Siciliano
  - Adrian Sluzarcyk
  - Jingjing Sun
  - Brian Teague
  - Rimchala Tharathorn
  - Zhen Xie
  - Liliana Wroblewska
- Princeton:
  - Craig Arnold
  - Ken Steiglitz
  - Hersch Rabitz
- Mt. Sinai:
  - Ihor Lemischka
- Caltech:
  - Frances Arnold
  - Jared Leadbetter
  - Yohei Yokobayashi
  - Cynthia Collins
- BBN:
  - Jake Beal
  - Fusun Yaman
  - Aaron Adler
- Duke:
  - Lingchong You
- ETH:
  - Kobi Benenson
- MIT:
  - Tom Knight
  - Gerald J. Sussman
  - Hal Abelson

Funding:
- DARPA BIOS
- NSF QuBIC
- NSF EMT
- NSF Career
- ONR
- DOE
- NIH
- NJ Spinal Cord

Thank you!

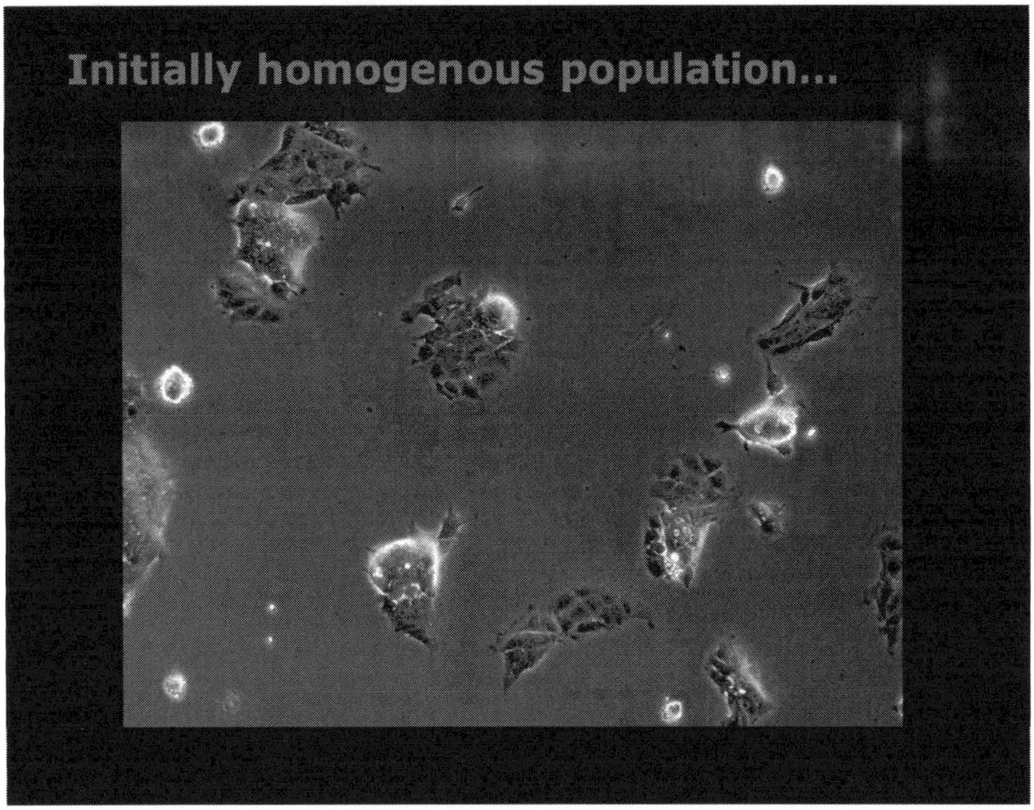

Control over population outcomes

Gata6: Endoderm
Nanog: Pluripotent

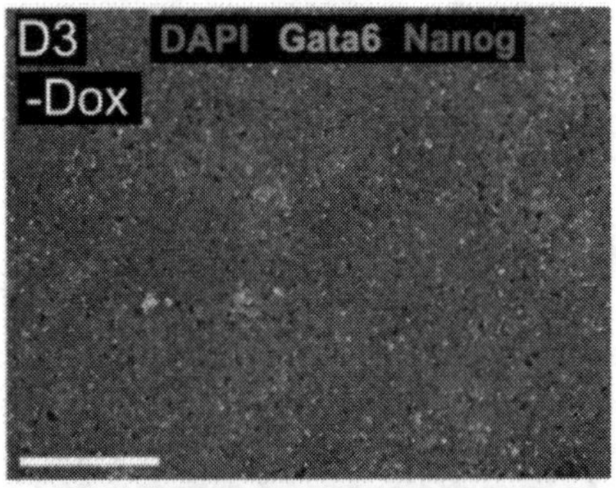

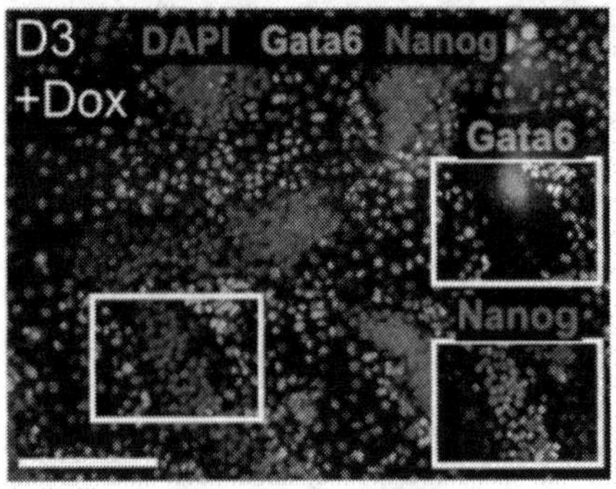

Control via
initial sub-population mix

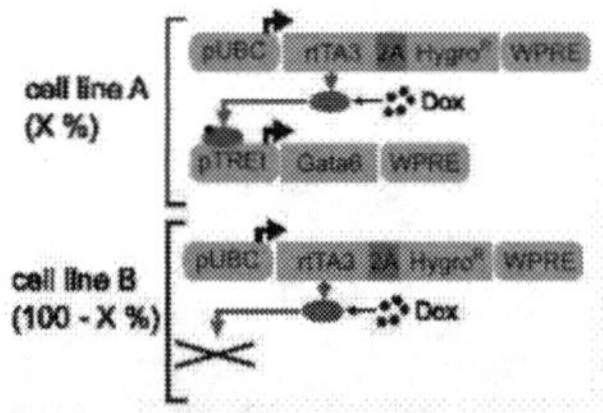

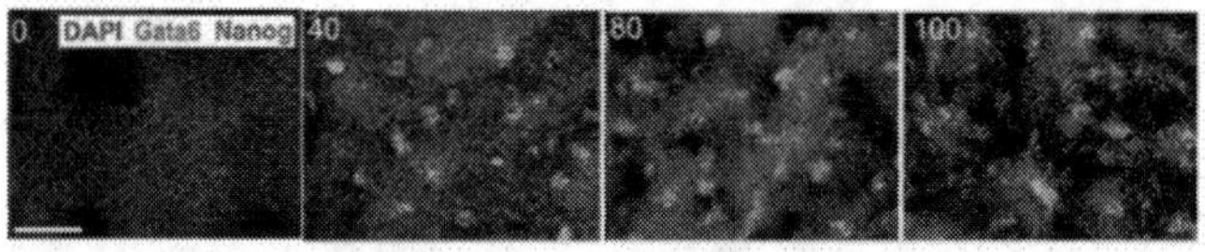

Control via
Dox induction

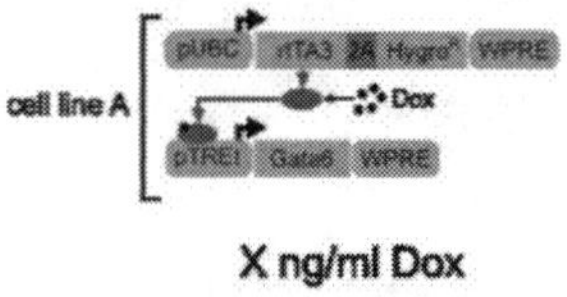

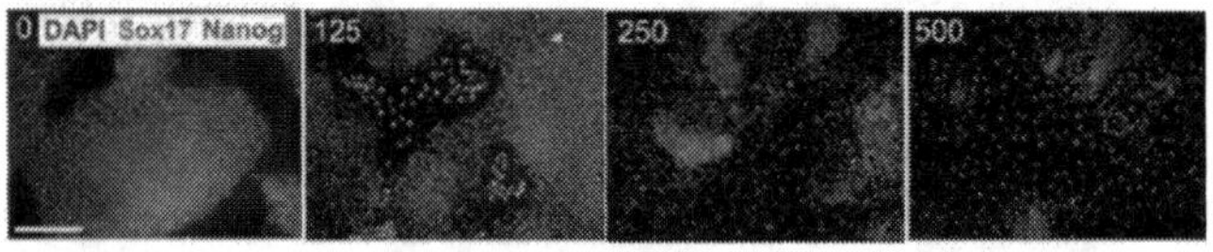

A revolutionary genome engineering technology began with…

curiosity about how bacteria fight the flu

Five Points…

- RNA-guided DNA cutting by CRISPR-Cas9
- Cas9 enables precision genome engineering
- Cas9's target search mechanism
- A CRISPR-associated DNA integrase: Cas1-Cas2
- A prudent path forward for genome editing

Five Points…

- **RNA-guided DNA cutting by CRISPR-Cas9**
- Cas9 enables precision genome engineering
- Cas9's target search mechanism
- A CRISPR-associated DNA integrase: Cas1-Cas2
- A prudent path forward for genome editing

CRISPRs: Hallmarks of acquired immunity in bacteria

Clusters of Regularly Interspaced Short Palindromic Repeats (CRISPRs)

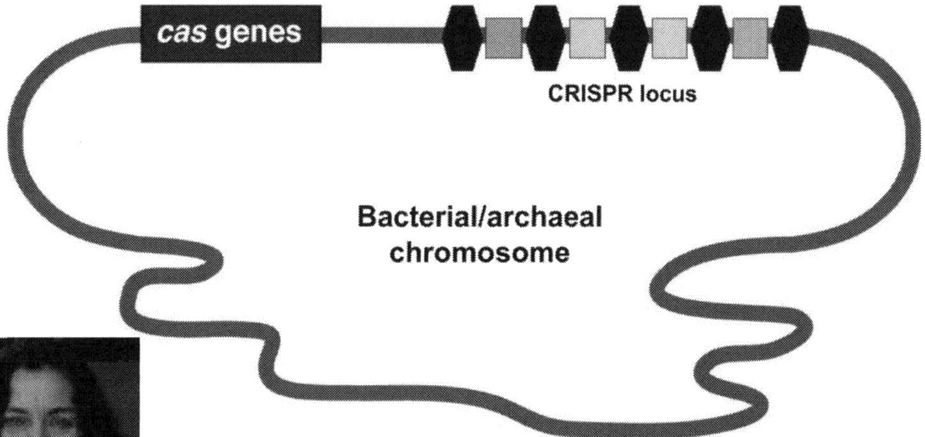

Jillian Banfield, UC Berkeley

Bolotin *et al.* 2005
Mojica *et al.* 2005
Pourcel *et al.* 2005

Three steps to acquire immunity in bacteria

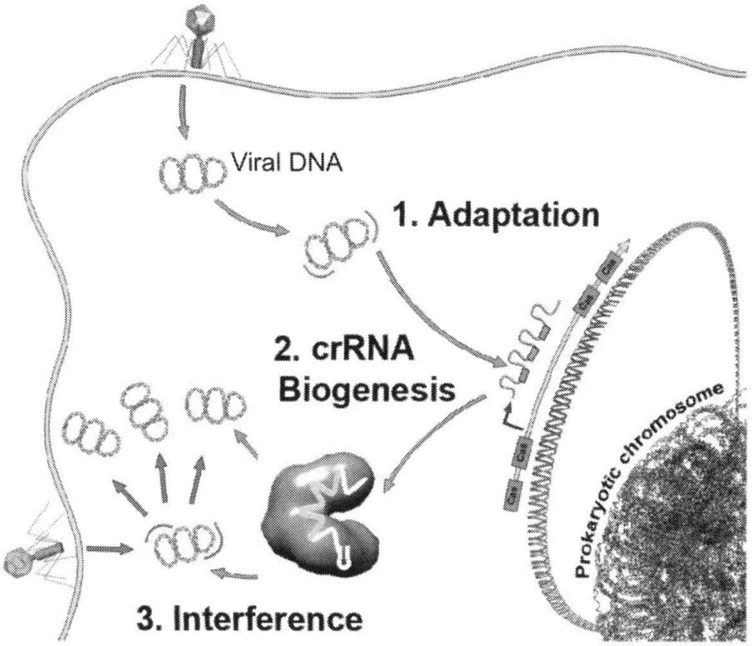

Reviews: Horvath & Barrangou (2010) *Science*; Wiedenheft et al. (2012) *Nature*

CRISPR-Cas9 Team

Emmanuelle Charpentier
Jennifer Doudna
Martin Jinek
Krzysztof Chylinski
Ines Fonfara
(Michael Hauer)

The Aha! Moment...

A programmable protein that cleaves DNA: Cas9

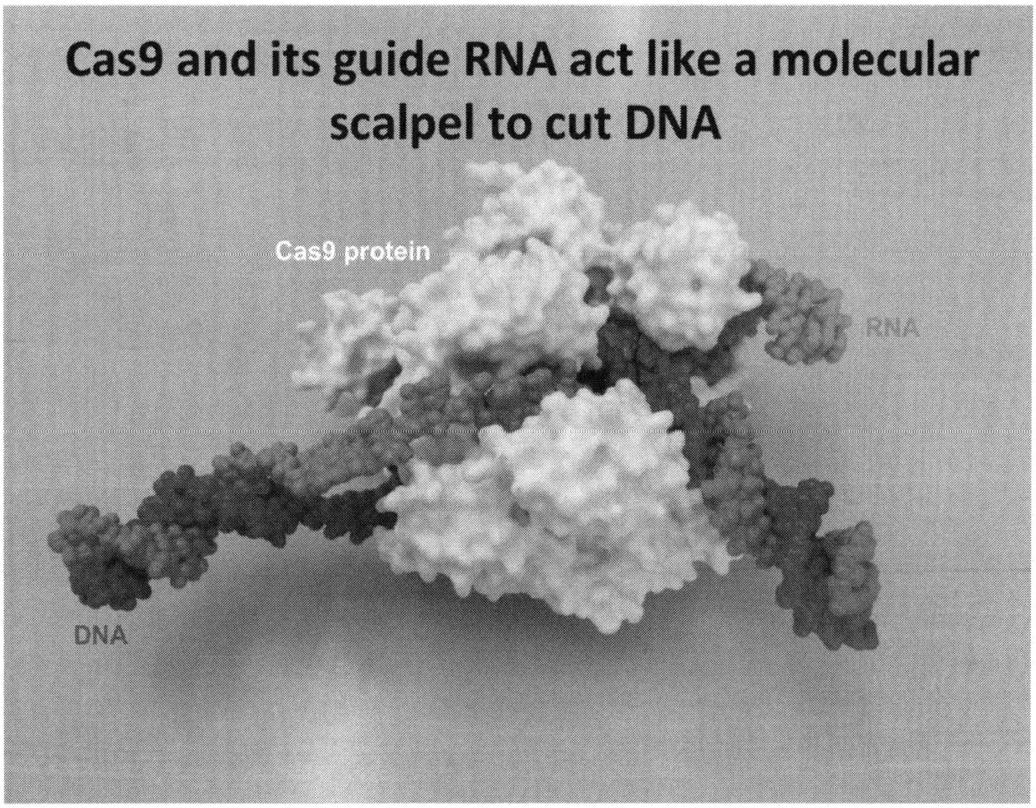
Cas9 and its guide RNA act like a molecular scalpel to cut DNA
Cas9 protein
RNA
DNA

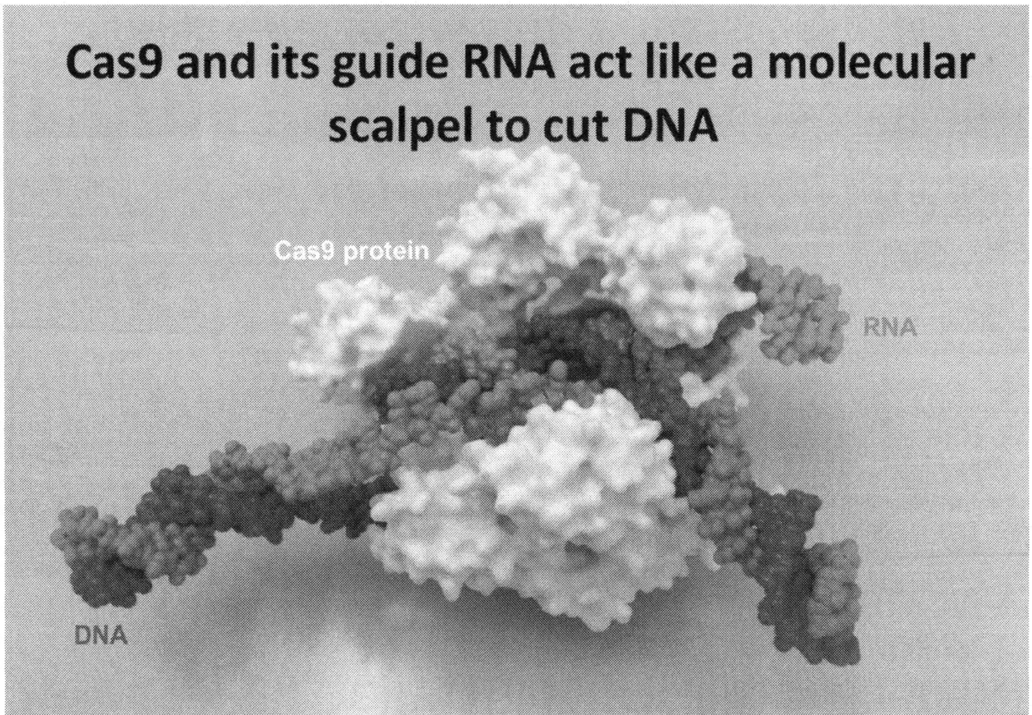
Cas9 and its guide RNA act like a molecular scalpel to cut DNA
Cas9 protein
RNA
DNA

Programming Cas9 with single-guide RNAs (sgRNAs)

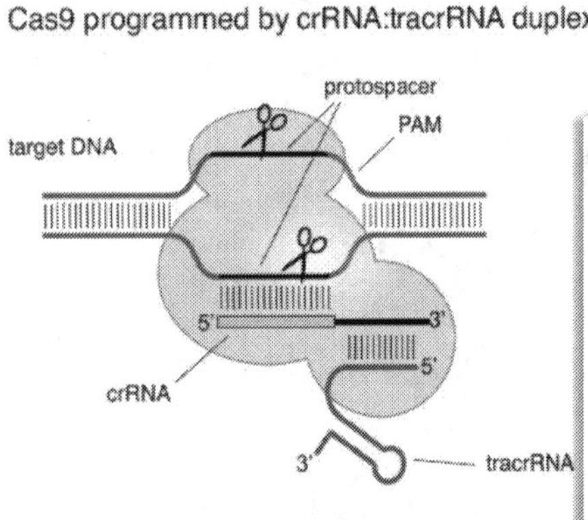

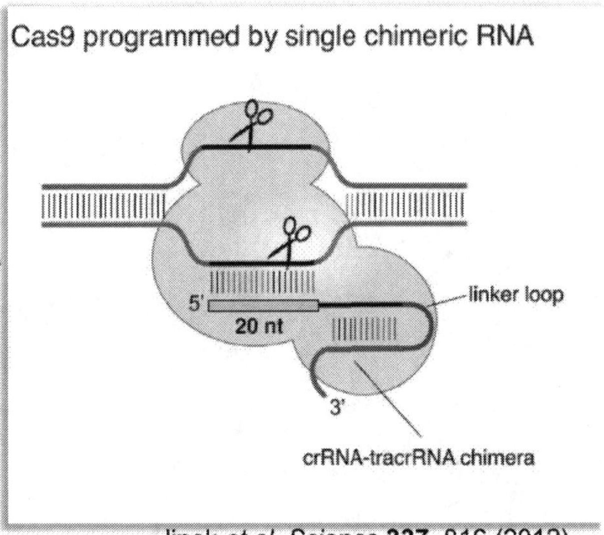

Jinek *et al. Science* **337**, 816 (2012)

Programmed Cas9 cleaves DNA at specified sites

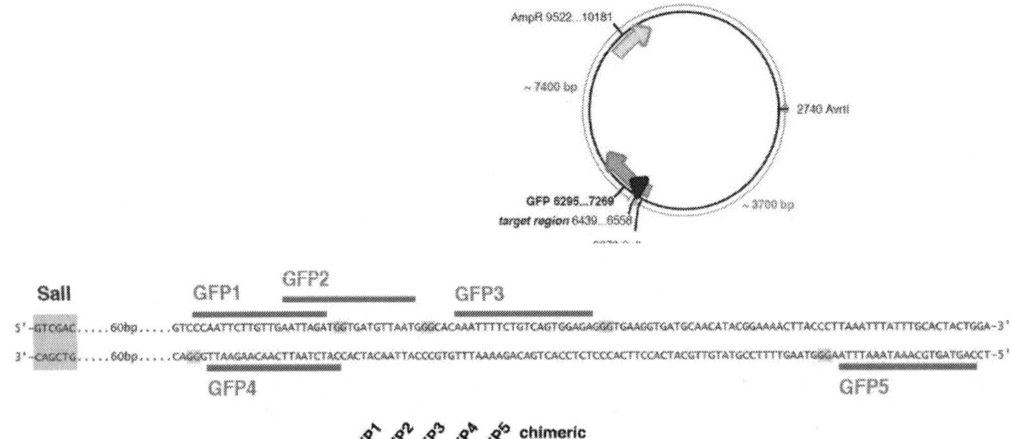

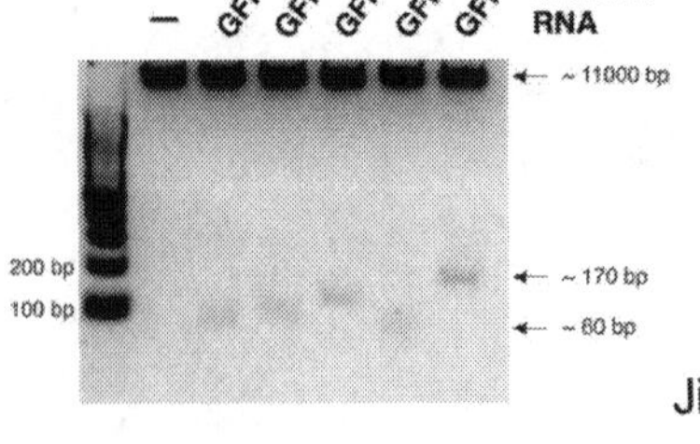

Jinek *et al. Science* **337**, 816 (2012)

Genome editing begins with dsDNA cleavage

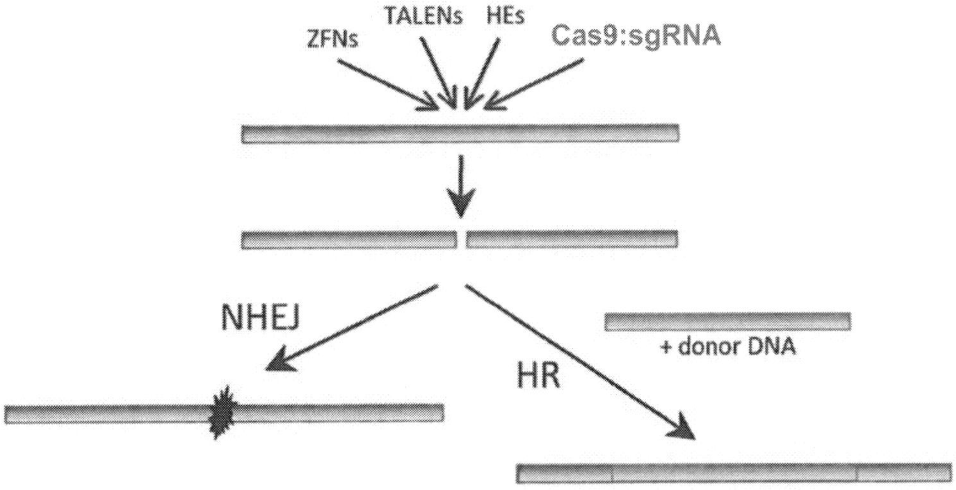

Five Points...

- RNA-guided DNA cutting by CRISPR-Cas9
- Cas9 enables precision genome engineering
- Cas9's target search mechanism
- A CRISPR-associated DNA integrase: Cas1-Cas2
- A prudent path forward for genome editing

Sciencexpress **Research Articles**

June 2012

Programmable DNA targeting by Cas9:gRNA

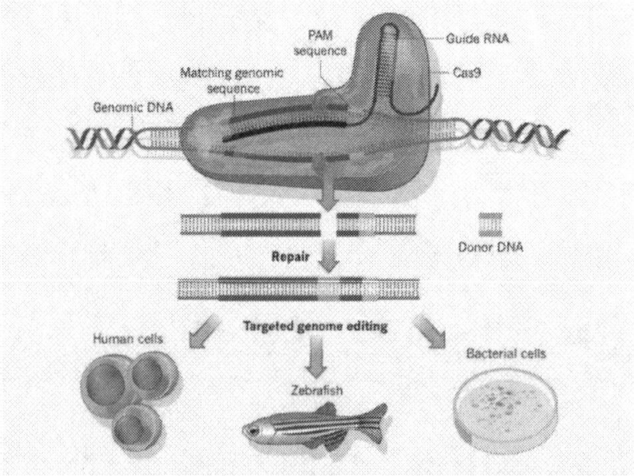

Charpentier, E. & Doudna, J. A. *Nature* **495**, 50–51 (2013)

- Specific genome editing in cells and organisms (Jan. 2013: Church; Doudna; Zhang)
- Robust transcriptional control with catalytically-inactive dCas9
- Live-cell imaging of genomic loci using fluorescently-tagged Cas9

Exponential growth of genome engineering!

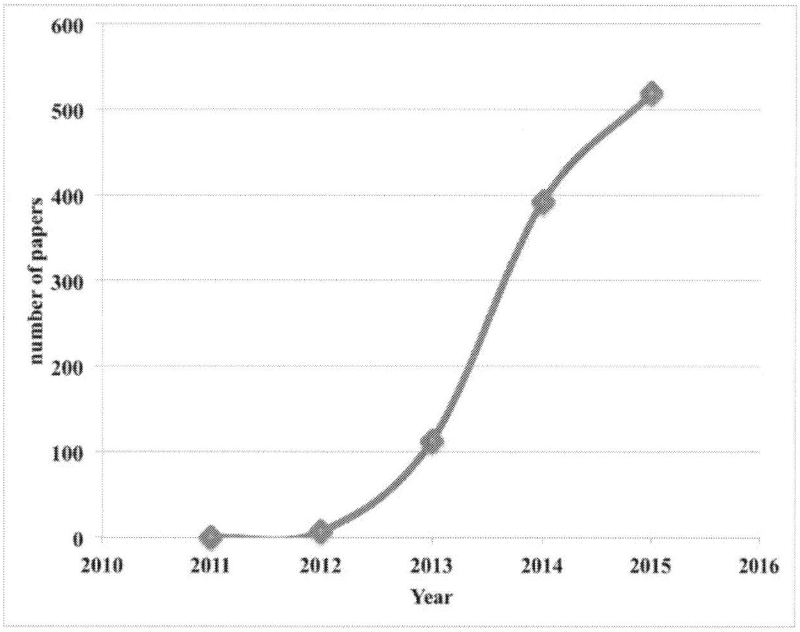

Courtesy of Jan Witkowski, Banbury Center

Five Points...

- RNA-guided DNA cutting by CRISPR-Cas9
- Cas9 enables precision genome engineering
- Cas9's target search mechanism
- A CRISPR-associated DNA integrase: Cas1-Cas2
- A prudent path forward for genome editing

Cas9:guide RNA binds to Phage λ DNA curtains

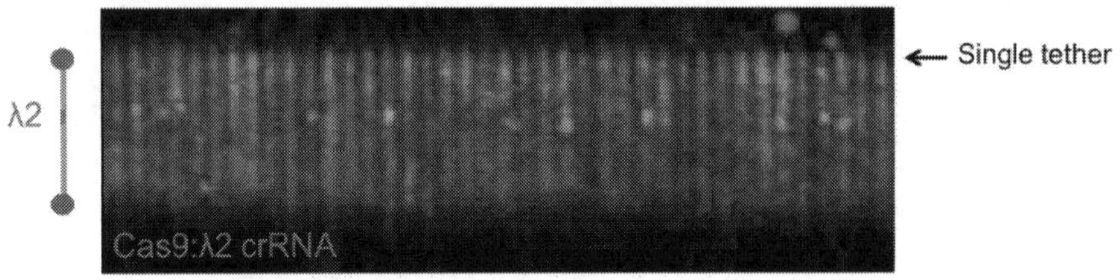

- DNA binding begins at PAM (NGG) motifs
- PAM contact triggers DNA unwinding, Cas9 catalytic activity
- High-affinity product binding, no substrate turnover

Sternberg et al. (2014) *Nature*

DNA target search and recognition by Cas9:gRNA

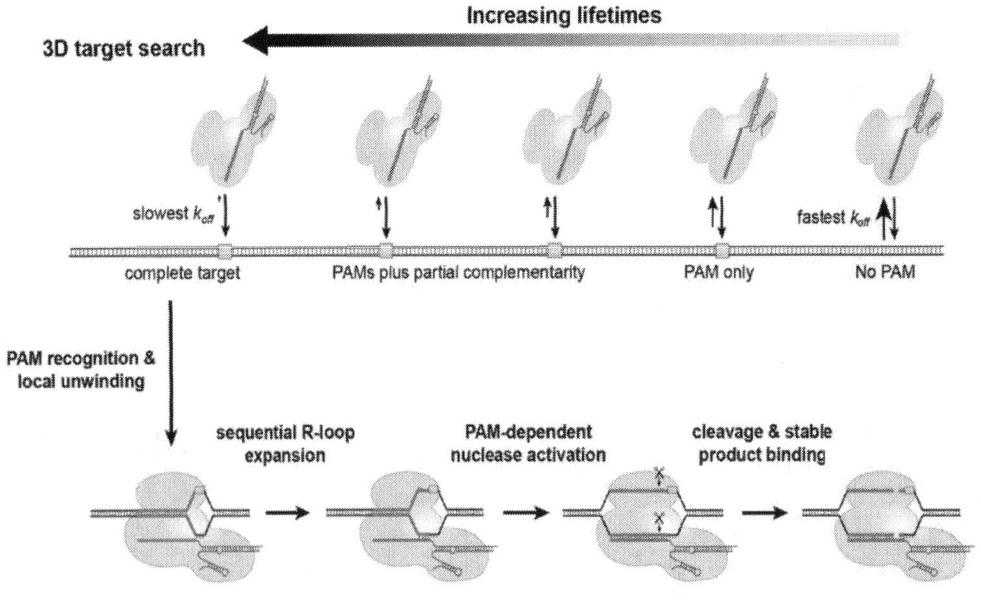

Sternberg et al. (2014) *Nature*

RNA-induced conversion of Cas9 into an active conformation for DNA surveillance

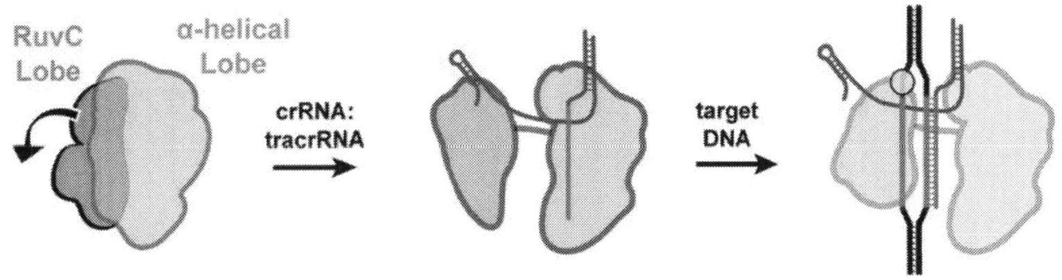

Jinek, Jiang, Taylor, Sternberg et al. (2014) *Science*

Jinek, Jiang, Taylor & Sternberg et al. (2014) *Science*
Nishimasu et al. (2014) *Cell*
Anders et al. (2014) *Nature*

Richard van der Oost, surfrender.com

RNA-induced conversion of Cas9 into an active conformation for DNA surveillance

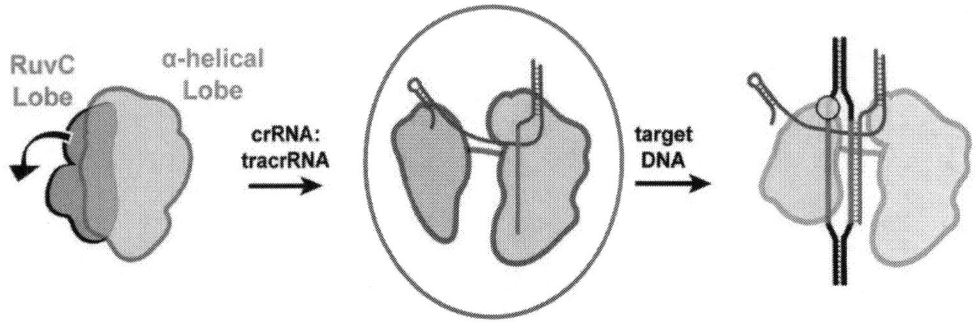

Jinek, Jiang, Taylor, Sternberg et al. (2014) *Science*

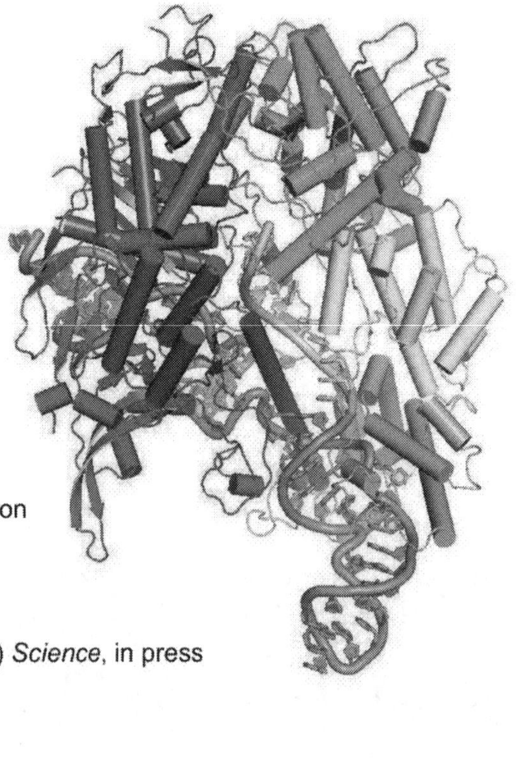

2.9 Å resolution

Jiang et al. (2015) *Science*, in press

In vivo labeling of Cas9 using HaloTag

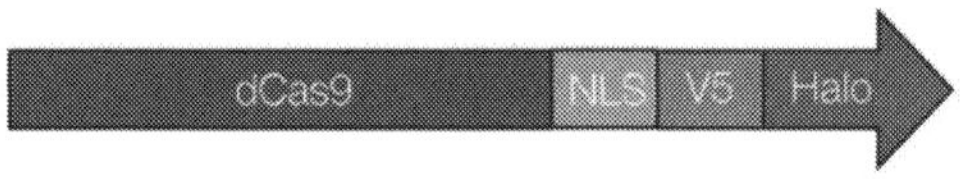

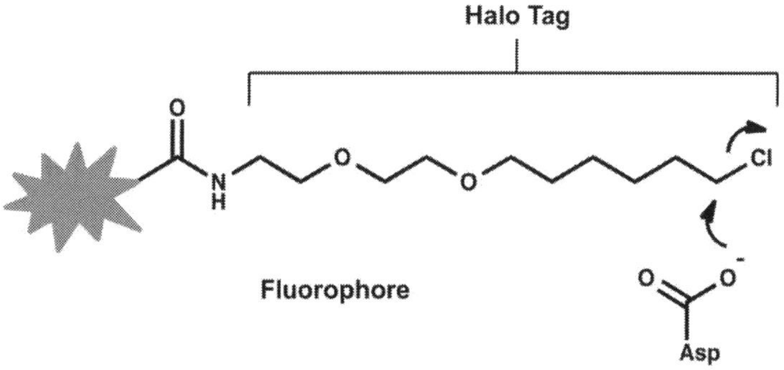

Los et al. (2008) *ACS Chem. Biol.*

Spencer Knight

Fast kinetics of Cas9 in Mammalian Cells

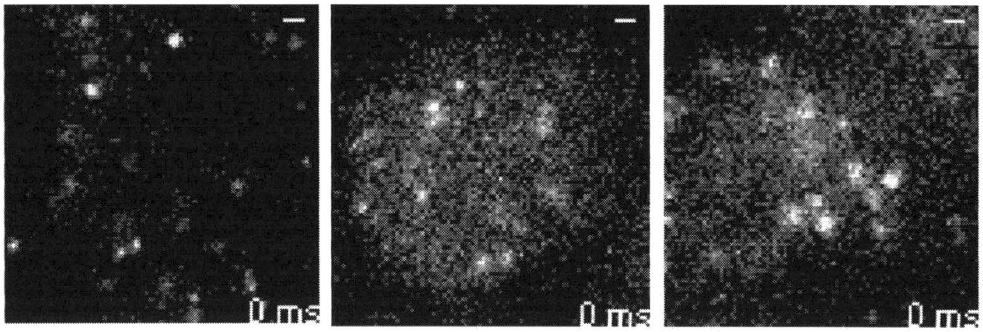

H2B　　Sox2　　dCas9

Spencer Knight, James Liu, Robert Tjian

3D Diffusion Dominates Cas9 searching

B2 SINE: 5′-GAGCTGGAGAGATGGCTCAGPAM

B2_0M: 5′-GAGCTGGAGAGATGGCGATTPAM

B2_13M: 5′-GAGGGGTAGAGATGGCTCAGPAM

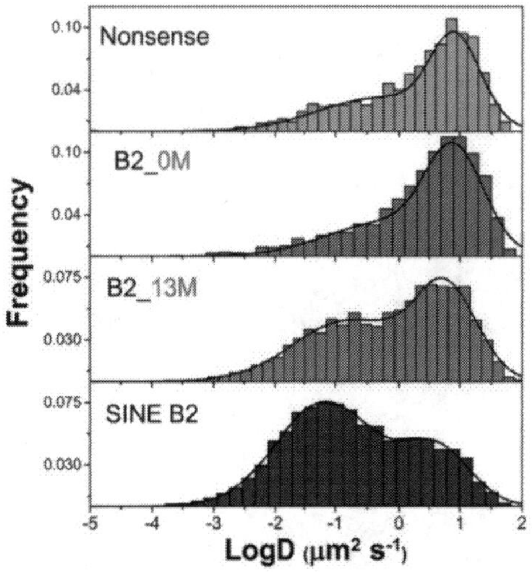

3D tracking with nonsense sgRNA
3 × 10^8 PAMS
3 × 10^5 5-nt seeds

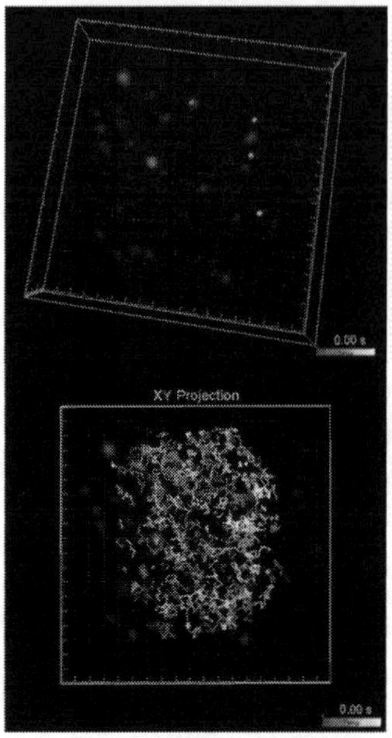

Cas9 Heterochromatin Sampling is Less Frequent

Five Points...

- RNA-guided DNA cutting by CRISPR-Cas9
- Cas9 enables precision genome engineering
- Cas9's target search mechanism
- **A CRISPR-associated DNA integrase: Cas1-Cas2**
- A prudent path forward for genome editing

How are new viral spacer sequences acquired?

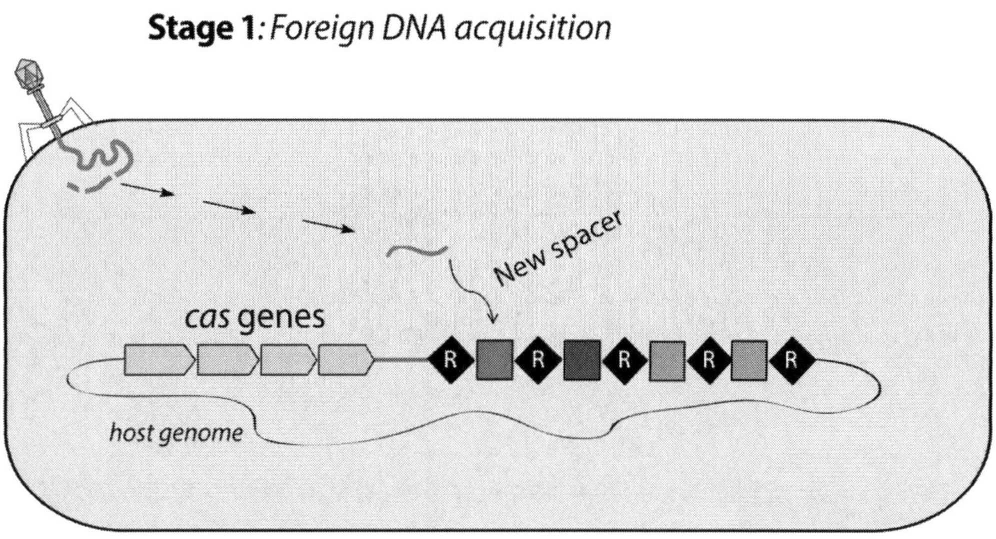

Cas1 and Cas2 are required for spacer acquisition

- Cas1 and Cas2 are nucleases found in all CRISPR–Cas systems.
- Overexpression leads to new spacer insertion into CRISPR locus.

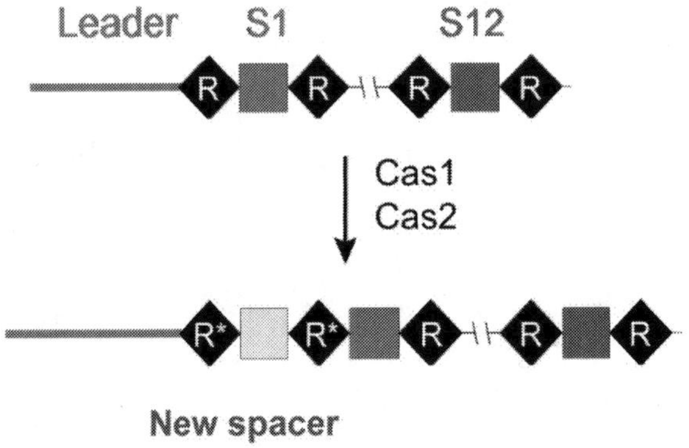

Yosef *et al.* (2012) *NAR*; Datsenko *et al.* (2012) *Nat. Comm.*; Swarts *et al.* (2012) P*LoS*

Cas1 and Cas2 function as a complex

E. coli **K-12 CRISPR locus**

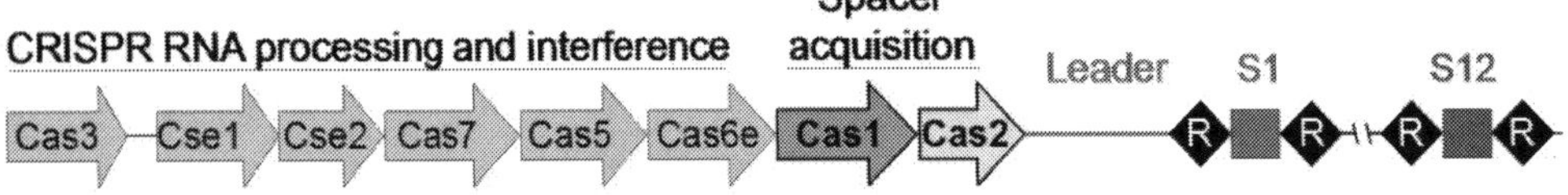

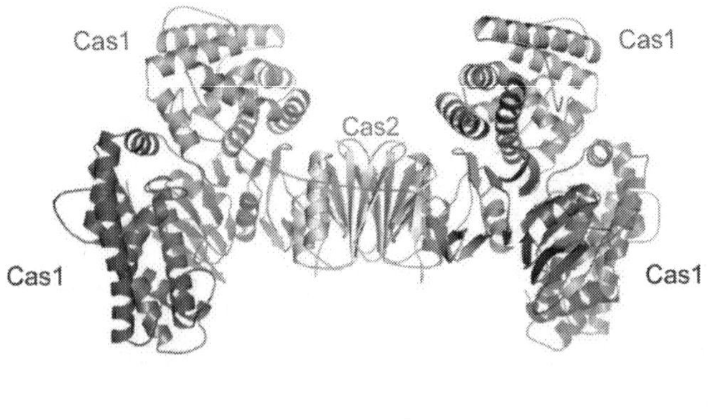

How does the Cas1–Cas2 complex insert spacers into the CRISPR locus?

Nuñez *et al.* (2014) *NSMB*

Establishing an *in vitro* protospacer integration assay

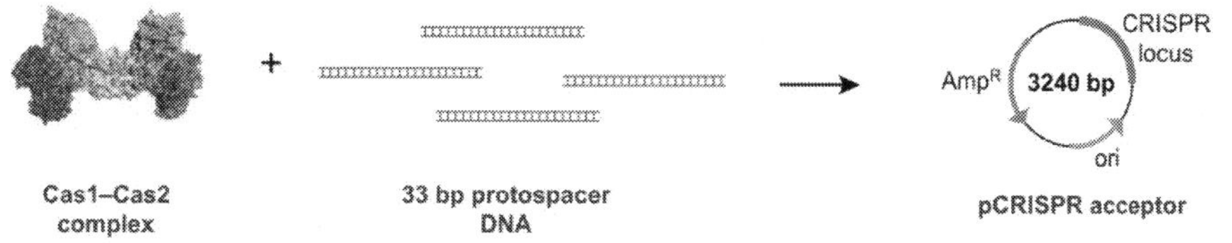

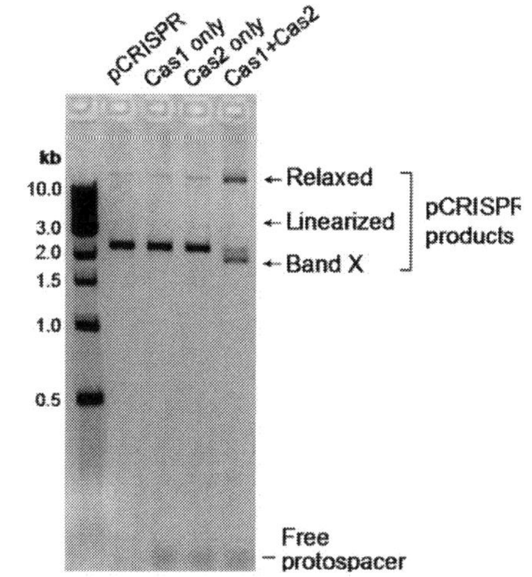

James Nuñez

Product formation is dependent on protospacers

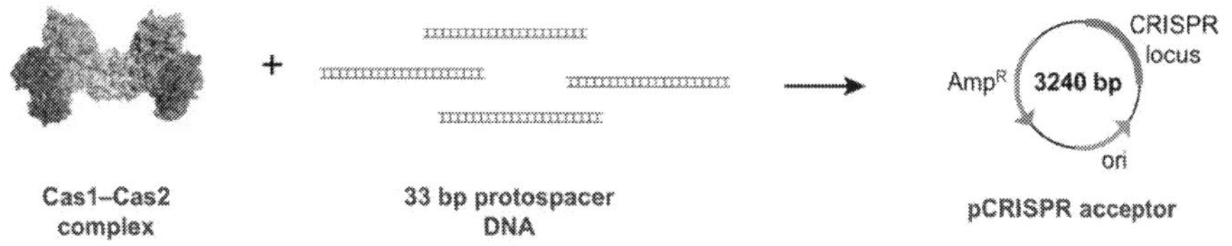

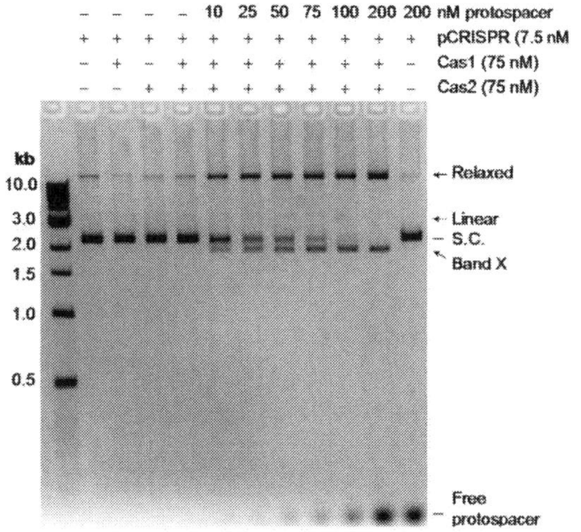

James Nuñez

Radiolabeled protospacers are integrated into pCRISPR

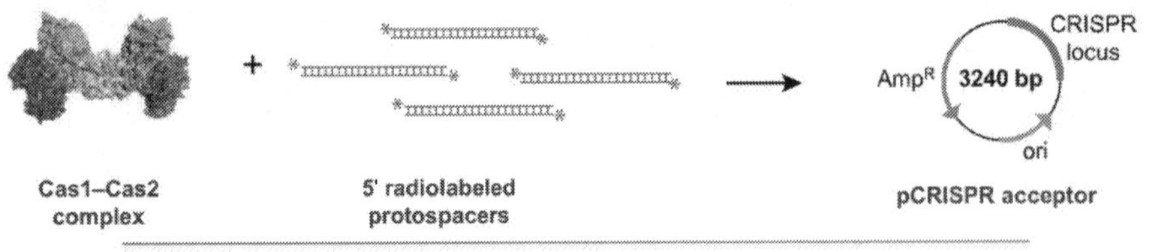

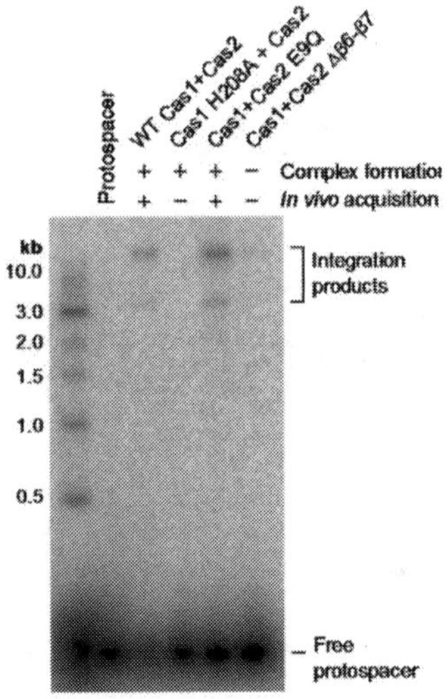

- Cas1 catalytic activity is required.
- Cas2 catalytic activity is not required.
- Cas1–Cas2 complex formation is required.
- Supercoiled plasmid target is required.

James Nuñez

Linear and relaxed integration products

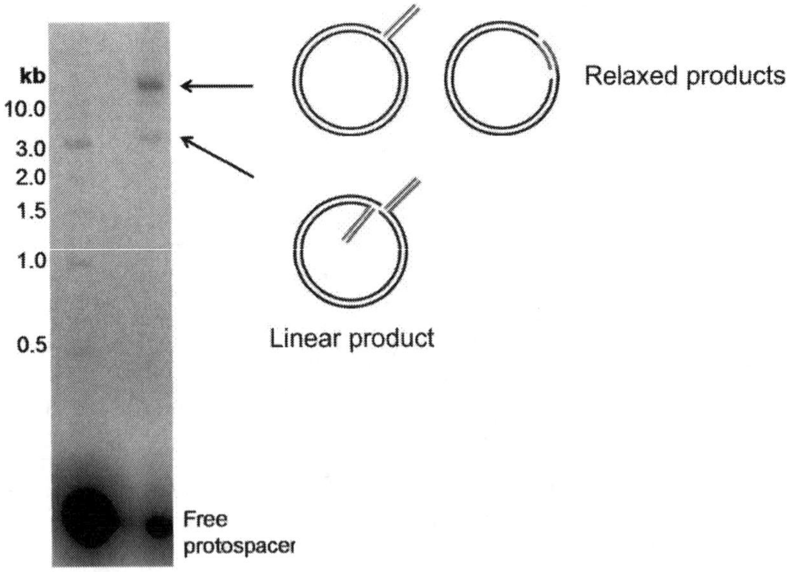

James Nuñez

Two possible mechanisms of DNA integration

Protein-covalent intermediate

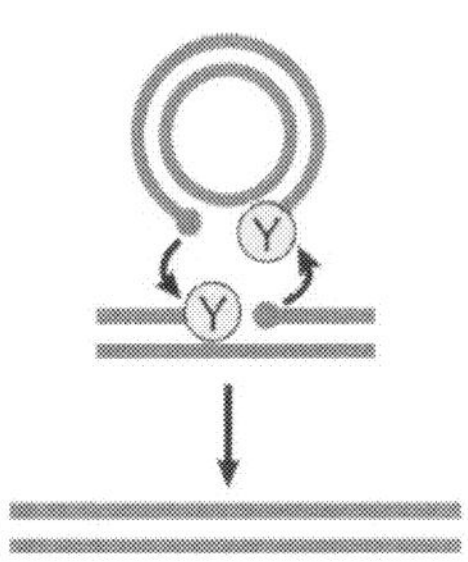

Examples: site-specific recombinases, Y- and S-transposases

Curcio & Derbyshire. *Nat. Rev. MCB*, 2004.

Direct nucleophilic attack

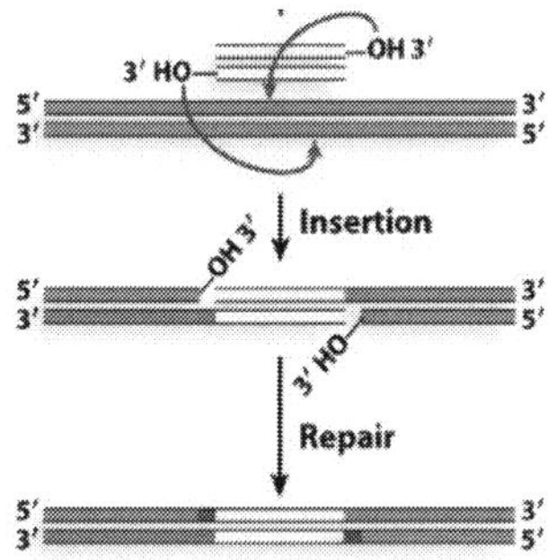

Examples: HIV-1 integrase, cut-and-paste transposases

Molecular Biology: Principles and Practice, 2012.

Integration proceeds via a 3'-OH nucleophilic attack

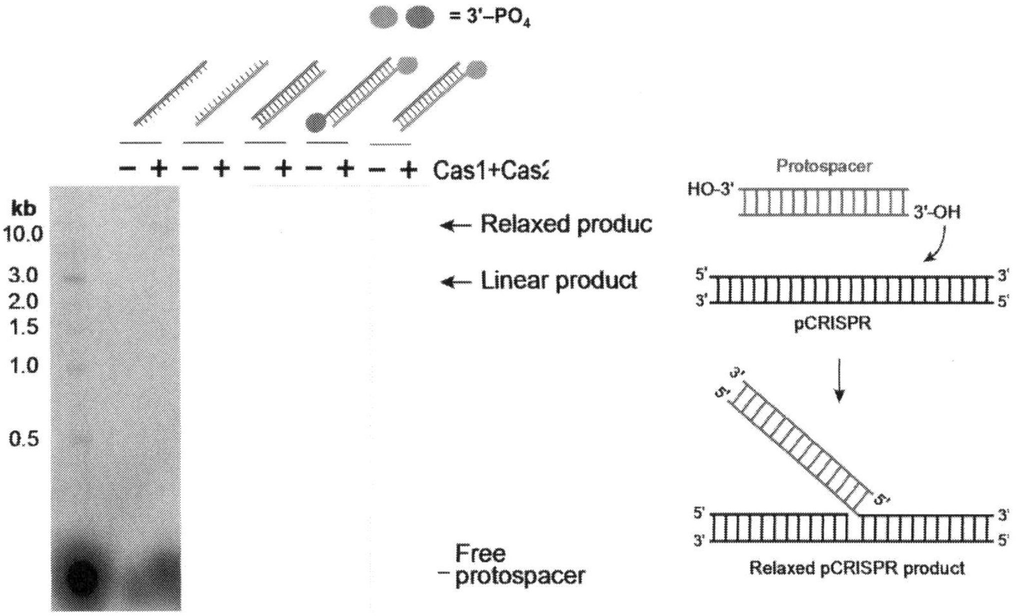

James Nuñez

Integration-seq to determine exact sites of integration

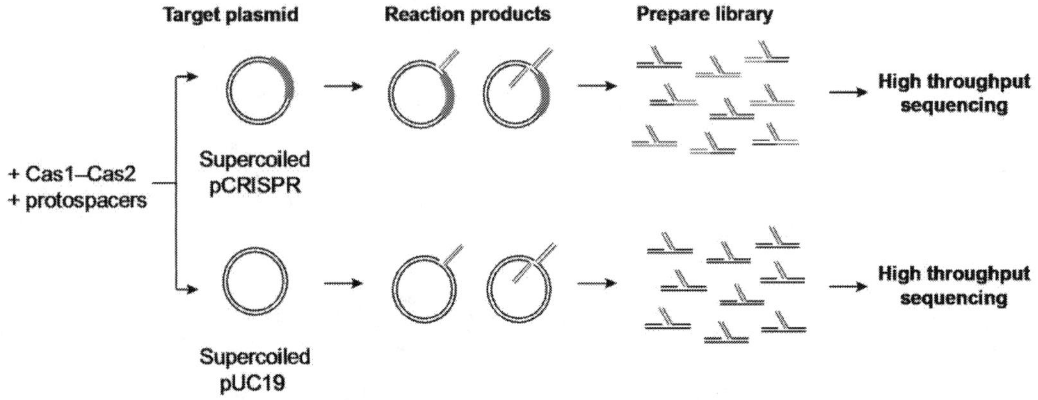

CRISPR repeats provide specificity for integration

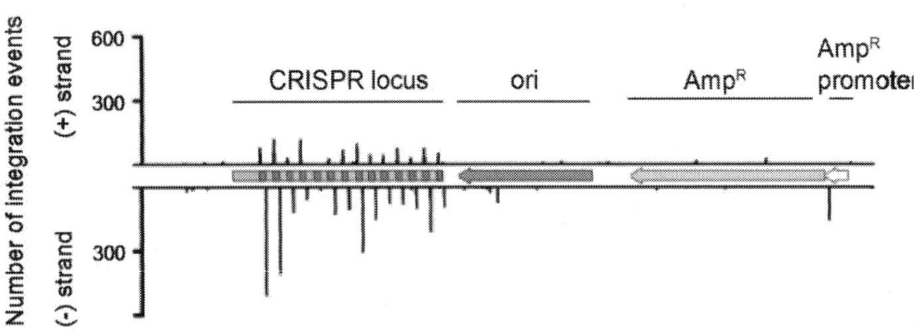

~71% of integration events are in the CRISPR locus

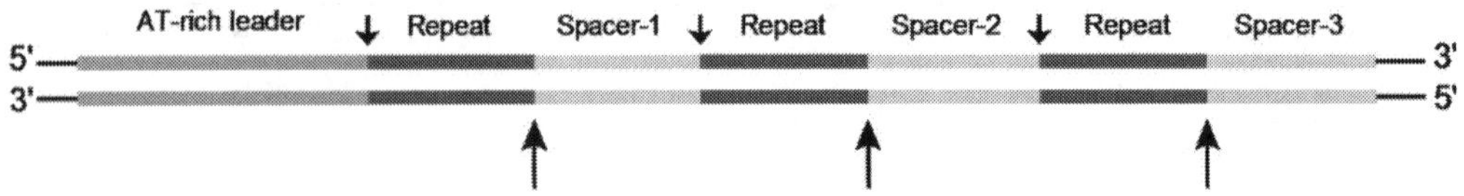

Repeat sequences can form DNA cruciforms

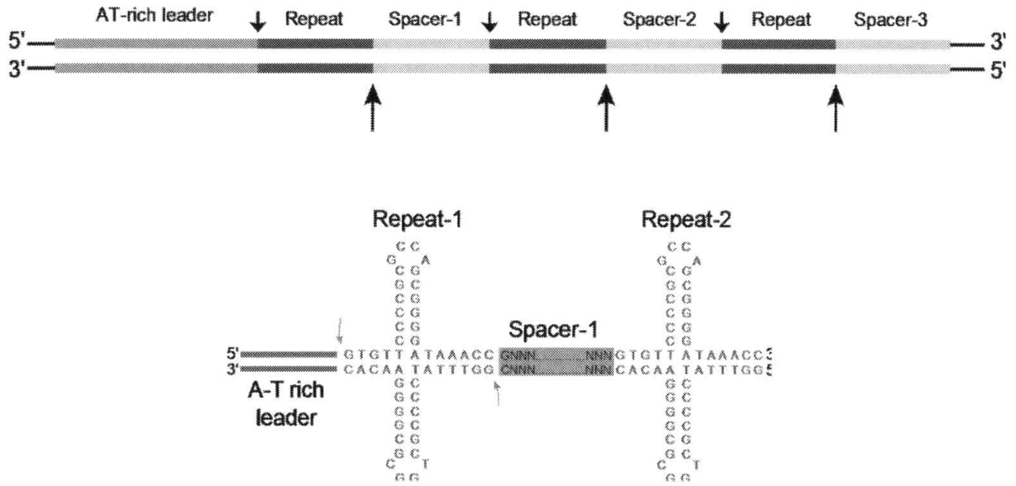

DNA cruciforms can form as a function of supercoiling, especially when found adjacent to AT-rich sequences.

Nonspecific integration site is similar to CRISPR array

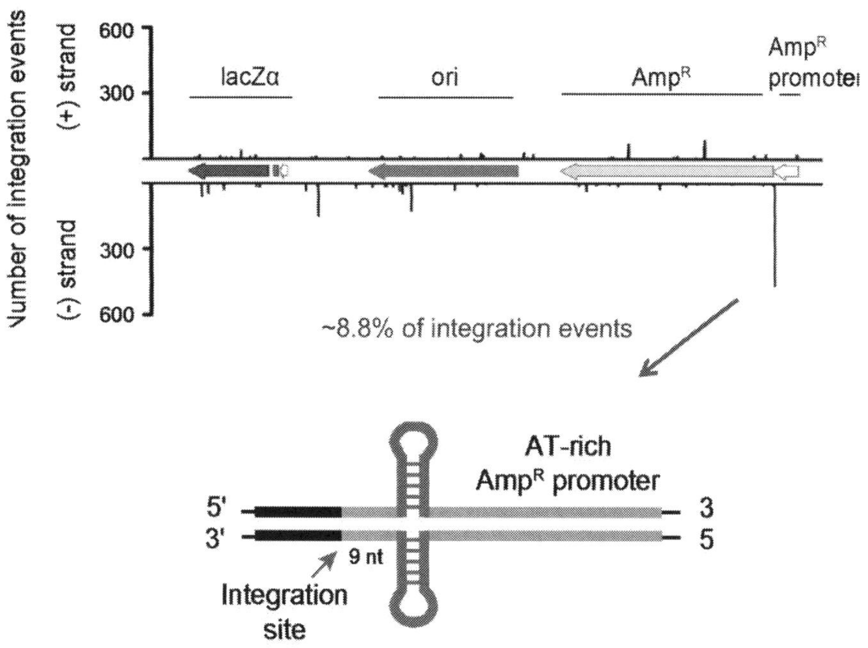

Mechanism of spacer integration

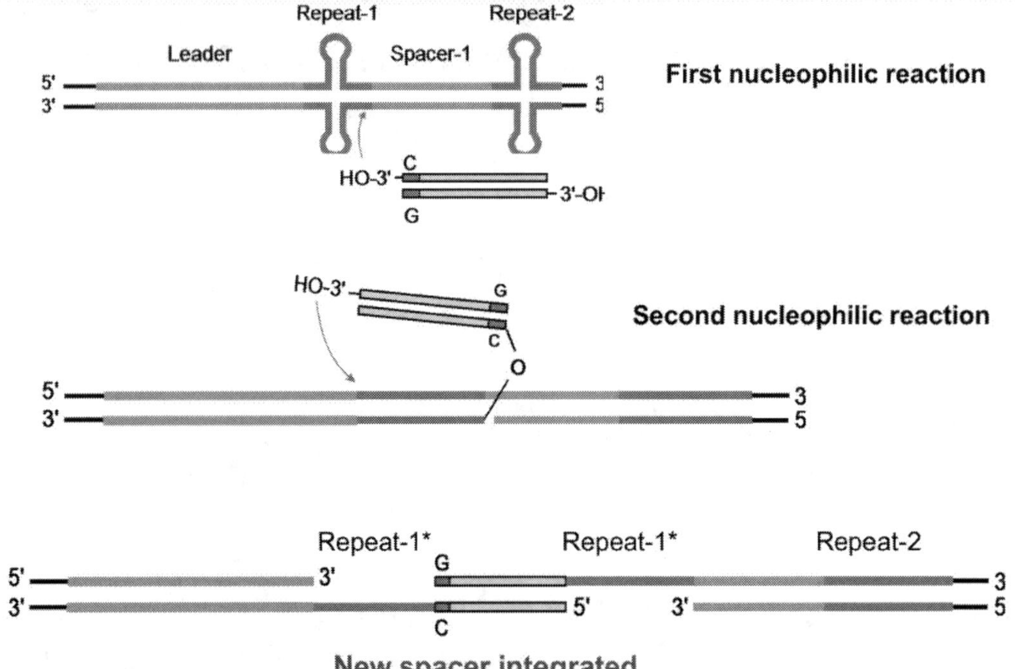

Nuñez *et al.* (2015) *Nature*

Mechanism of spacer integration

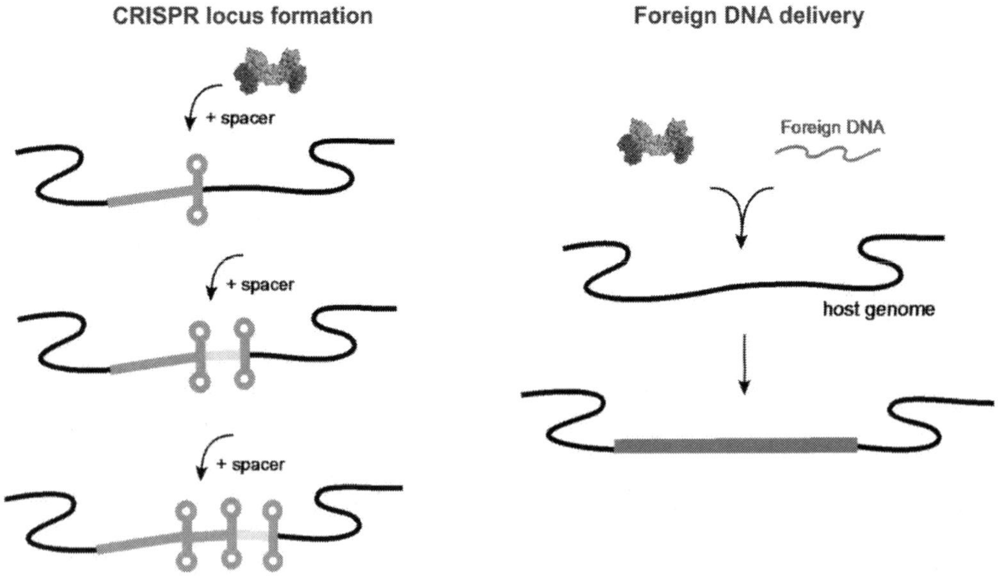

Five Points...

- RNA-guided DNA cutting by CRISPR-Cas9
- Cas9 enables precision genome engineering
- Cas9's target search mechanism
- A CRISPR-associated DNA integrase: Cas1-Cas2
- **A prudent path forward for genome editing**

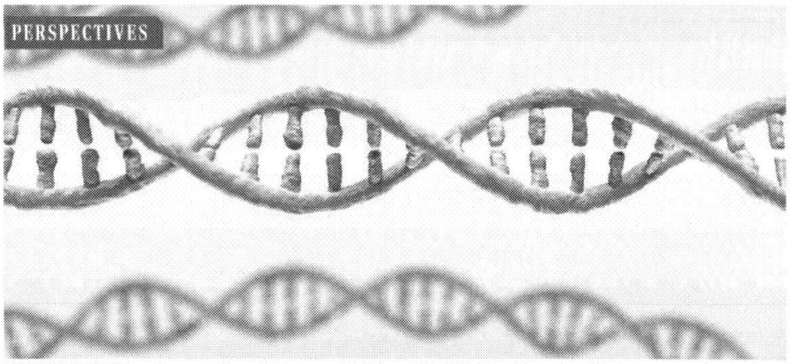

BIOTECHNOLOGY

amework for open discourse on the use of CRISPR-Cas9 technology to manipulate the human genome is urgently needed

Innovative Genomics Initiative

(IGI) @ UC Berkeley/UCSF

J. Doudna and J. Weissman, Co-directors
M. Botchan and K. Yamamoto, Administrative Directors
J. Corn, Scientific Director
M. Fenner, Program Director

August 2014

Acknowledgments

Krys Chylinski
Emmanuelle Charpentier, Braunschweig

Sy Redding
Eric Greene, Columbia University

Eva Nogales, UC Berkeley

Alan Engelman, Harvard Medical School

BILL & MELINDA GATES foundation

NSF

CRSB

hhmi | Howard Hughes Medical Institute

'Non-specific' binding is actually PAM binding

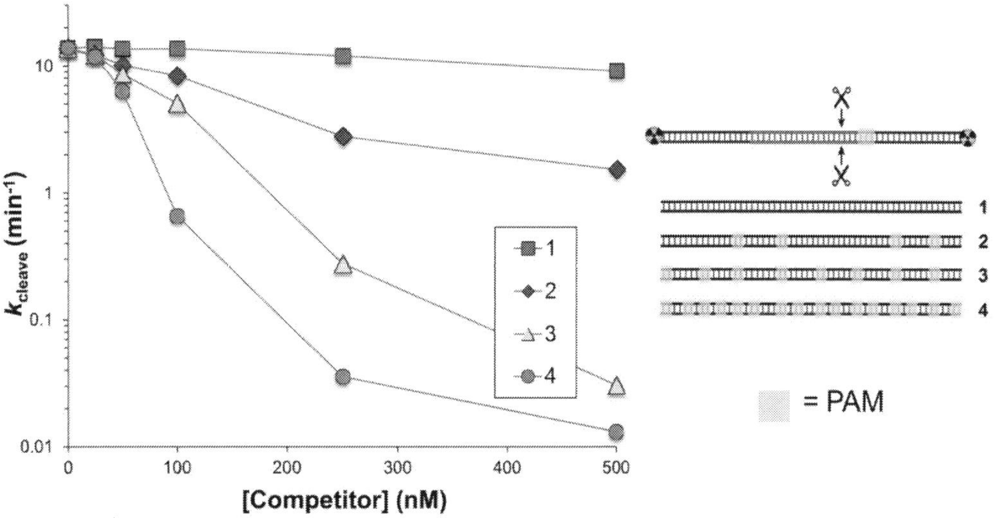

Increasing competition observed with an increasing number of PAM sites

dsDNA melting and RNA:DNA hybridization occurs only after PAM recognition

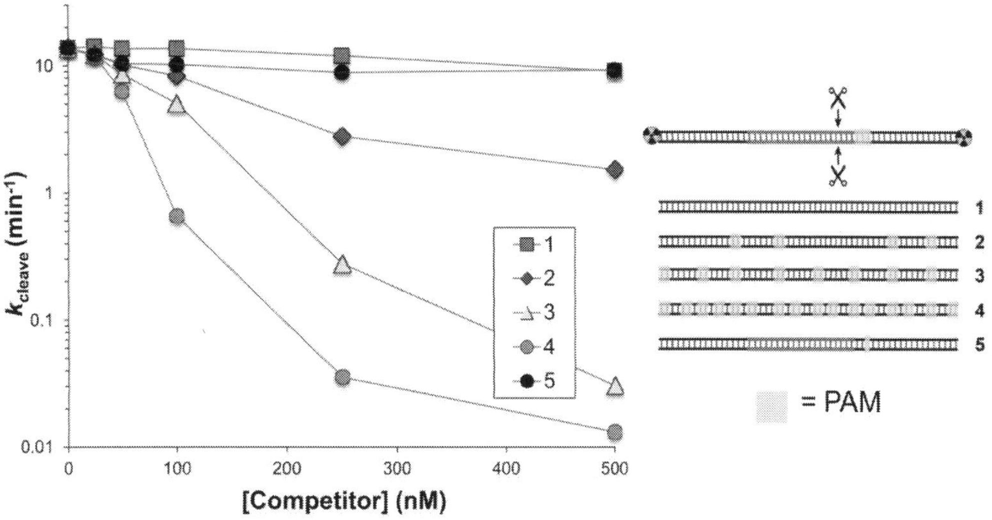

Single bp mutation in PAM converts DNA target into invisible substrate

Designing and verifying molecular circuits made of DNA

Erik Winfree

Computer Science & CNS & Bioengineering, Caltech
The DNA and Natural Algorithms (DNA) Group

Natural and artificial technology

DNA as an Engineering Material

Ned Seeman
(New York University)

Wood.
Glue.
Wool.
Silk.
Leather.
DNA.

single-stranded

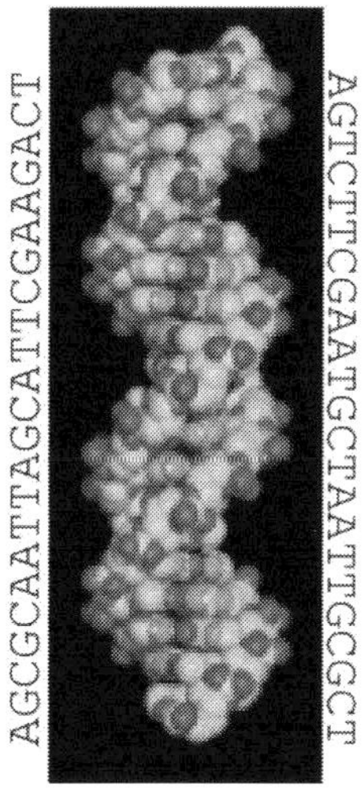

double-stranded

DNA as an Engineering Material

Ned Seeman
(New York University)

Wood.
Glue.
Wool.
Silk.
Leather.
DNA.

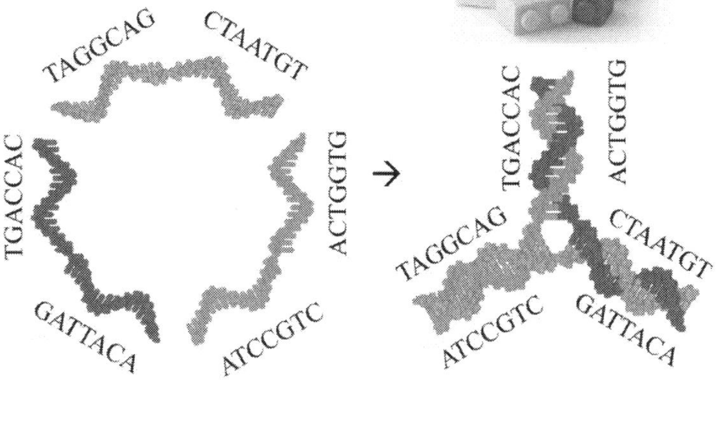

Seeman, *J. Theor. Biol.* 1982

DNA as a Computing Substrate

Len Adleman
(USC)

Gears.
Springs.
Fluids.
Vacuum tubes.
Transistors.
DNA.

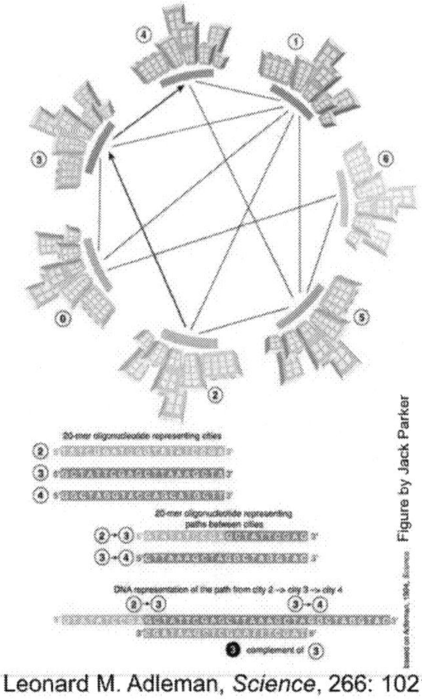

Leonard M. Adleman, *Science*, 266: 1021-1024 (1994)

Programmable information-based chemistry

RNA biology

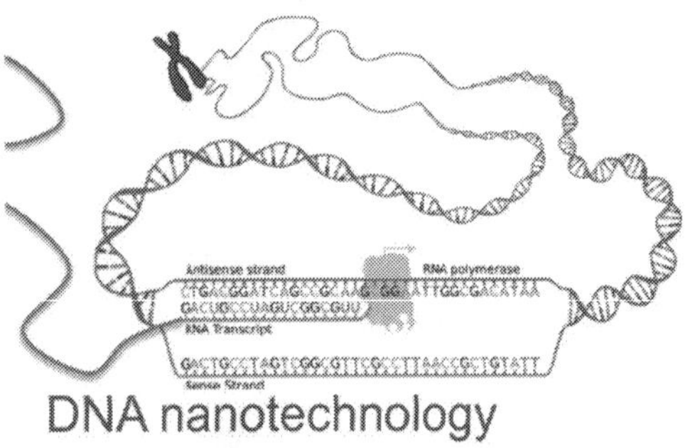

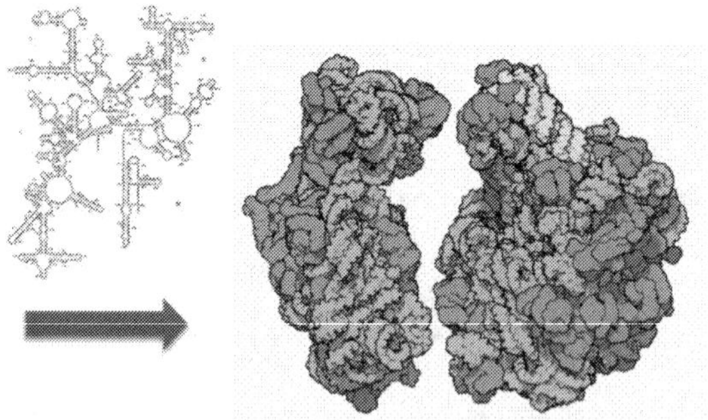

DNA nanotechnology

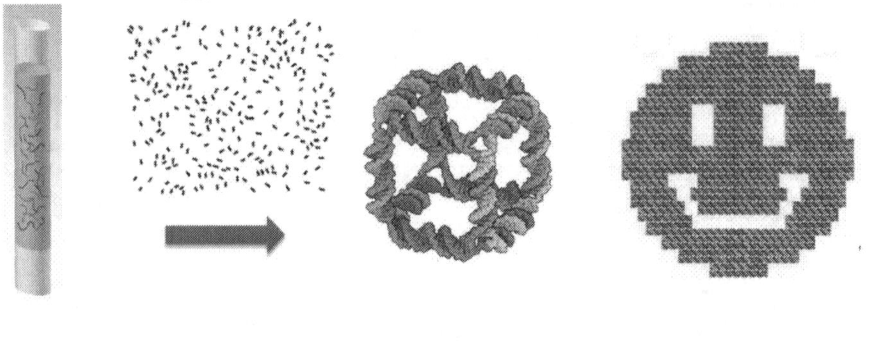

What kinds of dynamical behaviors are nucleic acid systems capable of?

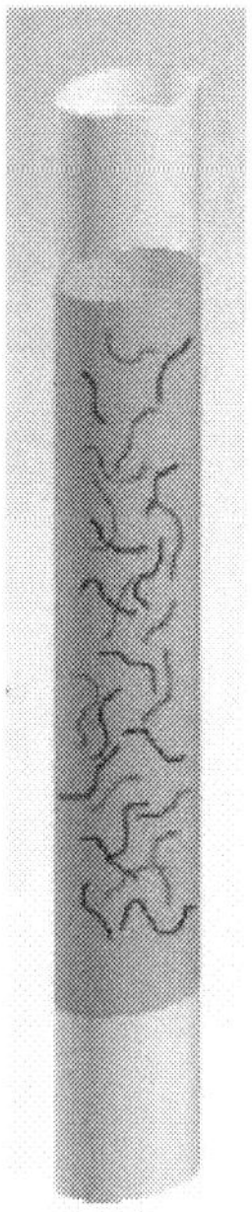

goes to completion...?

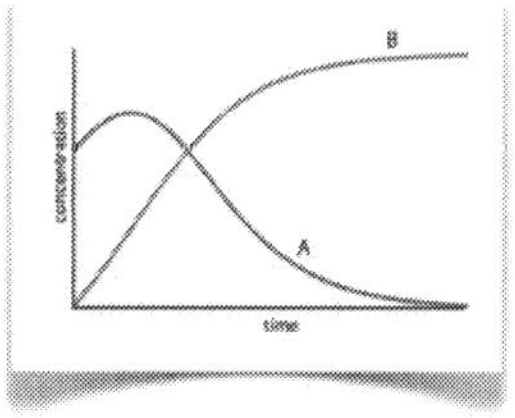

oscillates...?

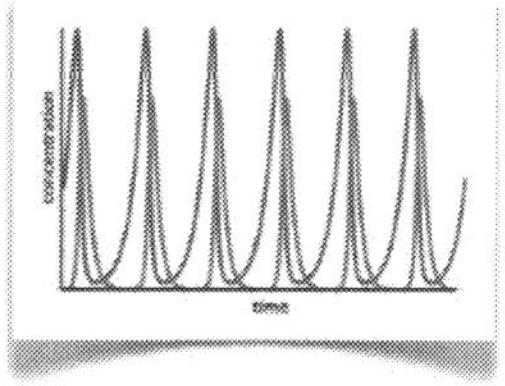

does something complex...?

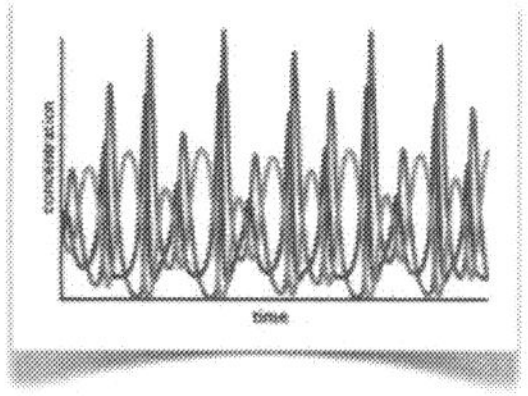

DNA hybridization

```
AGTCTTCGAATGCTAATTGCGCT

|||||||||||||||||||||||
TCAGAAGCTTACGATTAACGCGA
```

Toehold mediated strand displacement

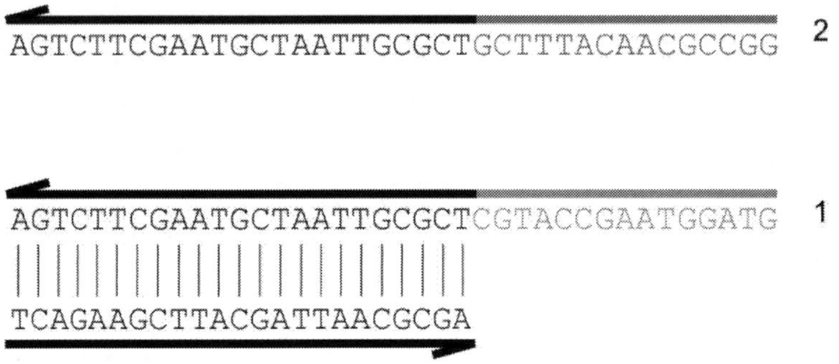

Toehold mediated strand displacement

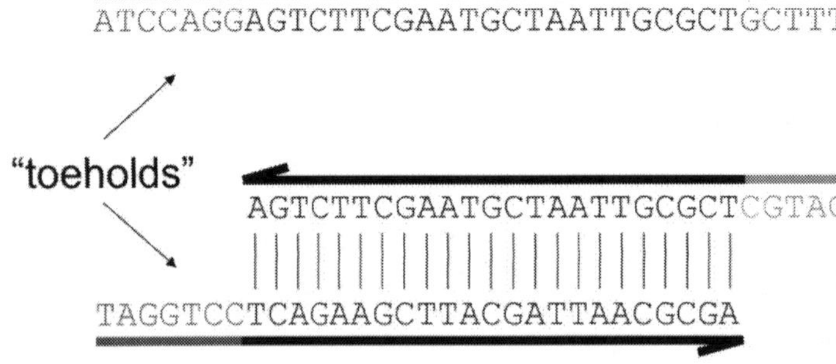

Yurke, Turberfield, Mills, Simmel, Neumann, **Nature** 406, 605 (2000)

Toehold mediated strand displacement

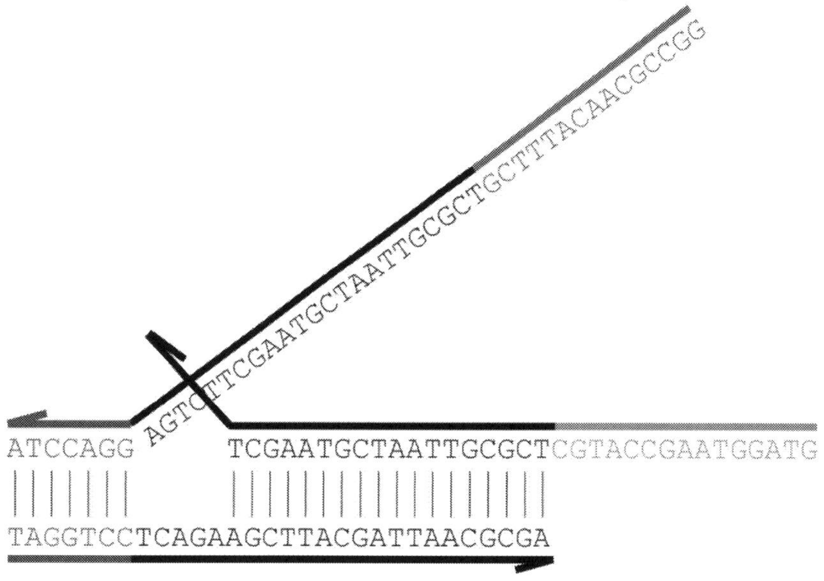

Yurke, Turberfield, Mills, Simmel, Neumann, **Nature** 406, 605 (2000)

Toehold mediated strand displacement

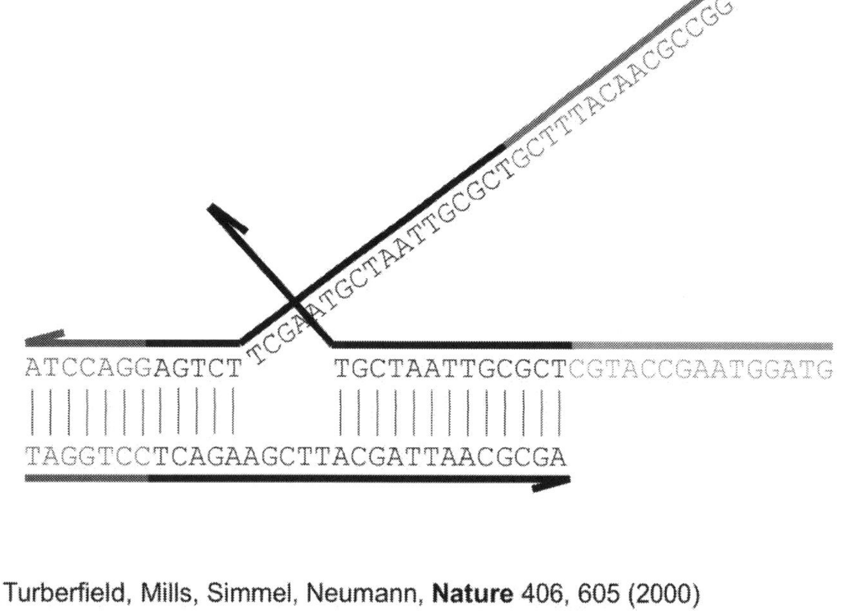

Yurke, Turberfield, Mills, Simmel, Neumann, **Nature** 406, 605 (2000)

Toehold mediated strand displacement

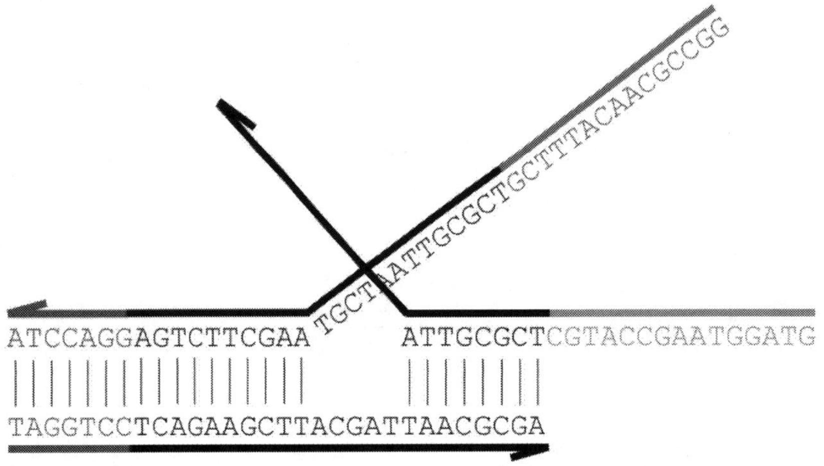

Yurke, Turberfield, Mills, Simmel, Neumann, **Nature** 406, 605 (2000)

Toehold mediated strand displacement

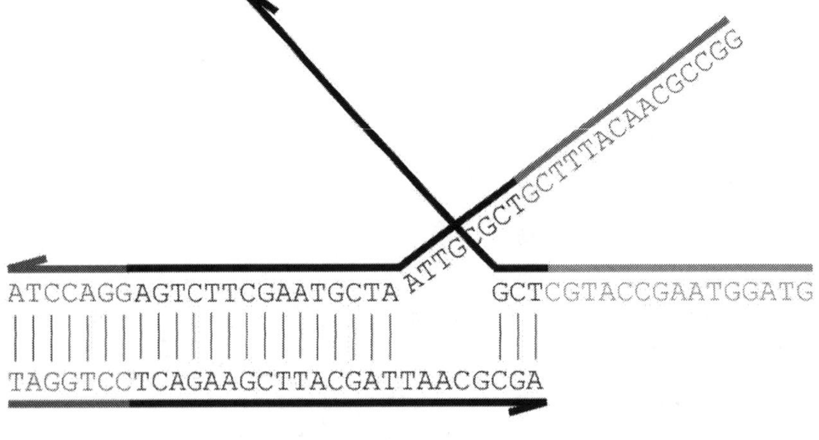

Yurke, Turberfield, Mills, Simmel, Neumann, **Nature** 406, 605 (2000)

Toehold mediated strand displacement

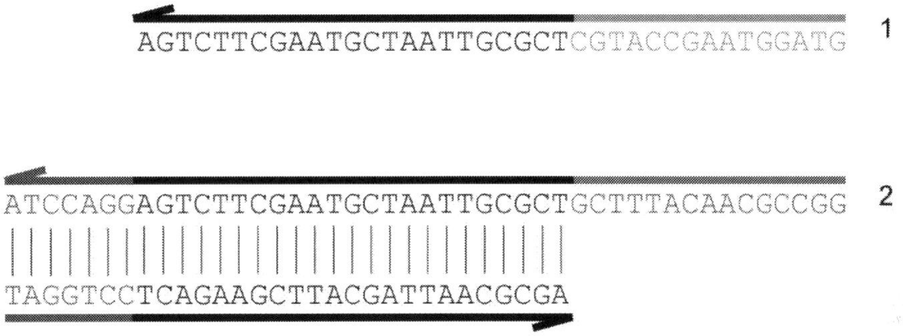

Yurke, Turberfield, Mills, Simmel, Neumann, **Nature** 406, 605 (2000)

Toehold mediated strand displacement

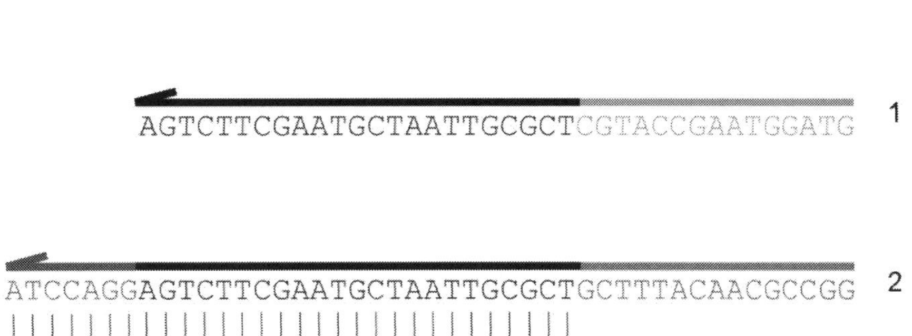

Yurke, Turberfield, Mills, Simmel, Neumann, **Nature** 406, 605 (2000)

Toehold mediated strand displacement

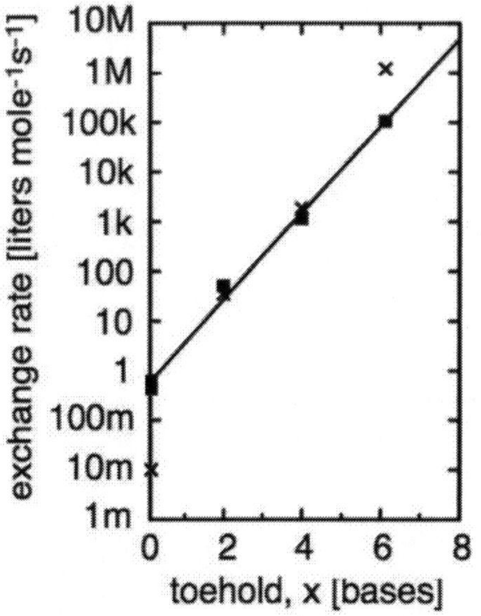

Bernard Yurke, Allen P. Mills, Genetic Programming and Evolvable Machines, 4: 111-122 (2003)

The programmable substrate: DNA strand displacement cascades

Logic: an AND gate

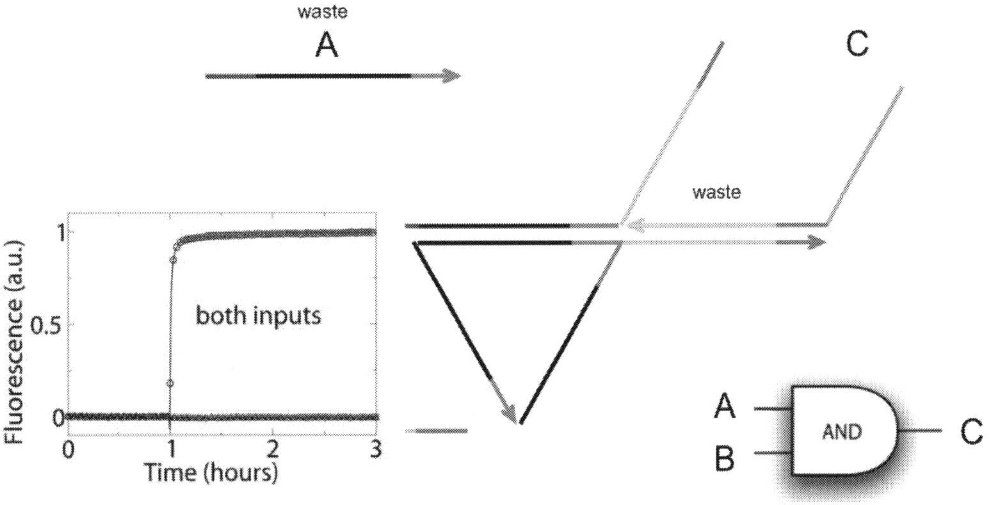

Seelig, Soloveichik, Zhang, Winfree, **Science**, 314: 1585-1587, 2006

A catalyst

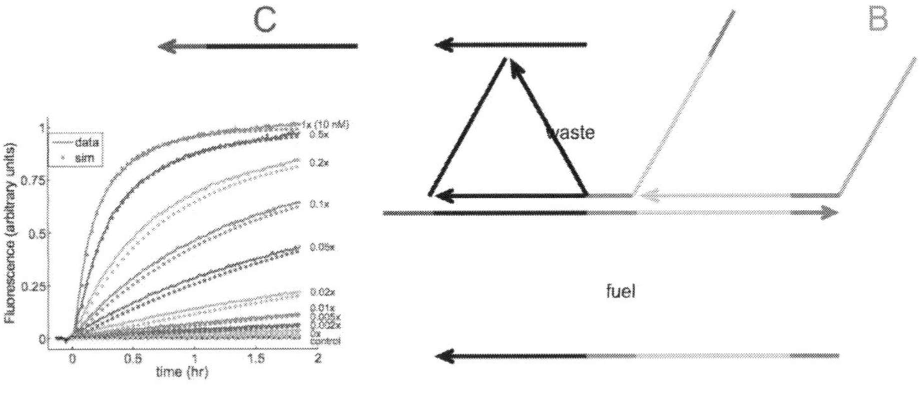

C + fuel1 + fuel2 → C + waste1 + waste2 + B

Zhang, Turberfield, Yurke, Winfree, **Science,** Nov 16, 2007

A square root circuit

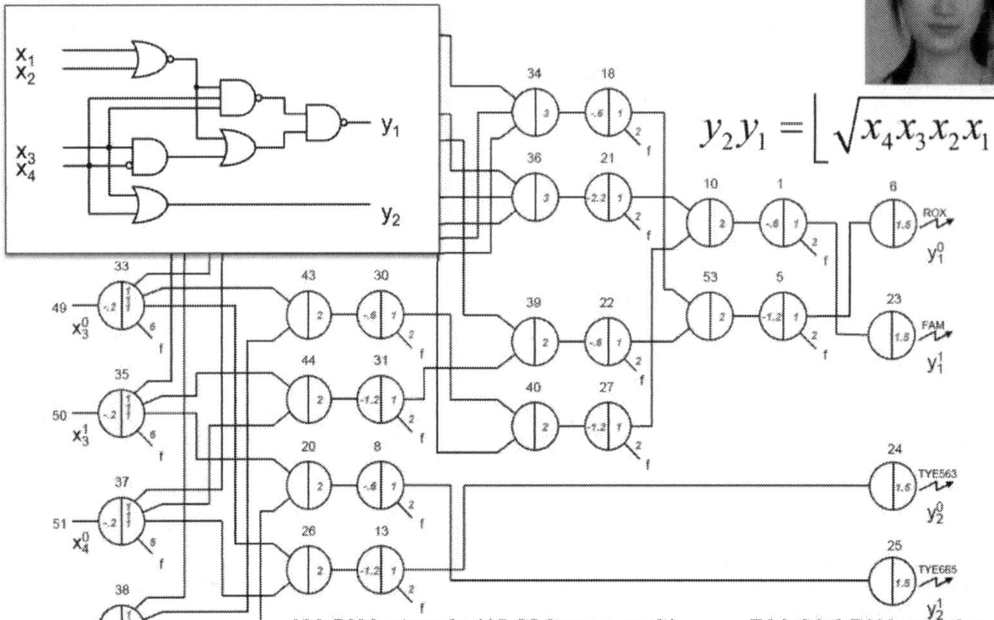

$$y_2 y_1 = \left\lfloor \sqrt{x_4 x_3 x_2 x_1} \right\rfloor$$

130 DNA strands (15-33 bases each) **74 initial DNA species**

Lulu Qian, Erik Winfree, *Science*, 2011

A square root circuit

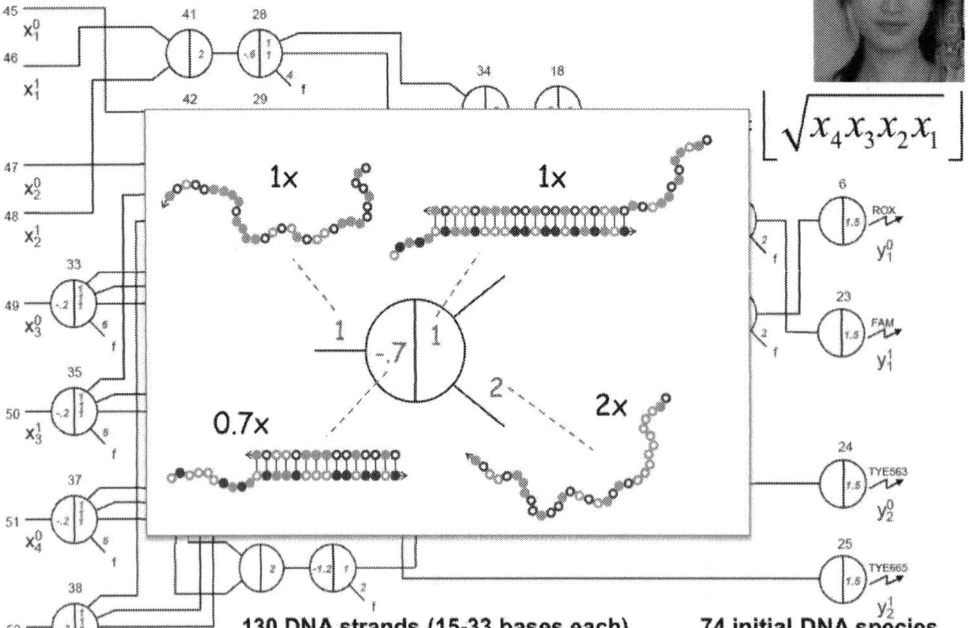

$$\left\lfloor \sqrt{x_4 x_3 x_2 x_1} \right\rfloor$$

130 DNA strands (15-33 bases each) **74 initial DNA species**

Lulu Qian, Erik Winfree, *Science*, 2011

A square root circuit

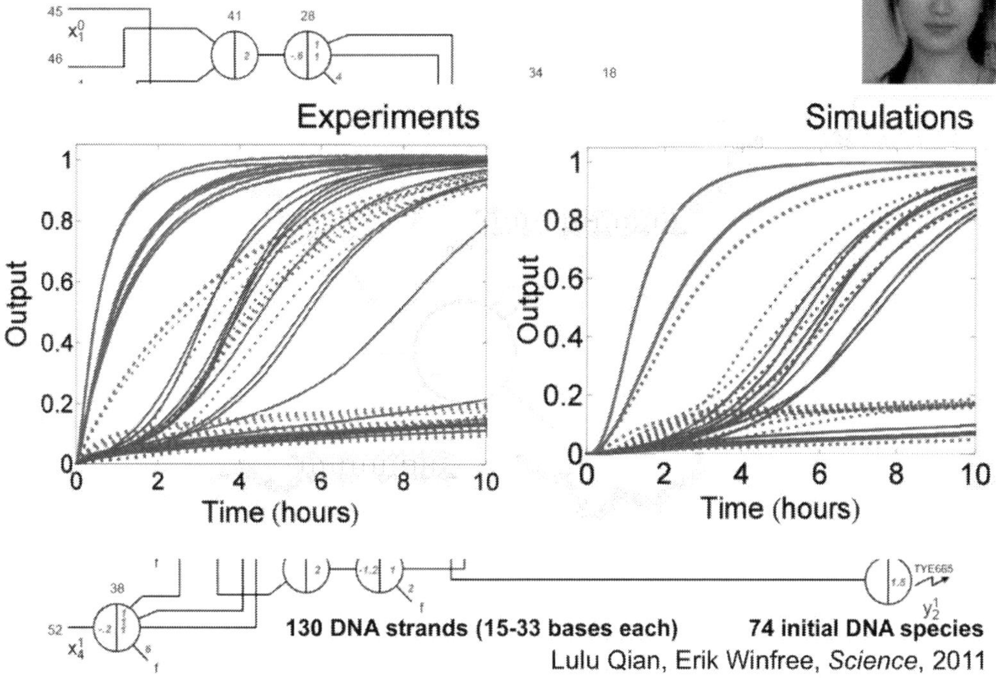

Lulu Qian, Erik Winfree, *Science*, 2011

Toss a mixture of DNA in a test tube... What can it do? What can't it do?

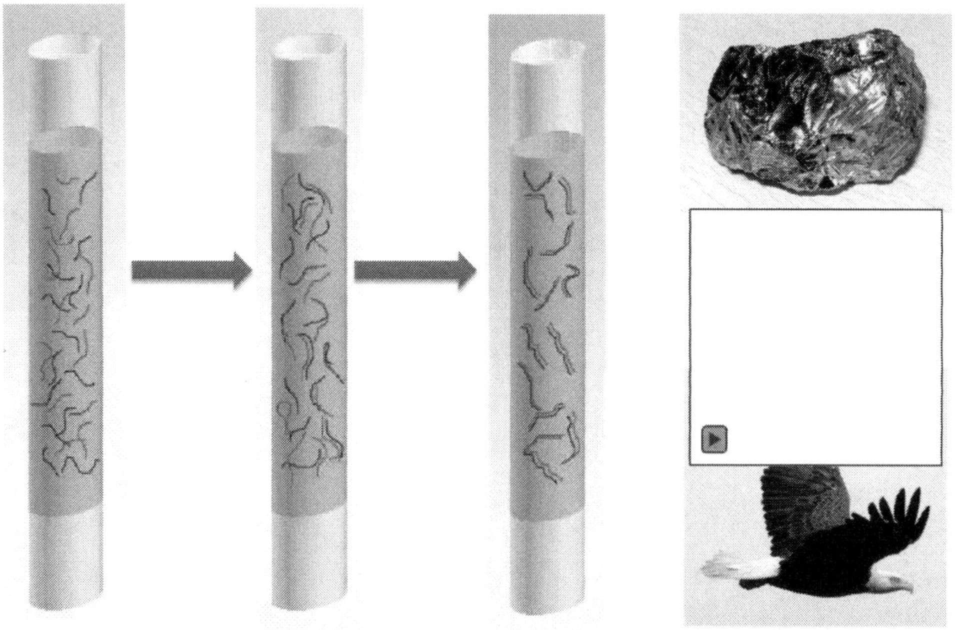

Implementing Formal Chemical Reaction Networks with DNA Strand Displacement Cascades

David Soloveichik, Georg Seelig, Erik Winfree, **LNCS,** 2009 and **PNAS**, 2010

Some behaviors of simple chemical systems (mass action, well-mixed)

stabilization | oscillation | chaos

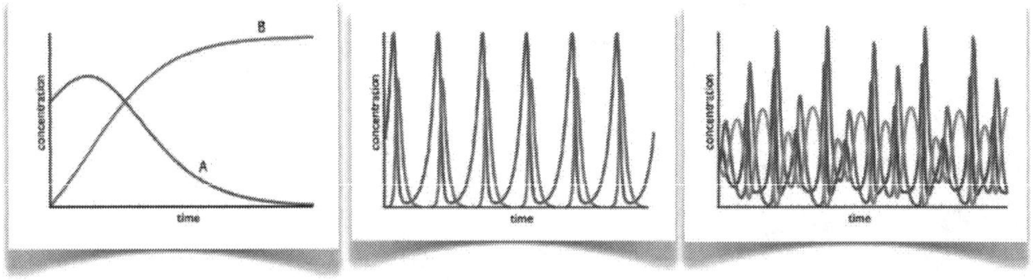

Stabilization:

$$A \longrightarrow B$$
$$A+B \longrightarrow B$$
$$A \longrightarrow 2A$$

Oscillation:

$$A \longrightarrow 2A$$
$$A+B \longrightarrow 2B$$
$$B \longrightarrow$$

Chaos:

$$A \xrightarrow{30} 2A$$
$$2A \xrightarrow{0.5} A$$
$$B+A \xrightarrow{1} 2B$$
$$B \xrightarrow{10}$$
$$A+C \xrightarrow{1}$$
$$C \xrightarrow{16.5} 2C$$
$$2C \xrightarrow{0.5} C$$

open systems: no conservation of mass/energy

Some behaviors of simple chemical systems

Chemical reaction networks (CRNs) as a programming language…
…use the formalism prescriptively, not descriptively…

- Mathematically equivalent to Petri Nets, Vector Addition Systems, etc.
  e.g. Petri (1966), Karp (1969), Goss & Peccaud (1998)
- Can implement arbitrary sequential digital logic circuits
  e.g. Magnasco (1997)
- Can implement (nearly) arbitrary dynamical system behaviors
  e.g. Korzuhin (1967), Klonowski (1983)
- Can can simulate Turing machine computations with arbitrarily small error
  e.g. Soloveichik, Cook, Winfree, Bruck (2008)

But does every formal chemical reaction network exist?

Domain-level model of DNA strand displacement

possible moves:
- **Hybridize** two complementary domains
- **Release** any strand held by only a short domain
- **Displace** a domain by an identical domain if this extends existing hybridization

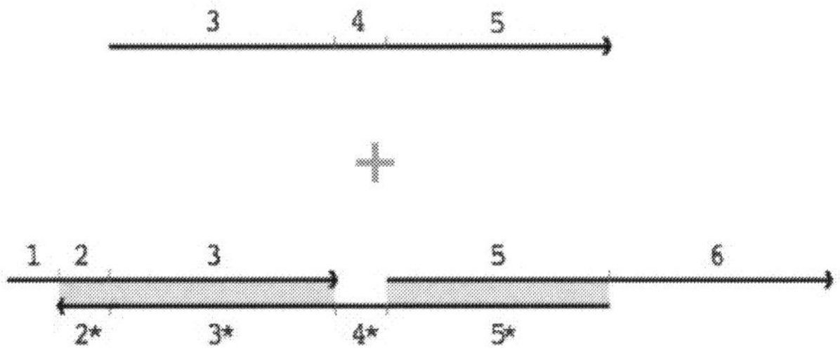

Soloveichik, Seelig, Winfree, PNAS, 2010

Domain-level model of DNA strand displacement

possible moves:

- **Hybridize** two complementary domains
- **Release** any strand held by only a short domain
- **Displace** a domain by an identical domain if this extends existing hybridization

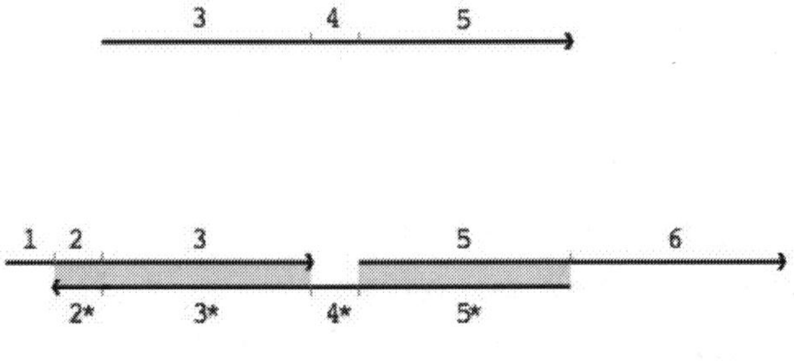

Soloveichik, Seelig, Winfree, PNAS, 2010

Domain-level model of DNA strand displacement

possible moves:

- **Hybridize** two complementary domains
- **Release** any strand held by only a short domain
- **Displace** a domain by an identical domain if this extends existing hybridization

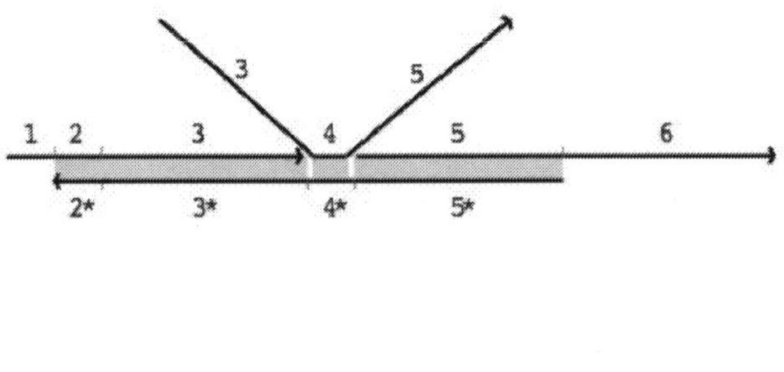

Soloveichik, Seelig, Winfree, PNAS, 2010

Domain-level model of DNA strand displacement

possible moves:

- **Hybridize** two complementary domains
- **Release** any strand held by only a short domain
- **Displace** a domain by an identical domain if this extends existing hybridization

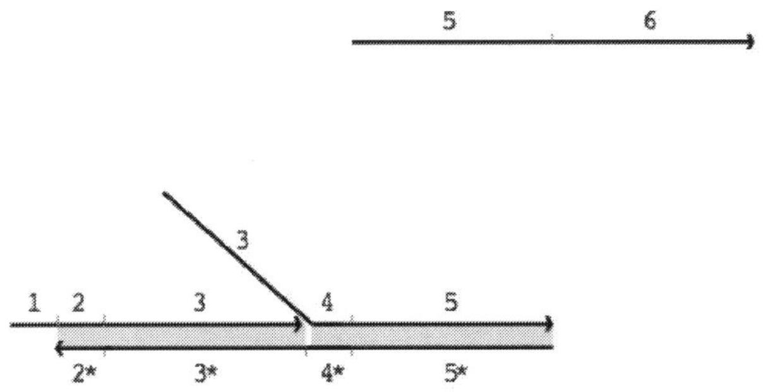

Soloveichik, Seelig, Winfree, PNAS, 2010

Domain-level model of DNA strand displacement

possible moves:

- **Hybridize** two complementary domains
- **Release** any strand held by only a short domain
- **Displace** a domain by an identical domain if this extends existing hybridization

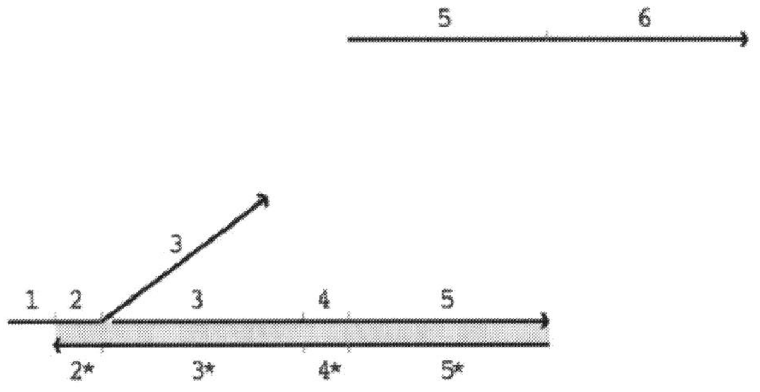

Soloveichik, Seelig, Winfree, PNAS, 2010

Domain-level model of DNA strand displacement

possible moves:

- **Hybridize** two complementary domains
- **Release** any strand held by only a short domain
- **Displace** a domain by an identical domain if this extends existing hybridization

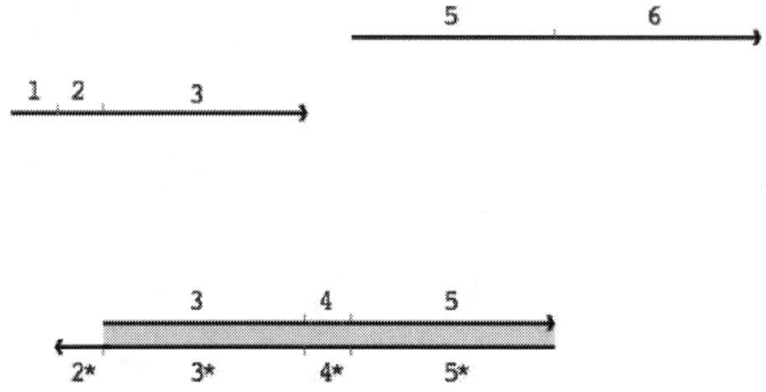

Soloveichik, Seelig, Winfree, PNAS, 2010

Compiling CRNs into DNA

CRN program:

$$A \xrightarrow{30} 2A$$
$$2A \xrightarrow{0.5} A$$
$$B+A \xrightarrow{1} 2B$$
$$B \xrightarrow{10}$$
$$A+C \xrightarrow{1}$$
$$C \xrightarrow{16.5} 2C$$
$$2C \xrightarrow{0.5} C$$

Initial conditions:

5 *A*'s
3 *B*'s
10 *C*'s

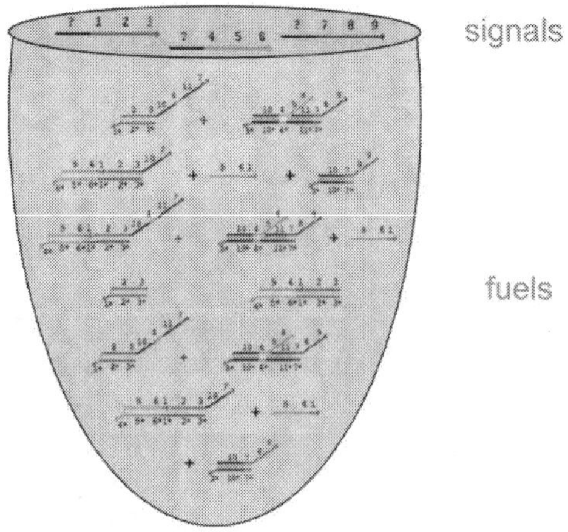

Soloveichik, Seelig, Winfree, PNAS, 2010

Programming Dynamical Behaviors in Chemical Systems using DNA Strand Displacement Cascades

Niranjan Srinivas, James Parkin, Georg Seelig, Erik Winfree, David Soloveichik
(in prep)

Compiling a CRN oscillator into a DNA system

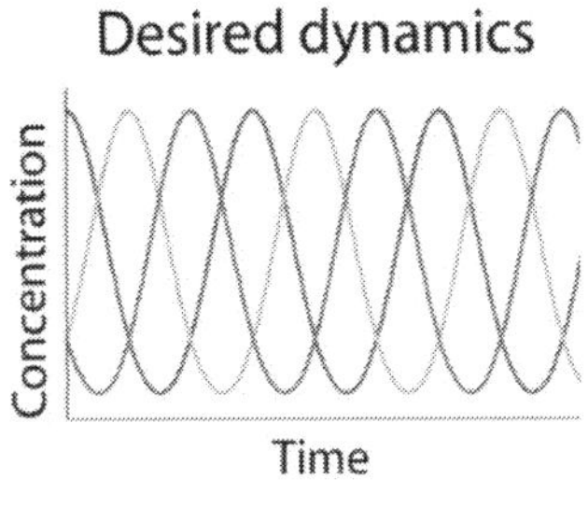

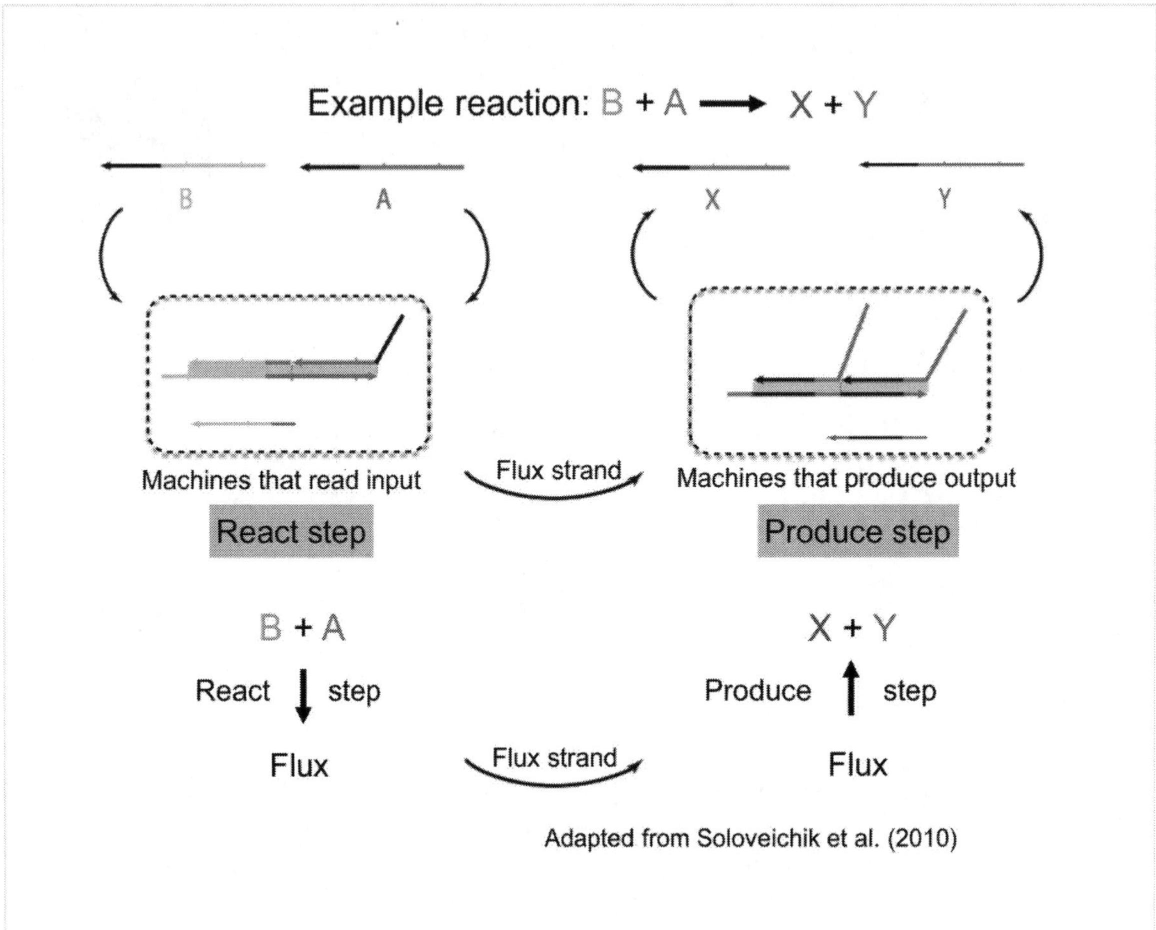
Example reaction: B + A ⟶ X + Y
B
A
X
Y
Machines that read input
Flux strand
Machines that produce output
React step
Produce step
B + A
React step
Flux
Flux strand
X + Y
Produce step
Flux
Adapted from Soloveichik et al. (2010)

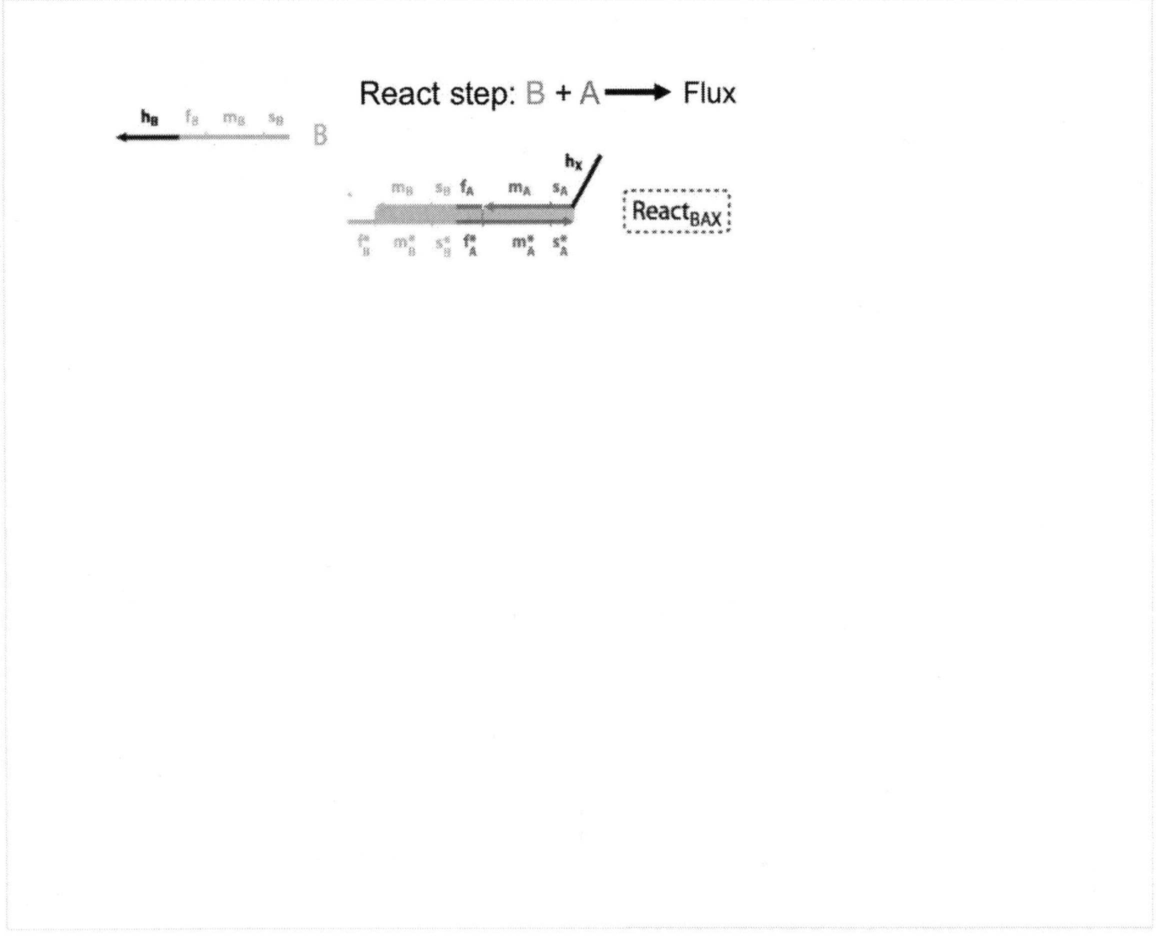
React step: B + A ⟶ Flux
hB fB mB sB
B
hX
mB sB fA mA sA
ReactBAX
fB* mB* sB* fA* mA* sA*

Produce step: Flux → X + Y

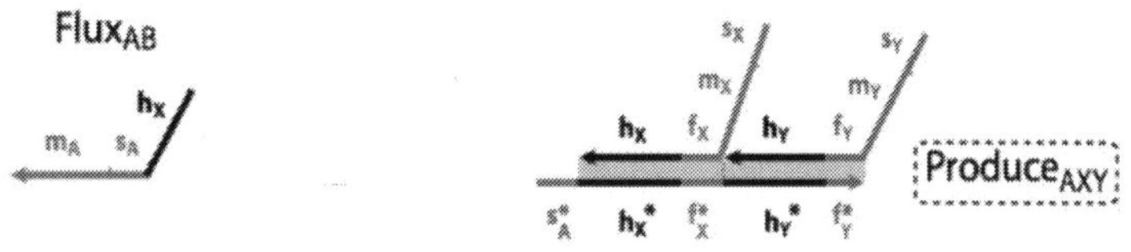

Example reaction: B + A → X + Y

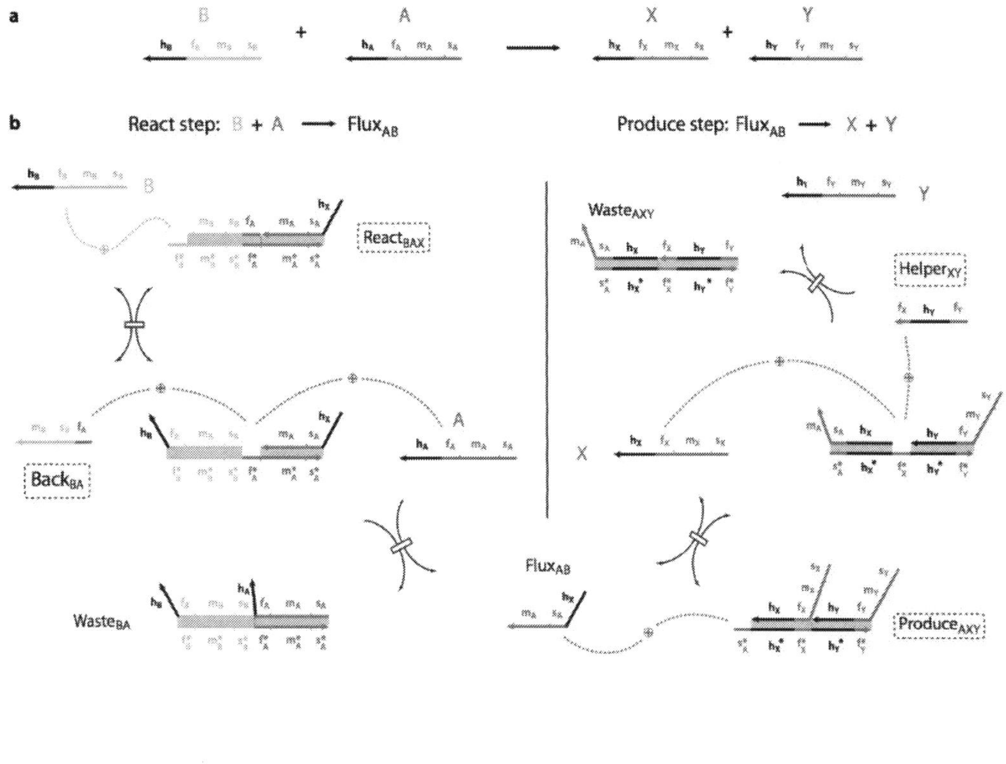

Compiling a CRN oscillator into a DNA system

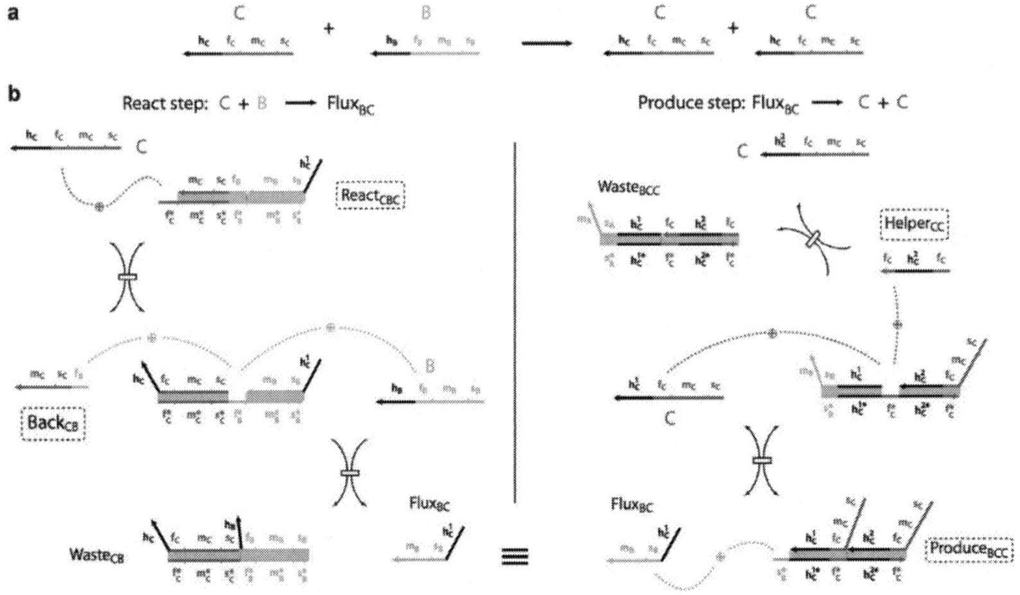

Signal strands with "thresholds" for tuning

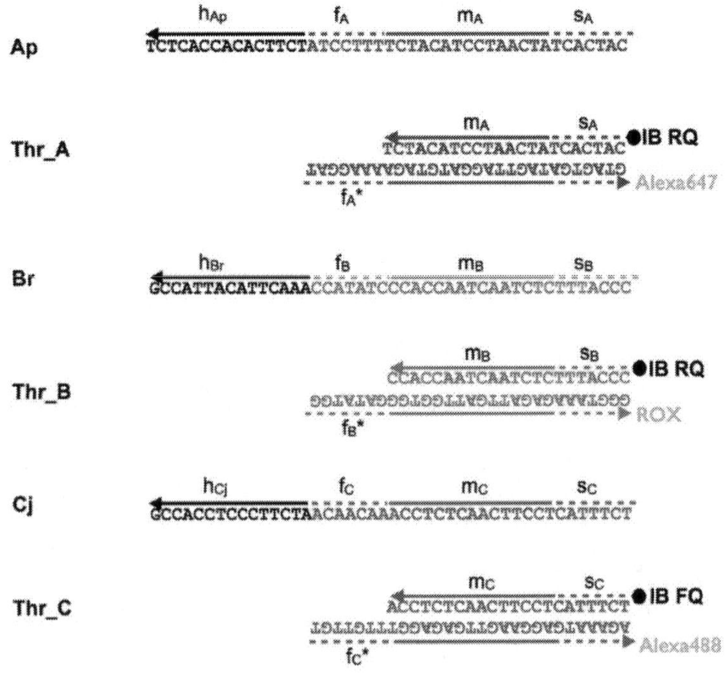

Fuels for B+A → B+B with "catalytic helpers" for tuning

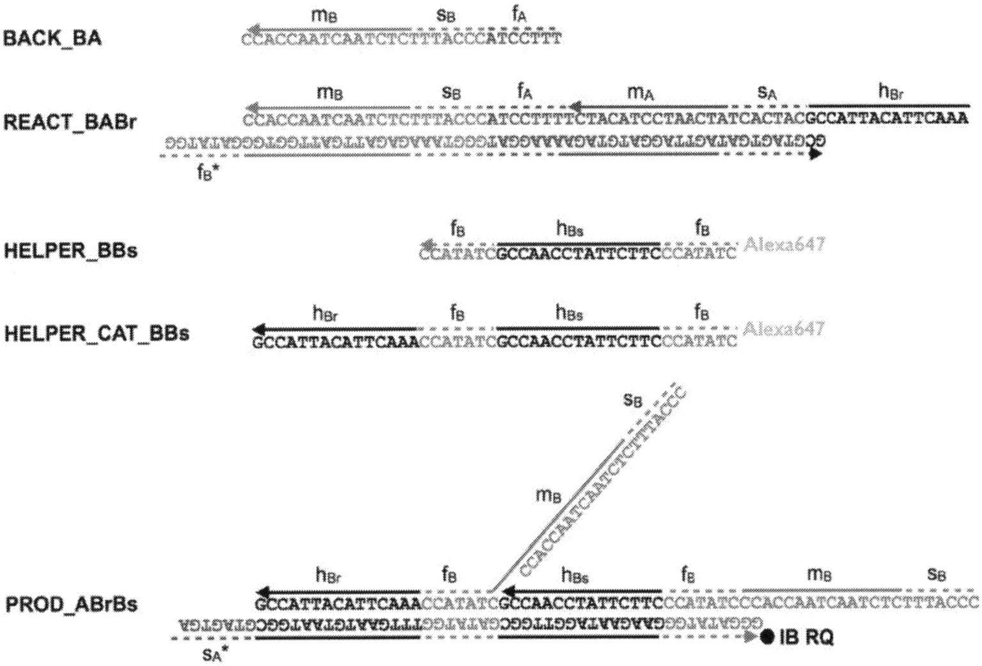

Fuels for C+B → C+C with "catalytic helpers" for tuning

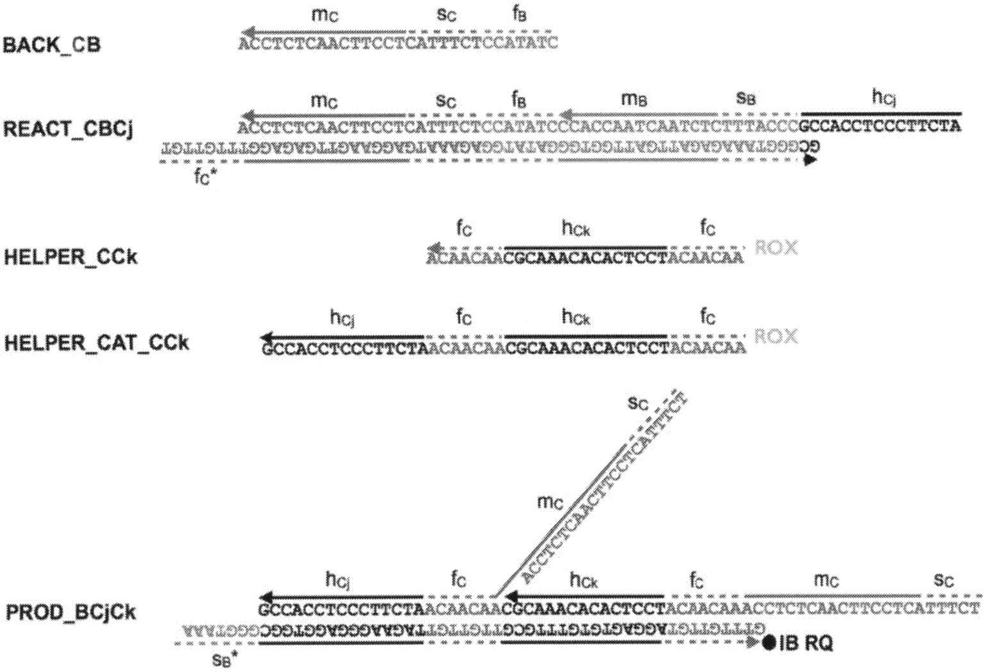

Fuels for A+C → A+A with "catalytic helpers" for tuning

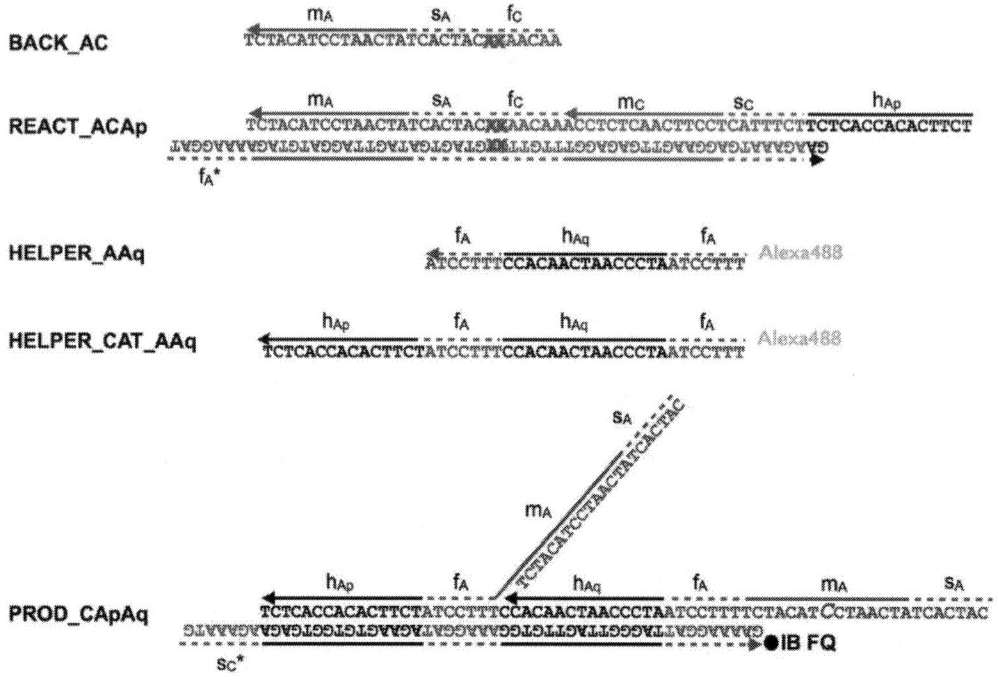

Critical design challenge: initial and gradual leak

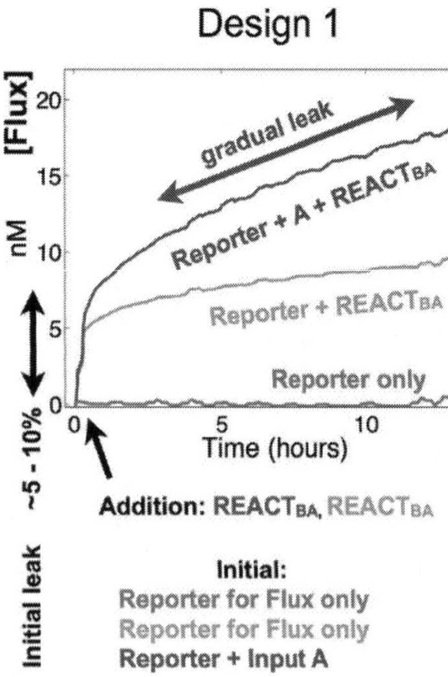

Testing the reaction modules independently

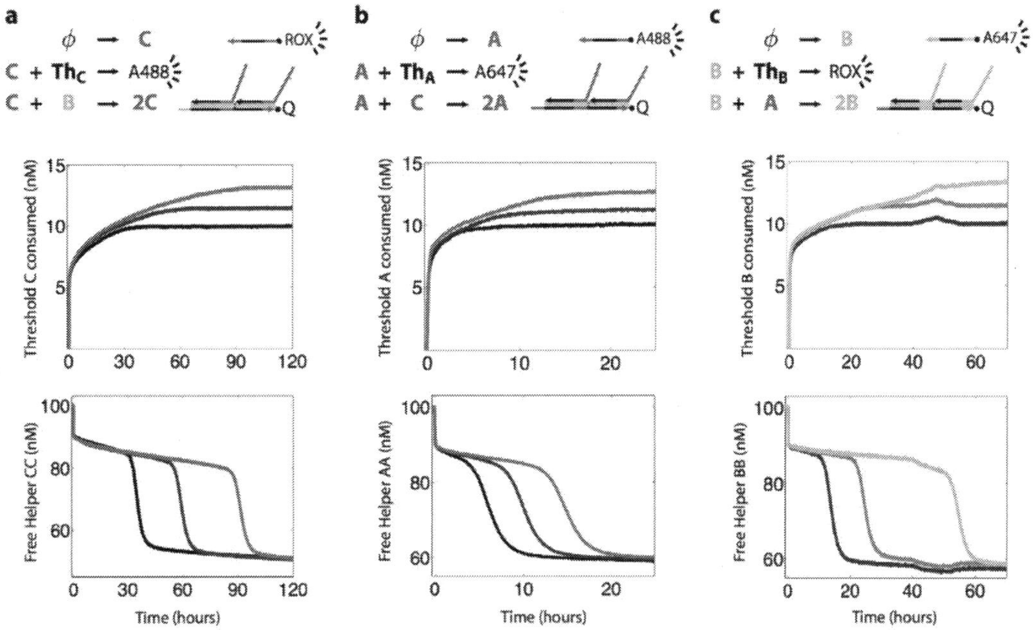

Simulation of the full system: What we expect to see

Ideal DNA implementation

Rate constants $2 * 10^5 \ /M/s$
(all strand displacement)

$([A],[B],[C])_0 = (30, 10, 10) \ nM$

Excess complexes: 300 nM

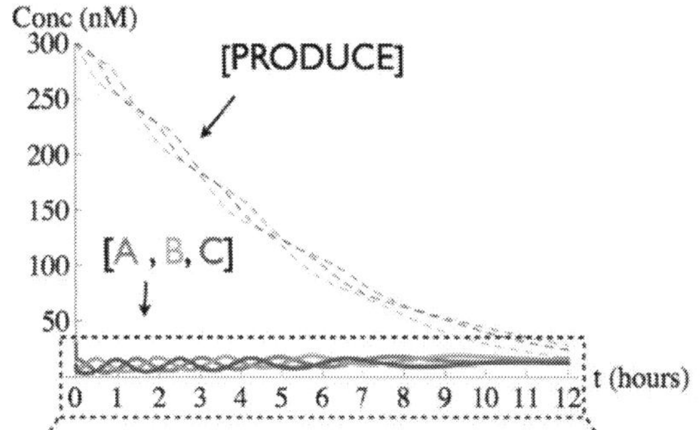

$([A],[B],[C])_0 = (60, 10, 10) \ nM$

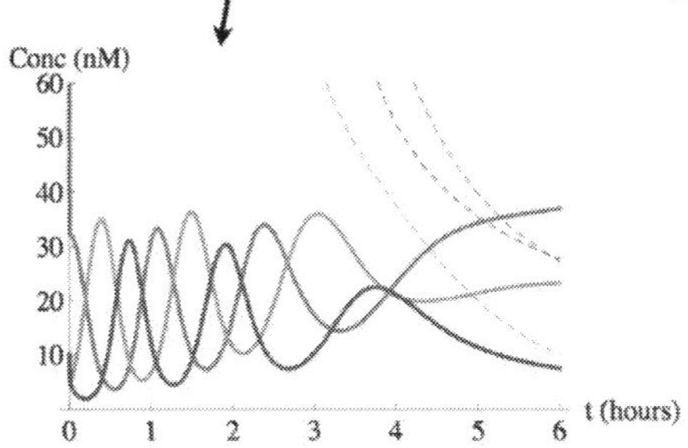

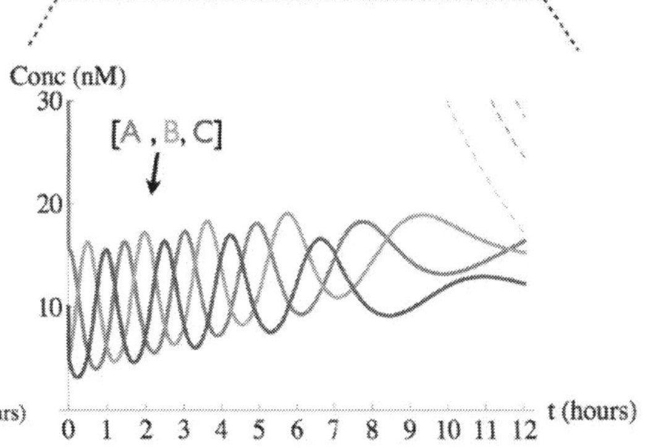

Initial experiments with the full system

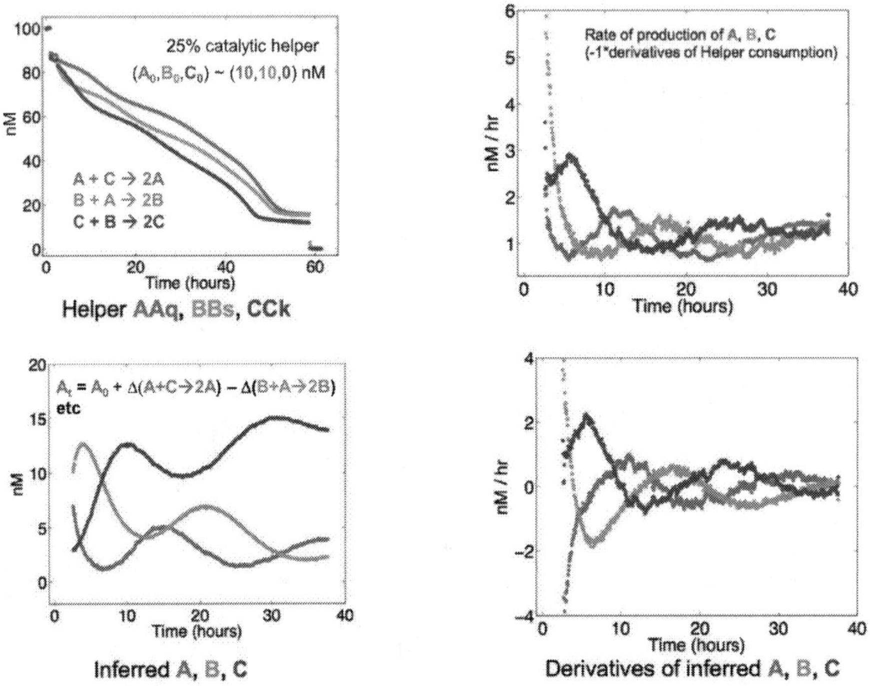

Oscillations from different initial conditions

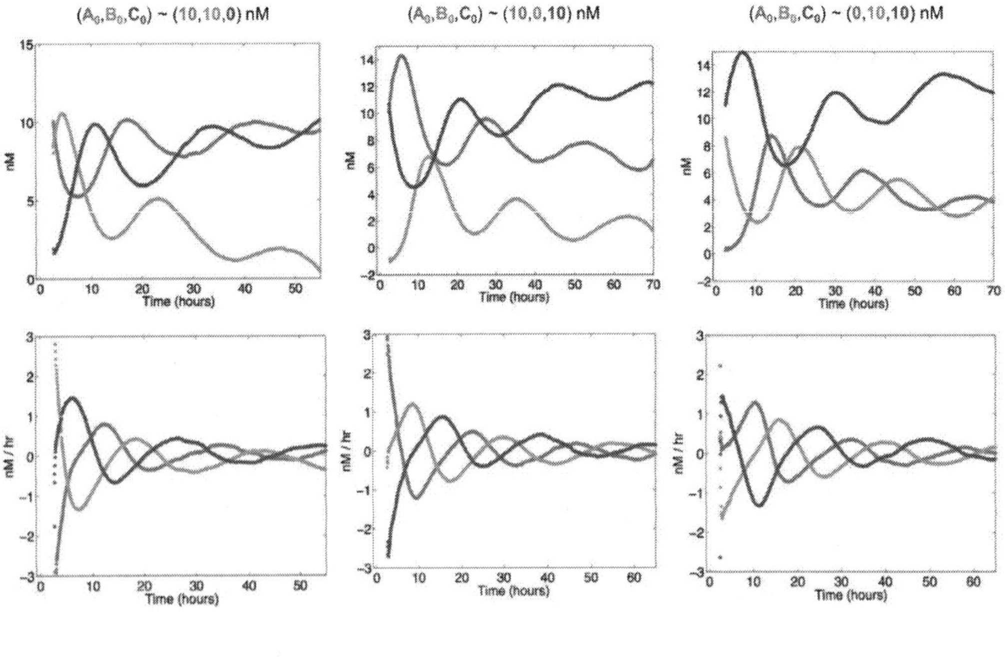

Oscillations from different initial conditions

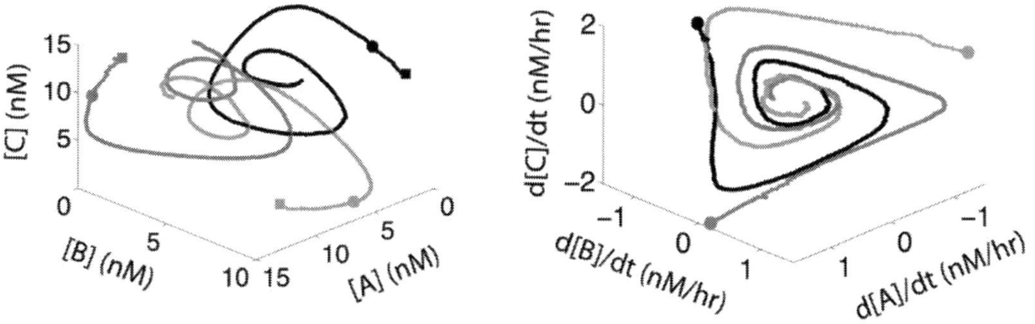

Compiling a CRN oscillator into a DNA system

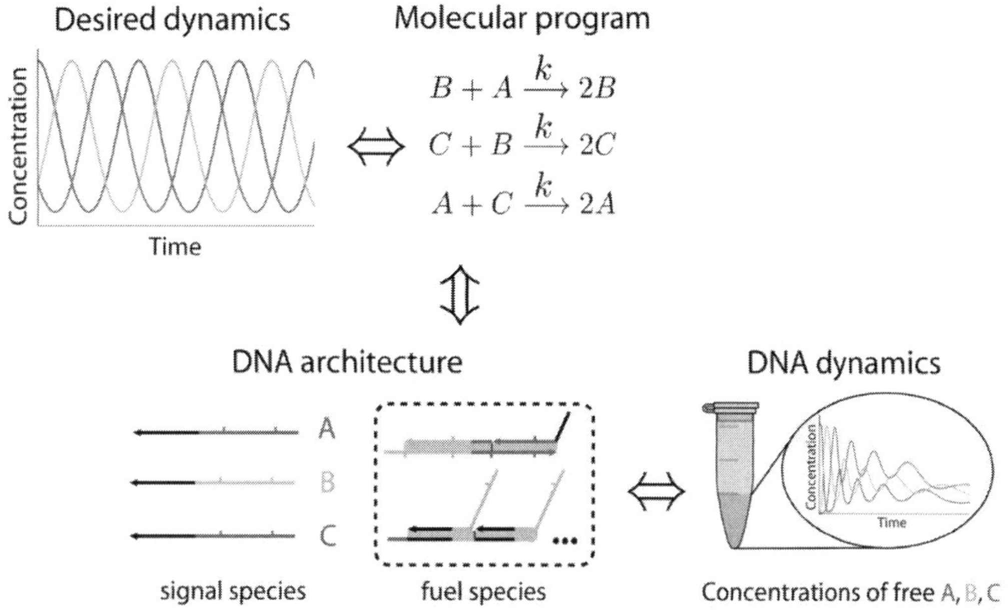

Are our design rules really useful?

Design orthogonal, iso-energetic toeholds

↓

Design all else - minimize spurious matches

↓

Custom heuristics for *in silico* verification

↓

Debugging experiments. Tuning rates.

Can we avoid this feedback step?

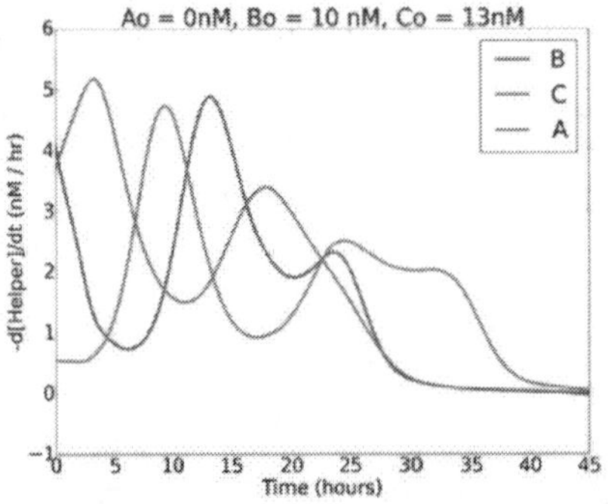

Data from first experiment with completely new sequence design **(literally done last week)**

StickyDesign (Evans & Winfree 2013), SpuriousC (Winfree, 2001), **Srinivas** et al., *in prep.*

Vision for molecular compilers: semantics and verification

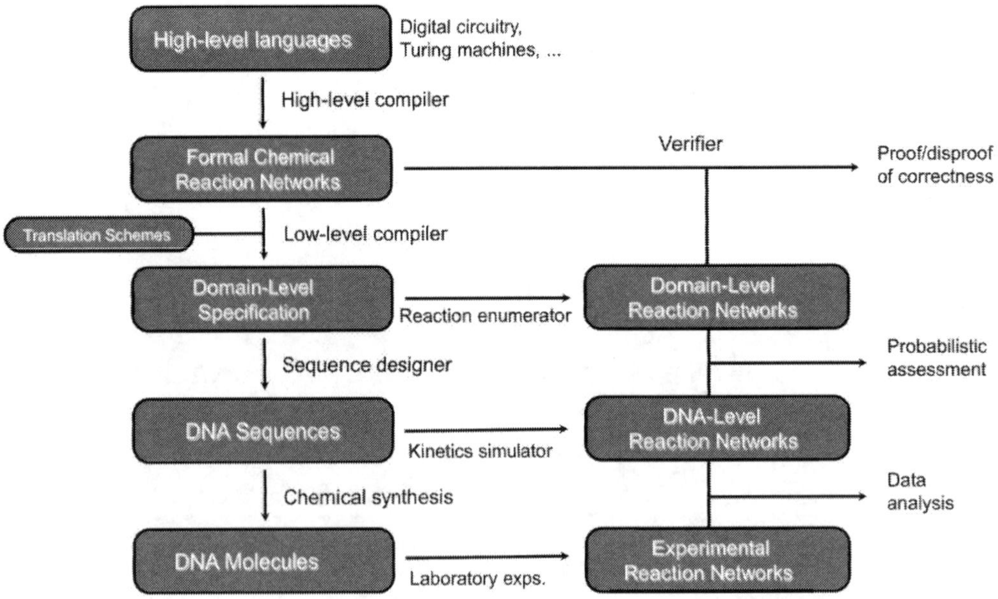

Toss a mixture of DNA in a test tube…
What can it do? What can't it do?

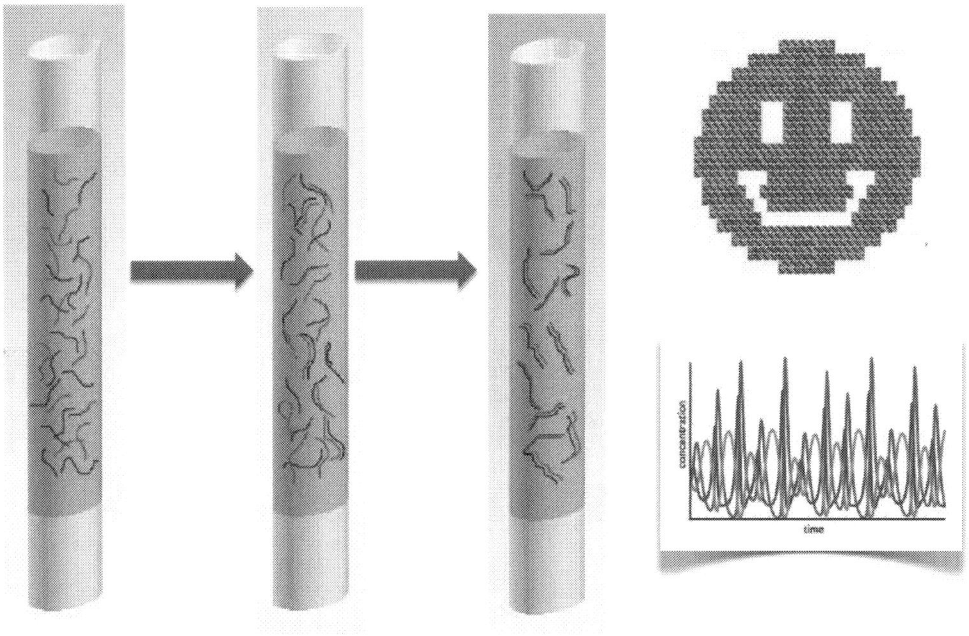

Thank you.

Powered DNA Strand Displacement Circuits

Rebecca Schulman

June 12, 2015

JOHNS HOPKINS UNIVERSITY

Department of Chemical and Biomolecular Engineering

Institute for NanoBioTechnology

Laboratory for Computational Sensing + Robotics, The Johns Hopkins University

Department of Computer Science

Biology points out fundamental challenges in materials design.

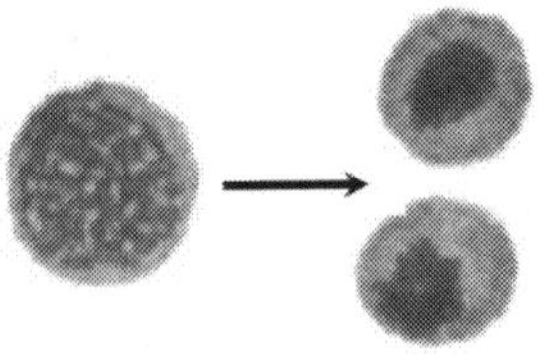

self-replication

self-healing

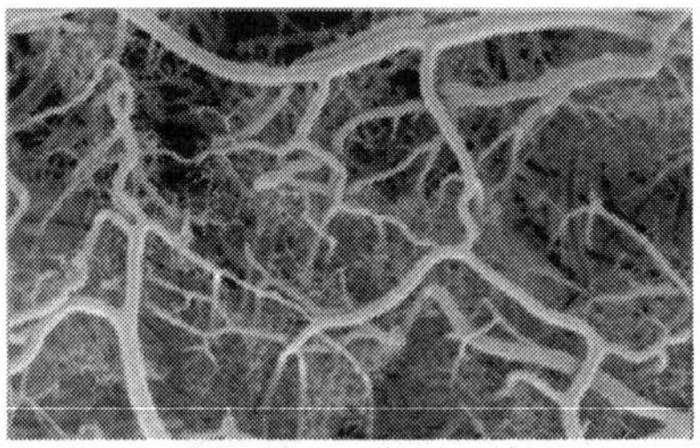

adaptive structure

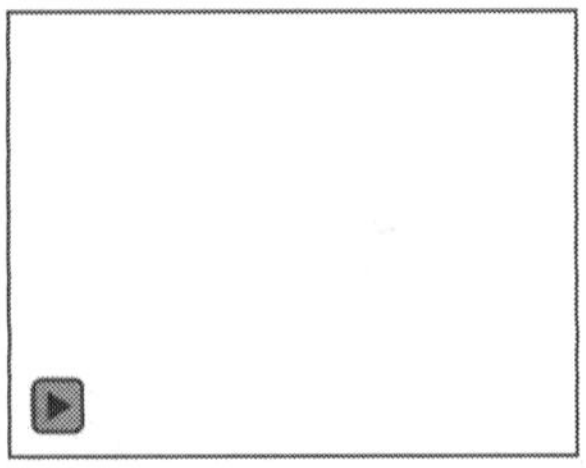

search and targeting

controlled dynamics and shape change

Biological systems combine complex structure and chemical reactions to achieve these features.

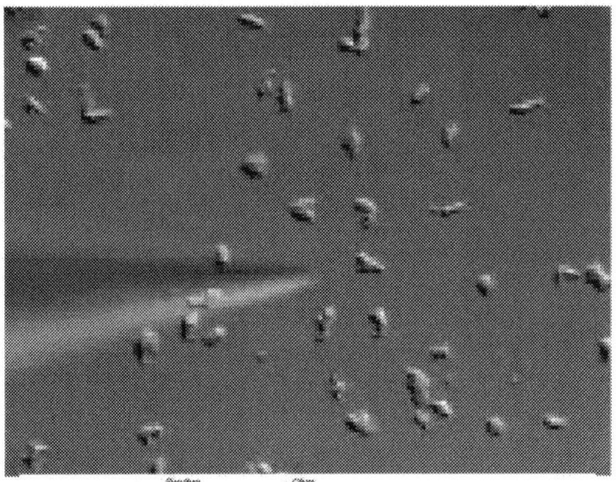

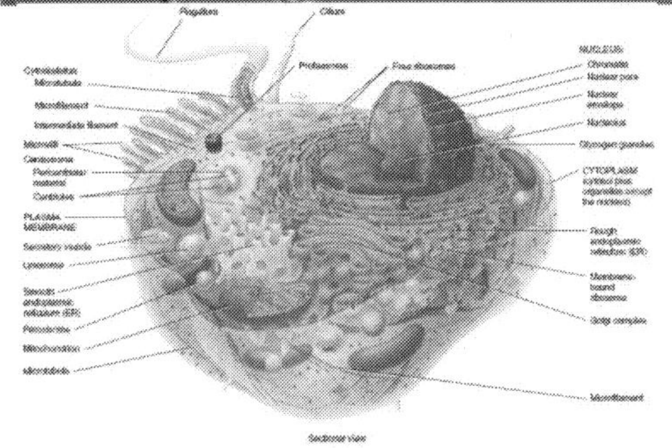

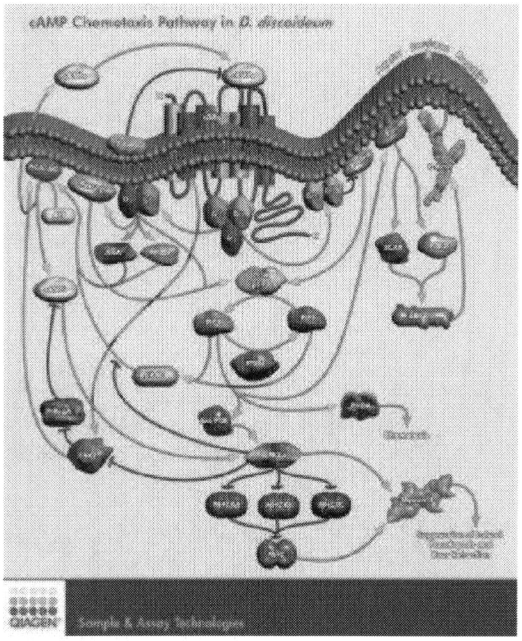

Networks of interactions direct biological growth and behavior.

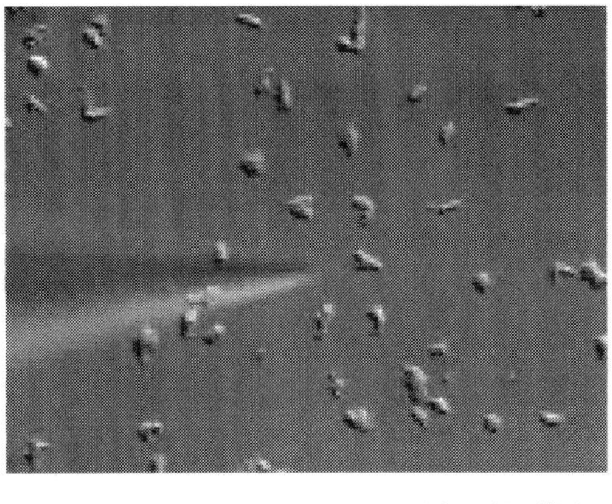

video: YouTube

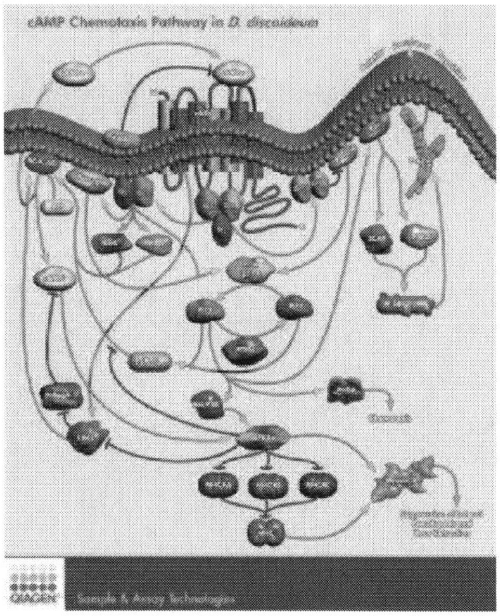

Our goal: recapitulate these integrated behaviors synthetically.

The molecular "circuit" is the essential control mechanism in the cellular world.

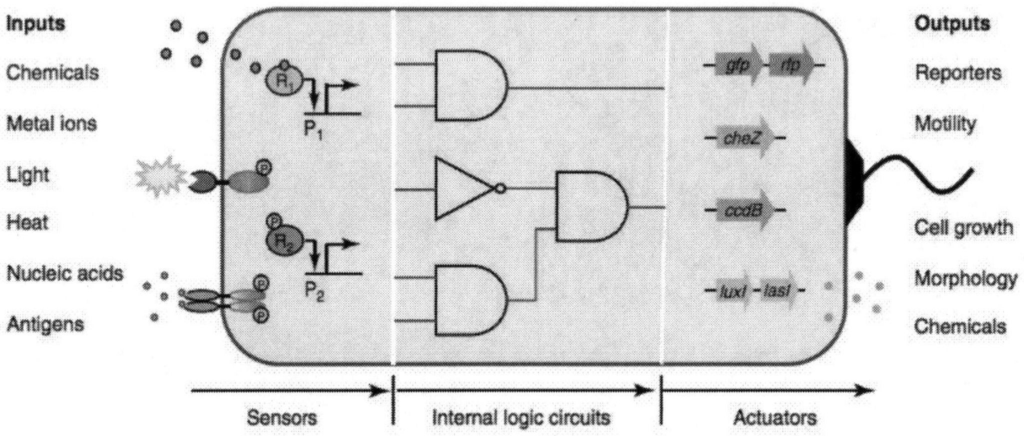

Wang and Buck 2011

Molecular signals are generally given as concentration levels.

In principle, logical control and response could also be executed by in vitro chemical reaction networks.

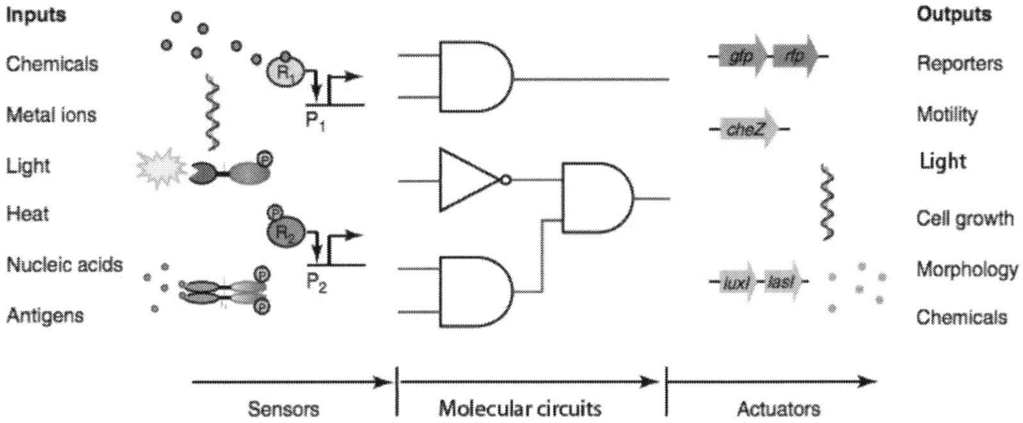

Could control cells, chemical reaction processes, therapeutics, etc.

In principle, logical control and response could also be executed by in vitro chemical reaction networks.

For many applications, in vitro circuits could offer important advantages over the use of whole cells.

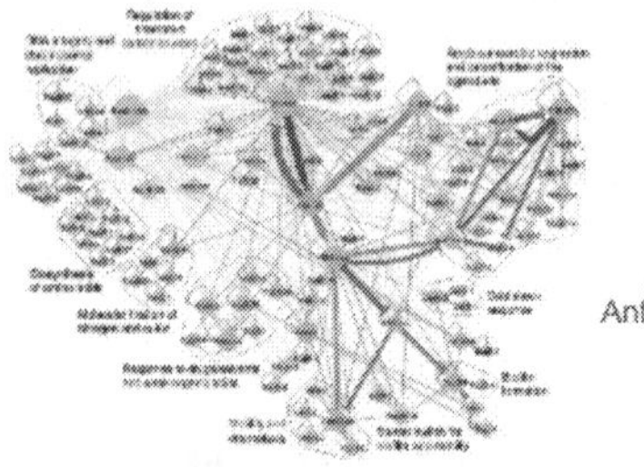

Martinez-Antonio et al 2008

Simplicity

Predictability in different environments

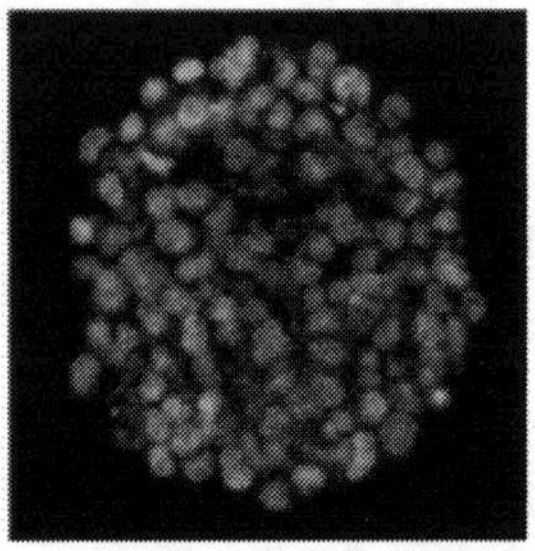

Boutin et al 2015

Whole cells don't need to be added

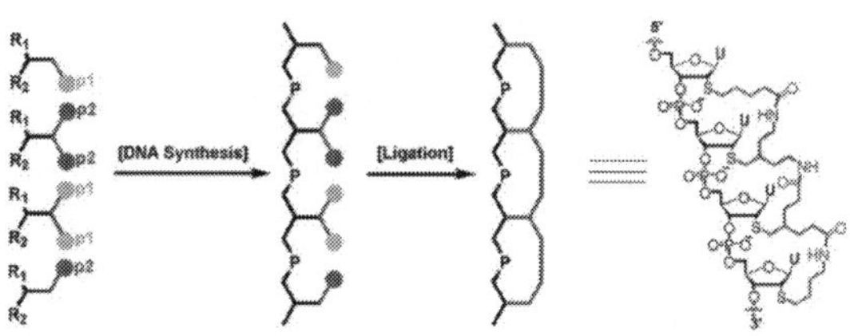

Zhu et al 2003

Expanded synthesis possibilities

Minimal *in vitro* transcriptional circuits could potentially be used to direct external processes.

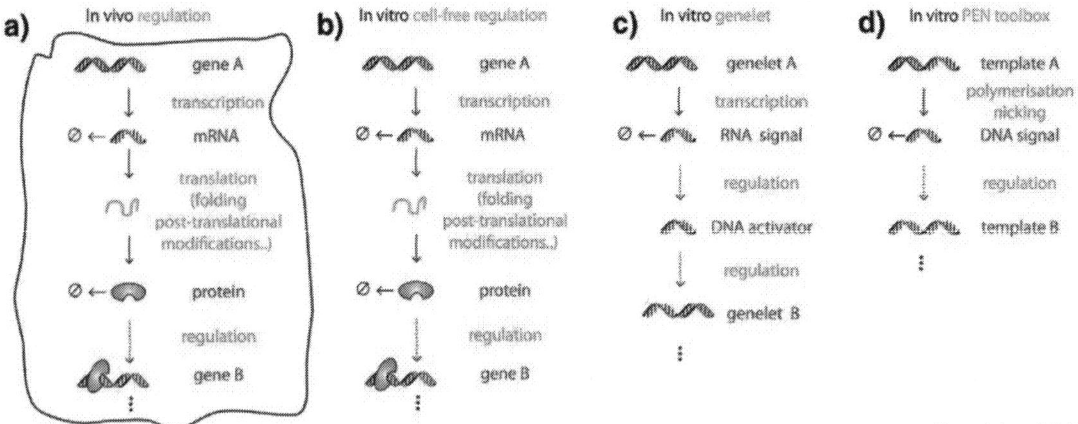

Rondelez 2013

Enzymes place many restrictions on physical conditions and DNA sequences and can be hard to produce and control.

Toehold-mediated strand displacement can produce reliable reaction processes.

Toehold-mediated strand displacement can produce reliable reaction processes.

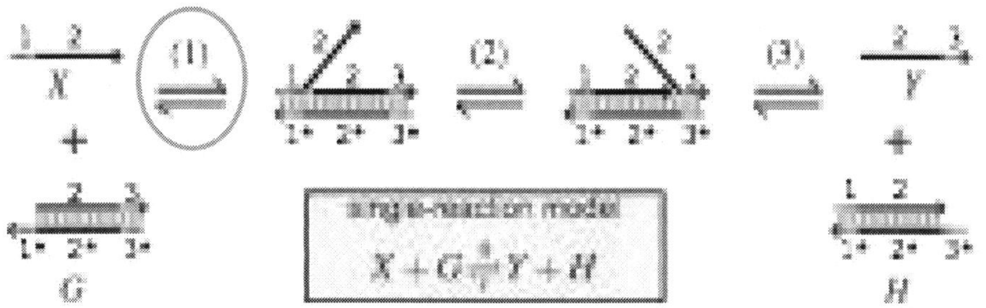

1. X binds to G transiently at a toehold.

Toehold-mediated strand displacement can produce reliable reaction processes.

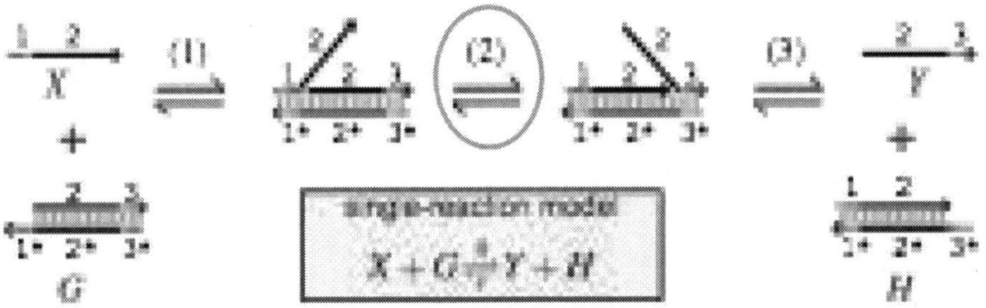

1. X binds to G transiently at a toehold.

2. The bases in domain 2 migrate back and forth randomly.

Toehold-mediated strand displacement can produce reliable reaction processes.

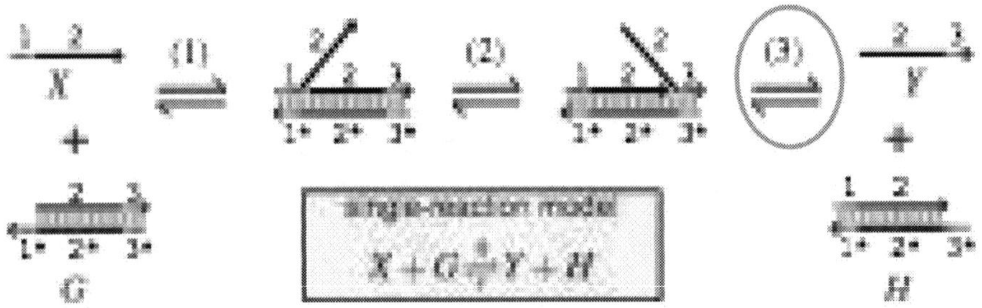

1. X binds to G transiently at a toehold transiently.

2. The bases in domain 2 migrate back and forth randomly

3. Y, bound to the complex transiently, can fall off so that X displaces Y.

Toehold-mediated strand displacement can produce reliable reaction processes.

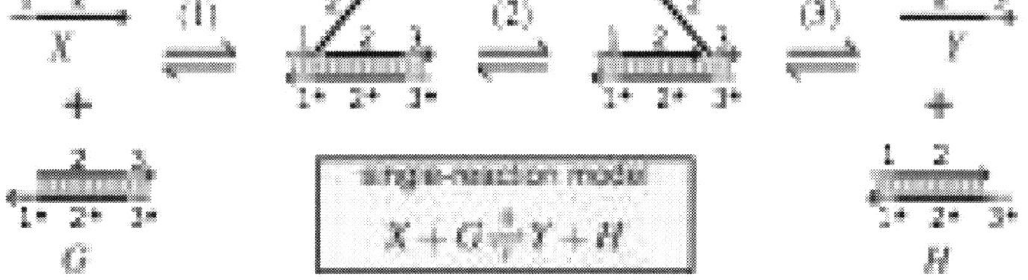

1. X binds to G transiently at a toehold transiently.

2. The bases in domain 2 migrate back and forth randomly

3. Y, bound to the complex transiently, can fall off so that X displaces Y.

Strand displacement circuits can designed in silico, and rapidly scaled.

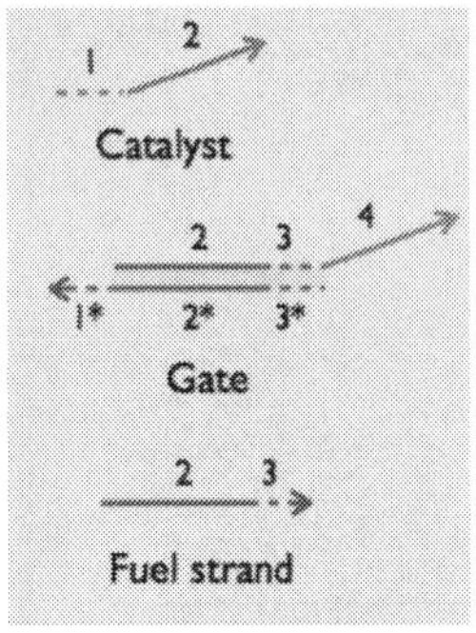

node motif

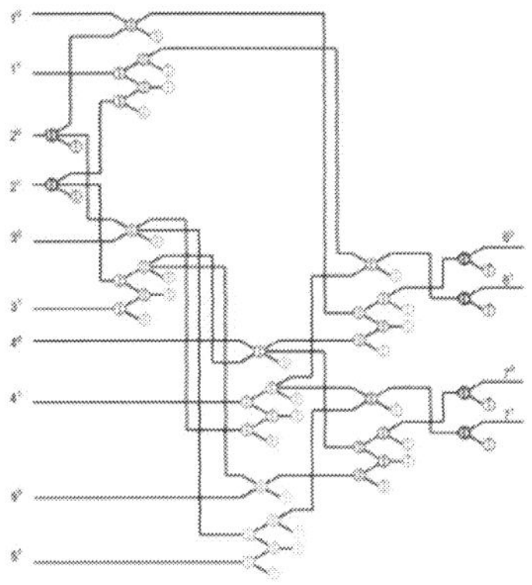

complex logic

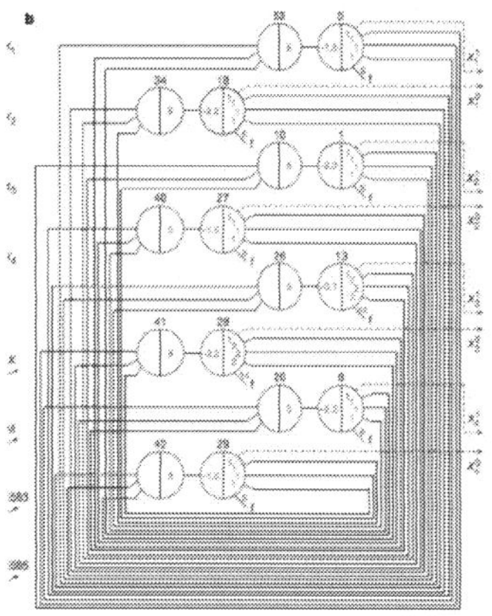

networks with feedback

Qian et al 2011, 2012
Seelig et al 2006

Rates and energetics of DNA strand displacement reactions are well-characterized and predictable.

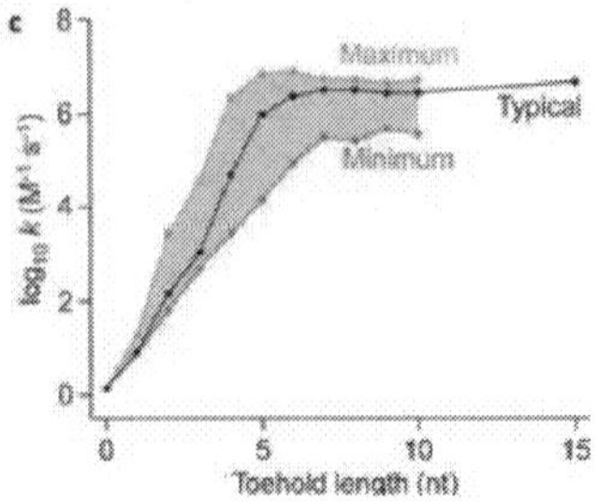

Zhang and Seelig 2011

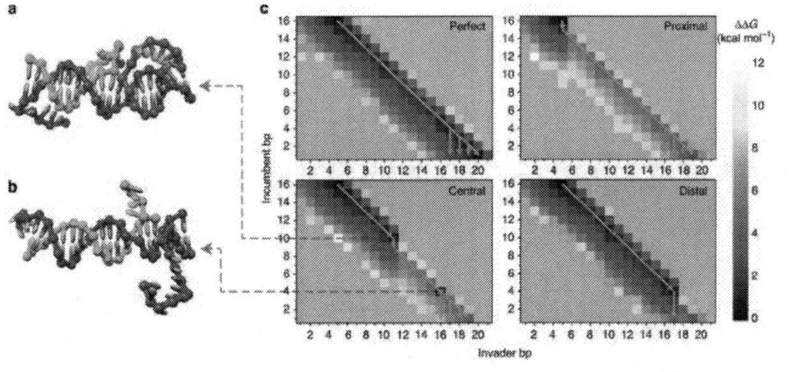

Machinek et al 2014

Networks of chemical reactions implemented using DNA can include more than a hundred species and can have complex functions.

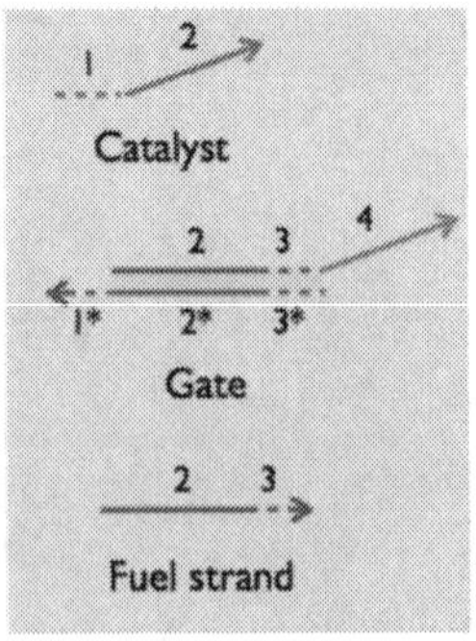

node motif

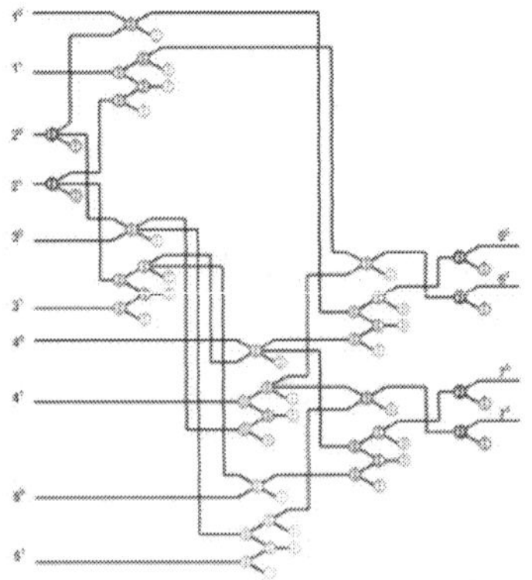

complex logic

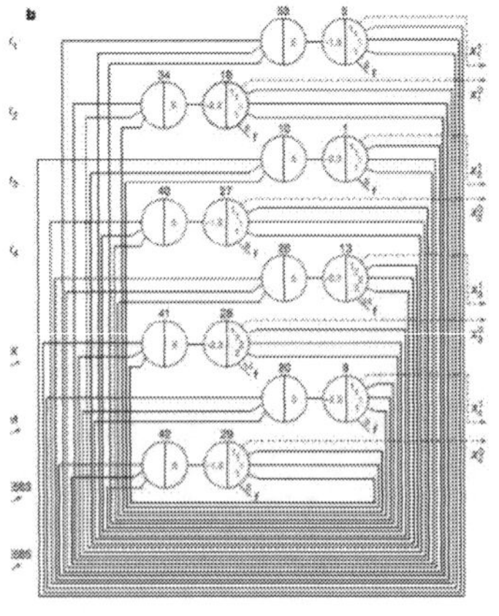

networks with feedback

Qian et al 2011, 2012
Seelig et al 2006

Most signals we would like to process come from dynamic processes with changing outputs.

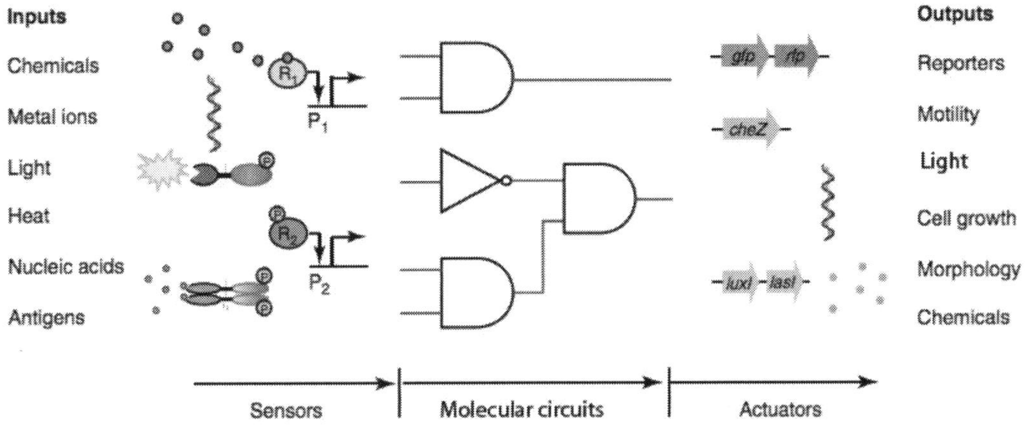

Some examples of control applications we might want in vitro circuits for:

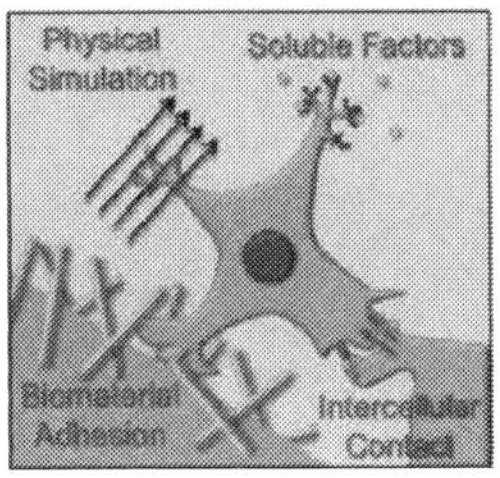

monitoring cell signals or other biomolecules

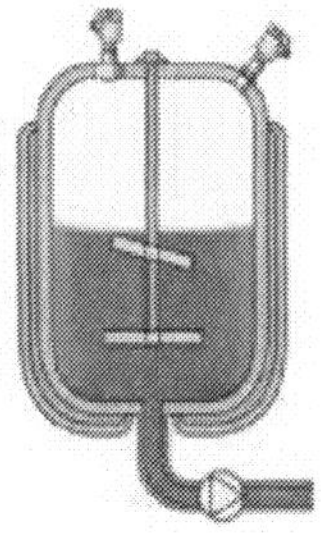

bioreactor control

Finio + Lipson, 2012

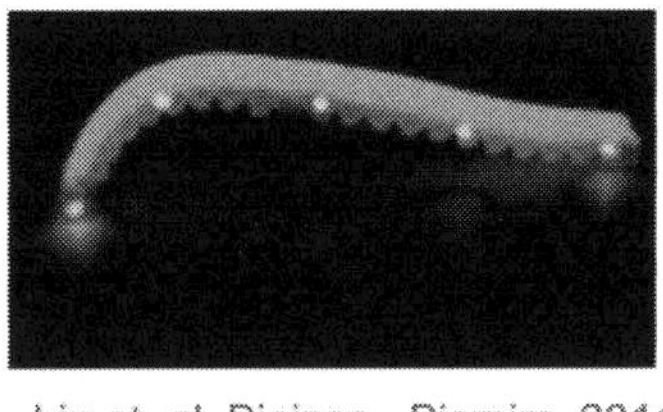

Lin et al, Bioinsp., Biomim. 2011

control of materials

All of these require ongoing, dynamic interaction.

Feedback control also requires dynamic response.

However, strand displacement circuits currently cannot respond to dynamic changes in output.

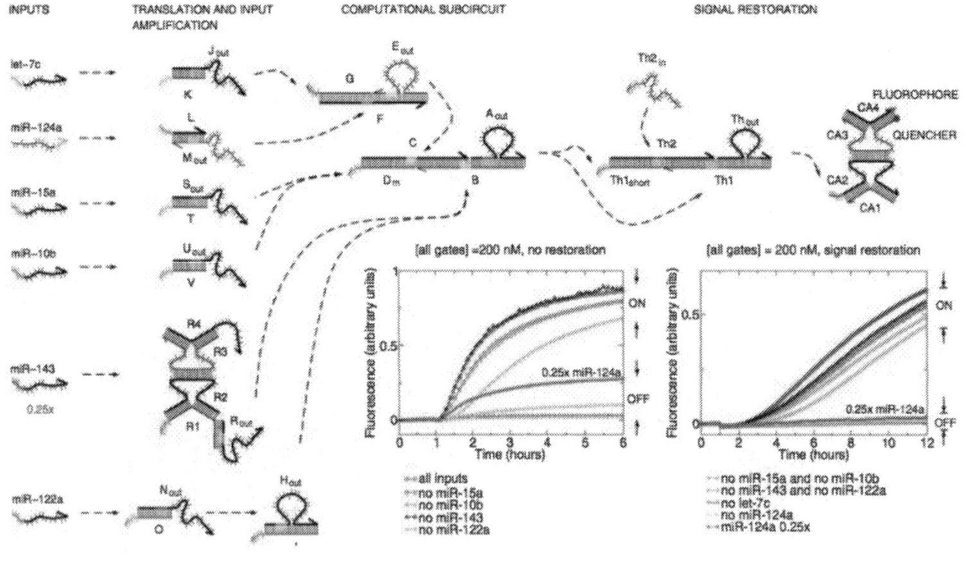

The logic gates are used up after a single computation!

Seelig et al 2006

"One time use" components fundamentally limit the application of current strand displacement systems.

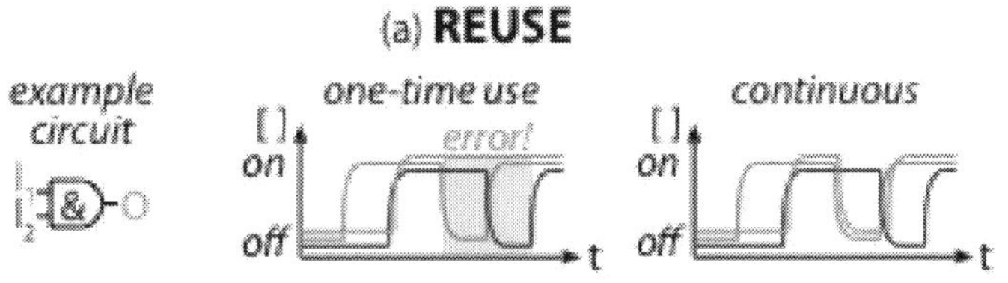

"Reversible" strand displacement circuits can respond dynamically but suffer from signal loss.

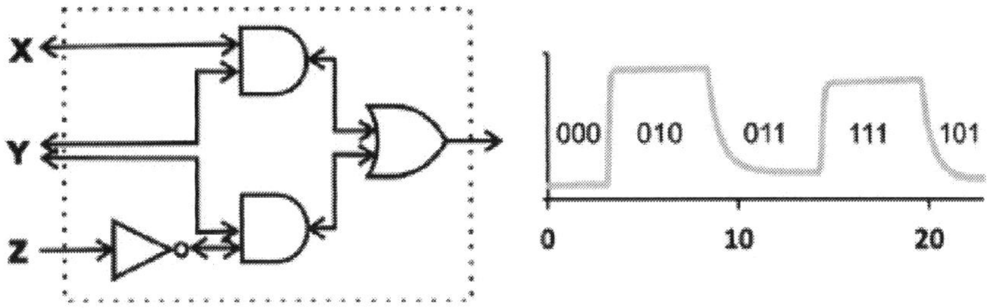

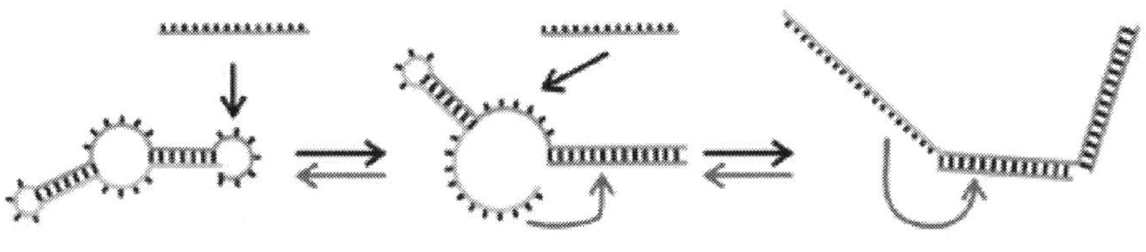

Genot et al 2011

"Reversible" strand displacement circuits can respond dynamically but suffer from signal loss.

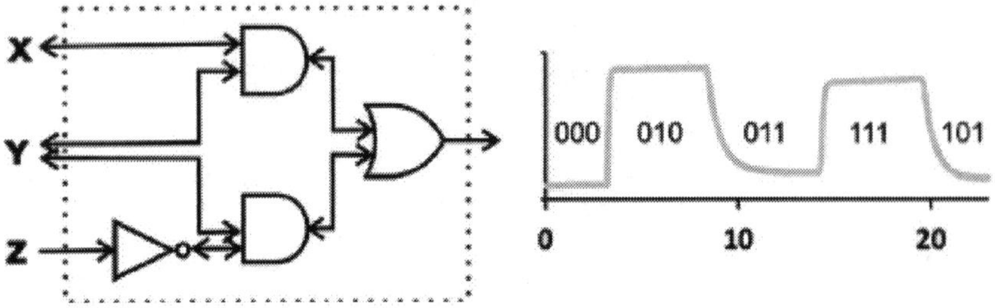

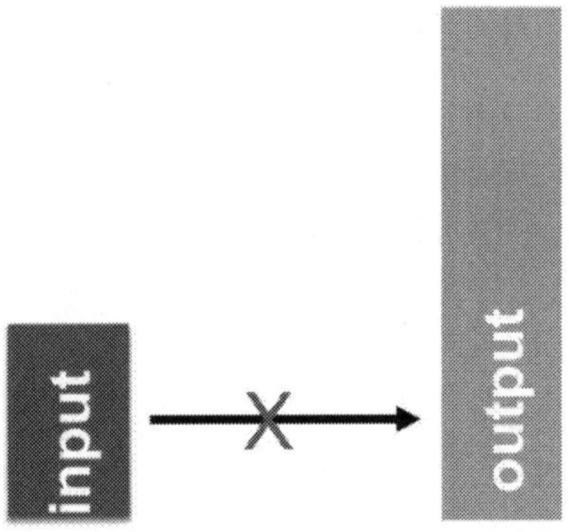

no energy input =
no signal restoration!

Genot et al 2011

The inability to respond dynamically makes monitoring or control impossible.

In vivo, biomolecular circuits are able to respond dynamically to changing inputs.

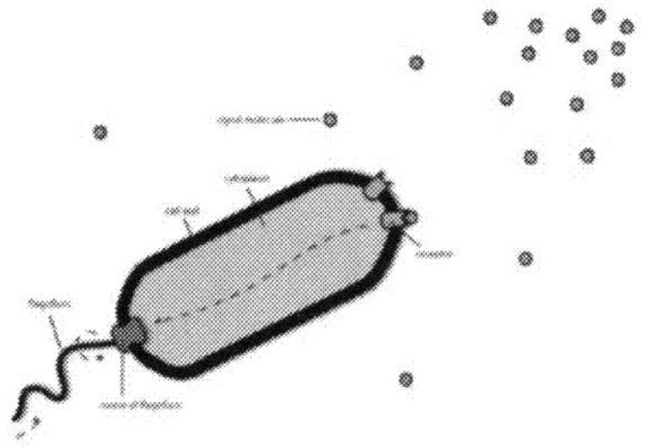

chemotaxis

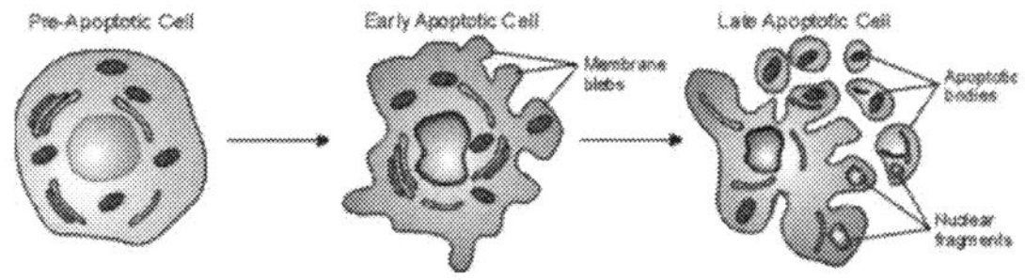

apoptosis

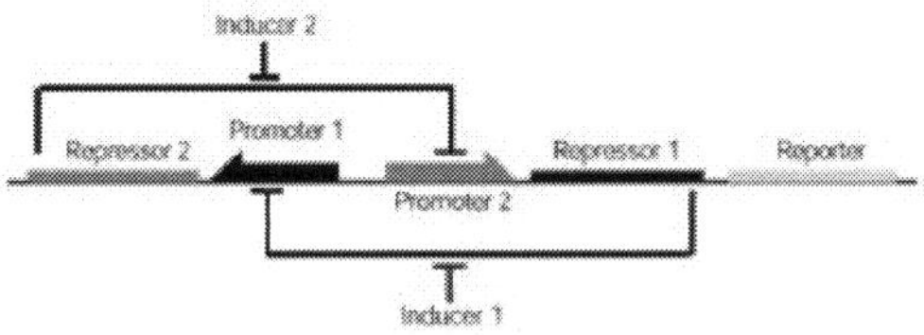

bistable switching

Likewise, electronic circuits respond to a series of clocked inputs.

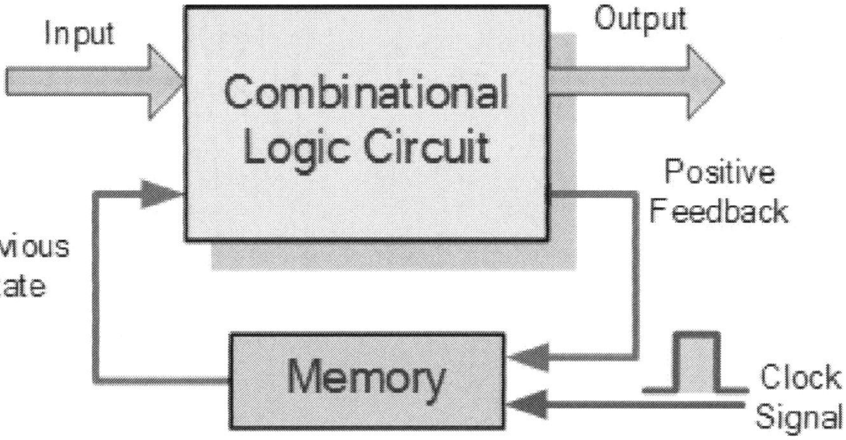

In both cases, a regulatory circuit operates on a driven, or powered input by controlling its outflow.

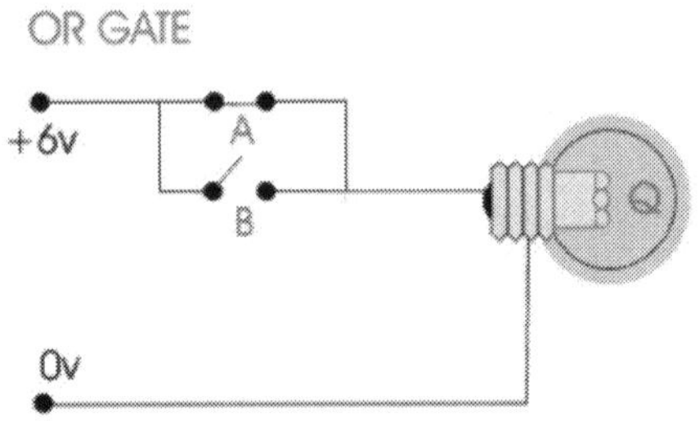

Electronic circuits and energy flux

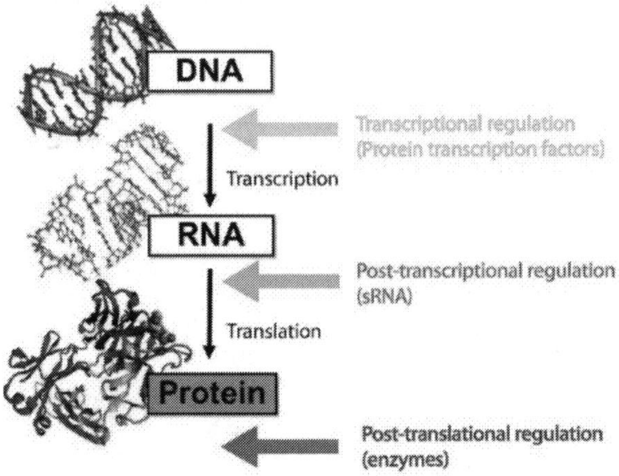

Biological circuits and energy flux

In biological systems, a slow "cycling" of component concentrations drives the system's dynamic response.

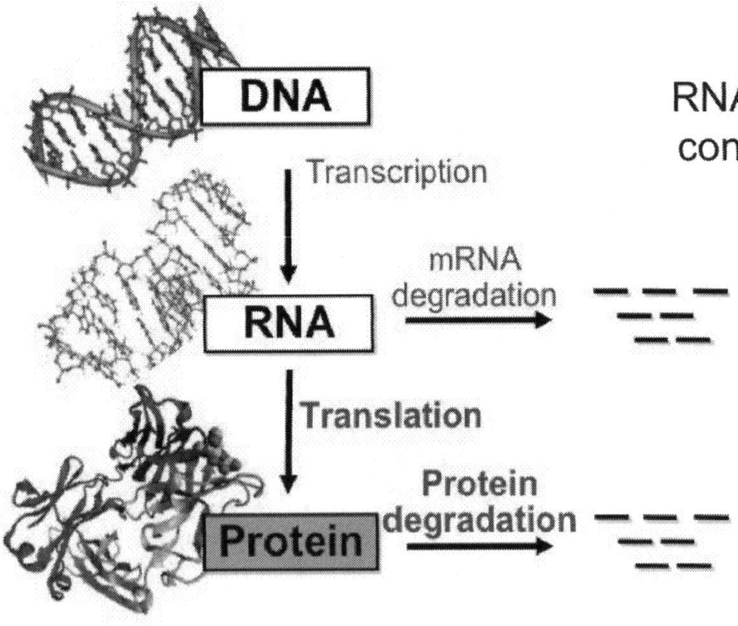

RNA and proteins are continually made and disassembled.

Biomolecular regulation (or logic) and production and degradation operate on different timescales.

Typical timescale for assembly

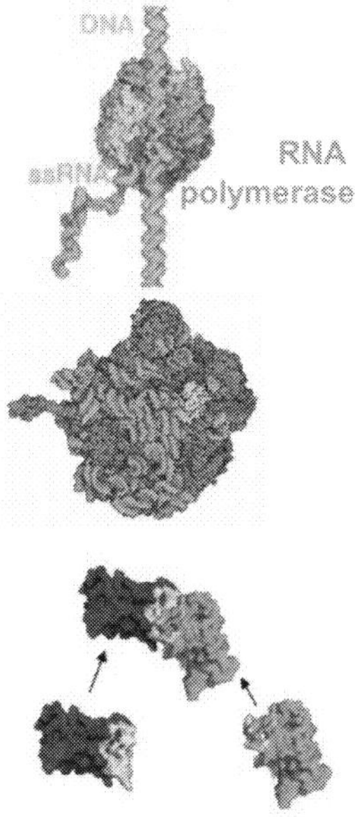

~ 1 to 30 minutes

~ 2 to 30 minutes

~ 1 microsecond - 1 second

Strand displacement reactions can also occur on very different, controlled time scales.

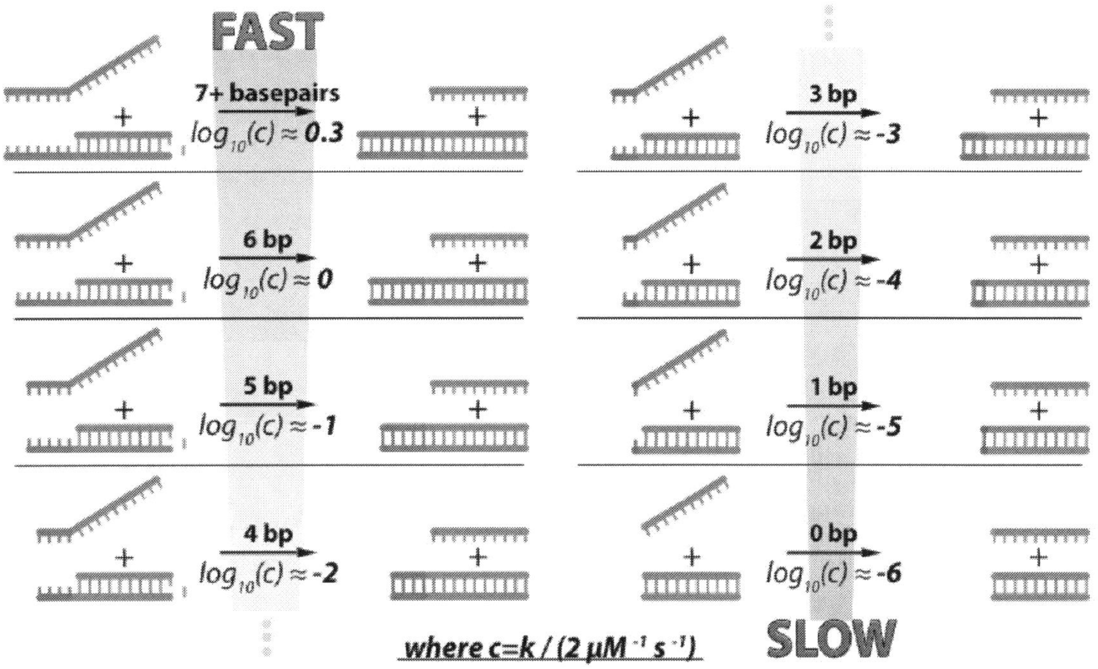

Computation, like biological signal processing, can occur on fast time scales.

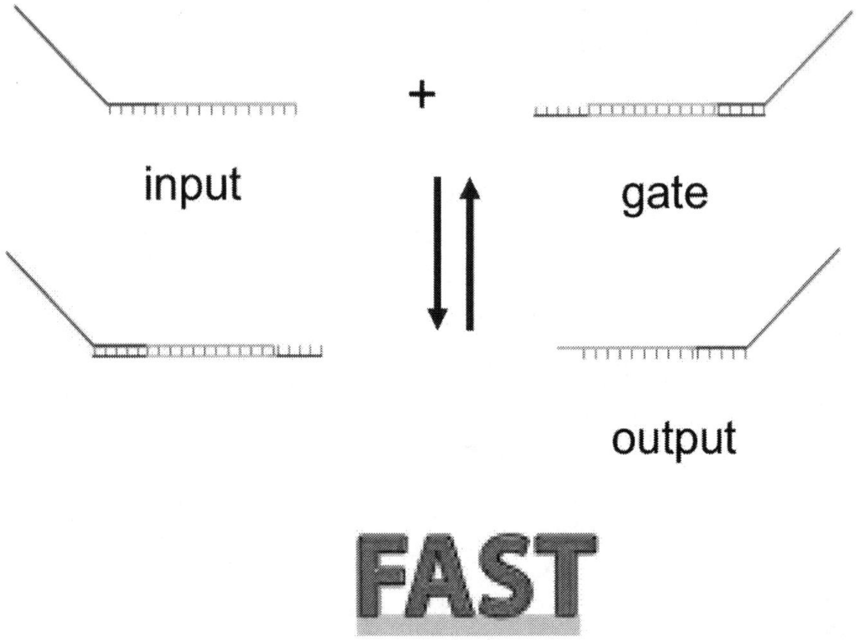

Qian et al 2011

A slow production system releases a signal output from a high energy precursor.

$$source \overset{slow}{\rightarrow} S$$

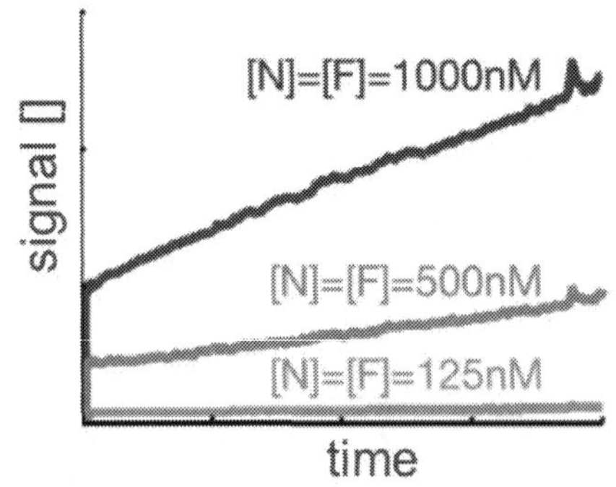

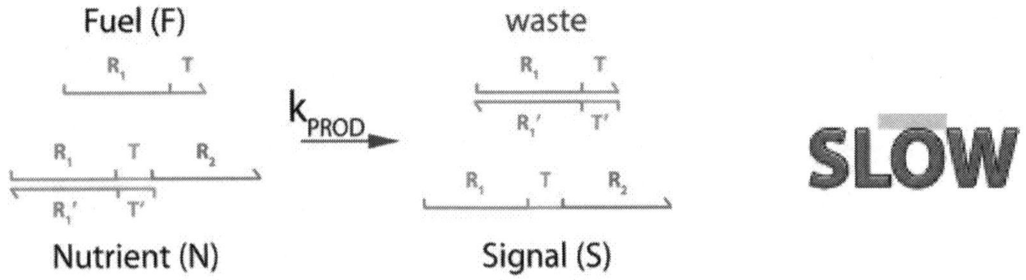

A slow degradation system *sequesters* a signal output, inactivating it as low-energy waste.

$$S \overset{slow}{\rightarrow} waste$$

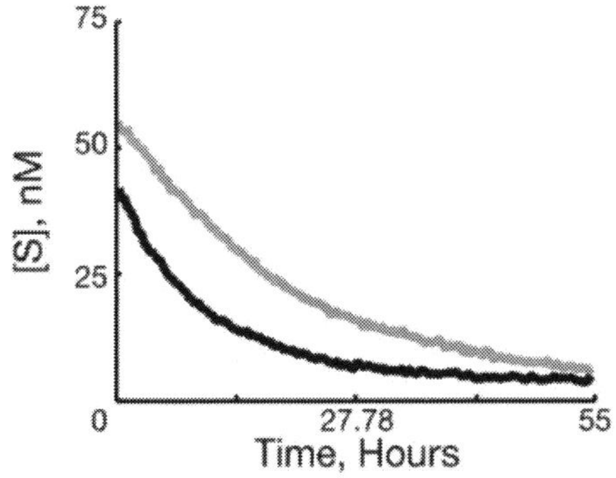

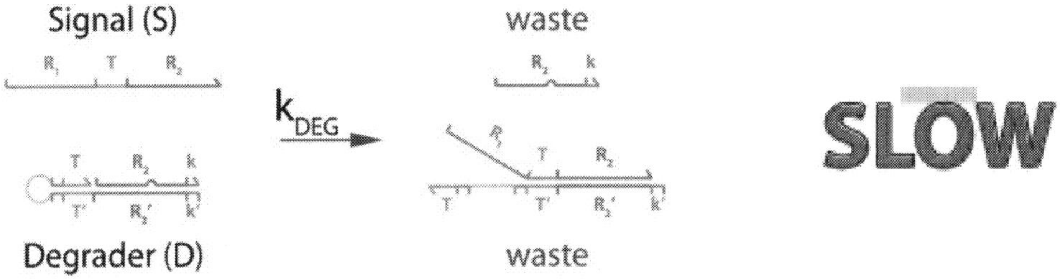

Together, a production and degradation system regulate the concentration of a given sequence.

$$source \overset{slow}{\rightarrow} S \overset{slow}{\rightarrow} waste$$

$$[S]_{ss} = \frac{k_f[N][F]}{k_d[D]}$$

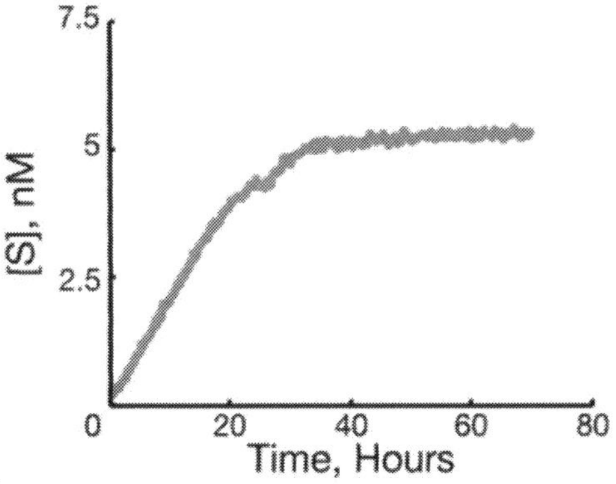

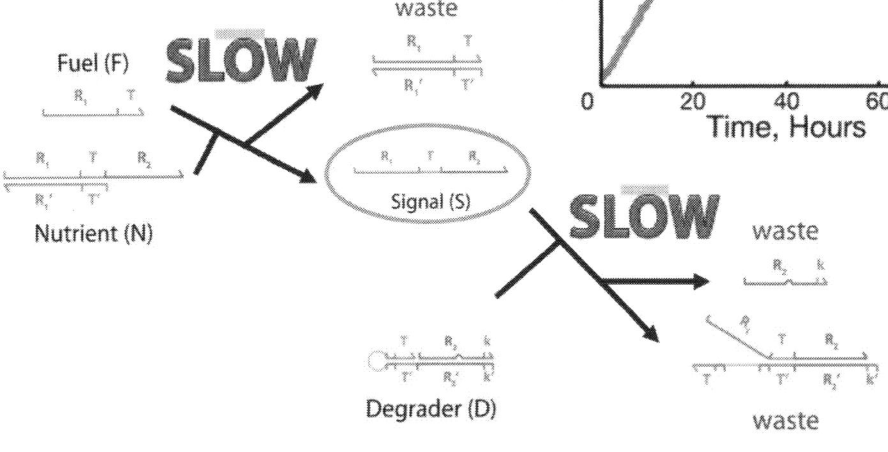

These systems together act to control energy flux through the system, enabling dynamic response.

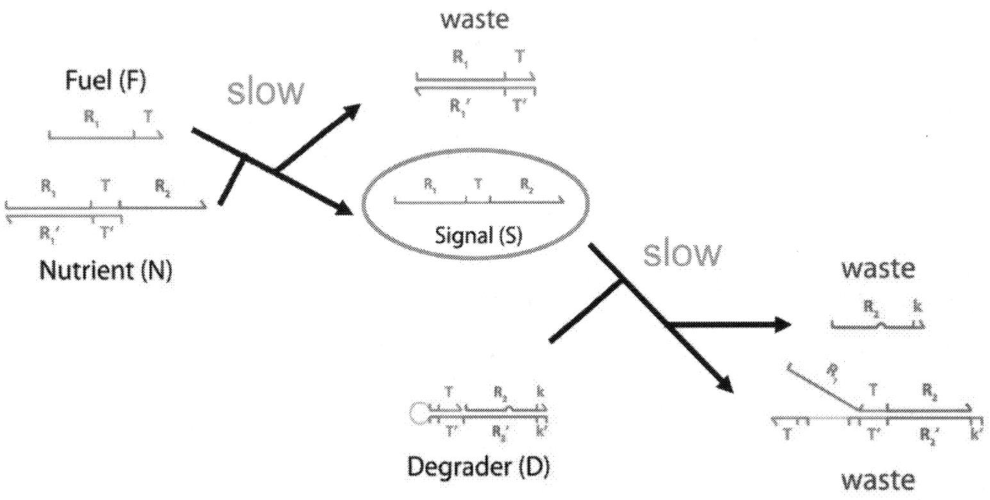

Example: An *in vitro* “sRNA” analog

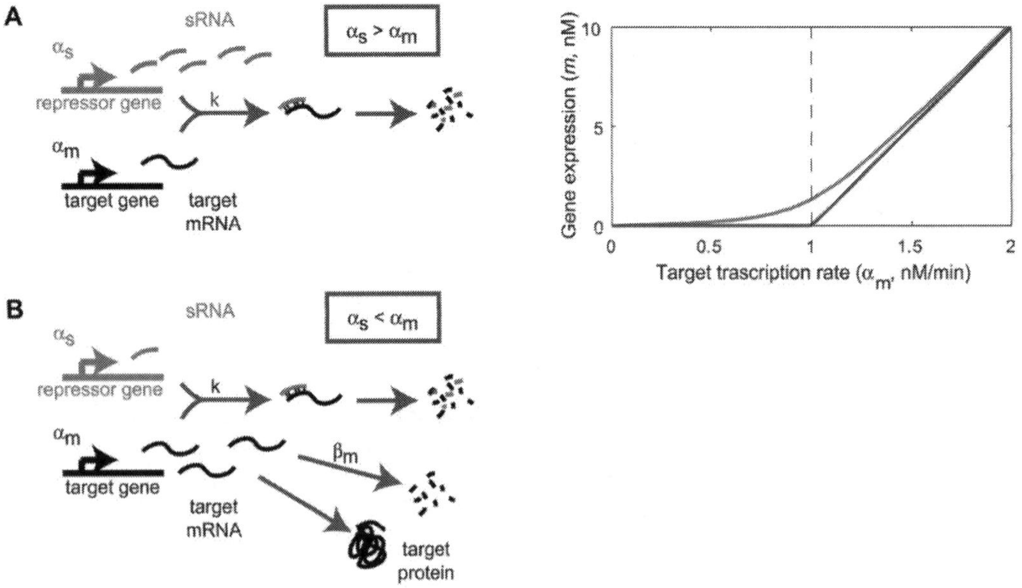

Levine et al, PLoS Biology, 2007

Example: An *in vitro* "sRNA" analog

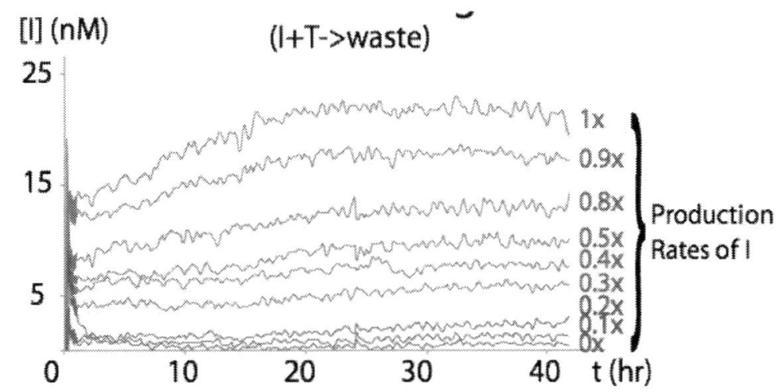

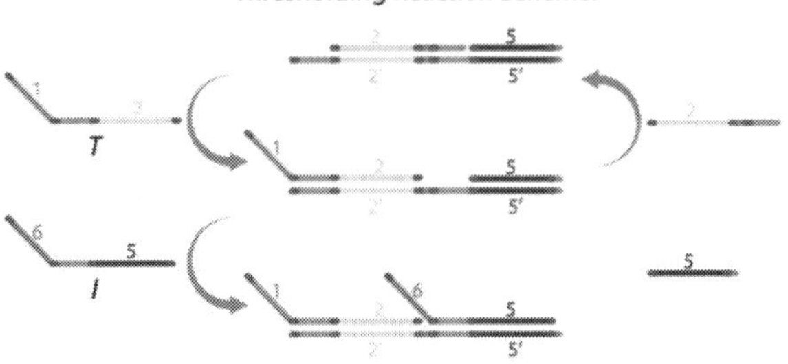

Example: An *in vitro* "sRNA" analog

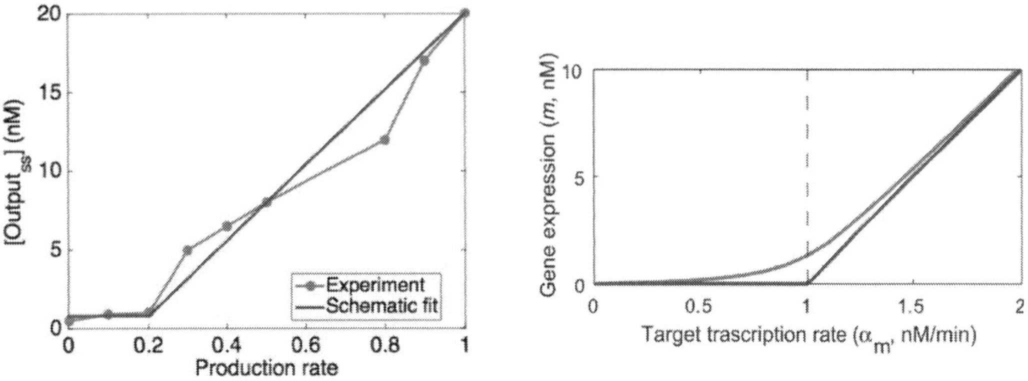

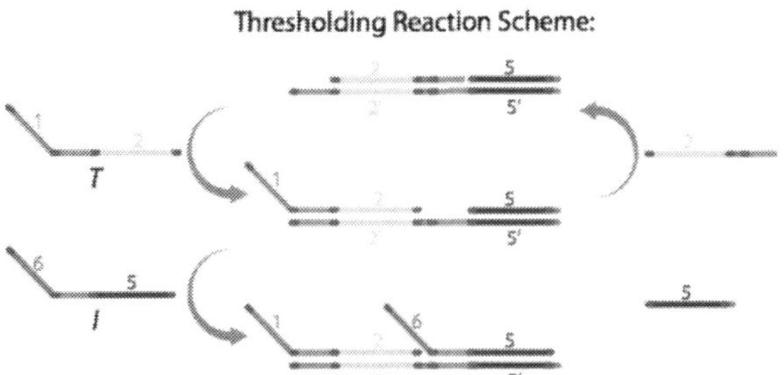

Fast reactions between "powered" species regulate their steady state values, performing computation.

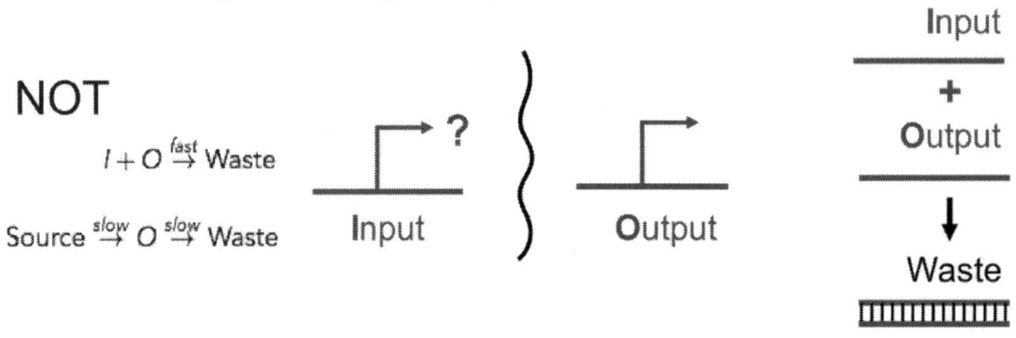

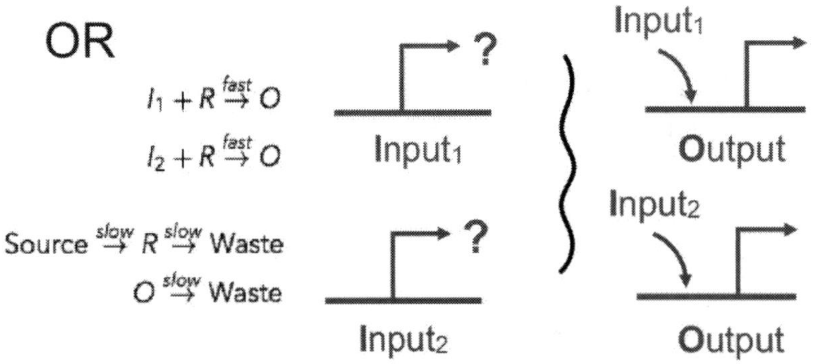

A powered strand displacement NOT gate can respond dynamically to a change in input values by changing its output value.

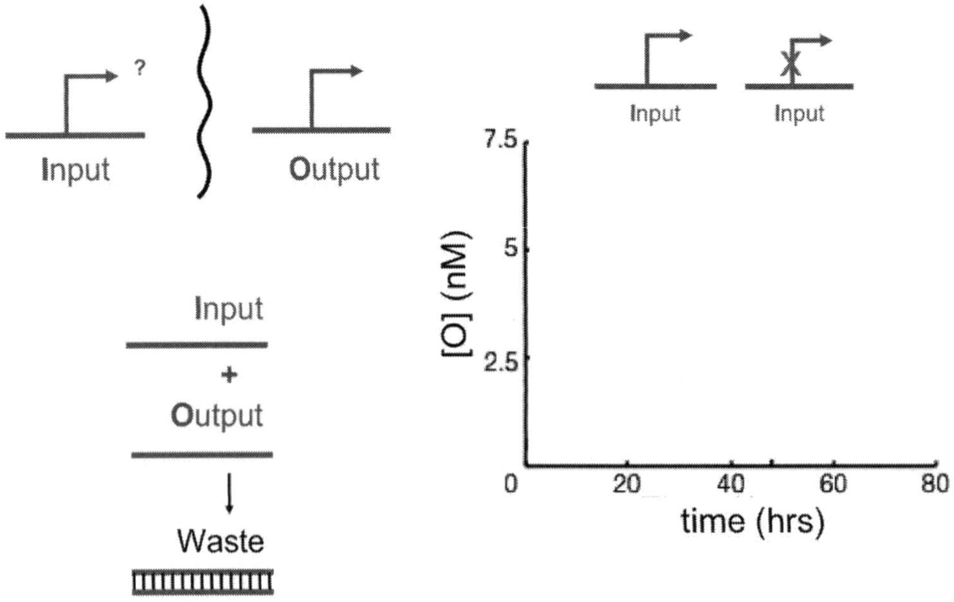

A powered logic circuit's response constantly approaches the value for the current input.

We have designed a full set of locircuits using this new powered approach.

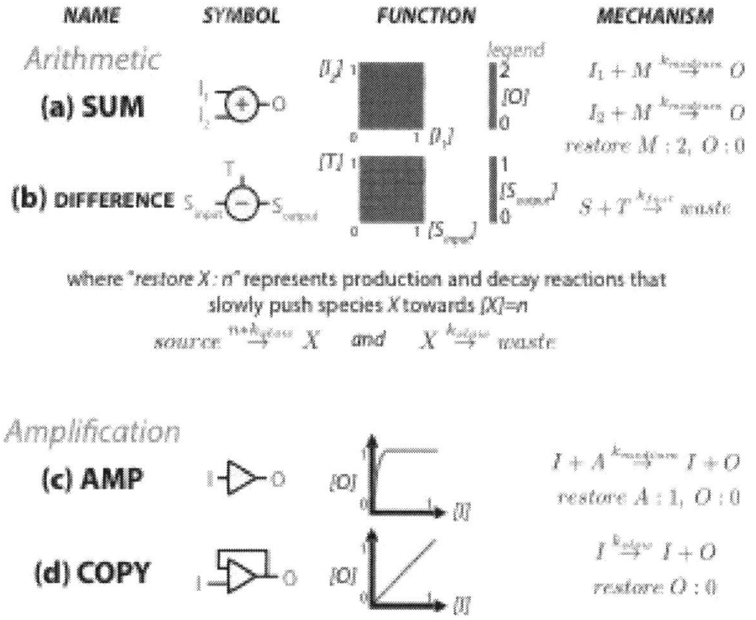

We have designed a full set of circuits using this new powered approach.

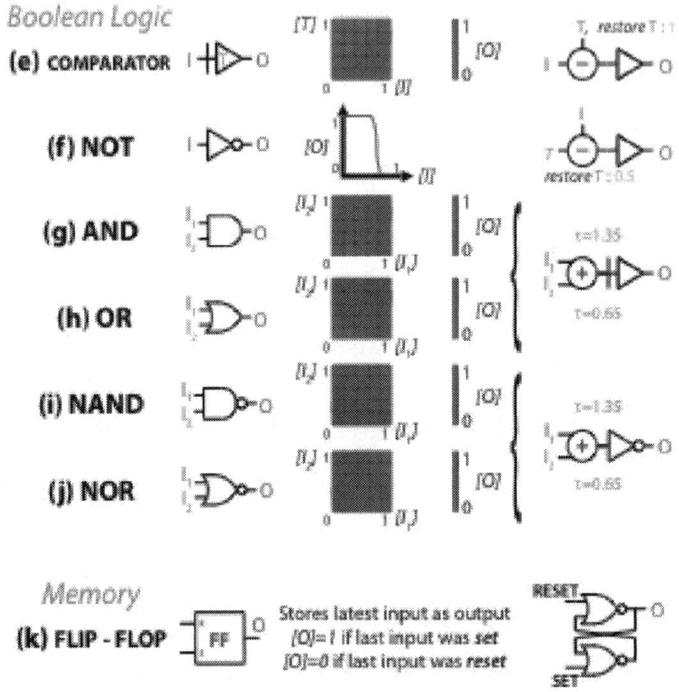

Larger circuits could be built using powered components to enable complex logic or memory.

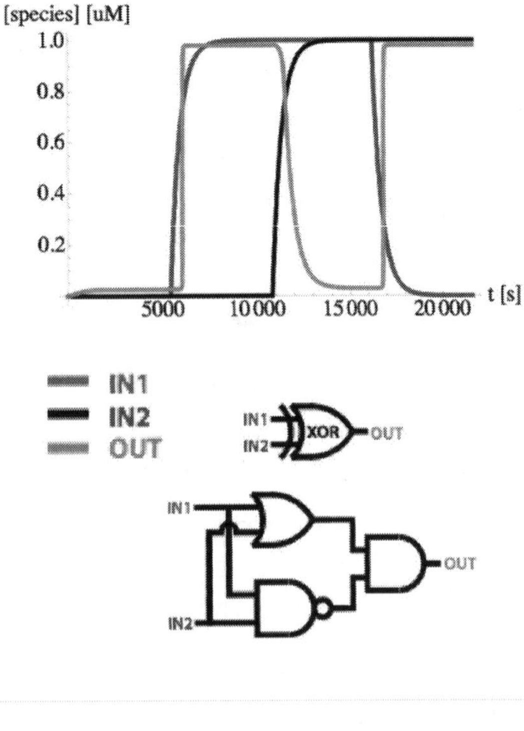

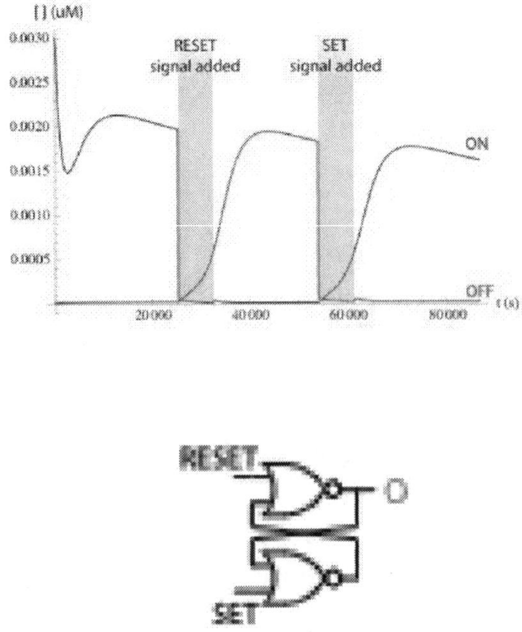

A signal restoration module makes it possible to build robust digital circuits.

$$\text{Source} \overset{slow}{\rightarrow} A \overset{slow}{\rightarrow} \text{Waste}$$

$$I + A \overset{medium}{\rightarrow} I + O$$

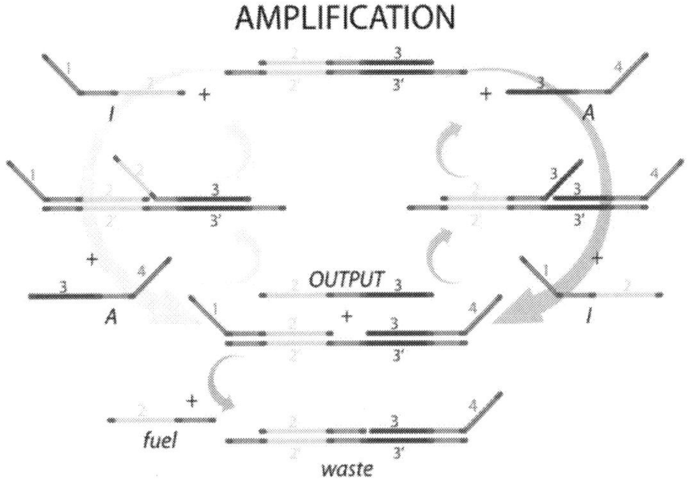

Powered strand-displacement systems are for more than logic: temporal control.

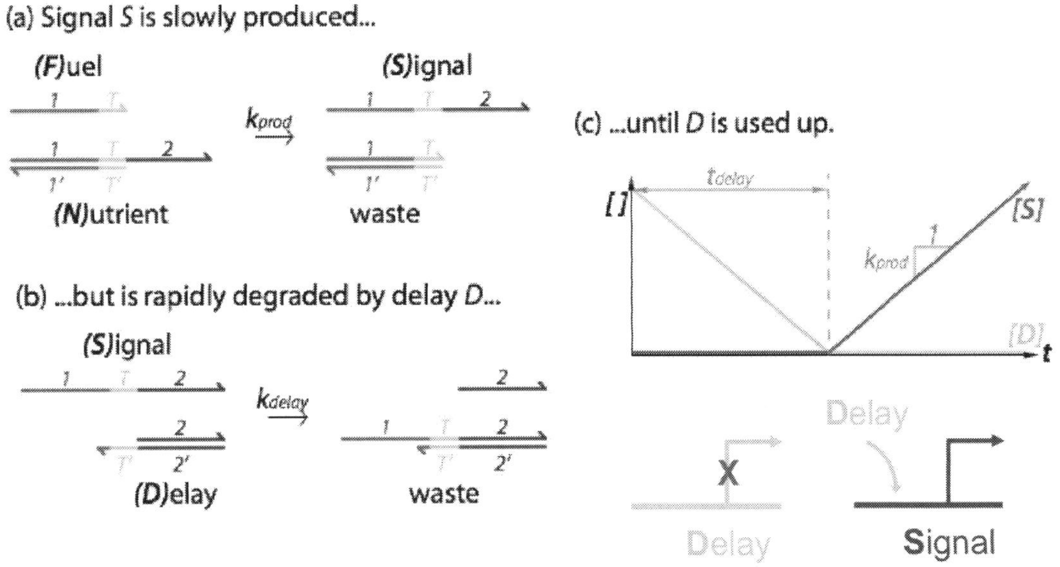

Scalise, Cangialosi et al, in prep.

Powered strand-displacement systems are for more than logic: temporal control.

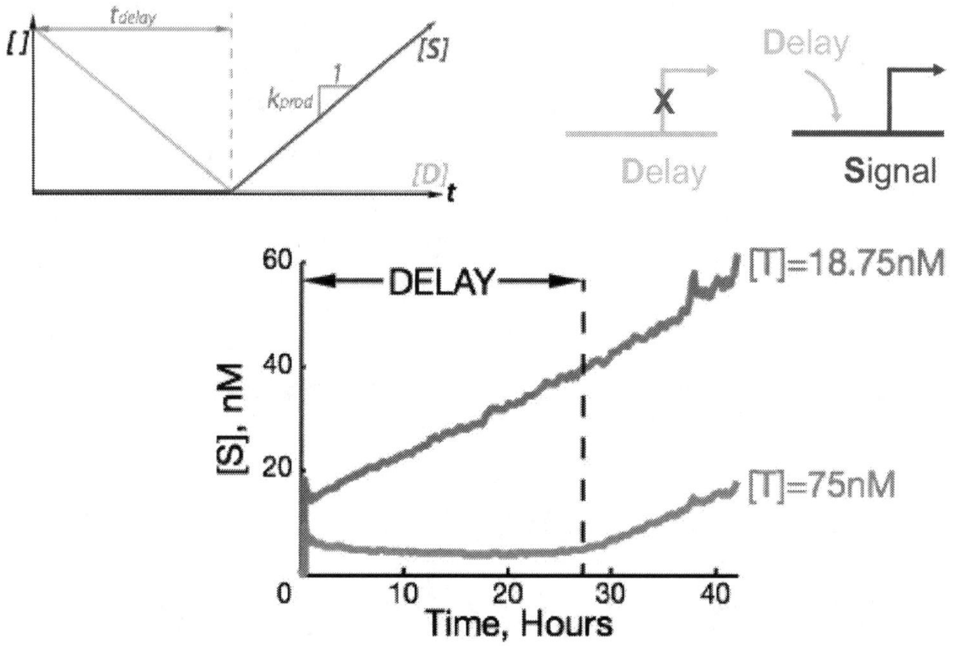

Scalise, Cangialosi et al, in prep.

Simulations of logic circuits show dynamic response to dynamic inputs.

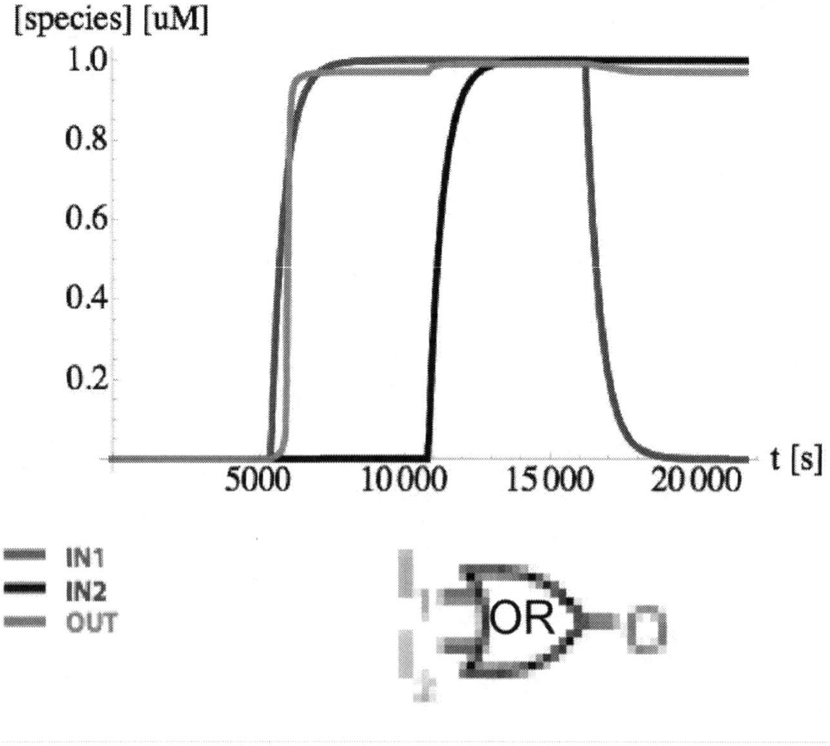

Powered circuits could also be useful for staged, stable pattern formation.

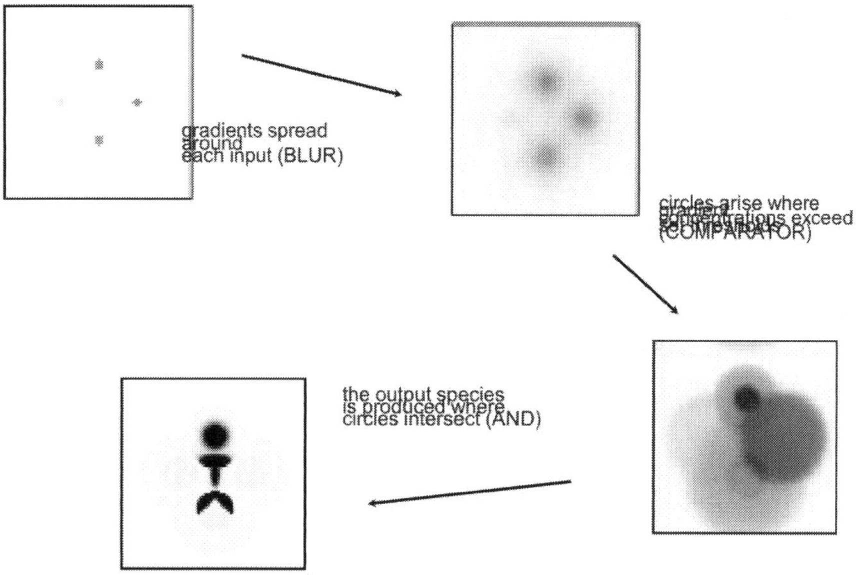

D. Scalise and R. Schulman
TECHNOLOGY (2014)

Powered circuits could also be useful for staged, stable pattern formation.

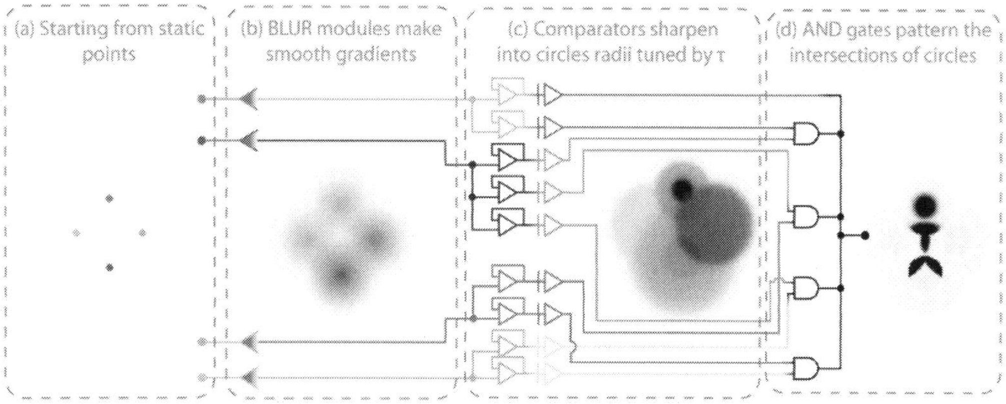

D. Scalise and R. Schulman
TECHNOLOGY (2014)

Powered circuits could also be useful for staged, stable pattern formation.

D. Scalise and R. Schulman
TECHNOLOGY (2014)

Aptamer exchange: Modular, reversible control of proteins and small molecules by powered DNA circuits

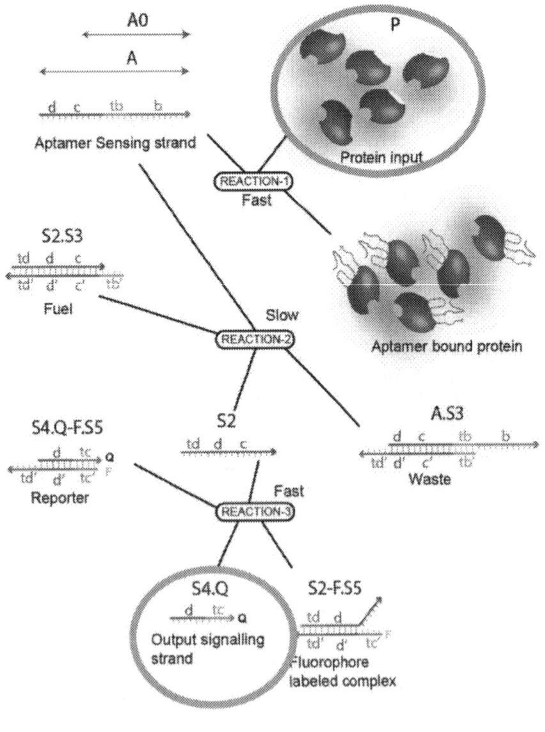

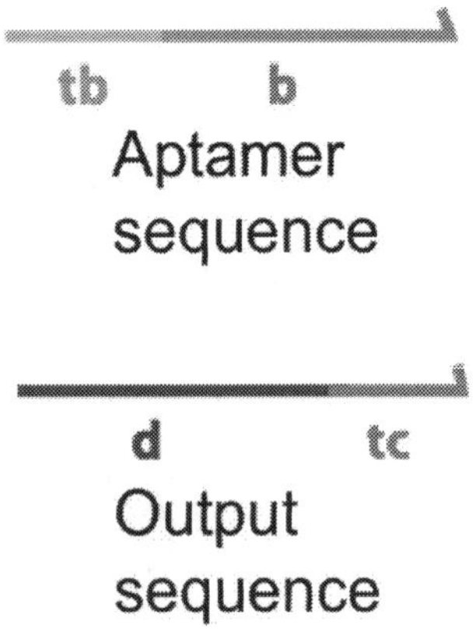

No aptamer <-> output sequence overlap.

Agrawal and Schulman, in prep.

Aptamer exchange: “Translators” have reliable, quantitatively controlled input and output values.

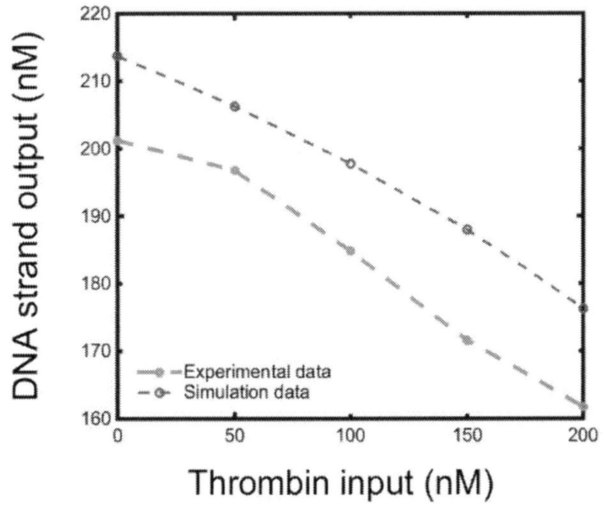

Agrawal and Schulman, in prep.

Traditional biosensors can either amplify a snapshot of a situation or monitor without amplification.

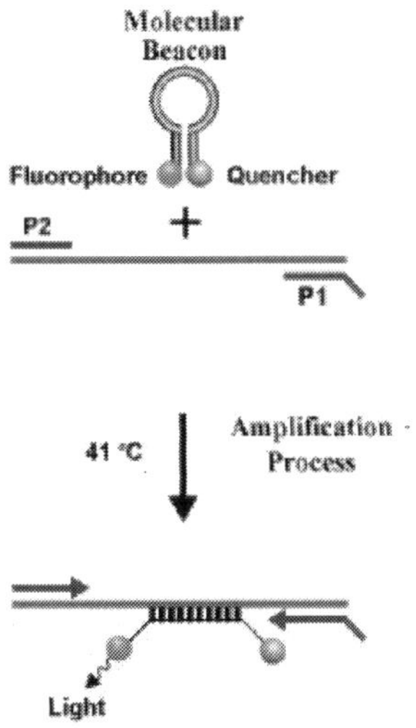

molecular beacon

Powered strand displacement circuits could make it possible to dynamically modify and amplify biomolecular signals.

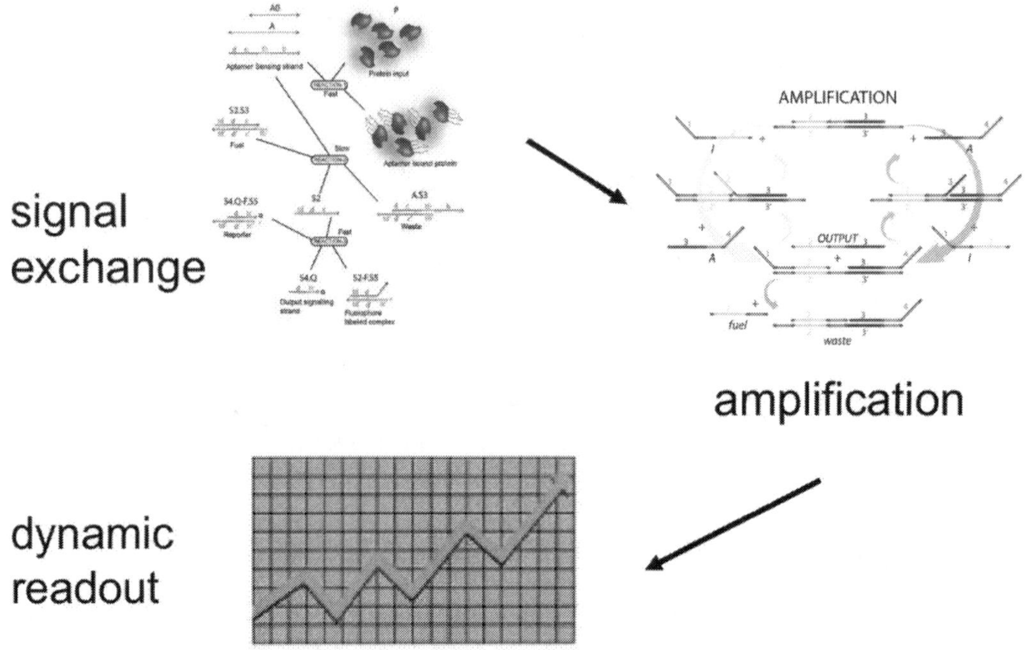

Powered strand displacement could enable feedback-controlled release for new control over therapeutics.

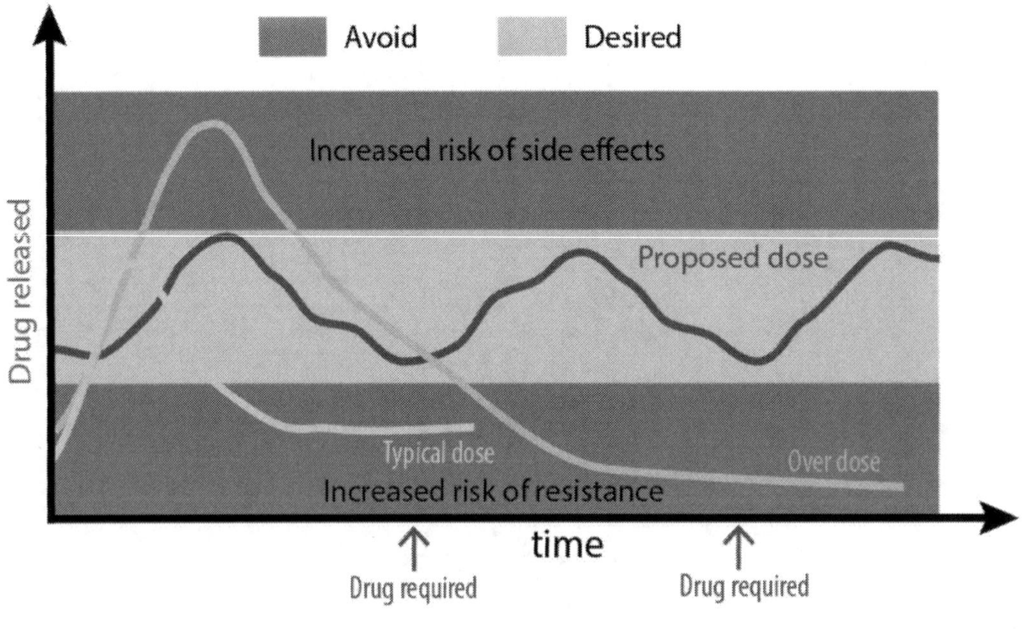

Summary

- Many potential applications of synthetic biology techniques could be accomplishing using in vitro molecular circuit for logic and control.
- DNA strand displacement can produce reliable scalable molecular circuits.
- The dynamic, robust response of a (molecular) circuit requires sustained energy flow through the system.
- Powered strand displacement circuits can enable dynamic molecular logic without enzymes.
- Reversible interaction with aptamers could enable these circuits to control a variety of molecular processes.

First, a word about the problem we will address.

We'll consider pattern transformation.

We can perform these logic operations in space, producing different outputs in different locations.

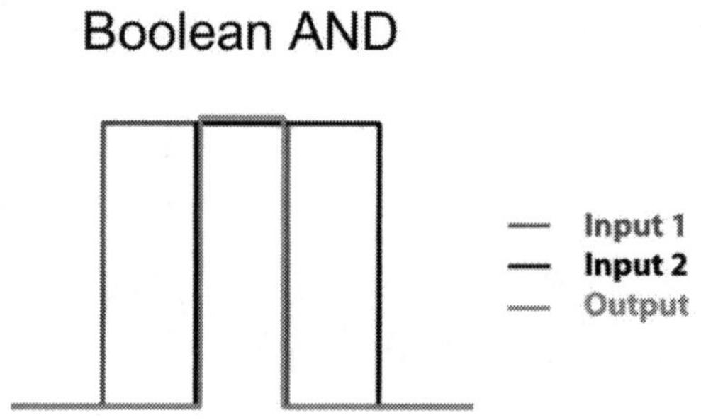

But diffusion processes "blur" the results of spatial transformations - the pattern disappears!

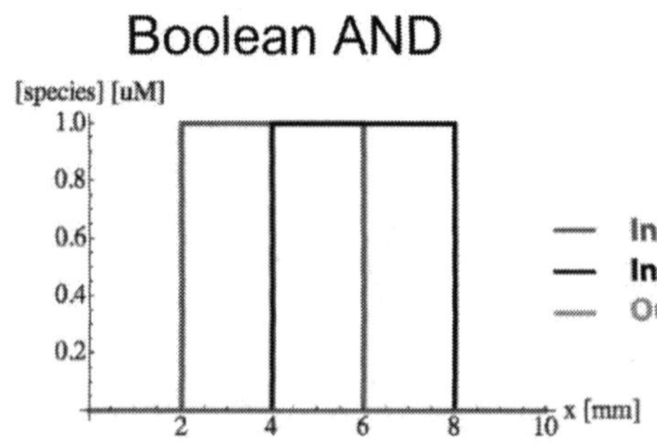

Biological networks are complex!

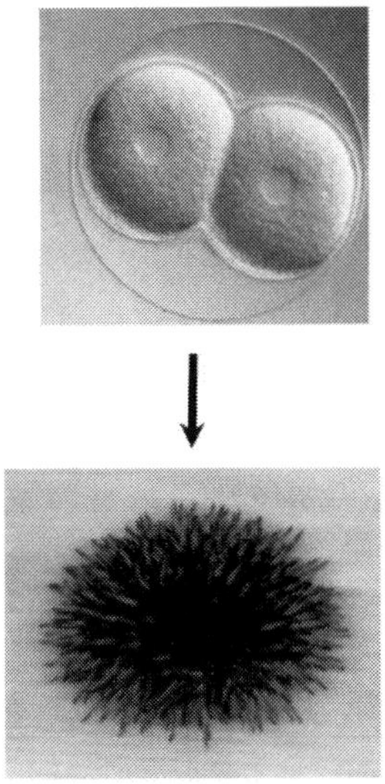

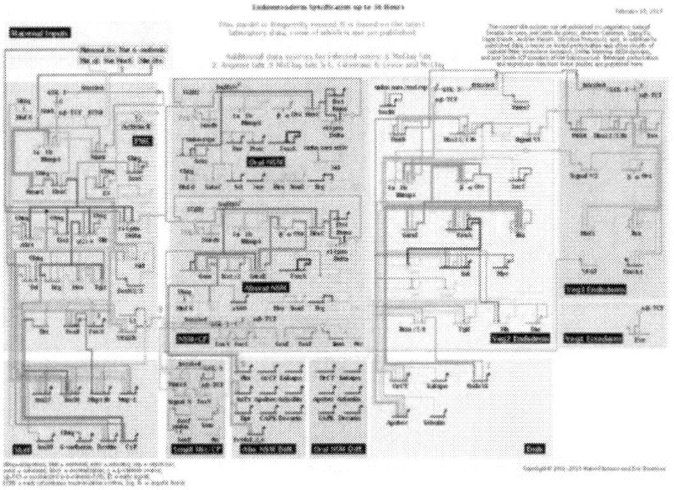

E. H. Davidson

Could more complex patterns arise from complex reaction-diffusion systems?

Patterned materials can produce arrangements of signalling molecules

- Limitations:
- Patterns require flows in the system
- 3-dimensional patterns are difficult to achieve

Mechanisms such as microfluidics are used to program the creation of patterns of growth factors and other molecules.

- Limitations:
- Patterns require flows in the system
- 3-dimensional patterns are difficult to achieve

Cells create patterns through combinations of reactions and diffusion processes.

Could we develop a biomimetic approach for controlled release based on synthetic reaction-diffusion processes?

Could we hope to build sufficiently complex synthetic networks synthetically?

What can we do with molecules?

- Any set of reactions with 0, 1 or 2 reactants are allowed, including those that do not obey "mass balance."

A + B-> C -> 2A

- We can set the diffusion rates of each molecule to any value.

A -> B B ->

Soloveichik et al, 2011, PNAS

- We can set the diffusion rates of each molecule to any value.

- We can use as many species as we'd like.

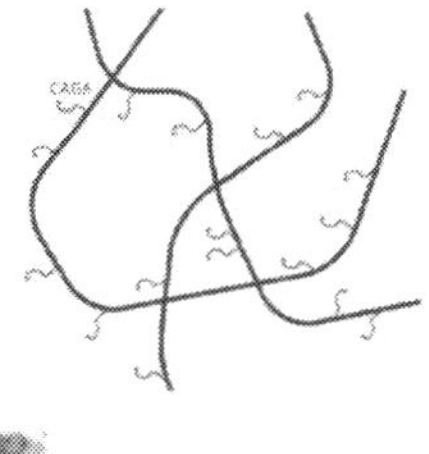

- We can use as many species as we'd like.

One important comment about these rules -- DNA, a real molecule, satisfies them.

Reactions and diffusion together produce dynamics described by reaction-diffusion equations, a coupled set of parabolic PDEs.

$$A + B \xrightarrow{k_1} C \qquad A + B \xleftarrow{k_2} C$$

$$\frac{\partial [A]}{\partial t} = D_A \nabla^2 [A] - k_1 [A][B] + k_2 [C]$$

$$\frac{\partial [B]}{\partial t} = D_B \nabla^2 [B] - k_1 [A][B] + k_2 [C]$$

$$\frac{\partial [C]}{\partial t} = D_C \nabla^2 [C] + k_1 [A][B] - k_2 [C]$$

Additional reactions introduce additive reaction rate terms.
Additional species introduce new equations.

Our vision: a modular series of transformations that can be combined to produce different final patterns.

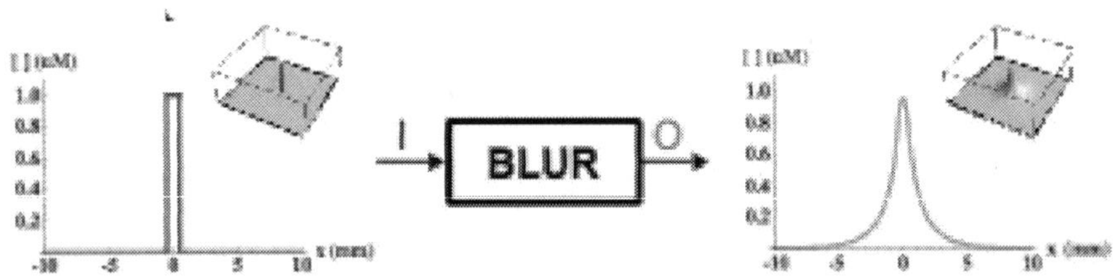

Each module takes a simple input pattern and creates a (potentially more complex) output pattern.

We could imagine pattern transformations that perform logic.

Input

$[I_1]_0 = \{0x, 1x\}$

$[I_2]_0 = \{0x, 1x\}$

Summing

$[M]_0 = 2x$

$I_1 + M \xrightarrow{k_{logic}} N$

$I_2 + M \xrightarrow{k_{logic}} N$

Thresholding

$[T]_0 = 1.3x$

$T + N \xrightarrow{k_{thresh}} waste$

$k_{thresh} >> k_{logic}$

Amplification

$A + N \xrightarrow{k_{logic}} N + O$

L. Qian and E. Winfree
Science (2011)

We could imagine pattern transformations that perform logic.

Input

$[I_1]_0 = \{0x, 1x\}$

$[I_2]_0 = \{0x, 1x\}$

Summing

$[M]_0 = 2x$

$I_1 + M \xrightarrow{k_{logic}} N$

$I_2 + M \xrightarrow{k_{logic}} N$

Thresholding

$[T]_0 = 1.3x$

$T + N \xrightarrow{k_{thresh}} waste$

$k_{thresh} >> k_{logic}$

Amplification

$A + N \xrightarrow{k_{logic}} N + O$

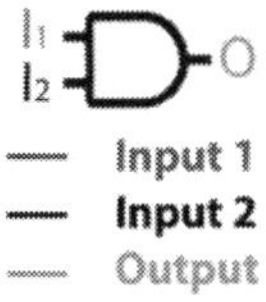

L. Qian and E. Winfree
Science (2011)

We can perform these logic operations in space, producing different outputs in different locations.

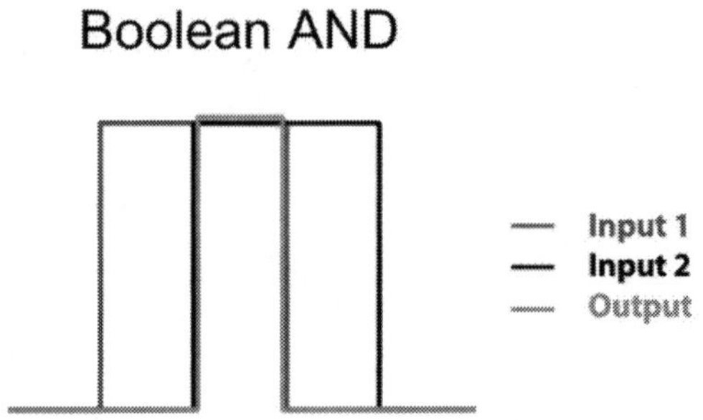

But diffusion processes "blur" the results of spatial transformations - the pattern disappears!

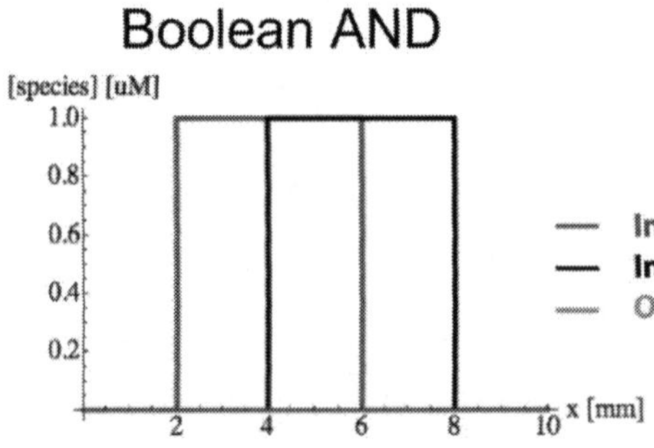

What is a “spatial pattern” of molecules?

Anv non-well-mixed solution is inherently out of equilibrium.

Video: Youtube

$$C(r,t) = \frac{M}{\sqrt{4\pi Dt}} \exp\left(-\frac{r^2}{4Dt}\right)$$

Without external energy, the species of a chemical "pattern" mix, dissolving the initial structure.

Energy-driven pattern formation processes are often inherently unstable or chaotic.

Our goal was a set of reactions that could maintain a pattern at steady state.

A constant, well-mixed inflow of high-energy, metastable "precursor" species can serve as a driving force.

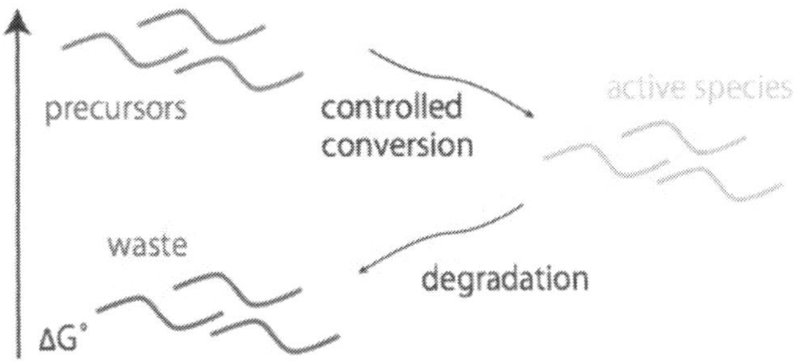

Steady state is achieved when production and decay are balanced.

Powered logic circuits enable spatial computation where each pattern is stable.

Cycling Reactions:

$$source \overset{k_{cycling}}{\rightarrow} Species \overset{k_{cycling}}{\rightarrow} waste$$

$$k_{cycling} << k_{logic}$$

Boolean AND

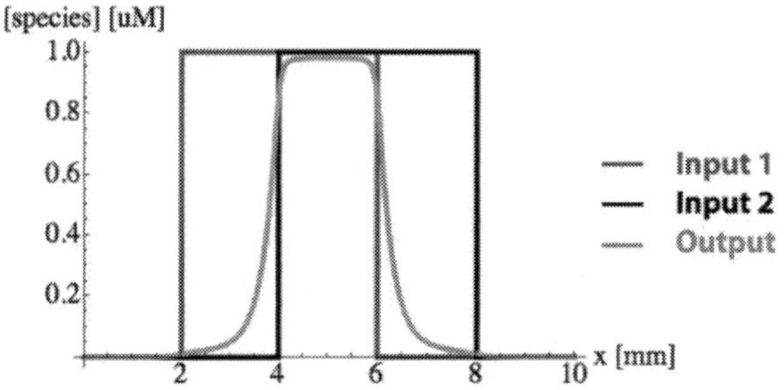

Powered logic gates can be used to assemble a full set of logic devices.

Comparator | Boolean NOT | Boolean OR

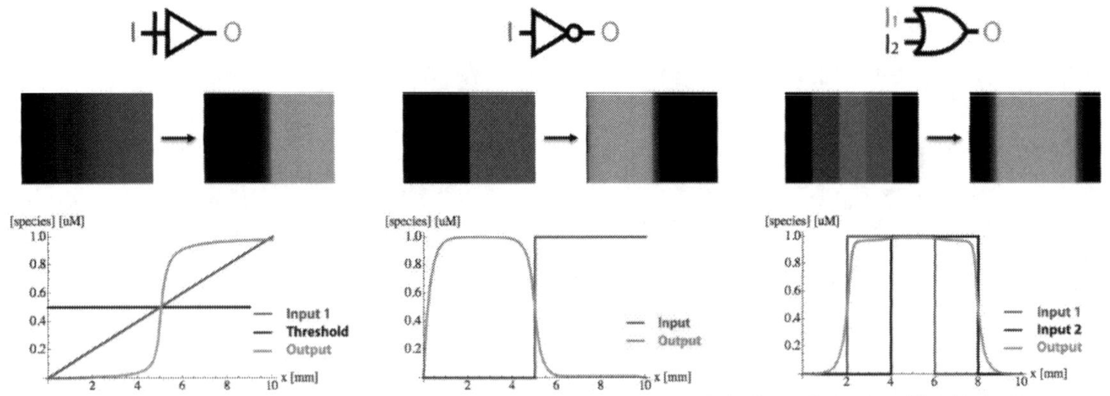

D. Scalise and R. Schulman *TECHNOLOGY* (2014)

We assume a single diffusion rate for all species.
Other rates would also work.

If a species is depleted by a module's reactions, it cannot serve as a reactant in multiple processes.

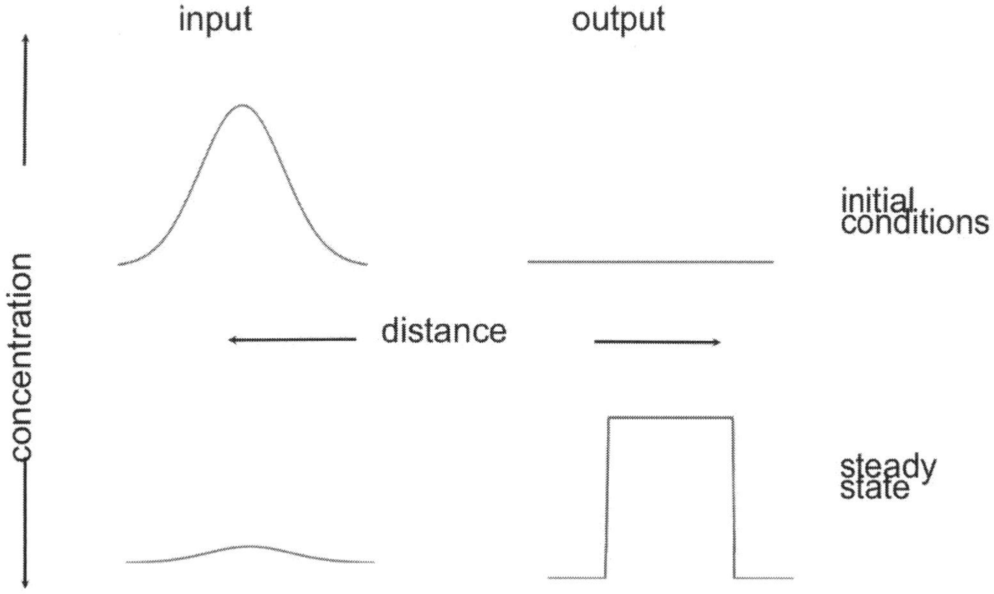

By "copying" a species in a catalytic process, we can ensure that reaction modules do not have side effects.

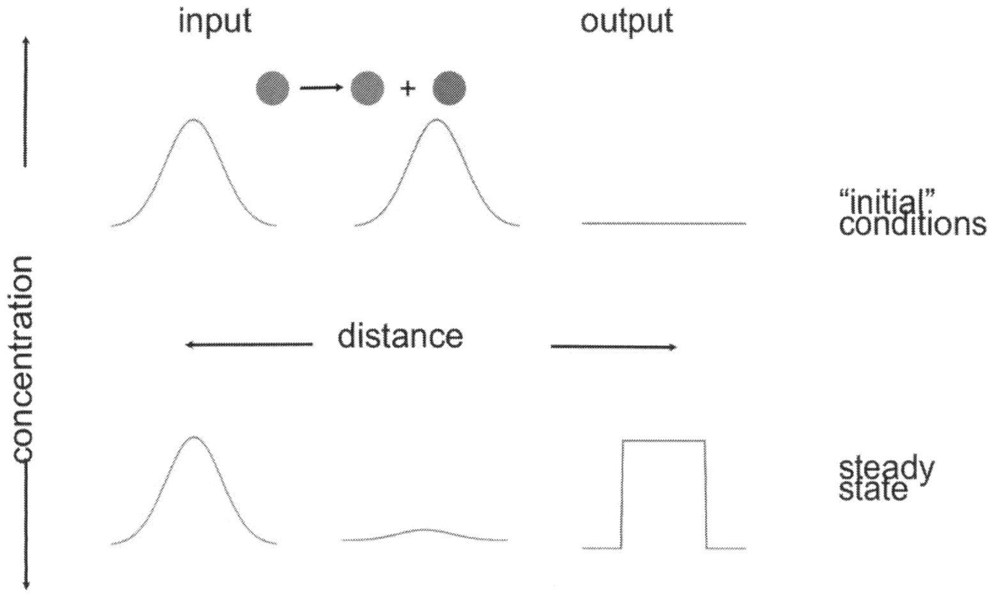

By "copying" a species in a catalytic process, we can ensure that reaction modules do not have side effects.

Copies input (I) onto output (O).

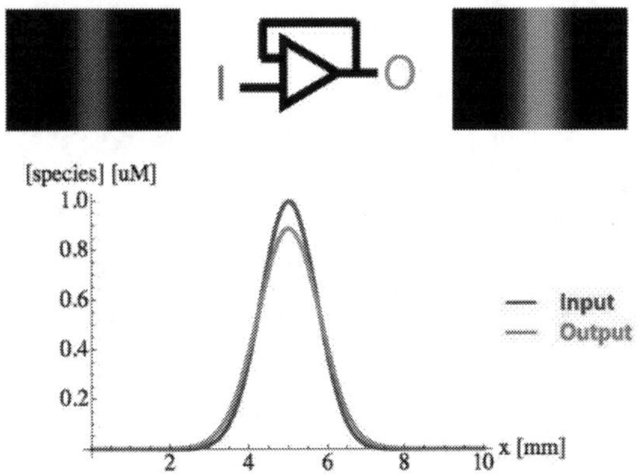

Reactions:

$$I \overset{k_{copy}}{\rightarrow} I + O$$

$$O \overset{k_{copy}}{\rightarrow} waste$$

PDEs:

$$\frac{\partial[I]}{\partial t} = 0$$

$$\frac{\partial[O]}{\partial t} = D_O \nabla^2[O] + k_{copy}[I] - k_{copy}[O]$$

D. Scalise and R. Schulman *TECHNOLOGY* (2014)

Module: BROADCAST

Transmits a blurry copy in all directions.

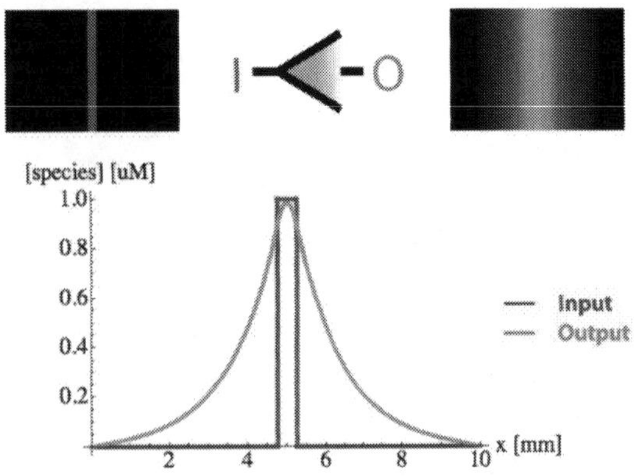

Reactions:

$$I \overset{k_{fast}}{\rightarrow} I + O$$

$$O \overset{k_{slow}}{\rightarrow} waste$$

PDEs:

$$\frac{\partial[I]}{\partial t} = 0$$

$$\frac{\partial[O]}{\partial t} = D_O \nabla^2[O] + k_{fast}[I] - k_{slow}[O]$$

D. Scalise and R. Schulman *TECHNOLOGY* (2014)

By composing these “modules”, we can create complex pattern transformations.

D. Scalise and R. Schulman *TECHNOLOGY* (2014)

By composing these “modules”, we can create complex pattern transformations.

D. Scalise and R. Schulman *TECHNOLOGY* (2014)

By composing these "modules", we can create complex pattern transformations.

D. Scalise and R. Schulman *TECHNOLOGY* (2014)

By composing these "modules", we can create complex pattern transformations.

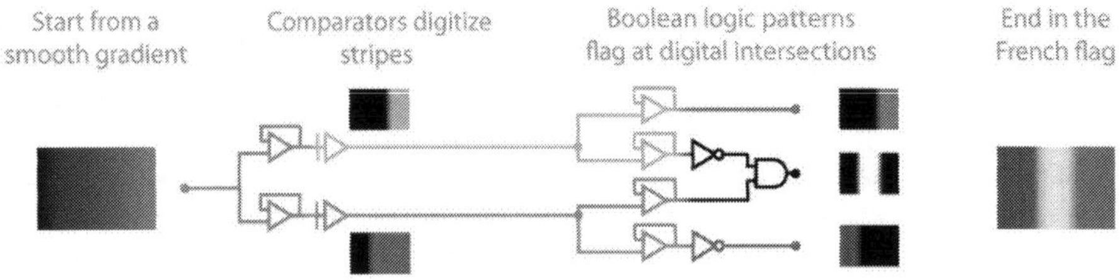

D. Scalise and R. Schulman *TECHNOLOGY* (2014)

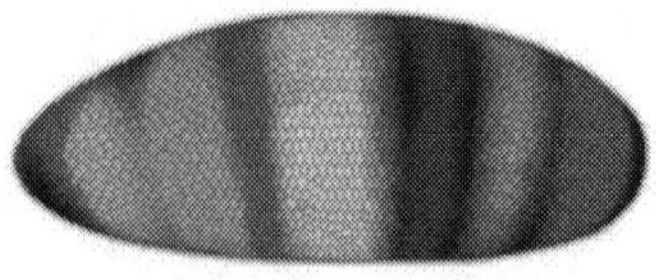

COPY modules within the circuit maintain modularity.

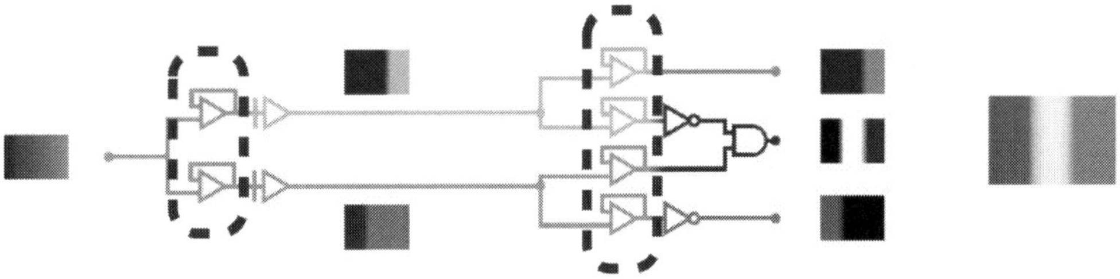

We can numerically simulate the pattern formation process using PDEs.

Input: 

D. Scalise and R. Schulman *TECHNOLOGY* (2014)

We can extend these ideas to two (or even three) dimensions to build complex pattern transformations.

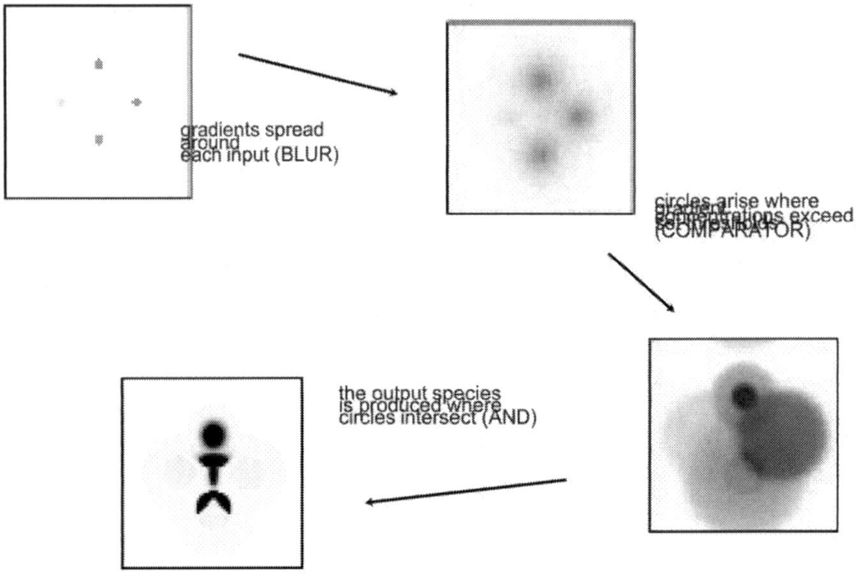

A BLUR module makes a spread out copy of the initial input.

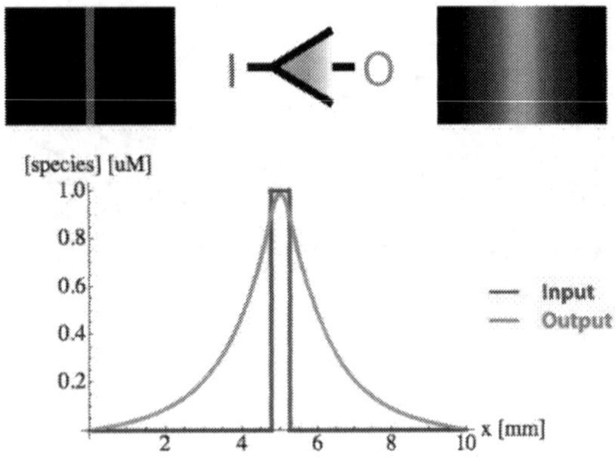

Reactions:

$$I \overset{k_{fast}}{\rightarrow} I + O$$

$$O \overset{k_{slow}}{\rightarrow} waste$$

PDEs:

$$\frac{\partial [I]}{\partial t} = 0$$

$$\frac{\partial [O]}{\partial t} = D_O \nabla^2 [O] + k_{fast}[I] - k_{slow}[O]$$

D. Scalise and R. Schulman *TECHNOLOGY* (2013)

A stick figure circuit is a stepwise diagram of the transformations involved.

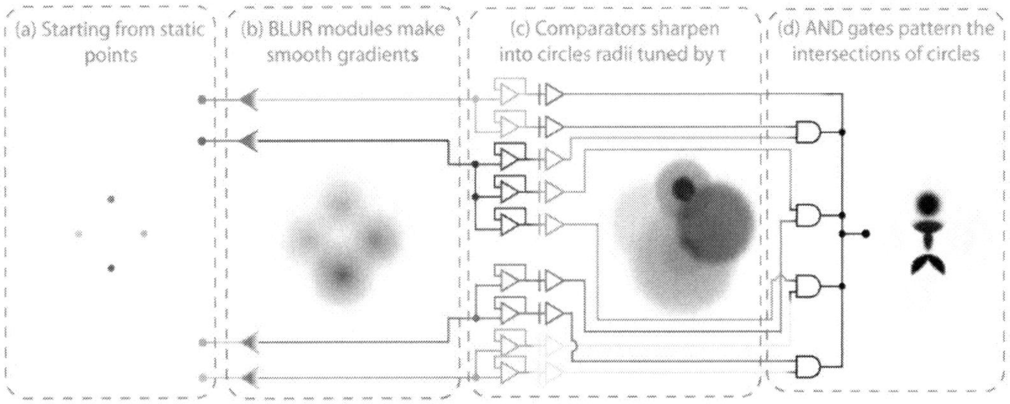

D. Scalise and R. Schulman *TECHNOLOGY* (2013)

Conversion of the circuit to a model or to the molecules is straightforward and algorithmic.

This looks complex, but is commensurate with both the complexity of biological networks and existing synthetic DNA systems:

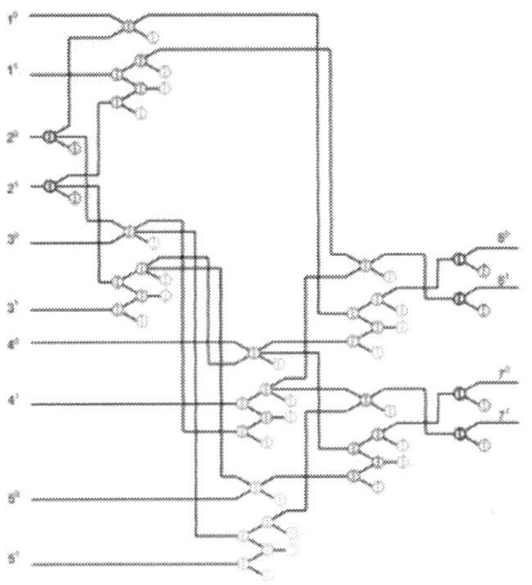

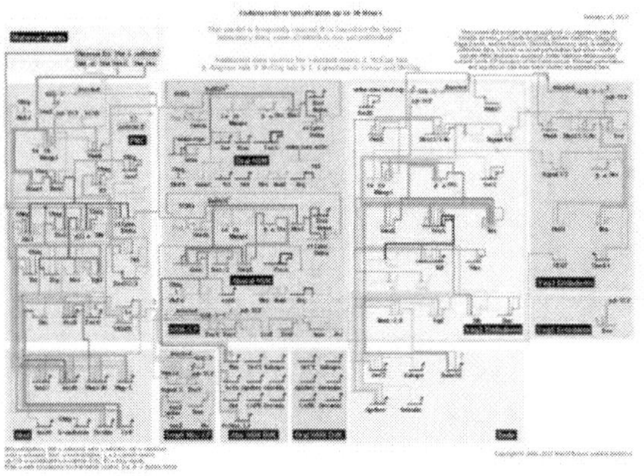

Numerical simulation of stick figure growth

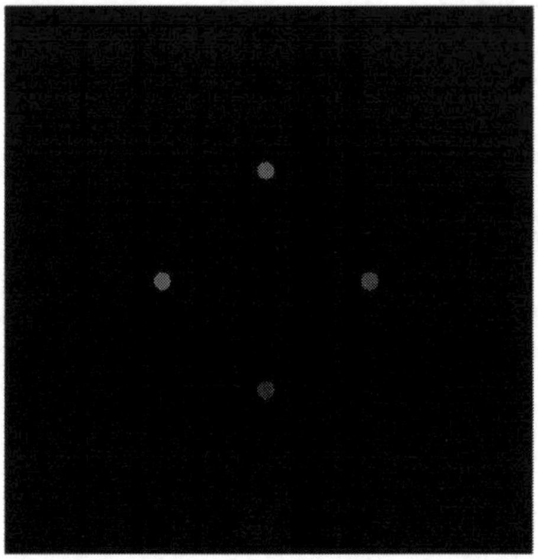

D. Scalise and R. Schulman
TECHNOLOGY (2013)

Thank you.

Schulman group
Deepak Agrawal
Angelo Cangeliosi
Josh Fern
Abdul Mohammed
Dominic Scalise
John Zenk

Ariana Cruz
Dylan Howie
Tyler Jorgenson
Seth Reinhart
Hirsh Shekhar
Jiaqi Wang
Kaiyuan Wang
Kathy Wang
Jiayao Wu

Collaborators
Elisa Franco
Deborah Fygenson
David Gracias
Vicky Nguyen
Georg Seelig
Denis Wirtz
Adam Woolley

Funding
DOE Basic Energy Sciences
NSF CCF
NSF NM (CAREER grant)
Turing Centenary Foundation

All concentrations are measured using a reversible "reporting" reaction:

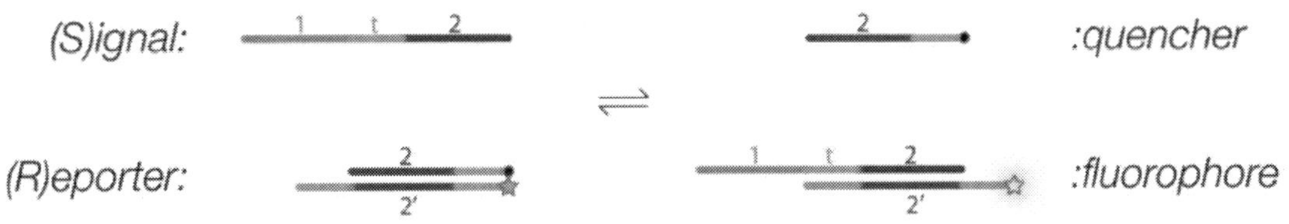

A signal restoration module makes it possible to build robust digital circuits.

$$\text{Source} \overset{slow}{\rightarrow} A \overset{slow}{\rightarrow} \text{Waste}$$

$$I + A \overset{medium}{\rightarrow} I + O$$

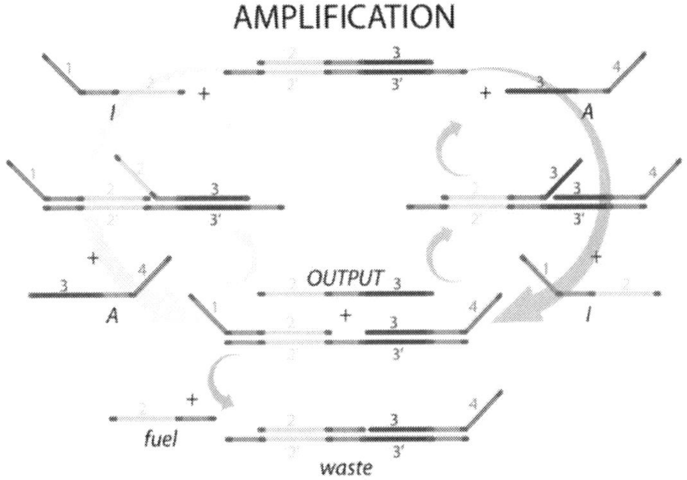

In vivo, biomolecular circuits are able to respond dynamically to changing inputs.

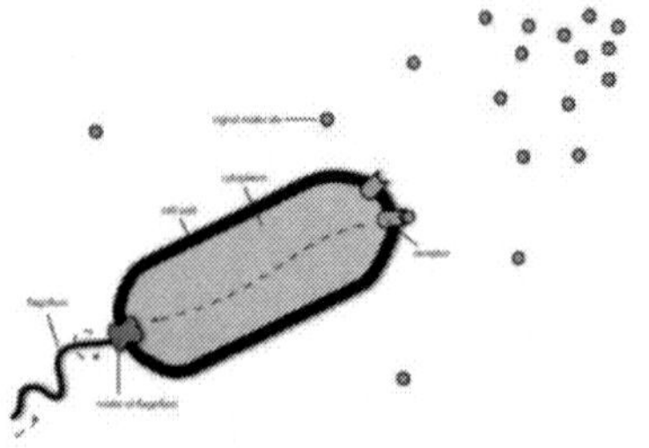

chemotaxis

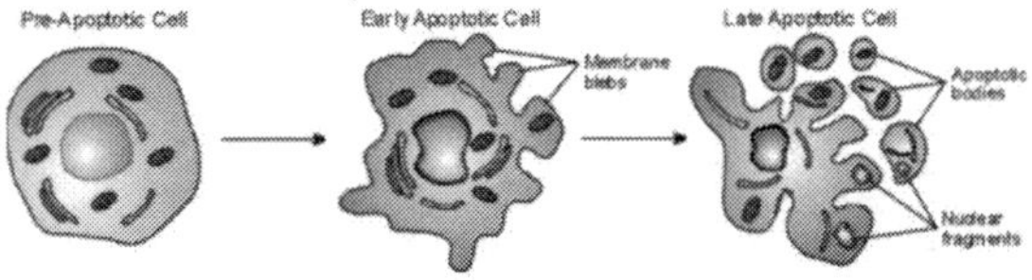

apoptosis

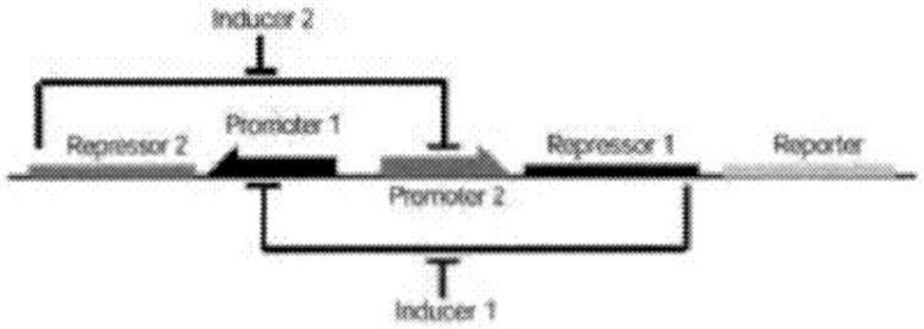

bistable switching

Synthetic DNA complexes that undergo conformational changes can implement designed networks of chemical reactions.

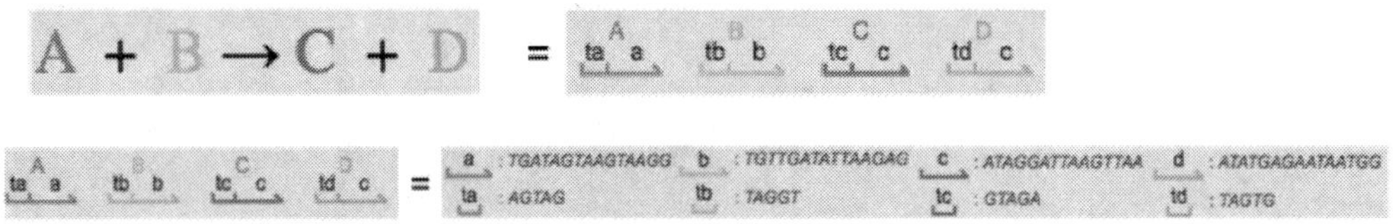

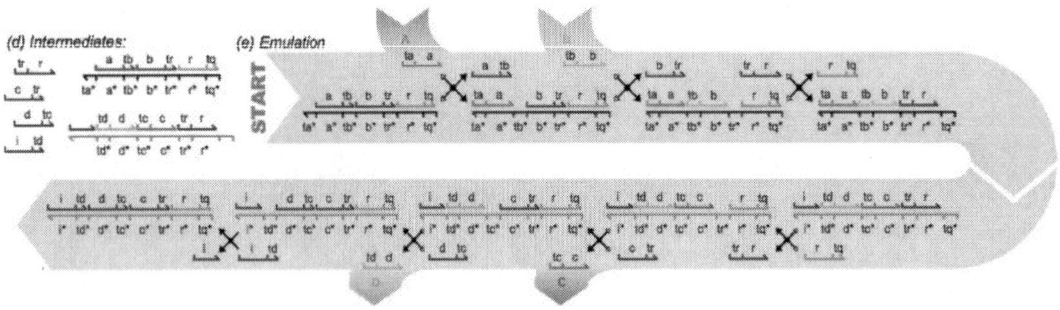

one reaction

Seelig et al 2013

Synthetic DNA complexes that undergo conformational changes can implement designed networks of chemical reactions.

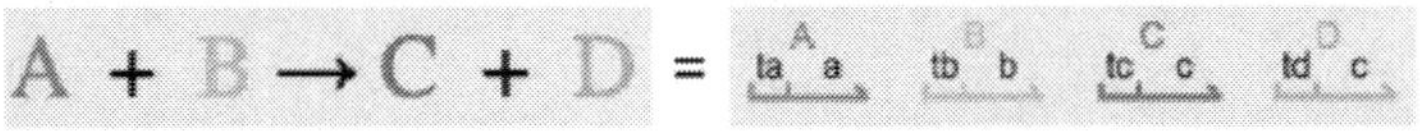

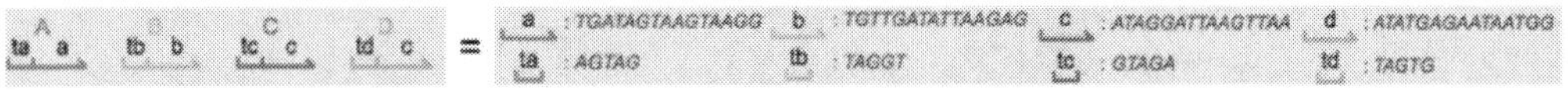

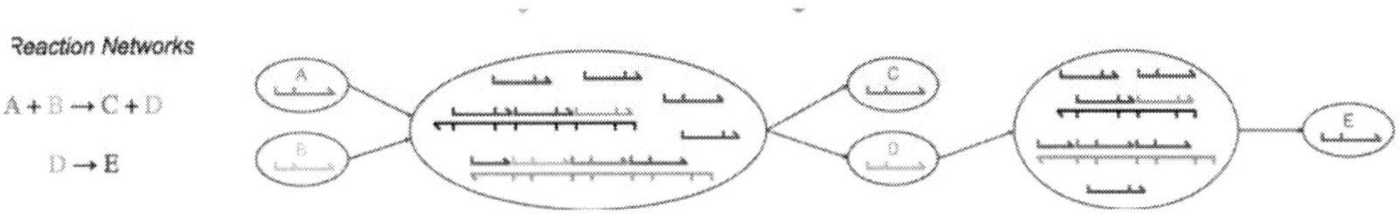

a network of reactions

Seelig et al 2013

Networks of chemical reactions implemented using DNA can include more than a hundred species and can have complex functions.

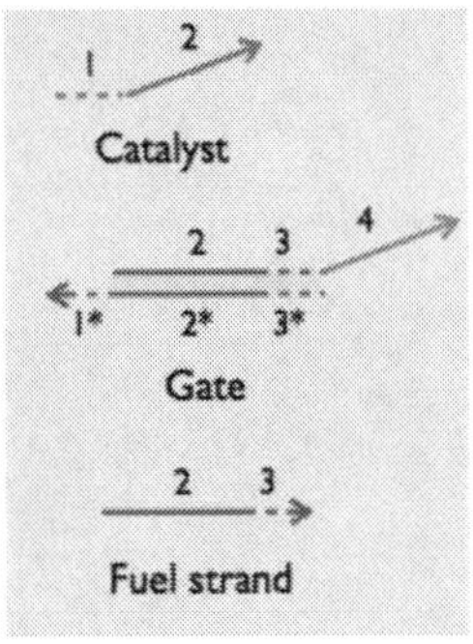

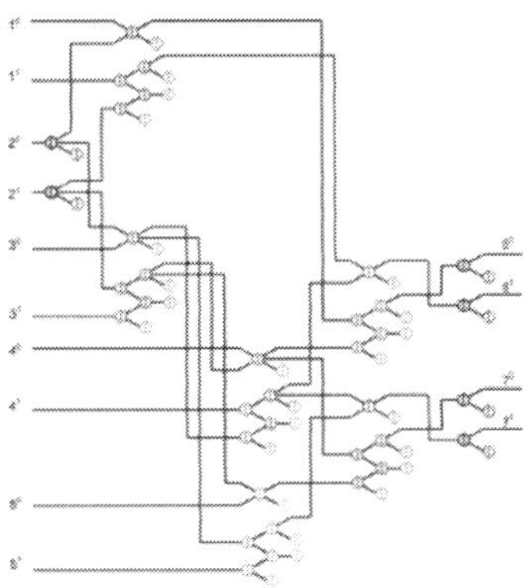

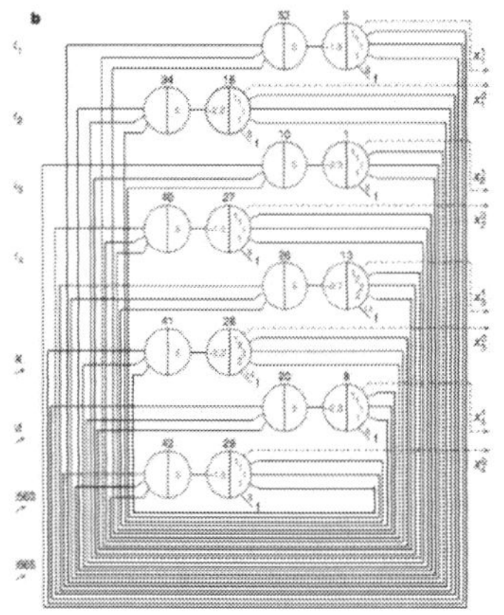

node motif

complex logic

networks with feedback

Qian et al 2011, 2012
Seelig et al 2006

Could we build nanoscale interconnects by self-assembling the connectors?

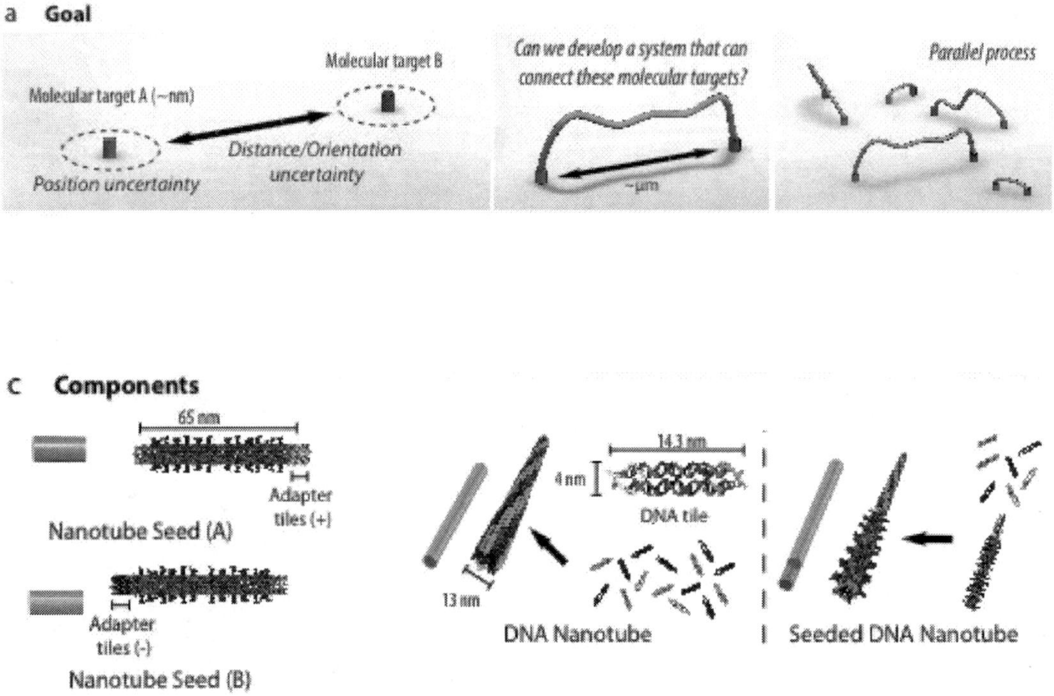

Conductive wires could be assembled by selectively attaching nano particles to the templates, then metallizing these structures.

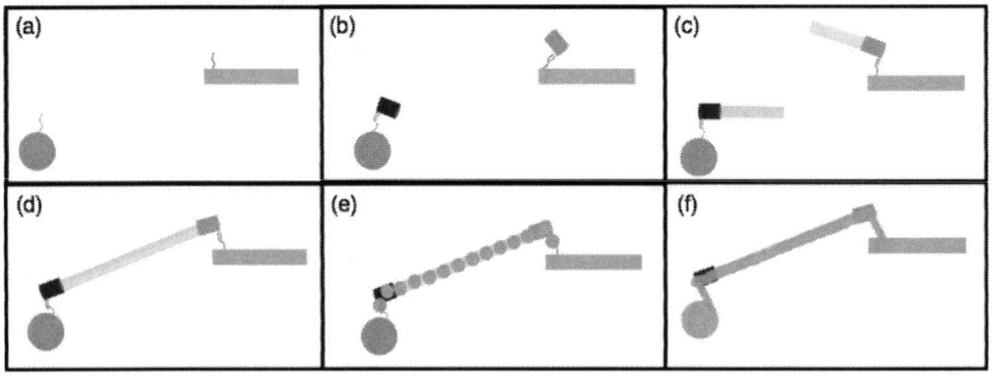

Also, build connections from cell surfaces, moving structures, even in vivo.

Joint work with Adam Woolley.

What kinds of patterns can be produced using modular transformations?

Observation #1: Any pattern can be described as a series of points.

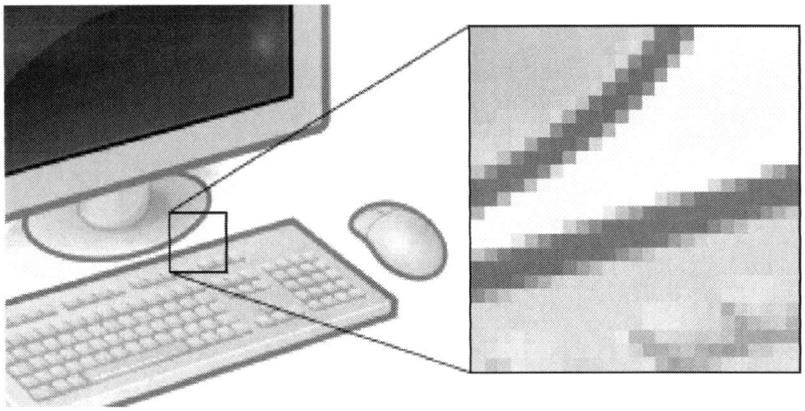

Observation #2 : If we can pattern a single pixel, we can pattern any image.

The reaction-diffusion process for each pixel can run in parallel.

An outline of a basic program to pattern a pixel in a prescribed location.

An outline of a basic program to pattern a pixel in a prescribed location.

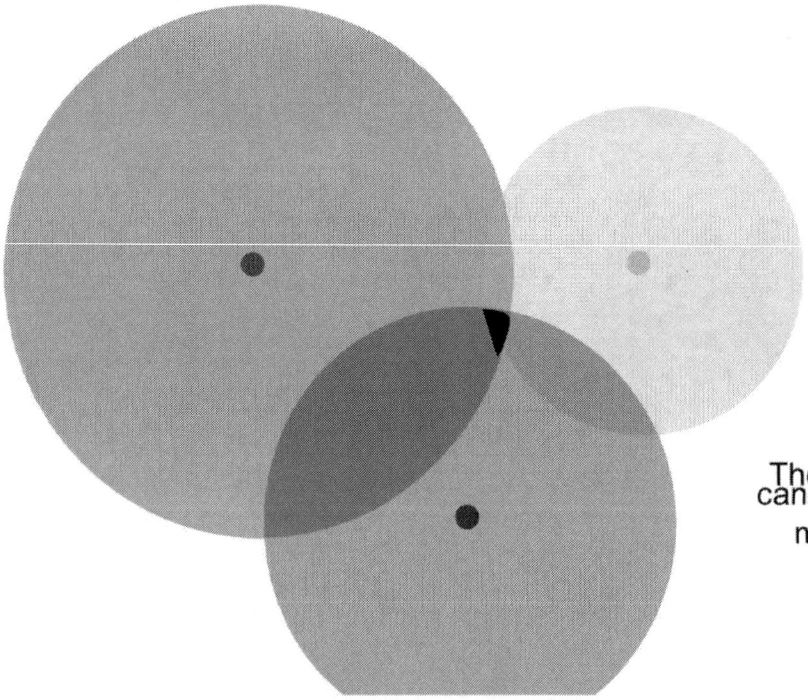

The same 3 points can serve as inputs (with COPY modules) for any set of desired pixels.

DNA nanotubes are seeded polymers that serve as wire templates.

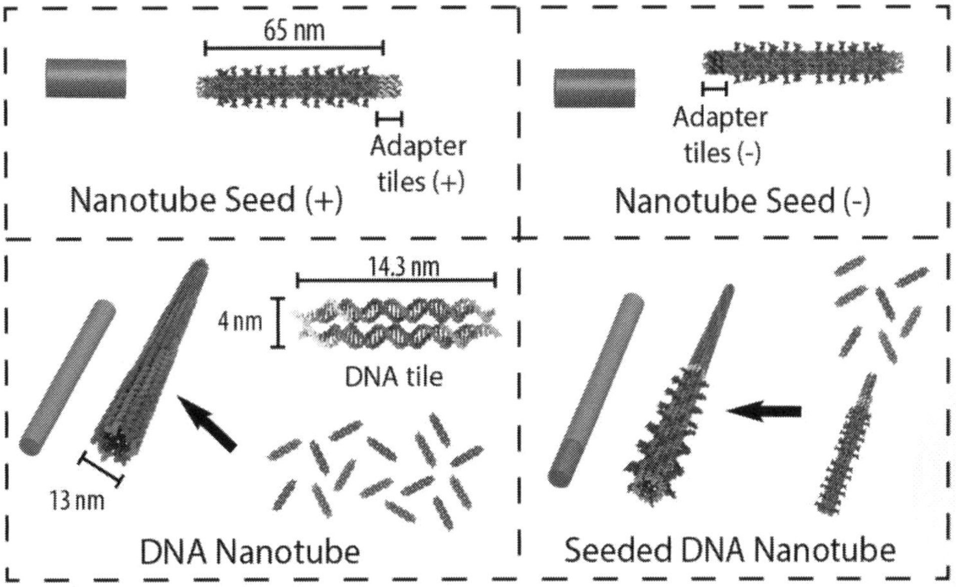

Mohammed *et al*, *Nano Lett.* (2013)

Nanotubes grow from either terminal then join to form a connection.

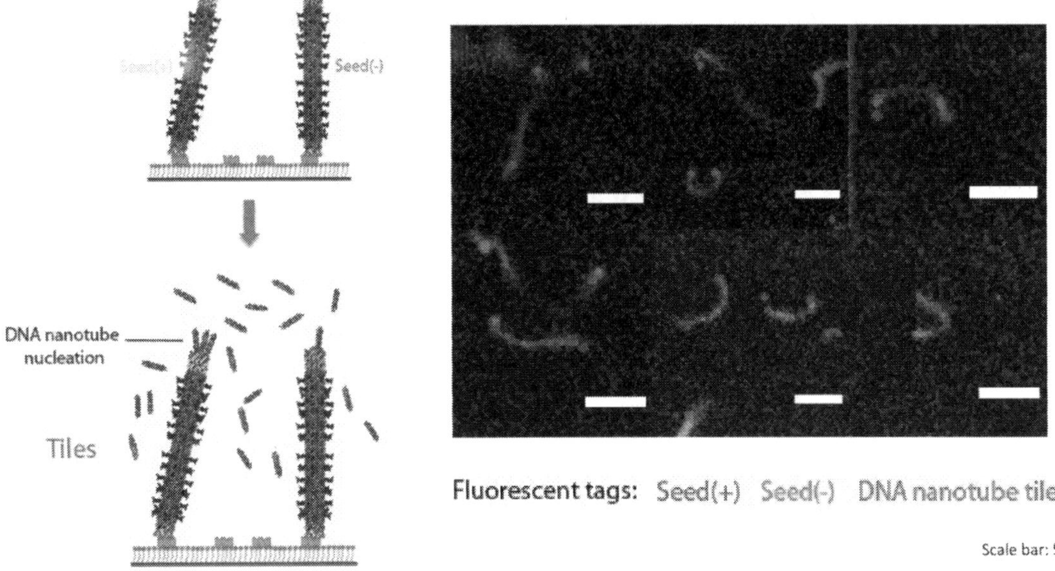

Fluorescent tags: Seed(+) Seed(-) DNA nanotube tiles

Scale bar: 5 µm

20 hours at 32°C

Yield of connections is >80% with no optimization.

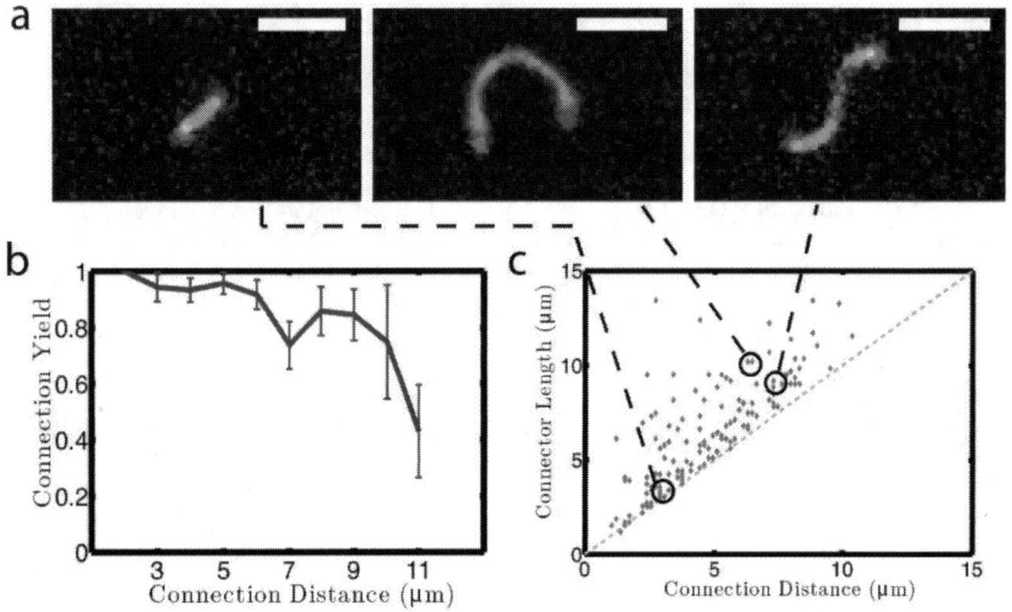

Time-lapse microscopy movies capture connection formation dynamics.

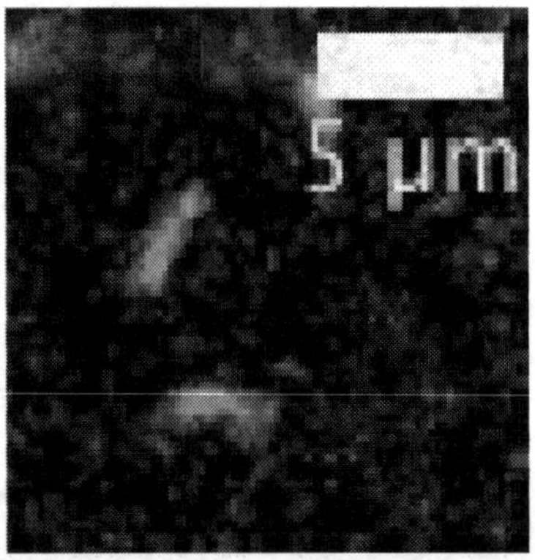

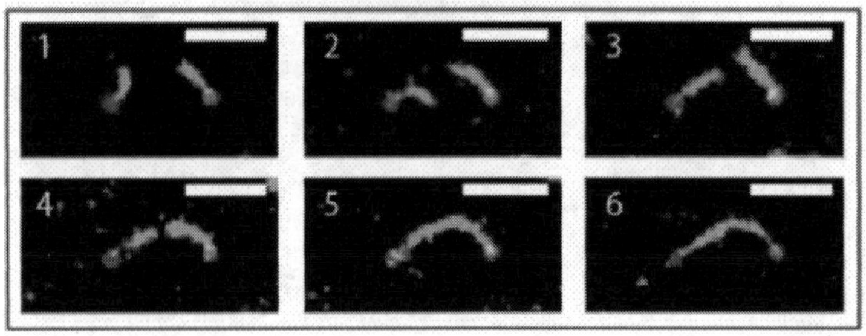

Most unconnected assemblies are either not long enough or the orientation of the endpoints make connection difficult.

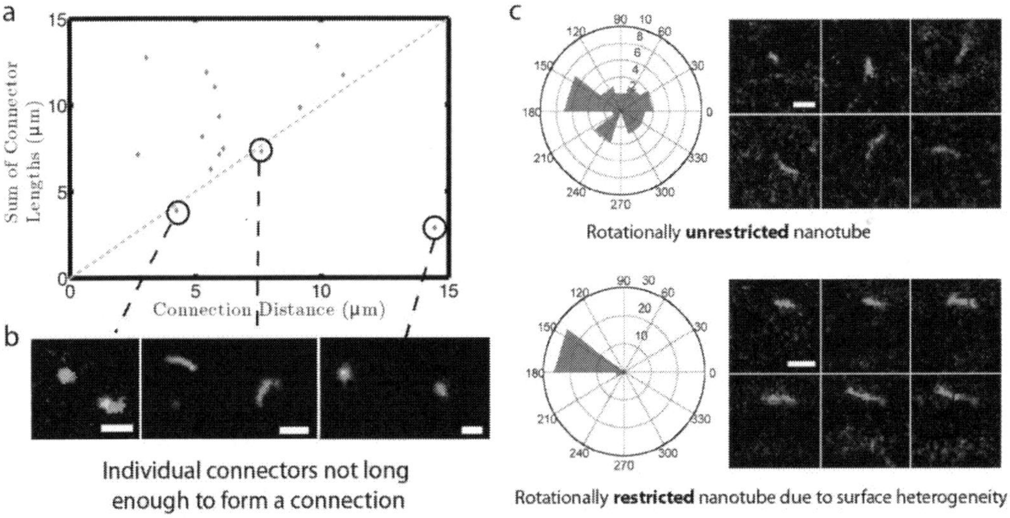

ɔes can produce 3-dimensional interconnects between siı

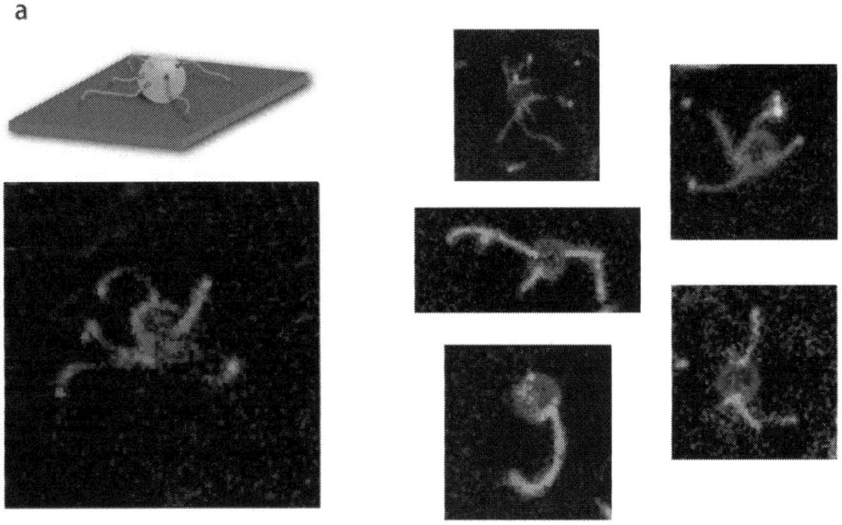

These images show (-) seeds blue on 3 micron beads and positive seeds (red) on the surface. Nanotubes (green) specifically connect these nanometer scale structures, but do not form elsewhere.

Could we use this interconnect assembly method to build terminals for nanoscale devices?

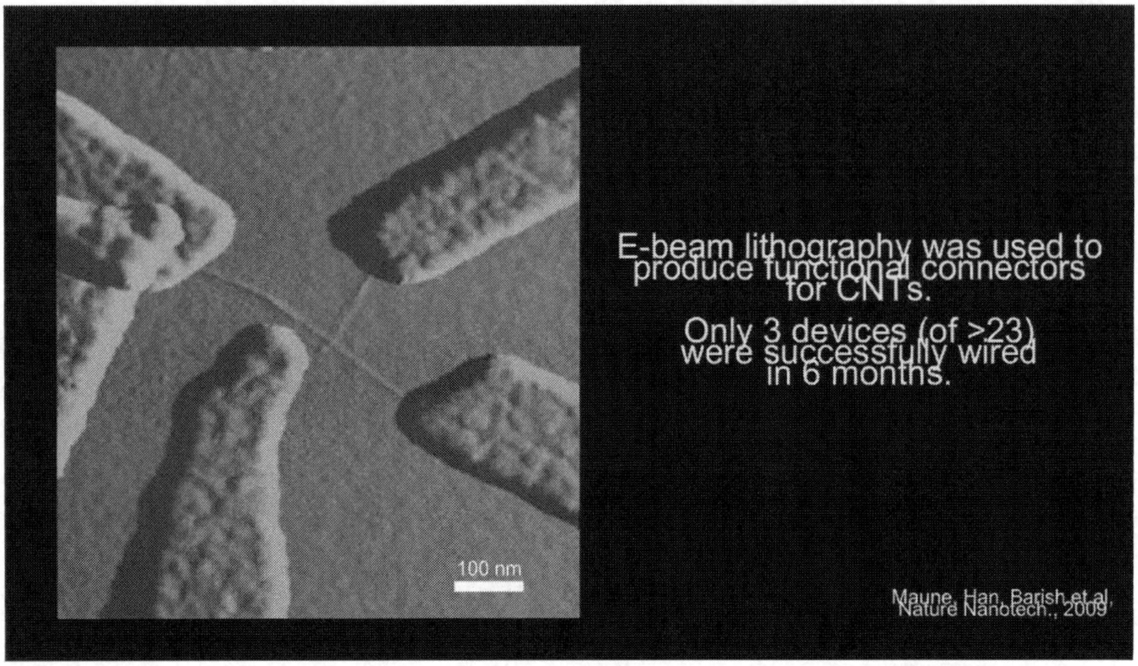

Could we build wires to specific ion channels, perhaps on nerves to wire cells, or rewire nerve tissue?

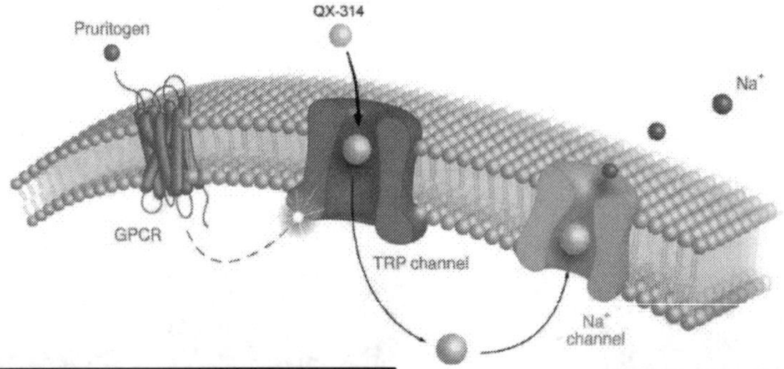

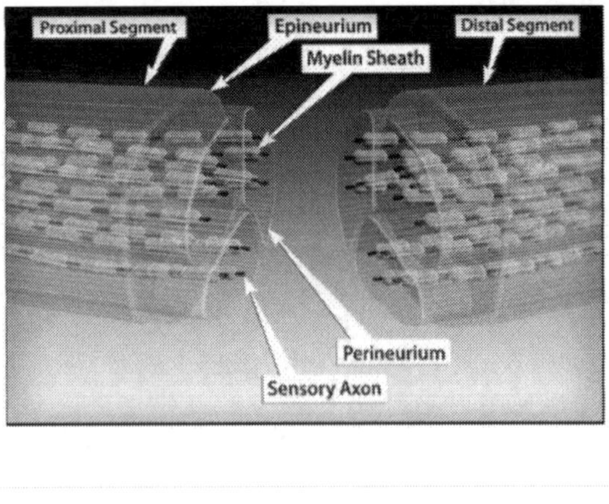

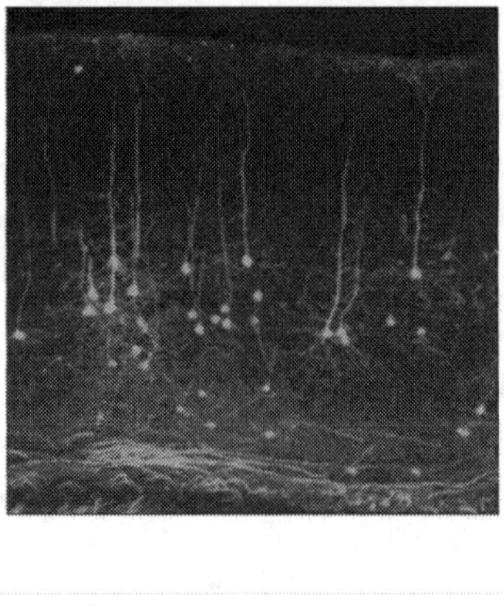

Circuit control could be used to self-assemble complex architectures, such as those made from nanotubes.

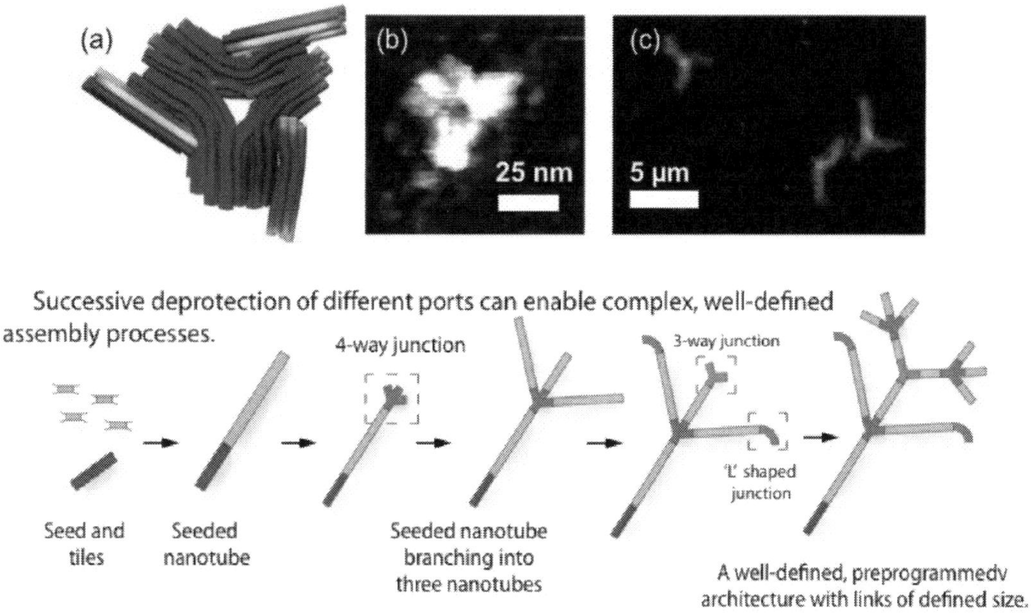

DNA crosslinked gels can be molded using standard assembly techniques.

Circuit control could be used to self-assemble complex architectures, such as those made from nanotubes.

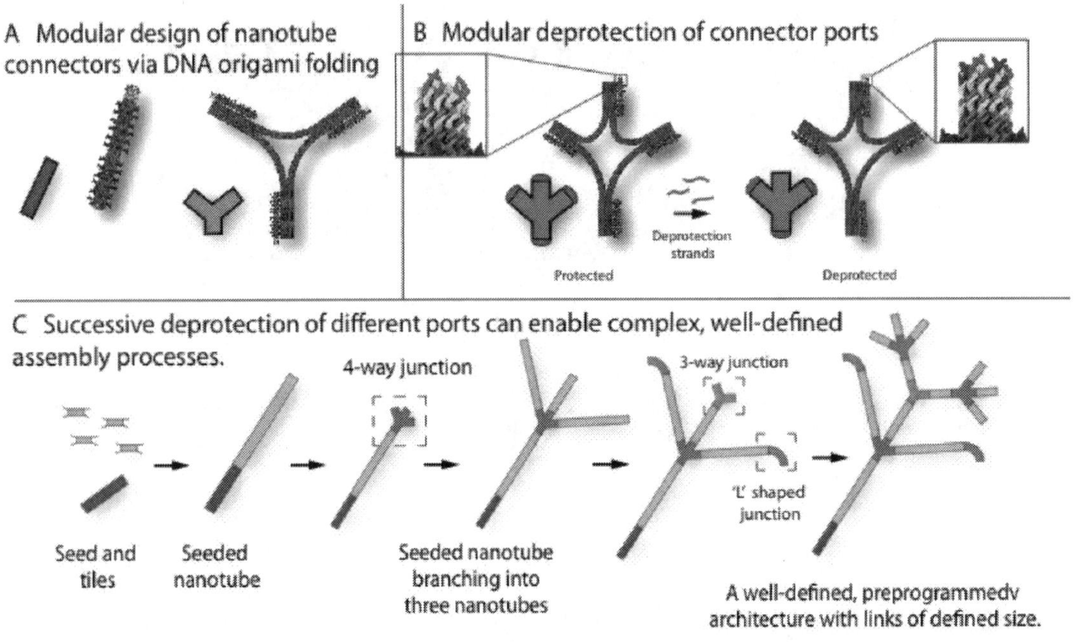

DNA nanostructures can template functional materials to make devices.

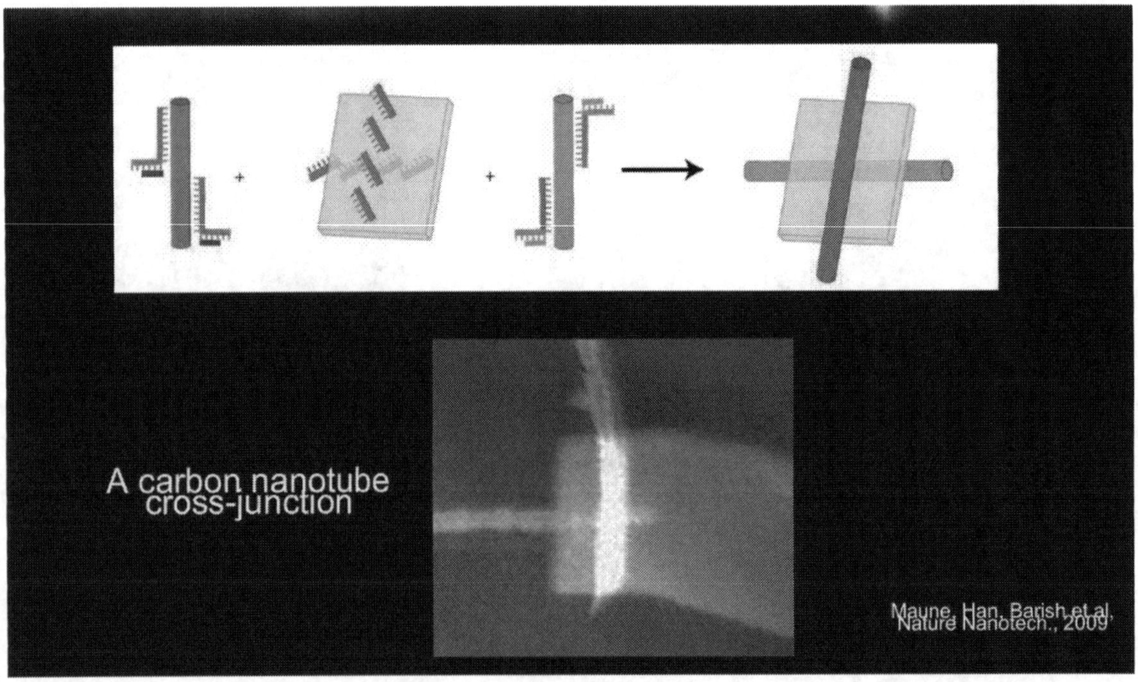

The difficulty of interfacing with these devices is a major limitation.

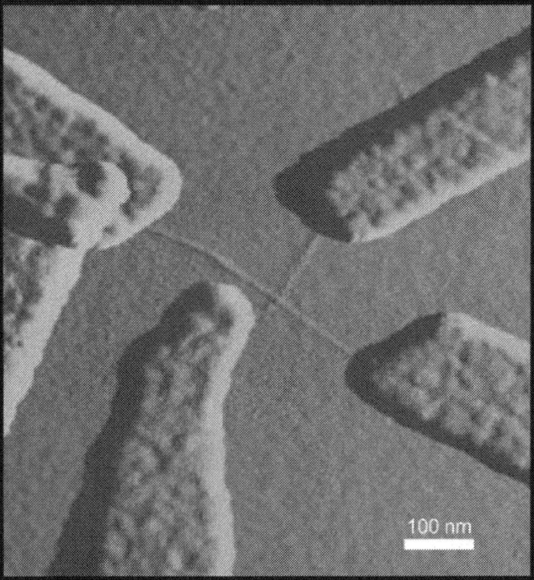

E-beam lithography was used to produce functional connectors for CNTs.

Only 3 devices (of >23) were successfully wired in 6 months.

Maune, Han, Barish et al. Nature Nanotech., 2009

We propose a new method for wiring nanoscale devices: self-organizing wires.

To assemble a self-guiding wire...

1. Chemically label the start and end sites.

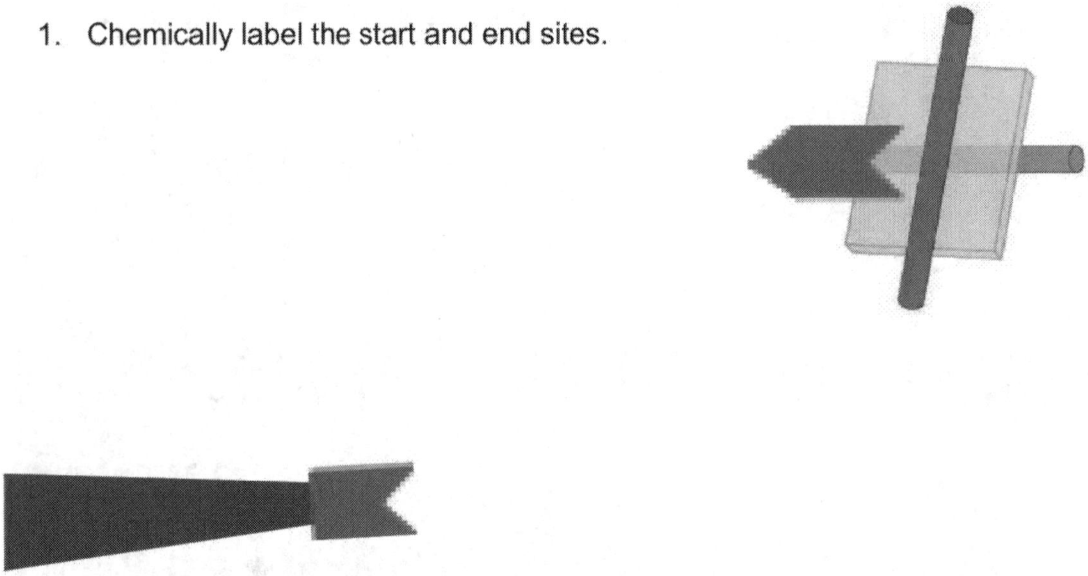

To assemble a self-guiding wire...

1. Chemically label the start and end sites.
2. Filament grows from the start site.

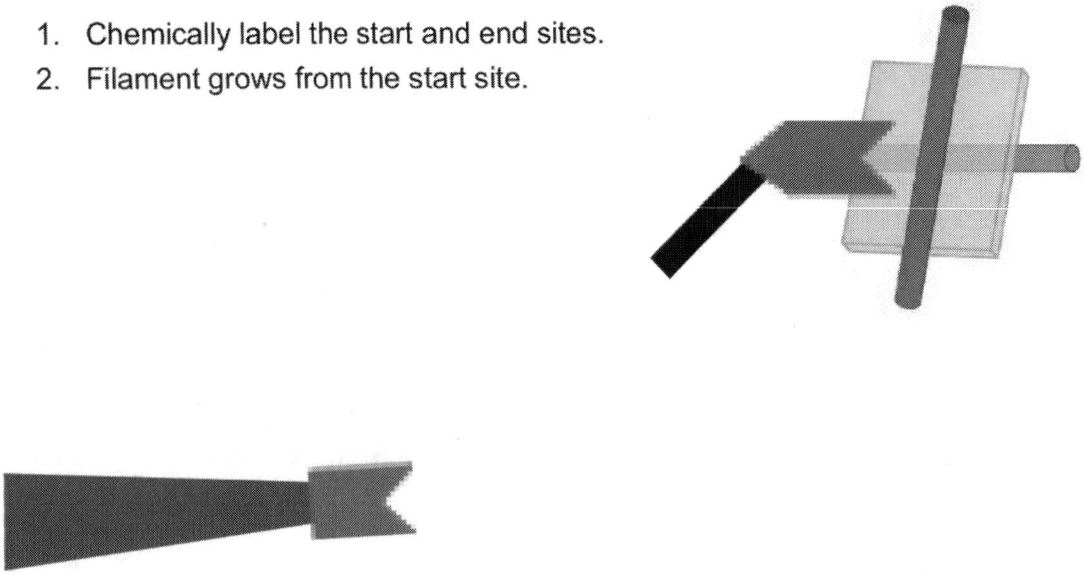

To assemble a self-guiding wire...

1. Chemically label the start
2. Filament grows from the s
3. As the filament grows, its

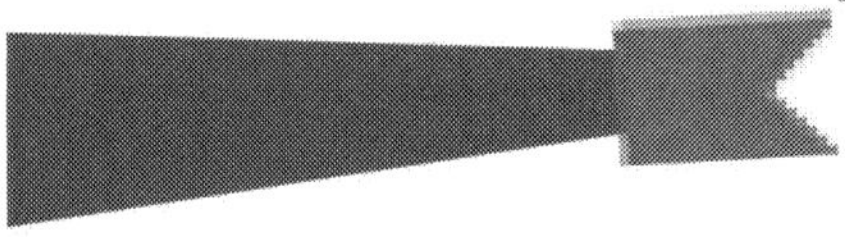

To assemble a self-guiding wire...

1. Chemically label the start and end sites.
2. Filament grows from the start site.
3. As the filament grows, its end diffuses.
4. Filament end attaches to end site.

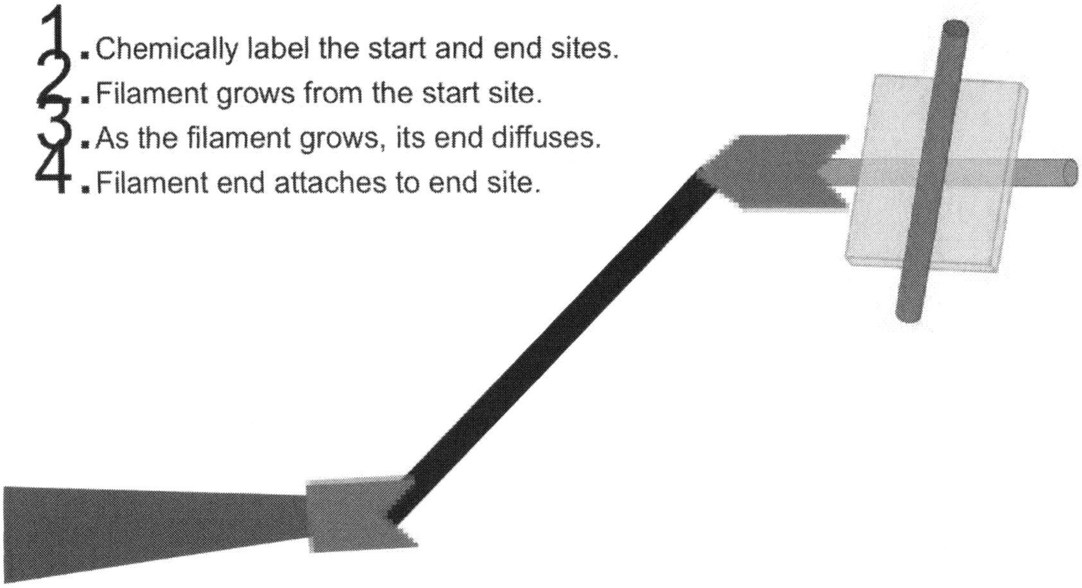

We can later functionalize the filament to produce a signal carrier.

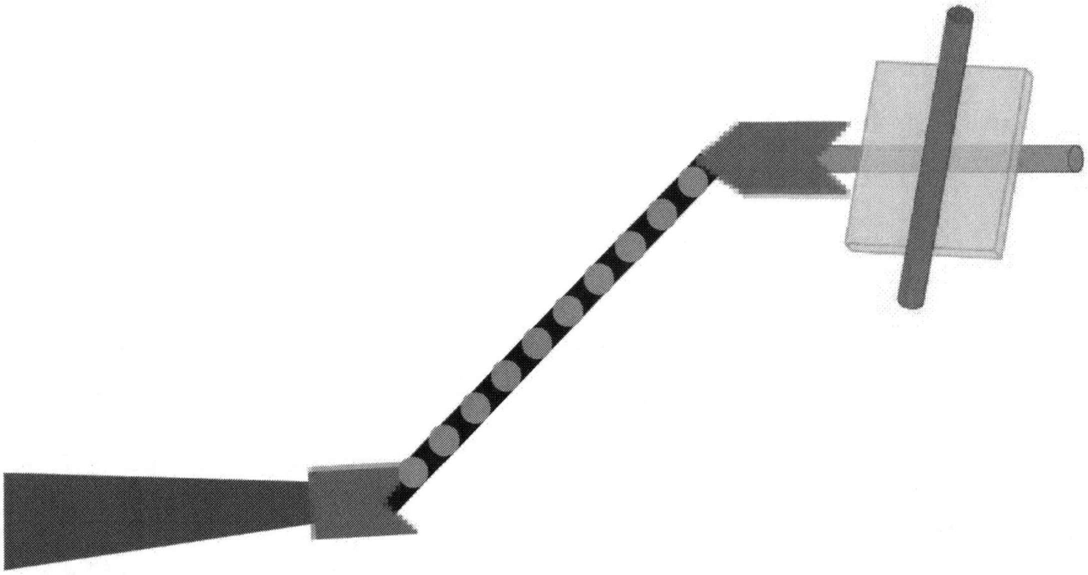

A nanotube seeds (image shown below) serve as templates for the growth of nanotubes

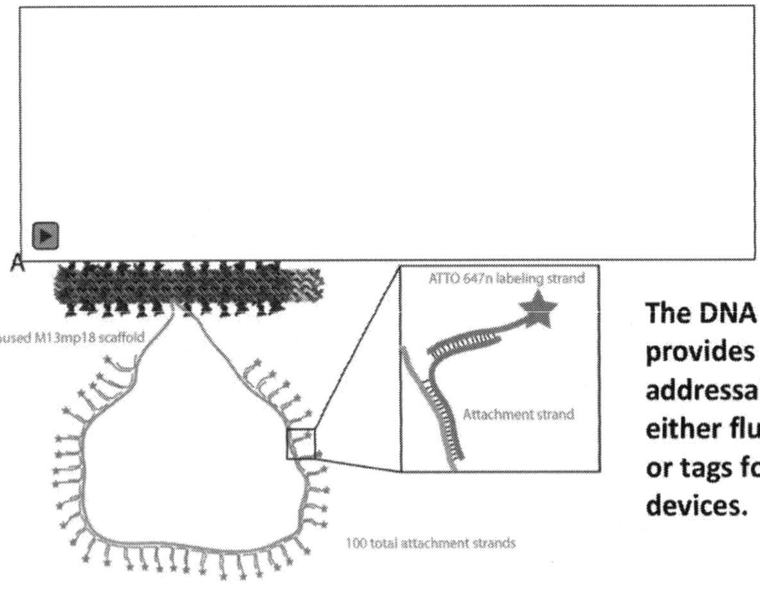

Approximate seed size" 80 nm length x10 nm radius

The DNA sequence on the scaffold provides several hundred bases of addressable binding locations for either fluorescent labels (shown) or tags for attachment to specific devices.

Reaction design, and illustration that nanotubes bind to seeds.

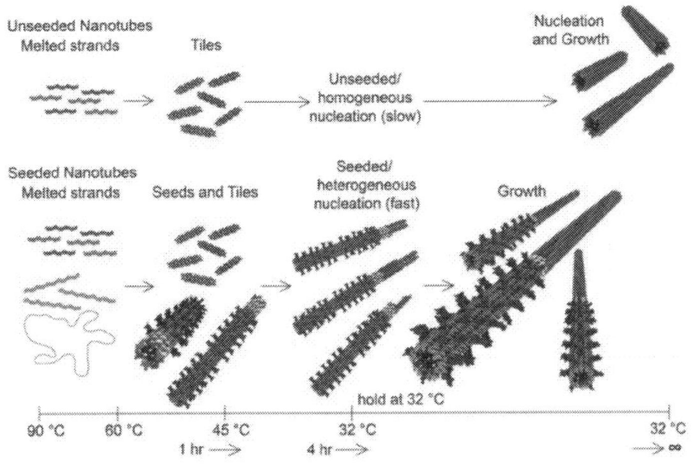

Schematic of how growth occurs

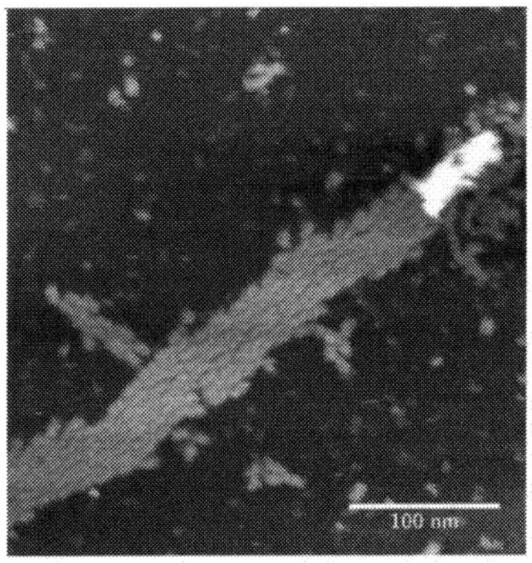

Seeded nanotube
AFM image

Nanotubes open on the mica surface but are tubes in solution.

Mohammed *et al*, *Nano Lett.* (2013)

Nanotubes grow only when seeds are present.

Unseeded nanotubes

Seeded nanotubes

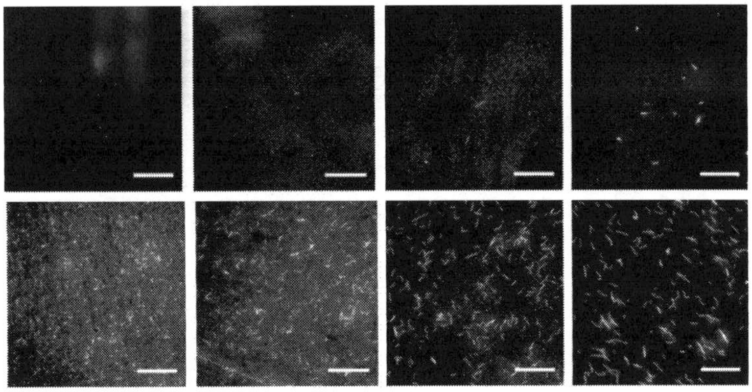

Scale bar: 20 µm — 2 hrs — 10 hrs — 25 hrs — 50 hrs

Mohammed *et al*, *Nano Lett.* (2013)

Nanotubes have seeds at their terminals.

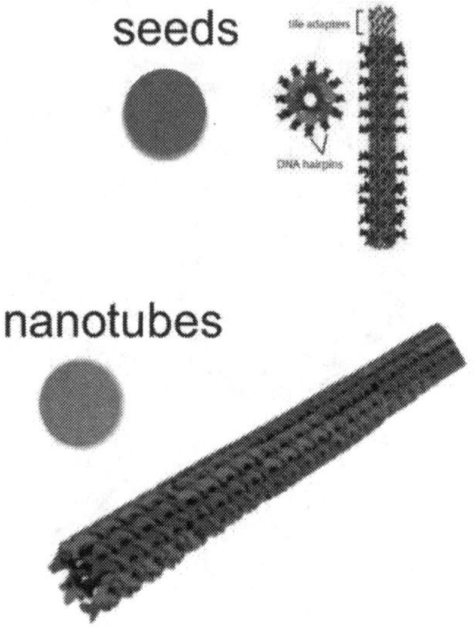

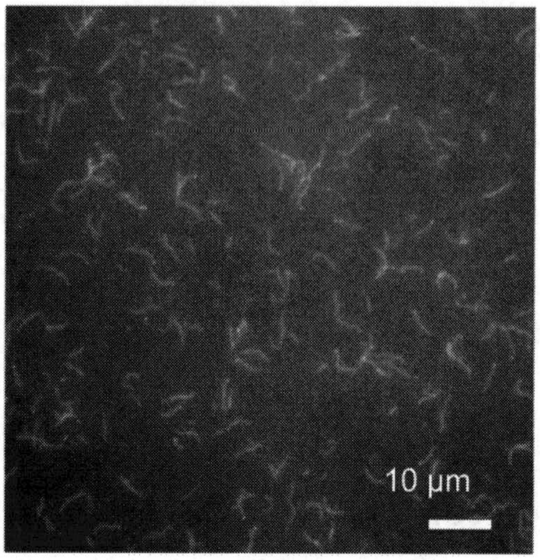

Molecular components:

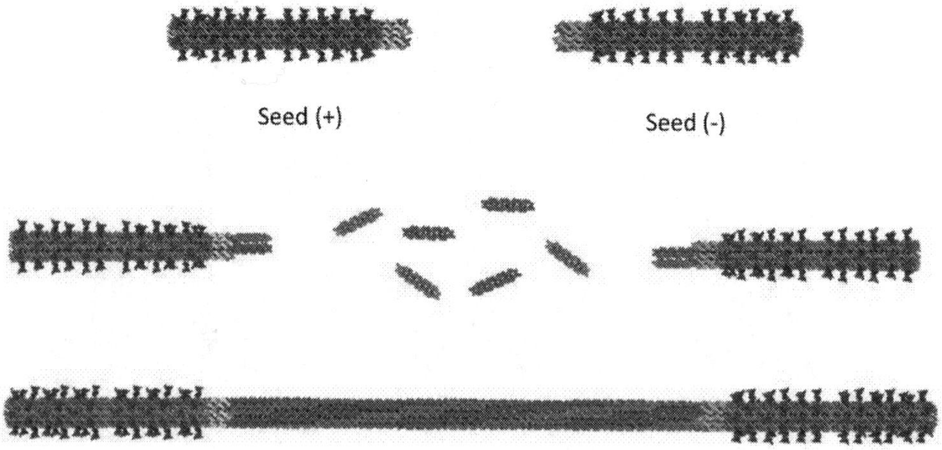

Seeds differ in their DNA sequences at their ends. Nanotubes from + and - seeds can join to form a stable structure, as established in the literature.

:achment of seeds to passivated surface and specific growth of nanotubes from these se

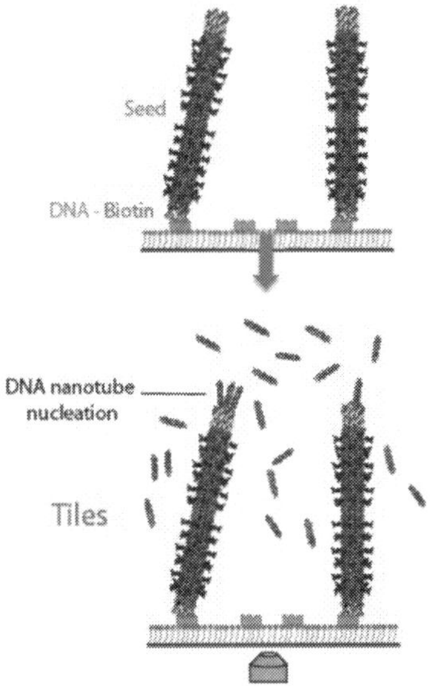

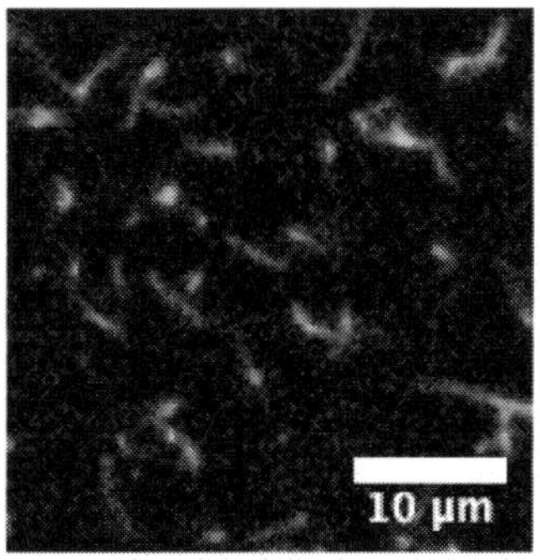

Nanotubes from (+) seeds

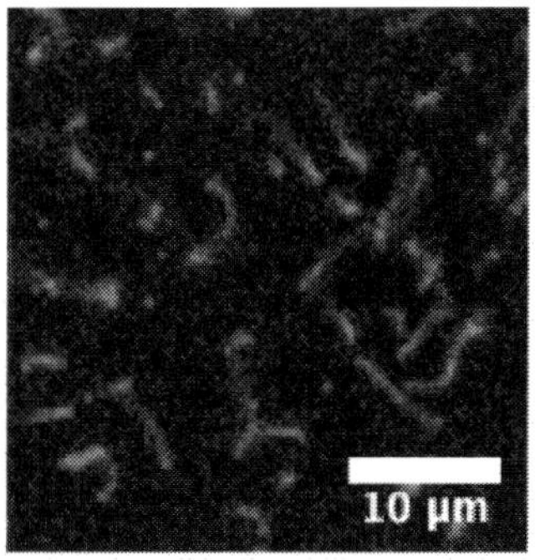

Nanotubes from (-) seeds

Fluorescent tags: Seeds DNA nanotube tiles

Fluorescence microscopy movies make it possible to resolve the dynamics of individual structures.

15 hours at 32°C

The material for interconnects: DNA nanotubes

2 types of tiles

REd tile

SEd tile

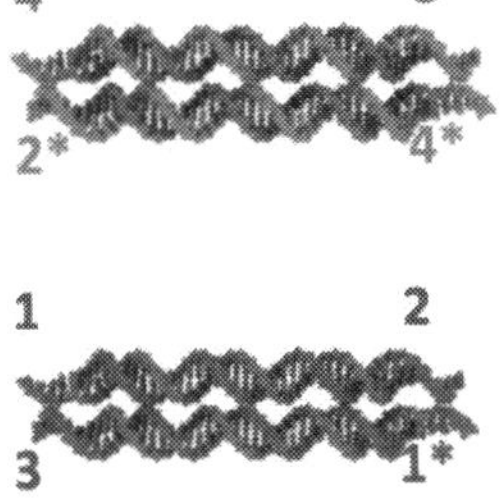

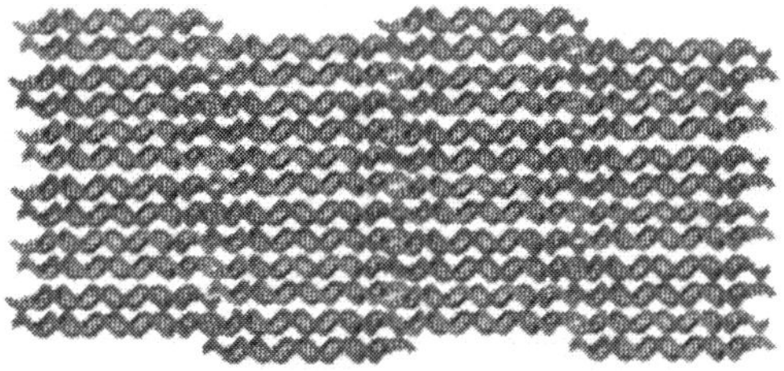

Sticky ends programmed to form diagonally stripped lattice

Design of a Toolbox of RNA Thermometers

Synthetic Biology: Evolution, Engineering & Design
June 10-13, 2015, Boston, USA

June 12, 2015

Shaunak Sen*, Rohit Satija, Dan Siegal-Gaskins, Richard M. Murray

1

Acknowledgements

Rohit Satija
Divyansh Apurva
Dan Siegal-Gaskins
Richard M. Murray

Jongmin Kim
Harry Choi
Clare Hayes
Anu Thubagere

Funding
FRTA, IIT Delhi
Department of Science and Technology, India
Gordon and Betty Moore Foundation, Caltech
NSF Molecular Programming Project, Caltech

Biomolecular sensors can have multiple uses

Example: Temperature-sensing applications

Large-volume biochemical reactors, where heat may be a convenient inducer

http://www.envirogen.com/pages/technologies/bioreactors/
accessed June 11, 2015

Design of temperature compensation in biomolecular circuits

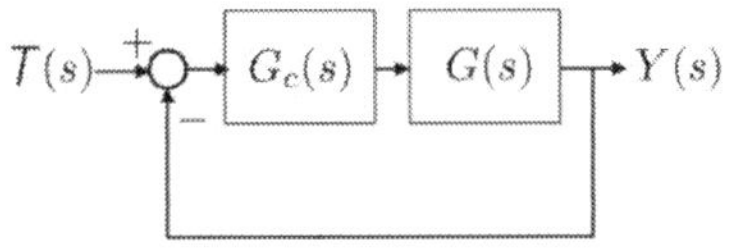

Spatiotemporal thermal control/ imaging in cells & tissues

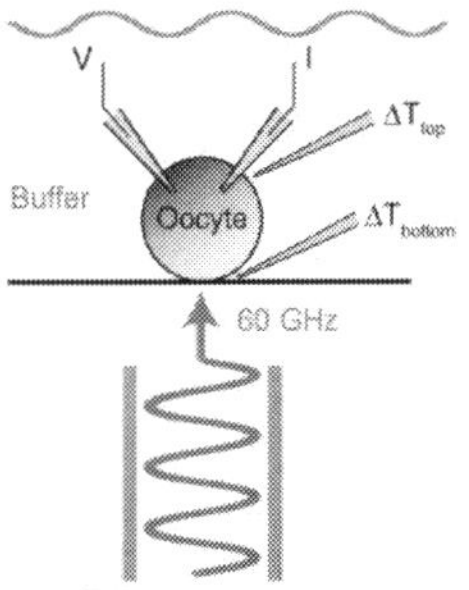

millimeter wave

Biophysical Journal Volume 104 June 2013 2622–2628
Mikhail G. Shapiro, Michael F. Priest,
Peter H. Siegel, and Francisco Bezanilla

RNA-based 'thermometers' can sense temperature

Natural Examples
Heat shock in *E. coli*,
Virulence

Basic Mechanism

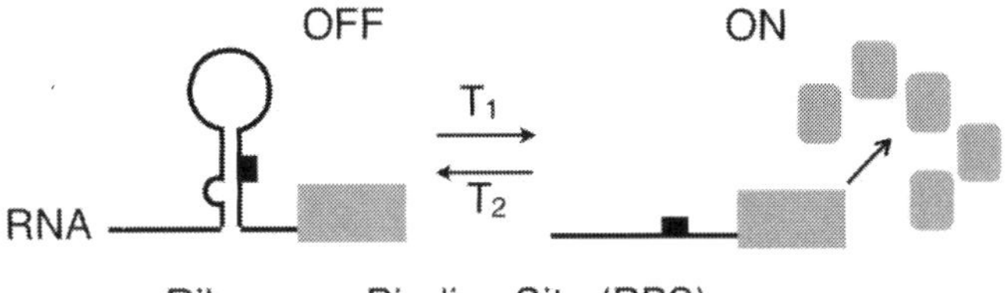

■ Ribosome-Binding Site (RBS)

Synthetic designs also made
Threshold temperature can be increased with a stronger secondary structure (longer stem, fewer bulges, smaller hairpin loop)

Interesting aspects of RNA thermometers

- Synthetic designs 'simpler' than natural counterparts

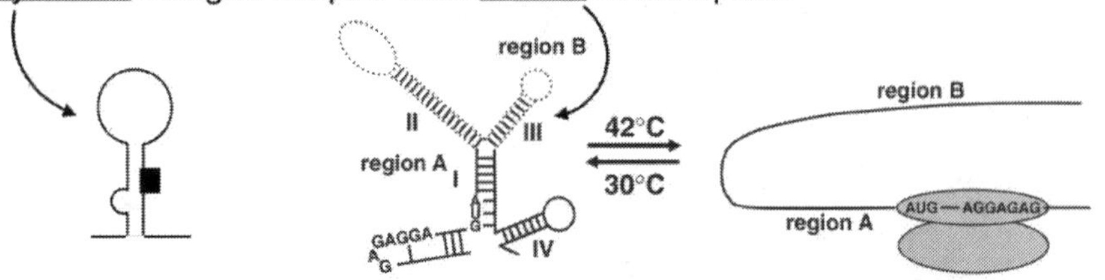

RNA thermometers FEMS Microbiol Rev **30** (2006) 3–16
Franz Narberhaus[1], Torsten Waldminghaus[1] & Saheli Chowdhury[1,2]

- Possible contribution of multiple structures to temperature response, not just of the Most Favoured structure

- 1bp sequence change can qualitatively change temperature response

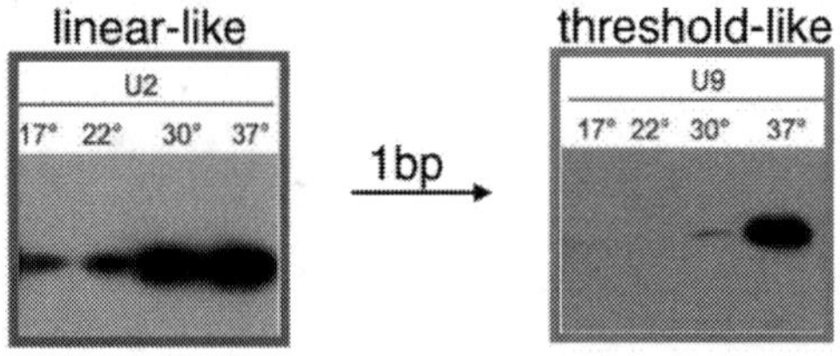

Design of simple synthetic RNA thermometers for temperature-controlled gene expression *in Escherichia coli*
Juliane Neupert, Daniel Karcher and Ralph Bock*
Nucleic Acids Research, 2008, Vol. 36, No. 19 e124
doi:10.1093/nar/gkn545

Goal is to design a toolbox of RNA thermometers with a variety of thresholds and steepnesses

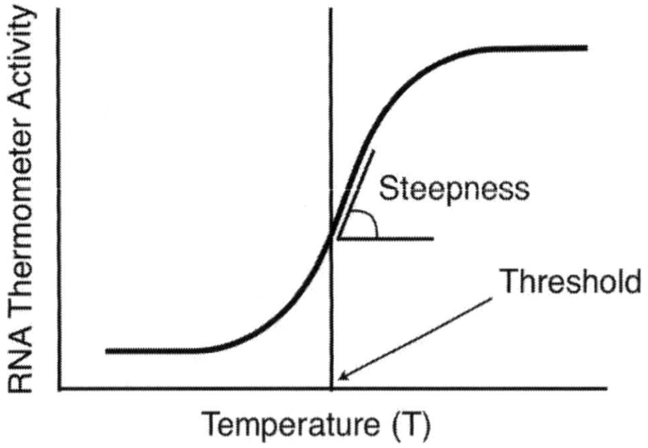

Temperature dependence of previously constructed RNA thermometers measured in cells

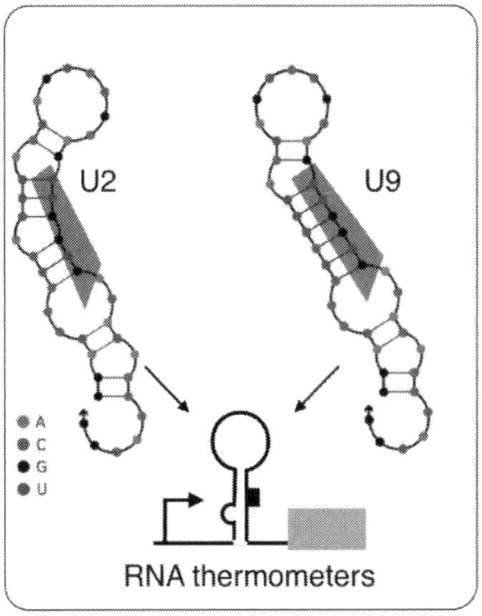

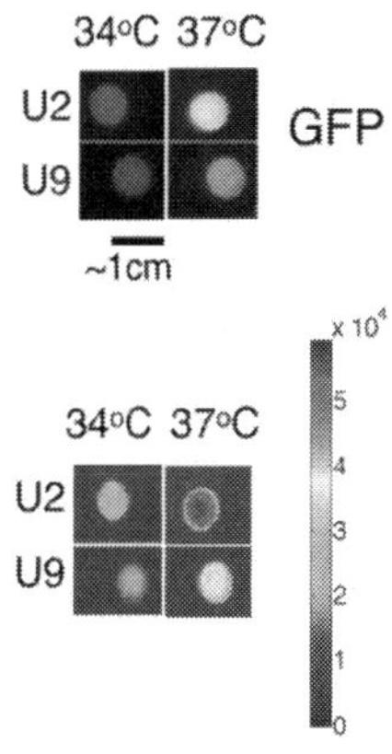

Cells, each disc is a colony

Temperature dependence of these RNA thermometers measured in cell extracts

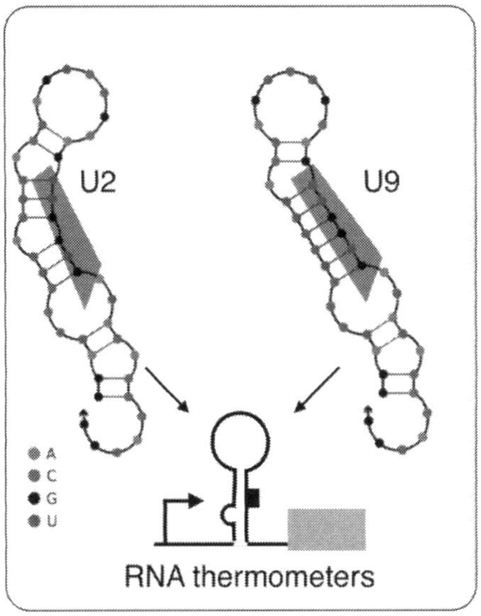

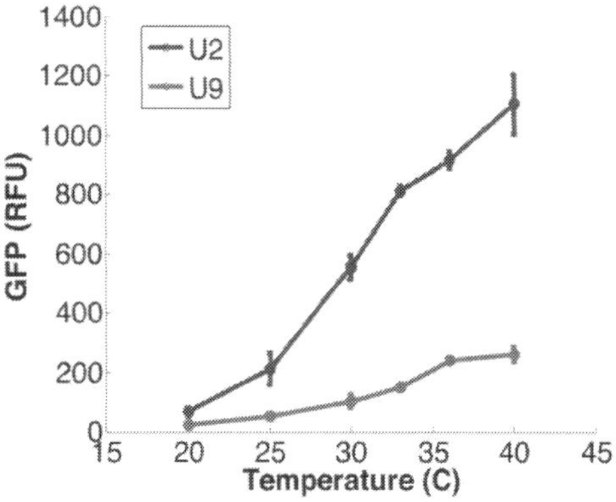

Cell extracts, t = 150 min

Temperature dependence measured consistent with a computational analysis

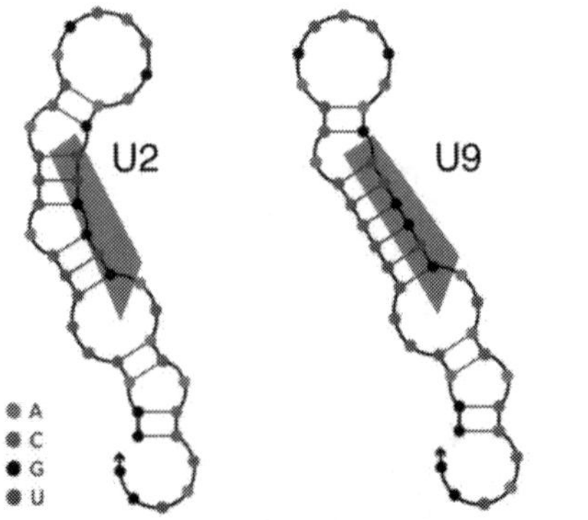

Free Energies
(Most Favoured Structure, 28°C)

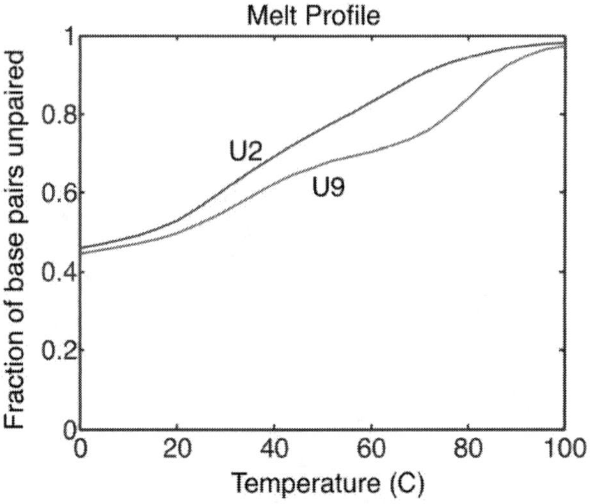

Computations using NUPACK nucleic acid package

RNA thermometer library designed by individually changing all base pairs involved in secondary structure pairing, RBS identical

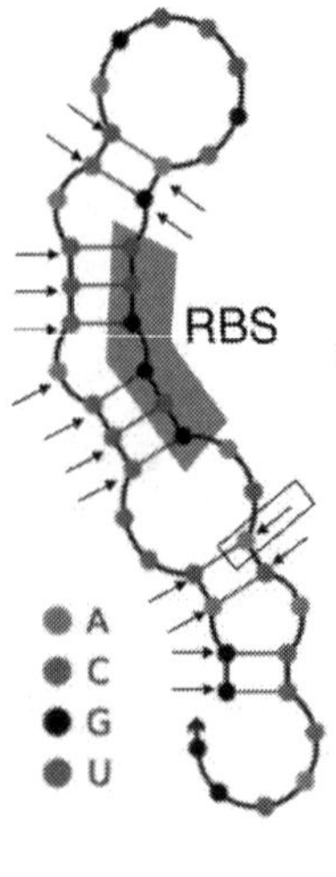

Sequence of U2

RBS

GGAUCCCUCA CUUACUAGUC UGCAGAAGGA GAUAUACCCA UGG

0 (deletion)

U9

0
C
... G ...
U

Library Size = 19 x 4 = 76

Computational analysis of library suggests a diversity of responses may be obtained

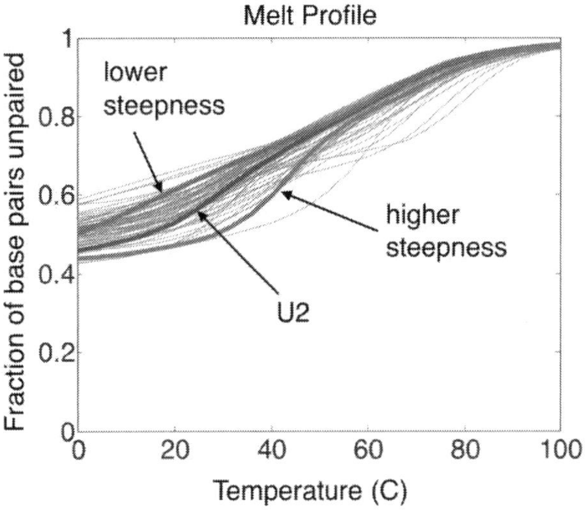

Computational analysis of library suggests a diversity of responses may be obtained

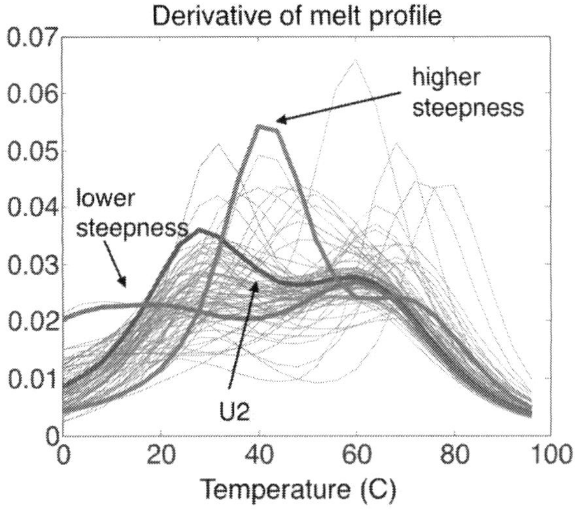

Preliminary measurements suggest a diversity of temperature responses

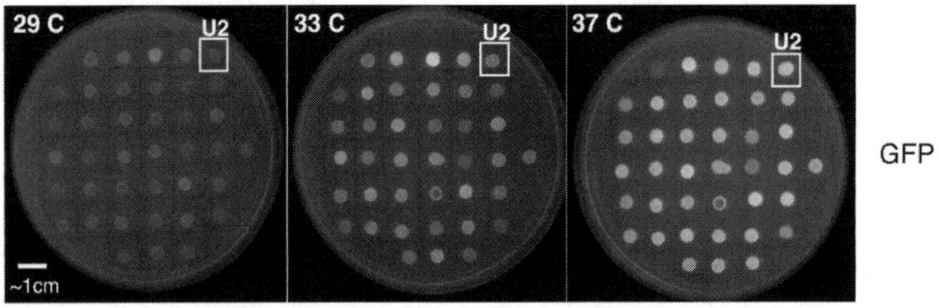

Each disc is a cell colony grown on a LB plate

Preliminary measurements suggest a diversity of temperature responses

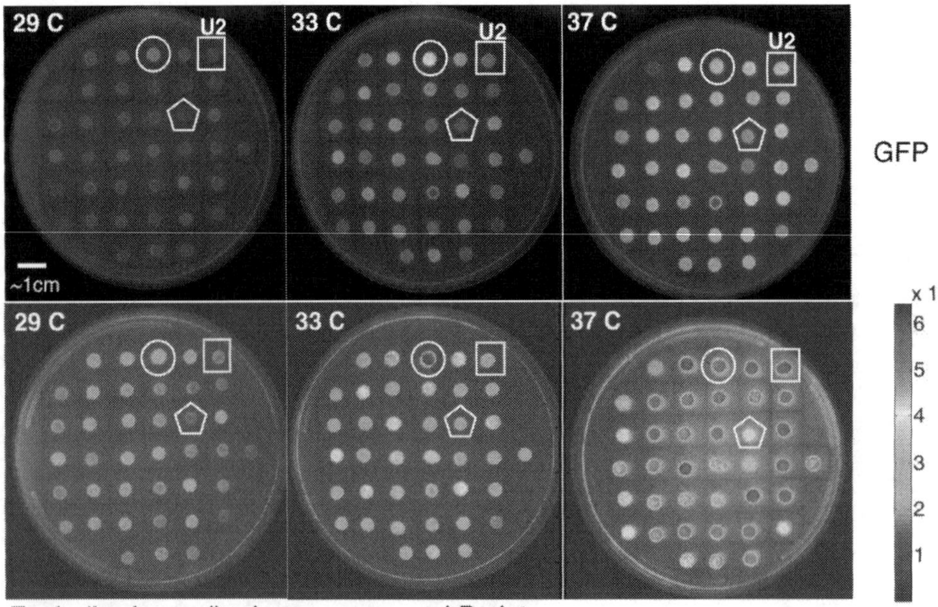

Each disc is a cell colony grown on a LB plate

Preliminary measurements suggest temperature responses with differing steepnesses

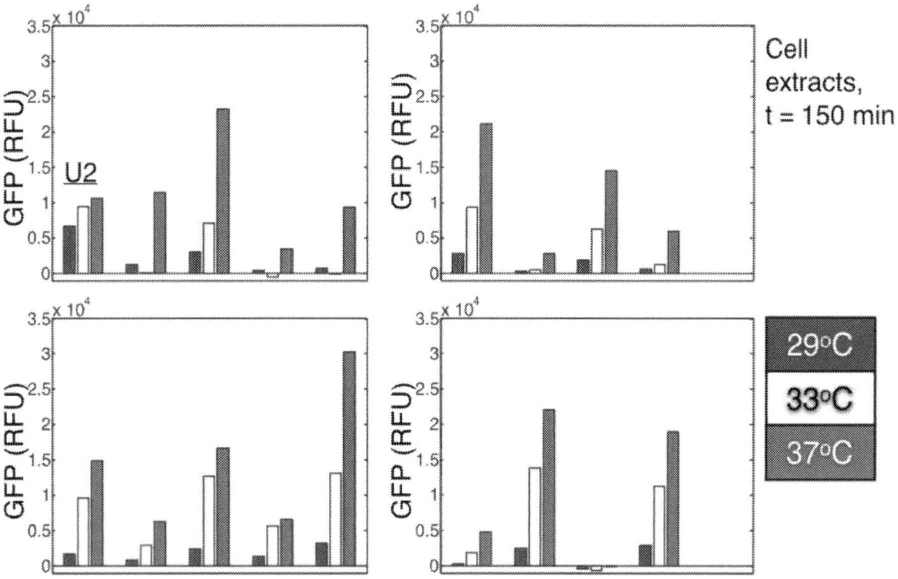

Summary

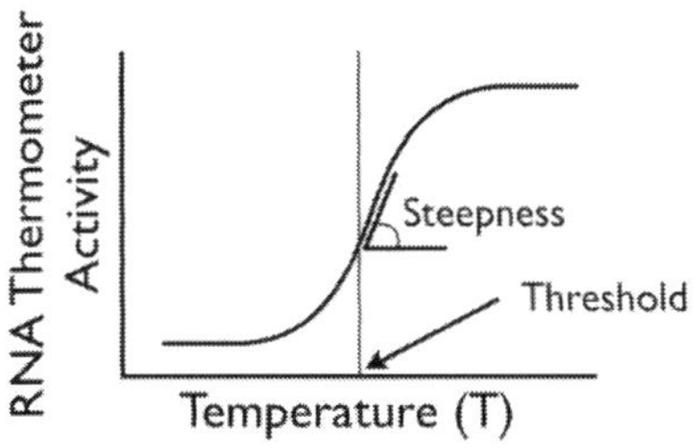

- Analyzed existing RNA thermometers, experimentally and computationally
- Designed RNA thermometer library, variety of temperature responses possible

- Measurements of library suggest existence of variety of temperature responses

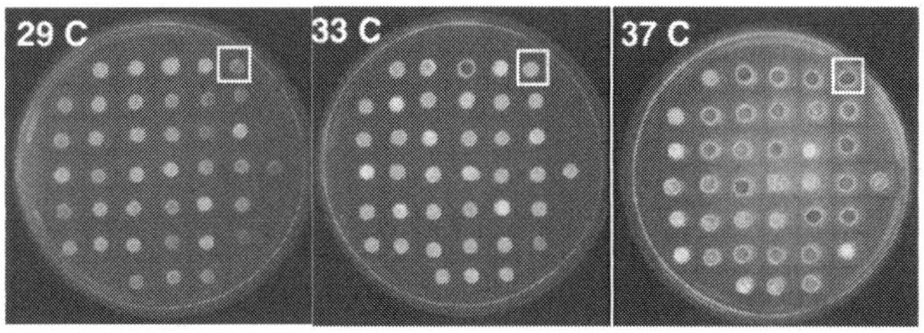

Future Work

- Characterisation of RNA thermometer activities for different temperatures (cells & cell extracts)
- Assessment of match between computations and measurements
- Combine RNA thermometers with other circuits

Example: Replace *u* in the Feedforward Loop with Temperature

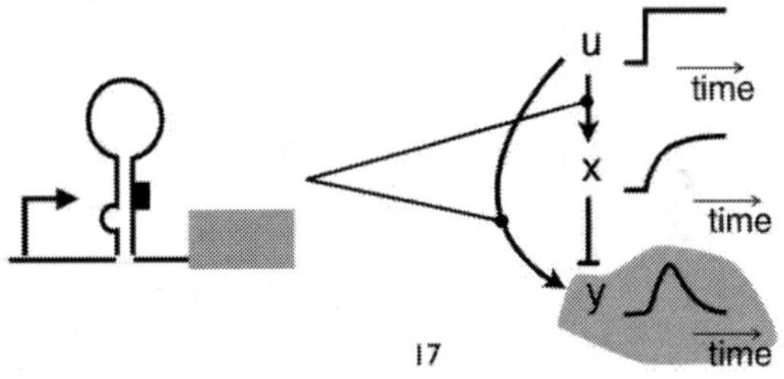

Acknowledgements

Rohit Satija
Divyansh Apurva
Dan Siegal-Gaskins
Richard M. Murray

Jongmin Kim
Harry Choi
Clare Hayes
Anu Thubagere

Thank you

Funding
FRTA, IIT Delhi
Department of Science and Technology, India
Gordon and Betty Moore Foundation, Caltech
NSF Molecular Programming Project, Caltech

Dynamic RNA Nanotechnology

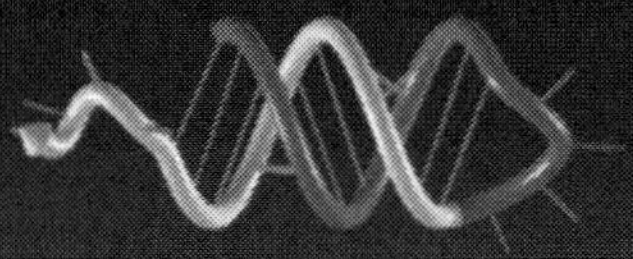

Niles A. Pierce
Biology & Biological Engineering
Engineering & Applied Science
Caltech

SEED
June 12, 2015
Boston

Life is orchestrated by programmable biomolecules

DNA GTGGTACAGGTGAATTTGGGTAGGCTAAATTGTCCATA
RNA GGGCUGUUUUUCUCGCUGACUUUCAGCCCCAAACAAAA
Protein MTYRLELNGKTLKGRTRTEAVDAATAEKVFKQYANDNG

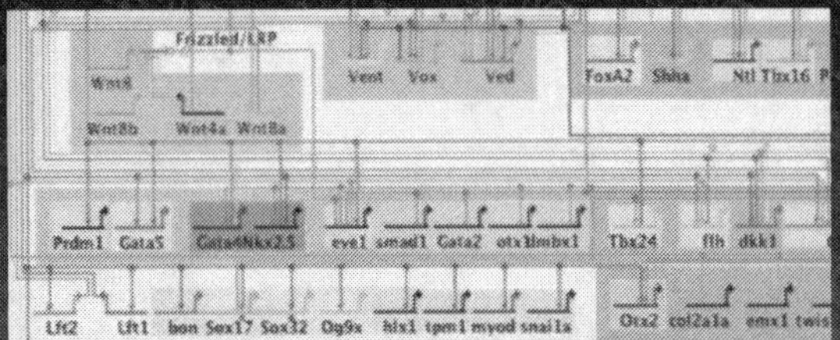

Image: Davidson Lab

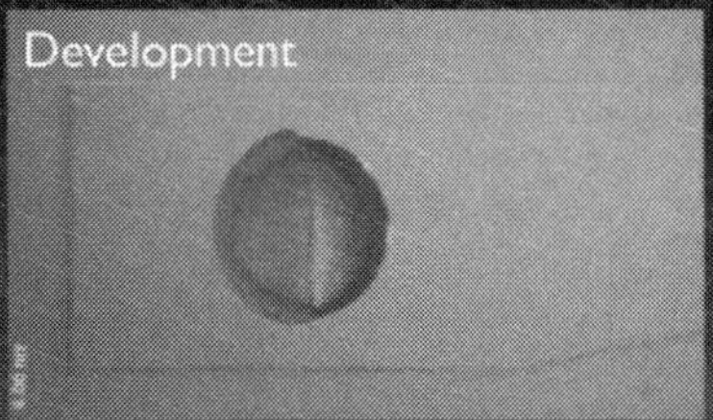

Movie: Megason Lab

Sequencing a genome provides a parts list but no manual

It remains challenging to interrogate and perturb the state of endogenous biological circuitry within intact organisms

Programmable Conditional Regulation

Exploit the very programmability that biological organisms exploit themselves

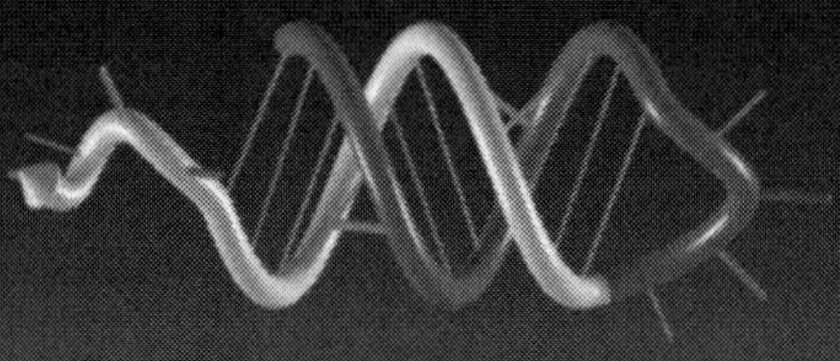

Engineer small conditional RNAs (scRNAs) that interact and change conformation to regulate the state of endogenous biological circuitry

Why Nucleic Acids?

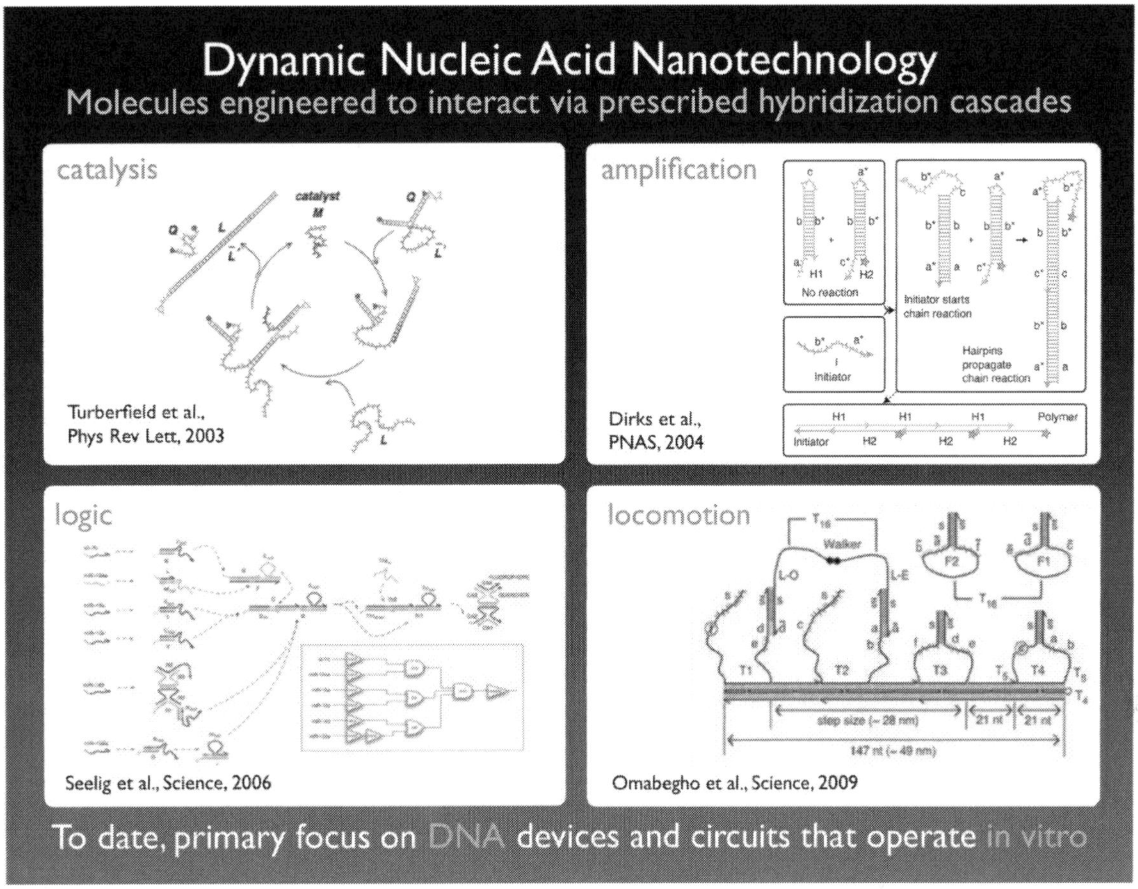
Dynamic Nucleic Acid Nanotechnology
Molecules engineered to interact via prescribed hybridization cascades
catalysis
Turberfield et al.,
Phys Rev Lett, 2003
amplification
Dirks et al.,
PNAS, 2004
logic
Seelig et al., Science, 2006
locomotion
Omabegho et al., Science, 2009
To date, primary focus on DNA devices and circuits that operate in vitro

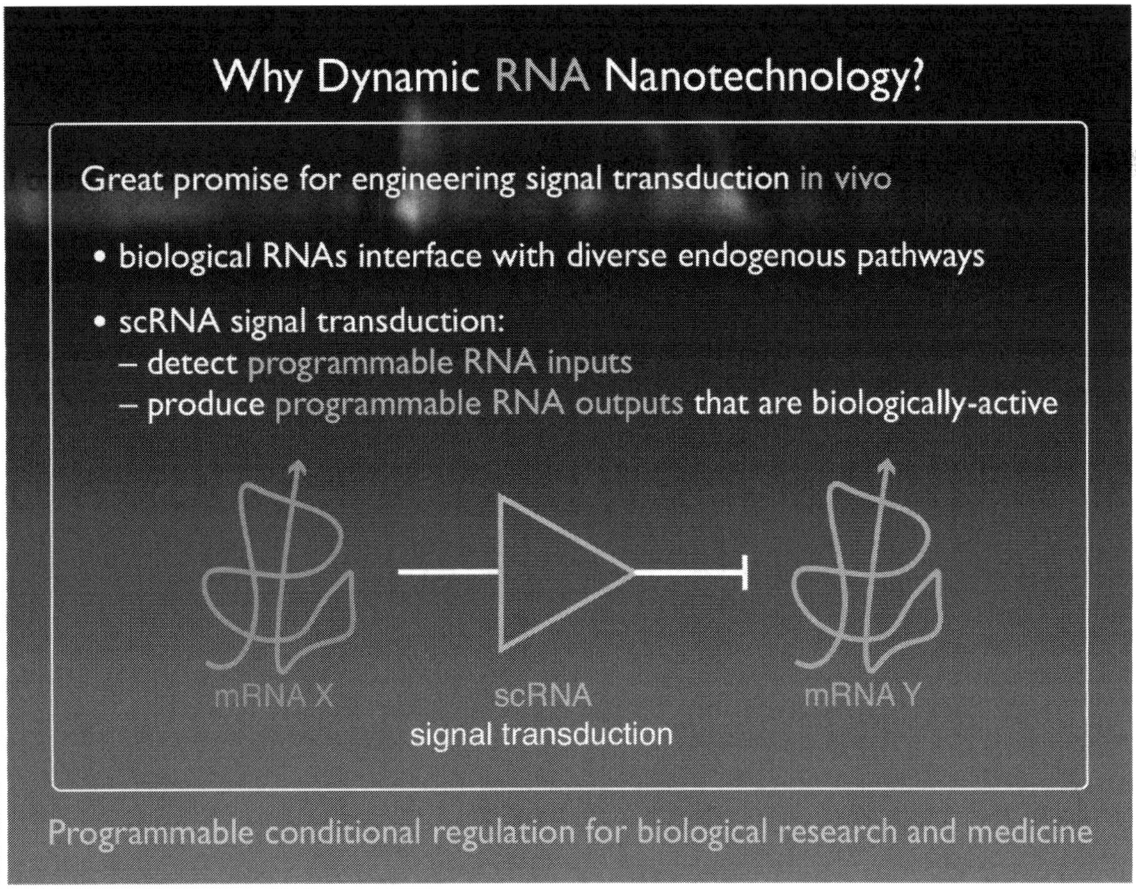
Why Dynamic RNA Nanotechnology?
Great promise for engineering signal transduction in vivo
• biological RNAs interface with diverse endogenous pathways
• scRNA signal transduction:
– detect programmable RNA inputs
– produce programmable RNA outputs that are biologically-active
mRNA X
scRNA
signal transduction
mRNA Y
Programmable conditional regulation for biological research and medicine

Nature: RNA Interference (RNAi)

Molecular logic: silence gene Y

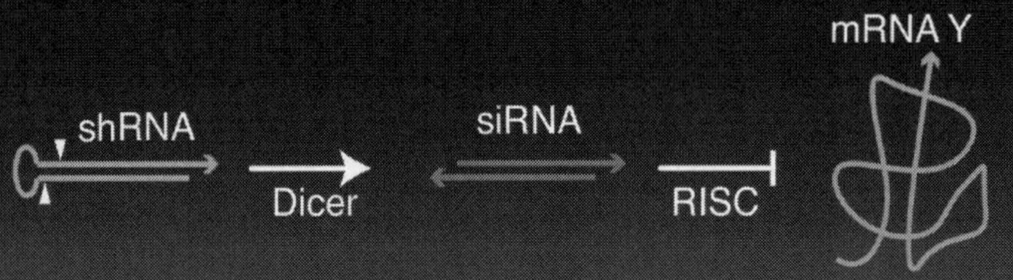

2006 Nobel Prize – Andrew Fire and Craig Mello

Good: program siRNA sequence to target gene of choice

Bad: siRNA constitutively active – difficult to confine silencing to specific locus and time

Goal: Conditional RNAi

Molecular logic: If gene X is transcribed, silence independent gene Y

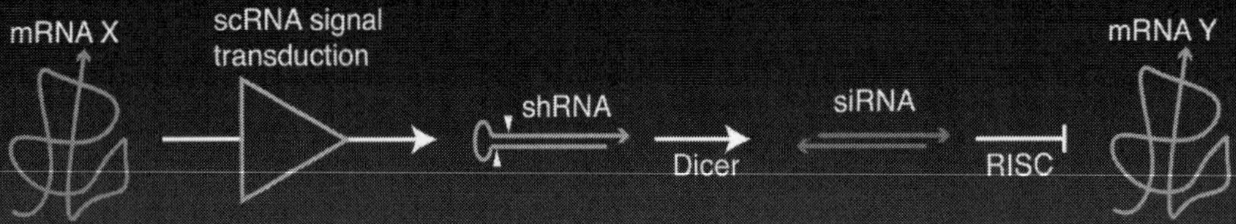

Programmable at two levels:

- sequence X controls scope of silencing
- sequence Y controls target of silencing

scRNA signal transduction:

- conditional shape change
- conditional sequence change

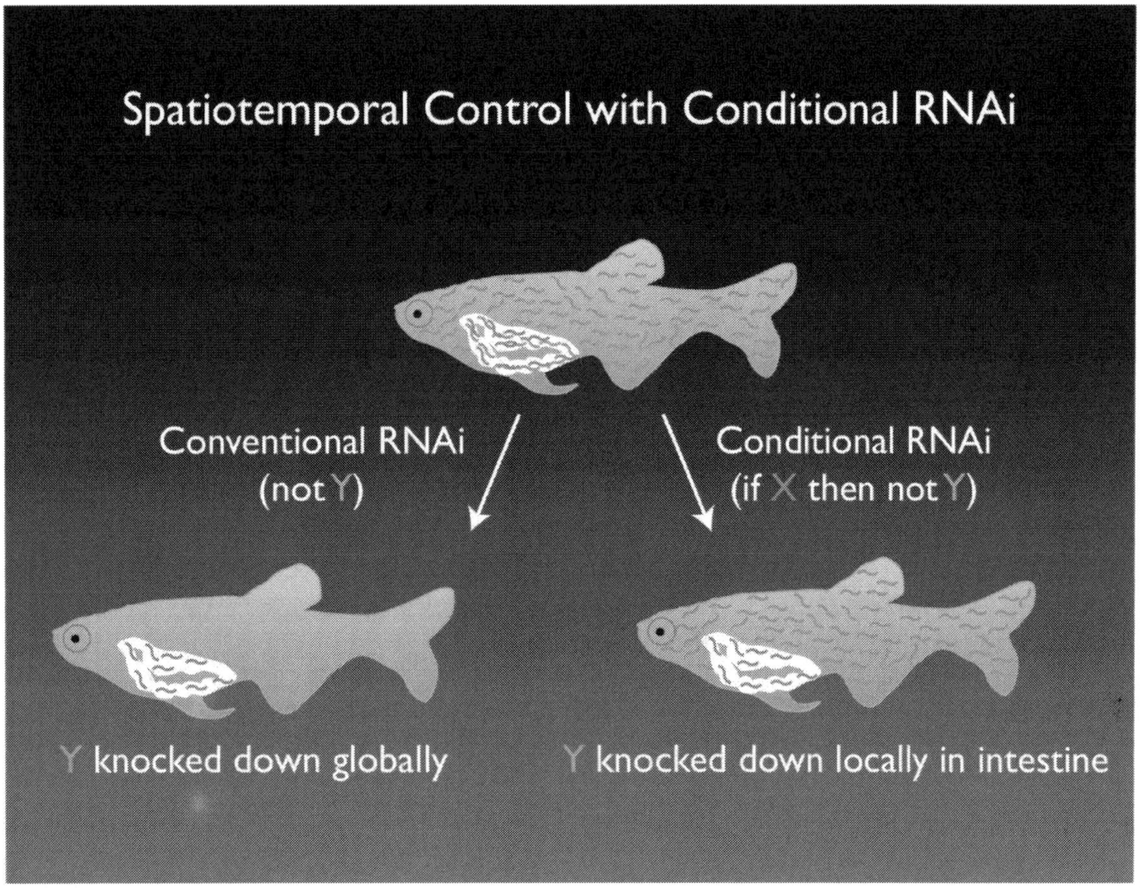
Spatiotemporal Control with Conditional RNAi
Conventional RNAi
(not Y)
Conditional RNAi
(if X then not Y)
Y knocked down globally
Y knocked down locally in intestine

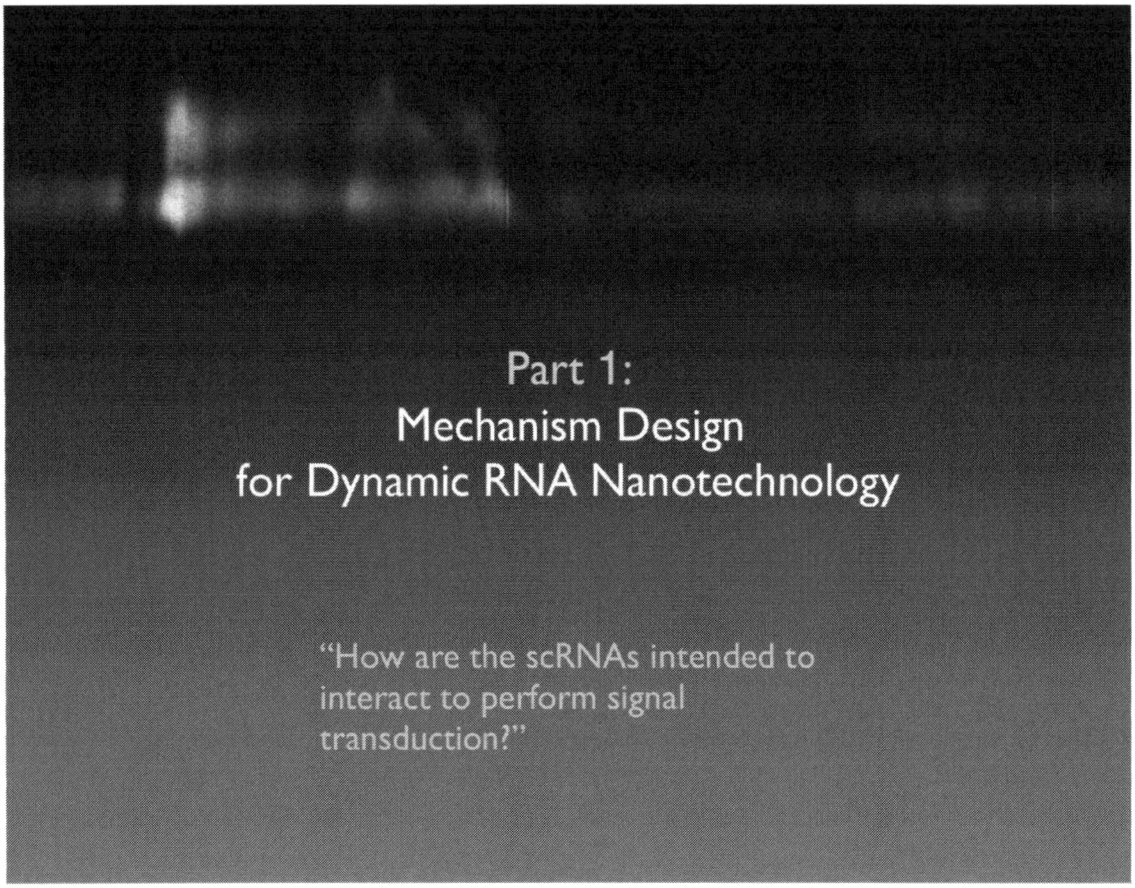
Part 1:
Mechanism Design
for Dynamic RNA Nanotechnology
"How are the scRNAs intended to interact to perform signal transduction?"

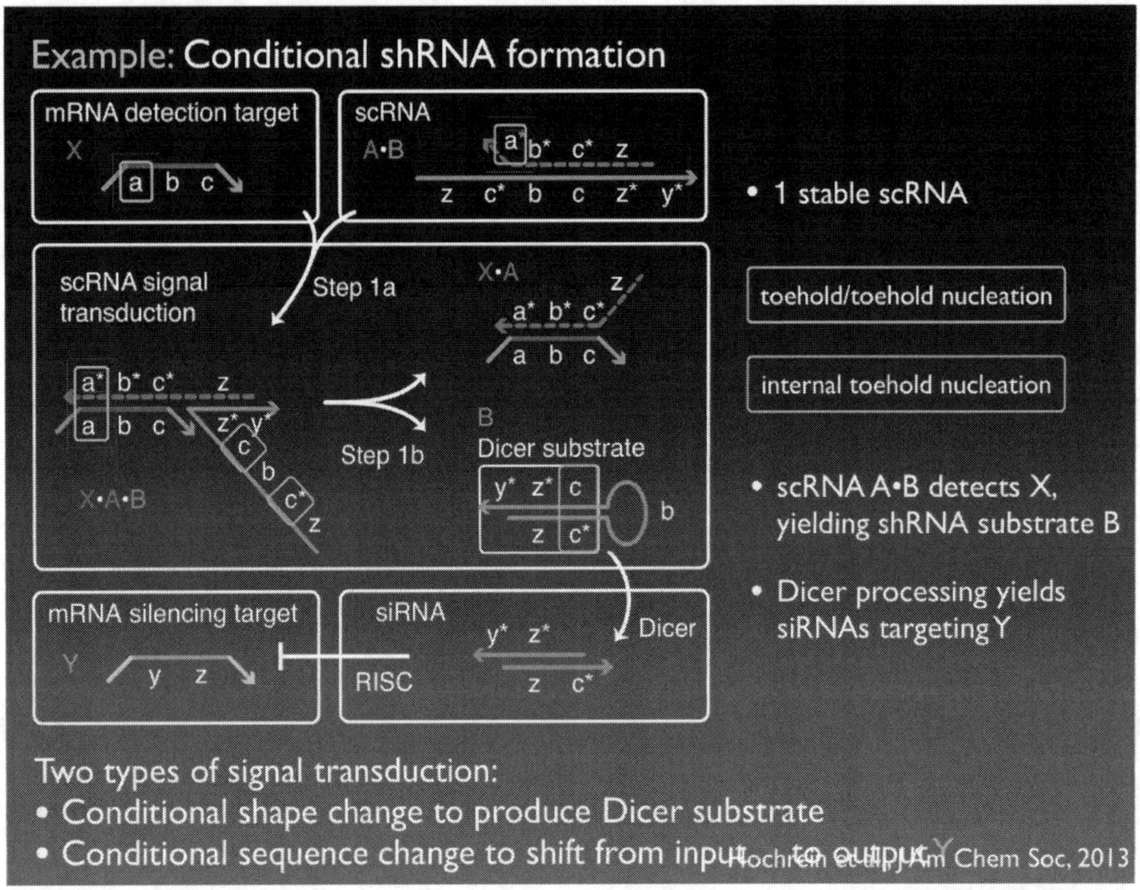
Example: Conditional shRNA formation
mRNA detection target
X
a b c
scRNA
A•B
a* b* c* z
z c* b c z* y*
scRNA signal transduction
Step 1a
X•A
z
a* b* c*
a b c
a* b* c* z
a b c z* y*
c
b
c*
z
X•A•B
Step 1b
B
Dicer substrate
y* z* c
z c*
b
Dicer
mRNA silencing target
Y
y z
siRNA
RISC
y* z*
z c*
• 1 stable scRNA
toehold/toehold nucleation
internal toehold nucleation
• scRNA A•B detects X, yielding shRNA substrate B
• Dicer processing yields siRNAs targeting Y
Two types of signal transduction:
• Conditional shape change to produce Dicer substrate
• Conditional sequence change to shift from input X to output Y
Hochrein et al., J Am Chem Soc, 2013

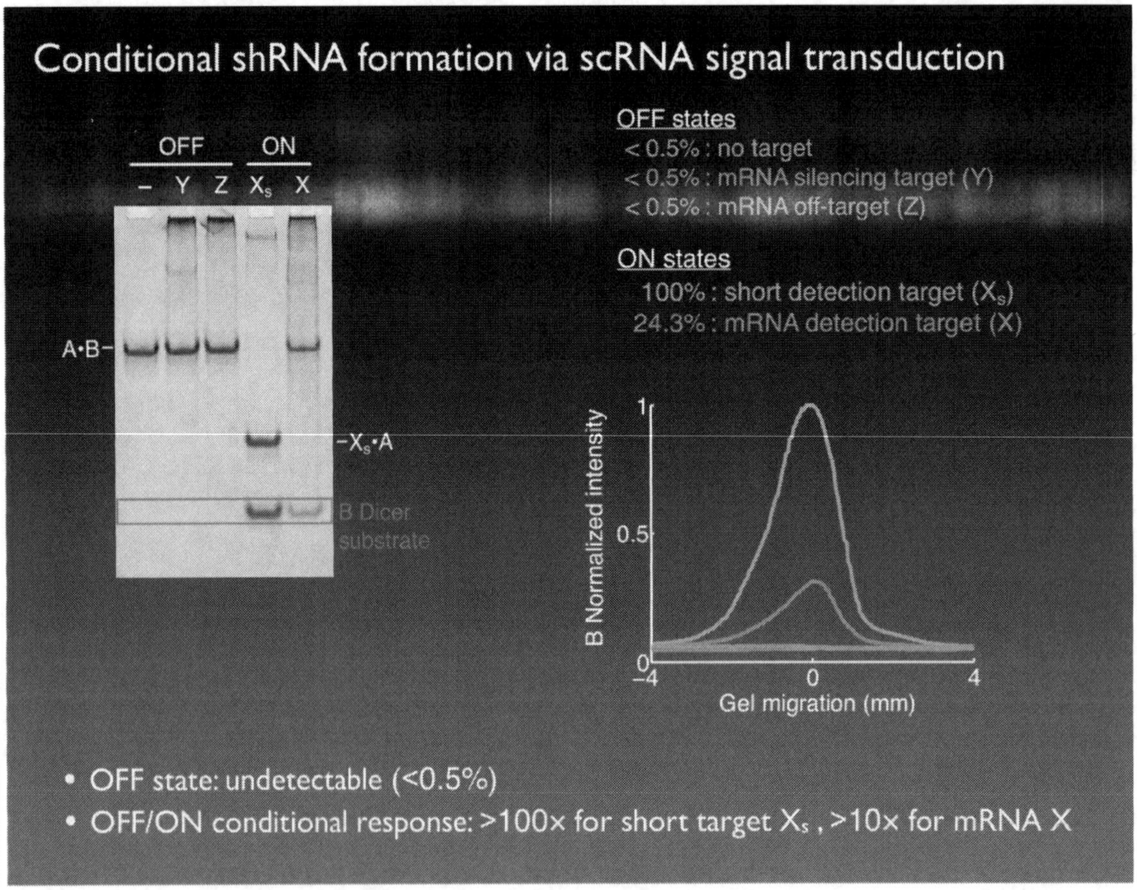
Conditional shRNA formation via scRNA signal transduction
OFF
ON
– Y Z Xs X
A•B–
–Xs•A
B Dicer substrate
OFF states
< 0.5% : no target
< 0.5% : mRNA silencing target (Y)
< 0.5% : mRNA off-target (Z)
ON states
100% : short detection target (Xs)
24.3% : mRNA detection target (X)
B Normalized intensity
1
0.5
0
–4
0
4
Gel migration (mm)
• OFF state: undetectable (<0.5%)
• OFF/ON conditional response: >100× for short target Xs , >10× for mRNA X

Conditional siRNA production via scRNA signal transduction

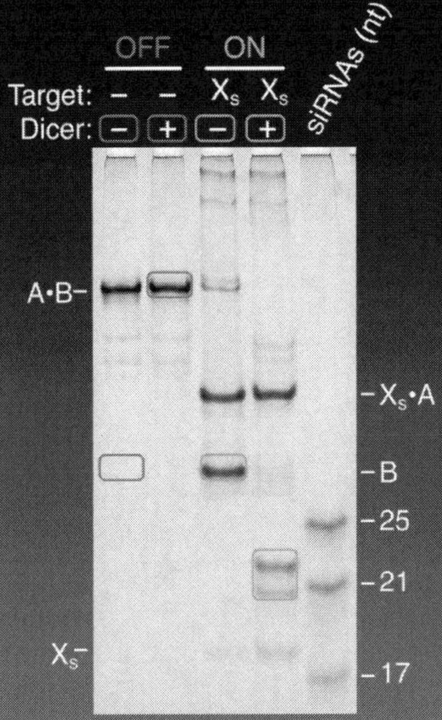

OFF state

- No detectable 'leakage' into substrate B
- No detectable Dicer processing of scRNA A•B

ON state

- Efficient formation of substrate B
- Efficient Dicer processing of B, yielding canonical 21- and 22-nt siRNAs (determined by mass spectrometry)

Hochrein et al., J Am Chem Soc, 2013

Design Elements for Dynamic RNA Nanotechnology

Design Alternatives	Mechanism				
	1	2	3	4	5
toehold/toehold nucleation	✓	✓	✓	✓	✓
loop/toehold nucleation		✓			
template/toehold nucleation				✓	
3-way branch migration	✓	✓	✓	✓	✓
4-way branch migration				✓	✓
spontaneous dissocation		✓		✓	
metastable reactants	✓			✓	✓
stable reactants		✓	✓		
catalytic production	✓				✓
non-catalytic production		✓	✓	✓	
scRNAs post-transcriptional	✓	✓	✓	✓	
scDNAs pre-transcriptional					✓

Hochrein et al., JACS, 2013

Principles of dynamic DNA nanotechnology appear well-suited for dynamic RNA nanotechnology

Rich design space for scRNA signal transduction

NUPACK

Part 2:
Sequence Design
for Dynamic RNA Nanotechnology

"What scRNA sequences will encode the desired signal transduction?"

Complex Design
Subsidiary design problem 1

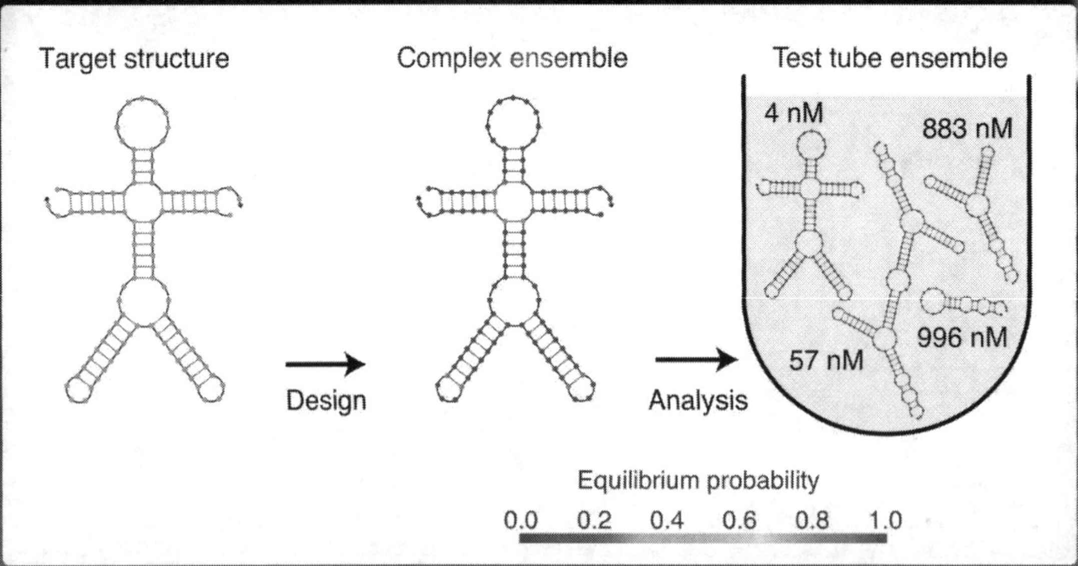

Complex design does not take into consideration:
- concentration of desired complex
- concentration of competing off-target complexes

Complex ensemble defect: average number of incorrectly paired nucleotides at equilibrium over the ensemble of the complex

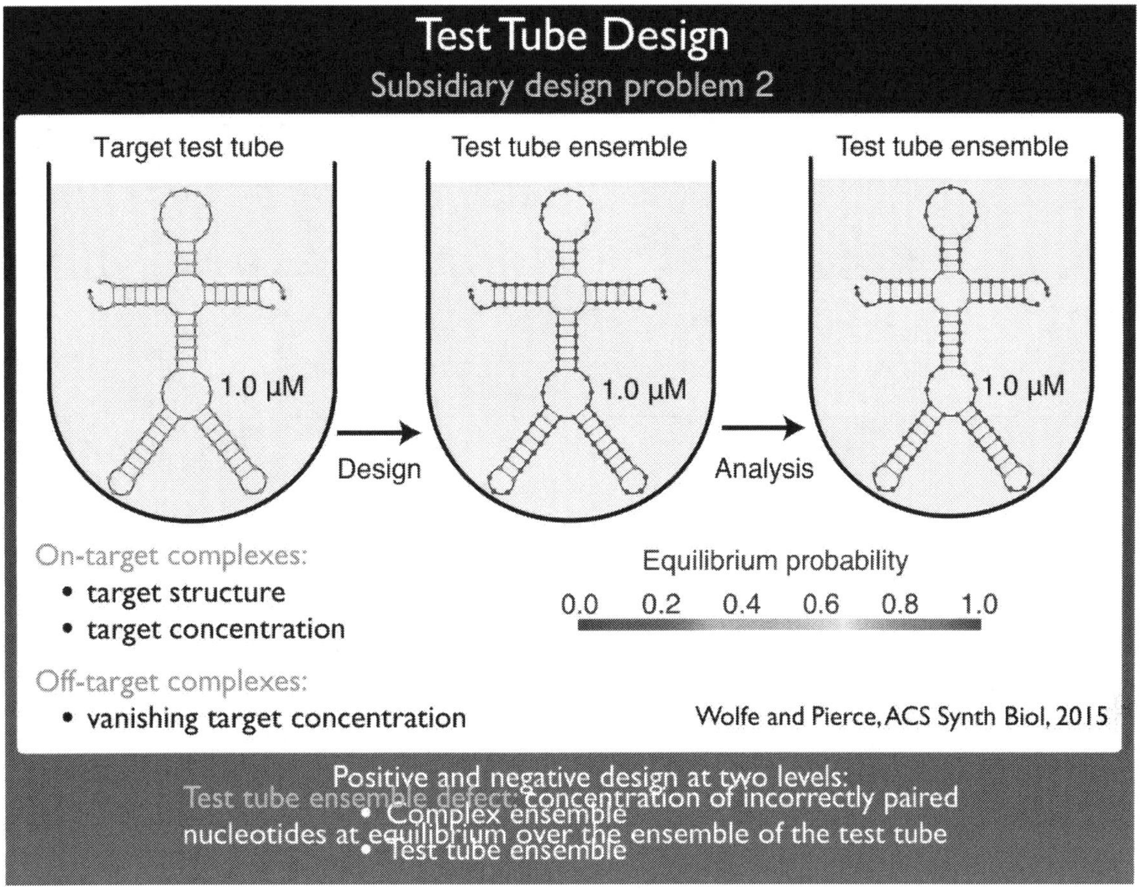
Test Tube Design
Subsidiary design problem 2
Target test tube
Test tube ensemble
Test tube ensemble
1.0 μM
1.0 μM
1.0 μM
Design
Analysis
On-target complexes:
• target structure
• target concentration
Off-target complexes:
• vanishing target concentration
Equilibrium probability
0.0 0.2 0.4 0.6 0.8 1.0
Wolfe and Pierce, ACS Synth Biol, 2015
Positive and negative design at two levels:
• Complex ensemble
• Test tube ensemble
Test tube ensemble defect: concentration of incorrectly paired nucleotides at equilibrium over the ensemble of the test tube

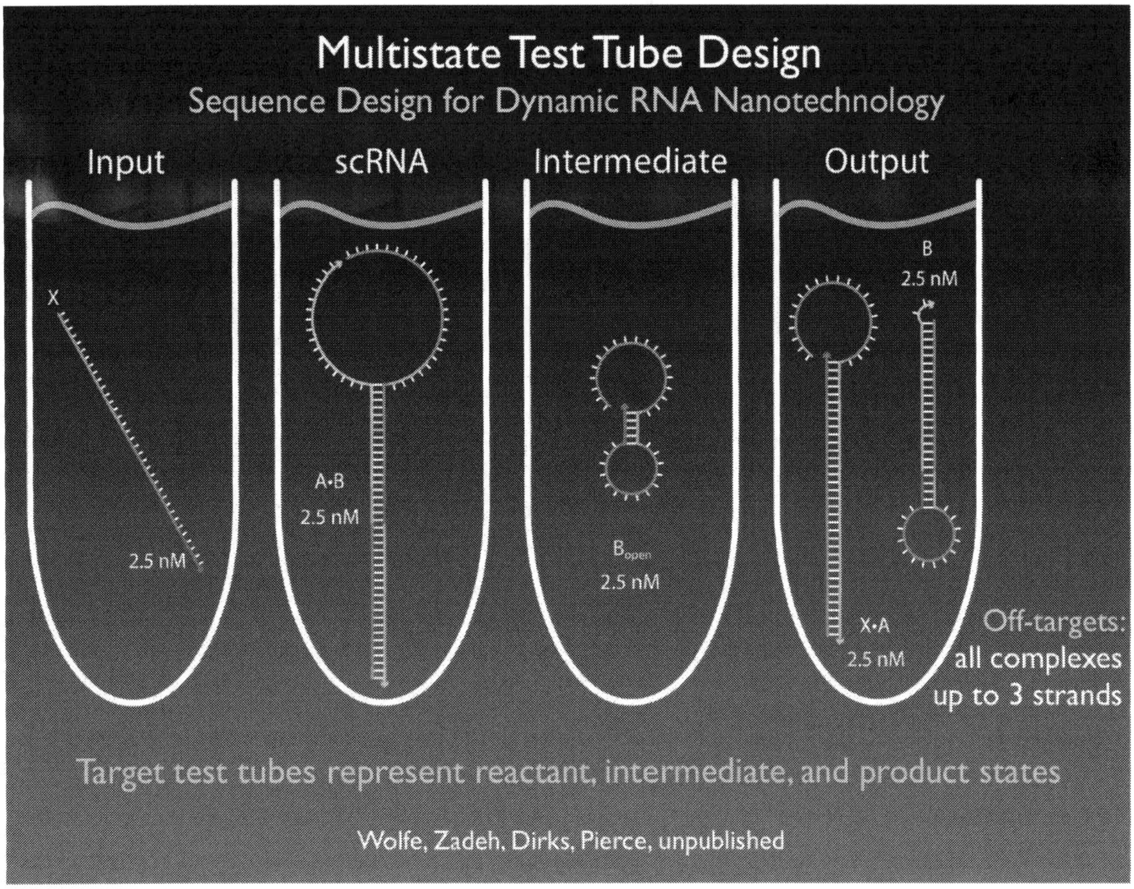
Multistate Test Tube Design
Sequence Design for Dynamic RNA Nanotechnology
Input
scRNA
Intermediate
Output
X
2.5 nM
A•B
2.5 nM
B open
2.5 nM
B
2.5 nM
X•A
2.5 nM
Off-targets:
all complexes
up to 3 strands
Target test tubes represent reactant, intermediate, and product states
Wolfe, Zadeh, Dirks, Pierce, unpublished

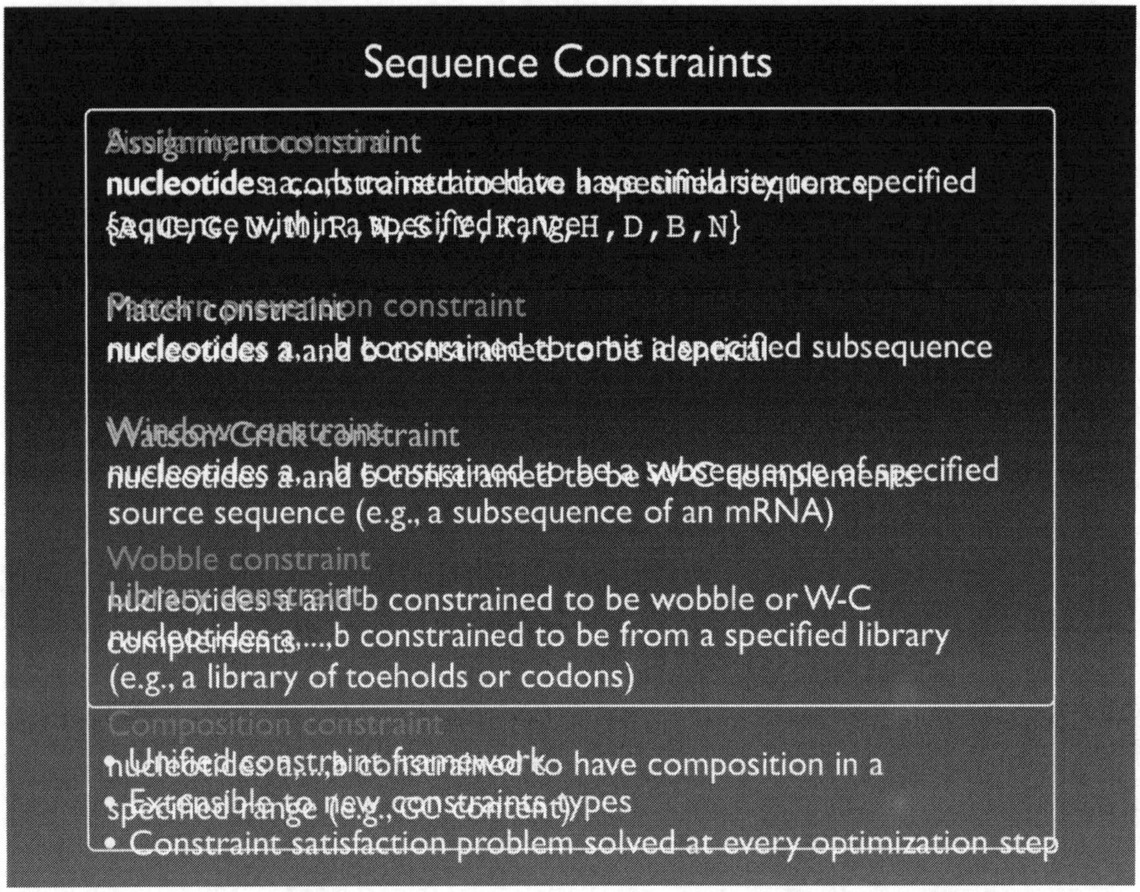
Sequence Constraints
Assignment constraint
nucleotide a constrained to have a specified value from
{A,C,G,U/T,R,Y,M,K,S,W,V,H,D,B,N}
Sequence constraint
nucleotides a,...,b constrained to have a specified sequence within a specified range
Pattern prevention constraint
nucleotides a,...,b constrained to prohibit a specified subsequence
Match constraint
nucleotides a and b constrained to be identical
Window constraint
nucleotides a,...,b constrained to be a subsequence of specified source sequence (e.g., a subsequence of an mRNA)
Watson-Crick constraint
nucleotides a and b constrained to be W-C complements
Wobble constraint
nucleotides a and b constrained to be wobble or W-C complements
Library constraint
nucleotides a,...,b constrained to be from a specified library (e.g., a library of toeholds or codons)
Composition constraint
nucleotides a,...,b constrained to have composition in a specified range (e.g., GC content)
• Unified constraint framework
• Extensible to new constraints types
• Constraint satisfaction problem solved at every optimization step

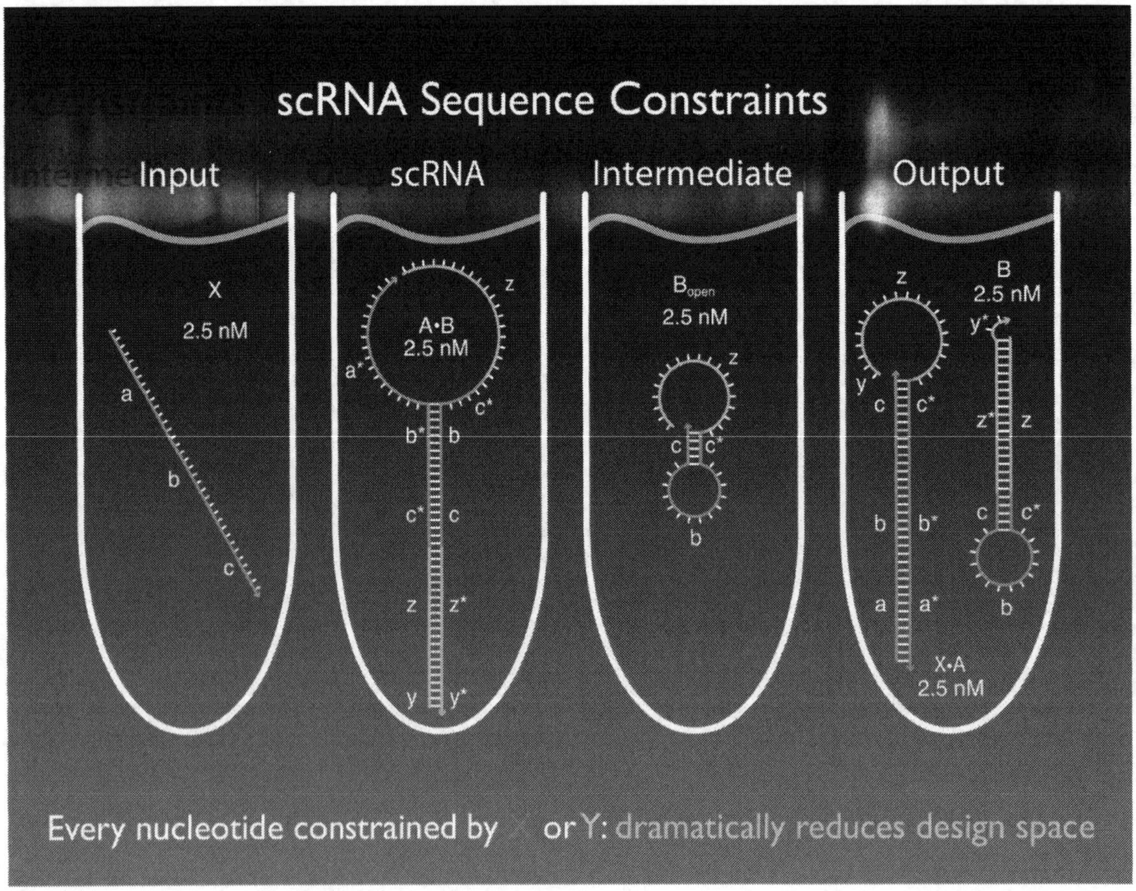
scRNA Sequence Constraints
Input
scRNA
Intermediate
Output
X
2.5 nM
a
b
c
A•B
2.5 nM
z
a*
c*
b*
b
c*
c
z
z*
y
y*
B_open
2.5 nM
z
c
c*
b
z
B
2.5 nM
y*
y
c
c*
z*
z
b
b*
c
c*
a
a*
b
X•A
2.5 nM
Every nucleotide constrained by X or Y: dramatically reduces design space

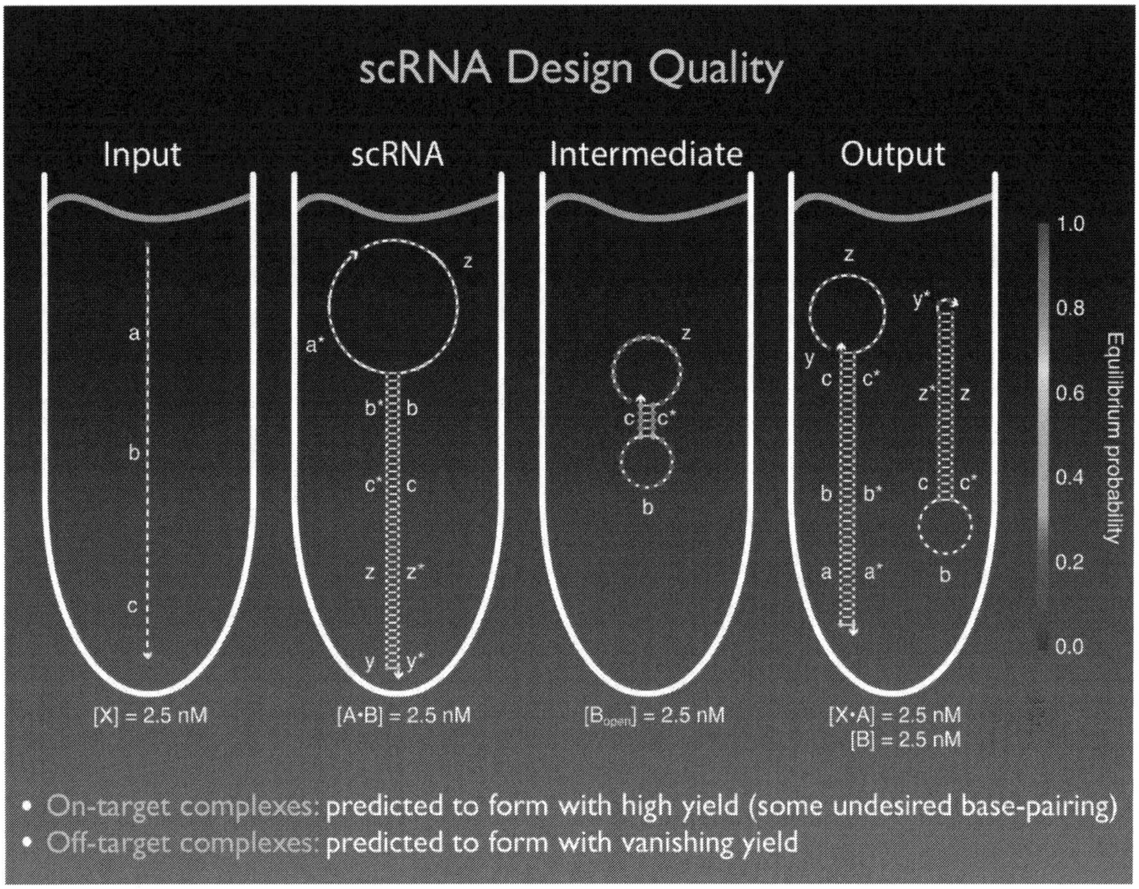
scRNA Design Quality
Input
scRNA
Intermediate
Output
a
b
c
z
a*
b* b
c* c
z z*
y y*
z
c c*
b
z
y
c c*
b b*
a a*
y*
z* z
c c*
b
1.0
0.8
0.6
0.4
0.2
0.0
Equilibrium probability
[X] = 2.5 nM
[A•B] = 2.5 nM
[B_open] = 2.5 nM
[X•A] = 2.5 nM
[B] = 2.5 nM
• On-target complexes: predicted to form with high yield (some undesired base-pairing)
• Off-target complexes: predicted to form with vanishing yield

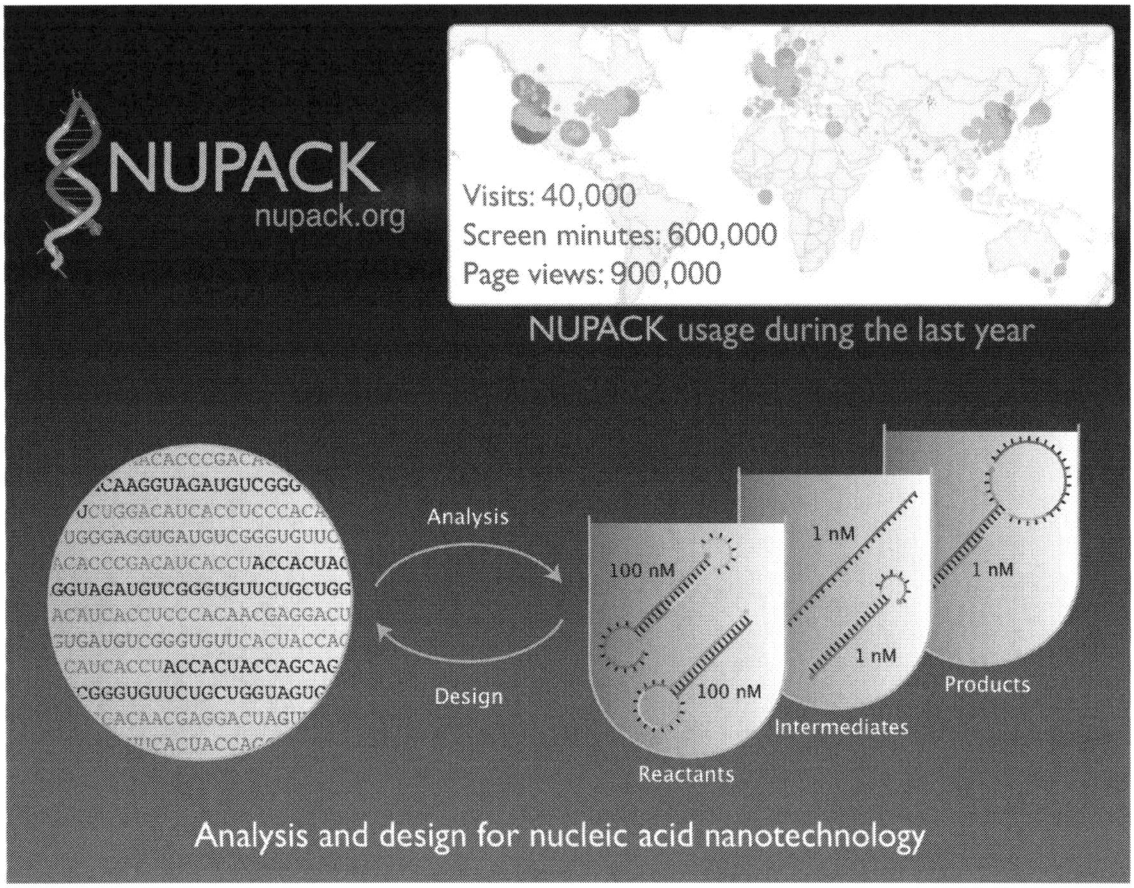
NUPACK
nupack.org
Visits: 40,000
Screen minutes: 600,000
Page views: 900,000
NUPACK usage during the last year
Analysis
Design
100 nM
100 nM
1 nM
1 nM
1 nM
Reactants
Intermediates
Products
Analysis and design for nucleic acid nanotechnology

Part 3:
Challenges and Opportunities
for Dynamic RNA Nanotechnology In Vivo

"Why is this hard? Why is it worth the effort?"

Practical Challenges

Will nature cooperate with dynamic RNA nanotechnology in vivo?

- native mRNA structure is difficult to predict
- sequestration and degradation of scRNAs by proteins
- off-target interactions with the transcriptome
- off-pathway interactions (including with undiscovered pathways)
- delivery of scRNAs into living cells and organisms

Mammalian cell lysate: intermediate test bed (eliminate delivery challenge)

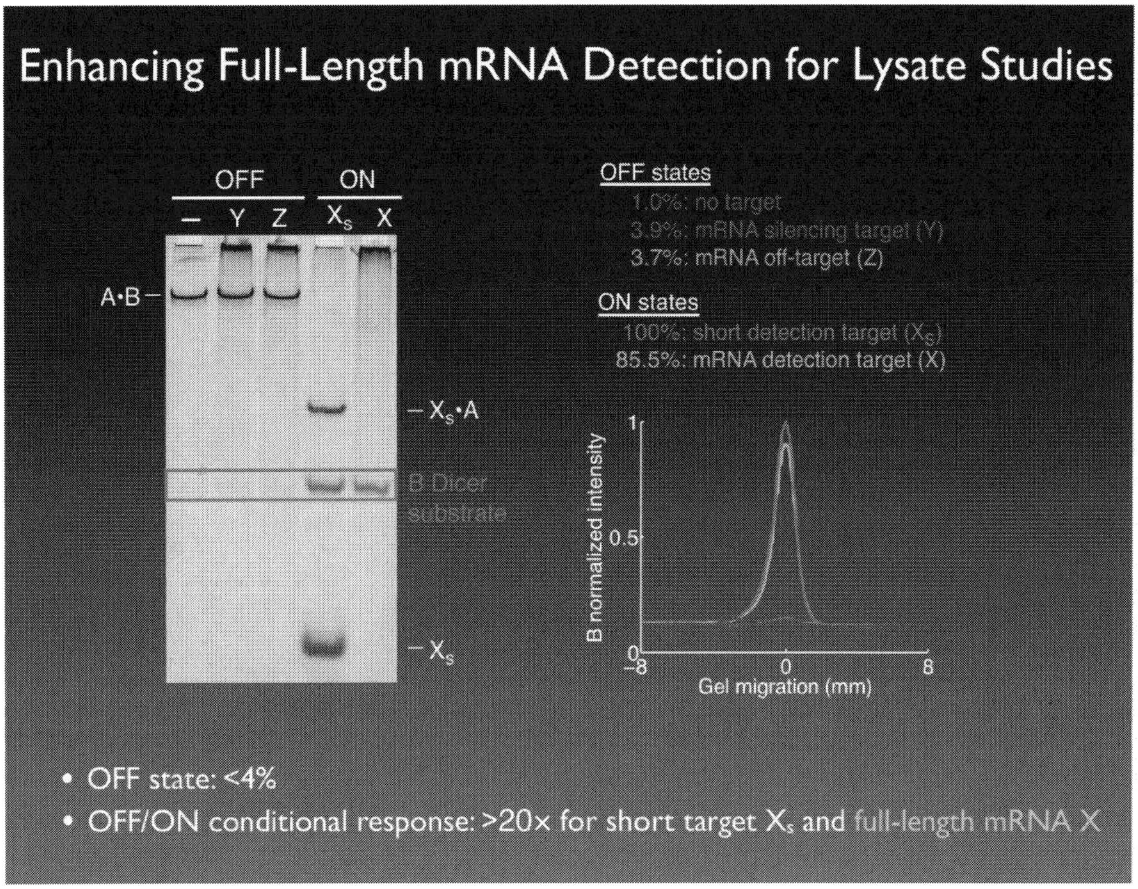
Enhancing Full-Length mRNA Detection for Lysate Studies
OFF
ON
– Y Z
X_s X
A•B –
– X_s•A
B Dicer substrate
– X_s
OFF states
1.0%: no target
3.9%: mRNA silencing target (Y)
3.7%: mRNA off-target (Z)
ON states
100%: short detection target (X_S)
85.5%: mRNA detection target (X)
B normalized intensity
1
0.5
0
–8
0
8
Gel migration (mm)
• OFF state: <4%
• OFF/ON conditional response: >20x for short target X_s and full-length mRNA X

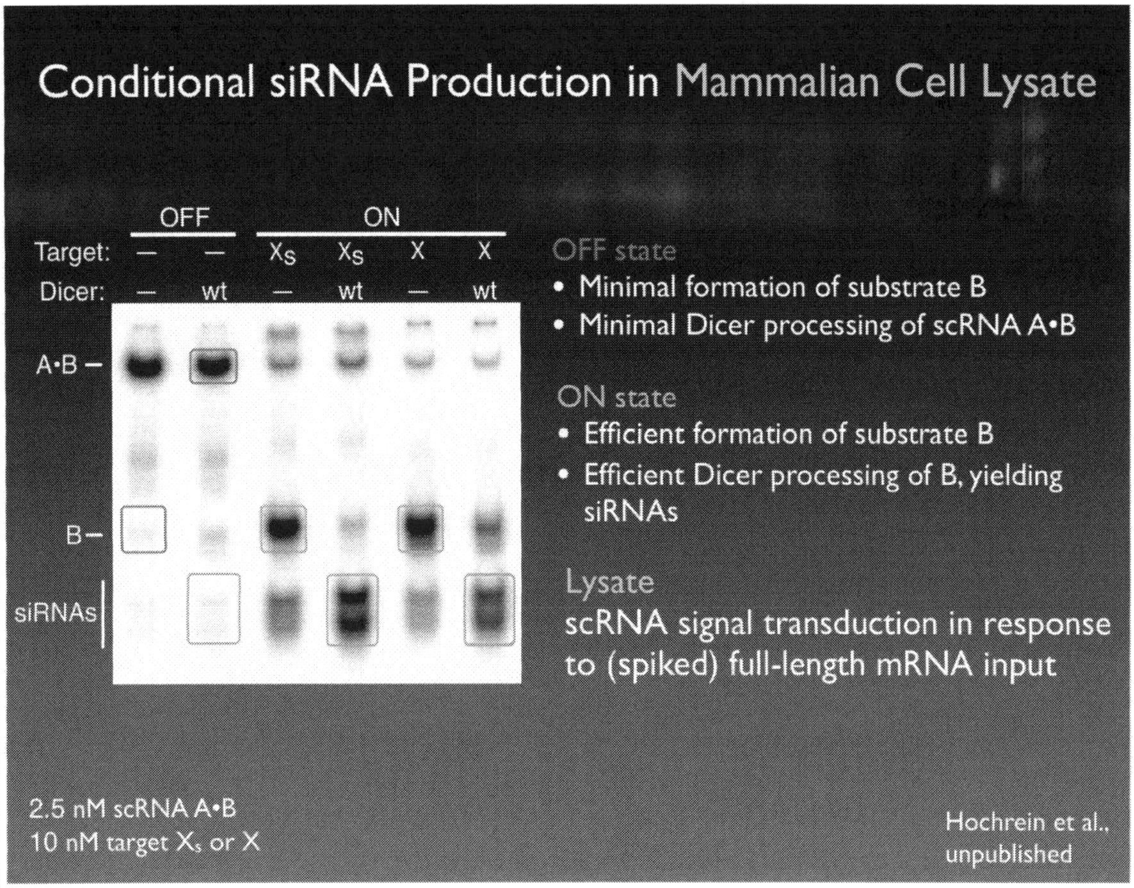
Conditional siRNA Production in Mammalian Cell Lysate
OFF
ON
Target: – – X_S X_S X X
Dicer: – wt – wt – wt
A•B –
B –
siRNAs
OFF state
• Minimal formation of substrate B
• Minimal Dicer processing of scRNA A•B
ON state
• Efficient formation of substrate B
• Efficient Dicer processing of B, yielding siRNAs
Lysate
scRNA signal transduction in response to (spiked) full-length mRNA input
2.5 nM scRNA A•B
10 nM target X_s or X
Hochrein et al., unpublished

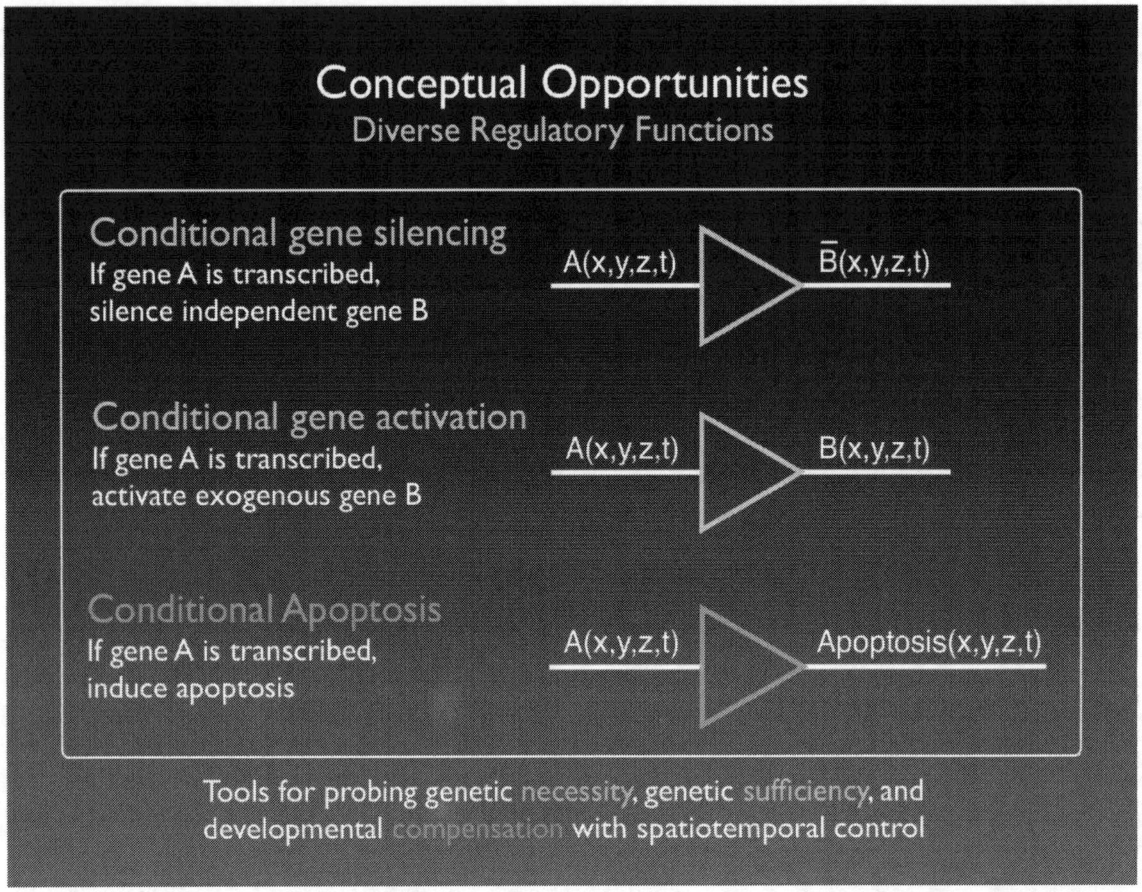
Conceptual Opportunities
Diverse Regulatory Functions
Conditional gene silencing
If gene A is transcribed, silence independent gene B
A(x,y,z,t)
B̄(x,y,z,t)
Conditional gene activation
If gene A is transcribed, activate exogenous gene B
A(x,y,z,t)
B(x,y,z,t)
Conditional Apoptosis
If gene A is transcribed, induce apoptosis
A(x,y,z,t)
Apoptosis(x,y,z,t)
Tools for probing genetic necessity, genetic sufficiency, and developmental compensation with spatiotemporal control

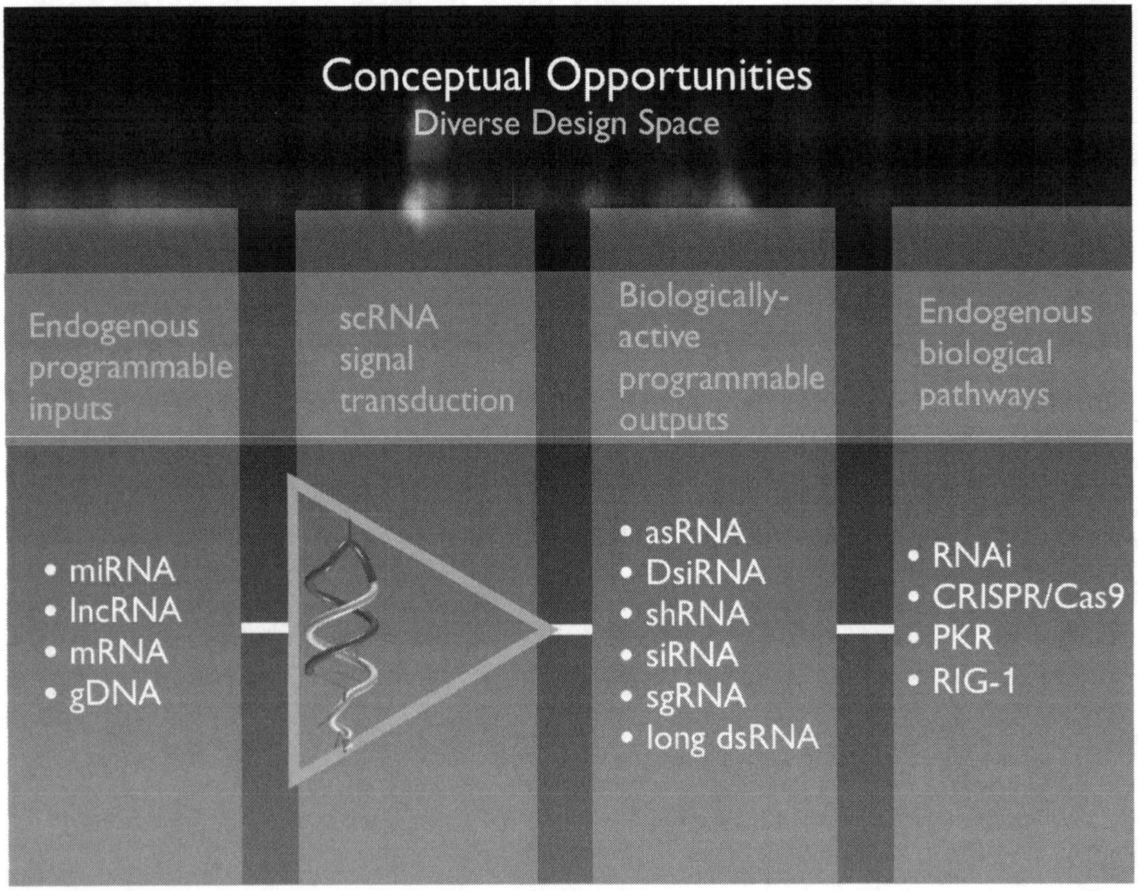
Conceptual Opportunities
Diverse Design Space
Endogenous programmable inputs
scRNA signal transduction
Biologically-active programmable outputs
Endogenous biological pathways
• miRNA
• lncRNA
• mRNA
• gDNA
• asRNA
• DsiRNA
• shRNA
• siRNA
• sgRNA
• long dsRNA
• RNAi
• CRISPR/Cas9
• PKR
• RIG-1

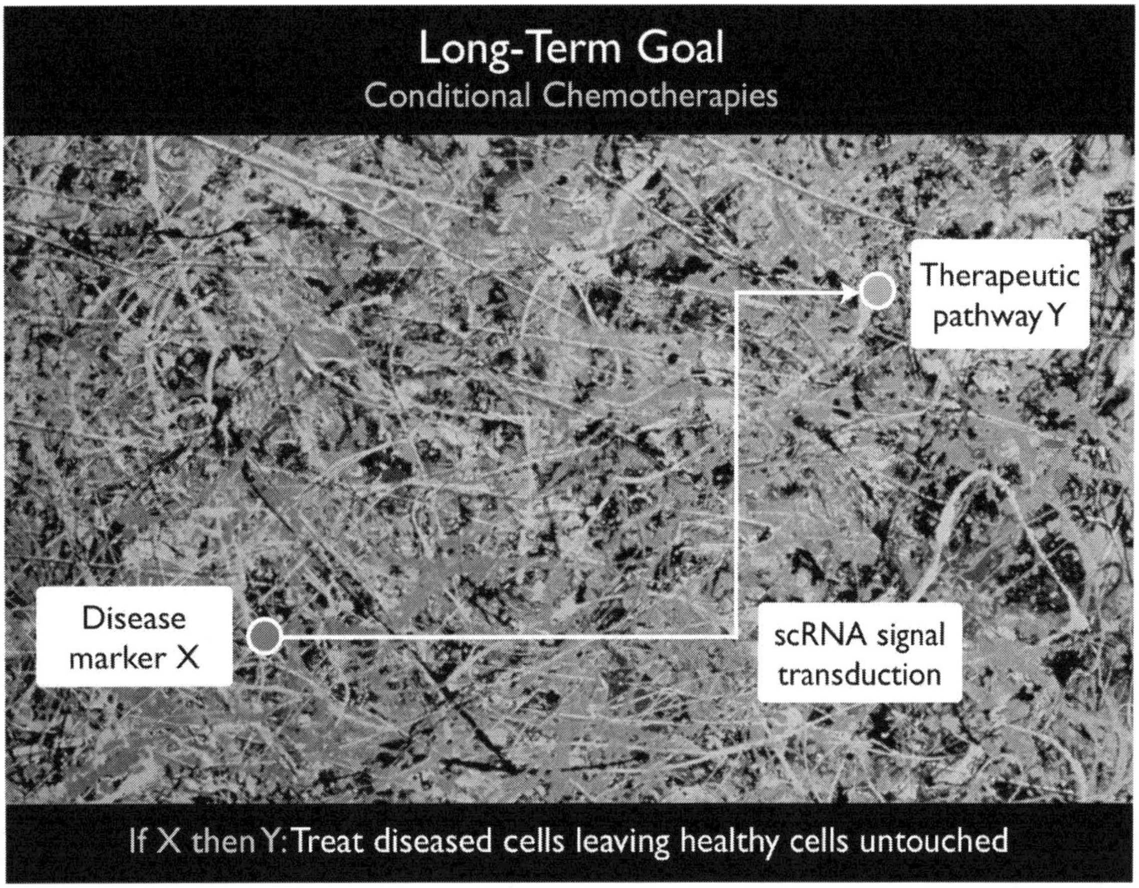
Long-Term Goal
Conditional Chemotherapies
Therapeutic pathway Y
Disease marker X
scRNA signal transduction
If X then Y: Treat diseased cells leaving healthy cells untouched

Engineering scRNA Signal Transduction
Towards Dynamic RNA Nanotechnology In Vivo
Dr. Lisa Hochrein
Dr. Maayan Schwarzkopf
Prof. Peng Yin (now at Harvard)
Mikhail Hanewich-Hollatz Graduate student
Dr. Robert Dirks (deceased)

Multistate Test Tube Design

Sequence Design for Dynamic RNA Nanotechnology

Dr. Brian Wolfe
(now at iRobot)

Nick Porubsky
Graduate student

Dr. Joseph Zadeh
(now at Airbnb)

Dr. Robert Dirks
(deceased)

Group Members

Dr. Lisa Hochrein
Dr. Maayan Schwarzkopf
Dr. Brian Wolfe
Dr. Harry Choi
Nick Porubsky
Mark Fornace
Mikhail Hanewich-Hollatz
Vikas Trivedi
Aneesh Acharya
Naeem Husain
Colby Calvert
Melinda Kirk

Funding

NIH
NSF
Moore Foundation
Beckman Institute
Pardee Foundation
Guggenheim Foundation

Collaborators

Prof. Peng Yin, Harvard
Prof. Julius Lucks, Cornell
Prof. Tatjana Sauka-Spengler, Oxford
Prof. Scott Fraser, USC
Dr. Rusty Lansford, USC
Prof. Erik Winfree, Caltech
Prof. Paul Rothemund, Caltech
Prof. Ebbe Andersen, Aarhus

Caltech

Bottom Up Construction of Dynamic Biomolecular Materials

UCR

Elisa Franco

SEED 2015

Leopold Green*, Jaimie Stewart*, Hari Subramanian*,
Vahid Mardanlou*, Jongmin Kim°, Rizal Hariadi†, Elisa Franco*
° Wyss Institute
† University of Michigan
*University of California, Riverside

Can we recapitulate the organization of cellular materials in a synthetic biomaterial?

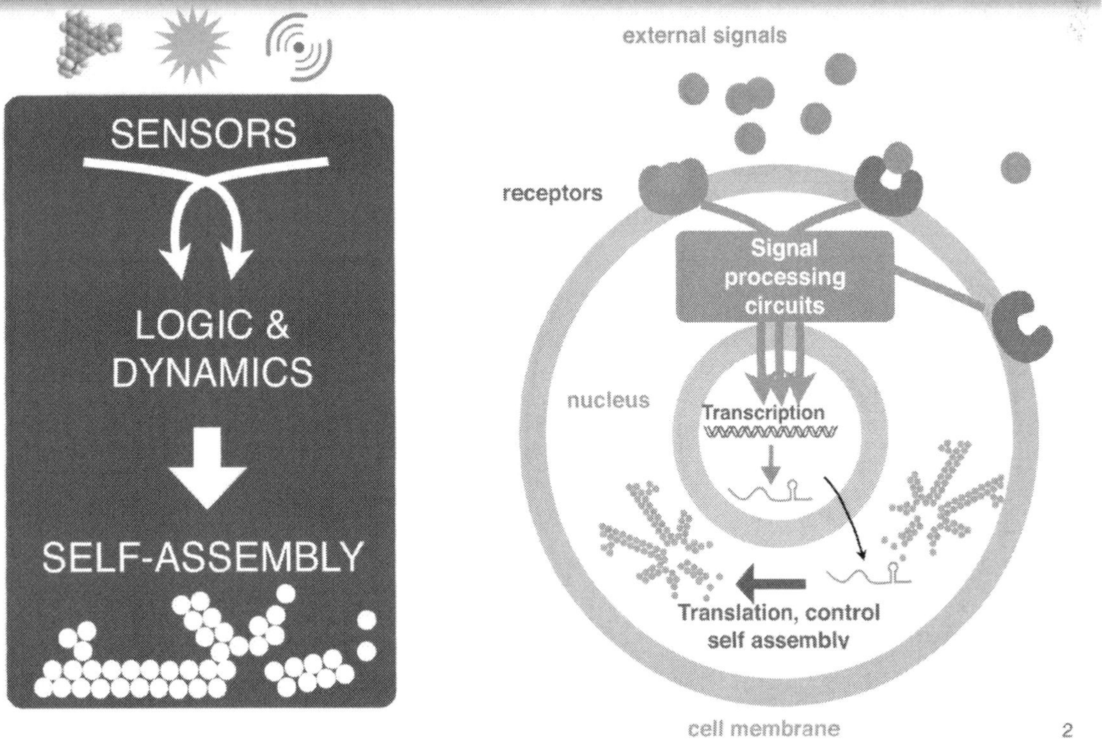

2

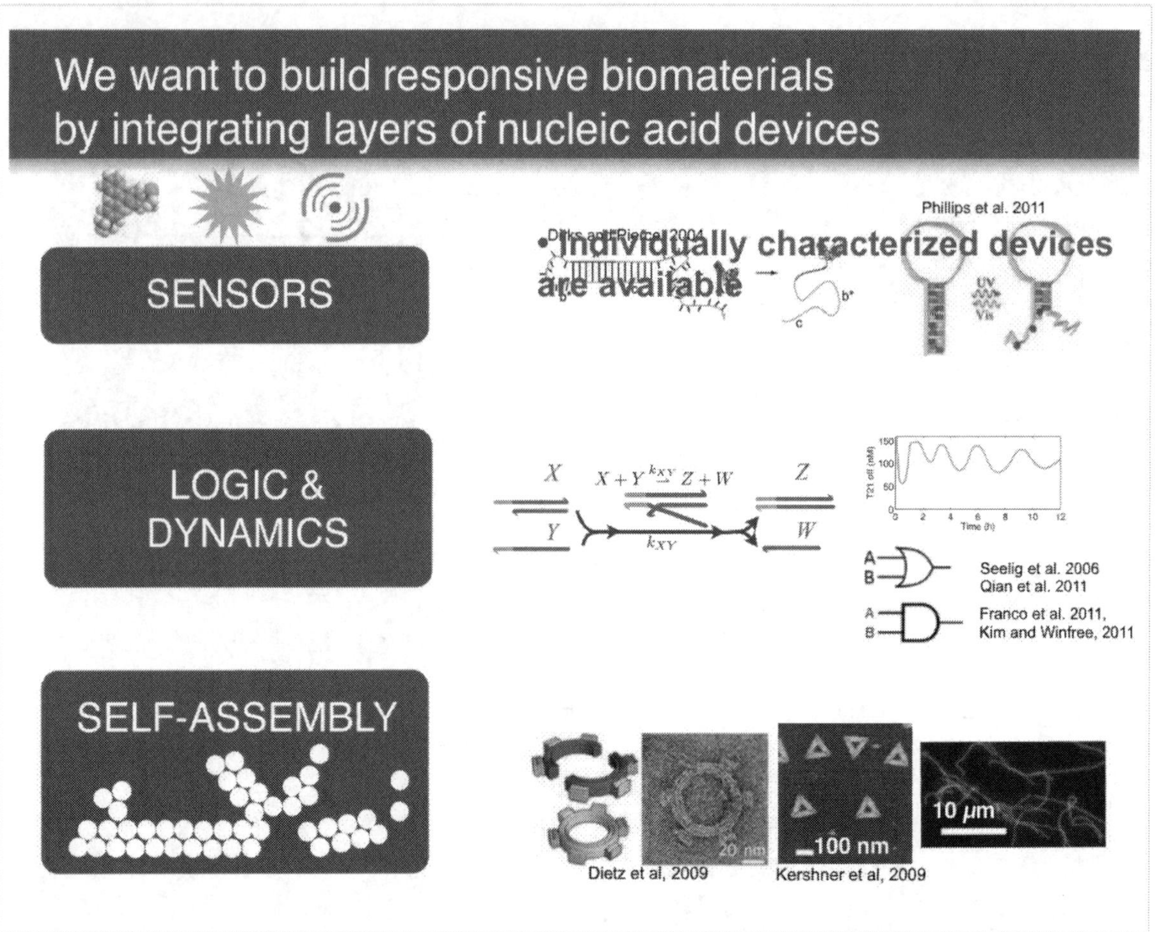
We want to build responsive biomaterials by integrating layers of nucleic acid devices
SENSORS
Phillips et al. 2011
• Individually characterized devices are available
UV
Vis
LOGIC &
DYNAMICS
X
$X + Y \xrightarrow{k_{XY}} Z + W$
Z
Y
k_{XY}
W
Time (h)
Seelig et al. 2006
Qian et al. 2011
Franco et al. 2011,
Kim and Winfree, 2011
SELF-ASSEMBLY
20 nm
100 nm
10 µm
Dietz et al, 2009
Kershner et al, 2009

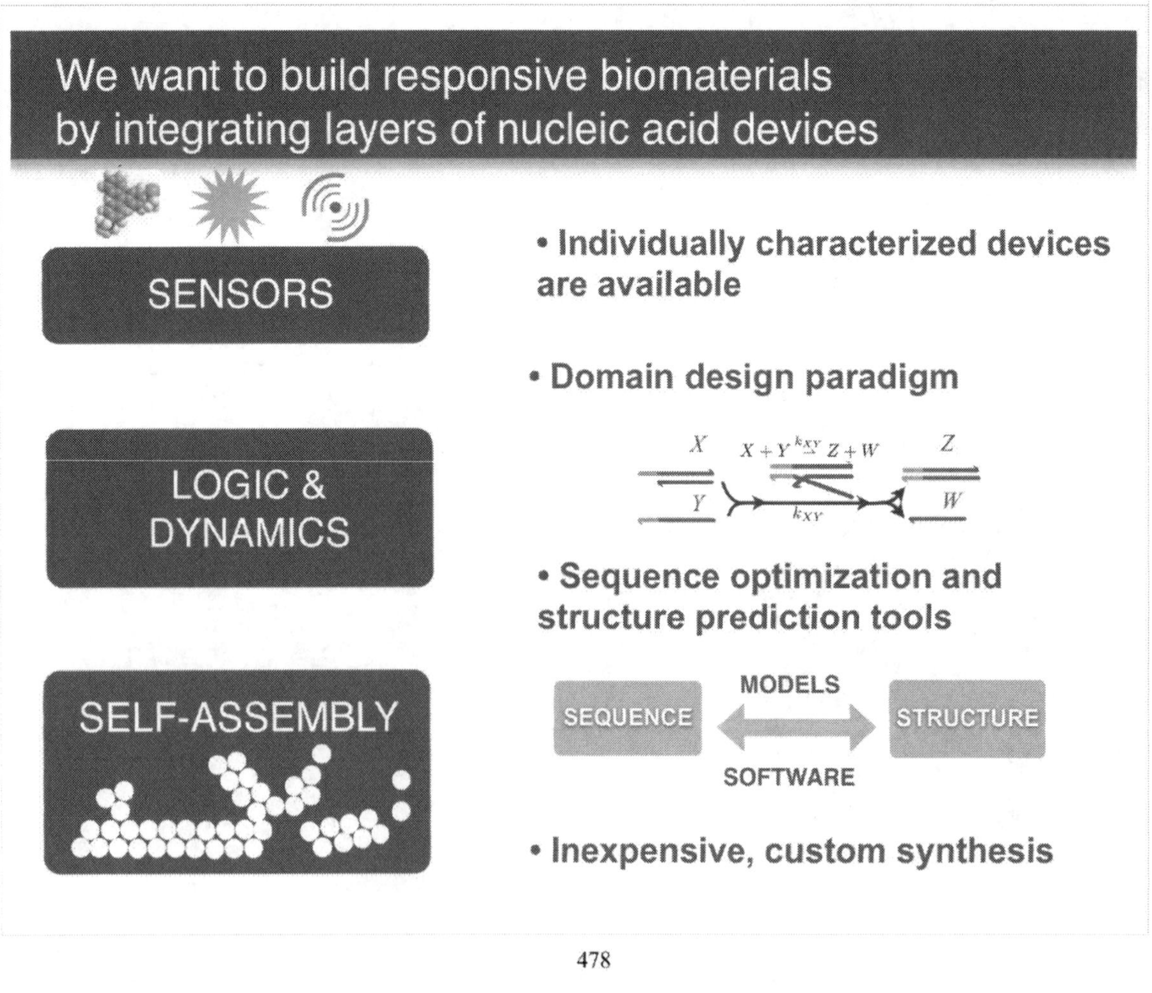
We want to build responsive biomaterials by integrating layers of nucleic acid devices
SENSORS
LOGIC &
DYNAMICS
SELF-ASSEMBLY
• Individually characterized devices are available
• Domain design paradigm
X
$X + Y \xrightarrow{k_{XY}} Z + W$
Z
Y
k_{XY}
W
• Sequence optimization and structure prediction tools
MODELS
SEQUENCE
STRUCTURE
SOFTWARE
• Inexpensive, custom synthesis

We want to build responsive biomaterials by integrating layers of nucleic acid devices

SENSORS

LOGIC & DYNAMICS

SELF-ASSEMBLY

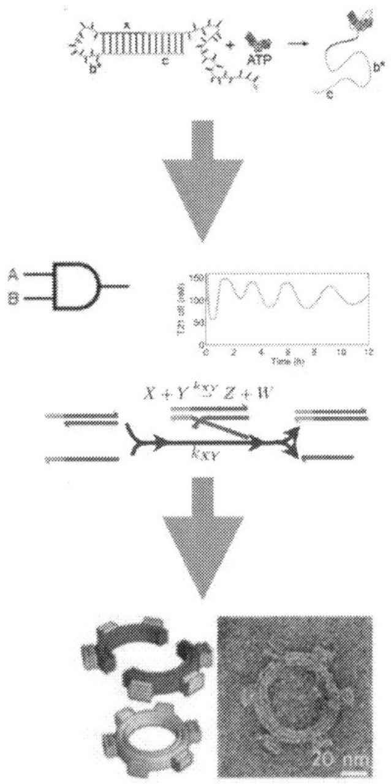

CHALLENGES

- Time scales
- Length scales
- Signal transmission
- Compatible environment

Solutions are portable to generic self-assembling, programmable matter

How do we build dynamical DNA structures?

DYNAMICS?

SELF-ASSEMBLY

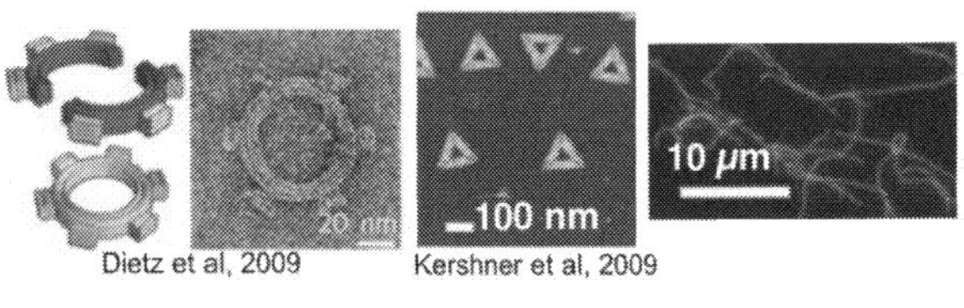

Dietz et al, 2009 Kershner et al, 2009

Dynamical tubular structures
Common motif in natural dynamic scaffolds, micron-scale

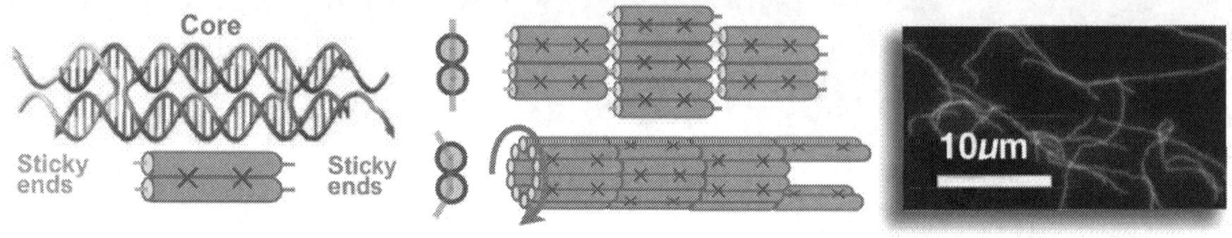

Fu and Seeman, 1993; Rothemund et al. 2004...

Add to the tiles an "access domain" for nucleic acid inputs

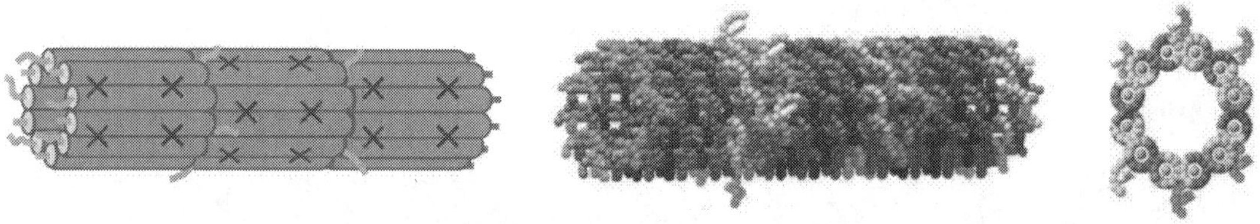

Dynamical tubular structures
Common motif in natural dynamic scaffolds, micron-scale

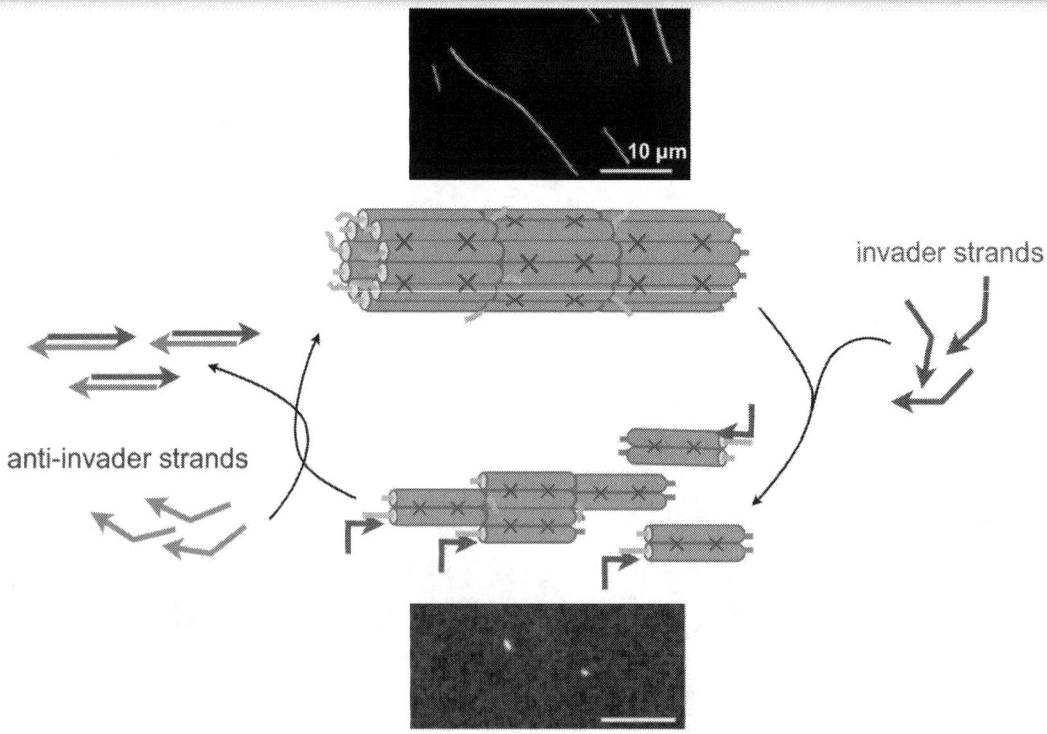

Methods: measuring nanotube length

Example: growth (end joining)

25˚C, TAE 12mM Mg++

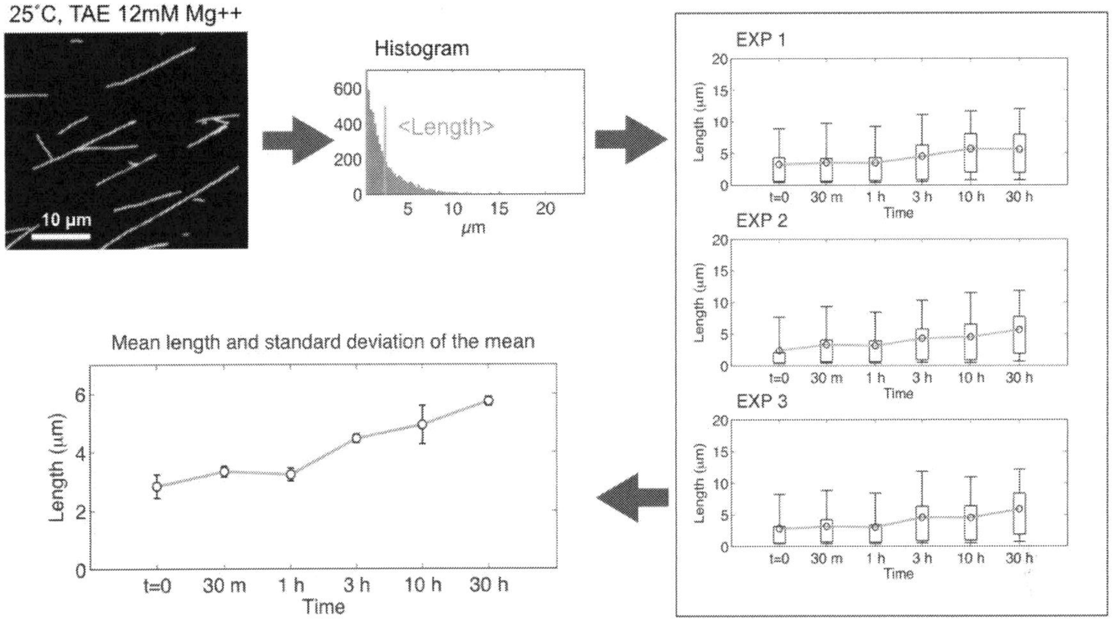

Results consistent with Ekani-Nkodo et al. 2004

9

Isothermal control of assembly and disassembly with DNA inputs

25˚C, TAE 12mM Mg++ , incubation time determines initial mean length

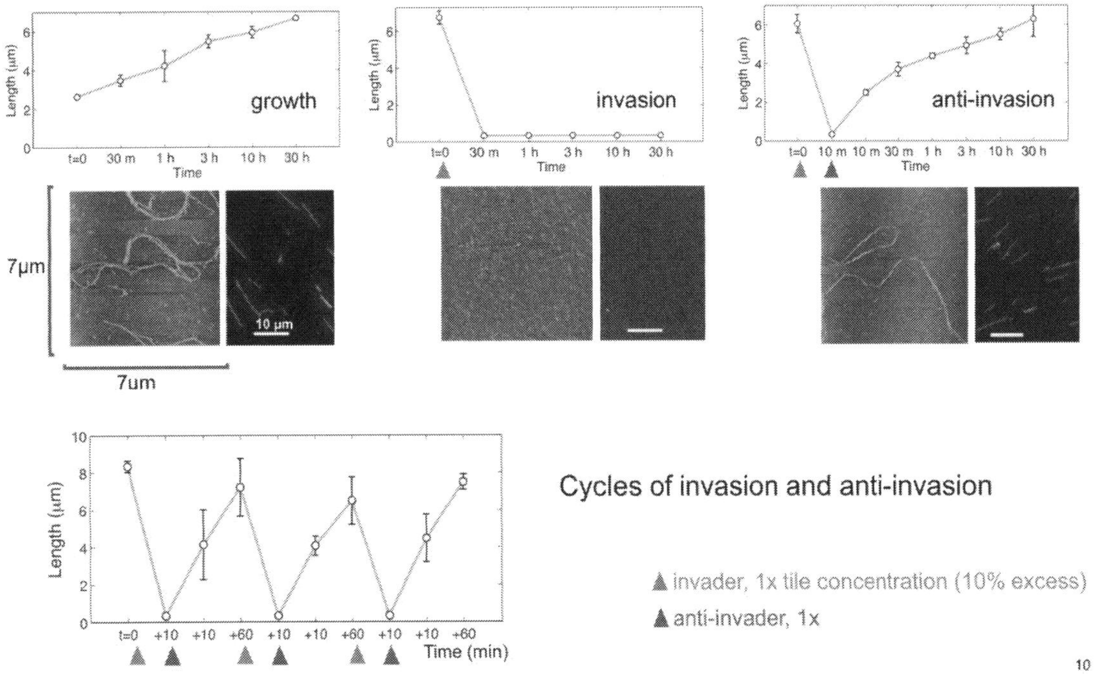

Cycles of invasion and anti-invasion

▲ invader, 1x tile concentration (10% excess)

▲ anti-invader, 1x

10

Assembly and disassembly rates can be modulated systematically

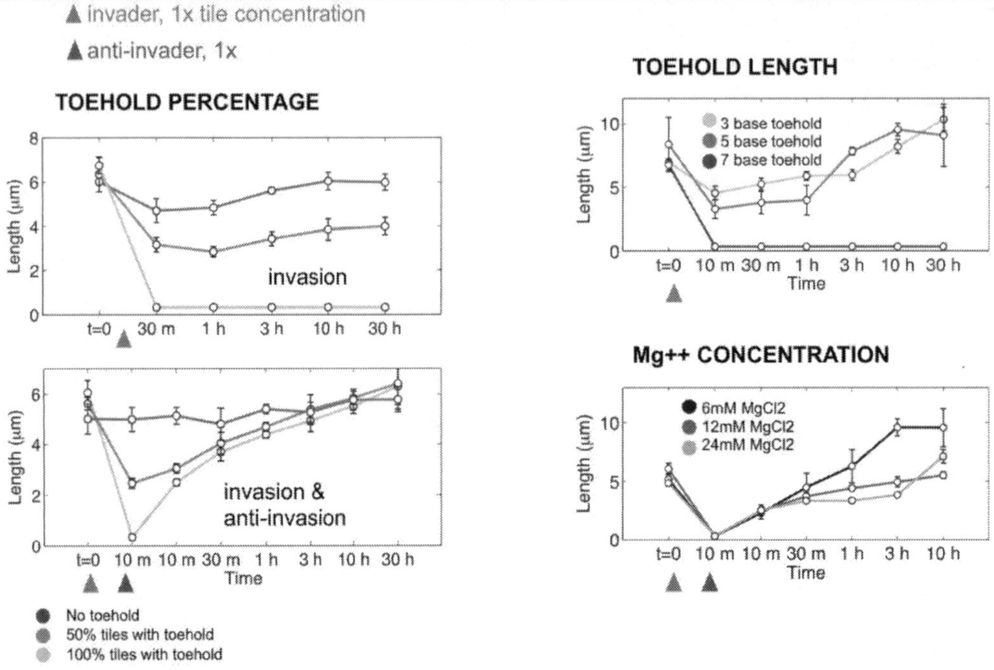

25˚C, TAE 12mM Mg++

11

Toehold position determines breakage mechanism

25˚C, TAE 12mM Mg++

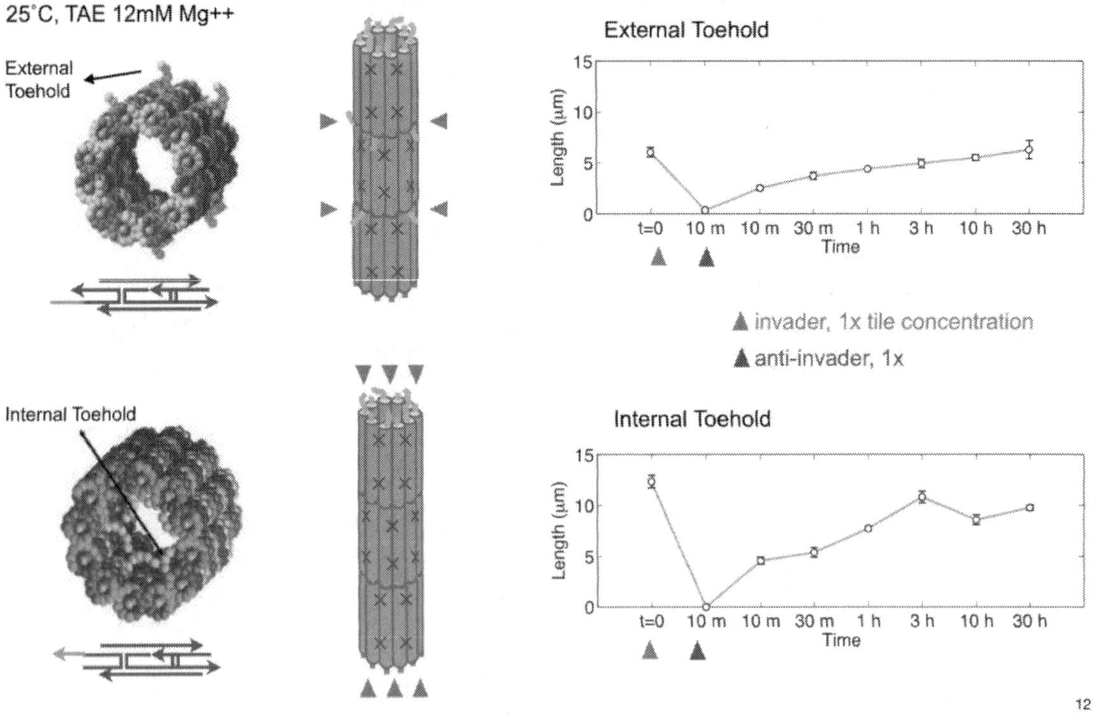

12

External vs internal toehold

External toehold invasion

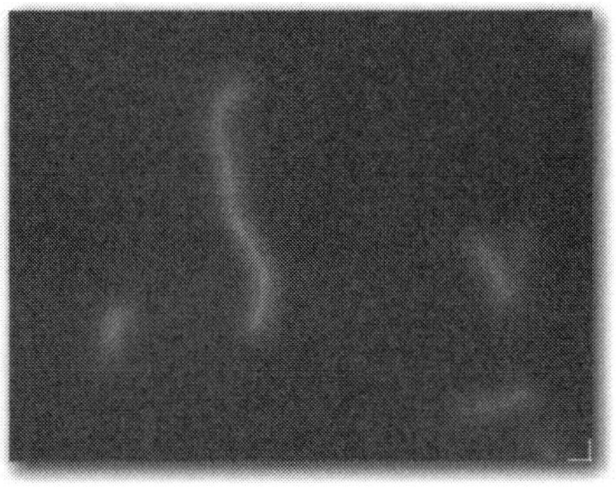

Internal toehold invasion

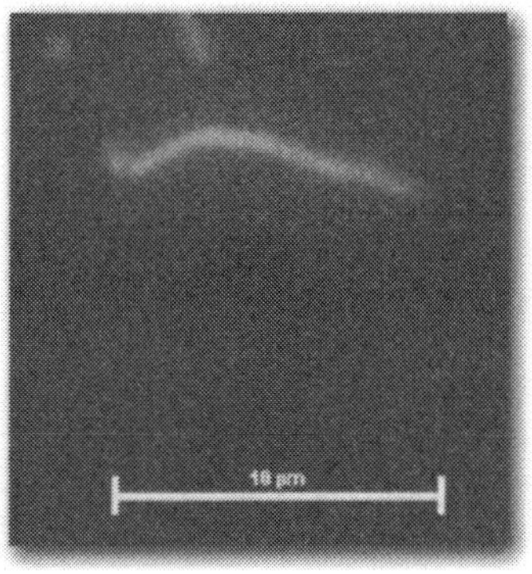

120 s after addition of invader
1 frame / 30 s

13

Anti-invasion

(Internal toehold)

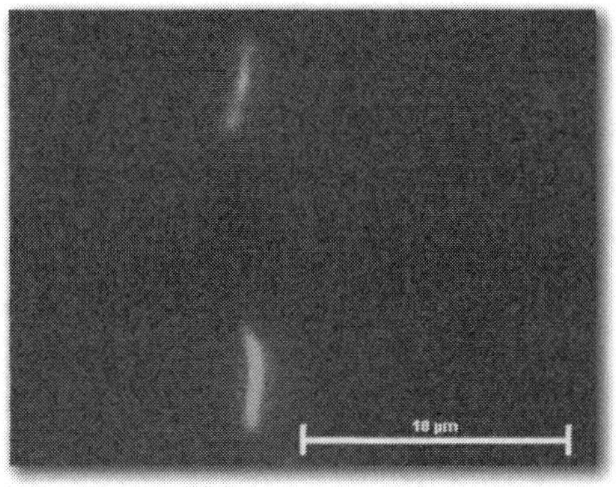

14

Macroscopic model of assembly and disassembly

GOAL: Estimate macroscopic breakage and joining rates in different conditions (PRELIMINARY)

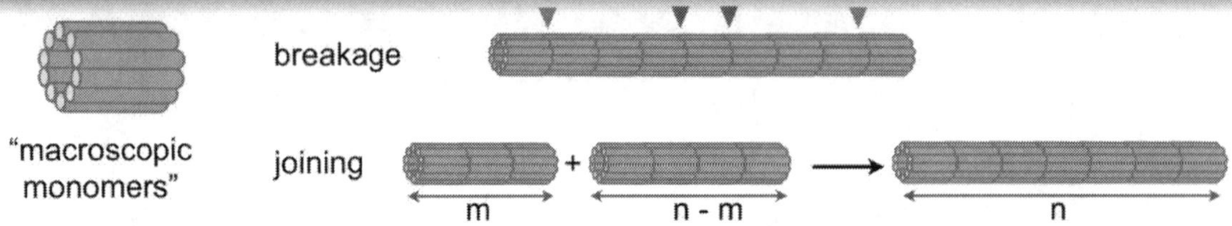

$$\frac{dL_n}{dt} = k_{break,end}(L_{n+1} - L_n) + \sum_i k_{break,int} L_i + \sum_m k_{join}(m, n-m) L_m L_{n-m} - \sum_m k_{join}(m,n) L_m L_n$$

$$k_{join}(m,n) = \frac{\alpha}{l_c}\left[\frac{1}{m}\ln\left(\frac{ml_c}{d}\right) + \frac{1}{n}\ln\left(\frac{nl_c}{d}\right)\right]$$

TEST: END JOINING $k_{break,end} \approx$ 2.45 10^{-5}/s $k_{break,int} \approx$ 2.5 10^{-6}/s $\alpha \approx$ 3.6 10^{-6} μm²/s

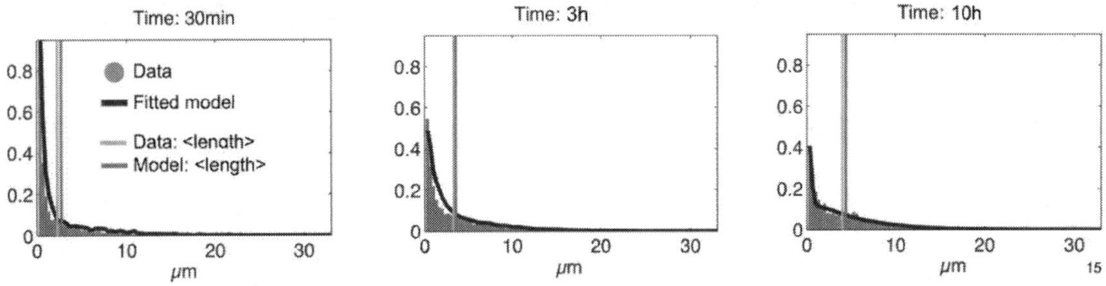

15

Can we direct nanotube growth with dynamical, autonomous circuits?

Synthetic biomaterial

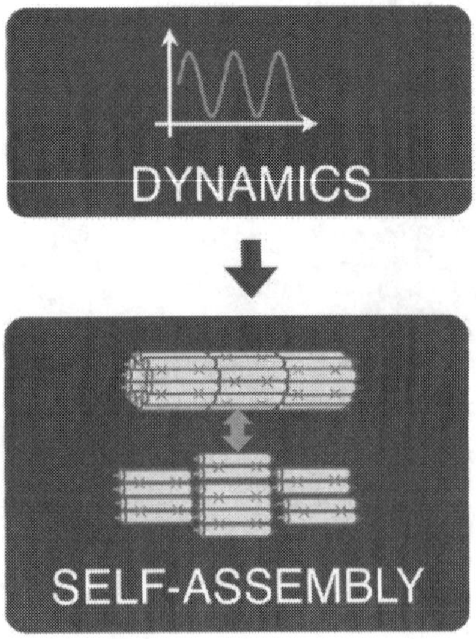

Cellular microtubules

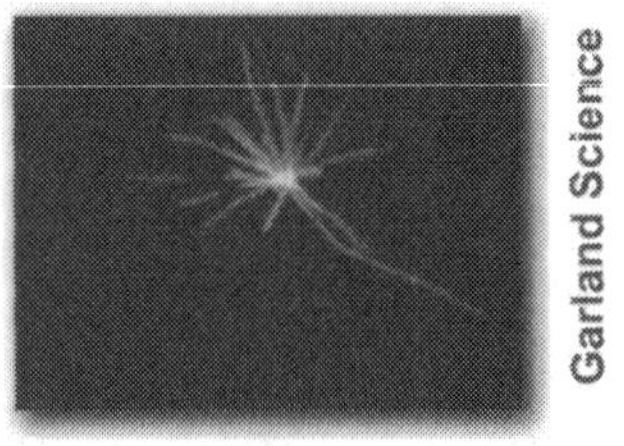

16

Can we direct growth with dynamical, autonomous circuits?

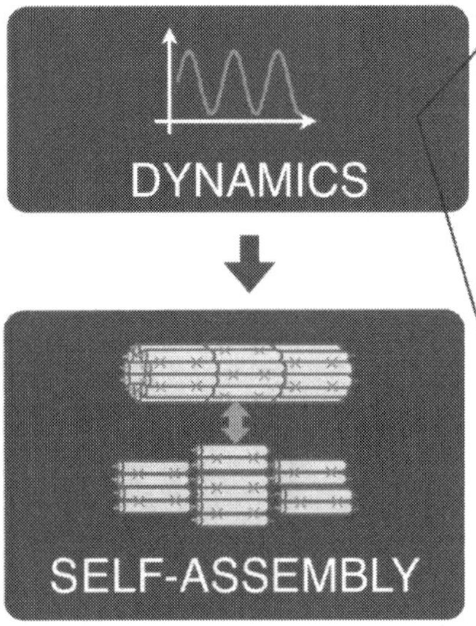

Test circuit: transcriptional oscillator

Challenges:

- Transcriptional environment
37 C, NTPs, Mg++, DNA binding enzymes
- Time scales
- Modularity (signal transmission)

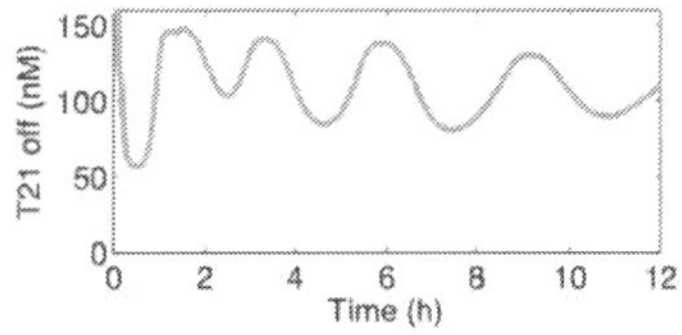

Kim and Winfree 2011, Franco et al 2011, Weitz et al 2014

17

Can we direct growth with dynamical, autonomous circuits?

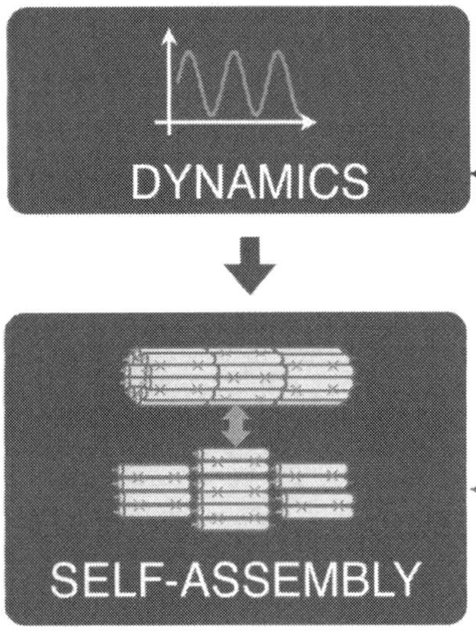

Challenges:

← **• Transcriptional environment**
37 C, NTPs, Mg++, DNA binding enzymes

- Time scales
- Modularity (signal transmission)

← **TAE, 12 mM Mg++ , 25˚C**

18

Redesign of sticky ends is necessary to make nanotubes compatible with transcriptional environment

1) Temperature/salt conditions

7 base sticky ends sufficient

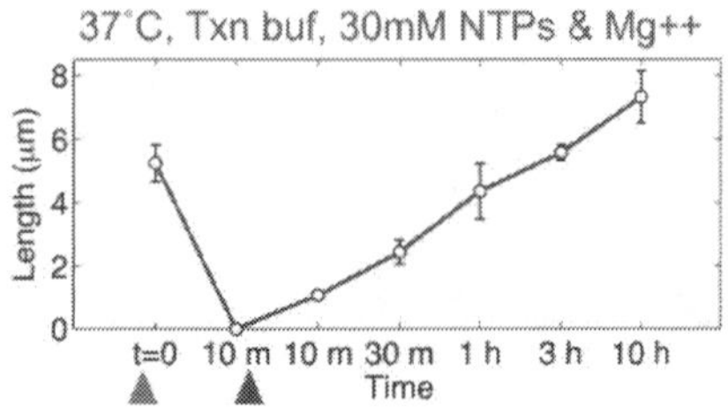

2) RNA Polymerase promotes dissolution of nanotubes

8 base sticky ends required

7 base

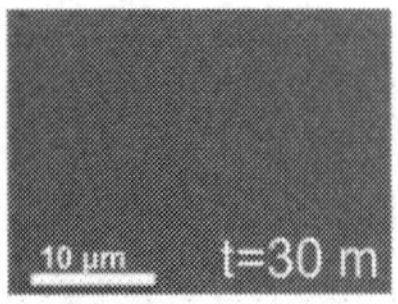

8 base

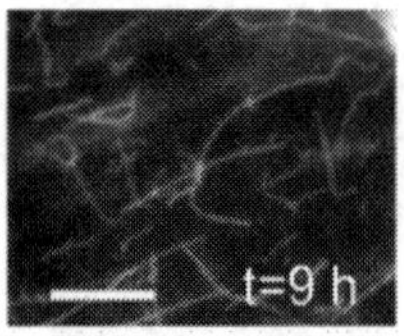

9 base

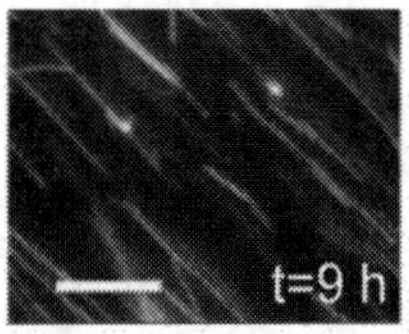

19

Controlling breakage and reassembly with a simple transcriptional system

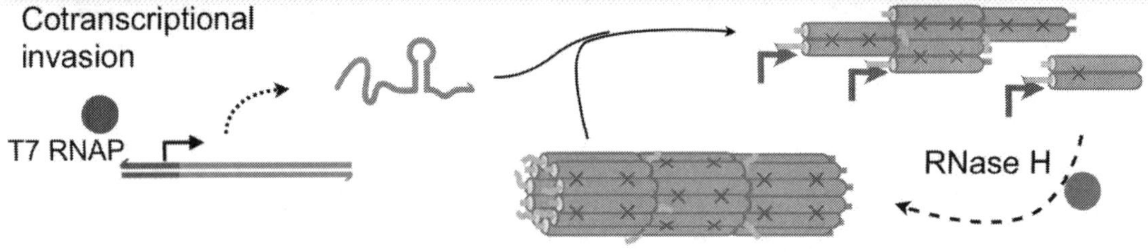

Initial population

Invasion 30 m

+ RNase H 5 h

Control (no invasion)

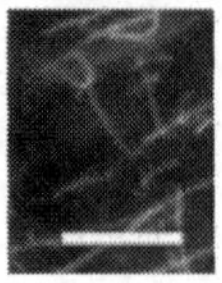

20

Can we direct growth with dynamical, autonomous circuits?

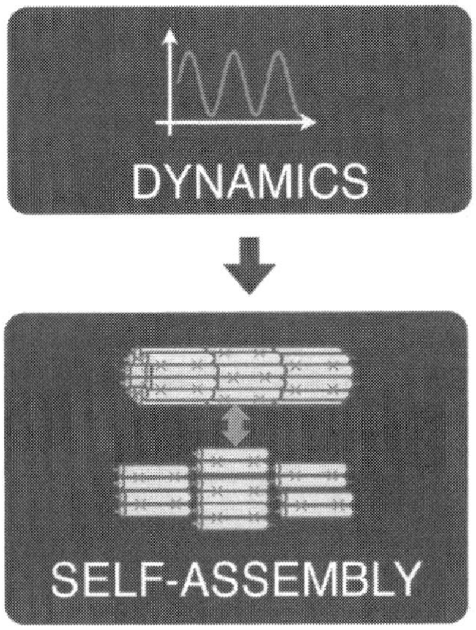

Challenges:

- Transcriptional environment
37 C, NTPs, Mg++, DNA binding enzymes

- **Time scales**

- Modularity (signal transmission)

21

Oscillator and nanotubes time scales are compatible

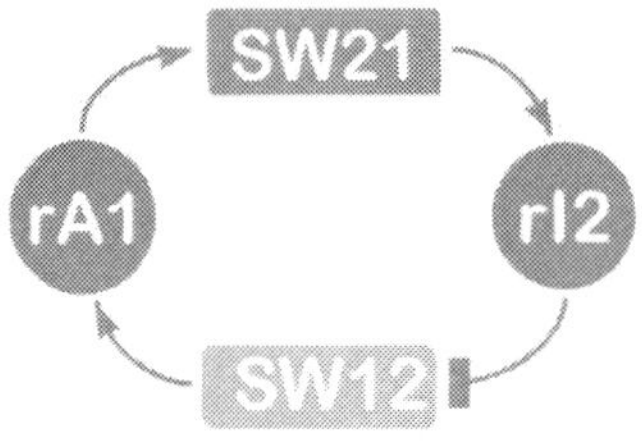

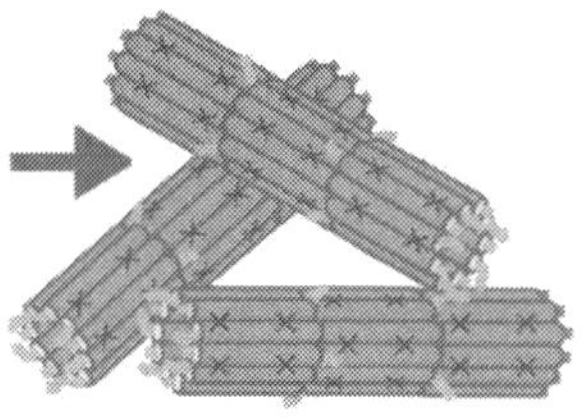

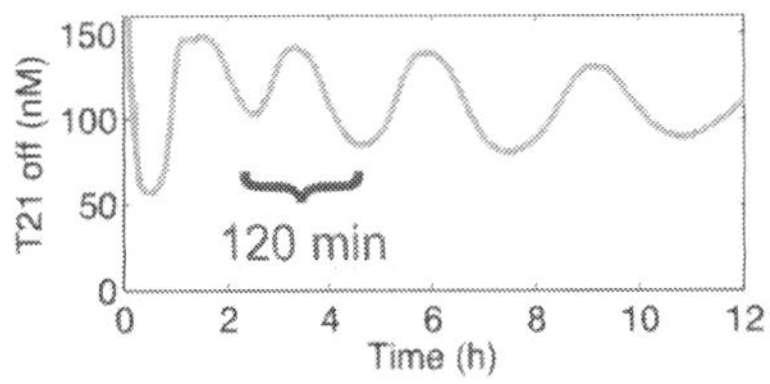

Tunable period
1-3 hours

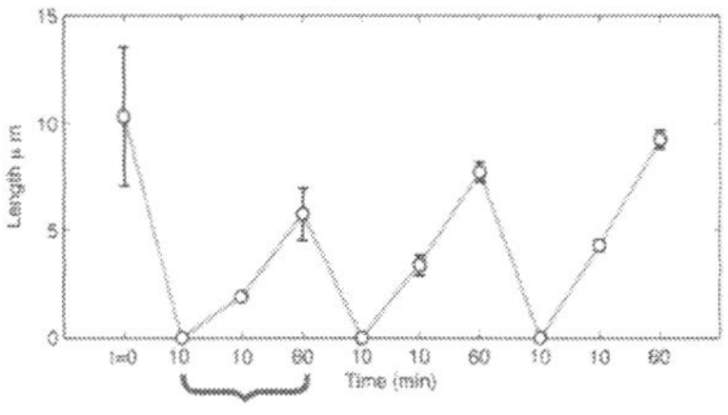

60 min

Growth speed -- concentration of monomers

22

Can we direct growth with dynamical, autonomous circuits?

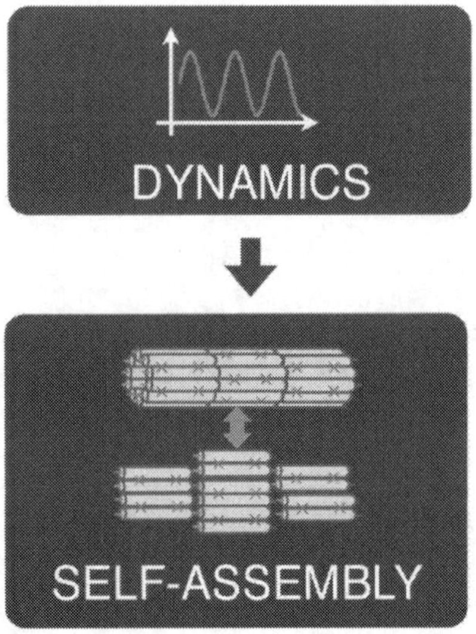

Challenges:

- Transcriptional environment
37 C, NTPs, Mg++, DNA binding enzymes
- Time scales
- **Modularity (signal transmission)**

23

Modularity is a well known problem in molecular circuits

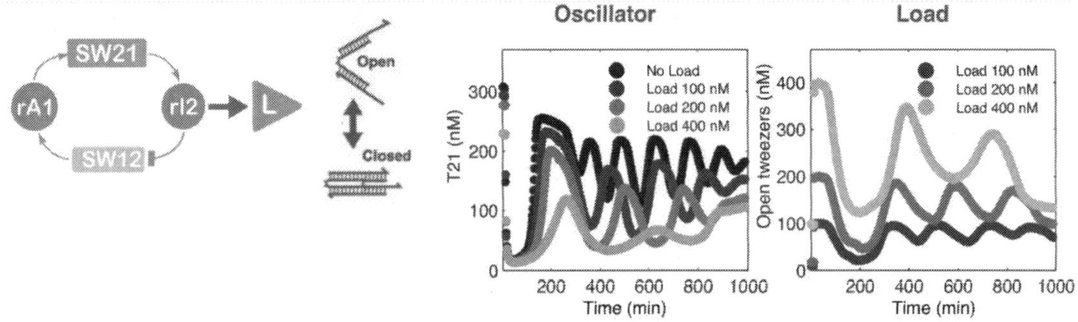

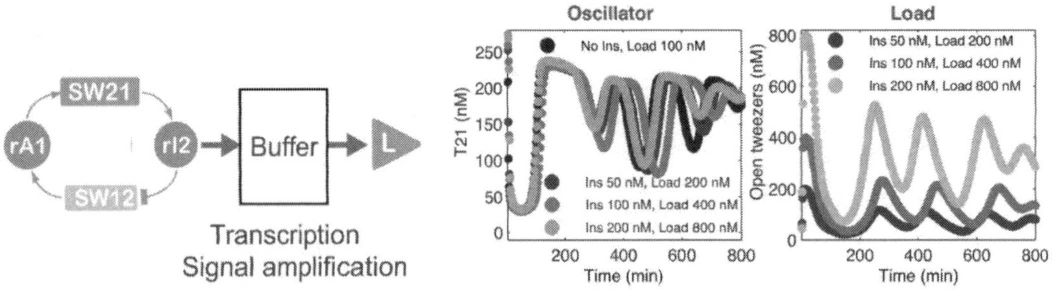

Franco et al. 2011 Retroactivity: Del Vecchio et al. 2008

24

Coupling the oscillator and the nanotubes

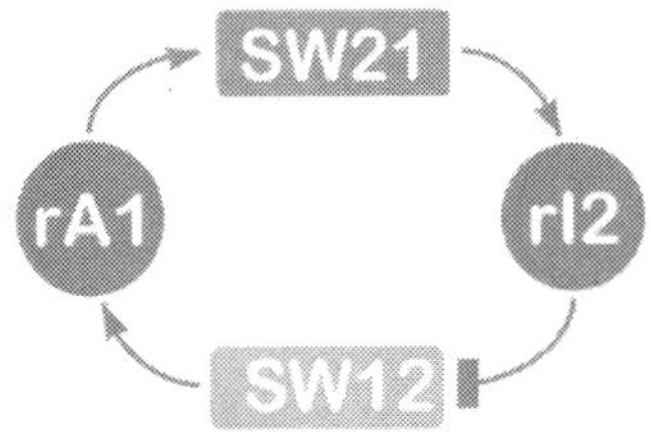

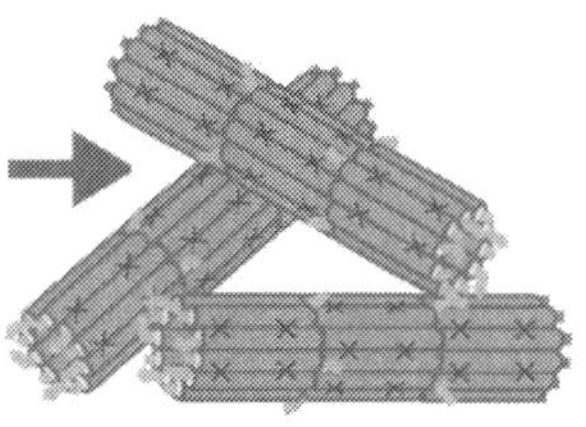

25

Enzyme sharing is another source of perturbation of the oscillator

PRELIMINARY

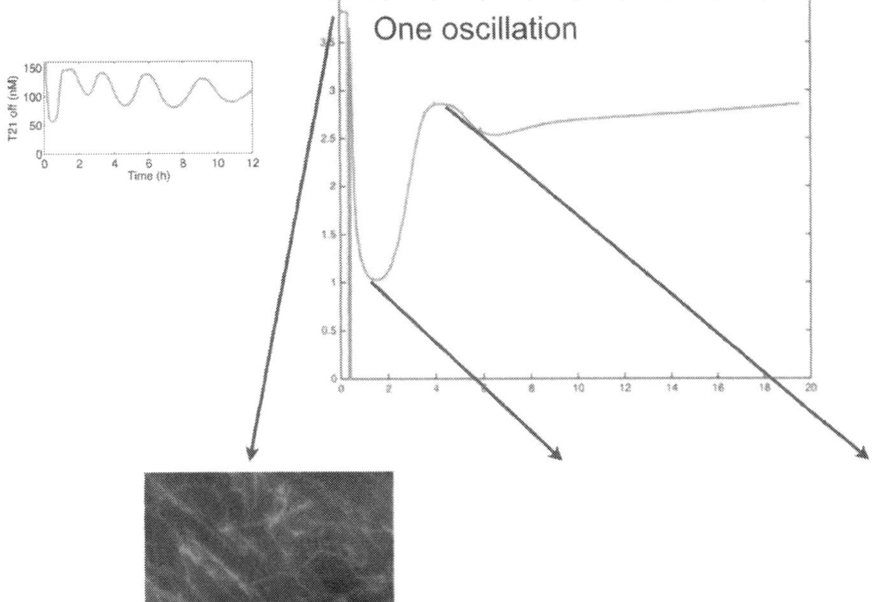

26

Outlook: nucleic acid active materials

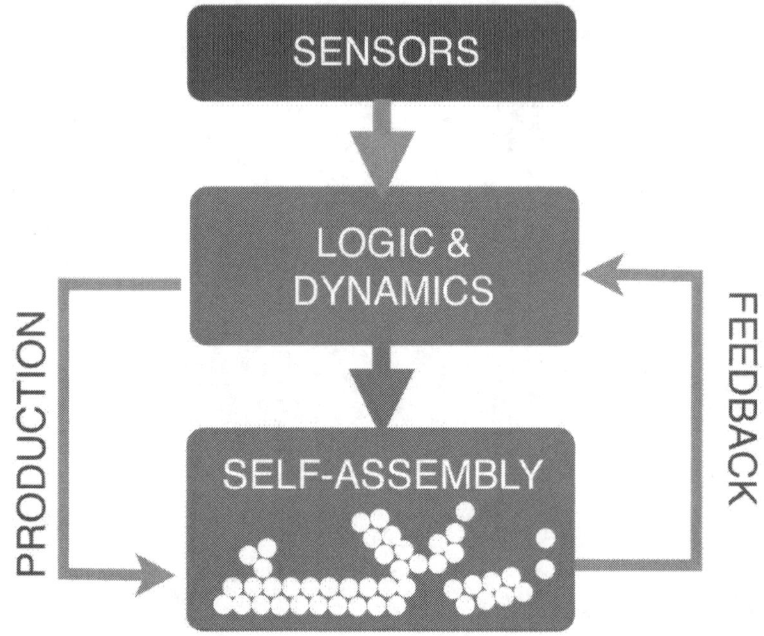

27

In progress: PRODUCTION AND CONTROL OF RNA NANOTUBES

Stewart et al. in preparation

RNA Tile

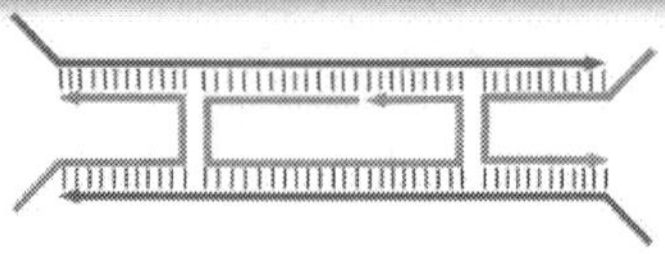

2) One pot anneal:

cf. Ko et al. 2010, DNA/RNA hybrid tiles

1) Transcribe + gel extract + anneal

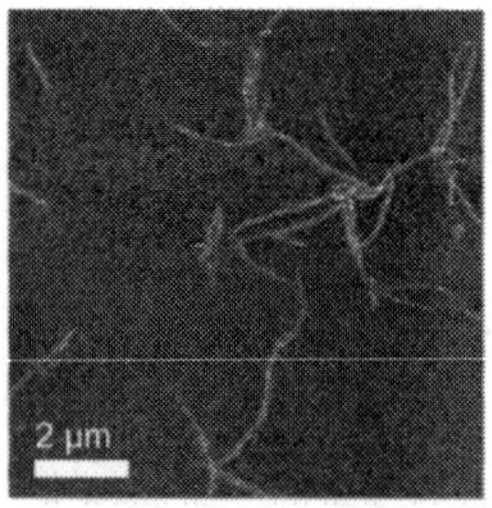

One pot invasion:

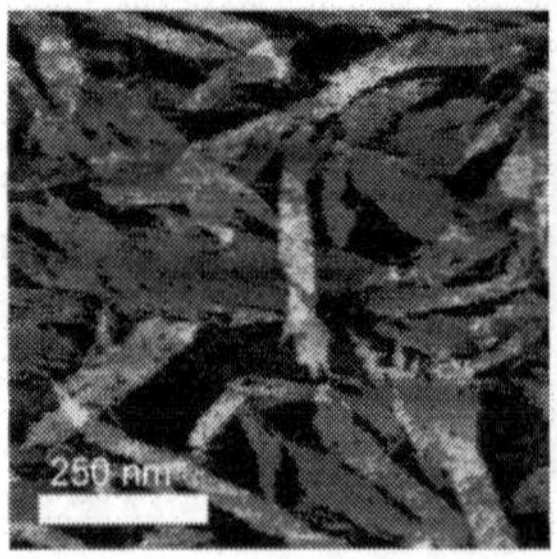

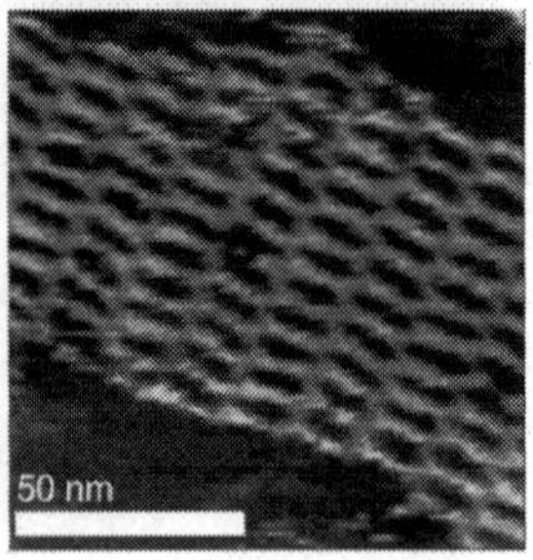

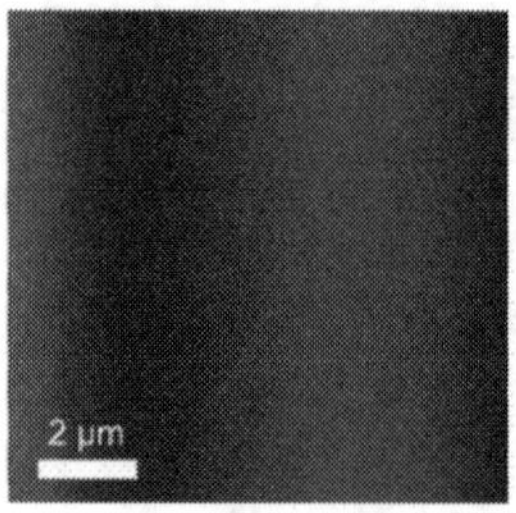

28

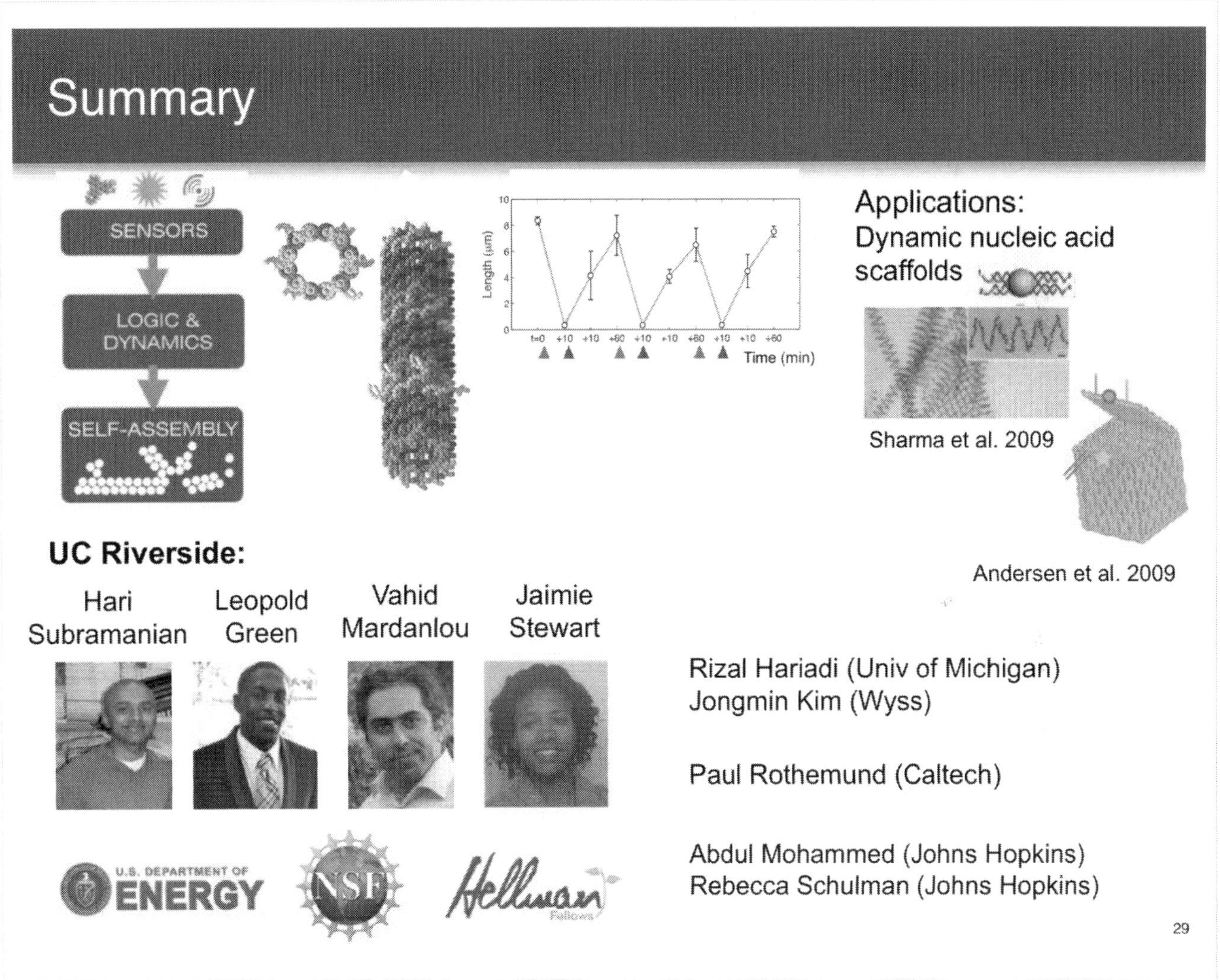
Summary
SENSORS
LOGIC & DYNAMICS
SELF-ASSEMBLY
Length (μm)
Time (min)
Applications:
Dynamic nucleic acid scaffolds
Sharma et al. 2009
Andersen et al. 2009
UC Riverside:
Hari Subramanian
Leopold Green
Vahid Mardanlou
Jaimie Stewart
Rizal Hariadi (Univ of Michigan)
Jongmin Kim (Wyss)
Paul Rothemund (Caltech)
Abdul Mohammed (Johns Hopkins)
Rebecca Schulman (Johns Hopkins)
U.S. DEPARTMENT OF ENERGY
NSF
Hellman Fellows
29

Reconstructing Anaerobic Microbiomes from the "Bottom Up": New Techniques to Decipher Interwoven Metabolism

Michelle A. O'Malley
Assistant Professor
Department of Chemical Engineering
University of California, Santa Barbara

Recalcitrant Biomass Hinders Production of Valuable Products from Cellulose

UCSB

Cellulosic biomass is an attractive renewable resource for biofuels, bioplastics, etc.

3 enzymatic components required:
- Endoglucanases
- Exoglucanases
- β-glucosidases

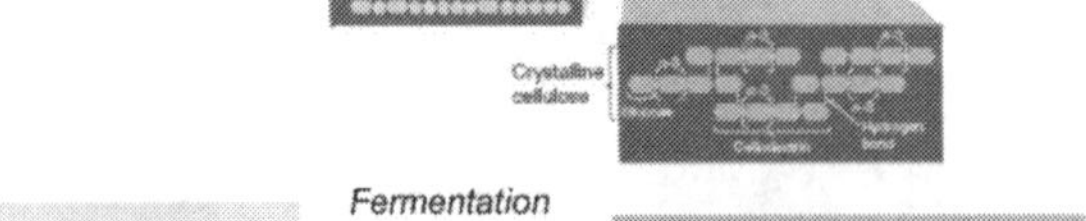

Cellulose-rich Feedstock → *Pretreatment, Enzymatic Hydrolysis* → Sugar → *Fermentation* → Value-added Product

Purified cellulases
(i.e. from *T. reesi, A. niger*)

Major Roadblocks:
- Recalcitrant nature of crude biomass (presence of lignin and other inhibitors)
- Costs associated with enzyme production and catalytic performance

2

Rubin, E. M. Nature 454 (2008) 841- 845.

What can we learn from nature?

UCSB

- Several *microbiomes* degrade biomass
- Anaerobes are promising understudied microbes
  - They populate the guts of animals, regulating immune function, nutrition, and other functions
  - Each microbe has a unique "job" within the community
  - New culture, sequencing, and proteomics tools allow us to understand and engineer these systems

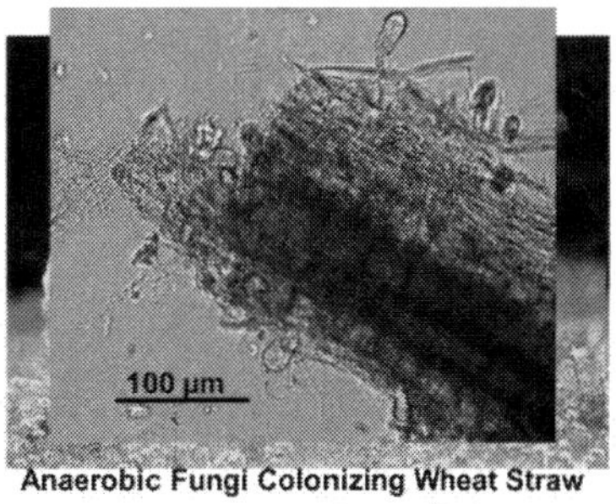

Anaerobic Fungi Colonizing Wheat Straw

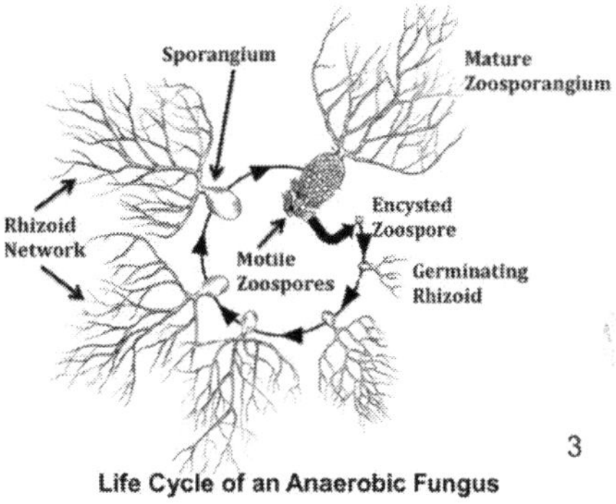

Life Cycle of an Anaerobic Fungus

Can we leverage natural partnerships to build novel consortia?

Haitjema CH, Solomon KV, Henske JK, Theodorou MK, O'Malley MA, *Biotechnology and Bioengineering*, 111(8): 1471-1482 (2014).
O'Malley MA, Theodorou MK, Kaiser CA. *Environmental Progress*, 31(1): 37-46 (2012).

3

Objectives

UCSB

Overall Goal: Accelerate production of renewable fuels & chemicals from waste materials

Approach: Learn from nature to build stable anaerobic consortia with novel functions

- Isolate gut microbes from their native microbiomes and integrate 'omics'-based tools to understand function
- Engineer anaerobic consortia that compartmentalize breakdown and bioproduction

4

Finding Anaerobes in the Muck...

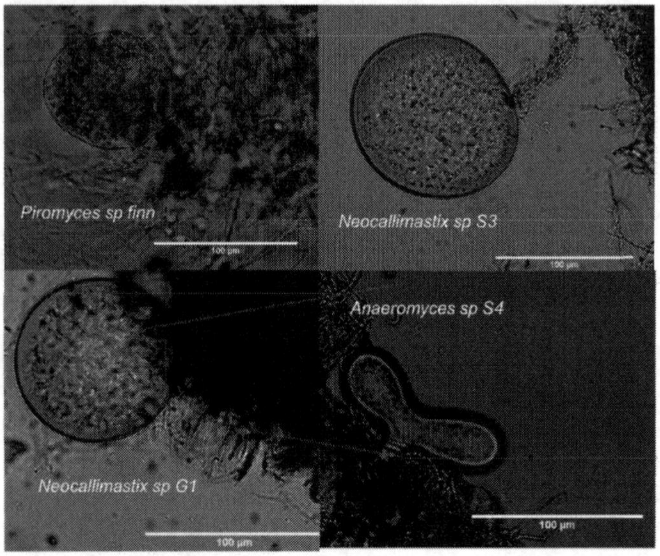

Fecal materials from Santa Barbara Zoo

↓

Roll tube isolation of unique strains

↓

Characterize biomass-degrading capacity

Solomon KV, Haitjema CH, Thompson DA, O'Malley MA, *Current Opinions in Biotechnology*, 28: 103-110 (2014).

5

Fungal Isolates Exhibit Substrate Preference in Batch Culture

- Fungi were extracted from a horse named "Finn" from Verrill Farm (Concord, MA)
- Fungal isolate renamed *Piromyces sp finn* and supported on reed canary grass
  - Full growth cycle is seen in 5 – 6 days
  - Active growth is characterized by fermentation gas accumulation
  - Pressure accumulation allows substrate preference to be determined

Finn
(courtesy of Verrill Farm)

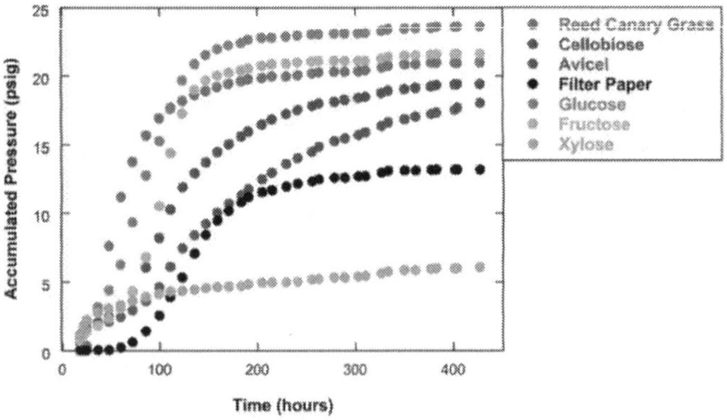

O'Malley MA, Henske JK, Zhong Y, Kaiser CA, *In preparation.*

6

RNA-Seq Reveals Biomass-degrading Machinery in Gut Fungi

UCSB

DNA —*transcription*→ (mRNA) —*translation*→ Protein

To identify the most powerful biomass-degrading enzymes, it is critical to examine the transcriptome from a growing organism.

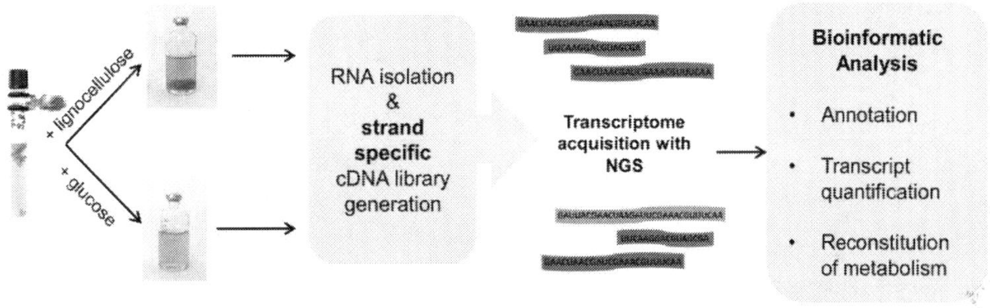

Hypothesis: Cellulosic substrates activate transcription of cellulolytic enzymes and cellulolytic complexes (cellulosomes) while sugars repress transcription

7

Over 27,000 unique transcripts *de novo* assembled by RNA-Seq

UCSB

The transcriptome of *Piromyces sp finn* contains more than 27,000 transcripts following a Poisson size distribution

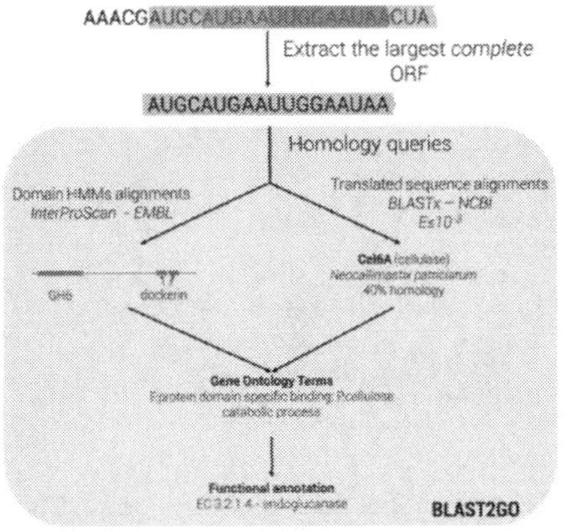

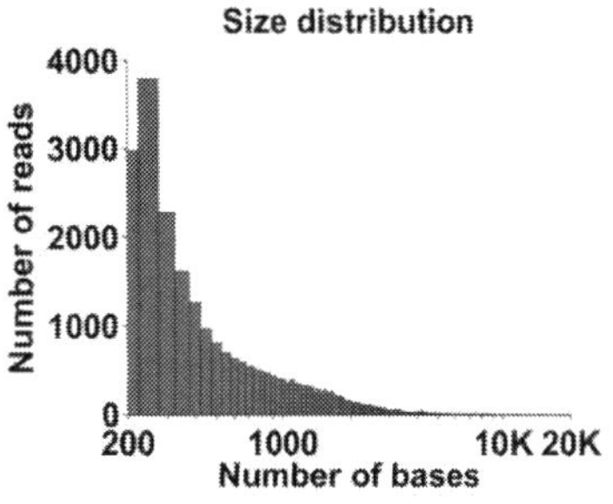

More than 30% of the transcriptome can be annotated with the majority being unique proteins (<70% homology)

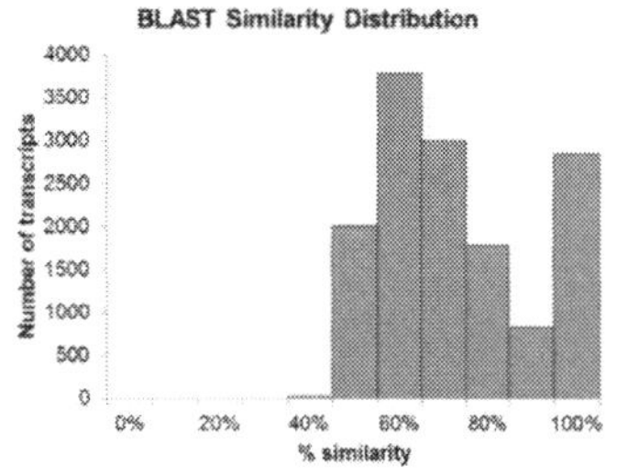

8

Solomon KV, Haitjema CH, Henske JK, Gilmore SP, Borges-Rivera D, Thompson DA, O'Malley MA, *In revision.*

Hundreds of Novel Enzymes are Identified from Anaerobic Gut Fungi

UCSB

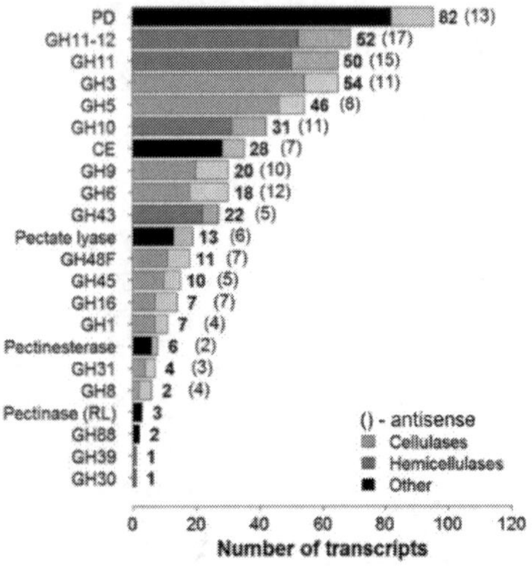

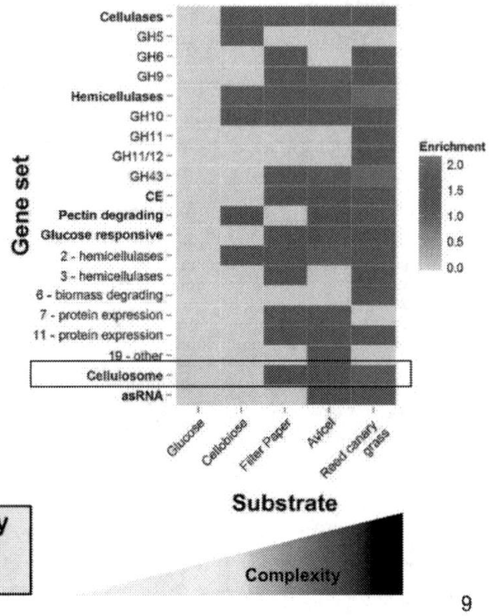

Gut fungi tune their enzymatic machinery based on substrate, and are enriched in xylan degrading enzymes

9

Solomon KV, Haitjema CH, Henske JK, Gilmore SP, Borges-Rivera D, Thompson DA, O'Malley MA, *In revision.*

Anaerobic Gut Fungi are Equal Opportunity Degraders

UCSB

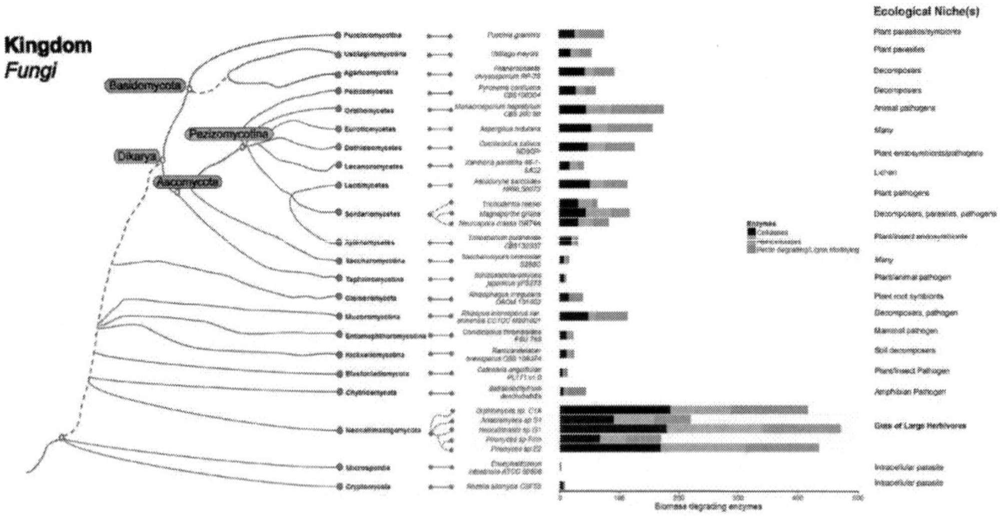

JGI

10

Solomon KV, Haitjema CH, Henske JK, Gilmore SP, Borges-Rivera D, Thompson DA, O'Malley MA, *In revision.*

The Fungal Transcriptome Rapidly Remodels in Response to Sugar Pulse

UCSB

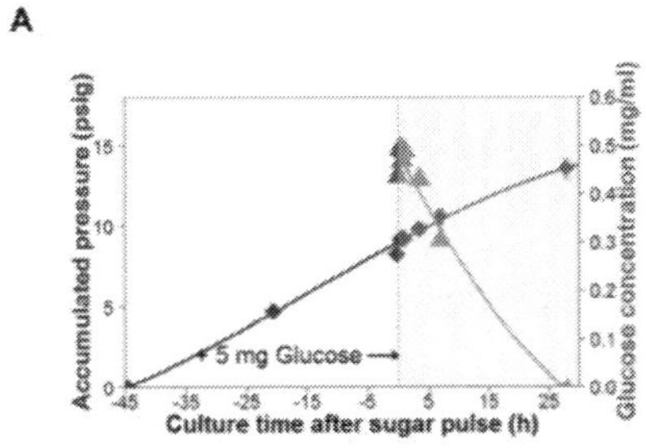

Majority of downregulated genes are associated with biomass breakdown

Are there new catalytic domains here?

NREL NATIONAL RENEWABLE ENERGY LABORATORY   Argonne NATIONAL LABORATORY

11

Solomon KV, Haitjema CH, Henske JK, Gilmore SP, Borges-Rivera D, Thompson DA, O'Malley MA, *In revision*.

Gut Fungal Enzymes vs. Standard Enzyme Cocktails

UCSB

Hypotheses:

1) Gut fungi display superior xylan degradation due to the wider array of endoxylanases (GH10, GH11)
2) Gut fungi more completely degrade crude biomass due to the diversity and abundance of PDs and β–glucosidases (GH5, GH9)

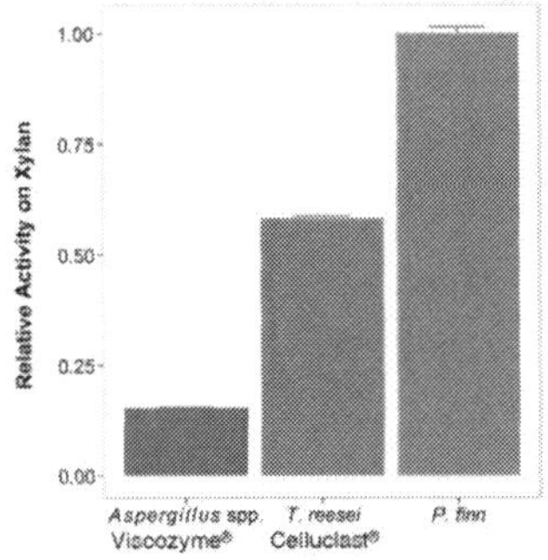

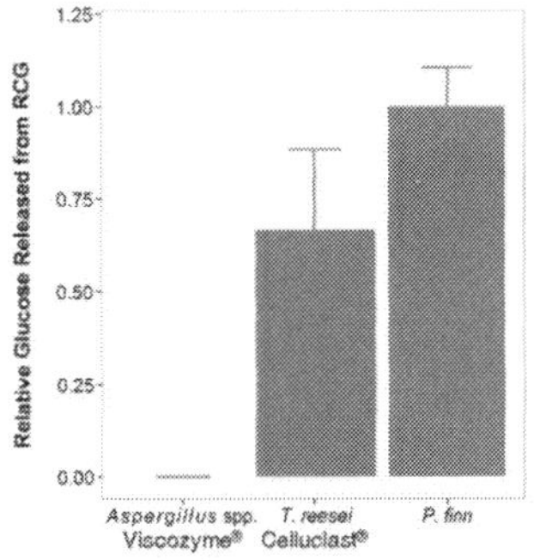

70% increased xylan activity   **50% more free glucose produced**

Enzymes secreted by gut fungi are powerful, even prior to optimization

12

Solomon KV, Haitjema CH, Henske JK, Gilmore SP, Borges-Rivera D, Thompson DA, O'Malley MA, *In revision*.

Objectives

Overall Goal: Accelerate production of renewable fuels & chemicals from waste materials

Approach: Learn from nature to build stable anaerobic consortia with novel functions

- ✓ Isolate gut microbes from their native microbiomes and integrate 'omics'-based tools to understand function
- ➢ Engineer anaerobic consortia that compartmentalize breakdown and bioproduction

13

Almost all Microbes on Earth Live within a Diverse, Dynamic Community

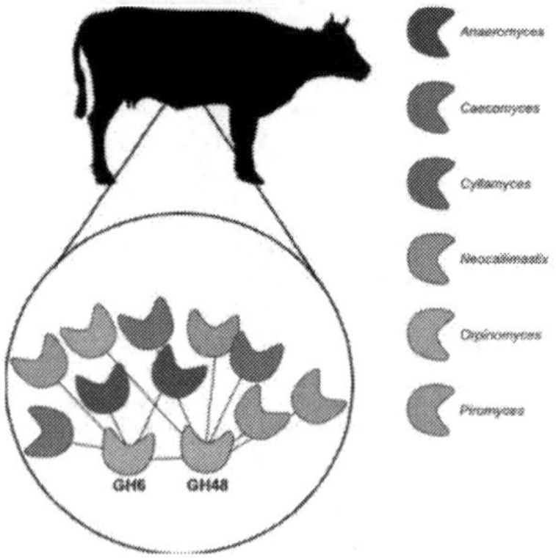

Within anaerobic gut communities:

- ➢ Fungi and bacteria degrade biomass
- ➢ Methanogens reduce H_2 to CH_4
- ➢ Protozoans regulate bacterial populations
- ➢ Some evidence of cross-reaction between secreted enzymes
- ➢ Unknown how nutrients flow within these systems, and what the "tipping point" is

Can we enhance fungal breakdown through the addition to microbial "neighbors"?

14

Anaerobic Syntrophy: Controlling H2 and CH4 Production

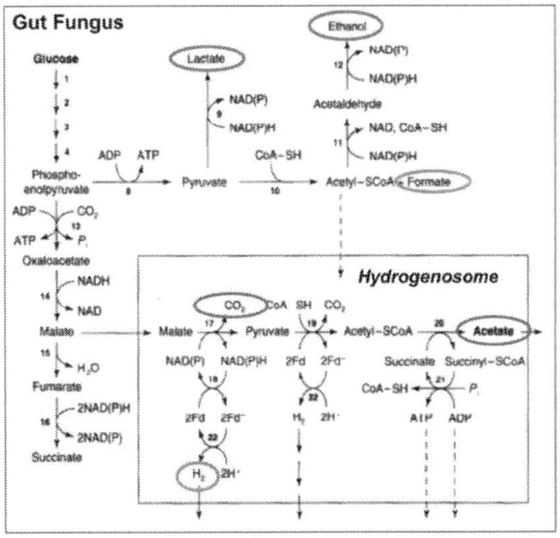

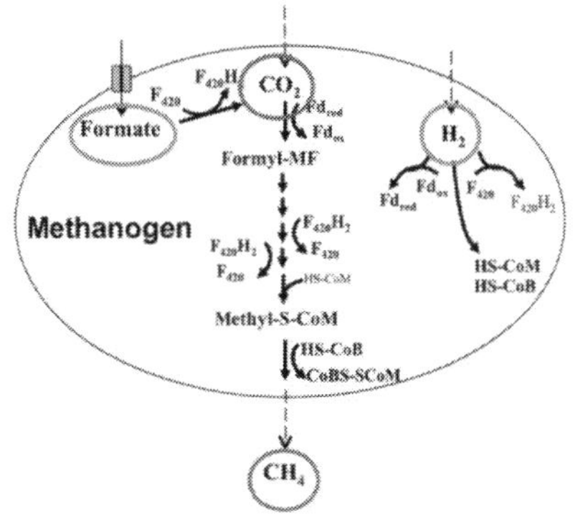

- ➢ Metabolism of anaerobic fungi and methanogens is linked
- ➢ Products depend on the microbes present in a culture
- ➢ Fungi – acetate, CO2, H2
- ➢ Fungi+Methanogens – CO2, methane

15

Jessica Sexton, Sean Gilmore, John Henske

Anaerobic Syntrophy: Controlling H2 and CH4 Production

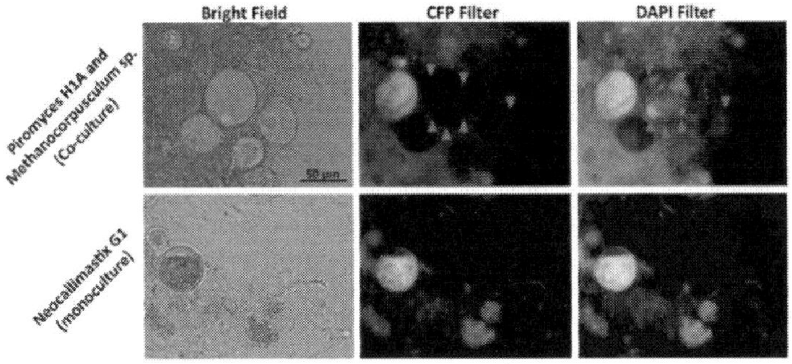

- ➢ Methanogens are co-extracted with fungi
- ➢ Cultures are stable 1+ years in continuous culture

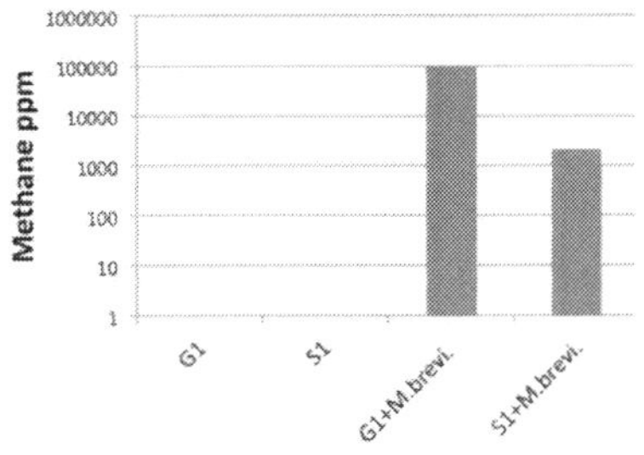

- ➢ Co-cultures produce methane compared to fungal monocultures

16

Jessica Sexton

Anaerobic Co-culture Accelerates Breakdown

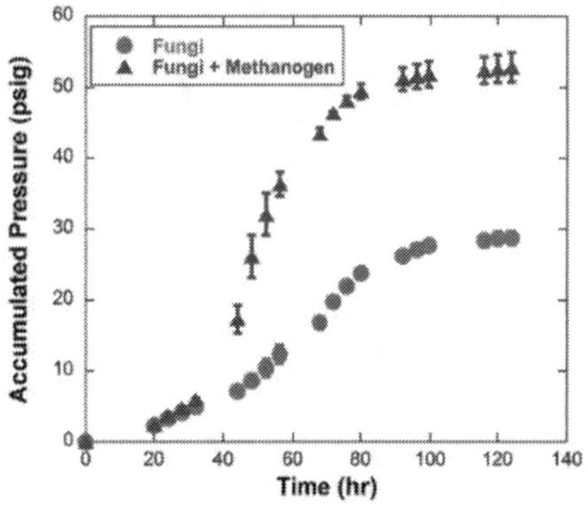

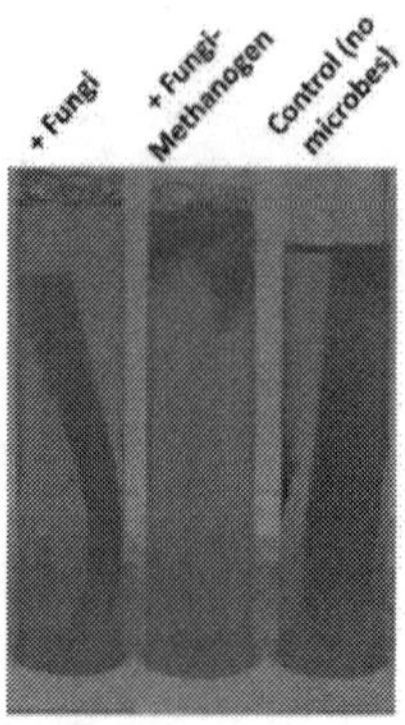

- Fungal/methanogen co-cultures **accelerate** biomass breakdown
- Metabolic underpinnings of the interaction are unknown
- Potential applications:
  - Synthetic systems for biogas production
  - Microbiome engineering to control carbon cycling

17

Jessica Sexton & John Henske

Mixing & Matching Gut Microbes to Tailor Degradation

FISH Analysis of Methanogen-Fungi Co-culture

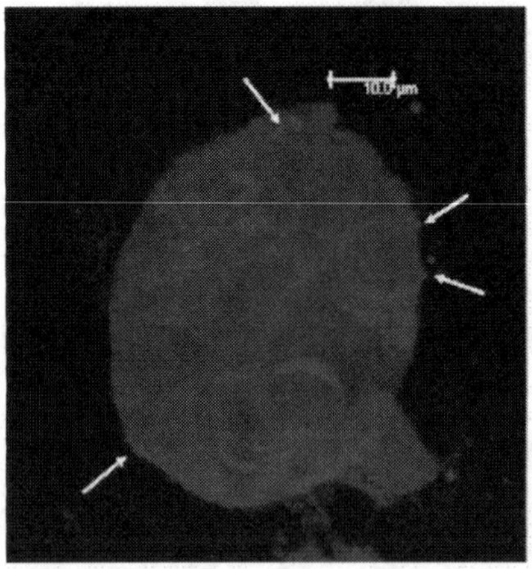

Leis, S. et al. Anaerobe 29 (2014) 34-43.

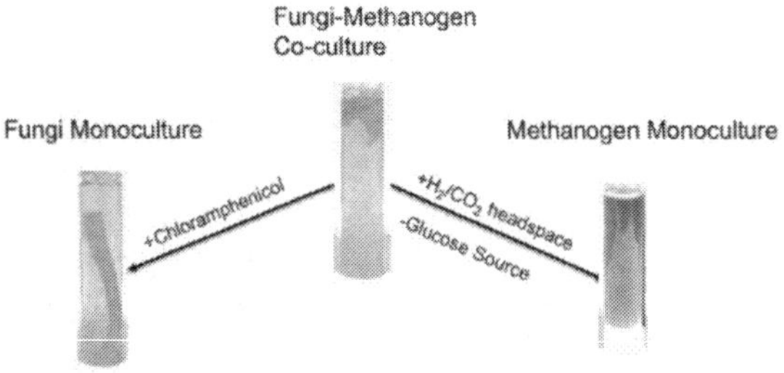

- Natural co-cultures can be separated into their components, then recombined to form new partnerships
- 7 novel methanogen genomes sequenced
  - *Methanosphaera cuniculi, Methanocorpusculum parvum, Methanosarcina spelaei, Methanobacterium bryantii,* ***Methanosphaera stadtmanae, Methanocorpusculum labreanum***

Can we tether fungi & methanogens together that normally wouldn't see each other?

18

Sean Gilmore, John Henske, Kevin Solomon

Where else can we go from here?

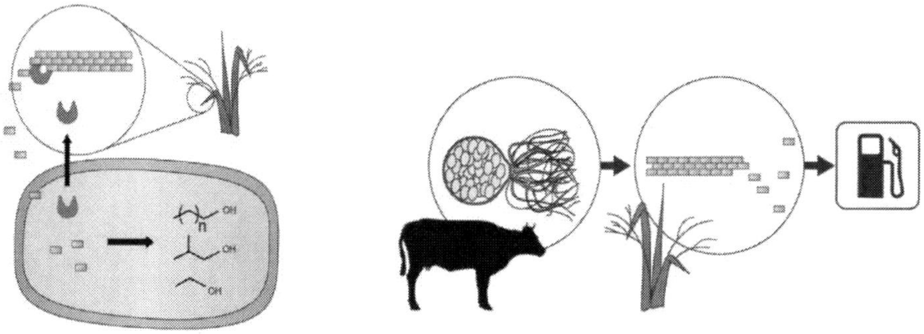

Transfer enzymes to model microbe for Consolidated Bioprocessing

Develop new genetic engineering strategies to control gut fungi or methanogens

Combine microbes in synthetic co-culture

19

"Compartmentalize" Biomass Breakdown and Product Conversion with Two Microbes

Goal: Metabolically link a model *facultative* anaerobe to released fungal sugars.

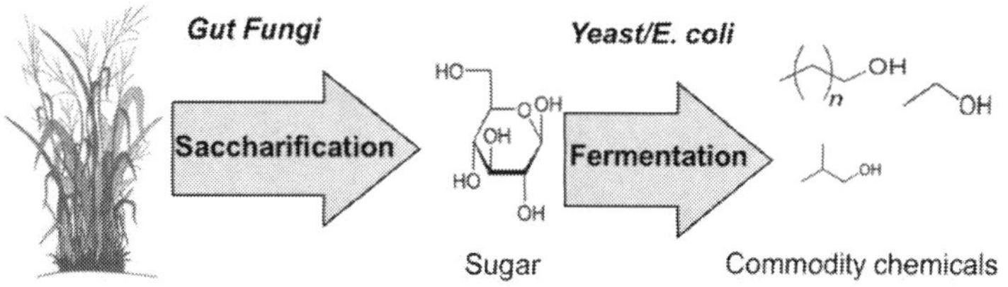

- ✓ Gut fungi likely produce more sugars than they need to grow and divide
- ✓ Model microbes like yeast and *E. coli* are easy to metabolically engineer
- ✓ "One pot" strategy to convert hydrolyzed sugars into value-added chemicals

20

John Henske

"Compartmentalize" Biomass Breakdown and Product Conversion with Two Microbes

UCSB

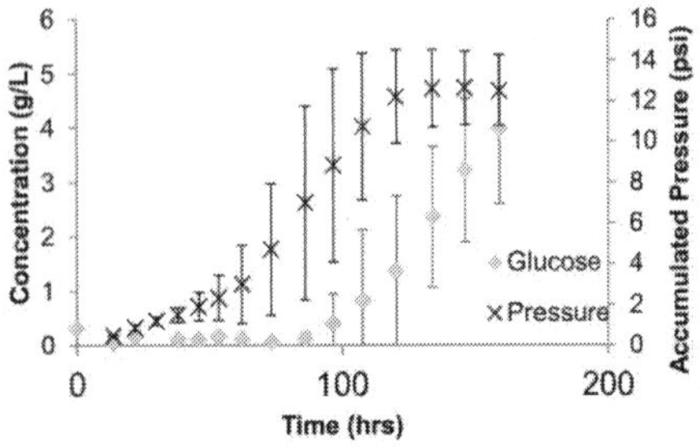

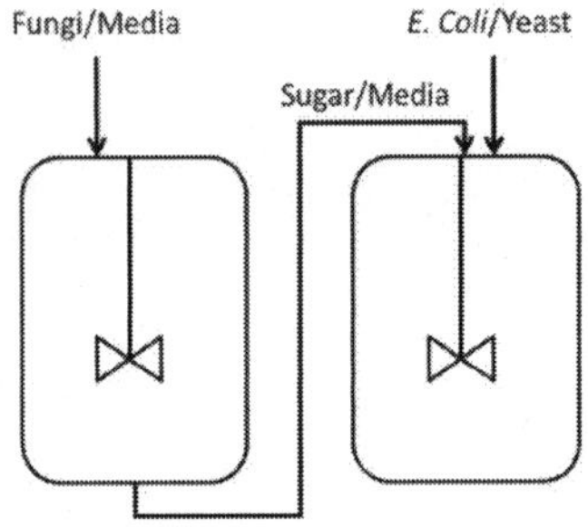

- Up to 5-7 g/L glucose accumulates in spent fungal cultures
- Both *E. coli* and *S. cerevisiae* can grow and divide in spent cultures

Current Targets:

- An anaerobic fluorescent protein confirms "hand-off" of carbon to the model microbe

Flavin-based Mononucleotide Fluorescent Protein (FbFP)

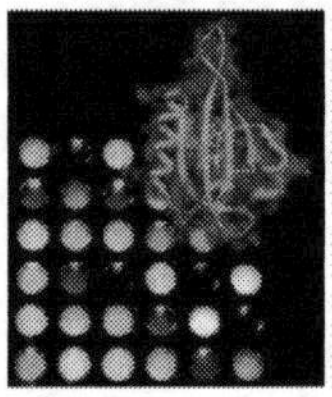

- Production of n-butanol in an anaerobic co-culture process

21

John Henske

Summary and Perspectives

UCSB

✓ Understudied anaerobes are powerful degraders of crude biomass

✓ Advances in 'omics' technologies make it possible to pinpoint nature's key enzyme cocktails

✓ Natural partnerships between anaerobes accelerates crude biomass processing

✓ By leveraging this understanding, we can engineer novel, stable microbial consortia for targeted value-added production

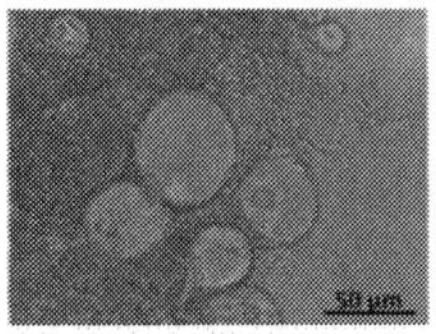

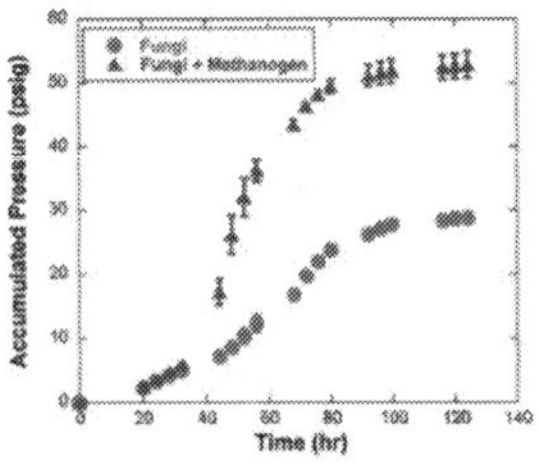

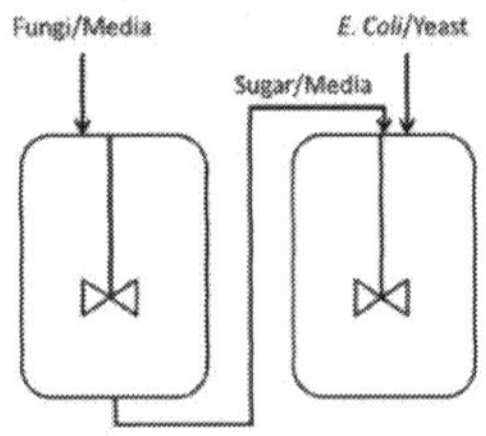

22

Acknowledgements

O'Malley Lab:
Dr. Kevin Solomon
Dr. Charles Haitjema
Dr. Jessica Sexton
Dr. Monica Rieth
Dr. Susanna Seppala
Dr. Nick Peng
Nikki Schonenbach
Jennifer Guerrero
John Henske
Sean Gilmore
Justin Yoo

Undergraduates:
Josie Tressel
Elisa Ovadia
Natalie Banakis
Erich Brodbeck
Sarah Maxel
Wen-Kang Chou
Charlotte Abrahamson

Broad Institute:
Prof. Aviv Regev
Dawn Thompson
Diego Borges-Rivera

Harper Adams (UK):
Mike Theodorou

Univ. of Delaware:
Prof. Kelvin Lee
Leila Choe

USDA:
Paul Weimer
Chris Odt

Verrill Farm
Santa Barbara Zoo

JGI:
Igor Grigoriev
Kerrie Barry

PNNL:
Aaron Wright
Sam Purvine
Scott Baker

Argonne NL:
Andrzej Joachimiak
Gyorgi Babnigg

NREL:
Steve Decker
Mike Himmel
Larry Taylor
Michael Resch

MIT:
Prof. Kristala Prather

UCSB:
Prof. Dave Valentine
Prof. Scott Shell

Funding and Support:

U.S. DEPARTMENT OF ENERGY, NREL, Argonne National Laboratory, MITSUBISHI RAYON, CNSI Challenge Grants, BIO-RAD, American Heart Association Learn and Live, JGI Joint Genome Institute, Hellman Fellows Fund, EMSL, USDA, ICB, BROAD INSTITUTE

Anaerobic Syntrophy: Controlling H2 and CH4 Production

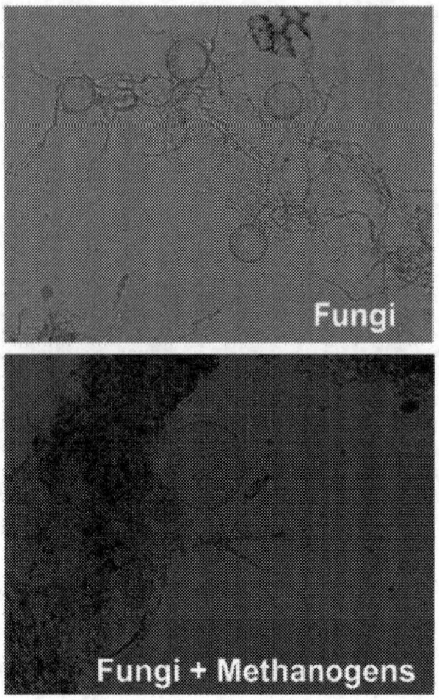

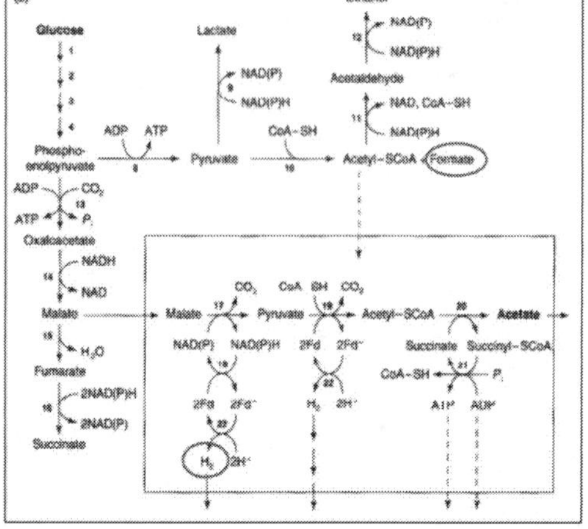

Biomass Degradation Pathways are Highly Tailored to Substrate

Functionality of enzymes needed increase with substrate complexity

Environment triggers changes in programming such as:
- Enzyme preferences
- Cellulosomal complex formation for optimal hydrolysis
- Antisense RNA regulation

Response is tuned by tightly conserved transcription factors
- **Negative regulators:** CreABC
- **Positive regulators:** XlnR, ACE1-2, ClbR, Clr1-2

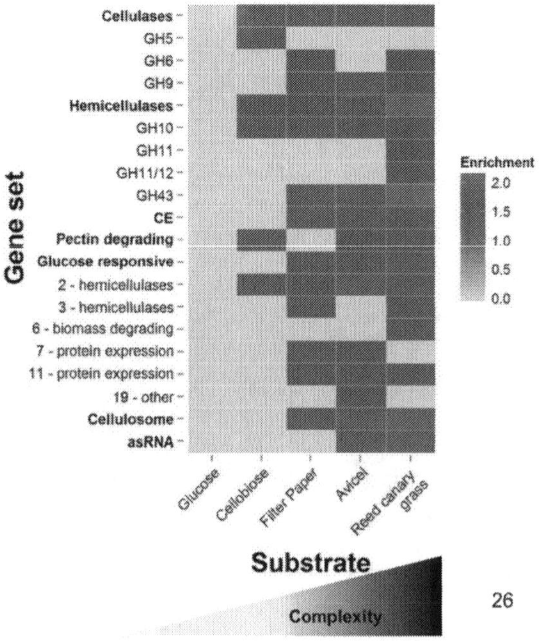

Solomon KV, Haitjema CH, Henske JK, Gilmore SP, Borges-Rivera D, Thompson DA, O'Malley MA, *In revision.*

Piromyces sp finn is Rich in Several GH Enzymes and Dockerin-tagged Proteins

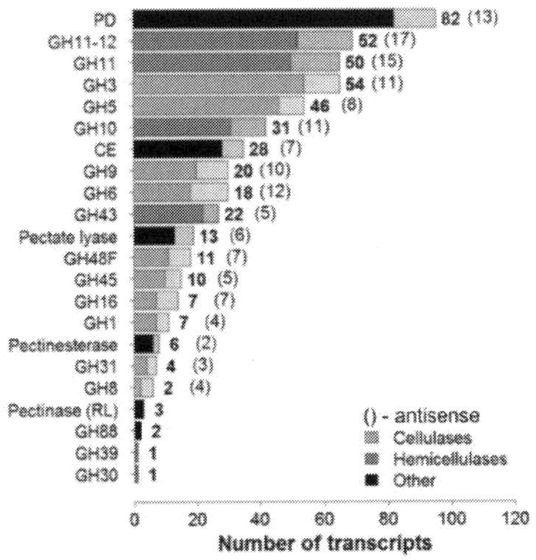

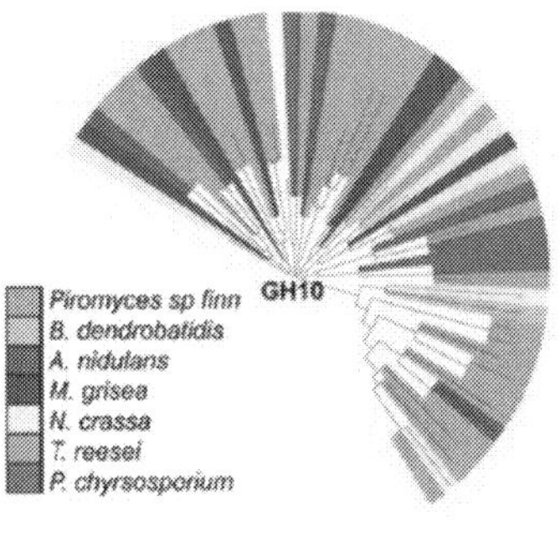

Gut fungi are rich in polysaccharide deacetylases, β–glucosidases (GH5, GH9), and endoxylanases (GH10, GH11)

Sequencing reveals hundreds of new enzymes within one strain!

27

Solomon KV, Haitjema CH, Henske JK, Gilmore SP, Borges-Rivera D, Thompson DA, O'Malley MA, *In revision.*

Xylan Degrading Machinery is Conserved Broadly Across Fungal Genera

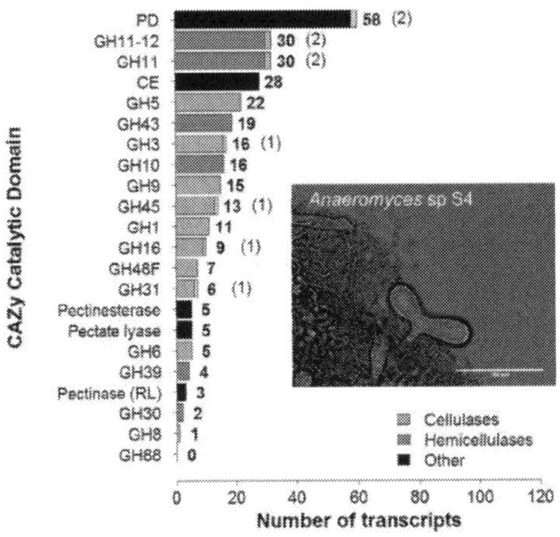

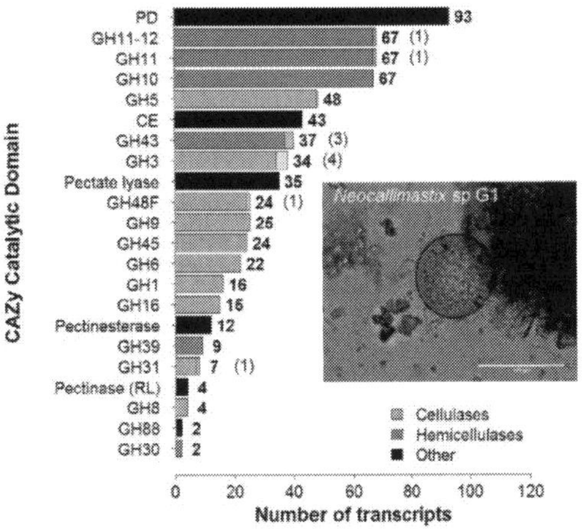

JGI Dr. Igor Grigoriev

Kevin Solomon & John Henske

Enzyme Discovery: Why Sequencing a Genome is Not Enough

UCSB

DNA —*transcription*→ (mRNA) —*translation*→ **Protein**

In order to identify the most powerful cellulolytic enzymes, it is critical to examine the transcriptome from a growing organism.

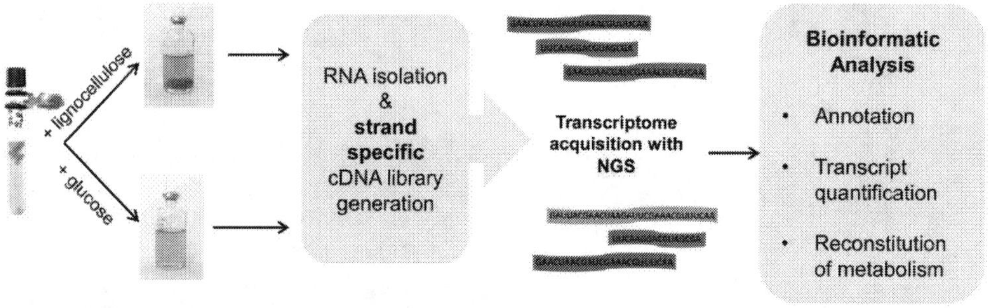

Hypothesis: Cellulosic substrates activate transcription of cellulolytic enzymes and cellulolytic complexes (cellulosomes) while sugars repress transcription

29

Cellulosomes Drive Targeted, Synergistic Cellulose Hydrolysis

UCSB

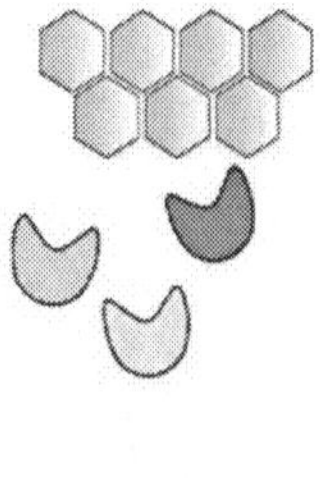

Aerobes Secrete Free Enzymes

- Enzymatic modules are fused to a protein scaffold via **dockerin-cohesin** interaction to form > **MDa complexes**
- Recent genomic sequencing of *Piromyces sp E2* by the DOE-JGI has revealed 360+ dockerin-tagged genes
- The structure and composition of fungal cellulosomes is unknown

30

Haitjema CH, et al, *Biotechnology and Bioengineering*, 111(8): 1471-1482 (2014).
Flint HJ, Bayer EA, et al. Nature Reviews Microbiology, 6 (2008) 121-131.

Famous Gut Fungi (Be careful what you wish for…)

UCSB

Los Angeles Times

LOCAL U.S. WORLD BUSINESS SPORTS ENTERTAINMENT HEALTH STYLE TRAVEL OPINION SHOP

BREAKING PHOTOS VIDEO CRIME OBITUARIES WEATHER TRAFFIC CROSSWORDS SUDOKU HOROSCOPES APPS

Horse manure yields secret to ethanol fuel

April 11, 2013 | By Geoffrey Mohan

Email Share +1 0 Tweet 0

Michelle O'Malley knows good horse poop when she sees it. While at MIT, the chemical engineer scooped up some manure from Finn, a grass-fed horse at a sustainable farm in Concord, Mass.

That offal has led to a potential breakthrough in turning grasses and nonfood crops into an alternative fuel in attempts to wean motorists from fossil fuels and stem man-made climate change.

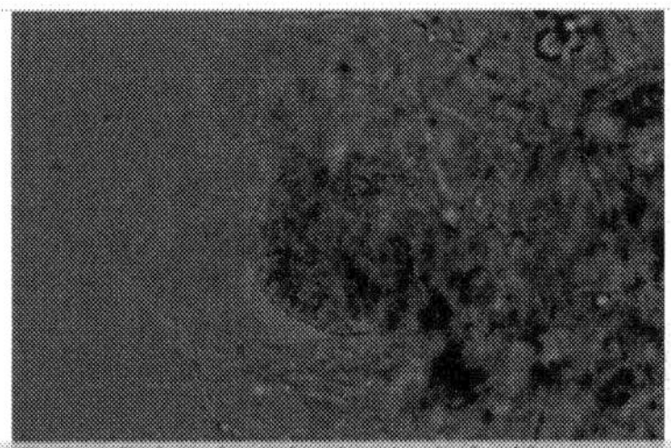

A UC Santa Barbara chemical engineer is using fungi from a horse's... (John Henske / UC Santa Barbara)

31

Over 27,000 unique transcripts *de novo* assembled by RNAseq

UCSB

The transcriptome of *Piromyces sp finn* contains more than 27,000 transcripts following a Poisson size distribution

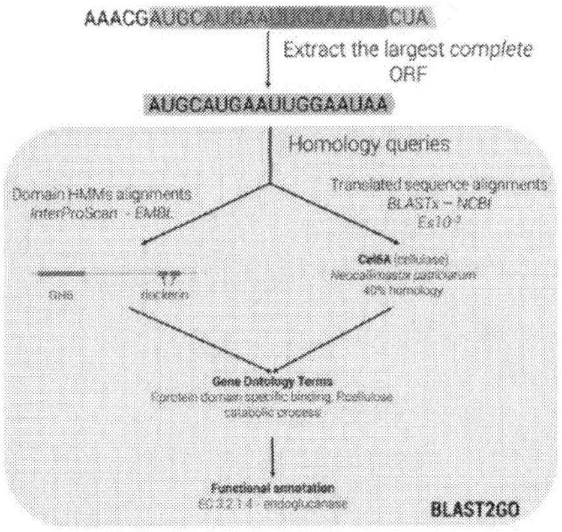

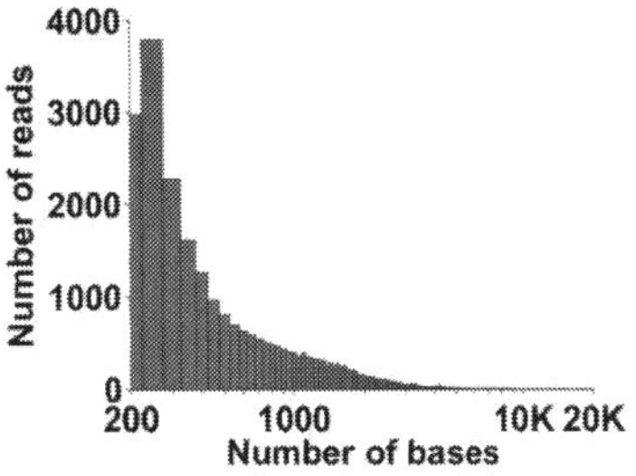

More than 30% of the transcriptome can be annotated with the majority being unique proteins (<70% homology)

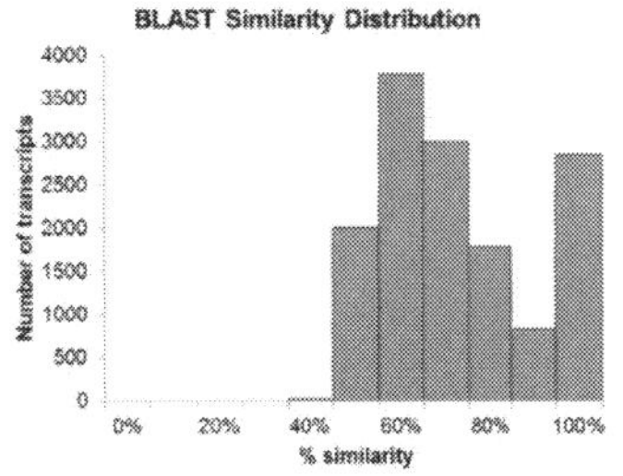

32

Kevin Solomon

A modular protein toolbox for RNA targeting

Daniel Martin-Alarcon

Katarzyna Adamala

Katriona Guthrie-Honea

Boyden Lab

synthetic neurobiology

mit media lab

State of the art

Specific RNA binding proteins (RBPs)

- MS2, PP7, and other viral coat proteins
- Lots of demonstrated applications
  - Imaging RNA molecules in vivo
  - Translational expression control
  - Cytosolic scaffolding
- Behave orthogonally from the cell and each other
- Bind to specific RNA hairpins

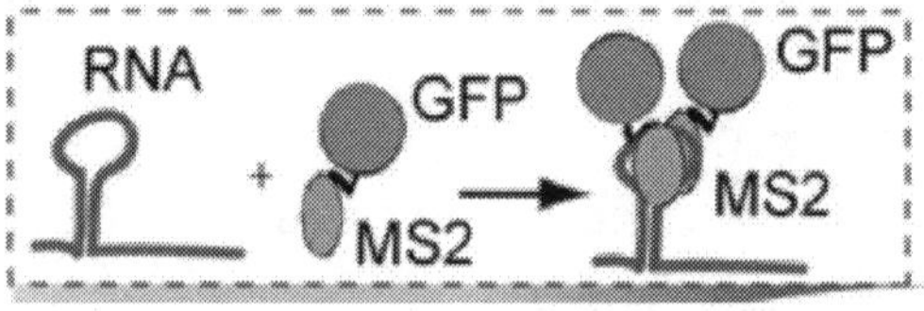

Ferguson, M.L., 2013. Measuring Transcription Dynamics in Living Cells Using Fluctuation Analysis. , 1042, pp.173–180.

2

Towards a general RBP

MIT BE
BIOLOGICAL ENGINEERING

A general RNA binding protein would

- Express in everything from bacteria to mammals
- Bind to arbitrary sequences
- Bind to targets long enough to be unique
  - (About 14 bp in the human transcriptome)

A general RBP would allow us to:

- Target RNA without cloning a binding sequence into it
- Target many sites in the same transcriptome with different effectors

3

PumHD protein

MIT BE
BIOLOGICAL ENGINEERING

The Pumilio Homology Domain:

- **8** binding units
- **3** key amino acids in each unit determine binding to one base.

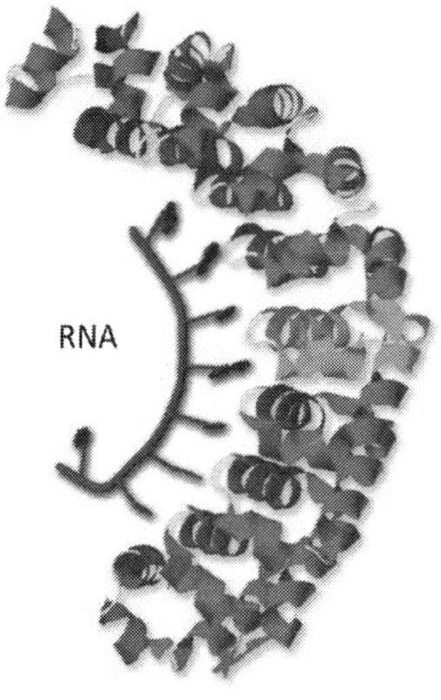

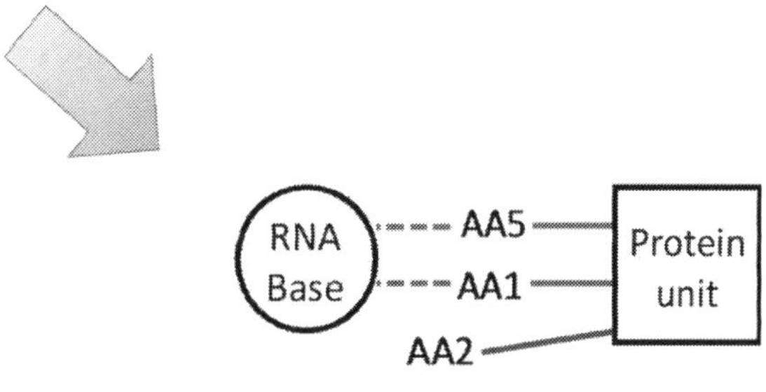

Human Pumilio1
in complex with RNA

4

PumHD protein

The binding code had been partially resolved.

We filled in the gaps and were the first to test binding to all bases at all positions in mammalian cells.

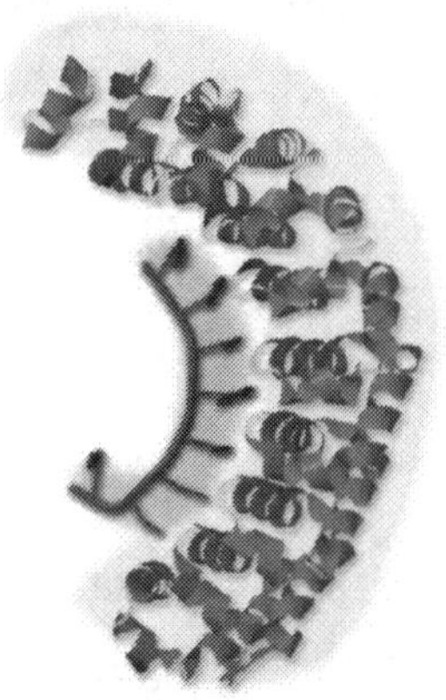

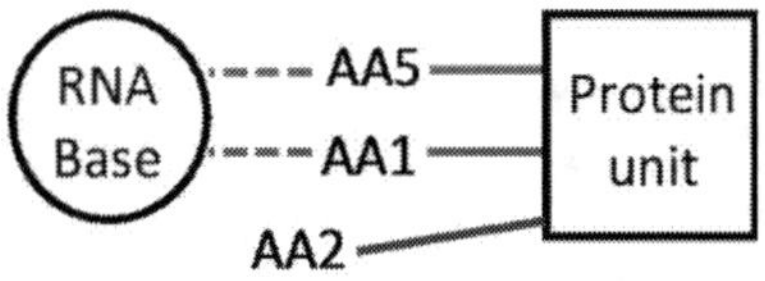

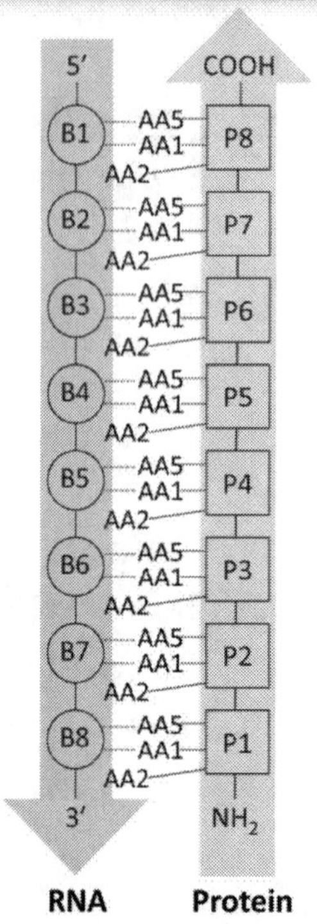

5

Prototyping PumHD mutants

We created a system for rapidly prototyping PumHD mutants.

- Adapted a Golden Gate cloning developed for TAL effectors
- Split GFP or split luciferase reconstitution allowed testing in live human cells

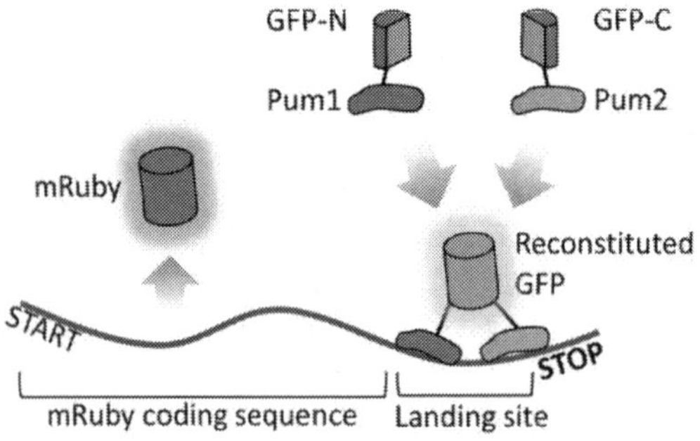

6

Prototyping PumHD mutants

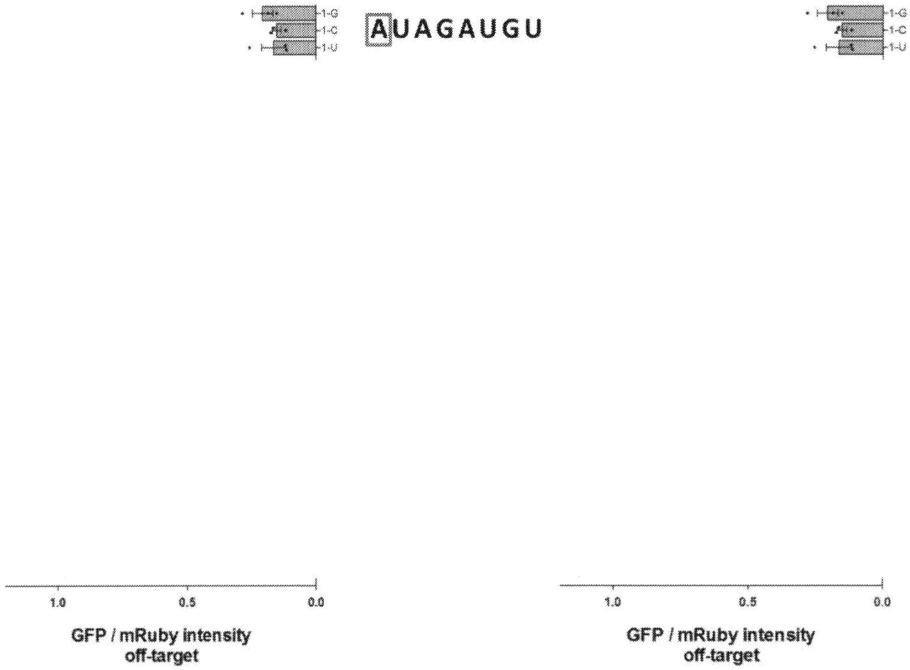

7

Pumby

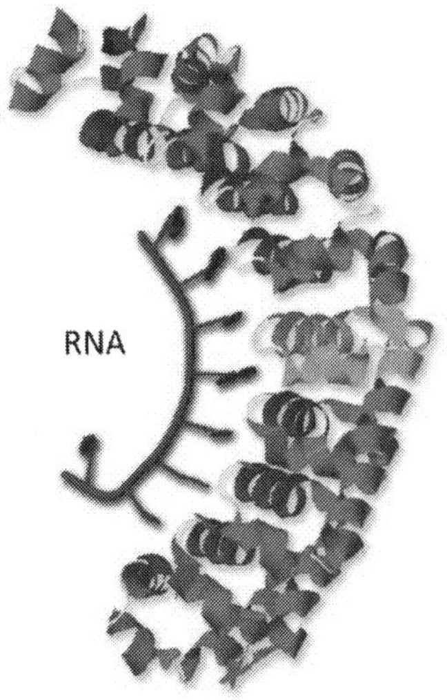

We discovered that a single, repeated unit of PumHD can create arrays of arbitrary length.

We call this system **Pum**ilio-**B**ased Assembl**y**, or **Pumby**.

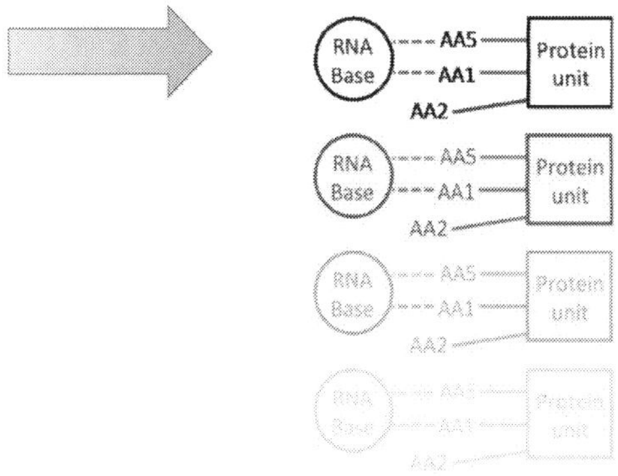

8

MIT BE
BIOLOGICAL ENGINEERING

Applications

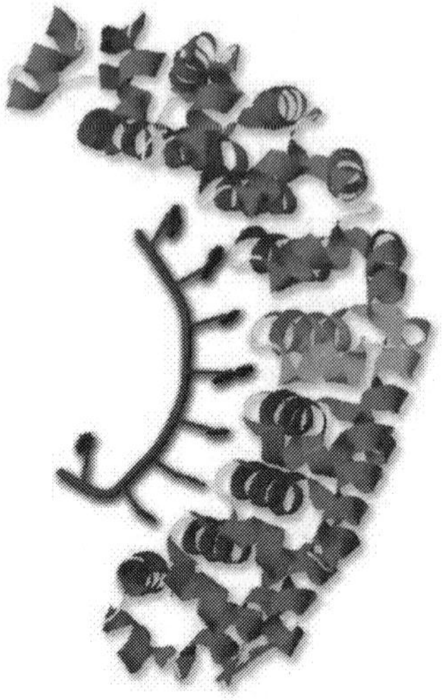

9

Application: Translation measurement

MIT BE
BIOLOGICAL ENGINEERING

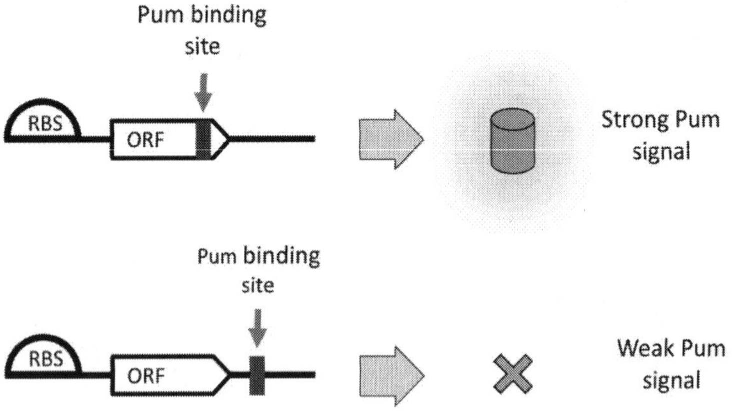

10

Application: Translation measurement

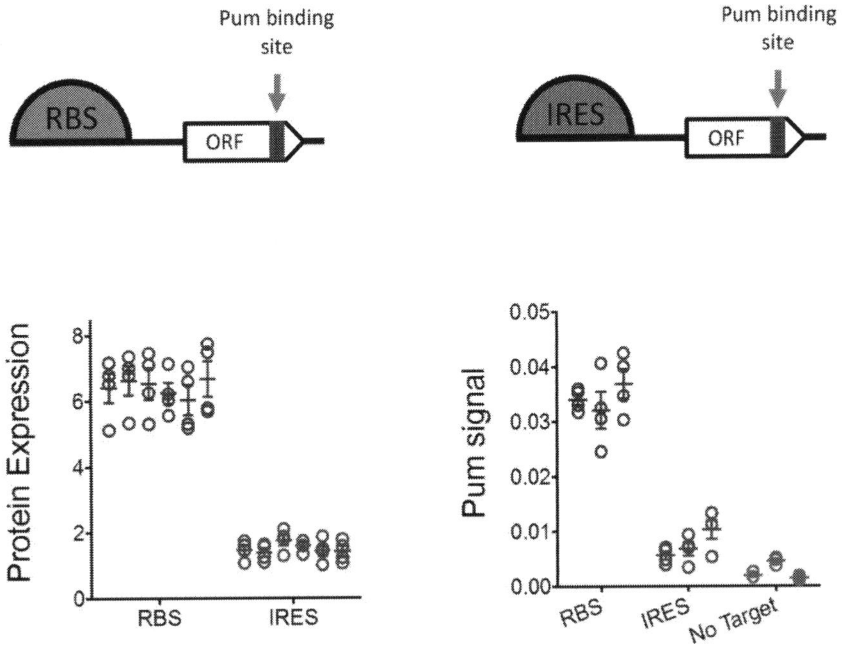

11

Application: Translation initiation

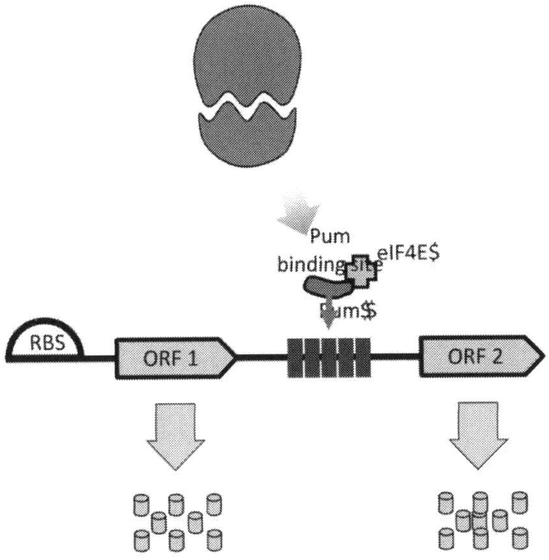

12

Application: Translation initiation

MIT BE
BIOLOGICAL ENGINEERING

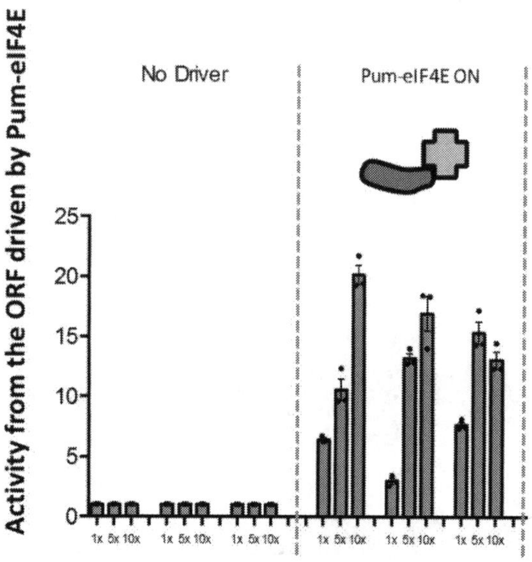

13

Orthogonality

MIT BE
BIOLOGICAL ENGINEERING

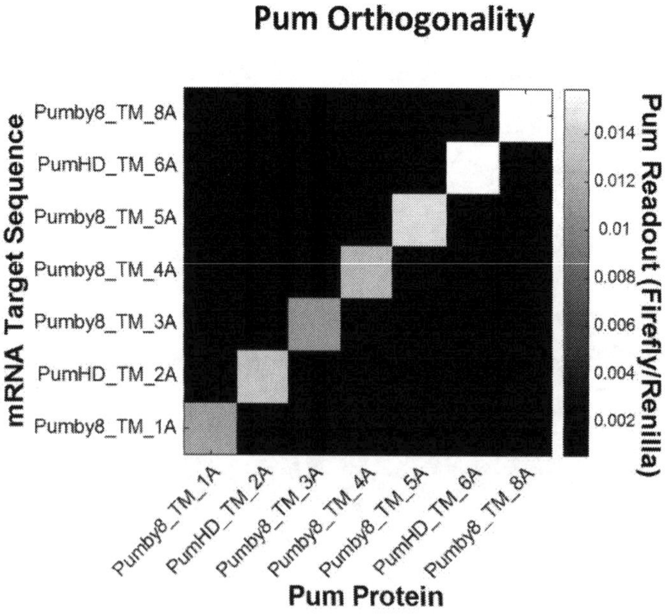

14

Summary

MIT BE
BIOLOGICAL ENGINEERING

Advantages of Pumby and PumHD

- Binding to arbitrary sequences
- Different binding lengths (Pumby only)
  - We tested 6 to 18 units.
- Good performance in fusions
- Inherently orthogonal to themselves

Future work

- Measuring single-base mismatch tolerance
- Applications in whole animals
- Translational cascades in genetic circuits
- Scaffolding for metabolic engineering

- [Your application here]

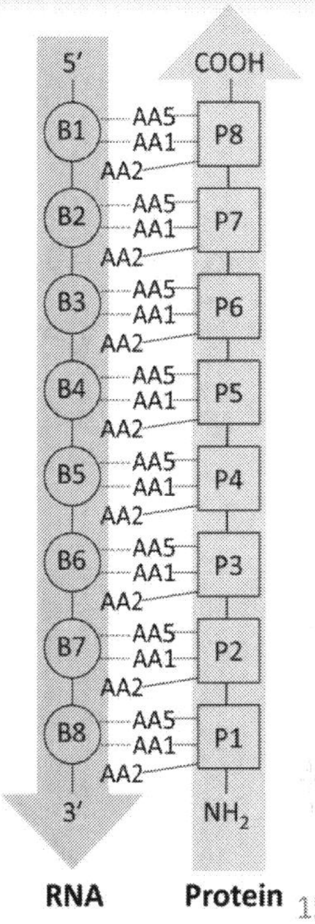

15

MIT

MIT BE
BIOLOGICAL ENGINEERING

Funding sources:

NIH 1R01NS075421, NSF CBET 1344219, Jeremy and Joyce Wertheimer, NIH BRAIN Initiative Grant 1U01MH106011, the New York Stem Cell Foundation-Robertson Award, NIH Director's Transformative Award 1R01MH103910 and NIH Director's Pioneer Award 1DP1NS087724, and the MIT Media Lab.

Thanks

Meet us at
Poster 29

Collaborators:

Katarzyna Adamala
Katriona Guthrie-Honea
Paul Reginato
Noah Jakimo
Jacob Becraft
Kiryl Piatkevich
Daniel Schmidt
Stuart Levine
Ed Boyden

synthetic neurobiology

mit media lab

Video Presentation in Original e-media

MIT BE
BIOLOGICAL ENGINEERING

Pumby

Binding of PumHD variants and Pumby variants to cognate and noncognate RNA as measured via fluorescence anisotropy of the FAM-labeled RNA target.

Protein	Active fraction	Cognate RNA	K_a	STDev K_a	K_d nM	Noncognate RNA
WT	0.35	UGUAGAUA	1.13E+10	± 1.71e+009	0.088	ACAUCUAU
PumHD-1	0.34	GUGCUCGC	8.50E+09	± 9.46e+008	0.118	CACGAGCG
PumHD-2	0.22	CAUGUCAG	2.77E+09	± 2.97e+008	0.362	GUACAGUC
Pumby-1	0.27	UGUAGAUA	7.44E+08	± 2.71e+008	1.343	ACAUCUAU
Pumby-2	0.26	CAUGUCAG	2.28E+09	± 1.71e+009	0.439	GUACAGUC

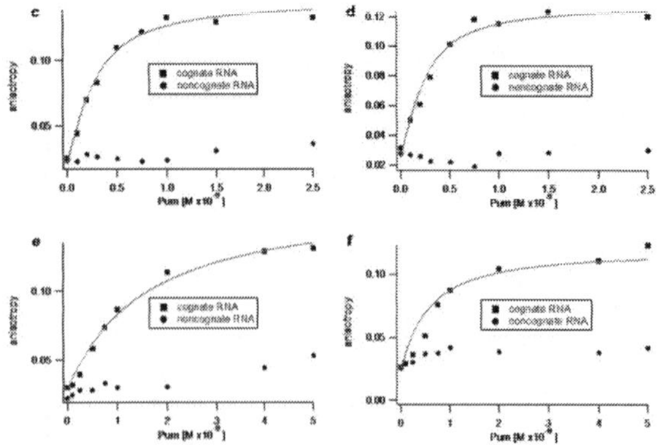

K_d nM
0.088
0.118
0.362
1.343
0.439

Prototyping Pumby chains

MIT BE
BIOLOGICAL ENGINEERING

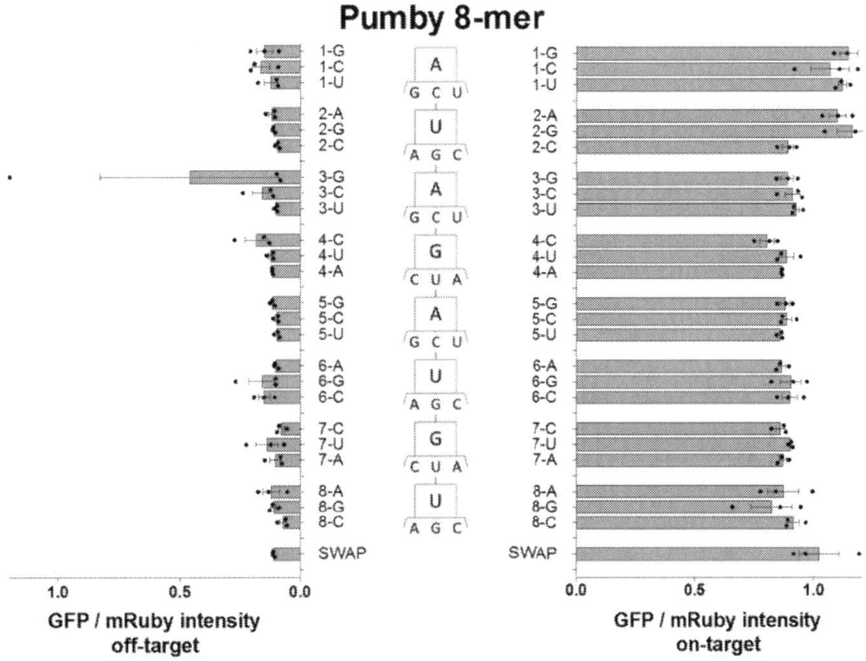

19

Prototyping Pumby chains

MIT BE
BIOLOGICAL ENGINEERING

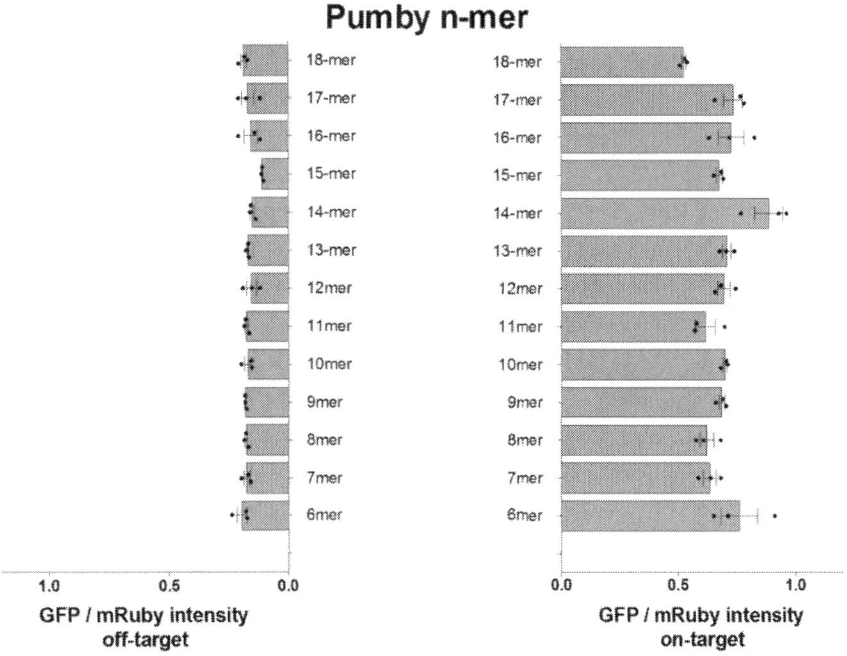

20

Prototyping Pumby chains

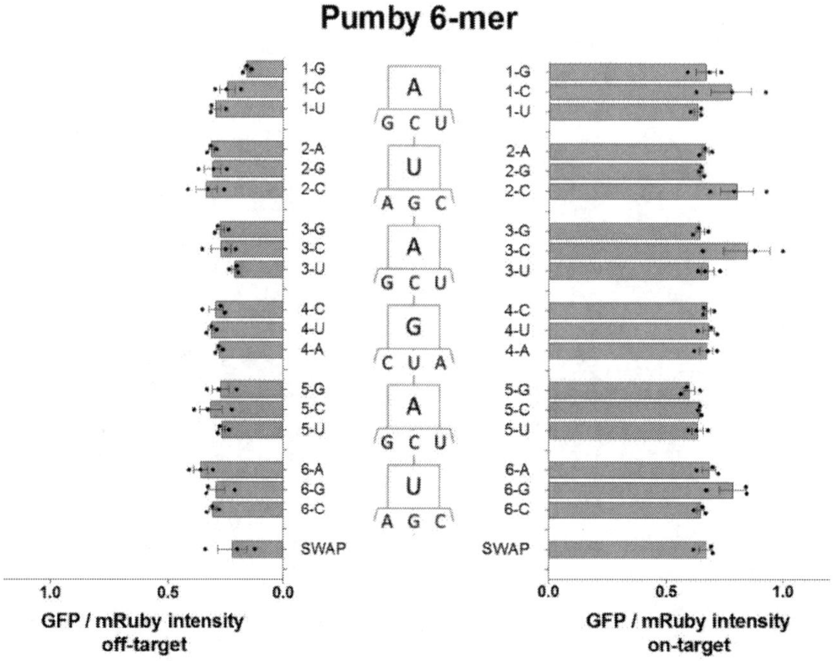

21

Application: Translation measurement

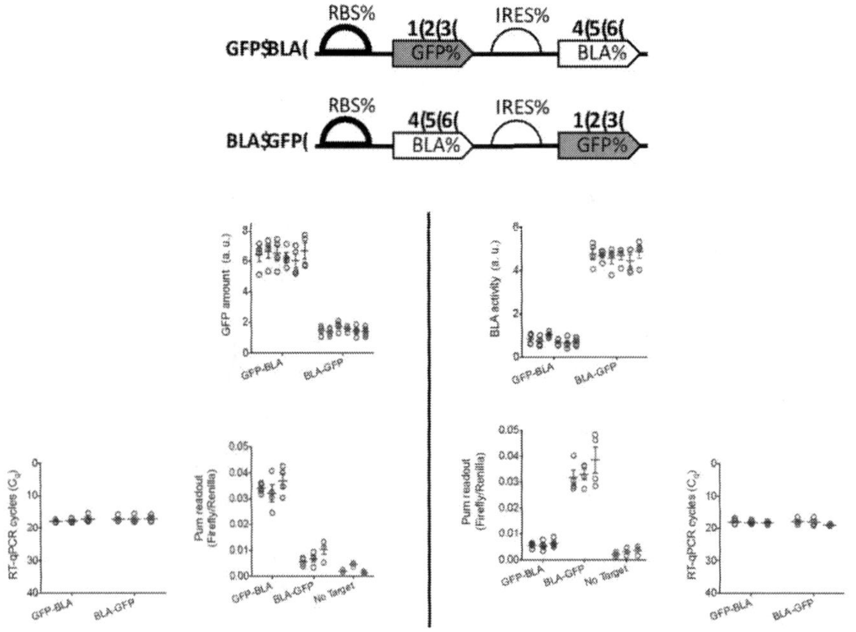

22

Application: Translation Initiation

MIT BE
BIOLOGICAL ENGINEERING

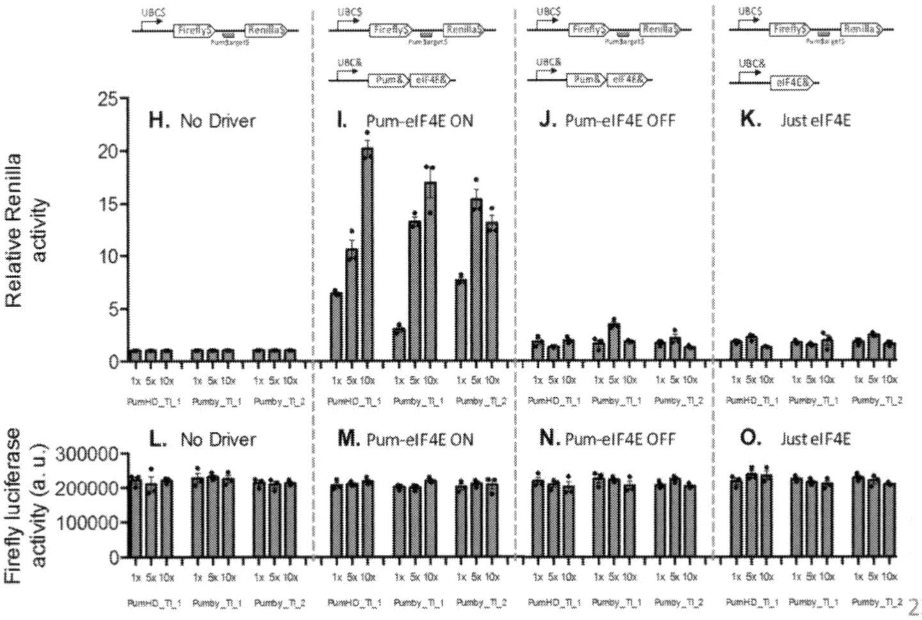

Bonus slides

MIT BE
BIOLOGICAL ENGINEERING

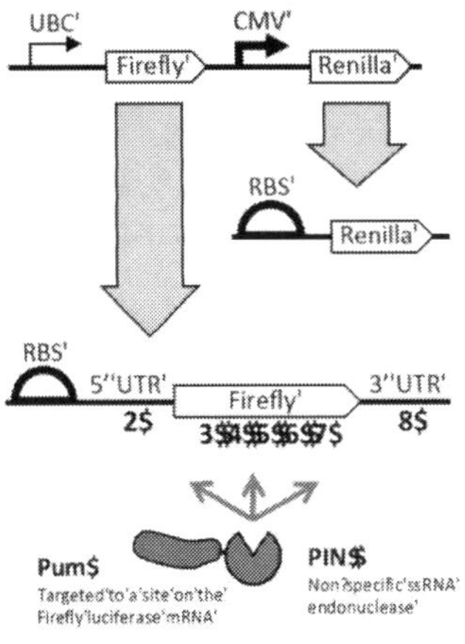

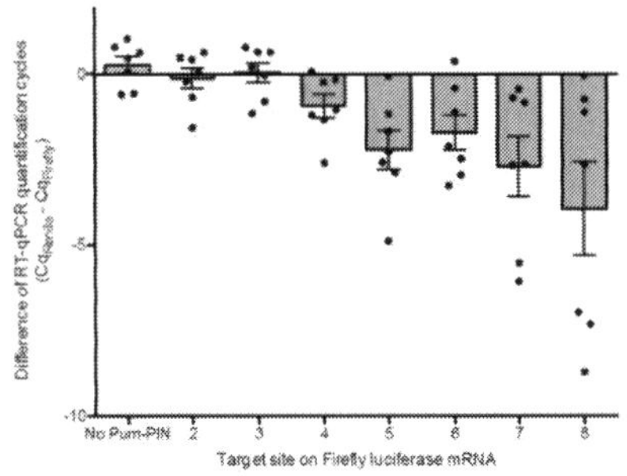

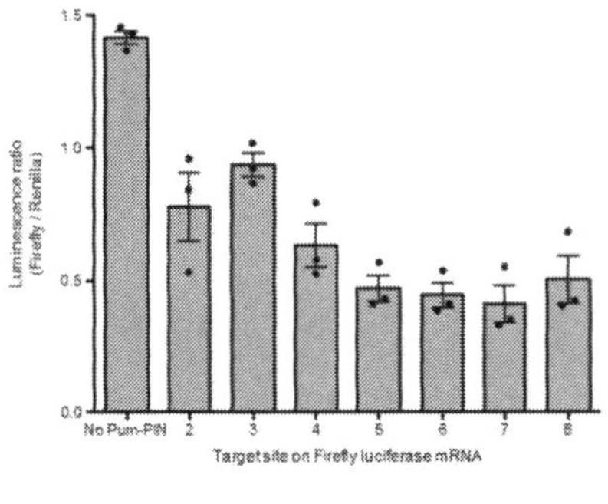

Bonus slides

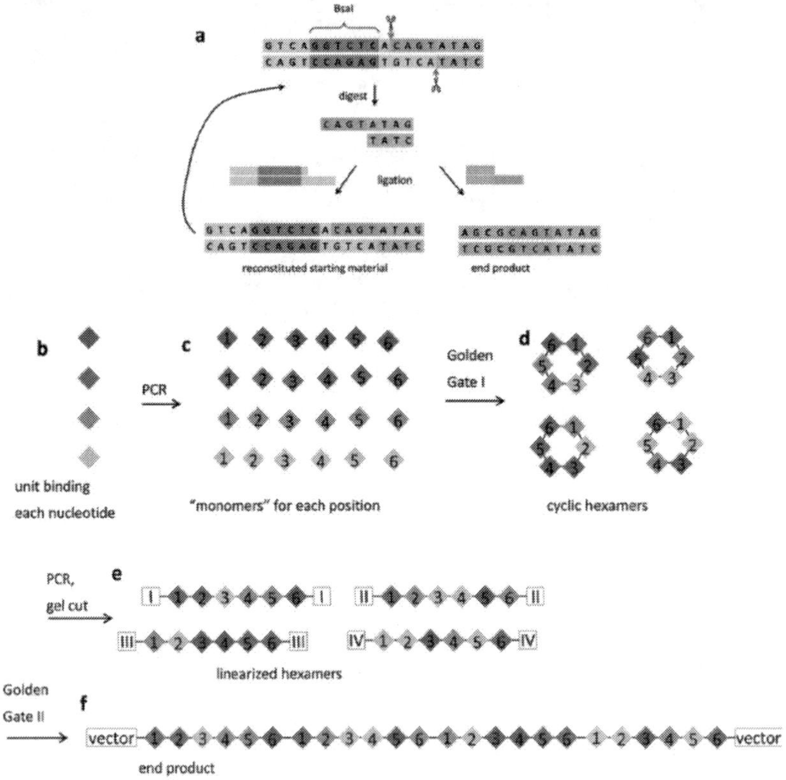

25

Designing Conservation Relations in Layered Synthetic Biomolecular Networks

Thomas P. Prescott
Department of Engineering Science, University of Oxford

Synthetic Biology, Engineering, Evolution, and Design (SEED)
Boston MA
Friday 12th June 2015

Mathematical modelling for systems design

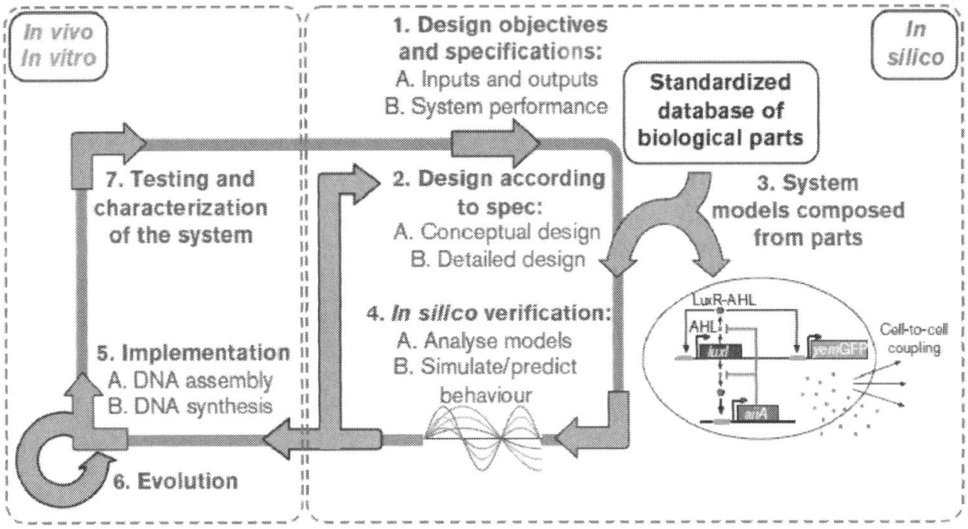

Taken from Arpino, Hancock, Anderson, Barahona, Stan, Papachristodoulou & Polizzi, *Microbiology* (2013)

Conservation Relations in Biomolecular Networks

$$\dot{x} = Sv(x)$$

The state values $x(t)$ lie in a subspace of R^n:

$$x(t) - x(0) \in \mathrm{Col}(S)$$

called the **stoichiometric compatibility class**.

Dual: vector λ such that $\lambda^T S = 0$ gives

$$\lambda^T \dot{x} = 0 \Rightarrow \lambda^T x(t) = K$$

for any $v(x)$. There are $n - \mathrm{rank}(S)$ independent such **conservation relations**.

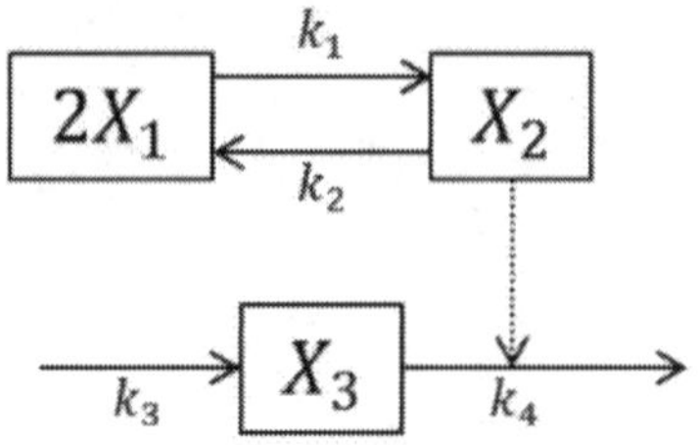

Concentration vector:
$$x = [x_1, x_2, x_3]^T$$

Flux vector:
$$v(x) = [k_1(x_1)^2, k_2x_2, k_3, k_4x_2x_3]^T$$

Stoichiometric matrix:
$$S = \begin{bmatrix} -2 & 2 & 0 & 0 \\ 1 & -1 & 0 & 0 \\ 0 & 0 & 1 & -1 \end{bmatrix}$$

Conservation Relations in Biomolecular Networks

$$\dot{x} = Sv(x)$$

The state values $x(t)$ lie in a subspace of R^n:

$$x(t) - x(0) \in \mathrm{Col}(S)$$

called the **stoichiometric compatibility class**.

Dual: vector λ such that $\lambda^T S = 0$ gives

$$\lambda^T \dot{x} = 0 \Rightarrow \lambda^T x(t) = K$$

for any $v(x)$. There are $n - \mathrm{rank}(S)$ independent such **conservation relations**.

Note that these conservation relations are
- Independent of the reaction rates $v(x)$.
- Linear in x

Aim of our work is to investigate the construction of new conservation relations that are
- A more general (nonlinear) conservation of the form $f(x(t)) = K$
- Tuneable based on the choice of reaction network parameters.

Tuning the dynamics of Genetic Regulation Networks

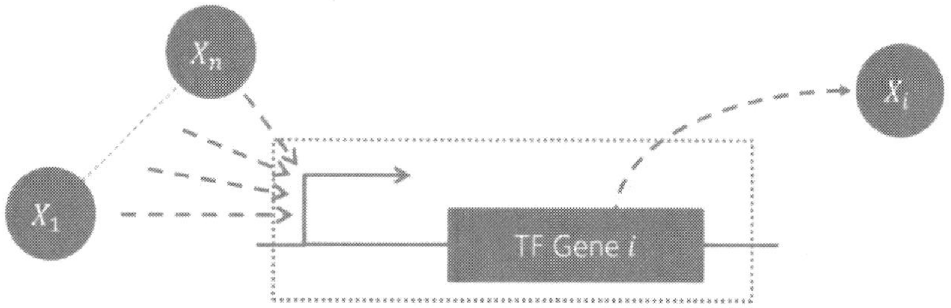

$$\dot{x}_i = \alpha_i - \delta_i x_i + R_i(x)$$

Tuning the dynamics of Genetic Regulation Networks

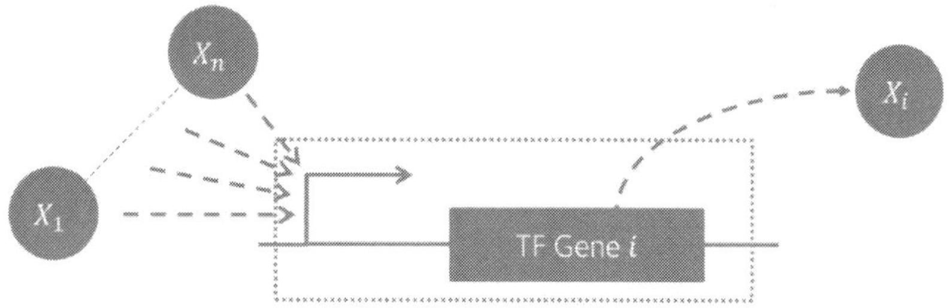

$$\dot{x}_i = \alpha_i - \delta_i x_i + R_i(x)$$

GRN determined by $R_i(x)$:

- TF X_j promotes expression of X_i may be modelled by
$$R_i(x) = \frac{\beta_{i,j} x_j}{\gamma_{i,j} + x_j}$$
- TF X_j promotes and X_k suppresses X_i may be modelled by
$$R_i(x) = \left(\frac{\beta_{i,j} x_j}{\gamma_{i,j} + x_j}\right)\left(\frac{\beta_{i,k}}{\gamma_{i,k} + x_k}\right)$$

Tuning the dynamics of Genetic Regulation Networks

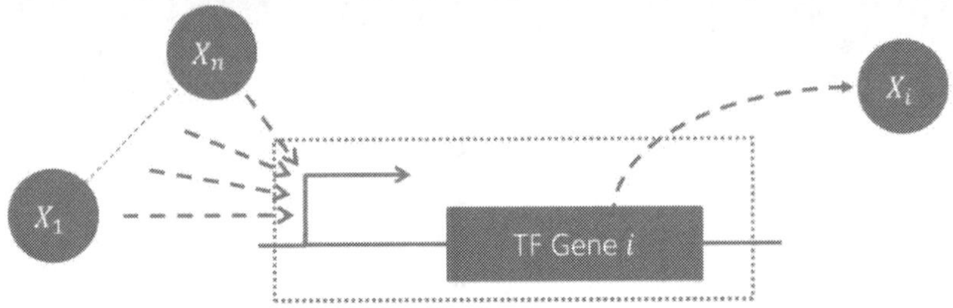

$$\dot{x}_i = \alpha_i - \delta_i x_i + R_i(x)$$

GRN determined by $R_i(x)$:

- TF X_j promotes expression of X_i may be modelled by
$$R_i(x) = \frac{\beta_{i,j} x_j}{\gamma_{i,j} + x_j}$$
- TF X_j promotes and X_k suppresses X_i may be modelled by
$$R_i(x) = \left(\frac{\beta_{i,j} x_j}{\gamma_{i,j} + x_j}\right)\left(\frac{\beta_{i,k}}{\gamma_{i,k} + x_k}\right)$$

A typical question for Synthetic Biology would be how to design this GRN (i.e. tune parameters, select the combinations of promoter/repressor/genes) to produce a specified performance.

Constraining a GRN state space

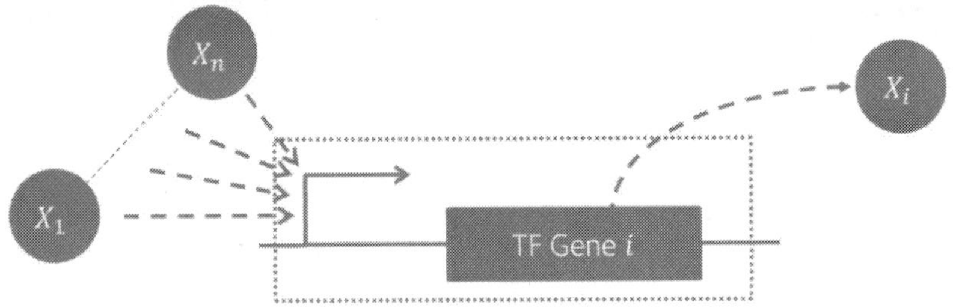

$$\dot{x}_i = \alpha_i - \delta_i x_i + R_i(x)$$

Introduce fast, direct interactions between transcription factors (and other intermediates, potentially). That is, overlay a ***fast*** Protein Interaction Network (PIN) layer with the GRN.

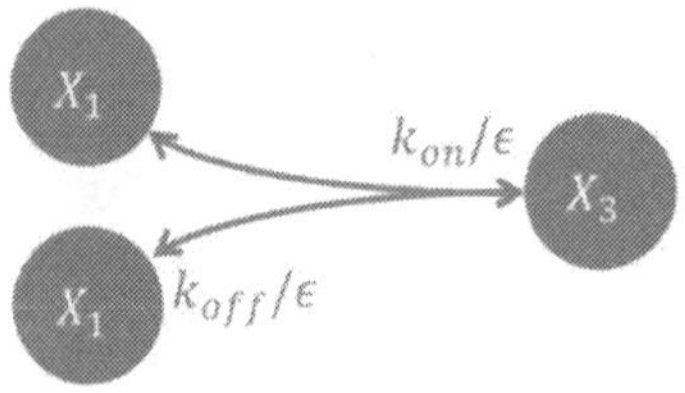

$$\dot{x}_1 = \alpha_1 - \delta_1 x_1 + R_1(x) - \frac{1}{\epsilon}\left[k_{on} x_1^2 - k_{off} x_3\right]$$

$$\dot{x}_3 = \alpha_3 - \delta_3 x_3 + R_3(x) + \frac{1}{\epsilon}\left[k_{on} x_1^2 - k_{off} x_3\right]$$

Constraining a GRN state space

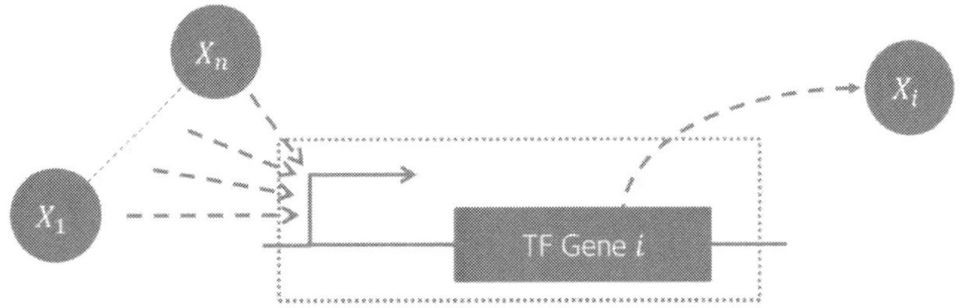

Nonlinear tuneable conservation relation

$$x_3 = \frac{k_{on}}{k_{off}} x_1^2$$

holds for all slow parameters α_i and δ_i, and also all regulation functions $R_i(x)$

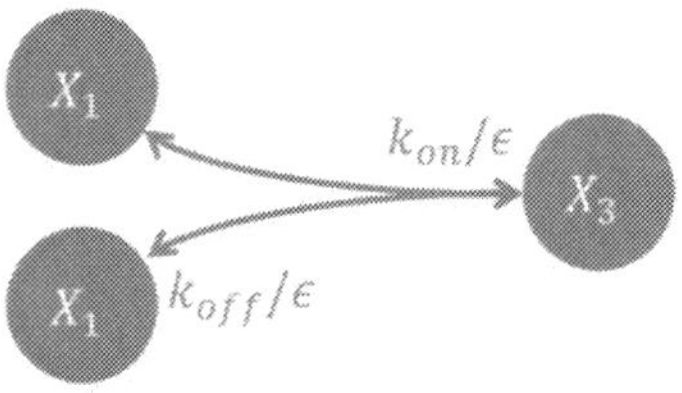

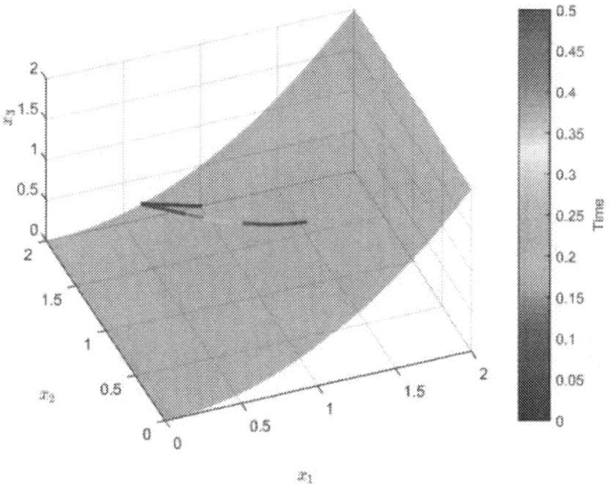

Previous Work: Timescale Separation

$$\dot{x} = Sv(x) = [S^s \quad S^f]\begin{bmatrix} v^s(x) \\ v^f(x)/\epsilon \end{bmatrix} = S^s v^s(x) + \frac{1}{\epsilon} S^f v^f(x)$$

Layered approach is to **replicate the state** and sum their contributions.

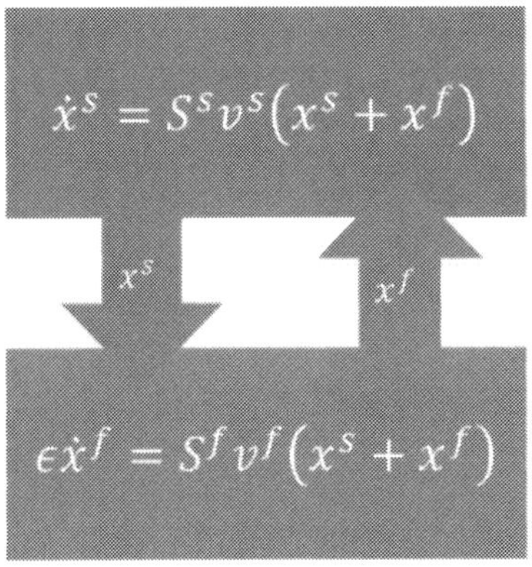

$$x = x^s + x^f$$

Previous Work: Timescale Separation

$$\dot{x} = Sv(x) = [S^s \quad S^f]\begin{bmatrix} v^s(x) \\ v^f(x)/\epsilon \end{bmatrix} = S^s v^s(x) + \frac{1}{\epsilon} S^f v^f(x)$$

Layered approach is to **replicate the state** and sum their contributions.

Now perform **quasi-steady state approximation** on the fast layer to produce a total of $r^f = \mathrm{rank}(S^f)$ new nonlinear, tuneable conservation relations

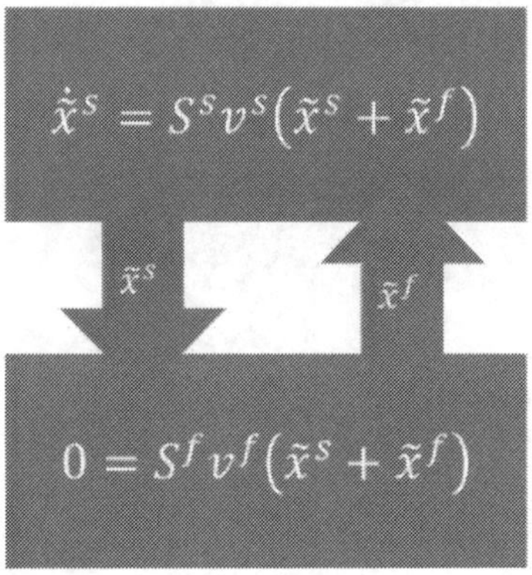

$$\tilde{x} = \tilde{x}^s + \tilde{x}^f$$

See Prescott & Papachristodoulou, J.Theor.Biol. (2014) for more details.

Fast reactions lead to new conservation relations

$$\dot{x} = Sv(x) = [S^s \quad S^f]\begin{bmatrix} v^s(x) \\ v^f(x)/\epsilon \end{bmatrix} = S^s v^s(x) + \frac{1}{\epsilon} S^f v^f(x)$$

Can we design a set of fast reactions to overlay on any arbitrary slow network such that a specified nonlinear state-space constraint $p(x) = 0$ holds – that is, such that $0 = S^f v^f(x)$ implies $p(x) = 0$?

$$0 = S^f v^f(\tilde{x})$$

Nonlinear Conservation Relations

$$0 = S^f v^f(\tilde{x})$$

- What possible conservation relations $p(x) = 0$ can be implemented by a fast reaction layer?
- Are there multiple fast layer implementations of the same specific nonlinear conservation relation? How should we choose which one to use?

Multiple Designs

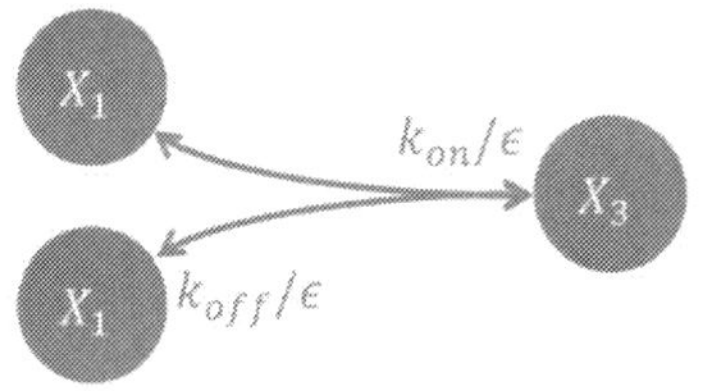

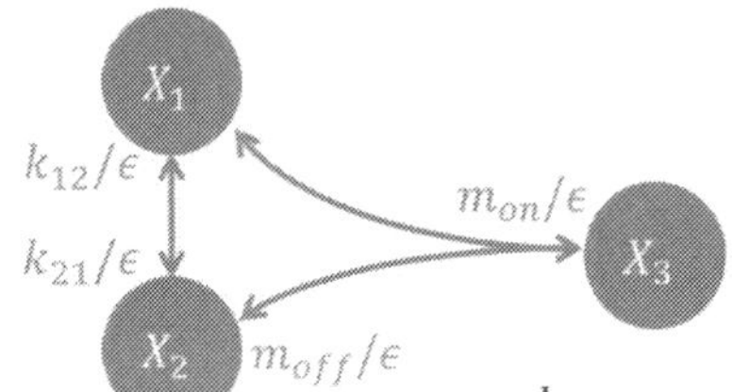

$$x_2 = \frac{k_{21}}{k_{12}} x_1$$

$$x_3 = \frac{k_{on}}{k_{off}} x_1^2 = \gamma x_1^2$$

$$x_3 = \frac{m_{on}}{m_{off}} x_1 x_2 = \gamma x_1^2$$

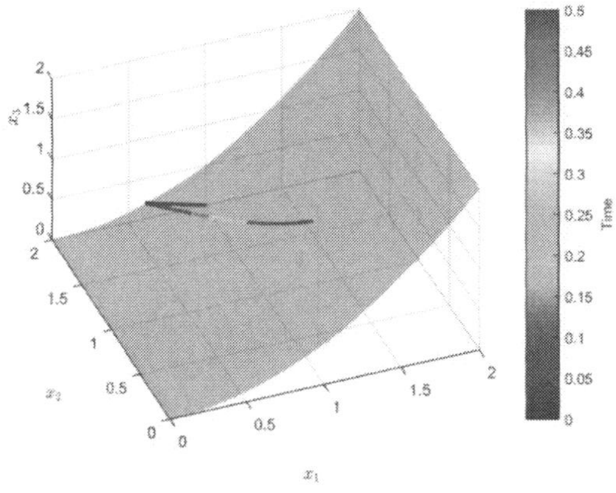

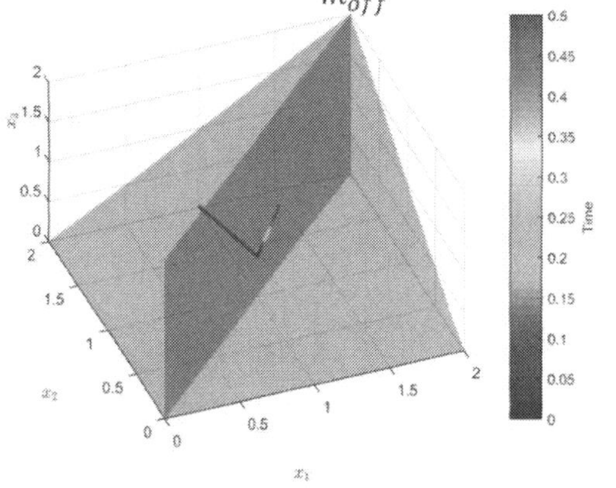

Multiple Designs

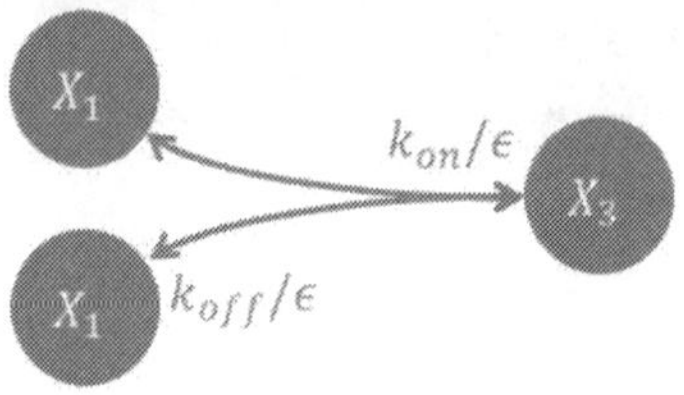

$$x_3 = \frac{k_{on}}{k_{off}} x_1^2 = \gamma^2 x_1^2$$

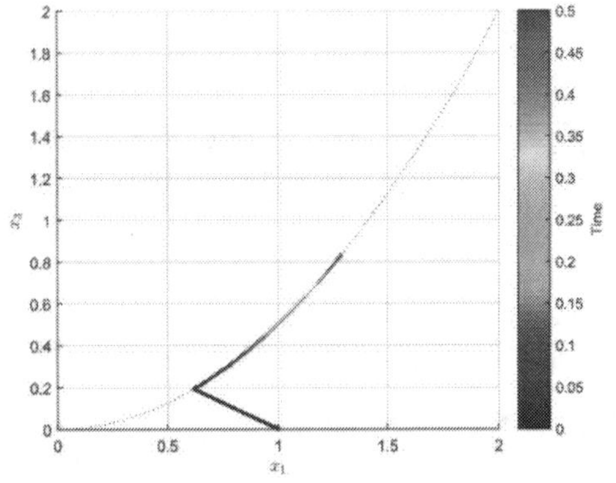

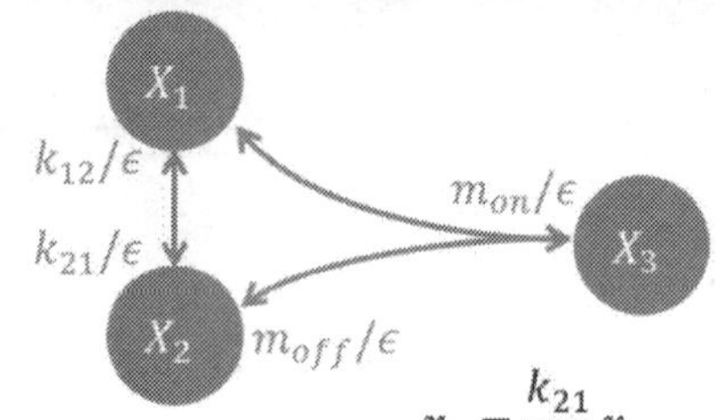

$$x_2 = \frac{k_{21}}{k_{12}} x_1$$

$$x_3 = \frac{m_{on}}{m_{off}} x_1 x_2 = \gamma x_1^2$$

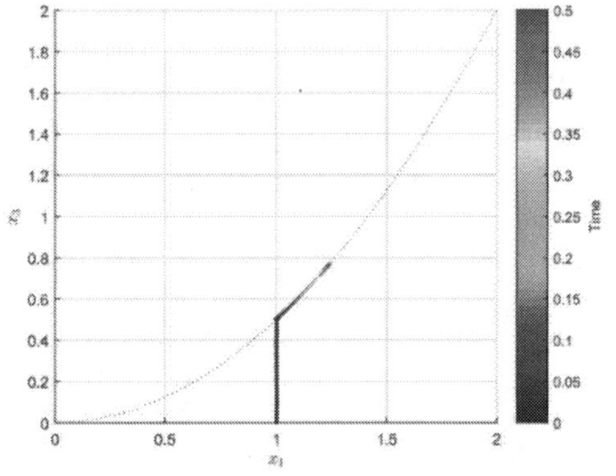

What about slow dynamics?

- The criterion for the optimal fast layer implementation is likely to be whatever makes the design of the dynamics (i.e. of the slow layer) easiest.
- The next task is to design the slow dynamics in the presence of a fast layer.

Can show (again TPP & AP, J.Theor.Biol., 2014) that the approximated system dynamics are

$$\dot{x} = \left(I + M^f(x)\right) S^s v^s(x)$$

where $M^f(x)$ is an easily-calculated matrix, which projects the isolated slow dynamics into the desired state space.

What about slow dynamics?

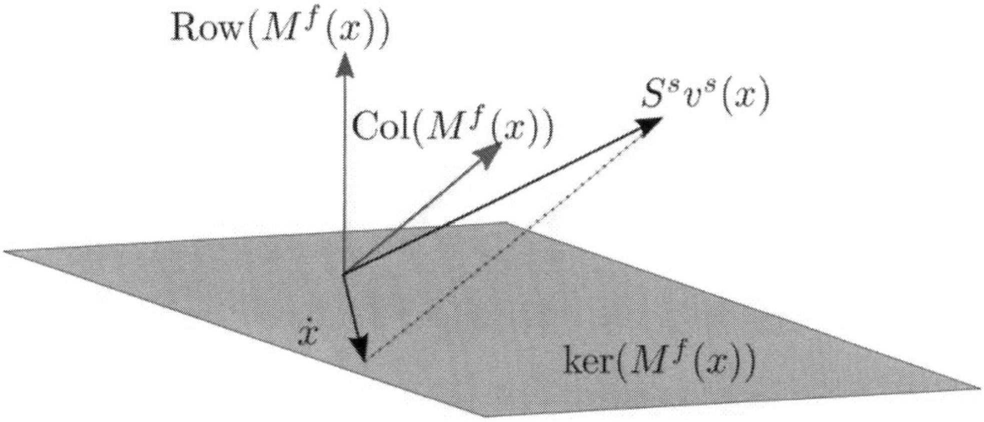

$$\dot{x} = \left(I + M^f(x)\right) S^s v^s(x)$$

$$0 = S^f v^f(x)$$

Thanks, SEED. ThEED.

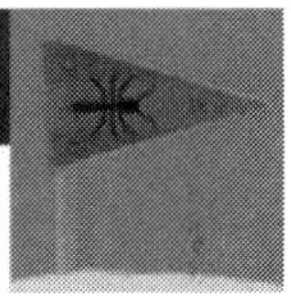

Thanks for help and funding are due also to

- The SySOS group of Prof Antonis Papachristodoulou at the Department of Engineering Science, University of Oxford.
- Engineering and Physical Sciences Research Council (UK)

EPSRC

Original dynamics:

$$\dot{x} = \frac{1}{\epsilon} S^f v(x) + S^s v(x)$$

Approximated dynamics:

$$\dot{\tilde{x}} = \left(I + M^f(\tilde{x})\right) S^s v(\tilde{x})$$

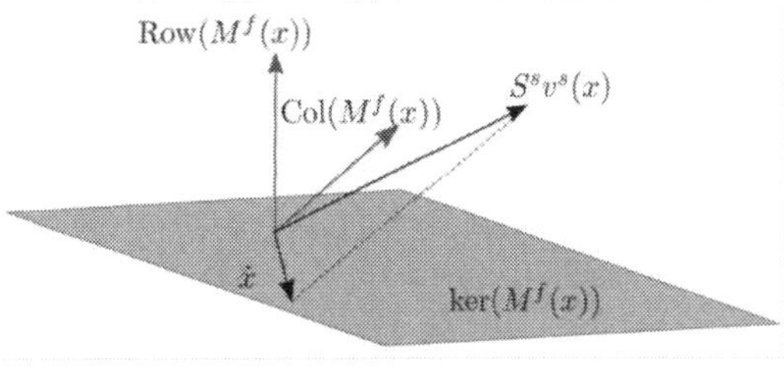

DESIGN AND OPTIMIZATION OF pH-RIBOSWITCH CIRCUITS FOR CELL SORTING APPLICATIONS

Hoang Long Pham, Adison Wong, Joey Ng, Matthew Chang

Department of Biochemistry, National University of Singapore
NUS Synthetic Biology for Clinical & Technological Innovation

pH-SENSING IN CELLULAR ENGINEERING

- Useful environmental sensor for cellular engineering.
- Examples: biopharmaceutical manufacturing, drug delivery.
- Shortage of well-characterized prokaryotic pH-sensor.

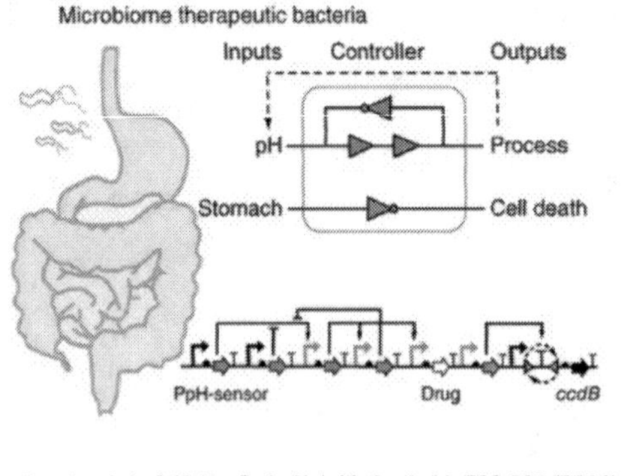

Brophy, J. A. & Voigt, C. A. *Nat. Methods* **11**, 508-520 (2014)

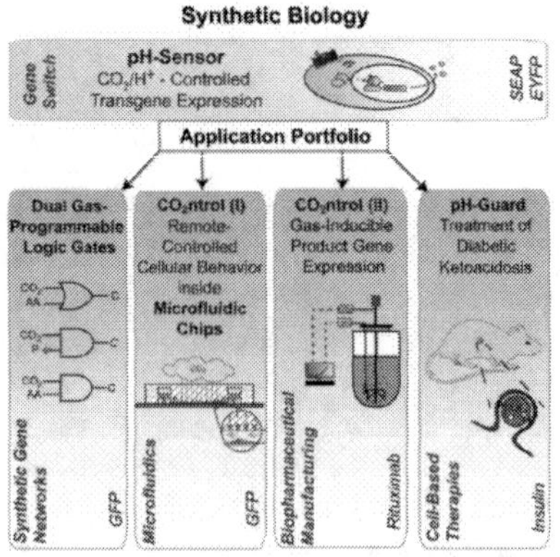

Auslander, D. *et al. Mol Cell* **55**, 397-408 (2014)

ENGINEERING NATURAL PH-RESPONSIVE ELEMENT

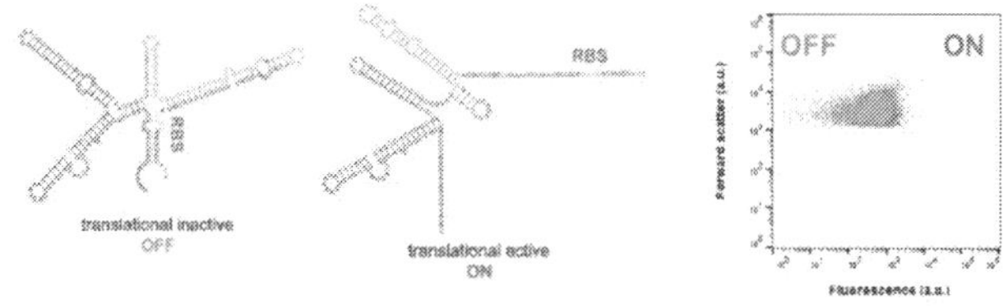

- Low output expression
- Low activation ratio

Gal Nechooshtan *et al.* *Genes Dev* **23**, 2650-2662 (2009)

- Systematic engineering workflow:
  - Part-based engineering
  - Device-based engineering
  - Circuit-based engineering

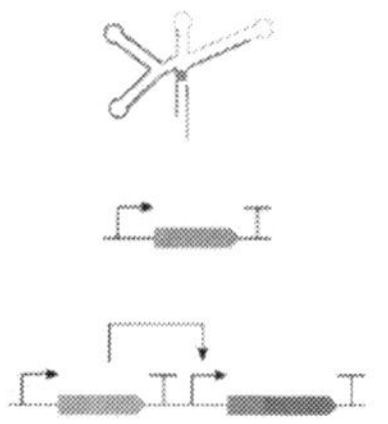

pHR: **pH-R**iboswitch
RFP: red fluorescent protein

PART ENGINEERING FOR IMPROVED SENSING PERFORMANCE

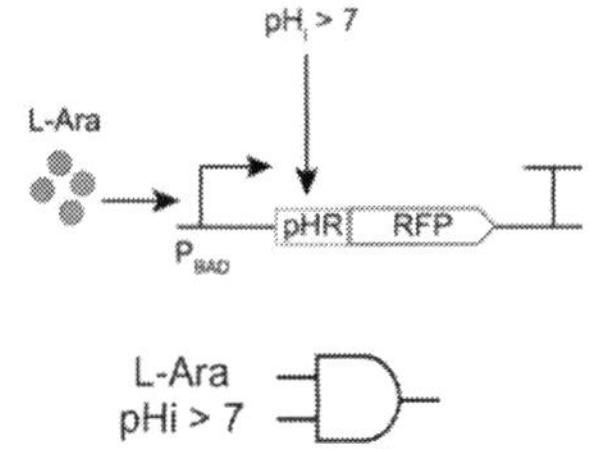

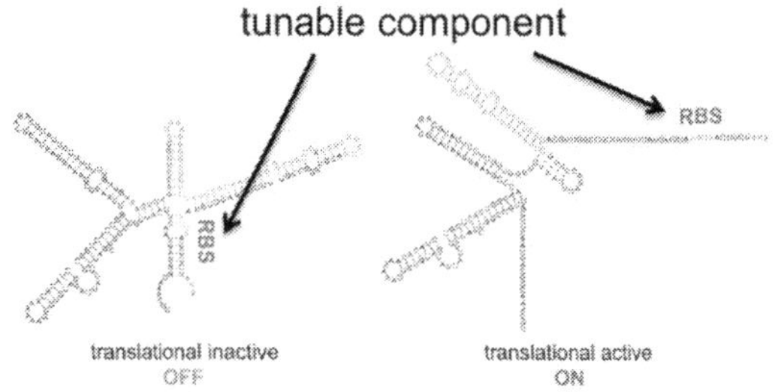

cRBS wt RBS wt

pHR wt UGCAAAGGGGAGTAACUUCA-172nt-GUGUAAGGAACUUCUaug

↓

UGCAAAGGGGANNNNNNNNNNNNNCUUCA-172nt-GUGUANNNNNNNNNNNNACUUCUaug

↓

pHR eng UGCAAAGGGGATTTCTCCTCTTTCUUCA-172nt-GUGUAAAAGAGGAGAAAACUUCUaug

cRBS eng RBS eng

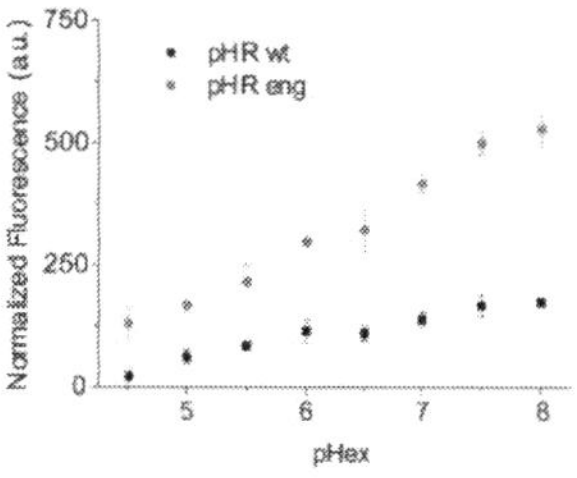

pHR: **pH-R**iboswitch
RFP: red fluorescent protein

DEVICE ENGINEERING FOR IMPROVED SENSING PERFORMANCE

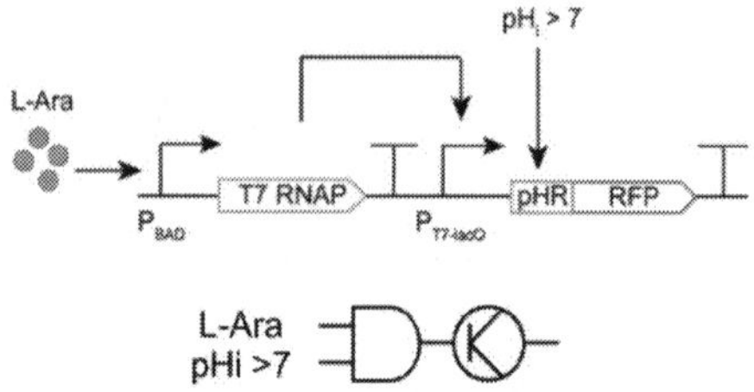

		In vitro speed (nt/sec)
	Wild type	240
T7 RNAP	Y639F	240
	F644A	170
	Q649S	160
	G645A	90
E. coli RNAP	Wild type	70
B. sublitis RNAP	Wild type	45

Makarova *et al.* *PNAS* **92**, 12250-12254 (1995)

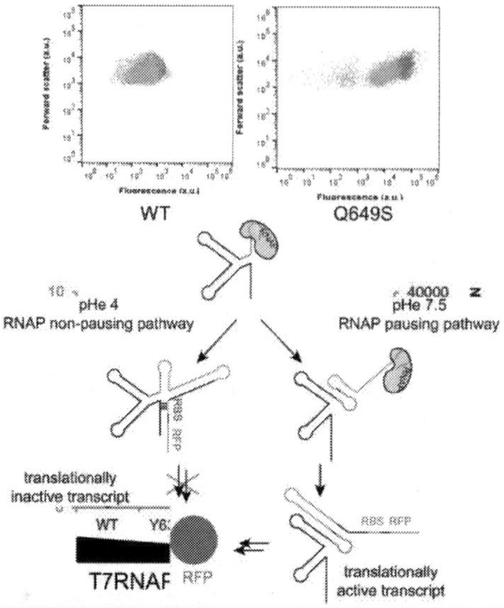

pHR: pH-Riboswitch
RFP: red fluorescent protein

Blue: OFF state
Red: ON state

CIRCUIT ENGINEERING FOR IMPROVED SENSING PERFORMANCE

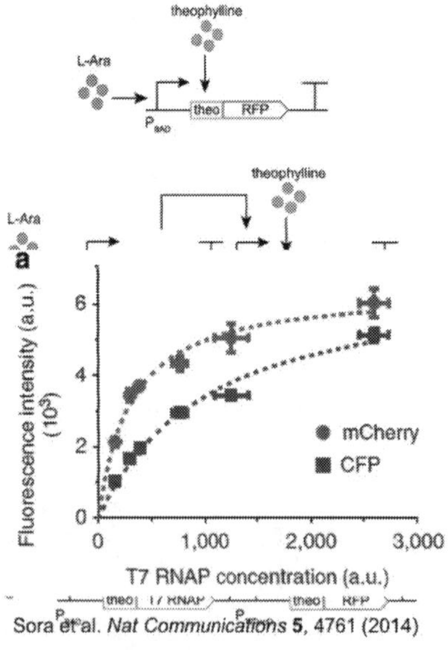

Sora et al. *Nat Communications* **5**, 4761 (2014)

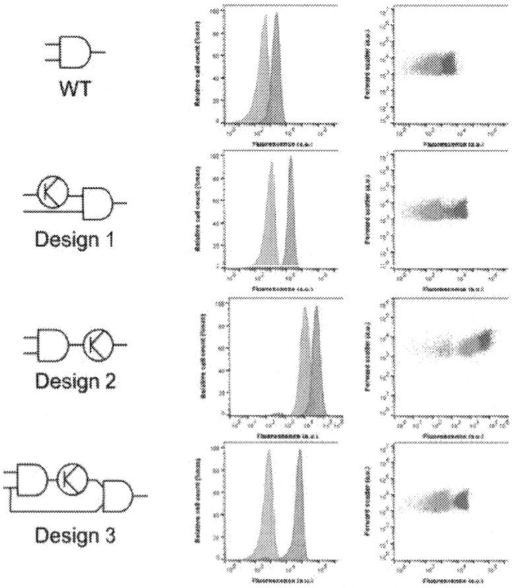

theo: theophylline-riboswitch
RFP: red fluorescent protein

Blue: OFF state
Red: ON state

CIRCUIT ENGINEERING FOR IMPROVED SENSING PERFORMANCE (cont.)

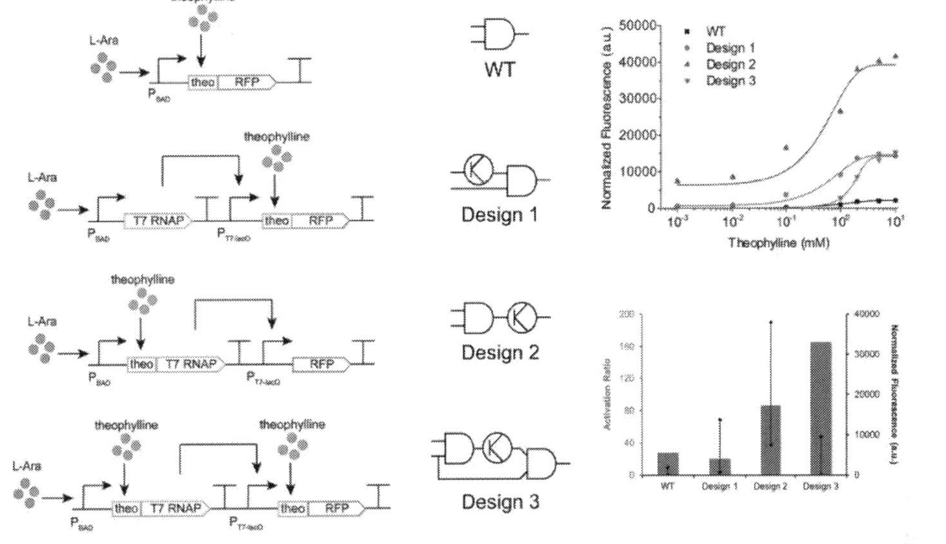

theo: theophylline-riboswitch
RFP: red fluorescent protein

OPTIMIZATION OF INTRACELLULAR pH SENSOR WITH LAYERED GENETIC CIRCUIT

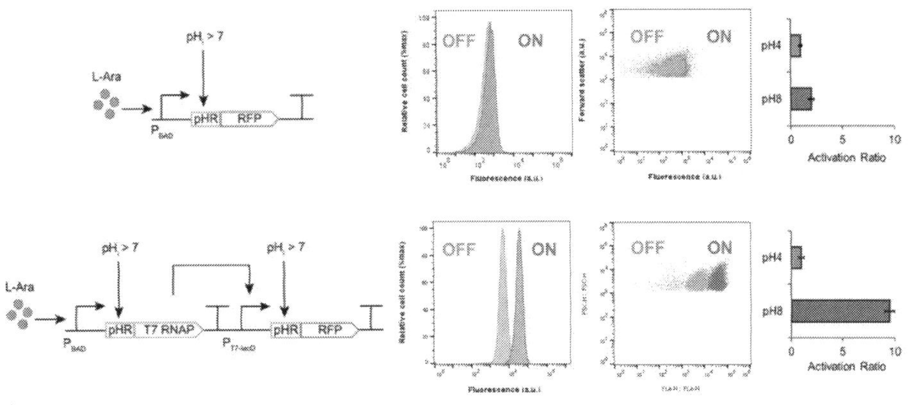

pHR: pH-Riboswitch
RFP: red fluorescent protein

INVERTED INTRACELLULAR pH SENSING DEVICE

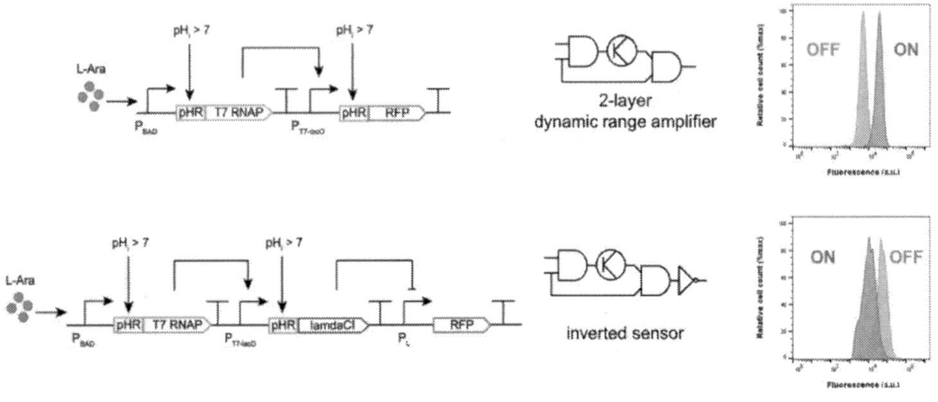

pHR: **pH-R**iboswitch
RFP: red fluorescent protein

FEEDBACK-REGULATED STRATEGY FOR DIRECTED EVOLUTION

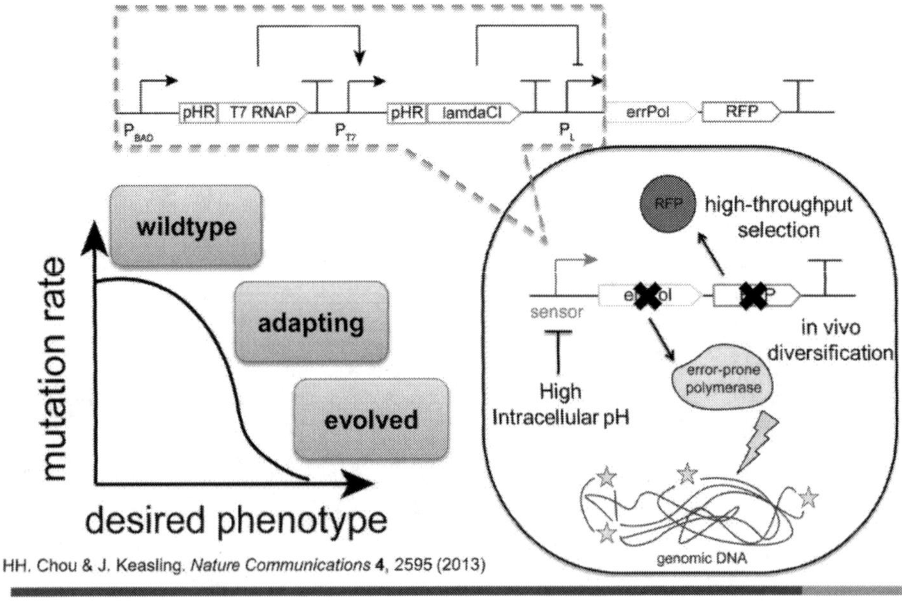

HH. Chou & J. Keasling. *Nature Communications* **4**, 2595 (2013)

SUMMARY

- Engineering workflow for improving riboswitch-based sensing performance:
  - Part-based engineering
  - Device-based engineering
  - Circuit-based engineering

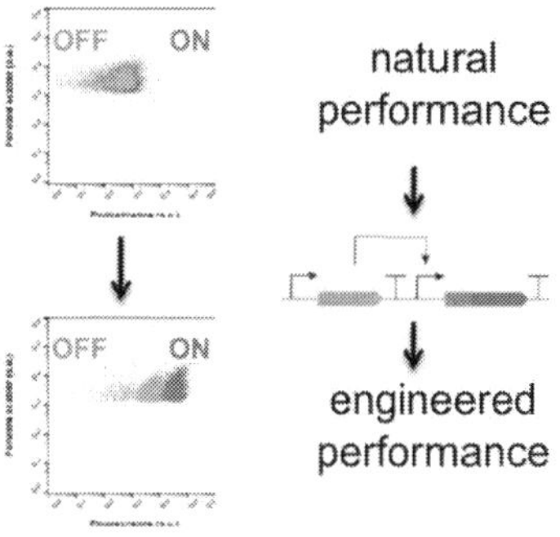

- Improvement:
  - Output expression
  - Fold activation ratio

- Future work: coupling of optimized pH-sensing circuits for feedback-regulated directed evolution of acid-tolerant phenotype.

ACKNOWLEDGEMENTS

NUS National University of Singapore

Chang lab at http://synbiolab.org/

U.S. AIR FORCE

National Environment Agency

Agency for Science, Technology and Research

NMRC National Medical Research Council Singapore

ThermoFisher SCIENTIFIC

KIAT Korea Institute for Advancement of Technology

NATIONAL RESEARCH FOUNDATION Prime Minister's Office, Republic of Singapore

life technologies™

NUS National University of Singapore

SUPPLEMENTARY INFORMATION

LABORATORY-SCALE DIRECTED EVOLUTION

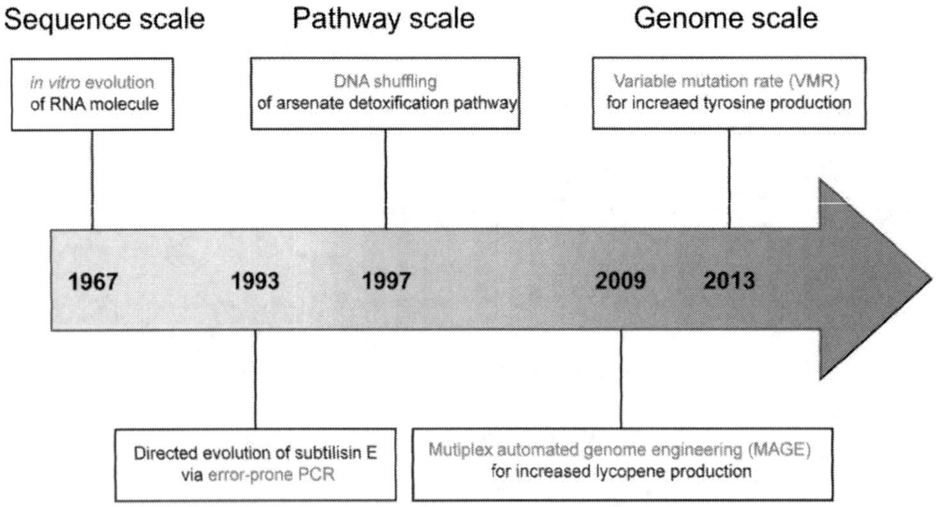

FUTURE OBJECTIVE

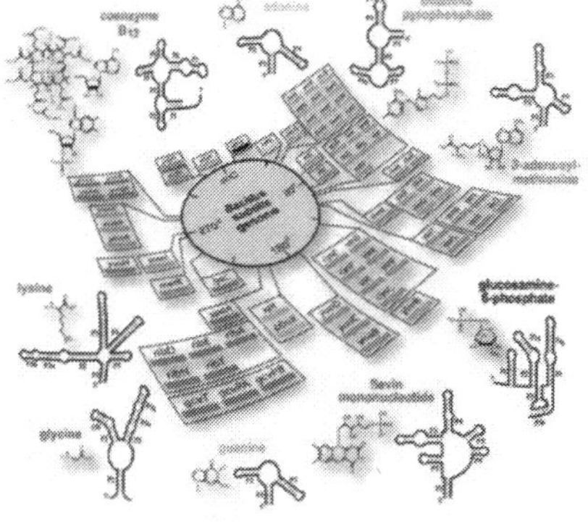

http://openwetware.org/wiki/CH391L/S13/Riboswitches

- Abundant in natural biological systems
- Versatile sensing mechanism
- Low sensitivity?
- Low activation ratio?

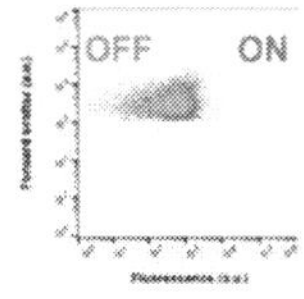

natural performance

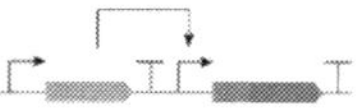

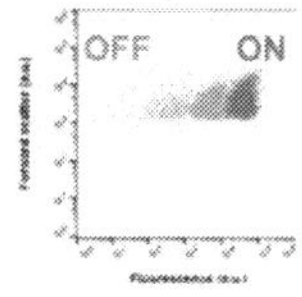

engineered performance

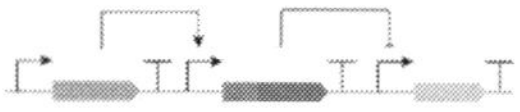

sorting applications

pH-SENSING IN NATURAL BIOLOGICAL SYSTEMS

- Constant environmental signal that significantly affects cellular physiology and behavior, e.g. chemotaxis, colonization, secretion.
- Examples: *H. pylori* gut colonization, cancer cell pH-homeostasis.

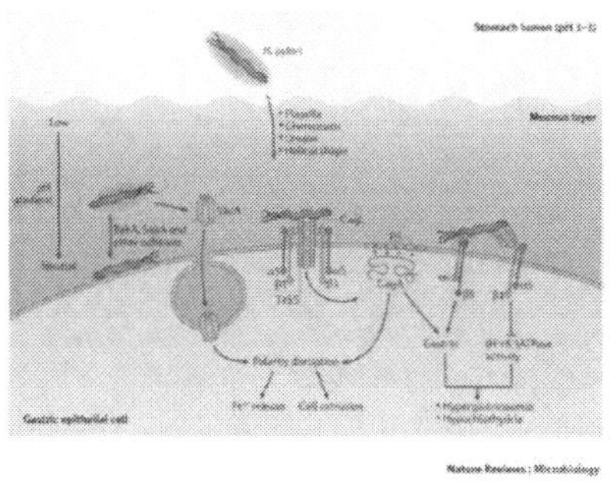

Bucker, R. *et al.* *Scientific reports* **2**, 994 (2012)

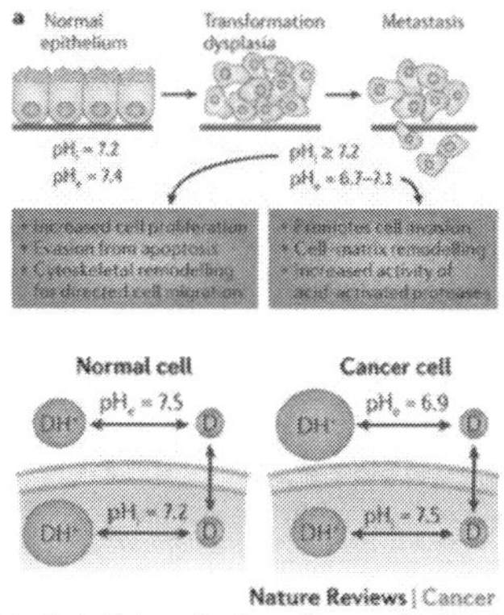

Nature Reviews | Cancer

Bradley A. Webb *et al.* *Nature Reviews Cancer* **11**, 671-677 (2011)

ADAPTATION OF BACTERIA IN ACIDIC ENVIRONMENT

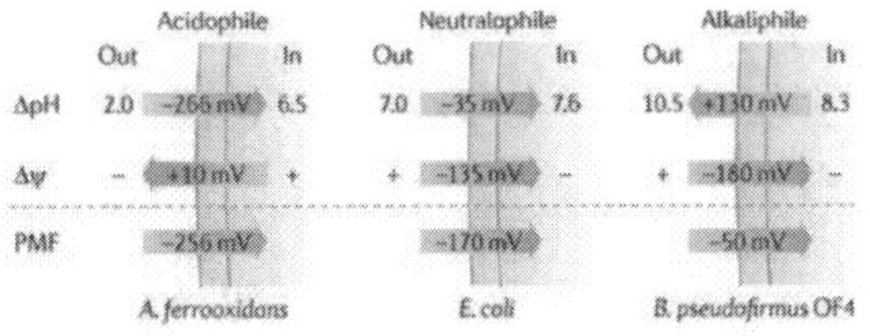

Intracellular pH (pHi) is an important signature of pH-homeostasis in bacteria

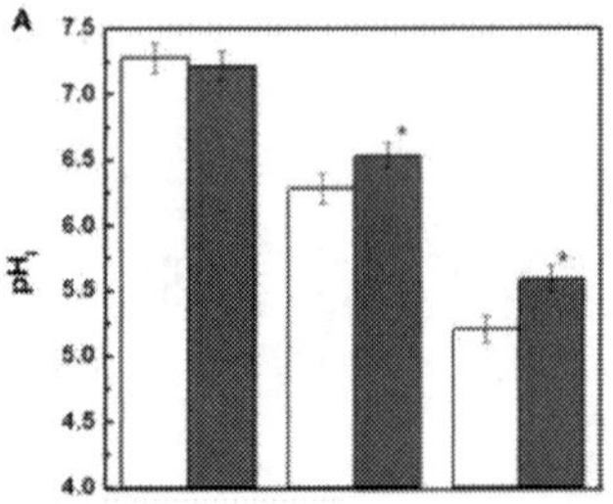

Acid-adapted strains show elevated pHi comparing to parental strain

INTRACELLULAR pH *E. coli*

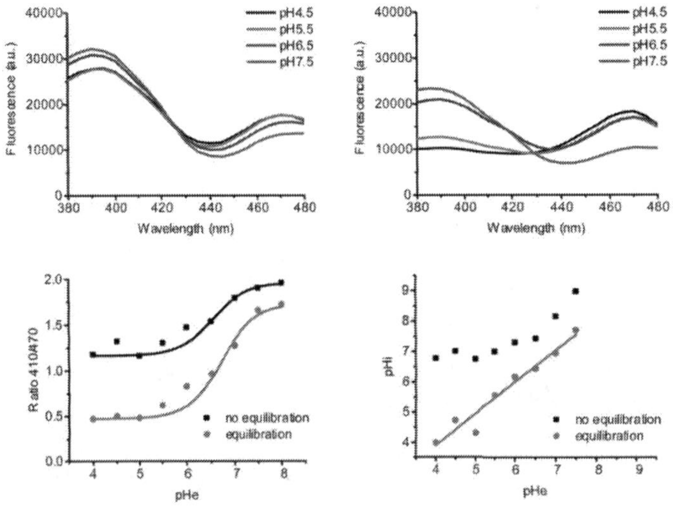

INDUSTRIAL NEEDS FOR ACID-TOLERANT MICROORGANISMS

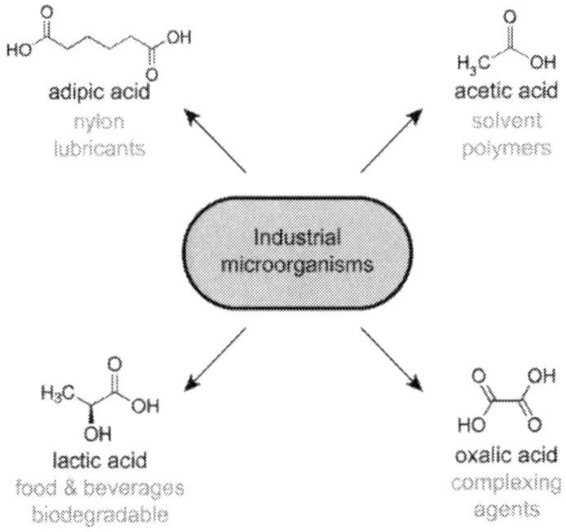

- Main producer of major organic acids products.
- Acid-tolerance is the key performance factor.
- Shortage of efficient high-throughput methods to develop robust microbial acid-tolerance strains for industrial organic acids production.

SYNTHETIC CIRCUIT FOR DIRECTED EVOLUTION OF ACID-TOLERANT STRAINS

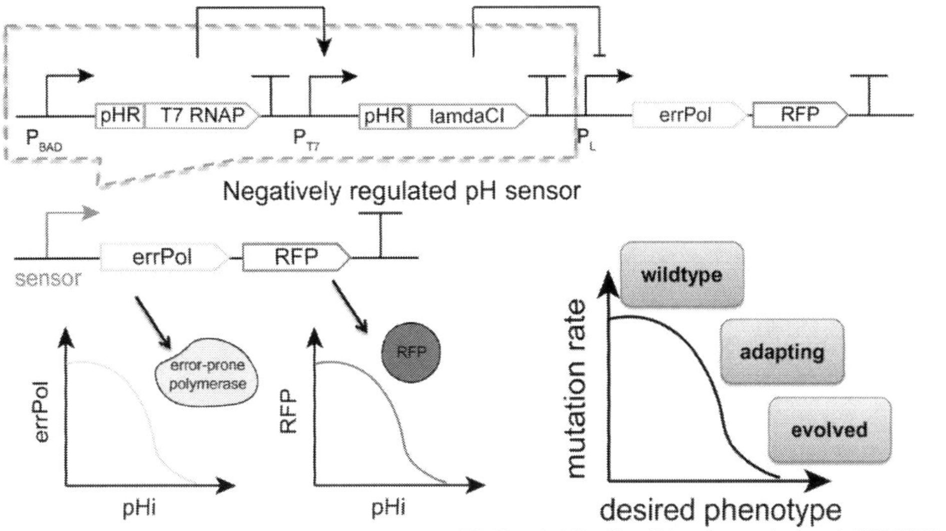

HH. Chou & J. Keasling. *Nature Communications* **4**, 2595 (2013)

AUTHOR INDEX

AUTHOR INDEX

AUTHOR INDEX

AUTHOR INDEX

AUTHOR INDEX

AUTHOR INDEX

Synthetic Biology Conference 2015 (SEED 2015)

Engineering, Evolution, and Design

Boston, Massachusetts, USA
10-13 June 2015

Volume 2 of 2

ISBN: 978-1-5108-1577-3

Printed from e-media with permission by:

Curran Associates, Inc.
57 Morehouse Lane
Red Hook, NY 12571

Some format issues inherent in the e-media version may also appear in this print version.

Printed by Curran Associates, Inc. (2015)

For permission requests, please contact AIChE
at the address below.

AIChE
120 Wall Street, FL 23
New York, NY 10005-4020

Phone: (800) 242-4363
Fax: (203) 775-5177

www.aiche.org

Additional copies of this publication are available from:

Curran Associates, Inc.
57 Morehouse Lane
Red Hook, NY 12571 USA
Phone: 845-758-0400
Fax: 845-758-2634
Email: curran@proceedings.com
Web: www.proceedings.com

TABLE OF CONTENTS

VOLUME 1

GENERAL SUBMISSIONS

VOLUME 2

POSTER SESSION

Enabling Selection in Directed Evolution of Enzymes via Cellular Engineering

Neda Hassanpour[1], Ehsan Ullah[1], Mona Yousofshahi[1], Nikhil U. Nair[2] and Soha Hassoun[1,2]
[1]Department of Computer Science
[2]Department of Chemical and Biological Engineering
Tufts University, Medford, MA

Directed Evolution of Enzymes

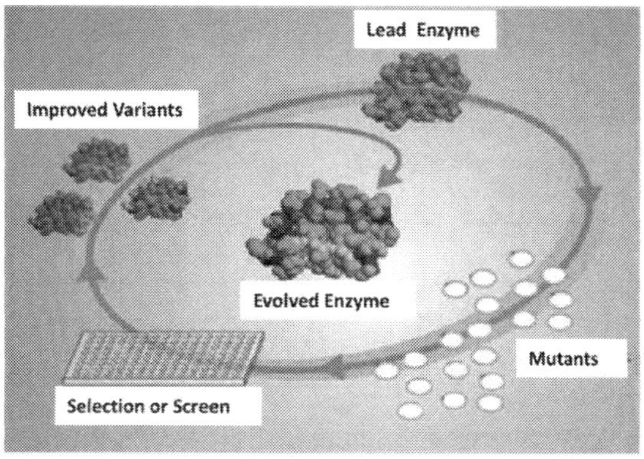

Google Image

2

Selection and Screen

- Selection is preferable to a screen
  - Screen: Link the desired property to some visual output
  - Selection: Link the desired property to survival of host cell
  - Selection has higher throughput than screen
- Selection scheme is hard to identify
  - There are currently no known high-throughput automated techniques for identifying appropriate selections

3

Problem Statement

Given a desired enzymatic reaction, *automatically* identify a consumption pathway, and couple it with cell survival

Computational Approach: Selection Finder (SelFi)

1. Construct a consumption pathway
2. Eliminate alternate carbon sources
3. Identify knockout targets
4. Rank the identified selection pathways

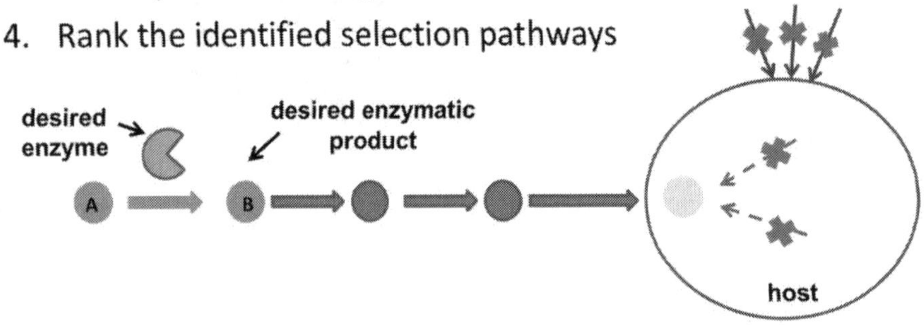

4

Step 1: Construct a Consumption Pathway

- Identify a set of possible pathways from the desired product to a metabolite in the cellular host
- Utilize a modified version of *ProbPath**

* M. Yousofshahi, K. Lee and S. Hassoun, Probabilistic Pathway Construction, *Metab . Eng.* 2011

5

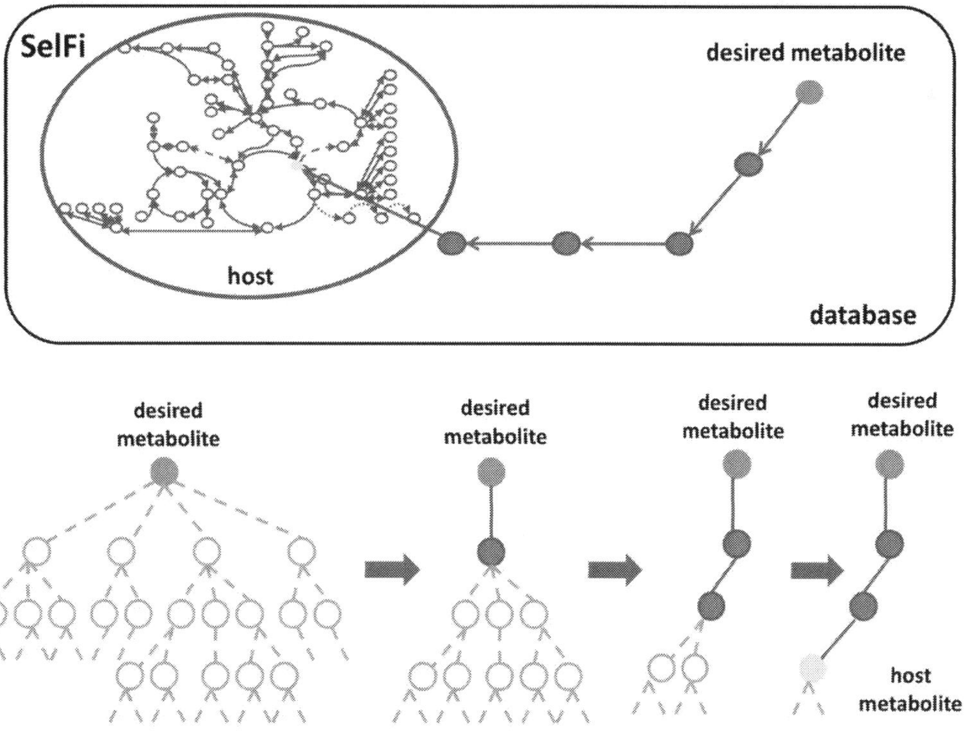

6

Step 2: Eliminate Alternate Carbon Sources

Restrict external carbon sources to maximize consumption flux

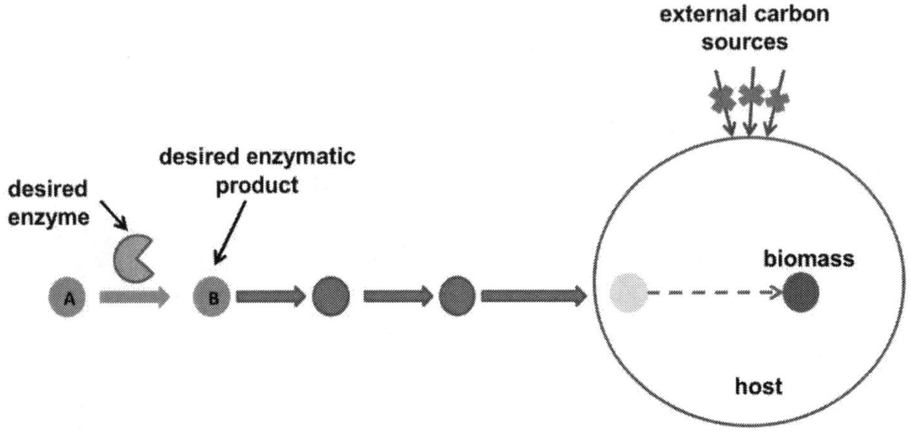

7

Step 3: Identify Knockout Targets

- Identify knockout targets
  - Increase the guaranteed minimum consumption flux
- Greedy method

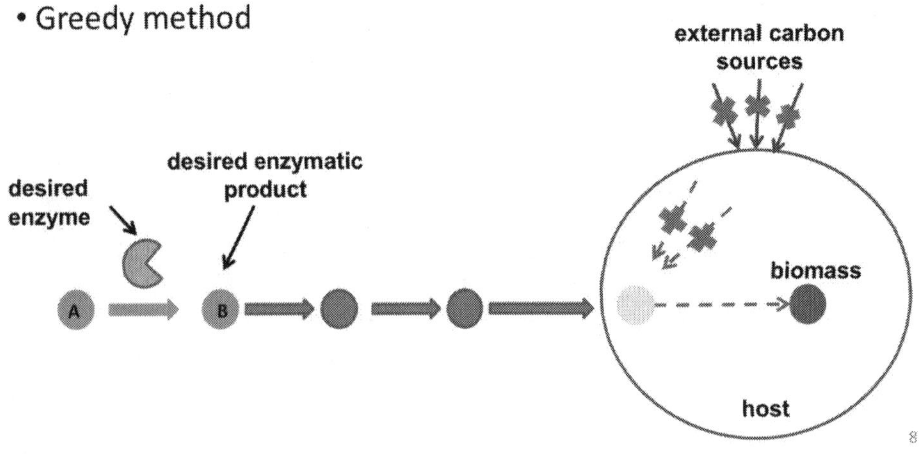

8

Step 4: Rank the Identified Pathways

- Score candidate pathways based on consumption rate, pathway length, and number of knockouts
- Higher guaranteed minimum flux, shorter pathways with the smallest number of knockouts are preferable

9

Computational Setup

- Host model: *i*AF1260 model of *E. coli* metabolism*
  - 1668 metabolites
  - 2382 reactions
- Guarantee ≥ 10% of wild type maximum biomass rate
- Desired product test cases
  - Xylitol
  - Aniline
  - Methanol
  - D-ribulose 1, 5-bisphosphate

* A. M. Feist et al., A genome-scale metabolic reconstruction for Escherichia coli K-12 MG1655 that accounts for 1260 ORFs and thermodynamic information, Mol. Syst. Biol. 2007

10

Results: Xylitol Test Case

Reduction of D-xylose to xylitol by Xylose Reductase (XR)*

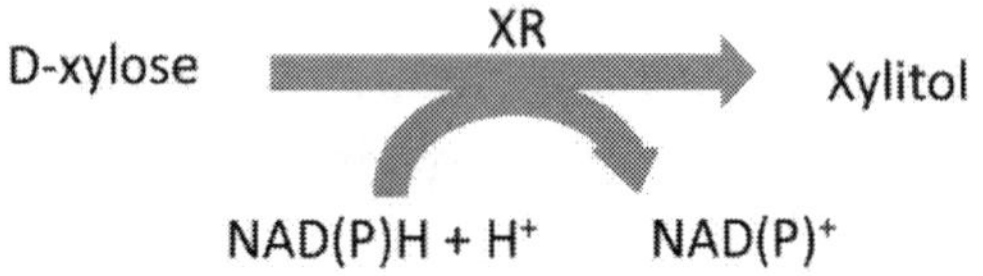

* N. U. Nair and H. Zhao, Evolution in Reverse: Engineering a D-Xylose-Specific Xylose Reductase, ChemBioChem 2008

11

Results: Xylitol Test Case

- Identified three selection pathways

* Pathway 1	xylitol + NAD^+ <=> D-xylulose + NADH + H^+
Pathway 2	xylitol + $NADP^+$ <=> L- xylulose + NADPH + H^+
Pathway 3	xylitol + NAD^+ <=> L- xylulose + NADH + H^+

- Identified three knockout targets

	Before knockouts	After first knockout	After second knockout	After third Knockout
Minimum consumption (mmol/sec)	0	21.8	57.66	95.34

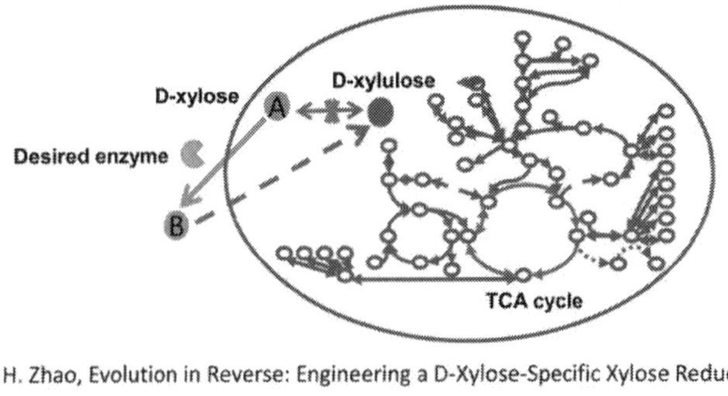

12

* N. U. Nair and H. Zhao, Evolution in Reverse: Engineering a D-Xylose-Specific Xylose Reductase, ChemBioChem 2008

Conclusion and Future Work

- Selection Finder (SelFi) is the first computational approach for constructing a selection pathway for enzyme directed evolution

- SelFi identifies selection pathways for various engineered enzymes
  - One of the identified pathways for xylitol was previously published in literature

- Future work: extend SelFi to non-KEGG molecules

13

Thank you

BACKGROUND

15

ProbPath

A probabilistic pathway construction algorithm

- Designed to identify synthesis pathways from the KEGG database
- Probabilistic reaction selection

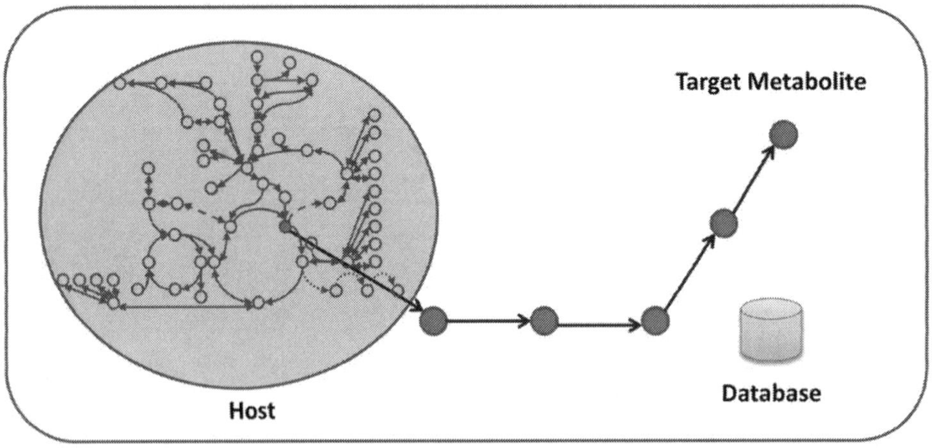

M. Yousofshahi, K. Lee and S. Hassoun, "Probabilistic Pathway Construction," Metabolic Engineering, vol. 13, p. 435–444, 2011

16

Flux Balance Analysis (FBA)

A mathematical approach for analyzing fluxes of reactions in a metabolite network

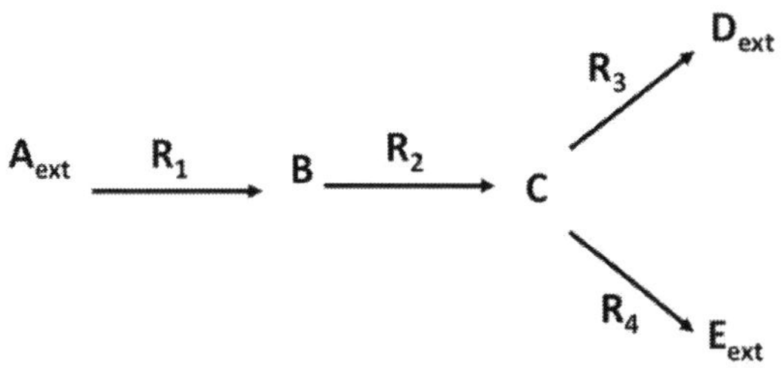

Objective Function: maximize v_3

Subject to:

$v_1 - v_2 = 0$
$v_2 - v_3 - v_4 = 0$
$v_1 \le 10$

17

Flux Balance Analysis (FBA)

A mathematical approach for analyzing fluxes of reactions in a metabolite network

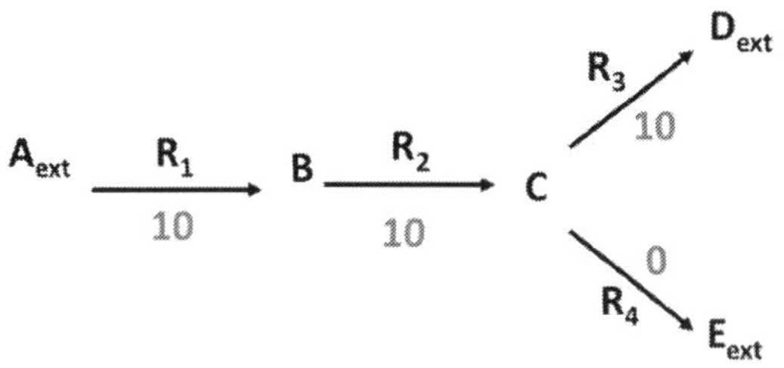

Objective Function: maximize v_3

Subject to:

$v_1 - v_2 = 0$
$v_2 - v_3 - v_4 = 0$
$v_1 \le 10$

18

ASF Algorithm – Step 2

- Synthesis pathway identification for cofactors
  - Identify supporting pathways from the host to the reactant-side cofactors along each selection pathway

19

ASF Algorithm – Step 1 & 2

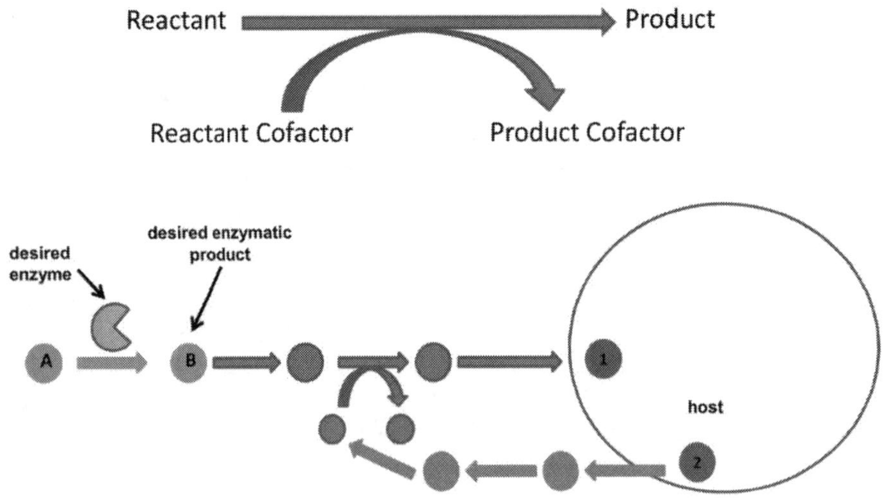

20

ASF Algorithm – Step 3

- Native reactant for desired enzymatic reaction
  - Raise of a special case
  - Selection pathway can not compensate knocked out carbon sources

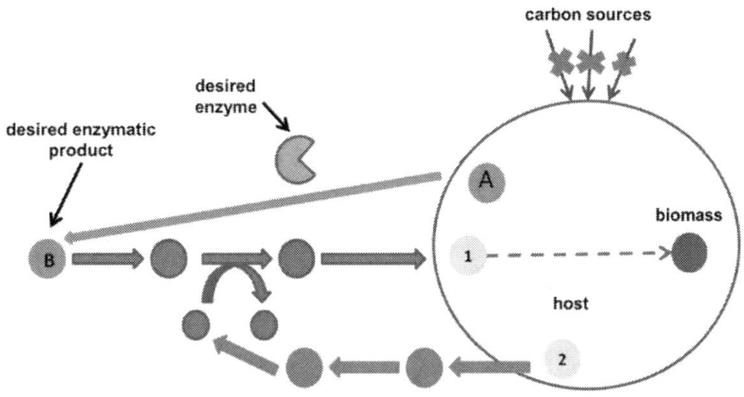

21

ASF Algorithm – Step 3

- Native reactant for desired enzymatic reaction
  - Identify the best carbon source to provide for the cell
  - Selection pathway is not coupled with cell survival

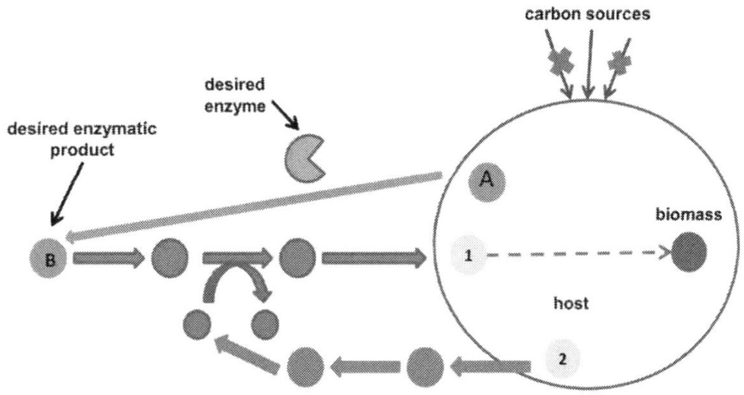

22

Future Work

- Flux of an identified consumption pathway is directly dependent on the demand of the end metabolite for cell growth
  - Aniline test case
  - Use of screen instead of selection
- Expanding the search space to contents of multiple databases
- Allowing non-database compounds by using structural similarity
  - Broader range of metabolites and enzymatic products

23

Quantifying and Modeling Compositional Context Effects in *E. coli*

Enoch Yeung

Engineering & Applied Sciences Division
California Institute of Technology

2015 SEED Conference

1

Challenges in Biological Circuit Design: Context Matters

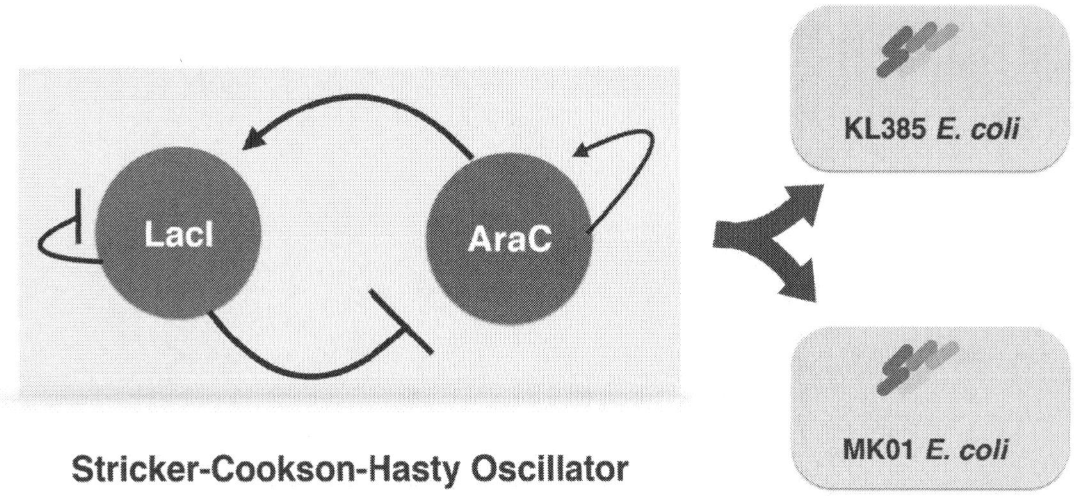

2

Challenges in Biological Circuit Design: Context Matters

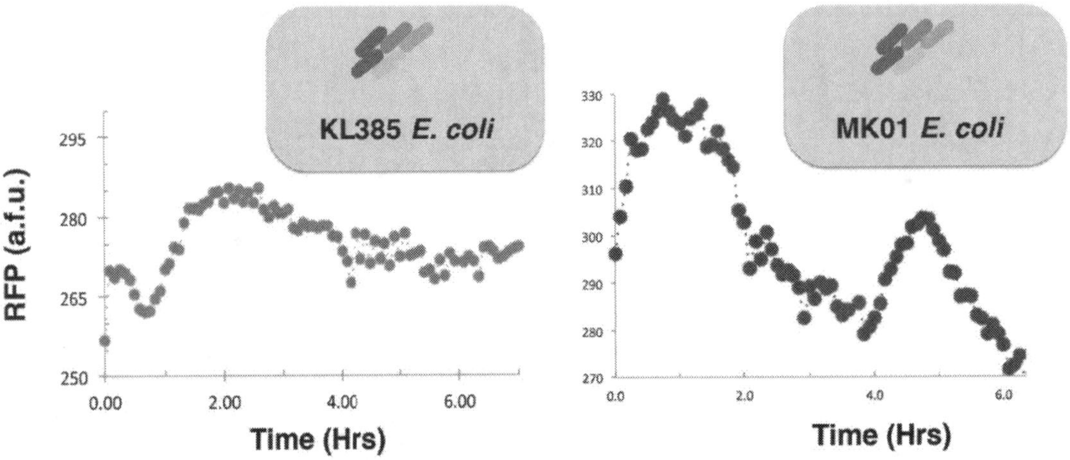

Biocircuit behavior can be affected by different types of context: environmental, host, and genetic.

3

From Biological Parts to Complex Biocircuits: Towards Model Guided Circuit Design

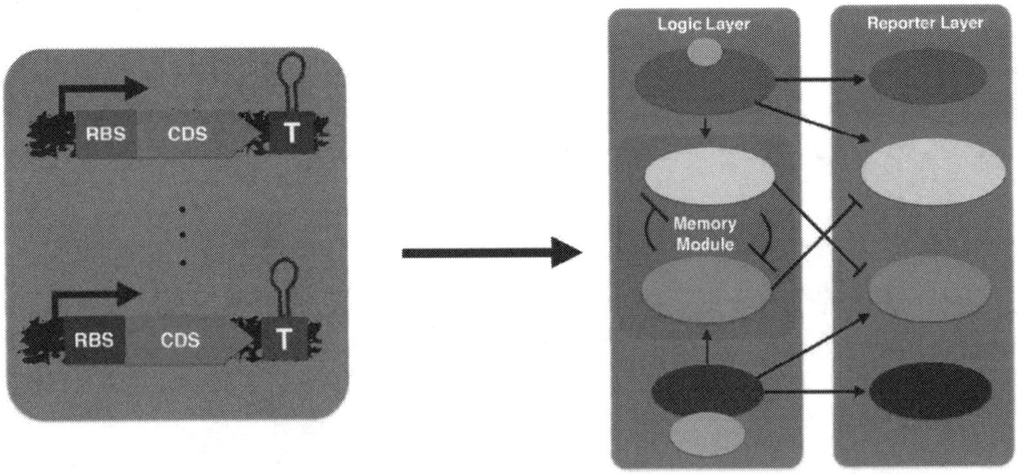

Biological Parts

"Complex" Biocircuit

4

From Biological Parts to Complex Biocircuits: Towards Model Guided Circuit Design

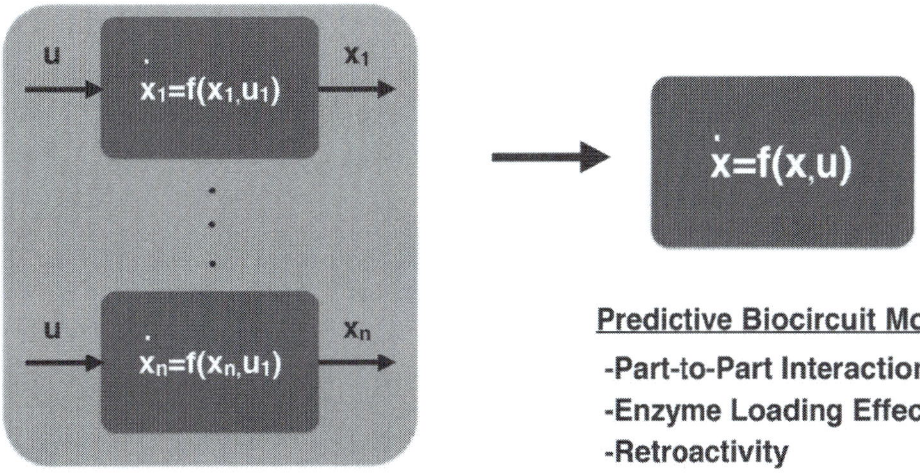

Biological Part Models

Predictive Biocircuit Model

- -Part-to-Part Interaction
- -Enzyme Loading Effects
- -Retroactivity
- -Host Effects

5

Does Gene Composition Affect Expression?

Genes of interest

Orientation

Convergent

Divergent

Tandem

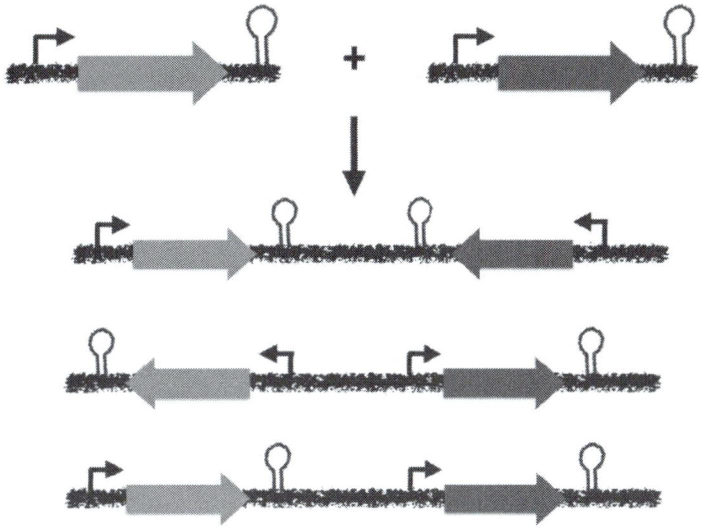

6

Does Gene Composition Affect Expression?

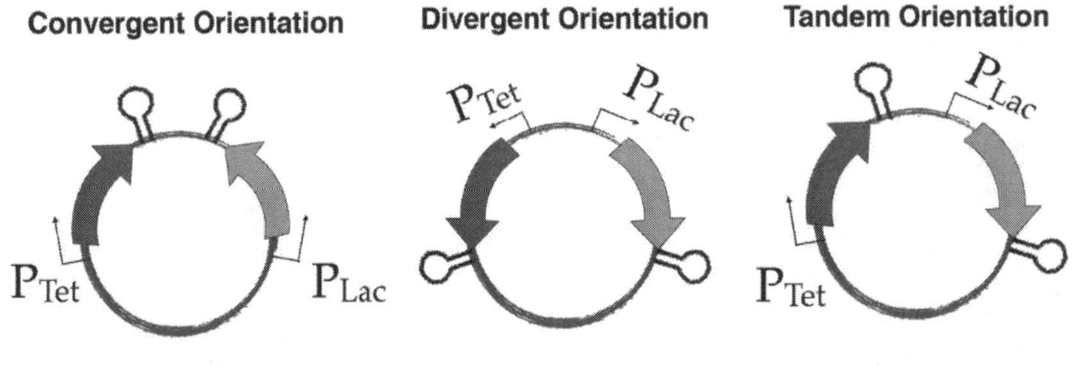

7

How does gene orientation affect expression?

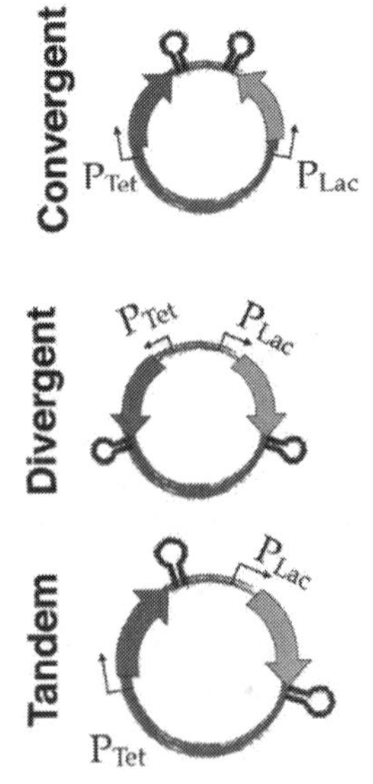

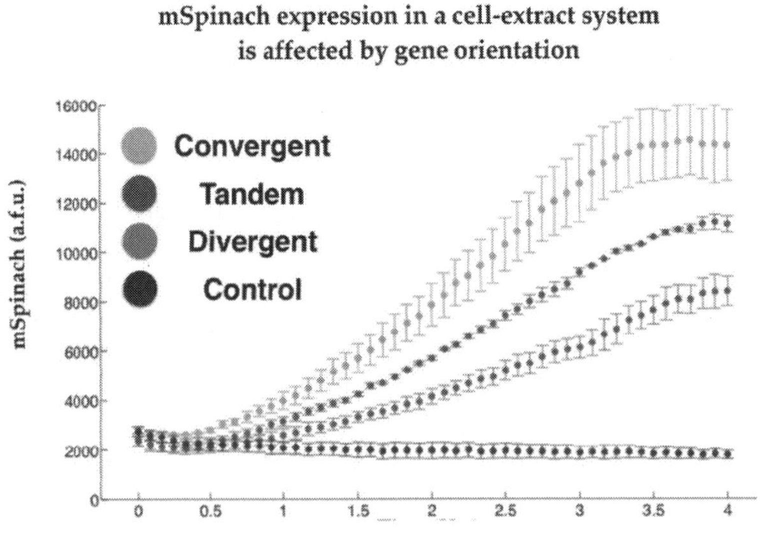

8

How does genetic context affect expression?

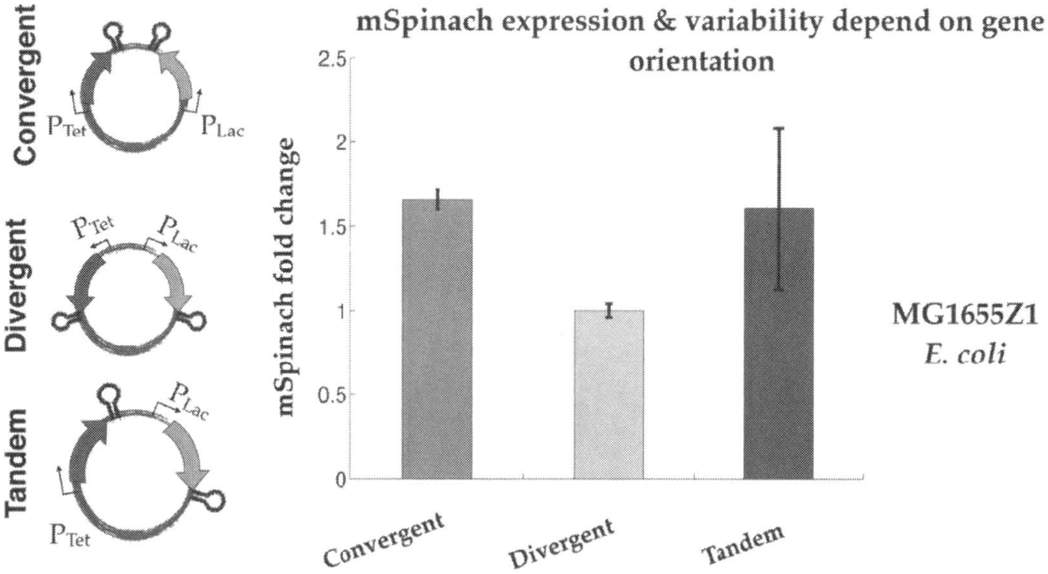

9

How does genetic context affect expression?

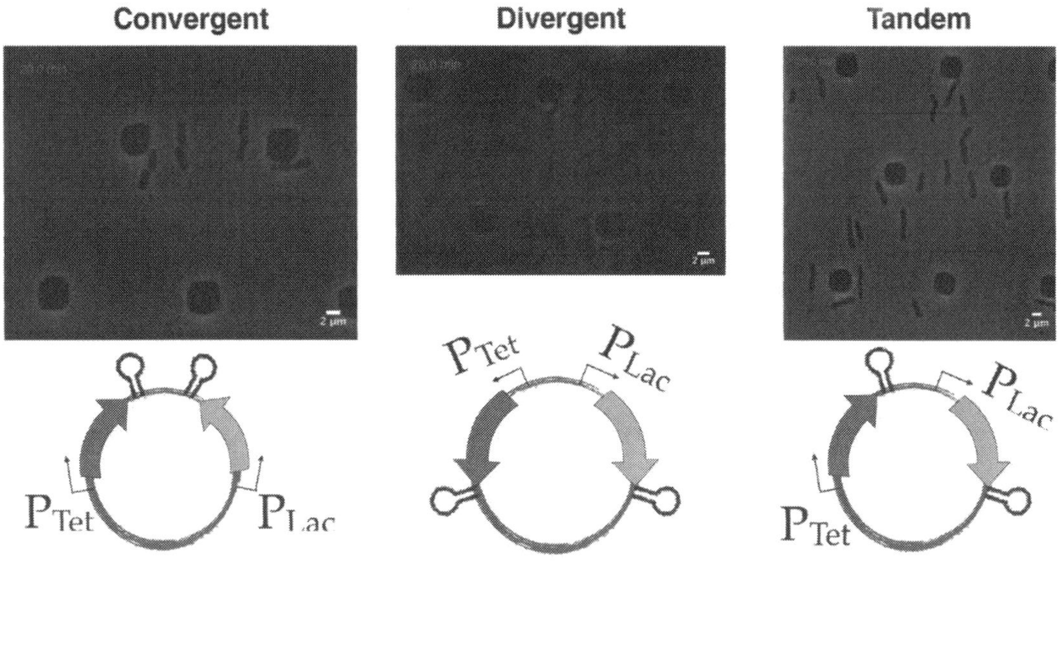

10

How does genetic context affect expression?

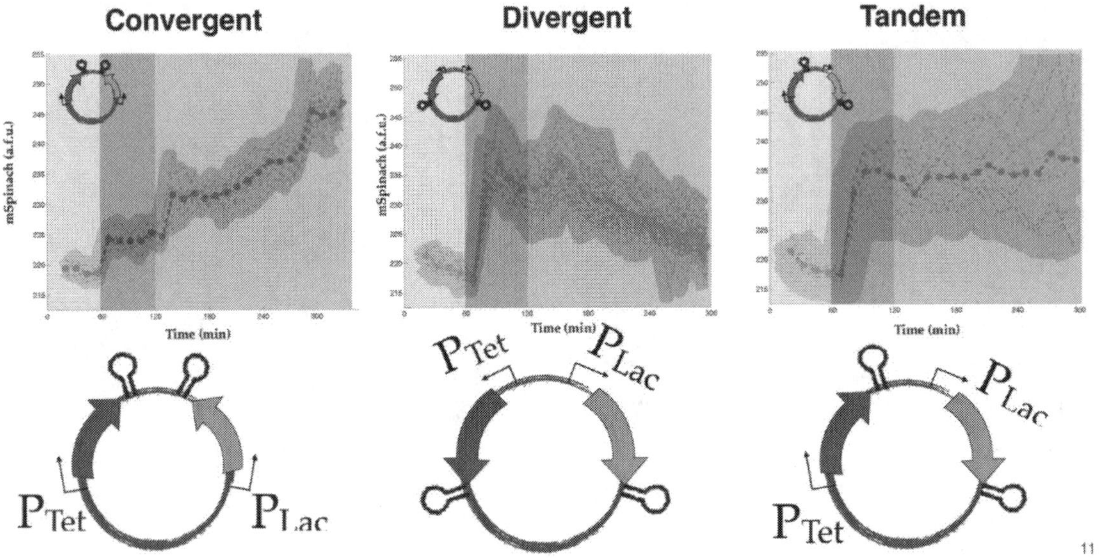

11

Does orientation affect translational expression?

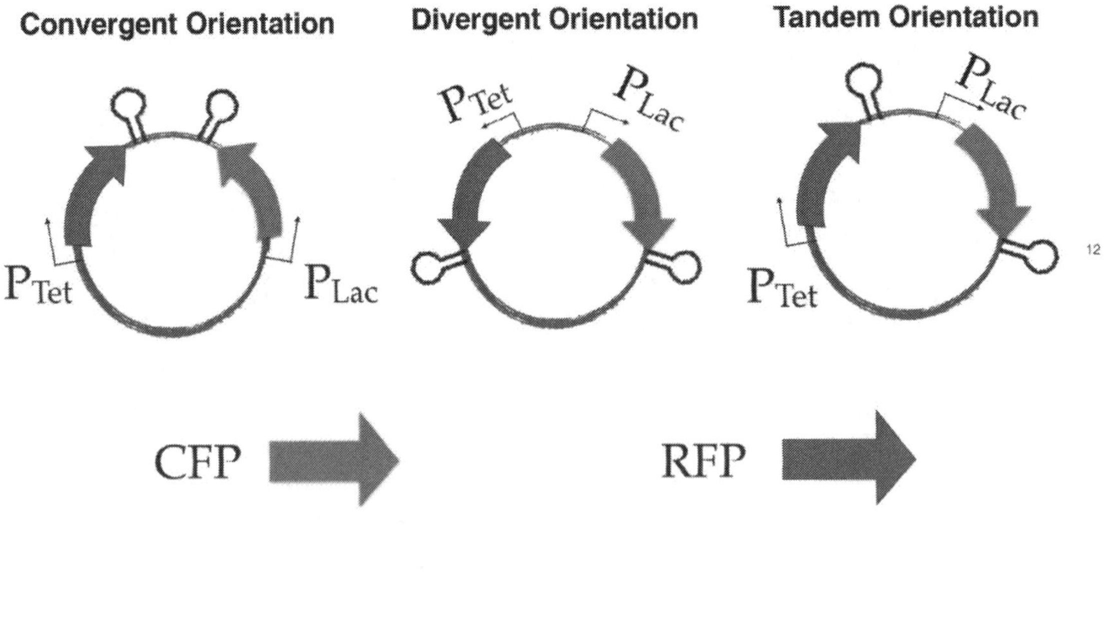

12

Does orientation affect translational expression?

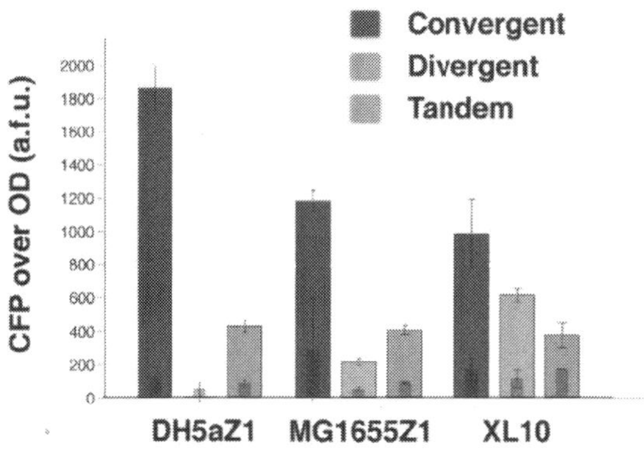

High (convergent) CFP expression is true across multiple strains of *E. coli*

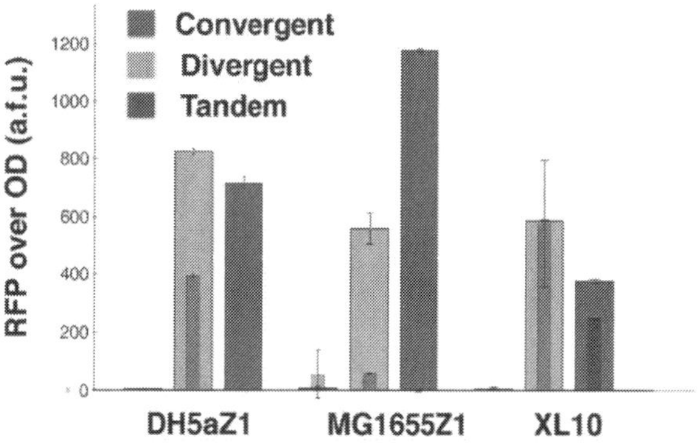

High (divergent & tandem) RFP expression is true across multiple strains of *E. coli*

Does orientation affect induction?

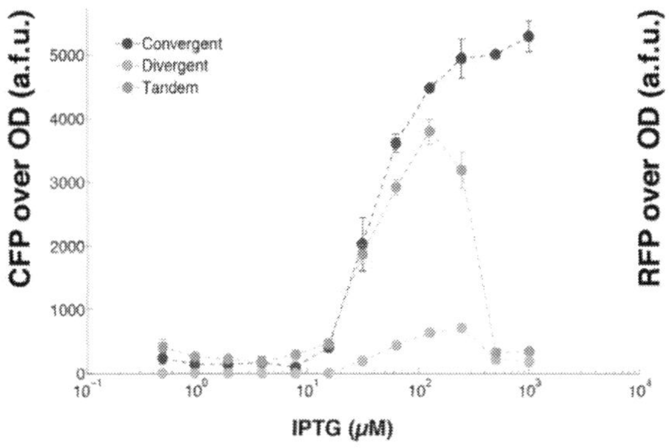

CFP is strongly inducible in the convergent orientation

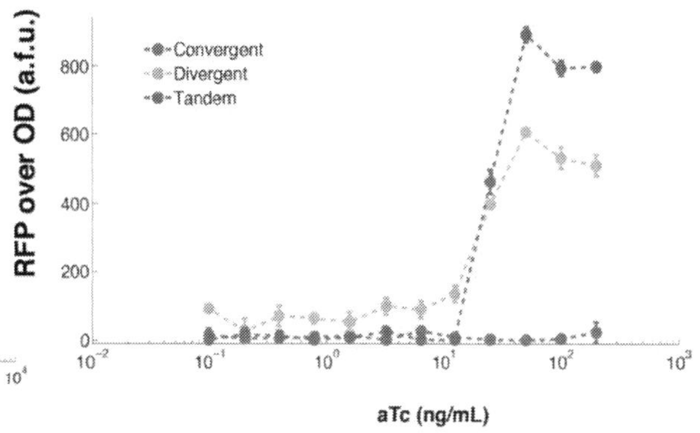

RFP is inducible in the divergent and tandem orientation

Genetic context modeling: supercoiling as a negative feedback mechanism for gene expression

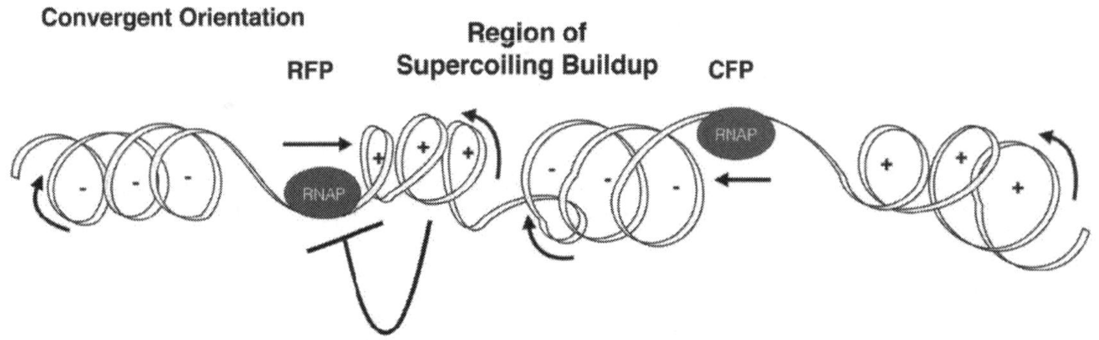

Propagation of positive and negative supercoiling in the intergenic region creates a kink, leading to supercoiling buildup.

Genetic context modeling: supercoiling as a negative feedback mechanism for gene expression

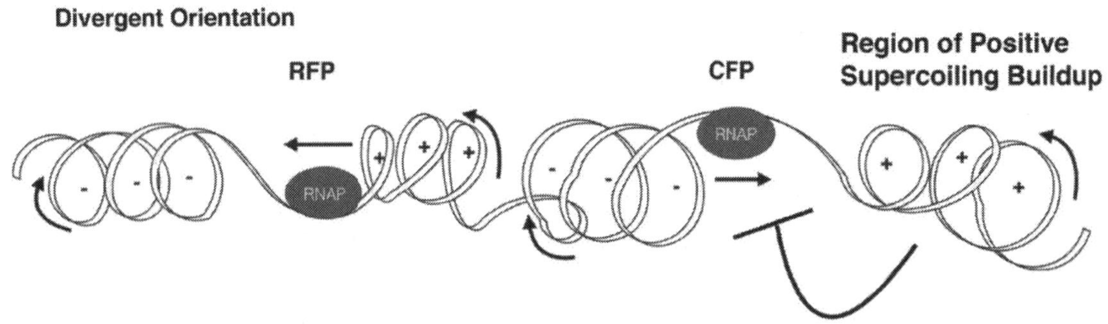

Positive supercoiling buildup downstream of CFP abolishes expression.

Genetic context modeling: supercoiling as a negative feedback mechanism for gene expression

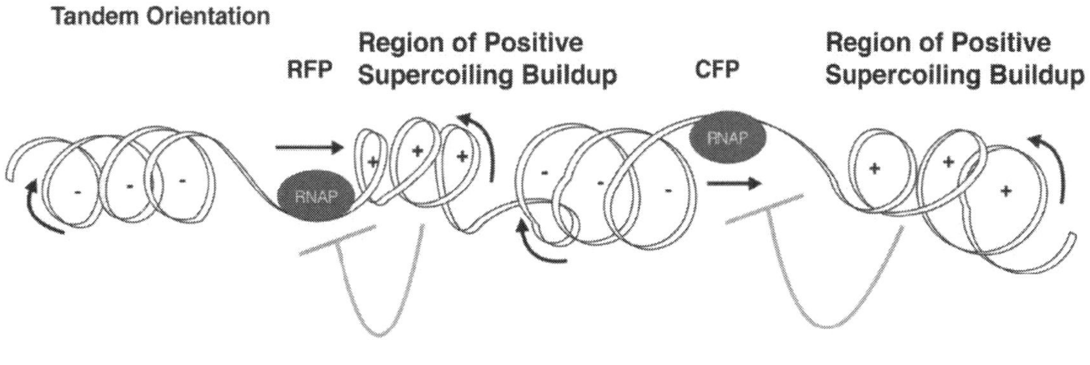

Propagation of supercoiling happens primarily in the downstream direction, thus reducing the torsional stress on the kink.

Genetic context modeling: supercoiling as a negative feedback mechanism for gene expression

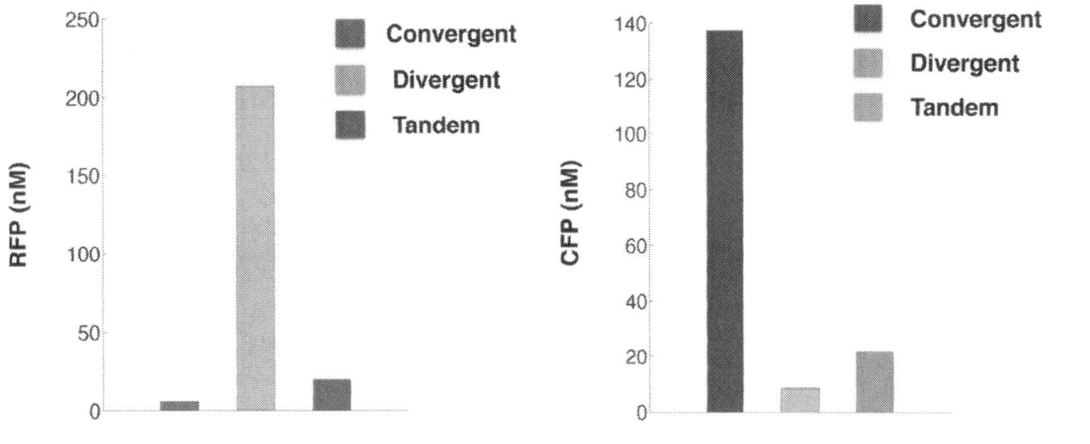

Yeung, E., Ng. A, Sun, Z. Z., Kim, J., Murray, R. M. "Quantifying the Effects of Compositional Context in a Cell-Free Transcription Translation System" Proceedings of the 2014 IEEE Conference on Decision and Control

Gyrase incubation reveals the effect of positive supercoiling

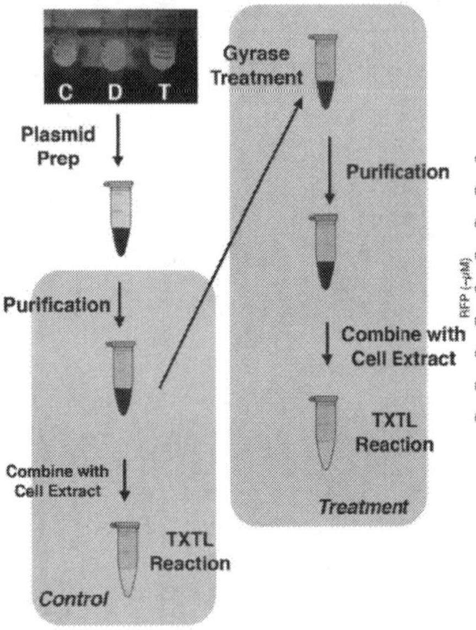

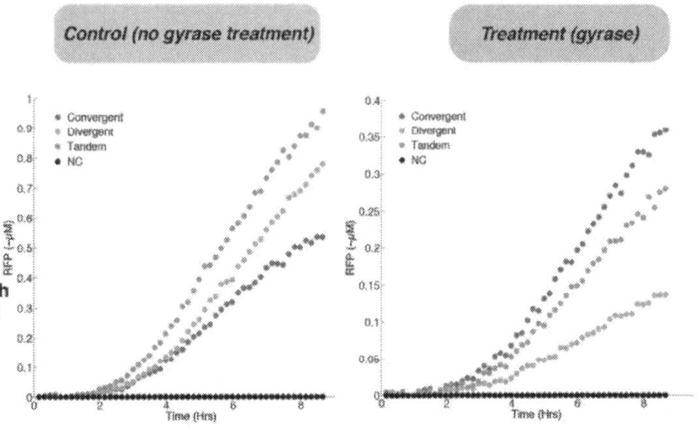

Gyrase treatment swaps the order of expression strengths!

Linear DNA controls match supercoiling model predictions

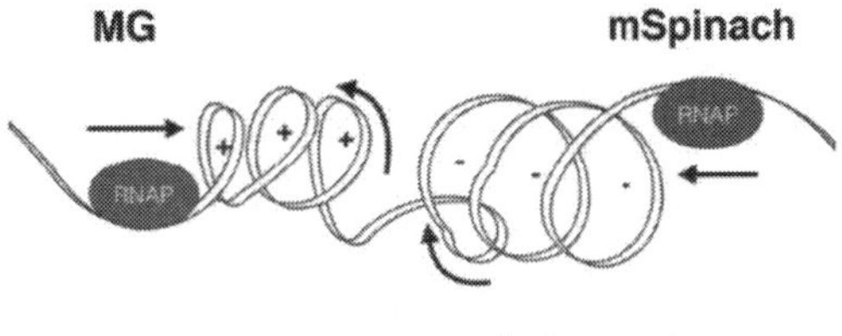

Convergent Orientation

-Convergent mSpinach expression is higher than divergent and tandem.

-Divergent and tandem expression are comparable to single gene expression.

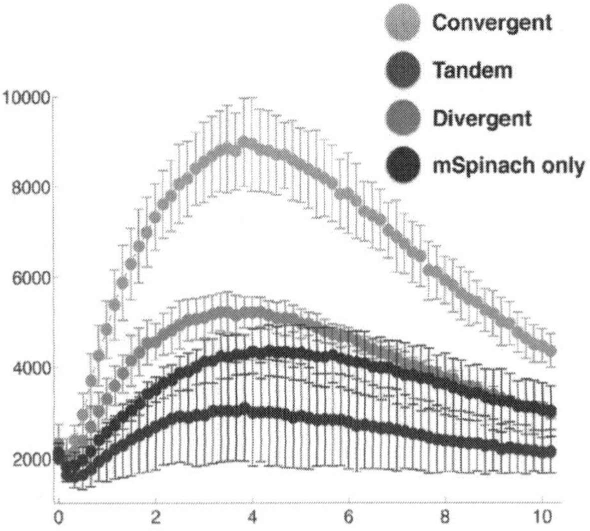

Genetic context modeling: supercoiling as an additional negative feedback loop

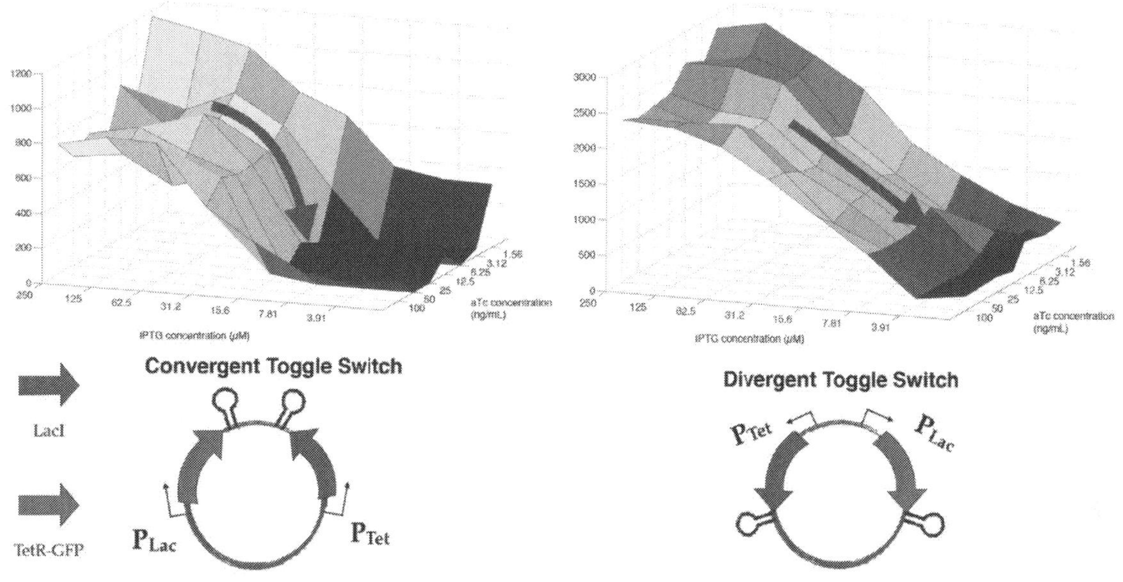

Genetic context modeling: supercoiling as an additional negative feedback loop

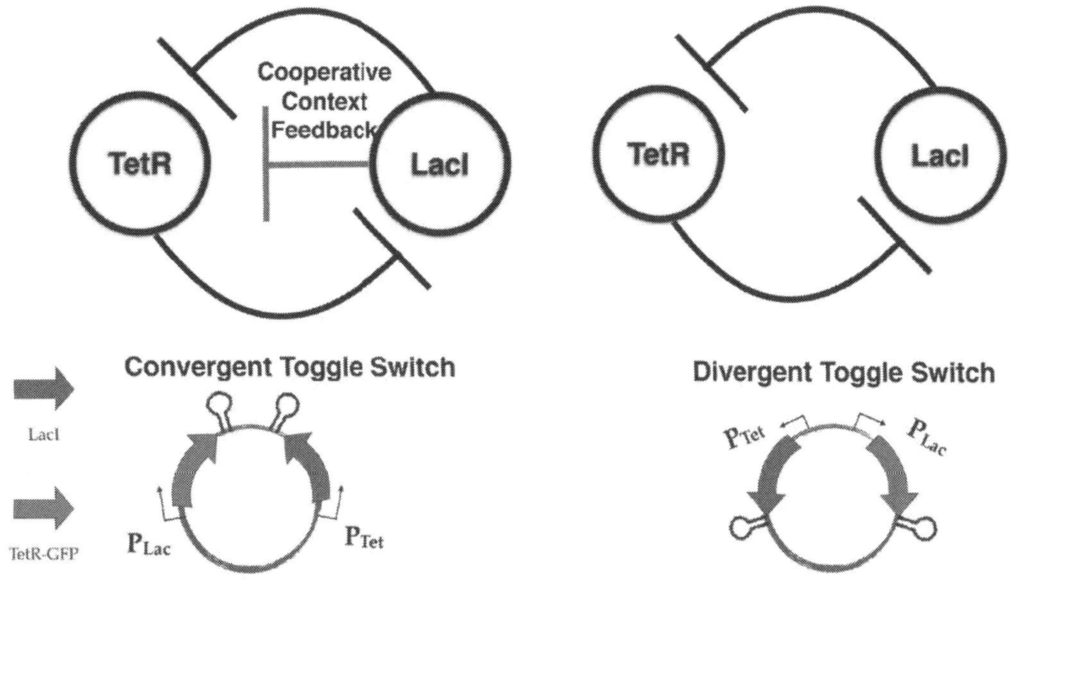

Conclusions & Acknowledgments

- Compositional context has a strong effect on gene expression. These effects are observed at the levels of both mRNA and protein, across multiple conditions and strains.
- Changing gene orientation can dramatically alter temporal response.
- Compositional context effects can alter gene induction, introducing leaks or abolishing response.
- Traditional models for compositional context fail to explain all the observed effects
- Supercoiling models provide a sufficient hypothesis for explaining these effects and they are supported by our control experiments.
- Compositional context can be used as an additional layer of genetic control to introduce feedback in biological circuits.

People:

- Richard M. Murray (Caltech)
- Ophelia Venturelli (UC Berkeley)
- David Younger (UWashington)
- Ania Baetica (Caltech)
- Jongmin Kim (Harvard)
- Andrew Ng (UC Berkeley/ UCSF)
- Kyle Martin (Caltech)
- Zachary Z. Sun (Caltech)
- Jin Park (Caltech)
- Adam Rosenthal (Caltech)

Funding Sources:

NSF DARPA

Charles Lee Powell Foundation

23

Steady-state RNA aptamer distributions reveal context-dependent logic

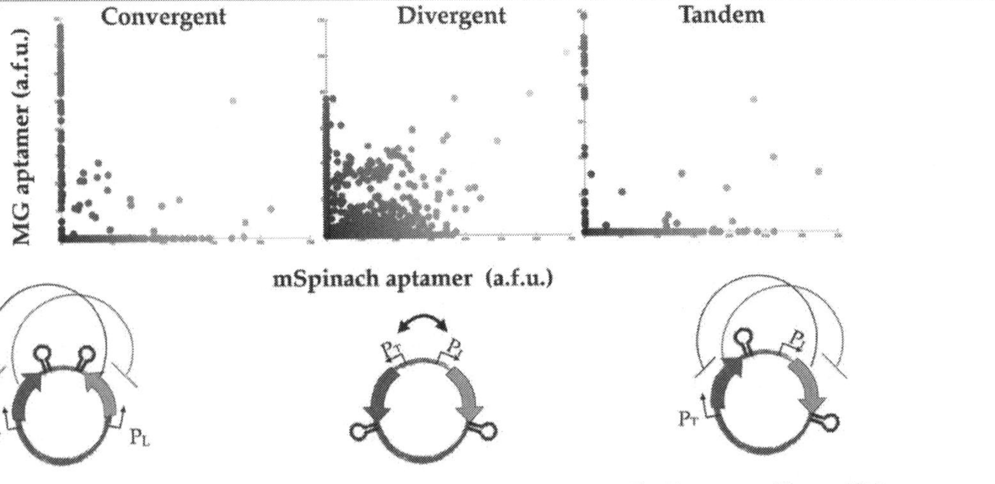

Kinked supercoiling creates *mutual negative feedback*, leading to XOR logic

Activation of one promoter triggers activation of the nearby promoter, leading to OR logic.

Positive supercoiling and R-loop complexes create *mutual negative feedback*, leading to XOR logic

24

Example System: Tandem reporters in the Hasty-Stricker oscillator can be used to program XOR logic

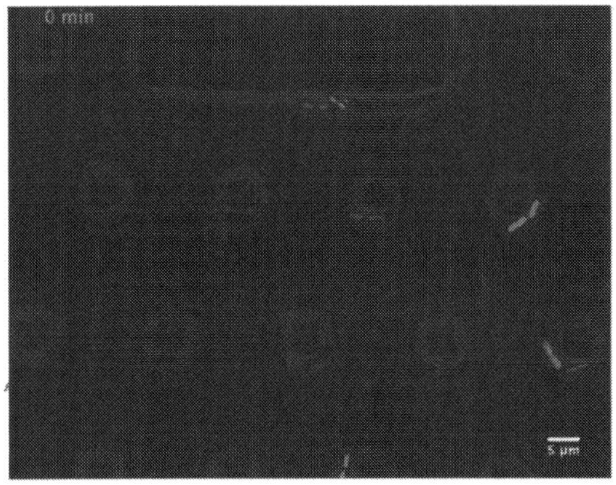

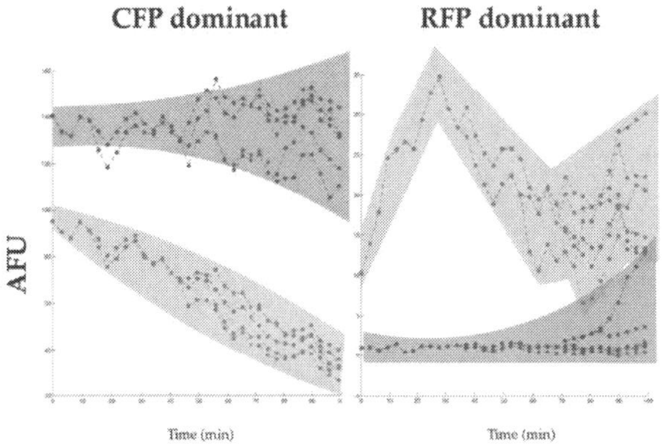

2

Exploiting Gene Orientation to Generate Bimodality

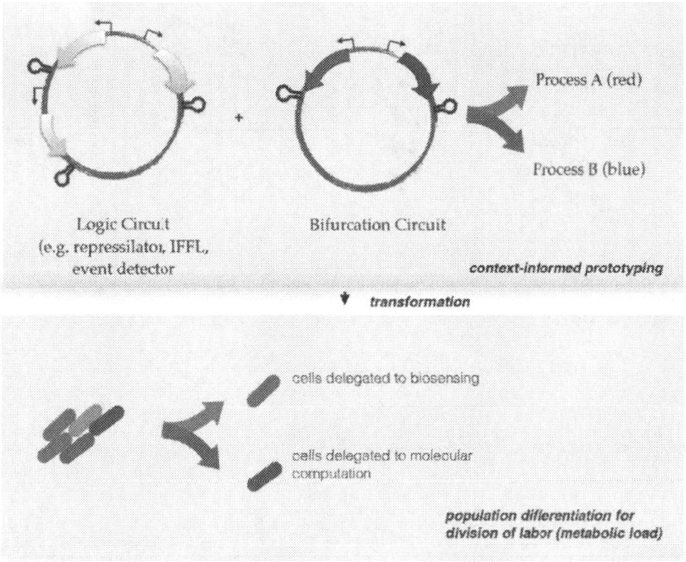

26

Supplementary Data: Coupling is present even in protein-based circuits

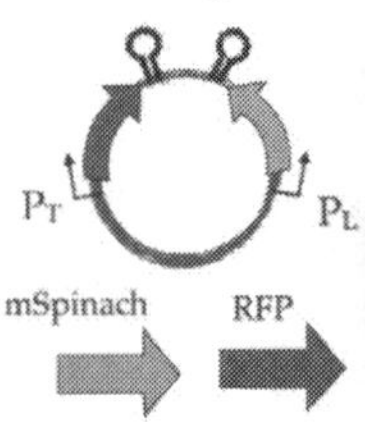

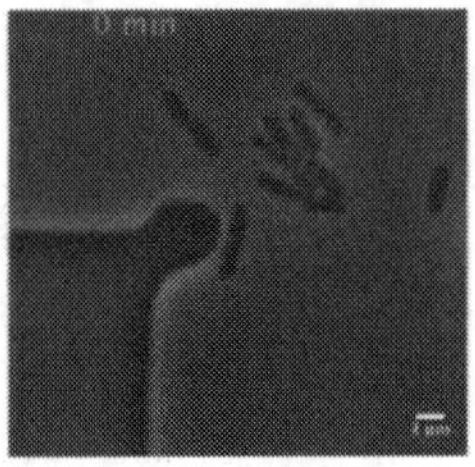

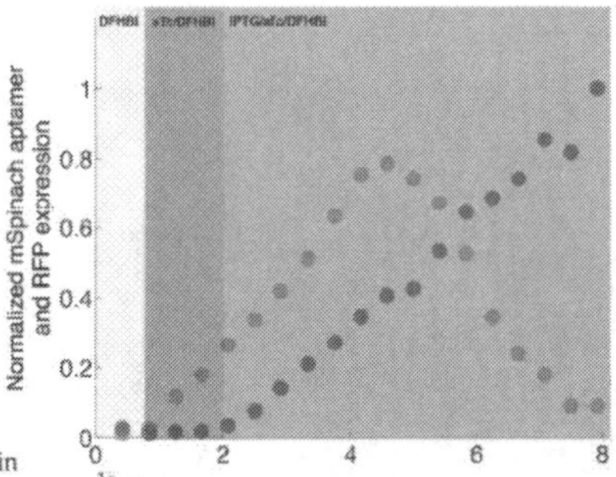

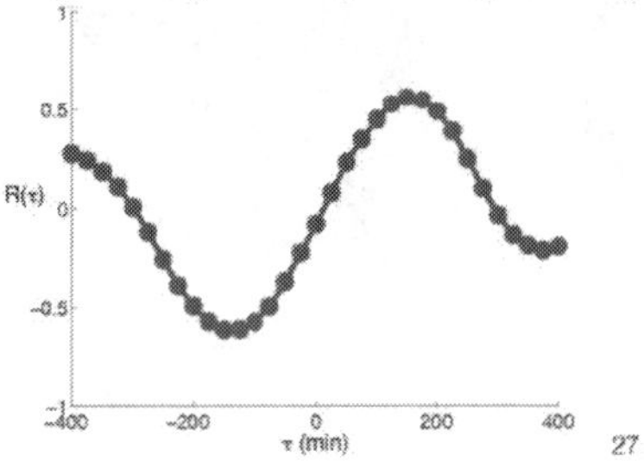

- mSpinach responds to aTc induction, over 500 AFU in response to induction of mRFP!
- aTc induction initially results in a small increase in RFP expression, but the expression picks up significantly once IPTG is added as well (the cross - correlation function takes the form of a repression circuit).
- Cells differentiate into two groups: mSpinach or RFP dominant

Enoch Yeung: eyeung@caltech.edu

Supplementary Data: Coupling is present even in protein-based circuits

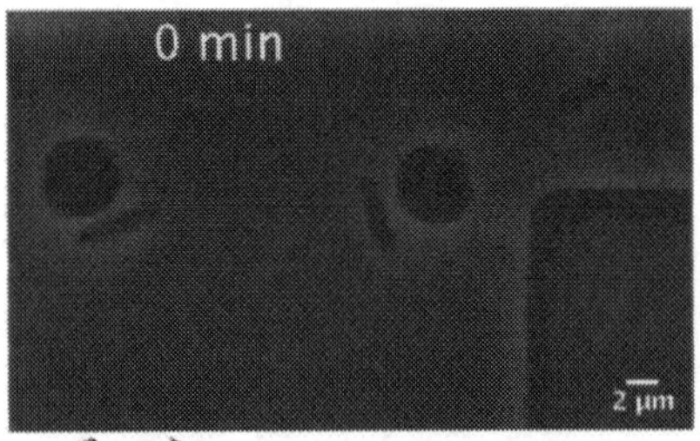

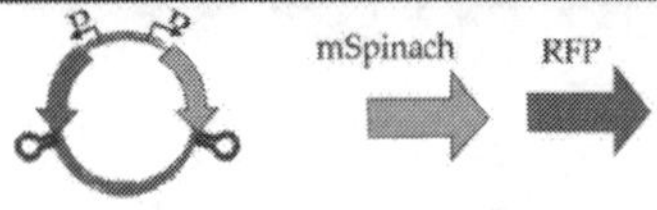

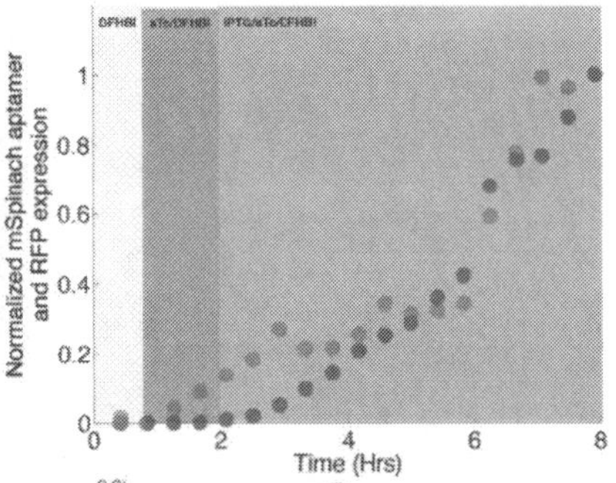

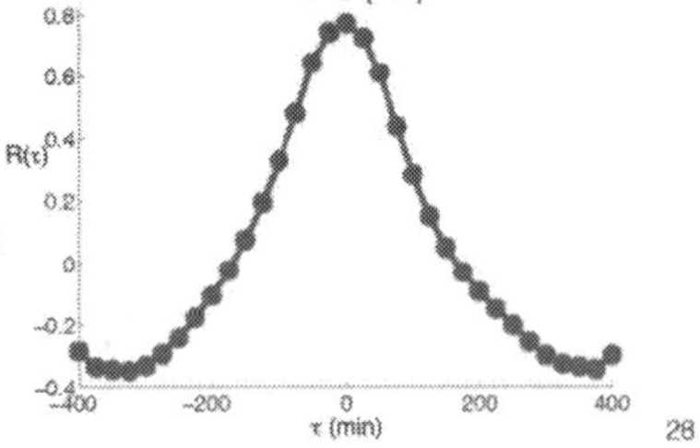

- mSpinach responds to aTc induction, faster response since it is transcription based.
- aTc induction initially results in a small increase in RFP expression, but the expression picks up significantly once IPTG is added as well (coupling?).
- mSpinach and RFP positively correlate in divergent orientation!

Enoch Yeung: eyeung@caltech.edu

Supplementary Data: Coupling is present even in protein-based circuits

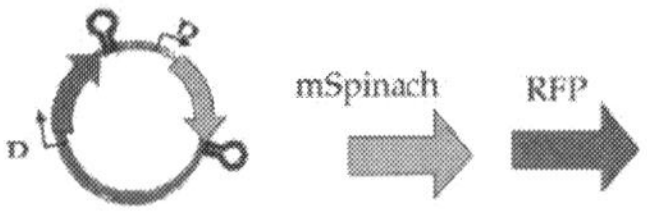

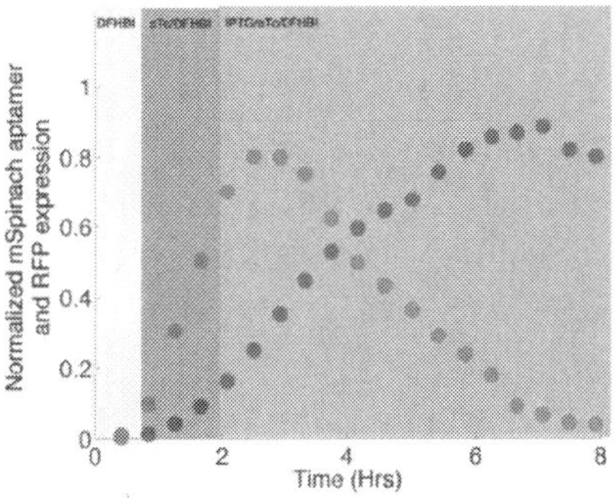

- mSpinach turns off as RFP turns on
- Response is very homogeneous => transcript length may eliminate differentiation aspect (RFP transcript is longer so more positive supercoiling is propragated into mSpinach construct)
- mSpinach and RFP anti-correlate (as in the convergent case)

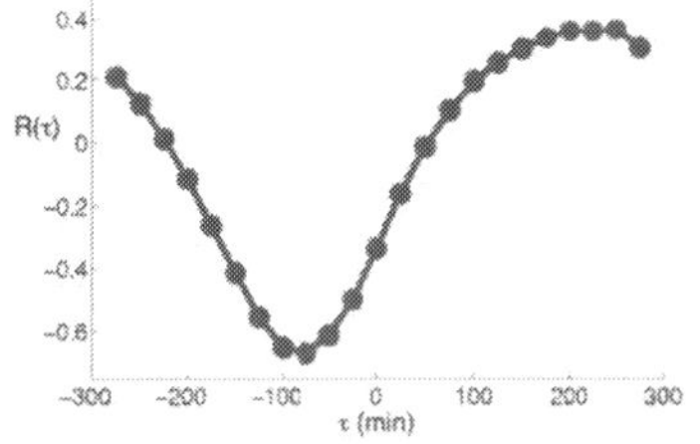

Compositional context as a fundamental source of gene regulation

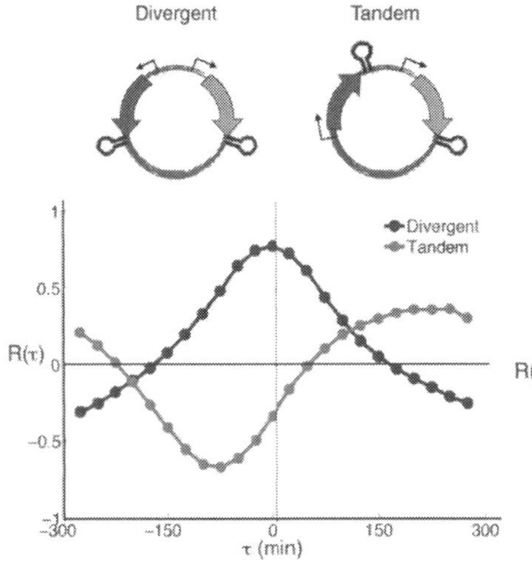

In vivo context data

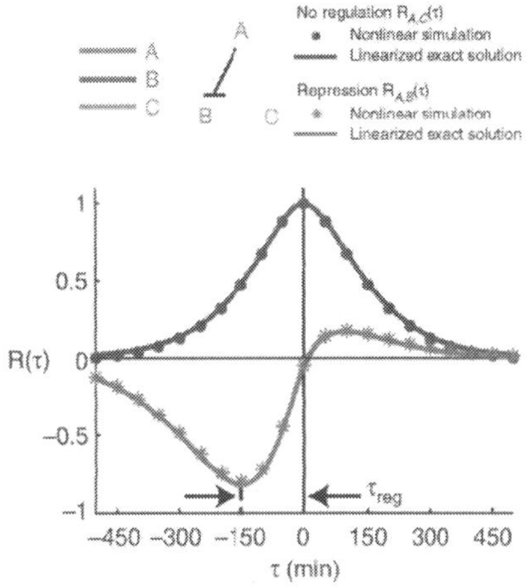

Simulation from Dunlop et. al., Nature Genetics 2008

Prosthetic Gene Networks for Biomedical Applications

Fussenegger, M., ETH Zurich

Originally presented on: 6/13/2015 8:30:00 - 8:55:00

Since Paracelsus' (1493-1541) definition that the dosing makes the drug the basic treatment strategies have lagely remained unchanged. We continue to use a precise prescribed dose of a small-molecule drug, a protein therapeutic or a therapeutic transgene to constitutively modulate or complement the activity of a disease-relevant target. However, this treatment concept does neither consider the metabolic dynamics nor the interdependence of the most important pathophysiologies of the 21st century such as obesity, diabetes and cardiovascular disorders. Synthetic biology-inspired prosthetic networks may act as metabolic prostheses that provide the dynamic interventions, the immediate pre-disease action and the multi-target capacity required to meet with the treatment challenges of the future. Prosthetic networks consist of synthetic sensor-effector gene circuits that (i) seamlessly operate in implanted designer cells, (ii) constantly sense, monitor and score metabolic disturbanes in peripheral circulation, (iii) process OFF-level concentrations of pathologic metabolites, and (iv) coordinate an adjusted therapeutic response in an (v) automatic and self-sufficient manner. We will present our latest generation of synthetic mammalian gene circuits and provide a few examples of prosthetic networks operating in animal models of prominent human diseases to highlight the challenges and impact of synthetic biology on future biomedical applications.

Controlling Organisms Electrically with Synthetic Biology

Caroline Ajo-Franklin
Staff Scientist, Berkeley Lab
cajo-franklin@lbl.gov
cafgroup.lbl.gov

MOLECULAR FOUNDRY

introduction

Electronically connecting living & non-living

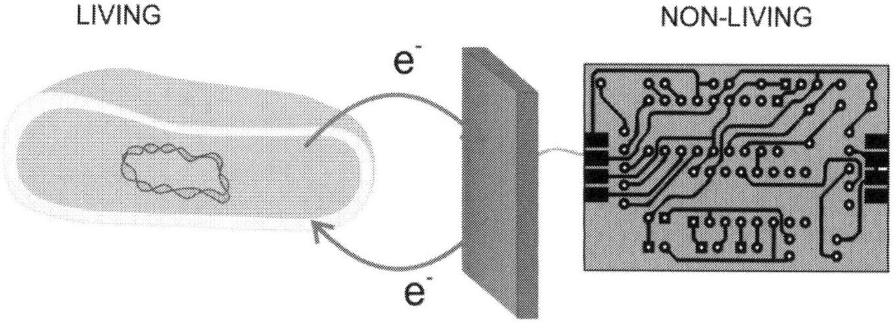

MOLECULAR FOUNDRY

introduction

Electron transfer regulates many aspects of cellular behavior

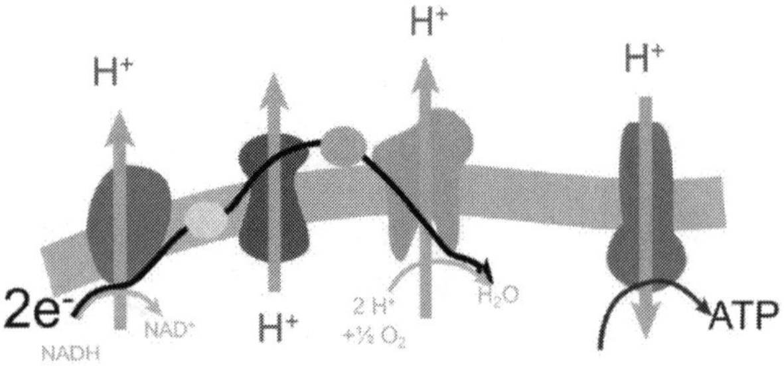

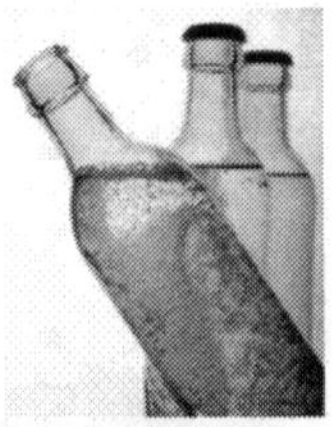

Goal: control behavior

MOLECULAR FOUNDRY

3

introduction

Existing methods of increasing e⁻ flux are non-specific, low yielding

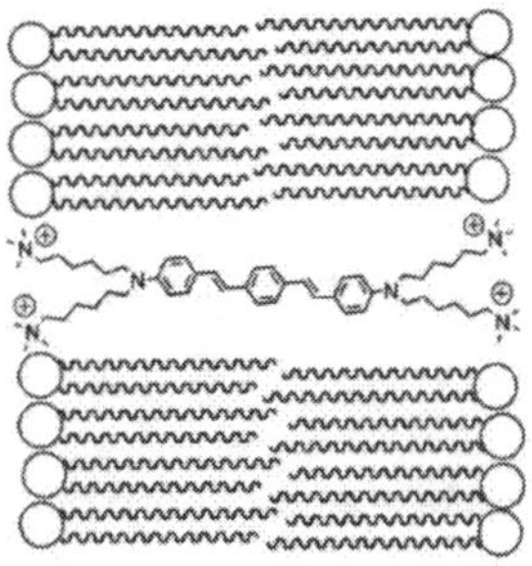

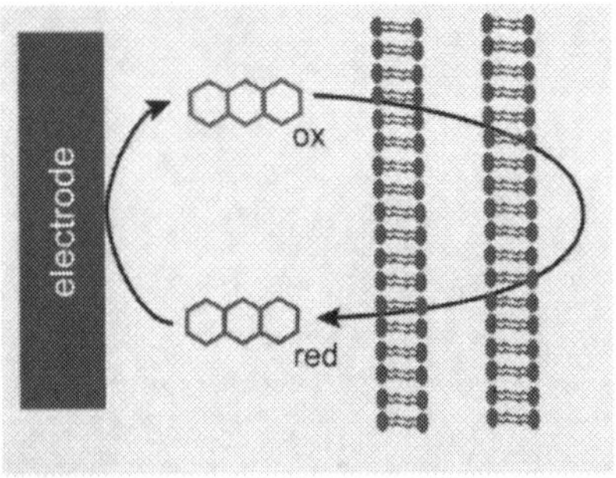

Log

Can we build a transmembrane 'wire' with molecular recognition?

CAF, Noy, *Advanced Materials* (2015).

MOLECULAR FOUNDRY

building

building

Mtr pathway moves e⁻ to/from electrodes

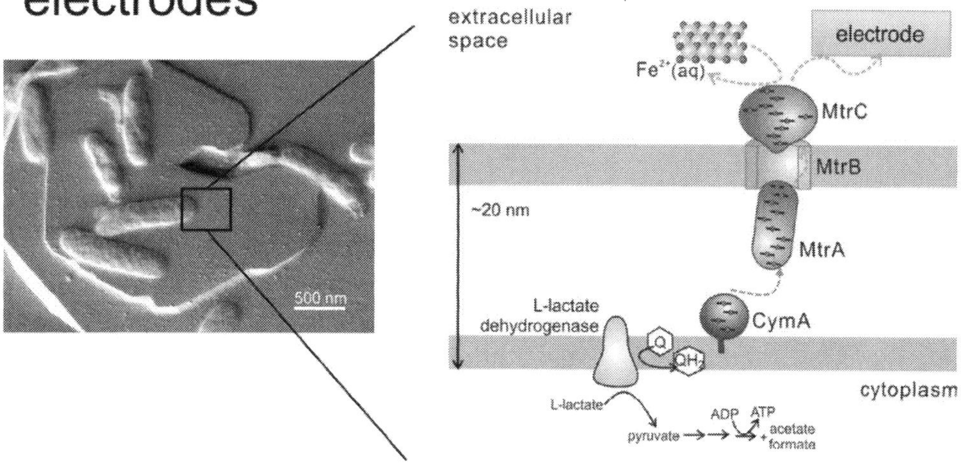

Bond, Gralnick, Fredrickson, Myers, Nealson, Richardson.

building

Mtr pathway moves e⁻ to/from electrodes

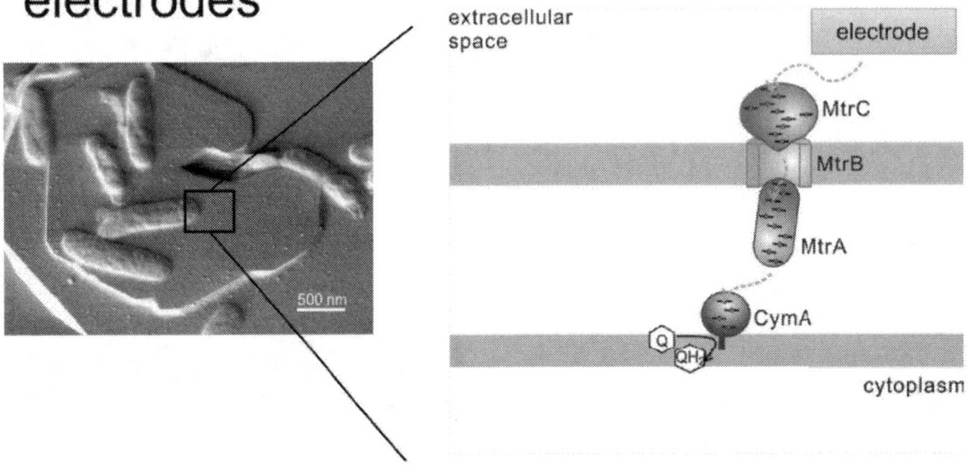

Ross, Gralnick, Bond. *PLoS One* (2012).

MOLECULAR FOUNDRY

7

building

Mtr proteins: difficult biosynthesis

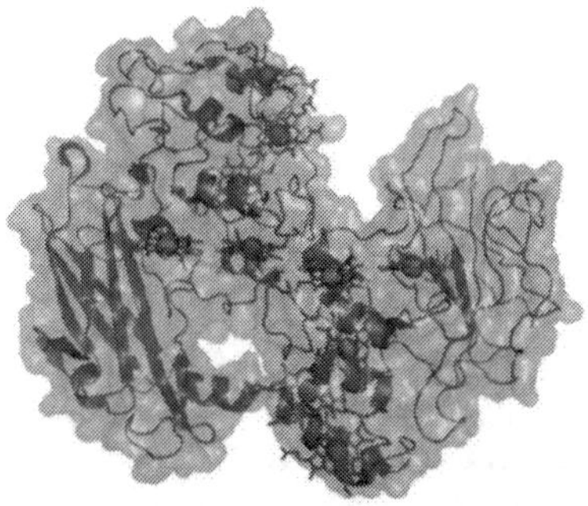

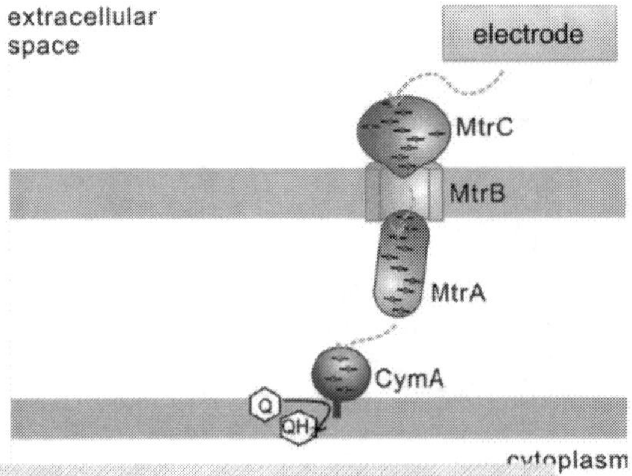

Can we use this electron transfer pathway to move e⁻ in & out of other organisms?

MOLECULAR FOUNDRY

8

building

Synthesizing an electron nanoconduit

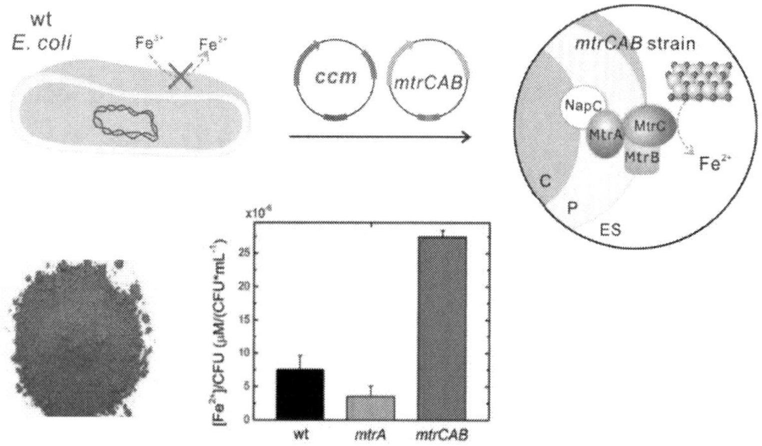

Jensen, *et al.*, CAF, *PNAS* (2010).

MOLECULAR FOUNDRY

building

e⁻ nanoconduits only: slow, inefficient, dead

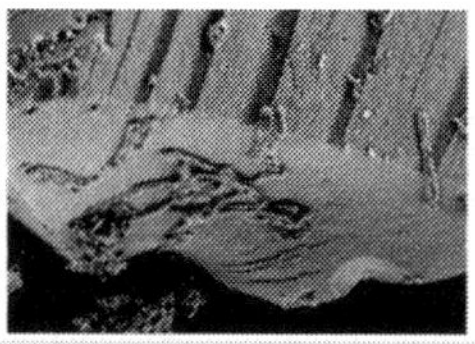

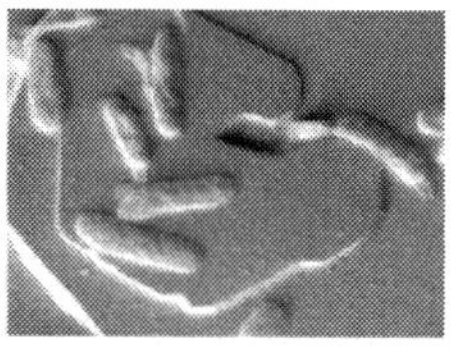

	e- coli v.1	*S. oneidensis*
rate	0.3 fA cell⁻¹	100 fA cell⁻¹
Coloumbic efficiency	~0.1%	~35%
cell survival	<10%	grow

MOLECULAR FOUNDRY

building

Too few e⁻ nanoconduits?

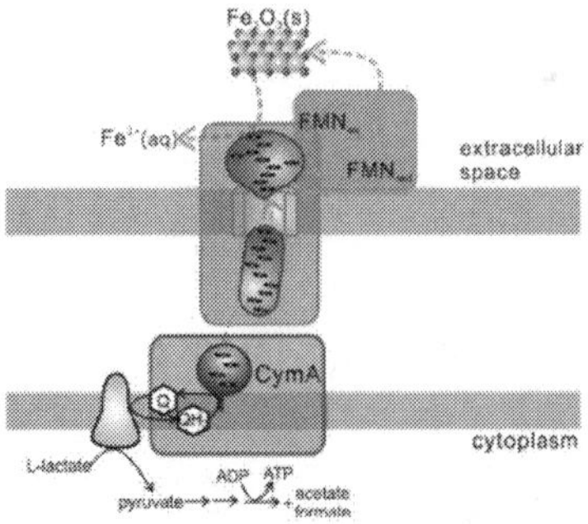

	e- coli v1	*S. oneidensis*
e⁻ exit	100% direct	20% direct 80% flavin-dependent
# e⁻ nanoconduits	10,000	72,000
e⁻ entry	NapC, other *E. coli* proteins	CymA Stc?, FccA?

Ross, et. al., Tien *Appl Environ Microbiol* (2009).
Brutinel, Gralnick. *Appl Microbiol Biotechnol* (2012).

MOLECULAR FOUNDRY

11

building

Hypothesis: need slow, tunable transcription

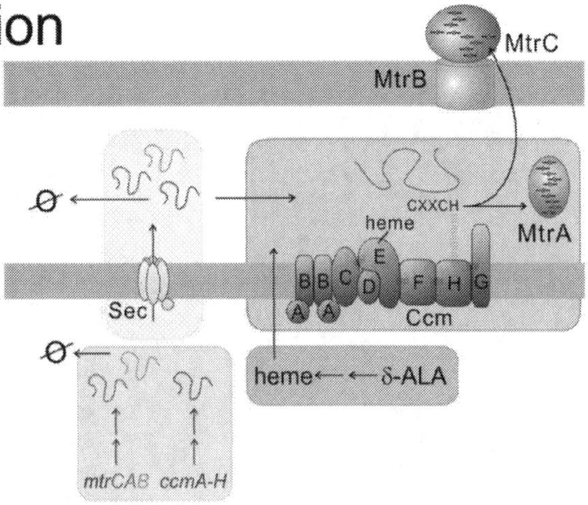

Goldbeck, *et al*, CAF, *ACS Synth Biol.* (2013).

MOLECULAR FOUNDRY

building

Hypothesis: need slow, tunable transcription

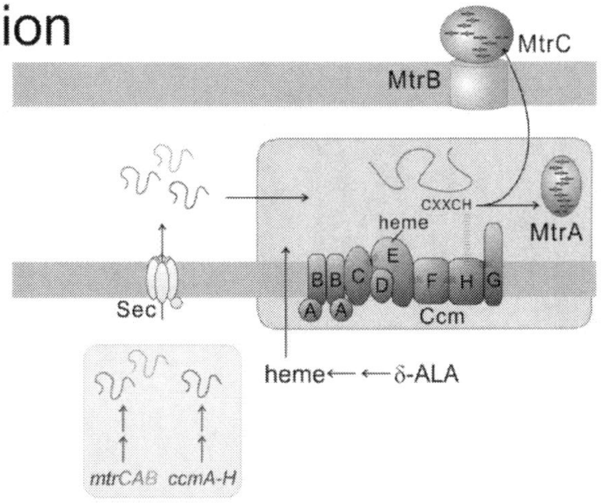

Goldbeck, *et al*, CAF, *ACS Synth Biol*. (2013).

MOLECULAR FOUNDRY

building

Tuning promoter activity affects protein production & cell growth

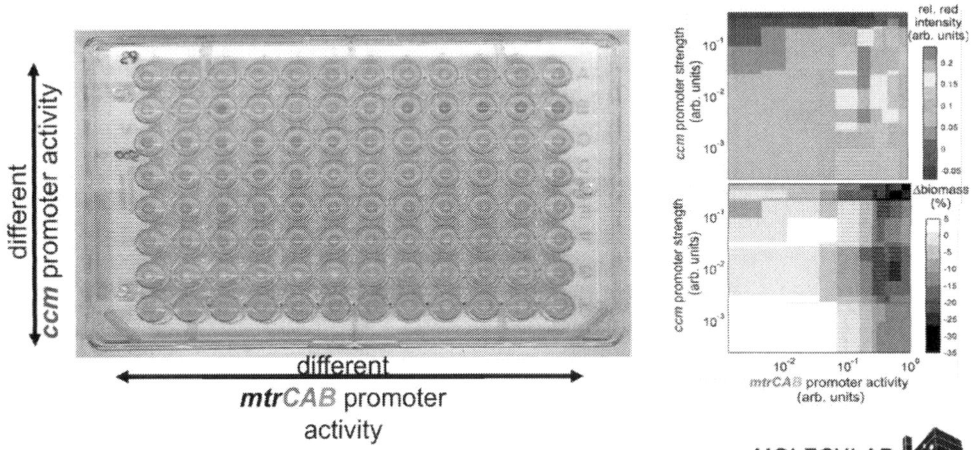

Goldbeck, *et al*, CAF, *ACS Synth Biol*. (2013).

MOLECULAR FOUNDRY

building

e^- *coli* v2 delivers measurable current

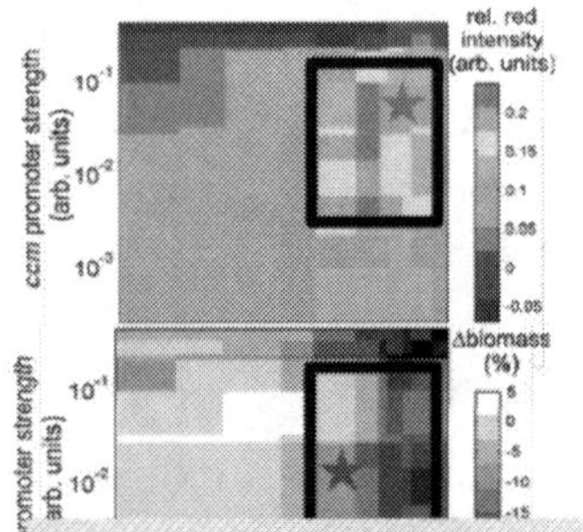

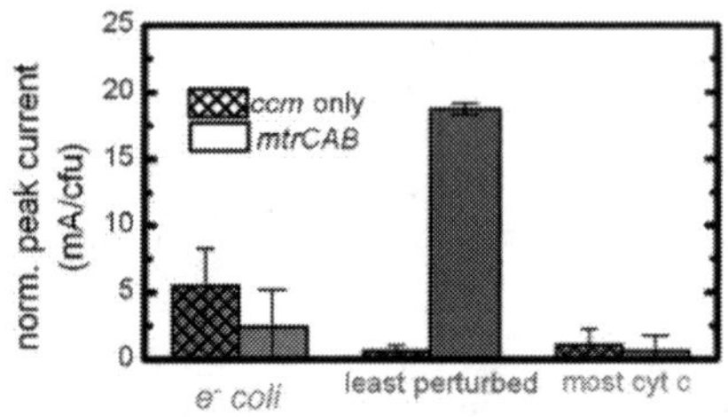

strong promoter ≠ most protein ≠ best function

Goldbeck, *et al*, CAF, *ACS Synth Biol.* (2013).

MOLECULAR FOUNDRY

building

Non-optimal entry & exit?

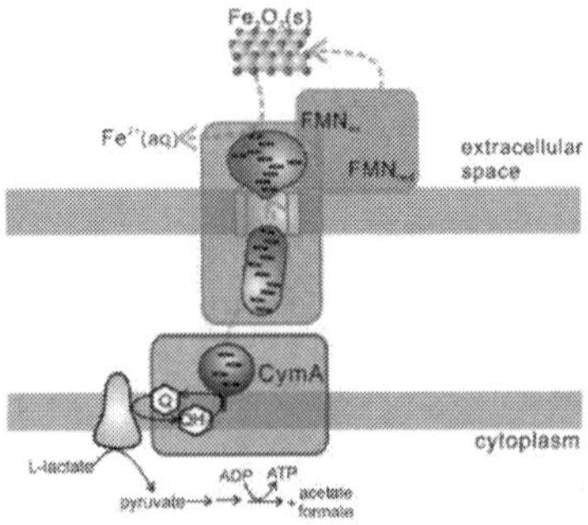

	e- coli v1	*S. oneidensis*
e^- exit	100% direct	20% direct 80% flavin-dependent
# e^- nanoconduits	10,000	72,000
e^- entry	NapC, other *E. coli* proteins	CymA Stc?, FccA?

Ross, et. al., Tien *Appl Environ Microbiol* (2009).
Brutinel, Gralnick. *Appl Microbiol Biotechnol* (2012).

MOLECULAR FOUNDRY

building

cymA, riboflavin increase Fe_2O_3(s) reduction & survival

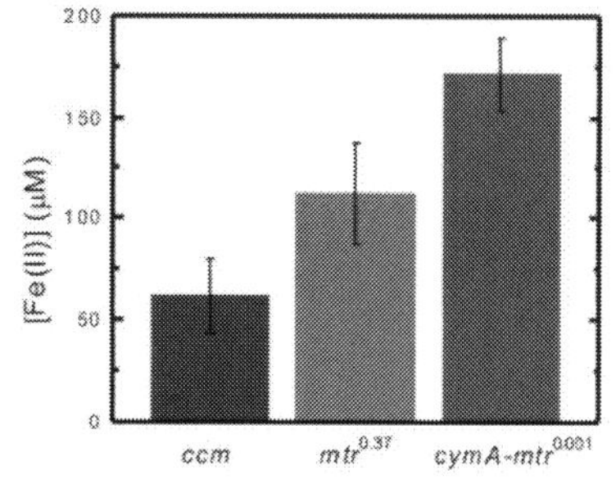

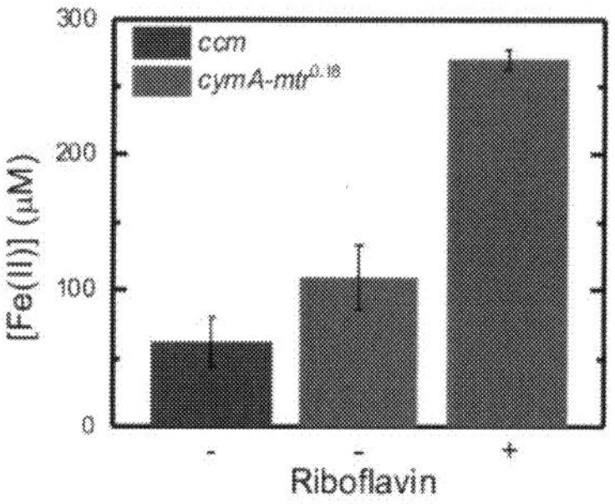

Jensen, *et al.*,CAF, in prep

MOLECULAR FOUNDRY

17

building

cymA-mtr E. coli make current

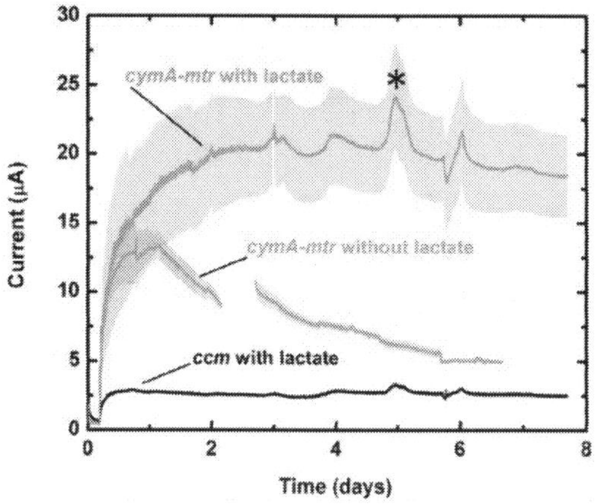

Current increase not due to:
# of viable cells,
cell attachment

TerAvest, Zajdel, CAF. *ChemElectroChem* (2014).

MOLECULAR FOUNDRY

18

building

e^- *coli* approaches electrical output of native organisms

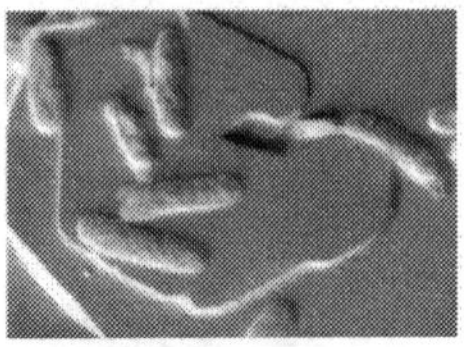

	e- coli v3.0	*S. oneidensis*
rate	~30 fA cell^{-1}	~100 fA cell-^{1}
Coloumbic efficiency	10%	~35%
cell survival	30-100%	grow

MOLECULAR FOUNDRY

19

using

Eliminating the need to redox balance

Autonomous sensing biorobots

Understanding mineralization

MOLECULAR FOUNDRY

20

using

Goal: Eliminate need for redox balance

1 mol glycerol : 3 mol biodiesel

glycerol → ethanol + CO_2 + H_2

glycerol → 1.5 ethanol + $2e^-$ + $2H^+$

glucose + $2H^+$ + $2e^-$ → succinate

Flynn, *et al.* , **mBio** (2010).

MOLECULAR FOUNDRY

using

e^- coli can get rid of electrons as current

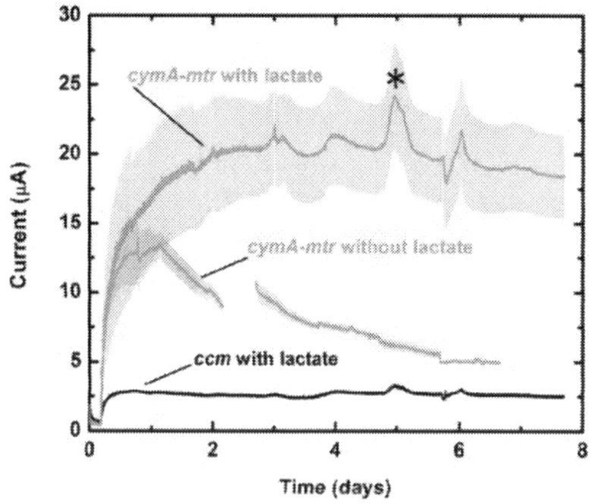

TerAvest, Zajdel & CAF, ChemElectroChem (2014).

MOLECULAR FOUNDRY

using

Electrode-interacting e^- *coli* are metabolically active for weeks

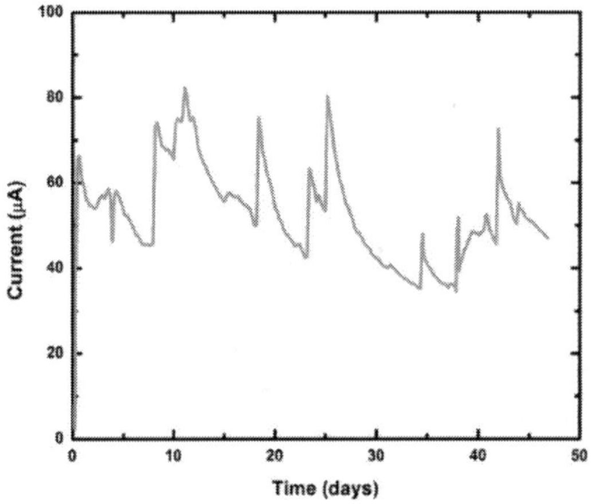

MOLECULAR FOUNDRY

23

using

Current production is linked to more oxidized metabolites

strain	[formate] (μM) 2 e^-	[ethanol] (μM) 2 e^-	[acetate] (mM) 4 e^-
ccm	44±13	64±0.6	0.95±0.05
cymA-mtr	n.d.	40±3	1.1±0.02

Current ~ Δe from metabolism

TerAvest, Zajdel & CAF, ChemElectroChem (2014).

MOLECULAR FOUNDRY

24

using

e$^-$ conduit changes *E. coli* metabolism

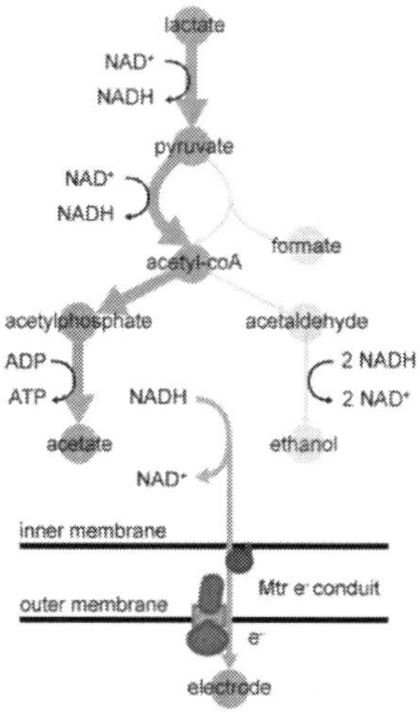

TerAvest, Zajdel & CAF, ChemElectroChem (2014).

MOLECULAR FOUNDRY

using

Can we supply electrons to the cytoplasm?

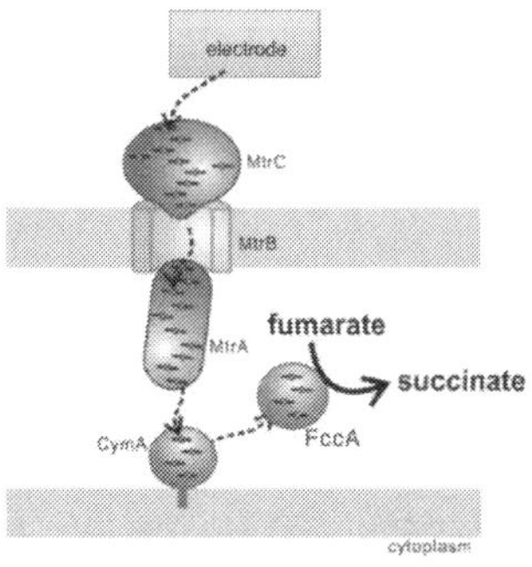

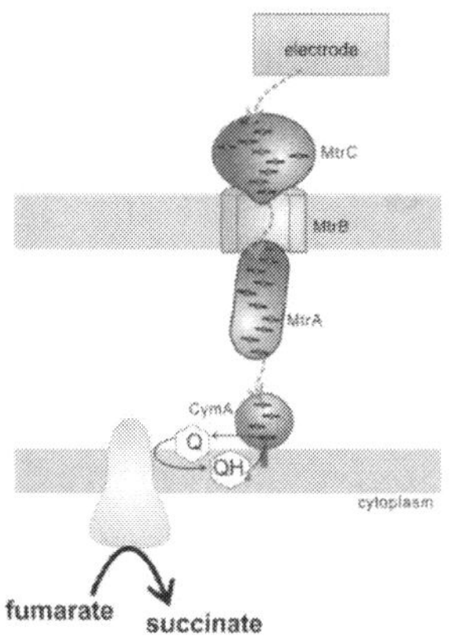

Ross, Gralnick, Bond. *PLoS One* (2012).

MOLECULAR FOUNDRY

using

Electrons enter e^- *coli* through e^- transfer pathway

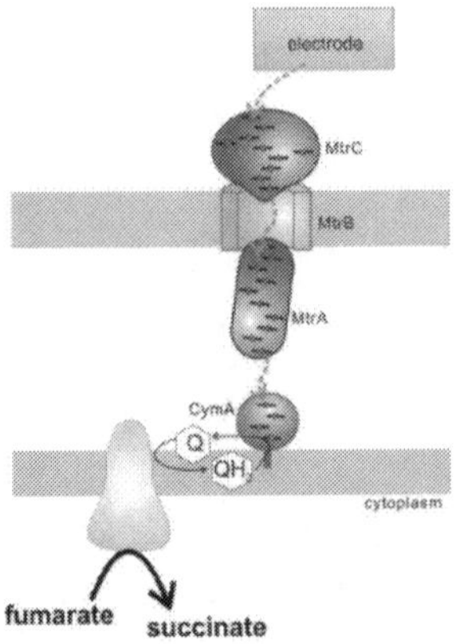

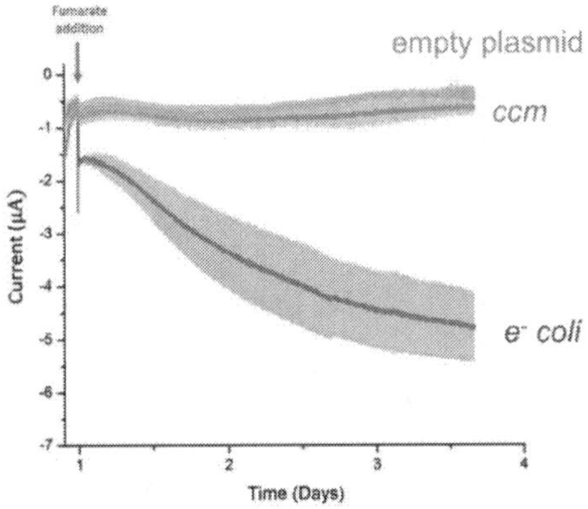

Baruch, TerAvest, CAF

MOLECULAR FOUNDRY

27

using

e^- are used in cytoplasmic biosynthesis

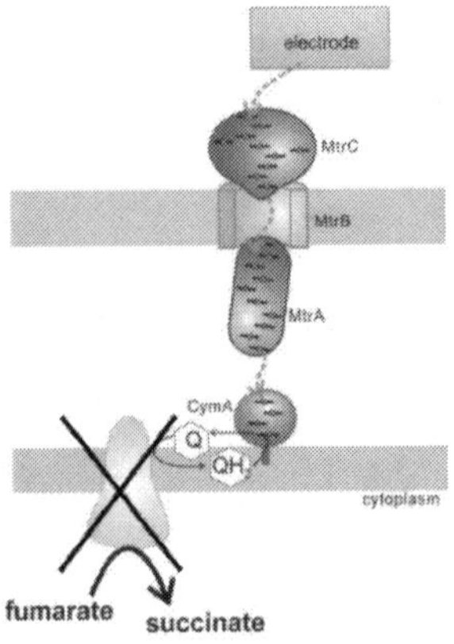

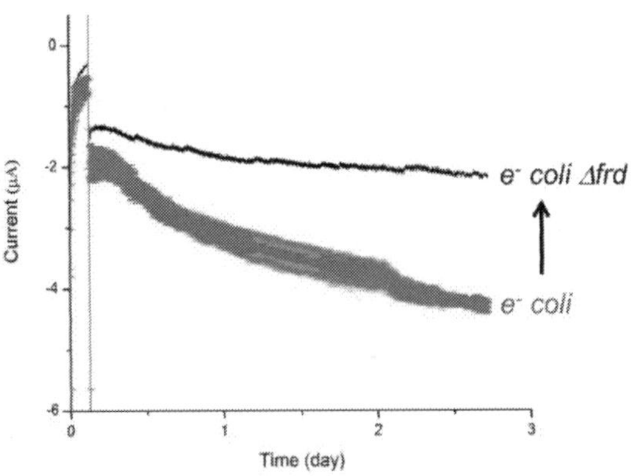

Baruch, TerAvest, CAF

MOLECULAR FOUNDRY

28

using

The electrode supplies reducing equivalents on demand to e^- *coli*

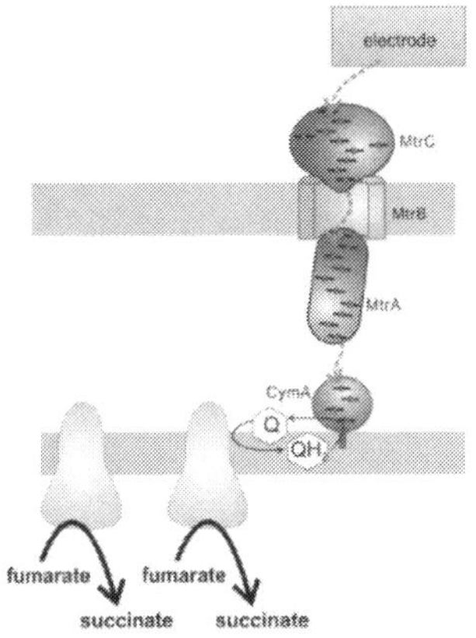

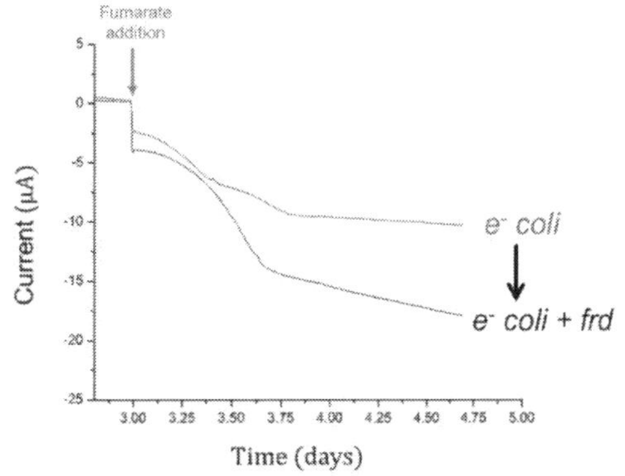

Baruch, TerAvest, CAF

MOLECULAR FOUNDRY

using

A cell-chip hybrid offers deployable sensing

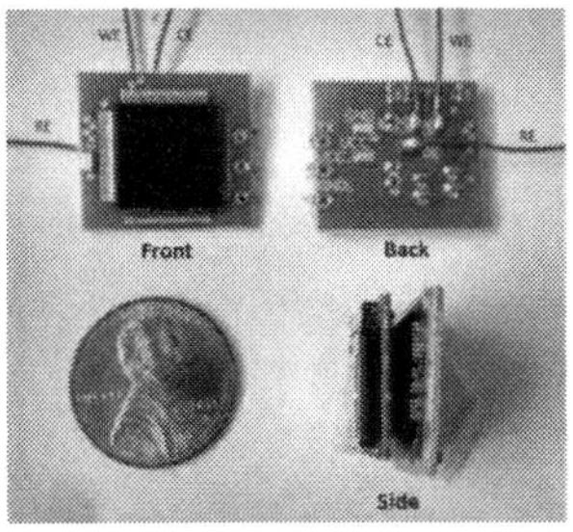

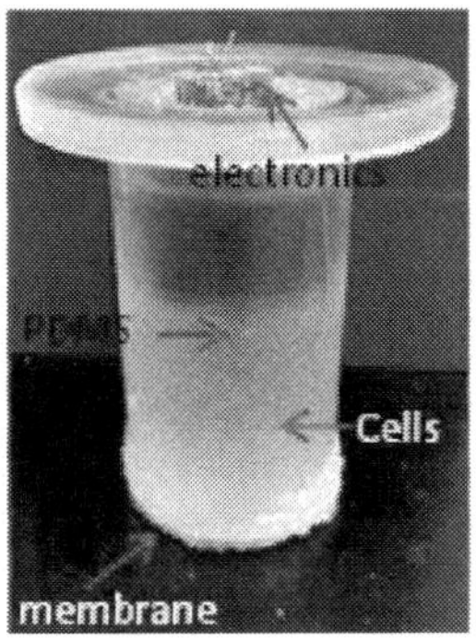

Zhou, Zajdel, TerAvest, Maharbiz, *submittted*

MOLECULAR FOUNDRY

using

Temperature sensing with cell-chip hybrid

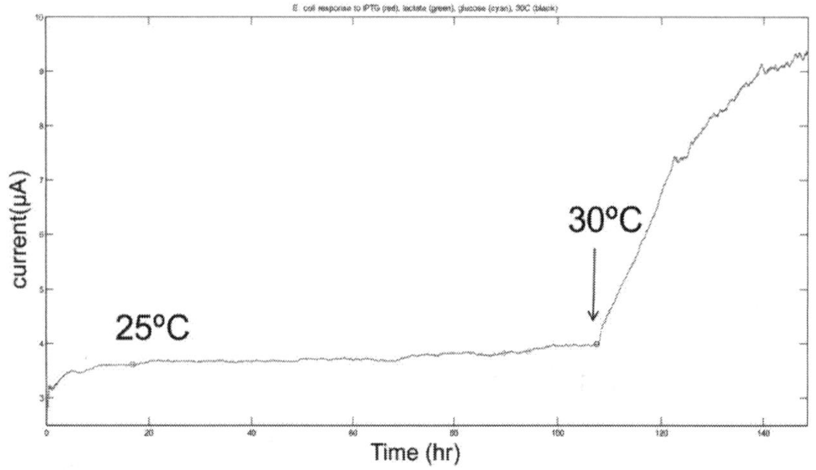

Zhou, Zajdel, TerAvest, CAF, Maharbiz

summary

Electron transfer pathways offer a **molecularly-defined route** to **efficiently** move electrons in & out of cells.

We have identified **general principles** for improved **synthesis & function** of electron transfer pathways in new hosts.

Moving e^- in & out of cells allows **readout and control** of intracellular state.

Want to go electronic? Come talk to me.

summary

acknowledgements

CAF lab
Dr. Michaela TerAvest
Dr. Tatsuya Fukushima
Dr. Moshe Baruch
Dr. Behzad Rad
Matt Hepler
Dr. Heather Jensen
Dr. Jenny Cappuccio
Cheryl Goldbeck
Greg Murray
Albert Shon, Nicole Beedle, Kostia Malley, Mark Kokish, Annie Li, Dimitri Umpriani

Collaborators:
Prof. Lars Angenent, Cornell
Prof. Michel Maharbiz, UC-Berkeley
Prof. Karina Sand, Univ. of Cohenhagen

Funding:
Office of Naval Research
Synthetic Biology Institute
LBL LDRD
DoE BES - Molecular Foundry
DoE EERE

MOLECULAR FOUNDRY

Towards the High-Throughput Construction of Fluorescent Biosensors

Stacy-Anne Morgan
Savage Lab
June 13, 2015

Energy Biosciences Institute

Berkeley
University of California

Overview

1. Background and motivations for metabolite-sensing fluorescent biosensors
2. Rational design of a fatty acid sensor
3. Transposon-mediated strategy for the high-throughput construction of biosensors

Measuring metabolism is a bottleneck in strain engineering

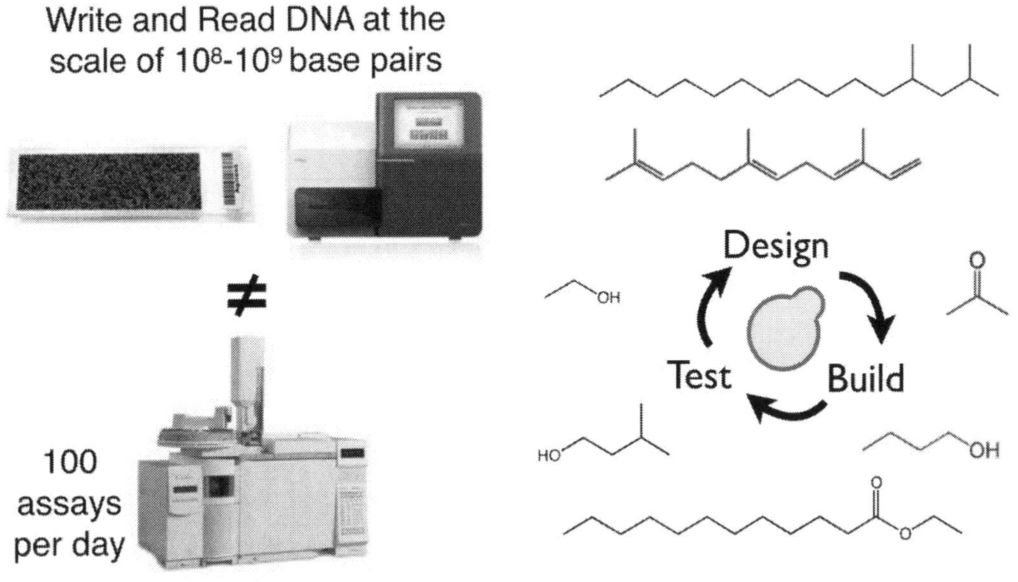

There are many ways to make fluorescent biosensors

Analyte → Measurable Output

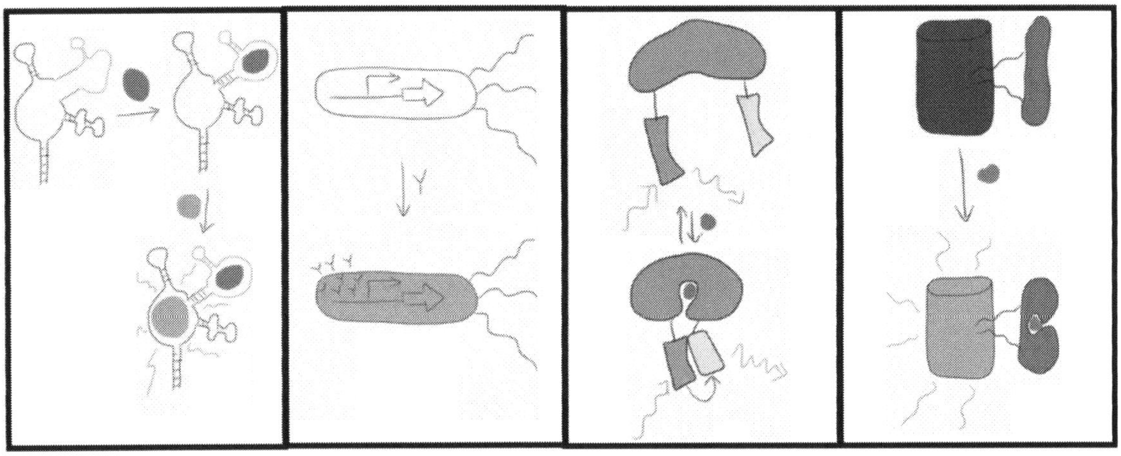

RNA-based biosensors	transcriptional reporters	FRET sensors	circularly-permuted GFP sensors

The chromophore of GFP exhibits switch-like behavior

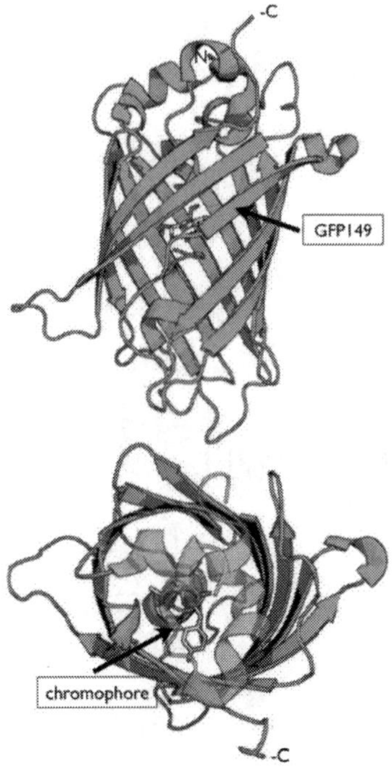

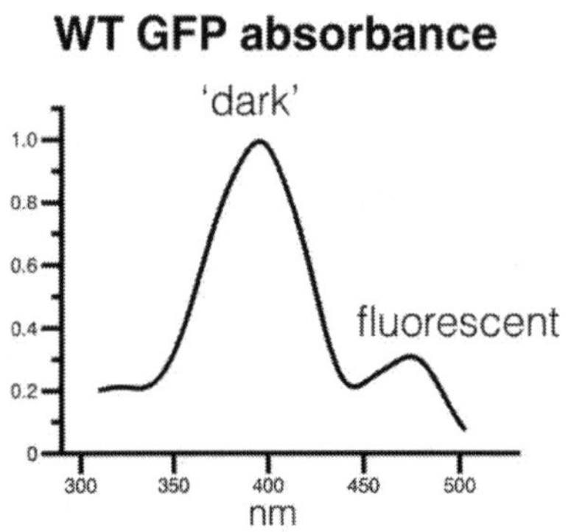

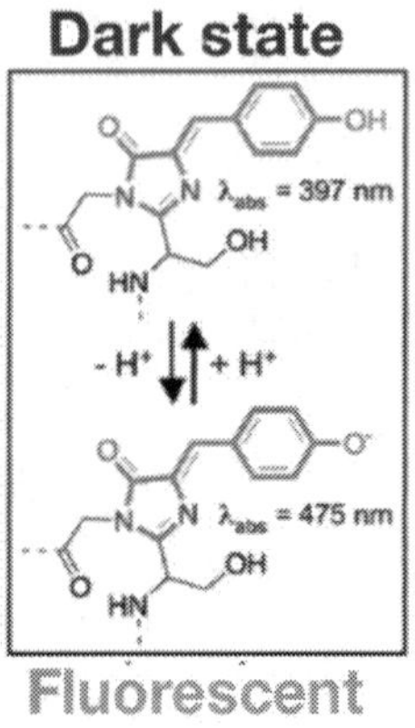

Emission properties of the GFP chromophore are sensitive to its local chemical environment

M. Chattoraj, B. A. King, G. U. Bublitz, S. G. Boxer, Proc. Natl. Acad. Sci. **93**, 8362–8367 (1996)., H. Niwa et al., Proc. Natl. Acad. Sci. **93**, 13617–13622 (1996)., R. Y. Tsien, Annu. Rev. Biochem. **67**, 509–544 (1998).

A ligand binding domain can modulate GFP fluorescence

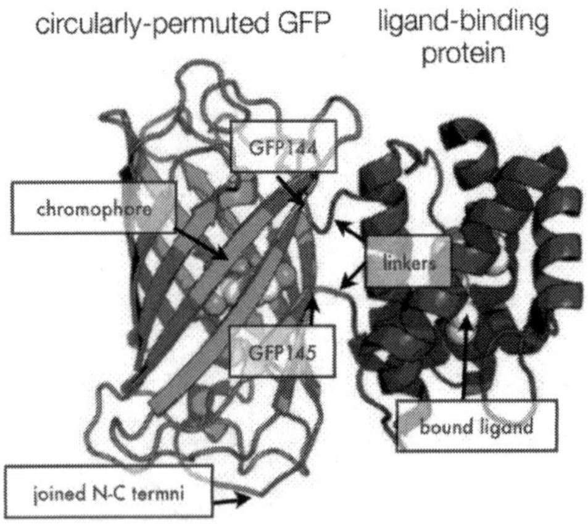

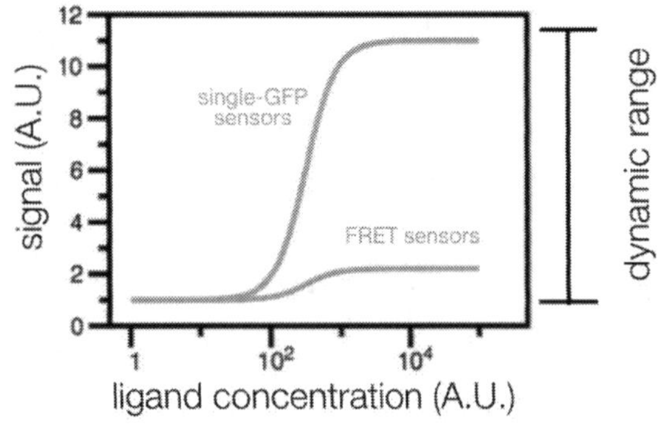

How do you construct an allosteric switch?

G. S. Baird, D. A. Zacharias, R. Y. Tsien, *Proc. Natl. Acad. Sci. U.S.A.* **96**, 11241–11246 (1999).

Overview

1. Background and motivations for metabolite-sensing fluorescent biosensors
2. **Rational design of a fatty acid sensor**
3. Transposon-mediated strategy for the high-throughput construction of biosensors

FadR acts as a sensor of fatty acid availability

- Repressor and activator of genes involved in fatty acid metabolism
- Antagonized by long chain fatty acid acyl-CoAs

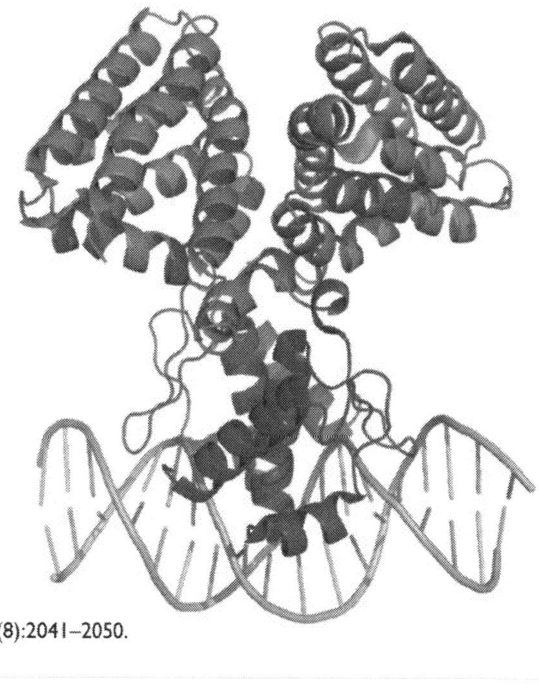

Van Aalten DMF, DiRusso CC, Knudsen J.EMBO Journal. 2001;20(8):2041–2050.

FadR acts as a sensor of fatty acid availability

- Repressor and activator of genes involved in fatty acid metabolism
- Antagonized by long chain fatty acid acyl-CoAs

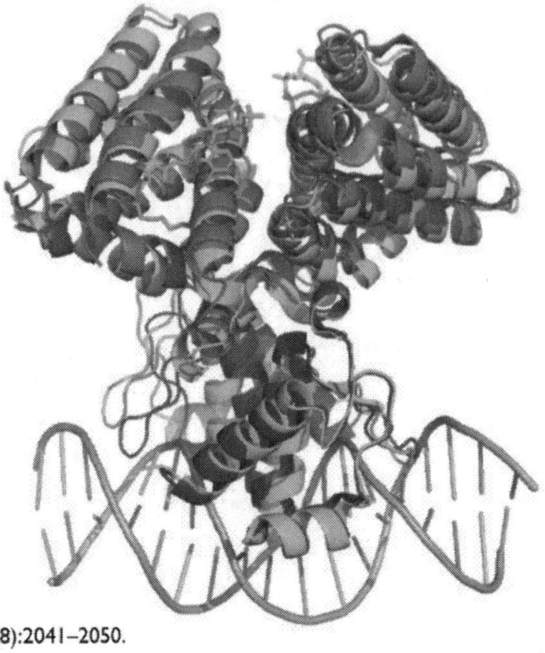

Van Aalten DMF, DiRusso CC, Knudsen J.EMBO Journal. 2001;20(8):2041–2050.

FadR acts as a sensor of fatty acid availability

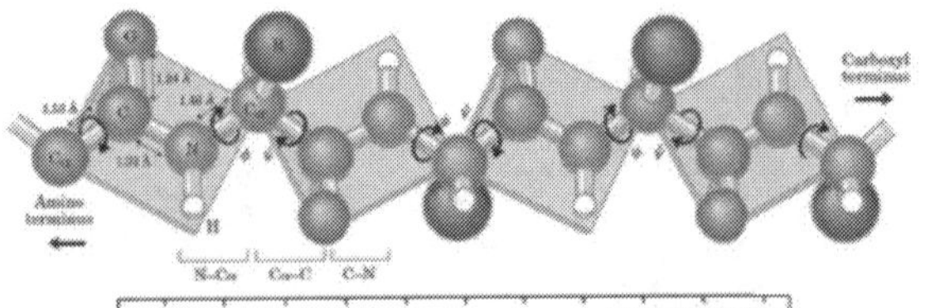

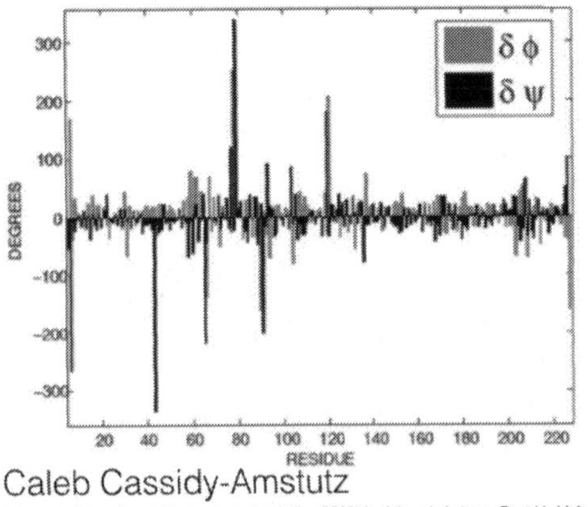

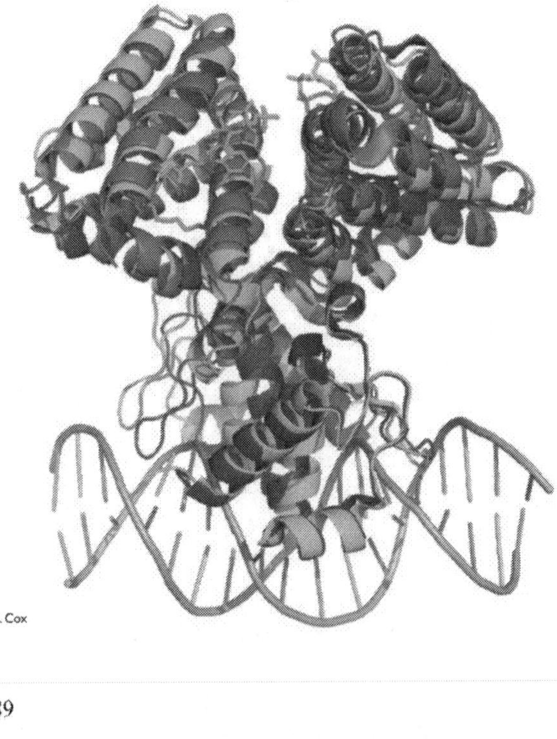

Caleb Cassidy-Amstutz

Lehninger Principles of Biochemistry Book (Jun 2008) by Albert Lehninger, David L. Nelson, Michael M. Cox

Van Aalten DMF, DiRusso CC, Knudsen J.EMBO Journal. 2001;20(8):2041–2050.

FadR acts as a sensor of fatty acid availability

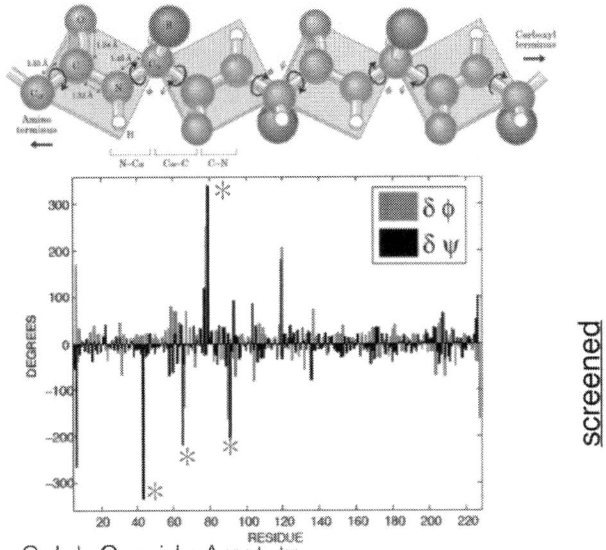

Caleb Cassidy-Amstutz

Lehninger Principles of Biochemistry Book (Jun 2008) by Albert Lehninger, David L. Nelson, Michael M. Cox

Van Aalten DMF, DiRusso CC, Knudsen J.EMBO Journal. 2001;20(8):2041–2050.

Screened Variants

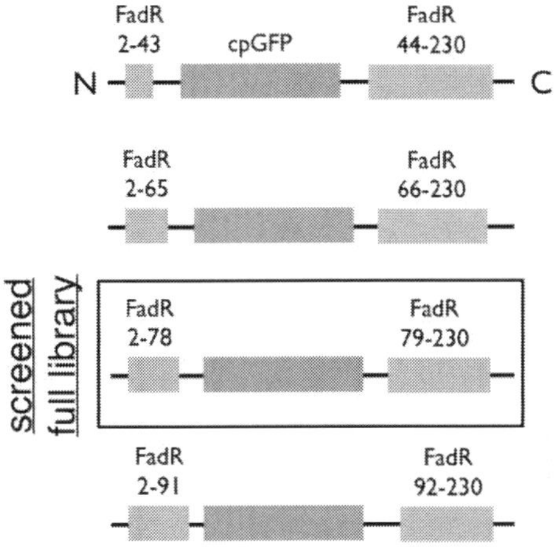

Construction of a fatty acid sensor

FadR from *E. coli*

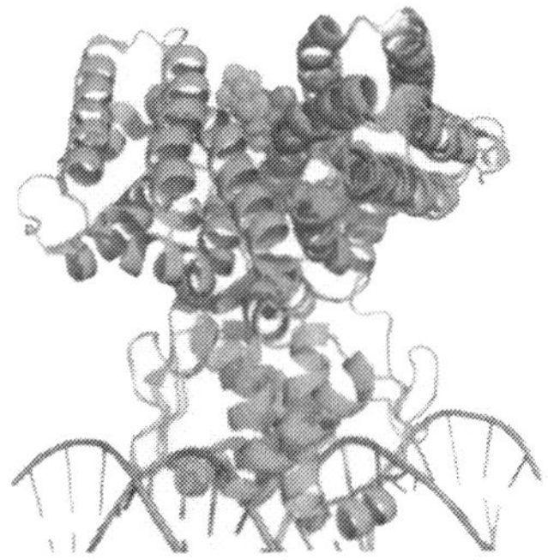

N-linker: Ser Ser Ala
C-linker: Thr Ser

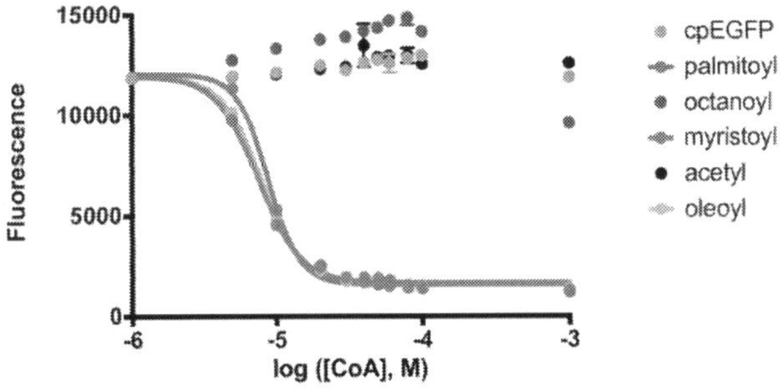

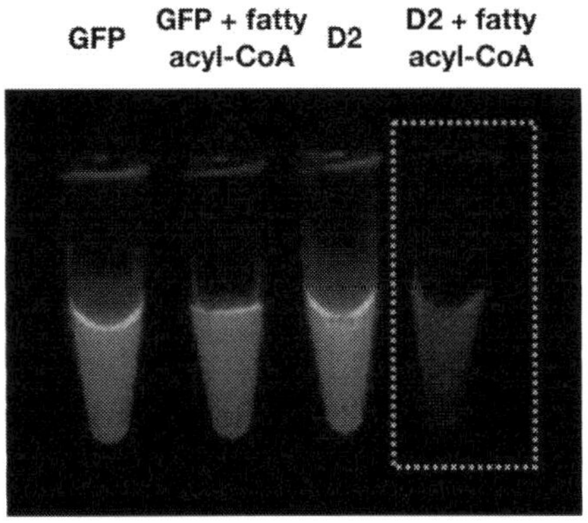

Conclusions

- Made a fatty acid biosensor with high dynamic range.
- Seems to affect culture viability. Employing known mutations to eliminate DNA binding.
- Turn-off sensor. Less desirable for *in vivo* work.

What did we learn?

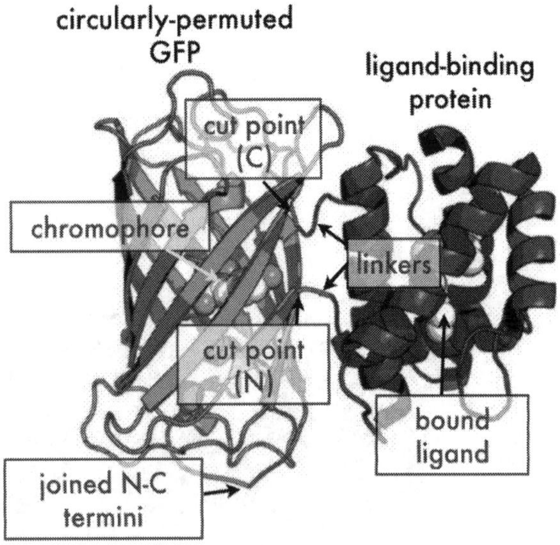

1. Topology is very important, hard to predict
2. Linkers important, hard to predict
3. Often, no x-ray structure information
4. Prefer turn-on sensors

Need: Big libraries and good way to screen

Constructing sensors that function well in vivo is not predictable

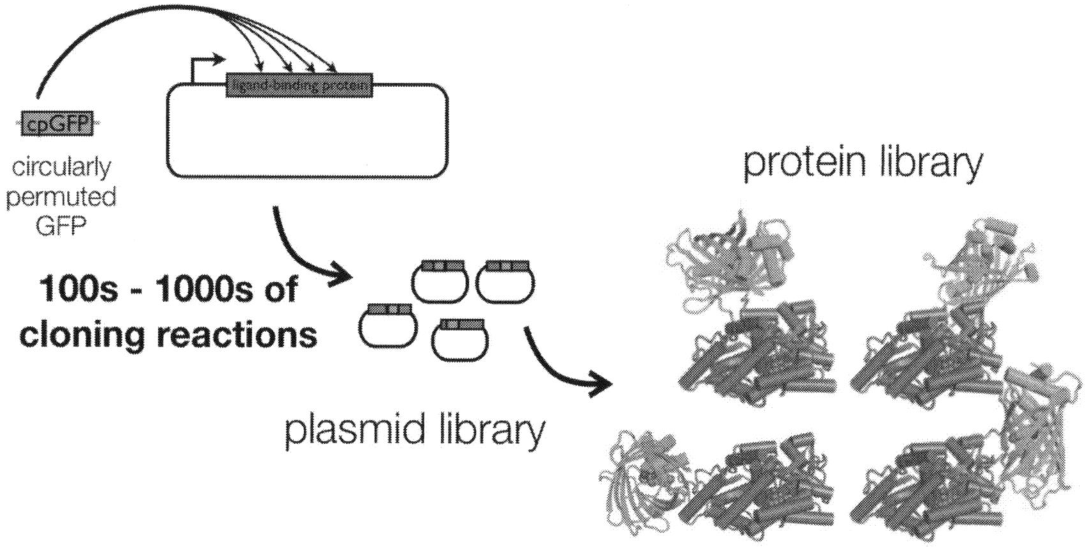

Sample all insertion sites and linkers in parallel to create the optimal biosensor

Overview

1. Background and motivations for metabolite-sensing fluorescent biosensors
2. Rational design of a fatty acid sensor
3. **Transposon-mediated strategy for the high-throughput construction of biosensors**

Transposons can create random domain insertion libraries

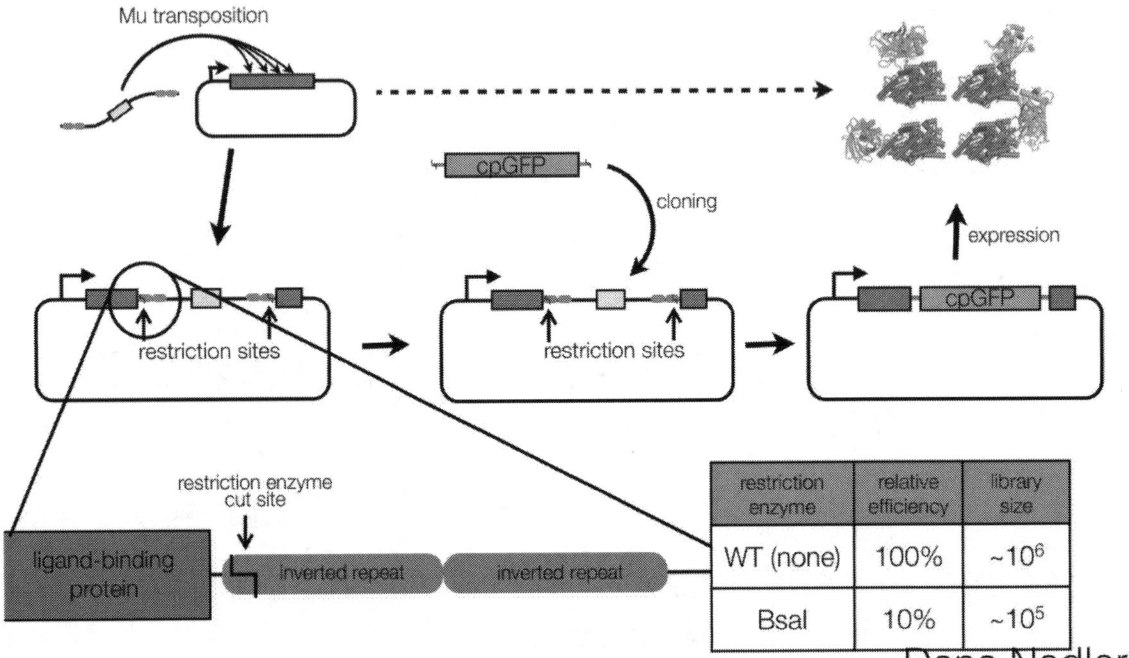

restriction enzyme	relative efficiency	library size
WT (none)	100%	~10^6
BsaI	10%	~10^5

Dana Nadler

Trehalose

- disaccharide containing two glucose units joined by a α,α-1,1-glycosidic linkage
- present in a variety of organisms
- energy source
- signaling molecule that directs/controls certain metabolic pathways
- can protect proteins and membranes from inactivation and denaturation caused by stress (desiccation, dehydration, heat, cold and oxidation)
- *Thermococcus litoralis* TMBP

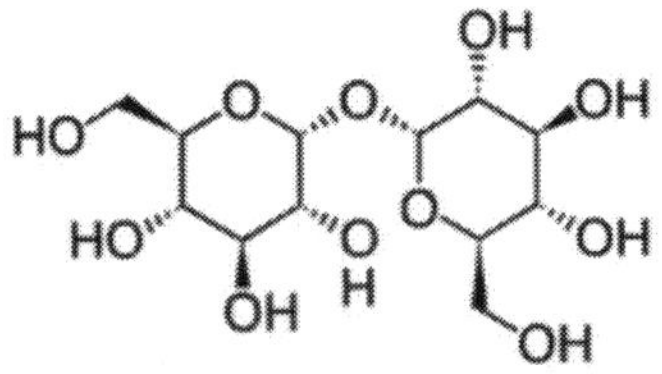

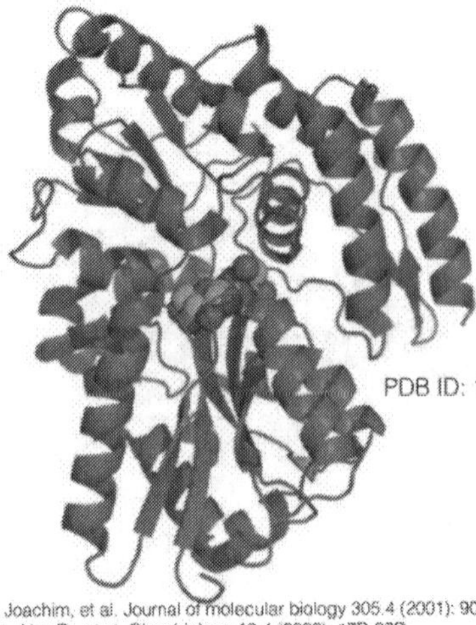

PDB ID: 1EU8

Diez, Joachim, et al. Journal of molecular biology 305.4 (2001): 905-915.
Elbein, Alan D., et al. Glycobiology 13.4 (2003): 17R-27R.

Summary of general strategy for trehalose-binding protein sensor

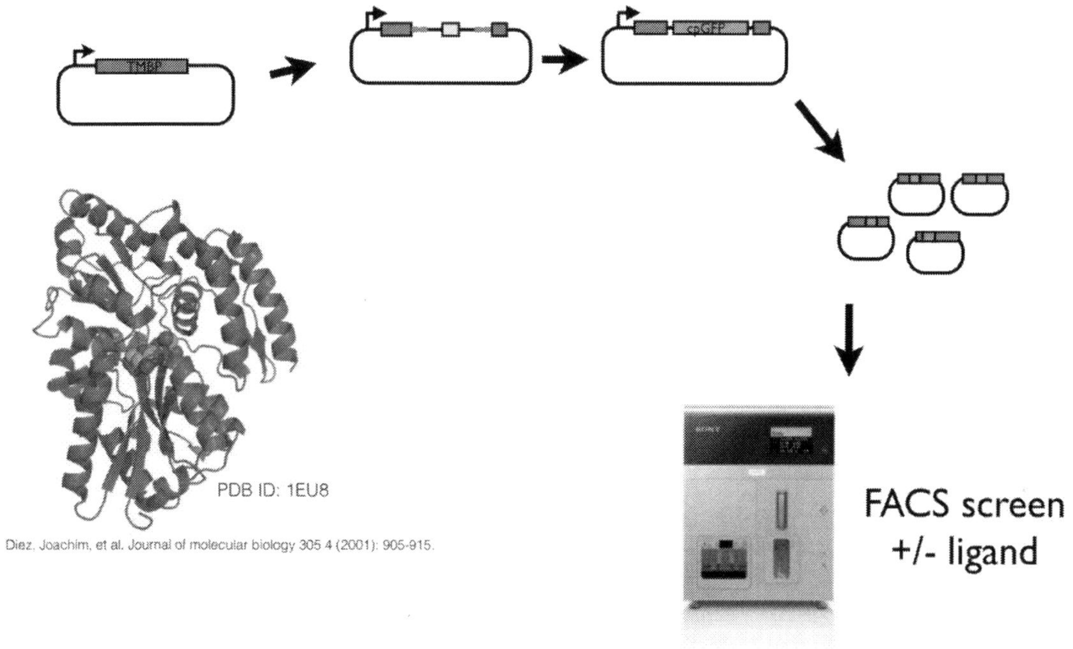

FACS can enrich for biosensing activity

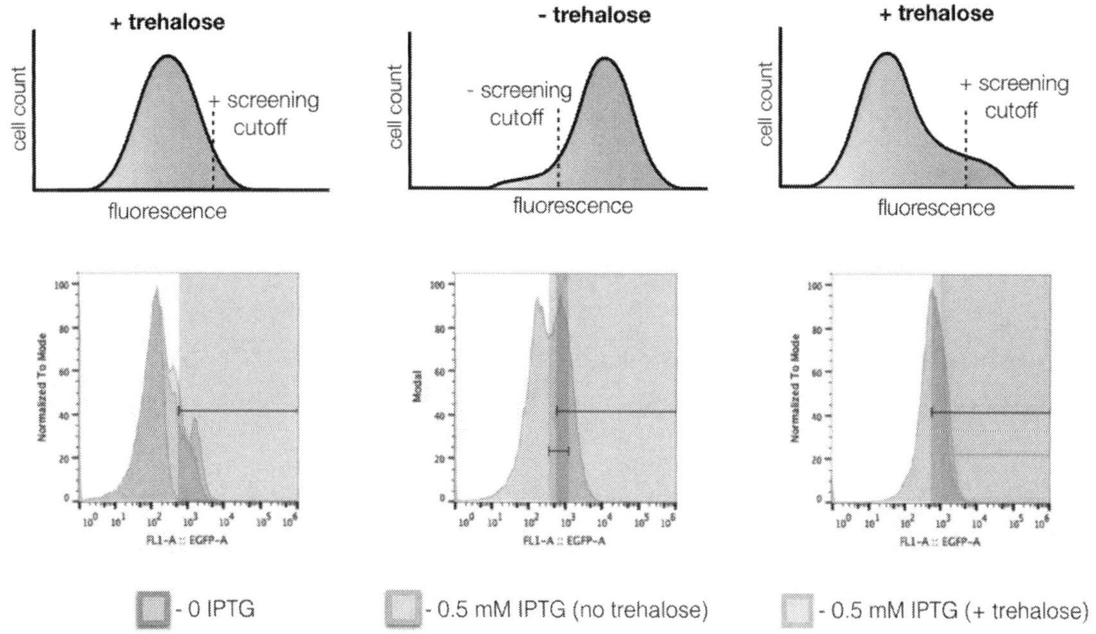

Sorting enriches for specific insertion sites

PCR assay

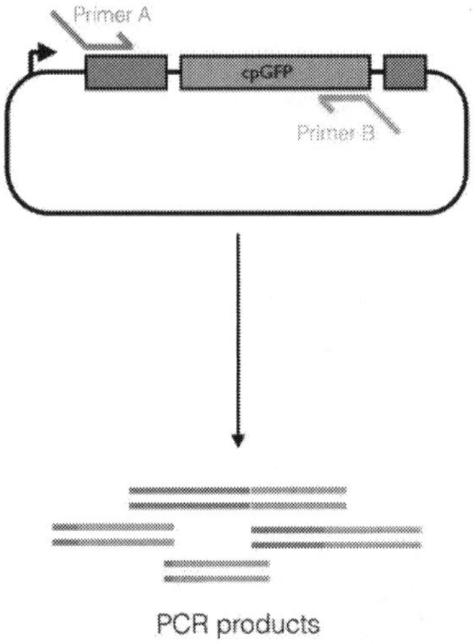

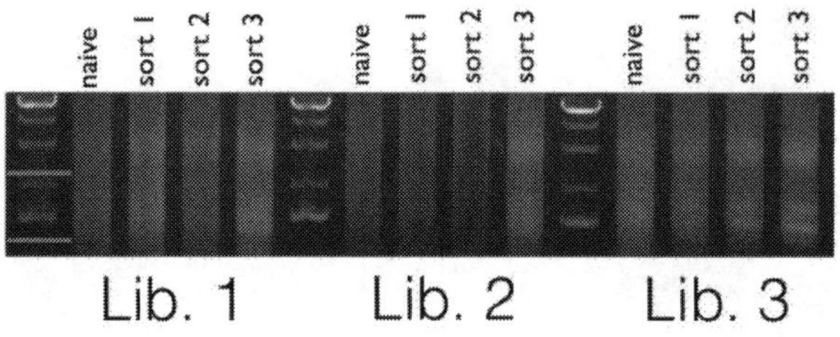

Screening of the TMBP libraries reveal functional sensors

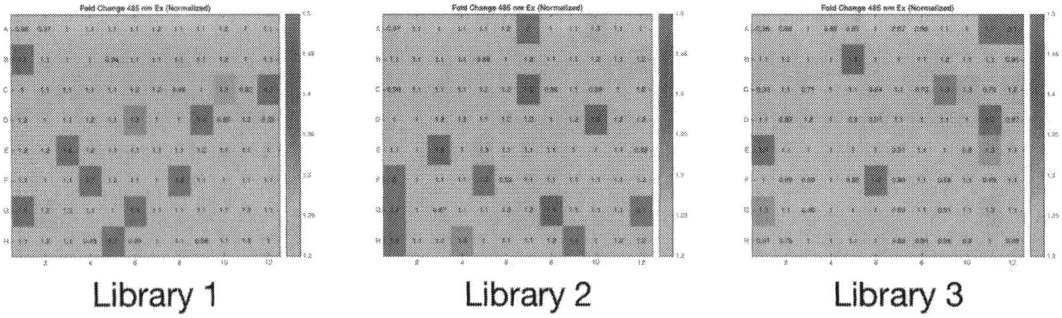

A1 - no colony
A2 - TMBP
A3 - cpEGFP

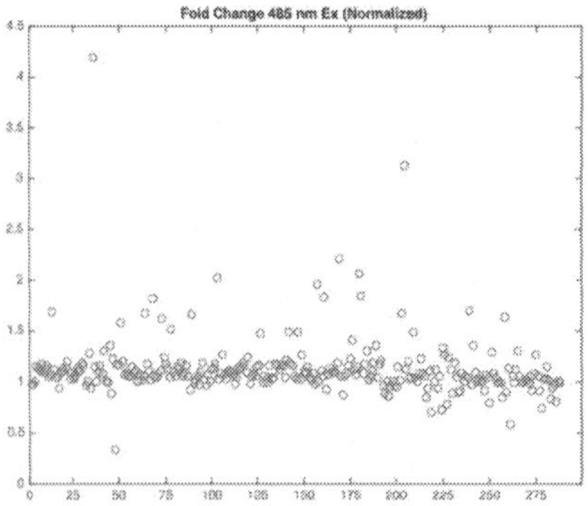

Sequencing confirms that the library is enriched at specific sites

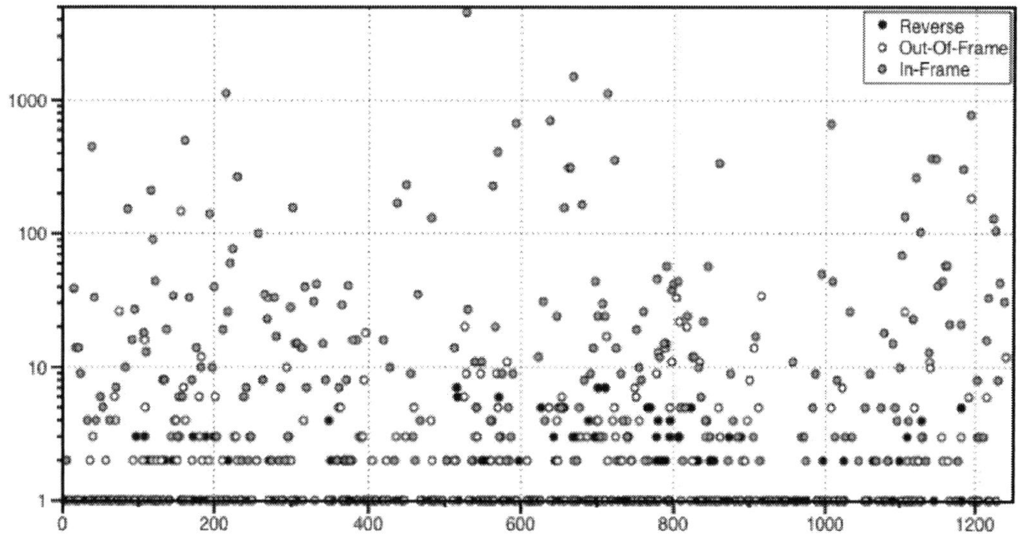

Sequencing confirms that the library is enriched at specific sites

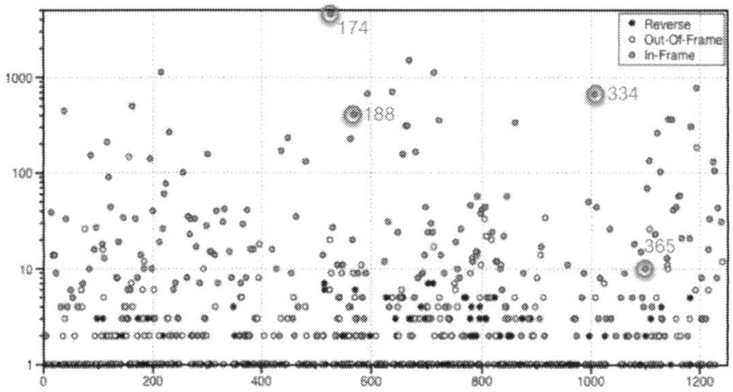

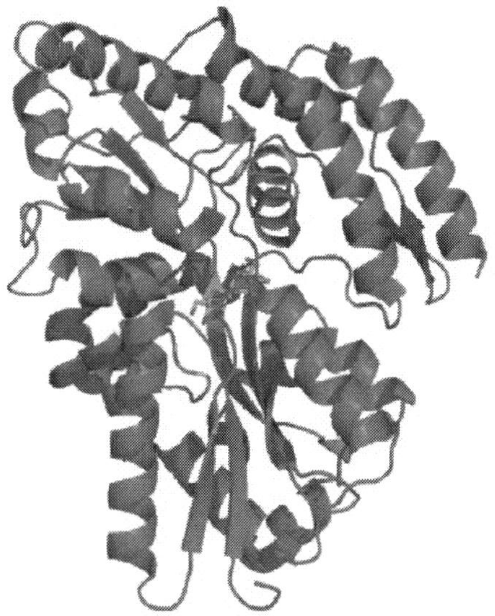

New sensor library contains variable linkers encoded by biased codons

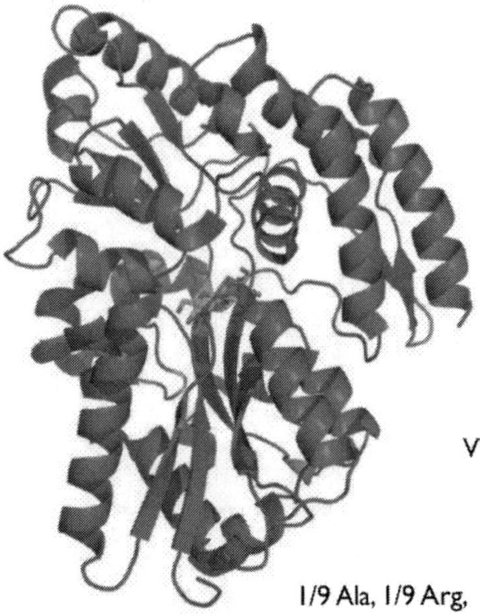

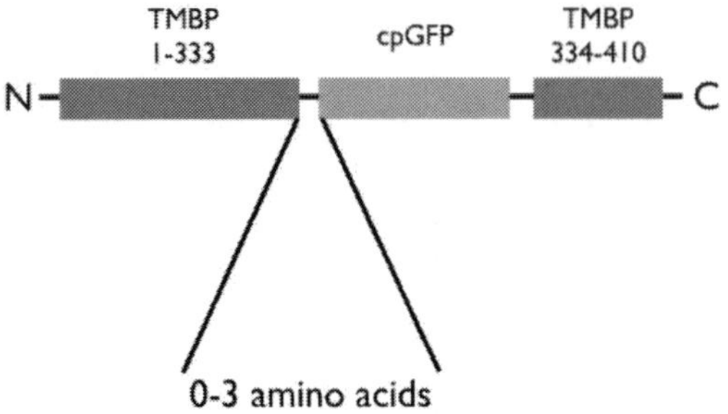

		1	2	3
	C	0.33	0.33	1
VVC codon:	G	0.33	0.33	0
	A	0.33	0.33	0
	T	0	0	0

1/9 Ala, 1/9 Arg, 1/9 Asn, 1/9 Asp, 1/9 Gly, 1/9 His, 1/9 Pro, 1/9 Ser, 1/9 Thr

The chosen insertion site is good

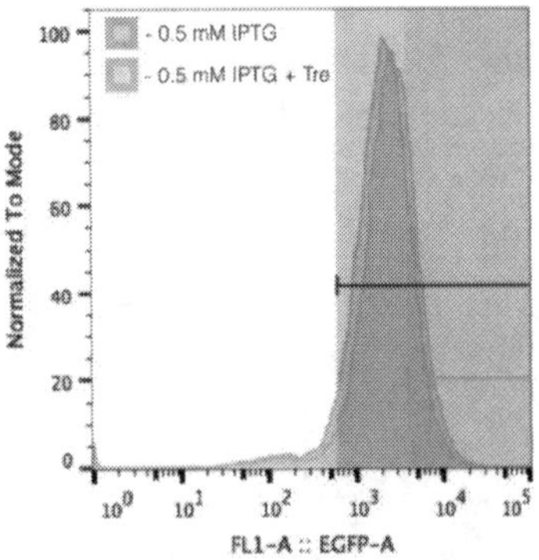

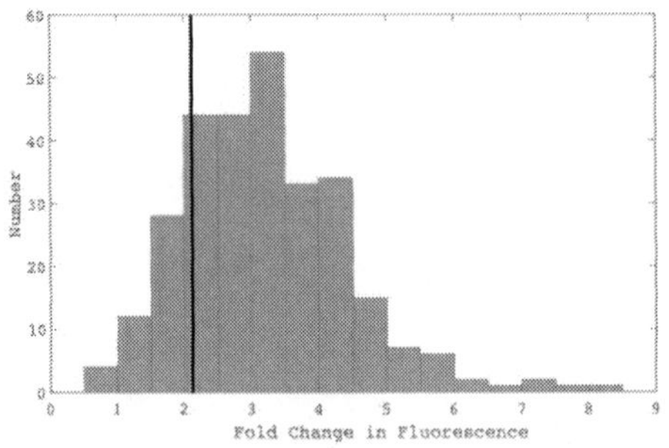

[IPTG]	0 mM	0.5 mM	
Sub	-	-	Tre
AX	0.01	16.13	23.89

in vitro characterization

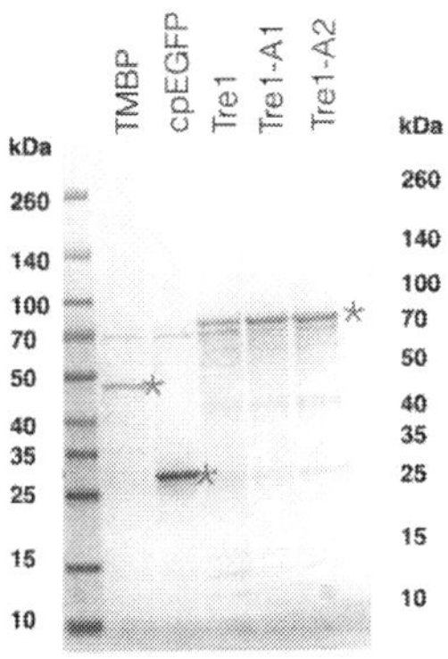

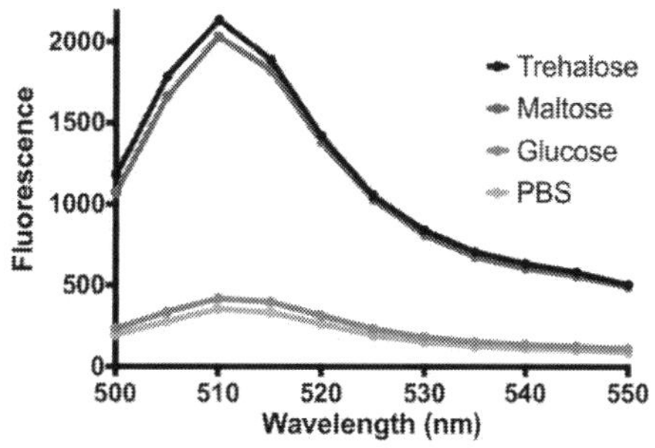

	K_d (nM)	
	Trehalose	Maltose
Tre1	10 ± 3	15 ± 5
Tre1-A1	53 ± 6	81 ± 3
Tre1-A2	31 ± 4	40 ± 6

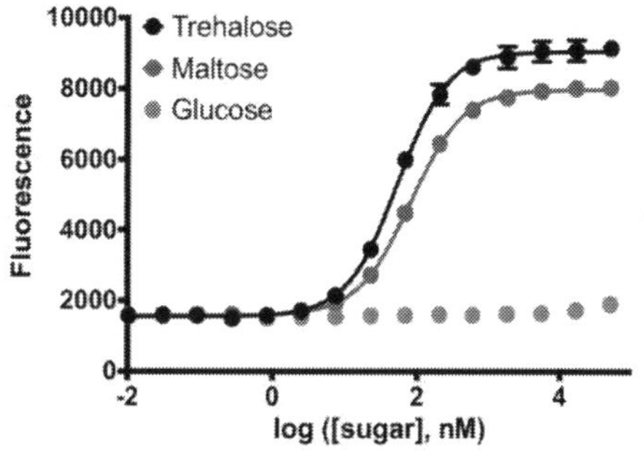

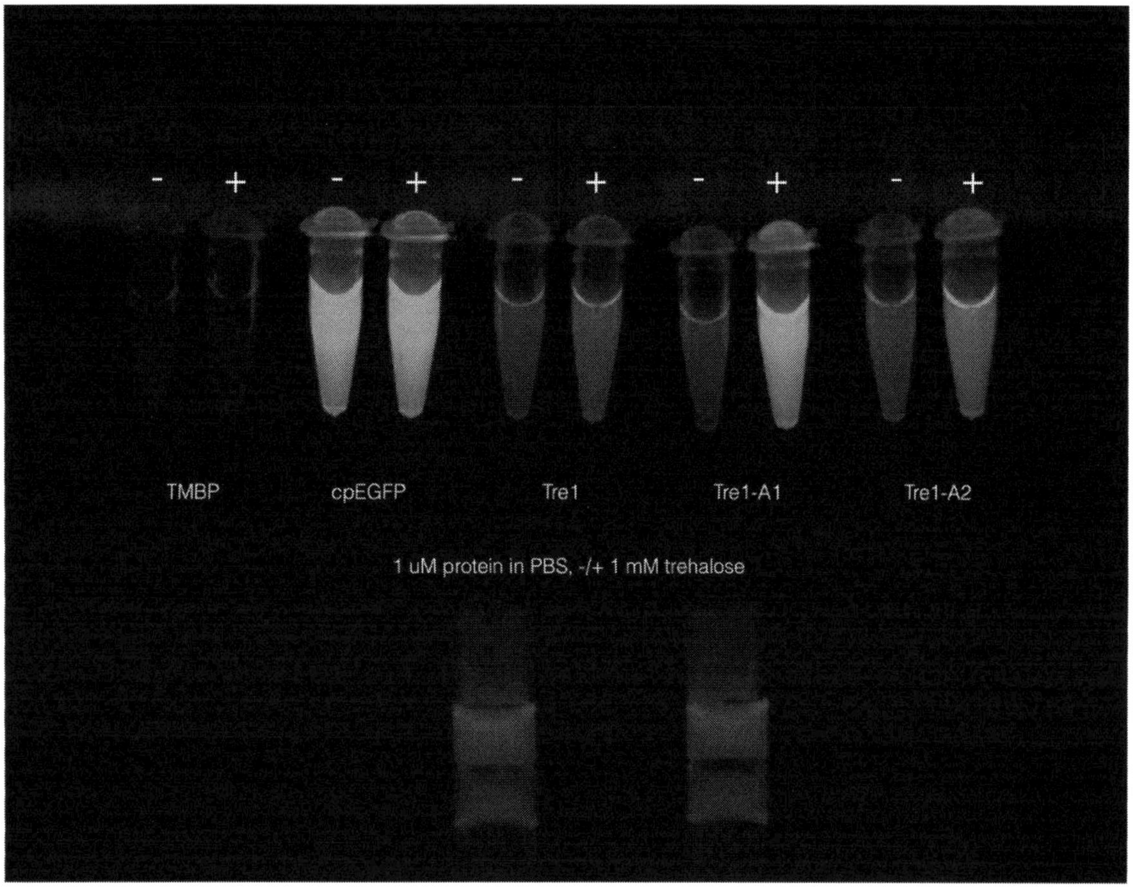

Sensors function well *in vivo*

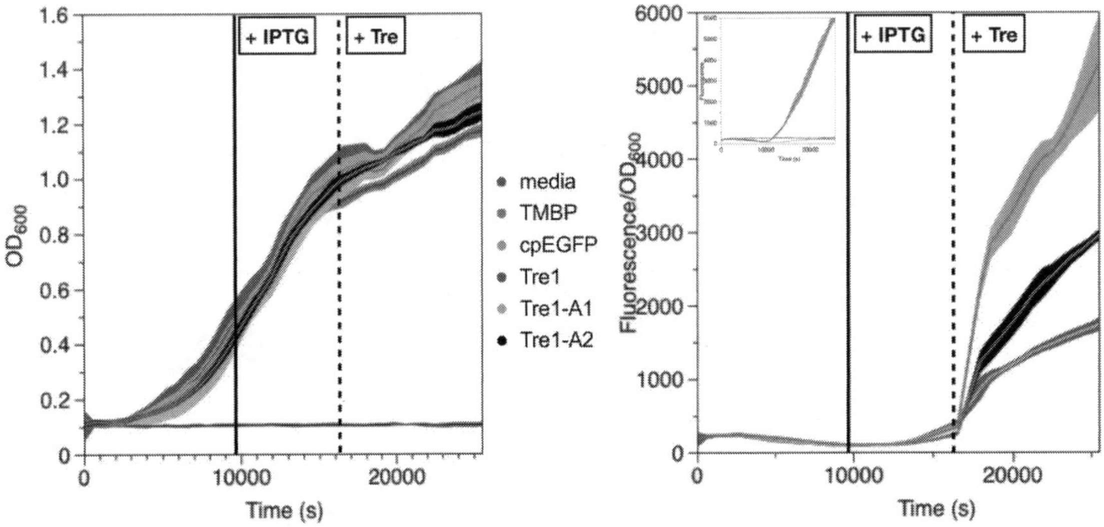

Sensors exhibit switching *in* vivo

EZ media
+ 1 mM trehalose

↕

EZ media
(wash)

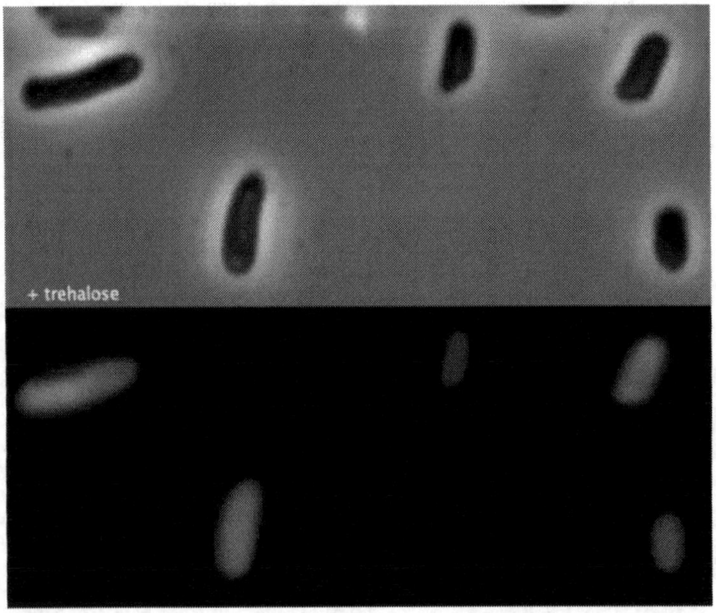

E. coli (BL21)

Conclusions

- Transposons are highly useful for certain types of protein engineering.
- We have built a screening platform for functional biosensors.
- We have isolated a novel high dynamic range sensor for trehalose.

Generalized Biosensor Construction Platform

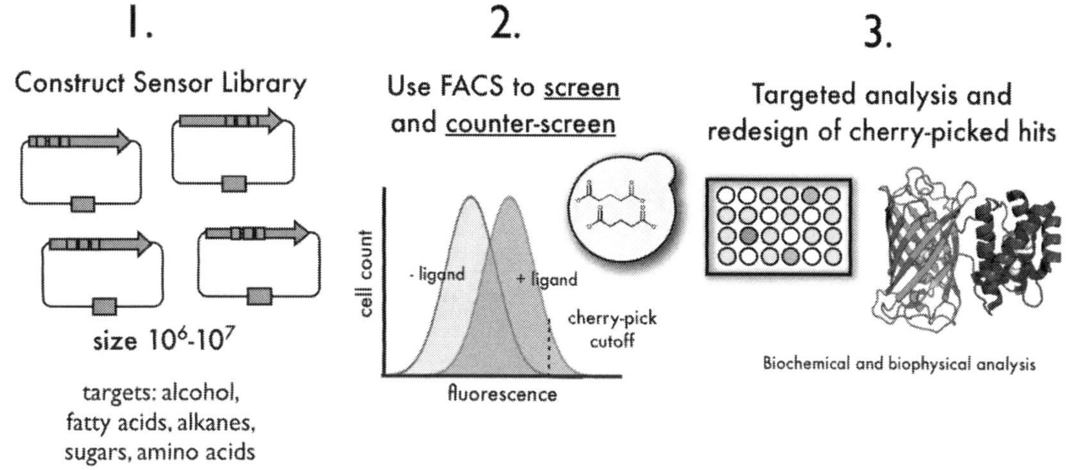

Interrogating disease and chemistry

Biosensor Construction

Cancer

Gene	Binding Target
mglB	glucose
glnBP	glutamine
pdhR	pyruvate
lldR	lactate

Biofuels

Gene	Binding Target
fadR	fatty acyl-CoA
psrA	free fatty acids
alkS	alkanes (C6-C12)
bmoR	alcohols (C2-C8)

Metabolic Disease

Gene	Binding Target
mglB	glucose
fadR	fatty acyl-CoA
lbp	phenylalanine
citA	glutarate

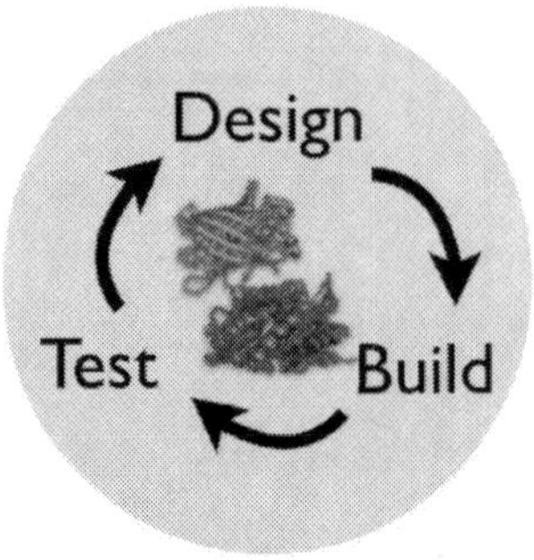

Biological questions

high-throughput single cell quantification

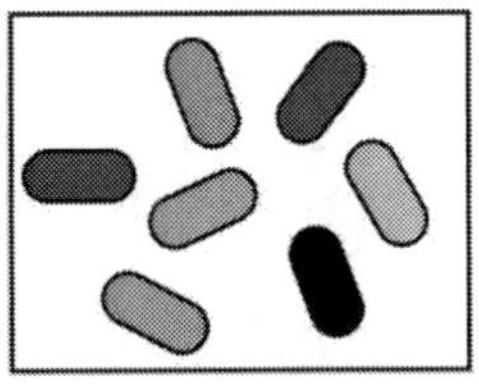

Whole animal atlas of metabolism

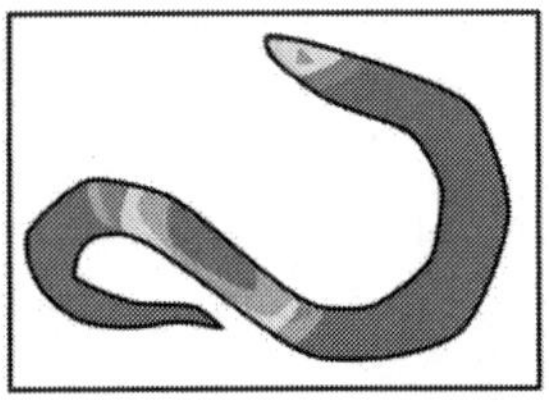

Acknowledgements

Savage Lab at Berkeley University of California

Dave Savage

Dana Nadler

Katie Kortright

Avi Flamholz

NIH DIRECTOR'S NEW INNOVATOR AWARD

Energy Biosciences Institute

march of dimes

Sensors function well *in vivo*

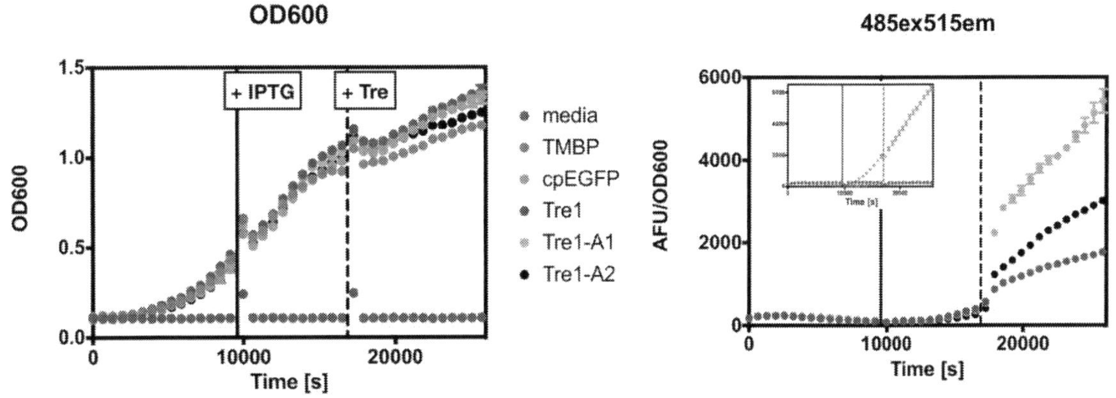

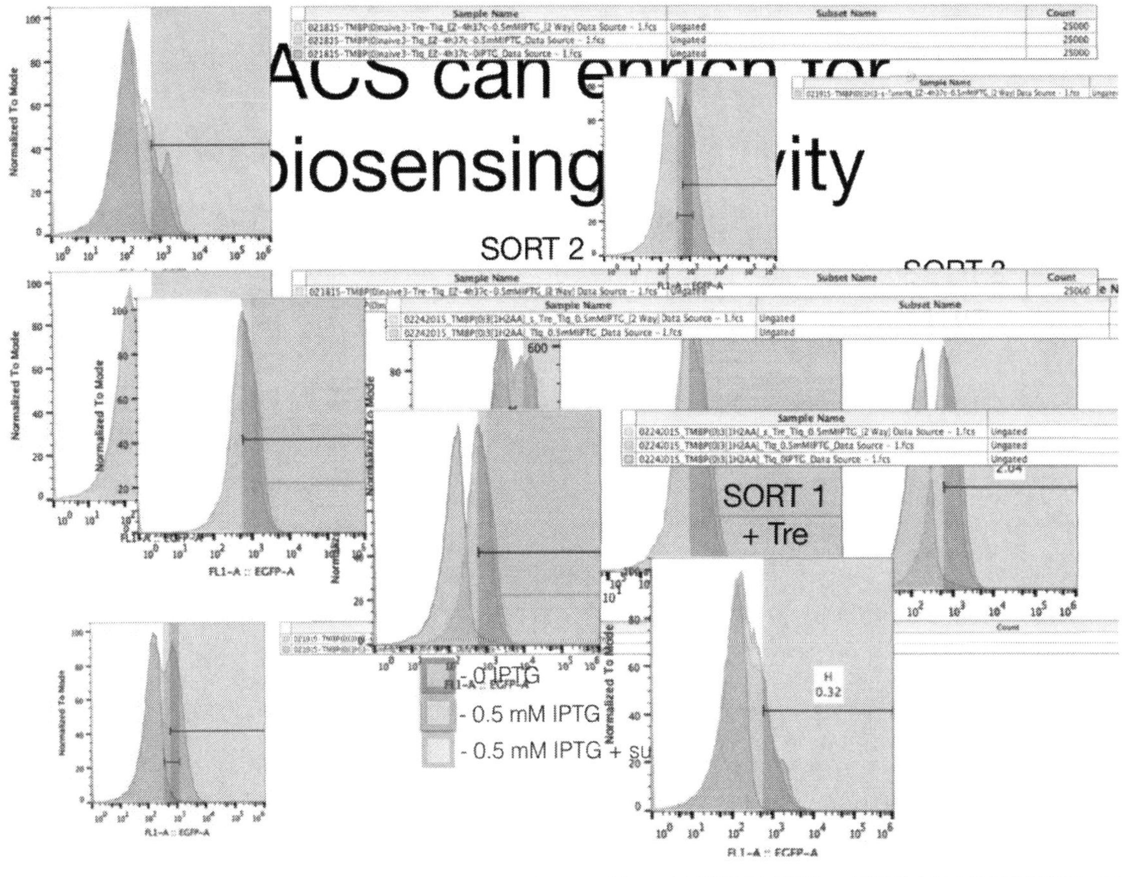

(0-3)VVC Sequencing

	Fold change	N-Linker	C-Linker
1	7.0	GDR	RSR
2	6.4	GDR	RSR
3	6.2	SGD	ANR
4	6.1	GGS	NSS
5	6.0	GDR	RSR
6	5.7	SNG	SNR
7	5.3	GSS	NDT
8	5.2	SNH	RGD
9	5.0	SSA	KNR
10	5.0	APG	HNH
11	4.7	GPP	SDR
12	4.6	NGG	HR
13	4.3	NNN	RSD
14	4.1	SGD	ANR
15	3.5	SAS	TTN
Tre1G09	2.1	GAS	ASP

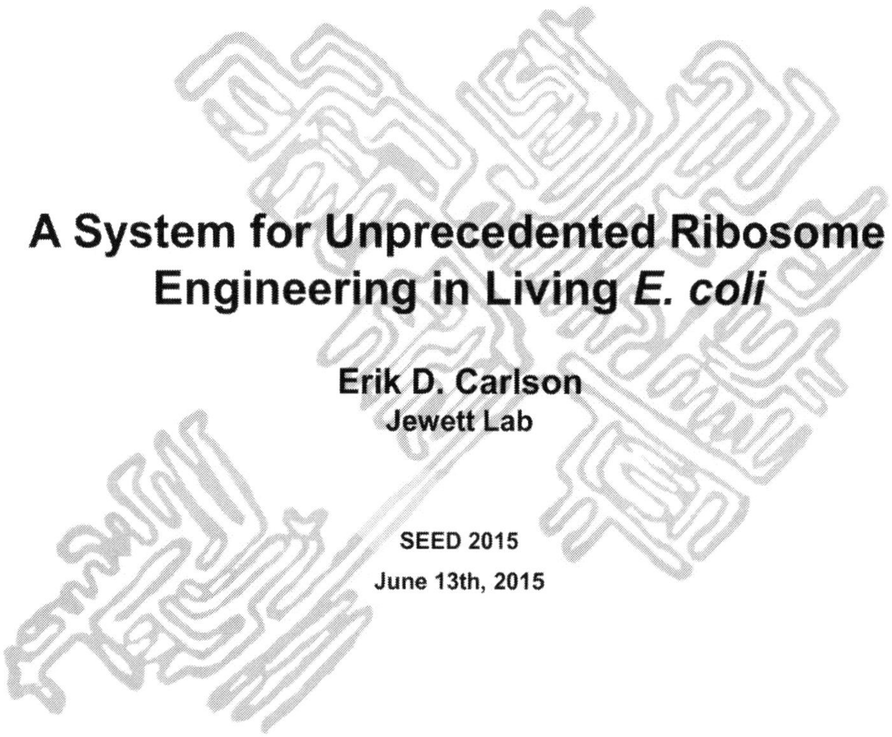

A System for Unprecedented Ribosome Engineering in Living *E. coli*

Erik D. Carlson
Jewett Lab

SEED 2015
June 13th, 2015

@SynBioCarlsons

The extraordinary synthetic capability of the cell has driven extensive efforts to engineer the translation machinery

- Translation is:
  - Accurate: 99.99%
  - Fast: ~20 amino acids/second in bacteria
- The ribosome is the cornerstone of translation

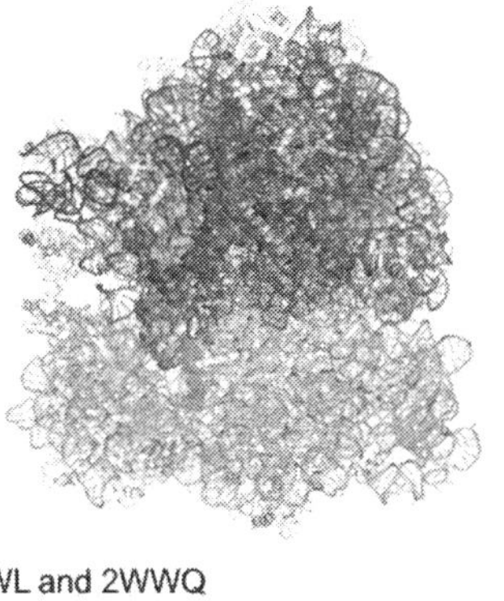

PDB 2WWL and 2WWQ

Venki Ramakrishnan group, MRC-LMB
http://www.mrc-lmb.cam.ac.uk/ribo

Making ~~proteins~~ a cake

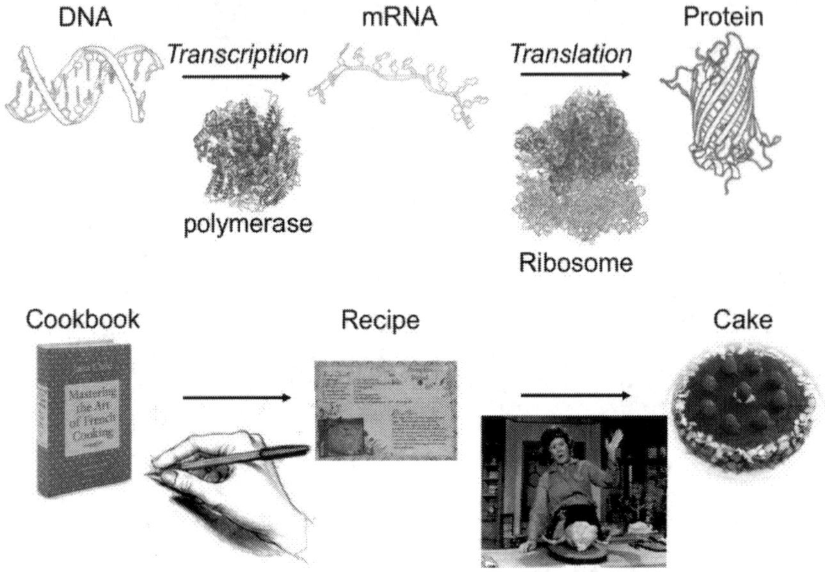

The chef determines what we can make

Engineering synthetic ribosomes makes possible new opportunities for synthetic biology

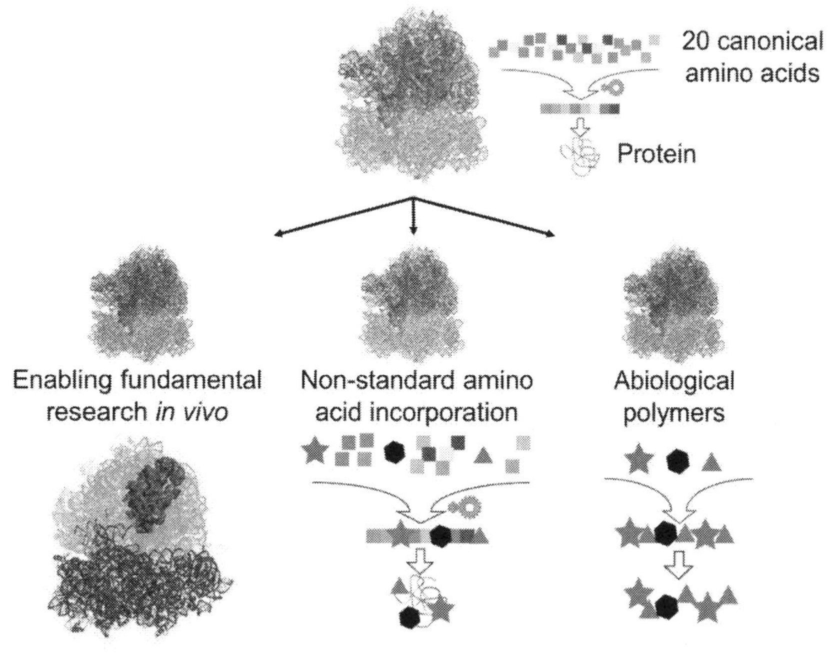

Nature's dual subunit ribosome: Composition and function of the *E. coli* ribosome

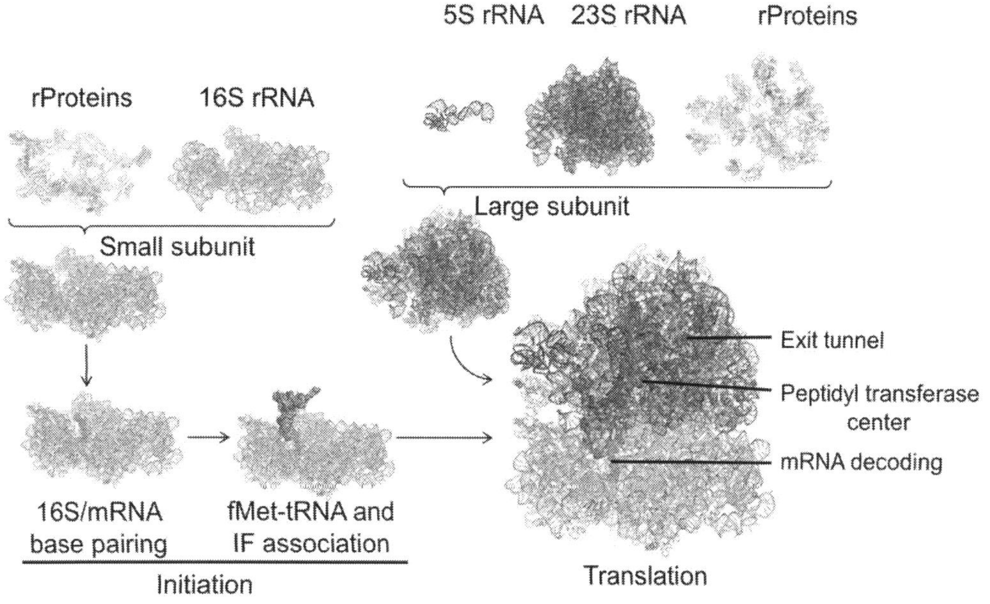

PDB 2WWL and 2WWQ

Nature's dual subunit ribosome: Composition and function of the *E. coli* ribosome

In all kingdoms of life:
- Two subunits
- Conserved functions
- Ratcheting of subunits

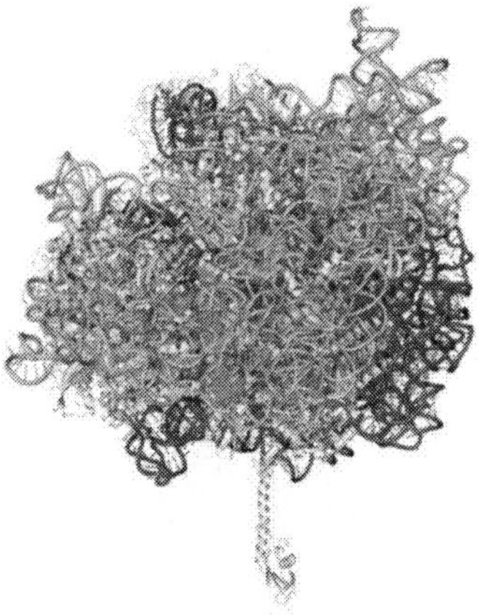

PDB 4GD2, 4GD1, 3R8T and 3R8S

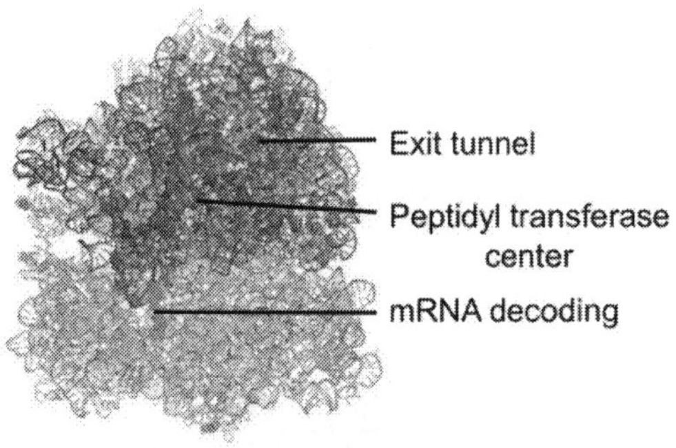

Translation

PDB 2WWL and 2WWQ

Engineering ribosomes is difficult

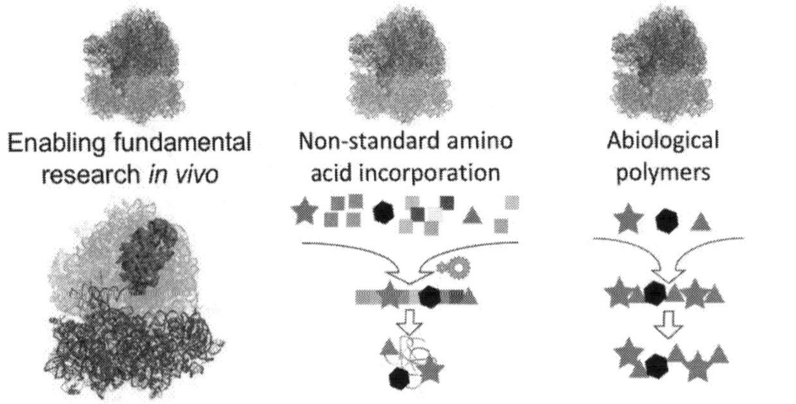

Cell vs. Engineer

The cell's objectives

- Growth and reproduction
- The ribosome must decode every mRNA in the cell to synthesize the proteome

The engineer's objectives

- Studying ribosome function
- New functionality to ribosome
  - Mutations altering function are often toxic in living cells

Mrs. MacGrady | Chef with many skills | Jiro | Extremely specialized

Arthur. Cookie Jar Group, WGBH.

Jiro Dreams of Sushi. Jiro Ono. Dir. David Gelb. Magnolia Pictures, 2011.

Mutating the 30S small subunit *in vivo*

E. coli

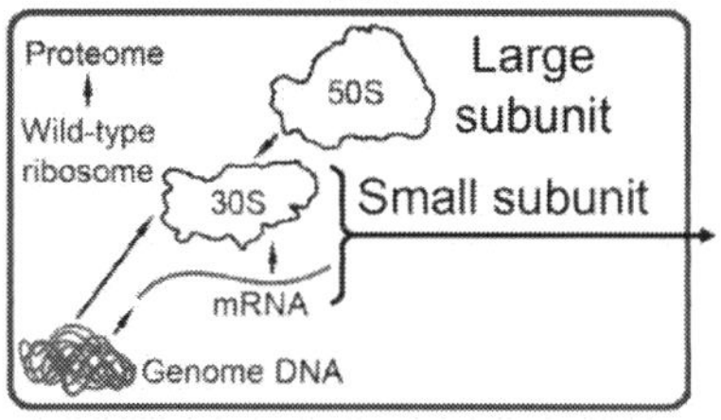

Previous state-of-the-art:

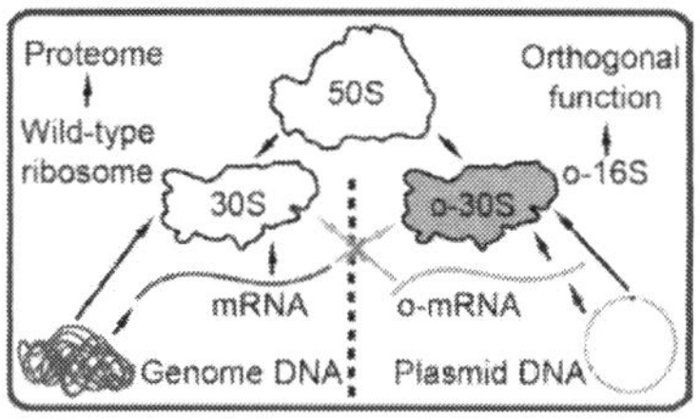

At translation initiation:

Shine-Dalgarno

wt mRNA 5′...UUC**AGGAGG**AGAUCU**AUG**...

wt 16S rRNA 3′AU**UCCUCC**ACUA...

anti-SD

Mutate ribosome binding site[1,2]

orthogonal SD

o-mRNA 5′...UUC**CACCAC**AGAUCU**AUG**...

o-16S rRNA 3′AU**GGUGUU**ACUA...

orthogonal anti-SD

1-Hui, A. and H. A. de Boer. *PNAS* **84** (14): 4762-4766, (1987).

2-Rackham, O. & Chin, J. W. *Nat Chem Biol* **1**, 159-166, (2005).

Mutating the 30S small subunit *in vivo*

E. coli

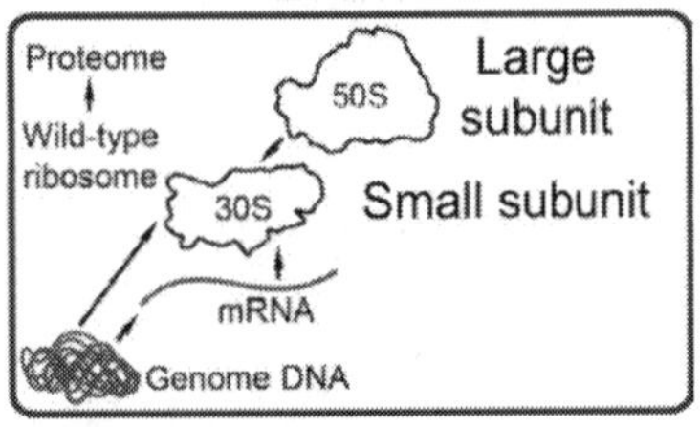

Previous state-of-the-art:

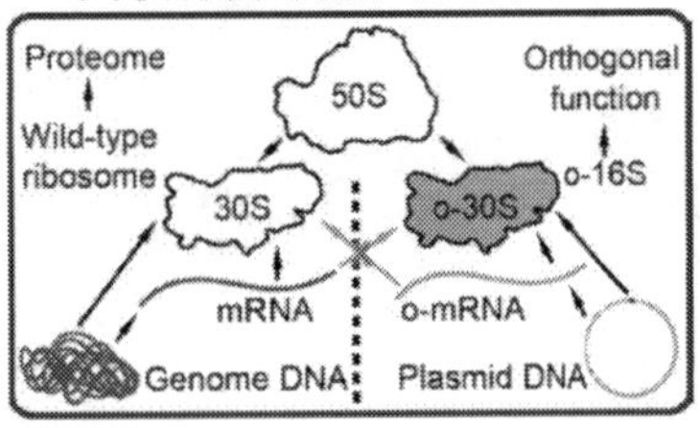

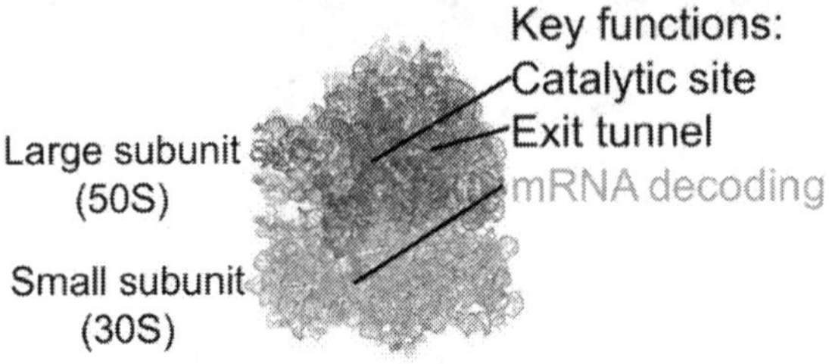

→Expanded functionality of 30S small subunit in *E. coli*[3,4]

Key limitation: 50S subunit not orthogonal

Ribosome's catalytic activity and the polypeptide exit tunnel are not engineerable

1-Hui, A. and H. A. de Boer. *PNAS* **84** (14): 4762-4766, (1987).
2-Rackham, O. & Chin, J. W. *Nat Chem Biol* **1**, 159-166, (2005).
3-Wang, K., Neumann, H., Peak-Chew, S. Y. & Chin, J. W. *Nat Biotech* **25**, 770-777, (2007).
4-Neumann, H., Wang, K., Davis, L., Garcia-Alai, M. & Chin, J. W. *Nature* **464**, 441-444, (2010).

A dedicated 50S subunit for the o-30S

E. coli

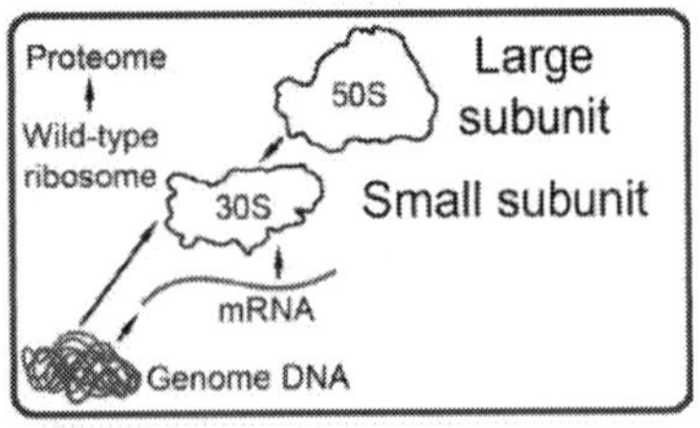

Previous state-of-the-art:

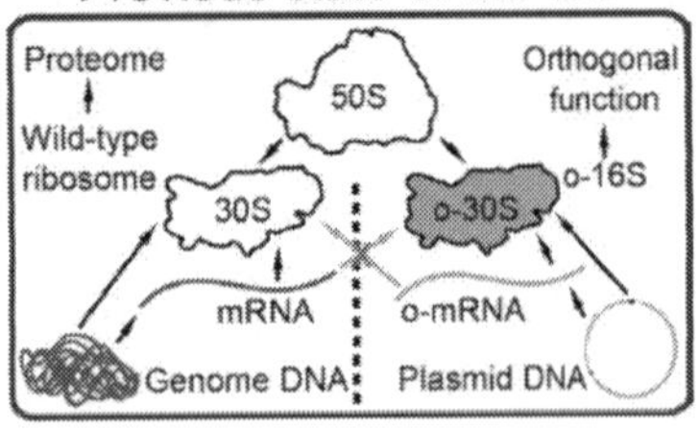

Ribo-T: A ribosome with tethered subunits

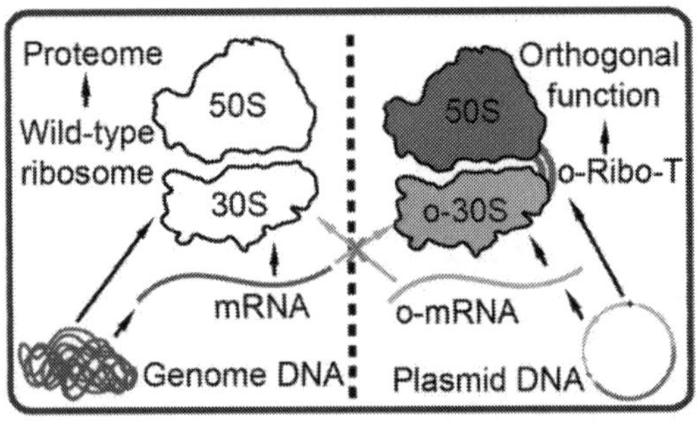

Connect 23S rRNA to 16S rRNA to form a single rRNA core molecule

How to connect core rRNA?

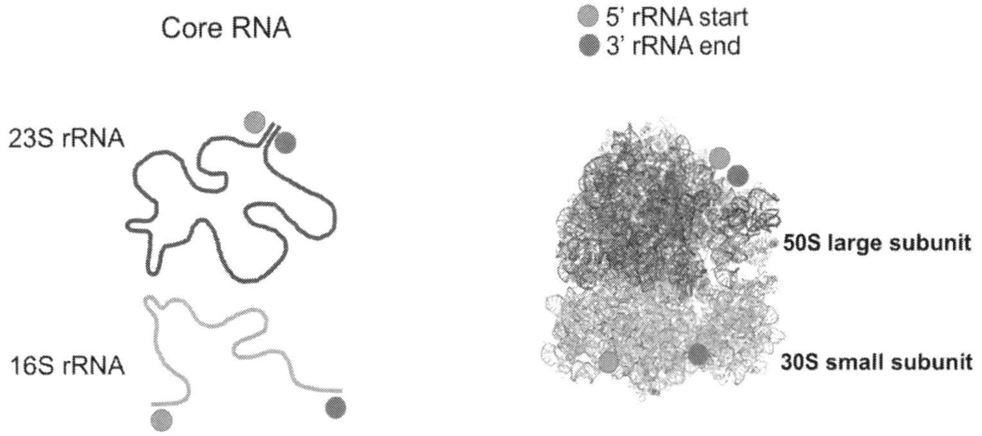

PDB 3R8T and 4GD2

How to connect core rRNA? Moving 23S 5' and 3' ends

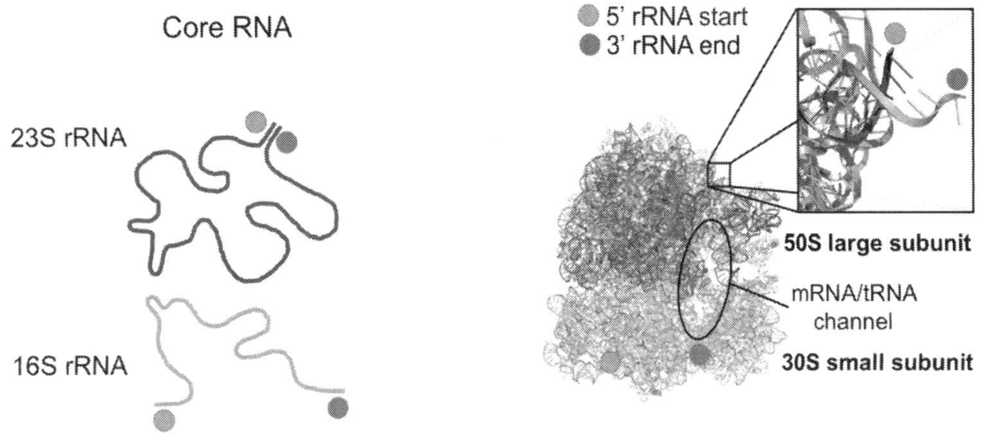

PDB 3R8T and 4GD2

How to connect core rRNA? Moving 23S 5' and 3' ends

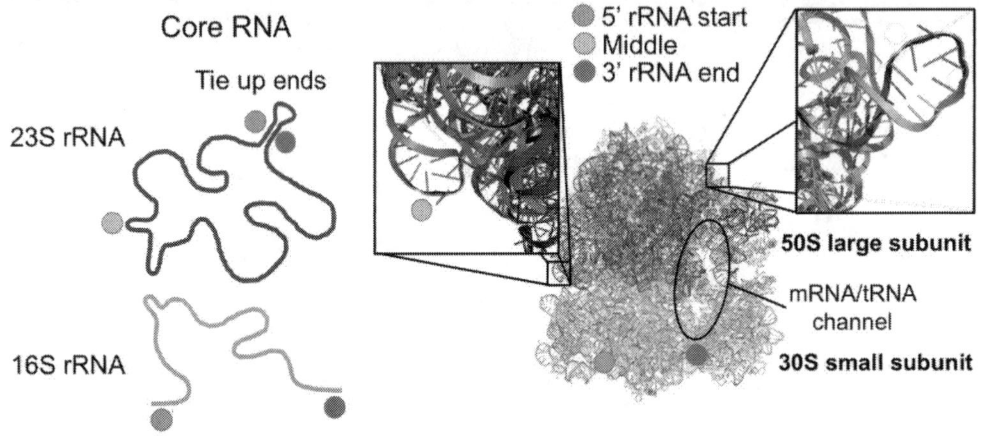

PDB 3R8T and 4GD2

How to connect core rRNA? Moving 23S 5' and 3' ends

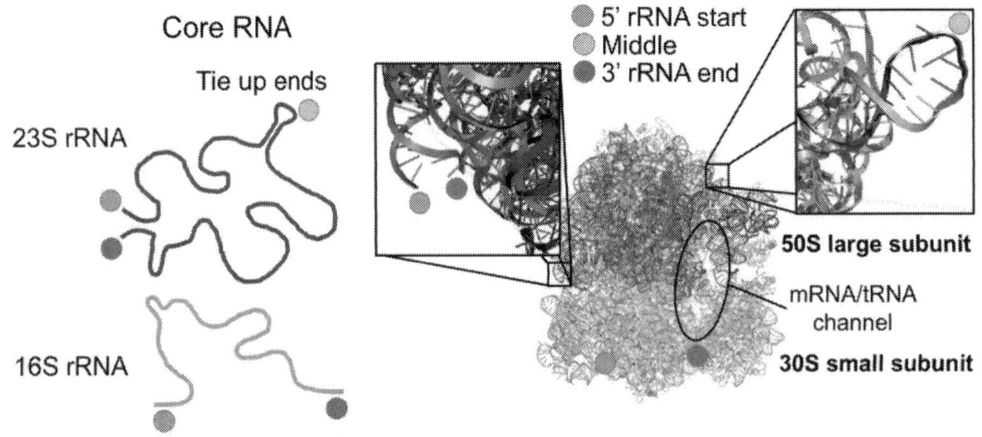

- Circular permutations can interfere with ribosome biogenesis
- 3 mutations known to be functional *in vivo*[1]
- ***Previous construction and testing methods[1] were tedious and not scalable***

[1]Kitahara, K. & Suzuki, T. *Mol Cell* **34**, 760-766, (2009)

PDB 3R8T and 4GD2

Finding the tether site: Can 23S circular permutations support cell growth?

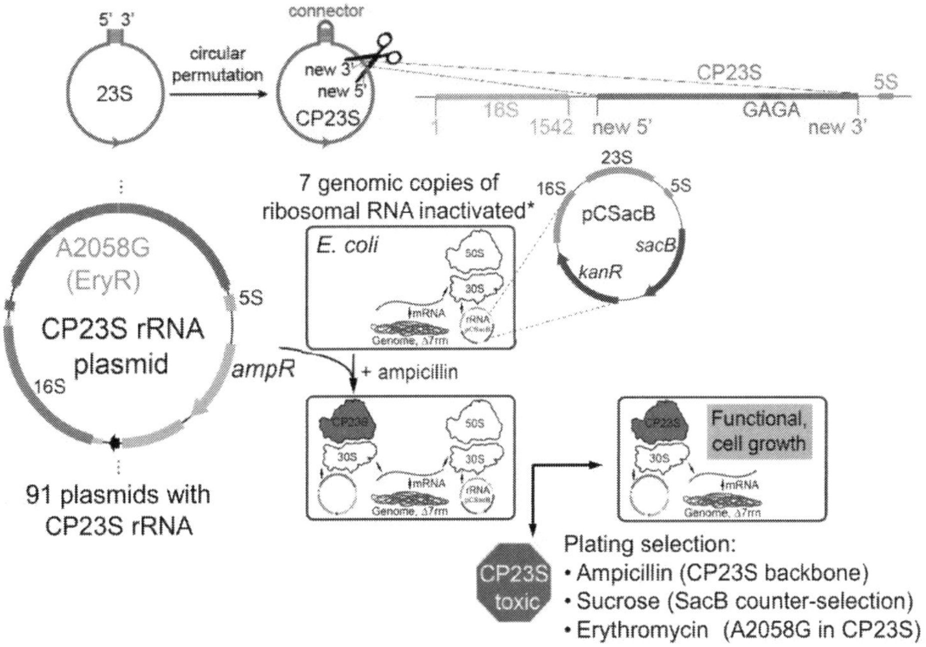

**E. coli* Squires strain, SQ171 from *Asai T et al. PNAS, 1999*

Finding the tether site: 23S rRNA circular permutation study

Results: **22** *circular permutations of 23S rRNA can fully support cell growth*

Detailed secondary structure map of 23S RNA

Viable mutation
Lethal mutation

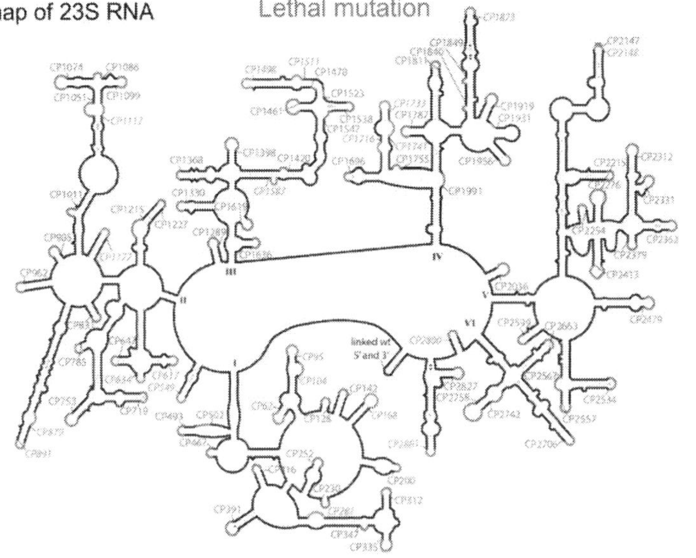

Finding the tether site: 23S rRNA circular permutation study

Results: **22** *circular permutations of 23S rRNA can fully support cell growth*

3D structure:

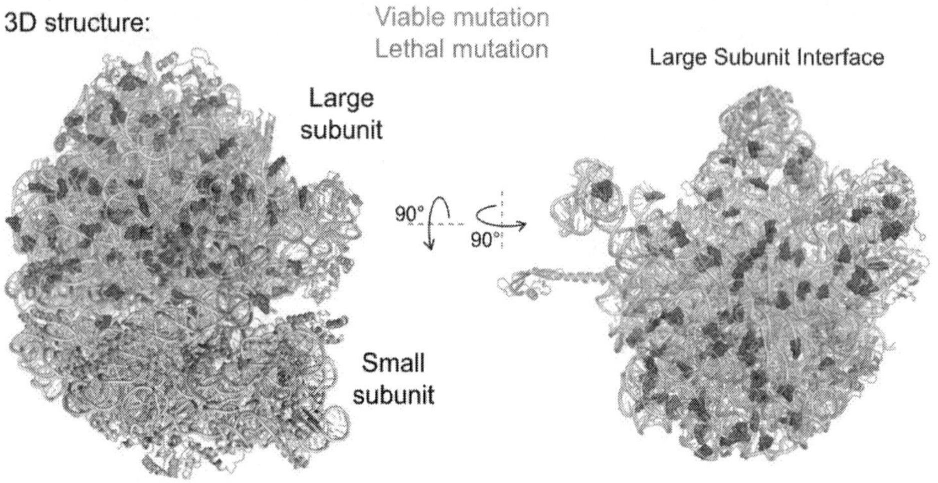

PDB 3R8T and 4GD2

Designing a ribosome with tethered subunits, Ribo-T

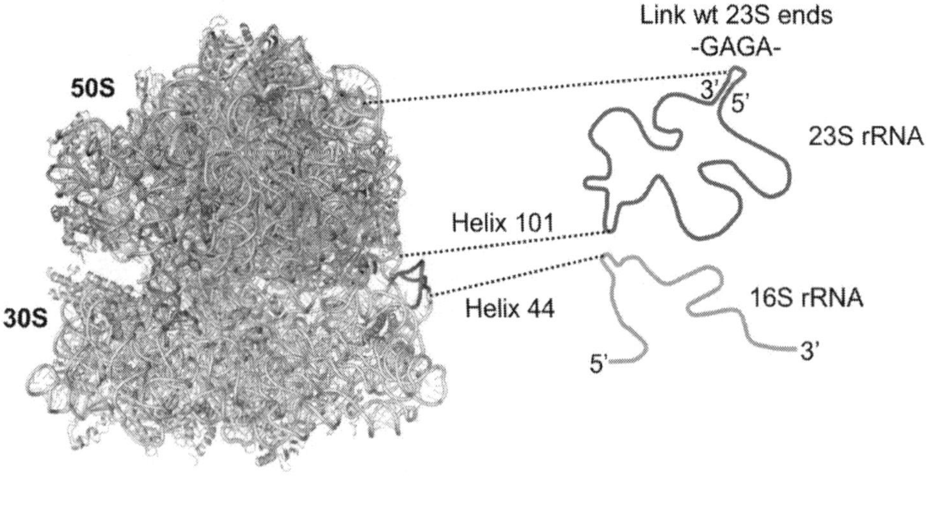

Designing a ribosome with tethered subunits, Ribo-T

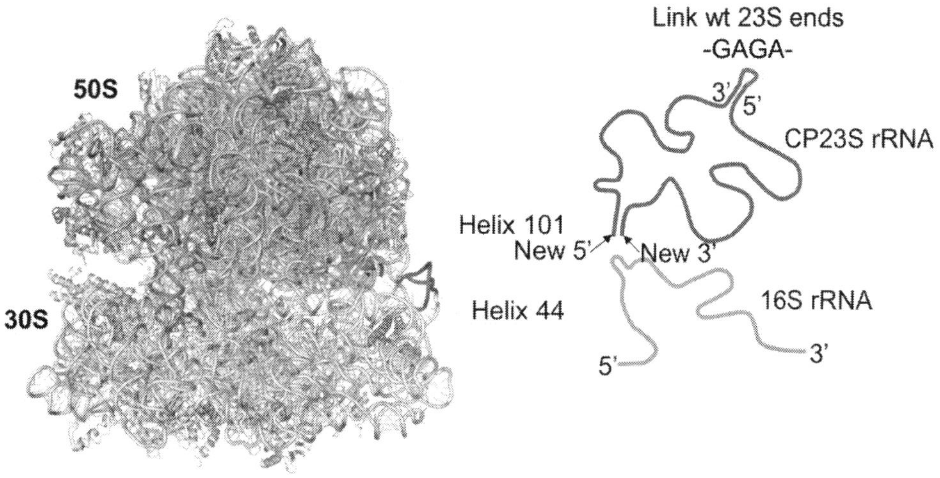

Designing a ribosome with tethered subunits, Ribo-T

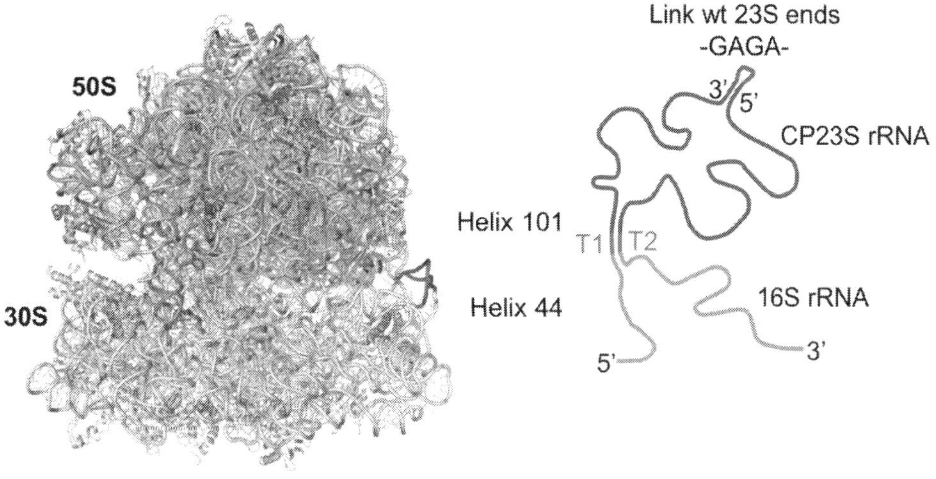

Selecting Ribo-T

Design:

Direct length is ~7 nucleotides long

→Test tether lengths ranging from 7-12 nucleotides

Build:

Built Ribo-T with tethers ranging from 7-12 nucleotides long (all As), small library of 36 combinations

Test:

Finding a functional Ribo-T:

- Replace all ribosomes in a living cell with Ribo-T
- Can it fully support cell growth?
  - Life or death→easy to screen

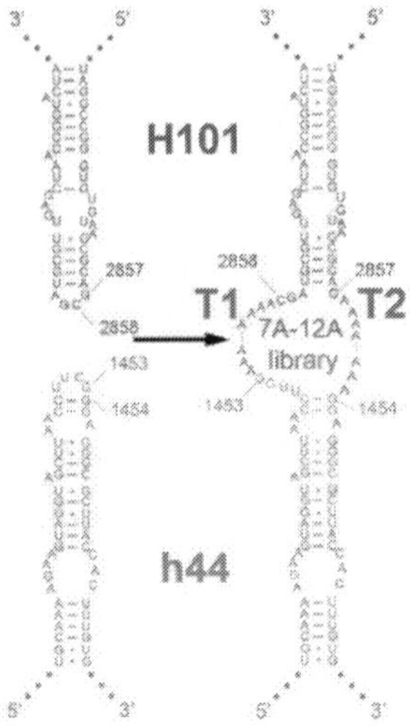

Can Ribo-T construct support cell growth?

7 genomic copies of ribosomal RNA inactivated* | Transform plasmid-based Ribo-T tether library | Remove wild-type, untethered ribosomes

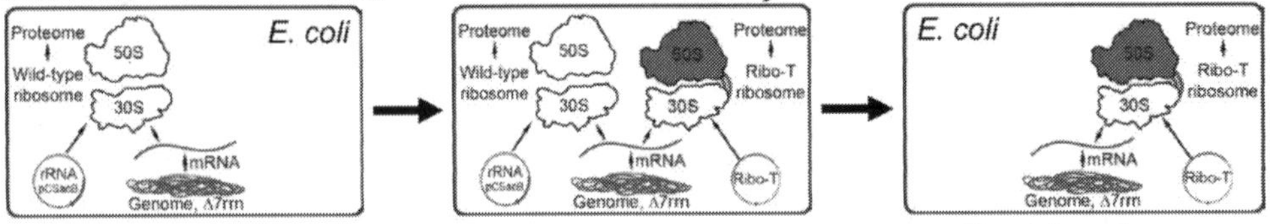

Doubling time 35±1 min

Doubling time 70±2 min

E. coli Squires strain, SQ171 from *Asai T et al. PNAS, 1999*

Ribo-T RNA expressed and stable? Extract total RNA, analyze size

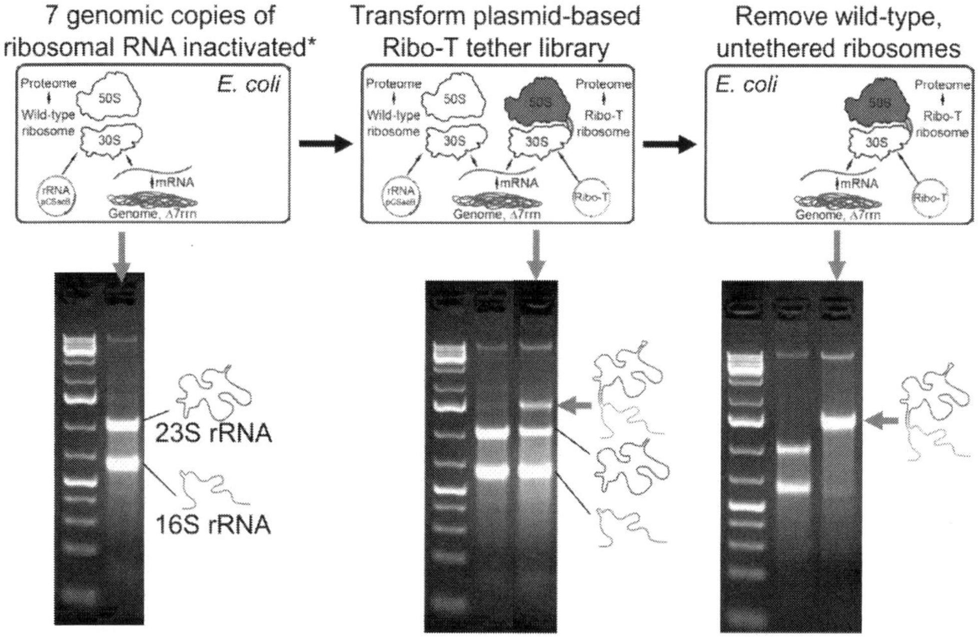

Ribo-T: A ribosome with tethered subunits

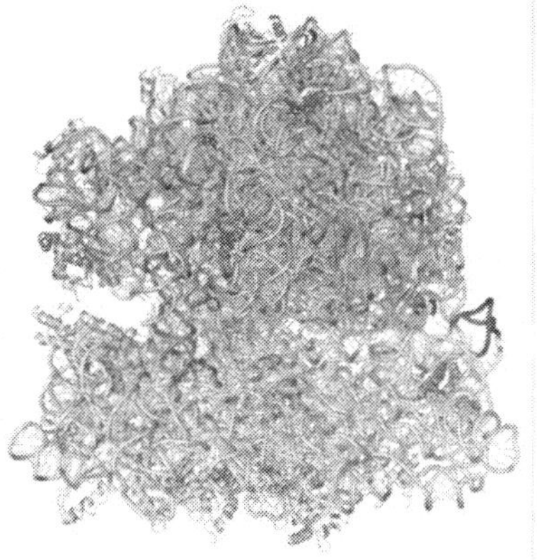

- Assembly
  - Assembled into a functional ribosome particle?

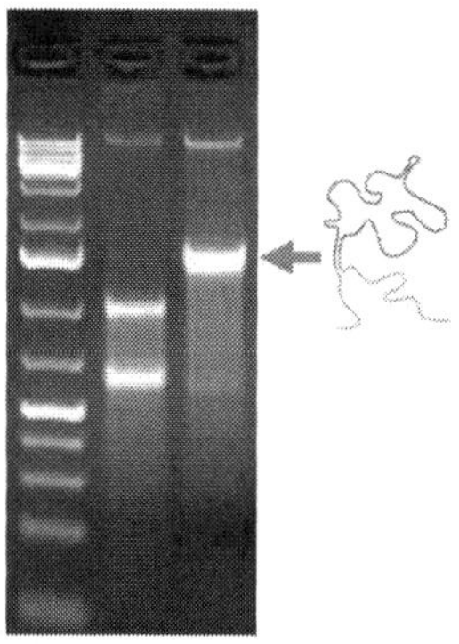

Sucrose gradient fractionation: analyze intact ribosome particles

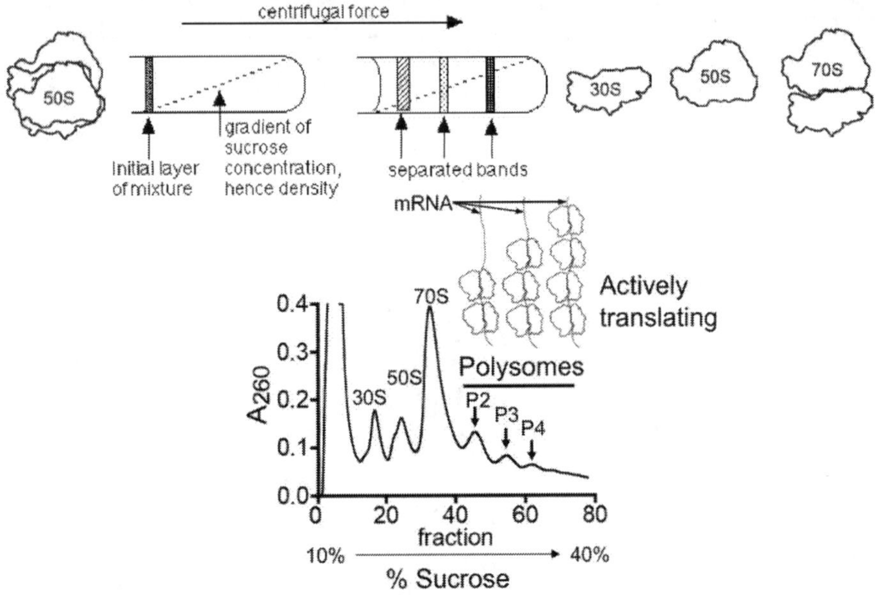

http://bricker.tcnj.edu/tech/BIOL311centrifugation.html

Sucrose gradient fractionation: analyze intact ribosome particles

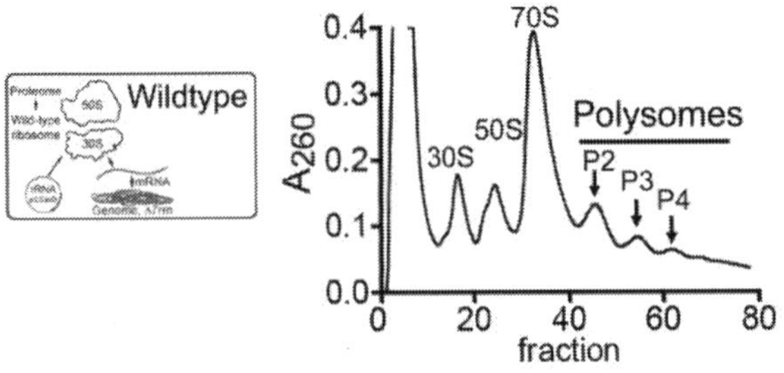

- Dominant 30S, 50S and 70S species
- Polysomes→actively translating ribosomes

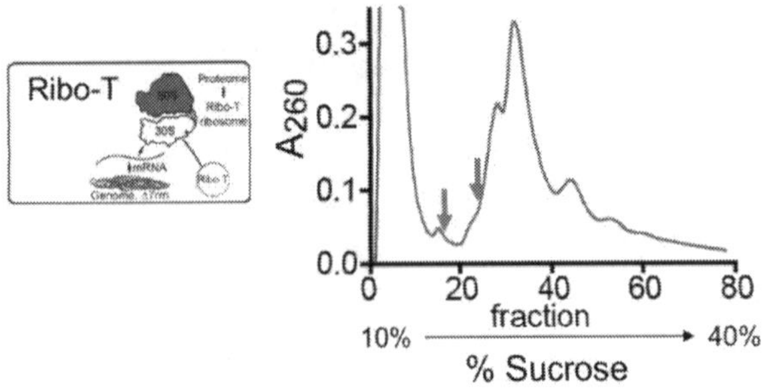

- No 30S and 50S peaks

Sucrose gradient fractionation: analyze intact ribosome particles

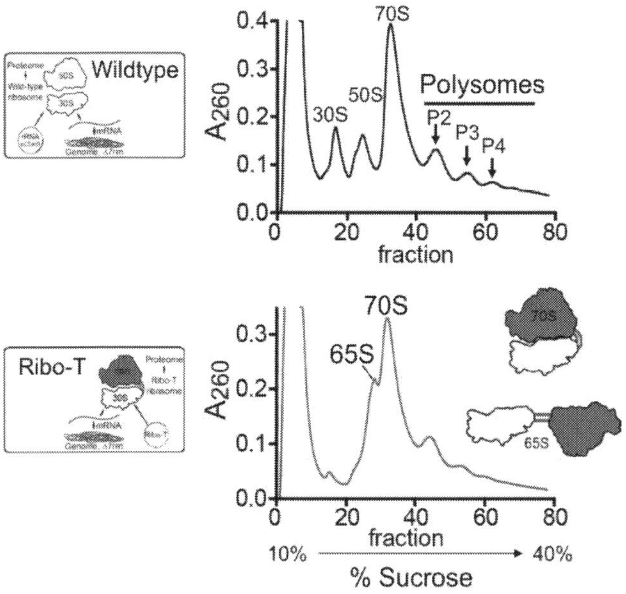

- Dominant 30S, 50S and 70S species
- Polysomes→actively translating ribosomes

- No 30S and 50S peaks
- Dominant 70S and 65S species

Sucrose gradient fractionation: analyze intact ribosome particles

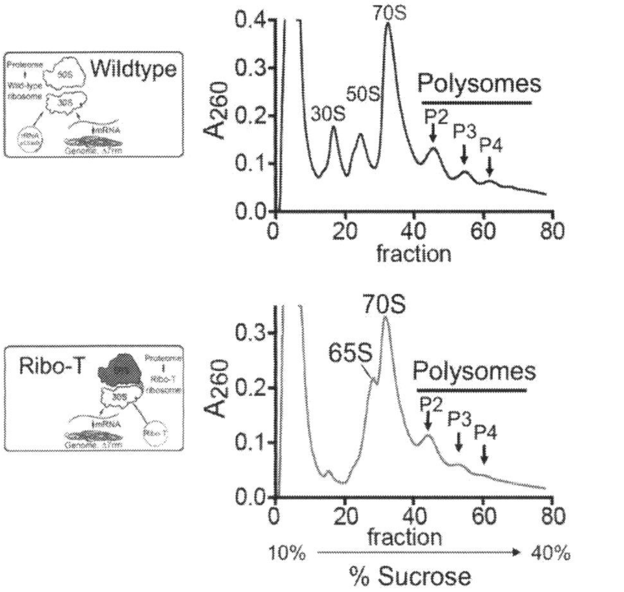

- Dominant 30S, 50S and 70S species
- Polysomes→actively translating ribosomes

- No 30S and 50S peaks
- Dominant 70S and 65S species
- Polysomes present

Ribo-T core rRNA in polysomes?

Ribo-T in polysome fractions
→actively translating *in vivo*

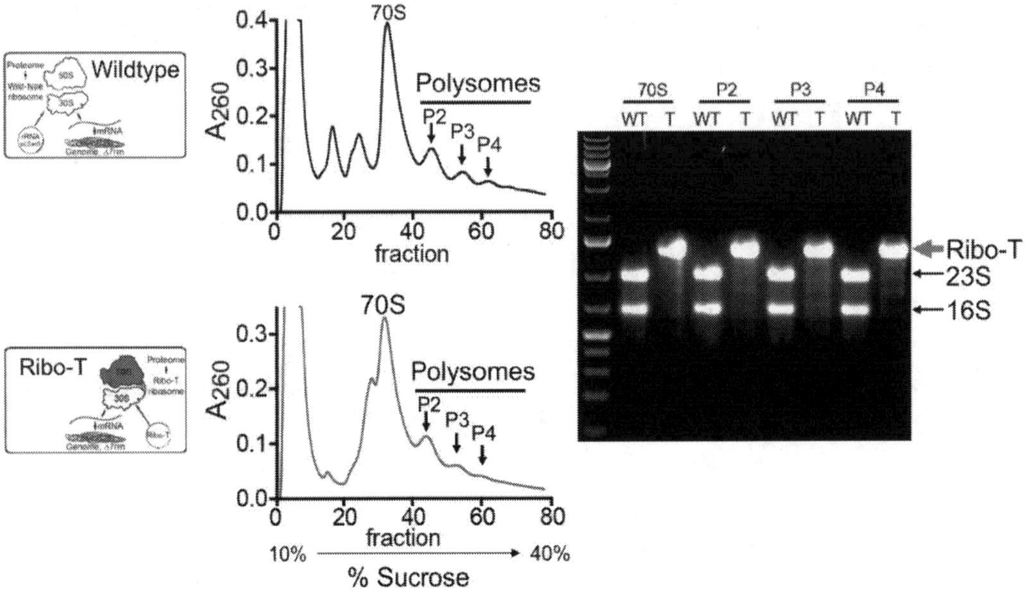

Ribo-T: A ribosome with tethered subunits

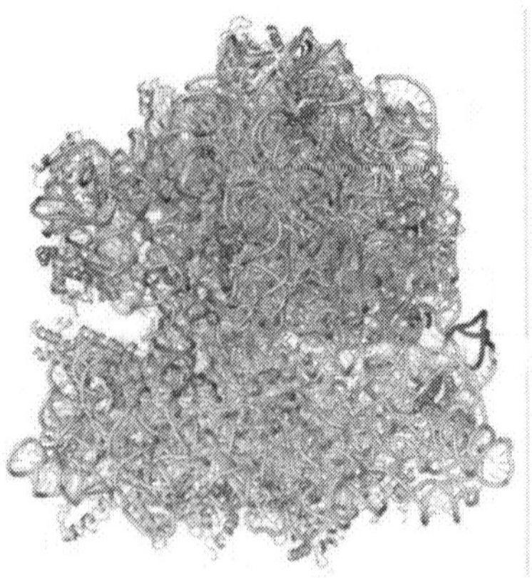

- Assembly
  - ✓ Assembled into a functional ribosome particle
- Composition
  - ✓ Single RNA core molecule
  - 5S rRNA
  - ribosomal proteins

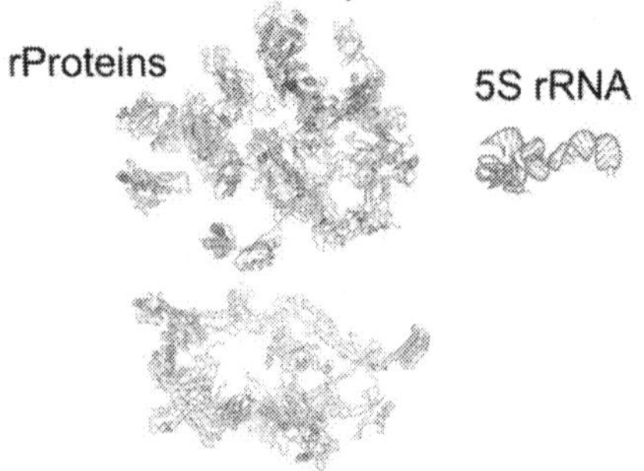

Ribo-T: A ribosome with tethered subunits

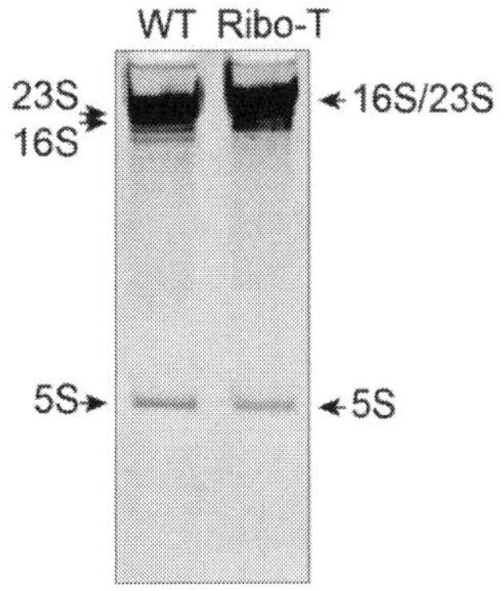

Denaturing polyacrylamide gel

- Assembly
  - ✓ Assembled into a functional ribosome particle
- Composition
  - ✓ Single RNA core molecule
  - ✓ 5S rRNA
  - ribosomal proteins

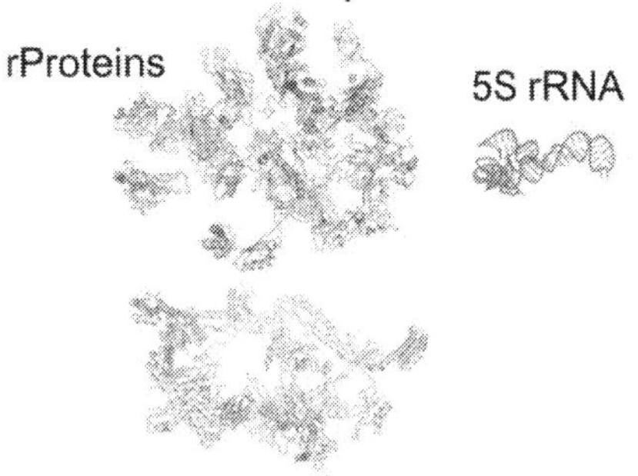

Ribo-T: A ribosome with tethered subunits

Tandem mass tagging with mass spec

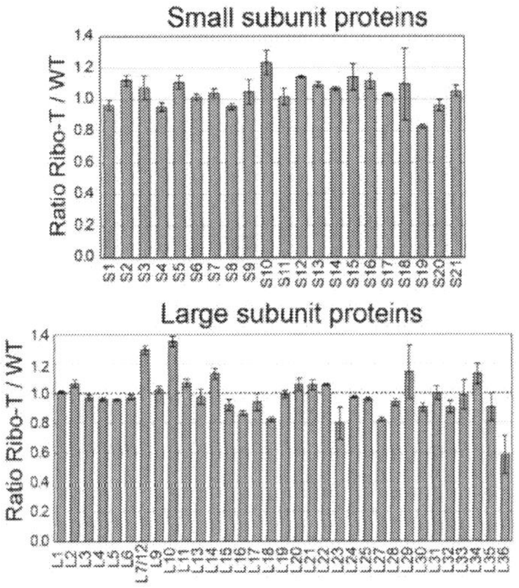

Mass spec by Ioanna Ntoi (NU, Kelleher)
**S22 not detectable in either sample*

- Assembly
  - ✓ Assembled into a functional ribosome particle
- Composition
  - ✓ Single RNA core molecule
  - ✓ 5S rRNA
  - ✓ ribosomal proteins

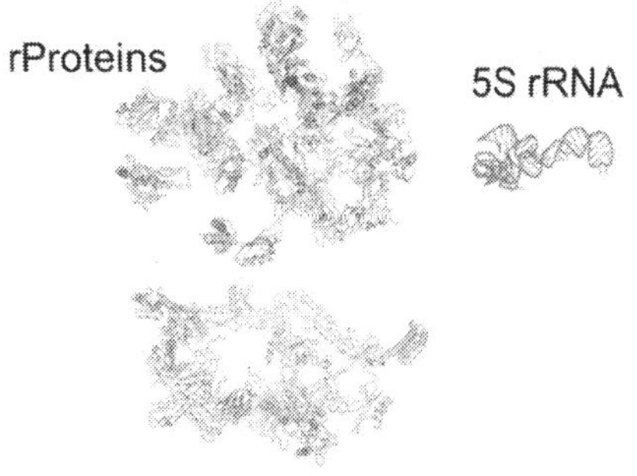

Ribo-T: A ribosome with tethered subunits

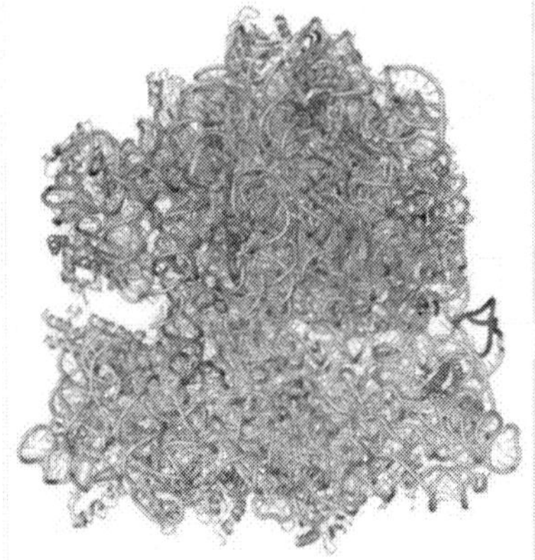

- Assembly
  - Assembled into a functional ribosome particle
- Composition
  - Single RNA core molecule
  - 5S rRNA
  - ribosomal proteins
- Utility
  - Modulating 23S rRNA function

Specialized functions for oRibo-T: accessing lethal mutations

23S rRNA mutations specializing ribosome function are often toxic *in vivo*
→ **requires oRibo-T**

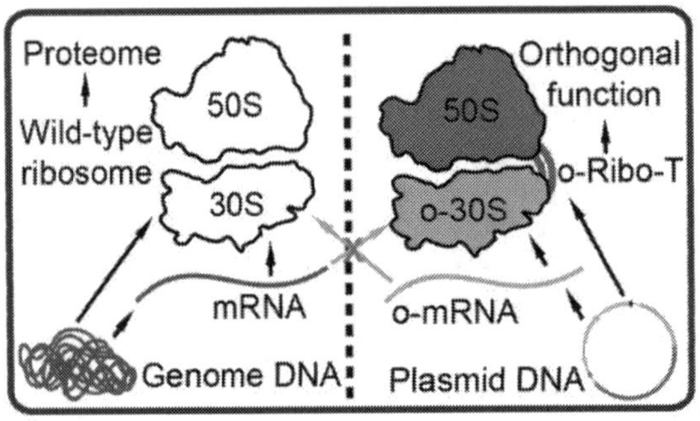

E. coli

Catalytic site and exit tunnel are primarily 23S ribosomal RNA

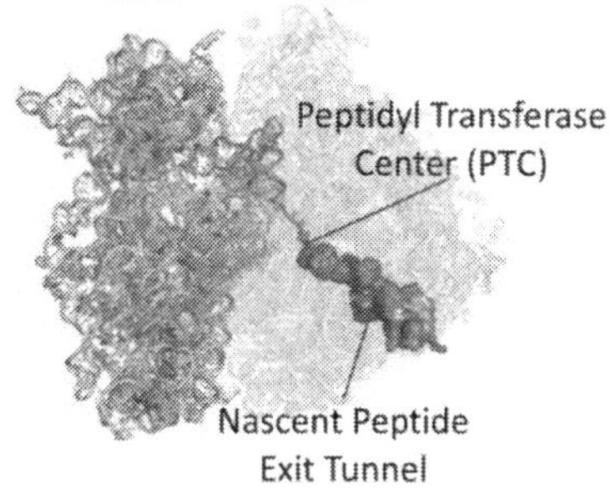

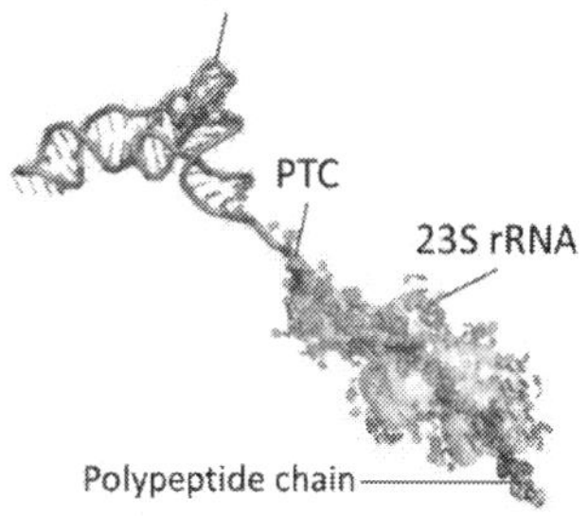

Shape, physicochemical, and dynamic properties are evolutionarily optimized for 20 natural amino acids and proteome expression

Leader peptide-induced stalling via large subunit interactions

- Leader polypeptide stalling used to regulate downstream gene expression
  - TnaC, ErmCL, SecM
- SecM stalling regulates SecA expression
  - Stalling sequence: ^{150}FSTPVWISQAQGIRAGP^{166}

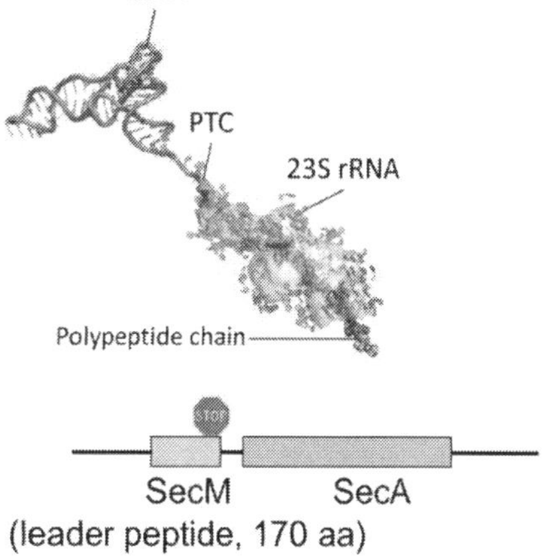

Gumbart et al., Biophys J 2012

Leader peptide-induced stalling via large subunit interactions

- Leader polypeptide stalling used to regulate downstream gene expression
  - TnaC, ErmCL, SecM
- SecM stalling regulates SecA expression
  - Stalling sequence: ^{150}FSTPVWISQAQGIRAGP^{166}
- 23S rRNA mutations can alleviate SecM stalling
  - ie A2058G (non-lethal)

What about other 23S mutations?

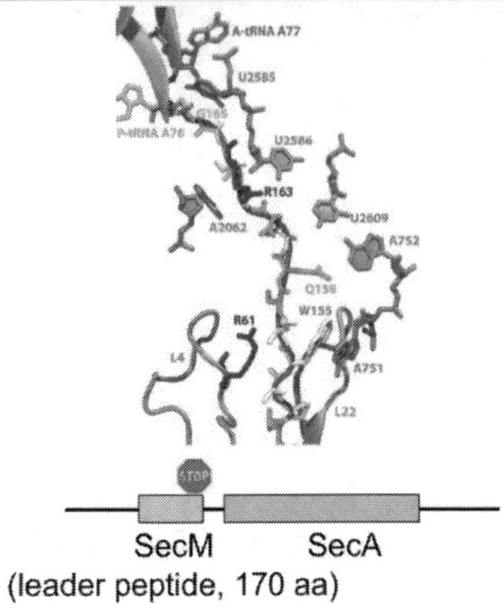

Gumbart et al., *Biophys J 2012*

Accessing greater mutation space with oRibo-T

- 23S residues A2451 and C2452 implicated in SecM stalling mechanism
  - A2451/C2452 mutations are toxic in wildtype ribosomes

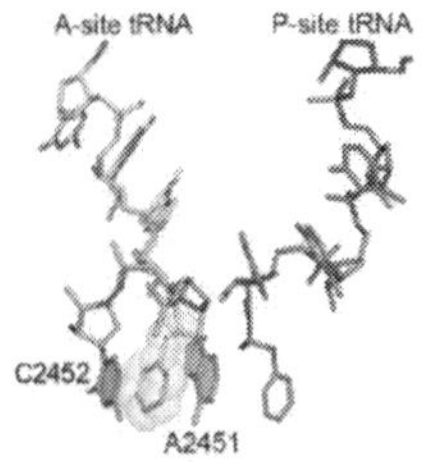

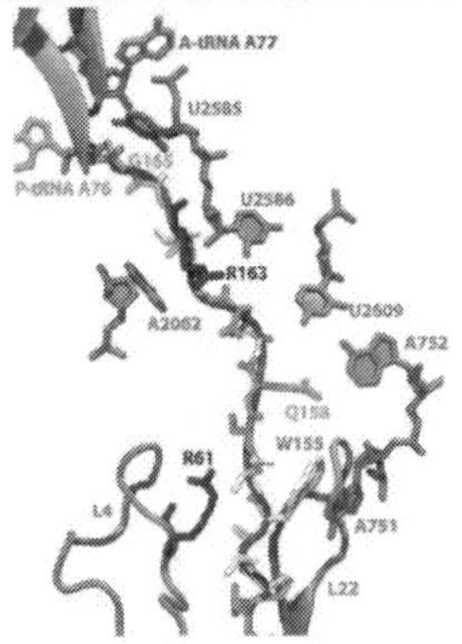

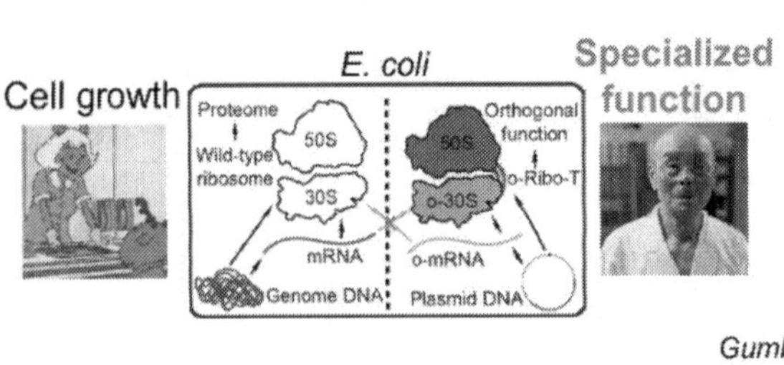

Gumbart et al., *Biophys J 2012*

Using oRibo-T to explore A2451/C2452 mutation effects on SecM stalling

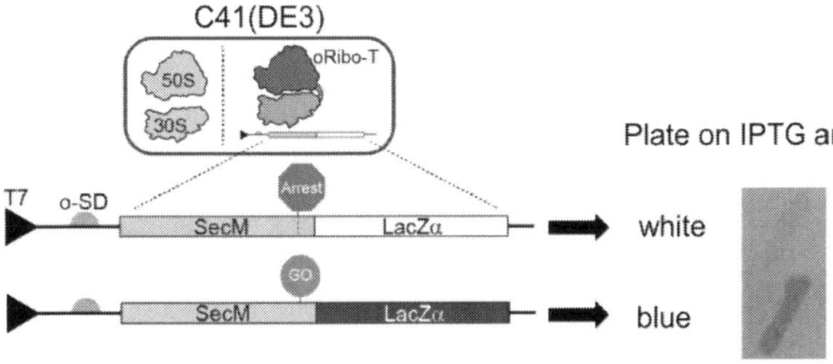

Nakatogawa and Ito, Cell 2002

Using oRibo-T to explore A2451/C2452 mutation effects on SecM stalling

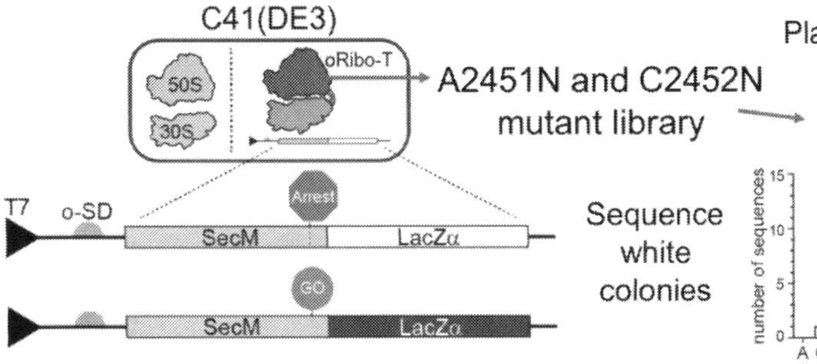

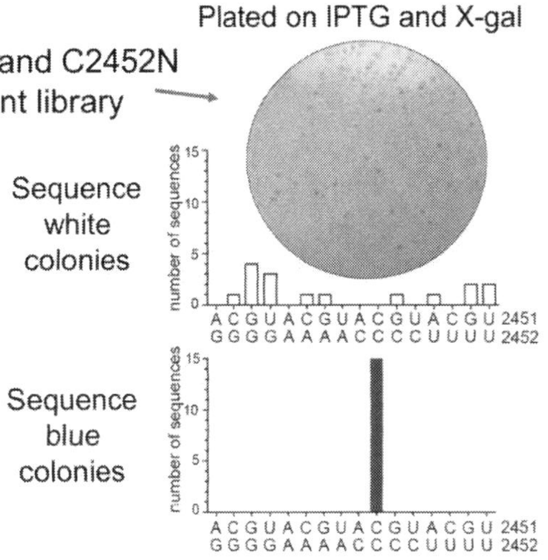

New oRibo-T function: SecM stalling alleviation with A2451C mutation

New oRibo-T function: SecM stalling alleviation

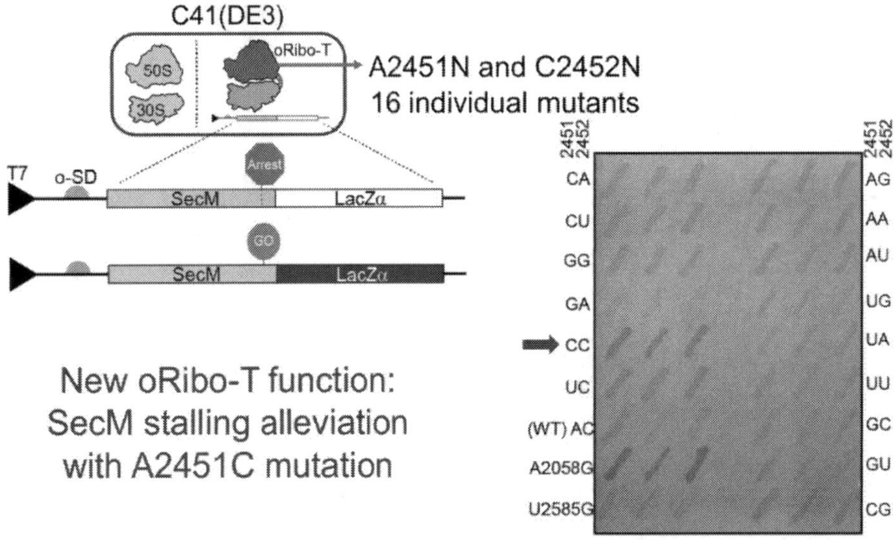

New oRibo-T function:
SecM stalling alleviation
with A2451C mutation

A System for Unprecedented Ribosome Engineering in Living *E. coli*

- We engineered a functional ribosome with tethered subunits
- *E. coli* can live with a tethered ribosome
- Full dissociation of ribosomal subunits is not a requirement for life or protein synthesis
- Ribo-T enables *in vivo* exploration of wildtype-lethal mutations altering ribosome function

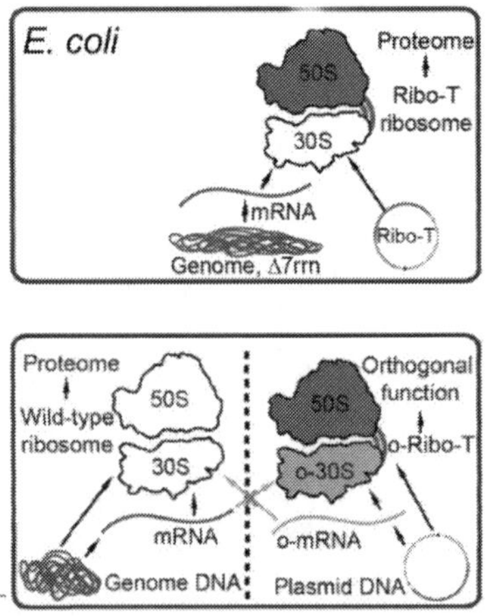

Enabling fundamental research *in vivo*

New synthesis capabilities

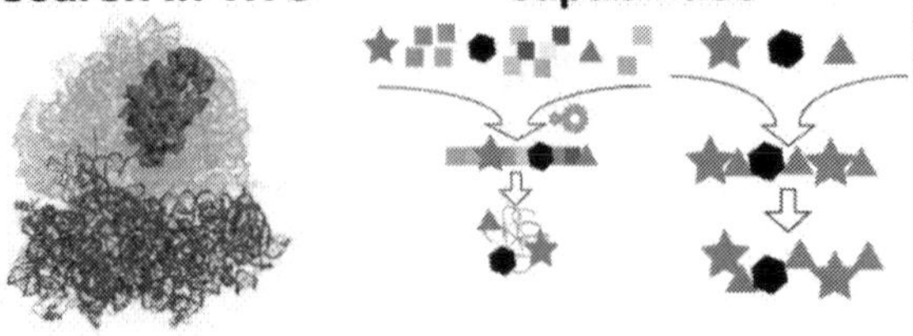

Thank you!
See you in Chicago for SEED 2016!

Thanks to:

Jewett lab

Mankin lab

Michael Jewett

Shura Mankin

Cedric Orelle

Dorota Klepacki (Mankin lab)
Ioanna Ntai (Kelleher lab)
Emm Fulk
Kirsten Swonger
Christina Burghard

Teresa Szal

Tanja Florin

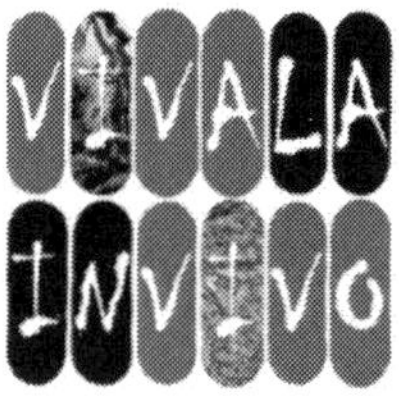

@SynBioCarlsons

SQ171/Ribo-T isolated genomic mutants

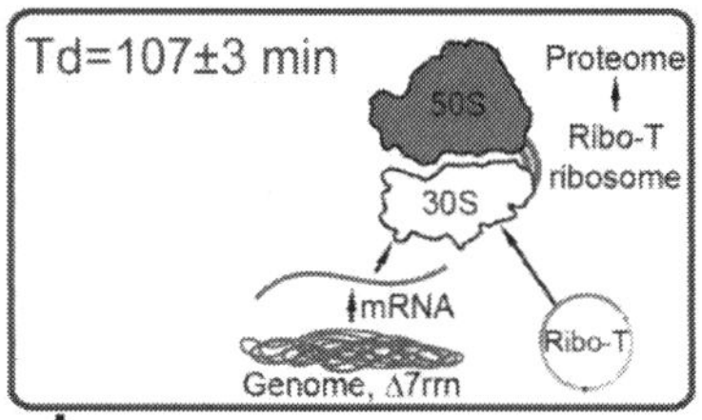

Serial passage 1 week:
Faster-growing strain isolated
Td = 70±2 min

Two genomic mutations found:

1: Non-sense mutation in putative Mg^{2+}/Co^{2+} efflux transporter

2: A549V mutation in S1 rProtein

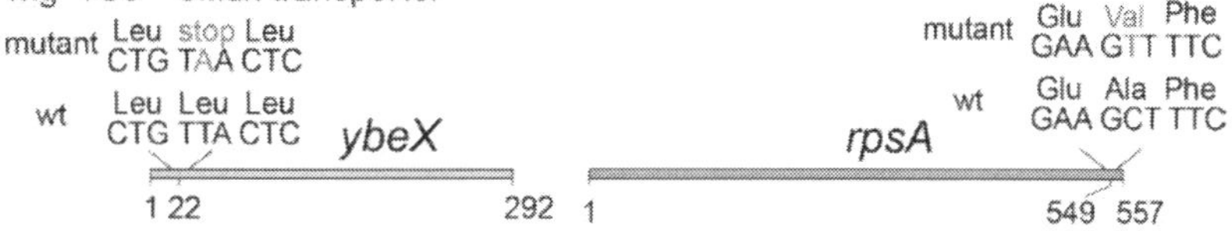

Quantifying tether clevage

No more than 7% cleavage observed in multiple preps

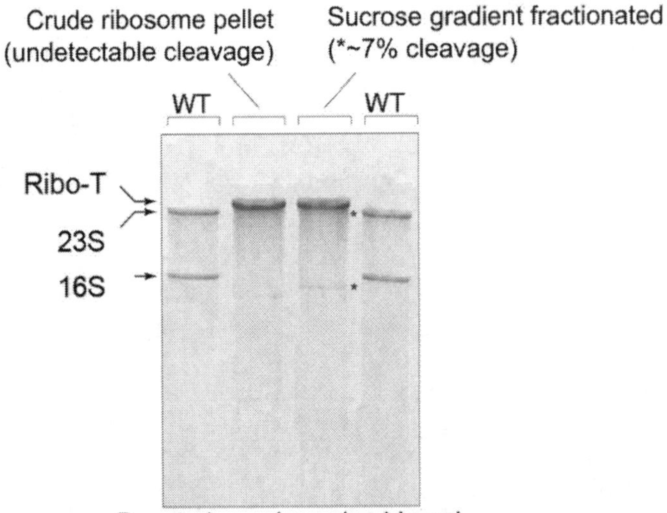

Denaturing polyacrylamide gel

Finding the tether site: 23S rRNA circular permutation study

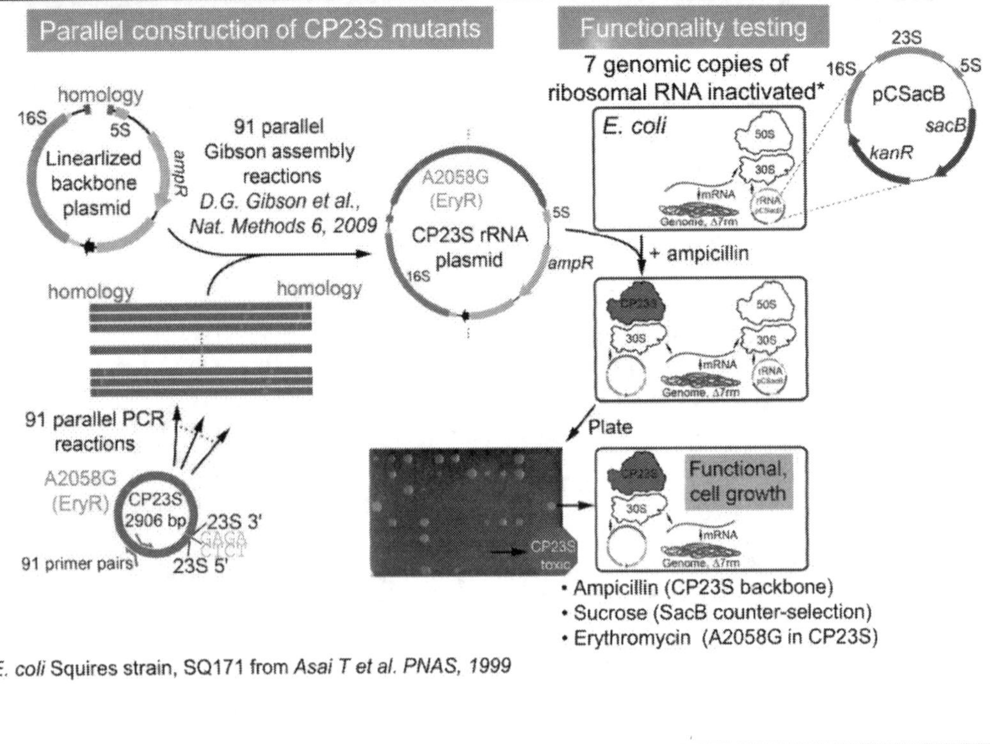

E. coli Squires strain, SQ171 from *Asai T et al. PNAS, 1999*

Ribo-T translation rate *in vivo*

7 genomic copies of ribosomal RNA inactivated*

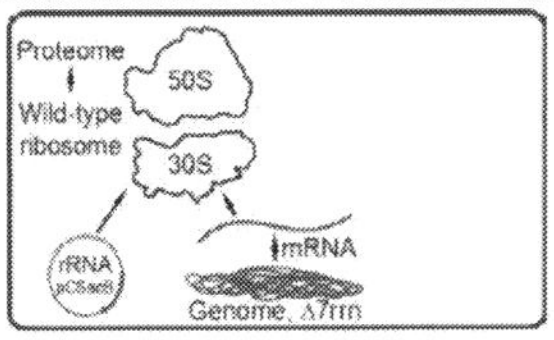

Doubling time
35±1 min

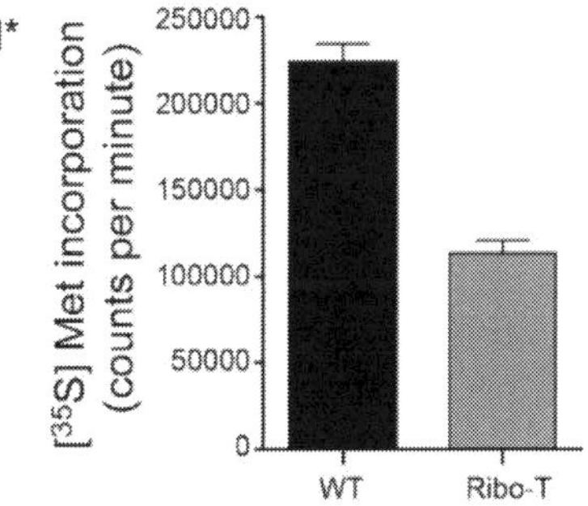

Remove wild-type, untethered ribosomes

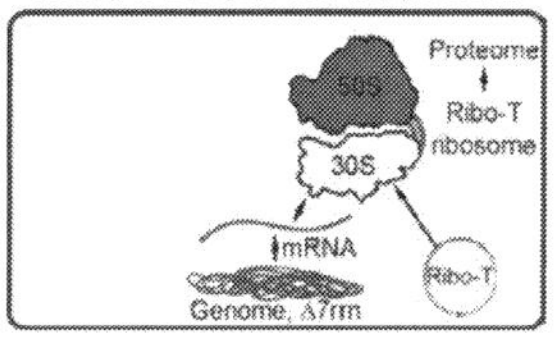

Doubling time
70±2 min

Ribo-T translation rate **50.5±3.5 %** of wild-type ribosomes

Ribo-T operon

Ribo-T transcript processed by 16S processing stems

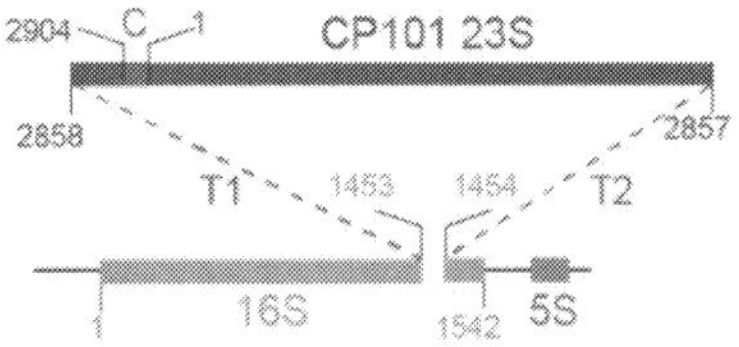

Ribosomes assembled into functional particles? Sucrose gradient fractionation

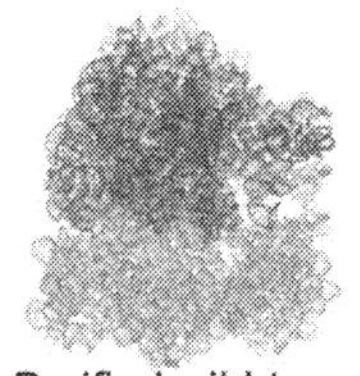

Purified wild-type ribosomes

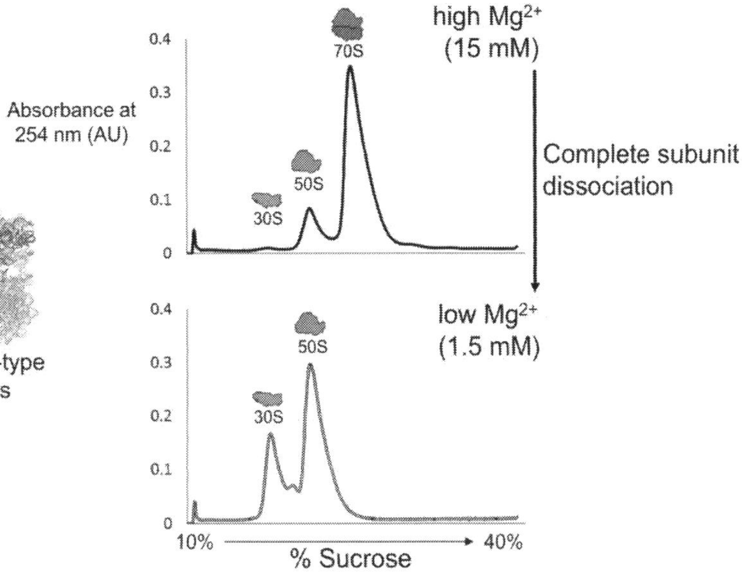

Ribosomes assembled into functional particles? Sucrose gradient fractionation

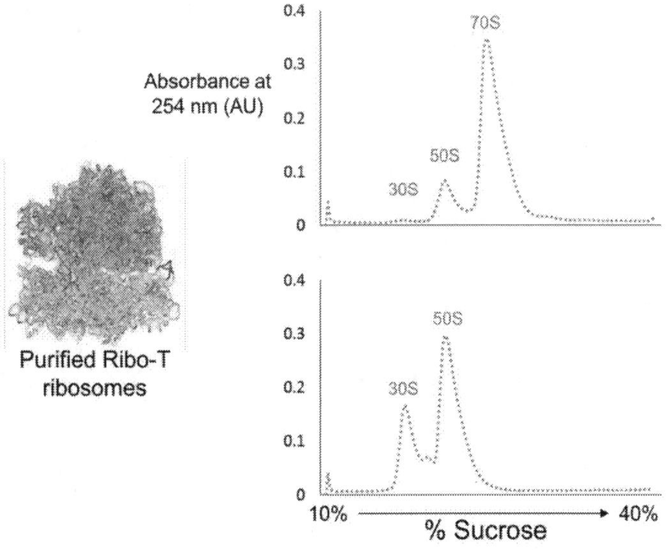

Ribosomes assembled into functional particles? Sucrose gradient fractionation

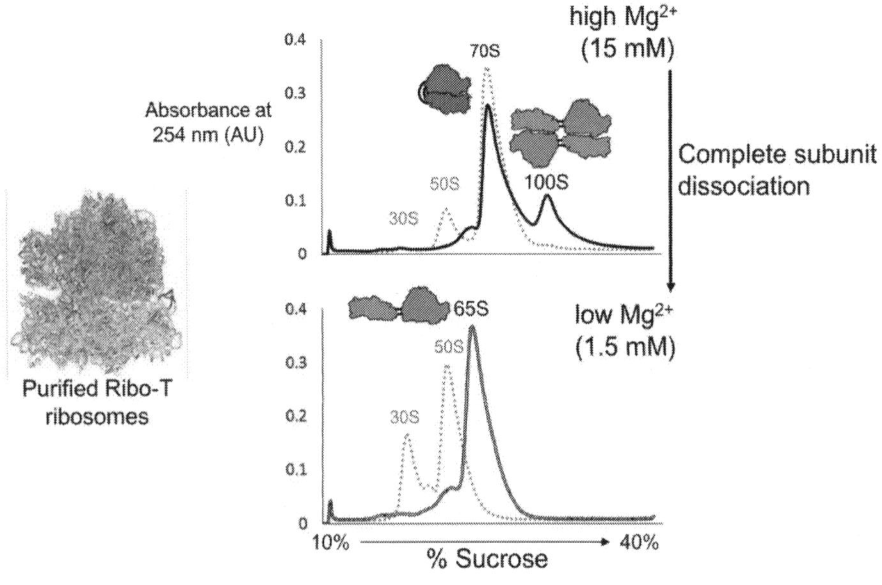

The selected tether in Ribo-T

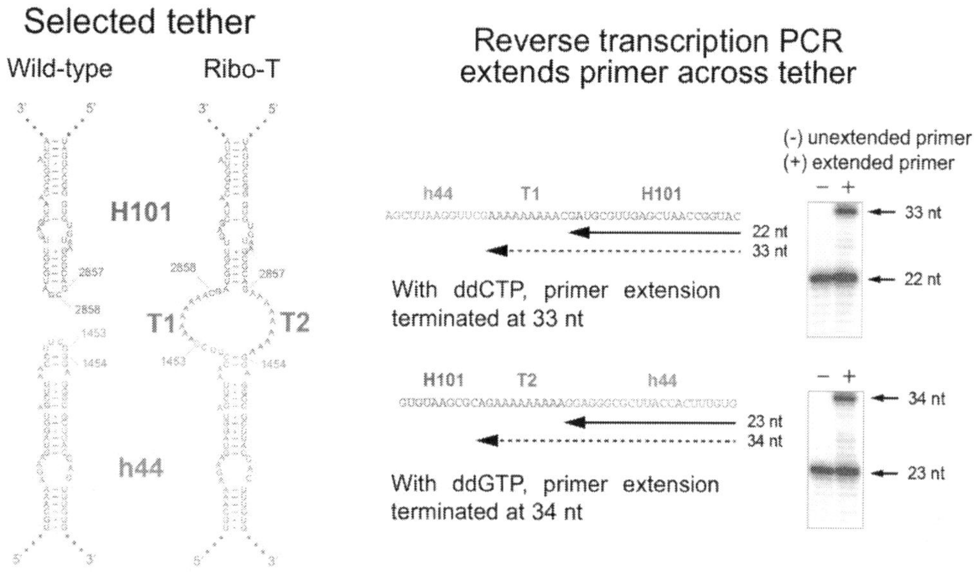

New oRibo-T function: SecM stalling alleviation

in vitro translation, Δribosome PURExpress

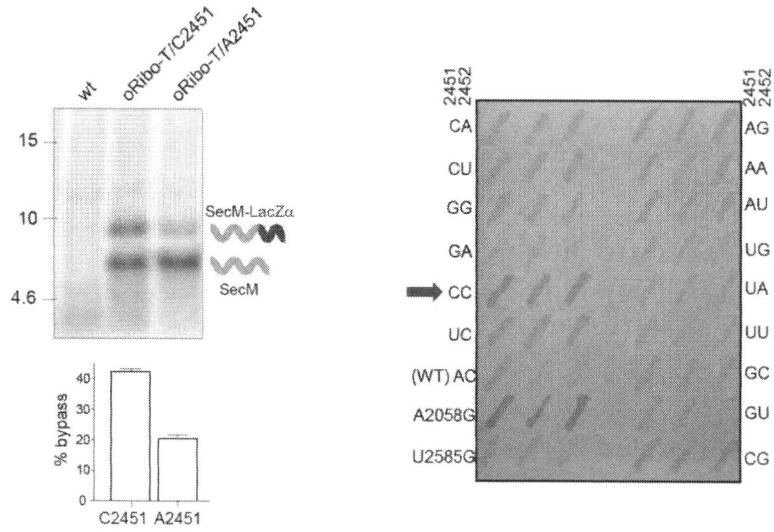

Ribo-T works as orthogonal system, oRibo-T

Orthogonal GFP mRNA expression in *E. coli* (POP2136 strain)

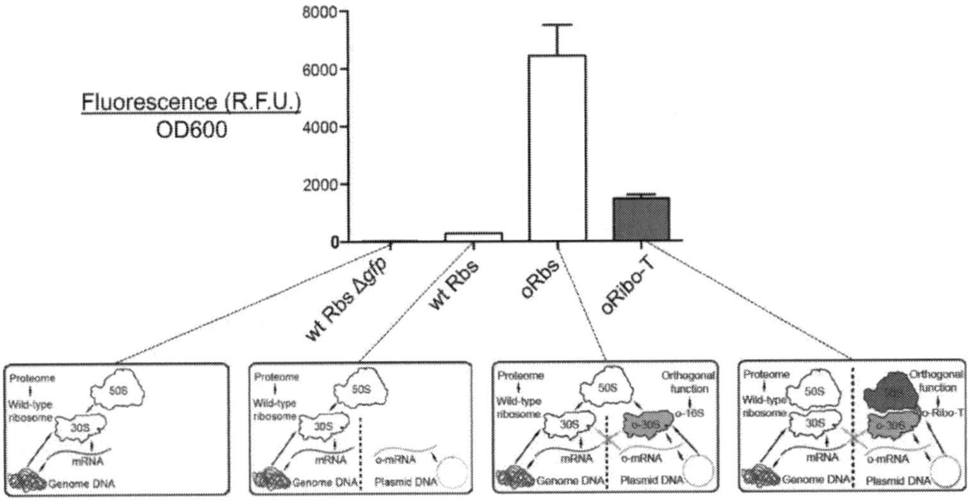

Testing A2451-C1452 library: New mutations found that alleviate secM stalling

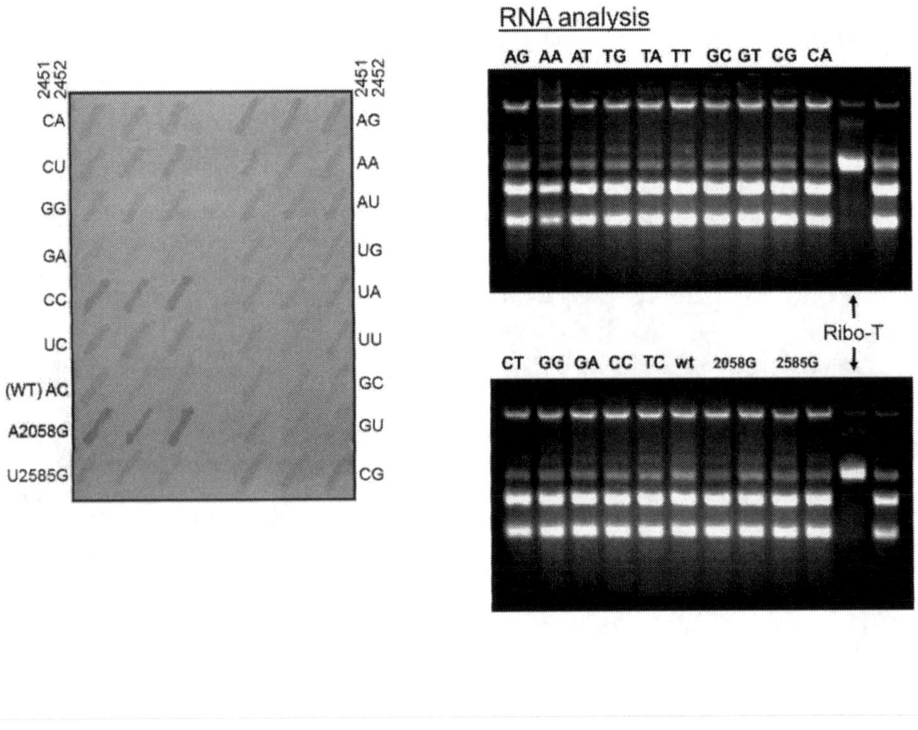

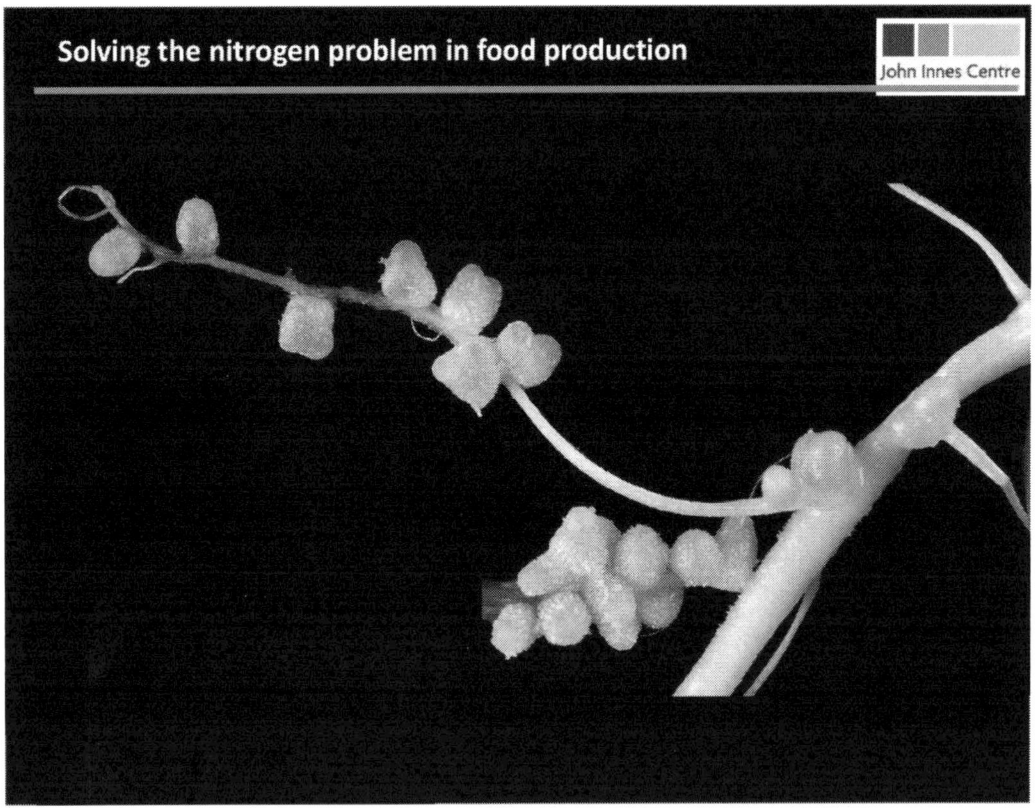
Solving the nitrogen problem in food production
John Innes Centre

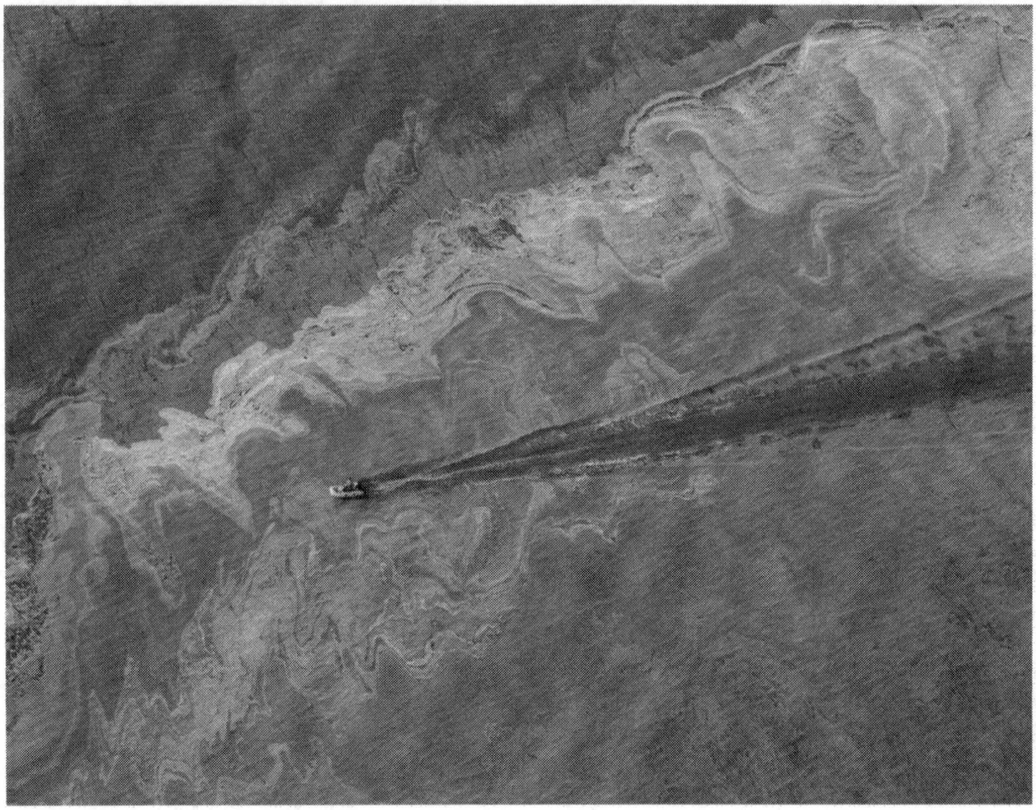

The fertiliser-use inbalance

John Innes Centre

Mueller et al 2012, *Nature*

Engineering the Nitrogen Symbiosis for Africa

ensa

BBSRC
bioscience for the future

BILL & MELINDA GATES foundation

Legumes are the prototype for self-fertilising crops

John Innes Centre

The developmental processes associated with nodulation

John Innes Centre

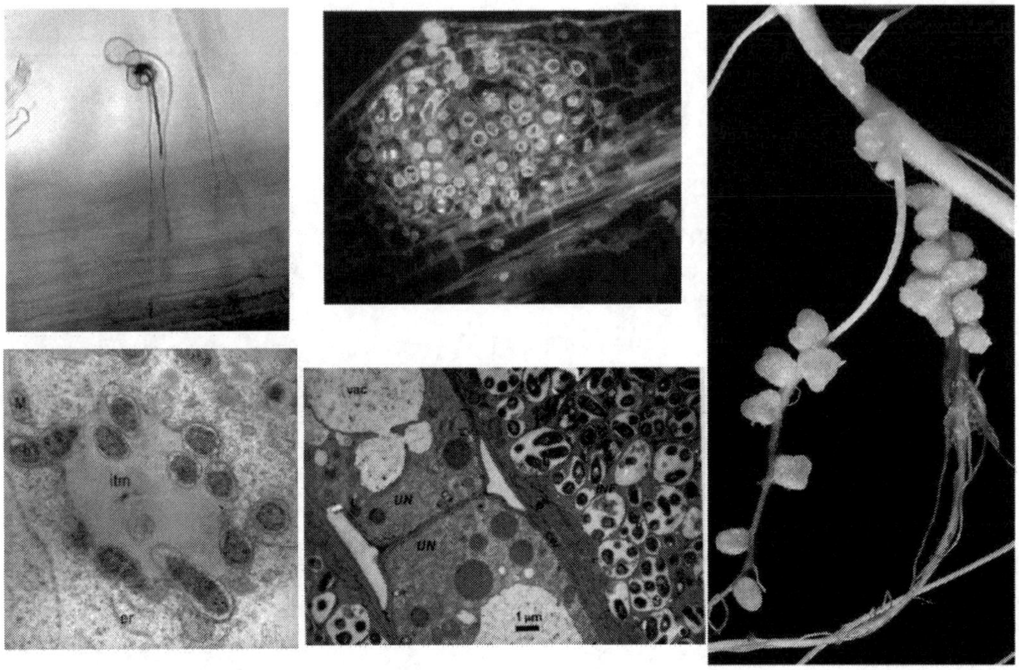

A complex engineering problem

John Innes Centre

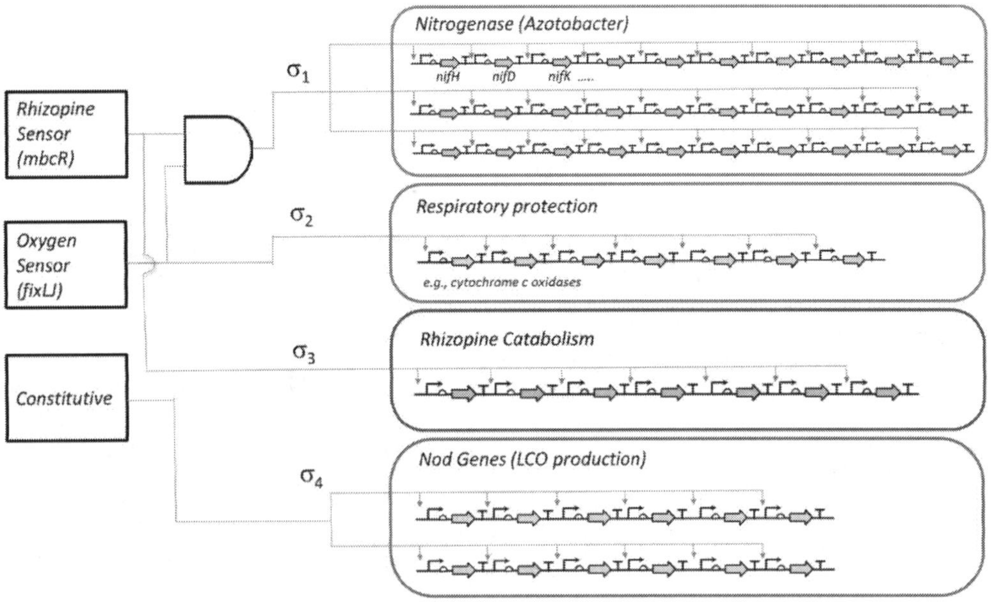

The early steps in symbiosis are a molecular dialogue

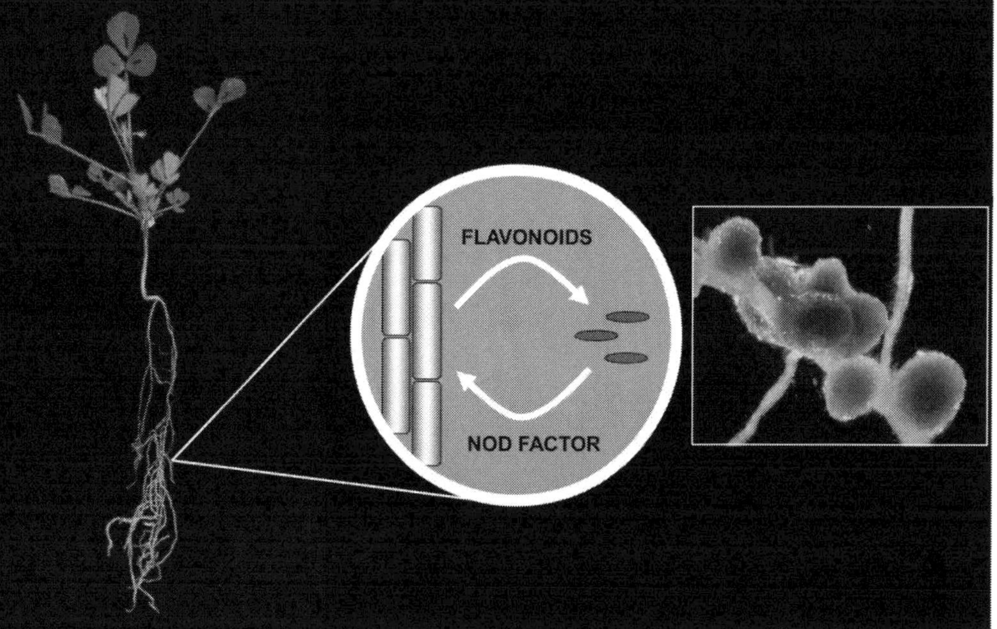

S. meliloti Nod factor

John Innes Centre

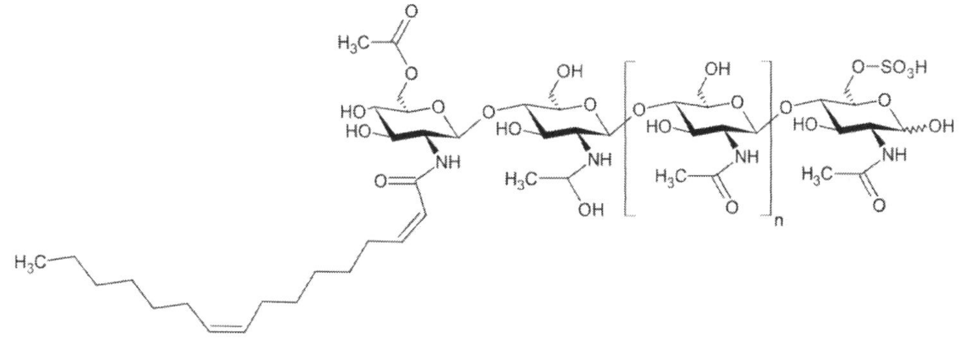

The symbiosis signalling pathway

John Innes Centre

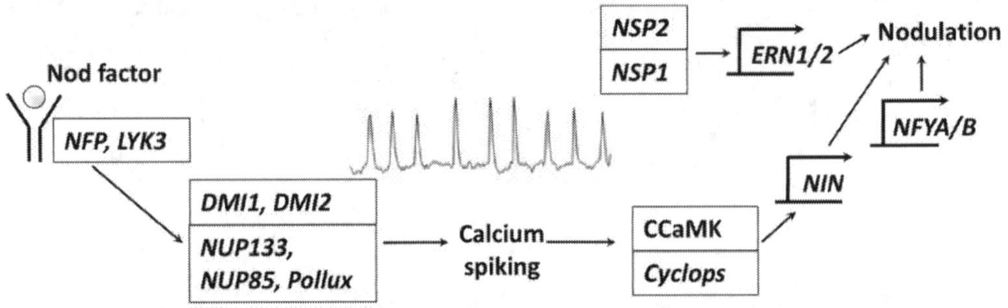

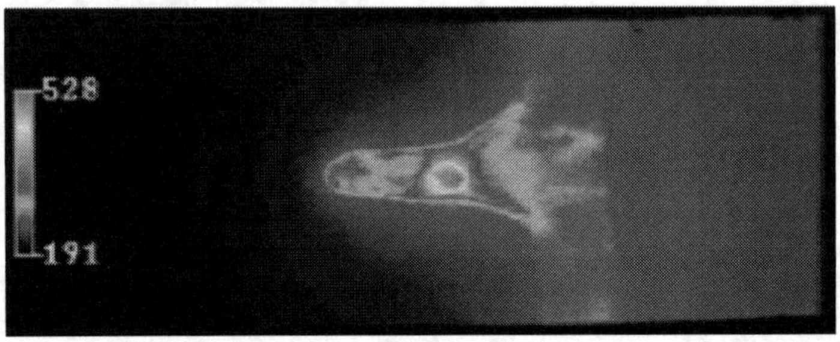

Engineering the symbiosis signaling pathway in cereals

John Innes Centre

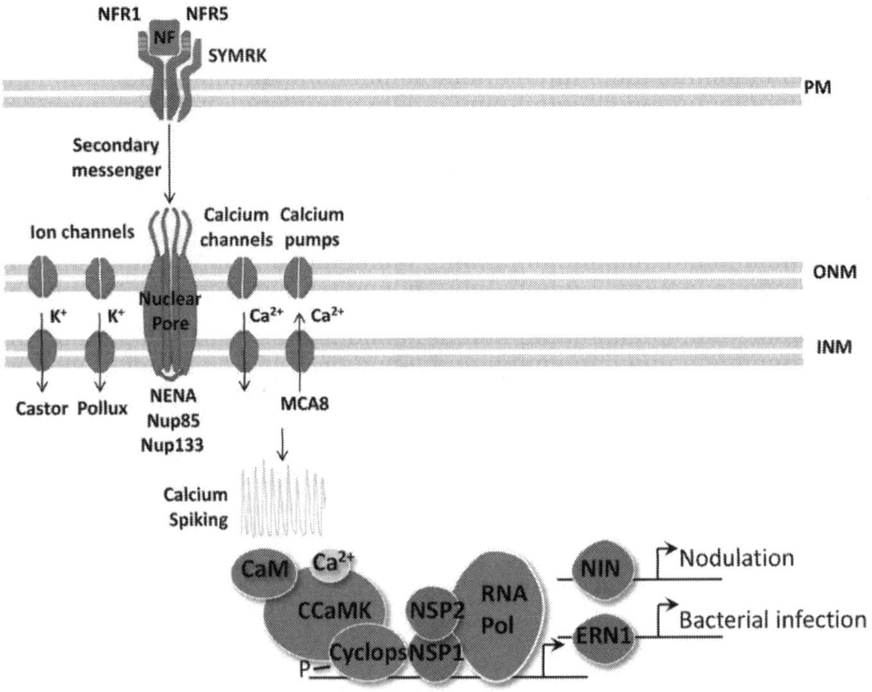

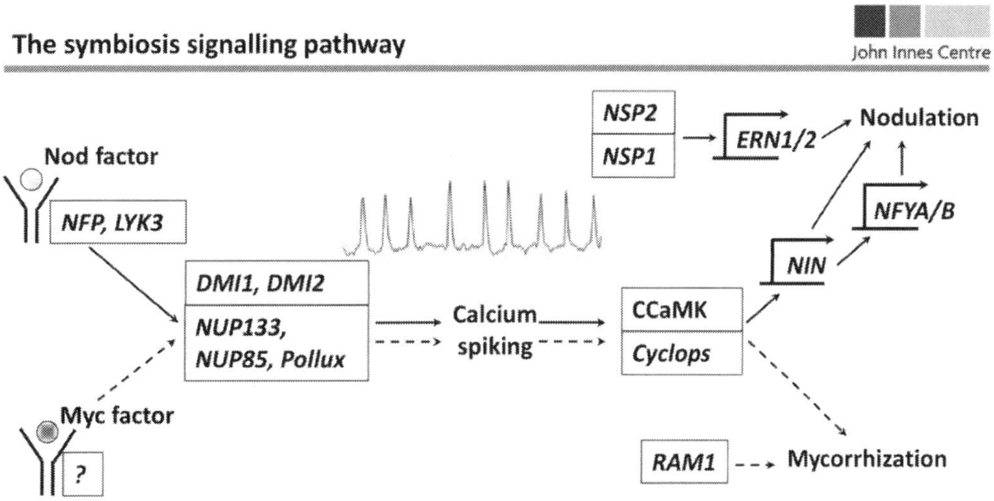
The symbiosis signalling pathway
John Innes Centre
Nod factor
NFP, LYK3
DMI1, DMI2
NUP133,
NUP85, Pollux
Calcium
spiking
CCaMK
Cyclops
NSP2
NSP1
ERN1/2
Nodulation
NIN
NFYA/B
Myc factor
?
RAM1
Mycorrhization

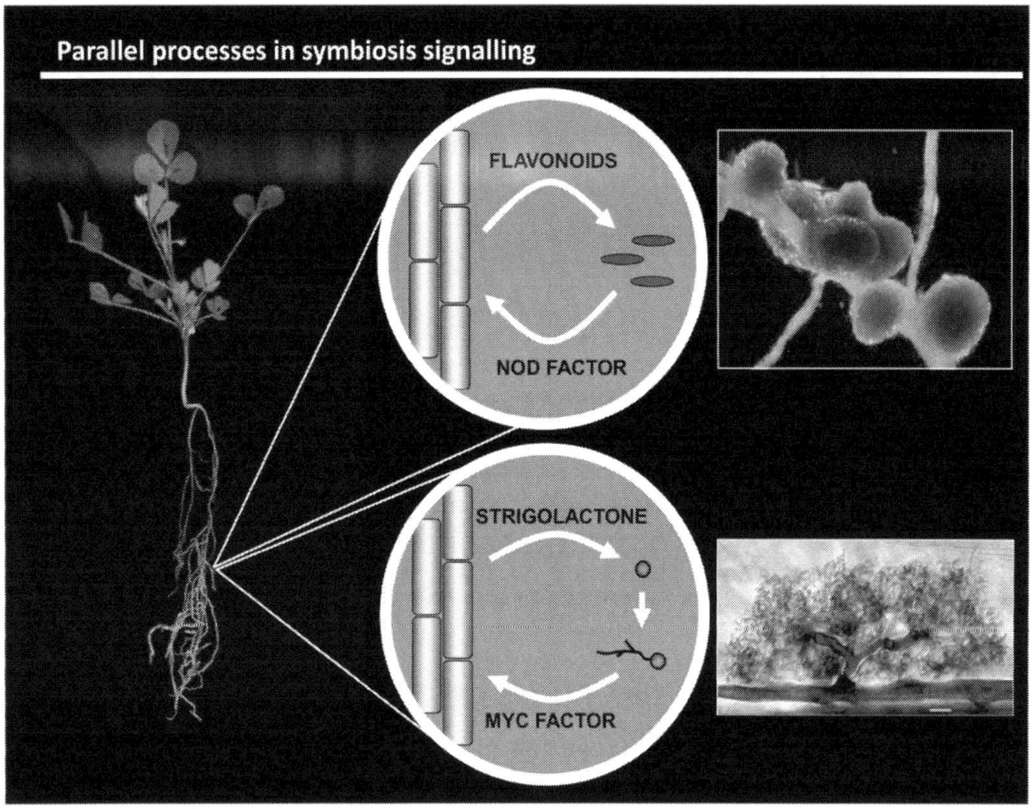
Parallel processes in symbiosis signalling
FLAVONOIDS
NOD FACTOR
STRIGOLACTONE
MYC FACTOR

Symbiotic N-fixation evolved from the mycorrhizal symbiosis

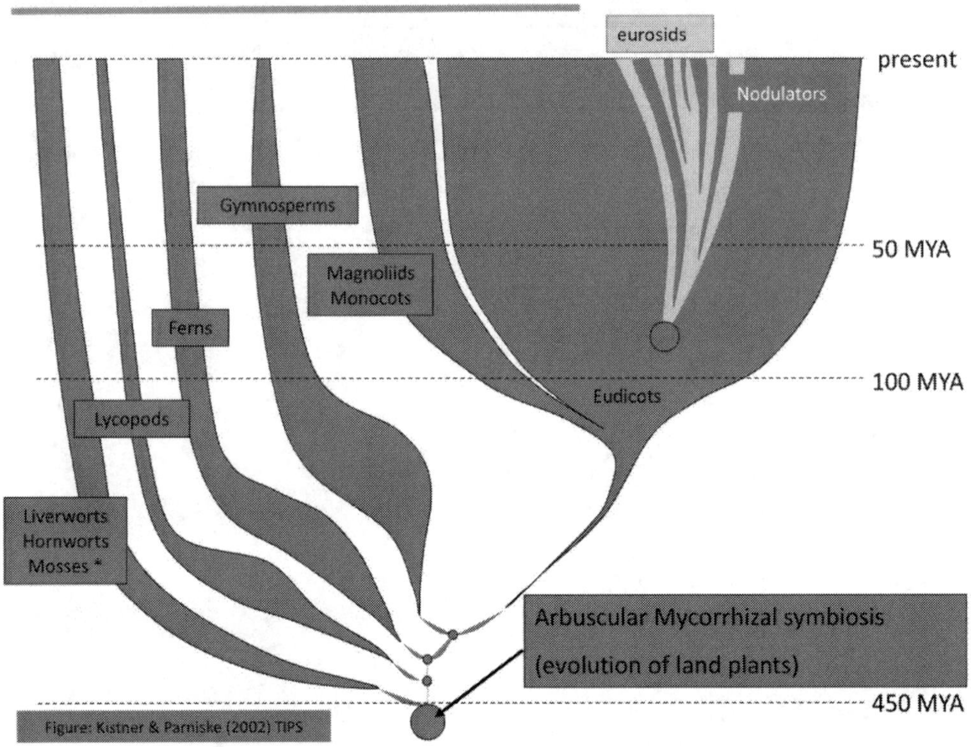

Rice symbiosis signalling mutants

John Innes Centre

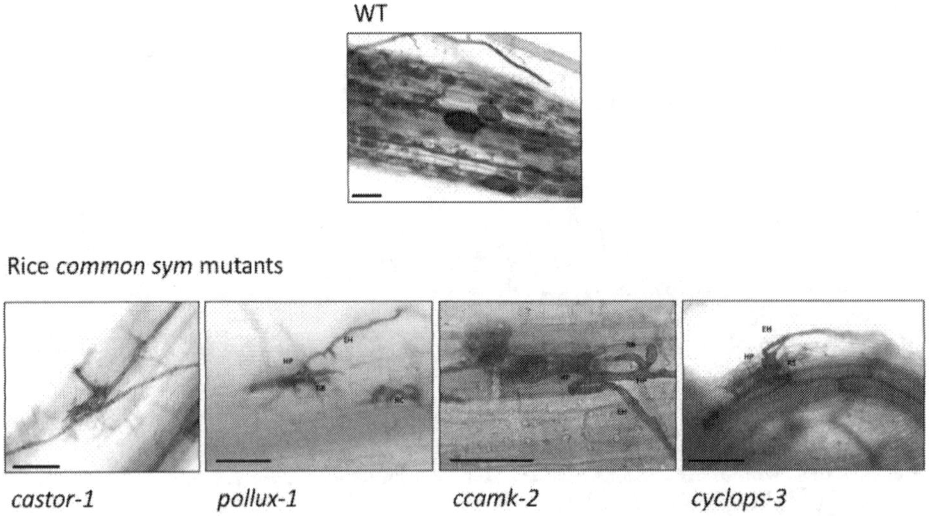

size bars = 50 μm

Gutjahr et al. 2008, Plant Cell

Cereals contain a functional symbiosis signalling pathway

John Innes Centre

Symbiosis signaling in cereals

John Innes Centre

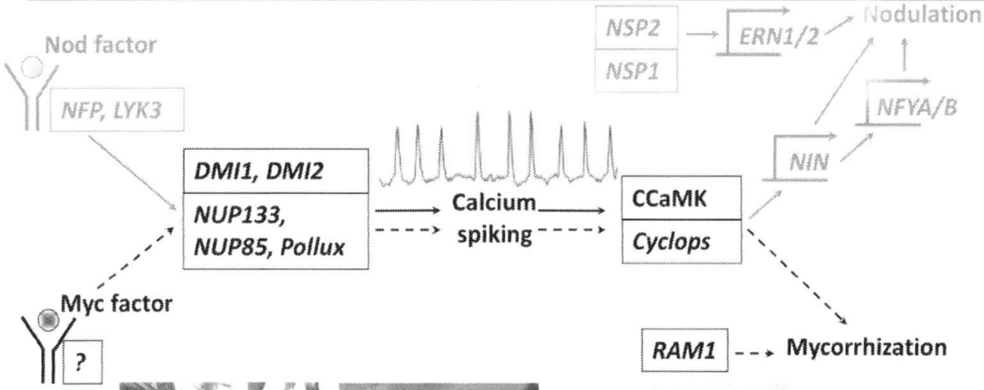

Plant recognition of mycorrhizal fungi

John Innes Centre

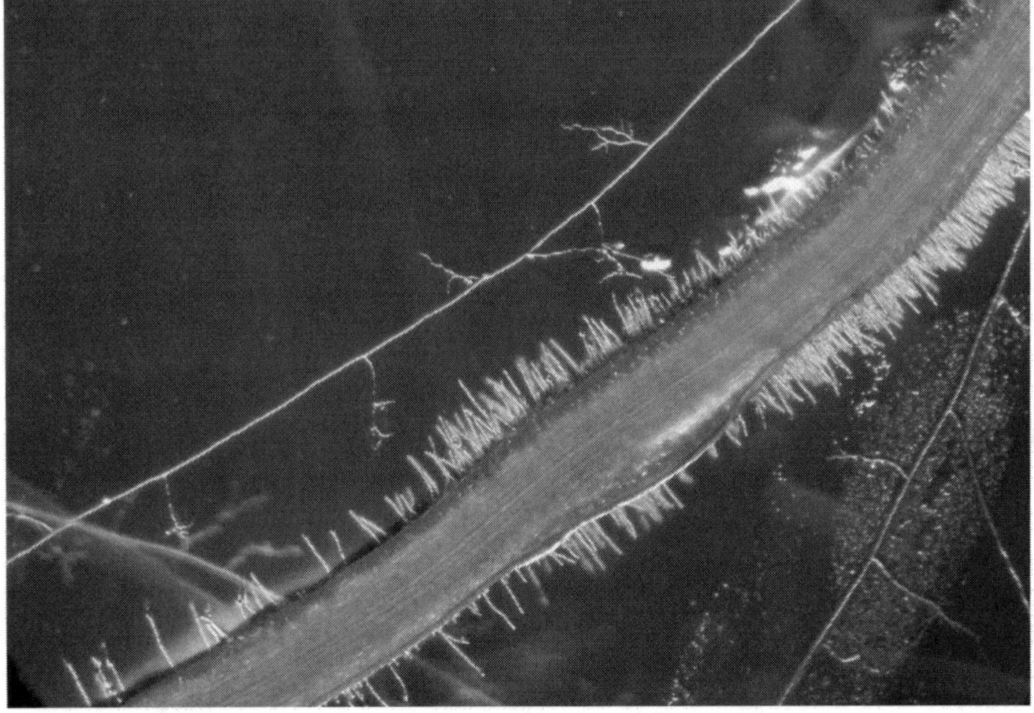

Mycorrhizal fungi produce LCOs and COs

John Innes Centre

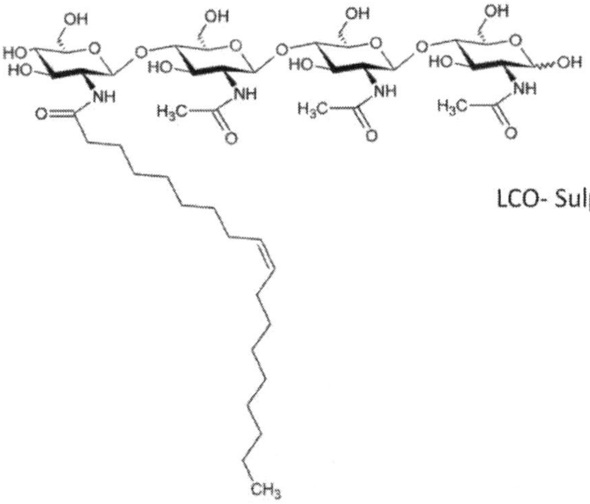

LCO- Sulphated and un-sulphated

Maillet et al 2011 *Nature*

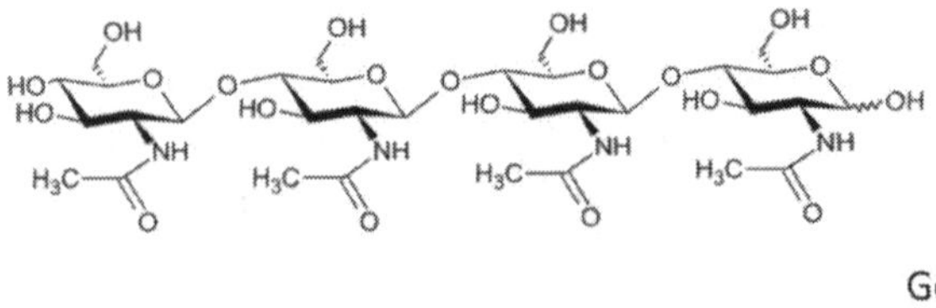

CO- CO4, CO5 most active

Genre et al., 2012 *New Phytologist*

Activation of calcium spiking in rice

John Innes Centre

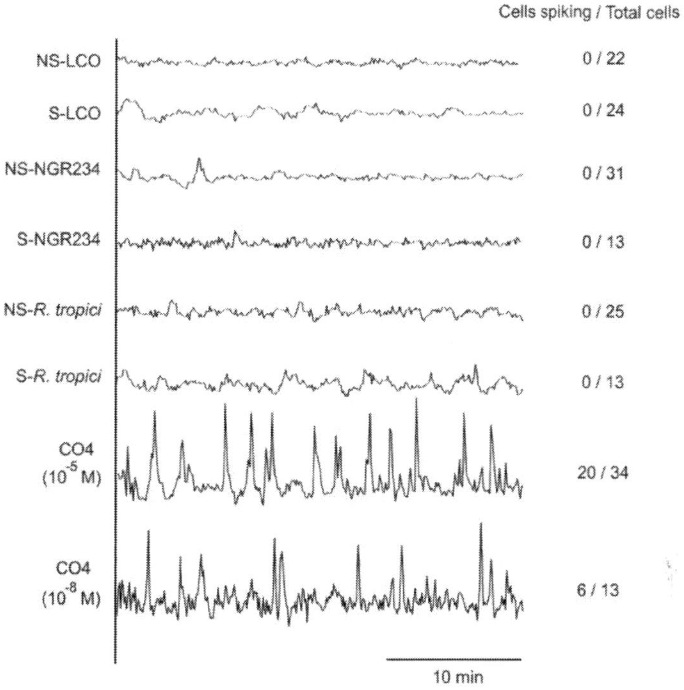

CO induced calcium spiking in rice

John Innes Centre

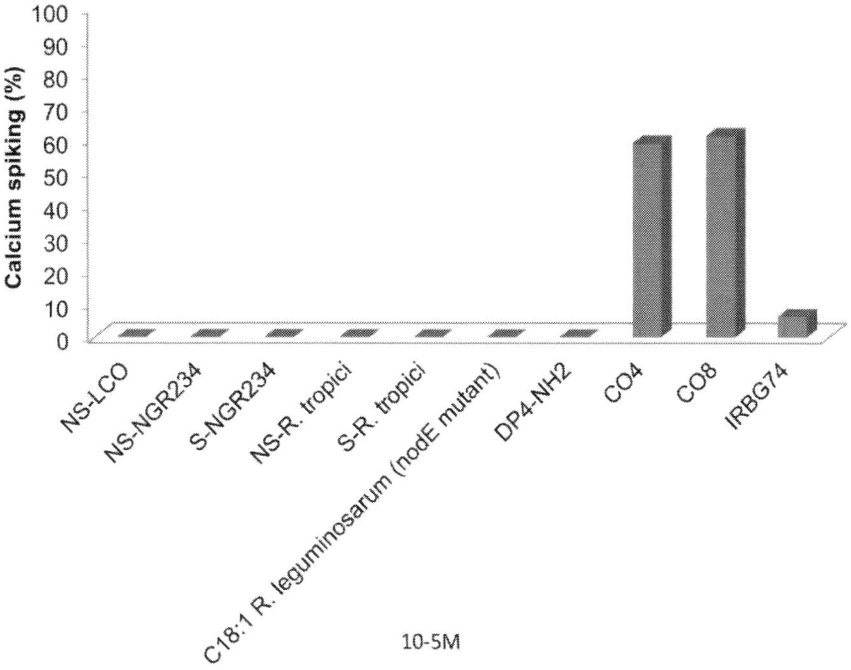

LCOs/COs induced calcium oscillations in Legumes and Non legumes

John Innes Centre

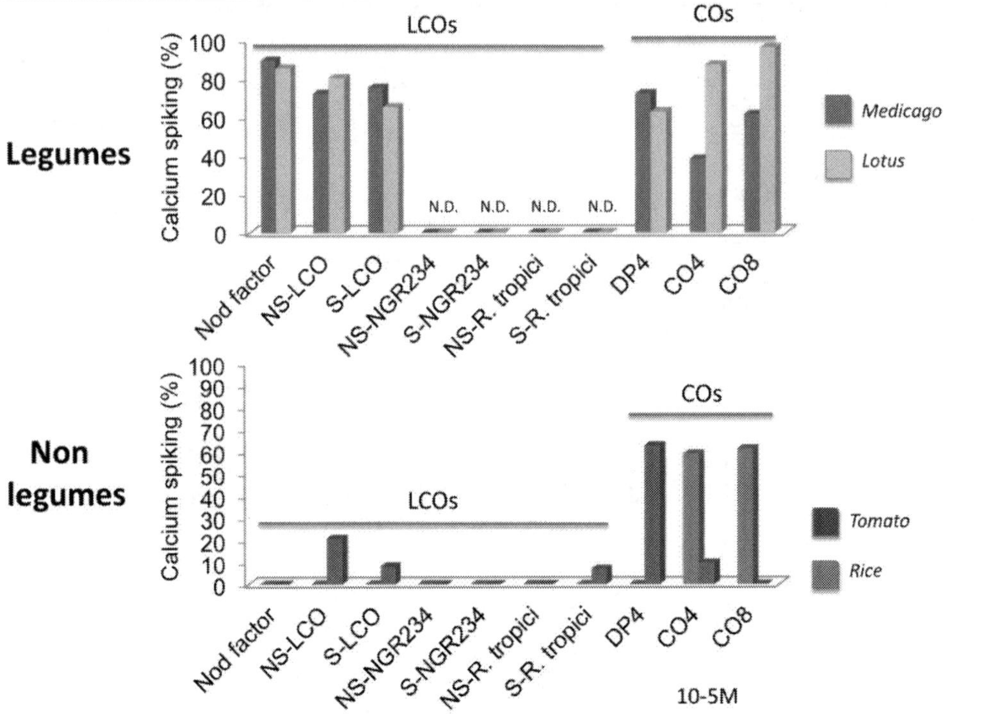

Symbiosis signaling in cereals

John Innes Centre

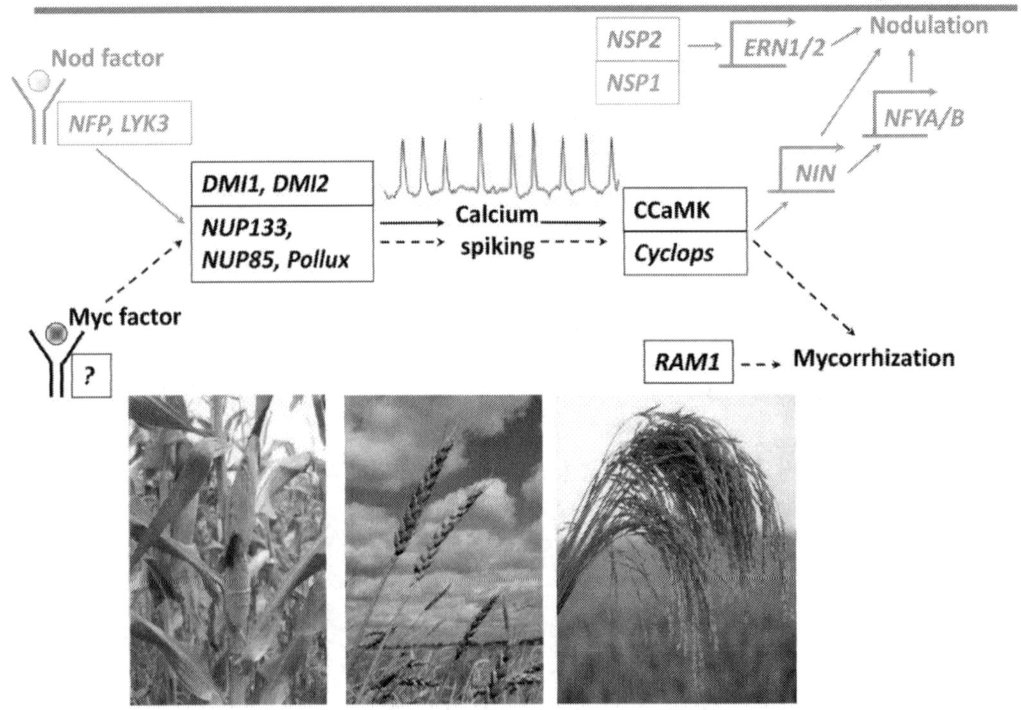

The Engineering design, build, test cycle

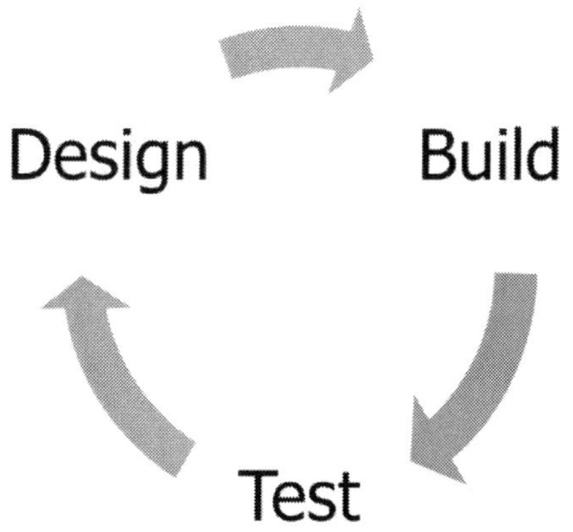

Golden Gate facilitates multi-gene vector construction

John Innes Centre

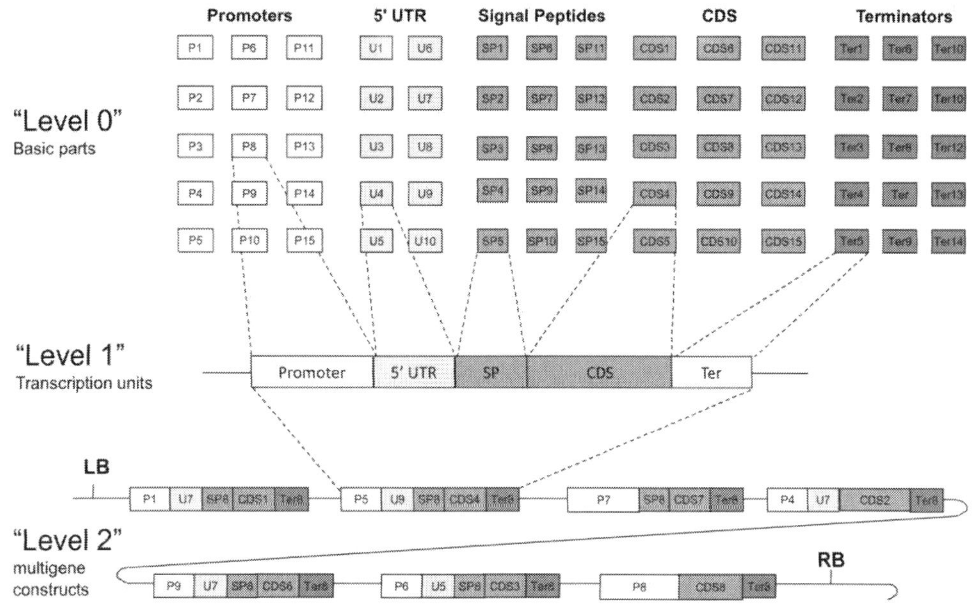

Weber 2011, *PloSone*

The Engineering design, build, test cycle

John Innes Centre

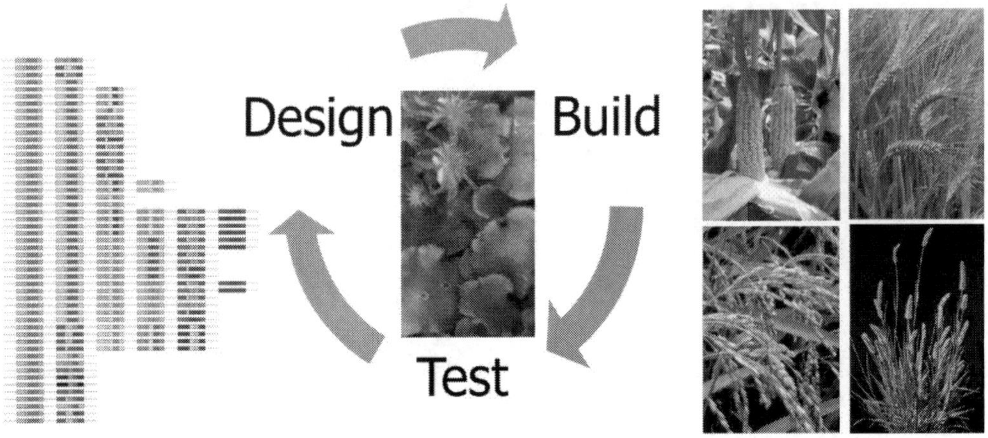

Engineering platforms in basal plants

John Innes Centre

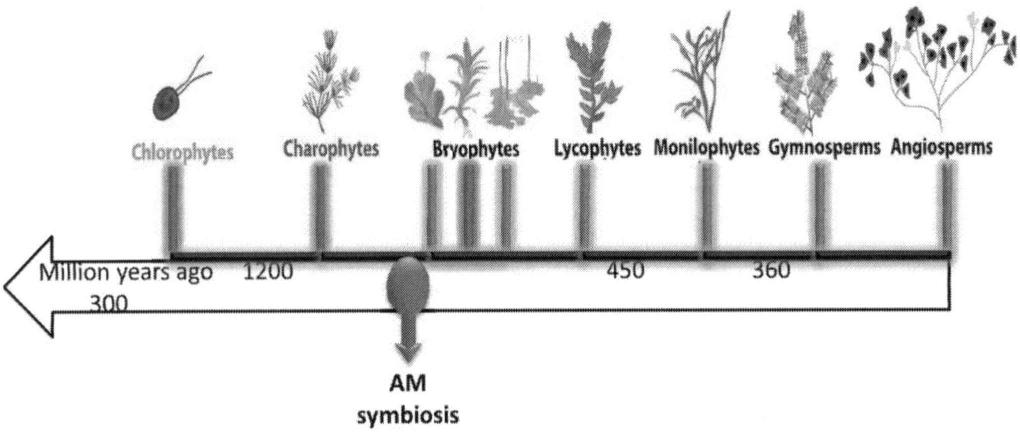

Mycorrhization in *Marchantia paleacea*

John Innes Centre

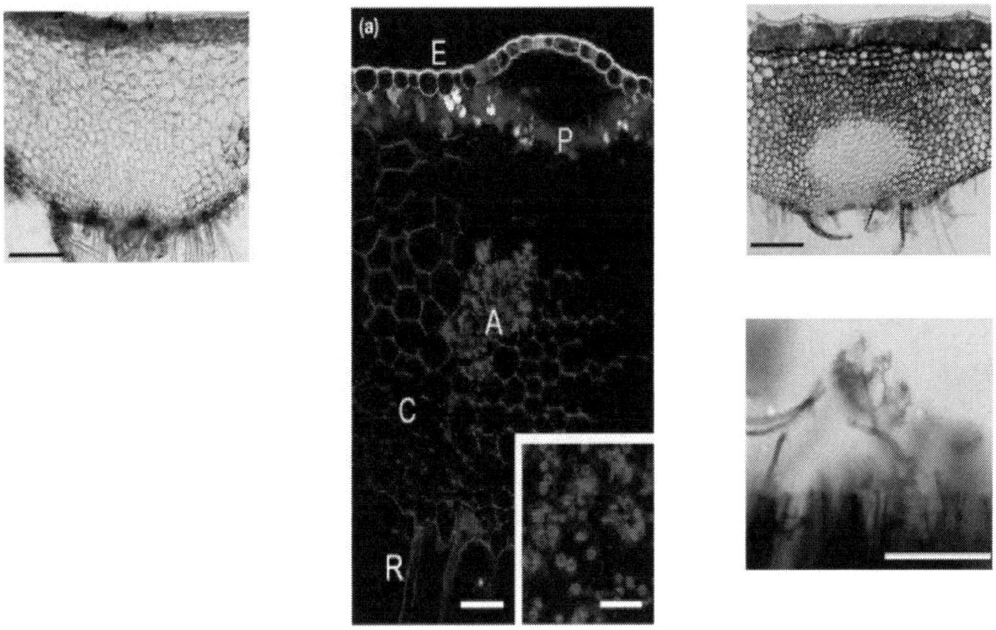

The Sym pathway in *Marchantia paleacea*

John Innes Centre

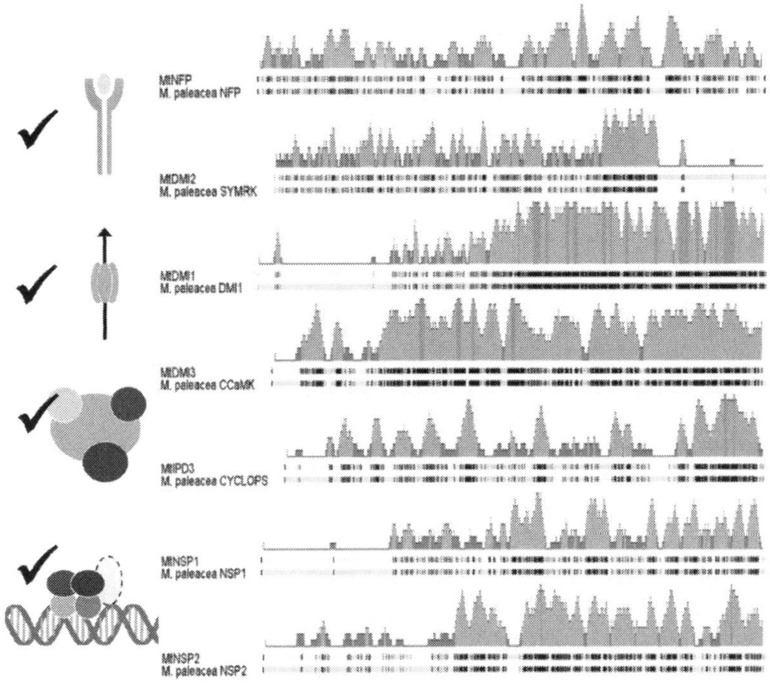

Symbiotic calcium oscillations in *M. paleacea*

John Innes Centre

1 second of video = 2 minutes of imaging

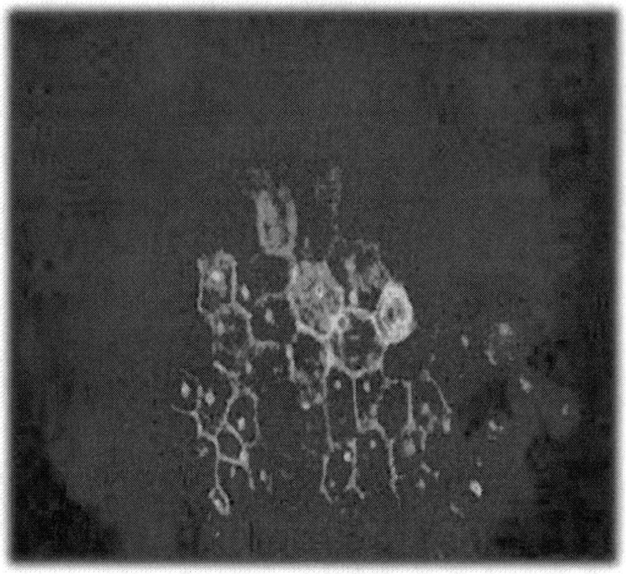

A simple proxy for engineering higher plants

John Innes Centre

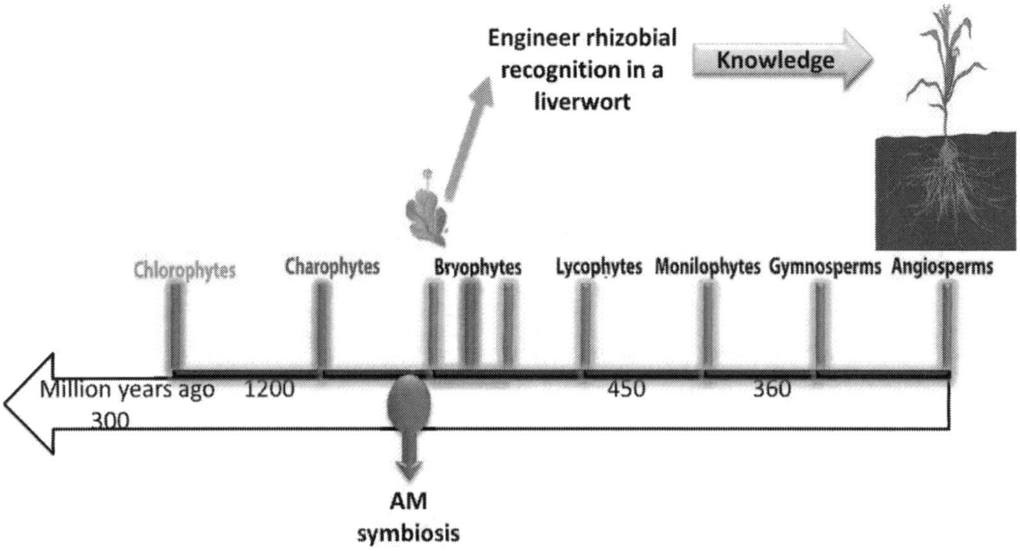

Acknowledgements

INRA Toulouse
Jean Dénarié
Clare Gough
Fabienne Maillet
Guillaume Bécard
Grenoble
Sébastien Fort

JIC
Giles Oldroyd
Christian Rogers
Jongho Sun
Feng Feng
Ben Miller
Andrey Korolev
Eleni Soumpourou
Andy Breakspear
Guru Radhakrishnan

BILL & MELINDA GATES *foundation*

Donald Danforth Plant Science Centre
Tom Brutnell
Hui Jiang
Quan Zhang

Boyce Thompson
Joyce Van Eck

BRACT
Wendy Harwood

Bill & Melinda Gates Foundation
Katherine Kahn

University of Aarhus
Jens Stougaard
Simona Radutoiu
Lene Madsen
Mickael Blaise
Dugald Reid
Ei-ici Murakami
Estelle Marchal

University of Wisconsin Madison
Jean-Michel Ané
Matt Crook

ENSA External Advisors
Allan Downie
Doug Cook
Michael Udvardi
Sharon Long

The JIC ENSA team

Funding:

BILL & MELINDA GATES *foundation* BBSRC bioscience for the future erc

One handful of soil:

10,000,000,000 microbes

100,000 different species

1. How do microbes interact with each other, and with multicellular organisms?

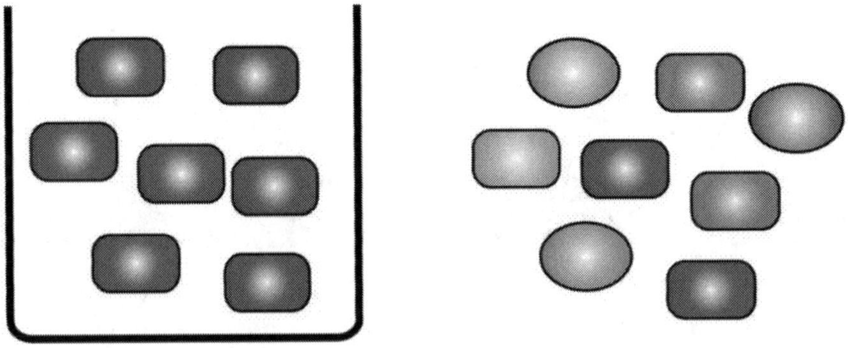

Microbes in the laboratory

Microbes in the real world

Natural products: Medicinal relevance

Natural products as drugs (1981-2002)

70% of new antibiotics

60% of new anticancer drugs

50% of new immunosuppressants

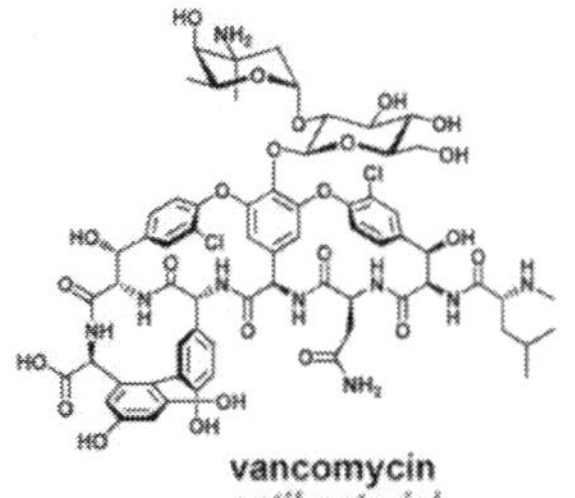

vancomycin
antibacterial

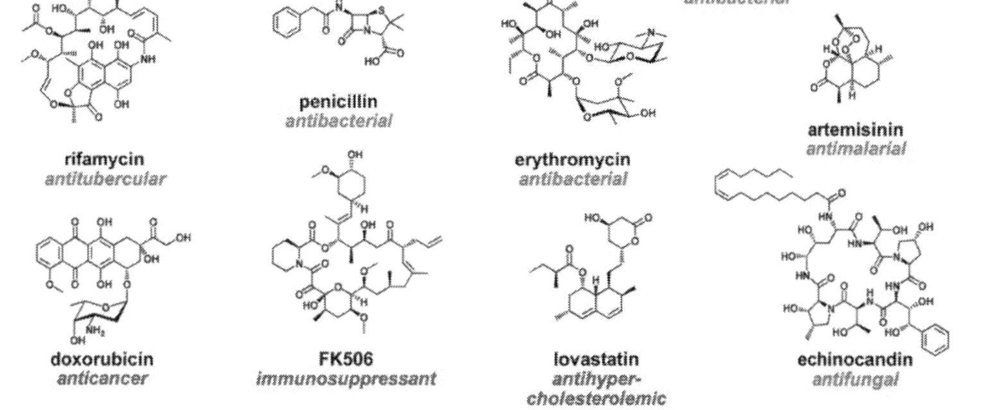

Newman et al, *J Nat Prod* **66**, 1022 (2003)

The metabolic cost of a natural product

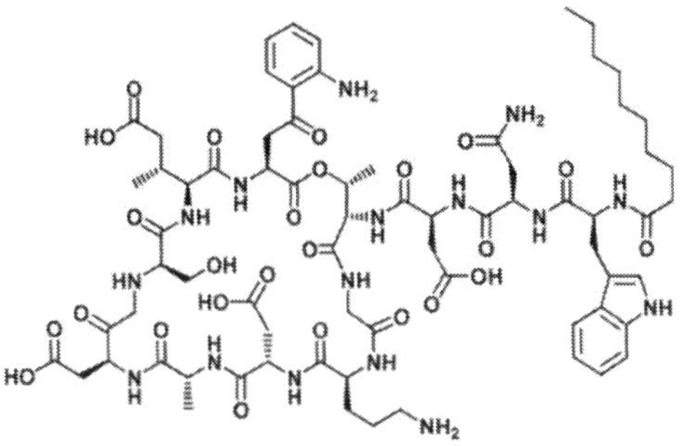

daptomycin

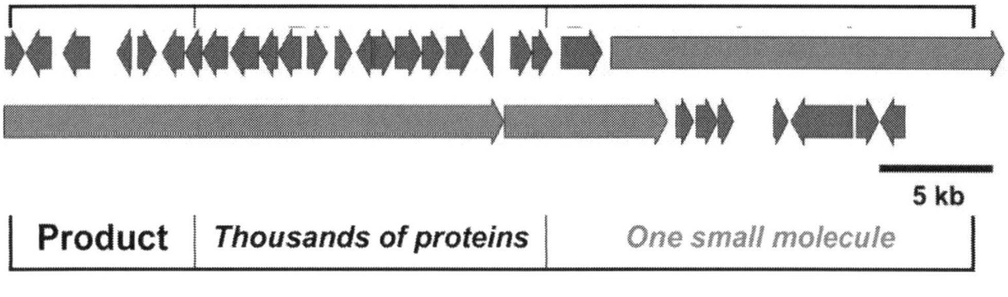

1. How do microbes interact with each other, and with multicellular organisms?

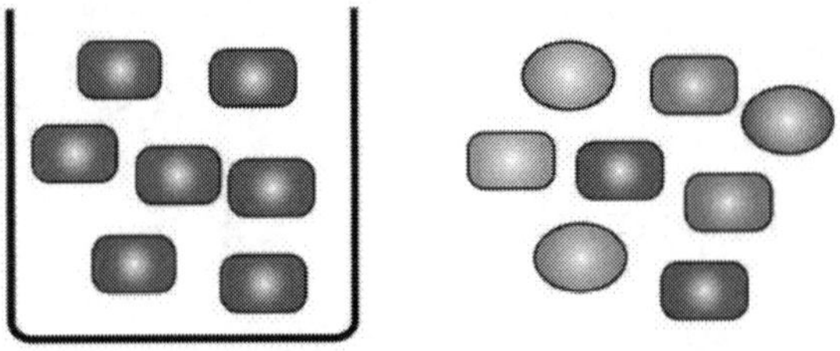

2. What are the natural roles of natural products?

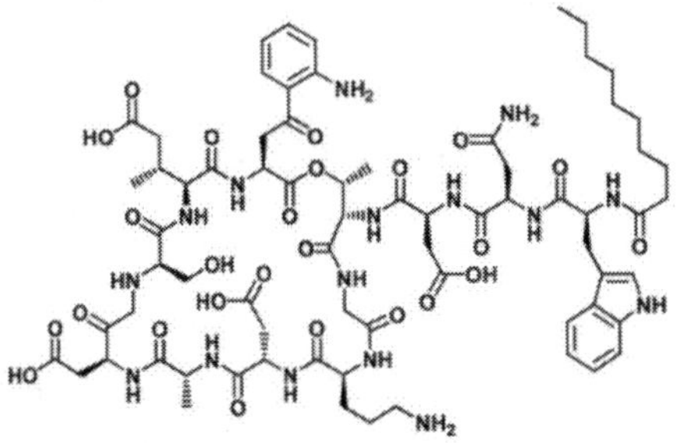

Natural products mediate interspecies interactions.

The Fischbach group

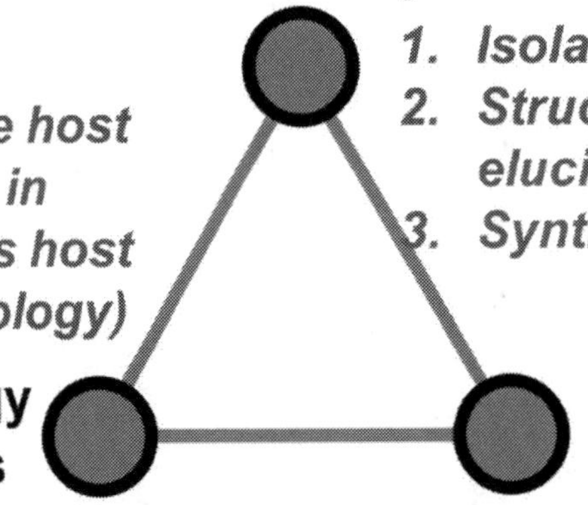

Chemistry:

1. *Isolation chemistry*
2. *Structure elucidation*
3. *Synthesis*

Microbiology & Genetics:

1. *KO in native host*
2. *Expression in heterologous host (synthetic biology)*

Bacillus
Bacteroides
Clostridium
Corynebacterium
Escherichia
Lactobacillus
Pseudomonas
Staphylococcus
Streptococcus
Streptomyces

Bioinformatics:

1. *Identify biosynthetic gene clusters*
2. *Predict structure of their small molecule product*

The old paradigm of product discovery

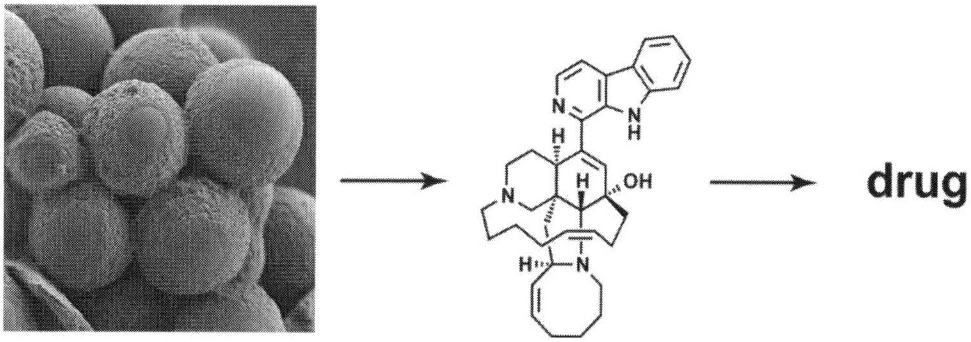

Connecting molecules to genes

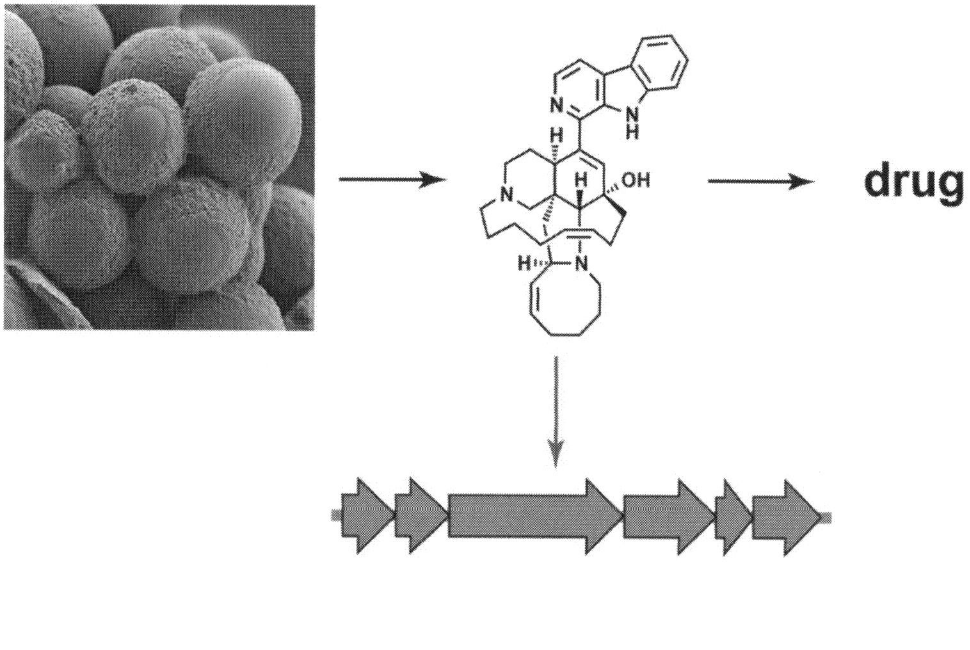

Connecting molecules to genes

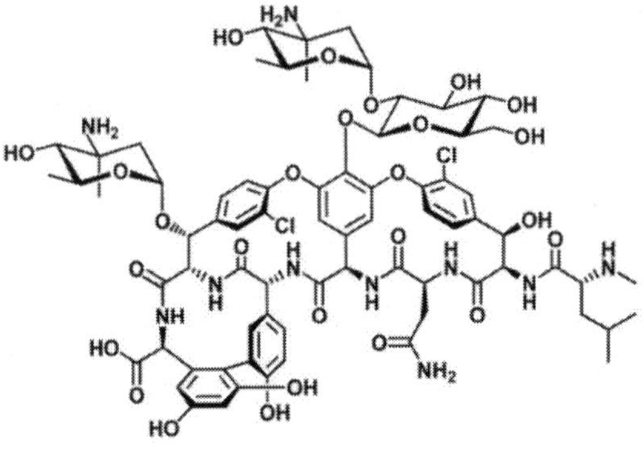

chloroeremomycin

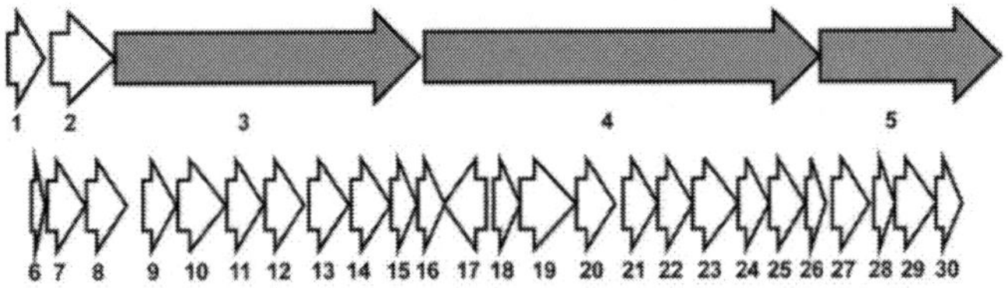

Van Wageningen *et al*, *Chem Biol* (1998) **5**, 155

Connecting molecules to genes

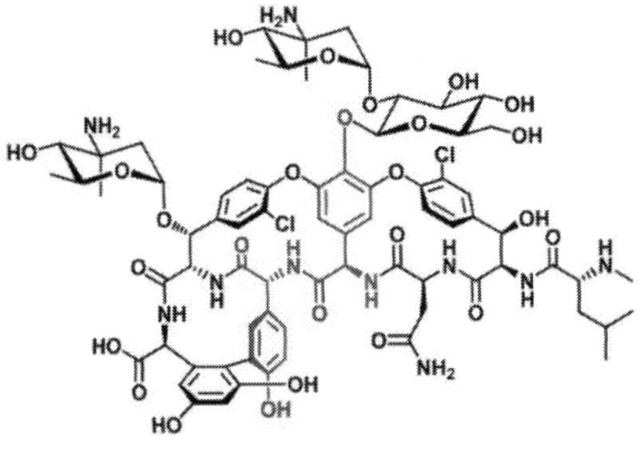

chloroeremomycin

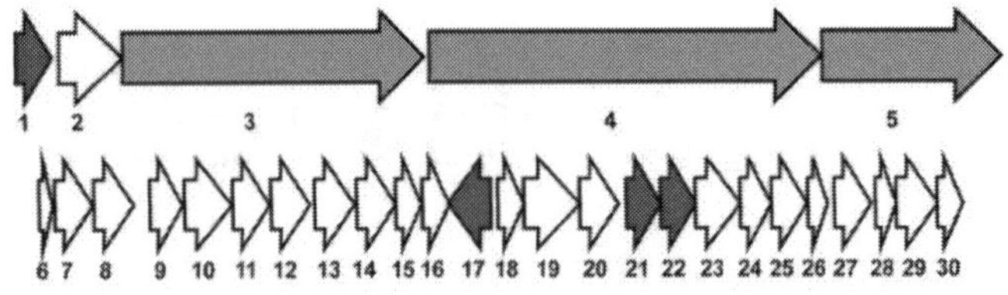

Van Wageningen *et al*, *Chem Biol* (1998) **5**, 155

Connecting molecules to genes

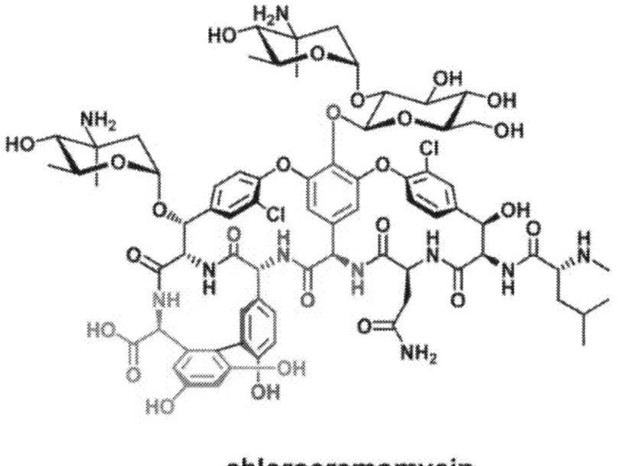

chloroeremomycin

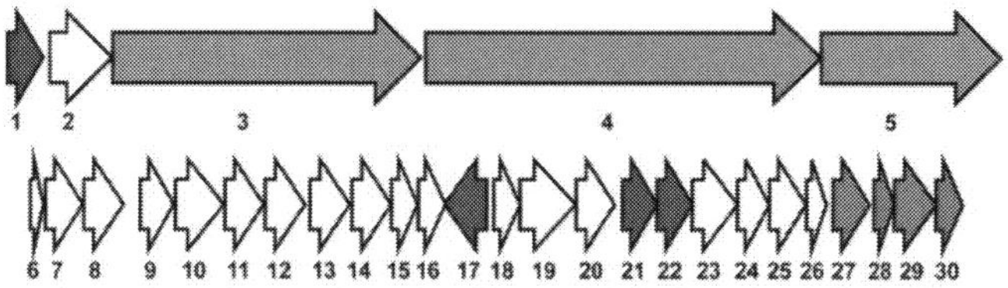

Van Wageningen *et al*, *Chem Biol* (1998) **5**, 155

Connecting molecules to genes

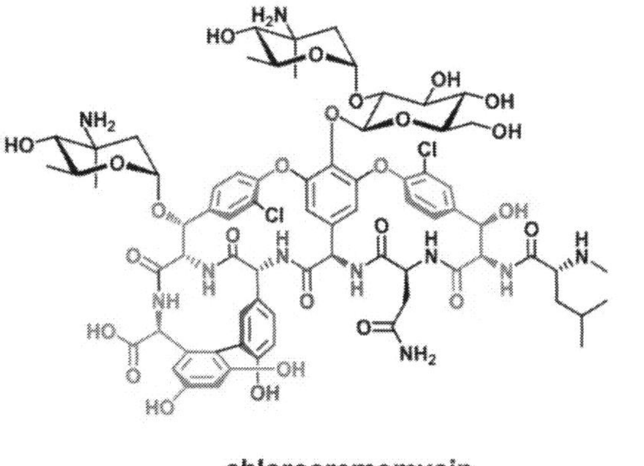

chloroeremomycin

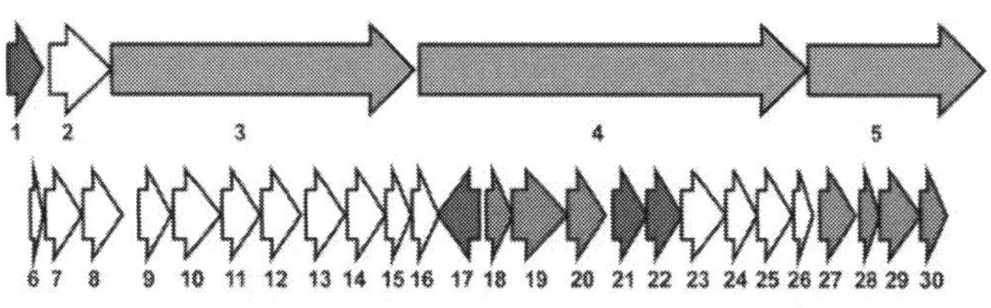

Van Wageningen *et al*, *Chem Biol* (1998) **5**, 155

Connecting molecules to genes

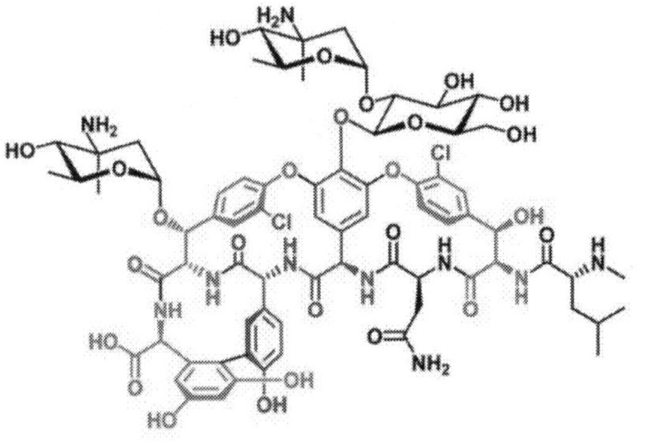

chloroeremomycin

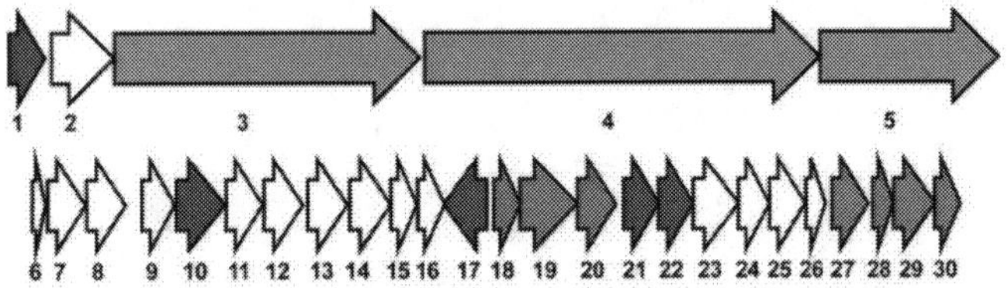

Van Wageningen *et al*, *Chem Biol* (1998) **5**, 155

Connecting molecules to genes

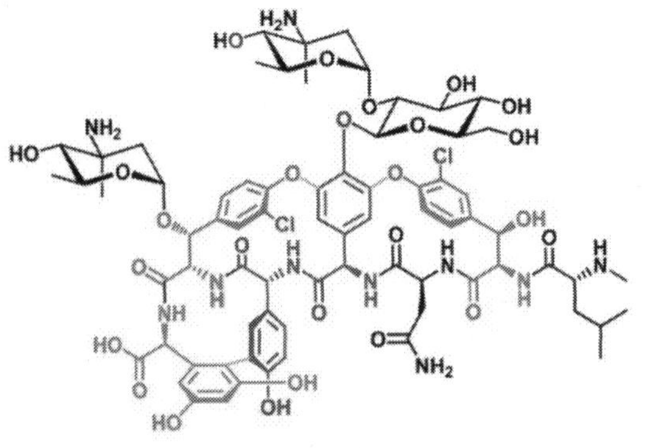

chloroeremomycin

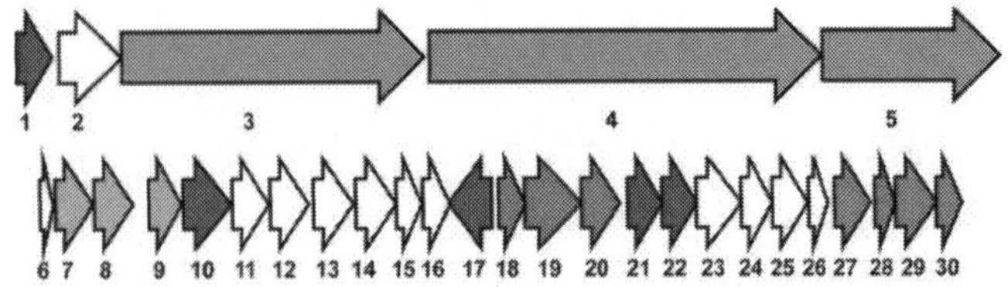

Van Wageningen *et al*, *Chem Biol* (1998) **5**, 155

Connecting molecules to genes

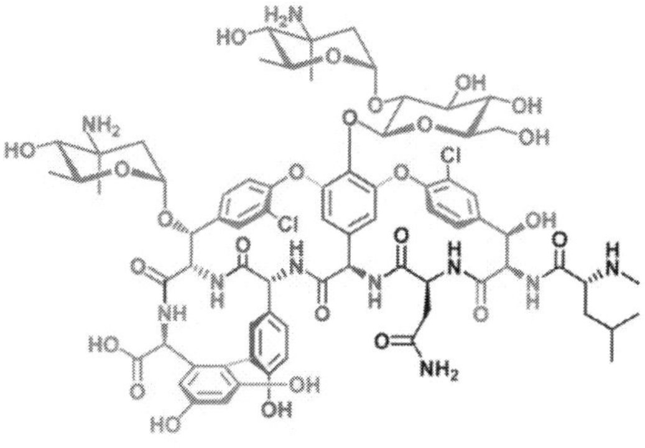

chloroeremomycin

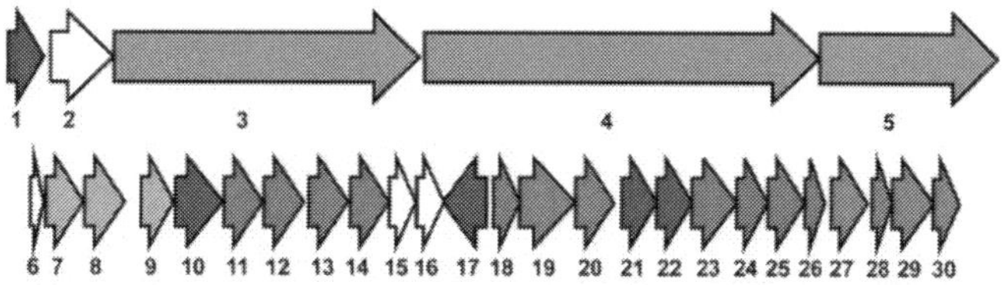

Van Wageningen *et al*, *Chem Biol* (1998) **5**, 155

Connecting molecules to genes

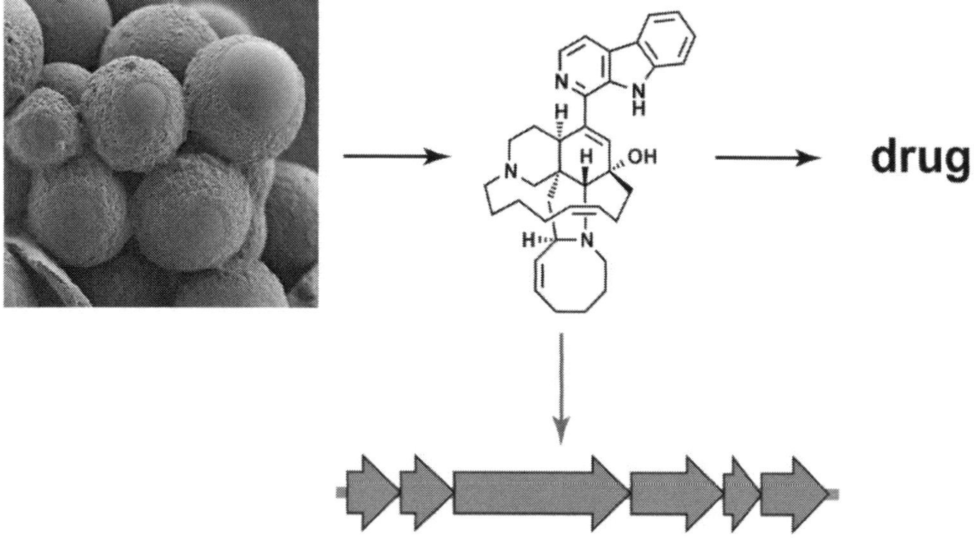

Connecting genes to molecules

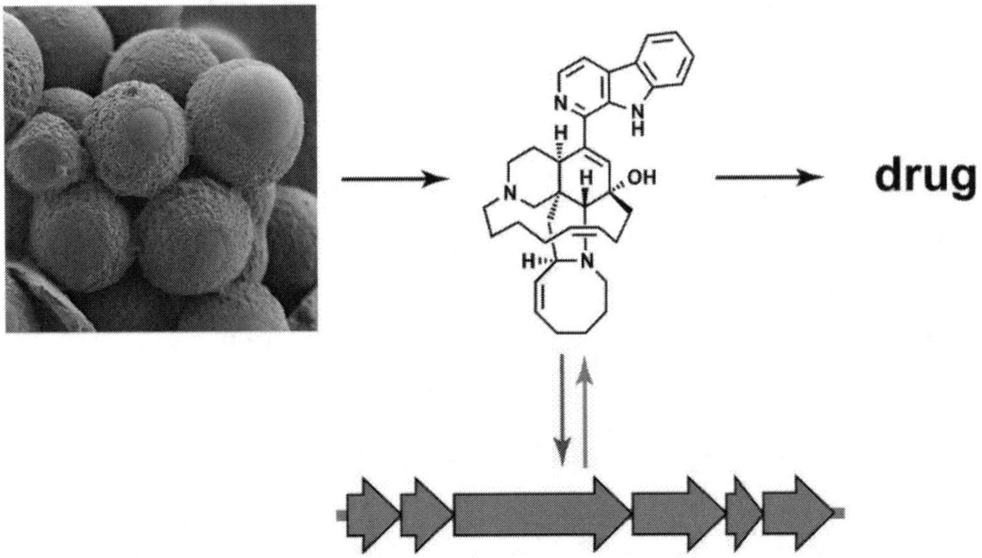

Connecting genes to molecules

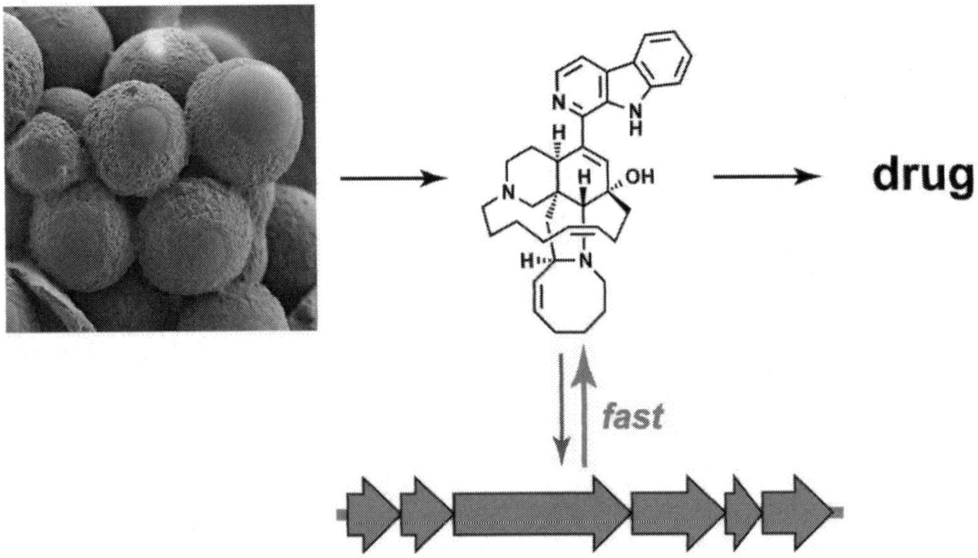

Connecting genes to molecules

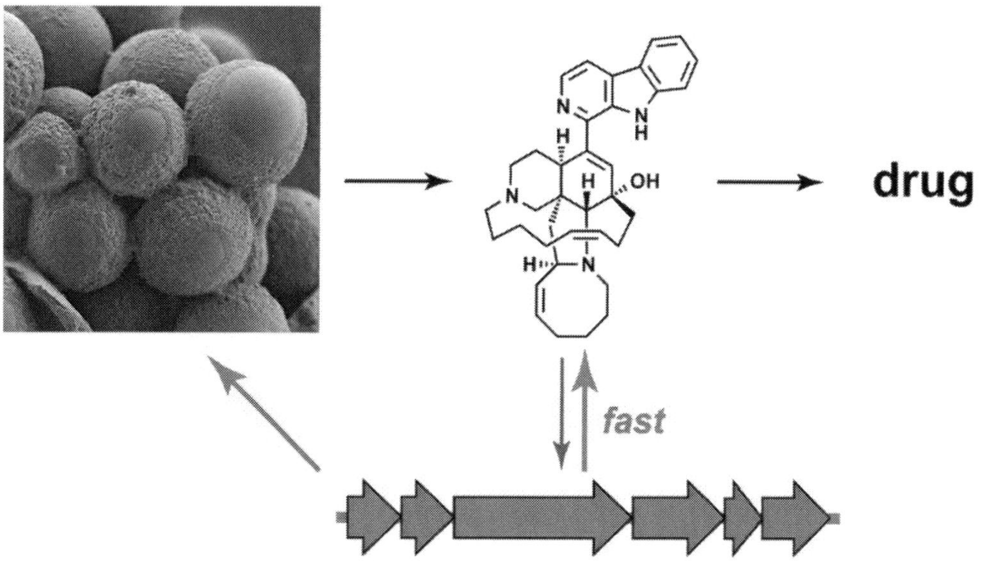

From genome to biosynthetic gene clusters

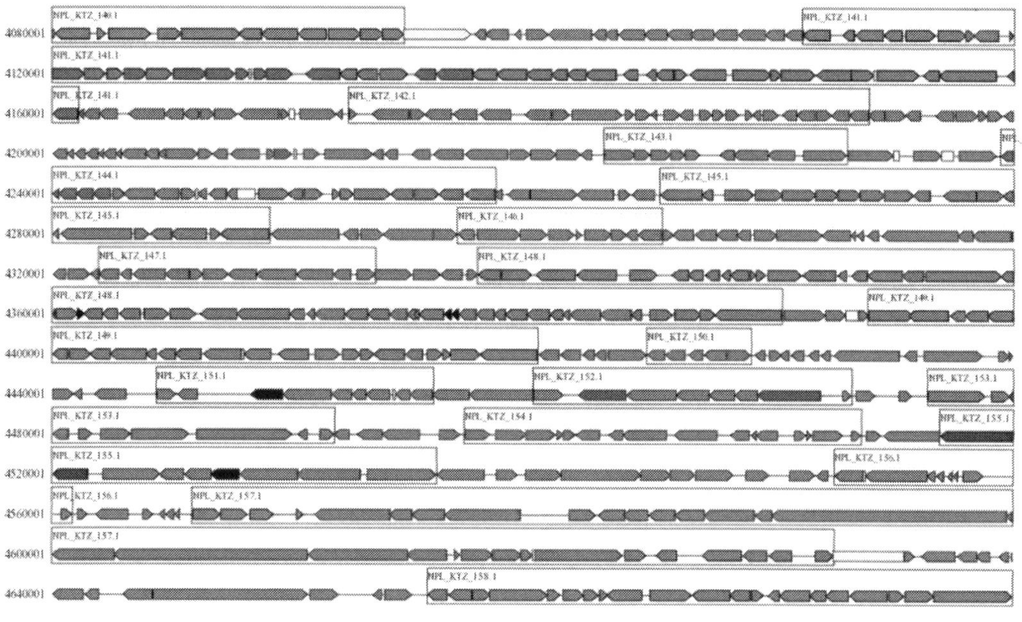

From gene cluster to molecule

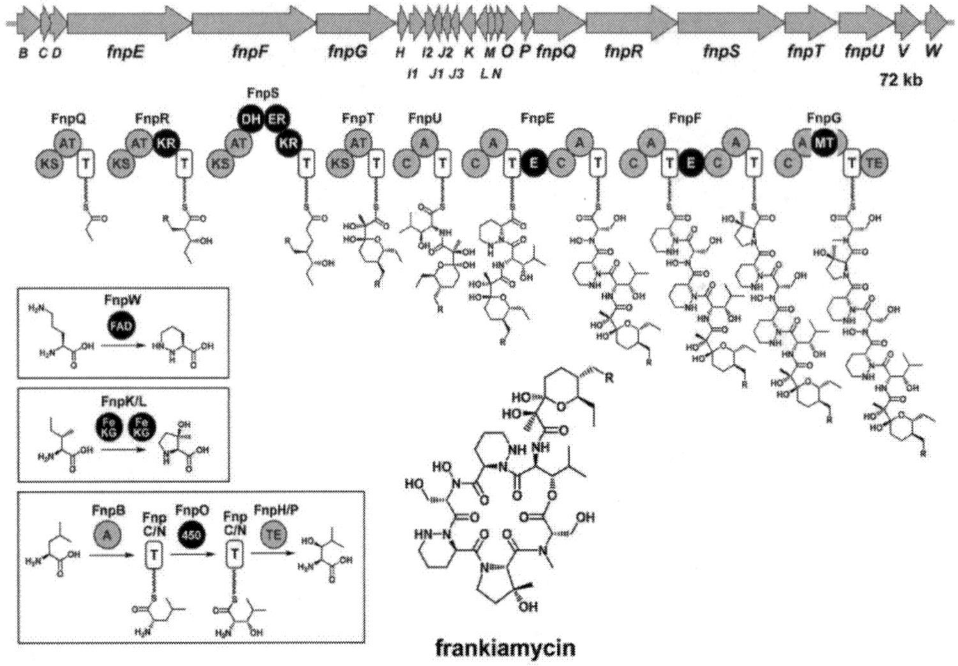

Genomics reveals cryptic natural products

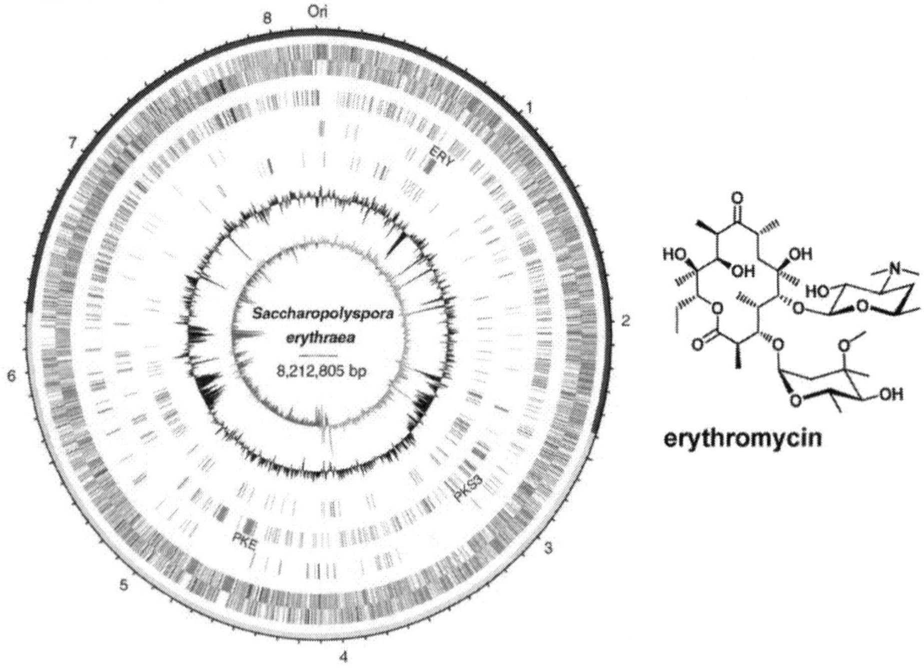

Oliynyk et al, *Nat Biotechnol* **25**, 447 (2007)

Genomics reveals cryptic natural products

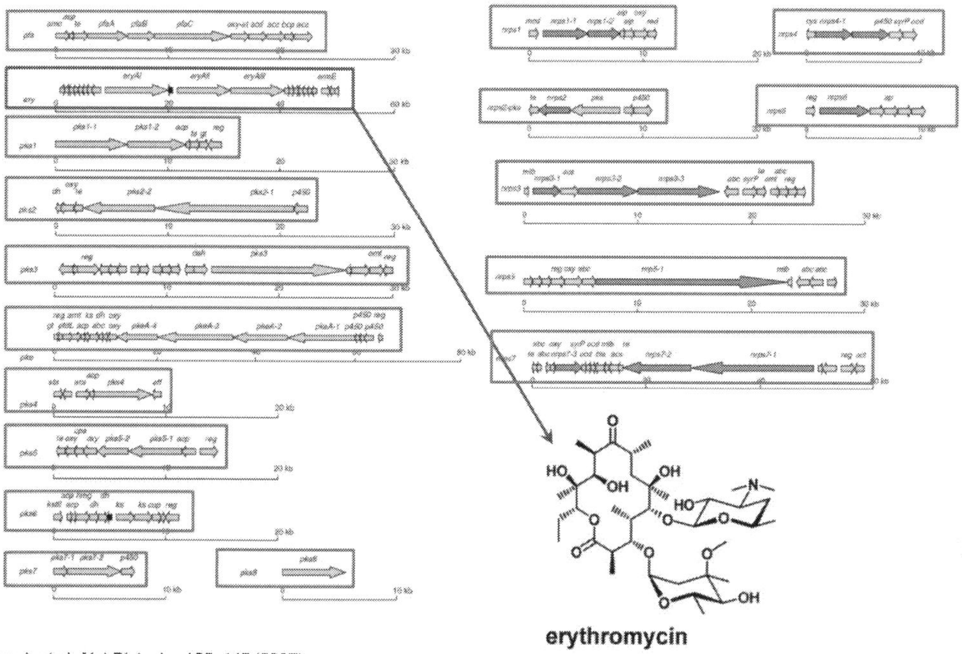

Oliynyk et al, *Nat Biotechnol* **25**, 447 (2007)

Automating gene cluster identification

Peter Cimermancic

Marnix Medema

Jan Claesen

Cimermancic, Medema, Claesen *et al*, *Cell* (2014) **158**, 412

Automating gene cluster identification

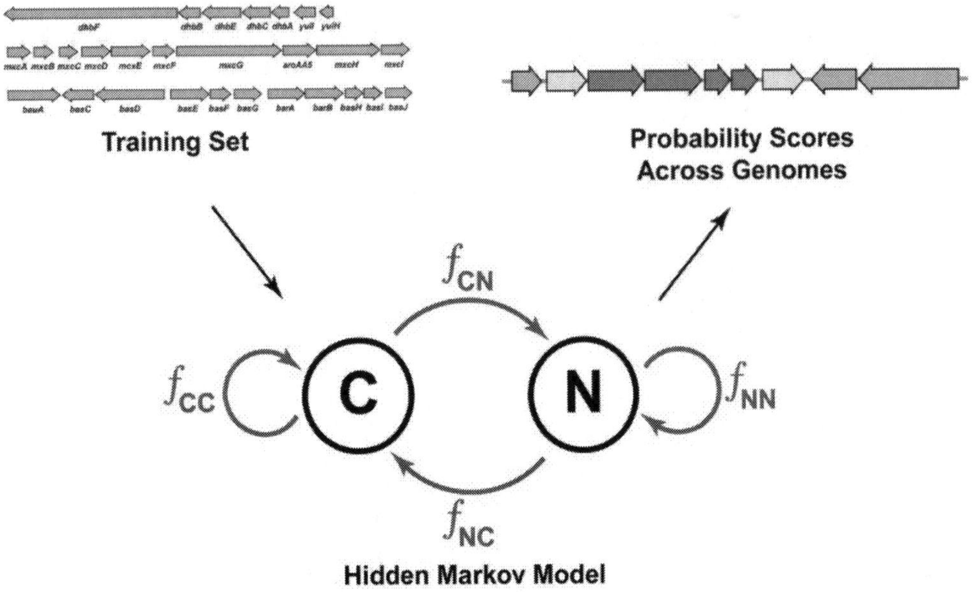

Cimermancic, Medema, Claesen *et al*, *Cell* (2014) **158**, 412

~30,000 putative biosynthetic gene clusters

Cimermancic, Medema, Claesen *et al*, *Cell* (2014) **158**, 412

A large family of unknown clusters

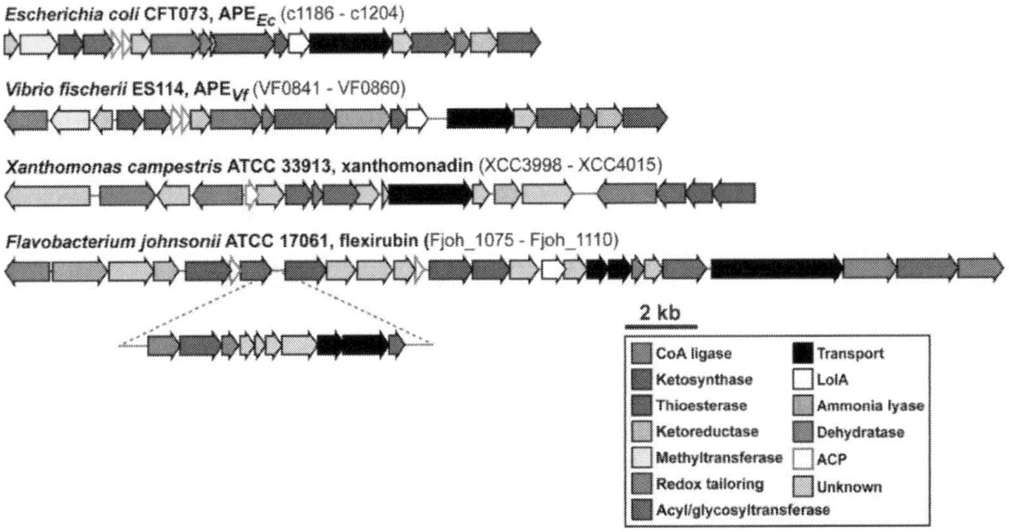

Cimermancic, Medema, Claesen *et al*, *Cell* (2014) **158**, 412

Unknown clusters encode a yellow pigment

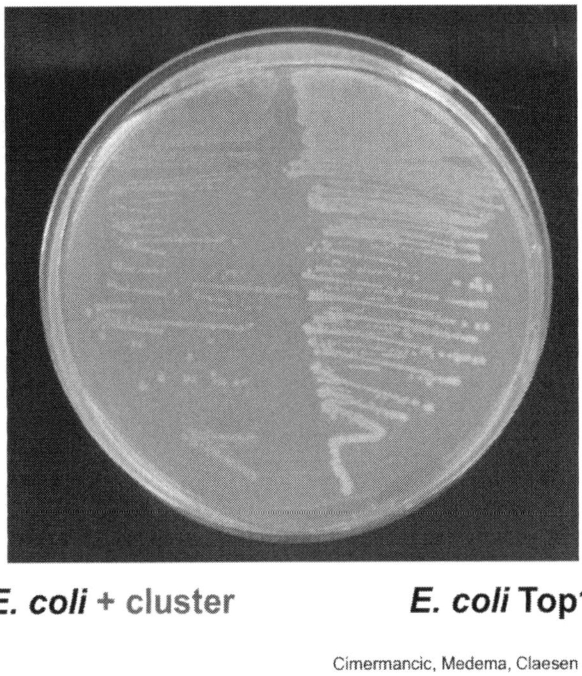

E. coli + cluster   *E. coli* Top10

Cimermancic, Medema, Claesen *et al*, *Cell* (2014) **158**, 412

Unknown clusters encode a yellow pigment

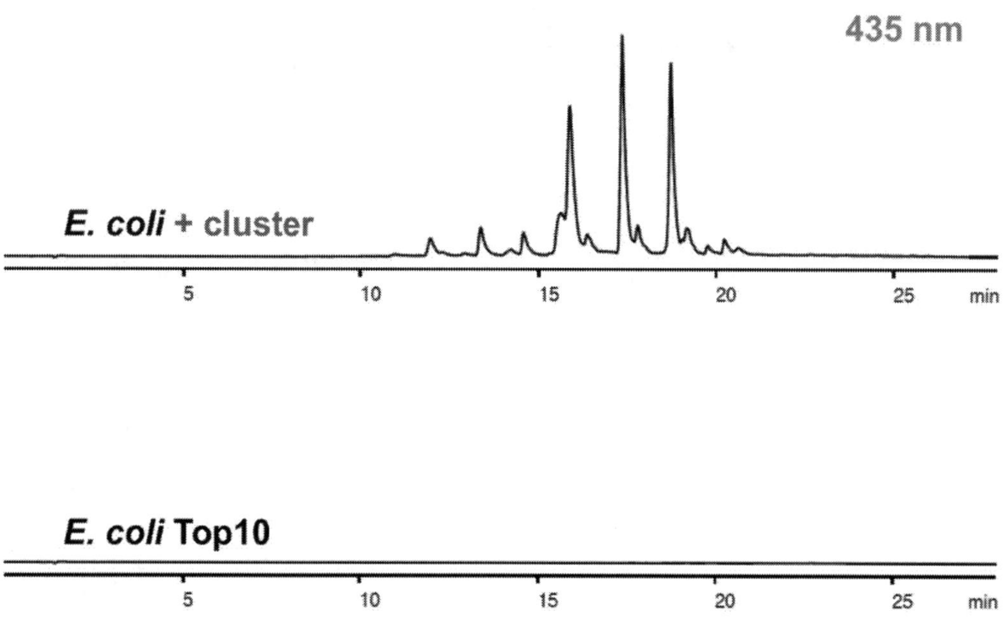

Cimermancic, Medema, Claesen *et al*, *Cell* (2014) **158**, 412

APEs: The largest family in the database

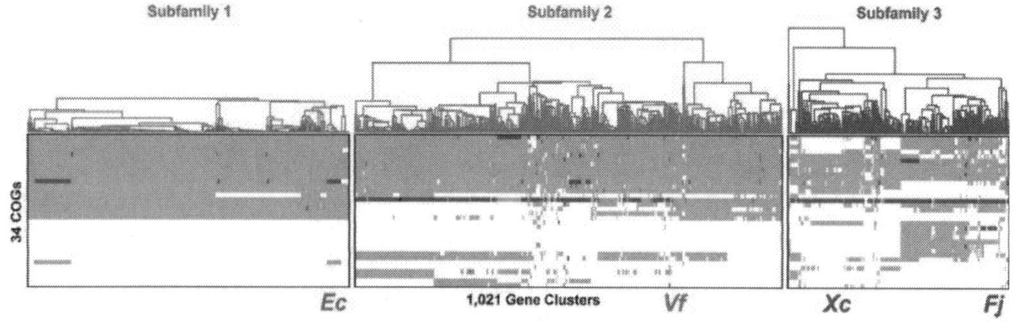

Cimermancic, Medema, Claesen *et al*, *Cell* (2014) **158**, 412

APEs: The largest family in the database

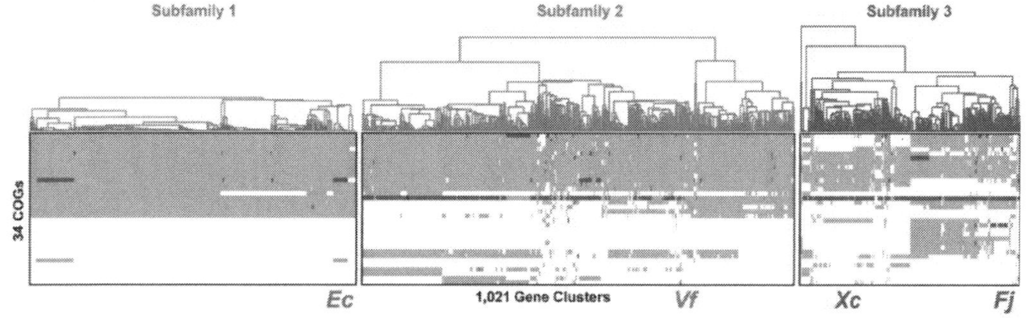

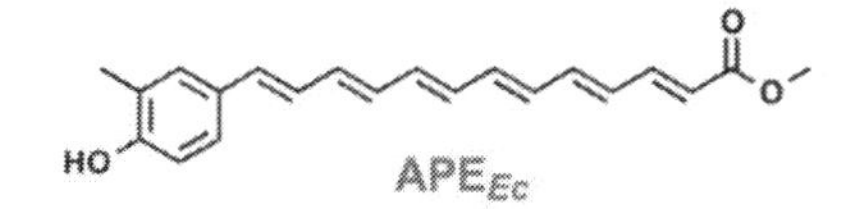

APE_{Ec}

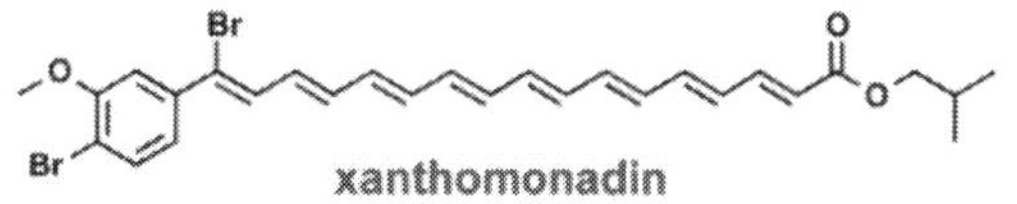

xanthomonadin

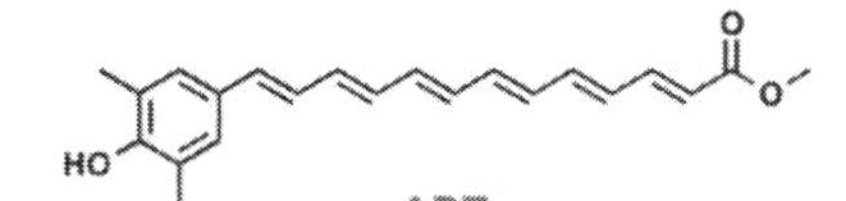

APE_{Vf}

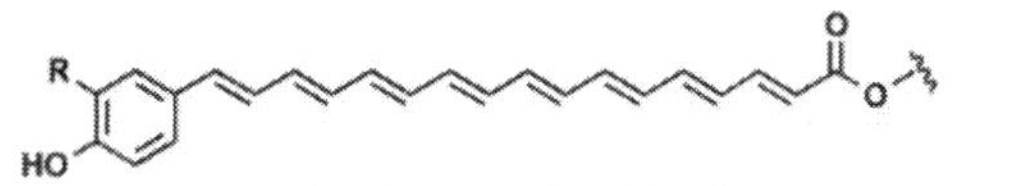

flexirubin (R = H, CH_3, Cl)

Cimermancic, Medema, Claesen *et al*, *Cell* (2014) **158**, 412

APEs: The largest family in the database

APEs (Gram-negative)

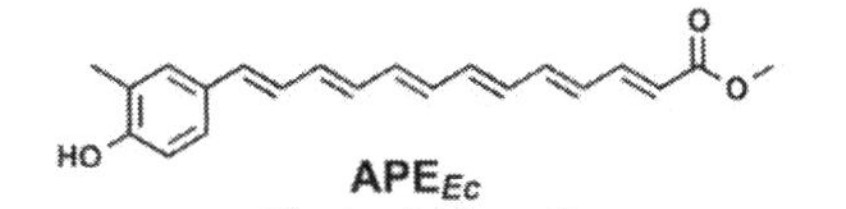

APE_{Ec}
Escherichia coli

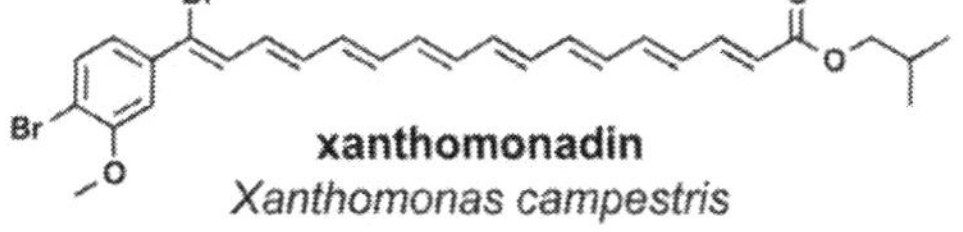

xanthomonadin
Xanthomonas campestris

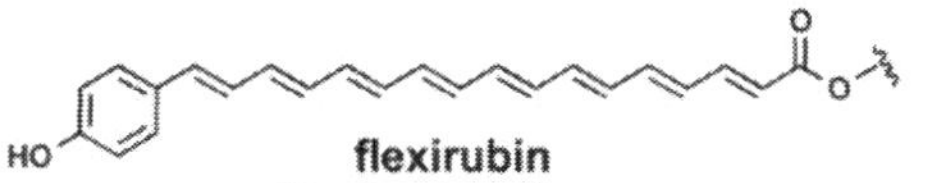

flexirubin
Flexibacter eligens

>1000 clusters

Carotenoids (Gram-positive)

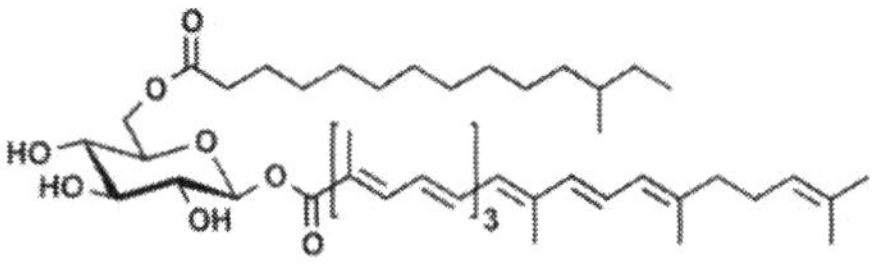

staphyloxanthin
Staphylococcus aureus

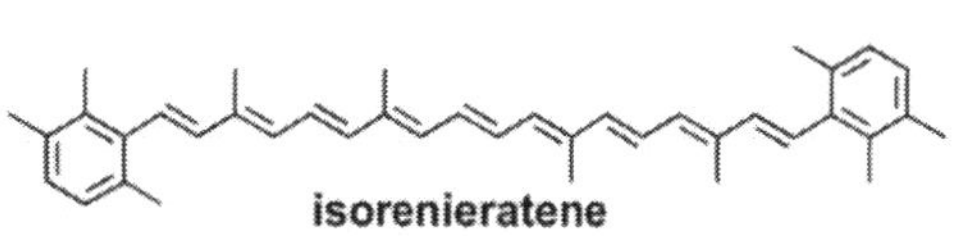

isorenieratene
Mycobacterium avium

870 clusters

Cimermancic, Medema, Claesen *et al*, *Cell* (2014) **158**, 412

Gene clusters in the human microbiome

Mohamed Donia

Donia *et al*, *Cell* (2014), **158**, 1402

Identifying gene clusters in the microbiome

2,430 human-associated bacterial genomes

↓ ***ClusterFinder***

44,000 predicted BGCs

↓ ***antiSMASH***

14,000 BGCs in known classes

↓ ***read recruitment from 752 metagenomes***

3,118 BGCs present in HMP metagenomic data

Donia *et al*, *Cell* (2014), **158**, 1402

Microbiome gene clusters: Facts and figures

2,430 human-associated bacterial genomes

↓ *ClusterFinder*

44,000 predicted BGCs

↓ *antiSMASH*

14,000 BGCs in known classes

↓ *read recruitment from 752 metagenomes*

3,118 BGCs present in HMP metagenomic data

Average gut: 599 gene clusters
Average oral cavity: 1,061 gene clusters

Saccharides predominate, but other classes well represented

Well-studied gene clusters are rare

Donia *et al*, *Cell* (2014), **158**, 1402

Thiopeptide antibiotics

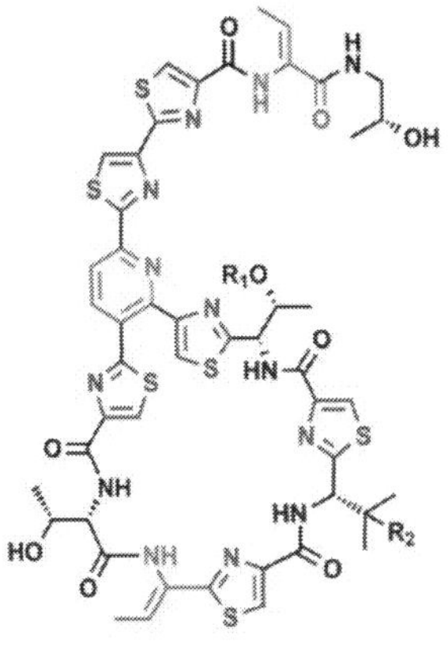

thiocillin
Bacillus cereus

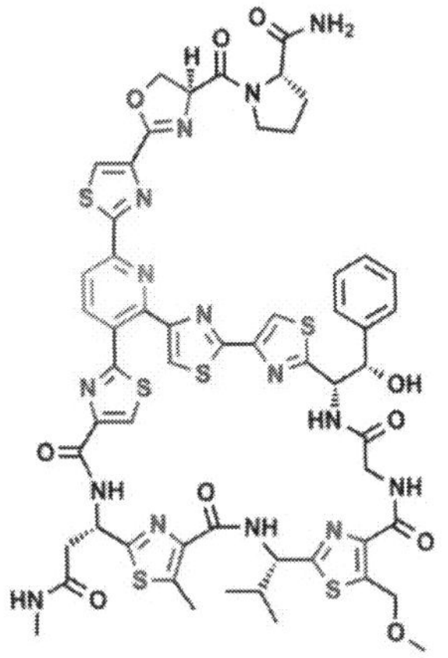

GE2270
Planobispora rosea

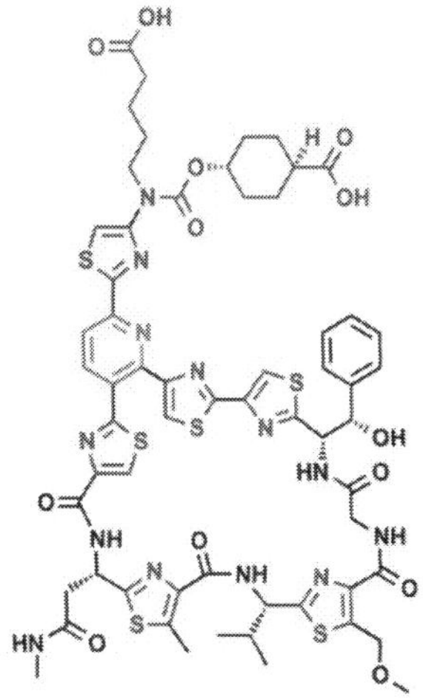

LFF571
Novartis, Phase II

Thiopeptides bind the 50S subunit

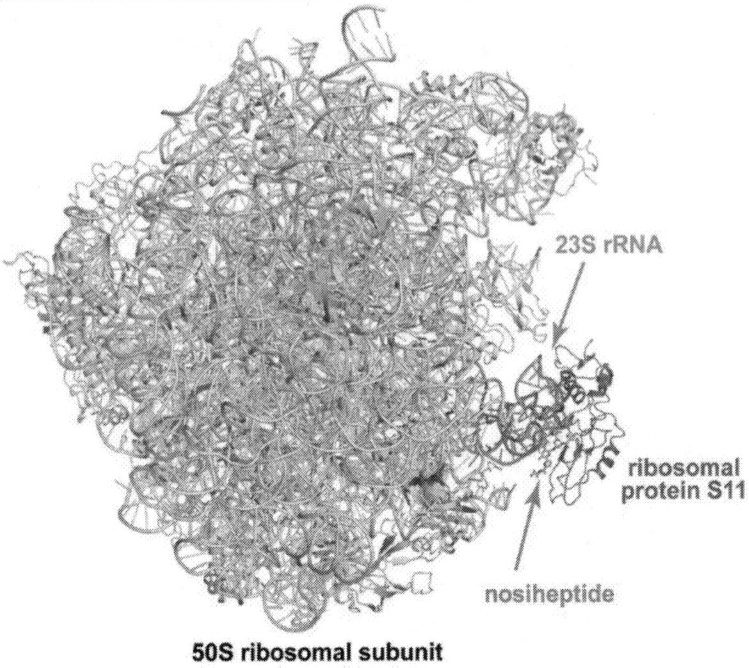

Harms *et al*, *Mol Cell* (2008) **30**, 26

Thiopeptides bind the 50S subunit

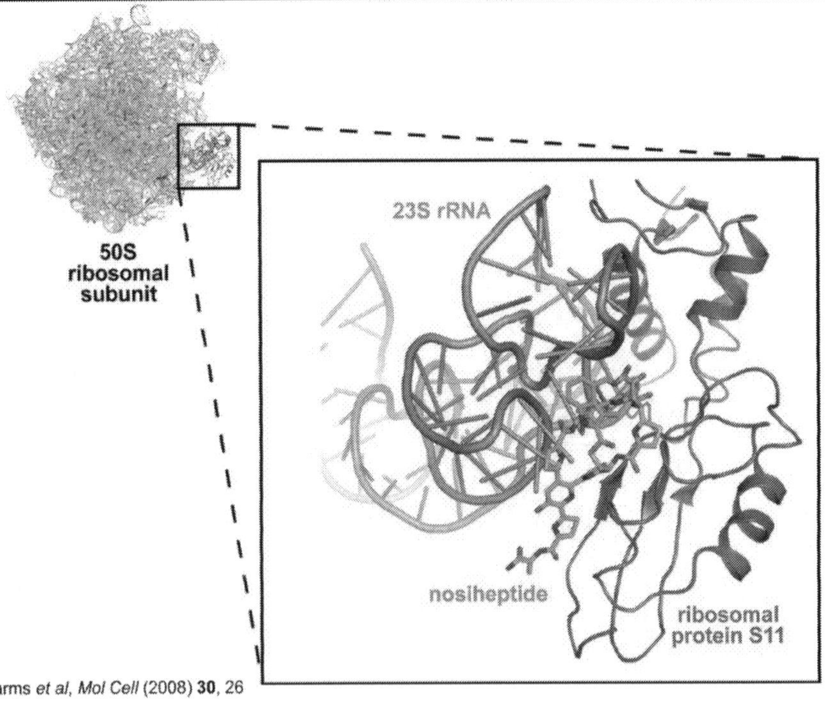

Harms *et al*, *Mol Cell* (2008) **30**, 26

Gene clusters for thiopeptide antibiotics

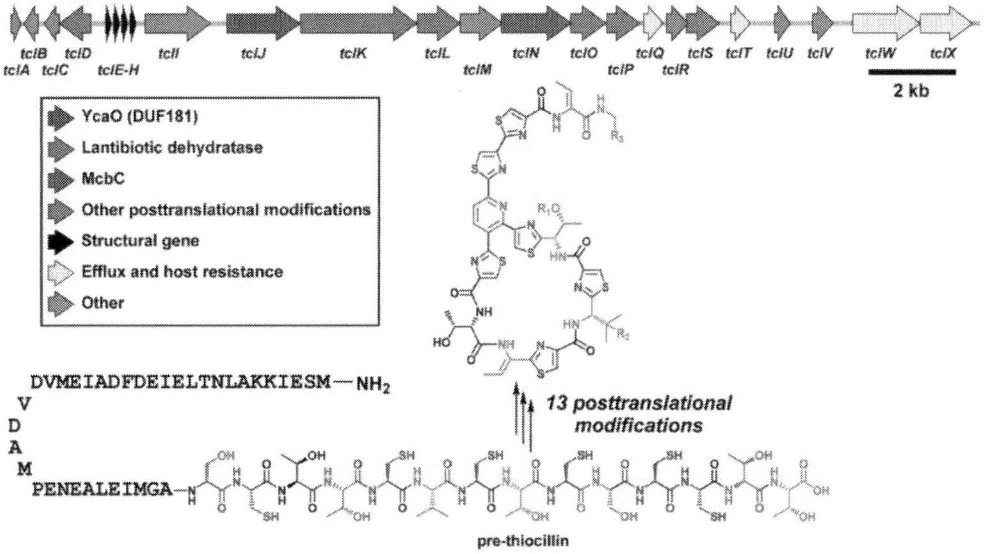

Wieland Brown *et al*, *Proc Natl Acad Sci* (2009) **106**, 2549

Gene clusters for thiopeptide antibiotics

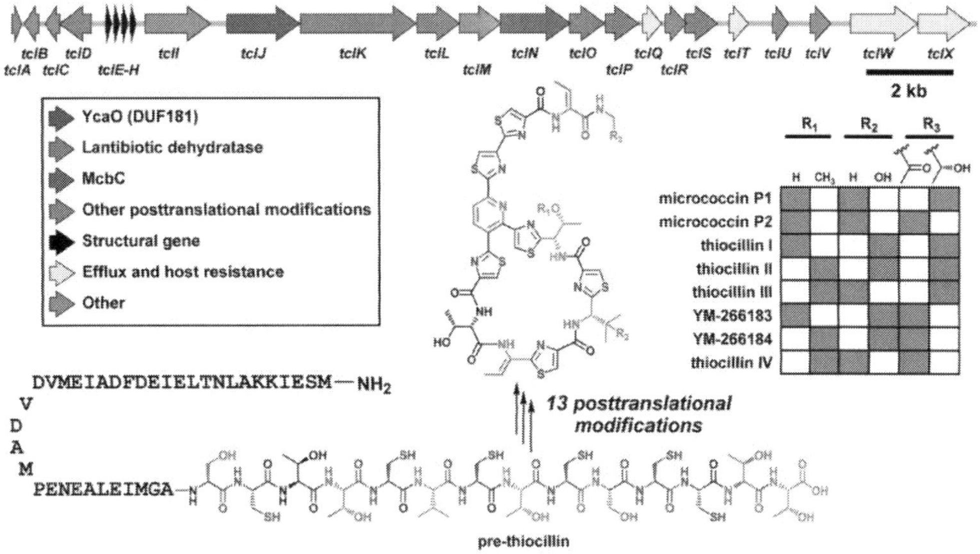

Wieland Brown *et al*, *Proc Natl Acad Sci* (2009) **106**, 2549

Thiopeptides from the human microbiota

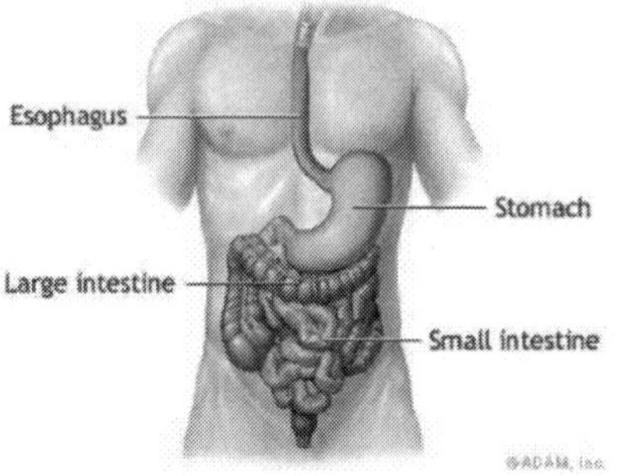

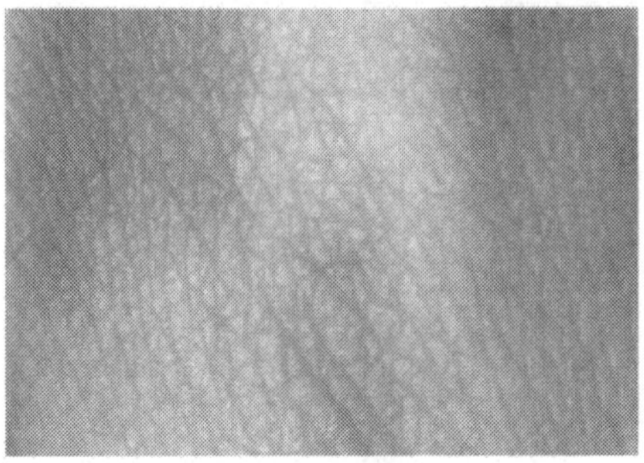

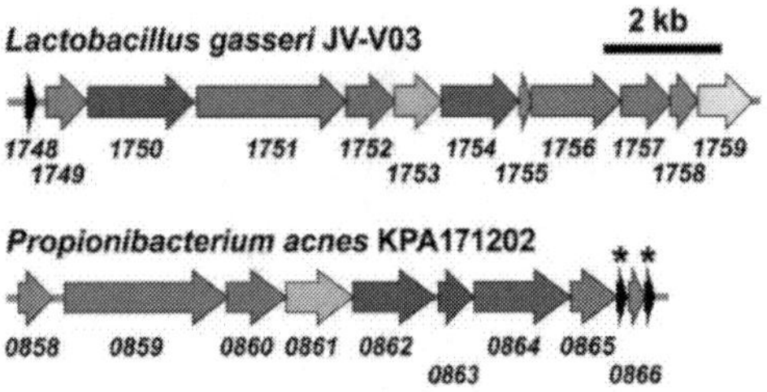

Donia *et al*, *Cell* (2014), **158**, 1402

Thiopeptides from the human microbiota

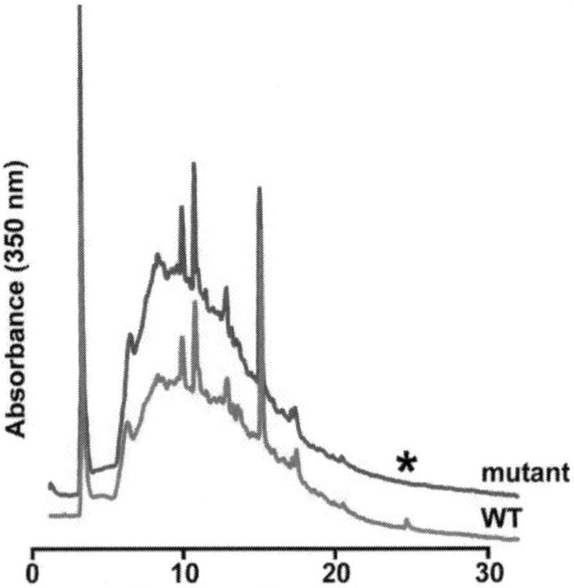

Donia *et al*, *Cell* (2014), **158**, 1402

Thiopeptides from the human microbiota

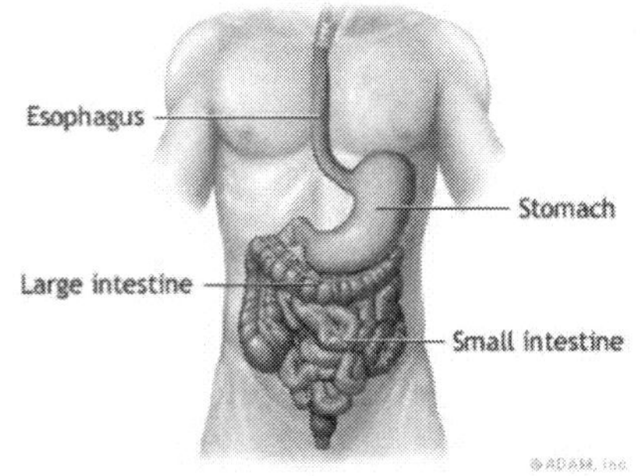

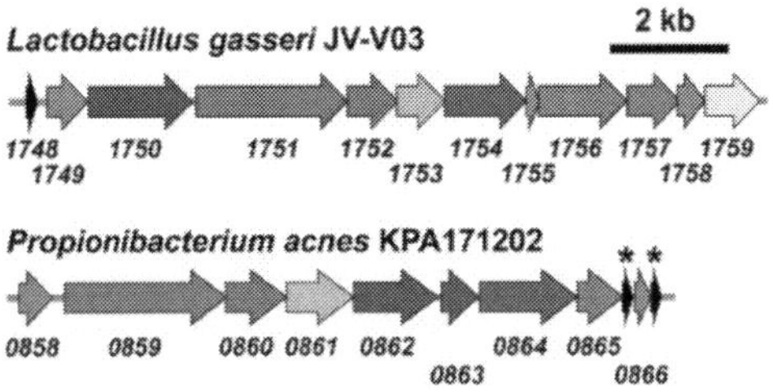

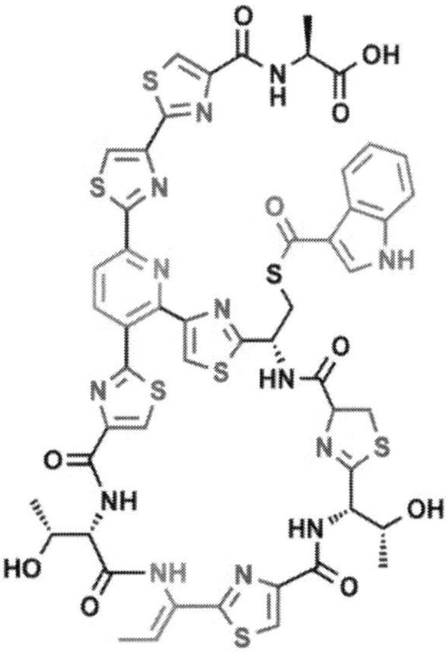

lactocillin
Lactobacillus gasseri

Donia *et al*, *Cell* (2014), **158**, 1402

Thiopeptides from the human microbiota

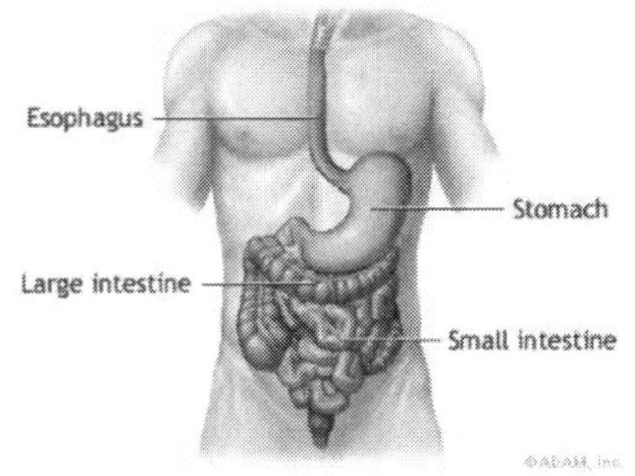

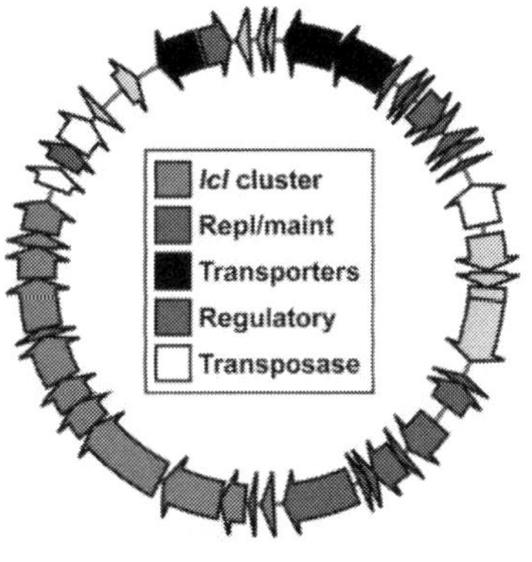

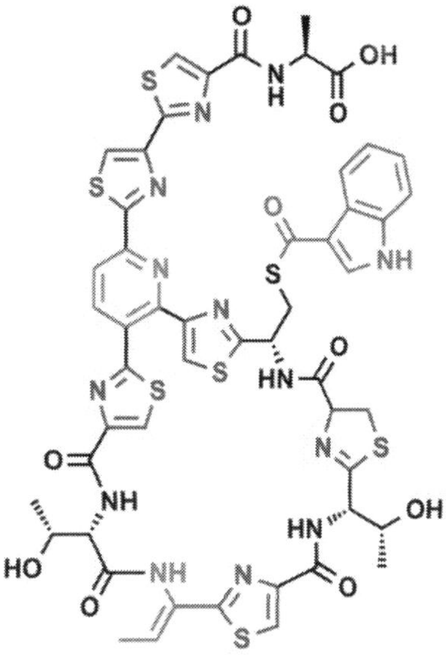

lactocillin
Lactobacillus gasseri

Donia *et al*, *Cell* (2014), **158**, 1402

Thiopeptides from the human microbiota

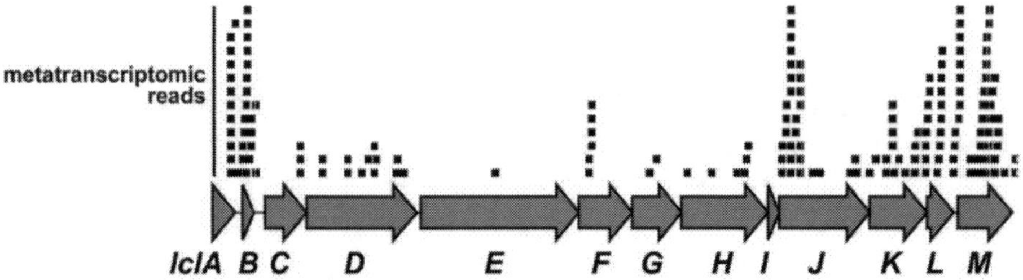

Donia *et al*, *Cell* (2014), **158**, 1402

Other thiopeptide clusters in the microbiota

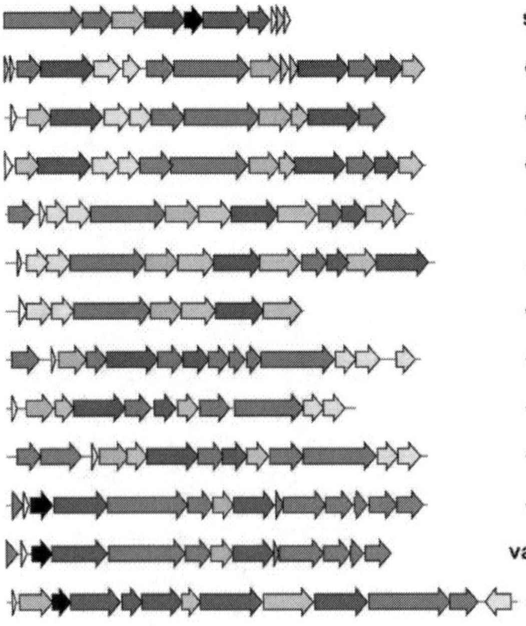

	Body Site	Prevalence in HMP samples
pac	skin	30%
sdo	oral	-
ot4	oral*	25%
ot5	oral*	8%
ljo	gut	-
gt1	gut*	9%
ot7	oral*	10%
sso	oral	-
ot3	oral*	1%
ot1	oral*	34%
ot2	oral*	1%
lcl	vaginal	-
ot6	oral*	22%

*Discovered first in metagenomic data

Donia *et al*, *Cell* (2014), **158**, 1402

A widely distributed gut gene cluster

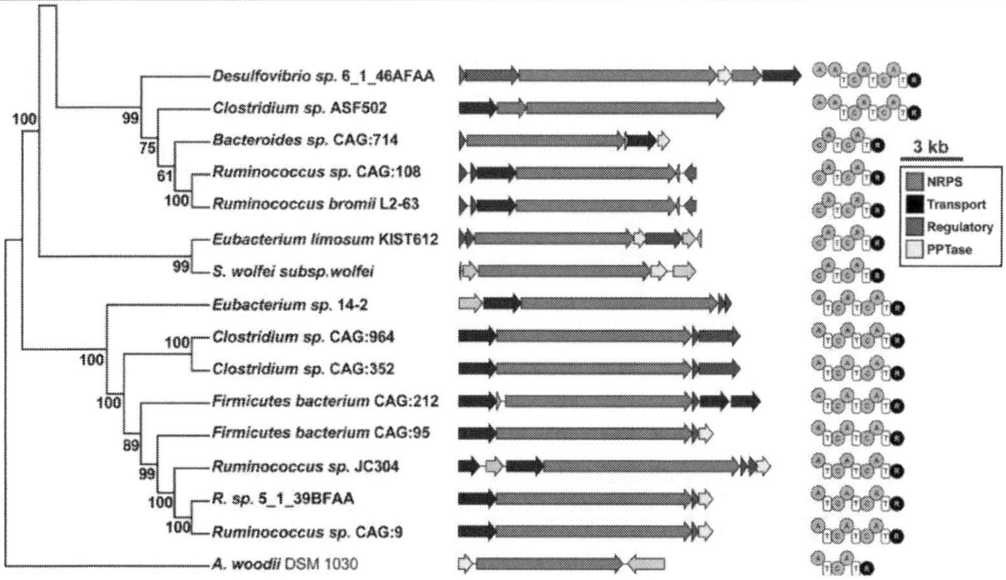

Ruminococcus **cluster present in 33% of gut WGS samples from HMP (n = 86)**

Donia *et al*, *Cell* (2014), **158**, 1402

Bile acid metabolism by gut bacteria

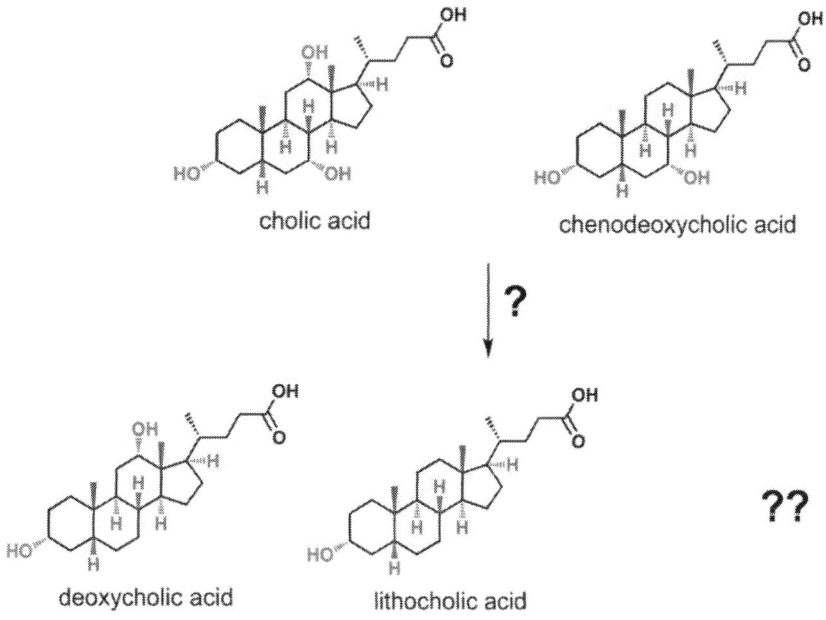

Yug Varma, Sloan Devlin, Jun Huh, Dan Littman

Bile acid metabolism by gut bacteria

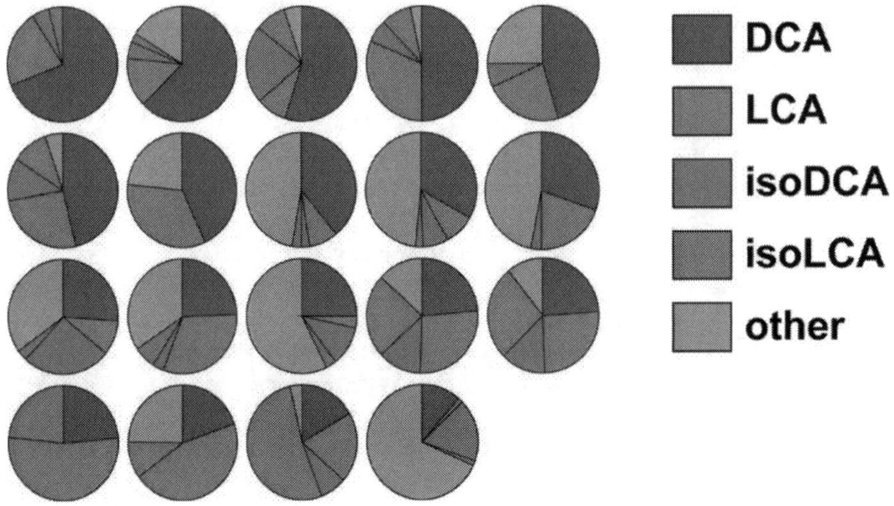

Yug Varma, Sloan Devlin, Jun Huh, Dan Littman

Bile acid metabolism by gut bacteria

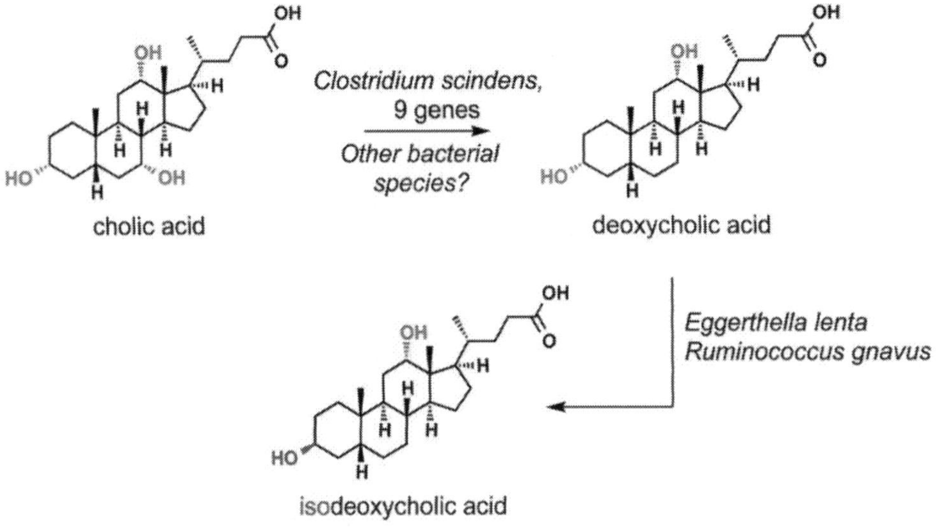

Yug Varma, Sloan Devlin, Jun Huh, Dan Littman

Microbiota: Largest endocrine organ?

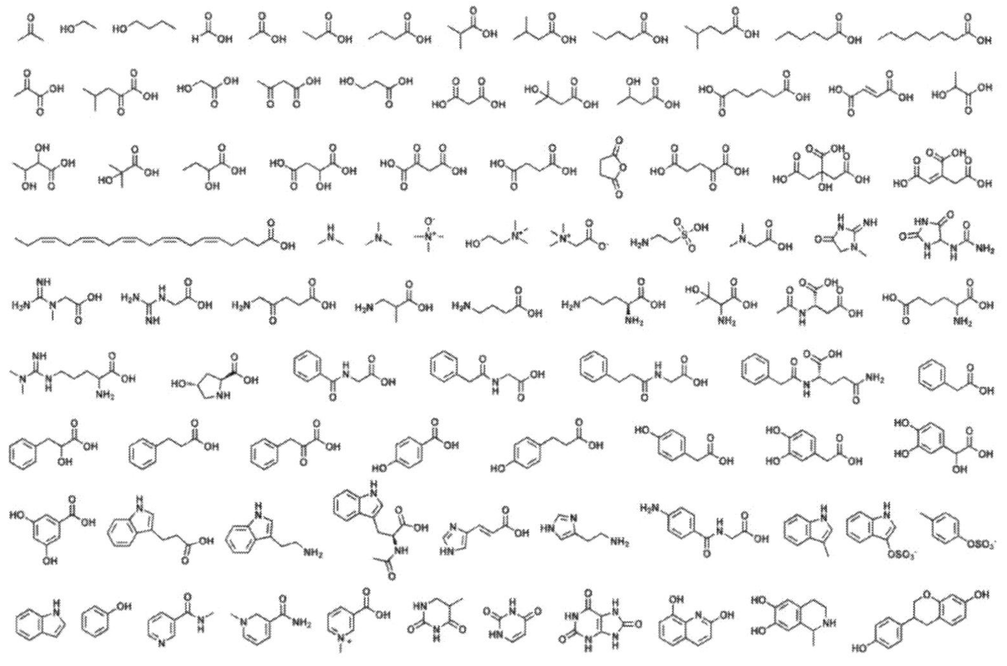

Acknowledgements

Fischbach Group
Jan Claesen
Sloan Devlin
Mohamed Donia
Amy Jacobson
Laurens Kraal
Aedan Liu
Colleen O'Loughlin
Tiffany Scharschmidt
Mao Taketani
Yug Varma
Brianna Williams
Peter Cimermancic
Marnix Medema (Max Planck Institute)
Laura Brown
Michael Zimmermann

Funding
Burroughs Wellcome Fund
David and Lucille Packard Foundation
DARPA
Glenn Foundation
NIH (New Innovator Award, NIAID, NIGMS, NIDDK)
Pfizer
W.M. Keck Foundation
UCSF Center For Systems and Synthetic Biology

Collaborators
Greg Barton
Bruce Birren
Jeff Bluestone
Rainer Breitling
Jon Clardy
Laurie Comstock
Joe DeRisi
Pieter Dorrestein
Larry Forney
Kenya Honda
Roberto Kolter
Katherine Lemon
Dan Littman
Averil Ma
Sarkis Mazmanian
Jacques Ravel
David Relman
Justin Sonnenburg
Eriko Takano
Peter Turnbaugh
Chris Voigt

Synthetic Biology Engineering of Biofilms As Nanomaterials Factories

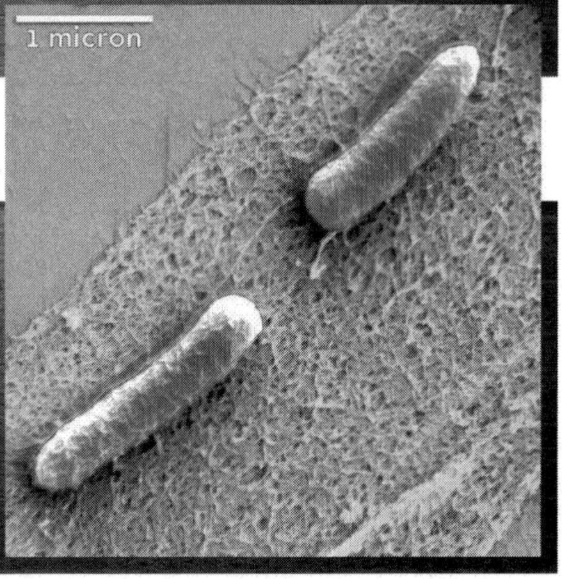

SEED Synthetic Biology, Engineering, Evolution & Design

Peter Q. Nguyen
Joshi Lab

HARVARD School of Engineering and Applied Sciences
WYSS INSTITUTE

---

1. Synthetic Biology for Genetically Encoded Nanomaterials.

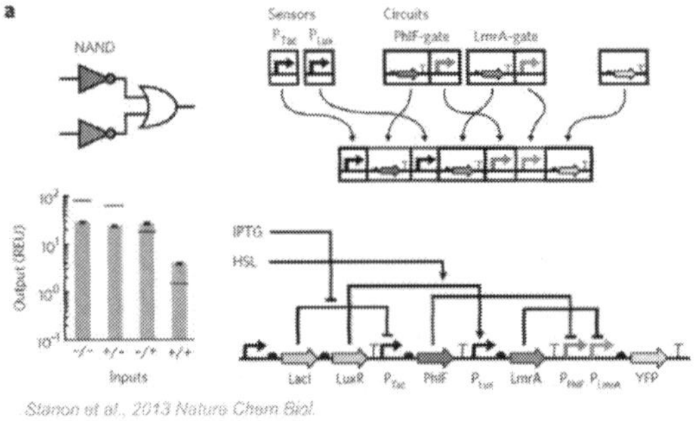

Stanton et al., 2013 Nature Chem Biol.

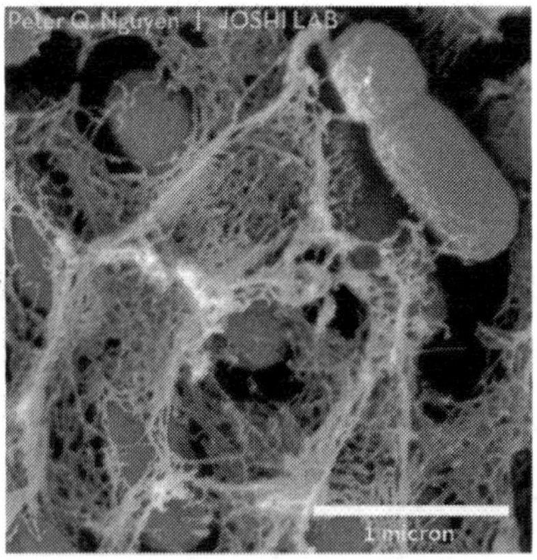

Nanoscale Engineering

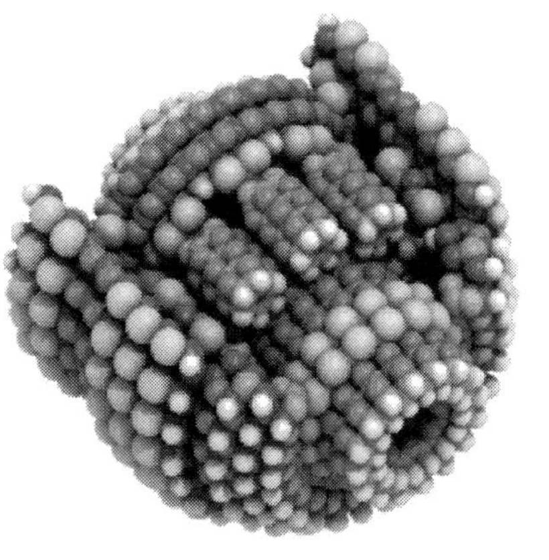

MarkIII(k) Planetary Gear, Nanorex, Inc.

WYSS INSTITUTE HARVARD School of Engineering and Applied Sciences

3

Nanoscale Engineering

Nanomanufacturing 'machines' for atomic assembly…

WYSS INSTITUTE HARVARD School of Engineering and Applied Sciences

4

(Bio)Nanoscale Engineering

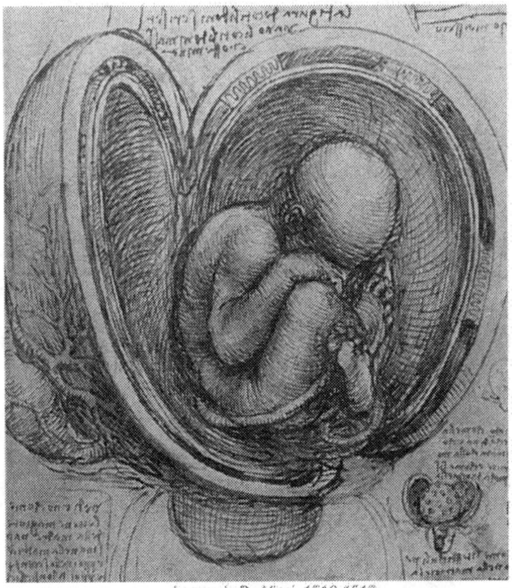

Leonardo Da Vinci, 1510-1512

...is old news to biology.

WYSS INSTITUTE HARVARD School of Engineering and Applied Sciences

5

SynBioNano

INPUT

OUTPUT

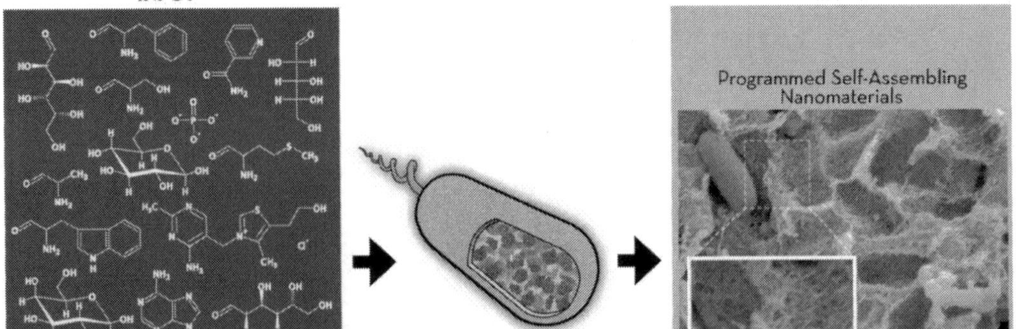

WYSS INSTITUTE HARVARD School of Engineering and Applied Sciences

6

Bioinspiration for Large-Scale Materials

Biofilm Macro-Architecture

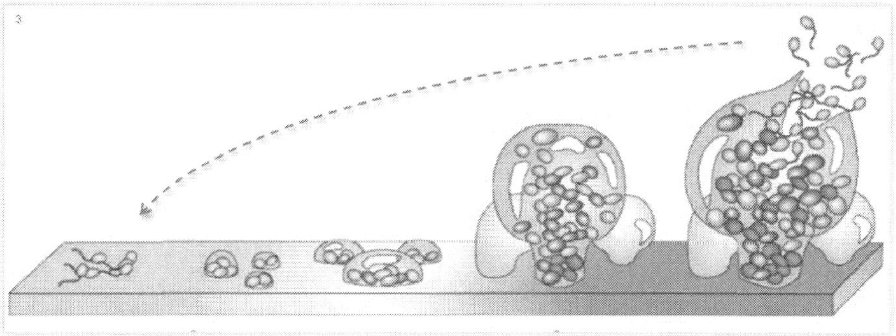

1,2. Nat. Rev. Microb. 2004. doi:10.1038/nrmicro821; 3. Nat. Rev. Drug Discov. 2003 doi:10.1038/nrmicro2980

8

Domesticating Nature

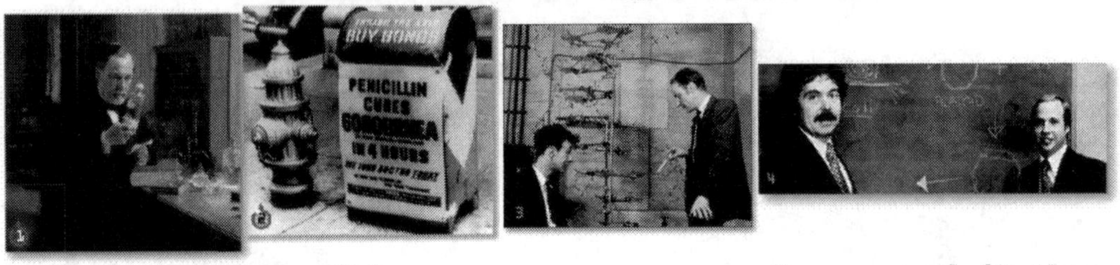

cteria are dangerous!!!

We understand how they work…

…We can exploit them.

Biofilms are bad!!!

We are starting to understand how they work…

…*Can* we exploit them?

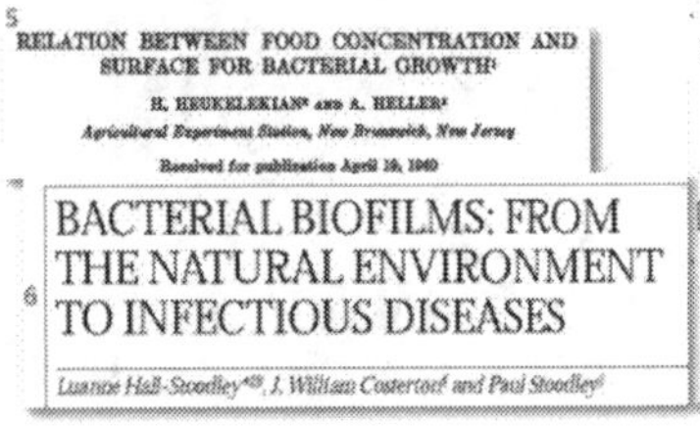
5 RELATION BETWEEN FOOD CONCENTRATION AND SURFACE FOR BACTERIAL GROWTH
H. HEUKELEKIAN AND A. HELLER
Agricultural Experiment Station, New Brunswick, New Jersey
Received for publication April 19, 1940

6 BACTERIAL BIOFILMS: FROM THE NATURAL ENVIRONMENT TO INFECTIOUS DISEASES
Luanne Hall-Stoodley, J. William Costerton and Paul Stoodley

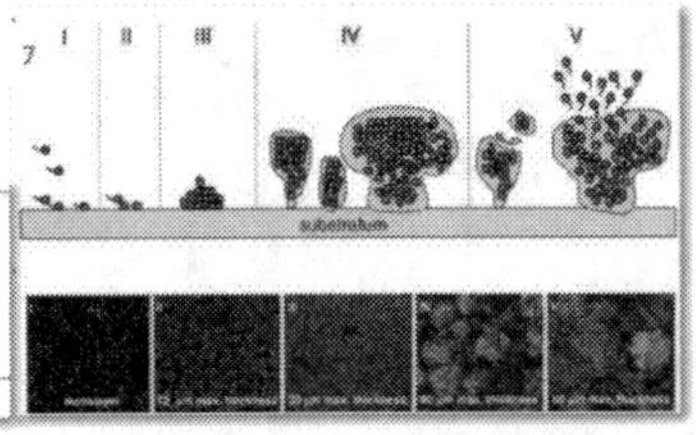

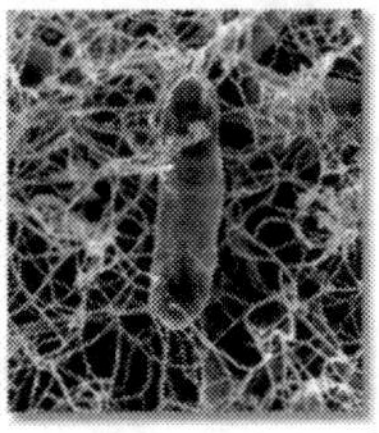

WYSS INSTITUTE HARVARD School of Engineering and Applied Sciences

1,2. Wikimedia Commons, 3. Time Magazine, Inc., 4. Genentech, 5. 1940 *J. Bacteriol.*, 6. 2004 *Nat. Rev. Microb.*, 7. 2008 Clin Rev Allergy Immunol. 8. Lower Lab, OSU.

9

2. Biofilms as Engineerable Bacterial 'Tissues'

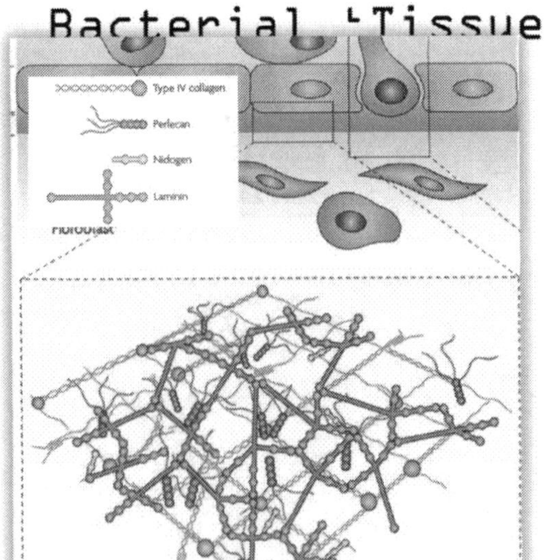

Sorokin. 2010 Nature Rev. Immunol.

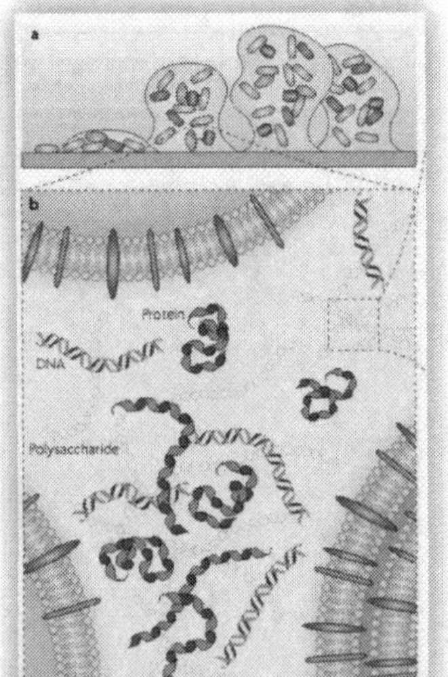

Flemming and Wingender. 2010 Nature Rev. Microb.

WYSS INSTITUTE HARVARD School of Engineering and Applied Sciences

11

Biofilm Nano-Architecture

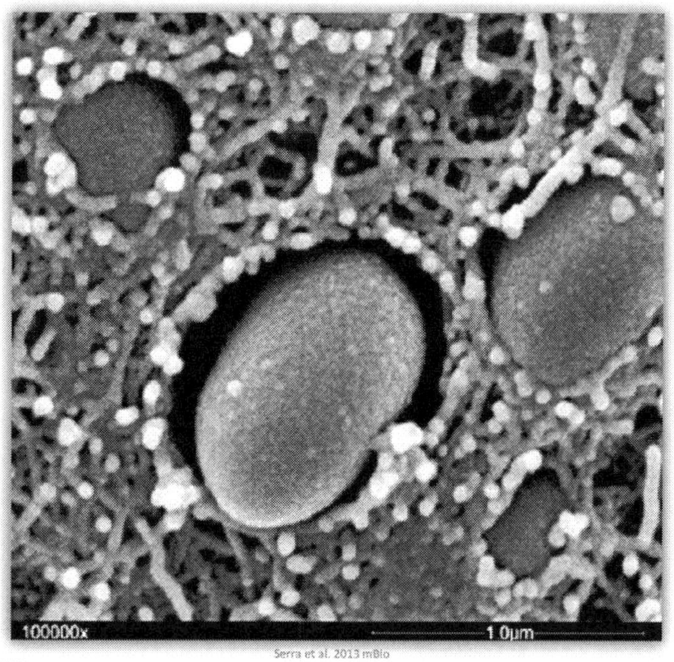

Serra et al. 2013 mBio

WYSS INSTITUTE HARVARD School of Engineering and Applied Sciences

12

Bacterial Extracellular Matrix Materials

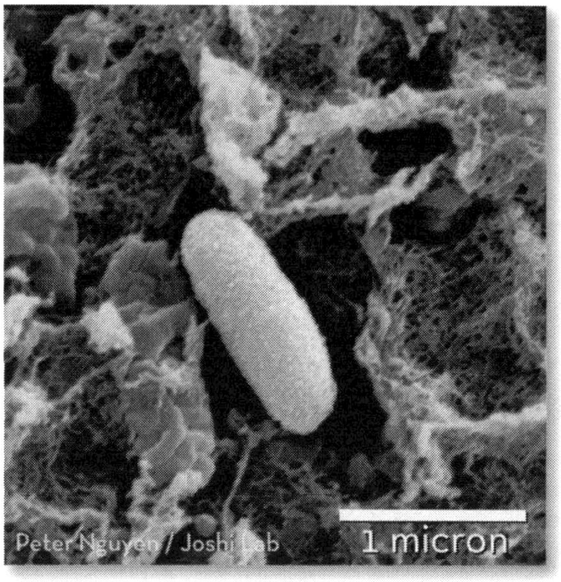

- Self-generating and self-renewing biomaterial.
- Vast nanoscale surface area.
- Programmable <u>modular</u> nanoscale scaffold will allow for rapid design and development.
- Multifunctionalization of scaffolds.
- Dynamic control of the biofilm by synthetic biology circuits.

WYSS INSTITUTE HARVARD School of Engineering and Applied Sciences

13

Curli Nanofibers

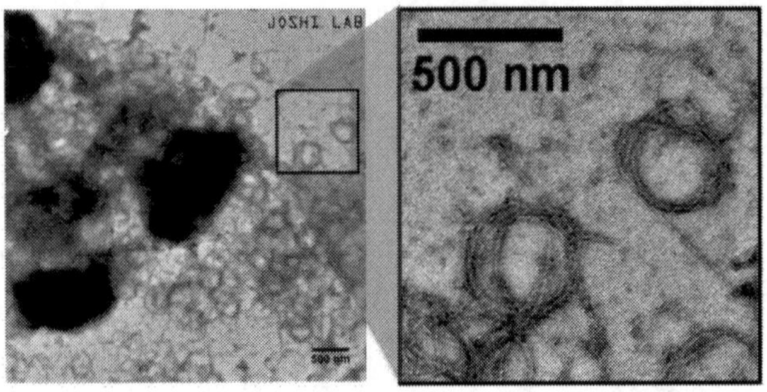

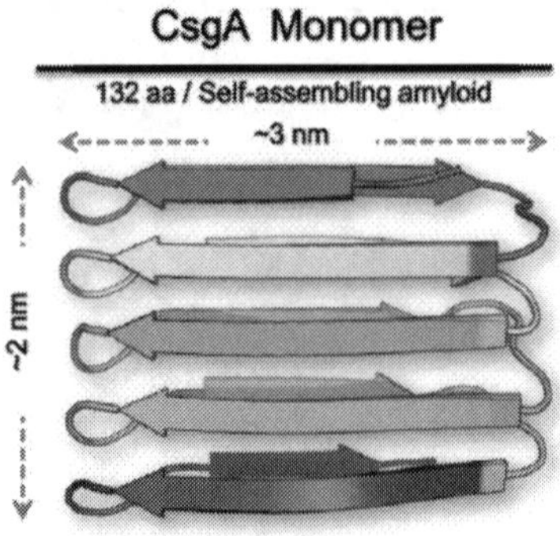

- Used by *E. coli* for biofilm adhesion and structure.
- Self-assembling amyloid protein monomer.
- Diameter of 3-7nm and tens of microns in length.
- Amyloid nanofibers are *really* tough:
  - Steel-like Strength
  - Silk-like Flexibility[1]

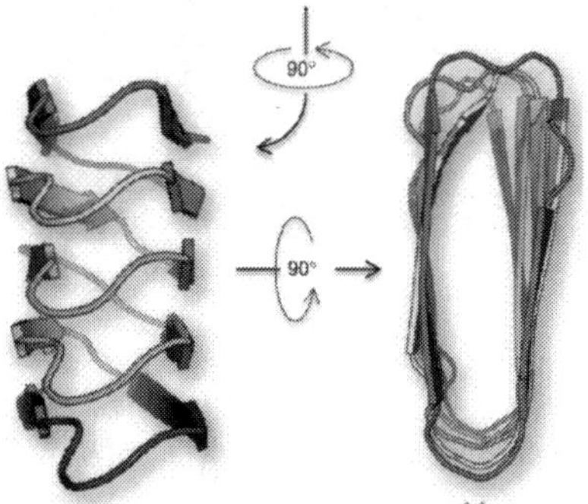

1. Proc Natl Acad Sci U S A 2006 Oct 24;103(43):15806-11.

Curli Biogenesis

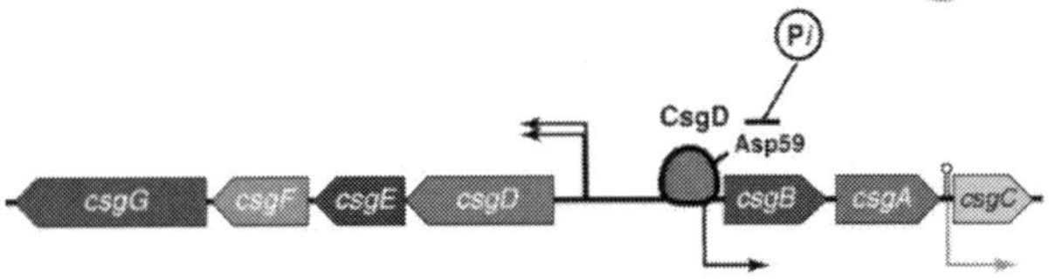

- Small set of genes (complete operon ~3kb).
- OM secretion does not require ATP.
- CsgA is secreted as an unfolded protein.
- CsgB acts as a membrane anchor and a nucleator.
- Facile assay methods.

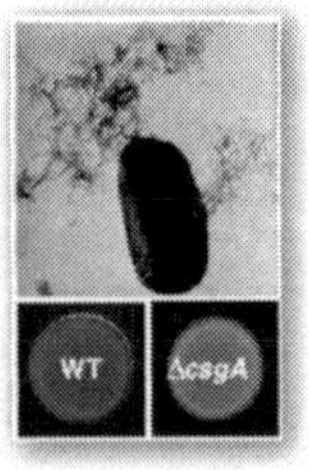

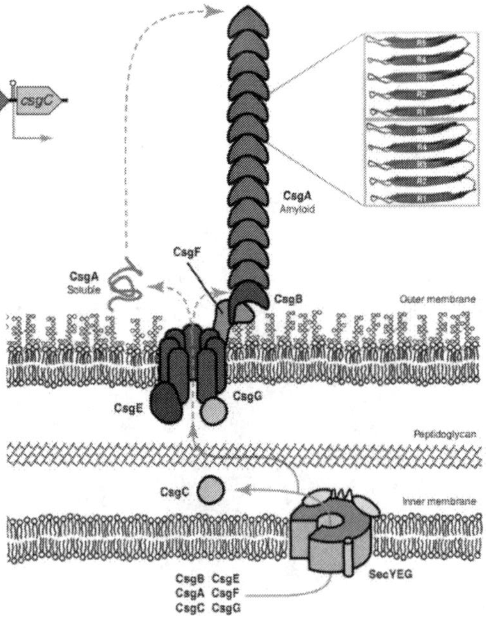

Trends in Microbiology February 2012, Vol. 20, No. 2

(Re)Conceptualizing Curli

- Rethink bacteria as programmable factories to produce bulk biomaterials.
- Repurpose self-assembling, hierarchically ordered protein biomaterials.
- Refactor amyloids as tough nanomaterials.

16

BIND – Biofilm Integrated Nanofiber Display

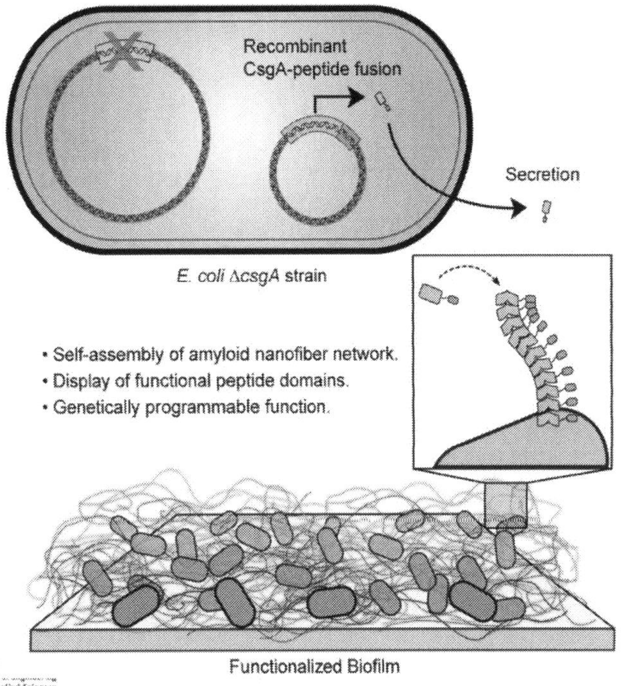

17

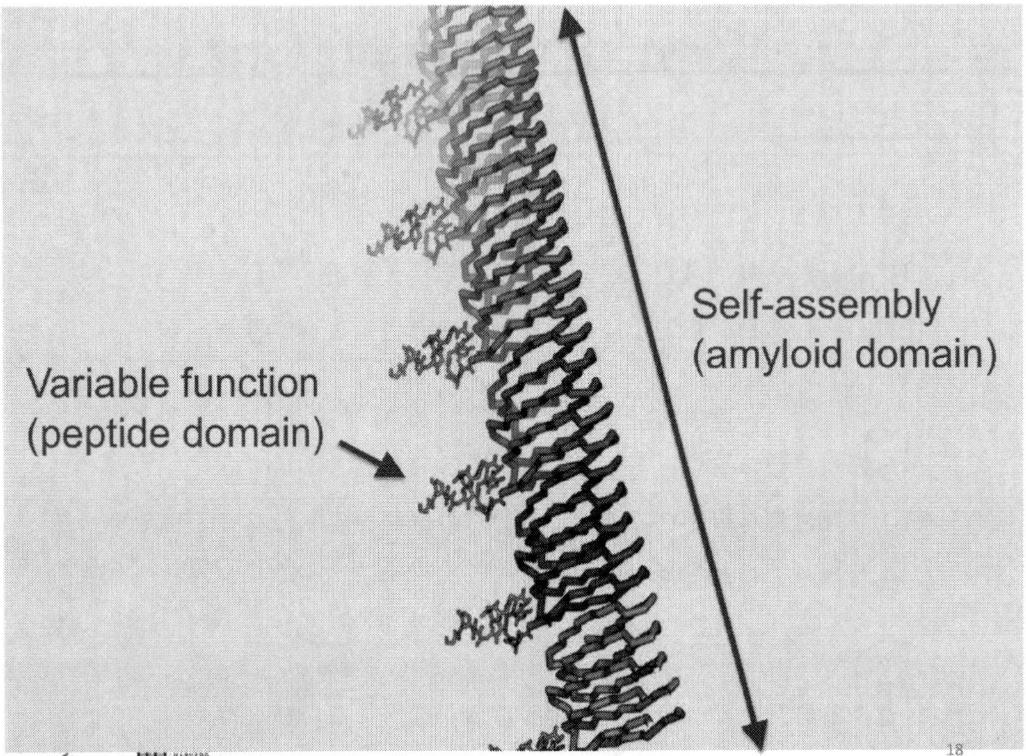

Protein Fusion Sites for CsgA

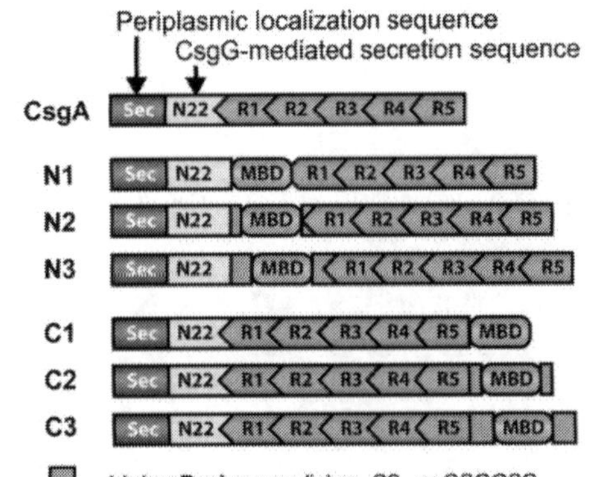

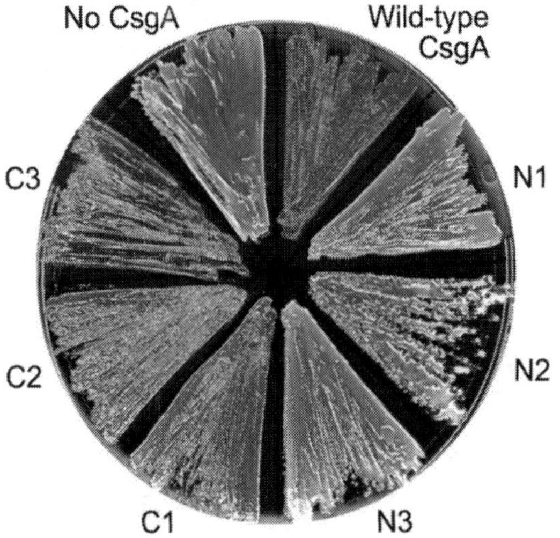

Nguyen et al., 2014 Nature Comm.

Peptide C3 Site Insertion Library

Peptide	Sequence	Length (aa)	Type	Function	Reference
HIS	HHHHHH	6	Affinity Tag	Affinity Tag	Bio/Technology 1988, 6(11): 1321.
GBP	EPLQLKM	7	Substrate Binding	Graphene edge binding	JACS 2011, 133: 14480.
FLAG	DYKDDDDK	8	Affinity Tag	Affinity Tag	Nature Biotech. 1988, 6: 1204.
CNBP	HSSYWYAFNNKT	12	Substrate Binding	Carbon nanotube binding	Nano. Lett. 2006, 6: 40.
A3	AYSSGAPPMPPF	12	Substrate Binding	Gold surface binding	Small 2005, 1(11): 1048.
Z8	LRRSSEAHNSIV	12	NP templating	ZnS quantum dot templating	J. Mater. Chem. 2003, 13: 2414.
E14	PWIPTPRPTFTG	12	NP templating	CdS quantum dot templating	J. Mater. Chem. 2003, 13: 2414.
CLP12	NPYHPTIPQSVH	12	Mineral templating	Hydroxyapatite nucleation	Langmuir 2011, 27: 7620.
QBP1	PPPWLPYMPPWS	12	Substrate Binding	Quartz/Glass binding	Bioinformatics 2007, 23: 2816.
SpyTag	AHIVMVDAYKPTK	13	Protein Display	General covalent capture/display of proteins	PNAS 2012, 109(12): E690.
BCCP	GLNDIFEAQKIEWH	14	Protein Display	Biotinylation tag	Prot. Sci. 1999, 8: 921.
MBD	KCTSDQDEQFIPKGCSK	17	Substrate Binding	Binding to stainless steel surfaces	Mol. Microb. 2006, 59(4): 1083.
CT43	CGPAGDSSGVDSRSVGPC	18	NP templating	ZnS quantum dot templating	JACS 2010, 132: 4731.
AFP8	DTASDAAAAAALTAANAKAAA ELTAANAAAAAAATAR	37	Substrate Binding	Ice crystal binding	JBC 1998, 273(19): 11714.
Mms6	GGTIWTGKGLGLGLGLGLGA WGPIILGVVGAGAVYAYMKSR DIESAQSDEEVELRDALA	59	NP templating	Magnetite NP templating	JBC 2003, 278(10): 8745.

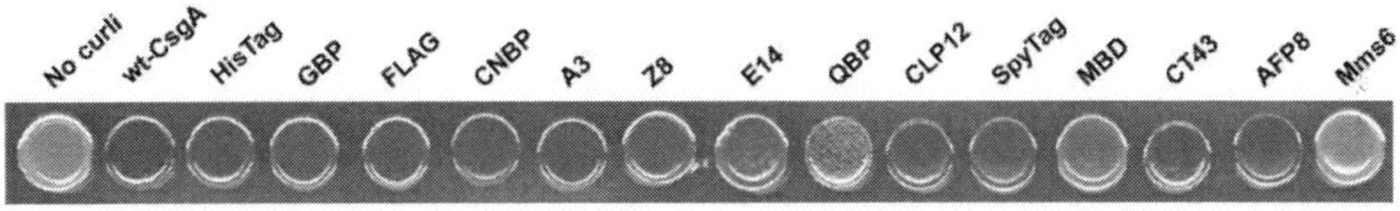

WYSS INSTITUTE · HARVARD School of Engineering and Applied Sciences

Nguyen et al., 2014 Nature Comm.

20

BIND Library Functional Testing

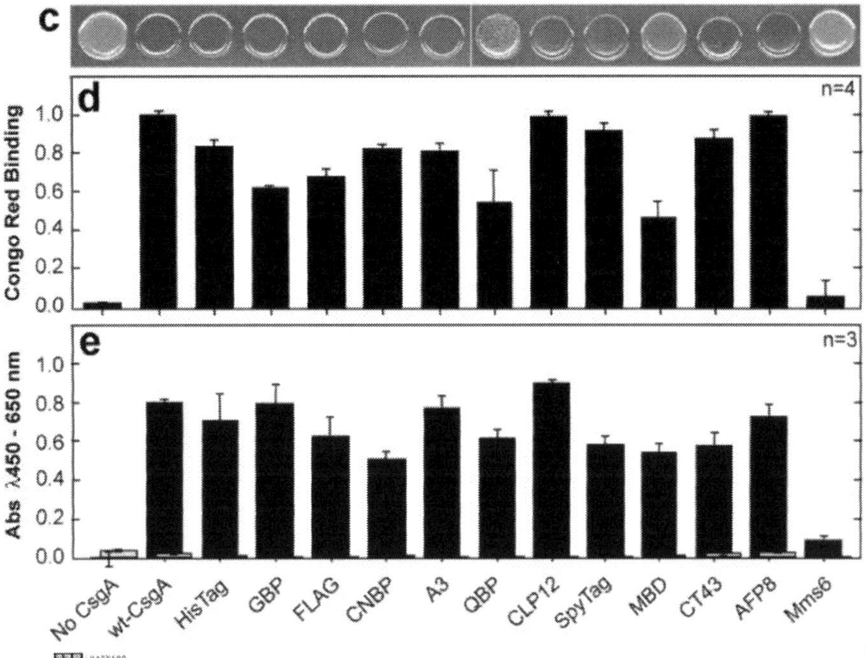

WYSS INSTITUTE · HARVARD School of Engineering and Applied Sciences

Nguyen et al., 2014 Nature Comm.

21

Congo Red Birefringence

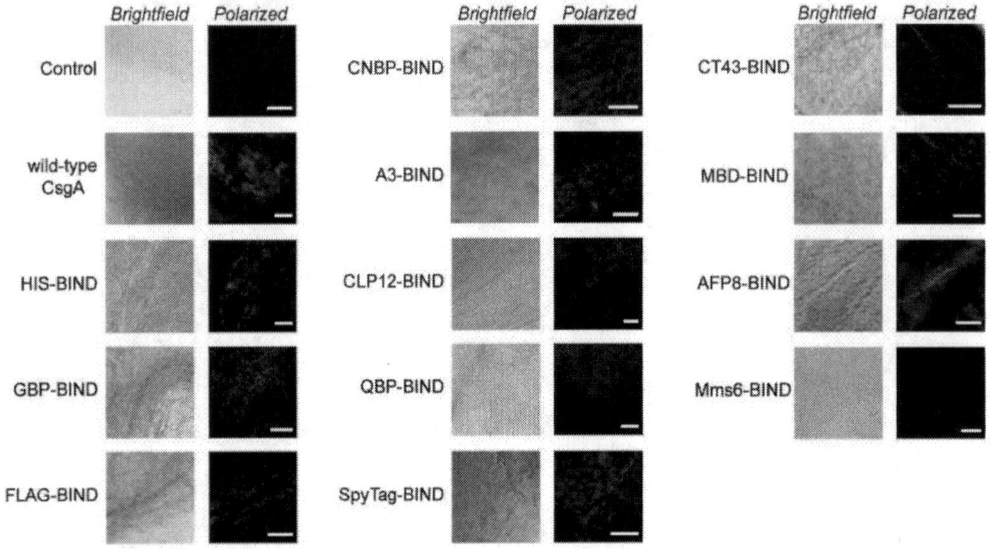

WYSS INSTITUTE HARVARD School of Engineering and Applied Sciences

Nguyen et al., 2014 Nature Comm.

22

MALDI-TOF Verification of Full Fusions

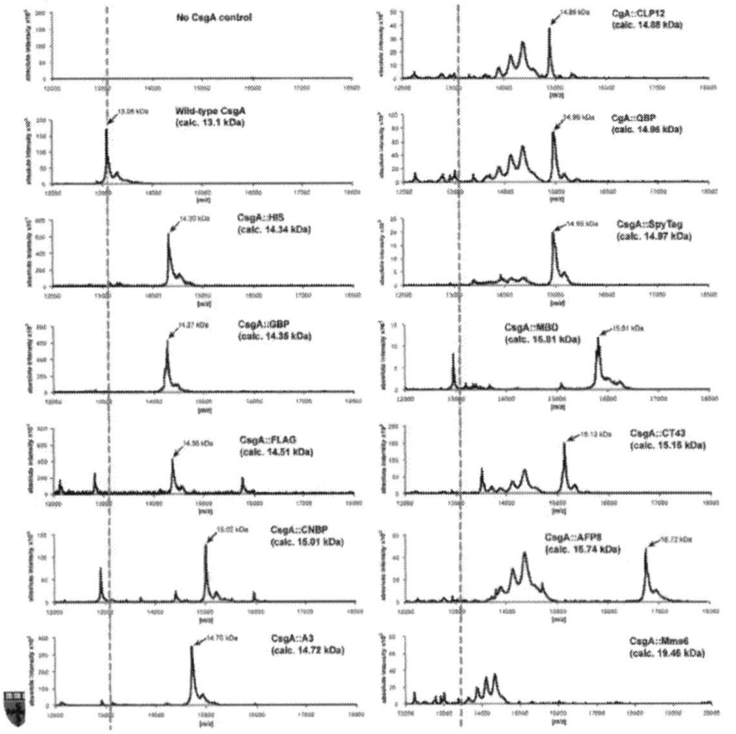

WYSS INSTITUTE

23

TEM Micrographs of BIND Ultrastructure

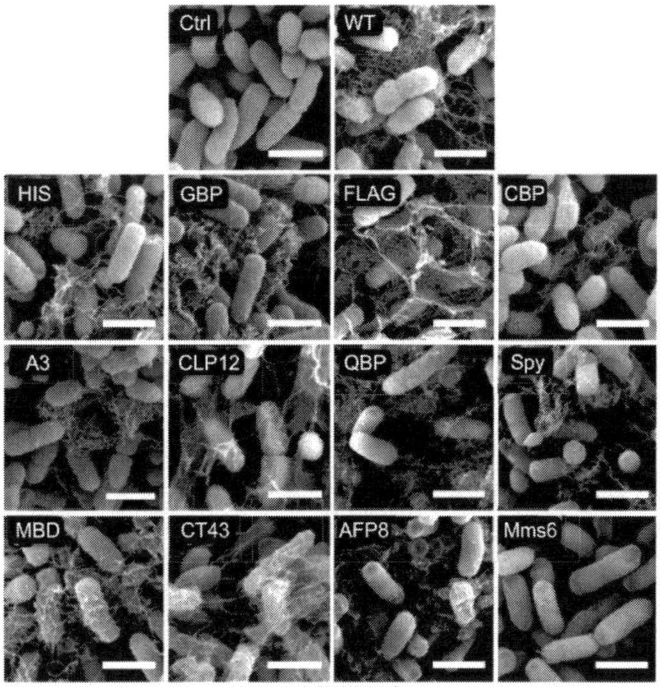

Wyss Institute

Nguyen et al., 2014 Nature Comm.

24

TEM Micrographs of BIND Ultrastructure

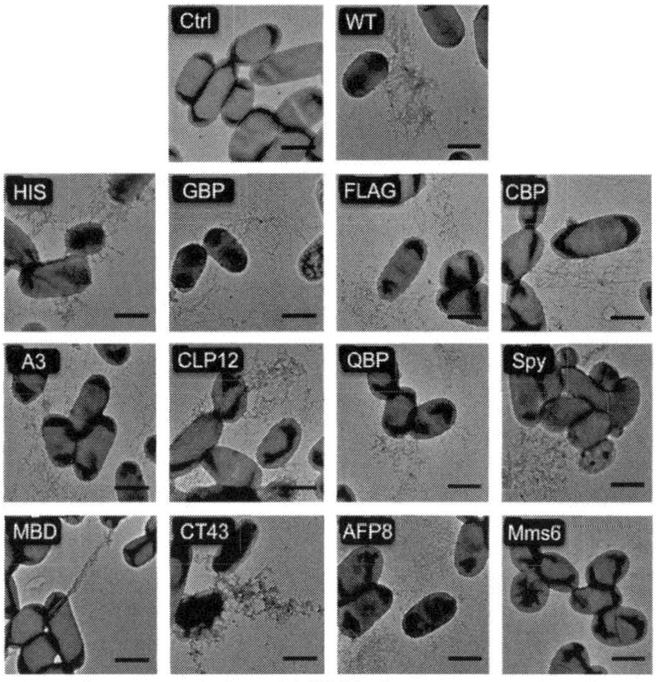

Wyss Institute

Nguyen et al., 2014 Nature Comm.

25

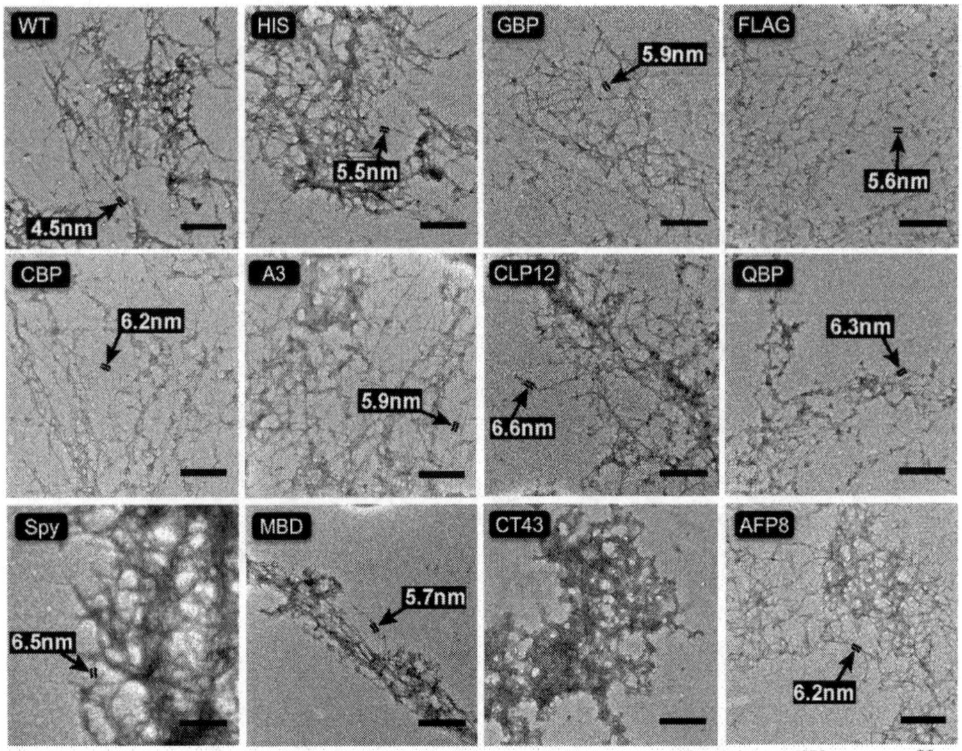

*All scale bars are 250 nm.

Nguyen et al., 2014 Nature Comm.

Wyss Institute | Harvard School of Engineering and Applied Sciences

26

Genetically Encoded BIND Nanomaterials

1. Nanoparticle Templating

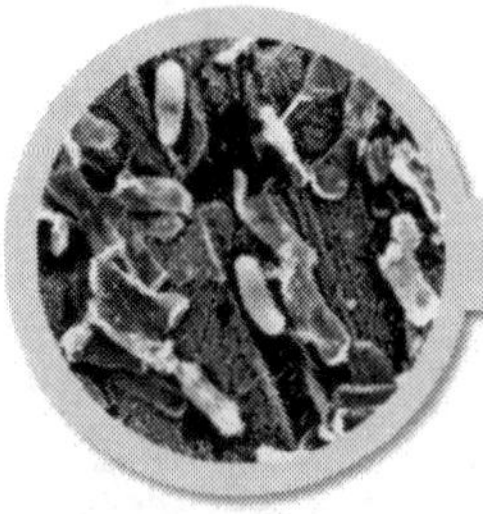

2. Material Adhesion

3. Protein Immobilization

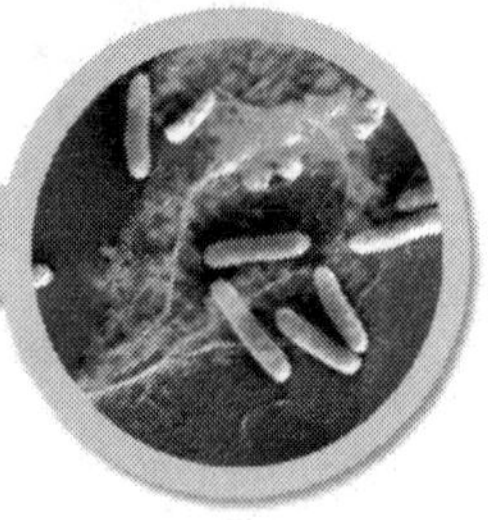

Wyss Institute | Harvard School of Engineering and Applied Sciences

27

Genetically Encoded BIND Nanomaterials

1. Nanoparticle Templating

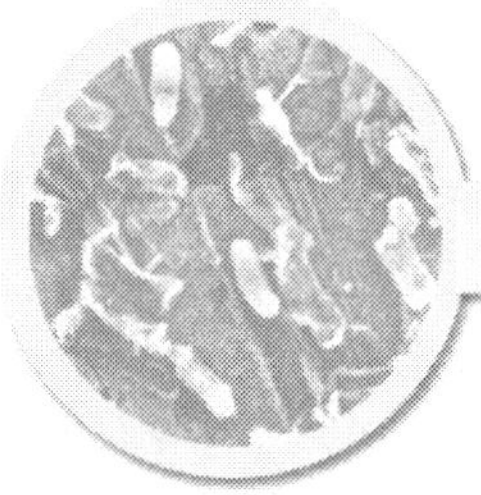

2. Material Adhesion

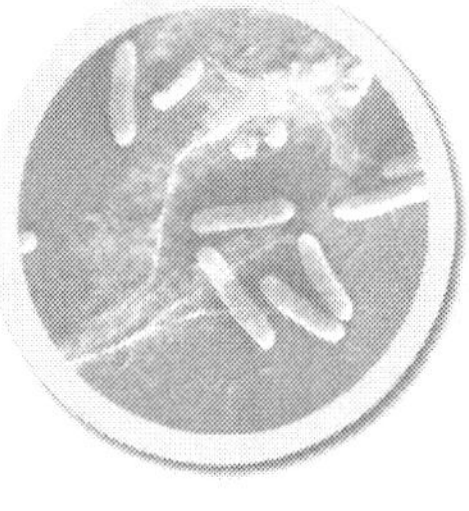

3. Protein Immobilization

Ag Nanoparticle Templating by A3-BIND

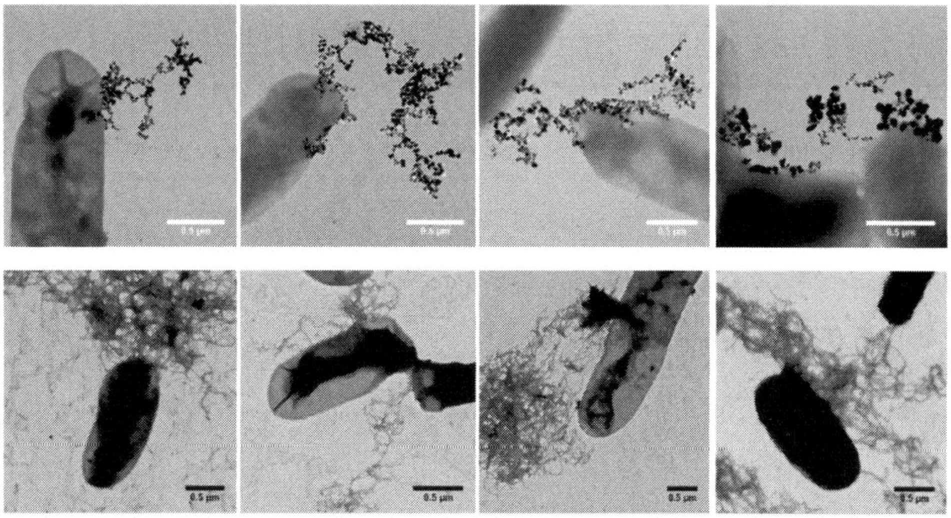

Nguyen et al., 2014 Nature Comm.

Ag Nanoparticle Templating by A3-BIND

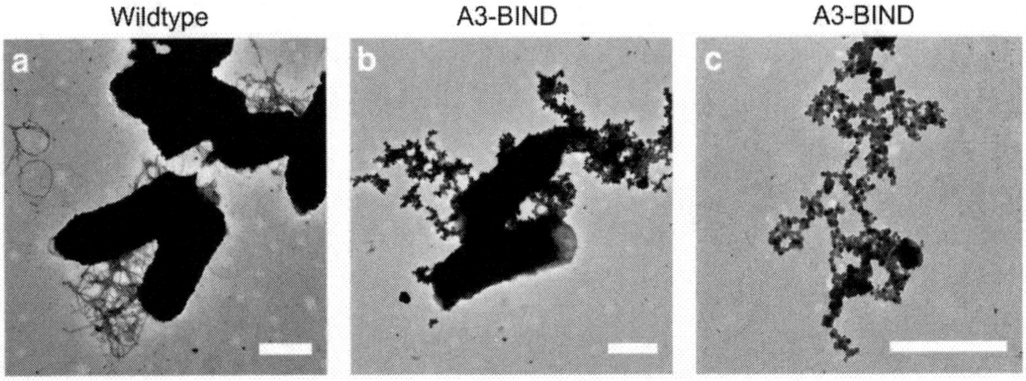

Nguyen et al., 2014 Nature Comm.

31

Genetically Encoded BIND Nanomaterials

1. Nanoparticle Templating

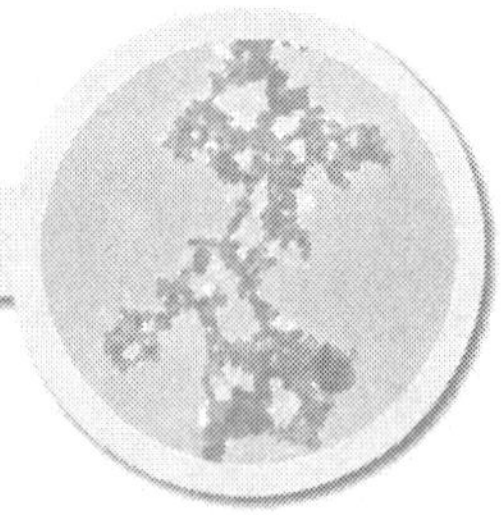

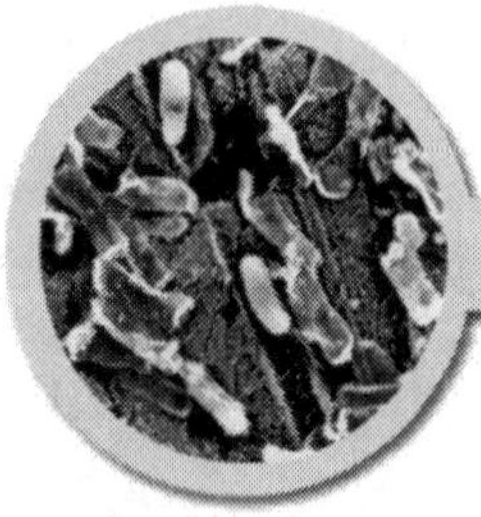

2. Material Adhesion

3. Protein Immobilization

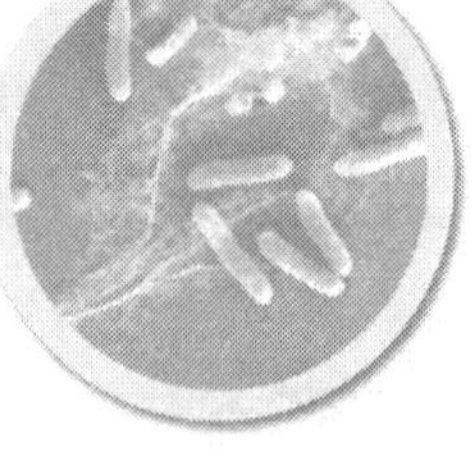

32

Adhesion to SS304L by MBD-BIND

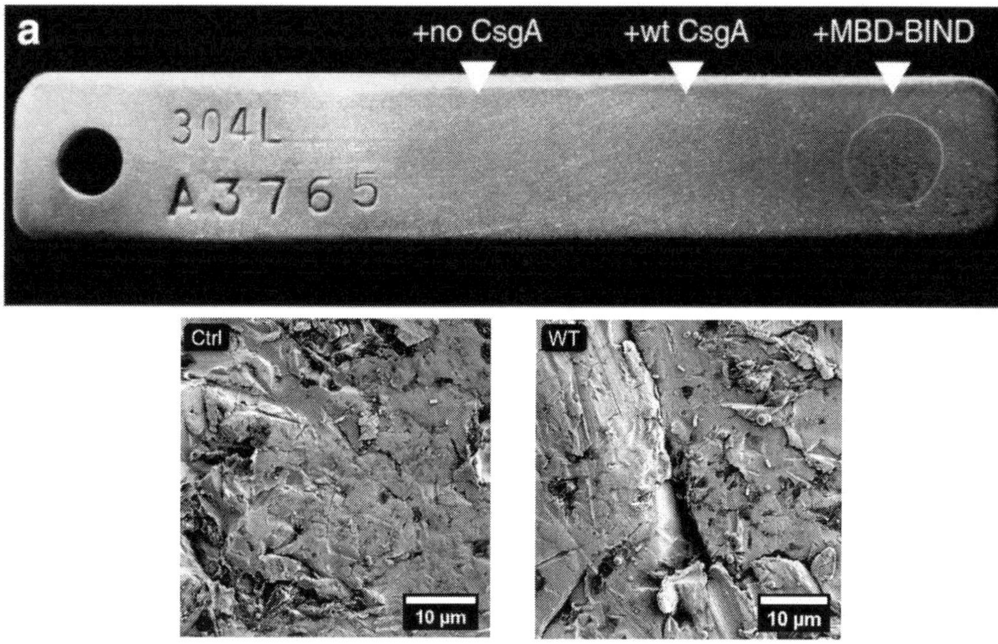

WYSS INSTITUTE HARVARD School of Engineering and Applied Sciences

Nguyen et al., 2014 Nature Comm.

Adhesion to SS304L by MBD-BIND

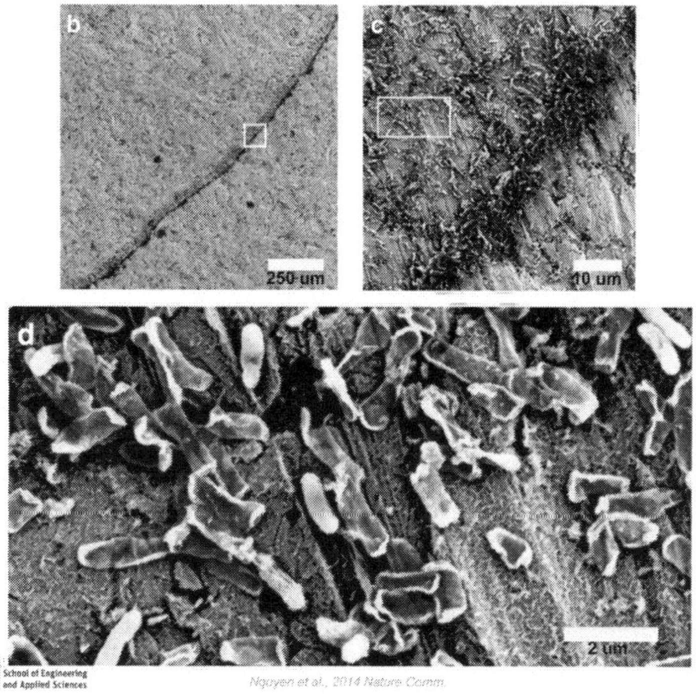

WYSS INSTITUTE School of Engineering and Applied Sciences

Nguyen et al., 2014 Nature Comm.

Genetically Encoded BIND Nanomaterials

1. Nanoparticle Templating

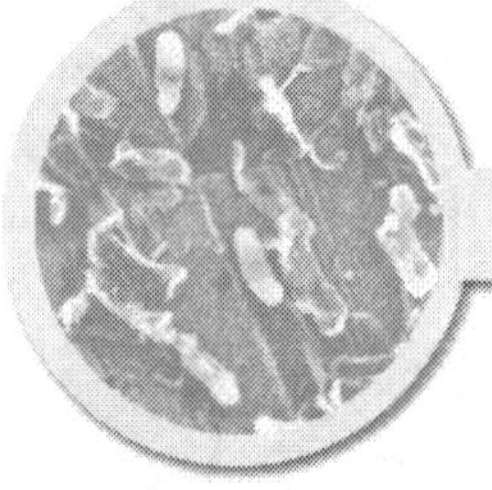

2. Material Adhesion

3. Protein Immobilization

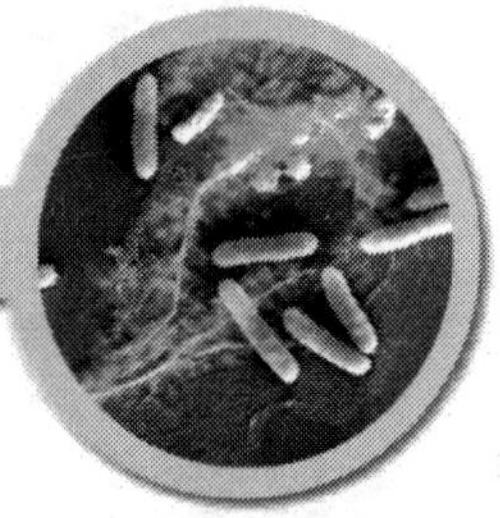

WYSS INSTITUTE HARVARD School of Engineering and Applied Sciences

35

Peptide C3 Site Insertion Library

Peptide	Sequence	Length (aa)	Type	Function	Reference
HIS	HHHHHH	6	Affinity Tag	Affinity Tag	Bio/Technology 1988, 6(11): 1321.
GBP	EPLQLKM	7	Substrate Binding	Graphene edge binding	JACS 2011, 133: 14480.
FLAG	DYKDDDDK	8	Affinity Tag	Affinity Tag	Nature Biotech. 1988, 6: 1204.
CNBP	HSSYWYAFNNKT	12	Substrate Binding	Carbon nanotube binding	Nano. Lett. 2006, 6: 40.
A3	AYSSGAPPMPPF	12	Substrate Binding	Gold surface binding	Small 2005, 1(11): 1048.
Z8	LRRSSEAHNSIV	12	NP templating	ZnS quantum dot templating	J. Mater. Chem. 2003, 13: 2414.
E14	PWIPTPRPTFTG	12	NP templating	CdS quantum dot templating	J. Mater. Chem. 2003, 13: 2414.
CLP12	NPYHPTIPQSVH	12	Mineral templating	Hydroxyapatite nucleation	Langmuir 2011, 27: 7620.
QBP1	PPPWLPYMPPWS	12	Substrate Binding	Quartz/Glass binding	Bioinformatics 2007, 23: 2816.
SpyTag	AHIVMVDAYKPTK	13	Protein Display	General covalent capture/display of proteins	PNAS 2012, 109(12): E690.
BCCP	GLNDIFEAQKIEWH	14	Protein Display	Biotinylation tag	Prot. Sci. 1999, 8: 921.
MBD	KCTSDQDEQFIPKGCSK	17	Substrate Binding	Binding to stainless steel surfaces	Mol. Microb. 2006, 59(4): 1083.
CT43	CGPAGDSSGVDSRSVGPC	18	NP templating	ZnS quantum dot templating	JACS 2010, 132: 4731.
AFP8	DTASDAAAAAALTAANAKAAA ELTAANAAAAAAATAR	37	Substrate Binding	Ice crystal binding	JBC 1998, 273(19): 11714.
Mms6	GGTIWTGKGLGLGLGLGLGA WGPIILGVVGAGAVYAYMKSR DIESAQSDEEVELRDALA	59	NP templating	Magnetite NP templating	JBC 2003, 278(10): 8745.

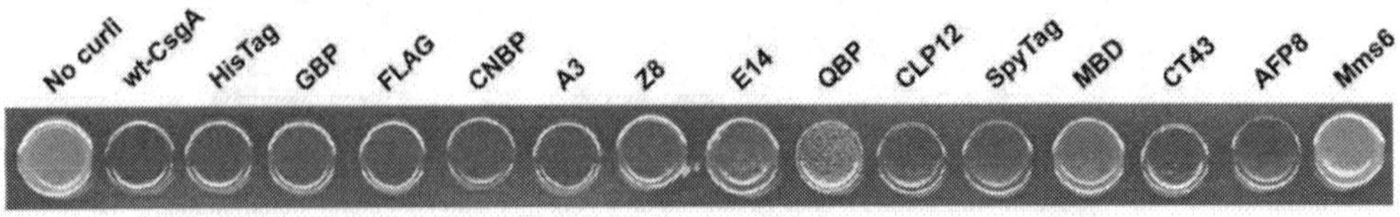

WYSS INSTITUTE HARVARD School of Engineering and Applied Sciences

Nguyen et al., 2014 Nature Comm.

36

Peptide C3 Site Insertion Library

Peptide	Sequence	Length (aa)	Type	Function	Reference
HIS	HHHHHH	6	Affinity Tag	Affinity Tag	Bio/Technology 1988, 6(11): 1321.
GBP	EPLQLKM	7	Substrate Binding	Graphene edge binding	JACS 2011, 133: 14480.
FLAG	DYKDDDDK	8	Affinity Tag	Affinity Tag	Nature Biotech. 1988, 6: 1204.
CNBP	HSSYWYAFNNKT	12	Substrate Binding	Carbon nanotube binding	Nano. Lett. 2006, 6: 40.
A3	AYSSGAPPMPPF	12	Substrate Binding	Gold surface binding	Small 2005, 1(11): 1048.
Z8	LRRSSEAHNSIV	12	NP templating	ZnS quantum dot templating	J. Mater. Chem. 2003, 13: 2414.
E14	PWIPTPRPTFTG	12	NP templating	CdS quantum dot templating	J. Mater. Chem. 2003, 13: 2414.
CLP12	NPYHPTIPQSVH	12	Mineral templating	Hydroxyapatite nucleation	Langmuir 2011, 27: 7620.
QBP1	PPPWLPYMPPWS	12	Substrate Binding	Quartz/Glass binding	Bioinformatics 2007, 23: 2816.
SpyTag	AHIVMVDAYKPTK	13	Protein Display	General covalent capture/display of proteins	PNAS 2012, 109(12): E690.
BCCP	GLNDIFEAQKIEWH	14	Protein Display	Biotinylation tag	Prot. Sci. 1999, 8: 921.
MBD	KCTSDQDEQFIPKGCSK	17	Substrate Binding	Binding to stainless steel surfaces	Mol. Microb. 2006, 59(4): 1083.
CT43	CGPAGDSSGVDSRSVGPC	18	NP templating	ZnS quantum dot templating	JACS 2010, 132: 4731.
AFP8	DTASDAAAAAALTAANAKAAA ELTAANAAAAAAATAR	37	Substrate Binding	Ice crystal binding	JBC 1998, 273(19): 11714.
Mms6	GGTIWTGKGLGLGLGLGLGA WGPIILGVVGAGAVYAYMKSR DIESAQSDEEVELRDALA	59	NP templating	Magnetite NP templating	JBC 2003, 278(10): 8745.

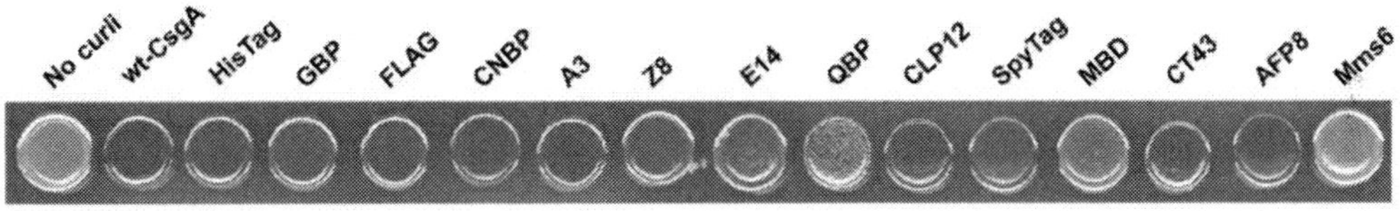

WYSS INSTITUTE · HARVARD School of Engineering and Applied Sciences

Nguyen et al., 2014 Nature Comm.

37

The SpyTag System

Peptide tag forming a rapid covalent bond to a protein, through engineering a bacterial adhesin

Bijan Zakeri[a,1], Jacob O. Fierer[a,1], Emrah Celik[b], Emily C. Chittock[a], Ulrich Schwarz-Linek[c], Vincent T. Moy[b], and Mark Howarth[a,2]

[a]Department of Biochemistry, University of Oxford, South Parks Road, Oxford OX1 3QU, United Kingdom; [b]Department of Physiology and Biophysics, University of Miami, Miller School of Medicine, P.O. Box 016430, Miami, FL 33101-6430; and [c]Biomedical Sciences Research Complex, University of St. Andrews, North Haugh, St. Andrews, Fife KY16 9ST, United Kingdom

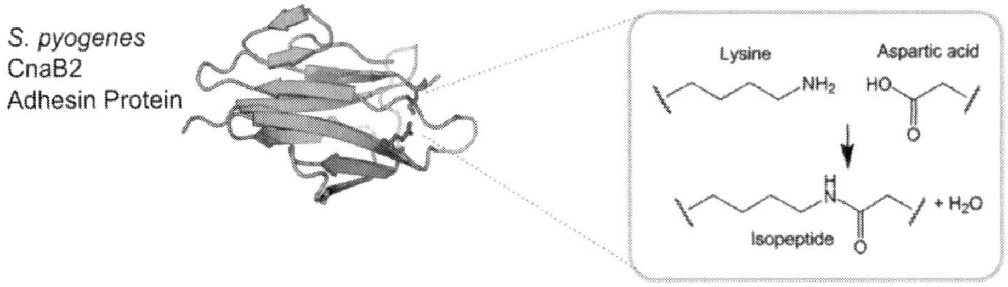

WYSS INSTITUTE · HARVARD School of Engineering and Applied Sciences

Zakeri et al., 2014 PNAS.

38

The SpyTag System

Peptide tag forming a rapid covalent bond to a protein, through engineering a bacterial adhesin

Bijan Zakeri[a,1], Jacob O. Fierer[a,1], Emrah Celik[b], Emily C. Chittock[a], Ulrich Schwarz-Linek[c], Vincent T. Moy[b], and Mark Howarth[a,2]

[a]Department of Biochemistry, University of Oxford, South Parks Road, Oxford OX1 3QU, United Kingdom; [b]Department of Physiology and Biophysics, University of Miami, Miller School of Medicine, P.O. Box 016430, Miami, FL 33101-6430; and [c]Biomedical Sciences Research Complex, University of St. Andrews, North Haugh, St. Andrews, Fife KY16 9ST, United Kingdom

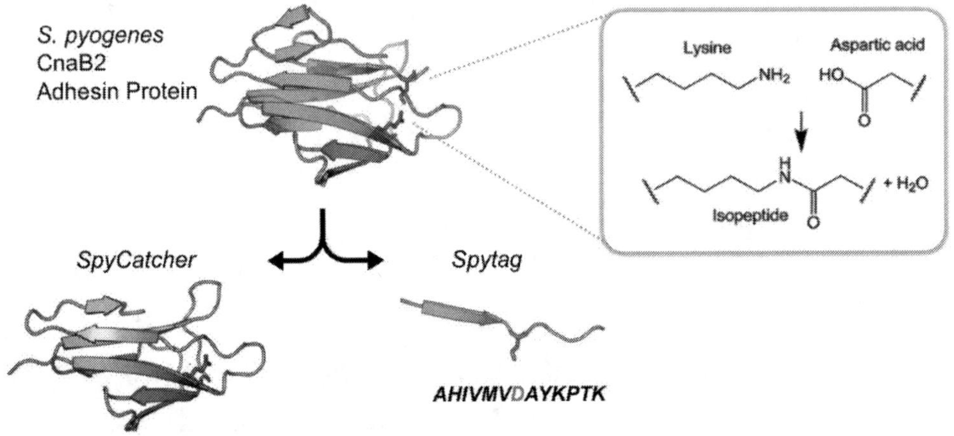

Zakeri et al., 2012 PNAS.

BIND for Protein Display

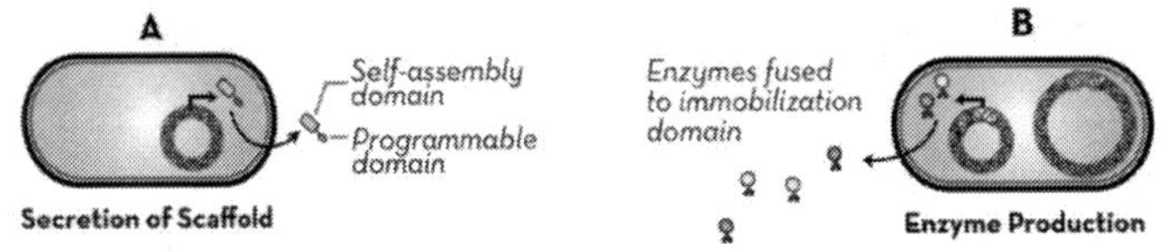

BIND for Protein Display

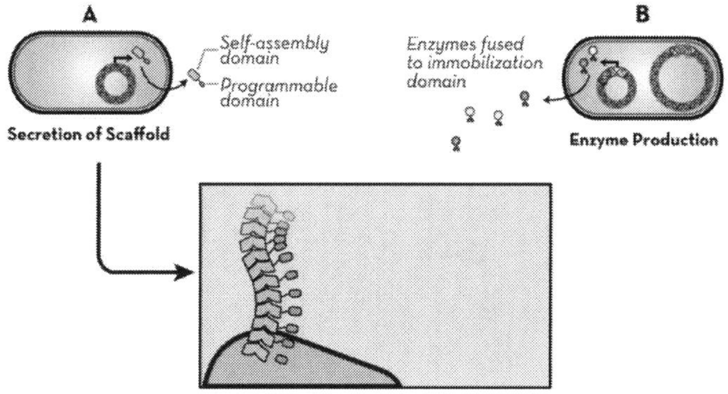

Wyss Institute | Harvard School of Engineering and Applied Sciences

41

BIND for Protein Display

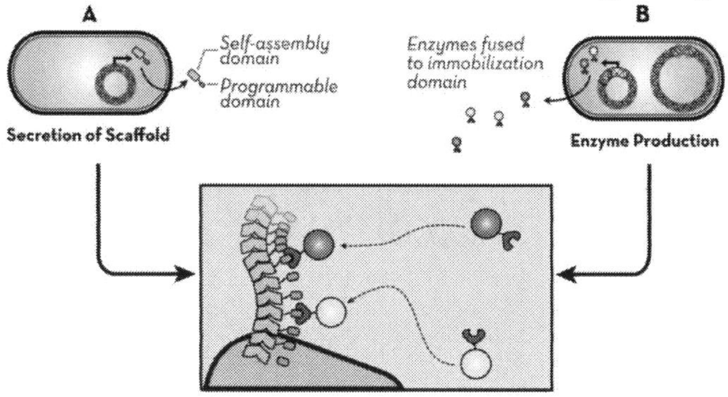

Wyss Institute | Harvard School of Engineering and Applied Sciences

42

BIND for Protein Display

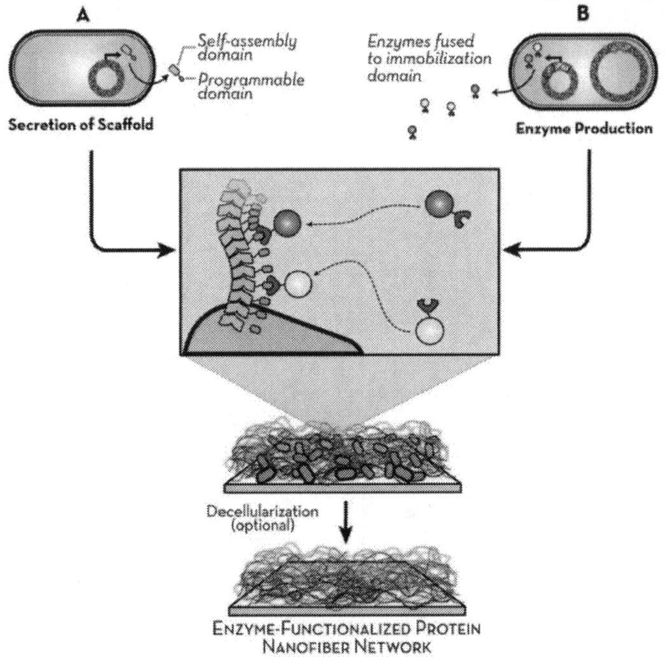

WYSS INSTITUTE HARVARD School of Engineering and Applied Sciences

43

SpyTag-BIND with GFP-SpyCatcher

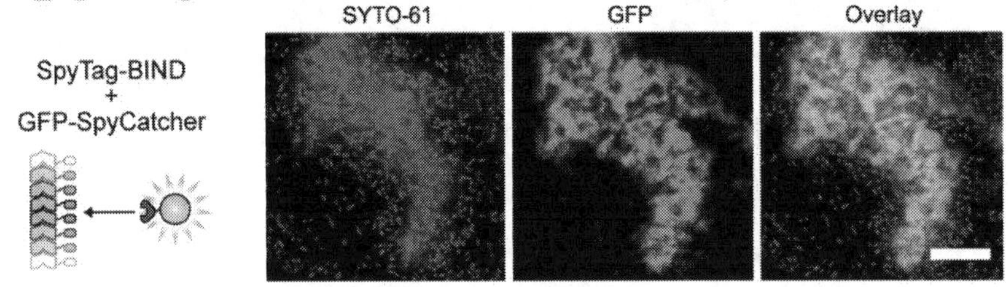

WYSS INSTITUTE HARVARD School of Engineering and Applied Sciences

Nguyen et al., 2014 Nature Comm.

44

SpyTag-BIND CLSM

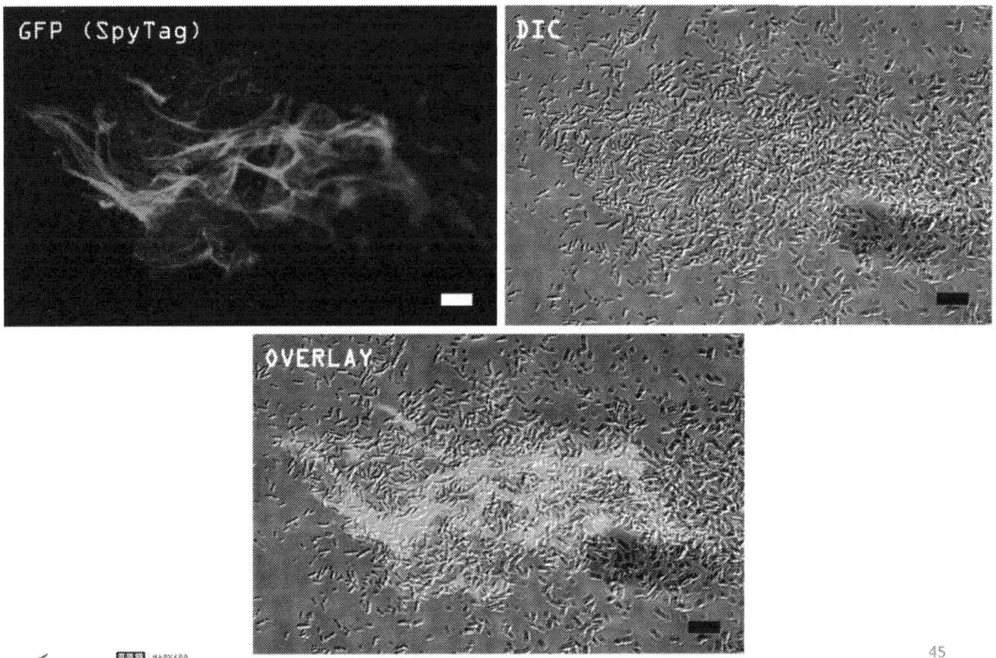

WYSS INSTITUTE HARVARD School of Engineering and Applied Sciences

Nguyen et al., 2014 Nature Comm.

45

Confocal and SEM Data Alignment

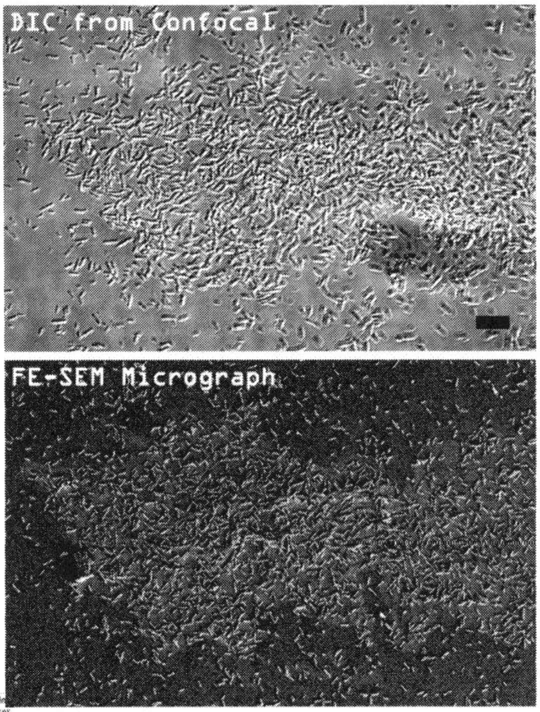

WYSS INSTITUTE HARVARD School of Engineering and Applied Sciences

46

Confocal and SEM Overlay – Region 1

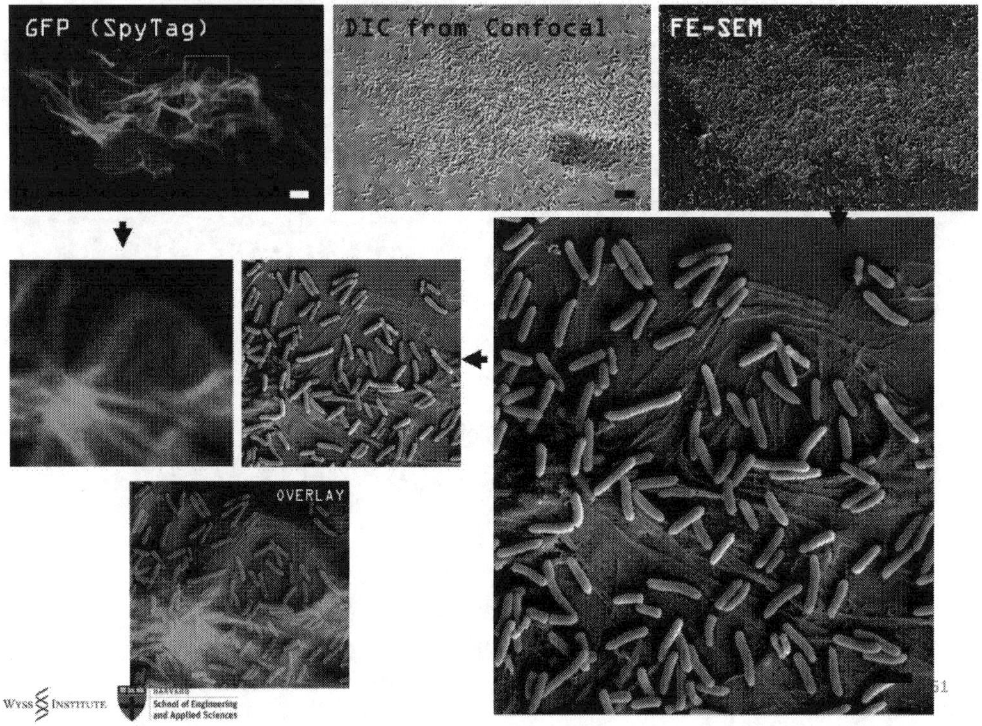

WYSS INSTITUTE HARVARD School of Engineering and Applied Sciences

51

Confocal and SEM Overlay – Region 1

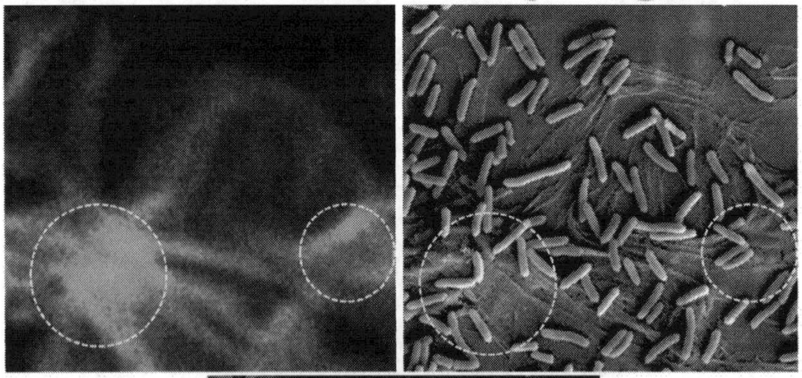

Highly fluorescent areas correlate with areas of high nanofiber density.

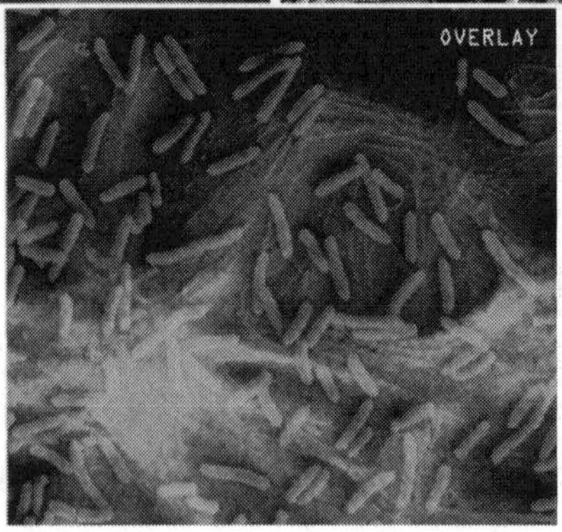

WYSS INSTITUTE HARVARD School of Engineering and Applied Sciences

52

Confocal and SEM Data Overlay – Region 2

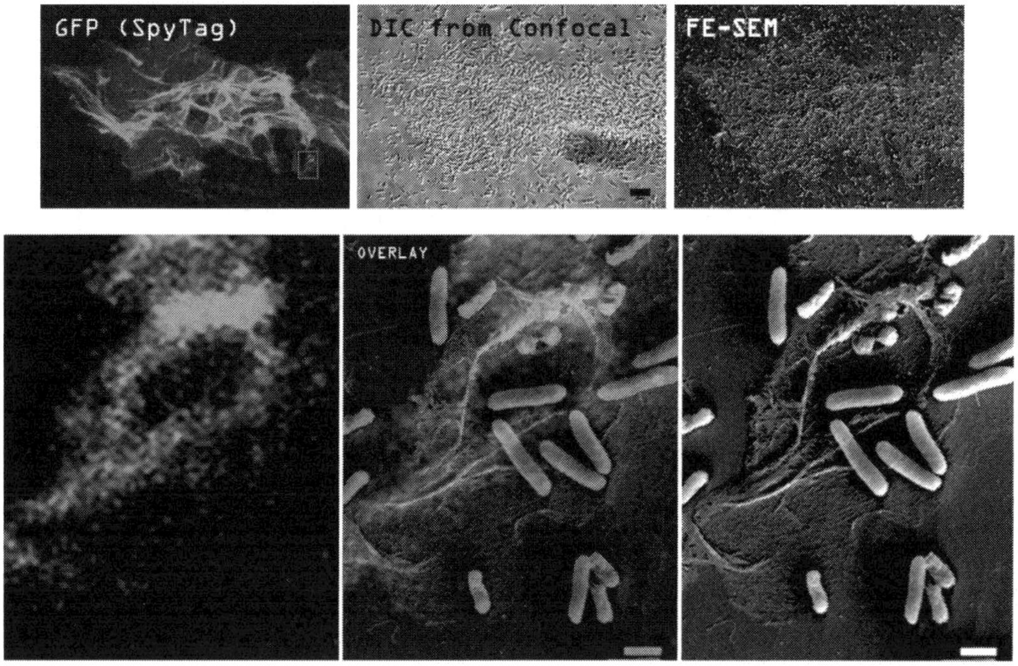

Multifunctionality of the BIND system

Enzymatic Biofilm Materials

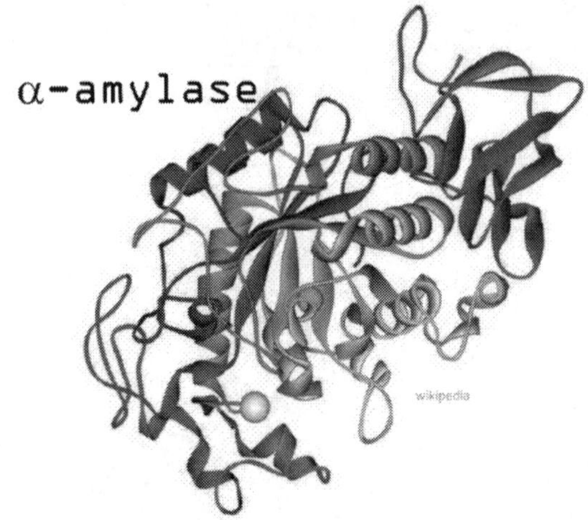

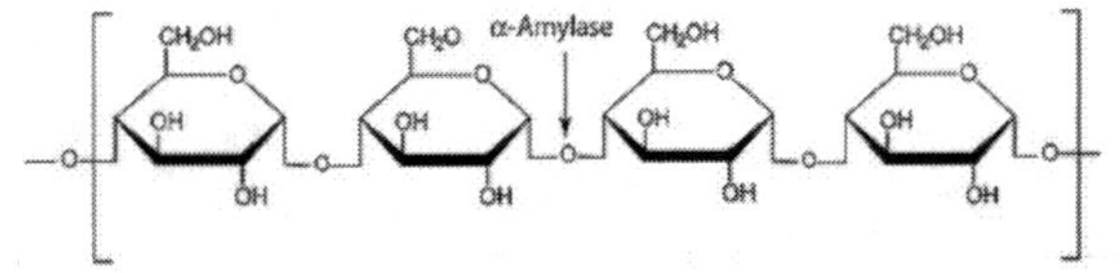

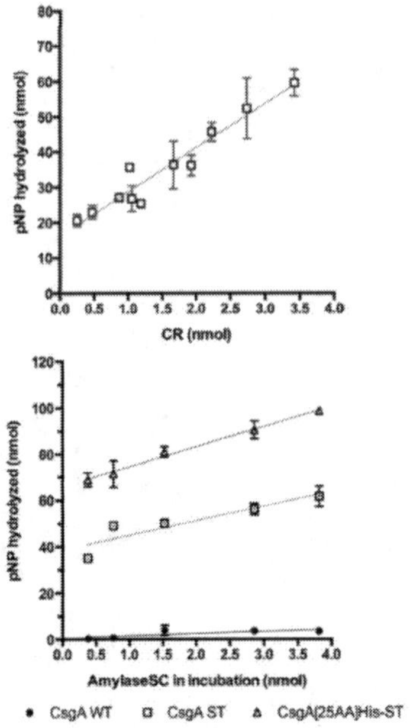

Botyanszki et al., 2015 Biotech. & Bioeng.

WYSS INSTITUTE | HARVARD School of Engineering and Applied Sciences

55

Current BIND Build

- BIND can be used to display various peptides on ECM amyloid nanofibers.
- Nanoparticle template has been demonstrated.
- Specific substrate binding has been demonstrated.
- A 2-component BIND covalent protein display has been constructed.
- BIND allows for multifunctional ECM scaffolds.

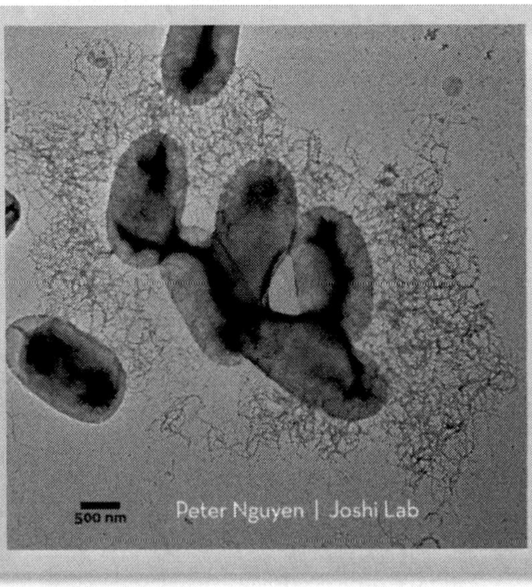

WYSS INSTITUTE | HARVARD School of Engineering and Applied Sciences

56

Ongoing Work

WYSS INSTITUTE | HARVARD School of Engineering and Applied Sciences

57

Synthetic Operon Design

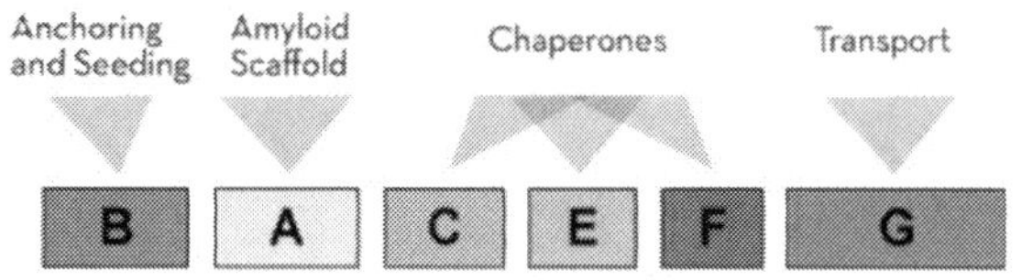

Nucleic Acids Research, 2013, Vol. 41, No. 9 e98

Spanning high-dimensional expression space using ribosome-binding site combinatorics

Lior Zelcbuch[1], Niv Antonovsky[1], Arren Bar-Even[1], Ayelet Levin-Karp[1], Uri Barenholz[1], Michal Dayagi[1], Wolfram Liebermeister[1,2], Avi Flamholz[1], Elad Noor[1], Shira Amram[1], Alexander Brandis[1], Tasneem Bareia[1], Ido Yofe[1], Halim Jubran[3] and Ron Milo[1,*]

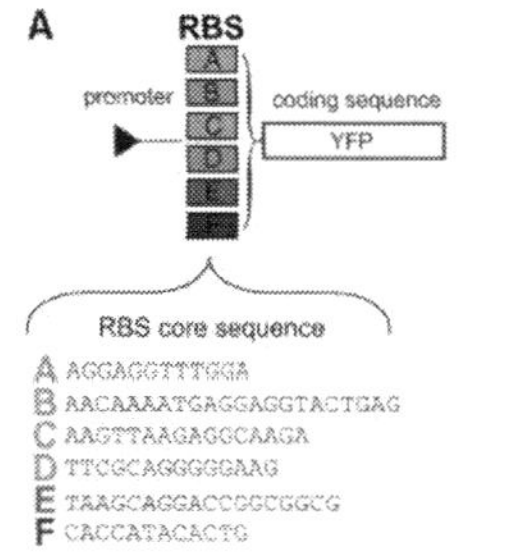

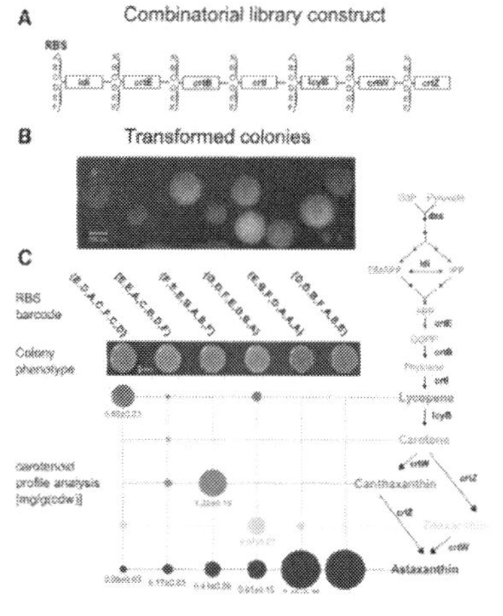

WYSS INSTITUTE | HARVARD School of Engineering and Applied Sciences

59

Curli Biogenesis

Growth Optimization
Biomass switches.

Systems Optimization
Tuning the system for optimal secretion dynamics.

Expression Optimization
RBS combinatorics of a synthetic curli Operon.

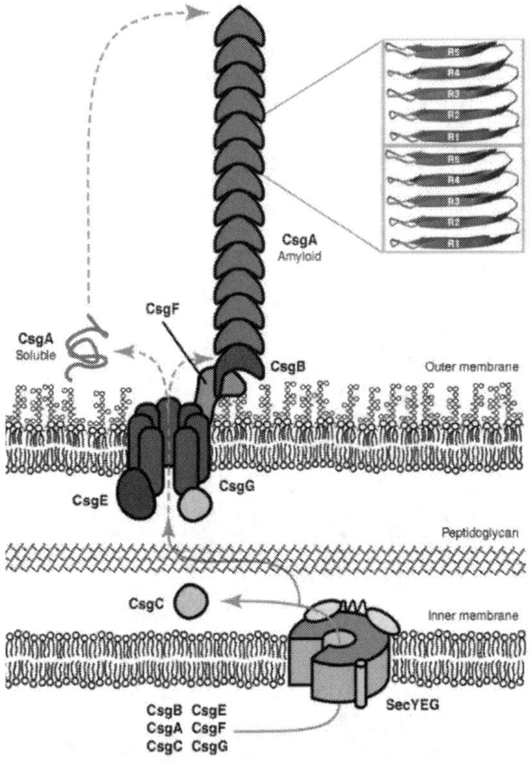

Trends in Microbiology February 2012, Vol. 20, No. 2

WYSS INSTITUTE HARVARD School of Engineering and Applied Sciences

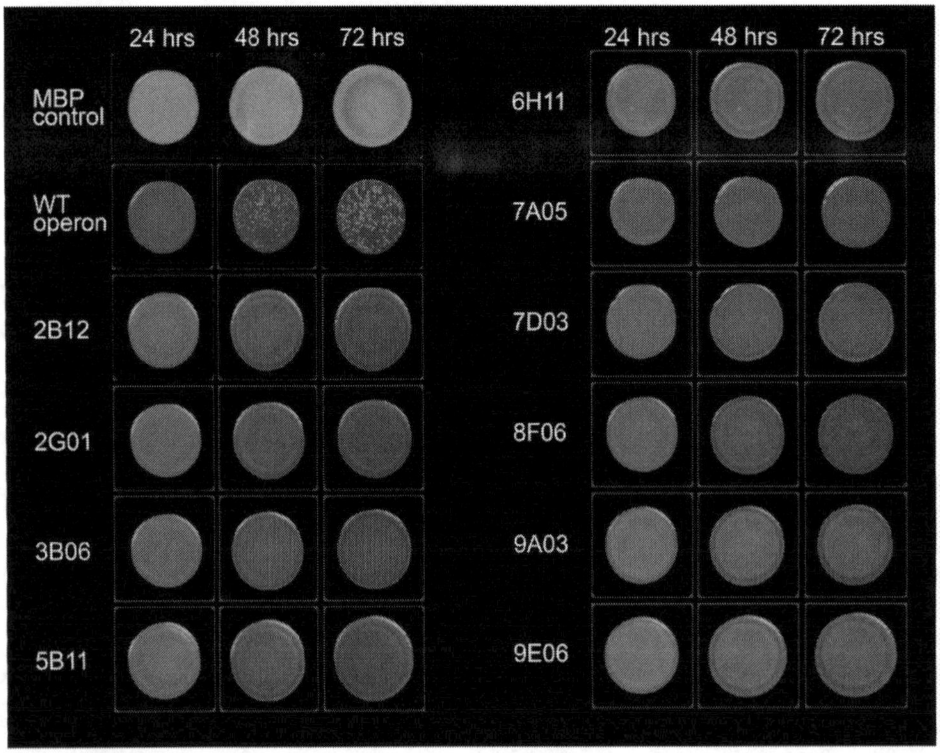

WYSS INSTITUTE HARVARD School of Engineering and Applied Sciences

Engineering Programmable Bacterial Tissues

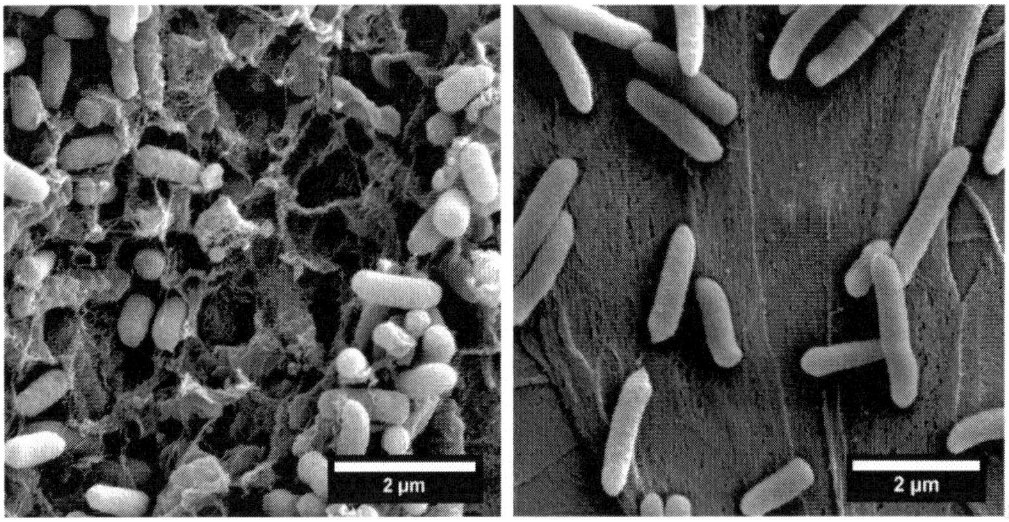

WYSS INSTITUTE HARVARD School of Engineering and Applied Sciences

62

Large Scale Nanomaterials

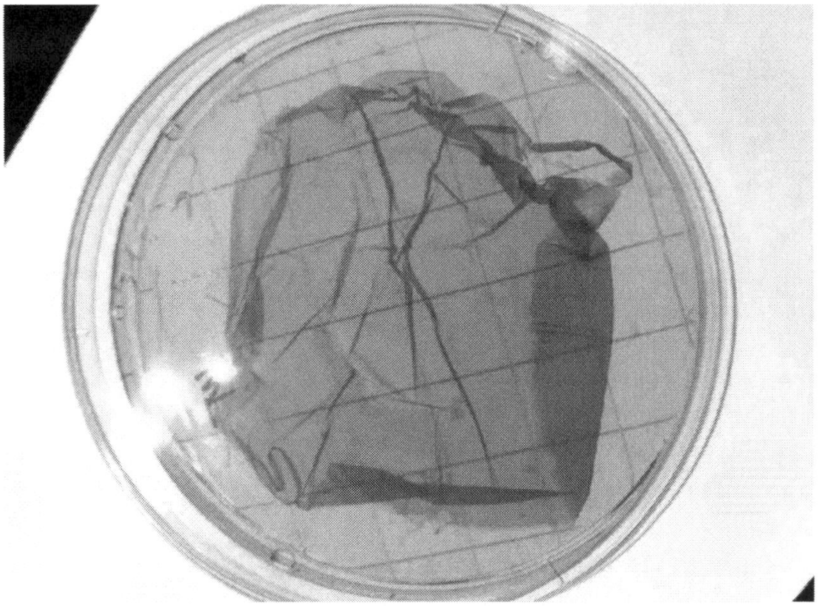

WYSS INSTITUTE HARVARD School of Engineering and Applied Sciences

63

Current Collaborations

SILVER
Dr. Avi Robinson-Mosher
Jernej Turnsek
Leon Liu

COLLINS
Dr. Keith Pardee

LOSICK
Jon Russell

LEWIS
Suman Gunasekaran

YIN
Tony Cho
Luvena Ong

BARSTOW
Dr. Buz Barstow

Wyss Institute — Harvard School of Engineering and Applied Sciences

64

Amyloid Nanomaterial Platforms

Lu Lab

nature materials — ARTICLES

Synthesis and patterning of tunable multiscale materials with engineered cells

nature nanotechnology — ARTICLES

Strong underwater adhesives made by self-assembling multi-protein nanofibres

Chao Zhong, Thomas Gurry, Allen A. Cheng, Jordan Downey, Zhengtao Deng, Collin M. Stultz and Timothy K. Lu

Hochschild Lab

A bacterial export system for generating extracellular amyloid aggregates

Viknesh Sivanathan & Ann Hochschild NATURE PROTOCOLS | VOL.8 NO.7 | 2013

Eisenberg Lab

JACS

Structure-Based Design of Functional Amyloid Materials

Dan Li, Eric M. Jones, Michael R. Sawaya, Hiroyasu Furukawa, Fang Luo, ...

PNAS

Designed amyloid fibers as materials for selective carbon dioxide capture

Dan Li, Hiroyasu Furukawa, Hexiang Deng, Cong Liu, Omar M. Yaghi, and David S. Eisenberg

Wyss Institute — Harvard School of Engineering and Applied Sciences

65

Summary - BIND as a SynBio Platform for Programmable Nanomaterials

- Thinking of microbes as living SynBio foundries for making materials. A factory in a test tube!
- Using functional amyloids as a green nanomaterial.
- Imagining protein-based large scale materials.

WYSS INSTITUTE HARVARD School of Engineering and Applied Sciences

66

Acknowledgements

Professor Neel S. Joshi
Zsofia Botyanszki
Richie P.K. Tay
Anna Duraj-Thatte, PhD
Glenna Meister, PhD
Noemie Courchesne, PhD
Martin Nussbaumer, PhD
Daniel Rubin, PhD
Rajiv Desai
Bom Praveschotinunt
Chris Wintersinger
Hannah Mayo

WYSS INSTITUTE

HARVARD
School of Engineering and Applied Sciences

- Prof. Matthew Chapman (U. Mi
- Prof. Anthony Hay (Cornell).
- Prof. Aaron White (U. Saskat

- PNM Platform.
- Everyone at the Wyss.
- SEED Organizers.

67

Evaluation of TALE- and Crispr/Cas9-Based Synthetic Transcription Factors for Heterologous Gene Expression in Saccharomyces Cerevisiae

Machens, F., University of Potsdam
Hochrein, L., University of Potsdam
Mueller-Roeber, B., University of Potsdam
Messerschmidt, K., University of Potsdam

Metabolic engineering in yeast, with the goal to introduce new biochemical pathways, calls for highly specific and orthogonal regulation of heterologous gene expression. Endogenous transcription factors and promoters are not suitable for this purpose, as they would strongly affect gene expression of the host. Synthetic transcription factors (synTFs) with programmable DNA-binding specificity provide one way to overcome this challenge and to ultimately construct complete gene regulatory networks. We aim to use TALE- and CRISPR/Cas9-based transcription factors in combination with synthetic promoters to control expression of a multitude of genes on circular expression yeast artificial chromosomes, which we call xYACs. Currently, we are characterizing different synTFs using a combination of flow cytometry based reporter gene assays, growth assays and host gene expression analysis. Our approach will finally yield a library of synTFs and corresponding synthetic promoters covering a wide range of expression strength and enable us to specifically modify expression levels of individual genes. The use of synTFs and their careful characterization will reduce negative effects on the host´s gene expression and overall growth fitness. In combination with a method allowing the modular assembly of xYACs and a blue and red light-sensitive induction system, our toolbox will allow easy and fast construction of new pathways in yeast.

Modular Design of Transcriptional Regulatory Parts By Insulating Operators from Their Cognate Promoters
Lou, C.,

Modular Design of transcriptional regulatory parts

Yeqing Zong, Haoqian Zhang, Xiangyu Ji, Qi Ouyang, Chunbo Lou

CAS Key Laboratory of Microbial Physiological and Metabolic Engineering, Institute of Microbiology, Chinese Academy of Sciences, Beijing, China Center for Quantitative Biology and Peking-Tsinghua Center for Life Sciences, School of Physics, Peking University, Beijng, 100871, China

Abstract

Predictably assembling genetic circuits requires a diversity of compositional DNA parts encoding defined functions. However, when assembling the parts, their defined functions are fragile, and usually interfered by other parts and their host cells. Several strategies have been developed to insulate some specific parts from their genetic context (*e.g.* RBS). Here, we describe a modular architecture to insulate the operators from their cognate promoters (recognized by T7 RNAP or sigma ECF11) in *E. coli*. In the modular architecture, the regulatory functions of different operators were described by a novel mathematical framework; meanwhile promoterâ??s activities were independent of the operators. More than 1000 transcriptional regulatory parts could be combinatorially constructed from a promoter and operator library, and their regulatory functions are easily predicted by a simple model without any fitting parameter. The versatile modular parts library is powerful toolbox to facilitate the construction of sophisticated genetic circuits.

Development of Modular, Self-Assembling Synthetic Protein Complexes As Screening Platforms for Enhanced Enzyme Discovery

Haydon, I., Novozymes
Hahm, J., Novozymes
Znameroski, E., Novozymes
Liu, F., Novozymes
Heu, T., Novozymes
Hasani, S., Novozymes, Inc.
Lamsa, M., Novozymes
Jones, A., Novozymes
Widner, W., Novozymes
Harris, P., Novozymes
Mullikin, R., Novozymes
Teter, S. A., Novozymes, Inc.
Lin, J.,

At the forefront of enzyme discovery, Novozymes continues to search for better ways to screen a growing diversity of glycosyl hydrolases in order to improve enzyme cocktails for second-generation biofuel production. To this end, we have successfully integrated a synthetic biology approach into our next generation of enzyme discovery platforms. Specifically, we have leveraged the available natural diversity of cellulosomal components to develop an artificial mini-cellulosome as the basis for a versatile cell-free method to screen combinations of multiple enzyme activities. In the process, we have engineered a highly thermostable scaffoldin comprised of distinct cohesin domains that stably bind cognate dockerins with high specificity for extended periods under lignocellulose hydrolysis conditions. These protein-protein interactions serve as respective binding modules for covalently-linked libraries of different enzyme classes. Our synthetic cellulosome approach has been adapted to an automated HTS system through which enzymes of synergistic quality are more efficiently identified compared to past conventional approaches. Finally, we have successfully employed this novel screening platform to discover promising new enzyme candidates tailored for improved hydrolytic performance on AFEX-PCS, an industrially-relevant biomass substrate.

Bridging the Gap: A Roadmap to Breaking the Biological Design Barrier

Beal, J., Raytheon BBN Technologies

This talk presents an analysis of an emerging bottleneck in organism engineering, and paths by which it may be overcome. Recent years have seen the development of a profusion of synthetic biology tools, largely falling into two categories: high-level "design" tools aimed at mapping from organism specifications to nucleic acid sequences implementing those specifications, and low-level "build and test" tools aimed at faster, cheaper, and more reliable fabrication of those sequences and assays of their behavior in engineered biological organisms. Between the two families, however, there is a major gap: we still largely lack the predictive models and component characterization data required to effectively determine which of the many possible candidate sequences considered in the design phase are the most likely to produce useful results when built and tested. As low-level tools continue to mature, the bottleneck in biological systems engineering is shifting to be dominated by design, making this gap a critical barrier to progress. Considering how to address this gap, we find that widespread adoption of readily available analytic and assay methods is likely to lead to rapid improvement in available predictive models and component characterization models, as evidenced by a number of recent results. Such an enabling development is, in turn, likely to allow high-level tools to break the design barrier and support rapid development of transformative biological applications.

High-Throughput Picodroplet-Based Analysis of Biosynthetic Libraries

Bricio, C., Imperial College London
Liu, X.,
Li, X.,
Rehak, M.,
Ellis, T., Imperial College London
Smith, C.,

Biosynthesis of high value chemicals by engineered microorganisms is an application of synthetic biology that offers both economic and environmental advantages. This application is increasing the need for high-throughput screening tools that can facilitate the detection of the best performance among a library of designed microbes. For this reason we are developing a high-throughput, miniaturised Mass Spectrometry (MS) tool for profiling synthetic designed libraries [(1)].

Combining microfluidics based picodroplet technology for cell encapsulation and sorting together with Mass Spectroscopy we aim to rapidly screen, identify and retrieve the best cell "hits" among synthetic metabolic pathway libraries. Based on this novel approach we will be able to determine which construct has the genetic combination that gives the best biosynthesis performance. To test this new tool we have designed three libraries of two synthetic metabolic pathways using molecular engineering techniques [(2,3,4)].

We chose two previously described synthetic pathways to produce non-natural amino acids [(5,6)] and focus on improving their level of expression. Various strategies have been explored such as the use of homologue genes from other organisms, varying the DNA copy number, transcription levels or translation activity. Then, using a pioneering picodroplet-based technology [(7)]that enables not only for the testing of up to 200,000 samples per day by MS, using miniaturised input volumes (400-700 pL) but also for retrieving identified 'hits' in a reproducible manner, we will select single cells, analyse their production of this non-natural amino acids and finally select and recover the best performing clones among the different profiles obtained for further studies.

This will enable new scientific breakthroughs, higher throughputs, lower screening costs, shorten design-build-test cycle and thus, be of interest to the current MS user base in the synthetic biology market and other sectors.

1. Joint project between Imperial College London and Sphere Fluidics Limited is jointly supported by a Technology Strategy Board Grant and the BBSRC.
2. Kittleson, J. T., Cheung, S., & Anderson, J. C. (2011). Rapid optimization of gene dosage in E. coli using DIAL strains. J Biol Eng, 5(1), 10.
3. Engler, C., Gruetzner, R., Kandzia, R., & Marillonnet, S. (2009). Golden gate shuffling: a one-pot DNA shuffling method based on type IIs restriction enzymes. PloS One, 4(5), e5553.
4. Gibson, D. G., Young, L., Chuang, R.-Y., Venter, J. C., Hutchison, C. a, & Smith,

H. O. (2009). Enzymatic assembly of DNA molecules up to several hundred kilobases. Nature Methods, 6(5), 343–5.

5. Mehl, R. a, Anderson, J. C., Santoro, S. W., Wang, L., Martin, A. B., King, D. S., … Schultz, P. G. (2003). Generation of a bacterium with a 21 amino acid genetic code. J. Am. Chem. Soc., 125(4), 935–9.
6. Zhang, K., Li, H., Cho, K. M., & Liao, J. C. (2010). Expanding metabolism for total biosynthesis of the non-natural amino acid L-homoalanine. Proc Natl Acad Sci USA, 107(14), 6234–9.
7. CA Smith, X Li, TH Mize, TD Sharpe, EI Graziani, C Abell, WT Huck (2013). Sensitive, High Throughput Detection of Proteins in Individual, Surfactant-Stabilized Picoliter Droplets Using Nanoelectrospray Ionization Mass Spectrometry. Anal. Chem. (85), 3812-3816.

*** Formatting Issue - Best Available Paper/Graphic ***

Synthetic Biology in a Chip: A Digital Microfluidic Platform for Cell Transformation, Culture and Expression

Gach, P., Sandia National Laboratories
Singh, A. K., Joint Bioenergy Institute
Shih, S. C. C., Joint BioEnergy Institute
Sustarich, J.,
Keasling, J. D., Lawrence Berkeley National Laboratory
Hillson, N. J., Joint BioEnergy Institute
Adams, P. D., Joint BioEnergy Institute

Â 1

Synthetic Biology in a Chip: A Digital Microfluidic Platform for Cell

Transformation, Culture and Expression

Philip C. Gach,1, 2 Steve C.C. Shih,1, 2 Jess Sustarich,1, 2 Jay D. Keasling,1,3,4,5,6,7 Nathan J. Hillson,1,3,8 Paul D. Adams,1,3,4 and Anup K. Singh1, 2 *

1 Technology Division, Fuels Synthesis Division, Joint Bioenergy Institute (JBEI), Emeryville, California 94608, United States
2 Applied Biosciences and Engineering, Sandia National Laboratories, Livermore,

California 94550, United States

3 Physical Biosciences Division, Lawrence Berkeley National Lab, Berkeley, California

94720, United States

4 Department of Bioengineering, University of California, Berkeley, California 94720, United States
5 Department of Chemical & Biomolecular Engineering, University of California,

Berkeley, California 94720, United States

6 QB3 Institute, University of California, Berkeley, California 94720, United States
7 Synthetic Biology Engineering Research Center, Emeryville California 94608, United

States

8 DOE Joint Genome Institute, Walnut Creek, California 94598, United States

* Corresponding Author email: aksingh@sandia.gov tel: (925) 294-1260

fax: 925-294-3020

Â 2

ABSTRACT

Synthetic biology experiments require optimization of pathways consisting of many genes and other genetic elements. Given the large number of alternatives available for each element, optimization of a pathway can require a large number of experiments. Currently, these experiments are done manually using fairly large amounts of costly reagents per experiments making the process very expensive, extremely slow and irreproducible. We have developed a digital microfluidic platform that uses aqueous droplets suspended in an oil phase as discrete reaction chambers (or incubators) to completely automate the molecular biology steps. It integrates all critical steps of transformation and culture in one chip including plasmid addition, transformation by
heat-shock, addition of selection medium, culture and phenotypic readout. The flexibility of digital microfluidics and peltier modules allows quick optimization of heat-shock parameters permitting transformation of a variety of plasmids and cell types. The chip was validated by introducing a variety of plasmids into *E. coli* including; plasmids containing genes for fluorescence proteins GFP, BFP and RFP, plasmids with selection markers for ampicillin or kanamycin, and plasmids built using golden gate assembly. Compatibility of this platform with multiple organisms was demonstrated by transforming DNA plasmids followed by culture of eukaryotic cells including; *S. cerevisiae* and *A. niger*. Employing the microfluidic device for heat-shock of GFP plasmid into *E. coli* cells gave excellent transformation efficiencies up to 4.4 Ã? 106 Â± 0.2
Ã? 106 CFU/Âµg DNA, similar to benchtop results, though requiring ~100-fold less reagent.
Reducing the duration of heat-shock steps afforded a throughput of 0.1 droplets/s, though with slightly compromised transformation efficiencies (9.0 Ã? 105 Â± 0.9 Ã? 105 CFU/Âµg

Â 3

DNA). This afforded the autonomous generation and processing of up to 100 discrete heat-shocked droplets in just seventeen minutes. The microfluidic device, because of its ability to integrate and automate molecular biology steps and use ~100-fold less reagents, will be of great use to researchers for optimization of synthetic biology pathways to produce biofuels, pharmaceuticals, and other chemicals or biochemicals.

KEY WORDS Digital Microfluidics, Synthetic Biology, Transformation, Cell Culture

Â 4

INTRODUCTION

Synthetic biology is a relatively new discipline that has demonstrated substantial advances in the development of pharmaceutical drugs,1 biofuels,2 chemicals3 and biomaterials.4 The key steps in synthetic biology involve design and characterization of standard biological parts (such as promoters, 5'UTR and terminator parts) and assembly of genes onto a plasmid to express novel pathways in host organisms for producing materials (including fuels, pharmaceuticals, chemicals and novel materials). The core set of technologies utilized to clone DNA has remained largely unchanged for the last 20 years. Successful synthetic biology experiments require a large number of genes or parts (and the variants/mutants for each gene) to be tested. Currently, the genes/parts are assembled manually using an appropriate combinatorial assembly strategy (BioBricks, SLIC/Gibson/CPEC, Golden-gate etc.). The assembled plasmids are then introduced into host organisms by transformation and the successful transformants are grown in liquid cultures in large enough amounts to quantify desired products. In summary, the overall process in designing, creating and optimizing a metabolic pathway involves many steps, each optimized by trial and error and consequently, a successful effort requires a large amount of time (months to years) and cost ($million in reagents and labor). Microfluidic devices have attracted significant attention towards developing high throughput screening platforms using greatly reduced amounts of reagents.6 These devices are capable of performing multiple reactions on a single platform through compartmentalizing reagents into discrete chambers or droplets. Digital microfluidics is a field of microfluidics in which individual aqueous droplets, separated by an immiscible phase such as oil or air, can be generated and manipulated on-chip. Advances in

Â 5

microfluidics have allowed very high throughput droplet generation, mixing multiple different reagents and using minimal quantities of reagents (down to femtoliter volumes). These technologies have been employed for automating several steps of the synthetic biology pipeline including synthesis of DNA,7, 8 cell transformation and cell sorting. Transformation of exogenous DNA into cells is a key aspect of this process, generally achieved through chemical heat-shock or electroporation. Electroporation and chemical heat-shock create porous cellular membranes to introduce DNA into cells and have been routinely employed in microfluidic devices. These devices have demonstrated equal to enhanced transformation efficiencies with much lower reagent volumes, compared to bulk methods.9, 10 Electroporation within microfluidic devices has been performed both in open channels11, 12 and in droplets.13-16 Chemical heat-shock has also been demonstrated in open microfluidic channels10, 17 and in droplets.18, 19 Unfortunately, only the gene introduction step is consistently executed on the device leaving other key procedures performed off-chip, including: DNA addition to cells, incubation, post- transformation addition of selection antibiotics, culture and analysis. A system capable of transformation and culture on a single device is necessary for in-line integration with existing gene assembly and cell sorting technologies thereby providing all

steps of standard synthetic biology on a single device.
To meet these requirements, we developed a microfluidic platform which affords droplet generation, mixing, heating and long-term storage. Droplet movement was achieved by digital microfluidics (DMF). This strategy for droplet manipulation is beneficial because it provides simple fabrication, biocompatibility, operational customization and consistent droplet volume generation. DMF has previously been

Â 6

demonstrated to be effective for DNA synthesis.20 In a proof-of-principle experiment, Au and coworkers demonstrated a DMF platform could transform bacteria with a fluorescent gene.19 However, this device did not possess integrated thermal elements; necessitating physical transfer of the device between a hot plate, ice bath and incubator and then moving the droplets to agar plates for culture. In this report, we describe a DMF device offering cell transformation and culture, which could seamlessly integrate with existing microfluidic tools to provide all steps of standard synthetic biology on a single device.
RESULTS AND DISCUSSION

Device construction and operation

Device requirements. Development of new genes in synthetic biology involves three main operations: 1) DNA synthesis 2) cell transformation and culture and 3) isolation of desired traits. Numerous methods have been developed for automating the synthesis of DNA,7, 8 cell transformation and sorting positive cells. Many studies have reported the use of microfluidic platforms for high-throughput addition of exogenous DNA to cells. However, these technologies do not allow for completely autonomous gene delivery, transformation and culture. A system capable of transformation and culture on a single device is necessary for in-line integration with gene assembly and cell sorting. To meet these requirements a microfluidic platform was developed which affords droplet generation, mixing, heating and long-term storage (Figure 1A).

Thermal regulation. Heat-shock is a common method for gene delivery into cells. It has the advantage of not requiring additional reagents or electrical shock to introduce DNA. Effective heat-shock generally requires three temperatures: 1) cold temperature (4Â°C) for DNA/cell mixing and recovery 2) hot temperature (37Â°C or 42Â°C) for heat-

Â 7

shock and 3) warm temperature (30Â°C or 37Â°C) for cell culture. Unlike previous systems, which employ temperature regulation over the entire device or reaction chamber,19 the current device has three discrete thermal zones to allow sequential operation of the three steps namely, DNA and cell addition at 4C, heat shock at 37 or 42c, and culture at 30 or
37. Three peltier modules attached to the bottom of the DMF chip provided the three thermal zones with exquisite temperature control. Peltier supply voltages were controlled

*** Formatting Issue - Best Available Paper/Graphic ***

by an Arduino board, which also afforded real-time temperature adjustments of device regions. The peltiers uniformly heat the glass slide and channels directly above the module after one-minute actuation. Tightly segmented temperature regions can be maintained on the device for a long period of time with this method (Figure 1C).

Droplet movement. DMF was employed for all aspects of droplet manipulation including: droplet generation, merging and movement. The DMF device was fabricated with chrome electrodes and SU-8 channels to afford a sturdy substrate that was also biocompatible (Figure 1B). This material is also non-reactive with the biocompatible fluorinated oil (HFE 7500) used for droplet segmentation. Channels were coated with Aquapel, which provided a hydrophobic surface ensuring aqueous droplets would not stick to the substrate. Aquapel is also resistant to degradation by the HFE 7500 and stable over several hours in contact with the fluorinated oil. A custom C++ program automated the control of an Arduino board, which provided the actuation voltages to the chrome electrodes on the device. Heat-shock on the DMF device consists of 6 primary steps: 1) generation of DNA and cell droplets 2) mixing DNA and cell droplets 3) DNA/cell droplet movement to heat-shock region 4) generation of media droplet 5) mixing of DNA/cell droplet with a media droplet and 6) DNA/cell/media droplet transfer to culture

Â 8

region (Figure 1D-K). Media storage on device consisted of a (10Ã?7Ã?0.175 mm) chamber, which supported up to 3 ÂµL reagent at a time. From these reservoirs, 235 Â± 11 nL droplets were generated at a rate up to 1 droplet/s. At this point, 54 individual 1150 Ã?1150 Âµm2 electrodes drove the droplets across the device using 250 ms electrode pulses.

Reagent Delivery. Two strategies were employed for reagent delivery to the reservoirs; pipetting prior to assembly or constant fluid addition during operation, as described previously.21 Experiments requiring few reactions could be prepared through manually pipetting 3 ÂµL liquid directly onto the reservoirs followed by channel filling with HFE 7500 and enclosing with an ITO-coated glass slide. This arrangement was beneficial for quick device reuse and was employed for generation of up to 7 individual 235 nL droplets per reservoir. Conversely, automated replenishment of reservoir reagents was utilized when large quantities of droplets needed to be generated. This configuration utilized a neMESYS syringe pump to continuously inject reagents directly into the reservoirs through a capillary imbedded in a channel on the device.

Droplet incubation. At the end of the DMF reaction region is a (50Ã?20Ã?0.175 mm) serpentine chamber for long-term culture of cells on the device. Droplets would naturally drift from the last electrode into this chamber due to convective flow. However, this movement was very slow on the order of minutes resulting in electrically induced droplet coalescence during high-speed operation. To overcome this limitation, a channel was constructed at the interface of the DMF electrodes and culture chamber. A

neMESYS syringe pump connected to the channel via a capillary provided a constant flow of oil (0.1 ÂµL/s) at this interface that propelled the droplet from the last electrode into the culture chamber. This addition eliminated droplet electrocoalescence at the

Â 9

culture chamber entrance. The constant addition of oil to the culture chamber had the additional benefit of replenishing oil lost to evaporation during prolonged experiments. During on-chip culture the oil flow rate was reduced to 0.01 ÂµL/s, which used only 0.86 mL oil for 24 hr cell culture.

E. coli Transformation

DMF heat-shock optimization. We tested the transformation efficiency in our DMF device by transforming *E. coli* cells (DH5Î±) with a GFP-expression plasmid. Unless otherwise stated, 5 individual 705 nL droplets were generated for each experiment. We started by performing chemical heat-shock following recommended protocols for the benchtop method, with minimal modifications to accommodate the droplet-based system. In this trial, 235 nL droplets containing cells and DNA were
merged and mixed at 4oC for 15 min, moved to 42oC for 45 s, allowed a 2 min recovery at 4oC, merged with a 235 nL droplet containing media and antibiotics (ampicillin) then relocated to a 37oC culture chamber for 1 day. Our device can potentially generate libraries of discrete purified DNA plasmids in individual droplets thereby eliminating the conventional requirement of cell plating on agar plates to achieve monoclonal colonies. All droplets possessed GFP positive cells following on-chip culture and afforded a fluorescence increase of 2072 Â± 290 relative fluorescence units (RFUs) (Figure 2C-D). However, this transformation protocol only afforded a transformation efficiency of 1.4 Ã? 106 Â± 0.1 Ã? 106 CFU/Âµg DNA (Figure 2F, 2H1), significantly lower than the standard
benchtop procedure under the same conditions (4.0 Ã? 106 Â± 0.4 Ã? 106 CFU/Âµg DNA) (Figure 2G, 2H1). Through simply changing the heat-shock temperature to 37oC and heat-shock duration to 10 s on the DMF device, the transformation efficiencies were

Â 10

increased to 2.3 Ã? 106 Â± 0.3 Ã? 106 CFU/Âµg DNA and 4.4 Ã? 106 Â± 0.2 Ã? 106 CFU/Âµg DNA, respectively (Figure 2H2, 2H3). These results are likely due to the much faster thermal change of the droplets within the DMF chip with respect to the a 50 ÂµL volume in a
centrifuge tube. These results are consistent with recommended heat-shock procedures for microtiter plates, where shorter heat-shock times are generally employed. These results demonstrate transformation efficiencies in our chip to be comparable to those achieved in benchtop heat-shock with the benefit of using ~100 times less reagent.

Improving heat-shock throughput. While recapitulating standard heat-shock procedures with the DMF chip produced high transformation efficiencies, this approach afforded extremely low throughputs (~1 droplet / hr). Unfortunately, this strategy would make experiments requiring large quantities of droplets infeasible. Time intensive steps

in traditional heat-shock include a 15 min DNA/cell mixing period, 2 min post heat-shock recovery and at least a 30 min outgrowth step. Several heat-shock protocols have demonstrated the length of various steps can be limited, though resulting in compromised transformation efficiencies. We adjusted the droplet duration at each chip region to efficiency investigate the effects of shorting each step on the transformation efficiency (Figure 2H, Supplementary Table 1). Employing ampicillin as a selection marker allows direct droplet transfer to the culture region, bypassing the outgrowth period, because ampicillin is a bacteriostatic thereby affording cells time to take up the DNA and replicate in itâ??s presence with no effect on the transformation efficiency. Reducing the DNA/cell mixing time and recovery period on the chip to 10 s each resulted in a mere 5 fold reduced transformation efficiency (9.0 Ã? 105 Â± 0.9 Ã? 105 CFU/Âµg DNA) (Figure
2H.5) and reduced fluorescence of droplet cultured on chip to 1570 Â± 180 RFUs. This is

Â 11

significantly better than reducing benchtop incubation times by the same factor which resulted in a 54-fold lower transformation efficiency (7.3 Ã? 104 Â± 1.3 Ã? 104 CFU/Âµg DNA). This is likely a result of the enhanced mixing and temperature change in the droplets that quickens DNA delivery and cell recovery. Although a 5-fold reduction in the transformation efficiency is significant, this approach allows heat-shock to be completed in just 30 s. Furthermore, by serially processing of the droplets we obtained a throughput of 360 droplets / hr. As a demonstration of large sample processing this protocol was used for preparing 100 individual droplets in just 17 min. Filled reagent reservoirs were maintained throughout operation by continuously supplying fluids with a syringe pump. Following 24 hrs culture all 100 of the heat-shocked droplets possessed GFP expressing *E. coli* cells. Efforts are underway to develop a larger DMF chip with larger number of electrodes operating in a linear and/or parallel manner to improve the throughput by 2 orders of magnitude.

Varying DNA concentrations. It has previously been shown that changing the quantity of DNA plasmids added to cells has a direct correlation to the number of transformants generated and resulting transformation efficiency. Â 22 We varied the plasmid DNA concentration in each droplet to examine its effect on *E. coli* transformation in the DMF chip. These experiments were performed on the DMF chip operating at the highest throughput (10 s DNA/cell mixing, 10 s heat-shock, 10 s recovery and with ampicillin selection marker). As the DNA concentration in each DNA/cell droplet was changed from 0.001ng/ÂµL to 1 ng/ÂµL the DMF device afforded transformation efficiencies ranging from 2.4 Ã? 106 Â± 0.7 Ã? 106 CFU/Âµg DNA to 1.9 Ã? 105 Â± 0.7 Ã? 105 CFU/Âµg DNA, respectively (Figure 2I). These colony counts equate to each heat-shocked droplet with

Â 12

0.001ng/ÂµL to 1 ng/ÂµL DNA generating 1.7 Â± 0.5 to 114 Â± 50 transformed cells,

respectively. Conversely, standard benchtop heat-shock afforded transformation efficiencies from 1.0 Ã? 107 Â± 0.2 Ã? 107 CFU/Âµg DNA to 9.4 Ã? 105 Â± 1.2 Ã? 105 CFU/Âµg DNA at DNA concentration of 0.001ng/ÂµL to 1 ng/ÂµL, respectively (Figure 2I). The benchtop and DMF systems each generate similar linear regression curves with increasing DNA concentrations, -0.76 and -0.79, respectively, with the benchtop approach affording 3.8 to 4.8 times greater transformation efficiencies though at significantly prolonged heat-shock durations and reagent consumption, as discussed above.
Importantly, all droplets cultured on the DMF chip for 24 hrs following heat- shock with 0.001ng/ÂµL to 1 ng/ÂµL DNA possessed GFP expressing *E. coli*. Transformation of a single DNA plasmid in each isolated droplet at these low plasmid concentrations offers the possibility of library screening without requiring multiple rounds of cell dilution and sorting; each droplet would contain a unique plasmid colony.
Using different plasmids. To demonstrate the versatility of platform, cells were transformed with plasmids containing genes for blue fluorescent protein (BFP) or red fluorescent protein (RFP). Cells transformed with BFP or RFP under identical conditions as the GFP plasmid above afforded slightly lower transformation efficiencies, 4.8 Ã? 105 Â±
0.8 Ã? 105 CFU/Âµg DNA and 5.5 Ã? 105 Â± 1.3 Ã? 105 CFU/Âµg DNA, respectively (Figure 3).
Previous reports have shown a direct correlation between transformation efficiency and plasmid size; it is more difficult for larger plasmids to transverse the created pores and enter cells.22 The small GFP plasmid (3.6 kb) likely could more easily enter the *E. coli* cells than the larger BFP plasmids (4.8 kb) and RFP plasmids (5.3 kb). Successful transformation of BFP and RFP plasmids could also be confirmed after 24 hr on-chip

Â 13

culture. Following imaging with a blue filter set or red filter set the BFP and RFP transformed cells afforded droplets with increased fluorescence, 620 Â± 130 and 2830 Â± 500 RFUs, respectively.
Kanamycin selection. To this point, ampicillin was used as the selection antibiotic because of its role as a bacteriostatic allowing immediate culture following heat-shock. However, during prolonged culture *E. coli* can secrete beta-lactamase that inactivates the ampicillin resulting in potential growth of non-selective cells. When this is the case researchers turn to other antibiotics, such as kanamycin, which is bacteriocidal and immediately primes cells for death.23 To ensure cell survival, an outgrowth step in antibiotic free media is required following heat-shock and prior to the addition of kanamycin. The outgrowth period on the DMF device was recapitulated by holding the droplet on the last DMF electrode for 30 min following heat-shock of *E. coli* with a GFP plasmid with a kanamycin resistance gene. Following outgrowth, a 235 nL droplet
containing 150 Âµg/mL kanamycin was added and the 940 nL droplet moved to the culture region. After 24 hr culture, droplets were imaged through a green filter set and a fluorescence increase of 1170 Â± 230 RFUs was obtained, demonstrating successful transformation and culture of GFP cells. Furthermore, droplets transferred to a kanamycin selective agar plate following outgrowth on-chip and cultured for 24 hrs afforded a transformation efficiency of 4.1 Ã? 105 Â± 0.6 Ã? 105 CFU/Âµg DNA (Figure

2H.6), slightly lower than GFP transformation with ampicillin selection.

Golden gate assembled DNA. To demonstrate the compatibility of the DMF heat- shock device with assembled DNA plasmids we examined the on-chip transformation of pProm1_BCD1-GFP expressing an ampicillin resistance gene into E. coli. DNA

Â 14

assembly was initially performed off-chip utilizing the golden gate assembly method24 and then added to the DMF device and 30 s serial droplet heat-shock performed, as described above. Following 24 hrs on-chip culture, all droplets possessed GFP positive cells and the droplets afforded a 890 Â± 210 RFU increase when fluorescently imaged with a green filter set. Successful on-chip transformation and culture show the DMF heat-shock device could seamlessly integrate with current microdevices working with other aspects of the synthetic biology pipeline.

Fungal Transformation

Yeast Transformation. Eukaryotic systems are frequently used for expressing proteins and pathways that are difficult to produce in bacterial systems with desired yields. Introduction of exogenous DNA into *S. cerevisiae* is routinely performed by heat- shock or electroporation. A common procedure for delivering DNA into *S. cerevisiae* is to heat-shock competent cells with DNA in a PEG solution at 30oC followed by culture on an agar plate.25 Flexibility of the DMF heat-shock device was demonstrated by adapting its operation to transform CEN.PK2 cells with a mCherry generating plasmid. The same heat-shock steps as the 30 s serial droplet transformation protocol was used except for changes in reagent use and operating temperatures. For on-chip culture, a 4
day incubation period was used to account for the longer replication time of yeast cells. Following 4 days incubation, all 5 of the cultured droplets possessed mCherry expressing yeast cells (Figure 4). Transformation efficiencies were assessed by transferring 3 droplets generated on the device to separate â??URA agar plates. 20 Â± 5 colonies were observed after 4 days culture on the agar plate corresponding to a transformation efficiency of 1.0 Ã? 104 Â± 0.2 Ã? 104 CFU/Âµg DNA. Bulk heat-shock following Zymo

Â 15

protocol afforded higher transformation efficiencies (1.0 Ã? 105 Â± 0.2 Ã? 105 CFU/Âµg DNA), likely a result of the increased cell/DNA incubation duration in concentrated PEG solution. Transformation efficiencies lower than those obtained with bacteria are common with yeast due to the difficulty of DNA transport through the cell membrane. Effective transformation in droplets is possible because the PEG helps confine the DNA to the cells following addition of culture media allowing continued gene delivery during the culture period.
Filamentous Fungi Transformation. Filamentous fungi, such as *A. niger*, is another important system for molecular biology and chemical synthesis.26 Â Although laborious

sample preparation was required to generate protoplasts,27 once prepared these cells can be transformed using similar protocols as with yeast heat-shock above. Transformation of *A. niger* protoplasts on the DMF device was examined by testing the incorporation of plasmid DNA expressing GFP and Hygromycin B resistance. A 30 min outgrowth step was included prior to the addition of Hygromycin B to afford sufficiently high gene expression required to retain cell viability in its presence. Five droplets were processed on the device and each droplet contained GFP positive cells following 5 days culture in the DMF chip (Figure 4). Various cell morphologies indicative of fungal growth were observed in the droplet cultures including individual mycelium (hyphal mats), mycelium clumps and individual *A. niger* spores all expressing GFP. Similar morphologies were observed along with larger conidiophores when *A. niger* cells were transformed using the traditional benchtop method with bilayer agar culture.27 Agarose hydrogels have previously been generated by DMF, which could potentially be adapted for recapitulating the bilayer agar culture of transformed *A. niger* cells.28

Â 16

Conclusions

A thermally controlled DMF platform was developed to afford cellular heat-shock and culture in discrete droplets on a single device. Micro peltier modules were attached to the bottom of a DMF chip to provide the different temperatures required during heat- shock. Automated DMF afforded the droplet generation, mixing and movement across the various thermal regions of the device. This high level of droplet control allowed employment of all pivotal heat-shock steps on a single device: cell/DNA mixing, heat-shock, media addition, culture and sometimes outgrowth. The flexibility of DMF and the micro peltier modules allowed easy alteration of the operating protocols based on the experimental requirements. A single device was utilized to optimize the heat-shock parameters of *E. coli* to afford high transformation efficiencies, high throughput sample processing and introduction of various DNAs. Likewise, the same device was used to transform *S. cerevisiae* and *A. niger* cells with exogenous DNA. Integrated gene delivery and culture on a single platform will be of great utility for screening gene variants and systematic investigation of DNA delivery strategies.

METHODS

Device assembly and operation. A digital microfluidic device was initially fabricated as described in the Supporting Information. Prior to operation the device was connected to tubing, peltiers and droplet drive electronics. Segmented control of device temperature was achieved by three peltier heaters/coolers attached to

the bottom of the device by thermal adhesive tape and controlled by an Arduino microcontroller (Modern Device, Providence, RI). A total of three peltiers were attached to the device: 1) a cooler

Â 17

module for DNA/ cell suspension mixing and cell recovery covered by an aluminum block 2) 2 mm peltier for heat-shock and 3) 24.5 mm peltier for cell culture (TE Technology, Inc., Traverse City, MI). Two techniques were employed for liquid addition of DNA, cell suspension, media and oil to the device; pipetting before assembly and continuous flow during operation. For the majority of experiments, 3.0 mL liquid was pipetted directly onto the device reservoirs and the channels were flooded with HFE 7500 then topped with an ITO glass slide and secured with Nd magnets (K&J Magnetics). For experiments requiring more reagents, the ITO glass slide was first bound to the device then ends of capillary tubes connected to a neMESYS syringe pump (Centoni GmbH, Korbussen, Germany) were glued into channels associated with the device reservoirs using 5 minuteÂ® Epoxy Gel (Devcon, Danvers, MA). Following assembly of either technique, a capillary was inserted into a fifth channel at the entrance of the culture chamber from which HFE 7500 was flowed continuously to remove droplets from the reaction chamber and replenish evaporated oil. Finally, the device was connected to a custom-built control board activated by a RBBB Arduino microcontroller (Modern Device, Providence, RI). A customized C++ program controlled the driving potentials of
the Arduino board to allow automation of droplet movement in the device.21
Microfluidic Transformation. *E. coli.* The transformation chip reservoirs were first loaded with all reagents using either a pipette or the syringe pump: 1) 3 ÂµL thawed chemically competent DH5Î± cells, 2) 3 ÂµL DNA solution (0.001 â?? 1.0 ng DNA /ÂµL) and
3) 3 ÂµL Luria-Bertani (LB) broth with selection antibiotics. A 235 nL droplet containing cells was first mixed with a 235 nL DNA droplet at 4oC. This droplet was then moved to a 37oC or 42oC chip region for 10 s or 45 s. The cell/DNA droplet was then merged with

Â 18

a 235 nL media droplet. For ampicillin selection, 300 Âµg/mL ampicillin was included with the media and droplet was immediately moved to the devices culture region. Conversely, kanamycin selection required 30 min outgrowth following media delivery prior to addition of a 235 nL droplet containing 200 Âµg/mL kanamycin. Once delivered to the culture region the droplets were incubated for at least 30 min at 37oC.

S. cerevisiae. A modified procedure from the Frozen-EZ Yeast Transformation II

Kit was employed for transforming CEN-PK cells with mCherry-expressing DNA (Zymo Research Corp., Irvine, CA). The transformation chip reservoirs were initially filled with reagents: 1) 3 ÂµL thawed chemically competent CEN-PK cells in EZ 2 solution, 2) 3 ÂµL mCherry DNA (20 ng/ÂµL) in EZ 3 solution (40-50% PEG) and 3) 3 ÂµL medium minus uracil. As per Zymo protocols for yeast transformation the entire chip was heated to 30oC. A 235 nL droplet containing competent cells in EZ 2 solution was first mixed with a 235 nL DNA droplet mixed with EZ 3 solution 30s. The heat-shocked droplet was then merged with a 235 nL droplet containing media minus uracil and moved to the devices culture region where it was incubated for 4 days at 30oC.

A. niger. Necessary reagents for *A. niger* transformation were first loaded: 1) 3 ÂµL thawed protoplast, 2) 3 ÂµL GFP DNA (20 ng/ÂµL) in 40% PEG 3) 3 ÂµL minimal medium with 1 M sorbitol and 4) 3 ÂµL minimal medium with 1 M sorbitol and 400 Âµg/mL Hygromycin B. A 235 nL droplet containing protoplast was first mixed with a 235 nL
DNA droplet at 4oC for 1 min. This droplet was then moved to a 30oC chip region for 30 min. The cell/DNA droplet was then merged with a 235 nL media droplet and afforded 30 min outgrowth at 30oC. At this point the droplet was merged with the 235 nL

Â 19

Hygromycin B droplet and moved to the devices culture region. Here the droplet was incubated for 5 days at 30oC.

Transformation efficiency *Off-chip culture*. To assess the transformation efficiency of *E. coli*. by the DMF chip, a 705 nL droplet was pipetted from the culture region following 30 min outgrowth. This droplet was immediately diluted with 50 ÂµL LB broth and spread on an LB agar plate containing antibiotics and incubated for 24 h at
37oC. Fluorescence images of the transformed colonies were taken using a FluorChemÂ®Q
MultiImage III system equipped with a CCD camera (Alpha Innotech, Santa Clara, CA). Transformation efficiencies were calculated from the average of 5 droplets and represent the number of colony forming units (CFUs) per Âµg plasmid DNA. These transformation efficiencies were directly compared to the conventional benchtop method performed following previous procedures.23

On-chip culture. The efficiency of DMF transformation and on-chip culture was initially evaluated by imaging droplets with a 4x objective following 1 to 5 days culture. A total droplet fluorescence value obtained by image analysis using ImageJ software was utilized to assess the quality of transformation with fluorescent protein generating DNA. High magnification images of cells were obtained by transferring droplets from the

device to a coverslip by a pipette. The droplet was covered by a second coverslip and then transferred to a microscope for imaging by a 100x objective. All images were taken with an inverted IX73 microscope (Olympus, Center Valley, PA) and iXON+ EMCCD camera (Andor Technology LTD, Belfast, UK).

Â 20

*** Formatting Issue - Best Available Paper/Graphic ***

ASSOCIATED CONTENT

Supporting Information

Details are supplied in the Supporting Information for the following: reagents, device fabrication, cell preparation, plasmid preparation and transformation protocols. This material is available free of charge via the Internet at http://pubs.acs.org.

AUTHOR INFORMATION

Corresponding Author

*E-mail: aksingh@sandia.gov

Notes

The authors declare no competing financial interest.

ACKNOWLEDGMENTS

This work conducted by the Joint BioEnergy Institute was supported by the Office of Science, Office of Biological and Environmental Research, of the U.S. Department of Energy under Contract No. DE-AC02-05CH11231. Sandia National Laboratories is a multi-program laboratory managed and operated by Sandia Corporation, a wholly owned subsidiary of Lockheed Martin Corporation, for the U.S. Department of Energyâ??s National Nuclear Security Administration under contract DE-AC04-94AL85000. The authors thank Dr. Anna Lechner and Dr. Leopold dâ??Espaux for assistance with yeast transformation. *A. Niger* protoplast and GFP plasmid were kindly provided by Dr. Saori Campen and Dr. Jinxiang Zhang.

Â 21

REFERENCES

[1] Paddon, C. J., and Keasling, J. D. (2014) Semi-synthetic artemisinin: a model for the use of synthetic biology in pharmaceutical development, *Nat Rev Microbiol 12*, 355-367.
[2] Rabinovitch-Deere, C. A., Oliver, J. W. K., Rodriguez, G. M., and Atsumi, S. (2013) Synthetic Biology and Metabolic Engineering Approaches To Produce Biofuels, *Chem Rev 113*, 4611-4632.

*** Formatting Issue - Best Available Paper/Graphic ***

[3] Keasling, J. D. (2008) Synthetic biology for synthetic chemistry, *Acs Chem Biol 3*, 64-76.
[4] Bryksin, A. V., Brown, A. C., Baksh, M. M., Finn, M. G., and Barker, T. H. (2014) Learning from nature - Novel synthetic biology approaches for biomaterial design, *Acta Biomater 10*, 1761-1769.
[5] Ben Yehezkel, T., Nagar, S., Mackranrs, D., Marx, Z., Linshiz, G., Shabi, U., and Shapiro, E. (2011) Computer-aided high-throughput cloning of bacteria in liquid medium, *Biotechniques 50*, 124-127.
[6] Huang, H. Y., and Densmore, D. (2014) Integration of microfluidics into the synthetic biology design flow, *Lab on a Chip 14*, 3459-3474.
[7] Lee, C. C., Snyder, T. M., and Quake, S. R. (2010) A microfluidic oligonucleotide synthesizer, *Nucleic Acids Res 38*, 2514-2521.
[8] Kersaudy-Kerhoas, M., Amalou, F., Che, A., Kelly, J., Liu, Y., Desmulliez, M. P. Y., and Shu, W. (2014) Validation of a Fully Integrated Platform and Disposable Microfluidic Chips Enabling Parallel Purification of Genome Segments for Assembly, *Biotechnology and Bioengineering 111*, 1627-1637.
[9] Nagamine, K., Onodera, S., Torisawa, Y. S., Yasukawa, T., Shiku, H., and Matsue, T. (2005) On-chip transformation of bacteria, *Anal Chem 77*, 4278-4281.
[10] Sha, J., Wang, Y., Wang, J., Ren, L., Tu, Q., Liu, W., Wang, X., Liu, A., and Wang, L. (2011) Capillary-composited microfluidic device for heat shock transformation of Escherichia coli, *J Biosci Bioeng 112*, 373-378.
[11] Wei, Z. W., Zhao, D. Y., Li, X. M., Wu, M. X., Wang, W., Huang, H., Wang, X. X., Du, Q., Liang, Z. C., and Li, Z. H. (2011) A Laminar Flow Electroporation System for Efficient DNA and siRNA Delivery, *Analytical Chemistry 83*, 5881-5887.
[12] Lin, Y. C., Jen, C. M., Huang, M. Y., Wu, C. Y., and Lin, X. Z. (2001) Electroporation microchips for continuous gene transfection, *Sensor Actuat B-Chem 79*, 137-143.
[13] Zhan, Y. H., Wang, J., Bao, N., and Lu, C. (2009) Electroporation of Cells in Microfluidic Droplets, *Analytical Chemistry 81*, 2027-2031.

Â 22

[14] Xiao, K., Zhang, M. Y., Chen, S. Y., Wang, L. M., Chang, D. C., and Wen, W. J. (2010) Electroporation of micro-droplet encapsulated HeLa cells in oil phase, *Electrophoresis 31*, 3175-3180.
[15] Qu, B., Eu, Y. J., Jeong, W. J., and Kim, D. P. (2012) Droplet electroporation in microfluidics for efficient cell transformation with or without cell wall removal, *Lab on a Chip 12*, 4483-4488.
[16] Luo, C., Yang, X., Fu, Q., Sun, M., Ouyang, Q., Chen, Y., and Ji, H. (2006) Picoliter-volume aqueous droplets in oil: electrochemical detection and yeast cell electroporation, *Electrophoresis 27*, 1977-1983.
[17] Hong, J. W., Chen, Y., Anderson, W. F., and Quake, S. R. (2006) Molecular biology on a microfluidic chip, *J Phys-Condens Mat 18*, S691-S701.
[18] Sha, J., Wang, Y. L., Wang, J. C., Liu, W. M., Tu, Q., Liu, A. J., Wang, L., and Wang, J. Y. (2011) Heat-shock transformation of Escherichia coli in nanolitre droplets formed in a capillary-composited microfluidic device, *Anal Methods-Uk 3*, 1988-1994.

[19] Au, S. H., Shih, S. C., and Wheeler, A. R. (2011) Integrated microbioreactor for culture and analysis of bacteria, algae and yeast, *Biomed Microdevices 13*, 41-50.
[20] Kim, H., Jebrail, M. J., Sinha, A., Bent, Z. W., Solberg, O. D., Williams, K. P., Langevin, S. A., Renzi, R. F., Van De Vreugde, J. L., Meagher, R. J., Schoeniger, J. S., Lane, T. W., Branda, S. S., Bartsch, M. S., and Patel, K. D. (2013) A Microfluidic DNA Library Preparation Platform for Next-Generation Sequencing, *Plos One 8*.
[21] Shih, S. C., Gach, P. C., Sustarich, J., Simmons, B. A., Adams, P. D., Singh, S., and Singh, A. K. (2014) A droplet-to-digital (D2D) microfluidic device for single cell assays, *Lab Chip 15*, 225-236.
[22] Hanahan, D. (1983) Studies on Transformation of Escherichia-Coli with Plasmids, *J Mol Biol 166*, 557-580.
[23] Ausubel, F. M. (2001) Current protocols in molecular biology, J. Wiley,, New York.
[24] Linshiz, G., Stawski, N., Goyal, G., Bi, C. H., Poust, S., Sharma, M., Mutalik, V., Keasling, J. D., and Hillson, N. J. (2014) PR-PR: Cross-Platform Laboratory Automation System, *Acs Synth Biol 3*, 515-524.
[25] Ito, H., Fukuda, Y., Murata, K., and Kimura, A. (1983) Transformation of Intact Yeast-Cells Treated with Alkali Cations, *J Bacteriol 153*, 163-168.
[26] Lubertozzi, D., and Keasling, J. D. (2009) Developing Aspergillus as a host for heterologous expression, *Biotechnol Adv 27*, 53-75.

Â 23

[27] Szewczyk, E., Nayak, T., Oakley, C. E., Edgerton, H., Xiong, Y., Taheri-Talesh, N., Osmani, S. A., and Oakley, B. R. (2006) Fusion PCR and gene targeting in Aspergillus nidulans, *Nat Protoc 1*, 3111-3120.
[28] Fiddes, L. K., Luk, V. N., Au, S. H., Ng, A. H., Luk, V., Kumacheva, E., and Wheeler, A. R. (2012) Hydrogel discs for digital microfluidics, *Biomicrofluidics 6*, 14112-1411211.

Â 24

*** Formatting Issue - Best Available Paper/Graphic ***

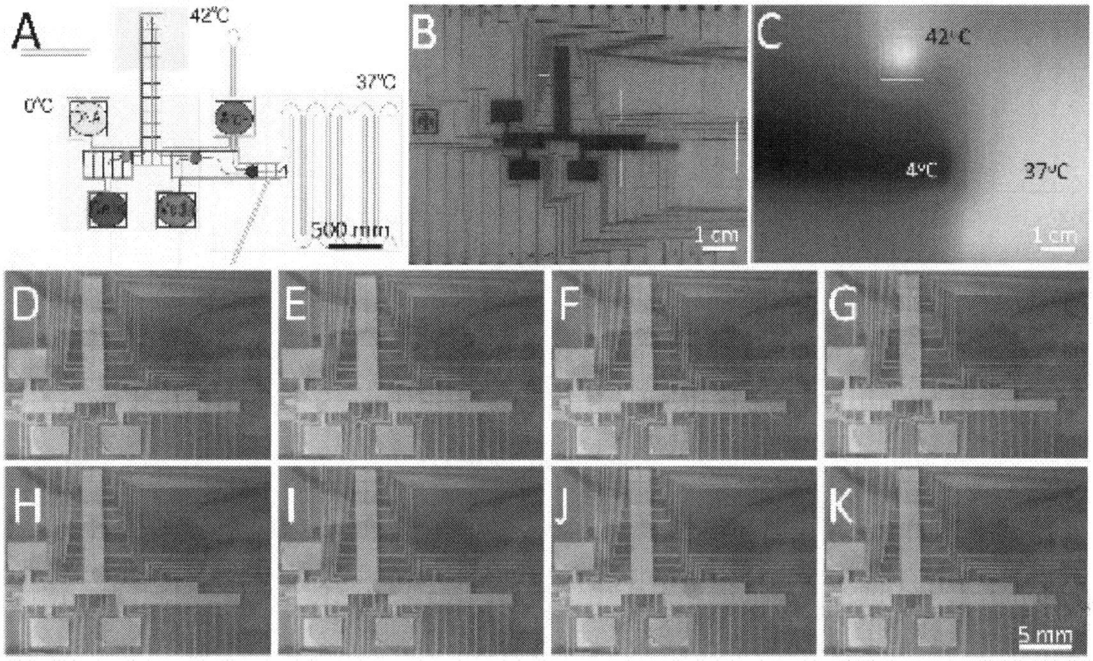

Figure 1. Schematic of the basic operations of the DMF transformation including: droplet generation, merging and relocation to thermally controlled regions (A). Brightfield (B) and infrared (C) images of the DMF device for cell transformation. Image sequence of on-chip DMF transformation. Images of filled reservoirs (D), generation of droplets containing DH5Î± cells or GFP plasmid (E) followed by mixing of droplets (F) at 4oC. Droplet transferred to 42oC region of device (G) then moved to 4oC device region during generation of droplet consisting of LB and AMP (H). Droplets are then merged (I) and allowed to recover at 4oC (J) then transferred to a 37oC culture chamber (K).

Â 25

*** Formatting Issue - Best Available Paper/Graphic ***

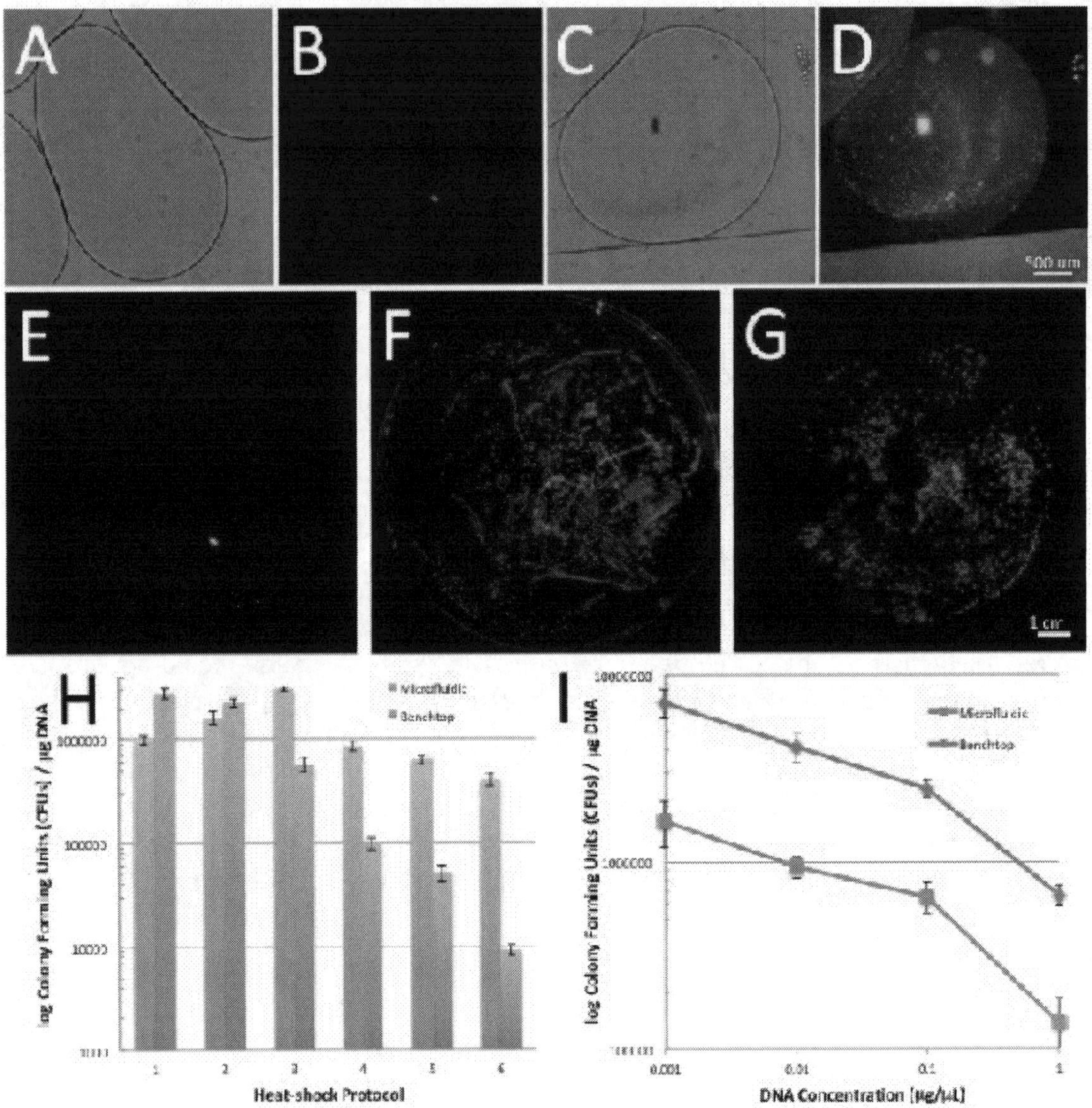

Figure 2. Brightfield and fluorescence images of a droplet containing DH5Î± cells and pRSET-EmGFP plasmid DNA without heat shock (A,B) and following transformation with in the DMF device (C,D). Fluorescence images of LB agar plates containing DH5Î± cells without heat shock (E) and following transformation with GFP plasmid DNA by the DMF device (F) and the conventional tube-based method (G). Plot of transformation efficiencies obtained under various heat shock procedures (H) and with different concentrations of GFP plasmid DNA (I).

*** Formatting Issue - Best Available Paper/Graphic ***

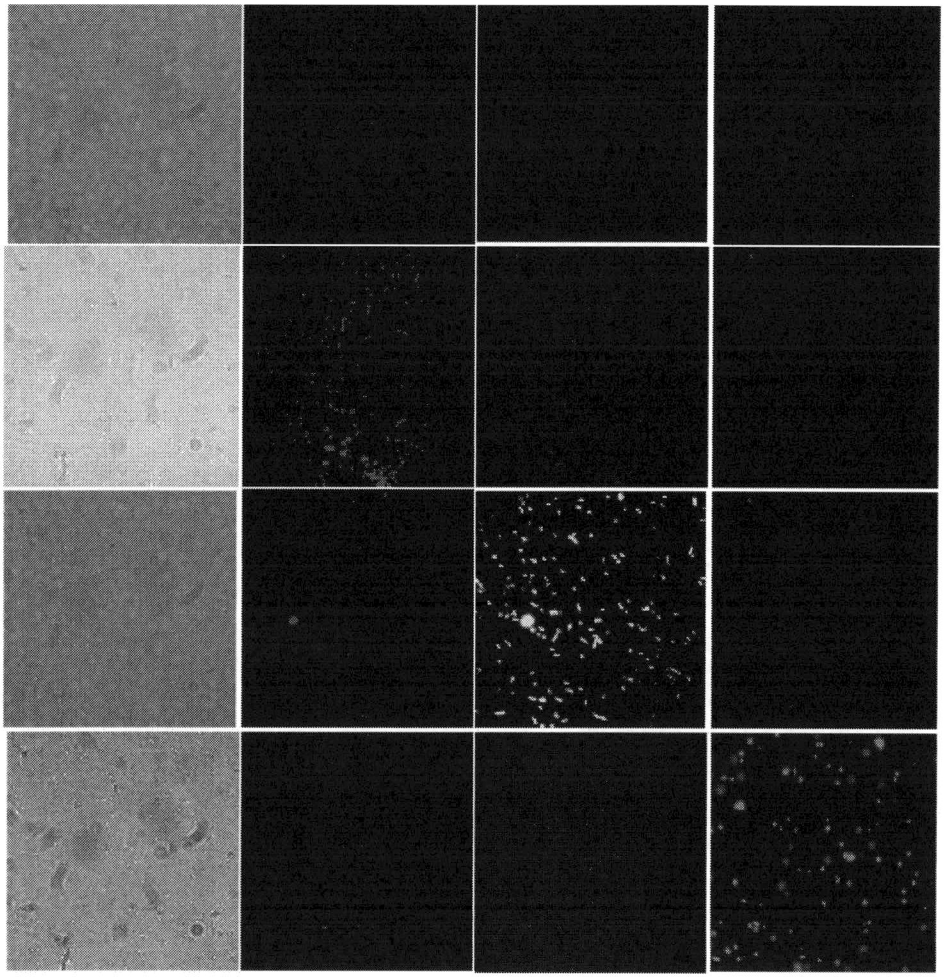

Â 26

BF' Blue' Green' Red'

Control'

BFP' GFP'

RFP

25'u

Figure 3. Brightfield and fluorescence images of DH5Î± cells transformed with blue fluorescence protein (BFP), green fluorescence protein (GFP) or red fluorescence protein (RFP) by the DMF device.

*** Formatting Issue - Best Available Paper/Graphic ***

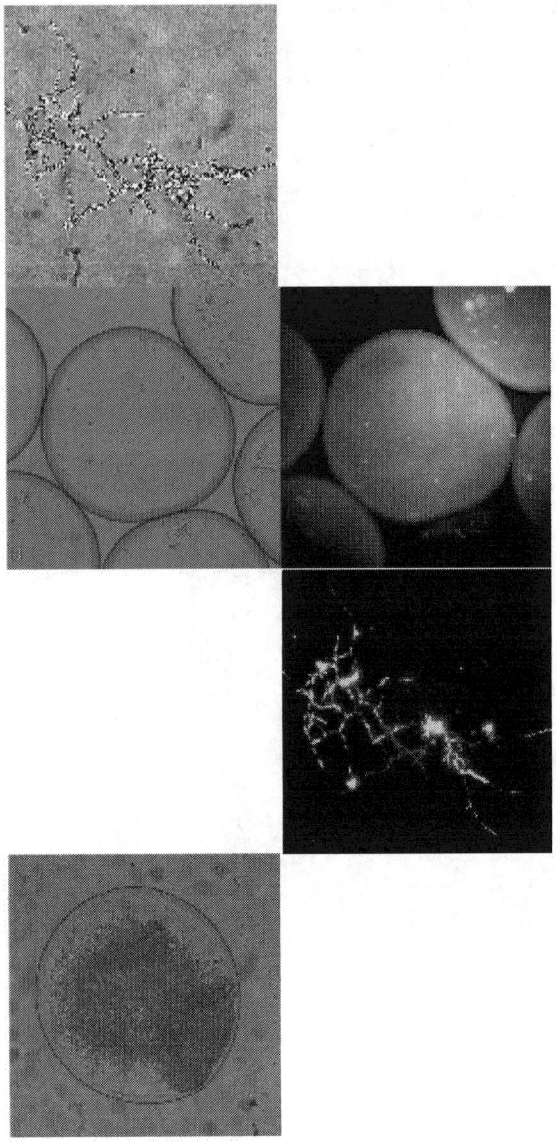

*** Formatting Issue - Best Available Paper/Graphic ***

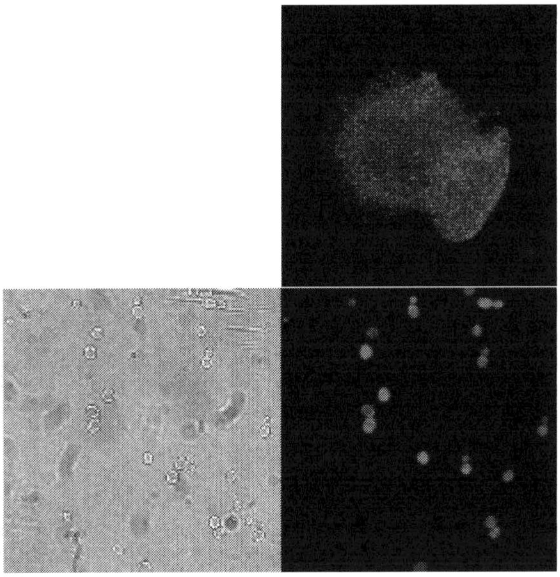

Â 27

Yeast#+#mCherry# A.#Niger#+#GFP##

m# m#

100#um#

Figure 4. Brightfield and fluorescence images of droplets containing CEN.PK cells transformed with mCherry (left) or *Asperagillus niger* cells transformed with GFP (right) following 5 days culture at 30oC on chip.

Â 28

Supporting Information For

Synthetic Biology in a Chip: A Digital Microfluidic Platform for Cell

Transformation, Culture and Expression

Philip C. Gach,1, 2 Steve C.C. Shih,1, 2 Jess Sustarich,1, 2 Jay D. Keasling,1,3,4,5,6,7 Nathan J. Hillson,1,3,8 Paul D. Adams,1,3,4 and Anup K. Singh1, 2 *

1 Technology Division, Fuels Synthesis Division, Joint Bioenergy Institute (JBEI), Emeryville, California 94608, United States
2 Applied Biosciences and Engineering, Sandia National Laboratories, Livermore,

California 94550, United States

3 Physical Biosciences Division, Lawrence Berkeley National Lab, Berkeley, California

94720, United States

4 Department of Bioengineering, University of California, Berkeley, California 94720, United States
5 Department of Chemical & Biomolecular Engineering, University of California,

Berkeley, California 94720, United States

6 QB3 Institute, University of California, Berkeley, California 94720, United States
7 Synthetic Biology Engineering Research Center, Emeryville California 94608, United

States

8 DOE Joint Genome Institute, Walnut Creek, California 94598, United States

* Corresponding Author email: aksingh@sandia.gov tel: (925) 294-1260
fax: 925-294-3020

Â 29

METHODS

Reagents and materials. Unless otherwise specified, general-use reagents were purchased from Sigma Aldrich. Fabrication reagents and supplies included SU-8-5, SU-

8-2025, SU-8-2075, S-1811 and SU-8 Developer from Microchem (Newton, MA), MF-321 positive photoresist developer from Rohm and Haas (Marlborough, MA), CR-4 chromium etchant was from OM Group (Cleveland, OH), and AZ-300T photoresist stripper from AZ Electronic Materials (Somerville, NJ).

Device fabrication. The digital microfluidic heat shock devices were fabricated using standard photolithography as described previously with moderate adjustments.1 The device consisted of a glass bottom, gold electrodes, dielectric layer, channel layer and an ITO cover. All photolithography was performed by exposing a photopolymer through a photomask (Fineline Imaging Inc., Colorado Springs, CO) by an OrielÂ® 500 W Hg exposure system (Newport Corporation, Irvine, CA). Gold electrodes were initially micropatterned on a glass slide by exposing, developing and etching a gold and S1811 coated glass slide (Telic Company, Valencia, CA). Following cleaning with acetone, IPA and O2 plasma (Harrick Plasma, Ithaca, NY): devices were coated with a 5 mm layer of SU-8 5. After selective removal of the SU-8 5 over the side contacts and another cleaning step, 150 mm channels composed of SU-8 2075 were fabricated over the dielectric layer. Following development, devices were rinsed with IPA and DI water, dried with N2 gas and hard-baked for 15 min. at 200Â°C. Prior to assembly, the DMF device and ITO coated glass slide (Delta Technologies, Stillwater, MN) were coated with 0.2 mm filtered Aquapel (TCP Global, San Diego, CA) for 15 min. A kimwipe was then used to remove

Â 30

the Aquapel and the device was rinsed with DI water, dried with N2 gas and kept at 21Â°C
for 1h.

Cell Preparation. *Chemically competent E. coli.* A common strain of *Escherichia coli (E. coli)* for laboratory cloning (DH5Î±) was acquired from the JBEI registry (https://registry.jbei.org) and made chemically competent using a standard CaCl2 procedure.2 Reagents and materials were kept chilled at 4oC during all aspects of cell preparation. Briefly, a bacterial cell colony in Luria-Bertani (LB) broth was cultured in a 37Â°C incubator at 200 rpm until early log phase was achieved (OD 600 = 0.3) measured with a spectrophotometer. At this point, the cell suspension was placed on ice for 15 min then centrifuged for 10 min at 3300 g. The medium was then replaced with 0.1 M CaCl2 in which the cells suspended on ice for 30 min. The suspension was again centrifuged for 10 min at 3300 g and the supernatant exchanged for a 0.1 M CaCl2 solution containing 15% glycerol. The cell suspension was then aliquoted into micro-centrifuge tubes and stored at -80oC for future use.

Chemically competent yeast. A commercially available kit (Frozen-EZ Yeast Transformation IITM) from Zymo Research Corp. (Irvine, CA) was used to prepare competent *Saccharomyces cerevisiae* (CEN.PK) cells obtained from the JBEI registry. Following the manufactures protocol, cells were grown at 30oC at 200 rpm until mid-log phase was reached (OD600 = 0.9). The cell suspension was then centrifuged at 500 g for 4 min and media replaced with EZ 1 solution (â?¤1% Tris and â?¤1% EDTA in water). Suspended cells were again centrifuged at 500 g for 4 min and the supernatant replaced with EZ 2 solution (1M sorbitol and 7% DMSO in water). Competent cells are then aliquoted into micro-centrifuge tubes and kept frozen at -70oC until needed.

Â 31

Protoplast preparation. Aspergillus niger protoplasts were prepared using a previous protocol with minor modifications.3 Briefly, 106 *A. niger* (ATCCÂ® 11414TM) conidia in 1 mL H2O were added to 100 mL YEPD broth and cultured in a 30oC
incubator while shaking at 150 rpm for 12 hrs. A Miracloth was used to filter the mycelia, which were then washed with stile H2O. Mycelia were digested in 40 mL Vinoflow solution (30 mg/mL Vino Taste Pro, 0.6 M Ammonium Sulphate, 50 mM Maleic acid, pH 5.5) at 30oC, shaking at 70 rpm for 4 hrs. At this point, the culture was filtered through a Miracloth and the filtrate centrifuged at 800 g for 10 min. The supernatant was then

removed and the pellet washed twice with 25 mL of ST solution (1 M sorbitol, 50 mM Tris, pH 8.0) then resuspended in 10 mL STC solution (1 M sorbitol, 50 mM Tris, pH 8.0, 50 mM CaCl2). This suspension was centrifuged at 800 g for 10 min and then supernatant replaced with fresh STC solution until a protoplast concentration of 1.2Ã?107 spores/mL was achieved. 40% PEG 4000 in STC solution is then gently mixed with this suspension (20% v/v), which is then mixed with DMSO (7% v/v). Protoplasts are then aliquoted into micro-centrifuge tubes and frozen at -80oC.

Plasmid Preparation. E. coli. Plasmid DNA was extracted from E. coli cells

expressing blue fluorescence protein (pBAD-mTag BFP2) 4 acquired from Addgene, green fluorescence protein (pRSET-EmGFP) from Life Technologies (Grand Island, NY) or red fluorescence protein (DsRed.T3)5 and kanamycin resistant green fluorescence protein obtained from the JBEI registry (https://registry.jbei.org) using QIA miniprep from Qiagen (Valencia, CA).
pProm1_BCD1-GFP DNA fragments used for Golden Gate assembly were prepared by PCR using primers and plasmids described elsewhere.6 Briefly, 50 ÂµL PCR

Â 32

reactions consisted of 2.5 ÂµL (2.5 ÂµM) of each forward and reverse primer, 1 ÂµL template, 1 ÂµL dNTPs (10 mM), 0.5 ÂµL high fidelity iProof phusion polymerase (BioRad; Hercules, CA), 10 ÂµL 5Ã?~ high fidelity phusion buffer, and 32.5 ÂµL deionized water.
Four 50 ÂµL PCR reactions (200 ÂµL total) were performed for each fragment amplified. The following PCR thermocycling conditions were used: denaturation at 98Â°C for 30 s, 38 cycles of denaturation at 98Â°C for 20 s, annealing at 68 Â° C for 30 s, and elongation at
72Â°C for 15 s each kb, and a final extension at 72Â°C for 10 min. Following PCR amplification the samples were digested and purified. At this point a modified Golden Gate DNA assembly method was performed. 1 ÂµL BsaI-digested BCD1-GFP fragment was mixed with 1 ÂµL BsaI-digested pProm1 promoter fragment, 1 ÂµL BsaI-digested vector backbone, 1 ÂµL T4 ligase enzyme (Thermo Scientific), 1 ÂµL of T4 ligase buffer and 5 ÂµL deionized water for 30 min at room temperature.

S. cerevisiae. pL211 containing a TEF promotor and ADH1, URA- and mCherry was provided by the JBEI registry (https://registry.jbei.org).

A. niger. GFP genes were cloned into an A. Niger/ E. coli Shuttle vector at JBEI

as described previously.

Supplementary Table 1. Heat-shock protocols

***** Formatting Issue - Best Available Paper/Graphic *****

Protocol name Description Antibiotic DNA - cell mixing Heat-shock Recovery Outgrowth Culture duration / duration / duration / duration / duration /

temperature temperature temperature temperature temperature

1 Long heat-shock at 42 oC Ampicillin 15 min / 4 oC 45 s / 42 oC 2 min / 37 oC 0 min 1 day / 37 oC

2 Long heat-shock at 37 oC Ampicillin 15 min / 4 oC 45 s / 37 oC 2 min / 37 oC 0 min 1 day / 37 oC

3 Quick heat-shock Ampicillin 15 min / 4 oC 10 s / 42 oC 2 min / 37 oC 0 min 1 day / 37 oC

5 Quick mixing and heat-shock Ampicillin 10 s / 4 oC 10 s / 42 oC 2 min / 37 oC 0 min 1 day / 37 oC

6 Quick mixing, heat-shock and recovery Ampicillin 10 s / 4 oC 10 s / 42 oC 10 s / 37 oC 0 min 1 day / 37 oC

7 Kanamycin resistance Kanamycin 10 s / 4 oC 10 s / 42 oC 10 s / 37 oC 30 min / 37 oC 1 day / 37 oC

8 Yeast heat-shock Ampicillin 10 s / 30 oC 10 s / 30 oC 10 s / 37 oC 0 min 4 days / 30 oC

9 A. niger heat-shock Hygromycin B 10 s / 4 oC 10 s / 30 oC 10 s / 37 oC 30 min / 30 oC 5 days / 30 oC

Â 33

References

[1] Shih, S. C., Gach, P. C., Sustarich, J., Simmons, B. A., Adams, P. D., Singh, S., and Singh, A. K. (2014) A droplet-to-digital (D2D) microfluidic device for single cell assays, *Lab Chip 15*, 225-236.
[2] Gannon, F., Neilan, J., and Powell, R. (1988) Current Protocols in Molecular-Biology - Ausubel,Fm, *Nature 333*, 309-310.
[3] Szewczyk, E., Nayak, T., Oakley, C. E., Edgerton, H., Xiong, Y., Taheri-Talesh, N., Osmani, S. A., and Oakley, B. R. (2006) Fusion PCR and gene targeting in Aspergillus nidulans, *Nat Protoc 1*, 3111-3120.
[4] Subach, O. M., Cranfill, P. J., Davidson, M. W., and Verkhusha, V. V. (2011) An Enhanced Monomeric Blue Fluorescent Protein with the High Chemical Stability of the Chromophore, *Plos One 6*.
[5] Sorensen, M., Lippuner, C., Kaiser, T., Misslitz, A., Aebischer, T., and Bumann, D. (2003) Rapidly maturing red fluorescent protein variants with strongly enhanced

*** Formatting Issue - Best Available Paper/Graphic ***

brightness in bacteria, *Febs Lett 552*, 110-114.
[6] Linshiz, G., Stawski, N., Goyal, G., Bi, C. H., Poust, S., Sharma, M., Mutalik, V., Keasling, J. D., and Hillson, N. J. (2014) PR-PR: Cross-Platform Laboratory Automation System, *Acs Synth Biol 3*, 515-524.

Resource Allocation and Whole Cell Response in Heterologous Gene Expression

Ceroni, F., Imperial College London
Stan, G. B., Imperial College London
Ellis, T., Imperial College London

The expression of exogenous proteins triggers physiological changes in the host cell, usually leading to decreased growth due to consumption of cellular resources. This loading-effect results in undesired cross-interactions and can cause many of the unpredictable behaviours characteristic of the failures common to synthetic biology. Working in standard *E. coli*strains, we show here that the 'load' of exogenous expression mainly impacts on the cell's translational resources and that this effect can be measured with a plate-based assay and predicted by a mathematical model of protein synthesis. We demonstrate that for the control of gene expression, riboswitches are less costly for the cell compared to equivalent transcription factor regulators, due to fewer resources being required for the production of RNA parts compared to proteins.

Extending from this we show preliminary RNAseq analysis of the *E. coli* transcriptome response to induced exogenous expression in the cell. Transcriptome profiling by this method provides a route to identify native genes whose expression changes in response to burden. We find that expression of many genes associated with general stress response, including ibpA, htpG, dnaK and lon, rapidly increase in the first 15 minutes of synthetic overexpression from an engineered plasmid. These native genes are thus useful biomarkers for host cell stress response to exogenous expression and their promoters can be exploited to engineer future burden-based feedback genetic devices.

*** Formatting Issue - Best Available Paper/Graphic ***

Spytag/Spycatcher: Programmable Protein Assembly for Synthetic Biology
Zakeri, B., MIT

Bijan Zakeri
02-Ââ? 06-Ââ? 15

SpyTag/SpyCatcher: programmable protein assembly for synthetic biology

Bijan Zakeri, Allen Y. Chen, Zhengtao Deng, Peter A. Carr, and Timothy K. Lu

Protein-Ââ? protein interactions in biology are generally weak and temporary. Yet many applications in synthetic biology require the stable and long-Ââ? term assembly of proteins post-Ââ? translationally. Conventional assembly methods that rely on non-Ââ? covalent interactions lead to the engineering of unstable and transient molecular architectures that are not ideal for most downstream applications. Here we present the application of the SpyTag/SpyCatcher covalent tagging system for genetically programmable and irreversible engineering of covalent protein architectures both *in vitro* and *in vivo*. We demonstrate covalent protein assembly for engineering living materials using biofilms and nanoparticles that bridge the biotic-Ââ? abiotic interphase. Furthermore, we demonstrate the intracellular assembly of a resistance marker for the maintenance of dual plasmids with a single origin, for applications in molecular steganography for DNA camouflage.

Hands-Free Control of Heterologous Gene Expression in Batch Cultures

Borkowski, O., INRA
Endy, D., Stanford University
Subsoontorn, P., Stanford University

Autonomous control of heterologous gene expression can simplify batch-culture bioproduction by obviating the need for monitoring growth condition. Existing technologies include the uses of special media, wasteful cell-cell communication or biochemical sensors designed only for specific synthetic pathways. There is a need for simple, resource-efficient and general-purpose autonomous control system responding to changing growth condition in batch cultures.

We exploited *E.coli* stationary-phase promoters as sensors for growth-phase transition in batch cultures. We used a recombinase switch to relay transcription signal between a stationary-phase promoter and an output heterologous gene. Specifically, transcription signal from the stationary-phase promoter triggers recombinase expression; recombinase enzyme then inverts a constitutive promoter that turns ON the output gene. We demonstrated that we could predict switching efficiency given measured promoter dynamic ranges and recombinase cassettes with standardized translation initiation elements. We showed that the recombinase switch amplifies signal fold change, lowers cell-to-cell signal variability and eliminates signal fluctuation.

We introduced an autonomous control system for heterologous gene expression. Our system uses stationary-phase promoters and recombinase switches to sense and relay endogenous signals during growth-phase transition. Unlike other existing technologies, our system does not require special media or resource-consuming production of cell-cell signaling molecules. Together, this work provides a simple, generic and efficient tool for batch-culture bioproduction.

Construction of Novel Malate Sensing Recombinant Escherichia coli By Engineering of Bacterial Two-Component System

Hong, S., University of Ulsan

In an attempt to develop a high throughput screening system for screening microorganisms which produce high amounts of malate, a MalKZ chimeric HK-based biosensor was constructed. Considering sequence the similarity among Escherichia coli MalK with Bacillus subtilis MalK and E. coli DcuS, the putative sensor domain of MalK was fused with the catalytic domain of EnvZ. The chimeric MalK/EnvZ (MalKZ) TCS induced the ompC promoter through the cognate response regulator, OmpR, in response to extracellular malate. Real-time quantitative PCR and GFP fluorescence studies showed increased ompC gene expression and GFP fluorescence as malate concentration increased. By using this strategy, various chimeric TCS based bacteria biosensors can be constructed, which may be used for the development of biochemical-producing recombinant microorganisms. This work was supported by a grant from the Next-Generation BioGreen 21 Program (SSAC, grant number: PJ011116) by RDA, and Basic Science Research Program by the Ministry of Education (NRF-2014R1A1A2054726).

Engineering Bacterial Microcompartments to Shield Toxicity during Protein Expression and Purification

Yung, M. C., Lawrence Livermore National Laboratory
Coleman, M. A.,
Bourguet, F. A.,
Carpenter, T. S., Lawrence Livermore National Laboratory

Engineering bacterial microcompartments to shield toxicity during protein expression and purification

M. C. Yung, M. A. Coleman, F. A. Bourguet, T. S. Carpenter
Bacterial microcompartments (BMCs) are â??organellesâ? that are natively produced by bacteria to organize and sequester enzymes in a biosynthetic pathway within the confines of a protein shell. BMCs are thought to shield the toxicity of pathway intermediates and increase reaction efficiency. As such, there is significant interest to engineer BMC shell proteins for biotechnology applications. Here, we express toxic proteins inside BMC shell proteins in *E. coli* as a platform for protein expression and purification. We demonstrate that BMCs can shield toxicity of these proteins during expression. Our results have broad utility to improve cytoplasmic expression of toxic proteins for protein purification applications.
(Work at LLNL conducted under Contract DE-AC52-07NA27344. LLNL-ABS-667012.)

Metabolic Engineering of Ralstonia Eutropha for Biosynthesis of Natural and Unnatural Polyhydroxyalkanoates from Renewable Resources

Park, S. J., Myongji University
Oh, Y. H., Korea Research Institute of Chemical Technology
Lee, S. Y., Korea Advanced Institute of Science and Technology (KAIST)

Metabolic engineering of *Ralstonia eutropha* for biosynthesis of natural and unnatural polyhydroxyalkanoates from renewable resources

Si Jae PARK1, Young Hoon OH2, Sang Yup LEE3

1Department of Environmental Engineering and Energy, Myongji University, Yongin-si, Gyeonggido 449-728, Republic of Korea, parksjdr@gmail.com

2Research Center for Biobased Chemistry, Korea Research Institute of Chemical Technology, Daejeon 305-600, Republic of Korea, yhoh@krict.re.kr

3Department of Chemical and Biomolecular Engineering, KAIST, Daejeon, 305-701, Republic of

Korea, leesy@kaist.ac.kr

Polyhydroxyalkanoates (PHAs) have been considered as promising environmentally friendly alternatives of petroleum-based polymers due to their material properties that are similar to those of chemically synthesized plastics. Since the cost effective production of PHAs is one of the most important factors for the commercialization of PHAs, much effort has been devoted to the development of microorganisms able to efficiently utilize cheap carbon sources such as sucrose and glycerol. For example, sucrose is one of the most abundant and least expensive carbon sources extracted from sugarcane and sugarbeet. Thus, development of microorganisms capable of utilizing cheap carbon sources can provide cost-competitiveness of fermentation-driven products.
Here, we report recombinant *Ralstonia eutropha* strains able to produce natural and unnatural PHAs from various renewable resources derived from biomass as a carbon source and detailed results will be presented in this presentation.

Acknowledgements

This research was supported by the Basic Science Research Program through the National Research Foundation of Korea (NRF) funded by the Ministry of Education

(NRF-2013R1A1A2058379) and by the Technology Development Program to Solve Climate Changes (Systems Metabolic Engineering for Biorefineries) from the Ministry of Science, ICT and Future Planning through the NRF (NRF-2012- C1AAA001-2012M1A2A2026556).

*** Formatting Issue - Best Available Paper/Graphic ***

Design and Construction of a Switchable Bistable Switch in Yeast

Yang, Y.,
Klavins, E.,

Design and Construction of a Switchable Bistable
Switch in Yeast

Yaoyu Yang, Eric Klavins Department of Electrical Engineering University of Washington, Seattle, WA

Abstractâ??The genetic bistable switch is a fundamental build- 1"

ing block for engineering synthetic living systems. In synthetic biology, researchers have designed and constructed rudimentary bistable switches in *E. coli* and in mammalian cells. However, it remains challenging to construct a reusable, modular, switchable bistable switch in eukaryotes. Here we describe a design for such

a bistable switch in *S. cerevisiae* (yeast), develop a predictive 2"

mathematical model, and present experimental results demon-

FBox'

GEV" 3" GAVNY"

strating the functionality, and in particular, the switchability, of the switch.

I. INTRODUCTION

G! E! V!

G. GAL4 DNA binding domain!

E. Estrogen Receptor!

V. VP16!

Y!

G! A! V! N!

G. GAL4 DNA binding domain!

A. Auxin degron IAA17.T2!

V. VP16!

N. Nuclear localization signal (2x SV40)!

Y. EYFP!

The genetic bistable switch is arguably a key building block for engineering synthetic living systems as it forms an addressable cellular memory unit and could be used for controlling cell states [1]. Researchers have designed and constructed bistable switches resulting in various designs and implementations from *E. coli* to mammalian cells over the past 15 years of synthetic biology [1], [3], [4]. However, currently there has not been a robust inducible bistable switch in yeast. Here, we introduce a design and implementation of a switchable bistable switch in yeast. Specifically, the switch can can be switched to a high state by adding Î²-estradiol and switched to a low state by exposure to the plant hormone auxin. Experimental results also show that the switch is able to maintain either its low or high state without the continued presence of these small molecules.

II. A BISTABLE SWITCH DESIGN IN YEAST

The design incorporates a positive feedback loop to achieve bistability and uses two other components to make the cir- cuit switchable via exposure to small molecules. The circuit diagram is shown in Fig. 1. The circuit consists of three components: GEV, a Î²-estradiol sensitive transcriptional acti-

***** Formatting Issue - Best Available Paper/Graphic *****

Fig. 1. The bistable switch design in yeast 1) Circuit diagram of the yeast bistable switch. 2) Construction scheme of GEV where it consists of three protein subdomains. GAL4 DNA binding domain, which recognizes pGAL1 promoter; the estrogen receptor domain interacts with Î²-estradiol, which makes the activator inducible by Î²-estradiol; VP16 is an transcriptional activation domain. 3) Construction scheme of GAVNY where it consists of four protein subdomains and a fluorescence tag. IAA17.T2 represents for T2 domain of Aux/IAA17 protein from *Arabidopsis thaliana* [2] and nuclear localization signal is responsible for brining the protein into nucleus to perform transcriptional control function.

each constitutively expressed through promoters pACT1 and pGPD. The GAVNY is fluorescent, as it is fused to an EYFP reporter. The functionality of this design is first explored by the following mathematical models.
III. MATHEMATICAL MODELS
This circuit can be modeled through a biochemical reaction network derived from the circuit design. Assuming mass action kinetics, we get the following reaction rate equation for GAVNY based on the assumption that other related species are in quasi-steady-state: vator developed and widely used in yeast [5], which activates promoter pGAL1 in the presence of Î²-estradiol; GAVNY, a novel auxin sensitive transcriptional activator that we built and

dA Î»a An + Î»b v

=

dt 1 + Ï□a An + Ï□b v

Î±f Î³af Au

â?? Î»f + Î³af Au

â?? Î´a A (1)

*** Formatting Issue - Best Available Paper/Graphic ***

activates pGAL1 forming the positive feedback loop; FBox from *Arabidopsis thaliana* [2], which degrades GAVNY when induced with auxin. We hypothesize that the circuit behaves as follows. When induced with Î^{2}-estradiol, GAVNY is produced by the activation of GEV and stays at a high state owing to the positive feedback loop even

*** Formatting Issue - Best Available Paper/Graphic ***

when the Î²-estradiol is removed; when induced with auxin, GAVNY is completely degraded by the FBox and stays at low state even when auxin is removed. The GEV and FBox in this circuit are

Here A represents the concentration of GAVNY, v represents the concentration of Î²-estradiol and u represents the concentra- tion of auxin. Î»a and Ï□a are parameters modeling the activation strength of GAVNY by itself, n is the Hill coefficient, Î»b and Ï□b are parameters responsible for the activation strength of GAVNY by GEV that induced by Î²-estradiol, Î±f , Î³af and Î»f are parameters that capture the degradation of GAVNY induced by auxin, Î´a models the basal degradation and dilution rate of GAVNY.

1"

200"

u"="0,"v"="1" u"="1,"v"="0" u"="0,"v"="0"

1" 5000"

100"nM" Î²:estradiol"

20"Î¼M"" auxin"

no"inducer"

0"

2" 200"

0" 300" 600" 900"

1000"

2"

0" 5" 17" 37"

u"="0,"v"="1" u"="0,"v"="0"

0"

0" 300" 600" 900"

t"

***** Formatting Issue - Best Available Paper/Graphic *****

5000" :

:

1000":

100"nM"

Î²:estradiol" no"inducer"

0" 5" 17" 37"

Time"(hr)"

Fig. 2. Numerical simulations demonstrating bistability and switchability..

1) The initial condition is A = 0, the system is first induced with Î²-estradiol shown as u = 0, v = 1 and the system is switched to the high state. Then at t = 300, Î²-estradiol is removed and auxin is added, simulated as u =

1, v = 0, the system is switched back to the low state. After t = 600, auxin is removed and there is no inducer around, simulated as u = 0, v = 0, and the system is still kept at the low state. 2) The system is first induced with Î²- estradiol and switched to the high state. Then Î²-estradiol is removed and the system is still kept at the high state. Parameters used in both simulations are n = 2, Î»a = 0.08, Ï a = 0.01, Î´a = 0.045, Î±f = 8, Î³af = 0.113, Î»f =

1, Î»b = 1, Ï?b = 1.

Analysis show that this system can exibit bistability under certain parameter constraints. In Fig. 2, we show simulation results that demonstrate the bistability and switchability. of this system.

IV. EXPERIMENTAL CONSTRUCTION AND RESULTS

The bistable switch was constructed following the design shown in Fig. 1. pACT1-GEV, pGAL1-GAVNY, and pGPD- FBox were first constructed on ampicillin resistant plasmids, then digested and integrated into the yeast genome sequen- tially.

The preliminary results for validating the functionality of this inducible bistable switch are shown in Fig. 3. The yeast cells were cultured in 30 C and fluorescence measurements were taken using a flow cytometer. The preliminary experi- mental results show that the circuit was induced to a high state with the addition of 100 nM Î²-estradiol and switched to a low state by adding 20 ÂµM auxin. The data also shows that the circuit was able to keep at the low state when auxin was diluted out and remained in the high state when Î²-estradiol was diluted out. These results demonstrate the bistability and switchability of this circuit.

*** Formatting Issue - Best Available Paper/Graphic ***

V. CONCLUSION

In this abstract, we presented an inducible bistable switch design and implementation in yeast. A design scheme for the inducible bistable switch was described and a mathematical model was developed to explore the functionality of this

Fig. 3. Experimental results demonstrating bistability and switchability.

1) The circuit was first induced with 100 nM Î²-estradiol at $t = 0$, the fluorescence measurement at $t = 5$ shows that the circuit switched to the high state. Also at $t = 5$, Î²-estradiol was diluted out and 20 ÂμM auxin was added, after 12 hrs, data shows that the circuit switched to the low state. Auxin was diluted out at $t = 17$ and data after 20 hrs shows that the circuit kept at low state. 2) The circuit was first induced with 100 nM Î²-estradiol at $t = 0$, data at $t = 5$ shows that the circuit switched to high state. At $t = 5$, Î²-estradiol was diluted out and later fluorescence measurements at $t = 17$, $t = 37$ show that the circuit kept at high state.

design. An experimental implementation was constructed and preliminary results were presented. The results showed a bistable switch in yeast that can be switched into high and low state using two different inducers Î²-estradiol and auxin. This switch is able to remain at either states for the time duration tested in our preliminary experiments. These results thus describe the first switchable synthetic bistable switch in *S. cerevisiae.*

REFERENCES

[1] T. S. Gardner, C. R. Cantor, and J. J. Collins. Construction of a genetic toggle switch in escherichia coli. *Nature*, 403(6767):339â??342, 2000.

[2] K. A. Havens, J. M. Guseman, S. S. Jang, E. Pierre-Jerome, N. Bolten, E. Klavins, and J. L. Nemhauser. A synthetic approach reveals extensive tunability of auxin signaling. *Plant Physiol.*, 160(1):135â??142, Sep 2012.

[3] B. P. Kramer, A. U. Viretta, D. M. Baba, D. Aubel, W. Weber, and M. Fussenegger. An engineered epigenetic transgene switch in mam- malian cells. *Nature biotechnology*, 22(7):867â??870, 2004.

[4] T. Lebar, U. Bezeljak, A. Golob, M. Jerala, L. Kadunc, B. PirsË?, M. StrazË?ar, D. VucË?ko, U. ZupancË?icË?, M. BencË?ina, et al. A bistable genetic switch based on designable dna-binding domains. *Nature communications*, 5, 2014.

[5] R. S. McIsaac, S. J. Silverman, M. N. McClean, P. A. Gibney, J. Macin- skas, M. J. Hickman, A. A. Petti, and D. Botstein. Fast-acting and nearly gratuitous induction of gene expression and protein depletion in saccharomyces cerevisiae. *Molecular biology of the cell*, 22(22):4447â??

4459, 2011.

The Joint Genome Institute's Synthetic Biology Internal Review Process

Simirenko, L., Joint Genome Institute
Harmon-Smith, M.,
Visel, A.,
Rubin, E., Joint Genome Institute
Hillson, N., Lawrence Berkeley National Labs

Serious concerns have been raised over the possible intentional use of synthetic biology approaches to engineer pathogenic organisms as well as the possible accidental environmental release of genetically engineered organisms. Scientists pursuing synthetic biology research must diligently consider issues such as these. As such, the U.S. Department of Energy (DOE) Joint Genome Institute (JGI) has developed a Synthetic Biology Internal Review process to assess, beyond technical and scientific merit, certain broader aspects (e.g. environmental, biocontainment, biosafety, or biosecurity) of the research proposals associated with the JGI's DNA synthesis program.

The JBEI Experiment Data Depot

Morrell, W., Sandia National Laboratory
Echols, N., Lawrence Berkeley National Laboratory
Birkel, G., Lawrence Berkeley National Laboratory
Vaino, I., Joint BioEnergy Institute
Hillson, N. J., Joint BioEnergy Institute
Garcia-Martin, H., Joint Bioenergy Institute

Synthetic Biology research draws on several disciplines and ties together diverse data sets. Collaboration efforts between researchers is often hampered by inaccessible or difficult-to-access data. Results are scattered across instruments, document folders, and email accounts. Discovering whom to ask for a copy of the needed data can be a frustrating and time-consuming process. Deciphering data in varied and exotic spreadsheet layouts and file formats can also add delays.

The Experiment Data Depot (EDD) is a database and associated set of software tools meant to ease the management and organization of "actionable" experimental data sets -- meaning data that can be fed directly into computational models and other analysis tools. The aim of the EDD is explicitly not to be a generic sample-tracking or laboratory information management system (LIMS). Rather, it is designed to be a repository of data in a Design-Build-Test-Learn workflow, focused on capturing information resulting from primarily the Test phase. Data is organized into Studies, with read/write access controls for supervisors, colleagues, and collaborators. Each Study collects metadata on the experiment setup and measurement protocols used, and links to strain records in the JBEI Inventory of Composable Elements (ICE). Data import from lab instruments - such as mass spectrometers and gene sequencers - is facilitated with custom extensions. The ability to inspect data and metadata is integrated into the user interface, along with exports to SBML, Excel, and CSV files for use in other software packages.

Produce Steviol Glycosides in Engineered Yeast

Mao, G., Conagen Inc.
Vick, J. E.,
Yu, O., Conagen Inc.

Steviol glycosides are natural products isolated from *Stevia rebaudiana* leaves, and are widely used as high intensity, low-calorie sweeteners. Naturally occurring steviol glycosides have the same base structure (steviol) and differ in the content of carbohydrate residues (e.g. glucose, rhamnose, and xylose residues) at the C13 and C19 positions of the diterpene steviol. *Stevia* extracts are commercially available, which typically contain stevioside (ST) and rebaudioside A (Reb A) as the primary compounds. Although these two compounds are much sweeter than sucrose on per unit base, they tend to have strong off-flavor taste, limiting their applications in foods and beverages. Other steviol glycosides, which demonstrate many desirable taste properties, typically are present in *Stevia* extracts at much lower concentrations.

The majority of the steviol glycosides are formed by several glycosylation reactions of steviol, which typically are catalyzed by the UDP-glycosyltransferases (UGTs) using uridine 5'-diphosphoglucose (UDP-glucose) as a donor of the sugar moiety. In plants, UGTs are a very divergent group of enzymes that transfer a glucose residue from UDP-glucose to steviol. In order to identify the key UGTs in the biosynthesis pathway of steviol glycosides, bioinformatics analysis was performed and selected UGT candidate genes were enzymatically characterized. This work identified several UGTs that demonstrated 1, 3-13-O-glucose, 1, 3-19-O-glucose, 1, 2-13-O-glucose, 1, 2-19-O-glucose and/or 1, 6-13-O-glucose glycosylation activity to produce 11 steviol glycosides including 6 known steviol glycosides (rebaudioside G, rebaudioside KA, rebaudioside A, rebaudioside E, rebaudioside D and rebaudioside M) and 5 novel steviol glycosides. Furthermore, we can produce these 11 steviol glycosides in large scale by using engineered UGT proteins and recombinant yeast cells. This recombinant yeast system has produced yields as high as 93% while providing a novel approach to successfully produce these desirable minor steviol glycosides at an industrially relevant scale.

Emergent Population-Level Genetic Oscillations in a Synthetic Bacterial Microconsortium

Chen, Y., Rice University
Kim, J.,
Hirning, A.,
Josic, K.,
Bennett, M.,

Emergent population-level genetic oscillations in a synthetic bacterial microconsortium

Ye Chen1,*, Jae Kyoung Kim2,*, Andrew J. Hirning1, Kresimir Josic3,4 and Matthew R. Bennett1,5

1 Department of Biosciences, Rice University, Houston, TX 77005, USA

2 Mathematical Biosciences Institute, The Ohio State University, Columbus, OH 43210, USA

3 Department of Mathematics, University of Houston, Houston, TX 77204, USA

4 Department of Biology and Biochemistry, University of Houston, Houston, TX 77204, USA

5 Institute of Biosciences and Bioengineering, Rice University, Houston, TX 77005, USA

* These authors contributed equally

To date, the majority of synthetic gene circuits have been constructed to operate within single, isogenic cellular populations. It has been proposed that synthetic microconsortia will provide a means of engineering novel population-level phenotypes that are difficult to obtain with single strains. While circuits that use multiple strains simultaneously to achieve population-level phenotypes have been reported, they generally use a single intercellular signaling molecule and hence the variety of their phenotypes is limited. Here, we use two genetically distinct populations of Escherichia coli and two different signaling mechanisms to engineer a bacterial microconsortium that exhibits robust oscillations in gene transcription. Specifically, we used two different bacterial quorum sensing systems to construct an â??activatorâ? strain and a â??repressorâ? strain that respectively up- and down-regulate gene expression in both strains. When co-cultured in a microfluidic device, the two strains form coupled positive and negative feedback loops at the population-level. The interacting strains exhibit robust, synchronized oscillations that are absent if either strain is cultured in isolation. We used a combination of mathematical modeling and targeted genetic perturbations to better understand the roles of circuit topology and regulatory promoter strengths in generating and maintaining these oscillations. We found that the dual-feedback topology was robust to changes in promoter strengths and fluctuations in the population ratio of the two strains. These findings demonstrate that one can program population-level dynamics through the genetic engineering of multiple cooperative strains.

Effects of Cell-to-Cell Variability in MarA Expression on Escherichia coli Antibiotic Resistance

Siu, Y.,
El Meouche, I., University of Vermont
Dunlop, M., The University of Vermont

Traditionally, bacterial populations are viewed as individual cells that share both genotypic and phenotypic identity. Consequently, when individual cells within a population demonstrate elevated antibiotic resistance levels this was assumed to be the result of genetic mutation. However, recent studies have successfully demonstrated that there exist disparities between genotypic identity and phenotypic identity. For instance, studies have shown that, as a mechanism to evade the lethal effects of the bactericidal drugs, bacteria cells are capable of switching between a regular drug susceptible state and a drug tolerant dormant state (persister) by stochastic phenotypic variation. In addition, recent work by our group has suggested that transient antibiotic resistance can stem from stochastic pulsing in genes related to antibiotic resistance. In particular, we are focusing on a transcription factor, MarA from *Escherichia coli*. MarA is known to play an essential role for the regulation of more than 40 downstream antibiotic resistance genes, such as the AcrAB-TolC efflux pump genes and *micF* gene. In our study, we aim to demonstrate that MarA expression is heterogeneous for each individual cells within an isogenic bacterial population. We hypothesize that differences in MarA expression may render individual bacteria with different antibiotic resistant levels. To achieve this goal, we are exposing wild-type *E. coli* cells to various antibiotics. Plasmids that carry a fluorescence protein gene controlled by a *marA* promoter are used as reporters for MarA expression. In an effort to attain a precisely controlled external environment, we are using a microfluidic chip to control addition of antibiotics and have coupled this with time-lapse microscopy. Our preliminary data indicate that there are differences in expression of fluorescence proteins between individual cells within a microcolony, which reflects noisy MarA expression. In addition, when antibiotics are added, cells demonstrate different antibiotic resistance levels, which correlated with the noisy MarA expression. Hence, the isogenic bacteria population could use this mechanism as a "bet-hedging" strategy, which could eventually lead to a new bacteria population with pre-adapted antibiotic resistance ability.

Rational Engineering of Erythropoietin for Targeted Red Blood Cell Production

Burrill, D. R., Harvard University
Vernet, A., Wyss Institute for Biologically Inspired Engineering
Graveline, A. R., Harvard University
Turczyk, B. M., Harvard University
Collins, J. J., Harvard University
Silver, P. A., Harvard Medical School
Way, J. C., Harvard University

In tuning the quantitative parameters of protein-protein interactions, evolution often employs combinations of weak physical contacts to achieve specificity. This natural phenomenon can be applied to the engineering of targeted, protein-based drugs with improved cell-type specificity relative to their existing counterparts. Our approach uses mutagenesis to weaken a signaling protein's affinity for its receptor on non-target cells, combined with its fusion to a cell-specific targeting element via a flexible linker. Targeting should be driven by the targeting element, and despite the mutation, subsequent binding of the mutated protein should occur due to its high local concentration at the cell surface. We have applied this approach to the development of an erythropoietin (EPO)-based therapeutic that targets human glycophorin A (huGYPA) on red blood cell precursors (RBCs). Recombinant EPO has been used for two decades to treat forms of anemia associated with end-stage renal failure, AIDS, chemotherapy, and hemoglobinopathies. Clinical use of EPO has recently decreased due to concerns over the drug's off-target effects, including tumor recurrence and platelet formation or activation, which may lead to coronary disease or thrombosis. Using a transgenic mouse strain that expresses huGYPA, we have demonstrated that our targeted version of EPO stimulates RBC production without activating platelet formation. We also show that our compound has a similar serum half-life to Aranesp, a long-acting EPO derivative that is used clinically. In patients, targeting EPO to RBCs should allow higher doses to fully restore RBC levels without increasing the risk of cardiovascular events or cancer progression. In addition, the targeted EPO developed here should have an extended serum half-life and reduced immunogenicity relative to existing EPO drugs. More broadly, this work establishes the utility of our approach as a general platform for the rational design of targeted therapeutics.

Grant acknowledgements: NIH NRSA Fellowship, Wyss Institute, DARPA

Rapid Microfluidic Assay for Bacterial Electrotransformation

Garcia, P. A., Massachusetts Institute of Technology
Ge, Z., Massachusetts Institute of Technology
Moran, J. L., Massachusetts Institute of Technology
Buie, C. R., Massachusetts Institute of Technology

Electroporation results from exposure of cells to electric fields of sufficient strength to disrupt the plasma membrane. The local trans-Ââ? membrane voltage (TMV) significantly increases during exposure of cells to external electric fields. When the local TMV exceeds a critical threshold, pores are created on the cell membrane, allowing transport of ions and macromolecules (e.g. DNA/RNA) across the membrane. There is vast empirical literature establishing protocols to increase the electrocompetency (capacity for DNA uptake by electroporation) of cells, a process that is currently time consuming and lacks real-Ââ? time feedback. Despite the tremendous need, there are currently no protocols for improving bacterial electrocompetency without relying upon time-Ââ? consuming empirical experimental processes.

Here, we present a rapid microfluidic platform capable of determining electric field thresholds for electroporation. In our microfluidic platform, a converging microchannel generates a linear gradient of electric field. Fluorescence-Ââ? encoded DNA plasmids or nucleic acid stains permeate cell membranes in regions where the electric field is sufficiently high. We correlate the fluorescent region of bacteria with the range of electric fields that results in electroporation. Analysis of the fluorescence images from a single experiment provides electroporation conditions that would otherwise require hundreds of experiments conducted in parallel.

Our microfluidic platform quantitatively characterizes the electric field required for successful electroporation of microorganisms, one of the parameters for determining electrocompetency. The assay is based on a PDMS microfluidic channel with an active electroporation region spanning 2-Ââ? 4 mm that enables the generation of high electric fields (5-Ââ? 15 kV/cm) at applied voltages of 1000-Ââ? 3000 V. This geometry has been optimized to yield a linear variation of electric field strength, facilitating correlation of cell location within the chip and local electric field.

Under typical operation, the inlet to the device is connected via flexible tubing and a hollow stainless steel needle to a syringe pump for controlled delivery of the cell-Â-â? dye mixture to the microfluidic device. The outlet is connected via a needle and tubing to an external reservoir, enabling the sample to be collected for downstream analysis. The needles serve as the electrical contacts required to deliver the electric field from the MicroPulserTM (Bio-Ââ? Rad, Hercules, CA) to the sample within the channel. The assay utilizes SYTOXÂ® (Life Technologies, Grand Island, NY), a dye which cannot penetrate the intact cell envelope and only fluoresces once it binds to intracellular DNA; therefore, it is an effective marker for determining electroporation conditions. We have utilized numerical simulations in COMSOLÂ® (Burlington, MA) to develop a â??rulerâ? that allows us to determine the electric field that correlates with the onset of electroporation. The simulations allow us to quantitatively determine the electric field range in regions of dye uptake.

We envision this rapid microfluidic platform as a tool to enable and optimize genetic transformation of
intractable or previously challenging microbial chassis. This platform systematically samples a continuous spectrum of electric fields; differing from the traditional trial and error approach with discrete steps. Results of this study will broaden the scope of bacteria available for applications in synthetic biology and genetic engineering.

Optimizing Glycoengineered E. coli with Synthetic Small Regulatory RNAs

Glasscock, C. J., Cornell University
DeLisa, M. P., Cornell University
Lucks, J. B., Cornell University

One of the most important modifications for many proteins is *N*-linked glycosylation, which modulates an array of therapeutic properties from pharmacokinetic activity to immunogenicity. Large-scale production of therapeutic glycoproteins is currently accomplished using eukaryotic cell culture platforms; however, recent advances in *Escherichia coli*-based glycoprotein synthesis could enable higher yields and better control over glycoform. These *E. coli*-based platforms involve heterologous, multi-enzyme pathways for glycan synthesis that require optimization for production-scale yields. The aim of this study is to optimize *E. coli* for increased glycan production with synthetic small regulatory RNAs (sRNAs). We constructed a panel of 41 synthetic sRNAs to repress the endogenous expression of *E. coli* genes that were predicted by a genome-scale flux balance model to reduce sugar precursor levels in the *N*-linked heptasaccharide biosynthetic pathway from *Campylobacter jejuni*. We then screened our sRNA panel for improved glycan production with a fluorescence-based assay of cell-surface displayed glycans and identified *E. coli* strains with over 50% increases in fluorescence intensity. Further work is anticipated to identify even greater improvements by screening multiplexed sRNA libraries with fluorescence-activated cell sorting (FACS)-based monitoring of glycan production. Parallel work is ongoing to optimize an engineered biosynthetic pathway that produces eukaryotic core mannose$_3$-*N*-acetylglucosamine$_2$ ($Man_3GlcNAc_2$) glycans in *E. coli*. Our approach of using sRNAs for metabolic flux optimization provides an effective strategy for increasing glycan yields and opens the door for more complex regulation of *E. coli*-based glycosylation by integration within genetic circuits.

An Ultra High Throughput Screening Platform for Discovery and Engineering of Full-Length Antibodies in Soluble and Secreted Forms

Fang, Y.,
Griswold, K. E., Dartmouth
Chu, T., Dartmouth
Crowley, A.,

Traditional display techniques for antibody engineering have two major limitations. First, the antigen binding affinities are normally detected with single-chain variable fragment (scFv) or fragment antigen-binding (Fab). However, the physicochemical properties and expressibilities of those fragments may not correlate with their cognate full-length formats. Second, high throughput screening or selections require high quality, soluble recombinant antigen, which can be challenging or even intractable with membrane proteins such as G-protein-coupled receptors (GPCRs) and ion-channel receptors.

To overcome these limitations, we have developed a novel high throughput screening platform for the discovery and engineering of full-length monoclonal antibodies (mAbs) in soluble, secreted forms. This state-of-the-art technology utilizes hydrogel microdroplets (GMDs) coupled with fluorescence-activated cell sorting (FACS). The GMDs provide a linkage between phenotype and genotype. Both bulk emulsion and microfluidic compartmentalization have been used to co-encapsulate antibody secreting cells (*Pichia pastoris or Saccharomyces cerevisiae*) and antigen bearing cells (cancer cell lines) in picoliter-volume hydrogels. Full-length antibodies secreted by recombinant yeast will bind with the cognate antigens on the surface of mammalian target cells within the same droplet, which can then be labeled with fluorophore conjugated secondary antibodies. Fluorescently stained GMDs can be sorted by FACS at rates of 2,000 events per second or more. It has enabled us to easily screen millions of clones per day, which is impossible with the conventional microtiter well plate- based method. Furthermore, this technology is very cost-efficient. The expenses for the entire setup (except for the FACS system) were estimated to be only several thousand dollars, while each round of GMDs production costs around 50 USD.

We successfully validated our approach by the enrichment and one-step isolation of anti-EGFR IgG secreting *Pichia pastoris* from a 1:10,000 mixture of *Pichia* cells secreting negative control anti-CCR5 IgG antibodies. We also applied the GMD-FACS screening platform to distinguish cancer cells with different surface expression levels of EGFR.

In general, we have presented an innovative high throughput screening platform which is efficient, robust, and cheap for discovery and engineering of secreted, full-length mAbs. The broad utility of this methodology has been demonstrated through the screening of a spike-in library and identification of different cancer cells. In addition, this platform can also be utilized to isolate and engineer affinity proteins such as F(ab')2, nanobodies, and various non-immunoglobulin scaffold proteins. Our GMD-FACS technique can be a valuable tool to facilitate antibody discovery in both academic and industrial settings.

Engineering E. coli Nissle to Degrade Toxic Compounds in Mouse Models of Metabolic Disease

Kotula, J. W., Synlogic, Inc.
Falb, D., Synlogic
Isabella, V., Synlogic, Inc.
Silva, A., Synlogic, Inc.
Anderson, C., Synlogic, Inc.
Millet, Y., Synlogic, Inc.
Miller, P., Synlogic, Inc.

The complex communities of microbes that live within the human gastrointestinal (GI) tract greatly impact human health through metabolic processes, their interactions with each other, and their interactions with the immune system. When functioning in concert with their hosts, the gut microbiota can supply nutrients, degrade toxic compounds, and prevent overstimulation of the immune system. The gut microbiota, as a whole, is so active that its combined metabolic output can be thought of as an additional organ. By manipulating these processes through genetic engineering, we can generate microbes that supply missing metabolites and degrade specific toxic compounds in order to relieve inborn metabolic disorders and supplement organ deficiencies.

Urea cycle disorders (UCD) are a collection of genetic conditions with an estimated incidence of 1:8,500 births, in which the patient is unable to convert nitrogen to urea, resulting in the toxic accumulation of ammonia systemically. This hyperammonemic state can lead to CNS impairment, affecting neurological function, impacting normal development and in some cases, can be fatal. Importantly, as much as 70% of excess ammonia in hyperammonemic patients accumulates in the GI tract, supporting the use of a modified gut microorganism-based therapeutic. We have engineered Nissle, a probiotic strain of *E. coli*, to reduce the concentration of blood ammonia in those suffering from hyperammonemia caused by UCD. Nissle has been modified to respond to the gut environment and activate a carefully optimized ammonia consumption pathway. Through *in vivo* studies, we have shown that our engineered Nissle strain can sequester the ammonia produced via digestion, thus preventing diffusion of ammonia into the blood. Lowering blood ammonia levels also decreased mouse mortality compared to controls. Our work demonstrates that engineered microbes can be introduced into the gut microbiota, sense their local environment in that compartment, and affect an intended response in order to treat metabolic disorders. This prototype provides a meaningful example of how the tools developed in synthetic biology can be translated into the clinic.

*** Formatting Issue - Best Available Paper/Graphic ***

A Biofilm Free E. coli Chassis for Use in Continuous Culture

Takahashi, C.,
Klavins, E.,

A Biofilm Free *E. coli* Chassis for Use in Continuous Culture

Chris N. Takahashi Department of Electrical Engineering University of Washington

Seattle, WA 98195

Email: cnt@u.washington.edu

Eric Klavins

Department of Electrical Engineering University of Washington Seattle, WA 98195

Email: klavins@u.washington.edu

a)

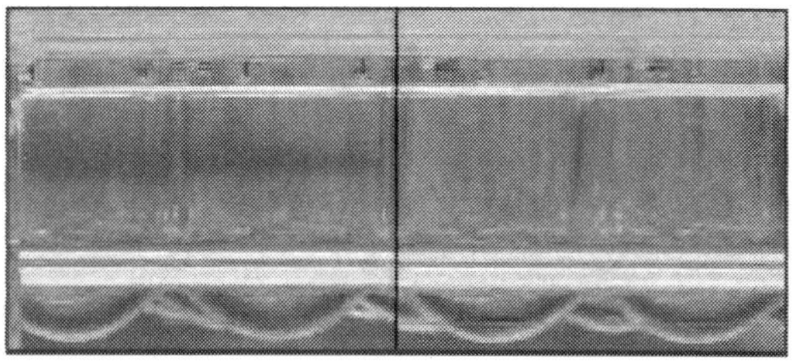

WT BF019 c)

*** Formatting Issue - Best Available Paper/Graphic ***

b)

WT BF019

selective pressures in directed evolutions. In continuous culture, substrate concentrations and metabolic byprod- ucts reach a steady state eliminating the drawbacks of batch culture. However, continuous culture selects strongly for phenotypes that are resistant to washout, such as flocculent yeast or adherent biofilms in *E. coli* [2]. Lab strains of yeast that are resistant to flocculation [3] are readily available, however a strain of *E. coli* that does not form biofilms in continuous culture has not yet been reported. Here, we have developed a five gene/operon *E. coli* knockout strain BF019 that is resistant to biofilm formation in continuous culture. We show that wild type *E. coli* can form disruptive biofilms in as little as 48 hours, while our novel strain remains biofilm free for at least two weeks, after which we ended the experiment. Strain BF019 and our previously reported [4] open-source turbidostat design [5] form a platform enabling characterization of long-term dynamics as well as directed evolution of better

WT BF019

Fig. 1. Comparison of wild type and BF019 a) Cross-sectional schematic of a continuous culture vessel. Wild-type *E. coli* have formed a biofilm causing the washout of planktonic cells. BF019 cannot form biofilms and therefore stays in a planktonic state. b) Wild type vs. BF019 biofilm formation on glass culture tubes used in continuous culture. c) A standard crystal violet stained plate assay [1] showing WT vs BF019. Wild type *E. coli* readily forms biofilms near the gas-liquid interface and on the bottom of the well (not shown), while BF019 does not form detectable biofilms.

Abstractâ??Continuous culture provides constant culture conditions enabling better characterization of synthetic circuits as well as constant selective pressures for directed evolution. When

*** Formatting Issue - Best Available Paper/Graphic ***

characterizing synthetic circuits over generational timescales (tens of minutes to hours), cellular metabolism impacts batch culture conditions, which can act as a disturbance to synthetic circuits and change

substrate utilizers and more tolerant strains required for consolidated bioprocessing.

REFERENCES

[1] J. H. Merritt, D. E. Kadouri, and G. A. Oâ??Toole, â??Growing and Analyzing Static Biofilms,â? in *Current Protocols in Microbiology* (R. Coico, T. Kowalik, J. Quarles, B. Stevenson, and R. Taylor, eds.), Hoboken, NJ, USA: John Wiley & Sons, Inc., Aug. 2011.

[2] J. D. Bryers, â??Biofilm formation and chemostat dynamics: Pure and mixed culture considerations,â? *Biotechnology and Bioengi- neering*, vol. 26, pp. 948â??958, Aug. 1984.

[3] W. S. Lo and A. M. Dranginis, â??FLO11, a yeast gene related to the STA genes, encodes a novel cell surface flocculin,â? *Journal of Bacteriology*, vol. 178, pp. 7144â??7151, Dec. 1996.

[4] C. N. Takahashi, A. W. Miller, F. Ekness, M. J. Dunham, and E. Klavins, â??A low cost, customizable turbidostat for use in synthetic circuit characterization,â? *ACS Synthetic Biology*, vol. 4, no. 1, pp. 32â??38, 2015.

[5] â??Klavins lab hardware.â? http://klavinslab.org/hardware.

RNA-Guided Gene Drives Can Efficiently Bias Inheritance in Wild Yeast
DiCarlo, J., Harvard Medical School, Boston University
Chavez, A.,
Dietz, S.,
Church, G. M., Harvard University
Esvelt, K.,

RNA-guided gene drives can efficiently bias inheritance in wild yeast

James E. DiCarlo1,2,3,7, Alejandro Chavez1,2,4,5,7, Sven L. Dietz1,2,4,6, Kevin M. Esvelt2,4*, and George M. Church1,2,4*
1 Department of Genetics,
2 Harvard Medical School, Boston, Massachusetts, USA.
3 Department of Biomedical Engineering, Boston University, Boston, MA 02215, USA
4 Wyss Institute for Biologically Inspired Engineering, Boston, Massachusetts, USA.
5 Department of Pathology, Massachusetts General Hospital, Boston, MA 02114, USA.
6 Department for Biosystems Science and Engineering, ETH ZÃ¼rich, Basel, Switzerland.
7 These authors contributed equally to this work.

*For correspondence: esvelt@sculptingevolution.org (K.M.E), gmc@harvard.edu (G.M.C.)

Abstract

Inheritance-biasing elements known as â??gene drivesâ?□ may be capable of spreading genomic alterations made in laboratory organisms through wild populations. We previously considered the potential for RNA-guided gene drives based on the versatile CRISPR/Cas9 genome editing system to serve as a general method of altering populations1. Here we report molecularly contained gene drive constructs in the yeast *Saccharomyces cerevisiae* that are typically copied at rates above 99% when mated to wild yeast. We successfully targeted both non-essential and essential genes and showed that the inheritance of an unrelated â??cargoâ?□ gene could be biased by an adjacent drive. Our results demonstrate that RNA-guided gene drives are capable of efficiently biasing inheritance when mated to wild-type organisms over successive generations.

Introduction

Gene drives have the potential to address diverse ecological problems by altering the traits of wild populations. As 'selfish' genetic elements, they spread not by improving the reproductive fitness of the organism, but by increasing the odds that they themselves will be inherited. Because this inheritance advantage can counteract the fitness costs associated with the drive itself or with adjacent genes carried along with it, they are theoretically capable of 'driving' unrelated traits through populations over many generations1.

*** **Formatting Issue - Best Available Paper/Graphic** ***

Inheritance-biasing is a common strategy in nature2. One elegant class of inheritance-biasing genes spread by cutting homologous chromosomes that do not contain them, thereby inducing the cellular repair process to copy them onto the damaged chromosome by homologous recombination (Fig. 1A). This process is known as 'homing'3. The best-known homing endonuclease gene is *I-SceI*, whose product cuts the gene encoding the large rRNA subunit of *S. cerevisiae* mitochondria. Nearly all natural homing endonucleases are embedded within self-splicing introns or inteins, which help to minimize disruption of the targeted gene. Most are capable of homing with extremely high efficiencies; I-SceI is correctly copied 99% of the time4.

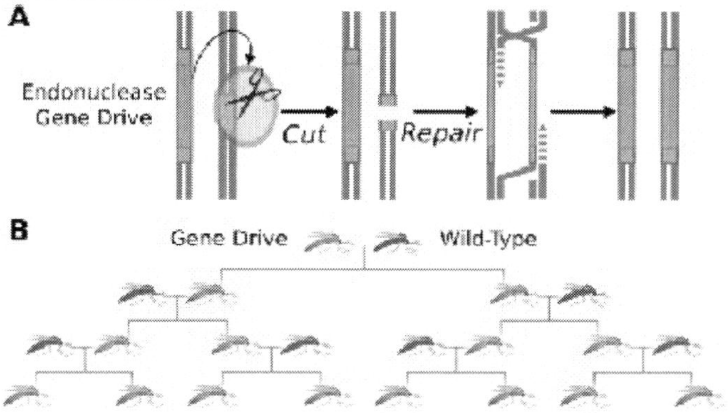

Figure 1. Mechanism and population-level effect of endonuclease gene drives.

(A) Homing endonucleases cut competing

alleles, inducing the cell to repair the damage by copying the endonuclease gene. (B) By converting heterozygous germline cells into homozygotes containing two copies (teal), gene drives increase the odds that they will be

inherited and consequently spread themselves and associated changes through wild populations (grey).

1

Austin Burt first suggested that homing endonucleases might be used to construct synthetic gene drives capable of altering wild populations of multicellular organisms in 2003 (Fig. 1B)5. Laboratories subsequently reported that *I-SceI* endonuclease genes inserted into mosquitoes6 or fruit flies7,8 exhibited homing in transgenic laboratory populations with an I-SceI recognition site inserted into the corresponding wild-type locus. However, the versatility of gene drives based on homing endonucleases is constrained by the difficulty of retargeting these enzymes to cleave useful sequences within native chromosomal loci9.

The recent development of the CRISPR nuclease Cas9, which cleaves target sequences specified by â??guide RNAâ?□ molecules, has enabled scientists to edit the genomes of diverse species10â??20. The question of whether Cas9 can be used to build efficient gene drives is highly relevant due to the potential for such constructs to alter any gene in any population that can be edited with Cas9, a group that may constitute every species tested to date21. We previously detailed the theoretical potential for RNA-guided gene drives to alter wild populations, including an evolutionary analysis, novel drive architectures, and containment measures robust to human error1. However, these new architectures and control strategies have not yet been validated. Indeed, whether Cas9 can bias inheritance in a eukaryotic organism at all remains unknown, raising the question of whether initiating public discussions and engaging in regulatory reform22 is truly necessary. We addressed these issues in the yeast *S. cerevisiae* by building gene drive constructs in haploid cells, mating them to wild-type haploid strains, and measuring the relative abundance of unmodified versus gene drive alleles in the resulting diploids and their sporulated haploid progeny over successive mating events.

Results

The inheritance-biasing efficiency of an endonuclease gene drive is determined by 1) the frequency of cutting and 2) the fraction of repair events that lead to the drive being copied onto the target chromosome. Only repair by homologous recombination (HR) results in drive copying and inheritance bias; the competing non-homologous end-joining (NHEJ) repair pathway ligates the broken ends together, preventing the drive from being copied and often generating mutations at the site of the break that prevent further cutting.

To prevent the accidental escape of our gene drives into wild yeast populations, we employed a method of molecular containment1 in which we split our Cas9 based gene drive system into two physically separate genetic parts: an episomally encoded Cas9 gene and a drive element encoding the guide RNA. This allowed us to avoid creating a self-sufficient inheritance-biasing cassette while still targeting wild-type yeast strains. Though simple, this form of molecular containment is not vulnerable to human error; even if drive-containing yeast were to escape into the wild, the required Cas9 episomal plasmid would rapidly be segregated away from the drive element, rendering the drive inoperative.

In order to directly measure the efficiency of CRISPR/Cas9 gene drives in yeast, we took advantage of the red color that builds up in yeast lacking functional copies of the *ADE2* gene23. If red *ade2â??* haploids are mated with cream-colored wild-type haploids, the resulting heterozygous diploids inherit one functional copy of *ADE2* and are therefore cream-colored. When these diploids undergo meiosis and reproduce via sporulation, half of the resulting haploids inherit the broken copy and are consequently red; the other half inherit the intact copy and are cream-colored (Fig. 2A).

But if the red haploids encode a functional gene drive, it will cut and replace the intact *ADE2* locus inherited from the wild-type parent, yielding red diploids (Fig. 2B). When these sporulate, all of the haploid offspring will inherit one of the two copies of the drive and will also be red. Thus, the

2

cutting efficiency of a gene drive that targets and replaces *ADE2* can be assessed by plating yeast cells and counting the fraction of colonies that are red.

We built a Cas9-based gene drive construct targeting *ADE2* by placing a guide RNA targeting the wild-type *ADE2* gene in place of the endogenous *ADE2* locus. We mated these red *ade2*::sgRNA haploids to wild-type yeast of the opposite mating type in the presence or absence of the Cas9 plasmid and plated on media that selects for diploids. Nearly all diploid colonies were red when the Cas9 plasmid was present, indicating highly efficient cutting of the *ADE2* copy inherited from the wild-type parent (Fig. 2C). We did not observe red diploid colonies in the absence of Cas9, demonstrating that the drive only functions in laboratory yeast populations encoding both Cas9 and guide RNA.

*** Formatting Issue - Best Available Paper/Graphic ***

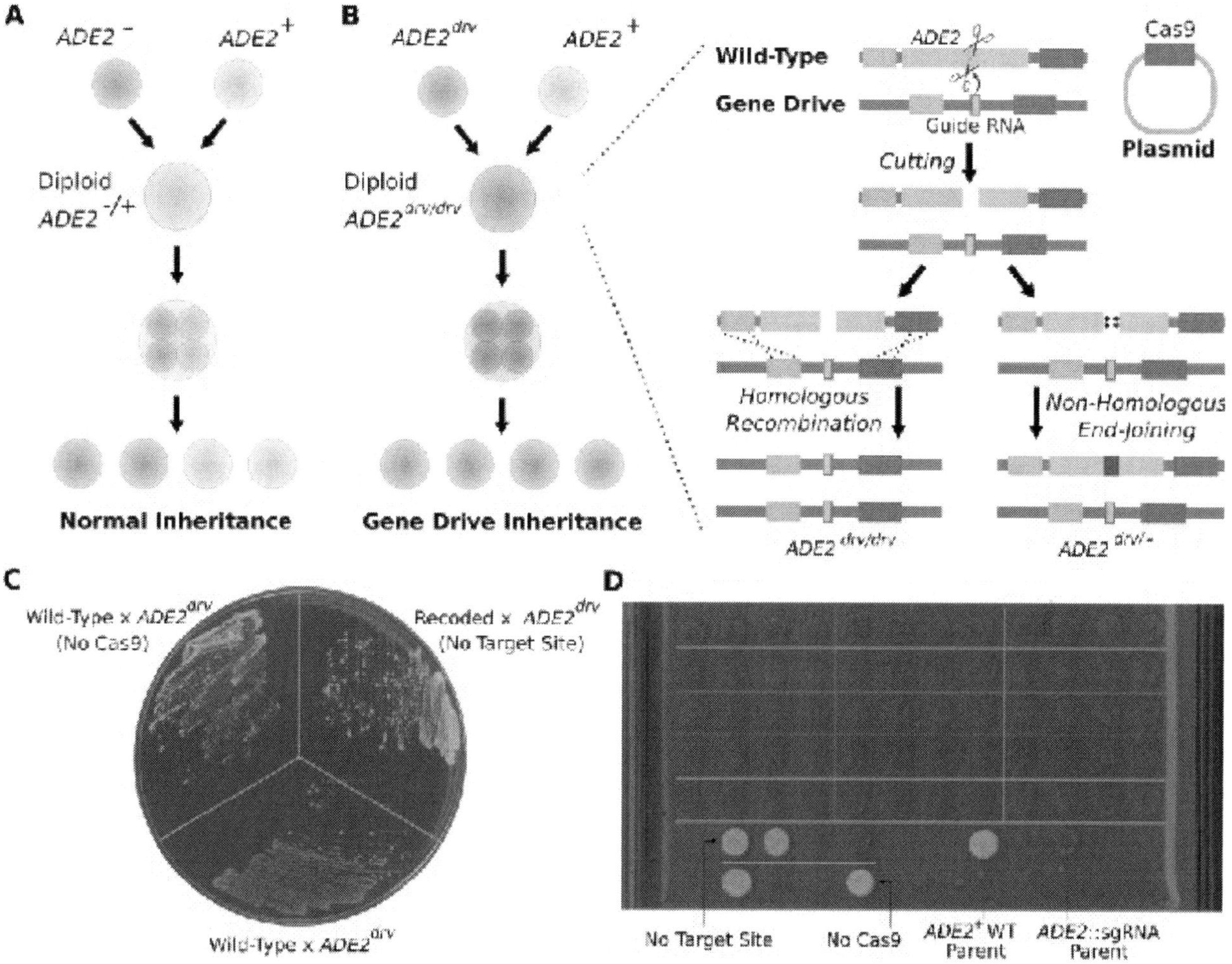

Figure 2. Biased inheritance of *ADE2* is readily visible in *S. cerevisiae.* (A) Mutations in *ADE2* generate a red phenotype on adenine-limiting media due to the buildup of red pigments. Mating a red mutant haploid to a wild-type haploid produces cream-colored diploids, which yield 50% red and 50% cream-colored progeny upon sporulation. (B) When haploids with a gene drive targeting *ADE2* mate with wild-type haploids in the presence of Cas9, cutting and subsequent replacement or disruption of *ADE2* produces red diploids that yield exclusively red progeny. (C) Diploids produced by mating wild-type and *ade2*::sgRNA gene drive haploids yield cream-colored colonies in the absence of Cas9 or when the target site is removed by recoding but uniformly red colonies when both are present, demonstrating Cas9-dependent disruption of the wild-type *ADE2* copy. (D) Spores from 15 dissected tetrads produce uniformly red colonies on

adenine-limited plates, confirming disruption of the *ADE2* gene inherited from the wild-type parent. In the absence of the target site or Cas9, normal 2:2 segregation is observed.

3

To verify that the *ADE2* alleles from wild-type parents were indeed lost, we sporulated the mated diploids and examined their resultant haploid progeny. Upon dissecting 18 *cas9+* diploids, we observed a perfect 4:0 ratio of red:cream haploids, confirming that all copies of the *ADE2* locus were disrupted. In contrast, 18 cream-colored *cas9*â?? diploids yielded a 2:2 red:cream ratio, indicating normal inheritance of the inactivated drive and the wild-type alleles (Fig. 2D).

To determine whether *ADE2* disruptions in red diploids were the result of successful copying of the drive element via HR, we sequenced the 72 haploids derived from dissected *cas9+* diploids. All sequenced colonies contained intact drives without additional mutations, indicating that drive mobilization was efficient and occurred at high fidelity. This represents the first example of a synthetic endonuclease gene drive that biases its own inheritance when mated to a wild-type eukaryote.

We next tested whether RNA-guided gene drives could be designed to bias the inheritance of not only the minimal drive element, but also any closely associated â??cargoâ?□ gene whose spread through an existing population may be desirable. As a proof of principle, we inserted the *URA3* gene *in cis* to the *ade2*::sgRNA minimal guide element. *URA3* allows laboratory modified yeast strains to grow in the absence of uracil supplementation (Fig. 3A). We mated these *URA3*-containing drive haploids to wild-type haploids in the presence of an episomal Cas9 plasmid, selected diploids (all of which were red), sporulated them, and dissected 18 tetrads. As was the case for the original *ADE2* gene drive, all of the sporulated haploid cells formed red colonies. Crucially, all grew normally when replica plated on uracil deficient media, indicating that *URA3* was efficiently copied with the drive (Fig. 3B).

While both *ADE2*-based gene drives are highly efficient at biasing inheritance in the laboratory, they would be unable to cut a chromosome lacking the non-essential *ADE2* gene or one with a

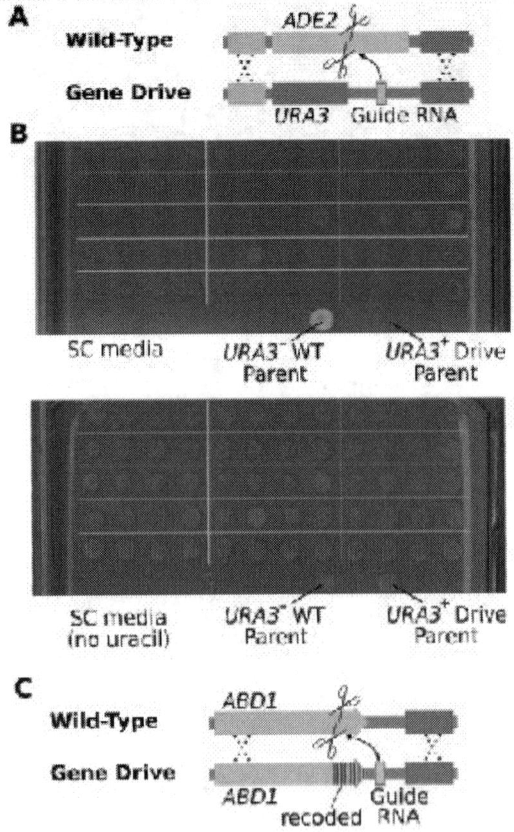

mutation in the targeted sequence. This is a particular problem for drives seeking to alter entire populations because any aberrant repair by non-homologous end-joining (NHEJ) following cutting will remove the target sites and thereby generate a drive- resistant allele. We previously postulated that gene drives that target and recode an essential gene could avoid drive resistance in large populations because error prone repair events that modify the target site would cause lethality1. Non-essential but nonetheless important genes could similarly be edited because mutants created by NHEJ events would still be less fit than the drive itself.

Figure 3. Gene drives and cargo genes remain intact upon copying and can spread by targeting both non-essential and essential genes. (A) The *ADE2*-targeting gene drive was modified to carry *URA3* as a cargo gene. (B) Diploids produced by mating wild-type *URA3*- haploid yeast with haploids encoding the gene drive carrying *URA*3 were sporulated and tetrads dissected to isolate colonies arising from individual spores. All of these grew when replica- plated onto plates lacking uracil, demonstrating that the drive successfully copied *URA3* in all diploids. (C) The

ABD1-targeting gene drive cuts and recodes the 3' end of the

essential *ABD1* gene.

4

To validate this strategy of essential gene recoding during drive insertion, we constructed a third gene drive targeting the essential *ABD1* gene (Fig. 3C)24. A haploid strain containing a recoded *ABD1* allele upstream of a guide RNA targeting the natural *ABD1* coding sequence was mated to wild- type cells in the presence of Cas9. We then selected diploids, sporulated the cells, dissected 18 of them, and sequenced the 72 resulting segregants. As expected, all of them contained the recoded *ABD1* locus and the guide RNAs, demonstrating the feasibility of gene drives based on essential gene recoding.

*** Formatting Issue - Best Available Paper/Graphic ***

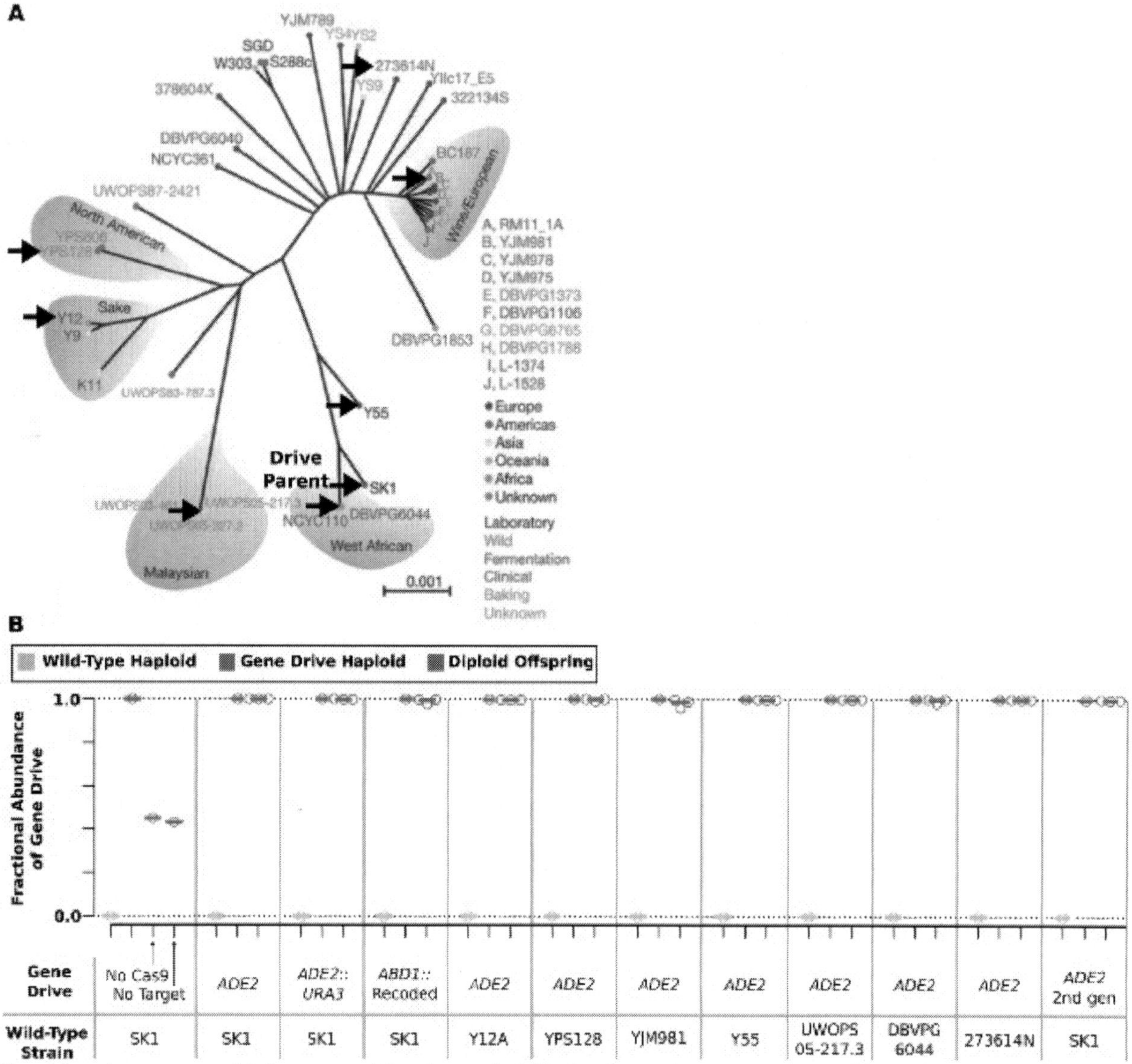

Because the rate at which a gene drive is copied likely depends on a variety of host factors, including the types of repair machinery available to the cell at the time of the cut, the chromatin state of the locus, and the degree of homology flanking the double-strand break generated by the drive, we were curious whether our gene drives could be copied from our laboratory strain into a diverse group of native *S. cerevisiae* strains. We correspondingly mated *ADE2* drive-containing haploids with 6 phylogenetically and phenotypically diverse wild-type strains of haploid *S. cerevisiae* (Fig. 4A)25. To quantitatively measure the efficiency of gene drive copying for each cross, we performed quantitative PCR on populations of all diploids using one set of primers specific to the drive and another set designed to amplify either wild-type or NHEJ-disrupted alleles.

Figure 4. The extent of inheritance-biasing in diverse yeast strains as measured by quantitative PCR. (A) A phylogenomic tree indicating the relationships between wild- type strains selected for gene drive testing. Adapted with permission from Macmillan Publishers Ltd: *Nature* 458:337-341, copyright 2009. (B) Quantitative PCR results depicting the relative abundance of wild-type and drive-containing alleles in diploids

arising from matings between SK1 haploids bearing gene drives and diverse wild-type haploid strains. â??No Cas9â?□ and â??No Targetâ?□ refer to the haploid cells containing the *ADE2* drive element mated to wild-type haploids in the absence of Cas9 or to an otherwise wild- type strain containing Cas9 that also has a mutation in the targeted sequence that blocks cutting. â??2nd genâ?□ refers to the haploid

progeny of an earlier mating.

5

The mean fraction of diploid chromosomes containing the *ADE2* gene drive was over 99% regardless of wild-type parent (Fig. 4B), attesting to the robustness of the drive in diverse backgrounds. Addition of the *URA3* cargo gene did not appreciably change this efficiency. The *ABD1* drive was copied at an equivalent rate.
A final question of importance concerns the efficiency and stability of the drive over successive copying events. Because *S. cerevisiae* reproduces mostly through asexual division and only a subset of the population sporulate in the laboratory even when induced to undergo meiosis26, experiments to determine the long-term population-level efficiency of gene drives are difficult to perform and interpret. As a surrogate for the efficiency of the drive element over multiple generations and hence how it may behave on population scales, we sought to measure the efficiency of successive copying events by the same drive. We correspondingly mated several haploid offspring of the first-round

ADE2 gene drive diploids to wild-type haploids containing the Cas9-expressing plasmid (Fig. 4B, right). All of the gene drive constructs biased inheritance at the same efficiency in the second generation as they did in the first, indicating a continued ability to spread through sexually reproducing populations over multiple generations so long as the fitness cost associated with the drive is less than the inheritance benefit in the relevant environment.

Discussion

Our discovery that Cas9 can bias inheritance in diverse wild yeast strains over successive generations at very high efficiency demonstrates that RNA-guided gene drives can function in eukaryotic organisms as predicted. By itself this does not guarantee the success of gene drives in other organisms as the rate of homologous recombination varies between species and is known to be particularly high in yeast. However, the fact that we observed inheritance biasing rates equal to or exceeding that of the natural *I-SceI* homing endonuclease gene is highly promising. Because a drive based on I-SceI was correctly copied after ~97% of successful genome cutting events in transgenic *Anopheles gambiae* mosquitoes that contained an I-SceI recognition site6 and Cas9 typically cuts more efficiently than does I-SceI7,27, our findings suggest that RNA-guided gene drives will be highly
effective at suppressing5,28,29 or spreading antimalarial alleles30,31 through populations of this important
malaria vector. Similarly, RNA-guided gene drives are likely to be more effective than those based on I-SceI, zinc-finger nucleases, or TALENs7,8,32 due to their ability to efficiently cut multiple sequences without requiring highly repetitive elements.

*** Formatting Issue - Best Available Paper/Graphic ***

Crucially, our gene drive that successfully targets and recodes the essential *ABD1* gene in diploid yeast demonstrates the feasibility of targeting and recoding genes important for fitness, a strategy expected to improve the evolutionary stability of gene drives1. We recommend that future efforts seeking to build gene drives intended for eventual release adopt this same approach. Targeting sites adjacent to both homology arms may also promote more efficient drive copying by biasing repair towards homologous recombination33 and increase the overall frequency of cutting27.

Our findings suggest that yeast may prove a useful platform for swiftly testing RNA-guided gene drive architectures before moving them into multicellular organisms. The power of yeast genetics and the ease of genome manipulation will facilitate combinatorial investigations into gene drive optimization. For example, studies might explore how biasing repair pathway choice34 affects the efficiency of copying for gene drives of various sizes. Because the factors involved in these pathways
are broadly conserved, these experiments could guide gene drive optimization in other organisms35.

6

While highly encouraging for potential gene drive applications, our results also sound a note of caution for subsequent experiments. That our drives were readily copied into a variety of yeast strains collected from all over the world underscores the potential for a single gene drive to affect very large populations. Poor flanking homology is not an effective barrier, at least in *S. cerevisiae*. Moreover, the *ADE2* gene drive took only two weeks to design, build, and test, suggesting that many laboratories are capable of building gene drives in yeast. Since yeast reproduce mainly through asexual division, gene drives would need to be considerably less costly to organismal fitness in order to spread in the wild than would a comparably efficient gene drive in an organism that always reproduces via mating. However, natural endonuclease gene drives such as *I-SceI* do exist within yeast. Whether our gene drives or the typical RNA-guided gene drive will constitute this level of burden is as yet unknown.

It is considerably more difficult to edit the genomes of multicellular model organisms such as Drosophila than the yeast genome, and still more difficult to alter those species for which gene drive applications are most likely to be relevant. However, a growing number of laboratories now make heritable alterations in more than a dozen sexually reproducing species. This confluence of factors demands caution. Because synthetic gene drives would alter the global environmental commons, the decision to deploy such a drive must be made collectively by society. Any accidental release could severely damage public trust in scientists. As demonstrated by numerous containment breaches involving pathogenic viruses and bacteria, physical methods of containment are always susceptible to human error and should not be exclusively relied upon whenever alternatives are available.

All scientists making heritable alterations with Cas9 should therefore employ non-physical containment methods sufficient to prevent the creation of an RNA-guided gene drive capable of spreading in the wild. Even scientists not intending to work with gene drives should consider taking precautions, since any unintended insertion of the cas9 gene and guide RNAs near a targeted site
could generate a gene drive. Fortunately, a simple and costless precaution is both available and already utilized for different reasons by many laboratories: *avoid delivering the* Cas9 *gene on a DNA cassette that also encodes a guide RNA*. As we have shown, guide RNAs alone cannot bias inheritance in the absence of Cas9 and consequently cannot spread through wild populations (Fig. 4B).

We therefore recommend that future gene drive experiments in yeast should also separate Cas9 from guide RNAs, for example by employing our episomal Cas9 plasmid, or employ another method of molecular containment. Gene drive experiments in species that always reproduce sexually pose greater risks and consequently should employ additional precautions such as targeting sequences not present in wild organisms or performing experiments in geographic areas where the organism in question cannot survive1. Working in genetic backgrounds that are less likely to escape the laboratory and mate *â??* such as wingless flies in the case of *Drosophila â??* may also be prudent. Should the scientific community deem it necessary, a voluntary peer review system designed exclusively for at-risk gene drive experiments, ideally conducted at the proposal stage, could further reduce the risk of accidental release.

In conclusion, our demonstration of diverse gene drive architectures enabling Cas9-mediated inheritance biasing in wild *S. cerevisiae* can guide efforts to build RNA-guided gene drives in other organisms and underscores the urgent need for precautionary control strategies, inclusive public engagement, and regulatory reform22 in advance of real-world applications.

7

Methods:

Plasmids and genomic cassettes

Gene drive cassettes were synthesized from gBlocks (Integrated DNA Technologies, Coralville, IA) and inserted into SK1 cells via Cas9-mediated genome modification as follows. Guide RNAs for each drive were cloned into p416-Cas9 containing plasmids with expression driven by the SNR52 promoter18. 60 base pair homology arms to the target locus were added on both ends of the gene drive cassette via PCR and 5 ug of PCR product was co-transformed with the p416-Cas9-gRNA plasmids. Correctly integrated gene drives were verified by sequencing and p416-Cas9-gRNA plasmids were cured using 5-Fluoroorotic Acid (FOA) selection.

To create the *URA3*-containing *ADE2* gene drive, the *ADE2* gene drive was cloned next to the *Candida albicans URA3* gene in the pAG60 plasmid. The entire URA3 cassette and gene drive were PCR amplified and inserted using Cas9-mediated genome modification into the *ADE2* locus of haploid SK1 cells.

The recoded C-terminus of the *ABD1* gene and corresponding gene drive were synthesized as a gBlock to remove homology and generate mutations in the seed sequence via synonymous changes. The TEF1 terminator was inserted at the 3â??end of the recoded *ABD1* gene between the gene and the gRNA as *ABD1* shares a terminator with the *VHC1* gene. The entire cassette was integrated into the haploid SK1 genome using Cas9-mediated genome modification.

The p416-Cas9-gRNA plasmid (conferring uracil prototrophy) is a variant of the previously described p414-Cas9-gRNA plasmid (conferring tryptophan prototrophy)18 (Addgene #43802). One or the other was used in each mating experiment. The pRS413 vector was transformed into select cell types to confer histidine prototrophy as a marker to select for diploid cells.

Strain genotypes:

*** Formatting Issue - Best Available Paper/Graphic ***

Strain Genotype

SK1 A MATa ho::LYS2 lys2 ura3 leu2::hisG his3::hisG trp1::hisG SK1 Î± MATÎ± ho::LYS2 lys2 ura3 leu2::hisG his3::hisG trp1::hisG Y12A MATa ho::HygMX ura3::KanMX

YPS128 MATa ho::HygMX ura3::KanMX

YJM981 MATa ho::HygMX ura3::KanMX Y55 MATa ho::HygMX ura3::KanMX UWOPS05-217.3 MATa ho::HygMX ura3::KanMX DBVPG 6044 MATa ho::HygMX ura3::KanMX

273614N MATa ho::HygMX ura3::KanMX

Yeast mating experiments

Haploid drive-containing SK1 yeast and haploid wild-type strains of the opposite mating type were mixed in equal amounts in YPAD liquid media and incubated overnight. The resulting diploids were washed in sterile water and plated on selective media for both parental genotypes. The chart below details the specific crosses:

8

MATa Genotype MATÎ± Genotype Selection

SK1 pRS414 â?? Cas9 SK1 ade2::gRNA (gene drive), pRS413 SC-histidine - tryptophan

SK1 ade2::gRNA + URA3 (gene drive), p414-Cas9

SK1 pRS413 SC-histidine - tryptophan

SK1 p414-Cas9 SK1 abd1::ABD1 recoded +gRNA (gene drive), pRS413

SC-histidine - tryptophan

*** Formatting Issue - Best Available Paper/Graphic ***

Y12A Hygromycin B

resistance (HygR)

YPS128

Hygromycin B resistance (HygR) YJM981

Hygromycin B

resistance (HygR) Y55 Hygromycin B resistance (HygR)

UWOPS05-217.3

*** Formatting Issue - Best Available Paper/Graphic ***

Hygromycin B resistance (HygR) DBVPG 6044

Hygromycin B

resistance (HygR)

273614N Hygromycin B resistance (HygR) SK1 ADE2:: ADE2 silently recoded genomic target seed sequence, p414- Cas9

SK1 ade2::gRNA (gene drive), p416-Cas9 SC-uracil+300 ug/mL Hygromycin B

SK1 ade2::gRNA (gene drive), p416-Cas9 SC-uracil+300 ug/mL Hygromycin B

SK1 ade2::gRNA (gene drive), p416-Cas9 SC-uracil+300 ug/mL Hygromycin B

SK1 ade2::gRNA (gene drive), p416-Cas9 SC-uracil+300 ug/mL Hygromycin B

SK1 ade2::gRNA (gene drive), p416-Cas9 SC-uracil+300 ug/mL Hygromycin B

SK1 ade2::gRNA (gene drive), p416-Cas9 SC-uracil+300 ug/mL Hygromycin B

SK1 ade2::gRNA (gene drive), p416-Cas9 SC-uracil+300 ug/mL Hygromycin B

SK1 ade2::gRNA (gene drive), pRS413 SC-histidine - tryptophan

SK1, p414-empty SK1 ade2::gRNA (gene drive), pRS413 SC-histidine - tryptophan

Sporulation and tetrad dissection

After mating in liquid YPAD and selection for diploids on selection plates, the selection plates were scraped into 10 mL selective media and grown overnight at 30Â°C. A fresh 5 mL YPAD culture was then inoculated to and OD=0.1 and grown 4-5 hours at 30Â°C. The entire culture was then washed twice in

10 mL water, inoculated into 2 mL of sporulation media(1% potassium acetate), and incubated at room-temperature for 3 days or until spores were visible. Sporulated cells were suspended in 50 ÂµL of a stock solution of zymolyase (50 Âµg/mL in 1M sorbitol) and incubated at 30C for 5 minutes, transferred to ice, diluted with 150 ÂµL cold H2O, microdissected using a Zeiss tetrad dissection microscope, and isolated spores grown on YPAD plates.

Selection for URA3 function

Dissected spores were grown in synthetic complete (SC) media and then spotted onto SC medium as well as SC medium without uracil. To enhance red color, all SC solid media used for plate images contained 0.5 X adenine hemisulfate (final concentration of 0.08 mM).

*** Formatting Issue - Best Available Paper/Graphic ***

9

Quantitative PCR

Candidate primer pairs were designed to amplify short regions specific to each drive or the wild-type sequence replaced by the drive, as well as the *ACT1* gene as a control. All sequences are included in the supplementary information. Genomic DNA was extracted using Method A as described in Looke et al.36

KAPA SYBR FAST qPCR Master Mix (2X) was used to perform the qPCR reaction along with

25 ng of genomic DNA. The amplification efficiency and relative specificity of each primer pair were measured by amplifying dilutions of genomic DNA from wild-type and drive haploids, respectively, and the best-performing and well-matched pairs selected for use (see below for all primers used). Quantitative PCR reactions were performed on genomic DNA isolated from each parental haploid as well as from diploids arising from three independent mating events. Three reactions (technical replicates) were performed per sample on a LightCycler 96 machine by Roche.

Calculations:

Results from three technical replicates were averaged for calculations. In order to directly calculate the ratio of alleles before PCR amplification, we first determined the efficiencies of the different primer pairs. Efficiencies were calculated from qPCR runs of serial dilutions (6 orders of magnitude) as:
Efficiency=10-1/slope
R2 values were higher than 0.99 in all cases except for one pair (ade2::URA3+sgRNA). The allelic ratios were calculated as:
xa . E Ct,a
= xb . E Ct,b
xa / xb = E Ct,b
/ E Ct,a
with xa and xb being the initial concentration of drive and wt DNA, Ea and Eb the efficiency of the respective primer pairs and
Ct,a and Ct,b the Ct values for each sample.

Figure 4B was generated using BoxPlotR37.

***** Formatting Issue - Best Available Paper/Graphic *****

Genome modification primers and gBlocks ADE2.sgRNA.ade2.1 insert.F

ADE2.sgRNA.ade2.1 insert.R

Sequence

TACGAACCGGGTAATACTAAGTGATTGACTCTTGCTGACCTTTTATTAAGAACTAAATGGt
ctttgaaaagataatgtatgattatgctttc

TAATAAGTGATCTTATGTATGAAATTCTTAAAAAAGGACACCTGTAAGCGTTGATTTCTAa
gacataaaaaacaaaaaaagcaccac

gRNA+CaURA3.ade2.F
TACGAACCGGGTAATACTAAGTGATTGACTCTTGCTGACCTTTTATTAAGAACTAAATGGa
gacataaaaaacaaaaaaagcaccaccg
gRNA+CaURA3.ade2.R
TAATAAGTGATCTTATGTATGAAATTCTTAAAAAAGGACACCTGTAAGCGTTGATTTCTAt
cgacactggatggcggcgttagtatc

ABD1.recode+gRNA
AGCCAGATGCCATTCAACAAGTTCTTCGTGCAGGAGATACCAAAGTGGATAGAACGTTTCA
GCCCAAAGATGCGTGAGGGGCTTCAGCGTAGCGACGGGCGTTACGGGGTGGAGGGTGACGA

GAAAGAGGCTGCTAGCTACTTTTACACGATGTTCGCTTTTAGAAAAGTTAAGCAATACATA
GAGCCTGAGTCAGTTAAACCAAATTGAACGGCTCCTCGCTGCAGACCTGCGAGCAGGGAAA
CGCTCCCCTCACAGACGCGTTGAATTGTCCCCACGCCGCGCCCCTGTAGAGAAATATAAAA
GGTTAGGATTTGCCACTGAGGTTCTTCTTTCATATACTTCCTTTTAAAATCTTGCTAGGAT
ACAGTTCTCACATCACATCCGAACATAAACAACCATGGGTATGACCGACCAAGCGACGCCC
AACCTGCCATCACGAGATTTCGATCCCACCGCCGCCTTCTATGAAAGGtctttgaaaagat
aatgtatgattatgctttcactcatatttatacagaaacttgatgttttctttcgagtata
tacaaggtgattacatgtacgtttgaagtacaactctagattttgtagtgccctcttgggc
tagcggtaaaggtgcgcatttttcacaccctacaatgttctgttcaaaagattttggtca
aacgctgtagaagtgaaagttggtgcgcatgtttcggcgttcgaaacttctccgcagtgaa
agataaatgatcTGAAGGGGATGAAAAGGAAGGTTTTAGAGCTAGAAATAGCAAGTTAAAA
TAAGGCTAGTCCGTTATCAACTTGAAAAAGTGGCACCGAGTCGGTGGTGCTTTTTTTGTTT
TTTATGTCT

ABD1.recode+gRNA.int.F tatgttgtgccattcgaaaccttaagaagtttggctgatgaatacggtttggaactagtaA
GCCAGATGCCATTCAACAAGTTC
ABD1.recode+gRNA.int.R gtaatacggccgaaatacagatgctttatagtagggttattgtttctattcatttttattA
GACATAAAAAACAAAAAAAGCACCACC
ADE2.gRNA tctttgaaaagataatgtatgattatgctttcactcatatttatacagaaacttgatgttt
tctttcgagtatatacaaggtgattacatgtacgtttgaagtacaactctagattttgtag

*** Formatting Issue - Best Available Paper/Graphic ***

tgccctcttgggctagcggtaaaggtgcgcatttttcacaccctacaatgttctgttcaa aagattttggtcaaacgctgtagaagtgaaagttggtgcgcatgtttcggcgttcgaaact tctccgcagtgaaagataaatgatcACTTGAAGATTCTTTAGTGTGTTTTAGAGCTAGAAA TAGCAAGTTAAAATAAGGCTAGTCCGTTATCAACTTGAAAAAGTGGCACCGAGTCGGTGGT GCTTTTTTTGTTTTTTATGTCT

ADE2.1.silent.seed.90mer. F
ADE2.1.silent.seed.90mer. F

TGATGTGCTAACGATTGAGATTGAGCATGTTGATGTTCCTACCCTGAAAAACCTGCAAGTA AAACATCCCAAATTAAAAATTTACCCTTC

GAAGGGTAAATTTTTAATTTGGGATGTTTTACTTGCAGGTTTTTCAGGGTAGGAACATCAA CATGCTCAATCTCAATCGTTAGCACATCA

ABD1.ver.F ATAGATAATGTTCCTGAATATGTTGTGCCA ABD1.ver.R TTACTACATATAGAAGTCTTGTAATACGGCCG ADE2.ver.F GCTACGAACCGGGTAATACTAAGTGATTG ADE2.ver.R CAGGTAATTATTCCTTGCTTCTTGTTACTGG

qPCR Primers Sequence

ade2.WT.qPCR.F TACGAACCGGGTAATACTAAGTGATTGACTC ade2.gRNA.qPCR.R CGCTAGCCCAAGAGGGCACTACA ade2.WT.qPCR.R TACCAACTGTTCTAGAATCCATACTTGATTGTTT URA3.genedrive.ade2.WT. qPCR.F URA3.genedrive.ade2.WT. qPCR.R URA3.genedrive. ade2.gRNA.qPCR.R

TACGAACCGGGTAATACTAAGTGATTGACTC CCTCCTAATATACCAACTGTTCTAGAATCCAT AAACTTCTCCGCAGTGAAAGATAAATGATC

ABD1_rec_qPCR.R CGAGGAGCCGTTCAATTTGGTTTAACTGAC
ABD1_rec_qPCR.F AGATGCGTGAGGGGCTTCAGC

*** Formatting Issue - Best Available Paper/Graphic ***

ABD1_WT_qPCR (JDwt1.4).F ABD1_WT_qPCR (JDwt1.3).R

GAAGGGGATGAAAAGGAAGC CGCTTTCCGGTTCGATATAC

C4td Cure for Type Two Diabetics
Chaudhary, D. M.,

Cure for Type 2 Diabetics Permanently for Ever

It sounds too good to be true: reversing type 2 diabetes through exercise and healthy eating.

No person suffering from diabetes has ever been cured of diabetes through pharmaceutical medications. It has never happened in the history of medicine. And yet type-2 diabetes can be cured at a rate of 90% - 95% with a living foods diet. Risks and challenges

Human Trial After the C2TD Injection produce

Massively Parallel High-Order Combinatorial Genetics By Combigem in Human Cells

Wong, A., Massachusetts Institute of Technology
Choi, G., Massachusetts Institute of Technology
Lu, T.,

Massively Parallel High-Order Combinatorial Genetics by CombiGEM in Human Cells

Alan S. L. W ong, Gigi C. G. Choi, Allen A. Cheng, Oliver Purcell, Timothy K. Lu

Synthetic Biology Group, MIT Synthetic Biology Center, Research Laboratory of Electronics, Department of Biological Engineering and Electrical Engineering & Computer Science, Massachusetts Institute of Technology, Cambridge, MA 02139, USA.
The systematic analysis of combinatorial genetic interactions, which play significant roles in regulating complex biological traits, has been limited in the throughput and order of complexity of genetic combinations that can be studied through current methods. To overcome these bottlenecks and accelerate the study of combinatorial genetics in human systems, we created the CombiGEM (Combinatorial Genetics *En Masse*) technology for the scalable and easy assembly of barcoded combinatorial genetic perturbation libraries. CombiGEM enables multiplexed quantification of all members in a given combinatorial genetic library by using next-generation sequencing technologies. The genetic elements included in CombiGEM libraries can be arbitrary, including gene expression/knockdown constructs, microRNAs, synthetic-biology circuit components, and programmable genome editing tools. We generated high-coverage combinatorial libraries comprising two-wise and
three-wise barcoded genetic components in a lentiviral delivery system for efficient and stable genomic integration in human cells. We have validated the entire pipeline for complex genotype-to-phenotype mapping, including combinatorial library assembly, library verification, pooled screening assays with barcode sequencing, computational analysis, and hit validation, in a relevant human cell model of disease. More broadly, our work establishes a powerful platform for the high-throughput profiling of multifactorial genetic combinations that regulate a myriad of phenotypes relevant to biomedical, biotechnology, and basic science applications.
This work was supported by the NIH New Innovator Award (DP2 OD008435), the Office of Naval Research, and the Ellison Foundation New Scholar in Aging Award, and the Croucher Foundation.

Illinois Biological Foundry for Advanced Biomanufacturing (iBioFAB)
Zhao, H., University of Illinois at Urbana Champaign

Illinois Biological Foundry for Advanced Biomanufacturing (iBioFAB)

Huimin Zhao, University of Illinois at Urbana-Champaign zhao5@illinois.edu
Inspired by the exponential growth of the microelectronic industry, synthetic biologists have been attempting to build biological foundries for rapid prototyping and manufacturing of biological systems for synthesis of bioproducts ranging from chemicals to materials to therapeutic agents. In this talk, I will introduce the Illinois Biological Foundry for Advanced Biomanufacturing (iBioFAB) that we recently established at UIUC, which consists of (i) a computational design framework that automates the design-build-test-analyze cycle and (ii) an integrated automated robotic system that automates DNA cloning, DNA assembly, heterologous expression, and product detection (Figure 1). This iBioFAB features a 6 degree of freedom articulated robotic arm that travels on a 5-meter-long track and transfers microplates among more than 20 instruments installed on the platform; a sophisticated sample tracking system; and libraries of parts and modules for a variety of platform organisms (see video: http://youtu.be/Hwb735qZ- IQ). To demonstrate its unprecedented power, the iBioFAB was recently used to synthesize up to 1000

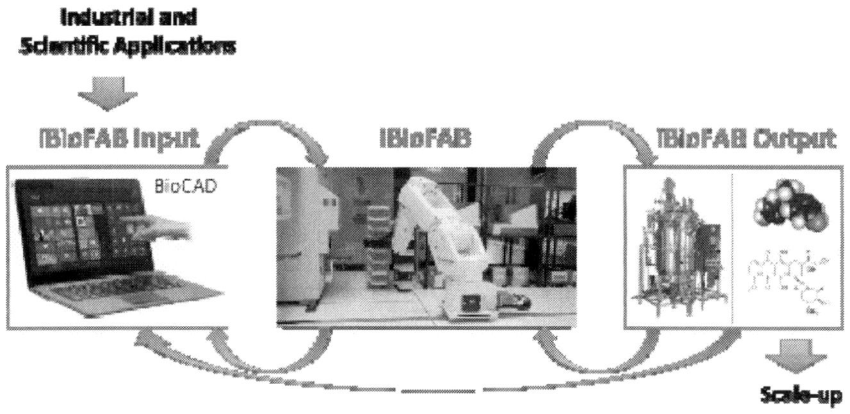

TALENs per day with dramatically reduced cost for large-scale genetic editing application. In addition, the iBioFAB has been used to automate the entire process of the RNA interference (RNAi)-assisted genome evolution (RAGE) method we recently developed for metabolic engineering of *Saccharomyces cerevisiae* (1). Through iterative cycles of creating a library of RNAi induced reduction-of-function mutants coupled with high throughput screening or selection, RAGE can continuously improve target trait(s) by accumulating multiplex beneficial genetic modifications in an evolving yeast genome. Previously, we used the manual RAGE method to improve acetic acid tolerance (1) and the furfural tolerance (2), two key traits for microbial production of chemicals and fuels from cellulosic materials. Recently, we augmented the genetic diversity created by the RAGE method by including a library of cDNA overexpression induced enhancement-of-function mutants and further developed a fully automated modified RAGE method using the iBioFAB. We then used this new method to dramatically improve the glycerol utilization rate and cellulase production level in addition to further increasing the acetic acid tolerance in *S. cerevisiae.*

Figure 1. Establishment of iBioFAB will create a new biomanufacturing paradigm.

1. T. Si, Y. Luo, Z. Bao, and H. Zhao. â??RNAi-Assisted Genome Evolution in *Saccharomyces cerevisiae* for Complex Phenotype Engineering.â? *ACS Synthetic Biology*, DOI: 10.1021/sb500074a.
2. H. Xiao and H. Zhao. â??Genome-wide RNAi Screen Reveals the E3 SUMO-protein Ligase Gene *SIZ1* as a Novel Determinant of Furfural Tolerance in *Saccharomyces cerevisiae.*â? *Biotechnology for Biofuels*, 7:78 (2014).

Combigem for Systematic and Massively Parallel Analysis of Drug-Gene Combinations for Cancer Therapeutics

Wong, A., Massachusetts Institute of Technology
Choi, G., Massachusetts Institute of Technology
Lu, T. K., Massachusetts Institute of Technology

CombiGEM for Systematic and Massively Parallel Analysis of Drug-Gene Combinations for Cancer Therapeutics

Alan S. L. Wong, Gigi C. G. Choi, Allen A. Cheng, Oliver Purcell, Timothy K. Lu
Synthetic Biology Group, MIT Synthetic Biology Center, Research Laboratory of Electronics, Department of Biological Engineering and Electrical Engineering & Computer Science, Massachusetts Institute of Technology, Cambridge, MA 02139, USA.

The chemosensitivity of cancer cells is under the influence of multiple cellular pathways and genetic factors. Combinatorial drug therapies can achieve higher efficacy than individual drugs since targeting multiple pathways can be synergistic. Understanding drug- gene interactions should enable us to devise optimized therapeutics for sensitizing cancer cells to drugs and reversing drug resistance. Conventional approaches for analyzing a large number of drug-gene combinations are low-throughput, and require huge effort and resources. Harnessing the power of synthetic biology and next-generation sequencing technologies, we developed CombiGEM (Combinatorial Genetics *En Masse*) as a powerful platform for high-throughput functional characterization of combinatorial genetic perturbations in human cells, which can be broadly applied in biomedical research. We applied
the CombiGEM technology to achieve scalable assembly of microRNA (miRNA) overexpression constructs. We generated high-coverage combinatorial miRNA libraries and performed systematic screens to identify combinatorial miRNA effectors that sensitized drug- resistant ovarian cancer cells to chemotherapy and/or inhibited cancer cell proliferation. We identified and validated a list of miRNA combinations that act synergistically to achieve anti- cancer phenotypes when expressed in combination. This effort revealed new insights into complex miRNA interaction networks, including previously unknown interactions between miRNAs that modulate drug resistance and cell growth phenotypes.

This work was supported by the NIH New Innovator Award (DP2 OD008435), the Office of Naval Research, and the Ellison Foundation New Scholar in Aging Award, and the Croucher Foundation.

Initial Steps Towards the Establishment of a Multigene Engineering Platform in Drosophila Melanogaster
Sarrion-Perdigones, A., Baylor College of Medicine

Initial steps towards the establishment of a Multigene Engineering Platform in

Drosophila melanogaster

Alejandro Sarrion-Perdigones1, Nicholas Abel2, Nick Matinyan3, Kristi L. Hoffman1, Audrey E. Christiansen1, Koen J.T. Venken1,2,3,4,5.

1 Verna and Marrs McLean Department of Biochemistry and Molecular Biology

2 Department of Pharmacology

3 Program in Integrative Molecular and Biomedical Sciences

4 Dan L. Duncan Cancer Center

5 Program in Developmental Biology

Baylor College of Medicine, TX, 77030

The application of synthetic biology principles for the analysis of biology and disease in multicellular organisms is a recently growing field. Currently there is no standard DNA assembly methodology to facilitate multigene engineering in *Drosophila melanogaster*. To overcome the existing limitations on the design and construction of new genetic combinations in this model organism and enable the exchange of genetic building blocks, we propose the establishment of a platform founded on the synergistic combination of two transgenesis technologies: GoldenBraid2.0, a synthetic assembly platform that facilitates multigene engineering via Type IIs restriction enzyme assembly, and P[acman], the high-capacity site-specific transgenesis platform developed for *Drosophila melanogaster*. Apart from providing an optimized cloning strategy that generates fully exchangeable genetic elements for multigene engineering, the platform will include the most advanced tools for genetic engineering in *Drosophila*. Our broad technological framework aims to establish the foundations of Synthetic Biology in the fly field and can be easily extrapolated towards other model organisms.
Here we present an initial group of elements characterized in *Drosophila* Schneider 2 (S2) cells. These elements include a strategically chosen group of functionally-tested, pre-made genetic parts to build frequently-used modules such as constitutive, tissue-specific and inducible expression cassettes as well as fluorescent protein and other reporter units common to *Drosophila* research. These elements are components of an

ever-growing open collection of DNA parts that will be available to any interested researcher. This platform, in combination with other shared resources in the fly community (e.g., Bloomington *Drosophila* Stock Center, BACPAC Resources, *Drosophila* Genomics Resource Center, and Addgene), will facilitate the expansion of complex synthetic biology advances in *Drosophila melanogaster*.

Unleashing the 21st Amino Acid: Removing the Sequence Constraints of Selenocysteine Incorporation

Thyer, R., UT Austin
Robotham, S. A., UT Austin
Brodbelt, J. S., UT Austin
Ellington, A. D., University of Texas at Austin

Selenocysteine is a rare, naturally occurring amino acid with a mosaic distribution across all domains of life. Unlike canonical amino acids, selenocysteine does not exist freely, and is directly synthesized on its tRNA which is pre-charged with serine. In bacteria, selenocysteine is co-translationally incorporated in response to specific opal stop codons, designated by the presence of a selenocysteine insertion sequence (SECIS) element which recruits the selenocysteine specific elongation factor and charged $tRNA^{Sec}$ needed to reassign the UGA codon. The SECIS element is a stem-loop RNA structure immediately following the UGA codon and forms part of the coding sequence in bacterial selenoproteins. With a low pKa and strong nucleophilicity, the site specific incorporation of selenocysteine has great potential for protein engineering, but the sequence constraints imposed by the adjoining SECIS element severely limit its use. A potential solution to this problem is to interface selenocysteine incorporation with the canonical translation machinery. We developed a novel genetic reporter system which depends on the formation of an essential selenyl-sulfhydryl bond to confer antibiotic resistance, allowing high-throughput screening of $tRNA^{Sec}$ variants compatible with the canonical elongation factor EF-Tu. From a $tRNA^{Sec}$ library containing a variable antideterminant region, which blocks interaction with EF-Tu, we isolated a $tRNA^{Sec}$ variant which is compatible with the canonical translation machinery and can suppress amber stop codons to incorporate selenocysteine with high efficiency. Using this evolved $tRNA^{Sec}$ we have produced new recombinant selenoproteins containing structural motifs such as selenyl-sulfhydryl and diselenide bonds.

Modified Transposons for Enhancing Protein Function

Nadler, D., UC Berkeley
Morgan, S. A., UC Berkeley
Kortright, K., UC Berkeley
Oakes, B., UC Berkeley
Flamholz, A.,
Savage, D., UC Berkeley

Multi-domain proteins are a common and essential component of the cellular proteome and contribute complex functions that single domains cannot achieve. The field of protein engineering uses sophisticated techniques to modify proteins for a given purpose, but generally focuses on single domains without consideration of the protein's overall topology. In this work, we set out to develop tools that enable the high-throughput sampling of different domain connections. A novel method of random transposon-mediated domain insertion was devised which efficiently created large plasmid libraries of domain fusions. As a proof of concept, we made fluorescent biosensors for small molecules by inserting circularly permuted GFP into ligand-binding proteins. We also demonstrated the generality of this method by constructing a pool of functional dCas9 variants with insertions of a PDZ protein-protein interaction domain. These successful protein-engineering efforts illustrate the advanced functions possible with multi-domain constructs and how this transposon-based insertion method can facilitate the creation and study of this new class of synthetic proteins.

Engineering Auto-Regulatory Genetic Circuits That Enable Modular and Programmable Functionalities for Biotechnological Applications

Lo, T. M.,
Chang, M. W., National University of Singapore

Engineering auto-regulatory genetic circuits that enable modular and programmable functionalities for biotechnological applications

Tat-Ming Lo1,2 and Matthew Wook Chang1,2

1Department of Biochemistry, Yong Loo Lin School of Medicine, National University of

Singapore, Singapore 117597

2Synthetic Biology Program for Clinical and Technological Innovation, National

University of Singapore, Singapore 117456

Microbes can be reprogrammed to demonstrate biotechnologically important functionalities such as biochemicals production. However, reprogramming microbes to perform complex and coordinated tasks often requires the use of additional exogenous chemicals such inducers, which can increase production costs, introduce host toxicity and incompatibilities with industrial scale-up. To overcome this challenge, we developed auto-regulatory genetic circuits by introducing novel sensors and feedback loops into microbial systems. Not only our engineered systems are independent of costly inducers, they are capable of regulating protein expression based on desired features such as substrate concentration and cell-density. One example is the use of our genetic circuit in auto-regulatory lignocellulosic bioconversion. In this system, the genetic circuit was engineered to regulate enzyme expression based on lignocellulose substrate cues and cell-density. In another example, we used our cell density-dependent circuit to facilitate product extraction in *Escherichia coli*. Macromolecular product extraction in *E. coli* often requires chemical, enzymatic, and/or mechanical lysis, thus increasing production costs. To aid in the extraction of biochemical products and simplify the product harvest process, we developed a genetic circuit that enabled cell density-dependent auto-regulatory lysis based on a transducer-switch scheme. As a proof of concept, we demonstrated the applicability of the lysis circuit to biotechnological applications by employing plasmid DNA extraction as a test bed. We envision that our engineered auto-regulatory circuits that enable programmable functionalities can be readily applied to a range of biotechnological applications.

Enabling Multiplex Genome Engineering in Yeast By RNAi and Crispr

Si, T., University of Illinois at Champaign-Urbana
Chao, R., University of Illinois at Champaign-Urbana
Zhao, H., University of Illinois at Urbana Champaign

Successful metabolic engineering practice often requires simultaneous manipulation of many genes. Whereas recombination-based genetic engineering (recombineering) is able to create combinatorial genetic diversity on a genome scale, such method is mainly limited to bacterial cells. Here we report a strategy that can generate multiplex modifications in the *Saccharomyces cerevisiae* genome, by combining the advances in full-length cDNA library construction, RNA-interference technology and CRISPR-Cas system. Briefly, upon introduction of specific double-stranded breaks (DSBs) in the repetitive sequences by CRISPR-Cas nucleases, both the overexpression and knockdown cassettes of every yeast gene can be integrated at high efficiency into the genomic loci of repetitive sequences. This process may be iteratively performed to accumulate dozens of genetic modifications in a single cell of an evolving yeast population. We fully automated this process through an integrated robotic platform, enabling generation of vast genetic diversity from which new or improved properties may emerge. We envision this new tool can greatly accelerate genome engineering in *S. cerevisiae* for basic and applied biological research and medicine.

Engineering Clostridium Thermocellum for High Ethanol Yield

Hon, S.,

Clostridium thermocellum, a gram-positive, thermophilic anaerobic bacterium, is a candidate organism for consolidated bioprocessing, due to its ability to solubilize cellulosic biomass and ferment said biomass into organic acids and ethanol. There have been several attempts to date to engineer *C. thermocellum* to produce ethanol at high yield; however, the maximum reported yield is 60% of theoretical.

Our group has previously engineered another thermophilic anaerobe, *Thermoanaerobacter saccharolyticum*, to produce ethanol at >90% theoretical yield. We have characterized and identified the genes that play a major role in producing the high ethanol yield, and now seek to transfer the key genes of this pathway into *C. thermocellum* to improve ethanol production.

In order to do this, we have improved our plasmid-based expression system to use in our organism. We have used this new plasmid to express *adhE*, a gene that encodes activities thought to play a role in the last two steps of ethanol production.

Creating a Riboswitch-Based Whole-Cell Biosensor for BPA

Shaffer, J., Pomona College
Velasquez, E.,
Liu, J., Pomona College

Detection of small organic molecules using riboswitches in whole-cell biosensors could allow for screening of environmental pollutants, such as bisphenol A (BPA), in an effective and economical manner. Riboswitches are RNA sequences typically embedded within the 5â?? untranslated region of mRNA that are capable of regulating gene expression and, when paired with a reporter gene, can signal the presence of small organic
molecules. In order to create a riboswitch in *Escherichia coli* that could bind BPA and, upon binding, translationally upregulate the expression of *gfp*, we investigated (1) the best method of engineering a novel BPA-activated riboswitch via dual genetic selection and (2) the effectiveness of using such a riboswitch to detect BPA.
Dual genetic selection was performed using a *tetA-gfp* reporter downstream of a mutagenized *thiM* riboswitch. Two *E. coli* libraries â?? Library 1 (3 x 104 members) and Library 2 (4 x 106 members) â?? were subjected to positive selection followed by negative selection. Library 1 was only subjected to one round of selection, whereas Library 2 was subjected to several rounds of selection under increasingly stringent selection conditions. These libraries were then screened for BPA-activated riboswitches via a GFP assay. Pre- selection sequencing of random members revealed that Library 1 and Library 2 were both highly diverse; no conserved sequences were identified within each library. Initial
analysis of post-selection hits suggests that multiple rounds of selection may be necessary. While more investigations must be conducted to verify these preliminary results and to determine the effectiveness of using a riboswitch to detect BPA, this technology serves as a promising method for detecting pollutants in the environment.

Self-Assembly through Toehold-Mediated Four-Way Branch Migration

Kishi, J.,

Self-assembly is a fundamental mechanism in the formation of complex and functional biological systems, and the ability of information-carrying DNA species to confer specific two- and three-dimensional shapes is a testament to the programmability and versatility of nucleic acids. This precise control of structure at the nanoscale could have profound implications for DNA probing and small molecule drug delivery. Here, we explore the possibility of utilizing toehold-mediated four-way branch migration to sequentially assemble complex structures, ranging from simple k-arm junctions to more intricate closed shapes.

ICE: A Distributed and Interconnected Biological Part Registry

Plahar, H., Joint BioEnergy Institute
Chen, J., Joint BioEnergy Institute, Lawrence Berkeley National Laboratory
Dmytriv, Z.,
Ham, T.,
Keasling, J. D., Joint Bioenergy Institute
Hillson, N. J., Joint BioEnergy Institute

Advances in synthetic biology research and genetic design automation, along with the availability of rapid and reliable DNA sequencing has resulted in a steady increase in the number of complex engineered biological devices. Maintaining information about the constituent parts of these devices, along with the ability to share them with other members in the synthetic biology community as well-characterized components with the goal of reusability is a challenge that is not fully addressed by any of the existing registry software platforms.

We have developed an open-source web-based platform for synthetic biology information management known as the Inventory of Composable Elements (ICE). It includes a distributed software platform to enable efficient sharing of biological elements across labs in the synthetic biology research community. Referred to as "Web of Registries", this feature aims to increase the efficiency of bioengineering by enabling scientists to publish and share datasets across multiple ICE installations in potentially different geographic locations. Thus, users from one ICE instance (e.g. the JBEI public repository) can query and access parts, sequences, strains and *Arabidopsis* seeds of interest stored in other connected ICE instances (e.g. the Synberc repository). ICE also integrates browser-based tools such as VectorEditor and PigeonCAD for sequence annotation, manipulation, analysis and visualization. Support for data formats such as GenBank, FASTA and the Synthetic Biology Open Language (SBOL), ensures compatibility with existing and emerging tools for synthetic biologists.

Quantitative Metabolic Modeling for Biofuel Production at the Joint Bioenergy Institute

Garcia Martin, H., Lawrence Berkeley National Laboratory
Alonso-Gutierrez, J., Joint Bioenergy Institute (JBEI)
Birkel, G., Lawrence Berkeley National Laboratory
Ghosh, A., Lawrence Berkeley National Laboratory
Morrell, W., Sandia National Laboratory
Echols, N., Lawrence Berkeley National Laboratory
Hillson, N. J., Joint BioEnergy Institute
Petzold, C., Joint BioEnergy Institute
Adams, P. D., Joint BioEnergy Institute
Keasling, J. D., Joint Bioenergy Institute
Lee, T. S., Joint BioEnergy Institute

The goal of the Joint BioEnergy Institute (JBEI) is to produce fundamental scientific discoveries to enable the development of commercially viable processes for large-scale conversion of lignocellulosic biomass into fuels.

The Quantitative Metabolic Modeling (QMM) directorate at the Fuels Synthesis Division is devoted to developing models of metabolism in order to improve biofuel production in a rationally directed fashion. We will describe how the QMM directorate fits in the general JBEI strategy and how we use experimental, computational and mathematical tools to achieve our goal. We will first show how the use of ^{13}C labeling data allows us to effectively constrain metabolic fluxes for genome-scale models and make quantitative predictions. Secondly, we will show how we used quantitative targeted proteomic data to guide metabolic pathway engineering and increase the production titers of limonene, a biofuel molecule. Finally, we will show how new web-based tools for standardized data storage (the Experiment Data Depot, EDD) and interactive flux visualization (Multiomics Visualization Tool, MvT) produce the necessary base to make these predictions possible.

Metabolic Engineering for the Production of Medium-Chain Fatty Acids and Derivatives in Baker's Yeast

Ling, H., National University of Singapore
Liu, R., National University of Singapore
Hee, K., National University of Singapore
Chang, M. W., National University of Singapore

Medium-chain fatty acids (MCFAs, with carbon chain length of 6 to12) and their derivatives are valuable compounds which can be used as food additives, anti-pathogenic agents, nutraceuticals, and drop-in fuels. Thus, they are very attractive target compounds for microbial bioproduction. In this study, we aimed to metabolically engineer baker's yeast to produce MCFAs and their derivatives. We enhanced the metabolic flux towards MCFA production by using combinatorial metabolic engineering approaches, including competitive pathway engineering, heterologous pathway reconstitution, and cofactor supply engineering. Blocking of competitive pathways to MCFA biosynthesis by multi-gene deletion led to significant accumulation of C8-C12 fatty acids as well as long-chain fatty acids. Further enhancement in MCFA production was achieved by introduction of a mammalian MCFA biosynthesis pathway into the MCFA-accumulating mutant. Subsequently, we demonstrated the conversion of MCFAs into valuable compounds by reconstituting and functionally characterizing a dietary lipid biosynthesis pathway in the MCFA-producing strain. Our findings could shed new light on metabolic engineering for developing microorganisms for the production of valuable fatty acids and derivatives. Additionally, the obtained MCFA-producing yeast strains could be applied to produce other value-added compounds.

Optogenetic Frequency Analysis of Bacterial Two-Component Systems and Synthetic Gene Circuits

Hartsough, L. A., Rice University
Groszman, K., Rice University
Tabor, J. J., Rice University

Synthetic biology is not yet a mature engineering discipline, due in part to a poor understanding of the dynamical input/output (I/O) properties of genetic components. Despite the fact that gene regulatory systems are inherently dynamic, most attempts to characterize them have been static. We have previously engineered green/red (CcaS/R) and red/far red (Cph8/OmpR) light-switchable two component systems (TCSs) with transcriptional outputs in *E. coli*[1,2]. In recent work, we developed quantitatively predictive mathematical models of the I/O dynamics of these signaling pathways by measuring their gene expression outputs in response to step changes in light input[3]. However, this characterization relied upon intuition to design non-linear models and iterative characterization and validation experiments. In contrast, the electrical engineering community has developed standardized, automatable, and scalable system identification (SI) methods that employ frequency analysis to construct linear transfer function (TF) models. These dynamical TF models are used to systematically characterize the performance of engineered components, reliably predict the behavior of those components when they are composed with others, and thereby design new systems with high predictability.

Here, we have combined frequency analysis, optogenetics, light-switchable TCSs and synthetic gene circuits to develop a standardized I/O characterization framework for transcriptional regulatory parts. Specifically, we subject *E. coli* populations to sinusoidally oscillating light inputs spanning a wide range of frequencies, and monitor the corresponding gene expression outputs with high temporal resolution. We then fit the sinusoidal response data to a small set of TF models, selecting one with the best fidelity for each system. We have validated this approach by characterizing both light-switchable TCSs and using the resulting TF models to predict gene expression dynamics in response to non-sinusoidal, time-varying inputs, with only 5% error (RMSE) over 12-hour experiments. This result demonstrates that frequency analysis can supplant our previous intuition-guided approach. Additionally, our TF models have yielded a better understanding of the TCSs' filtering characteristics, noise, and signal transduction limits, motivating characterization of other optogenetic components. We are also extending optogenetic frequency analysis to a wide range of genetic circuit components, including the Voigt TetR inverter family[4] and large sets of CRISPRi-based inverters, and using the data to create technical data sheets. Finally, we will use the resulting TF models to design larger genetic circuits with predictable dynamic properties. Our standard, scalable, and quantitatively rigorous approach will help synthetic biology transition into a mature engineering discipline.

[1] Tabor JJ, Levskaya A, & Voigt CA (2011). Multichromatic control of gene expression in *Escherichia coli*. *Journal of Molecular Biology*, 405(2), 315–24.

doi:10.1016/j.jmb.2010.10.038

[2] Schmidl SR, Sheth RU, Wu A, & Tabor JJ (2014). Refactoring and Optimization of Light-Switchable *Escherichia coli* Two-Component Systems. *ACS Synthetic Biology*, *3*(11), 820–831. doi:10.1021/sb500273n

[3] Olson EJ, Hartsough LA, Landry BP, Shroff R & Tabor JJ (2014). Characterizing bacterial gene circuit dynamics with optically programmed gene expression signals. *Nature Methods*, 1–11. doi:10.1038/nmeth.2884

[4] Stanton BC, Nielsen AK, Tamsir A, Clancy K, Peterson T, & Voigt CA (2014). Genomic mining of prokaryotic repressors for orthogonal logic gates. *Nature Chemical Biology*, 10(2), 99–105. doi:10.1038/nchembio.1411

Sentinels of Disease: Engineering Bacteria to Sense and Remember Salmonella Infection in the Mammalian Gut

Riglar, D. T., Wyss Institute for Biologically Inspired Engineering
Kerns, S. J., Wyss Institute for Biologically Inspired Engineering
Kotula, J., Harvard University
Way, J. C., Harvard University
Silver, P. A., Harvard Medical School

Enteric bacterial infections are a leading cause of global morbidity and mortality. Non-typhoidal *Salmonella* (NTS) species alone infect around 100 million people, and kill over 100 thousand of these, annually. Lack of accessibility to adequate medical care during severe infections is one of the major causes of death. The development of new, cheap and easy-to-administer diagnostics and therapeutics is therefore an important step towards better monitoring and treating this disease. We recently developed proof-of-concept engineered bacteria that sense the mammalian gut environment and remember its state for at least a week to be monitored by fecal sampling. Using 'Gut-on-a-Chip' microfluidic organ models and mouse models of *Salmonella* infection we have further developed this work to produce bacterial reporters that respond to markers of *Salmonella* infection *in vitro* and *in vivo*. This approach could provide a cheap and easy means to monitor infection and provides an important basis for the ongoing development of inducible therapeutics and other non-invasive diagnostic approaches using engineered bacteria.

Metabolic Engineering of Saccharomyces Cerevisiae for Production of Alkanes from Free Fatty Acids

Foo, J. L., National University of Singapore
Susanto, A. V., National University of Singapore
Leong, S. S. J., Singapore Institute of Technology
Chang, M. W., National University of Singapore

In view of concerns regarding depletion of petroleum resources and environmental issues arising from heavy reliance on fossil fuel for energy, much attention has been paid to developing microbial biosynthesis of biofuels for generating sustainable and renewable energy sources. Alkane is an ideal biofuel candidate because it is a major component of fossil fuel and is high in energy density. Although production of alkane in bacterial hosts has been extensively explored in the past few years, research on alkane biosynthesis in yeast lags far behind. Despite *Saccharomyces cerevisiae* being a robust industrial production host, biosynthesis of alkane from fatty acyl-CoA in *S. cerevisiae* was reported only recently. However, the yield was limited partly by low production of the aldehyde intermediate. Here, we report the engineering of *S. cerevisiae* for improved production of aldehyde from free fatty acid source for alkane biosynthesis.

Oxidoreductases that convert fatty acids to aldehydes are present in plants for signaling against biotic and abiotic stresses. We functionally expressed a plant oxidoreductase heterologously in *S. cerevisiae* to produce aldehydes from free fatty acids *in vivo*. Co-expression of aldehyde decarbonylases effected production of alkanes. The heterologous alkane production pathway was further optimized by gene deletions to increase the availability of free fatty acids to serve as substrate and selecting appropriate promoters for gene expression to improve metabolic flux. The engineered *S. cerevisiae* described here is a highly promising platform for alkane production from free fatty acids and the findings in this work will facilitate future development of alkane-producing yeast hosts.

Upgrading the Alkane Precursor Pool through Thioesterase Engineering

Ziesack, M.,
Dusel, B.,
Way, J. C., Harvard University
Silver, P. A., Harvard Medical School

In recent years microbially produced alkanes have become popular as direct drop-in fuel compounds. In 2013 Howard and colleagues introduced a highly modular alkane production pathway demonstrating that modification of the fatty acid precursor pool led to altered alkane production. However, they were unable to generate medium-length (C_5 to C_{11}), branched chain compounds. Here I propose to engineer thioesterase specificity to enable production of medium-length branched free fatty acids that can be converted into respective alkanes. In order to identify residues that are involved in specificity determination we expressed chimeras of *Cp*FatB1 (C_8) and *Cp*FatB2 (C_{14}) in *Escherichia coli* and measured their fatty acid profiles. We found that a region spanning 100 amino acids around the proposed binding pocket highly contributes to substrate specificity. We will use the Rosetta Enzyme Design Application to identify mutants that enable branched-chain activity in four different thioesterases (*Bs*TES, $C_{4/6}$; *Cp*FatB1: C_8; *Uc*FatB2: C_{12} and *Cp*FatB2: C_{14}). Mutated thioesterases will be expressed in a branched-chain fatty acid production strain and free fatty acids will be extracted and measured via GC-MS. This work will amplify the pool of microbial-derived fuel compounds and provide insight into enzyme specificity determination.

Optogenetic Regulation of Proteolysis Using an Engineered Light-Switchable Split-Tev Protease

Gerhardt, K., Rice University
Tabor, J., UC San Francisco

Cells use proteases and proteolytic cleavage to control diverse processes which include protein activation or inactivation by removal of inhibitory or catalytic domains, mediation of protein degradation by revealing stabilizing or destabilizing residues, and initiation of protein translocation by removal of peptide signal sequences or exposure of latent ones. Due to the potential for regulating such a diverse array of cellular processes, proteases and proteolytic cleavage are excellent candidates for synthetic control.

Here, we have engineered a synthetic light-switchable protease by fusing split halves of the NIa tobacco etch virus protease (TEVp) to the light sensing proteins Phytochrome B (PhyB) and Phytochrome Interacting Factor 6 (PIF6). PhyB and PIF6 form high affinity heterodimers under red light but rapidly dissociate under far-red light, making them ideal candidates for reconstituting otherwise inactive split-TEV fragments[1]. By coupling the catalytic activity of TEVp to the light-dependent activity of PhyB and PIF6 we are able to rapidly, photoreversibly, and quantitatively tune TEVp activity using time-varying ratios of red and far-red light. This result suggests that split TEV fragments rapidly become functional and non-functional upon dimerization and dissociation of PhyB and PIF6. We show that this light-switchable protease can be used to cleave and modify target proteins *in-vivo*, and that light-gated proteolytic cleavage can be engineered to initiate target protein degradation through the conserved N-end rule pathway in the model eukaryotic organisms *Saccharomyces cerevisiae* and *Dictyostelium discoideum*.

Using a custom multichromatic LED-tissue culture plate device, we have characterized the spectral, steady-state, and dynamic performance of the photoswitchable TEVp protein degradation system. We measured action spectra (activating and deactivating) and determined peak activating and deactivating wavelengths to be 648 nm and 721 nm, respectively, which correlate closely with in-vitro measured PhyB P_r/P_{fr} absorbance peaks spectra (660/730 nm). The steady-state dose response to 648 nm light intensity follows a Hill-like relationship with fold-change = 4.5, n = 1.39, k = 0.02 $\mu mol/m^2 * s$. Step-function response dynamics, $\tau_{ON,1/2}$ = 150 min, are fast compared to gene expression dynamics (cell-cycle time scales, 12 hrs in *D. discoideum*) but slow compared to reported half-times of N-end rule substrates (< 10 min). We have constructed a mathematical model that indicates dynamic range can be increased and target protein half-life under red light decreased by optimizing expression level of system components relative to host proteolysis components and we are currently implementing these optimizations in-vivo. Finally, our plate based characterizations will enable us to couple light-switchable TEVp with other optogenetic tools[2,3,4] without crosstalk for multi-channel control of protein production and degradation. This tool has many applications in science and biotechnology including improving control of dynamics and noise in eukaryotic synthetic circuits and the fabrication and modification of protein materials in vitro.

[1] Wehr M.C. *et al.*, Nat Methods. 2006 Dec;3(12):985-93

[2] Renicke C. *et al.*, Chem Biol. 2013 Apr 18;20(4):619-26.

[3] Tabor J.J. and Levskaya A, J Mol Biol. 2011 Jan 14;405(2):315-24.

[4] Muller K. *et al.*, Nucleic Acids Res. 2013 Jul;41(12):e124.

Cell-Free Systems for Synthetic Biology

Jewett, M. C., Northwestern University

Cell-free systems provide a valuable platform for understanding, using, and expanding the capabilities of natural systems. They reduce complexity, remove structural barriers, and do not require the maintenance of cell viability. As a consequence, cell-free systems provide an unprecedented and otherwise unattainable freedom of design to modify and control biological systems. In this presentation, I will describe both top-down and bottom-up efforts to design, construct, and modify biological catalysts involved in protein synthesis. In one example, I will describe our efforts to establish the first crude extract based cell-free protein synthesis (CFPS) platform based on an *Escherichia coli* strain lacking release factor 1 (RF1) and capable of high-level batch expression (>1.5g/L) of pure proteins harboring non-standard amino acids (nsAAs). Our yields outperform the best expression of proteins with single or multiple nsAAs *in vivo* by >10x, which widens the aperture to new diverse applications in functional biomaterials and medically relevant proteins. In another example, I will describe a bottom-up approach for constructing minimal cells *in vitro*. Towards this objective, we show significant progress in building a synthetic ribosome, the most complex of the macromolecular catalysts needed to boot-up self-replication. This goal has been precluded for decades because conventional *E. coli* ribosome reconstitutions are non-physiological, and ribosomes reconstituted with *in vitro* transcribed ribosomal RNA (rRNA) are essentially non-functional. To move beyond previous limitations, we developed an integrated synthesis, assembly, and translation method (called iSAT) that enables efficient one-step co-activation of rRNA transcription, assembly of transcribed rRNA with native ribosomal proteins into *E. coli* ribosomes, and synthesis of functional protein by these ribosomes in a ribosome free S150 extract. A novel feature of iSAT is that it mimics co-transcription of rRNA and ribosome assembly as it occurs *in vivo*. We also developed a ribosome synthesis and evolution method (termed RISE) for the selection of variant ribosomes *in vitro*. We have used RISE, the first purely *in vitro* method for ribosome engineering, to evolve the large ribosomal subunit for altered functionalities. We anticipate that iSAT and RISE will aid studies of ribosome biogenesis, make possible minimal cells, and catalyze a new paradigm for the synthesis and evolution of abiological polymers. Taken together, our complementary approaches are enabling a deeper understanding of why nature's designs work the way they do and opening new frontiers for harnessing a dramatically expanded genetic code for manufacturing novel therapeutics and synthesizing genetically-encoded materials.

Engineering Light- and Pharmacologically Triggered Extracellular Biological Switches

Hörner, M., Faculty of Biology, University of Freiburg
Hotz, N., Centre for Biological Signalling Studies (BIOSS), University of Freiburg
Kaufmann, B., Centre for Biological Signalling Studies (BIOSS), University of Freiburg
Gübeli, R., Spemann Graduate School of Biology and Medicine (SGBM), University of Freiburg
Christen, E., Faculty of Biology, University of Freiburg
Büning, H.,
Grimm, D.,
Zurbriggen, M. D., Faculty of Biology, University of Freiburg
Weber, W., Spemann Graduate School of Biology and Medicine (SGBM), University of Freiburg

Synthetic molecular switches are the key components to obtain control of mammalian cells. While the majority of the developed switches focuses on intracellular processes we now transfer concepts of synthetic biology out of the cell. In this context, we (i) engineered pharmacologically switchable viral vectors enabling precise control of the transfer of genetic information into cells and (ii) a synthetic extracellular matrix the mechanical properties of which can be modulated by light.

For the chemical switch controlling viral infectivity we used adeno-associated viral particles (AAV) due to their high relevance in gene-therapeutic approaches. We engineered the system in such a way that infection of the target cells occurs only in the presence of the rapamycin structural analog AP21967.

The development of the light-tunable extracellular matrix was realized by combining light-gated molecular switches from the field of optogenetics with cell-compatible polymers from material sciences. To this aim, we covalently coupled the cyanobacterial phytochrome Cph1 to an eight-arm polyethylene glycol forming a biohybrid hydrogel. Illumination with red light triggers dimerization of Cph1, thereby increasing the number of crosslinks within the hydrogel and thus enhancing its stiffness. Vice versa, far-red light illumination induces Cph1 monomerization and decreases hydrogel stiffness. Due to the fast switching properties of Cph1, the stiffness can be modulated reversibly within seconds. By incorporation of RGD cell adhesion motifs our hydrogel serves as suitable matrix for primary cells as well as cell lines. Since variations in matrix stiffness can modulate cellular signaling pathways and even direct cell differentiation, we propose that our engineered light-tunable extracellular matrix will be a unique tool for studying matrix-cell interactions.

Regulation of Biomass Degrading Enzymes in Anaerobic Gut Fungi and Their Application in Synthetic Co-Culture Systems

Henske, J. K., University of California, Santa Barbara
Solomon, K. V., University of California, Santa Barbara
Gilmore, S. P., University of California, Santa Barbara
Sexton, J. A., University of California, Santa Barbara
Theodorou, M. K., Harper Adams University
O'Malley, M. A., University of California, Santa Barbara

Regulation of biomass degrading enzymes in anaerobic gut fungi and their application in synthetic co- culture systems

John K. Henske, Kevin V. Solomon, Sean P. Gilmore, Jessica Sexton, Michael K. Theodorou, Michelle A.
Oâ??Malley

To support renewable technologies, it is necessary to develop more efficient methods to extract sugars from crude plant biomass (lignocellulose). Plants contain cellulose that depolymerizes into fermentable sugars for microbial biofuel production. However, in crude biomass, cellulose is trapped within lignin, hemicellulose and other biopolymers that complicate its hydrolysis. To address this issue, one can turn to nature, particularly to microbes that routinely degrade plant biomass. Many large herbivores, such as cows and horses, harbor a consortium of microbes in their digestive tracts that convert recalcitrant biomass into sugars. Within this consortium, anaerobic gut fungi are the primary colonizers of plant material, and represent a rich source of biomass degrading enzymes. We have isolated several novel strains of gut fungi from herbivores at the Santa Barbara Zoo to characterize their ability to release sugars from crude biomass. We have used transcriptomic analysis to identify specific enzymes required for the breakdown of plant material including cellullases (GH5, GH9, GH48), hemicellulases (GH10, GH11, GH43), and accessory enzymes (Polysaccharide deacetylases). Through examining the regulatory pattern of these enzymes during growth on a variety of carbon sources, those that are most important for degradation of crude plant material can be identified. We have also used transcriptomics to identify sensor proteins and transcription factors that may be responsible for the global regulation of these enzymes. Through examination of putative regulatory proteins and the environmental stimuli that trigger their response, we can develop methods to control gut fungal metabolism and the production of biomass degrading enzymes.

While tools to engineer gut fungi directly are severely underdeveloped, another way to incorporate them into industrial processes is to create co-culture systems. Natively, gut fungi maintain a syntrophic relationship with archaeal methanogens by which the fungi produce CO 2 and H2 that the methanogens convert into methane. This relationship results in enhanced biomass breakdown by the fungus. In a synthetic system gut fungi are used for their degrading power to release sugars from biomass (~5 g/L released from cellulosic substrates). This excess sugar is then used to fuel production of a value-added chemical in a model microbe, such as *S.* cerevisiae or *E.* coli. We have used the production of Flavin- based fluorescent proteins (FbFPs) to quantify growth in both systems and production of n-butanol in *E. coli* to assess the ability to produce a fuel molecule in this system. If methanogens are incorporated into this synthetic system, it is expected that the amount of sugar available for value-added chemical production will increase. By coupling the lignocellulose-degrading capabilities of the gut fungi with the production capacity of model microbes, many different products may be generated directly from biomass.

Quantitative Gene Circuits and Computationally Designed Transcriptional Sensors in Plants

Morey, K.,
Prasad, A.,
Baker, D., University of Washington
Medford, J.,
Antunes, M., Colorado State University

Plants are an attractive system for application of synthetic biology. Due to the long life cycle of plants, rigorous quantitative characterization of genetic components, assembly and testing of complex gene circuits, can take several months to years. For this reason, there is currently a very limited number of characterized genetic parts available for plants. We have developed methodology that allows rapid testing and quantitative characterization of a large number of synthetic genetic parts in isolated plant cells. We used this methodology to select genetic components for assembly of complex gene circuits and predict their function in transgenic plants. We have developed quantitatively tuned genetic circuits, such as positive feedback loops and toggle switches, which can be used to precisely control traits in plants with external inputs. In addition, we have developed a computationally redesigned transcriptional sensor in plants. Computationally redesigned proteins are engineered to be destabilized in the absence of the ligand and stabilized by ligand binding, leading to transcriptional activation in plants. These advances are allowing synthetic biology to take root in plants.

Domesticating the Genome of a Naturally Transformable Bacterium

Renda, B. A., University of Texas at Austin
Suarez, G. A., University of Texas at Austin
Dasgupta, A.,
Leon, D.,
Barrick, J. E., University of Texas at Austin

As the field of synthetic biology moves towards more ambitious genome engineering efforts, the need for chassis organisms better suited for genome modification is becoming more apparent. The bacterium *Acinetobacter baylyi* ADP1 could provide an ideal next-generation synthetic biology chassis due to its natural transformability. We conducted a 1000-generation adaptive evolution experiment in ADP1 to study the stability of this strain. We observed the gradual loss of transformability over time and, furthermore, observed the emergence of a high-aggregation phenotype. The genomes of evolved clones were sequenced to determine the genetic basis of these phenotypic changes. We found that
the activity of ADP1's sole IS-element family IS*1236* was the main driver behind these changes and also observed the loss of a 49-kb prophage region that had been previously used as a landing site for integrating heterologous genes into the chromosome. We deleted all copies of IS*1236* from the ADP1 genome to create an IS-less strain (ADP1-ISx) with increased genetic stability. This strain showed a much greater decrease in the frequency of gene-inactivating mutations than expected from the rates of spontaneous IS*1236* insertions. Unexpectedly, ISx also has improved transformability and a reduced rate of autolysis. In sum, our work demonstrates the utility of 'domesticating' the genomes of microbial strains by removing selfish DNA elements and has generated a new bacterial chassis (ADP1-ISx) with improved transformability and genomic stability for future genome engineering efforts.

Characterization of Crispri Dynamics Using Optogenetics and Mathematical Modeling

Castillo-Hair, S. M., Rice University
Igoshin, O., Rice University
Tabor, J. J., Rice University

Characterization of CRISPRi Dynamics Using Optogenetics and Mathematical Modeling

Castillo-Hair SM1, Igoshin OA1,2,3, Tabor JJ1,2
1. Department of Bioengineering, Rice University, Houston, TX 77005, United States
2. Department of Biosciences, Rice University, Houston, TX 77005, United States
3. Center for Theoretical Biophysics, Rice University, Houston, TX 77005, United States
CRISPR interference (CRISPRi) can be used to repress transcription from virtually any promoter in bacteria, yeast, or mammalian cells [1-3], and to construct multilayered synthetic gene circuits [4,5]. In this system, inactivated nuclease dCas9 forms a complex with a 102-nucleotide long single guide RNA (sgRNA), which is then directed to a segment of DNA complementary to a 20-nucleotide region in the sgRNA. If the DNA segment is a promoter or the coding region of a gene, dCas9 will then block transcription initiation or elongation, effectively acting as a transcriptional repressor. Though the steady state response from sgRNA to output transcription rate has begun to be studied, a thorough characterization of CRISPRi dynamics has not been performed. A quantitative understanding of CRISPRi dynamics would enable researchers to create precise, time varying perturbations in a wide range of natural gene networks, and establish design principles for engineering analog gene circuits.
Here, we study CRISPRi repression dynamics in *E. coli* using our previously developed CcaS/CcaR optogenetic system [6]. We use light-induced sgRNA expression to find that the dose-response curve follows a power-law-like scaling, and that it is very sensitive to small levels of sgRNA, in agreement with previous work [4]. Under typical operating conditions, repression is found to occur within minutes, with expression from the target decreasing within one cell cycle, whereas de-repression requires 2-3 cell cycles. Using a combination of mathematical modeling and experiments, we find that fast repression occurs because sgRNA:dCas9:DNA forms quickly, whereas slow de-repression occurs due to a need for stable sgRNA:dCas9 complexes to be diluted and new DNA to be synthesized. We then use the mathematical model to identify specific system re-designs that allows us to tune and accelerate de- repression dynamics. Finally, we use a combination of modeling and experiments with competing sgRNAs to study how competition for dCas9 (i.e. dCas9 loading) affects system performance. This work should inform future optimizations to CRISPRi system that will allow the reliable and

predictable construction of large analog circuits in single cells.

References

[1] Lei S. Qi, *et al.* Repurposing CRISPR as an RNA-guided platform for sequence-specific control of gene expression. *Cell* **152**, 1173â??1183 (2013).
[2] Luke A. Gilbert, et al. CRISPR-Mediated Modular RNA-Guided Regulation of Transcription in
Eukaryotes. *Cell* **154**, 442-451 (2013).
[3] Luke A. Gilbert, et al. Genome-Scale CRISPR-Mediated Control of Gene Repression and Activation.

Cell **159**, 647-661 (2014).

[4] Alec AK. Nielsen, Christopher A. Voigt. Multi-input CRISPR/Cas genetic circuits that interface host regulatory networks. *Mol. Syst. Biol.* **10**: 763 (2014).
[5] Samira Kiani, et al. CRISPR transcriptional repression devices and layered circuits in mammalian cells.

Nature Methods **11**, 723-726 (2014)

[6] Sebastian R. Schmidl, et al. Refactoring and Optimization of Light-Switchable *Escherichia coli* Two- Component Systems. *ACS Synth. Biol.* **3**, 820â??831 (2014)

The Effect of Cellular Noise on MSC Cell Fate Decisions
Lam, A., University of Utah
Deans, T., University of Utah

The Effect of Cellular Noise on MSC Cell Fate Decisions

Alec H. Lam1, Tara L. Deans, PhD1

1Department of Bioengineering, University of Utah, Salt Lake City, UT, USA

Variation in gene expression level, termed as cellular noise, is a ubiquitous characteristic among all cell populations. Despite this inherent stochasticity, cells such as mesenchymal stem cells (MSCs) still have the ability to undergo important cell fate decisions. To investigate the role of noise in cell fate decisions, we are examining the differentiation efficiency of MSCs in the presence of various amounts of exogenous noise. To accomplish this, we are building genetic circuits to control fluctuations of noise in MSCs and assessing the differentiation outcome based on this noise. We have found that cellular noise is an important mechanistic factor that influences a MSC differentiation potential. Further, these results suggest that cellular noise is a tool that can be used to control cell differentiation.

A Rule-Based Library for the Verification and Modification of Synthetic DNA Sequences

Oberortner, E., Lawrence Berkeley National Laboratories
Deutsch, S., Lawrence Berkeley National Laboratory
Cheng, J. F., Lawrence Berkeley National Laboratories
Hillson, N. J., Joint BioEnergy Institute

Many synthetic biology research groups design complex DNA sequences *in silico* according to biological discoveries and rules, but order the physical sequence from DNA synthesis vendors. Before the physical synthesis process starts, at latest, the designed sequences must be verified against rules of DNA synthesis methods and technologies. If a sequence violates a rule, then corrective actions are performed manually that depend on the type of rule and the designers' knowledge. Moreover, manual sequence modifications are hard to reenact, error-prone, time-consuming, and circumvent fully automated workflows.

Various algorithms have been developed to verify sequences against rules *in silico*. Biological advances have shown that rules can be mainly categorized into two types: *repeats* and *GC content*. Both types can occur in the entire sequence (i.e. *globally*) or in specific regions (i.e. *locally*). Direct repeats denote *k*-mers (e.g. polymers, dimers, trimers) and inverted repeats indicate potential undesired secondary structures (e.g. hairpins). Here, we demonstrate the Sequence Polishing Library (SPL) and its integration into an automated workflow at the DOE Joint Genome Institute. Currently, the SPL imports sequence information from GenBank, FASTA, FASTQ, or SBOL file formats. Besides finding only exact matches of sequence repeats, the SPL can find repeats to a configurable degree of similarity. Therefore, SPL programmatically integrates the BBMap bioinformatics library (http://sourceforge.net/projects/bbmap).

For the specification of rules, we develop a yet simple but expressive language, which also supports the specification of corrective actions. The SPL differentiates between human, recommended, and automated corrective actions. Per default, human corrective actions must be performed since which action to perform depends on where and what type of violation occurs. In order to modify sequences in an automated fashion, information is needed about the sequences' structure. For example, if a coding sequence contains repeats, then organism-specific codon tables can be consulted to modify the DNA sequence. If a promoter violates a rule, then the SPL can recommend alternative promoters to the designer. Such structural information can be specified, for example, using the GenBank or SBOL format.

The SPL can be invoked programmatically via an Application Programming Interface (API) and utilized manually via a web-based User Interface (UI). Both interfaces provide expressive feedback about rule violations and corrective actions. Therefore, we believe that synthetic biology researchers as well as DNA synthesis vendors can benefit from utilizing the SPL, contributing to the automation of internal and cross-organizational workflows.

CX1: A Web-Based Application for Augmenting Open Synthetic Biology Laboratory Automation

Bates, M., Autodesk
Silverman, D., Autodesk
Berliner, A., Autodesk
Hessel, A.,

Individual experimentation done at the laboratory bench is the foundation of classical life science R&D, but this craftsman approach has substantial barriers to entry and limits information sharing, reproducibility, and scaling. Digital design, process management and automation, common in many areas of product development and manufacturing including pharmaceutical development, are now finding their way into everyday bioscience. Collectively, they have the potential to dramatically accelerate the iterative design, build, test, learn (DBTL) cycle of synthetic biology. Recently Transcriptic, Inc., has leveraged advances in robotics and software to automate the build and test components of the cycle and make them available via the cloud. However, many researchers still lack the computer programming expertise required to implement the laboratory operations to augment or replace their routine bench work. Working with Transcriptic and their Autoprotocol language, we have developed a web-based application called CX1 to provide an easy-to-use framework for designing, visualizing, controlling, and distributing JSON-based Autoprotocol constructs. The aim of CX1 is to augment openly available synthetic biology lab automation. To demonstrate the utility of the CX1 application, a process development pipeline has been implemented to direct the assembly and characterization of a synthetic PhiX174 bacteriophage from design through *Escherichia coli* plaque assay.

Evolving Around the Genetic Switches and Circuits
Umeno, D., Japan Science and Technology Agency (JST)

Application for Speaker (SEED2015)

Evolving around the genetic switches and circuits.

Daisuke Umeno

Graduate School of Engineering, Chiba University, Japan,

E-mail address: umeno@faculty.chiba-u.jp

PRESTO, Japan Science and Technology Agency (JST), Japan

Abstract

Directed evolution is a powerful strategy to construct complex yet functional systems ranging from genetic switch, regulatory circuits, and metabolic networks. In the last SEED meeting, we developed and tested various types of positive/ negative selection systems to widely meet the demand for improving switching stringency, PoPS values, and rapidity in on/off transition. In this meeting, we would like to show various applications of our systems ranging from new induction systems to rapidly tuning their device specificities, as well as their gathering into multi-input genetic circuits.

1. Evolving Switches.

To create new induction system that strictly meet the demand from industrial sides â?? self cloning, use of safe and cheap induction reagents, high induction level, and high stringency-, we created plasmid-encoded regulatory systems based on Bet-operon. Due to the need to minimize the interactions with chromosome-encoded Bet operons, the prototype Bet-switches were implemented on high-copy vectors and under highly active promoters, resulting in the â??always-offâ?? behaviour. Random mutagenesis of BetI genes, followed by ON-/ OFF- selection in the presence/ absence of choline, resulted in the isolation of dozens of switches with different properties (PoPS, stringency, IC50, â?¦.etc). Multiple round of directed evolution also (and quickly) isolated various types of Bet- switches with (1) altered ligand specificity and/or (2) reversed phenotypes (where choline acts as co- repressor). In this presentation, we also show the directed evolution of various known switches to
â??betterâ?? behaveâ?? in various context.

2. Two-input gates and circuits.

Even the complex and multi-membered circuits ultimately turn-on and -off the gene expression. The switches we created was assembled into the prototype logic gates and subjected to the
randomization and systematic on/ off selection in various input states. Almost all of the starter
circuits malfunctioned, but we obtained rapidly obtained a series of two-input gates (AND, imply, N-imply, NAND, â?¦.etc) with decent fidelity/ robustness.

Engineering a Synthetic Regulon in Saccharomyces Cerevisiae for Efficient Xylose Utilization

EndalurGopinarayanan, V., Tufts University
Nair, N. U., Tufts University

Saccharomyces cerevisiae is one of the most widely used microbes for generating biofuels using lignocellulosic feedstock. Since this yeast cannot utilize xylose, the second most abundant sugar in lignocellulose, several metabolic engineering approaches have been attempted in the recent past for efficient xylose consumption. However, growth on xylose is still poor and economically unfeasible for biofuel production. We hypothesize that further enhancement in xylose utilization and biofuel production can be made by addressing the issue from a regulatory perspective rather than a metabolic perspective. Therefore, our work aims at developing a regulon (genome-wide regulatory infrastructure) to enhance the growth and biocatalytic fitness of this yeast for enhanced respiratory growth and advanced biofuel (non-ethanolic) production using xylose. To achieve this, we developed a protein sensor-actuator system that regulates the expression of aerobic xylose metabolism based on its concentration. The well-characterized galactose sensor protein, Gal3p, was engineered for increased affinity towards xylose through semi-rational protein engineering. Targets for mutagenesis were identified using *in silico* docking studies and high-throughput screens were developed based on G418 resistance and GFP fluorescence. In these screens, expression of *kanMX* and *gfp*, placed under *GAL* promoters, were triggered by a regulatory cascade initiated by the binding of Gal3p to xylose. Initial selection was performed in G418 selection plates and the strains that showed resistance were further quantitatively screened for increased fluorescence. The results reveal further scope for increasing xylose sensitivity in Gal3p mutant variants. Future work involves up- and down-regulating genes and pathways required for enhanced aerobic growth on xylose under the control of this synthetic regulatory system. Such a strain will act as a platform to engineer advanced biofuels using lignocellulosic biomass.

Facilitating Unconventional Yeast Engineering for Biorenewables Production

Shao, Z., Iowa State University
Cao, M., Iowa State University
Gao, M., Iowa State University

Many of the 1800 known yeast species have highly unusual metabolic, biosynthetic, physiological and fermentative capacities that are not possessed by model yeasts such as *Saccharomyces cerevisiae.* To date, strides have primarily made on transferring genetic elements attributed to the desired traits from native sources to model hosts. However, as the outcomes of long-term natural evolution in particular environments, many high-performing functions are granted by a network of genes through a complex hierarchy of regulations, many of which have not been clearly elucidated. When the predicted relevant genes were incorporated into model hosts, many functions were impaired due to the alteration of physiological backgrounds, general missing of desired regulations, lack of mechanisms for cofactor recycling, or simply expression imbalance. On the other hand, nowadays, we are no longer retarded by genome sequencing, and if we could establish generic design rules, hence shortening tool development cycle for unconventional microbes, we can directly leverage, engineer and continuously improve the treasures granted by nature.

Here we report our recent success in establishing the first stable expression platform for engineering *Scheffersomyces stipitis*, one of the species with the highest native capability for xylose fermentation. The current expression plasmid is extremely unstable, being lost within 2 days of cultivation. With the perfect integration of *in silico* prediction and high throughput screening, we were able to isolate all eight centromere sequences from *S. stipitis* genome in a monthly manner. The identified centromeres significantly enhanced the stability of the episomal vector, resulting in 3-fold increase in lactic acid production. Isolating a functional centromere in an efficient manner by far is still a missing piece from literature. Immunoprecipitation with specific kinetochore protein antibodies is the predominant method, but suffers from the extensive experiment procedures and the prerequisite of antibody that needs to be developed for each species. We are currently investigating the generalizability of our hybrid approach in rapid centromere identification in a broad spectrum of unconventional yeasts. In contrast to the previous developments in this field that were fairly scattered and short of systematicness, we are ambitious to elucidate design principles and establish platform technologies to enable rapid functional modifications of a series of high-potential yeast species.

Green, A. A., Arizona State University
Kim, J.,
Teichmann, M.,
Silver, P. A., Harvard Medical School
Collins, J. J., Massachusetts Institute of Technology
Yin, P., Harvard Medical School

Synthetic RNA Networks for Sophisticated *In Vivo* Computation

Alexander A. Green1,2*, Jongmin Kim2, Mario Teichmann2, Pamela A. Silver2,3, James J. Collins2,4,5, Peng Yin2,3
1Center for Molecular Design and Biomimetics, The Biodesign Institute, Department of Chemistry and Biochemistry, Arizona State University, AZ 85287, USA. 2Wyss Institute for Biologically Inspired Engineering, Harvard University, Boston, MA 02115, USA. 3Department of Systems Biology, Harvard Medical School, Boston, MA 02115, USA. 4Institute for Medical Engineering and Science, Department of Biological Engineering, and Synthetic Biology Center, Massachusetts Institute of Technology, Cambridge, MA

02139, USA. 5Broad Institute of MIT and Harvard, Cambridge, MA 02142, USA. *E-mail:

alexgreen@asu.edu.
Synthetic gene networks have generally employed well-characterized natural parts rewired to provide new functionality; however, the limited number and crosstalk between such parts has hindered the construction of more complex networks that can operate robustly in living cells. We have recently developed a new type of synthetic biological device called the toehold switch that addresses these limitations by providing dozens of orthogonal parts each with a wide dynamic range of gene expression. These validated new parts provide us with the opportunity to construct much more complex *in vivo* synthetic networks. Here, we present a new molecular programming strategy that employs long information-processing strands of RNA to evaluate the status of networks of short *de-novo*-designed RNAs being transcribed within the cell. These synthetic RNA computational networks, or ribocomputers, are enabled by sets of highly orthogonal toehold switches and self-assembling RNA complexes that act as conditional activators of gene expression. Ribocomputers can compute expressions combining AND, OR, and NOT operations for universal combinatorial logic in *E. coli*, and they exploit *in silico* RNA sequence design to greatly reduce the time and effort required to construct new functional systems. These systems can efficiently evaluate multi-input AND and OR logic operations using a single computational layer and routinely provide ON/OFF ratios greater than 20. Furthermore, multiple ribocomputers can be active within a cell at the same time providing multi-output capabilities. All these features make ribocomputers an

important new addition to the synthetic biology toolkit and highlight the strengths of programmable RNA-based regulation for realizing sophisticated *in vivo* logic.

High-Throughput Screening of Membrane Transport Proteins in Synthetic Lipid Vesicles

Romero, P. A., University of California, San Francisco
Sarhan, M.,
Abate, A.,

Membrane transport proteins mediate the exchange of ions, small molecules, or macromolecules across the cell membrane and play critical roles in many biological systems. Engineered transporters have myriad applications that include biosensing, metabolic engineering, and experimental tools for probing neural function. Nevertheless, transporter engineering has been limited by the lack of robust and quantitative high-throughput assays suitable for screening large transporter libraries. Cell-based assays are difficult due to the toxicity of heterologously expressed membrane proteins, the confounding effects of endogenous transporters, and the inability to precisely control the intracellular environment. We present a new method for high-throughput transporter screening in synthetic lipid vesicles. Microfluidic flow focusing produces double emulsion templates that, through a novel dewetting mechanism, are converted into monodisperse giant unilamellar vesicles. The vesicles can be loaded with cell-free expression systems to express the desired transporters in the membrane, where they function. This allows the transporter variants to be assayed and sorted at high throughput using commercial fluorescence activated cell sorters. Synthetic lipid vesicles enable high-throughput screening of membrane transport proteins, and represent a robust platform for the directed evolution of transporters with new biochemical properties.

Geneart® Strings™ DNA Libraries – the Fast and Economic Way of DNA Library Production

Netuschil, N., Thermo Fisher Scientific / GeneArt AG
Bauer, C., Thermo Fisher Scientific
Santl, T., Thermo Fisher Scientific / GeneArt AG
Greifenegger, N., Thermo Fisher Scientific / GeneArt AG
Kranz, C., Thermo Fisher Scientific / GeneArt AG
Liss, M., Thermo Fisher Scientific
Trefzer, A.,

GeneArt® Strings™ DNA Libraries - the fast and economic way of DNA library production

Nikolai Netuschil, Claudia Bauer, Tobias Santl, Norbert Greifenegger, Christian Kranz, Michael Liss, Axel Trefzer

Thermo Fisher Scientific / GeneArt AG, Josef-Engert-Str. 11, 93053 Regensburg, Germany

Thermo Fisher Scientific offers an extensive portfolio around *de novo* gene synthesis which consistently keeps up-to-date with the latest demands and applications. Particularly, our Directed Evolution products deliver top-of-the-line quality, showcasing our technology leadership in the market. Current processes for DNA library production in which diversity is introduced into a sequence verified template are time and labor intensive and thus make degenerated libraries products at the upper end of the customer value continuum.

Here we report the development of GeneArt® Strings™ DNA Libraries, a low cost library product that will allow our customers to obtain libraries much faster and at significantly reduced cost.

By combining our longstanding expertise in the production of GeneArt® Combinatorial Libraries with our very successful low cost gene synthesis product GeneArt® Strings™ DNA Fragments, we are able to offer an economical directed evolution solution to price sensitive academic and industrial customers alike: GeneArt® Strings™ DNA Libraries allow for the introduction of all IUPAC defined ambiguous nucleotides into the customer sequence while providing the advantages of GeneArt® Strings™ DNA Fragments of short production time, low price, flexible sequence design and full gene optimization options.

Cell-Free Metabolic Engineering for Butanol Production

Karim, A., The University of Texas at Austin
Jewett, M. C., Northwestern University

Cell-free biological systems have emerged as one approach to decouple the cell's physiological and evolutionary objectives from the engineer's process objectives to enable high yielding biomolecular transformations. In this work, we used a crude cell lysate approach to reconstitute the 17-step *Clostridium* 1-butanol biosynthetic pathway in extracts of *Escherichia coli*. We prove the ability to mix 5 crude lysates each with selectively over-expressed enzymes to activate the entire 1-butanol production pathway *in vitro* with high productivities and yields. We additionally demonstrate the ability to quickly optimize these cell-free systems by modulating the physicochemical environment to improve yields and productivities. Furthermore, we establish for the first time the ability to combine cell-free protein synthesis of enzymes with cell-free metabolic engineering to activate the metabolic pathway towards 1-butanol. This work provides a proof-of-concept that not only can we perform chemical synthesis by utilizing cellular metabolism in crude lysates but also that we can construct biosynthetic pathways by using crude lysates for both catalyst synthesis and utilization in one pot. This opens the door for novel biosynthetic pathway prospecting and rapid prototyping of metabolism by cell-free synthetic biology.

Reengineering for Robust Photorespiratory Bypass to Improve Photosynthesis and Crop Production

South, P., USDA-ARS
Ort, D., USDA-ARS

Reengineering for robust photorespiratory bypass to improve photosynthesis and crop production.

Paul F. South1,2 Berkley J. Walker1,2 and Donald R. Ort1,2

1USDA-ARS Photosynthesis Research Unit, 2Institute for Genomic Biology, University of Illinois

Over the past 50 years improvements in grain crop yield potential has maintained food supplies as demand has increased. As climate changes globally and human population increases traditional methods of crop improvement have become less effective in adapting and improve agricultural production on ever limiting land and water availability. Improving photosynthetic efficiency has been a long standing goal toward increasing output and crop yield that has to date only played a minor role in crop improvements. At 25Â°C and current CO2 levels roughly
25% of Rubisco activity is the fixation of oxygen instead of carbon dioxide resulting in the
conversion of RuBP to one molecule of phosphoglycerate and one molecule of glycolate. C3 plants recover the carbon in glycolate through the C2 photorespiratory pathway. The C2
pathway uses energy in the form of ATP and reducing equivalents and loses fixed CO2 resulting in a reduction in photosynthetic efficiency by 30%. Synthetic biology has provided new opportunities in altering photorespiratory metabolism to improve photosynthetic efficiency.
Indeed metabolic bypasses to photorespiration have been generated and have demonstrated improvements in growth. Using the Golden Gate synthetic biology approach we have
assembled a series of multigene constructs that contain alternate metabolic pathways to bypass
photorespiration. In addition, we designed a screen based approach to test a range of standardized parts (promoters, terminators) in model plants *Arabidopsis thaliana* and *Nicotiana tabacum*. We have successfully transformed in large multigene constructs and have demonstrated detectable gene expression. Using a fluorescence based screen we tested single construct transformed lines for rescue of photorespiratory deficient plants in Arabidopsis. Our results indicate that large multigene constructs containing a metabolic bypass to

photorespiration can rescue changes in fluorescence caused by low CO_2 stress in deficient plants. Determining robust photorespiratory bypass constructs can provide insight into next generation crops and our utilization of standard parts and fluorescent screening provide a new tool kit for plant synthetic biology to engineer improvements in photosynthetic efficiency.

Metabolic Engineering of Saccharomyces Cerevisiae for the Overproduction of Short Branched-Chain Fatty Acids

Yu, A. Q., National University of Singapore
Juwono, N. K. P., National University of Singapore
Leong, S. S. J., Singapore Institute of Technology
Chang, M. W., National University of Singapore

Short branched-chain fatty acids (sBCFAs, C4-6) are versatile platform chemicals in high demands because they can be converted to a wide range of valuable products in the manufacturing, pharmaceutical and food industry. For example, their esters are used for the production of plastics, plasticizers, surfactants, textile auxiliaries, fragrances and flavorings. Currently, sBCFAs are mainly synthesized chemically, which can be costly and causes pollution. In order to develop a more economical and environmentally friendly route for sBCFAs production, we engineered *Saccharomyces cerevisiae*, a model eukaryote of industrial significance, for the overproduction of sBCFAs. *S. cerevisiae* natively produces the sBCFAs isobutyric acid, isovaleric acid, and 2-methylbutyric acid from branched-chain amino acids via the Ehrlich pathway. However, the natural production levels of these sBCFAs are very low in *S. cerevisiae.* In order to utilize *S. cerevisiae* as an efficient cell factory for high-level production of our target sBCFAs, we optimized the endogenous Ehrlich pathway for sBCFA biosynthesis. We achieved overproduction of sBCFA in yeast by overexpressing pathway genes, knocking out genes in competitive pathways, reengineering native transporters and optimizing growth conditions. Our engineered yeast can thus serve as a platform strain for biosynthesis of valuable sBCFA-derived chemicals.

Combinatorial Engineering of Saccharomyces Cerevisiae for Terminal Alkene Production

Chen, B., National University of Singapore
Lee, D. Y., National University of Singapore
Chang, M. W., National University of Singapore

Biological production of terminal alkenes has garnered a significant interest due to their industrial applications such as lubricants, detergents and fuels. Here, we engineered the yeast *Saccharomyces cerevisiae* to produce terminal alkenes via a one-step fatty acid decarboxylation pathway and improved the alkene production using combinatorial engineering strategies. In brief, first, we screened and characterized eight fatty acid decarboxylases (OleT) to enable and enhance alkene production in *S. cerevisiae*. We then developed a fatty acid-overproducing strain to boost the precursor availability, which could enhance the metabolic flux and resulted in a higher production titer. We then improved the enzyme cofactor accumulation through cofactor genetic engineering. We then enhanced the cell growth in rich medium and tuned the enzyme expression by optimizing the combinations of the promoters and plasmids. Finally, we further increased the alkene production by optimizing the culturing conditions in bioreactors. This study represents the first report of terminal alkene biosynthesis in the yeast *S. cerevisiae*, and we envision that the abovementioned combinatorial engineering approaches could provide insights into devising engineering strategies to improve the production of fatty acid-derived biochemicals in *S. cerevisiae*.

Control of the Asymmetric Segregation of Proteins at Division Thanks to a Self-Assembling Peptide

Decrulle, A. L.,
Lindner, A. B.,

Asymmetric division may give rise to daughter cells with different fates. It often results from the unequal distribution of cellular components due to either random segregation of diffusive molecules or as a result of a localizing mechanism, placing molecules in specific locations within the cell prior to division. Unequal distribution of cellular components plays an important mechanistic role in a wide range of systems such as stem cells differentiation and renewal, bacterial sporulation control, antibiotic persistence, diauxie, and aging.

So far, most of asymmetric division studies were mostly observational and at best measured the degree of asymmetry using single-cell microscopy. We go one step further and use a synthetic approach in order to actively control the distribution of specific cellular components. By fusing proteins to a ionic self-assembling peptide (ELK16) we took advantage of the nucleoid macromolecular crowding effect to localize slowly diffusing, big assemblies at E. coli's pole.

We were then able to observe their asymmetric segregation during several divisions using time-lapse microscopy and microfluidic systems.We show that fused enzymatic activity is retained in vivo. As example, inducing asymmetry of the aminoglycoside-resistance enzyme (APH(3')-IIIa), resulted in polar segregation where only cells containing the enzyme cluster grew and survived multiple generations in the presence of the antibiotics (e.g. Kanamycin). In this work we created and characterized a useful tool to control the segregation of proteins inside E. coli, enabling the study of the fitness landscape governed by asymmetric division both at the single cell and at the population level. Moreover this might be used as a tool to cluster enzymes in order to favor metabolic channeling as recently shown as well as a way to differentiate monoclonal cell population.

An Attempt of Transplantation of Carotenoid Genes from Eukaryote to Prokaryote

Tsuge, K., Keio University
Hasebe, M., Keio University
Togashi, T., Keio University
Itaya, M., Keio University
Tomita, M., Keio University
Nakada, T.,
Ito, T.,

Bioproduction of useful chemicals by living organisms would have some problems with productivity. For one of the possible solutions, transplantation of the entire pathway to a well-known host cell, such like *Escherichia coli*, by gene transfer have been considered. Although the genome projects reveals all of gene set in one organism, however, it is empirically obvious that only gathering and transferring gene set for the relevant metabolic pathway isn't enough to perform the pathway in heterologous host. To elucidate possible problem with gene expression beyond the barrier of species, especially between eukaryote and prokaryote, a series of astaxanthin, carotenoid, biosynthesis genes from algae were tried to transfer to *E. coli* as a case study. *Haematococcus lacustris* (=*H. pluvialis*) NIES-144, which is one of the famous algae for astaxanthin production, was rendered for a donor. cDNA of the strain sampled at astaxanthin production phase was subjected to Next Generation Sequencer (NGS) not only to determine the sequence of carotenoid genes but also to evaluate relative expression abundance of these genes. Identified 8 genes for astaxanthin were then assembled into a plasmid with designed artificial operon by OGAB (Ordered Gene Assembly by *Bacillus subtilis*) method according to the gene expression abundance. In parallel with this, we also constructed an allelic carotenoid operon plasmid using bacterial genes obtained from *Erwinia*. Recombinant *E. coli* of alga operon didn't produce any astaxanthin, while that of bacterial one did. To identify possible problematic gene(s) in the algae operon, we performed to construct a series of chimeric operons comprised from gene(s) of algae and bacteria. Finally we identified that only lycopene synthesis step that is comprised from 3 genes have problem, while other 5 reaction steps than lycopene synthesis well work in *E. coli*. In the presentation, possible reason for defect of lycopene synthesis will be discussed.

Mammalian Genetic Circuit Construction Utilizing the Q System

Shimpi, A.,
Deans, T., University of Utah

It is highly desirable to have additional transcriptional regulatory tools that can be applied independently of existing systems in mammalian cells. The Q system is a binary expression system that was originally discovered in *Drosophila.* The regulatory parts have been moved upstream of the fluorescent molecule, GFP, and has been demonstrated to work in *Drosophila*, *C. elegans*, and zebrafish. Here the regulatory parts are introduced as a new component for future genetic circuits in mammalian cells. The regulatory component of the Q system contains two genes of interest QF and QS along with the necessary binding sites (QUAS). QF acts as a transactivator for genes downstream of its binding site, while QS inhibits the binding of QF and prevents transcription. The predictable binary manner of the system makes the system a prime candidate for inclusion into the synthetic biology repertoire. To demonstrate the Q system's candidacy it was introduced to mammalian cells to determine functionality. Plasmids containing QF and binding sites upstream of a reporter gene were found to function as predicted. However, the repression predicted for QS was absent and thus deemed unfit for inclusion in genetic circuits. Additionally, QF and QUAS were coupled with existing components to predictably control gene expression through genetic circuits such as Boolean logic gates and an oscillatory system. The Q system demonstrates its candidacy for inclusion into the synthetic biology repertoire to advance the construction of genetic circuits in a controlled and predictable manner.

Large-Scale Determination of Causative Mutations for Enhanced Chemical Tolerance in Evolved Strains of Escherichia coli

Lennen, R. M., University of Wisconsin-Madison
Herrgård, M. J., Technical University of Denmark
Nielsen, A. T.,
Feist, A. M., University of California, San Diego
Sommer, M.,
Mohamed, E. T. T.,

Overcoming product toxicity remains one of the largest barriers toward achieving economically viable cell factory processes. Even relatively non-toxic products impose significant stresses at the economically-relevant titers ultimately required for commodity chemical production, which are often in excess of 100 g/L. We have utilized a robotic platform to rapidly evolve multiple parallel populations of E. coli K-12 MG1655 for enhanced growth in the presence of toxic concentrations of 11 chemicals representing diverse functional classes that have significant interest as polymer precursors, biofuels, and bulk chemical intermediates. These include dicarboxylic acids, diamines, monocarboxylic acids, diols, and aromatic acids. Whole genome resequencing of over 200 strains isolated from nearly 90 independently evolved populations has enabled the determination of the genetic basis of tolerance to each chemical. The mutations in evolved strains were biased toward loss-of-function mutations, which were reconstructed in combinations and shown to be major drivers of tolerance in many cases. Additional combinations of probable gain-of-function mutations were reconstructed using oligonucleotide-mediated recombineering and a conjugation-based genome shuffling approach. Based on successful reconstructions of tolerance phenotypes, novel mechanisms of tolerance have been inferred, and numerous new gene targets conferring increased tolerance have been discovered. Cross-compound screening of all evolved isolates has also revealed specific sets of mutations that confer tolerance toward multiple classes of compounds. Ongoing work includes continuing to explore the mechanistic basis for tolerance and testing tolerant mutants for endogenous production capability.

Expanding Synthetic Biology's Toolkit By in Vitro Evolution: Characterization of a Fast and Uniquely Unnatural Hammerhead Ribozyme

Koussa, J., New York University Abu Dhabi
Jijakli, K., New York University Abu Dhabi
Kimizialtin, S., New York University Abu Dhabi
Szostak, J., Howard Hughes Medical Institute and Massachusetts General Hospital
Salehi-Ashtiani, K., New York University Abu Dhabi

Ribozymes, and more specifically hammerhead ribozymes, have been successfully used in synthetic biology approaches to artificially control gene expression via *trans-* or *cis-*acting mechanisms. A typical Hammerhead ribozyme consists of a highly conserved catalytic core of around 15 nucleotides flanked by three based-paired stems and two loops. This structural arrangement includes specific non- Watson-Crick interactions that prime the ribozyme for activity. Furthermore, kissing-loop interaction of loop II with loop I, which often occurs in naturally occurring ribozymes, mediates enhanced catalytic activity in natural instances of these ribozymes. In this study, we describe characterization of an atypical hammerhead ribozyme obtained through *in vitro* evolution that features an unusually short stem-loop II structure and a high catalytic activity. The minimal stem-loop II structure and the retaining of full functionality of this structure motivated us to characterize this molecule further. Extensive mutational analyses and biochemical assays evaluating the effects of point mutations on catalysis were done for this purpose. Our analyses showed sensitivity of catalytic rates to specific sequences of stem-loop II and stem I structure. We carried out *de novo in silico* simulations to obtain plausible 3D structural models that can explain our observed biochemical experiments, providing a deeper insight on the structural arrangement as well as a plausible three-dimensional folding pattern that would account for the high catalytic activity, or lack of, in certain mutants. Molecular dynamics simulations to further refine a plausible structure for this ribozyme are ongoing. This study has further been complemented by multiple genome-wide scans for naturally occurring RNA molecules sharing this specific RNA motif. These scans have yielded no significant hits in 108 eukaryotic and prokaryotic genomes searched so far, suggesting a lack of natural occurrences of this ribozyme. This apparent under-representation in extant genomes, suggest that this specific hammerhead ribozyme might have been under negative selective pressures throughout evolution. These characteristics underline a positive potential in using this specific ribozyme motif in synthetic biology approaches. The ease of activating/deactivating this ribozyme through its peripheral structures sets forth a solid, promising ground, to generation of an effective tool for efficient modulation of gene expression.

Genome Engineering of Genomically Recoded Escherichia coli Enhances Site-Specific Incorporation of Non-Standard Amino Acids in Cell-Free Protein Synthesis

Martin, R. W., Northwestern University
Hong, S. H., Northwestern University
Kwon, Y. C., Northwestern University
Des Soye, B. J., Northwestern University
Jewett, M. C., Northwestern University

Site-specific incorporation of nonstandard amino acids (NSAAs) into proteins and biopolymers by amber suppression makes possible new chemical properties, new structures, and new functions. However, competition between release-factor 1 (RF1) and orthogonal tRNAs have limited the technology. Here, we describe the development of a high yielding cell-free protein synthesis (CFPS) platform from crude extracts of genomically recoded Escherichia coli lacking release factor 1. Because this recoded strain has not been previously optimized for CFPS, we exploited multiplex automated genome engineering (MAGE) to design and construct synthetic genomes that, upon cell lysis, lead to improved extract performance. To stabilize template DNA, messenger RNA, amino acids, and protein products in CFPS, we targeted the deletion of 15 potential negative effectors. More than 50 strains were generated and tested in CFPS, allowing us to catalogue the systems impact by making numerous gene deletions both individually and in combinations. The protein synthesis activities of our most productive cell extracts were more than four-fold greater as compared to the extract from the parent strain, achieving more than 1.6 mg/ml of superfolder green fluorescent protein (sfGFP). We also show high efficiencies for producing modified sfGFP containing *p*-acetylphenylalanine at single and multiple positions. Our work has implications for using whole genome editing for CFPS strain development, expanding the chemistry of biological systems, and cell-free synthetic biology.

How to Best Control Gene Expression in Cell Populations in Real-Time

Fiore, G., Telethon Institute of Genetics and Medicine
Perrino, G., University of Naples Federico II
di Bernardo, M., University of Naples Federico II
di Bernardo, D., Telethon Institute of Genetics and Medicine

How to best Control Gene Expression in Cell Populations

in real-Ââ? time.

Gianfranco Fiore*, Giansimone Perrino*, Mario di Bernardo, Diego di Bernardo

Automatic control is an engineering approach to regulate the behaviour of a system in an automated and precise manner. Controllers are used in everyday appliances from thermostats to microwaves, as well as, in complex and large systems such as airplanes and power-plants. The design principle is based on negative feedback, where the quantity to be measured (e.g. the temperature) is measured and compared to the desired value to yield a â??control errorâ? . The controller changes some physical process in the system (e.g. switches the boiler on or off) to reduce the control error to zero. Thanks to advances in molecular biology and biotechnology, control engineering methods have been recently proposed to regulate gene expression in living cells using a variety on computational and experimental approaches. Due to its very recent application to biological systems, it is not yet clear what control strategy works best. Here we provide a computational and experimental comparison of the two main control strategies proposed in literature: proportional-integral (PI) control and model predictive control (MPC). We then propose and test an innovative control strategy, Zero Average Dynamics (ZAD) control, which has been extensively used in electrical power converters but never in biological systems. We used a microfluidics-based control strategy we recently presented [Menolascina F. et al, PLOS Comp. Biolo. 2014; Fiore G. et al, Chaos, 2013; Menolascina F. et al, Automatica, 2011] to control in real-time the expression of a fluorescent reporter protein from the inducible *GAL1* promoter in yeast *Saccharomyces cerevisiae*. In this set-up the computer can monitor cell fluorescence via a time-lapse fluorescent microscope and either feed cells with galactose, thus inducing expression from the *GAL1* promoter, or glucose to repress gene expression. The control tasks we investigated are: (i) â??set-pointâ? control in which the protein has to be maintained at 50% of its maximum value in galactose, for 1500 min and (ii) â??tracking-controlâ? in which the protein has to maintain a value of 75% for 500 min, then 50% for another 500 min and 25% for the remaining 500 min. Experiments confirm the numerical simulation, proving that MPC and ZAD strategies can achieve successfully the regulation of gene expression in living cells for both set-point and tracking control, whereas the PI strategy has a worse performance, at least for the â??trackingâ? control task. The ZAD strategy however is much less computationally expensive than the MPC and does not require a detailed mathematical model of the system to be controlled. In conclusion our work shows that it is now possible to achieve precise regulation of gene expression in eukaryotic cells thus making possible to investigate cellular behaviour in completely new ways.

Introducing Synbis – the Synthetic Biology Information System

Bultelle, M. A., Imperial College
Sainz de Murieta, I., Imperial College
Kitney, R. I., Imperial College

1. Introduction

An increasing concern of Synthetic biology is the rational design of pathways. The goal is to build a new pathway or modify an existing one, so that it is endowed with a given dynamic and may operate under a given set of constraints. Typically, the rational design cycle involves an in-silico design and modelling phase on series of CAD software (at part level and/or circuit level [1,2]), followed by construction of the circuit and a testing phase. Depending on the results, the process may be repeated several times until the goal is met.
Such iterative approach has proven its worth and become mainstream in several fields of engineering. In the case of synthetic biology, rational design requires a large catalogue of well characterised chassis, plasmids and bioparts (fundamental parts like promoters or RBS but also common devices such as logic gates, oscillators, pulse generators â?¦).
Furthermore, such a catalogue must not only be available online, but it must also be presented in a user- friendly manner and support widely-used data standards to integrate with existing and future CAD tools [3]. Although several well-known repositories exist (the iGEM parts registry [4], JBEI [5] or the virtual parts repository [6] to name a few), we are not aware of any existing repository that meets all those requirements.
Acknowledging this crucial need, the Centre for Synthetic Biology and Innovation at Imperial College (CSynBI) has launched a multi-faceted effort to support the automated characterisation of biological parts â?? focusing in particular on:
1. The support for existing data standards and development of a complementary, robust data standard for the acquisition of experimental data.
2. The construction of a common IT-spine to track and store all data as they are processed and curated.
3. A robust dissemination strategy, enabling public access to high quality biopart information.
In this paper we present the work conducted on points 2 and 3, and introduce SynBIS (the Synthetic Biology Information System).

2. Biopart Characterisation and SynBIS

SynBIS is the IT spine that has been designed at CSynBI to enable biopart- characterisation efforts on a scale that is difficult-to-impossible for human experimentalists to achieve. It was first developed with

constitutive promoters, and now supports the characterisation of other fundamental bioparts such inducible promoters or RBS.

Practically, SynBIS deals with characterisation in term of pipeline â?? characterisation is considered *â??a sequence of reliable, validated process aimed at generating high quality experimental data, extracting from them their most important features, and disseminating the resultsâ??*. Human intervention is kept to a minimum (curation mainly) and automation is used when possible (especially during data acquisition and data analysis). Currently, CSynBI pipelines are made of three steps.
1. **Data acquisition -** Acquisition has been automated as per [7] â?? using a set of validated protocols , whose purpose it is to improve reproducibility and increase the throughput. Plate reader and flow cytometry data are typically acquired.
2. **Data Processing -** Experimental data are processed to estimate the relationship between the input and output of the part. Although the information depends on the type of part, processing is modularised by using libraries of models.
3. **Dissemination -** To make it more usable, all the information on the biopart is compiled into a datasheet, which is then uploaded to dedicated website (with it is own APIs).
Practically, **SynBIS** logs three types of information as all datasets involved (controls included) proceed down the pipeline:

1. Custodial information

a. **Who ?** Staff / Algorithm Version in Charge b. **When ?** Time stamps

c. **Where ?** Institutions Involved

2. **Analysis Information**

a. **Input /Output** of each analysis step

b. Corresponding Metadata

3. Human Curation Information

a. Curator Decision b. Curator reason(s)

SynBIS data-model has been organised around the end-product of the pipeline: the data-sheet of the biopart/device. Datasheets are mainstays of engineering and not new to Synthetic Biology (the datasheet for F2620 [8] is a classic example) and there have been proposals to establish a minimal template [9]. Our work builds upon these earlier works by systematically describing bioparts with a black box approach including:

ï?· Identity (name, sequence, URI)
ï?· Input and Output (including crosstalk)
ï?· Qualitative relations between outputs and inputs
ï?· Quantitative relations between outputs and inputs (raw data and models)

SynBIS places a big emphasis on the experimental context (including the chassis, medium, plasmid, reporter, assay protocol and experiments settings). Since the influence of the context on parts behaviour is poorly understood, SynBIS will host several datasheets (one per context) for the same part.
SynBISâ?? template has been designed so **datasheets are human as well as machine readable.** The human readable part of the datasheet displays a small subset of all the data collected and generated as part of the characterisation process. The machine readable part makes all these data available to a user.

3. Data Dissemination

SynBIS currentlys host characterisation information on constitutive and inducible promoters:

ï?· 20 promoters from the Anderson Collection
ï?· 2 variants of promoters from the Anderson Collection
ï?· 6 wild-type E-coli promoters
ï?· A handful of inducible promoters (e.g XylF/XylR system)

More promoters are being constructed and characterised. The SynBIS team plans to have several dozens more promoters available by June 2015.
Al the data hosted on SynBIs are available for download and use under CC-BY Creative Common license and can be accessed in two different manners.
A website has been trialled with selected partners since June 2014 (http://synbis.bg.ic.ac.uk) and is now public access. It can be searched by name, sequence and according to the characteristic function of the biopart (typical requests when it comes to circuit design).
SynBIS also implements an API to enable programmatic access from external applications. This way the power users can not only access individual datasheets (and related data) but also run bulk queries. The API comprises two types of RESTful web services:

ï?· Inbound: input of new curated datasheet information into SynBIS
ï?· Outbound: retrieval of datasheet information.
o XML interface: provides the complete description of a datasheet formatted following the SynBIS database structure.
o SBOL interface: provides the basic information which can be encoded with SBOL-Core (the current stable release).

Whichever the method of access, all the information acquired and generated as part of the

characterisation of the biopart is available. SynBIs only uses free and open interchange data standards that are common or we believe will become common in the synthetic biology community.These include human readable formats such as SBOL and XML (serialisation of our datasheets), and SBML for computer models of biological processes. SynBIS also uses a novel (binary) data standard called DICOM-SB that has been developed at Imperial College, London for the purpose of storing raw experimental data in an efficient manner [10].
To integrate all these standards, SynBIS uses a useful property of SBOL version 2.0 [11]. Specifications of SBOL 2.0 (still in development) allow for the listing of the URIs of external resources in the annotation field of bioparts, thus offering an elegant method to link â??the essential information for synthetic DNA sequencesâ?? (SBOL) ,raw experimental data (DICOM- SB), processed data (SBML) and the datasheet (XML).

[1] MacDonald, J. T., Barnes, C., Kitney, R. I., Freemont, P. S., & Stan, G. B. V. (2011). Computational design approaches and tools for synthetic biology.*Integrative Biology*, *3*(2), 97-108. [2] Arpino, J. A., Hancock, E. J., Anderson, J., Barahona, M., Stan, G. B. V., Papachristodoulou, A.,

& Polizzi, K. (2013). Tuning the dials of synthetic biology. *Microbiology*, *159*(Pt 7), 1236-1253.

[3] Kitney, R., & Freemont, P. (2012). Synthetic biologyâ??the state of play. *FEBS letters*, *586*(15),

2029-2036.

[4] http://parts.igem.org/Main_Page

[5] https://public-registry.jbei.org/

[6] http://sbol.ncl.ac.uk:8081/

[7] Hirst CD, Ainsworth C, Kelwick RJR, Freemont PS, Kitney RI, Baldwin GS. An automated platform for the characterisation of regulatory biological parts in synthetic biology. In Submission. 2014.

[8] Canton, B., Labno, A., & Endy, D. (2008). Refinement and standardization of synthetic biological parts and devices. *Nature biotechnology*, *26*(7), 787-793.

[9] Arkin, A. (2008). Setting the standard in synthetic biology. *Nature biotechnology*, *26*(7), 771-773 [10] A DICOM EXTENSION SUPPORTING THE DATA ACQUISITION PROCESS IN SYNTHETIC BIOLOGY

[11] Roehner, N., Oberortner, E., Pocock, M., Beal, J., Clancy, K., Madsen,

C., ... & Myers, C. J.

(2014). Proposed data model for the next version of the Synthetic Biology Open Language. *ACS*

synthetic biology.

Heterologous Expression and Characterization of Hybrid Two-Component Systems from Gut Bacteria

Daeffler, K.,

Two component systems (TCSs) are the primary signal transduction modality by which bacteria sense and respond to the environment. Classical TCSs are comprised of an inner membrane bound sensor histidine kinase (SK) and a cytoplasmic transcription factor known as a response regulator (RR). In the presence of an extracellular input signal, the SK phosphorylates the RR, which then binds to a promoter, activating transcription. Recently, a new TCS subfamily called the hybrid TCSs (HTCSs) has been identified wherein the SK and RR are fused into a single membrane-bound protein. While the functional implications of this architecture are poorly understood, HTCSs are found extensively throughout the genus *Bacteroides*, which comprise approximately 50% of the human gut microbiome. Additionally, HTCSs have been shown to sense diverse host- and diet-derived carbohydrates, which are linked to host physiology and disease. Hindering their further characterization and use for synthetic biology, no full-length HTCS has been expressed or characterized outside the native organism. Here, we demonstrate that a HTCS can be expressed and function in *E. coli*, though with considerable toxicity and a weaker transcriptional response than is observed in the native organism. Through systematic expression of truncation variants of several human gut-derived HTCSs, we have identified an N-terminal predicted signal peptide targeting the Sec translocon as the major source of toxicity. Complete removal of this domain results in a dramatic reduction in toxicity of the overexpressed protein but is accompanied by a loss of ligand sensitivity in the fructose sensor BT1754. We are currently working to optimize protein expression and enhance promoter output through protein and promoter engineering as well as identify the ligands sensed by HTCSs without known inputs. This work has implications in understanding the dynamics of the human gut microbiome, understanding the functional implications of the HTCS architecture, and engineered novel diagnostics and therapeutic agents for the human gut.

Implementation of Synthetic Metabolic Pathways for the Production of 2,4-Dihydroxybutyric Acid

Walther, T.,
Auriol, C., Toulouse White Biotechnology Center
Irague, R., Toulouse White Biotechnology Center
Topham, C., Toulouse White Biotechnology Center
Andre, I., Toulouse White Biotechnology Center
Remaud-Simeon, M.,
François, J. M., Toulouse White Biotechnology

Implementation of Synthetic Metabolic Pathways for the Production of 2,4-Dihydroxybutyric Acid

Thomas Walther1-4, ClÃ©ment Auriol1-4, Romain Irague1-4, Christopher M. Topham1-4, ClÃ©mentine Dressaire1-4, Audrey Baylac1-4, Florence Calvayrac1-4, HÃ©lÃ¨ne Serrano-Bataille1-4, Julie Fredonnet1-4, HÃ©lÃ¨ne Cordier1-4, Isabelle AndrÃ©1-4, Marc Maestracci5, Robert Huet5, Magali Remaud-Simeon1-4, Jean Marie FranÃ§ois1-4

1UniversitÃ© de Toulouse; INSA, UPS, INP; LISBP, 135 Avenue de Rangueil, 31077 Toulouse, France ;

2INRA, UMR792 IngÃ©nierie des SystÃ¨mes Biologiques et des ProcÃ©dÃ©s, Toulouse, France ; 3CNRS, UMR5504, Toulouse, France ; 4TWB, 3 rue des Satellites, Canal Biotech Building 2, 31400 Toulouse, France; 5Adisseo SA, Antony Parc II, 10 place General de Gaulle, 92160 Antony, France.

2,4-Dihydroxybutyric acid (DHB) is a molecule with considerable potential as a versatile chemical synthon. Notably, it may serve as a precursor for chemical synthesis of the methionine analogue 2-hydroxy-4-(methylthio)butyrate, thus, targeting a considerable market in animal nutrition. However, petrochemical synthesis of DHB is not economically viable, and no natural metabolic pathways exist for the biochemical production of DHB. We have therefore conceived three synthetic metabolic pathways for the synthesis of DHB starting from the natural metabolites malate or homoserine.

(I)*

ATP ADP

malate

CoASH, ATP ADP, Pi

(IV)

glyoxylate

Acetyl CoA

(V)
malyl-Pi
malyl-CoA

NAD(P)H NAD(P)+, Pi

(II)*
malate

(VI)*

NAD(P)H NAD(P)+, CoASH

(VII)* (VIII)*

semialdehyde

NAD(P)H
2-oxogluta glutamate rate
NAD(P)H NAD(P)+

(III)*

NAD(P)+

homoserine
→2-oxo-4- →hydroxybutyrate

2,4-dihydroxybutyrate
(DHB)

Synthetic metabolic pathways for the biosynthesis of 2,4-dihydroxybutyric acid. (*) Previously unreported enzymatic activities. (I) malate kinase, (II) malate semialdehyde dehydrogenase, (III) malate semialdehyde reductase, (IV) malyl-CoA synthetase, (V) malyl-CoA-lyase, (VI) malyl-CoA reductase, (VII) homoserine transaminase, (VIII) 2-oxo-4-hydroxybutyrate reductase.

The two pathways departing from malate proceed via activation of malate by phosphorylation or acylation to yield malylphosphate or malyl-CoA, respectively. These compounds are then subject to two successive rounds of reduction to yield DHB via the common intermediate malate semialdehyde.
Synthesis of DHB from homoserine requires deamination of homoserine to yield 2-oxo-4- hydroxybutyrate followed by the reduction of the latter to obtain DHB.
The pathways employ several previously unreported enzymatic activities. Those activities were obtained by rational design based on structural and mechanistic knowledge, and/or by screening of candidate enzymes acting on sterically similar cognate substrates.
Each of the individual pathways has its particular thermodynamic and physiological requirements which were met by rational strain design. The pathways were expressed in genetically optimised Escherichia coli strains, and production of DHB from glucose was demonstrated.

Precision Expression Control in Bacteroides Thetaiotaomicron, a Human Commensal Bacterium

Mimee, M., Massachusetts Institute of Technology
Tucker, A. C., Massachusetts Institute of Technology
Voigt, C. A., University of California, San Francisco
Lu, T. K., Massachusetts Institute of Technology

The human microbiota plays an integral role in diverse human physiological processes including metabolism, development and immunity. Manipulation of the native microflora using genetically engineered microbes is a promising avenue for cellular therapies and diagnostics. However, dominant members of the microbiota are difficult to robustly and reliably engineer. Here, we describe tools for engineering *Bacteroides thetaiotaomicron,* a relatively abundant species of bacteria that resides in the human gastrointestinal tract. We identify a library of constitutive promoters and ribosome binding sites that span a 10,000-fold range in expression. This library is used to construct a recombinase-based memory gate that stores permanent memory on the *B. thetaiotomicron* chromosome. Additionally, we describe a series of fully orthogonal, sugar-inducible promoters able to conditionally elicit up to a 100-fold difference in gene expression. Finally, we demonstrate programmable knockdown of gene expression using CRISPRi targeted to both synthetic constructs and endogenous genes implicated in antimicrobial peptide resistance and carbohydrate metabolism. Together, these biological parts serve as a fundamental resource for the study and development of engineered commensal organisms that can positively impact human health.

Engineering Bacteriophages for Specific Bacterial Targeting

Citorik, R. J., Massachusetts Institute of Technology
Mimee, M., Massachusetts Institute of Technology
Lu, T. K., Massachusetts Institute of Technology

Engineering Bacteriophages for Specific Bacterial Targeting

Robert J. Citorik, Mark Mimee, Timothy K. Lu

Interest in bacteriophages has seen a rebirth in recent years as the result of increasing antibiotic resistance in bacterial infections, the recognition of their important role in the human microbiome, and the development of new tools and approaches for engineering designer phages. In the past year, we demonstrated the creation of sequence-Ââ? specific antimicrobials by using phages as delivery vehicles for RNA-Ââ? guided nucleases. Ongoing work has focused on expanding the application of this technology to target new bacteria. As we gain experience in optimizing the system for different targets, we aim to create a platform that can be translated into a programmable system for specific removal of bacteria to alleviate infection and dysbiosis in microbial communities.

Logic Gates for Engineering Programmable Interactions in Microbial Consortia

Beliaev, A.,
McClure, R., Pacific Northwest National Laboratory
Bernstein, H. C., Pacific Northwest National Laboratory

The current state of synthetic biology relies heavily upon cultivation of single species of model microorganisms in tightly controlled laboratory environments. Such, often highly engineered, domesticated strains suffer from poor stability and resilience in natural settings due to fluctuating environmental conditions and competition from communities of indigenous microbes. To address these challenges, we have developed a generalizable framework for design and engineering of autonomous synthetic microbial consortia capable of operating safely within unpredictable and dynamic natural environments. At the center of the proposed platform lies photoautotroph-heterotroph interactive partnership. Such associations, which naturally assemble to take advantage of services carried out by individual members, are ubiquitous in nature and mediate key ecological processes such as energy capture, carbon fixation, and nutrient cycling. Genetically tractable photoautotrophic organisms, such as cyanobacteria, can sustain and drive an engineered consortium through production of O_2, organic C, and, in some cases, fixed inorganic or organic N. While metabolite exchange provides ways for functional compartmentalization, a tight control and a high level of communication and interaction between members of the consortium are required. As demonstrated in single-species models, computational logic gates provide a controllable and modular approach to programming desired cellular behavior. Applying logic gates so that output responses are sustained rather than transient to communities of bacteria, as opposed to single cells, will allow for more complex interactions and offers a novel way to precisely engineer synthetic multi-species systems. This presentation discusses opportunities for implementing logic-based controls into metabolically interdependent phototrophic consortia for improved communication, functional stability, and controllable output.

Engineering of a High Lipid Producing Yarrowia Lipolytica Strain

Kamineni, A., Novogy Inc

Microbial lipid production has received interest both for the synthesis of specialty lipids and for the production of fuels and chemicals from various feedstocks. Microbial lipids are produced by many oleaginous organisms including the well characterized yeast *Yarrowia lipolytica*. Lipid yield in oleaginous organisms can be increased through the up-regulation and/or down-regulation or deletion of genes involved in the lipid pathway. In this study, we describe a strain-engineering strategy applying combinatorial screening of overexpression and deletion genetic targets that led to the construction of a high lipid-producing yeast biocatalyst. The resulting strain NS432 that combines three genetic modifications selected for their effect on lipid production reached 76% fatty acid content and 0.21 g lipid per g glucose yield in batch glucose fermentation. In fed-batch glucose fermentation NS432 produced 92 g/L lipid at a productivity of 0.68 g.L-1.hr^{-1}, which is 3.7 and 3.2-fold higher in titer and productivity, respectively, over previously reported *Y. lipolytica* engineered biocatalysts.

De Novo Designed Toehold Repressors
Carlson, P. D., Cornell University
Lucks, J. B., Cornell University

Title: *De novo* designed Toehold Repressors

Paul Carlson, Julius Lucks

Abstract: RNA regulators of gene expression are becoming increasingly important components of the synthetic biology toolbox. RNA is a versatile and designable molecule, and is one of the most powerful substrates for engineering gene expression at our disposal. Moreover, the ability to propagate information directly via RNA-RNA interactions creates the basis for the bottom-up design of genetic networks to implement logic or carry out complex patterns of gene expression. An important class of such regulators is RNA-mediated repressors, which repress downstream gene expression in response to a *trans*-acting antisense RNA. Rational mutagenesis and sequence mining have been successful in creating libraries of orthogonal RNA repressors that function on the transcriptional and translational levels. However, the construction of large networks will require additional regulators with enhanced dynamic range and reduced crosstalk. Inspired by recent work demonstrating the successful forward design of RNA-mediated translational activators called toehold switches, we sought to use a similar approach to create *de novo* designed RNA translational repressors. Using the Nucleic Acid Package (NUPACK), we successfully inverted the structural logic of toehold activators to create toehold repressors capable of attenuating gene expression up to 98% (50-fold repression). In this design, the structure of the toehold repressor frees the ribosome binding site (RBS) and start codon, allowing translation to proceed in a default ON state. Binding of a designed *trans*-acting antisense RNA leads to structural rearrangement and formation of a hairpin, which occludes the RBS and start codon, preventing translation in the OFF state. This approach shows promise for creating a large library of independent regulators with large dynamic range.

*** Formatting Issue - Best Available Paper/Graphic ***

Engineering Transcriptional Regulator Effector Specificity Using Computational Design and in Vitro Rapid Prototyping: Developing a Vanillin Sensor

de los Santos, E. L. C., California Institute of Technology
Meyerowitz, J. T., California Institute of Technology
Mayo, S. L., California Institute of Technology
Murray, R. M., California Institute of Technology

California Institute of Technology
1200 E. California Blvd
Pasadena, CA 91125
February 18, 2015
To the Editors of ACS Synthetic Biology and Organizers of the SEED2015 Conference:
On behalf of my fellow co-authors, I would like to submit for your consideration a manuscript entitled â??Engineering Transcriptional Regulator E?ector Speci-
ficity using Computational Design and In Vitro Rapid Prototyping: Developing a Vanillin Sensorâ?   for a presentation at the SEED conference, as well as publi- cation in ACS Synthetic Biology. This paper is a continutation of the research presented at last yearâ??s SEED conference under the title, â??Engineering Tran-
scriptional Regulator E?ector Specificity Through Rational Design and Rapid Prototypingâ?   . The work describes the engineering of qacR, a transcriptional regulator to create functional variants responsive to a new e?ector. A copy of
this manuscript and the supplemental information has been uploaded to the bioRxiv preprint server.
Results covered include:

â?¢ Computational protein design to generate a library of qacR mutants that respond to vanillin, a growth-inhibiting small molecule;

â?¢ Rapid screening using an in vitro transcription-translation (TX-TL) sys- tem;

â?¢ Further in vitro charactetrization of hits from the TX-TL screen

â?¢ In vivo characterization of qacR mutants

The further *in vitro* characterization, and *in vivo* characterization are new re- sults from further work on the system in the last year. We believe that this work fits well within the scope of the SEED2015 conference and the readers of ACS Synthetic Biology and would be of interest to those in attendance.
Thank you for your consideration.
Emmanuel Lorenzo de los Santos, for the authors

Engineering Transcriptional Regulator Effector

*** Formatting Issue - Best Available Paper/Graphic ***

Specificity using Computational Design and *In Vitro*

Rapid Prototyping: Developing a Vanillin Sensor

Emmanuel L.C. de los Santos,? Joseph T. Meyerowitz, Stephen L. Mayo, and

Richard M. Murray

Division of Biology and Biological Engineering, California Institute of Technology, Pasadena, CA, USA

E-mail: emzodls@caltech.edu
KEYWORDS: synthetic biology, cell-free systems, *in vitro* biological circuit prototyping, computational protein design, protein engineering

---

?To whom correspondence should be addressed

1

Abstract

The pursuit of circuits and metabolic pathways of increasing complexity and ro- bustness in synthetic biology will require engineering new regulatory tools. Feedback control based on relevant molecules, including toxic intermediates and environmental signals, would enable genetic circuits to react appropriately to changing conditions. In this work, variants of qacR, a tetR family repressor, were generated by computational protein design and screened in a cell-free transcription-translation (TX-TL) system for responsiveness to a new targeted e?ector. The modified repressors target vanillin, a growth-inhibiting small molecule found in lignocellulosic hydrolysates and other indus- trial processes. Promising candidates from the *in vitro* screen were further characterized *in vitro* and *in vivo* in a gene circuit. The screen yielded two qacR mutants that re- spond to vanillin both *in vitro* and *in vivo*. We believe this process, a combination of the generation of variants coupled with *in vitro* screening, can serve as a framework for designing new sensors for other target compounds.

Introduction

The utility of a synthetic genetic circuit for real world applications is dependent on the ability to e?ectively trigger the circuit. While we can control the expression of target genes with

*** **Formatting Issue - Best Available Paper/Graphic** ***

transcriptional regulators, triggers for these transcriptional regulators are limited to a small number of molecules and other inputs (e.g. light) (*1*). As a consequence, most synthetic circuits right now are limited to proof-of-principle demonstrations without being extendable to real world applications. In order to design synthetic control circuits for real world applications such as metabolic engineering or biofuel production, signals from these pathways, such as the level of a toxic metabolic intermediate, need to be transmitted into existing transcriptional control machinery. This work develops a framework to use a combi- nation of sequence generation by computational protein design and rapid prototyping using a cell-free transcription-translation (TX-TL) system to switch e?ector specificity of existing
2

Is there a sensor?

Computationally aided selection of

No sequences for

screen (35,577,057,600 sequences)

Rapid *in vitro* screening of mutants (27 mutants)

Yes

Optimize to desired specifications

No

Is there a hit?

Yes

in vivo testing and *in vitro* verification of hits

characterization

(2 mutants)

(2 mutants)

*** Formatting Issue - Best Available Paper/Graphic ***

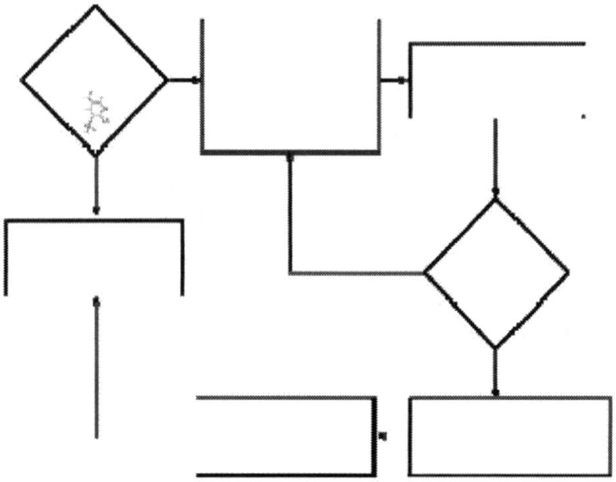

Figure 1: **Workflow for generating novel sensors.** The *in vitro* TX-TL platform allows for the rapid screening of sequences selected with the help of computational protein design. Hits from the *in vitro* screen are then verified by further *in vitro* testing. *In vivo* testing and characterization can then be performed to see if they meet the desired specifications. Further refinement of the hits through directed evolution or further computational design can be performed until specifications necessary are achieved. Numbers in parenthesis are the number of sequences considered by the computational algorithm, or the number of mutants assayed at the specified step for vanillin.

transcriptional regulators to respond to targeted small molecules of interest (Figure 1).

The tetR family is a large family of transcriptional regulators found in bacteria. They are named after the tetR repressor, which controls the expression of tetA, an e?ux pump for tetracycline (*2*). They contain two domains, a helical-bundle ligand-binding domain and a helix-turn-helix DNA-binding domain. In the absence of their inducing molecule, tetR repressors bind to DNA, preventing the transcription of downstream genes. Inducer binding to the ligand-binding domain causes a conformational change in the DNA binding domain that causes dissociation from the DNA, allowing transcription of downstream genes. The tetR transcriptional regulation machinery has been used in the design of synthetic circuits, including the repressilator (*3*) and the toggle switch (*4*).

QacR is a tetR-family repressor found in *S. aureus* that controls the transcription of qacA, an e?ux pump that confers resistance to a large number of quaternary anionic com-

3

pounds. The protein has been studied because it is induced by a broad range of structurally dissimilar compounds (*5*). Structural examination of qacR in complex with di?erent small molecules has shown that qacR has two di?erent binding regions inside a large binding pocket. While qacR has multiple binding modes for various inducers, in all cases for which there are structures, binding of the inducer causes a tyrosine explusion which moves one of the helices and alters the conformation of the DNA binding domain, rendering qacR unable to bind DNA (*6* â??*8*). Crystal structures of inducer-bound forms of qacR and the qacR-DNA complex coupled with a definitive structural mechanism for qacR induction make it the ideal starting point for computational design of new transcriptional regulators. In this work, we describe our e?orts to apply our framework to engineer qacR to sense vanillin, a phenolic growth inhibitor that is a byproduct of lignin degradation performed during the processing of biomass into intermediate feedstock in biofuel production (*9*).

*** Formatting Issue - Best Available Paper/Graphic ***

Results and discussion

Computationally Aided Selection of QacR Mutant Sequences

We created a computational model of vanillin to place into a crystal structure of QacR (PDB ID: 1JTO). A computational protein design algorithm was used to find potential vanillin binding sites close to the location of the tyrosine expulsion in the binding pocket of qacR (Figure 2A-B) while being in the proximity of amino acid positions that allowed for favorable pi-stacking and hydrogen bonding interactions. We used targeted ligand placement (*11*) to find potential binding positions for vanillin by defining an idealized binding site for the molecule. The algorithm yielded four potential binding positions for vanillin (Figure 2C). Computational protein sequence design was then used to select amino acid residues at the positions around the potential vanillin binding sites. In order to minimize the possibility of steric clashes in the protein, calculations that considered both the DNA-bound state and the ligand-bound state using a multi-state design algorithm were also performed (*14*). Finally,

4

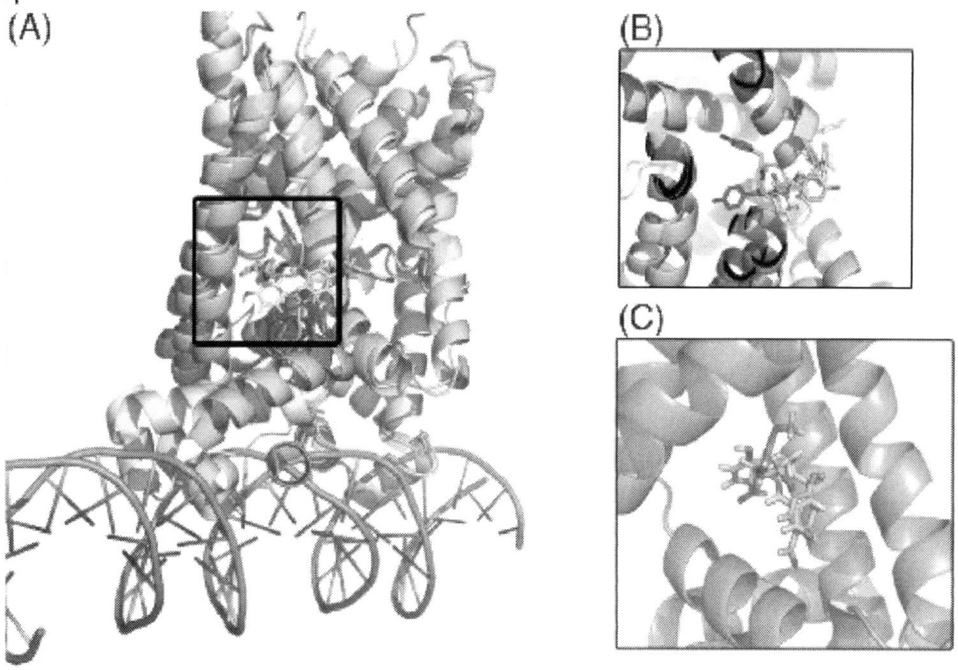

Figure 2: **Computationally Aided Selection of qacR mutants. (A)** Overlay of PDB structures of the non- ligand bound (cyan, PDB ID: 1JTO) and ligand bound (yellow, PDB ID: 3BQZ) conformations of qacR. A conformational shift in the binding pocket occurs upon entry of the small molecule causing the protein to dissociate from DNA. **(B)** A closer look at the binding pocket of qacR, the binding of the ligand causes the displacement of three tyrosine residues shown. **(C)** Computational model for potential vanillin binding sites.
Vanillin is shown as a di?erent color in each of the four sites. A protein design algorithm was asked to suggest mutations for amino acids close to the potential binding sites to support the placement of vanilin in these sites.

we also ran calculations that included an energy bias to favor the wild-type residue. The lowest energy sequences from these four calculations (single-state biased, single-state non- biased, multi-state biased, and multi-state non-biased) were analyzed, and used as a guide to compile a set of ten mutants for *in vitro* testing.

In Vitro Screening of Generated Sequences

We first decided to validate function of the wild-type protein. This was done by placing green fluorescent protein (GFP) downstream of the qacA promoter sequence (PQacA). While we observed a hundred-fold decrease in fluorescence in cells containing plasmids encoding the wild-type qacR gene in addition to PQacA -GFP, addition of berberine, a native qacR in-
5
ducer, yielded no observable di?erence in fluorescence (Figure S1). We hypothesized that the inducer was not getting into the cells due to the di?erences in cell wall between gram-positive and gram-negative bacteria. Because of this, we decided to use an *in vitro* transcription-translation (TX-TL) system to test the mutants (*15*). We first tested the wild type protein in our TX-TL system. We observed an increase in GFP fluorescence as we increased the con- centration of plasmid encoding pQacA-GFP from 2 nM to 8 nM (Figure 3A). The addition of plasmid encoding the qacR repressor to the system resulted in a decrease in fluorescence. However, we detected high autofluorescence from berberine. We switched to dequalinium, a previously characterized non-fluorescent qacR inducer (*6*). The addition of dequalinium resulted in an increase in fluorescence in the TX-TL system up to 85% of the fluorescence of the positive control, in which there was no DNA encoding the qacR repressor (Figure 3B). These results demonstrated a functional wild-type qacR repressor in TX-TL. After validating the function of wild-type protein in TX-TL, we used the system to look at the functionality of the qacR mutants. None of the initial ten mutants showed any repression of GFP fluorescence. We analyzed the ligand bound and DNA-bound computational models of one of the qacR mutants that contained only three amino acid substitutions from a qacR mutant that was previously shown to be functional by Peters *et al.* (*8*). The computational model showed the potential for some mutations to cause steric clashes in the DNA bound state. We created a second library reverting either the 50th and 54th positions (A50F/W54L) or the 119th position (Y119L) back to their wild-type identity (Table S1).

In order to determine if any of the mutants of our library warranted further character- ization, we performed a rapid screen of 17 qacR mutants in TX-TL (Figure S2). Plasmids containing DNA that encoded each of the qacR variants or the wild-type qacR sequence were placed into a TX-TL reaction containing either water, dequalinium or vanillin. QacR activity was monitored by a plasmid encoding GFP downstream of PQacA. Two of the mu-
tants, qacR2 and qacR5, displayed an increase in fluorescence in the presence of vanillin and
6
dequalinium over water (Figure 4). We focused on these two mutants for further *in vitro* and *in vivo* characterization.

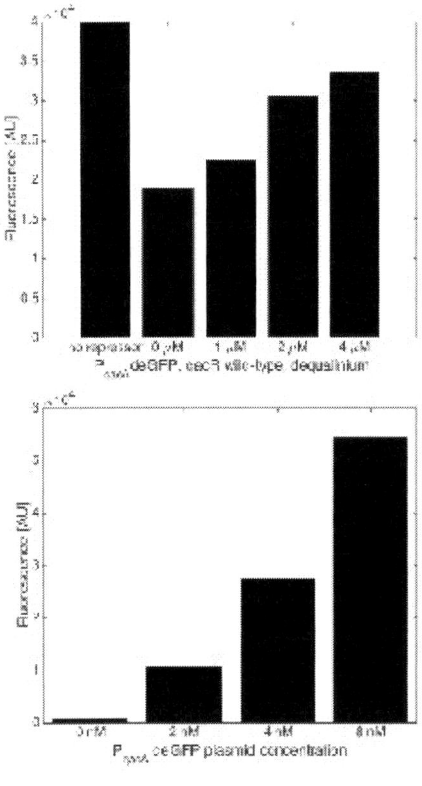

(a) (b)

Figure 3: **Validation of TX-TL Screening.** (A) GFP signal after three hours of a TX-TL reaction. Plasmid encoding GFP downstream of the native qac promoter was added to the TX-TL platform. Higher concentrations of plasmid yielded more GFP signal. (B) Response of wild-type qacR to dequalinium. DNA encoding GFP and wild-type qacR was added to the TX-TL system. Increasing fluorescent signal is observed with increasing concentrations of dequalinium. The highest fluorescent signal is observed when there is no repressor in the system, demonstrating the ability of TX-TL to test for qacR repression and de-repression.

Further *in vitro* testing of qacR2 and qacR5

In order to verify the response of qacR2, and qacR5 to vanillin, we performed more extensive TX-TL tests on the mutants. TX-TL reactions were set up with a constant amount of reporter (PQacA-deGFP) plasmid and either no repressor (water), or plasmids encoding wild- type qacR, qacR2, or qacR5. Reactions were incubated for 75 minutes at 29 C to produce the repressor protein. This initial reaction was then added to solution containing dequilinium, vanillin, or water. We monitored the rate of GFP production between the first and third hours of the reaction, where the rate of protein production appeared linear.
Figure 5 shows the ratio of GFP fluorescence between the case where there is no repressor, and each of the repressors tested with the di?erent inducers. The wild-type qacR is able to inhibit the production of GFP to around 15% of its maximum value. The mutants are less e?cient at repressing the production of GFP. Three times and four times more repressor
7

***** Formatting Issue - Best Available Paper/Graphic *****

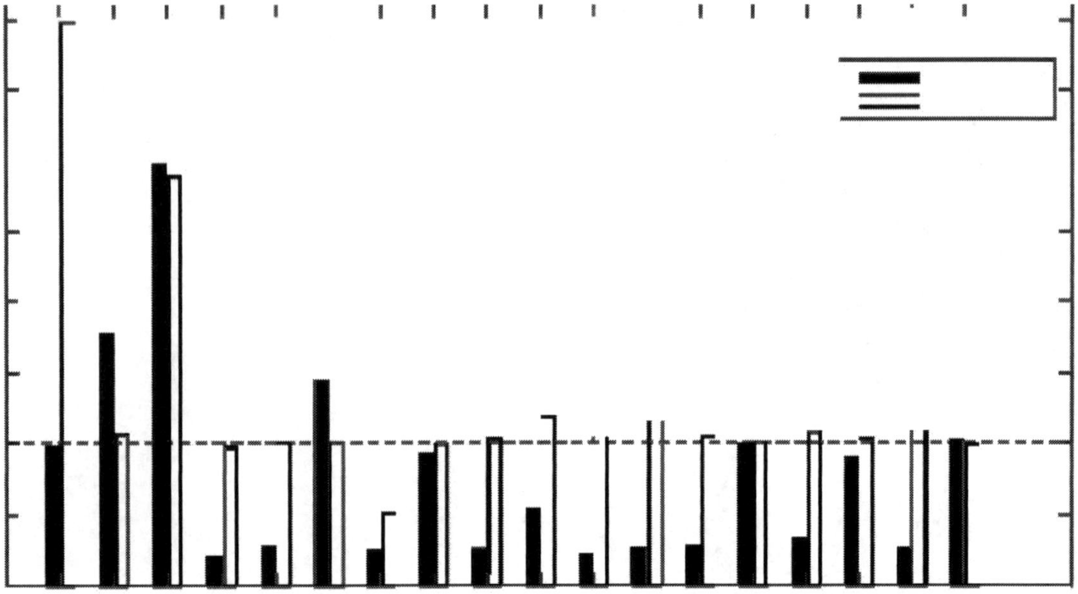

4

3.5

vanillin dequalinium

3

2.5

2

1.5

1

0.5

0 wt 1 2 3 4 5 6 7 8 9 10 11 12 13 14 15 16 17

qacR variants

Figure 4: ***In Vitro* TX-TL screen of qacR mutants found potential candidates for further testing.** Fold change in maximum fluorescence between water and inducer for qacR mutants. Seventeen qacR mutants were screened using TX-TL. Plasmids containing DNA encoding each of the qacR variants were placed into the system along with water, dequalinium (native qacR inducer) and vanillin. To monitor qacR response, a plasmid encoding GFP downstream of the native qacA promoter was also added to the system. qacR2 and qacR5 were seclected for further characterization. qacR1 was not selected due to low absolute signal (Figure S2)

DNA was added to the reactions of qacR2 and qacR5 respectively. In spite of the additional DNA, we do not observe the same level of repression that we see with the wild-type protein. Wild-type qacR is well induced by the native inducer, we observed full derepression at the dequalinium concentration used. Induction of qacR2 and qacR5 with dequalinium is also observed, although to a lesser degree than the wild-type protein. Resource limitations due to the increase in repressor DNA added may be contributing to the lower induction level observed. QacR2 and qacR5 display a response to vanillin at the concentration we tested, while no response to vanillin was detected for the wild-type protein. The mutations introduced to the protein decrease the ability of the mutants to repress DNA. This could be due to protein instability, or due to a weaker protein-DNA interaction. However, these mutations also increase the sensitivity of the mutants to vanillin, allowing their response to

8

be detectable in our *in vitro* platform.

We assumed that the maximum amount of GFP fluorescence that can be achieved for a specific inducer condition was when there is no repressor present. This takes into account potential inhibition of the TX-TL reaction by the inducer. The factors that can a?ect the ability of the particular repressor to reach this the no repressor case are resource limitations due to additional load from the production of the repressor DNA, and response of the repres- sor to the inducer in the reaction. We expect that resource limitations would have a negative e?ect on the ability of the repressor to reach the maximum fluorescence level. Conversely, response to repressor should have a positive e?ect in reaching the maximum fluorescence level.

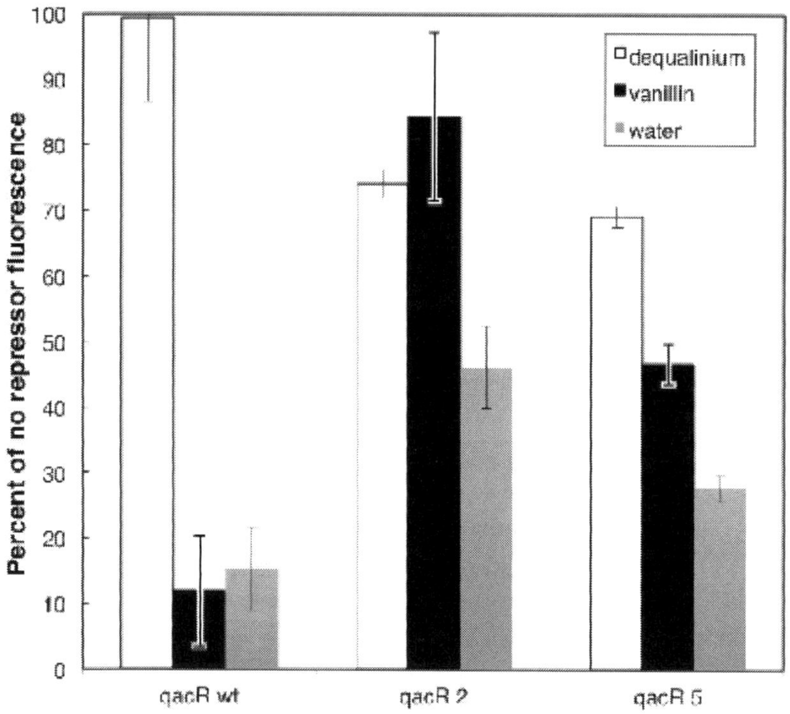

Figure 5: ***In vitro* testing of qacR2 and qacR5.** Ratio of the rate of GFP production between TX-TL reactions with and without repressor DNA. 10 $\hat{A}\mu$M of dequalinium and 5 mM of vanillin was used to induce the production of GFP for each of the qacR variants tested.

9

In Vivo Testing of qacR2 and qacR5

In order to further characterize the qacR mutants, and to see if we could detect vanillin in a more complex system, we decided to test the *in vivo* response of the qacR variants to vanillin. Plasmids containing genes that encode the wild-type qacR sequence, qacR2 or qacR5 downstream of Ptet and GFP downstream of PQacA were cloned into DH5?Z1 cells (Figure 6). For each of the qacR variants, we compared di?erences in fluorescence signal across increasing vanillin concentrations. We tested di?erent repressor concentrations by varying the amount of anhydrous tetracycline (aTc) in the system. Similar to the *in vitro* experiments, and in order to get an idea for the maximum fluorescence the system could achieve, we grew cells that only contained GFP downstream of PQacA without any repressor. Cells that were grown in higher aTc concentrations had a lower measured optical density (OD), indicating a slower growth rate. We hypothesize that this is due to the toxicity of the qacR repressor to the *E. coli* strain. Since qacR is not a native protein, it is possible that qacR is binding to locations in the *E. coli* genome. Interestingly, the di?erences in optical density measurements become less pronounced with increasing vanillin concentration, suggesting that vanillin may provide a mitigating e?ect to this toxicity. In order to account for di?erences in OD, fluorescence measurements were normalized to OD.

The lowest OD measurements were observed for cells encoding the wild-type qacR at 12 ng/mL aTc where very little growth was observed for cells expressing the wild-type protein. At this aTc concentration, all of the cells expressing repressor exhibited lower optical densities when compared to cells that were only expressing fluorescent protein. The di?erences in optical density are less pronounced at lower aTc concentrations. When no aTc is present in the system, cells at the higher vanillin concentrations had lower ODs. At higher aTc concentrations, cells at higher vanillin concentrations had higher ODs. This implies that both the vanillin concentration and the expression of the repressor have an e?ect on cellular growth. The optical densities for the cells at di?erent aTc and vanillin concentrations are shown in Tables S2-S5.

10

Figure 7A shows the e?ect of increasing the aTc concentration on the fluorescence of cells in the absence of vanillin. Similar to the *in vitro* tests, fluorescence was normalized to the no repressor case. Increasing the aTc concentration decreased the fluorescence of cells in the absence of vanillin, confirming that the qacR mutants are able to repress the expression of GFP at higher protein concentrations.

The response of wild-type qacR, qacR2, and qacR5 to increasing vanillin concentrations is shown in figure 7B. The response curves for each protein are plotted for the minimum aTc concentraion such that maximum GFP repression is observed. This corresponds to aTc concentrations of 4, 8, and 12 ng/mL for wild-type qacR, qacR2, and qacR5 respectively. This is consistent with the *in vitro* data that more qacR2 and qacR5 DNA was required to repress the expression of GFP. Similar to the *in vitro* tests, we expect the ability of the cell to reach the maximum fluorescence level to be dependent on its response to inducer, and toxicity from vanillin and qacR. Indeed, cells expressing the qacR mutants exhibited an increase in fluorescence with increasing vanillin levels demonstrating that they are capable of sensing vanillin. The mutants exhibit a marked increase in sensitivity to vanillin. QacR2 displays a response that goes from approximately 20% of the fluorescence of the cells not expressing any repressor to matching the fluorescence of the non-repressed cells at 1 mM vanillin. QacR5 saturates at around 40% of the fluorescence of the non-repressed cells. This correlates with the *in vitro* data that show qacR2 achieving close to the non-repressed fluorescence, with qacR5 less sensitive to vanillin (Figure 5).

*** Formatting Issue - Best Available Paper/Graphic ***

Framework Enables Engineering of Sensors through Rational Reduc- tion of Design Space

The framework developedâ?? a combination of sequence generation using computationally-aided design, preliminary screening with TX-TL, and *in vitro* and *in vivo* validationâ??can be used for other small molecule targets potentially facilitating the design of more actuators in synthetic circuits. While it is possible that the computational model of vanillin binding was 11

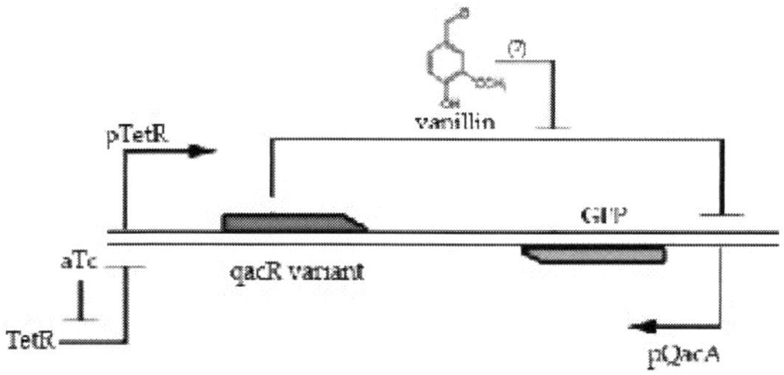

Figure 6: Circuit layout for *in vivo* tests. Genes encoding GFP under the control of the native qac promoter, and our QacR designs under the control of a tet-inducible promoter were

placed in a single plasmid and transformed into DH5?Z1 cells. qacR levels were controlled using aTc for varying vanillin concentrations. Candidate designs that are responsive to vanillin should show an increase in fluorescence with increasing vanillin concentrations

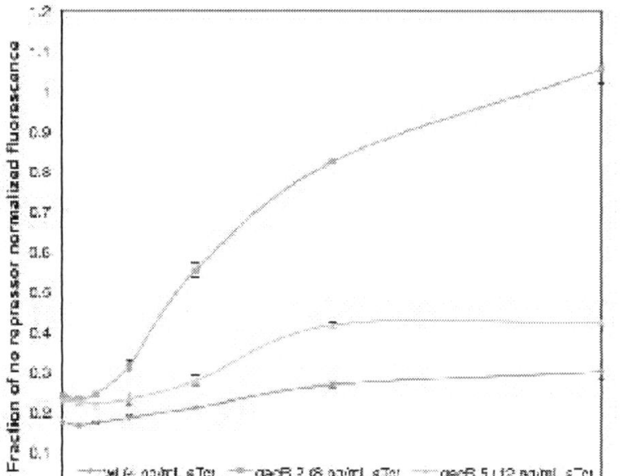

*** Formatting Issue - Best Available Paper/Graphic ***

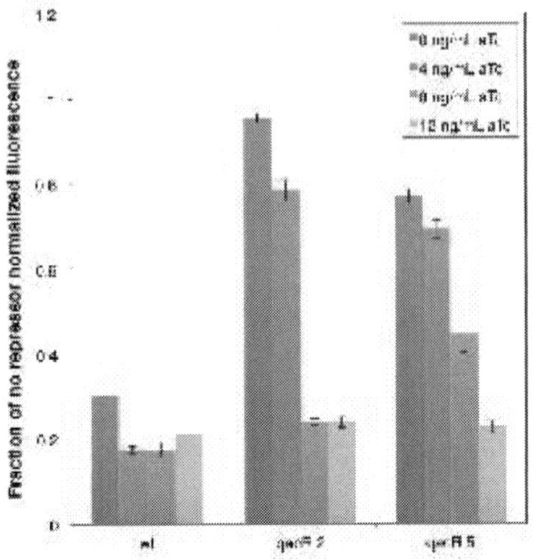

(a) (b)

Figure 7: ***In vivo* response of qacR to vanillin.** Cells expressing GFP without any repressor were used as a control to normalize for di?erences in fluorescence due to aTc and vanillin levels. (a) All of the proteins are able to repress the expression of GFP. The wild-type protein is able to inhibit the expression of GFP at lower aTc concentrations, while higher aTc concentrations are necessary for the mutants to achieve a similar level of repression. (b) QacR mutants respond to vanillin in a concentration dependent manner.

inaccurate, the computational design provided value in drastically reducing the number of sequences to test into a figure that was experimentally tractable. Without the computational design to reduce the size of the design space, we would not have had e?ective starting points to attempt the engineering of a vanillin sensor.

12

The use of the *in vitro* cell-free system in a preliminary screen provides many advantages. It allows the screening of more mutants in a shorter amount of time. The simpler system also reduces the number of variables to consider. Complicating factors such as cell membrane permeability and cell growth do not need to be considered during this part of the screen. Repressors whose native inducers cannot enter the target organism can be used as starting points with the cell-free system. Finally, we can use this framework to target molecules that are known to be toxic to cells and measure engineering results in a cell-free context.

As a result of this process, we now have functional vanillin sensors that can be used in a feedback circuit that dynamically responds to vanillin. QacR2 and qacR5 can be used as a starting point for a synthetic circuit that responds to vanillin concentrations. While we only tested the protein in *E. coli* recent work has developed a process that facilitates the transfer of prokaryotic transcription factors into eukaryotic cells, increasing the flexibility of the molecules for use in metabolic engineering (*16*). By linking a vanillin sensor to the expression of a gene that can mitigate the toxic e?ect of vanillin, such as an e?ux pump or an enzyme which converts vanillin to a less toxic molecule, we can design a dynamic feedback circuit and potentially improve metabolic yield. It remains to be seen whether the sensors developed will have the required dynamic range or sensitivity for a functional feedback circuit; however, if a better sensor is needed these proteins can be used as a starting point for directed evolution in order to obtain a sensor with the desired properties.

*** Formatting Issue - Best Available Paper/Graphic ***

Materials and Experimental Methods

Computationally Aided Selection of Mutant Sequences

An *in silico* model of vanillin was constructed using the SchrÃ¶dinger software suite. Partial charges for vanillin were computed using Optimization in Jaguar version 7.6 (*10*) using HF/6-311G** as the basis set. Vanillin rotamers were chosen by looking at the ideal angles for the carbon hybrid orbitals. A model of an idealized vanillin binding pocket was designed

13

by looking at the protein data bank for proteins that bound small molecules similar to vanillin. Models of vanillin in the qacR binding pocket were generated using the Phoenix Match algorithm (*11*).

Monte Carlo with simulated annealing (*12*) and FASTER (*13*) were used to sample conformational space. A backbone independent conformer library with a 1.0 Ã? resolution was used for the designed residues (*11*). Designed residues were chosen by compiling a list of amino acid residues within 15 Ã? of vanillin. Potential residues for each site were selected by chemical intuition and context within the structure. Computational models of qacR with vanillin present were scored using the PHOENIX forcefield with the inclusion of an additional geometry bias term that favored pi-stacking and hydrogen bonding interactions (*11*).

Cell Free *in vitro* Transcription-Translation System and Reactions

The transcription-translation reaction consists of crude cytoplasmic extract from BL21 Rosetta 2 *E. coli* (*15*). Preliminary tests were done with plasmids and inducers at the specified concentrations. For the initial screen, the qacR mutants were downstream of a T7 promoter. TX-TL reactions were run with 2 nM of the plasmid encoding the qacR variant, 0.1 nM plasmid encoding T7 RNA polymerase, and 8 nM plasmid encoding PQacA-deGFP. Vanillin was added at a concentration of 2.5 mM and dequalinium was added at 10 ÂμM.

For the *in vitro* tests to further characterize the hits. Plasmids encoding qacR2 or qacR5 downstream of a tet-responsive promoter were used along with a plasmid encoding deGFP downstream of a qac-responsive promoter. Plasmids were prepared using the Macherey-Nagel NucleoBond Xtra Midi/Maxi Kit. Plasmid DNA was eluted in water and concentrated by vacufuge to the desired concentration. TX-TL reactions were set up as follows: 5 ÂμL of bu?er, 2.5 ÂμL of cell extract and 1.5 ÂμL repressor DNA at a specific concentration was mixed and incubated at 29 C for 75 minutes to facilitate the production of repressor DNA. This mix was then added to a mixture of 1 ÂμL deGFP plasmid and 1 ÂμL of an inducer stock. Measurements were made in a Biotek plate reader at 3 minute intervals using excita-

14

tion/emission wavelengths set at 485/525 nm. Stock repressor plasmid concentrations were 0.24 ÂμM , 0.73 ÂμM , and 0.97 ÂμM for qacR wild-type, qacR2, and qacR5, respectively. The deGFP plasmid concentration was approximately 0.40 ÂμM. Inducer concentrations were 5 mM for vanillin, and 10 ÂμM for dequalinium.

Experimental conditions were done in triplicate and the error bars are the error propa- gated from the standard deviation of the means.

*** Formatting Issue - Best Available Paper/Graphic ***

Cell Strain and Media

The circuit was implemented in the *E.coli* cell strain DH5?Z1, a variant of DH5? which contains a chromosomal integration of the Z1 cassette (*17*). All cell culture was done in optically clear M9ca minimal media (Teknova M8010).

Genes and Plasmids

DNA encoding the qacR genes was constructed using overlap extension PCR. Plasmids used contained chloramphenicol resistance with a p15a origin of replication.

In vivo experiments

Cells were grown in at least two consecutive overnight cultures in M9ca minimal media. On the day of the experiment, overnight cultures were diluted 1:100 and grown for 5 hours to ensure that the cells were in log phase. Cells were then diluted 1:100 into fresh media at the specified experimental condition. Cells were grown in these conditions at 37 C for
12 15 hours in Axygen 96 well plates while shaking at 1100 rpm. Endpoint fluorescence was measured by transferring the cells to clear bottomed 96-well microplates (PerkinElmer, ViewPlate, 6005182) . GFP was read at 488/525 with gain 100.
Analysis of the data was done by taking fluorescence readings for each independent well. Experimental conditions of the qacR proteins were done in triplicate and repeats were
15
averaged. Error bars shown are the error propagated originating from the standard deviation of the mean.

Acknowledgement

The authors thank Jongmin Kim and Jackson Cahn for reading the manuscript. This research was conducted with supprot from the Institute for Collaborative Biotechnologies through grand W911NF-09-0001 from the U.S. Army Research O?ce. Additional support was granted in part by the Benjamin M. Rosen Bioengineering Center, the Gordon and Betty Moore Foundation through Grant GBMF2809 to the Caltech Programmable Molecu- lar Technology Initiative, and DARPA through the Living Foundries Program.

References

[1] Purnick, P. E. M., and Weiss, R. (2009) The second wave of synthetic biology: from modules to systems. *Nature reviews. Molecular cell biology 10*, 410â??422.
[2] Ramos, J. L., Martinez-Bueno, M., Molina-Henares, A. J., TerÃ¡n, W., Watanabe, K., Zhang, X. D., Gallegos, M. T., Brennan, R., and Tobes, R. (2005) The TetR family of transcriptional repressors. *Microbiology and Molecular Biology Reviews 69*, 326â??+.
[3] Elowitz, M. B., and Leibler, S. (2000) A synthetic oscillatory network of transcriptional regulators. *Nature 403*, 335â??338.
[4] Gardner, T. S., Cantor, C. R., and Collins, J. J. (2000) Construction of a genetic toggle switch

in Escherichia coli. *Nature 403*, 339â??342.
[5] Grkovic, S., Hardie, K. M., Brown, M. H., and Skurray, R. A. (2003) Interactions of the QacR multidrug-binding protein with structurally diverse ligands: implications for the evolution of the binding pocket. *Biochemistry 42*, 15226â??15236.
16
[6] Schumacher, M. A., Miller, M. C., Grkovic, S., Brown, M. H., Skurray, R. A., and Bren- nan, R. G. (2001) Structural mechanisms of QacR induction and multidrug recognition. *Science 294*, 2158â??2163.
[7] Schumacher, M. A., and Brennan, R. G. (2003) Deciphering the molecular basis of mul- tidrug recognition: Crystal structures of the Staphylococcus aureus multidrug binding transcription regulator QacR. *Research in Microbiology 154*, 69â??77.
[8] Peters, K. M., Brooks, B. E., Schumacher, M. A., Skurray, R. A., Brennan, R. G., and Brown, M. H. (2011) A single acidic residue can guide binding site selection but does not govern QacR cationic-drug a?nity. *PLoS ONE 6*, e15974.
[9] Klinke, H. B., Thomsen, ABâ?? and Ahring, B. K. (2004) Inhibition of ethanol-producing yeast and bacteria by degradation products produced during pre-treatment of biomass. *Applied Microbiology and Biotechnology 66*, 10â??26.
[10] Bochevarov, A. D., Harder, E., Hughes, T. F., Greenwood, J. R., Braden, D. A., Philipp, D. M., Rinaldo, D., Halls, M. D., Zhang, J., and Friesner, R. A. (2013) Jaguar: A high-performance quantum chemistry software program with strengths in life and materials sciences. *International Journal of Quantum Chemistry 113*, 2110â??2142.
[11] Lassila, J. K., Privett, H. K., Allen, B. D., and Mayo, S. L. (2006) Combinatorial methods for small-molecule placement in computational enzyme design. *Proceedings of the National Academy of Sciences of the United States of America 103*, 16710â??16715.
[12] Kuhlman, B., Dantas, G., Ireton, G. C., Varani, G., Stoddard, B. L., and Baker, D. (2003) Design of a novel globular protein fold with atomic-level accuracy. *Science 302*, 1364â??1368.
[13] Allen, B. D., and Mayo, S. L. (2006) Dramatic performance enhancements for the FASTER optimization algorithm. *Journal of Computational Chemistry 27*, 1071â??1075.
17
[14] Allen, B. D., and Mayo, S. L. (2010) An e?cient algorithm for multistate protein design based on FASTER. *Journal of Computational Chemistry 31*, 904â??916.
[15] Sun, Z. Z., Hayes, C. A., Shin, J., Caschera, F., Murray, R. M., and Noireaux, V. (2013) Protocols for implementing an Escherichia coli based TX-TL cell-free expression system for synthetic biology. *Journal of visualized experiments : JoVE* e50762.
[16] Stanton, B. C., Siciliano, V., Ghodasara, A., Wroblewska, L., Clancy, K., Trefzer, A. C., Chesnut, J. D., Weiss, R., and Voigt, C. A. (2014) Systematic Transfer of Prokaryotic Sensors and Circuits to Mammalian Cells. *ACS synthetic biology*
[17] Lutz, R., and Bujard, H. (1997) Independent and tight regulation of transcriptional units in Escherichia coli via the LacR/O, the TetR/O and AraC/I1-I2 regulatory elements. *Nucleic Acids Research 25*, 1203â??1210.
18

Supporting information for:

*** Formatting Issue - Best Available Paper/Graphic ***

Engineering Transcriptional Regulator Effector Specificity using Computational Design and *In Vitro* Rapid Prototyping: Developing a Vanillin Sensor

Emmanuel L.C. de los Santos,? Joseph T. Meyerowitz, Stephen L. Mayo, and

Richard M. Murray

Division of Biology and Biological Engineering, California Institute of Technology, Pasadena, CA, USA

E-mail: emzodls@caltech.edu
KEYWORDS: synthetic biology, cell-free systems, *in vitro* biological circuit prototyping, computational protein design, protein engineering

---

?To whom correspondence should be addressed

S1

Table S1: QacR Mutants Tested

Protein	Amino Acid Position																		
	50	54	57	58	61	86	89	90	93	96	99	102	116	119	120	126	154	157	161
qacR-wt	F	L	E	E	W	S	T	E	Y	Q	I	F	M	L	E	A	N	N	T
qacR1	F	L	Q	L	Y	S	T	Q	Y	M	Q	S	Q	Y	Q	A	M	Q	M
qacR2	F	L	Q	L	Y	S	T	Q	Y	Q	Q	F	Q	Y	Q	A	M	L	M
qacR3	F	L	Q	L	Y	S	T	Q	Y	Q	I	S	Q	Y	Q	A	M	L	M
qacR4	F	L	Q	L	Y	S	T	Q	Y	M	Q	Q	M	Y	Q	A	M	Q	M
qacR5	F	L	E	E	Y	S	T	Q	Y	M	Q	S	Q	Y	Q	A	N	N	T
qacR6	F	L	E	E	Y	S	T	Q	Y	Q	Q	F	Q	Y	Q	A	N	N	T
qacR7	F	L	E	E	Y	S	T	Q	Y	Q	I	S	Q	Y	Q	A	N	N	T
qacR8	F	L	Q	L	W	S	T	Q	Y	M	Q	S	Q	Y	Q	A	M	Q	M
qacR9	A	W	Q	L	Y	S	T	Q	Y	M	Q	S	Q	L	Q	A	M	Q	M
qacR10	A	W	Q	L	Y	S	T	Q	Y	Q	Q	F	Q	L	Q	A	M	L	M
qacR11	A	W	Q	L	Y	S	T	Q	Y	Q	I	S	Q	L	Q	A	M	L	M
qacR12	A	W	Q	L	Y	S	T	Q	Y	M	Q	Q	M	L	Q	A	M	Q	M
qacR13	A	W	E	E	Y	S	T	Q	Y	M	Q	S	Q	L	Q	A	N	N	T
qacR14	A	W	E	E	Y	S	T	Q	Y	Q	Q	F	Q	L	Q	A	N	N	T
qacR15	A	W	E	E	Y	S	T	Q	Y	Q	I	S	Q	L	Q	A	N	N	T

***** Formatting Issue - Best Available Paper/Graphic *****

qacR15	A	W	E	E	Y	S	T	Q	Y	Q	I	S	Q	L	Q	A	N	N	T
qacR16	A	W	Q	L	W	S	T	Q	Y	M	Q	S	Q	L	Q	A	M	Q	M
qacR17	A	W	E	E	W	S	T	Q	Y	Q	I	F	M	L	Q	A	N	N	T

S2

Table S2: OD600 of Cells at 4 ng/mL aTc

vanillin (mM)	GFP			wild-type			qacR2-1			qacR5-1		
1000	0.456	0.44	0.42	0.75	0.72	0.71	0.8	0.82	0.84	0.6	0.57	0.54
500	0.62	0.63	0.63	0.78	0.77	0.78	0.83	0.83	0.82	0.72	0.71	0.72
250	0.718	0.75	0.72	0.82	0.83	0.82	0.88	0.84	0.84	0.78	0.77	0.75
125	0.775	0.78	0.76	0.82	0.81	0.83	0.85	0.89	0.86	0.79	0.78	0.8
62.5	0.783	0.81	0.8	0.84	0.86	0.85	0.88	0.91	0.89	0.84	0.82	0.83
31.25	0.823	0.83	0.82	0.84	0.85	0.85	0.87	0.86	0.86	0.85	0.82	0.83
0	0.835	0.86	0.82	0.85	0.86	0.85	0.89	0.91	0.9	0.86	0.84	0.84
no aTc	0.866	0.85	0.85	0.88	0.9	0.88	0.92	0.92	0.91	0.9	0.83	0.85

vanillin (mM)	GFP			wild-type			qacR2-1			qacR5-1		
1000	0.52	0.49	0.48	0.66	0.66	0.68	0.78	0.79	0.78	0.6	0.6	0.59
500	0.63	0.64	0.63	0.73	0.73	0.75	0.83	0.84	0.83	0.71	0.7	0.72
250	0.78	0.74	0.74	0.75	0.76	0.79	0.89	0.85	0.85	0.76	0.77	0.77
125	0.76	0.8	0.78	0.74	0.77	0.75	0.86	0.89	0.88	0.8	0.79	0.79
62.5	0.84	0.8	0.81	0.75	0.77	0.77	0.87	0.88	0.89	0.81	0.82	0.85
31.25	0.79	0.82	0.81	0.75	0.74	0.77	0.89	0.9	0.89	0.81	0.8	0.79
0	0.84	0.86	0.85	0.78	0.8	0.77	0.93	0.91	0.88	0.85	0.84	0.85
no aTc	0.82	0.85	0.81	0.84	0.85	0.85	0.87	0.91	0.9	0.86	0.85	0.84

vanillin (mM)	GFP			wild-type			qacR2-1			qacR5-1		
1000	0.45	0.45	0.41	0.28	0.26	0.27	0.7	0.72	0.72	0.35	0.37	0.37
500	0.67	0.64	0.63	0.27	0.24	0.23	0.7	0.7	0.71	0.46	0.44	0.46
250	0.72	0.72	0.71	0.11	0.12	0.11	0.55	0.52	0.52	0.49	0.51	0.53
125	0.8	0.77	0.8	0.09	0.09	0.09	0.23	0.23	0.24	0.41	0.43	0.47
62.5	0.87	0.82	0.81	0.08	0.07	0.08	0.14	0.15	0.15	0.38	0.38	0.42
31.25	0.81	0.83	0.79	0.08	0.07	0.07	0.11	0.11	0.11	0.32	0.33	0.39
0	0.84	0.83	0.82	0.08	0.08	0.07	0.09	0.09	0.09	0.31	0.33	0.37
no aTc	0.8	0.79	0.8	0.84	0.86	0.83	0.89	0.88	0.88	0.83	0.8	0.82

vanillin (mM)	GFP			wild-type			qacR2-1			qacR5-1		
1000	0.48	0.45	0.49	0.05	0.06	0.05	0.27	0.26	0.26	0.15	0.15	0.14
500	0.69	0.67	0.67	0.06	0.07	0.05	0.23	0.23	0.24	0.17	0.17	0.19
250	0.76	0.76	0.75	0.04	0.05	0.05	0.08	0.08	0.08	0.11	0.12	0.12
125	0.82	0.79	0.79	0.04	0.05	0.05	0.05	0.05	0.05	0.07	0.08	0.08
62.5	0.81	0.82	0.8	0.04	0.04	0.04	0.04	0.05	0.05	0.06	0.06	0.06
31.25	0.82	0.84	0.82	0.04	0.04	0.04	0.05	0.04	0.05	0.06	0.06	0.06
0	0.84	0.87	0.84	0.04	0.04	0.04	0.05	0.04	0.05	0.06	0.05	0.05
no aTc	0.82	0.81	0.84	0.84	0.84	0.83	0.89	0.87	0.94	0.81	0.82	0.83

Table S3: OD600 of Cells at 4 ng/mL aTc Table S4: OD600 of Cells at 8 ng/mL aTc Table S5: OD600 of Cells at 12 ng/mL aTc

S3

*** Formatting Issue - Best Available Paper/Graphic ***

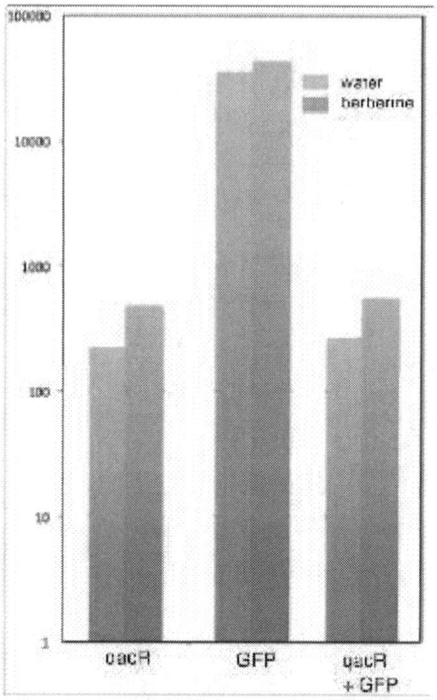

Figure S1: **Initial qacR induction test.** Fluorescence of cells encoding wild-type qacR was compared in the presence and absence of berberine, a native qacR inducer. While we observed repression upon the addition of qacR, we did not observe induction when berberine was added.

qacR mutants screen

4

x 10

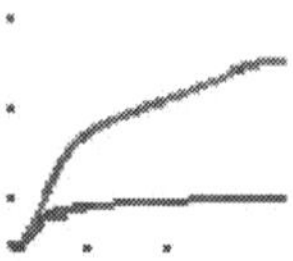

6

5

4

3

2

1

0

*** Formatting Issue - Best Available Paper/Graphic ***

4

x 10

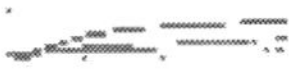

6

qacR wt

qacR 6

4

x 10

6

5

4

3

2

1

0

4

x 10

*** Formatting Issue - Best Available Paper/Graphic ***

6

qacR 1

qacR 7

4

x 10

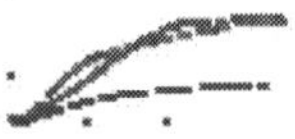

6

5

4

3

2

1

0

4

x 10

6

qacR 2

qacR 8

4

*** Formatting Issue - Best Available Paper/Graphic ***

x 10

.

.

.

6

5

4

3

2

1

0

4

x 10

.

.

.

6

qacR 3

qacR 9

4

x 10

.

.

.

6

***** Formatting Issue - Best Available Paper/Graphic *****

5

4

3

2

1

0

4

x 10

6

qacR 4

qacR 10

4

x 10

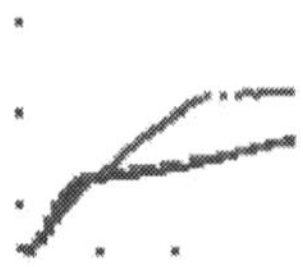

6

5

4

3

2

*** Formatting Issue - Best Available Paper/Graphic ***

1

0

4

x 10

.

.

.

6

qacR 5

qacR 11

5

4

3

2

1

4

x 10

.

.

.

6

5

4

3

*** Formatting Issue - Best Available Paper/Graphic ***

2

1

qacR 12

water vanillin dequalinium

5

4

3

2

1

4

x 10

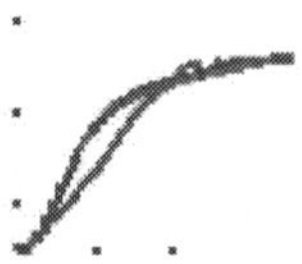

6

5

4

3

2

1

qacR 13

5

4

3

*** Formatting Issue - Best Available Paper/Graphic ***

2

1

4

x 10

.

.

.

6

5

4

3

2

1

qacR 14

5

4

3

2

1

4

x 10

.

.

.

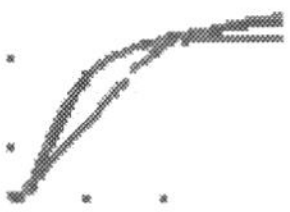

*** Formatting Issue - Best Available Paper/Graphic ***

6

5

4

3

2

1

qacR 15

5

4

3

2

1

4

x 10

.

.

.

6

5

4

3

2

1

qacR 16

*** Formatting Issue - Best Available Paper/Graphic ***

5

4

3

2

1

4

x 10

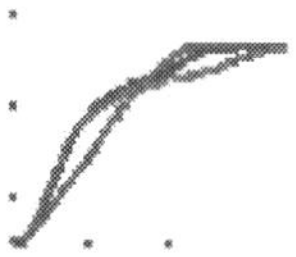

6

5

4

3

2

1

qacR 17

0 0 200 400 600

0 0 200 400 600

0 0 200 400 600

0 0 200 400 600

Time (min)

0 0 200 400 600

0 0 200 400 600

***** Formatting Issue - Best Available Paper/Graphic *****

Figure S2: **Timetrace of qacR mutant screen.** GFP fluorescence time traces for TX- TL reactions set up to screen for qacR mutants that were sensitive to vanillin. Reactions contained DNA encoding a qacR variant, T7 RNA polymerase, a fluorescent reporter and either water, dequilinium or vanillin.

S4

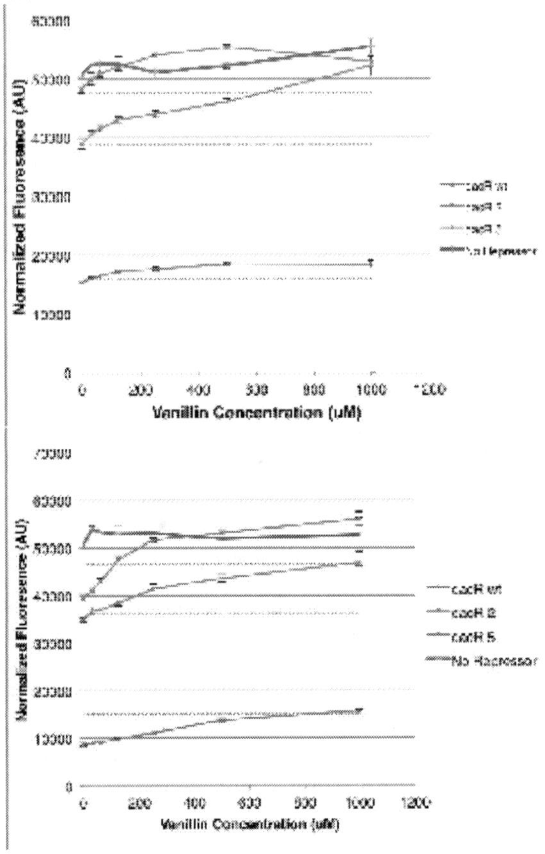

(a) 0 ng/ml aTc (b) 4 ng/ml aTc

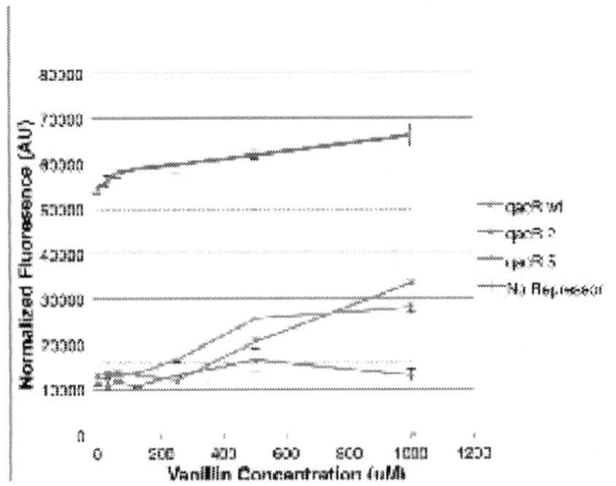

*** Formatting Issue - Best Available Paper/Graphic ***

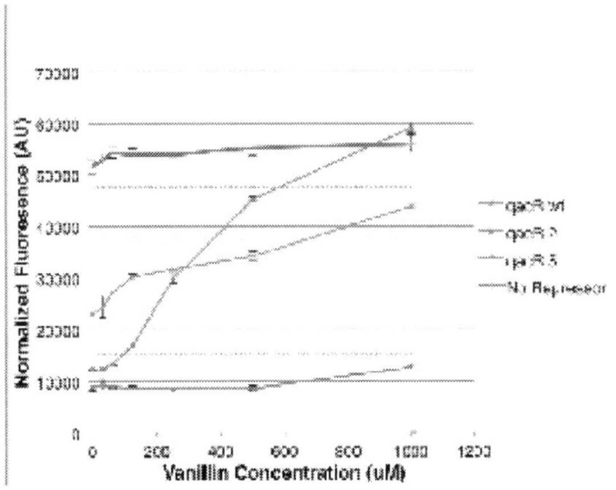

(c) 8 ng/ml aTc (d) 12 ng/ml aTc

Figure S3: **Vanillin dosage response curves.** GFP normalized to optical density for varying vanillin concentrations. The dotted lines represent the fluoresence level for cultures without aTc in order to observe the decrease in fluorescence due to background expression of the protein
S5

Epathoptimize: A Combinatorial Approach for Transcriptional Balancing of Metabolic Pathways

Jones, J. A., Rensselaer Polytechnic Institute
Vernacchio, V., Rensselaer Polytechnic Institute
Lachance, D., Rensselaer Polytechnic Institute
Lebovich, M., Rensselaer Polytechnic Institute
Fu, L., Rensselaer Polytechnic Institute
Shirke, A., Rensselaer Polytechnic Institute
Schultz, V., Rensselaer Polytechnic Institute
Cress, B., Rensselaer Polytechnic Institute
Linhardt, R. J., Rensselaer Polytechnic Institute
Koffas, M. A. G., Rensselaer Polytechnic Institute

Recent advancements in synthetic biology give scientists the ability to balance a variety of factors involved in gene regulation to improve product titers, yields, and productivities. Using a library of isopropyl β-D-1-thiogalactopyranoside (IPTG)-inducible mutant T7 promoters of varied strength, a combinatorial method was developed for transcriptional balancing of the violacein pathway. Violacein is a potent bioactive compound that has been shown to have various antitumor, antifungal, and anticancer properties. The violacein pathway is also an excellent model for exploratory metabolic engineering efforts into pathway regulation and control due to many colorful intermediate and side products allowing for easy analysis and strain comparison. Upon screening approximately 4% of the initial library, several high-titer mutants were discovered that resulted in up to a 63-fold improvement over the control strain. With further fermentation optimization, titers were improved to 1829±46 mg/L, a 2.6-fold improvement in titer and a 30-fold improvement in productivity over previous literature reports. Future applications of this technology towards improving the microbial production of biofuels, pharmaceuticals, and commodity chemicals could expedite the transition from the laboratory to the marketplace.

Complex Design, Simple Implementation: The Development of Red Light-Regulated Nuclear Protein Shuttling and Enzyme-Free DNA Assembly

Beyer, H. M., University of Freiburg
Juillot, S.,
Herbst, K., University of Freiburg
Gonschorek, P.,
Samodelov, S. L.,
Mueller, K.,
Roemer, W., University of Freiburg
Schaefer, E., University of Freiburg
Nagy, F.,
Straehle, U.,
Weber, W., University of Freiburg
Zurbriggen, M. D., University of Freiburg

Protein shuttling in and out of the nucleus represents a key step in controlling cell fate and function. Here we demonstrate a red light-inducible and far-red light-reversible synthetic system for controlling nuclear localization of proteins in mammalian cells and zebrafish. We synthetically reconstructed and validated the red light-induced Arabidopsis thaliana phytochrome B nuclear import mediated by phytochrome-interacting factor 3 in a non-plant environment. Based on this principle we induced nuclear import and activity of target proteins by the spatiotemporal projection of light patterns using a smartphone display. A synthetic transcription factor was translocated into the nucleus of mammalian cells and zebrafish to drive transgene expression. We describe the first plant phytochrome-based optogenetic in vivo application in vertebrates and expand the toolbox of available light-regulated molecular devices.

The development of such complex cellular networks requires suitable molecular cloning methods facilitating the implementation of the underlying design. Here, we describe AQUA (advanced quick assembly), a simple and versatile seamless assembly approach. It does not require any kits, enzymes or preparations of reagents and is the simplest assembly cloning protocol to date. We demonstrate the applicability and versatility of AQUA Cloning in selected proof-of-principle applications. These include (i) targeted insertion-, deletion- and site-directed point-mutagenesis, (ii) combined cloning and protein expression, (iii) combinatorial cloning, and (iv) the one step, one pot de novo assembly of multiple DNA fragments into a single circular plasmid encoding a complex light- and chemically-regulated Boolean A NIMPLY B logic operation.

The Directed Evolution of Halogenases for Late-Stage Functionalization of Bioactive Molecules

Payne, J.,
Andorfer, M., University of Chicago
Durak, L., University of Chicago
Kunze, K., University of Chicago
Poor, C.,
Wang, D. S.,
Lewis, J.,

Halogenation is a ubiquitous and vital process in the development of pharmaceuticals; an estimated quarter of all pharmaceuticals and agrochemicals are halogenated, and halogenation plays an integral role in the synthesis of many more. Despite this, the prevalent methods of halogenation suffer from a lack of site-selectivity and harsh reaction conditions, making the production of halogenated drug candidates difficult and environmentally unfriendly. To this end I have been working on the development of enzyme catalysts which perform site-selective halogenations of drug candidates in environmentally sustainable conditions. My preliminary work focused on improving the expression of natural halogenating enzymes in *E. coli*, particularly one called RebH, and exploring the scope and selectivity of wild-type RebH on a range of substrates on the preparative scale. My work now is focused on the directed evolution of RebH to further enhance its activity and expand its substrate scope with two primary goals: first, the halogenation of biologically active substrates, with a focus on bacterial biofilm inhibitors, in order to alter their biological properties; and second, the enantioselective halogenation of prochiral symmetrical compounds. Through the introduction of mutations in the genetic code for RebH, a variety of slightly altered RebH mutants are obtained and screened for changes in activity. Improved variants are characterized and further altered, thus recapitulating the process of evolution at a much accelerated pace in our laboratory. This research will both elucidate the role various mutations have in altering the activity of RebH and produce RebH variants useful for the development of novel drug candidates.

Real-Time Optogenetic Control of Intracellular Protein Concentration in Microbial Cell Cultures

McClean, M.,
Melendez, J.,
Oakes, B., UC Berkeley
Patel, M.,
Noyes, M.,

Real-time light-driven temporal control of gene expression and protein concentration in *S. cerevisiae*

Justin Melendez, Benjamin Oakes, Michael Patel, Marcus Noyes and Megan McClean
Department of Biomedical Engineering, University of Wisconsin, Madison

Studies of signaling and transcriptional networks are limited by the tools currently available to systematically perturb and interrogate such networks. Standard genetic tools such as deletion, overexpression, and mutation are effective for identifying network components but less suitable for understanding kinetics and identifying feedback interactions. Here we describe the development of a system for real-time light-inducible control of gene expression in *S. cerevisiae*. Controlling transcription enables fine-tuned temporal modulation of protein levels useful for interrogating genetic networks. Light-induction is achieved through fusion of the *Arabidopsis thaliana* proteins cryptochrome 2 (CRY2) and its interaction partner (CIB1) to appropriate DNA-binding and activation domains (1). CRY2 and CIB1 naturally dimerize on blue-light exposure, bringing the DNA-binding domain and the activation domain together to drive expression of the desired gene. We characterize the kinetics, off-target effects, and dose-response of this induction system and develop improved fusion proteins to allow for specific and controllable induction.
To implement real-time control we developed an integrated robotics system. Cells from a steady-state chemostat culture are continuously imaged for fluorescence. Information from this imaging is used to adjust the inducing light levels to vary the concentration of the fluorescently-tagged protein. By controlling the light-intensity, pulse frequency, and pulse duration we are able to maintain a steady- state protein concentration of our choosing. By modulating the inducing light we can produce dynamic protein level perturbations such as pulses and oscillations. This tool allows us to flexibly perturb protein levels to interrogate the kinetics and structure of genetic networks.

(1) Kennedy, MJ et al 2010 *Nature Methods* 7: 973-975

New Evidence for the Effect of Human Chromatin on Cas9-Mediated Editing

Davis, R.,
Haynes, K., Arizona State University

The CRISPR/Cas9 system is a popular genome engineering tool that has the potential to be developed into a safe gene therapy, where errors in a patient's genome can be precisely corrected with no off-target effects. By simply modifying the targeting sequence of the guide RNA, one can direct Cas9 to virtually any gene to generate mutations, deletions, or small insertions. However, bioengineers have yet to fully understand how the Cas9/RNA complex, a bacterial macromolecule, interacts with the complexities of a human genome that is packaged with nuclear proteins in a structure known as chromatin. So far, previous studies have only provided direct evidence that DNA methylation does not interfere with Cas9-mediated genome editing. However, genome-wide binding studies suggest Cas9 access to target DNA may be reduced by chromatin. The impact of the multiprotein complexes that make up the bulk of chromatin remains unclear. Using an inducible, transgenic cell line, we have directly measured chromatin inhibition of Cas9 editing efficiency at a reporter transgene that is stably integrated into a chromosome. Upon addition of doxycycline, the transgenic cells express a fusion protein that induces the accumulation of histone methylation, Polycomb complexes, and nucleosome compaction at a luciferase transgene. This switchable chromatin system allows us to compare the editing efficiency of a specific luciferase-targeted Cas9/gRNA complex in the presence and absence of compacted chromatin. We observed a decrease in editing efficiency at specific sites when chromatin is induced at the Cas9/gRNA target, luciferase. These results provide the first direct evidence that closed chromatin interferes with Cas9 function at a single site in the human genome.

Construction and Characterization of a Refactored Actinorhodin Biosynthetic Cluster

Nath, S., Lawrence Berkeley National Laboratory
Hoover, C., Lawrence Berkeley National Laboratory
Louie, K., Lawrence Berkeley National Laboratory
Bowen, B. P., Lawrence Berkeley National Laboratory
Petzold, C., Joint BioEnergy Institute
Kim, J.,
Northen, T., Joint BioEnergy Institute
Rubin, E., Joint Genome Institute
Keasling, J. D., Joint Bioenergy Institute
Hillson, N. J., Joint BioEnergy Institute
Deutsch, S., Lawrence Berkeley National Laboratory

Next-generation sequencing technologies have enabled single genomes as well as complex environmental samples (metagenomes) to be sequenced at an ever-increasing pace. Advanced bioinformatics analysis of the resulting genomics and transcriptomics data is revealing an unprecedented catalogue of proteins, and pathways that encode a myriad of novel catalytic activities and a vast diversity of novel compounds of potential value for environmental and biomedical applications.

Characterizing the products of predicted biosynthetic pathways remains challenging as (i) Clusters originate from diverse organisms many of which are difficult to culture and manipulate, (ii) Biosynthetic clusters are often under transcriptional or translational repression by unknown regulatory mechanisms, (iii) Biosynthetic clusters are often large and complex in terms of sequence composition, (iv) Molecule production may require the presence of unknown substrates or co-factors.

In order to systematically characterize the products of biosynthetic clusters we are building a scalable pipeline for pathway refactoring which includes automated design, efficient DNA construction and assembly, sequence verification, strain generation and compound detection.

As a proof-of-principle for our pipeline we designed, built and tested a fully refactored actinorhodin pathway rACT, which was synthesized from scratch using synthetic promoters and terminators, *in-silico* calculated RBSs, and optimized coding sequences. In total rACT comprised 22 Kb of synthetic sequence, which was organized as 5 contiguous operons that did not mimic the natural cluster architecture in terms of orientation, order or operon composition. The rACT DNA design was carefully screened for sequence features that might compromise synthesis efficiency and any such features present in the original design were removed by codon juggling. The resulting 'polished' DNA design was partitioned to optimize the assembly junctions, and synthesized using standard methods. Construct assembly was performed using a two-step process involving Gibson followed by yeast Gap-repair cloning. The integrity and sequence composition of rACT was verified using Pacific Biosciences sequencing technology.

The rACT cluster was integrated into a *Streptomyces coelicolor* strain lacking the native actinorhodin cluster (delACT strain). The resulting rACT strain was cultured and compared to the delACT strain as well as to wildtype *S. Coelicolor*. Samples were collected at days 2, 4 and 6 of culture in triplicates and subjected to transcriptomics, proteomics and metabolomics analyses. Actinorhodin as well as all of its metabolic intermediates were detected by LC-MS in the rACT and wildtype strains but not in the delACT strain. Interestingly, actinorhodin production was detected earlier in the rACT strain than in the wildtype strain showing a clear de-coupling of growth and molecule production.

Our design strategy, build pipeline and full OMICS analysis will be presented in detail as we seek to further refine and scale our genomes to molecules pipeline.

Automated Statistical Design of Experiments for Metabolic Engineering

Olson, B., Lawrence Berkeley National Laboratory
Dehal, P., Lawrence Berkeley National Laboratory

The growing field of metabolic engineering for natural product production typically involves transplant of a heterologous pathway comprising multiple genes into a host organism. A common challenge in metabolic engineering is balancing enzyme expression levels in this pathway to maximize product production. Finding this optimal balance requires designing, building and testing many strains with different combinations of genetic components (genes, promoters, ribosome binding sites, etc). Since pathways can contain non-linear interactions, the number of potential strains grows exponentially with the length of the pathway. Software tools make it easy to design constructs to cover the entire space of potential strains, however, the cost of building and testing constructs means that only a small fraction of the design space can be experimentally validated, even in a high-throughput environment. Thus effective Design of Experiments (DoE) is key to pathway optimization given limited experimental resources.

We approach this DoE challenge as a combinatorial optimization problem. The goal is to find a high target-producing strain out all potential strains in the pathway design space, while minimizing the total number of constructs which must be actually built and tested. The field of combinatorial optimization offers many algorithms for approaching such a problem. We propose an Estimation of Distribution Optimization (EDO) framework with an iterated regression model of the target metabolic pathway using proteomics and metabolomics data. This approach leverages multiple rounds of experiments to narrow in on the goal strain. In each round we first build a statistical model from the previous round's omics data to predict target product production for any potential strain in the design space. We then sample from this model to generate the DoE for the next round. After each round, the model accuracy increases and is able to more effectively direct the next round of experiments.

Previous approaches to guide metabolic engineering show that statistical models are able to capture non-obvious interactions between pathway components, but fail to realize the potential of machine learning as a tool for automated DoE. Previous methods, which employ PCA and linear regression, are limited in their ability to capture arbitrary complex interactions in high-dimensional data sets and require significant manual interpretation. Machine learning algorithms allow us to train arbitrarily complex regression models and the EDO framework provides automated, statistically sound, DoE.

We test the effectiveness of this approach on several existing metabolic engineering data sets. Our toolkit predicts high-yield strains for different target natural products on distinct pathways and suggests additional experiments which are likely to yield even higher production levels. These predictions are computed without the need for manual interpretation, however, the toolkit also provides visualizations of high-dimensional omics data for qualitative representations of the pathway model.

Probabilistic Design of RNA Based Genetic Components: Next Generation Rgr Devices

Voje, W. Jr., University of Washington
Gander, M. W., University of Washington
Klavins, E.,
Carothers, J.,

The current disconnect between designing biological systems and realizing their construction within biological chassis necessitates multiple iterations within the design-build-test engineering framework. It has been demonstrated that by considering genetic components at the level of RNA in terms of structure and free energy it is possible to predict integrated part performance. Recently, Cas9 proteins have been repurposed and extended to be modular transcriptional regulators. To construct genetic networks based on the Cas9 platform it is necessary to both robustly and efficiently express guide RNA (gRNA) and have that gRNA function in an orthogonal and predictable manner. Previous work has efficiently expressed gRNA from robust and well characterized pol II promoters by flanking gRNA with self-cleaving ribozymes (RGRs). Our work will demonstrate how probabilistic non-comparative RNA structure prediction methods as well as kinetic based RNA folding-path simulations can inform the design and implementation of the original RGR framework as well as develop more sophisticated expression devices. Finally, the aforementioned tools are used to develop rules for the predictable and orthogonal implementation of Cas9 based genetic systems.

Synthetic DNA Delay Circuits for Temporal Coordination

Scalise, D., Johns Hopkins University
Howie, D.,
Schulman, R., Johns Hopkins University

Biological systems use chemical circuits to coordinate elaborate sequences of events, from the timing of cell divisions inside the embryo, to the sustained rhythms of circadian clocks. Similar control over the timing of events in synthetic systems remains challenging, in part because of the complexity of designing *de novo* gene regulatory networks. Here we demonstrate a simple molecular circuit, consisting of only three species of synthetic DNA complexes in buffer, which automatically releases a target sequence of DNA into solution after a tunable delay period. The delay time can be specified on the order of hours to days, and multiple orthogonal delay circuits can be combined to coordinate the release of different sequences at independent times. This bare-bones circuit facilitates precise timing of target events inside of *in vitro* DNA networks without external stimulation.

An optogenetic system for nuclear translocation of viral nanoparticles and tunable gene delivery

Gomez, E. J., Rice University
Gerhardt, K., Rice University
Judd, J., Rice University
Tabor, J. J., Rice University
Suh, J., Rice University

The use of viruses to deliver nucleic acids for various applications, ranging from fundamental biological studies to clinical translation, is an active area of research recently fueled by the European approval of an adeno-associated virus (AAV) based product for human use. AAV is a promising vector able to deliver genes, parts of genetic circuits, RNAi, and most recently CRISPR/Cas. However, *in vivo* viral gene delivery still suffers from low gene expression at the target site and leaky off-target expression.

With the hopes of improving targeting and efficiency of gene delivery, vectors have been designed to be stimulus-responsive against a variety of inputs, including pH, temperature, and enzyme concentration in the extracellular environment. Although promising, engineering vectors to be responsive to exogenous stimuli, rather than endogenous inputs, may render the delivery process more controllable both temporally as well as spatially.

Additionally, mainly extracellular steps of the gene delivery process have been manipulated so far. Once inside the cell, gene vectors are challenged by several obstacles like sequestration in endosomes, ubiquitination and clearance by autophagosomes, cytosolic trafficking, and perinuclear accumulation. This makes nuclear localization a formidable rate-limiting step and a major determinant of effective gene delivery.

To overcome this problem, we have engineered AAV controllable by externally applied light. The virus is genetically engineered to display surface-exposed Phytochrome interacting factor 6 (PIF6) moieties capable of binding to Phytochrome B (PhyB). In HeLa cells constitutively expressing PhyB tagged with a C-terminal nuclear localization sequence (NLS), we are able to exert control over virus nuclear localization and gene delivery by varying the intensity, duration, and wavelength of light. Red light activates PhyB-NLS to bind to AAV-PIF6 and promote nuclear accumulation, and far-red light deactivates PhyB. A transfer function has been established relating the ratio of red:far-red light to the overall gene delivery by the virus, and spatial control of light is being investigated to enable patterned gene expression.

Combining optogenetic tools with viral gene delivery can lead to the development of predictable and controllable delivery platforms, which will be broadly useful in gene therapy and mammalian synthetic biology research. Other light-sensitive proteins such as the Light-Oxygen-Voltage (LOV) domains are also under investigation for use in gene delivery. Future applications include organized gene expression from regulated patterns of light in 2D and in 3D biomaterials for regenerative medicine applications.

Crispathbrick: Modular Combinatorial Assembly of Type II-a Crispr Arrays for dCas9-Mediated Multiplex Transcriptional Repression in E. coli

Cress, B., Rensselaer Polytechnic Institute
Toparlak, D., Rensselaer Polytechnic Institute
Guleria, S.,
Lebovich, M., Rensselaer Polytechnic Institute
Stieglitz, J. T.,
Englaender, J. A., Rensselaer Polytechnic Institute
Jones, J. A., Rensselaer Polytechnic Institute
Linhardt, R. J., Rensselaer Polytechnic Institute
Koffas, M. A. G., Rensselaer Polytechnic Institute

Programmable control over an addressable global regulator would enable simultaneous repression of multiple genes and would have tremendous impact on the field of synthetic biology. It has recently been established that CRISPR/Cas systems can be engineered to repress gene transcription at nearly any desired location in a sequence-specific manner, but there remain only a handful of applications described to date. In this work, we report development of a cloning procedure called "CRISPathBrick," enabling rapid modular assembly of natural type II-A CRISPR arrays capable of simultaneously repressing multiple target genes in *E. coli*. Iterative incorporation of spacers into this CRISPathBrick feature facilitates the combinatorial construction of arrays, from a small number of synthetic, user-defined DNA parts, which can be utilized to generate a suite of complex phenotypes. We demonstrate assembly and functionality of these arrays in *E. coli* by repressing genomic and plasmid-based reporter genes, as well as endogenous virulence factors and biosynthetic genes, using a single plasmid in multiple divergent strains. We show that significant repression is achieved in engineering hosts like BL21, K-12, and DH5alpha, in addition to wild-type probiotic strain Nissle 1917 and virulent strain K5. The value of addressable, orthogonal master transcriptional regulators like dCas9 cannot be overstated; thus, we have developed a two-plasmid toolkit for facile construction and characterization of CRISPR arrays. The first plasmid, pCRISPathBrick, possesses all components necessary to achieve repression of both heterologous and endogenous targets in *E. coli*, guided by the genetic program encoded in its readily expandable CRISPR array. The second plasmid, pCRISPReporter, is compatible with pCRISPathBrick and contains a transcriptionally-insulated multiple cloning site, where genomic promoters regions and genes of interest can be inserted to create a translational fusion with any desired reporter. With these two plasmids, we demonstrate facile characterization of synthetic spacers, quantified in terms of repression of the target protein-reporter fusion. Finally, we show that CRISPathBrick can be used to repress endogenous metabolic targets in *E. coli*, mediating up to 2.5-fold improvement in production of the heterologous plant natural product naringenin (from 7.6 mg/L to 18.9 mg/L). We believe that the tools developed here will be incredibly valuable for scientists in many disciplines, from systems and synthetic biology to metabolic engineering and the basic sciences.

New Promoters to Improve Heterologous Protein Production in Penicillium chrysogenum

Polli, F., University of Groningen
Meijrink, B., DSM Biotechnology Center
Bovenberg, R. A. L., DSM Biotechnology Center
Driessen, A. J. M.,

Filamentous fungi such *as Aspergillus* and *Penicillium* are used as hosts for expression and secretion of heterologous proteins and secondary metabolites. Studies have shown the promiscuous nature of fungi in recognizing transcriptional and translational elements present in genes from other fungal species. However, only few promoters have been characterized so far in *Penicillium chrysogenum.* To address this need, a set of heterologous and homologous promoters were tested in a reporter system.. We combined available synthetic biology tools such as BioBricks and *in vivo* homologous recombination in *Saccharomyces cerevisiae* in order to engineer twelve *Penicillium* and *Aspergillus* promoter pathways driving the expression of GFP while constitutive expression of DsRed was used as an internal strandard. These pathways were tested using the BioLector technology in fermentations with a *P. chrysogenum* strain and classified in strength and expression profile. This catalog of promoters will be used in future synthetic biology projects to tune the expression of target genes.

Small Molecule–Triggered Cas9 Protein with Reduced Off-Target Cleavage

Davis, K. M., Harvard University
Liu, D. R., Howard Hughes Medical Institute

The RNA-guided endonuclease Cas9 enables simple and efficient genome editing. We have developed Cas9 nucleases that are activated by the cell-permeable small molecule 4-hydroxytamoxifen (4-HT) by inserting a 4-HT-responsive intein at specific positions in Cas9. These intein-Cas9s enable small-molecule control of genome editing, with Cas9 activity being controlled at the post-translational level. In human cells, these conditionally active intein-Cas9s demonstrate improved genome-editing specificity over wild-type Cas9.

Optimizing heterologous pathways in mammalian cells via competitive splicing

Richardson, R., Harvard University
Hicks, W.,
Way, J. C., Harvard University
Silver, P. A., Harvard Medical School

The application of synthetic biology in mammalian cells offers great promise for developing new tools for exploring biology and novel therapeutics to treat human disease. However, the dearth of tools for regulating gene expression quantitatively limits the phenotypic landscape that can be explored with engineered genetic circuits. This challenge is particularly evident with efforts to engineer metabolite production, for which the relative expression level of enzymes can impact yield of the final product or result in the formation of unwanted or toxic byproducts. Similarly, from a developmental perspective, nuances in expression level among multiple genes often lead to drastic changes in cell fate. I address this technological gap by engineering alternative 3' splice site selection as a mechanism for regulating and optimizing the relative expression level of multiple genes from a single, polycistronic operon. In doing so, not only have I develop a novel strategy for modulating gene expression, but I will also gain significant insight into the rules that govern RNA splicing in mammalian cells. This technology will specifically be applied to engineer essential amino acid production in human cells and to improve the efficiency of induced pluripotent stem cell conversion during cellular reprogramming, demonstrating the far-reaching capabilities and impact of this modular platform.

A General Method for Sensing Small Molecules in Eukaryotes

Feng, J.,

Biosensors for small molecules can be used in applications that range from metabolic engineering to orthogonal control of transcription. We describe a method to create biosensors starting with a computationally designed ligand-binding domain (LBD) that, in principle, can be generated for any target molecule. The LBD is fused to either a fluorescent protein or a transcriptional activator and is destabilized by mutation such that the fusion accumulates only in cells containing the target ligand. We illustrate the power of this method by developing biosensors for digoxin and progesterone. Addition of ligand to cells expressing a biosensor activates transcription in yeast, mammalian cells and plants, with a dynamic range of up to ~100-fold.

We use the biosensors to improve the biotransformation of pregnenolone to progesterone and to regulate CRISPR activity. In concert with computational LBD design approaches, this method should enable the generation of biosensors for a broad range of molecules.

Enabling Rapid Protein and Circuit Engineering with High-Throughput Microfluidic Devices

Maerkl, S. J.,

We developed several methods and microfluidic approaches to enable the rapid engineering and quantitative characterization of proteins and genetic circuits *in vitro*. Protein engineering remains challenging, primarily due to the cumbersome experimental processes currently required for expressing, purifying and quantitatively characterizing proteins. We developed a novel solid-phase gene synthesis approach allowing the combinatoric assembly of protein variants. We directly coupled our gene synthesis method with high-throughput on-chip protein expression and characterization, allowing us to drastically reduce the time required for a complete design-test cycle. We applied our novel protein-engineering platform to the generation of hundreds of Zn-finger transcription factors and the precise quantitation of the binding specificity of each variant.

In order to streamline genetic network engineering we have recently demonstrated that genetic networks can be implemented and characterized *in vitro* using a microfluidic chemostat. This chemostat approach allows complex networks such as oscillators to be run *in vitro*. We have previously shown that several biologically relevant regulatory mechanisms can be implemented on this *in vitro* platform and we built and characterized the first *in vitro* genetic oscillator. Over the last year we focused on testing whether we can functionally implement genetic networks that are known to work *in vivo* in order to assess whether genetic networks can be designed or modified *in vitro* and consequently transferred *in vivo*. We have successfully implemented the Elowitz repressilator *in vitro* and performed a comprehensive analysis of the system that would be difficult to achieve *in vivo*. We additionally designed and implemented novel 3 node oscillators, and the first 5 node oscillators *in vitro* and we will present data on how these systems behave *in vivo*.

Controlling Multiple Gene Expression in Saccharomyces Cerevisiae By Optogenetic Switches

Hochrein, L., University of Potsdam
Machens, F., University of Potsdam
Mueller-Roeber, B., University of Potsdam
Messerschmidt, K., University of Potsdam

Our group is interested in the regulated production of small to large cohorts of proteins and peptides and the establishment of a toolkit for the modular assembly of circular expression yeast artificial chromosomes (xYAC) in *S. cerevisiae*. To simplify induction of protein production we will provide a reliable and easy to use induction system: Light-induced TALE-transcription factors based on two different light-sensing systems from *Arabidopsis thaliana*. While the CRY2/CIB1 system is blue light-sensitive and has no need for an exogenous cofactor, the red light-sensing PhyB/PIF3 interaction depends on a chromophore, which is not available in yeast. To avoid the exogenous supply of the chromophore, yeast will be engineered to express the necessary cofactor biosynthesis genes along with the red light-sensing proteins. In order to engineer two autonomous and non-interfering light-dependent gene switches, we will combine the red and the blue light-sensing systems on one single xYAC. With this approach it will be possible to easily regulate gene expression of a large number of proteins in a reversible, specific, dose-dependent and orthogonal manner totally independent of each other just by applying light of different wavelength. Results on the optimization of chromophore biosynthesis, evaluation of the individual light-induction systems, and the parallel use of red and blue light-controlled gene expression within our xYAC platform will be presented.

Engineering Caspases with Altered Specificities
Hill, M.,

Engineering Caspases with Altered Specificities

Maureen Hill, Derek MacPherson, Peng Wu, Olivier Julien, Sergey Savinov, James A. Wells and Jeanne A. Hardy

Caspases are proteases that play important regulatory roles in apoptotic cell death, neurodegeneration and inflammation. They are amongst the most specific of all proteases, and are therefore excellent targets for engineering new specificity. Unfortunately, because they are heterotetramers, they are not amenable to traditional engineering approaches like phage display and are difficult to tackle by rational design due to their flexible active sites. Such characteristics prompted our lab to develop a GFP-based reporter of caspase activity that has enabled us to perform a directed evolution screen and evolve caspases with new functions.

In our first application of this technology, we engineered caspase-7, which recognizes and cleaves the amino acid sequence DEVD, to recognize and cleave the new sequence VEID, which is recognized by caspase-6 but not caspase-7. Simple substitution of the homologous residues in caspase-6 dramatically diminished activity, but did not alter the specificity, indicating that a more holistic evolutionary procedure would be needed. The caspase-7 variants with the most strongly altered specificity displayed kinetic values similar to wild-type caspase-6 and mimicked caspase-6 cleavage patterns with protein substrates. Using *N-terminomics* we assessed the substrate specificity profile for the most successful engineered caspase. We found that had completely altered the specificity using our directed evolution approach. The resulting WebLogos for the engineered caspase were completely superimposable with caspase-6 and unlike the parent scaffold, caspase-7. Our seven crystal structures of evolved caspases bound to either DEVD or VEID substrates allowed us to observe how the new residues in the active site contributed to the change in specificity. In particular, F282 plays a critical role by arching inward to create a small pocket around the newly introduced isoleucine at the P2 peptide position. This unexpected finding underscores why rational design is insufficient for altering substrate recognition.

Engineering Emergent Properties in Synthetic Microbial Communities

Bernstein, H. C., Pacific Northwest National Laboratory
Carlson, R. P., Montana State University
Beliaev, A. S., Pacific Northwest National Laboratory
Fredrickson, J. K., Pacific Northwest National Laboratory

Traditionally, engineers and synthetic biologists exert control over microbial metabolism by adding or removing genes and pathways from a single host. However, new paradigms have been realized though the 'consortia' concept of engineering and compartmenting metabolic function within assigned community members. Beneficial emergent properties, such as enhanced productivity and/or stability, can be engineered into synthetic or thoughtfully constructed microbial communities. This presentation focuses on how emergent properties have been realized from community-engineering. Binary consortia were either metabolically engineered (via genome reduction) or naturally prone to cooperate in a 'producer-consumer' motif. These systems ranged from synthetic *Escherichia coli* co-cultures engineered for mutualistic exchange-detoxification of acetic acid to artificial wild-type assemblies of phototrophic cyanobacteria and heterotrophic counter parts. These distinct binary cultures demonstrated enhanced biomass productivity and resistance to metabolite feed-back inhibition. For example, the effect of oxygen availability on metabolically engineered *E. coli* systems was found to be a controlling factor for acetic acid exchange and subsequent phenotype-specific spatial patterning in biofilms and biomass productivity in chemostats. Similarly, oxygen gradients were imposed on a constructed consortium consisting of wild-type photoautotroph (*Thermosynechococcus elongatus*) and chemoheterotroph (*Meiothermus ruber*), which responded with an increased resistance to the growth inhibition effect of high oxygen-tension. This research lends critical insight to a grand-challenge for biological engineers and synthetic biologists aiming to capitalize on knowledge gained from model systems and phenomenological observation to rationally design microbial communities for controllable outputs.

Construction of Gamma-Aminobutyric Acid Producing Recombinant Escherichia coli By Introduction of Synthetic Scaffolds

Hong, S., University of Ulsan

Escherichia coli were engineered for the direct production of gamma-aminobutyric acid from glucose by introduction of synthetic protein scaffold. In this study, three enzymes consisting GABA pathway (isocitrate dehydrogenase, glutamate synthase and glutamate decarboxylase) were connected via synthetic protein scaffold. By introduction of scaffold, 0.92 g/L of GABA was produced from 10 g/L of glucose while no GABA was produced in wild type E. coli. The optimum pH and temperature for GABA production were 4.5 and 30oC, respectively. When competing metabolic network was inactivated by knockout mutation, maximum GABA concentration of 1.3 g/L was obtained from 10 g/L glucose. The recombinant E.coli strain which produces GABA directly from glucose was successfully constructed by introduction of protein scaffold. This work was supported by a grant from the Next-Generation BioGreen 21 Program (SSAC, grant number: PJ011116) by RDA, and Basic Science Research Program by the Ministry of Education (NRF-2014R1A1A2054726).

A 3-D Printed Hardware Platform for Quantitatively Characterizing the Performance Features of Diverse Optogenetic Transcriptional Regulatory Systems in E. coli and S. Cerevisae

Olson, E. J., Rice University
Gerhardt, K. P., Rice University
Ramakrishnan, P.,
Castillo-Hair, S. M., Rice University
Ekness, F.,
Hartsough, L. A., Rice University
Landry, B. P.,
Tabor, J. J., Rice University

There has recently been rapid development of optogenetic tools for optically regulating gene expression in microbes. These tools exhibit diverse spectral responses, light sensitivities, photoconversion and dark reversion rates, and chromophore requirements, which significantly impact their suitability for different experiments. Literature data on these performance characteristics is limited as different studies have performed varied measurements, resulting in only partial assessments. Furthermore, data are of variable quality and are reported in different arbitrary units. Altogether, this makes direct comparisons of the performance of these tools very difficult.

Here, we address this problem by developing a framework to quantitatively assess these and other performance metrics using a custom hardware device termed the LED Plate Apparatus (LPA). The LPA is a 3D-printed device incorporating a programmable array of through-hole LEDs which holds a 24-well microplate, providing an isolated light environment for each well. The two LEDs per well are reconfigurable, enabling them to be easily swapped for LEDs of any desired wavelength. The LPA can be programmed with complex light sequences using the Light Program Interface (LPI), a custom webtool that produces files which are loaded onto the LPA using a SD card. We are using the LPA and LPI to perform characterization experiments on many previously published but incompletely-characterized light-inducible systems, such as pDusk and BphS/MrkH in *E. coli* and PhyB/PIF6 and Cry/CIB in *S. cerevisae*. Our data include the action spectra (both activating and deactivating), steady-state light dose response, and step-function response dynamics. Using these characterization datasets, we are producing datasheets on the performance metrics of the current generation of *E. coli* and *S. cerevisae* optogenetic tools.

Furthermore, we are using these data to produce predictive mathematical models of the light-to-gene-expression signaling of each system that will enable a) easy selection of appropriate fluorescent protein reporters to combine each tool with and b) the design of multi-LED light input signals which can generate desired gene expression signals *in vivo*, even when programming multiple systems simultaneously. Our approach should be readily compatible with optogenetic proteolysis systems and mammalian cells. This work dramatically lowers the "barrier to entry" for scientists and engineers wishing to perform optogenetic experiments in their lab by opening up access to an easy-to-use standardized suite of hardware, software, and wetware tools.

Feedback Control of Synthetic Genetic Circuits for the Mitigation of Resource Competition

Hamadeh, A.,
Del Vecchio, D.,

A recurring challenge in the design of genetic circuits is context dependence, the fact that the behavior of a functional module is influenced by the state of the wider cellular milieu with which it interacts. One key player in context dependence is the scarcity of shared cellular resources, especially those required for transcription and translation during gene expression. Because of competition for these limited resources, the behavior of modules becomes coupled in subtle ways, preventing circuits from working as expected.
This poster proposes a classical feedback control approach to mitigate the steady state effects of the competition for resources necessary for gene expression. In particular, we analyze and compare the ability of several inhibitory feedback regulation architectures to reduce the interdependence between different gene expression processes due to resource limitations.

On the Engineering of Effective Biological Computations

Beal, J., Raytheon BBN Technologies

Engineering biological cells to perform computations has a broad range of important potential applications, including precision medical therapies, biosynthesis process control, and environmental sensing. Implementing predictable and effective computation, however, has been extremely difficult to date, due to a combination of poor composability of available parts and of insufficient characterization of parts and their interactions with the complex environment in which they operate. I demonstrate how this situation can be improved by quantitative signal-to-noise analysis of the relationship between computational abstractions and the variation and uncertainty endemic in biological organisms. This analysis takes the form of a delta-SNR function for each computational device, which can be computed from measurements of a device's input/output curve and expression noise. These functions can then be combined to predict how well a circuit will implement an intended computation, as well as evaluating the general suitability of biological devices for engineering computational circuits. Applying signal-to-noise analysis to current repressor libraries shows that no library is currently sufficient for the engineering of many-layer circuits with significant SNR in their outputs. The analysis also indicates key targets to remedy this situation and vastly improve the range of computations that can be used effectively in the implementation of biological applications.

Piace: Parallel Integration and Chromosomal Expansion of Biofuels Pathways in E. coli

Goyal, G.,
Gutierrez, J., Joint Bioenergy Institute
Kang, A.,
Keasling, J. D., Lawrence Berkeley National Laboratory
Lee, T. S., Joint BioEnergy Institute
Hillson, N. J., Joint BioEnergy Institute

Robust fermentation of biomass-derived sugars into biofuels demands the reliable microbial expression of underlying metabolic pathways. Previous studies have shown that plasmid-based expression systems, which may suffer from instability issues (despite selective pressure), can result in inconsistent and highly variable biofuel production rates and titers. An established mitigation strategy, Chemical Induced Chromosomal Expansion (CIChE; Tyo et al. 2009), exploits recA-mediated copy number expansion of a pathway post integration at a single chromosomal locus, and offers the gene copy numbers of plasmids while maintaining pathway stability even without the addition of antibiotics (a significant process cost-savings). Innovating beyond CIChE, we are developing PIACE (Parallel Integration and Chromosomal Expansion) to facilitate the simultaneous and independently-adjustable expansions of multiple pathways at separate chromosomal loci. PIACE offers a set of BglBrick-compatible suicide vectors, each of which site-specifically integrates into a distinct chromosomal locus via a single homologous cross-over and has high efficiency for transformation and chromosomal integration of long (> 3kb) pathways that reside in super-coiled sequence-validated plasmids. As a proof of principle for PIACE, we have integrated rfp and gfp at two separate chromosomal loci, expanded rfp and gfp independently in the presence of heat-curable plasmid pRedi2RecA, and demonstrated that both expansions are stable over 50 cellular generations in the absence of antibiotic selection pressure. We are now applying PIACE to the chromosomal stabilization and optimization of the isopentenol fuel pathway in E. coli DH1. PIACE affords an efficient, tunable method for improving the robustness, productivity, and titers of engineered metabolic pathways.

The Joint Genome Institute's Synthetic Biology Biosecurity DNA Screening Process

Simirenko, L., Joint Genome Institute
Rubin, E., Joint Genome Institute
Hillson, N., Lawrence Berkeley National Labs

Concerns have been raised that individuals with malicious intent could exploit DNA synthesis technology in order to obtain genetic elements from organisms and toxins that would otherwise be difficult to obtain. In response to these concerns, the U.S. Department of Health and Human Services issued the *Screening Framework Guidance for Providers of Synthetic Double-Stranded DNA*. The voluntary *Guidance* outlines the U.S. government's recommendations for screening double-stranded DNA to ensure that existing regulations[1] are followed, encourage best practices in addressing biosecurity concerns, and reduce the risk that individuals with ill intent may exploit the application of nucleic acid synthesis technology to obtain genetic material derived from or encoding Select Agents or Toxins, or agents on the EAR's Commerce Control List (CCL).

In accordance with the *Guidance*, the U.S. Department of Energy Joint Genome Institute has developed a DNA screening pipeline to screen all sequences that are synthesized through it's synthetic biology program. The screening method detects "sequences of concern" of at least 200 nucleotides in length on either DNA strand, and the resultant polypeptides from translations using the three alternative reading frames on each DNA strand (or six-frame translation). Sequences are aligned to the sequences in GenBank's nucleotide and protein databases rather than a curated database, to ensure that it automatically adapts as new sequences are added to GenBank. A "Best Match" approach is used to determine whether a query sequence is unique to Select Agents or Toxins, or CCL-listed agents, toxins or genetic elements, and to minimize false positives from closely related organisms or highly conserved "house-keeping genes" which do not pose a biosecurity threat.

---

[1] Select Agent Regulations (SAR) and, for international orders, the Export Administration Regulations (EAR)

Dynamic Control of ERG9 Expression for Improved Amorpha-4,11-Diene Production in Saccharomyces Cerevisiae

Yuan, J., National University of Singapore
Ching, C. B., Singapore Institute of Technology

Background

To achieve high-level production of non-native isoprenoid products, it requires the metabolic flux to be diverted from the production of sterols to the heterologous metabolic reactions. However, there are limited tools for restricting metabolic flux towards ergosterol synthesis. In the present study, we explored dynamic control of *ERG9* expression using different ergosterol-responsive promoters to improve the production of non-native isoprenoids.

Results

Several ergosterol-responsive promoters were identified using quantitative real-time PCR (qRT-PCR) analysis in an engineered strain with relatively high mevalonate pathway activity. We found mRNA levels for *ERG11*, *ERG2* and *ERG3* expression were significantly lower in the engineered strain over the reference strain BY4742, indicating these genes are transcriptionally down-regulated when ergosterol is in excess. Further replacement of the native *ERG9* promoter with these ergosterol-responsive promoters revealed that all engineered strains improved amorpha-4,11-diene by 2~5-fold over the reference strain with *ERG9* under its native promoter. The best engineered strain with *ERG9* under the control of P_{ERG1} produced amorpha-4,11-diene to a titer around 350 mg/L after 96 h cultivation in shake-flasks.

Conclusions

We envision dynamic control at the branching step using feedback regulation at transcriptional level could serve as a generalized approach for redirecting the metabolic flux towards product-of-interest

Resource Limitations Lead to Hidden Interactions in Gene Networks

Qian, Y.,
Huang, H. H.,
Del Vecchio, D.,

Protein production in gene networks relies on the availability of resources necessary for transcription and translation, which are found in cells in limited amount. As various genes in a gene network compete for a common pool of resources, including RNA polymerase and ribosomes, a hidden layer of interactions among genetic circuits arises. In this work, we develop a general modeling framework using deterministic reaction rate equations to account for these interactions. We verified the model with a two-stage activation cascade and a one-stage repression circuit in experiments with *E coli*. Resource limitations introduce (1) a feed-forward loop in the interaction graph of an activation cascade, which shows counterintuitive negative or biphasic steady-state responses, and (2) a positive auto-regulation in a repression circuit, which leads to bistability confirmed by hysteresis behavior.

Development and Experimental Validation of a Mechanistic Model of *in vitro* DNA Recombination*

Jack Bowyer, Jia Zhao, Susan Rosser, Sean Colloms, Declan Bates

I. INTRODUCTION

Synthetic Biology is a highly interdisciplinary field with the aim of establishing engineering protocols for the construction of synthetic biological circuits. One of the first synthetic biological devices to be built was the genetic toggle switch in *E. coli* [1]. Such transcriptional memory devices have paved the way for the creation of bi-stable genetic switches based on DNA recombination. These site-specific recombinases (SSRs) mediate distinct recombination events that give rise to two stable DNA states. Hence, DNA recombination has huge potential as a tool for DNA sequence assembly with numerous potential applications including biological data storage [2], DNA-based counting systems [3] and the assembly of metabolic pathways [4].

Attempts to experimentally design and build synthetic systems using recombinases have thus far been hindered by a lack of validated computational models that capture the mechanistic basis of DNA recombination. The predictive capabilities of such models could be exploited by Synthetic Biologists to reduce the number of iterative cycles required to align experimental results with design performance requirements. Here, we develop and validate the first detailed mechanistic model of DNA recombination, with a focus on how efficiently recombination can occur, and the model features required to replicate and predict experimental data.

II. A MECHANISTIC MODEL OF DNA RECOMBINATION

Recombinases can enable precise DNA manipulation both *in vitro* and *in vivo* [8]. These enzymes, known as integrase and excisionase, catalyse two recombination events termed integration and excision. Integrase alone is sufficient to mediate the integration reaction between an attachment site encoded within the host chromosome, attB, and an attachment site on the phage chromosome, attP. The phage genome is inserted into the host chromosome and is flanked by the newly formed attachment sites attL and attR. The excision reaction is mediated in the presence of integrase and a recombination directionality factor (RDF) also known as excisionase, restoring the independent substrates as well as the original attB and attP sites (Figure 1). A pre-integration state consisting of attB and attP is referred to as the BP state,

*Research supported by EPSRC via a DTA studentship to J. Bowyer

J. Bowyer and D.G. Bates are with the Warwick Integrative Synthetic

Biology Centre and School of Engineering, University of Warwick

(e-mail: J.E.Bowyer@warwick.ac.uk, D.Bates@warwick.ac.uk)

J. Zhao and S Colloms are with the Institute of Molecular, Cell and

Systems Biology, University of Glasgow

(e-mail: j.zhao.1@research.gla.ac.uk, Sean.Colloms@glasgow.ac.uk)

S. Rosser is with Synthetic and Systems Biology, School of Biological

Sciences, University of Edinburgh (e-mail: Susan.Rosser@ed.ac.uk)

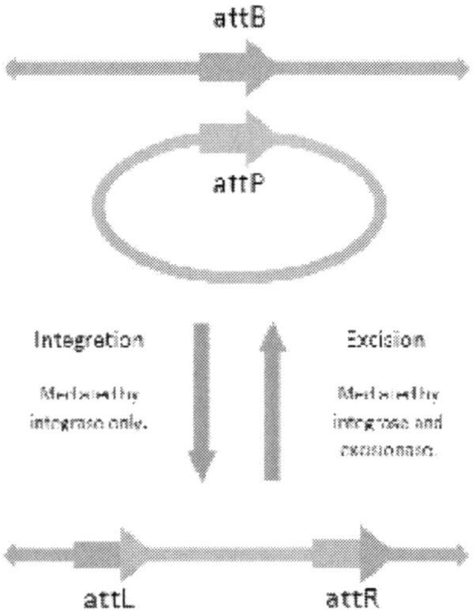

whereas a post-integration/pre-excision state containing attL and attR is the LR state. DNA recombination efficiency in switching between these two states is dependent on the concentrations of integrase and RDF in the system.

Figure 1. Schematic diagram of phage integration and excision. The phage genome attachment site, attP, is integrated into the host chromosome attachment site, attB. Integration gives rise to attL and attR, each formed of half of attB and attP. Excision restores attB and attP, removing the integrated phage genome from the host chromosome.

An extensive review of the experimental literature was carried out to synthesise current knowledge of the mechanistic basis of recombination, see [5-7] and references therein. Many structural features of the DNA recombination reaction network have widespread experimental support in the literature. Integrase forms dimers in solution [9] with one dimer bound to attB and attP necessary to mediate the integration reaction [10], which is unidirectional [11]. RDF does not bind directly to DNA attachment sites [12]. One integrase dimer and an additional RDF monomer bound to both attL and attR is necessary to mediate excision, restoring attB and attP [13]. However, we were unable to find consensus regarding three significant biological details:

â?¢ The directionality of the excision reaction.
â?¢ RDF dimerisation and subsequent tetramerisation in solution.
â?¢ Integrase monomer binding to DNA substrates.

By constructing a variety of models based on the well established properties of the reaction network that differed only according to the inclusion/exclusion of these three features, we were able to examine which model best matched our experimental data and hence investigate the most plausible representation of the true biological structure (Figure 2). We found that including unidirectional excision
[14], monomeric RDF that neither dimerises nor tetramerises in solution [15] and integrase monomer binding [16] resulted in optimal data fitting and prediction capabilities in our model. These results were determined via global optimisation of the parameters for each candidate model against the data (see Results section). Early results revealed that our models consistently produced greater recombination efficiency than that observed in our data. Hence we included a plausible mechanism for the sequestration of integrase: the formation of a dysfunctional integrase dimer at a rate kintX (Figure 2, top). Our modelling investigation focuses on DNA recombination mediated by Ï?C31 serine integrase and its RDF, gp3, *in vitro*. The network consists of twenty-two distinct biological entities and twenty-eight reaction constants. The rate of change in concentration of each entity is governed by a corresponding ordinary differential equation (ODE) in the model. We apply mass action kinetics to the biological interactions in order to derive the ODEs that comprise the model. Analysing the efficiency with which the system switches from BP state to LR state involves determining the total register of the system in either state. This equates to summing each of the ODEs that govern the network entities in the DNA state of interest (see Table I for the full list of ODEs). Model simulations are therefore the numerical solutions to the following equations:

ð ??ð ·ð ¿ð ¿ð ¿ð ¿ð ¿
ð ??ð ?? = ð ??ð ¿ ð ·ð µð µ ð ¼4 â?? ð ??ð ¿ ð ·ð ¿ð ¿
ð ¼4 ð ??4 , (1)

ð ??ð ·ð µð µð ¿ð ¿ð ¿
ð ??ð ?? = â?? ð ??ð ¿ ð ·ð µð µ ð ¼4 + ð ??ð ¿
ð ·ð ¿ð ¿ ð ¼4 ð ??4 , (2)

where DLRtot , DBPtot represent the total concentration of DNA in LR state and BP state respectively, DBP I 4 represents the concentration of the protein:DNA complex consisting of four integrase monomers bound to DNA in the BP state, DLR I 4 R4 represents the concentration of the protein:DNA complex consisting of four integrase monomers and four excisionase monomers bound to DNA in the LR state and kR represents the rate of recombination.

Experiments performed *in vitro* allow the initial concentrations of the recombinases and substrates to be quantified exactly and therefore provide the initial conditions with which to solve the system of ODEs. Hence, we do not require knowledge of integrase and excisionase expression levels, as was necessary for the simple model of *in vivo* recombination dynamics developed in [2].

III. RESULTS â?? EXPERIMENTAL VALIDATION

We compared the ability of our model to replicate and predict *in vitro* experimental data with a simple mathematical model of DNA recombination previously developed in Bonnet *et al.* [2]. Using a subset of our steady state data for a variety of different initial

concentrations of integrase and gp3, the parameters of both models were optimised using a Genetic Algorithm to minimise the difference between simulations and data values. The predictive power of both models was then assessed by evaluating their ability to match a different subset of experimental time course data. As the model of [2] was originally developed for *in vivo* data, we adapted it in order to simulate our *in vitro* system dynamics
by removing the parameters representing recombinase expression levels and adding our sequestration mechanism. As shown in Figure 3A, this simple model is unable to provide a reasonable replication of the experimental data. The more detailed model, however, exhibits strong predictive power, accurately reproducing the experimental recombination efficiencies across all initial concentrations of integrase and gp3 (Figure 3B). Similar results were seen for time-course data (to be included in the full paper). The parameter space within which both models are optimised is large, since evidence regarding the numerical values of many reaction rate constants is currently lacking in the literature. Future work will focus on further investigation of model parameters to ensure their biological plausibility.

IV. CONCLUSIONS

We have developed the first detailed mechanistic model of *in vitro* DNA recombination. The predictive power of the model was validated against a large set of experimental data on recombination efficiencies for different initial concentrations of integrase and gp3. The proposed model sheds light on a number of mechanistic features of DNA recombination for which there is currently no consensus in the experimental literature, and will be a valuable design tool for Synthetic Biologists working on the construction of recombinase-based genetic circuitry, potentially producing significant reductions in development times. Future work will extend our modelling investigation to the *in vivo* system in order to examine model performance under cellular biological conditions.

V. REFERENCES

[1] T. Gardner, C. Cantor and J. Collins. Nature. 2000 Jan

20;403(6767):339-42.

[2] J. Bonnet, P. Subsoontorn and D. Endy. Proc Natl Acad Sci

USA. 2012 Jun 5;109(23):8884-9. doi: 10.1073/pnas.1202344109. [3] A. Firedland, T. Lu, X. Wang, D. Shi, G. Church and J.

Collins. Science. 2009 May 29;324(5931):1199-202. doi:

10.1126/science.1172005.

[4] L. Zhang, G. Zhao and X. Ding. Sci Rep. 2011;1:141. doi:

10.1038/srep00141. Epub 2011 Nov 3.

[5] P. Ghosh, L. Wasil and G. Hatfull. PLoS Biol. 2006 Jun;4(6):e186. [6] P. Ghosh, N. Pannunzio and G. Hatfull. J Mol Biol. 2005 Jun

3;349(2):331-48. Epub 2005 Apr 7.

[7] P. Ghosh, L. Bibb and G. Hatfull. Proc Natl Acad Sci U S A. 2008

Mar 4;105(9):3238-43. doi: 10.1073/pnas.0711649105. [8] N. Grindley, K. Whiteson and P. Rice. Annu Rev

Biochem. 2006;75:567-605.

[9] P. Ghosh, N. Pannunzio and G. Hatfull. J Mol Biol. 2005 Jun

3;349(2):331-48. Epub 2005 Apr 7.

[10] T. Miura, Y. Hosaka, Y. Yan-Zhuo, T. Nishizawa, M. Asayama, H.

Takahashi and M. Shirai. J Gen Appl Microbiol. 2011;57(1):45-57. [11] P. Fogg, S. Colloms, S. Rosser, M. Stark and M. Smith. J Mol Biol.

2014 Jul 29;426(15):2703-16. doi: 10.1016/j.jmb.2014.05.014. Epub

2014 May 22.

[12] P. Ghosh, L. Bibb and G. Hatfull. Proc Natl Acad Sci U S A. 2008

Mar 4;105(9):3238-43. doi: 10.1073/pnas.0711649105.

[13] A. Groth and M. Calos. J Mol Biol. 2004 Jan 16;335(3):667-78.

[14] B. Swalla, E. Cho, R. Gumport and J. Gardner. Mol Microbiol. 2003

Oct;50(1):89-99.

[15] R. Keenholtz, S. Rowland, M. Boocock, M. Stark and P. Rice.

Structure. 2011 Jun 8;19(6):799-809. doi: 10.1016/j.str.2011.03.017. [16] M. Smith, R. Till, K. Brady, P. Soultanas, H. Thorpe and M. Smith.

Nucleic Acids Res. 2004 May 11;32(8):2607-17. Print 2004.

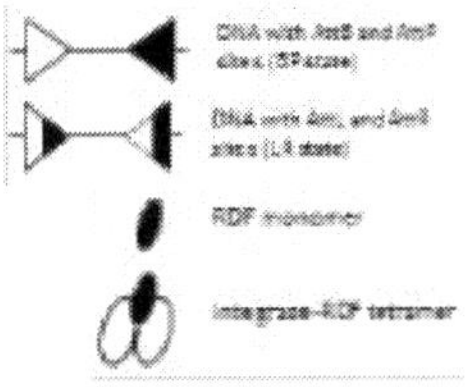

Integration Excision

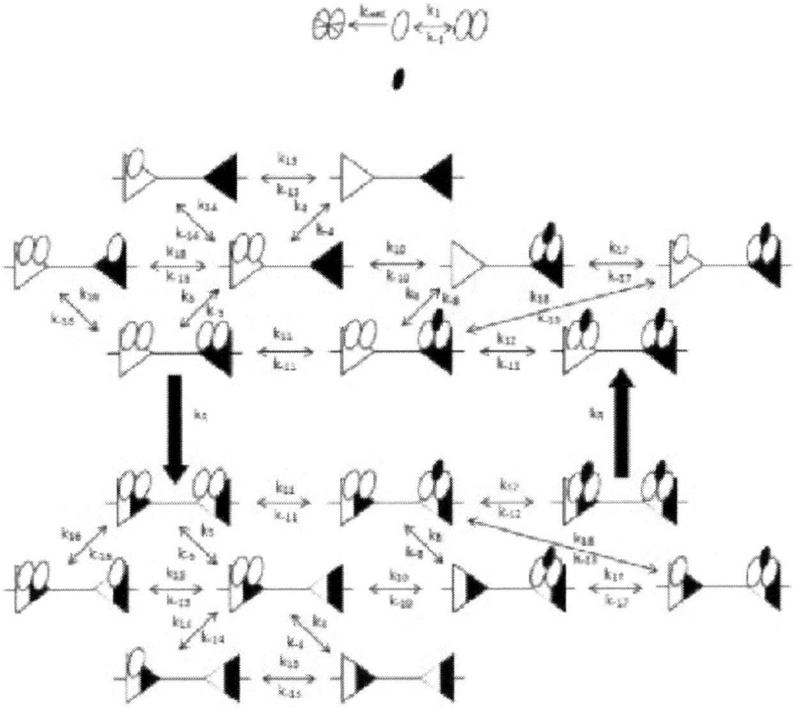

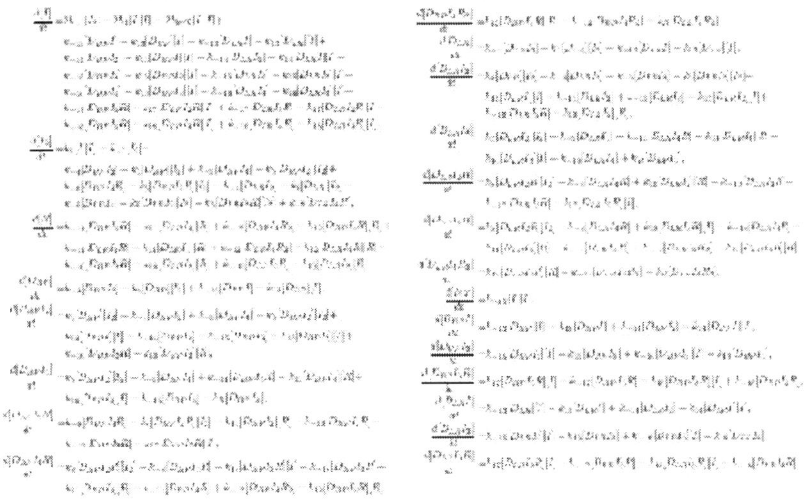

Table 1. Model Equations arising from the application of mass action kinetics to the network reactions shown in Figure 2. The model comprises

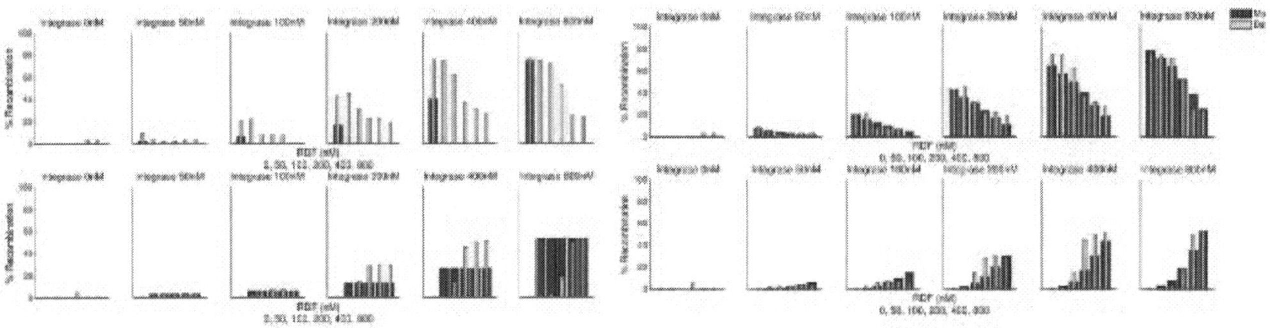

22 ODEs. Concentrations are represented by square brackets. DNA is denoted by D with the subscripts BP and LR representing the BP and LR states respectively. Integrase and RDF are denoted by I and R respectively, with numerical subscripts representing the number of monomers present / bound. The dysfunctional integrase dimer is denoted by I2X.

(A) (B)

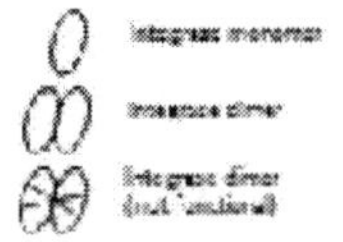

Figure 2. The DNA recombination reaction network used to construct the model. The network is based on the mechanisms underlying DNA recombination that have been verified in the current experimental literature. We model the dynamics of Ï?C31 integrase and its RDF, gp3. Reactions and their rate constants are depicted by arrows and their corresponding numbered k. The rate of recombination is kR.

Figure 3. A) Data fitting / prediction results for the model of [2]. B) Data fitting / prediction results for our model. In both A) and B) the top row of bar graphs represents the integration (BP-LR) reaction and the bottom row of bar graphs represents the excision (LR-BP) reaction. The wider blue bars represent model simulations and the thinner cyan bars represent data.

Transcriptional Bursting in Confined Cell-Free Reactions
Caveney, P. M., University of Tennessee Knoxville
Norred, S. E.,
Chin, C., University of Tennessee Knoxville
Retterer, S. T., Oak Ridge National Laboratory
Collier, C. P., Oak Ridge National Laboratory
Simpson, M. L., Oak Ridge National Laboratory

Transcriptional Bursting in Confined Cell-Free Reactions

Patrick M. Caveney1,2, Sarah E. Norred1,2, Charles Chin1,2, Scott T. Retterer1,2, Charles P. Collier2,

Michael L. Simpson1,2,3

1Bredesen Center, **University of Tennessee Knoxville**, 2Center for Nanophase Materials Sciences, **Oak Ridge**

National Laboratory, 3Department of Materials Science and Engineering, **University of Tennessee Knoxville**

Abstract:

Transcriptional bursting, bursts of transcription followed by periods of inactivity, is widely observed across all life1,2,3,4,5,6,7. This phenomenon is important for many cell fate decisions such as proviral latency in HIV8,9,10,11,12. The widely studied two-state (or random telegraph) model is 13,14,15,16,17,18 characterized by the transitions between ON and OFF, kON and kOFF respectively

Many mechanisms have been proposed including: supercoiling19, transcription factor kinetics20, chromatin remodeling21,22, and transcriptional reinitiation23,24,25. However, these studies neglect spatial correlations caused by *in vivo* crowding (40%26) and confinement (~10 femtoliters)27,28 that may contribute to transcriptional bursting29,30. We present experimental and simulation evidence of transcriptional bursting in confined cell-free protein synthesis reactions absent the proposed molecular mechanisms of bursting. And demonstrate that the magnitude and dynamics of bursting may be manipulated by control of confinement.

References:

1. Blake, William J., Mads KÃ¦rn, Charles R. Cantor, and James J. Collins. "Noise in eukaryotic gene expression." *Nature* 422, no. 6932 (2003): 633-637.

2. Skupsky, Ron, John C. Burnett, Jonathan E. Foley, David V. Schaffer, and Adam P. Arkin. "HIV promoter integration site primarily modulates transcriptional burst size rather than frequency." *PLoS computational biology* 6, no. 9 (2010): e1000952.
3. Taniguchi, Yuichi, Paul J. Choi, Gene-Wei Li, Huiyi Chen, Mohan Babu, Jeremy Hearn, Andrew Emili, and X. Sunney Xie. "Quantifying E. coli proteome and transcriptome with single-molecule sensitivity in single cells." *Science* 329, no. 5991 (2010): 533-538.
4. Dar, Roy D., Brandon S. Razooky, Abhyudai Singh, Thomas V. Trimeloni, James M. McCollum, Chris D. Cox, Michael L. Simpson, and Leor S. Weinberger. "Transcriptional burst frequency and burst size are equally modulated across the human genome." *Proceedings of the National Academy of Sciences* 109, no. 43 (2012): 17454-17459.
5. Hensel, Zach, Haidong Feng, Bo Han, Christine Hatem, Jin Wang, and Jie Xiao. "Stochastic expression dynamics of a transcription factor revealed by single-molecule noise analysis." *Nature structural & molecular biology* 19, no. 8 (2012): 797-802.
6. Levine, Joe H., Yihan Lin, and Michael B. Elowitz. "Functional roles of pulsing in genetic circuits." *Science* 342, no. 6163 (2013): 1193-1200.
7. Carey, Lucas B., David Van Dijk, Peter MA Sloot, Jaap A. Kaandorp, and Eran Segal. "Promoter sequence determines the relationship between expression level and noise." *PLoS biology* 11, no. 4 (2013): e1001528.
8. Weinberger, Leor S., John C. Burnett, Jared E. Toettcher, Adam P. Arkin, and David V. Schaffer. "Stochastic gene expression in a lentiviral positive-feedback loop: HIV-1 Tat fluctuations drive phenotypic diversity." *Cell* 122, no. 2 (2005): 169-182.
9. Weinberger, Leor S., and Thomas Shenk. "An HIV feedback resistor: auto-regulatory circuit deactivator and noise buffer." *PLoS biology* 5, no. 1 (2006): e9.
10. Weinberger, Leor S., Roy D. Dar, and Michael L. Simpson. "Transient-mediated fate determination in a transcriptional circuit of HIV." *Nature genetics* 40, no. 4 (2008): 466-470.
11. Dar, Roy D., Brandon S. Razooky, Abhyudai Singh, Thomas V. Trimeloni, James M. McCollum, Chris D. Cox, Michael L. Simpson, and Leor S. Weinberger. "Transcriptional burst frequency and burst size are equally modulated across the human genome." *Proceedings of the National Academy of Sciences* 109, no. 43 (2012): 17454-17459.
12. Weinberger, Ariel D., and Leor S. Weinberger. "Stochastic fate selection in HIV-infected patients." *Cell* 155, no. 3 (2013): 497-499.
13. Kepler, Thomas B., and Timothy C. Elston. "Stochasticity in transcriptional regulation: origins, consequences, and mathematical representations."*Biophysical journal* 81, no. 6 (2001): 3116-3136.
14. Simpson, Michael L., Chris D. Cox, and Gary S. Sayler. "Frequency domain chemical Langevin analysis of stochasticity in gene transcriptional regulation."*Journal of theoretical biology* 229, no. 3 (2004): 383-394.
15. Paulsson, Johan. "Models of stochastic gene expression." *Physics of life reviews* 2, no. 2 (2005): 157-175.
16. Pedraza, Juan M., and Johan Paulsson. "Effects of molecular memory and bursting on fluctuations in gene expression." *Science* 319, no. 5861 (2008): 339-343.
17. Elgart, Vlad, Tao Jia, and Rahul V. Kulkarni. "Applications of Littleâ??s Law to stochastic models of gene expression." *Physical Review E* 82, no. 2 (2010): 021901.
18. Jia, Tao, and Rahul V. Kulkarni. "Post-transcriptional regulation of noise in protein

distributions during gene expression." *Physical review letters* 105, no. 1 (2010): 018101.
19. Chong, Shasha, Chongyi Chen, Hao Ge, and X. Sunney Xie. "Mechanism of transcriptional bursting in bacteria." *Cell* 158, no. 2 (2014): 314-326.
20. Bialek, William, and Sima Setayeshgar. "Physical limits to biochemical signaling." *Proceedings of the National* Academy of Sciences of the United States of America 102, no. 29 (2005): 10040-10045.
21. Raj, Arjun, Patrick Van Den Bogaard, Scott A. Rifkin, Alexander Van Oudenaarden, and Sanjay Tyagi. "Imaging individual mRNA molecules using multiple singly labeled probes." *Nature methods* 5, no. 10 (2008): 877-879.
22. Raser, Jonathan M., and Erin K. O'Shea. "Control of stochasticity in eukaryotic gene expression." *Science* 304, no. 5678 (2004): 1811-1814.
23. Golding, Ido, Johan Paulsson, Scott M. Zawilski, and Edward C. Cox. "Real-time kinetics of gene activity in individual bacteria." *Cell* 123, no. 6 (2005): 1025-1036.
24. Zenklusen, Daniel, Daniel R. Larson, and Robert H. Singer. "Single-RNA counting reveals alternative modes of gene expression in yeast." *Nature structural & molecular biology* 15, no. 12 (2008): 1263-1271.
25. Blake, William J., Mads KÃ¦rn, Charles R. Cantor, and James J. Collins. "Noise in eukaryotic gene expression." *Nature* 422, no. 6932 (2003): 633-637.
26. Zimmerman, Steven B., and Stefan O. Trach. "Estimation of macromolecule concentrations and excluded volume effects for the cytoplasm of Escherichia coli." *Journal of molecular biology* 222, no. 3 (1991): 599-620.
27. Tan, Cheemeng, Saumya Saurabh, Marcel P. Bruchez, Russell Schwartz, and Philip LeDuc. "Molecular crowding shapes gene expression in synthetic cellular nanosystems." *Nature nanotechnology* 8, no. 8 (2013): 602-608.
28. Fowlkes, Jason D., and C. Patrick Collier. "Single-molecule mobility in confined and crowded femtolitre chambers." *Lab on a Chip* 13, no. 5 (2013): 877-885.
29. Meyer, B., E. Agliari, O. BÃ©nichou, and R. Voituriez. "Exact calculations of first-passage quantities on recursive networks." *Physical Review E* 85, no. 2 (2012): 026113.
30. van Zon, Jeroen S., Marco J. Morelli, Sorin TÇ?nase-Nicola, and Pieter Rein ten Wolde. "Diffusion of transcription factors can drastically enhance the noise in gene expression." *Biophysical journal* 91, no. 12 (2006): 4350-4367.

A Barcoding Strategy Enabling Higher-Throughput Library Screening By Microscopy

Chen, R., Imperial College London

Dramatic progress has been made in the design and build phases of the design-build-test cycle for engineering cells. However, the test phase usually limits throughput as many outputs of interest are not amenable to rapid analytical measurements. For example, phenotypes such as motility, morphology, and subcellular localization can be readily measured by microscopy, but analysis of these phenotypes is notoriously slow. To increase throughput, we developed microscopy-readable barcodes (MiCodes) composed of fluorescent proteins targeted to discernible organelles. In this system, a unique barcode can be genetically linked to each library member, making possible the parallel analysis of phenotypes of interest. As a first demonstration, we MiCoded a set of synthetic leucine zippers (SYNZIPs) to allow an 8x8 matrix to be tested for specific interactions in a pool. A novel microscopy-readable two-hybrid fluorescence localization assay for probing candidate interactions in the cytosol was also developed using a bait protein targeted to the peroxisome and a prey protein tagged with a fluorescent protein. This work introduces a generalizable, scalable platform for making microscopy amenable to higher-throughput library screening experiments, thereby coupling the power of imaging with the utility of combinatorial search paradigms.

Engineering artificial mechanosensitive cells by combining cell-free expression and ultrathin double emulsion template

Jin Woo Lee1,*, Kenneth K.Y. Ho1,*, Allen P. Liu1,2,3,4

1Mechanical Engineering Department, 2Biomedical Engineering Department, 3Cellular and

Molecular Biology program, 4Biophysics program, University of Michigan, Ann Arbor, MI 48109.
* contributed equally to this work
Assembling biological parts into a biologically functional system using a bottom-up *in vitro* reconstitution approach offers the possibility of designing artificial cells with the ability of sensing and responding to external stimuli. Artificial cells are defined as the encapsulation of biologically active material in a biological or synthetic membrane. Here, we describe a robust and general method to produce artificial cells for the purpose of mimicking one or more behaviors of a cell. A microfluidic double emulsion system is used to encapsulate a mammalian cell-free expression system that is able to express membrane proteins into the bilayer or soluble proteins inside the vesicles. The development of a robust platform that allows the assembly of artificial cells is valuable in understanding subcellular functions and emergent behaviors in a more cell-like environment as well as for creating novel signaling pathways to achieve specific cellular behaviors. As a test bed, we are engineering mechanosensitive vesicles to respond to the external physical environment. The platform development and inclusion of additional information processing components could potentially open up new applications in biosensing and medicine.

Structure and Assembly of the Cyanobacterial Carboxysome Matrix

Niederhuber, M., Harvard Medical School
Polka, J.,
Giessen, T.,
Lambert, T., Harvard Medical School
Silver, P. A., Harvard Medical School

Intracellular microcompartments are potentially useful tools for metabolic regulation and optimization. The cyanobacterial Î²-carboxysome is a natural microcompartment used to concentrate bicarbonate and carbon dioxide, while excluding oxygen, in the vicinity of the carbon fixing enzymes ribulose-1,5- bisphosphate carboxylase (RuBisCO) and carbonic anhydrase (CA). Our lab has previously shown that in *Synechococcus elongatus* Î²-carboxysomes assemble in an inside-to-outside manner, with RuBisCo aggregations budding off existing carboxysomes prior to encapsulation by an outer protein shell. Based on previous observations and known protein structures, it is hypothesized that this initial protein aggregation (termed the carboxysome matrix) has an ordered lattice-like structure and is a key aspect of Î²-carboxysome assembly, guiding both the final size and number of carboxysomes within the cell. Here we describe our current efforts to elucidate the spatial and temporal pattern of internal carboxysome matrix assembly using super resolution microscopy (SIM) and *in vitro* techniques. To this end we have established a Î²-cyanobacterial line expressing a fluorescently tagged copy of the inner matrix structural protein CcmM for *in vivo* imaging alongside fluorescently tagged RuBisCo and carboxysome shell proteins. We have also purified the short isoform of CcmM (M35), RuBisCo small and large subunits, and shell components for characterization with electron microscopy. We hope that by developing an improved picture of carboxysome self-assembly using these techniques, we will provide a useful foundation for producing synthetic protein-based microcompartments and engineering improved photosynthetic efficiency in plants and cyanobacteria.

Viral Based Design: Remodeling the Hippo Pathway from Tumor Suppression to Oncogenesis

Shanzer, M., The Weizmann Institute of Science
Reuven, N., The Weizmann Institute of Science
Shaul, Y., The Weizmann Institute of science

During the process of evolution viruses have learned to tailor cellular functions to their own needs. Due to the frugal nature of viral design, viral regulatory proteins are small, multipurpose, and efficient. For this reason, viruses curate a treasure-trove of molecular information for the synthetic biologist, providing a glimpse into a natural process of ?reverse-engineering? of cellular functions. The polyomavirus middle T antigen (PyMT) induces oncogenic transformation by hijacking cellular signaling effectors. Activation of the tyrosine kinase, c-Src, is a key step, imperative for viral oncogenesis. However, the underlying mechanism of Src activation by PyMT remains elusive. We addressed this fundamental question by examining the virus-host interaction between PyMT and the Hippo tumor suppressor pathway. The Hippo pathway is a major regulator of cellular proliferation. PyMT physically interacts with the Hippo pathway effector Taz (WWTR1). In proliferating cells Taz is nuclear and coactivates the pro-oncogenic TEAD transcription factors. Taz phosphorylation by Hippo pathway core kinase, Lats, leads to Taz nuclear exclusion and a consequent downregulation of TEAD coactivation. We found that Taz was required for transformation by PyMT in a counter-intuitive manner. Despite the cells undergoing oncogenic proliferation, Taz was exclusively cytoplasmic in the presence of PyMT, an observation expected to have rendered Taz deprived of its reported pro-proliferative function. Indeed, PyMT markedly inhibited TEAD coactivation by Taz, an event associated with tumor suppression. We resolved the molecular basis of the enigmatic Taz cytoplasmic retention by PyMT by demonstrating that PyMT activated the Lats tumor suppressor kinase leading to Taz nuclear exclusion. Next, we searched a role for cytosolic Taz in PyMT oncogenesis and identified Shp2, a phospho-tyrosine phosphatase, as an important player. Taz regulates Shp2 subcellular localization via physical interaction. PyMT induced cytosolic localization of Taz and Shp2. We demonstrate how PyMT designs a new network in oncogenesis where, counter-intuitively, the tumor suppressor Hippo pathway plays a critical role. We propose that understanding viruses is of great value in designing sophisticated synthetic modules.

Adaptation of Bacteriophage on Synthetic Transcription Machinery

Garry, D., UT Austin
Ellington, A., University of Texas at Austin

With the emergence of synthetic biology, it remains unclear if synthetic parts (i.e. altered activity, substrate specificity, orthogonality, etc.) can significantly alter the evolutionary trajectory of the host organism. In this study, T7 bacteriophages lacking their wild-type RNA polymerase (T7Δ1) were passaged on an engineered T7 RNA polymerase designed to recognize an orthogonal promoter. T7 RNA polymerase plays a key step in the life cycle to the bacteriophage as it controls expression of 17 critical promoters. We hypothesize that massive changes in the regulatory architecture would emerge to better accommodate the orthogonal RNA polymerase. After 100 passages on the orthogonal polymerase, we observed multiple changes in multiple promoters and the evolved strains had a higher overall fitness than the ancestor, reaching that of wild-type T7 bacteriophage. Additionally, the evolved strains were able to accommodate insertion of the orthogonal RNA polymerase back into their genomes, whereas the ancestor T7Δ1 was not. This provides an example of how simple design strategies may frequently require further evolutionary optimization. Our work raises the question whether synthetic biology can be used for laboratory speciation events and the development of orthogonal viruses.

Cellulose Production in *Escherichia coli*

Gizem Buldum, Alexander Bismarck, Athanasios Mantalaris

Abstractâ?? Over the last 30 years, attention has been paid to bacterial cellulose (BC). BC exhibits unique physical, chemical and mechanical properties when compared to plant-based cellulose, including high purity and biocompatibility. It is currently being used in the production of a wide variety of products such as paper, tissue- engineering products, electronics, biomedical devices, textiles, and food. Although *Acetobacter xylinum* is the most efficient producer of BC, it has many weaknesses such as long doubling time compared to most other bacteria, low productivity and its susceptibility to culture conditions. These limit the cost-effective production of BC. *E.coli* is an industrially attractive microorganism because it is a well- characterized host for protein expression and it has a significantly high growth rate. Herein, *E. coli* C41 (DE3) was selected as the host organism for the expression of bacterial cellulose synthase operon (*bcs*) of *A.xylinum*. The expression system was created using the pET-Duet1 carrying *bcs* operon (*bcs*ABCD) and pCDF carrying upstream of the bcs operon. The upstream region of the bcs operon contains two genes: *cmc*ax (encoding endo-Î²-1,4-glucanase) and *cpp*Ax (encoding cellulose complementing protein). All *bcs* genes were successfully transferred and expressed in *E.coli* C41 (DE3). The cells were cultured in LB medium at 37Ë?C and 180 rpm before IPTG induction. Following 0.05 mM IPTG induction, culture temperature shifted to 23Ë?C. After 24 hour culturing, cellulose production was obtained with a yield of 0.026 g/L. Cellulose samples were analyzed by FTIR spectra and scanning electron microscopy. SEM images clearly showed the shape and size distribution of the microfibrils. The microfibrils have lengths of more than 800-1000Âµm and diameters of the fibres are between 10â??20 Âµm. The expressions of *bcs* proteins were analysed by SDS page. The system stability was also monitored by plasmid stability test. The percentage of plasmid harboring cells was obtained 98% of the total viable cells even after 24-hr culturing. This proves that the system has been created for BC production is consistent. This study is novel in functional BC production in *E. coli* by the transfection of the cellulose production ability of *A. xylinum* and a big step to create valuable strains for the effective production of BC. BC productivity is affected by many factors such as the composition of the culture medium, environmental factors (pH, temperature, dissolved oxygen content) and the type of cultures used (culture flasks, agitated fermenters). In order to increase the yield of production in *E. coli*, culture conditions should be optimised by further studies. [1-3].

Keywordsâ?? Bacterial cellulose, *E.coli*, biopolymer, recombinant expression system

G.Buldum is with Imperial College London, Department of Chemical Engineering, London SW7 2AZ, UK (phone: +44 (0)20 7594 1224; e-mail: g.buldum11@ imperial.ac.uk).
A.Bismarck was with Imperial College London, SW7 2AZ, UK. He is now with Institute of Materials Chemistry and Research, Faculty of Chemistry, University of Vienna, WÃ¤hringer Strasse 42, 1090 , Vienna , Austria (phone:
+43 (1) 4277 71301; e-mail: alexander.bismarck@univie.ac.at) A.Mantalaris is with the Imperial College London, Department of
Chemical Engineering, SW7 2AZ, London UK (phone: +44 (0)20 7594 5601;

e-mail: a.mantalaris@ imperial.ac.uk).
REFERENCES
[1] K-Y. Lee, G. Buldum, A. Mantalaris, A. Bismarck, More than meets the eye in bacterial cellulose: biosynthesis, bioprocessing, and applications in advanced fiber composites. *Macromol Biosci*, 2014, 14(1):10â??32.
[2] S. Kawano, K. Tajima, Y. Uemori, H. Yamashita, T. Erata, M. Munekata, M. Takai, Cloning of cellulose synthesis related genes from Acetobacter xylinum ATCC23769 and ATCC53582: Comparison of cellulose synthetic ability between strains. *DNA Res*. 9, 2002, 149â??156.
[3] P. Ross, R. Mayer, M. Benziman, Cellulose biosynthesis and function in bacteria. *Microbiol Rev*, 1991, 55:35â??58.

Crispr-Based Genome Editing Tools: New Applications and Streamlined Workflows

Chesnut, J.,

CRISPR-Cas9 systems provide a platform for high efficiency genome editing that can lead to innovative applications in cell engineering. However, the delivery of Cas9 and synthesis of guide RNA (gRNA) remain two steps that limit overall efficiency and general ease of use. Here we describe novel methods for rapid synthesis of gRNA and for delivery of Cas9 protein/gRNA complexes into a variety of cells through liposome-mediated transfection or electroporation. We will present a streamlined cell engineering workflow that goes from gRNA design to analysis of edited cells in as little as three days and results in highly efficient genome editing and biallelic knockout of multiple genes in hard-to-transfect cells. The reagent preparation and delivery to cells requires no plasmid manipulation so is amenable for high throughput, multiplexed genome-wide cell engineering.

Further, we will show data using lentivirus-based CRISPR delivery for high-throughput screening of mammalian cell populations. Until recently RNAi has been the main tool for performing loss of function studies in mammalian cell populations. Using CRISPR technology, we are now able to perform complete gene knock-out studies which hold promise for clearer phenotypes and fewer false readouts as what is currently seen with the variable knock-down of expression using RNAi. We are creating gene family-specific arrayed libraries of CRISPR-lenti particles. These new tools enable high throughput, arrayed gene knockout screens using various cell types, including our proprietary CellSensor lines that give a fluorescent readout of various signaling pathway activities.

These two CRISPR-based gene-editing platforms represent the latest in the rapid evolution of editing tools for mammalian genomes by simplifying and increasing the cell engineering workflow and providing a pre-designed, ready to use platform for efficient compound screening in mammalian cell lines.

Development and Characterization of a Nucleoside Kinase That Accepts Unnatural Deoxyribonucleoside P, a Part of an Expanded DNA

Matsuura, M. F., The Westheimer Institute for Science and Technology
Shaw, R. W., Foundation for Applied Molecular Evolution
Moussatche, P., Firebird Biomolecular Sciences LLC
Fajardo, D. S., Firebird Biomolecular Sciences LLC
Moses, J. D., The Westheimer Institute for Science and Technology
Benner, S. A., The Westheimer Institute for Science and Technology

One of the more ambitious long-term goals of synthetic biology seeks to construct living cells that accept and use biopolymers different from DNA, RNA, and proteins. For example, the Romesberg group at The Scripps Research Institute reported last year that a strain of *Escherichia coli* could maintain an unnatural nucleotide pair whose structure greatly deviates from the standard Watson-Crick geometry. Since the Scripps group could not obtain kinases to convert
fed nucleosides intracellularly to the triphosphates needed for DNA replication, they instead altered the *E. coli* to have transport systems that transported their nucleoside triphosphate analogs from the growth medium into the cell directly.

Our approach is different. We seek to construct a host cell that creates the triphosphates of components of an unnatural genetic system inside of the cell. This would allow the relatively inexpensive (and much more stable) nucleosides to be added to the growth medium, where we have now shown that natural nucleoside transporters bring them inside of the cells. This strategy requires (a) that the structure of the added nucleotides not differ too greatly from the structures of the standard nucleotides, so that (b) we can engineer standard nucleoside and nucleotide kinases to accept them in a synthetic metabolic pathway that makes the triphosphates intracellularly.

Over the past five years, we have developed a set of alternative nucleotides that create a 12 letter artificially expanded genetic information system (AEGIS) (**Fig. 1**). AEGIS pairs are joined by alternative patterns of hydrogen bond donor and acceptor groups. In my work, I have examined one of these AEGIS components, 2-amino-8-(1â??-?-D-2â??-deoxyribofuranosyl)-imidazo[1,2-a]-1,3,5-triazin-4(8H)-one, trivially called **P**. DNA polymerases, RNA polymerases, and reverse transcriptases have all been developed to copy and transcribe DNA and RNA containing **P** and its pair, 6-amino-5-nitro-3-(1â??-?-D-2â??-deoxyribofuranosyl)-2(1H)-pyridone (trivially called **Z**).

A nucleoside kinase from *Drosophila melanogaster* (*Dm*dNK) was engineered to accept unnatural nucleoside d**P** to produce unnatural nucleoside monophosphates d**P**MP. This is the first phosphorylation step to synthesize the corresponding nucleoside triphosphate d**P**TP, a substrate for DNA polymerases.

In this study, the activity of *Dm*dNK Q81E mutant was measured in a three-step coupled-enzyme assay [1] and its kinetic parameters were calculated. These results showed that the Q81E
mutation allowed the engineered kinase to accept d**P** as a substrate. The Michaelis constant

(KM), the turnover rate (kcat), and specificity constant (kcat /KM) for d**P** were similar to those for standard nucleosides dA, dT, dC.

To formally complete the construction of an artificial metabolism *in vitro*, I am also developing nucleoside monophosphate kinases (NMPK) to phosphorylate d**P**MP to give d**P**DP. I will describe work adding these to nucleoside diphosphate kinase (NDPK), which I have already shown phosphorylate d**P**DP to give d**P**TP, the third phosphorylation step to synthesize triphosphate substrates for DNA polymerases. Once transferred into living cells, these enzymes should allow the formation of d**P**TP *in vivo*, and represent an example of a standard problem-solution in synthetic biology.

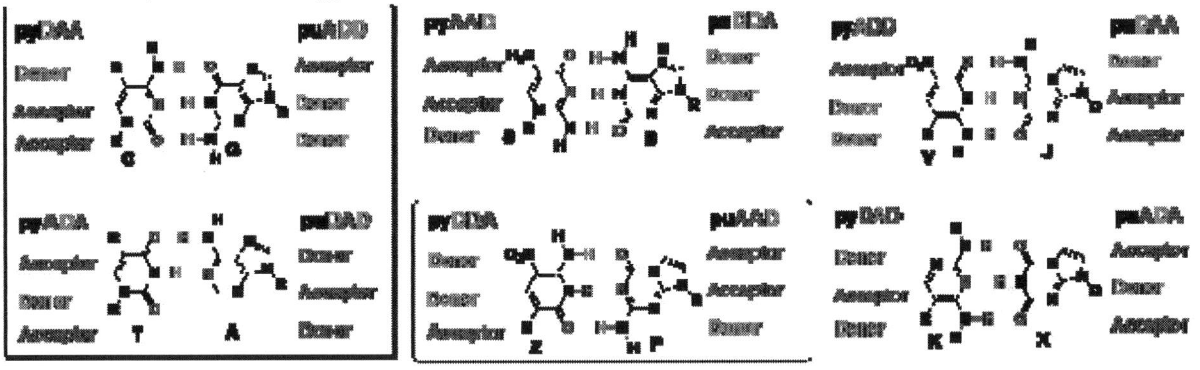

Figure 1. The expanded genetic alphabet is created by shuffling hydrogen bond donor and acceptor groups within the Watson-Crick pairing geometry. Standard base are in the box on the left. The **Z**:**P** pair is boxed in the center. The similarity between **Z** and **P** and natural nucleotides allows us to solve the metabolism problem that the Scripps group left unsolved. Also, once the **Z**:**P** pair is in place, the strategy allows further expansion to add three more base pairs.

Reference

[1] Gerth, M. L.; Lutz, S. (2007) Mutagenesis of non-conserved active site residues improves the activity and narrows the specificity of human thymidine kinase 2. *Biochem. Biophys. Res. Commun.*, *354*, 3: 802-807.

2-Quinoxalinecarboxylic Acid Production in Escherichia coli

Kim, C. Y., National University of Singapore

Quinoxaline, also known as benzopyrazine, is an important pharmaceutical and industrial compound. Despite their simple chemical structure, quinoxaline and quinoxaline derivatives display a myriad of useful therapeutic properties including antibacterial, antiviral, antiparasitic, and antitumor activities. They also possess unique electronic and luminescent properties and are used in the production of organic semiconductors, light emitting diodes, and solar panels. The most common method for large scale production of quinoxaline compounds is condensing benzene-1,2-diamine with 1,2-diketones, by heating for several hours in organic solvent. It is therefore desirable to develop a biological method for quinoxaline preparation that does not require use of solvents and does not produce toxic waste products. We are engineering *E. coli* to produce 2-quinoxalinecarboxylic acid. In particular, we are using the native Sec pathway of *E. coli* to export the biosynthesized 2-quinoxalinecarboxylic acid.

Genome Engineering to Support Rational Design for Metabolic Engineering in Bacillus Subtilis

Planson, A. G., INRA
Mangan, M., INRA
Noirot, P., INRA
Dervyn, E., INRA
Jules, M., INRA

Genome engineering to support rational design for metabolic engineering in *Bacillus subtilis* Anne-GaÃ«lle Planson, Michael Mangan, Philippe Noirot, Etienne Dervyn, Matthieu Jules Micalis Institute, INRA, AgroParisTech, Jouy-en-Josas, France Genome engineering is a rapidly growing field of microbial research but still remains limiting when
developing *Bacillus subtilis*-based cell factories. *B. subtilis* is the model organism for the Gram- positive bacteria, is GRAS (Generally Regarded As Safe) and is heavily used in industry. We successfully designed and developed *B. subtilis* chassis strains lacking functions such as sporulation, antibiotic resistance, mobile DNA elements, prophages, isozymes while retaining deletion making functions. This latter work represents a proof-of-concept for strain development for biotechnologies. To develop adapted chassis strains it is critical to be able to modify the genome "*Ã la carte*". We are currently developing a set of novel recombineering methodologies based on single strand annealing proteins (SSAP) in *B. subtilis*. Such recombinases mediating oligonucleotide-targeted mutagenesis will serve to introduce point mutations in multiple *loci*, enabling simultaneous modification of several chromosomal targets. We are also developing a CRISPR-Cas system for *B. subtilis* using the inactivated Cas9 as an interference system for gene expression modulation. Altogether this will result in an integrated toolbox for synthetic biology and metabolic engineering. These new methods for rapid introduction of multiple point mutations combined with the interference system should lead to the development of robust and high-performing bacterial cell factories.

Isolation and Charactization of Novel Terpene Synthases from Diverse Metagenomic Libraries

Crook, N., The University of Texas at Austin
Yoneda, A., Washington University in Saint Louis
Dantas, G., Washington University in Saint Louis

Nature produces a stunning diversity of molecular functionality, of which humanity has only exploited a small fraction for industrial and energy applications. To date, chemical space has been confined to that enabled by petroleum feedstocks or abundant natural products, in turn putting strong constraints on achievable material properties. Terpenoids are one of the most abundant and diverse classes of biomolecules produced in nature, collectively enabling a variety of therapeutic, energy, and cosmetic applications. However, this class of molecules remains recalcitrant to synthetic chemistry approaches, necessitating the use living systems. Although there has been recent outstanding progress in the high-level microbial production of several key terpenoids, much of the remaining molecular diversity is locked within production hosts which produce small product titers and are expensive to cultivate. This limitation is primarily due to a lack of diverse biological parts, as current databases of protein function are currently populated by proteins which conform to well-understood sequence-function relationships. Interestingly, despite this known diversity, recent genomics investigations have shown a vast untapped reservoir of bacterial terpene synthases residing in the genomes of uncultivable organisms living in the soil, indicating a vast array of terpenoids waiting to be discovered and offering a wider palette of biological parts for the production of these compounds in robust microbial hosts. However, in order to validate these predictions and expand our catalogs of useful biocatalysts, techniques for high-throughput enzyme discovery, characterization, and annotation are urgently needed. In this work, we exploited the toxicity of a key node in the terpenoid biosynthesis pathway in order to develop a high-throughput approach for discovery of terpene synthase activity. We first develop optimized culture conditions and screening protocols in order to maximize enrichment of terpene synthases. Then, we experimentally mine hundreds of gigabases of genetic material isolated from diverse habitats in order to discover novel biocatalysts with the ability to produce new classes of terpenoid molecules in *E. coli*. After biochemical characterization of terpenoid products generated by these novel enzymes, we then perform bioinformatics analysis to generate improved predictors of terpene synthase function and use these predictors in a focused approach to experimentally investigate terpene synthase activity in sequenced microbial genomes.

Anti-Virulent Disruption of Pathogenic Biofilms Using Synthetic Quorum-Based Technology

Yew, W. S.,

The rapid emergence of multi-drug resistant bacteria has accelerated the need for novel therapeutic approaches to counter life-threatening infections. The persistence of bacterial infection is often associated with quorum-sensing-mediated biofilm formation. Thus, the disruption of this signaling circuit presents an attractive anti-virulence strategy. Quorum-quenching lactonases have been reported to be effective disrupters of quorum-sensing circuits. However, there have been very few reports of the effective use of these enzymes in disrupting bacterial biofilm formation. Here, we present the use of Synthetic Quorum-Based Technology to disrupt biofilm formation in a clinically relevant *A. baumannii* S1 strain through the use of an engineered quorum-quenching lactonase. *Acinetobacter baumannii* is a major human pathogen associated with serious hospital-acquired infections, and its virulence is attributed predominantly to the persistence of its biofilms. The engineered lactonase treatment achieved significant *A. baumannii* S1 biofilm reduction. This study also showed the possibility of using engineered quorum-quenching enzymes in future treatment of biofilm-mediated bacterial diseases.

Decoupling Growth and Metabolism in Escherichia coli

Chubukov, V.,
Mukhopadhyay, A., Joint Bioenergy Institute
Sauer, U., ETH Zurich
Keasling, J. D., Joint Bioenergy Institute

Microbial chemical production involves an intrinsic tradeoff between biomass formation and product formation. Theoretically, the ideal resolution of this tradeoff is production during stationary phase, in which growth is limited but carbon is nevertheless abundant. However, microbes typically respond to such conditions by shutting down central metabolism. Decoupling growth and metabolism is therefore one of the central problems in host engineering for synthetic biology. In this study, we characterize the metabolism of the model bacterium *Escherichia coli* under a variety of stationary phase conditions and show significant variation in metabolic activity that strongly depends on the limiting nutrient. We find an unusual phenotype of high glucose uptake rates under conditions of magnesium limitation, suggesting a lack of intrinsic barriers to enhancing metabolic activity under other starvation conditions. Based on quantitative metabolomics, we suggest a role for phosphoenolpyruvate in regulating glucose uptake. We further focus on the condition of nitrogen limitation, where a key regulatory interaction is the inhibition of the glucose uptake enzyme PtsI by accumulation of alpha-ketoglutarate. We find that by manipulating the expression of PtsI, we can enhance metabolic rates under nitrogen starvation, and we investigate the effect of this enhancement on the productivity of several biofuel pathways. We will also discuss how high-throughput and single-cell based screening and selection approaches will be used to further enhance stationary phase productivity for renewable biochemical and biofuel production.

In Vivo Characterization of Stress-Responsive Promoters in the Normal and Inflamed Murine Gastrointestinal Tract

Kotula, J. W., Synlogic, Inc.
Millet, Y., Synlogic, Inc.
Isabella, V. M., Synlogic, Inc.
Falb, D., Synlogic, Inc.
Anderson, C., Synlogic, Inc.
Silva, A., Synlogic, Inc.
Miller, P., Synlogic, Inc.

Research in synthetic biology has led to the generation of various biological systems using specific promoter sequences for the activation of gene expression in response to particular environmental or chemical stimuli. Those systems will likely be key in the development of synthetic therapeutic organisms requiring a precise spatiotemporal control of their activation in patients suffering from the targeted medical syndromes. Thus far, however, all of the studies describing those biological systems have been performed *in vitro*, and additional information on their behavior in animal disease models is needed to confirm their utility for potential therapeutic applications. Using the well-known probiotic *E. coli* Nissle strain carrying promoter-fluorescent protein reporter constructs, we conducted the first comprehensive *in vivo* study of low-oxygen, oxidative stress and nitrosative stress-responsive promoters in the gastrointestinal tract of healthy mice and murine IBD (Inflammatory Bowel Disease) models. Our data are providing valuable insights into the kinetics, fine localization, and level of those promoters' activities along the mouse GI tract and how they differ between a normal and inflamed gut. In parallel, our studies are providing important information regarding the environmental stresses synthetic therapeutic organisms would face in the context of IBD, a knowledge that will play a critical role in engineering more efficient microbial therapeutic agents.

Mining for novel regulatory noncoding RNA in bacteria

Jeremy I. Marcus1, Soha Hassoun1, **Nikhil U. Nair**2

1Department of Computer Science

2Department of Chemical & Biological Engineering

Tufts University, Medford, MA 02155
Produced during futile attempts by the RNA polymerases to clear the promoter region and transition to transcriptional elongation phase, the synthesis of â??abortiveâ? RNAs have been known for decades in all kingdoms of life. However, until the recent identification of abortive RNA in the bacteriophage T7 with transcriptional antitermination function, they had been considered only as â??junkâ? RNAs. Functional abortive RNAs have not, however, been identified to exist endogenously in more complex biological systems such as bacteria. In this work, we describe an *in silico* predictive approach to identify possible new regulatory noncoding RNAs in bacteria and evaluate whether these abortive RNAs might be able to serve biological functions in *E. coli*. Abortive RNAs generated from 3,780 transcriptional units (obtained from RegulonDB) were used as query sequences within their respective transcription units to search for possible binding sites. Sites that fell within known regulatory features, based upon data obtained from RegulonDB and WebGeSTerDB, were then ranked based upon the standard free energy of annealing (reported by UNAfold) of the transcript to the binding site. Matches with a sufficiently high standard free energy of annealing were then explored further as likely sites for abortive RNA- mediated regulation. We will showcase one such result â?? the uncovering of a potentially novel abortive RNA-mediated regulation of an alcohol dehydrogenase in *E. coli*. We also discuss the potential application of using such functional abortive RNAs in a design-based approach to regulating gene expression in synthetic systems. Future experimental validation of this and other loci may reveal the pervasive physiological role of this new class of noncoding RNAs in *E. coli*.

Programmable Control of Acetate Metabolism Escherichia coli

Moser, F.,

Acetate production in Escherichia coli fermentations reduces the total yield and quality of recombinant protein products. Acetate production occurs due to metabolic overflow in the presence of excess glucose, is increased when oxygen levels are low, and is facilitated by the phosphate acetyltransferase (*pta*) gene. We have engineered a genetic program that measures levels of glucose, oxygen, and acetate in the cell, combines the information into two Boolean AND gates, and produces an output that reduces acetate accumulation in batch cultures of E. coli. The output is composed of three modules, each regulating the expression of *pta* at different levels of the central dogma. A CRISPRi guide RNA module reduces the transcription of the gene, an interfering RNA that reduces translation, and a LON protease that degrades the enzyme. Together, these outputs enable real-time control of *pta* expression in response to conditions that lead to acetate formation in *E. coli*.

Multiplexed Screens of Olfactory Receptor Binding

Jones, E. M., University of California Los Angeles
Kosuri, S., University of California, Los Angeles

Animal olfaction is extremely sensitive and precise mechanism for chemical detection with the ability to distinguish very dilute compounds from complex mixtures. This sense is mediated by molecules binding to one or more olfactory receptors (ORs), which are a subtype of G protein-coupled receptors (GPCRs) that mediate signal transduction by stimulating cyclic AMP (cAMP) production. Current methods for monitoring OR activation are either qualitative or rely on luminescent reporters not practical for screening thousands of receptors against thousands of ligands. Advancements in next generation sequencing and genome editing techniques allow us to develop a novel platform to measure multiplexed responses from a library of engineered human cells each clonally expressing unique ORs on the genome using landing pad integration. Receptor activation is linked to a cAMP responsive reporter expressing a barcode that identifies each OR and measure quantitatively using RNAseq. Apart from chemical sensing, the system enables us to pursue a variety of applications such as: deciphering functional differences in ORs among populations, screening drugs that interact with G protein coupled receptors, and engineering receptors with novel specificities.

Replicable Single-stranded DNA Origami

Dongran Han, Cameron Myhrvold, Peng Yin
Department of Systems Biology, Harvard Medical School
Wyss Institute for Biologically Inspired Engineering at Harvard

DNA has been used to create a variety of complex nanoscale shapes and devices since the birth of DNA nanotechnology in 1982. One main challenge is to create complex nanostructures from a single-strand of DNA, like proteins which typically fold from a single polymer. Here we report the design and synthesis of single-stranded DNA origami structures (ssOrigami) containing up to 4,000 nucleotides, which can fold into designed shapes. Such ssOrigami can be easily replicated by polymerases *in vitro* and *in vivo*. ssOrigami structures can also be used as a template for amplification by PCR.

In previous studies, multiple DNA strands were successfully designed to self-assemble into complex structures with or without the help of a long scaffold strand. However, biological macromolecules, such as mRNAs and proteins, typically fold from a single polymer into a well-defined structure. The ability to fold DNA nanostructures in this manner would enable robust assembly and even replication of such structures from a unimolecular template. However, folding complex nanostructures from a single molecule of single-stranded DNA (ssDNA) still remains challenging for the field of DNA nanotechnology, due to the complexity of the topology in most existing strategies and the difficulty of large-scale clonal production of suitable DNA sequences.

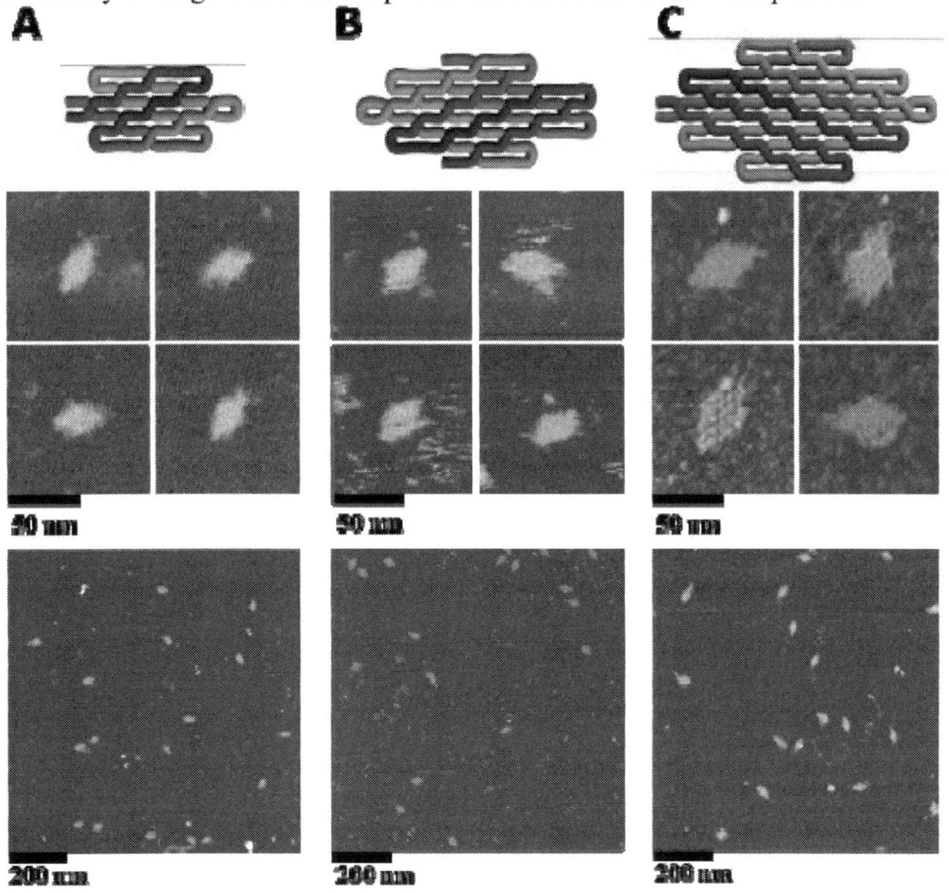

Figure 1. ssOrigami structures with different sizes.

In our study, ssOrigami structures with different sizes ranging from 1,000 to 4,000 nucleotides (nt) were assembled. Three diamond shape ssOrigami containing 1k, 1.6k and 2.3k nt were demonstrated (Figure. 1
A-C), and they folded with high yield.
More complicated and larger ssOrigami structures were also achieved (Figure 2). While the folding path is less obvious than simple diamond shapes, these complex ssOrigami structure were also successfully assembled with good quality. ssOrigami of up to 4,000 nt in size were demonstrated (Figure 2 C, D), which are larger than 99.99% of all proteins in the human proteome and three times the size of the largest catalytic rRNA (16S rRNA). This result will greatly broaden our vision and understanding about the self-assembly capability of macro-biomolecules, especially when such a procedure can happen during a short (2 hour)
annealing step.

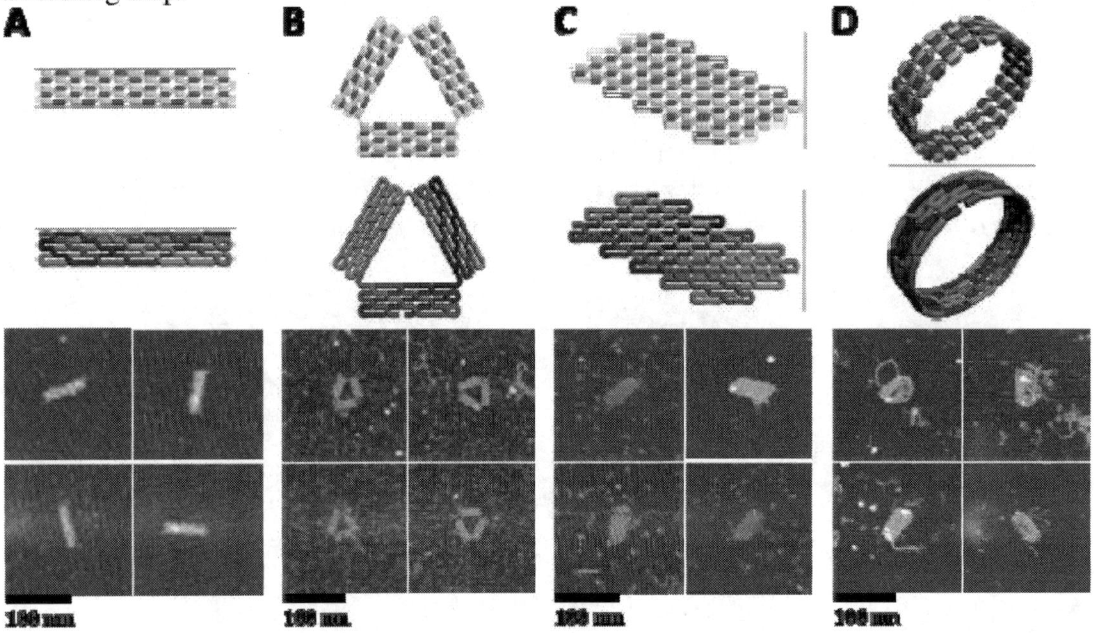

Figure 2. ssOrigami structures with different shapes.

Unlike multi-stranded DNA systems such as DNA origami and DNA bricks, which contain hundreds of distinct components and assemble with undesirable defects and heterogeneity, ssOrigami is a homogenous,
â??pureâ?? system. Additionally, ssOrigami can be easily replicated and amplified by polymerases *in vitro* and *in vivo,* which can greatly decrease the cost of DNA nanotechnology, especially when large amount of DNA nanostructures are required. Finally, ssOrigami structures can be used as a template for amplification by PCR.

Simplexes Characterize the Coupling in Gene Expression Due to Competition for Rnap and Ribosomes

Gyorgy, A., MIT
Del Vecchio, D.,

The limited availability of shared cellular resources couples the expressions of any two genes. This, among other factors, renders the behavior of gene circuits context-dependent. As a result, our ability of predicting the emergent behavior of a complex gene network from that of the composing modules remains limited, posing a challenge in systems and synthetic biology. To overcome this limitation, here we characterize the coupling among protein concentrations due to competition for the shared transcriptional/translational machinery in gene networks with arbitrary topology.

In particular, we show that the set of protein concentrations that are simultaneously realizable is given by the intersection of simplexes. Exploiting these simplexes, we further characterize how changing the concentration of one protein affects the concentration of other proteins, even in the absence of regulatory linkages among them. The fact that the vertices of these simplexes are straightforward to measure experimentally renders this approach easy to employ in practice.

This simplex-based method aids the rational and predictable design of complex biocircuits as follows. First, it provides a simple tool to check whether the desired behavior of a large gene circuit is realizable. If the transcriptional/translational resources are insufficient, the simplexes aid the partitioning of circuits into smaller components (subnetworks), each realizable within a single cell. Second, we can predict how tuning various biochemical parameters affect the strength of interdependence among protein concentrations. Therefore, we can mitigate the coupling among components, facilitating the modular design of complex gene circuits. Finally, standard pathway optimization problems, such as maximizing the concentration of protein complexes or maximizing the rate at which a reaction occurs, can be formulated as convex optimization problems. These problems can be naturally interpreted in the context of economics, giving rise to concepts such as the price and utility of proteins.

Metabolic Engineering of Saccharomyces Cerevisiae for Production of Fatty Acid Short-Chain Alkyl Esters

Teo, W. S., National University of Singapore
Chang, M. W., National University of Singapore

Biodiesel is a mixture of fatty acid short-chain alkyl esters obtained via transesterification of vegetable oils with an alcohol. Since methanol is a cheap alcohol, biodiesels are more commonly found as methyl esters. As a more renewable and scalable biodiesel production method, microbes have been genetically engineered to produce fatty acid ethyl esters from biomass. While fatty acid (methyl and ethyl) esters are useful alternative diesel fuels, fatty acid esters with longer chain alcohol moiety, and with branching have better fuel properties. Here, we engineered yeast to produce fatty acid short-chain esters, including ethyl, isobutyl and isoamyl esters using endogenously synthesized fatty acids and alcohols. Two wax ester synthases were identified, cloned and expressed. Both enzymes were found to catalyse the formation of fatty acid short-chain esters, with different alcohol preference. In order to boost the ability of yeast in producing the aforementioned esters, multiple gene disruptions were carried out to increase flux towards fatty acyl-CoAs. In addition, as native production of isobutanol is insufficient and rate-limiting in yeast, isobutanol pathway genes targeted into the mitochondria were overexpressed. By combining these engineering strategies, we successfully engineered yeast metabolism to produce fatty acid short-chain esters, including ethyl, isobutyl and isoamyl esters

Engineered Antibiotic Marker-Free Probiotic Dual Expression Chassis

Koh, E., National University of Singapore
Hwang, I. Y., National University of Singapore
Chang, M., National University of Singapore

Probiotics present an attractive strategy for long-term delivery of therapeutic molecules *in vivo* as they may colonise as part of the human microbiota. Synthetic biology genetic circuits that facilitate the expression of these therapeutic molecules are usually stabilized in cells on plasmid by antibiotic resistance genes as selection markers. However, antibiotic resistance genes are not suitable for *in vivo* application as constant antibiotic selection pressure is required for plasmid retention. Moreover, the presence of antibiotic resistance genes introduces the risk of potential horizontal transfer of resistance to the native microbiome. In this study, we engineered an antibiotic marker-free dual expression system in a probiotic strain, *Escsherichia coli* Nissle. The essential genes involved in alanine and aspartate metabolisms were deleted from the bacterial genome to create a double auxotrophic *E. coli* strain. The auxotrophic phenotypes could be rescued by the balanced co-expression of the deleted genes on two separate plasmids to achieve plasmid selection and retention. To further enhance the dual plasmid system, a regulatory switch was implemented on the two co-expressed plasmids to ensure their stability. Our study demonstrates that the co-expressed plasmids could be maintained in the probiotic expression chassis without antibiotic selection. This dual expression system selected for auxotrophic complementation can be used for gastrointestinal delivery of therapeutic molecules *in vivo* with a high degree of biosafety.

Developing a DNA Polymerase-Based Biosensor for Neural Recording

de Paz, A. M., Northwestern University
Cybulski, T. R., Rehabilitation Institute of Chicago
Boyden, E. S.,
Church, G. M., Harvard University
Kording, K. P., Northwestern University
Tyo, K. E. J., Northwestern University

Progress in neural recording is critical to better understand the brain and to evolve treatments for brain-related illnesses. Further advances in mapping brain activity are limited by the low spatiotemporal resolution of current technologies. Our objective is to create a molecular recording device that combines the rapid writing capability of DNA polymerases with the robust information storage capability of DNA to generate dense spatiotemporal data on neural activity. Such a recording device, combined with the advent of highly parallel next generation sequencing technologies, has the potential to simultaneously monitor neurons at time scales matching mental activity1-3.

Ca2+ fluctuations on the millisecond timescale are indicative of neuron firing. We aim to engineer a DNA polymerase-based biosensor that responds to high calcium by writing errors into DNA. By coupling the nucleotide misincorporation rate of a DNA polymerase with levels of Ca2+ in the environment we can read out the relative Ca+2 concentration at a particular time by sequencing the copied DNA1-3.

As a first step toward building a Ca2+ biosensor, we are interested in understanding the metal cation-polymerase fidelity landscape to determine strategies for engineering DNA polymerase recorders. In order to screen metal-dependent fidelity properties of a multitude of DNA polymerases, we have developed a novel medium-throughput method for rapidly identifying changes in DNA polymerase error rate across a wide spectrum of metal cation and DNA sequence context conditions. Unlike former methods for determining polymerase fidelity, which rely on bacterial transformations4 or gel electrophoresis5, this *in vitro* technique is scalable (100 conditions/day) and allows for standardization of fidelity changes. Additionally, since DNA polymerase properties are difficult to study in their full complexity, this screen will allow us to decouple the singular and combinatorial effects of environment and sequence context on inducing replication errors.

Using this screen, we will identify polymerase candidates, environmental conditions and DNA template sequences that facilitate recording metal cation time course information via changes in fidelity. We expect these results to catalyze synthetic biology efforts toward a high-performance metal cation sensing and storage system based on incorrect nucleotides being incorporated into DNA. This first demonstration of a divalent metal recording platform will inform future development of a Ca2+-specific biosensor for neural recording applications.

Works Cited

1 Kording, K. P. Of toasters and molecular ticker tapes. *PLoS Comput Biol* **7**, e1002291, doi:10.1371/journal.pcbi.1002291 (2011).

2 Zamft, B. M. *et al.* Measuring cation dependent DNA polymerase fidelity landscapes by deep sequencing. *PLoS One* **7**, e43876,

doi:10.1371/journal.pone.0043876 (2012).
3 Glaser, J. I. *et al.* Statistical analysis of molecular signal recording. *PLoS Comput Biol* **9**, e1003145, doi:10.1371/journal.pcbi.1003145 (2013).

4 Bebenek, K. & Kunkel, T. A. Analyzing fidelity of DNA polymerases. *Methods Enzymol* **262**, 217-232 (1995).

5 Creighton, S., Bloom, L. B. & Goodman, M. F. Gel fidelity assay measuring nucleotide misinsertion, exonucleolytic proofreading, and lesion bypass efficiencies. *Methods Enzymol* **262**, 232-256 (1995).

*** Formatting Issue - Best Available Paper/Graphic ***

MinMCell: Designing Minimal Microbial Cells

Andrea PatanÃ¨*, Andrea Santoro*Â°, Jole Costanza**, Giuseppe Nicosia*

* Dept of Mathematics & Computer Science - University of Catania, Italy
**Center for Genomic Science of IIT@SEMM, Istituto Italiano di Tecnologia, Milan, Italy
Â° Scuola Superiore di Catania, Catania, Italy

A minimal cell can be viewed as the kernel of life. A minimal microbial cell would be a perfect framework to understand how the minimal parts (e.g., genes, pathways, networks, compartments) work together to make a living cell. The long-term goal is to try to understand which are the minimal biochemical objects required for life which will provide unexpected (and probably counterintuitive) insights into how cells function. More pragmatically, in this research work we developed a suite of tools and algorithms, *MinMCell*, to infer a minimal microbial cell; to this purpose we tackled three different organisms: *E. coli iJO1366*, *S. cerevisiae iTO977* and *M. genitalium iPS189*.

The goal of this research has been to identify a minimal set of essential genes needed to the cell
to divide and grow in pure culture.

In order to discover a minimized genome the designed approach is able to maintain the essential
genes and the lethal pairs of genes (genes encoding an essential function) and concurrently to erase
(knock-out) the non-essential genes.

We used the public databases [1-4] and the literature [5-10] in order to classify the essential genes and the lethal genes of the three analyzed organisms. The genes to knock-out have been identified by an ad-hoc algorithm for genome-scale re-design based on *Pareto optimality*.

For the *E. coli* in anaerobic condition, we obtained a minimal strain with 857 knock-out genes
with a biomass production of 0.554h-1. Moreover, inspecting the obtained Pareto Front it is possible to determine the strain with a maximum value of biomass (1.0334937h-1, basically the biomass value of the *wild type E. coli*) and a high number (but not the maximal one) of genes off 479; hence it is possible to re-design a bacterium to achieve the same biomass performance of the *wild type E. coli* erasing the
11,171% of the redundant and non-essential genes.

Overall, the designed method identified 35413 distinct feasible strains (the biomass values belong
to the following feasible set (0.05h-1, 1.0334937h-1]) and 28 Pareto optimal strains (non-dominated strains with respect the maximization of biomass and number of knock-out genes).

Similar results have been obtained for the other two organisms. For the *S. cerevisiae*, in anaerobic and carbon-limited conditions, we discovered a minimal set of 424 non-

essential genes that can be
knocked-out in order to the cell to divide and grow (biomass value of 1.800568e-13h-1). While for the

M. genitalium, in anaerobic condition, the algorithm detected 39 knout-out genes with a biomass value

of 0.707h-1.
The designed method doesnâ??t disrupt the essential protein-coding genes, lethal genes, genes that
encode proteins with potential redundant essential functions and genes encoding proteins of unknowns functions. For the three organisms the disruption of some genes accelerated bacterial growth. A comparative genome analysis completes the present study in order to detect the common essential genes.

References

[1] http://ecoliwiki.net/colipedia/index.php/Essential_genes
[2] http://www.genome.wisc.edu/resources/essential.htm
[3] http://www.ecogene.org/old/topic.php?topic_id=5
[4] http://www-sequence.stanford.edu/group/yeast_deletion_project/deletions3.html
[5] A. Joyce et al, *Experimental and computational assessment of conditionally essential genes in*

Escherichia coli, J Bacteriol 188:8259-71, 2006

[6] T. Baba et al, *Construction of Escherichia coli K-12 in-frame, single-gene knockout mutants: the*

Keio collection, Mol Syst Biol 2:2006.0008, 2006

[7] M. Lajoie et al, *Probing the limits of genetic recoding in essential genes*, Science 342:361-3, 2013 [8] J. Glass et al, *Essential genes of a minimal bacterium*, PNAS 103(2):425-430, 2005
[9] K. Jin et al, *PhenoM: a database of morphological phenotypes caused by mutation of essential genes in Saccharomyces cerevisiae*, Nucleic Acids Res D687-94, 40, 2012
[10] M. Lluch-Senar et al, *Defining a minimal cell: essentiality of small ORFs and ncRNAs in a genome-reduced bacterium*, Mol Syst Biol 11(1):780, 2015
2

Engineering of an NADPH/NADP+ redox sensor in *Saccharomyces cerevisiae*

Jie Zhang,1 Thomas P. B. Pihl,1 Michael K. Jensen,1 Jay D. Keasling1,2,3,4,5,*

1 The Novo Nordisk Foundation Center for Sustainability, Technical University of Denmark, Denmark

2 Joint BioEnergy Institute, Emeryville, CA, USA

3 Physical Biosciences Division, Lawrence Berkeley National Laboratory, Berkeley, CA, USA
4 Department of Chemical and Biomolecular Engineering & 5 Department of Bioengineering University of California, Berkeley, CA, USA

* E-mails: jdkeasling@lbl.gov

Abstract

Last two decades have seen an enormous progress in metabolic engineering as an enabling technology for a biosustainable society. By introducing a heterologous pathway into a microbial host and rewiring the carbon fluxes, it is often achievable that a new compound can be produced with improved titer, rate or yield. However, the improvement may be largely limited by non-optimal flux distribution or imbalanced cofactors, including NADPH, a reducing equivalent for the biosynthesis of many economically important chemicals. Due to the technical challenges, analysis of this cofactor can only be performed at a low throughput. To facilitate high-throughput screening of strains with increased reducing equivalent capacity, we developed an NADPH/NADP+ redox sensor in the yeast *Saccharomyces cerevisiae* by exploiting the native oxidative stress defense system, namely, the yeast transcription factor Yap1, and the promoter of its target gene, *TRX2*. When coupled with yeast enhanced green fluorescent protein (yEGFP), our sensor- reporter can generate an output of up to 10-fold increase induced by diamide, a disulfide generating oxidant, as well as by genetic modifications that hamper NADPH production. In this study, we built a sensor-actuator circuit to demonstrate the capability to select strains with better NADPH regeneration capacity. In principle, this approach can be scaled up for screening of a larger mutant library to identify novel target for metabolic engineering. The sensor can also be applied to drive the expression of
NADPH-generating enzymes to rescue the cofactor deficiency.

A Dicom Extension Supporting Data Acquisition in Synthetic Biology

Sainz de Murieta, I., Imperial College
Bultelle, M. A., Imperial College
Kitney, R. I., Imperial College

Usage of standard parts is a common approach successfully adopted in several engineering disciplines. Different biology initiatives have mainly focused on maximising the public availability of bioparts without a strong focus on data quality. To improve the quality of data, the Centre for Synthetic Biology and Innovation at Imperial College has developed an automated characterisation pipeline for biological parts, that comprises:
1. A robust data standard for the acquisition of experimental data.
2. A common IT-spine to track and store all data as they are processed and curated.
3. A robust dissemination strategy, enabling public access to high quality biopart information. The Synthetic Biology Information System (SynBIS) has been developed to implement the points 2
and 3 above. It will centralise biological knowledge from the characterised part data, such that only high quality information is disseminated.
This paper presents our work with regard to point 1: the development of a robust data standard for the acquisition of experimental data. In order to this standard to successfully support our characterisation pipeline, we pursued the following objectives:

ï?· Enable the encoding of good quantitative and qualitative information.
ï?· Empower reproducibility of the same results when the experiment is repeated in a different location.
ï?· Support the representation of modular data.
ï?· Optimise the storage of massive amounts of characterisation data.
ï?· Provide services oriented to the automation and exchange of information between data acquisition points and central repositories.

DICOM-SB

We undertook a review of existing standards which might be augmented for synthetic biology. The result was that the existing standard, Digital Imaging and Communications in Medicine (DICOM), was able to represent experimental data, whilst complying with the objectives above. DICOM is an extensive standard supported by a large industrial consortium (NEMA), including the main medical hardware manufacturers. Given its long term success and proven industry orientation, we decided to take advantage of that work and experience and develop our synthetic biology data acquisition standard as an extension of DICOM. The Synthetic Biology extension for DICOM (DICOM-SB, working title) provides a framework that allows the integration of wet lab experimental data acquisition modalities into a common data model. It enhances the basic architecture inherited from DICOM, allowing to encode new synthetic biology data acquisition modalities not present in the general standard (*NEMA PS3 / ISO 12052*, n.d.).
DICOM encodes data objects as a series of items (or data elements), such that each item

is identified by a predefined *attribute* (also called *tag*). Each attribute is related to a data type (e.g. integer, float, character, string, etc.) which in DICOM is called *Value Representation* (VR). Finally, the value to be represented is encoded at the end (last bytes) of each item, preceded by the total items length.
Since the amount of DICOM attributes is so extensive, building consistent objects that include all the required information can become a tedious task if they have to be searched and chosen one by one. In order to ease this task, DICOM clusters the attributes describing properties of one entity or
group of entities into the same *Information Module*. This way, when designing a DICOM object to encode a certain data structure, modules will be the minimal blocks that will be combined to assemble them. The module specification also determines what attributes are mandatory (and thus must always be completed) and which ones can be left incomplete. Finally, the different

Information Entities (IE's) are built from the Information Modules, and Information Object
Definitions (IOD's) are aggregated from IE's.

Each and every data object must implement a standard IOD, and their entities are related following a *hierarchical information model*. We have defined the DICOM-SB Hierarchical Information Model, comprising the following Information Entities:

ï?· Biopart is the object of study, and thus this entity is located at the top of the hierarchy.

ï?· Each biopart can be studied by a *characterization experiment*, whose objective is to perform all the experimental procedures required to generate a standard biopart datasheet.
ï?· Each characterisation experiment comprises a set of assays / modalities that are repeated over a number of days. Each single repeat of a modality, performed with a concrete equipment and following a specific protocol constitutes a *series*.
ï?· Each series can include a number of *modalities*, encoding the information relative to a specific data acquisition activity within a concrete equipment.
The experimental measurements for each data acquisition activities are stored in the corresponding attributes of each modality. Our current protocol uses two different types of modalities:

ï?· Cell population measurements: a plate reader provides pairs of correlated time series of

GFP intensity and total amount of cells per wave.

ï?· Single cell measurements: a flow cytometer provides a series of individual GFP

measurements per cell.
The remaining modality fields, as well as the rest of the higher entities in the data model, store the metadata required to classify, process, analyse and disseminate the experimental measurements.

Once a new data model (IOD) is set, DICOM must offer a service to allow the communication of between different data acquisition points and repositories. In a data acquisition context like ours, there is a need of (at least) a service that stores the acquired IOD into a data repository. For that purpose we have developed a web service that implements a *DICOM Message Service Element* (DIMSE): the *Store* service. The combination of an IOD and the corresponding DIMSE service creates the *Service Object Pairs* (SOP's).

Discussion

A standard like DICOM-SB can provide very important advantages when supporting data acquisition processes in synthetic biology.
One mayor advantage is optimized data storage. Instead of using text-based representations like most standards (e.g. XML, SBML, SBOL), DICOM encodes data in binary format. While this doesn't make a difference when dealing with text strings or characters, it offers significant savings when dealing with numbers: binary representations allow to encode up to 256 different numbers per byte; conversely, text based representations use 1 byte per digit, meaning up to 10 numbers per byte. In sum: DICOM offers up to a 25:1 downscale just by using binary encoding without data compression. This feature becomes especially interesting when dealing with data intensive modalities, such as flow cytometries (up to 50000 events per file), microarrays, etc.
Communication is another very relevant feature, especially when applied to a data acquisition process. The fact that a standard data representation is always packed within a corresponding communication service facilitates the automated distribution of results between measuring equipment and repositories. Medical imaging has benefited from this automation for over the last decade thanks to DICOM being embedded in most of their commercial modalities. The adoption of a similar standard by the synthetic biology modality suppliers would be a key milestone towards the achievement of those automation standards. We have trialled DICOM-SB with the Poh lab at NTU in Singapore. Based on the results, we are confident about the potential benefits and its acceptance by industry.
Far from competing with other synthetic biology standards, we think DICOM-SB complements what is currently available in the area. For example, SBML is widely accepted as the standard to encode simulation models; SBOL is very powerful providing qualitative descriptions of bioparts (Galdzicki
et al., 2014, 2012) and their interactions (Roehner et al., 2014), as well as sequence annotations. But neither of them offer a solution to encode experimental data as DICOM-SB does. We aim to empower the coexistence of complementary standards, therefore we use both DICOM-SB and SBOL within our SynBIS platform. We integrate them using a property of SBOL 2.0 (Roehner et al.,
2014) that allows the annotation of bioparts using links to resources encoded following a different standard (in our case, DICOM-SB).

References

Galdzicki, M., Clancy, K.P., Oberortner, E., Pocock, M., Quinn, J.Y., Rodriguez, C.A., Roehner, N., Wilson, M.L., Adam, L., Anderson, J.C., Bartley, B.A., Beal, J., Chandran, D., Chen, J., Densmore, D., Endy, D., GrÃ¼nberg, R., Hallinan, J., Hillson, N.J., Johnson, J.D., Kuchinsky, A., Lux, M., Misirli, G., Peccoud, J., Plahar, H.A., Sirin, E., Stan, G.-B., Villalobos, A., Wipat, A., Gennari, J.H., Myers, C.J., Sauro, H.M., 2014. The Synthetic Biology Open Language (SBOL) provides a community standard for communicating designs in synthetic biology.
Nat. Biotechnol. 32, 545â??550. doi:10.1038/nbt.2891

Galdzicki, M., Wilson, M., Rodriguez, C.A., Pocock, M.R., Oberortner, E., Adam, L., Adler, A., Anderson, J.C., Beal, J., Cai, Y., Chandran, D., Densmore, D., Drory, O.A., Endy, D.,
Gennari, J.H., GrÃ¼nberg, R., Ham, T.S., Hillson, N.J., Johnson, J.D., Kuchinsky, A., Lux,
M.W., Madsen, C., Misirli, G., Myers, C.J., Olguin, C., Peccoud, J., Plahar, H., Platt, D., Roehner, N., Sirin, E., Smith, T.F., Stan, G.-B., Villabos, A., Wipat, A., Sauro, H.M., 2012. Synthetic Biology Open Language (SBOL) Version 1.1.0.

NEMA PS3 / ISO 12052, n.d. , Digital Imaging and Communications in Medicine (DICOM) Standard. Rosslyn, VA, USA.

Roehner, N., Oberortner, E., Pocock, M., Beal, J., Clancy, K., Madsen, C., Misirli, G., Wipat, A.,
Sauro, H., Myers, C.J., 2014. Proposed Data Model for the Next Version of the Synthetic Biology Open Language. ACS Synth. Biol. doi:10.1021/sb500176h

From Bacillus Subtilis Gene Expression Decomposition to Synthetic Biology

Borkowski, O., INRA
Mäder, U., Institute for Microbiology, EMAU
Sauveplane, V., Center for Systems and Synthetic Biology and Biotechnology
Goelzer, A.,
Dervyn, E., INRA
Müntel, J.,
Schaffer, M.,
Calabre, M., INRA
Aymerich, S., INRA
Noirot, P., INRA
Becher, D., EMAU
Nicolas, P., INRA
Jules, M., INRA
Fromion, V.,

We aimed at decomposing gene expression into distinct and well characterized genetic elements (gene location, TATA box, transcription start, translation initiation region, etc.) that could be rationally assembled to control gene expression across growth conditions. To address this issue, we revisited the contribution of replication, transcription and translation to global mechanisms allowing bacteria to modulate abundance of single proteins with respect to the growth rate. Growth rate has long been regarded as a global variable modulating macromolecular cellular contents but the growth-rate-dependent variation of the transcription machinery abundance was recently shown to clearly influence gene expression.

Regarding translation, we demonstrated that translation efficiency strongly depends on growth rate and that cells can differentially translate individual transcripts as function of the growth rate without dedicated regulators. The mechanism underpinning this translational regulation relies on the selective titration of the *free* (untranslating) ribosome fraction by the transcript-specific translation initiation regions. We identified this mechanism by developing a knowledge-based, nonlinear, mathematical model of translation, followed by successive cycles of predictions and experimental validations using a combination of specific and global quantitative approaches. Similar modi operandi were adopted to characterize the intrinsic properties of replication and transcription. Integrating knowledge from these three fundamental cellular processes allowed us to acquire genome-wide the quantitative parameters that characterize the native *B. subtilis* genetic sequences. We then rationally designed a library of synthetic constructs combining a selection of different genetic elements giving rise to any desired gene expression profile across growth conditions.

The intrinsic properties of these hard-wired regulatory layers of gene expression open novel avenues for synthetic biology, particularly in the potential it offers to bypass the need for multiple specific regulators to modulate complex synthetic circuits.

Synthetic Yeast Based Cell Factories for the Production of Plant Natural Colors

Coumou, H. C., Technical University of Denmark
Petersen, P. D., Chr. Hansen A/S
Mortensen, M. M.,
Frandsen, R. J. N.,
Thrane, U.,
Okkels, F. T.,
Mortensen, U. H., Technical University of Denmark

The vast majority of today's consumers prefer natural over chemically synthesized pigments in their food products and the global market for natural colors is increasing and expected to reach $2.3 billion by 2019. Among the industry's preferred natural food colors are anthocyanins, which are pigments extracted from edible fruits and vegetables, such as grape, elderberry, red radish and black carrot. Being polyphenols, increasing attention has been paid to their health benefits as potent antioxidants.

Anthocyanins are a class of flavonoids that are derived from shikimic acid and acetyl-CoA. They are extensively glycosylated, acylated and/or methylated resulting in immense diversity; and up to date more than 600 compounds have been identified. These decorations with sugar, methyl, aromatic and aliphatic groups influence the molecular properties of anthocyanins, such as their solubility in water and color shade.

The complex structures of anthocyanins are difficult to mimic by chemical synthesis and their production is therefore limited to extraction from the natural resource, typically resulting in low yields, variable quality and supply, and high production costs. To serve the increasing need for plant natural food colors, it is desirable to develop cheap and reliable production methods. Our aim is therefore to transfer the production of anthocyanins to a microbial host.

The baker's yeast *Saccharomyces cerevisiae* is an intensively applied cell factory for the production of renewable fuels and chemicals. Its genetics and physiology have been extensively studied and recent publications show successful metabolic engineering of acetyl-CoA metabolism and the shikimate pathway, greatly improving yields of different classes of plant natural products. Furthermore, yeast supports heterologous production of e.g. membrane bound cytochrome p450s and glycosyltransferases and synthesis of the flavonoid naringenin, the precursor of anthocyanins, has been demonstrated previously. Therefore, we selected *S. cerevisiae* as microbial host for the production of anthocyanins.

To exploit yeast as cell factory for the biosynthesis of a wide range of natural food colors, we are applying synthetic biology tools to develop an anthocyanin production screening platform. This platform uses a combinatorial approach based on mating a library of gene candidates for decorating enzymes to strains containing the synthetic pathways for the production of defined anthocyanin scaffolds. Novel biosynthetic production systems are constructed by shuffling genes originating from different plant species, expanding the biochemical space of anthocyanins and steering required molecular properties for utilization in the food industry.

Modular Engineering of Bacterial Two-Component Systems to Encode Novel Signaling Properties

Sheth, R. U., Rice University
Tabor, J. J., Rice University

Humans are populated by trillions of bacteria that sense and respond to host-, microbe-, and diet- derived molecules. A major goal of synthetic biology is to engineer microbes with custom sense-respond behaviors for therapeutic and diagnostic applications. Two-component systems (TCSs) are the primary means by which bacteria sense the environment. In a canonical TCS, a histidine kinase phosphorylates its cognate response regulator, eliciting a transcriptional response. However, natural TCSs contain specific steady-state and dynamical input/output (I/O) properties that have evolved to permit the organism to best adapt to different environmental signals. To leverage and multiplex the tens of thousands of TCSs known from sequence for useful applications, a generalizable strategy for designing desired TCS signaling properties is needed.

Here, we exploit the modular protein domain architecture of TCSs to program diverse signaling properties. Using our previous *E. coli* light-switchable TCSs, we identify domains and conserved mutations to create new signaling modules that can increase or decrease pathway activity, and show that this approach is generalizable across different TCSs. Furthermore, using these elements we engineer simple feedback and feedforward loops to encode ultrasensitivity, acceleration, pulsing, delay, and memory. We are applying this framework to engineer probiotic bacteria that can sense and remember exposure to diagnostic profiles of disease-associated human gut metabolites and respond with therapeutic molecules.

Engineering Acyl Carrier Protein to Enhance Production of Shortened Fatty Acids

Xueliang Liu1,2# Wade M. Hicks1,3#*, Pamela A. Silver1,3, and Jeffrey C. Way1*

1 Wyss Institute for Biologically Inspired Engineering; 2 School of Engineering and Applied

Sciences, Harvard University; 3 Department of Systems Biology, Harvard Medical School

Keywords: acyl carrier protein; protein engineering; thioesterase; free fatty acid; biofuels

Abstract

We engineered the specificity of the acyl-carrier protein (ACP), an essential enzyme of fatty acid metabolism, to alter the *E. coli* lipid pool and enhance production of medium-chain fatty acids as biofuels precursors. A homology model of the *S. elongatus* ACP was constructed showing a hydrophobic pocket harboring the growing acyl chain. Amino acids within the pocket were mutated to increase steric hindrance to the acyl-chain, and gas chromatography-mass spectrometry analysis performed on cellular extracts identified mutants that, when over- expressed in *E. coli*, increased the proportion of shorter-chain lipids; I75W and I75Y showed the strongest effects. We further demonstrated increased production of lauric acid in *E. coli* also expressing the C12-specific acyl-ACP thioesterase from *Cuphea palustris*.

Introduction

With the continuous rise in global energy needs and adverse climate changes, development of cleaner and renewable alternatives to fossil fuels has become paramount. Microbial synthesis of biofuels is an attractive, renewable alternative to fossil fuels.1â??3 Organisms naturally synthesize large quantities of fuel-like hydrocarbons in the form of lipids,

which are used in cell membranes and other molecules. In microbes, the end products of fatty acid metabolism are long acyl-chains consisting mostly of 16-18 carbons. When extracted for fuels, these long chain carbon molecules remain solid at room temperature and lack favorable physical properties such as higher volatility and lower viscosity. Such properties are characteristic of medium-length (8-12) carbon chains used ubiquitously in

fuels for vehicles and jets.
Previous work on the biological synthesis of medium length fuel precursors has employed thioesterase enzymes with medium-length chain specificity to release free fatty acids (FFA) from intermediates in fatty acid synthesis.4â??7 Here we employ a complementary strategy
to bias FFA synthesis toward shorter chains by engineering acyl-carrier protein (ACP), an essential enzyme and key component of fatty acid metabolism. In fatty acid synthesis in bacteria and plants, ACP is attached to the acyl chain and presents it to the other enzymes during successive cycles of elongation and reduction (Figure 1).8â??11 ACP is a small (~9kB), acidic (pI =
4.1) protein abundant in the cytoplasm, constituting about 0.25% of all soluble proteins in *E.*
coli.8 The structure of ACP is highly conserved even among variants with low sequence similarity. There are, in general, four alpha helices with the major helices I, II and IV running parallel to each other enclosing a hydrophobic pocket that harbors the acyl-chain and minor helix III running perpendicular to these (Figure 2). The acyl chain is connected to a 4- phosphopantetheine modification at a conserved serine and enters the hydrophobic cavity between helices II and III. Roujeinikova et al. solved the structures of *E. coli* ACP attached to C6, C7, and C10 fatty acids.12 In each case, the distal end of the fatty acid terminates in a deep pocket within the protein near Ile72 (corresponding to Ile75 of the *S. elongatus* ACP), with the phosphopantetheine group also entering the pocket to varying degrees. Acyl chains up to 8 carbons are fully bound within the pocket, with the thioester bond sequestered in the core of the protein.8,12â??14 We therefore hypothesized that the size of ACPâ??s hydrophobic pocket serves as a molecular ruler that influences the composition of lipid lengths in a cell. As the acyl chain grows to a length of around 16, the thioester bond becomes fully solvent exposed, which may facilitate cleavage by phosphate:acyl-ACP acyltransferases for downstream phospholipid processing.
To test our molecular ruler hypothesis, we have mutated residues at the base of the hydrophobic pocket to contain bulkier hydrophobic side chains. We found that over-expressing certain mutant ACPs increased production of medium chain fatty acids. Our findings could be potentially advantageous for microbial production of transportation biofuels based on artificial metabolically engineered pathways.

Results and Discussion

To test our hypothesis of ACP acting as a molecular ruler for growing acyl-chains, we constructed mutants of ACP designed to decrease the acyl-chain pocket size (Figure 2). Variants of the cyanobacterial (*S. elongatus*) ACP were expressed in an *E. coli* host. We chose *S. elongatus* ACP due to its potential compatibility with recently discovered enzymes of the cyanobacterial alkane biosynthesis pathway15, which could enable microbial synthesis of fatty alcohol or alkanes. To determine which hydrophobic residues of *S. elongatus* ACP lined the inner, acyl-chain pocket, we constructed a structural homology model using the published crystal structure of *E. coli* ACP bound to a C10 fatty acyl chain (2FAE) as a template (Figure 2). We constructed a number of single amino acid mutants by exchanging small hydrophobic side chain residues, such as

isoleucine or leucine, with bulkier hydrophobic side-chains such as phenylalanine, methionine, tyrosine or tryptophan. ACPs initially fold into an inactive apo state. Conversion to the active holo state is achieved through post-translational modification whereby
4â??-phosphopantetheine is transferred from co-enzyme A (CoA) to a specific serine residue on the apo-ACP (Ser39 on *S. elongatus* ACP).8,16 ACP over-expression may reduce the CoA pool leading to toxic accumulation of apo-ACP, which inhibits sn-glycerol-3-phosphate acyltransferase lipid metabolism and suppresses cellular growth.16,17 To verify functional expression of recombinant ACPs, we measured culture growth kinetics over 15 hours. Compared to controls, cells overexpressing WT *E. coli* ACP (Ec-ACP), WT *S. elongatus* ACP (Se-ACP), or mutant Se-ACPs all showed suppressed growth at low levels of induction and worsened at higher induction levels (Figure 3). The decrease in culture density after 15 hours was as large or larger for Se-ACP mutants than for WT Se-ACP (Figure S1). Growth suppression by the point mutants suggests that these recombinant cyanobacterial ACPs were expressed and properly folded.

To analyze the effect of mutant Se-ACPs on lipid pools, we used gas chromatography mass spec (GC-MS) to characterize fatty acid methyl esters (FAMEs) derived from lipid pools in
Se-ACP overexpressing cells. We compared ratios of FAME peak areas for each sample to remove the influence of growth kinetics and cell densities. We detected peaks for FAMEs derived from the naturally most abundant palmitic acid (C16) and the shorter, less abundant myristic acid (C14) and quantified these peaks in all sample spectra. The ratios of C14 and C16 peak areas were calculated and compared across controls and Se-ACP point mutants. For all uninduced samples, the C14:C16 ratio is around 0.1 (Figure 4A). After induction, only the I75W and I75Y Se-ACP mutants demonstrated a statistically significant increase in the C14:C16 ratio relative to WT Se-ACP, 3- and 2.7-fold respectively ($p<0.05$, two-tailed student-t test) (Figure
4B), indicating that their lipid pools had shifted toward shorter acyl chains. Mutants that replaced Leu49 or Ile57 did not increase the proportions of shorter fatty acids compared to over- expressing WT ACPs. The side chain of isoleucine 75 is positioned in the hydrophobic pocket close to the terminus of the acyl-chain, more so than residues 49 and 57 (Figure 2A).12 We suggest that mutations at position 75 more greatly enhance steric hindrance inside the ACP pocket (Figure 2B and 2C). The I75W and I75Y mutations seem sufficient to bias the fatty acid pool toward a higher proportion of shorter chains, however, mutating Ile75 to phenylalanine or methionine did not cause nearly as significant shifts in lipid pool chain-length composition (Figure 4). This could be because tyrosine and tryptophan have bulkier hydrophobic side chains compared to methionine and phenylalanine and therefore create a smaller acyl-chain pocket within ACP. Indeed, our homology models indicate that the Tyr75 and Trp75 sidechains protrude roughly two carbon-carbon bond distances further into the hydrophobic acyl-chain pocket than an isoleucine at the same position (Figure 2B and 2C; only I75W shown). Therefore, I75W and I75Y Se-ACP mutants may directly hinder elongation from C14 to C16 in fatty acid synthesis and skew the fatty acid pool towards shorter chain lengths. C16 lipids are still dominant, possibly due to the presence of WT Ec-ACP, and/or selectivity at the level of phospholipid formation, or an inherent preference of ACP for C16 and C18 chain lengths

being altered to C14 and C16. A certain level of C16 lipids may be necessary for cell membrane
integrity and survival. This may also explain the stronger growth defect found in I75W and I75Y mutant Se-ACP strains compared to the control strain overexpressing WT Se-ACP (Figure 3, Figure S1).

To explore the potential to further skew cellular lipids towards short chain lengths, particular those shorter than 14 carbons long, we introduced secondary point-mutations in addition to the Se-ACP I75W or I75Y mutations (Table 1). Here amino acids with small hydrophobic side-chains such as isoleucine, valine or alanine were exchanged for a bulkier methionine, a polar glutamine, or a hydrophilic arginine and assayed lipid content for changes to the C14:C16 ratio. Our results show that double mutant Se-ACPs do not significantly increase the C14:C16 ratio beyond either single I75W or I75Y mutation alone (Figure S2). Chains shorter than C14 were not detected at reliable quantities by GC-MS most likely due to their naturally low abundance as transient intermediates in FFA synthesis rather than accumulated end products in membranes.

As an additional control, Se-ACP serine 39 residue, which is post-translationally modified with 4-phosphopantethene, was mutated to alanine (S39A), thereby generating an inactive, obligate apo-ACP. Overexpressing this inactive ACP resulted in similarly low C14:C16 ratio compared to WT (Figure 4). Furthermore, cellular growth was strongly suppressed by over- expressing this mutant, confirming that an overabundance of apo-ACP induces toxicity.

A 24 hour time-course measuring lipid pool composition shows that the highest C14:C16 ratio occurs around 5 hours post-induction (Figure 5). Longer induction times result in a decreased C14:C16 ratio for all strains, particularly for Se-ACP I75W and I75Y mutants, which become indistinguishable from WT controls by 24 hours. This highlights the importance of growth phase on lipid composition. During exponential growth phase when cells are actively dividing and building new membranes, fatty acid metabolism is highly active, and an abundance
of mutated ACPs with reduce pocket sizes likely biases the fatty acid pool toward shorter acyl- chains. It may be that as cell growth slows membrane synthesis proceeds with greater fidelity. Alternatively, short-chain fatty acids may be degraded or â??repairedâ?? and actively replaced with fatty acids of the correct length, which would be more apparent in stationary phase when new C14 fatty acids are not being added to membrane lipids.

We next tested the effect of mutant ACPs on production of shorter chain FFAs, fatty alcohols and alkanes. To over-produce FFAs, a thioesterase that specifically produces 12-carbon chains (*CpFatB1* from *Cuphea palustris*)6 was co-expressed with *Se*-ACP. We hypothesized that by expressing our mutant ACPs, increased cellular concentration of shorter chain acyl-ACPs would serve as substrates to the medium chain-specific thioesterase and further increase the yield of medium chain FFAs. To test this hypothesis, we expressed the C12-specific thioesterase along with WT or mutant *Se*-ACP and measured FFA production by GC-MS analysis of fatty acid ethyl esters (FAEE) derived from produced FFAs (Figure 6).

In conjunction with expressing the C12 thioesterase, strains over-expressing I75W or I75Y mutant ACPs significantly increased medium chain FFA yields compared to those expressing either WT Ec- or Se-ACP (Figure 6). Double mutants based on I75W or I75Y

(e.g. I75W+I57M) also produced similar amounts of medium chain FFAs as the single mutants alone. Similar to our hypothesis on ACP functionality, C12-specific and other medium-chain thioesterases likely act as molecular rulers specifically recognizing fatty acid chains extruded from ACP. We propose that overexpressing mutant Se-ACPs enhances thioesterase-dependent C12 FFA production simply by increasing the C12-acyl-ACP substrate pool. Empty vector and obligate apo-ACP (S39A) controls both produced less FFA than the I75W or I75Y mutants but more than the WT *Ec*- or *Se*-ACP controls. Over-expression of active holo-ACPs could burden cells by depleting CoA pools through excess 4-phosphopantetheine modification of ACP.16,17 If
so, we may be underestimating the enhanced medium-chain FFA synthesis observed for the
I75W and I75Y mutants.

We have shown that ACP, an essential enzyme in fatty acid metabolism, can be modified by site-directed mutagenesis to skew cellular lipid pools towards smaller acyl-chain lengths. Specifically, expressing mutant ACPs enhanced the level of C14 fatty acids in membrane lipids, and by co-expressing mutant ACPs with chain-length specific thioesterases we have demonstrated the ability to tune the length of fatty acid production. Taken together, these results support our hypothesis that bacterial ACPs function as a molecular ruler during fatty acid synthesis. Other enzymes involved in fatty acid synthesis also likely function as molecular rulers and engineering modified acyl-chain specificity has been similarly achieved. For example, FabB and FabF catalyze elongation of fatty acid chains (Figure 1), and have a clearly defined pocket that should accommodate carbon chains up to about 18.18 Val et al.
engineered the FabF pocket to accommodate a maximum of 6 carbons.19 Similarly, the cyanobacterial aldehyde decarbonylase solved structure20,21 contains electron density corresponding to a C18 fatty acid or aldehyde; Khara et al. modified this enzyme to have specificity for medium-chain substrates. The C8, C12 and C14-specific plant-derived Acyl-ACP thioesterases apparently also function as molecular rulers, although the underlying structural mechanisms have not been identified. Since FFAs are hydrophilic, they are not ideal fuel molecules. Instead, FFAs can act as precursors to further enzymatic modification for in transformation into highly desired fuel molecules such as fatty alcohols and alkanes. Engineering such enzymes (e.g. aldehyde decarbonylases, acyl-ACP reductases, and carboxylic acid reductases) towards shorter carbon chain substrate recognition will likely be key to tailoring biofuel formulations. To achieve the ultimate goal of efficient biofuel synthesis, it may be necessary to engineer the length specificity of several enzymes â?? most such enzymes have evolved to handle chains of 16-18 carbons, but shorter chains are desired in fuels. This
technology could help to optimize biofuel yield and molecular makeup, which would benefit the goal of developing energy sources alternative to fossil fuels.

Methods

Homology Modelling

The structural model of *Se*-ACP harboring a decanoyl-chain was obtained by homology to the published x-ray crystal structure of the *E. coli* decanoyl-ACP (2FAE) using SWISS-MODEL.12

Strain Construction

Double stranded DNA encoding *E. coli* and *S. elongatus* ACP genes were synthesized as gBlocks (Integrated DNA Technologies) and cloned into the pCDF-Duet vector by Gibson Assembly.22 Single and double amino acid mutations of the *Se*-ACP gene were incorporated during DNA synthesis. An empty pCDF-Duet-1 vector (Millipore) without the ACP gene was included as control. Plasmids were sequence-verified and transformed into *E.coli* BL21(DE3). For free fatty acid (FFA) production, the C12 thioesterase gene (*CpFatB1* from *Cuphea palustris*) was cloned into pET-Duet-1 vector (Millipore) and transformed into strains harboring the plasmids carrying the ACP variants. For fatty alcohol and alkane production, in one pathway the *S. elongatus* acyl-ACP reductase and aldehyde decarbonyase were cloned into pCOLA-Duet-1 and the pACYC-Duet-1 vectors, respectively. The plasmids were transformed into the strains expressing ACP (in the pCDF-Duet-1 vector) by electroporation. In an alternative pathway, a carboxylic acid reductase (*Mycobacterium marinum*), its maturation factor phosphopantetheinyl transferase (*Bacillus subtilis*), and the petF ferredoxin gene were cloned into the pET-Duet vector with the C12 thioesterase. An aldehyde decarbonylase (*Prochlorococcus marinus*) was cloned into pRSF-Duet-1 vector. The pET-Duet and pRSF-Duet vectors were transformed into the ACP-expressing strains by electroporation. The sequences of these plasmids are given in Supplementary Materials.

Growth Kinetics Assay

ACP expressing strains in triplicates were inoculated from single colonies representing independent transformants into LB medium, grown overnight to saturation and back-diluted into M9 minimal media containing 0.4% glucose. The cultures were grown to mid-exponential phase (OD~0.4), dispersed into 96-well plates, induced with various concentrations of IPTG and left to grow shaking at 37Ë?C in a plate reader (BioTek NEO). The optical densities (OD) of the cultures were recorded every 5 minutes over 15 hours by the plate reader. The growth curves, as well as the final OD after 15 hours were compared among the strains to quantify growth suppression by ACP over-expression.

Analysis of Cellular Lipid Composition

ACP expressing strains in triplicates were inoculated in LB and grown overnight and back- diluted into M9 minimal media containing 3% glucose. The cultures were grown to an optical density of 0.4, induced with 1mM IPTG, and grown for 6 more hours at 37Ë?C. For the time course experiment, the cultures were left to grow for up to 24 hours. After growth, the cells were pelleted and resuspended in 1:1 methanol:chloroform with 2% glacial acetic acid for lysis, hydrolysis of membrane lipids, and solubilization of fatty acids into the organic phase. An octanoate (C8 fatty acid) was added into the mixture as an internal standard. After vigorous mixing by vortexing, the organic phase was

transferred by glass pipettes into glass vials, and the chloroform solvent was evaporated by nitrogen. The vials were then treated with methanol containing 1.25M HCl at 50 50Ë?C for 15 hours to catalyze methylation of the fatty acids. The reaction was quenched by adding 5ml of 100mg/ml sodium bicarbonate. 0.5ml hexane was added and the mixture was vortexed vigorously before the hexane phase containing the fatty acid methyl esters (FAME) was extracted and subsequently analyzed on a GC-MS (Agilent 6890/5975).23 First a standard set of FAMEs with varying chain lengths was run on the GC-MS
in scan mode to determine the identity of each fatty acid peak based on the elution time for each fatty acid and comparison of its fragment profile to those in the NIST database (via Agilent ChemStation software). Fatty acid peaks from the extracted cell samples were also identified using scan mode. To quantify peak areas, the background was minimized by using Selective Ion Mode (SIM) whereby the elution times were used to determine fatty acid identity and only the most dominant mass peaks pertaining to each fatty acid methyl ester were counted. To compare the proportions of different chain lengths in each sample, the ratio of the peak areas for 14-carbon chain to 16-carbon chain was taken.

Analysis of Free Fatty Acid (FFA)

ACP and C12 thioesterase expressing strains in triplicates were grown in M9 minimal media containing 3% glucose and induced with IPTG as described above. After 6 or 24 hours of growth, 5 microliters of each culture (cells and media, as medium chain FFA may be secreted) was transferred to wells of a new 96-well plate for high-throughput spectrometric determination of FFA concentration using the Roche Free Fatty Acid Kit (Product Number 11383175001). To specifically quantify lauric acid, cultures of ACP plus thioesterase-expressing cells were lysed and extracted with chloroform. The FFA was ethylated and run on the GC-MS to determine the spectrum of chain lengths.

Author Information

Corresponding Authors

*EMAIL: wade.hicks@wyss.harvard.edu
*EMAIL: jeff.way@wyss.harvard.edu

Author Contributions

XL and WMH contributed equally to this work.

References

(1) Lennen, R. M., and Pfleger, B. F. (2013) Microbial production of fatty acid-derived fuels and chemicals. *Curr. Opin. Biotechnol.* 1â??10.
(2) Choi, Y. J., and Lee, S. Y. (2013) Microbial production of short-chain alkanes. *Nature*. (3) Dellomonaco, C., Clomburg, J. M., Miller, E. N., and Gonzalez, R. (2011) Engineered
reversal of the Î²-oxidation cycle for the synthesis of fuels and chemicals. *Nature 476*, 355â??9.
(4) Zheng, Y.-N., Li, L.-L., Liu, Q., Yang, J.-M., Wang, X.-W., Liu, W., Xu, X., Liu, H., Zhao, G., and Xian, M. (2012) Optimization of fatty alcohol biosynthesis pathway for selectively enhanced production of C12/14 and C16/18 fatty alcohols in engineered Escherichia coli. *Microb. Cell Fact.*

11, 65.

(5) Lennen, R. M., and Pfleger, B. F. (2012) Engineering Escherichia coli to synthesize free fatty acids. *Trends Biotechnol. 30*, 659â??67.
(6) Torella, J. P., Ford, T. J., Kim, S. N., Chen, A. M., Way, J. C., and Silver, P. a. (2013) Tailored fatty acid synthesis via dynamic control of fatty acid elongation. *Proc. Natl. Acad. Sci. U. S. A. 110*, 11290â??5.
(7) Voelker, T. A., and Davies, H. M. (1994) Alteration of the specificity and regulation of Alteration of the Specificity and Regulation of Fatty Acid Synthesis of Escherichia coli by Expression of a Plant Medium- Chain Acyl-Acyl Carrier Protein Thioesterase.
(8) Chan, D. I., and Vogel, H. J. (2010) Current understanding of fatty acid biosynthesis and the acyl carrier protein. *Biochem. J. 430*, 1â??19.
(9) Crosby, J., and Crump, M. P. (2012) The structural role of the carrier protein â?? active controller or passive carrier â? 1111â??1137.
(10) Nguyen, C., Haushalter, R. W., Lee, D. J., Markwick, P. R. L., Bruegger, J., Caldara-Festin, G., Finzel, K., Jackson, D. R., Ishikawa, F., Oâ??Dowd, B., McCammon, J. A., Opella, S. J., Tsai, S.-C., and Burkart, M. D. (2014) Trapping the dynamic acyl carrier protein in fatty acid biosynthesis. *Nature 505*, 427â??31.
(11) Masoudi, A., Raetz, C. R. H., Zhou, P., and Pemble IV, C. W. (2013) Chasing acyl carrier protein through a catalytic cycle of lipid A production. *Nature* 1â??7.
(12) Roujeinikova, A., Simon, W. J., Gilroy, J., Rice, D. W., Rafferty, J. B., and Slabas, A. R. (2007) Structural studies of fatty acyl-(acyl carrier protein) thioesters reveal a hydrophobic binding cavity that can expand to fit longer substrates. *J. Mol. Biol. 365*, 135â??45.
(13) Chan, D. I., Stockner, T., Tieleman, D. P., and Vogel, H. J. (2008) Molecular dynamics simulations of the Apo-, Holo-, and acyl-forms of Escherichia coli acyl carrier protein. *J. Biol. Chem. 283*, 33620â??9.
(14) Roujeinikova, A., Baldock, C., Simon, W. J., Gilroy, J., Baker, P. J., Stuitje, A. R., Rice, D. W., Slabas, A. R., Rafferty, J. B., and Sheffield, S. (2002) X-Ray Crystallographic Studies on Butyryl-ACP Reveal Flexibility of the Structure around a Putative Acyl Chain Binding Site *10*,
825â??835.
(15) Schirmer, A., Rude, M. a, Li, X., Popova, E., and del Cardayre, S. B. (2010) Microbial biosynthesis of alkanes. *Science 329*, 559â??62.

(16) Carey, M. R. (1995) The Unmodified (Apo) Form of Escherichia coli Acyl Carrier Protein Is a Potent Inhibitor of Cell Growth. *J. Biol. Chem. 270*, 22229â??22235.
(17) Magnuson, K., Jackowski, S., Rock, C. O., and Cronan, J. E. (1993) Regulation of fatty acid biosynthesis in Escherichia coli. *Microbiol. Rev. 57*, 522â??42.
(18) White, S. W., Zheng, J., Zhang, Y.-M., and Rock. (2005) The structural biology of type II
fatty acid biosynthesis. *Annu. Rev. Biochem. 74*, 791â??831.
(19) Val, D., Banu, G., Seshadri, K., Lindqvist, Y., and Dehesh, K. (2000) Re-engineering ketoacyl synthase specificity. *Structure 8*, 565â??566.
(20) Jia, C., Li, M., Li, J., Zhang, J., Zhang, H., Cao, P., Pan, X., Lu, X., and Chang, W. (2015) Structural insights into the catalytic mechanism of aldehyde-deformylating oxygenases. *Protein Cell 6*, 55â??67.
(21) Khara, B., Menon, N., Levy, C., Mansell, D., Das, D., Marsh, E. N. G., Leys, D., and Scrutton, N. S. (2013) Production of propane and other short-chain alkanes by structure-based engineering of ligand specificity in aldehyde-deformylating oxygenase. *Chembiochem 14*, 1204â??
8.
(22) Gibson, D. G., Young, L., Chuang, R.-Y., Venter, J. C., Hutchison, C. a, and Smith, H. O. (2009) Enzymatic assembly of DNA molecules up to several hundred kilobases. *Nat. Methods 6*,
343â??5.
(23) Lennen, R. M., Braden, D. J., West, R. a, Dumesic, J. a, and Pfleger, B. F. (2010) A process for microbial hydrocarbon synthesis: Overproduction of fatty acids in Escherichia coli and catalytic conversion to alkanes. *Biotechnol. Bioeng. 106*, 193â??202.

Figures

Figure 1

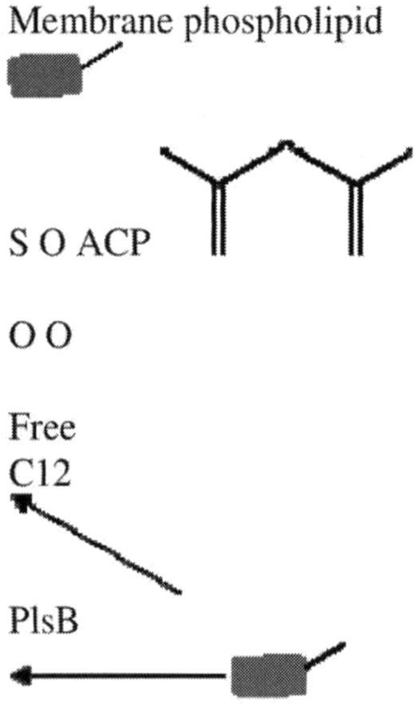

S

FabB FabF

S

Fatty
Acid

ACP

thioesterase

FabI

1+2n

O

ACP

O O

FabG

1+2n

ACP

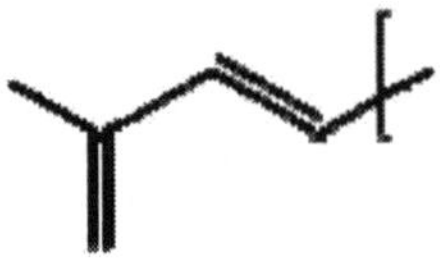

S]

1+2n

O

FabA FabZ

ACP

S

O OH

]

1+2n

Figure 1 Overview of Fatty acid Synthesis. Fatty acid synthesis proceeds through iterative cycles of elongation. In each cycle, the acyl-chain is extended by 2-carbons using a malonyl- ACP as a carbon donor (by FabB or FabF) and

subsequently reduced into a saturated chain (by FabG, FabA, FabZ and FabI). From the first 2-carbon malonyl-ACP to the final length fatty acid processed through this cycle, the hydrophobic acyl-chain is attached to and shielded by the ACP instead of existing in a free form.

Figure 2

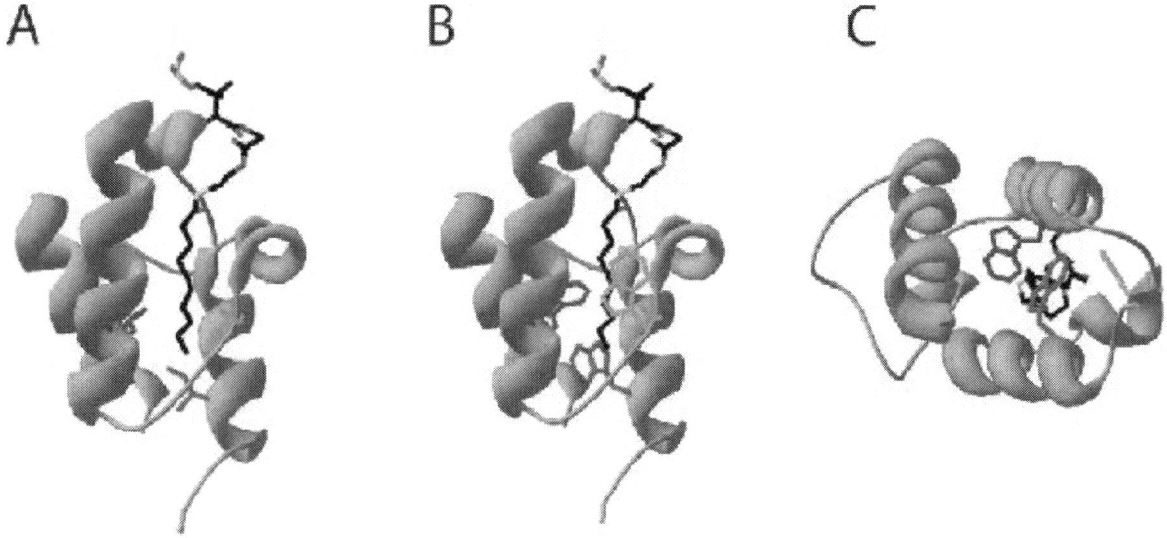

Figure 2 Se-ACP Structural Homology Models with WT and Mutant Residues. A. Homology model of Se-ACP bound to a C10 acyl-chain is shown. Highlighted in blue (residue 49), green (residue 57) and red (residue 75) are small hydrophobic amino acids lining the WT ACP pocket,

Leu, Ile, and Ile, respectively. Each residue was separately mutated to a bulkier hydrophobic amino acid: methionine, tyrosine or tryptophan in order to induce steric hindrance and favor shorter-chain fatty acid synthesis. B. For simplicity, a homology model with all three residues of interest mutated to tryptophan shows how each side chain might be positioned when mutated separately. Trp75 (red) extends closest to the acyl-chain terminus. C. Looking up through the axis of the acyl chain from the bottom perspective of the ACP pocket, Trp75 (red) is more directly in line with the acyl chain, as compared to the other mutant residues. This positioning likely introduces the most direct steric hindrance to the acyl chain, resulting in increased production of shorter chain lipids.

Figure 3

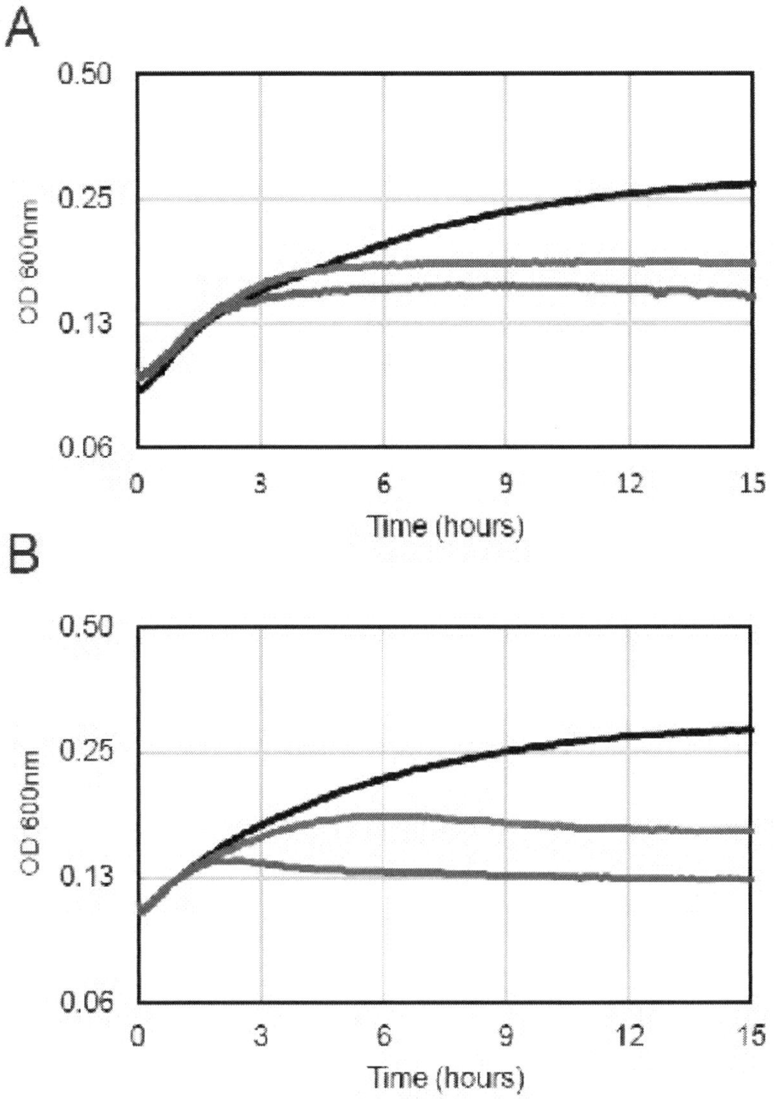

Figure 3 Growth Suppression by ACP expression. Growth of *E. coli* is suppressed by induction of Se-ACP expression increasing from 0mM (black), 0.5mM (blue) to saturation at
1mM (red) IPTG. The growth defect is likely due to inhibition of phospholipid metabolism by apo-ACP. Se-ACP I75W expression (**B**) shows similar growth suppression compared to WT (**A**), indicating proper folding and functionality of ACP. All mutant ACPs show similar growth curves (data not shown). Representative growth curves are shown.

Figure 4

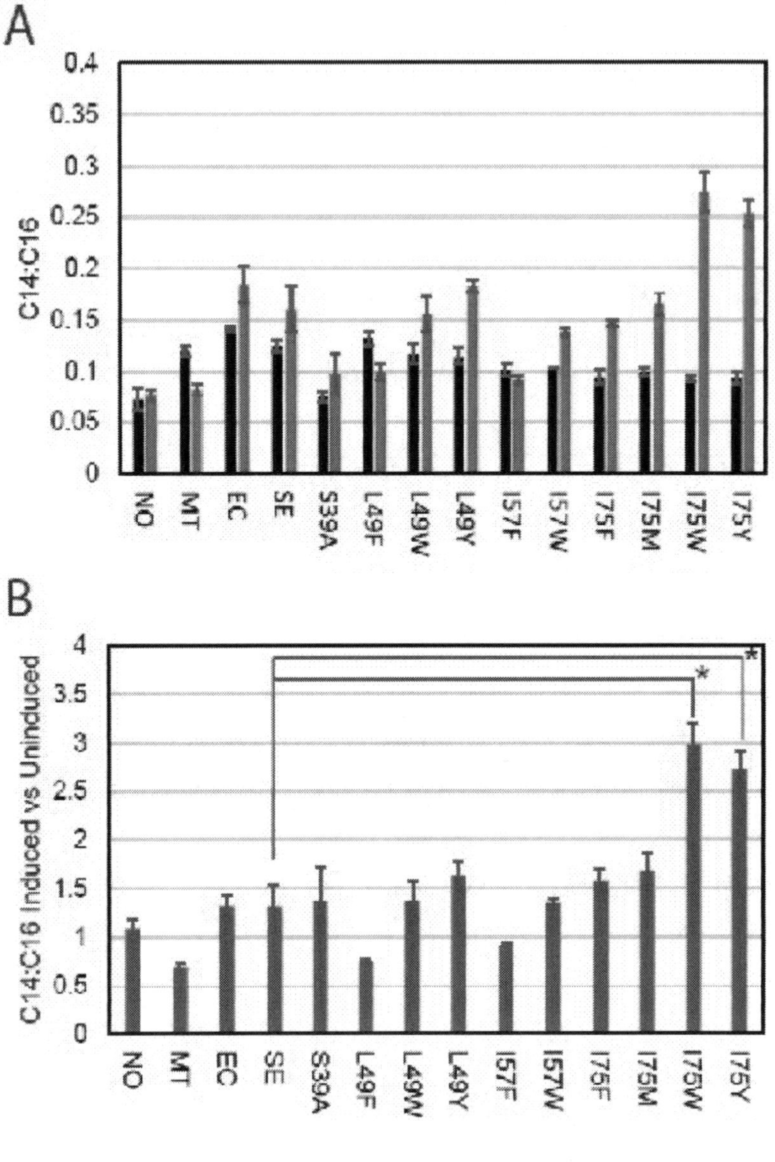

Figure 4 GC-MS analysis of Cellular Lipids in Single ACP Mutants. A. Ratios of C14 to C16

GC-MS peak areas for uninduced (black) and induced (red) strains. **B.** Fold changes of induced vs. uninduced C14:C16 ratios. The I75W and I75Y mutants have significantly increased C14:C16 ratios as compared to expressing WT Se-ACP. Data represents triplicate biological measurements. Error bars are standard error of the mean (S.E.M). * $p<0.05$, two-tailed student-t test.

Figure 5

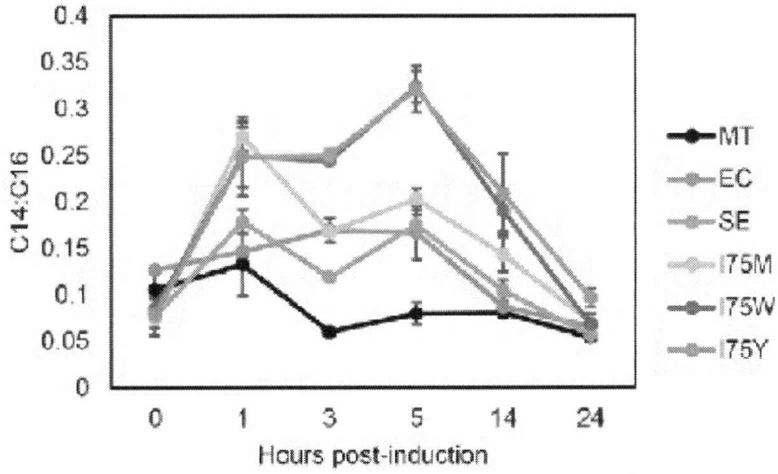

Figure 5 Time Course of C14:C16 Ratios Se-ACP I75W and I75Y both demonstrate the highest C14:C16 cellular lipid ratio at 5 hours after induction during the growth phase. As the cell cultures saturate past 14 hours, the ratios decrease to the baseline of around 0.05-0.1. Data represent triplicate biological measurements. Error bars are S.EM.

Figure 6

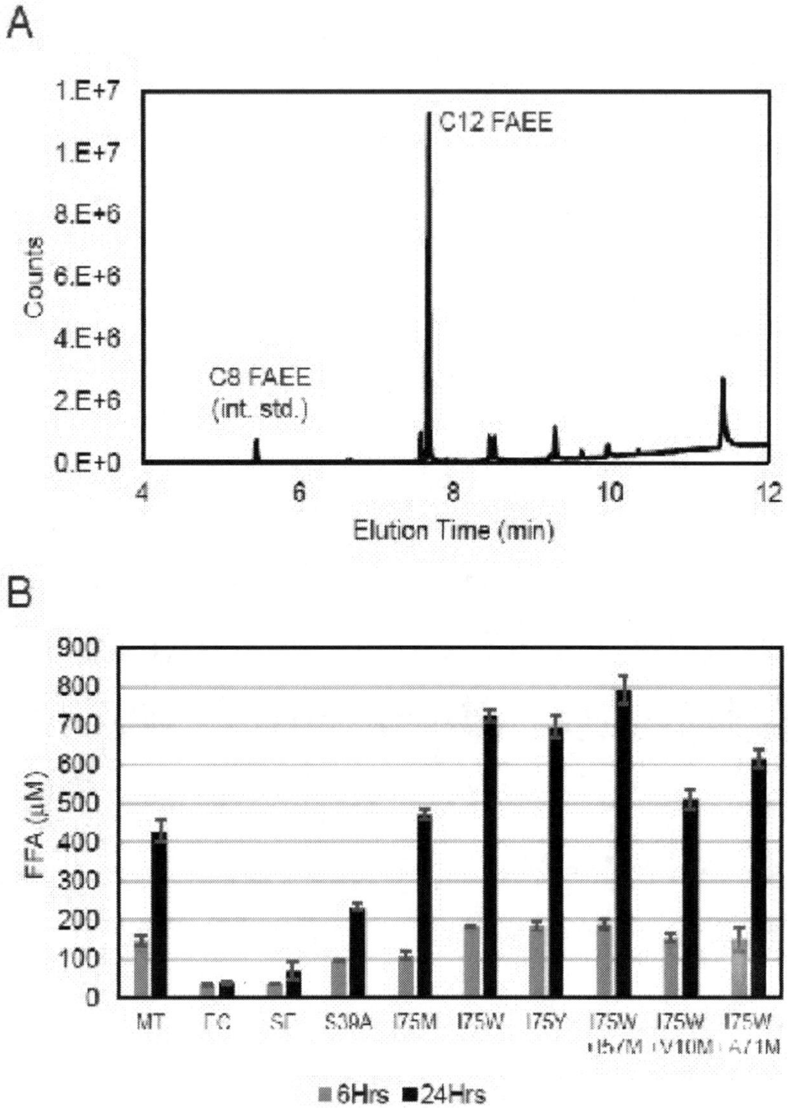

Figure 6 Free Fatty Acid Production by C12 thioesterase. A. Representative GC-MS trace of FAEEs derived from cell cultures shows thioesterase specificity toward 12-carbon acyl-chains. B. FFA concentrations measured from cell

cultures at 6 hours (blue) and 24 hours (black) post- induction of both the C12 thioesterase and the indicated ACP. The Se-ACP I75W and I75Y mutants and their derivatives yield more FFA than controls. WT Ec-ACP and Se-ACP controls yield the least FFA even compared to the empty vector (MT) and apo-ACP (S39A) controls. Data represent triplicate biological measurements. Error bars are S.E.M.

Supporting Information

Figure S1

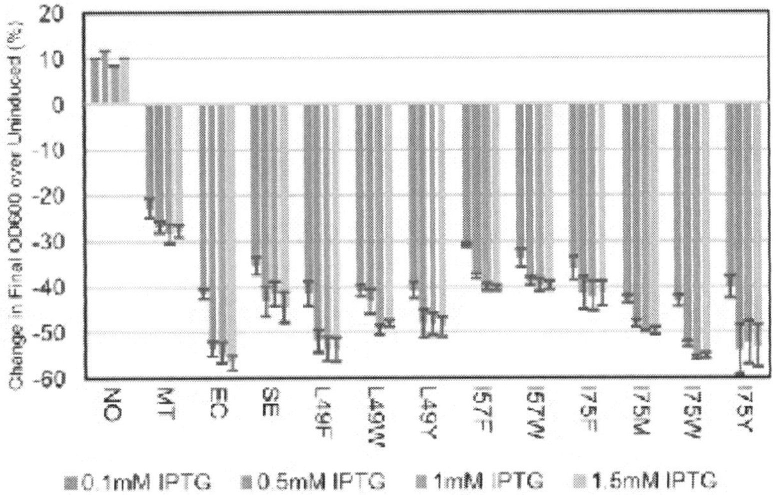

Figure S1 Growth Suppression by ACP expression. Shown are changes in culture OD of *E. coli* strains induced to overexpress various ACPs versus their uninduced condition (0mM IPTG). Culture densities were measured after 15 hours of growth in M9 minimal media with 0.4% glucose. When overexpressed, most mutants show equal or stronger growth suppression vs. WT Se-ACP. Data represent triplicate biological measurements.

Error bars are S.E.M.

Figure S2

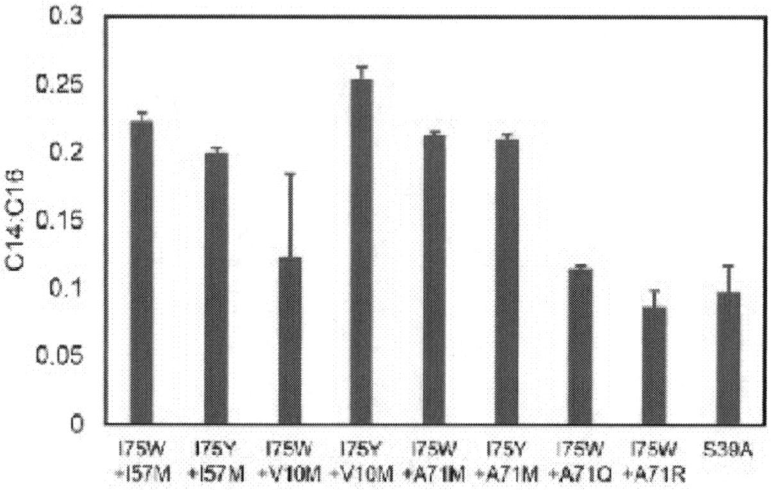

Figure S2 GC-MS analysis of cellular lipids from double mutants. Combining the single Se- ACP I75W or I75Y point mutants with a second set of residues mutated to methionine does not significantly change the C14:C16 ratio from that observed for the single point mutants alone. Mutating this second set of residues to arginine (A71R) or glutamine (A71Q) reduced the C14:C16 ratio to WT Se-ACP levels. Data represent triplicate biological measurements. Error bars are S.E.M.

Realising Efficient and Robust Synthetic Biology Systems Using Systems and Control Engineering

Tomazou, M. III, Imperial College London
Ang, J.,
Stan, G. B., Imperial College London

Synthetic biology systems are as good as their reliability in terms of robustness and performance characteristics within a given host and an ever-changing environment. Progression to larger scale designs and higher order functionalities requires solid foundations for the design and implementation of robust and efficient biological systems. The lessons learned by the advancements made in more established engineering disciplines point to the fact that adapting and implementing systems and control theory principles to synthetic biology is crucial for meeting these robust performance requirements. To that effect, we study and work towards implementing and combining synthetic feedback control mechanisms within cells. In particular we focus on the design, implementation and combination of synthetic biology feedbacks at various levels: (a) The first level focuses on the synthetic circuitry level, where feedback auto-regulation can be mediated directly through transcriptional regulation and implemented through various network topologies. (b) The second level focuses on engineering feedback in metabolic processes for optimizing the production of certain metabolites but more importantly, for balancing the flux of nutrients and ensuring a dynamically balanced allocation of cell resources when part of the native metabolic flux becomes diverted by synthetic branches. (c) At the population level we are working towards an extracellular environment sensor-actuator feedback control mechanism that can robustly keep extracellular concentrations around a desired setpoint. We study the above both from both a theoretical and an experimental viewpoint and consider combinations of these various feedbacks with the aim to deliver more resilient chasses for operating synthetic systems and to consequently reduce the host-imposed uncertainty on synthetic systems behaviour.

Engineered Probiotics That Detect and Kill the Human Pathogen Vibrio Cholerae

Mao, N., Boston University
Cameron, D. E., Massachusetts Institute of Technology
Collins, J. J., Massachusetts Institute of Technology

New synthetic biology techniques have enabled us to engineer the human microbiota for therapeutic purposes. In this project, we have engineered the probiotic lactic acid bacteria *Lactococcus lactis* to detect *Vibrio cholerae*, the causative agent of the infectious disease cholera. We used the CqsS sensor domain from the *V. cholerae* CAI-1 quorum-sensing system to create a hybrid two component system in *L. lactis* that can directly detect the *V. cholerae* CAI-1 signal. Incorporation of a TetR-based signal inverter into the detection circuit allowed us to couple CAI-1 recognition to activation of the output module, where a combination of several antimicrobial peptides will be used to specifically kill *V. cholerae*. We have tested the circuit *in vitro* and in co-culture conditions and will carry out *in vivo* tests in a *V. cholerae* mouse infection model in the near future. This novel design uses the pathogen's own sensing system to detect and kill it.

Design Principles of Protein Organelles

Savage, D.,

Cells use structure to catalyze and facilitate the chemical reactions of metabolism. This principle is embodied by bacterial microcompartments, a class of protein-based organelles that are found in nearly 30% of bacterial genomes. More specifically, microcompartments are capsid-like structures composed of thousands of proteins that self-assemble to form an outer shell and encapsulated cargo proteins that form a short metabolic pathway within the inner lumen. Despite knowledge of the component proteins, it is less well understood how these structures self-assemble and function *in vivo*. Here, we describe our efforts to answer these questions using a combination of synthetic expression in heterologous hosts, modeling of the reaction-diffusion process, and biochemical reconstitution of assembly. These results suggest general themes for how encapsulation can improve pathway performance and specific methods for employing microcompartments and other capsid-forming proteins as a means of engineering metabolism and biomaterials.

Development and Characterization of the Peroxisome for Metabolic Engineering

Russ, Z. N., UC Berkeley
DeLoache, W. C., UC Berkeley
Dueber, J. E., U.C. Berkeley

Engineered metabolic pathways often suffer from undesirable interactions with the production host's native cellular processes. Evolution has solved the problem of metabolic crosstalk by segregating distinct cellular functions into membrane-bound organelles. By improving the modularity and efficiency of protein import into the yeast peroxisome, determining the natural chemical permeability of the peroxisomal membrane, and characterizing both the rate and capacity of protein import, we have transformed the peroxisome into a useful option for compartmentalization and sequestration of proteins and pathways. With this new, characterized system, we demonstrate successful compartmentalization of a model pathway and inducible, rapid sequestration of essential proteins. These advances will serve as the basis for turning the peroxisome into a well-characterized synthetic organelle for use in limiting crosstalk in pathways and protein sequestration.

Paper-Based Synthetic Gene Networks

Pardee, K., Wyss Institute, Harvard University
Green, A., Arizona State University
Ferrante, T.,
Cameron, E., Massachusetts Institute of Technology
DaleyKeyser, A.,
Donghia, N.,
Yin, P., Harvard Medical School
Collins, J. J., Massachusetts Institute of Technology

SEED 2015 abstract:

Synthetic gene networks are incredibly powerful tools, with wide-ranging uses in reprogramming and rewiring organisms. To date, there has not been a way to harness the vast potential of these networks beyond the constraints of a laboratory or *in vivo* environment. Here, we present an *in vitro* paper-based platform for synthetic biology that promises to provide both a new venue for synthetic biologists to operate, and a much-needed medium for the safe deployment of engineered gene circuits out of the lab. It enables the simple, sterile and abiotic distribution of synthetic biology-based technologies for the clinic, global health and industry. Based on commercially available cell-free transcription and translation systems, bacterial and mammalian components can be freeze-dried onto paper and other porous substrates to create poised synthetic gene networks that are stable for long-term storage at room temperature and are activated by rehydration. The resulting engineered materials have the transcription and translation properties of a cell and can host genetically-encoded tools. We demonstrate this technology with small molecule and RNA actuation of genetic switches, the characterization of novel gene circuits, and the construction of paper-based sensors for glucose and mRNAs, including antibiotic resistance genes and strain-specific Ebola virus sensors. Moreover, the rational design and rapid prototyping elements of the system provide a powerful combination for accelerating research. For greater practical use, gene circuits were enhanced with colorimetric outputs for detection by the naked eye, as well as with the fabrication of a low cost, electronic optical interface for quantification and possible automation of reactions. These low cost, paper-based synthetic gene networks have the potential to bring bio-based sensors, counters, timers and simple logic to portable devices.

Modelling Population Cell Cycle Heterogeneity to Deconvolute Population-level Measurements

Melchior du Lac Andrew Scarpelli Joshua Leonard Declan G. Bates

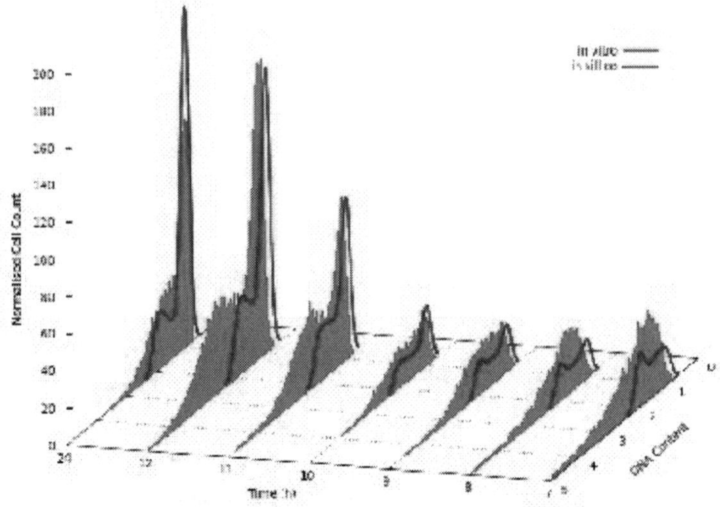

I. INTRODUCTION

Synthetic biology uses mathematical modelling, combined with *in silico* simulations to guide the design and optimisation of synthetic genetic circuits (SGC). These circuits are usually modelled as a system of ordinary differential equations (ODE). For example, the repressillator and the toggle switch, two of the founding SGCâ??s, were successfully designed using this method [1][2]. The models employed in these studies did not, however, predict the level of variations in the oscillatory signals or the desynchronisation observed between even small number of generations when expressed in *Escherichia Coli* (*E.Coli*). Even if the molecular mechanisms and biochemistry of the transcription and translation mechanism controlling these SGCâ??s are quite well understood, biological noise dic- tates these non-deterministic expression dynamics [3]. This noise is a combination of the probabilistic nature of the biochemical reactions, called intrinsic noise, and extrinsic noise, that refers to variations in other cellular components that may affect the transcription and translating machinery of the gene or system of interest [4]. For example, in the scope of synthetic biology, the latter noise could arise from cell variability of gene copy numbers from different plasmid copy numbers or from asymmetric cell duplication [5][6][7]. While the effects of intrinsic noise is being increasingly well formalised through stochastic modelling, extrinsic noise is more difficult to elucidate because of the vast number of potential sources and their uncertain influence on the system of interest [3].

To obtain a quantitative understanding of any biological system, intracellular concentrations over time courses are required [8]. However, most -omics technologies sample pop- ulations instead of individual cells. At the population level, the accumulation of biological noise leads isogenic cells to quickly become phenotypically heterogneic, a property that is undesirable for the accurate estimation of expression levels [9]. To reduce the influence of extrinsic noise, basic microbiolog- ical practices dictate that single celled organisms be sampled during exponential growth of culture, to assure that individuals have a similar molecular profile [7]. However, even under these stringent experimental conditions, cells are seldom homoge- neous [10]. This makes the output of such measurements a statistical property of the predominant molecular state of the population, that cannot be used as a reliable estimation of discrete intracellular concentrations [11][8].

Although the origins of heterogeneity are diverse and pre-

Fig. 1. Histogram of the *in vitro* (red) and *in sillico* (blue) DNA distributions along the stationary phase of growth in batch culture. The DNA content is normalised by number of chromosomes, and Gaussian spread has been applied to the simulation results using a CV of 0.4.

dominantly still to be elucidated, in some cases, they depend on known biological mechanisms [10][7], such as the cell cycle [12][13]. Indeed, many genes are differentially expressed based on the growth cycle state of the cell [14]. To deconvolve the dynamics of time course data and formalise the influence of cell cycle extrinsic noises on measured sample expression, it is necessary to understand cell cycle heterogeneity in populations of cells. The term deconvolution, in this case, is used not as a strict mathematical term, but to illustrate that population measurements are the sum of individual expression. Using an individual based simulation method to represent the inherent heterogeneity of bacterial populations, combined with a real- istic growth model, we simulate the cell cycle dynamics of a population of cells in conditions of restricted and exponential growth.

II. DRIVING INSTEAD OF DICTATING GROWTH

Because of the multi-dimensional nature of extrinsic noise, it is necessary in the context of modelling frameworks to make assumptions on the state of the cells, and usually takes the form of population â??steady stateâ? [7]. Although this is not equivalent to the systems theory definition, it does imply time invariant properties of the population, where cells synthesise a particular molecule at a constant differential rate [15]. In such a case, the population is said to be in steady state *vis-a-vis* the molecule of interest. It is experimentally achieved by growing cells exponentially and in conditions of unrestricted growth (also called balanced growth), where Malthusian growth theory can be applied [16][17]. This established method has been designed to drive populations to behave in a predictable manner in order to reduce the inherent heterogeneity between individual cells [17]. For example, mathematical formalisation of the age distribution in exponentially growing populations, has enabled successful models of population chromosomal distributions for rapidly growing cultures [12]. However the use of Malthusian growth to describe population dynamics and the underlying cellular processes, is not only restricted to exponential experimental conditions, but many examples in the literature show that this is not a robust method to describe slower growth rates [18][19][17]. Worse, the very concept of population steady state is increasingly being challenged [20]. Other growth models such as the Monod equation that prac- tically describes the growth dynamics under nutrient limiting conditions, have no mechanistic biological basis, and so are of limited use in understanding cell growth and the downstream influence on the cellular properties [6].

Since most growth models have limitations in accurately representing the dynamics of population growth, instead of dictating the dynamics of individual cell growth using for example the Monod equation, we drive individual cell growth by population level measurements. Optical density (OD) is a standard method that determines the turbidity of a sample and is used as proxy for the relative changes of mass of a sample. Although the data reported is relative, if one were to determine the true volumetric concentration based on a particular OD, one could normalise OD data that would return time course volumetric changes of a population [8]. Using this estimate of volumetric population concentration, we distribute the volume changes among all members of a population that we are simulating. We term this the â??injection methodâ? , because we are driving the growth of the cell by injecting volume to match the population volumetric concentration. OD data is also used to extract growth growth rates and cell cycle parameters that are of interest to us. However because of we are using the injection method, we cannot infer them from growth models. To perform this without making any assumption of the dynamics of the growth, we use a method qualified as â??instantaneous growth rateâ? [21]. This implies smoothing OD data using the spline interpolation method, enabling the growth dynamics to be accurately extracted from directly measured experimental data [21].

III. MECHANISTIC CELL CYCLE MODEL

Evidence suggests that the dynamics of the cell cycle is tightly correlated with the mass of a cell. Indeed, cells have a constant mass at initiation of DNA replication [22]. This observation formalises why bacterial cells are larger at higher growth rates while being smaller at lower growth rates, and also explains overlapping rounds of replication, where one chromosome may have several origin of replication open if

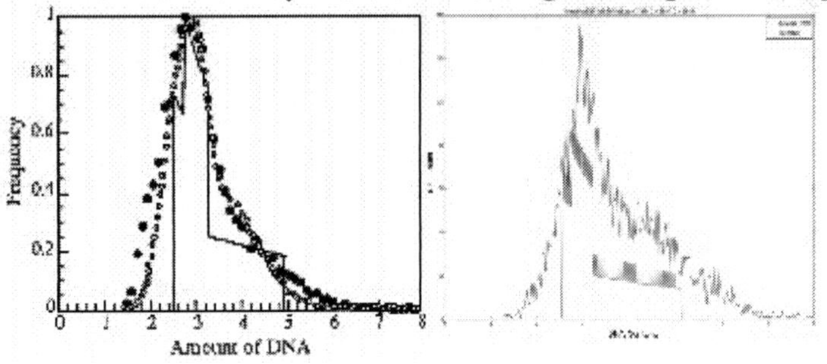

Fig. 2. Comparison between Keasling et al. Monte Carlo simulation fit to flow cytometry data (left) and injection method growth (right) for a doubling rate of 27 minutes. For both the x-axis represents the DNA content in chromosome numbers and both the y-axis represents the relative cellular concentration. The solid line and the green line both represent the traditional method of deducing DNA population distributions, based on the theoretical age distribution of a population. The noise associated with the green distribution arises from the discrete solution to a continuous function. The solid circles are the simulation results of the Keasling et al. paper and the open circles are experimental data while the blue line is the simulation results of our injection method.

the bacteria is dividing fast enough [22]. We use the Cooper and Helmsetter mechanistic model of the cell cycle that implements the above observation [23], and has been used to model exponentially growing populations [13].

To simulate the inherent heterogeneity of bacterial cell populations, agent based simulations are becoming a more popular method to capture innate social and self emergent behaviours of many aspects of biology [24]. By explicitly representing all members of a populations, simulating them in parallel and implementing some type of communication between agents, the sum of the individuals affect on the system dictates the system behaviour. For our purpose the cells do not communicate, and thus our method can be characterised as individual based.

IV. VALIDATING THE INJECTION METHOD

It has been reported that Monte Carlo simulations of ex- ponentially growing bacterial populations successfully repro- duced the DNA distribution of a bacterial populations [13]. To illustrate that our growth method is valid, we reproduced their simulation conditions using our growth method [13]. We generated a dummy growth curve whoâ??s OD increased with a constant rate and translates to a doubling rate of 27 minutes, as per the authors simulation. The results are presented in Figure 2. It shows that the DNA distribution in the simulation spans from 1.5 to 7 chromosomes, similar to the Keasling et al. simulation and their flow cytometry fit, and that the distribution is skewed to the right. Both are also plotted with the output of the traditional model of the cell cycleâ??s DNA distribution (solid line and green line). It uses the theoretical age distribution, depicted by a probability density function, of a population in exponential growth and returns the relative abundance of the cells with particular DNA concentrations [13].

V. PRELIMINARY RESULTS

To show that our method can replicate the right dynamics of the cell cycle regardless of the growth conditions, we have obtained time course flow cytometry data from the exponential phase to the late stationary phases of growth. To calibrate the data, we arrested the duplication of DNA and fit the corresponding peaks using Gaussian functions and deduced their spread ($CV = 0.4$). Our initial time course simulation results (Fig. 1) demonstrates the shift from one distribution in the early stationary phase, constituted mostly of cells with two chromosomes, to two distinct ones in the late stationary phase (24h), constituted of mostly one chromosome per cell and a smaller peak at two chromosomes. However the spread and distribution, as seen *in vitro*, are mostly wrong. We believe that the reason for this are badly conditioned parameters of the cell cycle. In this simulation we used the parameters from the literature that are not only different from the strain of bacteria (TOP10) used in the flow cytometry experiment, but are also extracted during the exponential part of the growth curve [19]. We are currently testing if we can infer more accurate cell cycle parameters from the DNA distribution [19]. Assuming that we can simulate the cell cycle reliably using this method, we plan on extending it to model the effect of plasmid dynamics on SGC functionality. Different plasmid types have different partitioning patterns and understanding the heterogeneic influence of the plasmid copy number and replication would, in much the same way as described for chromosomal genes, lead to differentially expressed genes based on the cell cycle and growth conditions [5][6]. Our method could reliably predict this pattern of heterogeneity and help understand the noise associated with measured expression profiles for plasmids genes as well as chromosomal genes.

REFERENCES

[1] T. S. Gardner, C. R. Cantor, and J. J. Collins, â??Construction of a genetic toggle switch in escherichia coli,â? *Nature*, vol. 403, no. 6767, pp. 339â??

342, 2000.

[2] M. B. Elowitz and S. Leibler, â??A synthetic oscillatory network of transcriptional regulators,â? *Nature*, vol. 403, no. 6767, pp. 335â??338,

2000.

[3] M. B. Elowitz, A. J. Levine, E. D. Siggia, and P. S. Swain, â??Stochastic gene expression in a single cell,â? *Science*, vol. 297, no. 5584, pp. 1183â??

1186, 2002.

[4] P. S. Swain, M. B. Elowitz, and E. D. Siggia, â??Intrinsic and extrinsic contributions to stochasticity in gene expression,â? *Proceedings of the National Academy of Sciences*, vol. 99, no. 20, pp. 12 795â??12 800, 2002.

[5] I. Golding, J. Paulsson, S. M. Zawilski, and E. C. Cox, â??Real-time kinetics of gene activity in individual bacteria,â? *Cell*, vol. 123, no. 6, pp. 1025â??1036, 2005.

[6] J. L. Snoep, M. Mrwebi, J. M. Schuurmans, J. M. Rohwer, and M. T. De Mattos, â??Control of specific growth rate in saccharomyces cerevisiae,â? *Microbiology*, vol. 155, no. 5, pp. 1699â??1707, 2009.

[7] S. Cooper, *Bacterial growth and division: biochemistry and regulation of prokaryotic and eukaryotic division cycles*. Elsevier, 1991.

[8] B. Volkmer and M. Heinemann, â??Condition-dependent cell volume and concentration of escherichia coli to facilitate data conversion for systems biology modeling,â? *PLoS One*, vol. 6, no. 7, p. e23126, 2011.

[9] J. M. Buscher, D. Czernik, J. C. Ewald, U. Sauer, and N. Zamboni, â??Cross-platform comparison of methods for quantitative metabolomics of primary metabolism,â? *Analytical chemistry*, vol. 81, no. 6, pp. 2135â??

2143, 2009.

[10] M. E. Lidstrom and M. C. Konopka, â??The role of physiological het- erogeneity in microbial population behavior,â? *Nature chemical biology*, vol. 6, no. 10, pp. 705â??712, 2010.

[11] D. Dubnau and R. Losick, â??Bistability in bacteria,â? *Molecular microbi- ology*, vol. 61, no. 3, pp. 564â??572, 2006.

[12] K. Skarstad, H. B. Steen, and E. Boye, â??Escherichia coli dna distri- butions measured by flow cytometry and compared with theoretical computer simulations.â? *Journal of bacteriology*, vol. 163, no. 2, pp. 661â??

668, 1985.

[13] J. Keasling, H. Kuo, and G. Vahanian, â??A monte carlo simulation of the escherichia coli cell cycle,â? *Journal of theoretical biology*, vol. 176, no. 3, pp. 411â??430, 1995.

[14] P. Sobetzko, A. Travers, and G. Muskhelishvili, â??Gene order and chro- mosome dynamics coordinate spatiotemporal gene expression during the bacterial growth cycle,â? *Proceedings of the National Academy of Sciences*, vol. 109, no. 2, pp. E42â??E50, 2012.

[15] D. B. Kell, H. M. Ryder, A. S. Kaprelyants, and H. V. Westerhoff, â??Quantifying heterogeneity: flow cytometry of bacterial cultures,â? *An- tonie van Leeuwenhoek*, vol. 60, no. 3-4, pp. 145â??158, 1991.

[16] T. R. Malthus, *An essay on the principle of population*. Cosimo, Inc.,

2013, vol. 1.

[17] T. Akerlund, K. NordstroÂ¨ m, and R. Bernander, â??Analysis of cell size and dna content in exponentially growing and stationary-phase batch cultures of escherichia coli.â? *Journal of Bacteriology*, vol. 177, no. 23, pp. 6791â??6797, 1995.

[18] R. Allman, T. Schjerven, and E. Boye, â??Cell cycle parameters of escherichia coli k-12.â? *Journal of bacteriology*, vol. 173, no. 24, pp.

7970â??7974, 1991.

[19] O. Michelsen, M. J. T. de Mattos, P. R. Jensen, and F. G. Hansen, â??Precise determinations of c and d periods by flow cytometry in escherichia coli k-12 and b/r,â? *Microbiology*, vol. 149, no. 4, pp. 1001â??

1010, 2003.

[20] F. C. Neidhardt, â??Bacterial growth: Constant obsession withdn/dt,â?

Journal of bacteriology, vol. 181, no. 24, pp. 7405â??7408, 1999.

[21] H. Zhang, Y. Liu, B. Liu, and P. Gao, â??A novel approach for estimating growth phases and parameters of bacterial population in batch culture,â? *Science in China Series C*, vol. 49, no. 2, pp. 130â??140, 2006.

[22] W. Donachie, â??Relationship between cell size and time of initiation of dna replication,â? 1968.

[23] C. Helmstetter, S. Cooper, O. Pierucci, and E. Revelas, â??On the bacterial life sequence,â? in *Cold Spring Harbor symposia on quantitative biology*, vol. 33. Cold Spring Harbor Laboratory Press, 1968, pp. 809â??822.

[24] G. An, Q. Mi, J. Dutta-Moscato, and Y. Vodovotz, â??Agent-based mod- els in translational systems biology,â? *Wiley Interdisciplinary Reviews: Systems Biology and Medicine*, vol. 1, no. 2, pp. 159â??171, 2009.

Low-Cost Healthcare Diagnostics By Directed Evolution of Peptide Receptors in Yeast

Tyo, K. E. J., Northwestern University
Adeniran, A., Northwestern University
Stainbrook, S.,

Low cost, point-of-care (POC) diagnostics for specific proteins would have imminently useful applications in resource-poor healthcare, as well as counter-bioterrorism (detecting pathogens), and environmental monitoring (detecting contamination). However, existing technologies for protein detection are either too expensive or too bulky for portability (mass spec) or not suitable for continuous monitoring and have limited analyte possibilities (lateral flow immunoassays, i.e., pregnancy tests). To date, the majority of Synthetic Biology biosensing parts detect analytes in the cytoplasm and nucleus, which preclude most proteins that can not cross the membrane.

We have developed a directed evolution approach for tuning sensitivity and specificity of G-protein coupled receptors (GPCRs). Nature has already evolved mechanisms for sensitive and specific detection of peptides and proteins using GPCRs. In particular the yeast GPCR, Ste2p, detects the 13 amino acid peptide called a-factor. Evolving Ste2p to recognize different peptide sequences would be a relatively small evolutionary step.

Using a FACS-based high throughput screen, we have evolved receptors that can detect several different peptide sequences, including sequences with homology to important clinical biomarkers for renal failure and tuberculosis. We have shown that we can alter the sensitivity and specificity of a receptor by selecting mutant receptors that respond to low concentrations of ligand (sensitivity) and not respond to non-target ligands until the concentration is very high (specificity). Evolved receptors can discriminate peptides with only one amino acid variation.

This approach should have many useful applications in Synthetic Biology, including: healthcare diagnostics, novel cell-to-cell communication (as cells can be easily engineered to produces any number of peptide sequences), and to reveal sequence-activity relationships (SARs) for GPCRs, a class of drug-target receptors that have been difficult to characterize because of difficulties crystalizing transmembrane receptors.

Cell-cell communication in yeast using auxin biosynthesis and auxin responsive CRISPR transcription factors

Arjun Khakhar1*, Nicholas J Bolten2*, Jennifer Nemhauser3, Eric Klavins2
1. University of Washington Department of Bioengineering 2. University of Washington Department of
Electrical Engineering 3. University of Washington Department of Biology

Abstract

A true engineering framework for synthetic multicellular systems requires a programmable means of cell- cell communication. Such a communication system would enable complex behaviors, such as pattern formation, division of labor in synthetic microbial communities, and improved modularity in synthetic circuits. However, it remains challenging to build synthetic cellular communication systems in eukaryotes due to a lack of molecular modules that are orthogonal to the host machinery, easy to reconfigure, and scalable. Here, we present a novel cell-to-cell communication system in *Saccharomyces cerevisiae* (yeast) based on CRISPR transcription factors and the plant hormone auxin that exhibits several of these features. Specifically, we engineered a *sender* strain of yeast that converts indole-3-acetamide (IAM) into auxin via the enzyme iaaH from *Agrobacterium tumefaciens*. To sense auxin and regulate transcription in a *receiver* strain, we engineered a reconfigurable library of auxin degradable CRISPR transcription
factors (ADCTFs). Auxin-induced degradation is achieved through fusion of an auxin sensitive degron (from IAA co-repressors) to the CRISPR TF and co-expression with an auxin F-box protein. Mirroring the tunability of auxin perception in plants, our family of ADCTFs exhibits a broad range of auxin sensitivities. We characterized the kinetics and steady state behavior of the sender and receiver independently, and in co-cultures where both cell types were exposed to IAM. In the presence of IAM, auxin is produced by the sender cell and triggers de-activation of reporter expression in the receiver cell. The result is an orthogonal, rewireable, tunable, and arguably scalable cell-cell communication system for yeast and other eukaryotic cells.

Introduction

Multicellular systems in nature are capable of incredible feats of distributed computation and self- organization. Examples range from division of labor in filamentous algae1, to the exquisite sensitivity of the adaptive immune system2, to morphogenesis and development of tissues, organs, and multicellular
organisms. Computer scientists have shown that cells are in principle capable of computing a wide variety of functions3, generating complex morphologies4, and of making decisions5,6. Experimentally, synthetic multicellular systems have been built to

regulate populations7, synchronize oscillations8, form patterns9–11, implement logic functions through distributed computation3, and cooperate to solve problems12. However, a scalable suite of cell-cell communication modules has yet to emerge. In particular, in *Saccharomyces cerevisiae*, strategies that use components of native signal transduction pathways can lead to crosstalk and undesirable phenotypes such as growth arrest9,13,14. Such systems are not obviously portable to other eukaryotes, are difficult to reprogram, and require significant changes to the host cell to function correctly7. Here, we describe progress toward building an engineering framework for yeast cell-cell communication that is orthogonal to yeast (and many other eukaryotic cells except plants15), modular, and tunable.

Orthogonality is crucial for rationally engineering cell-cell communication. Auxin, a plant hormone, does not have measurable effects on laboratory strains of yeast16,17 when grown in standard conditions. Our receiver cells use elements of the *Arabidopsis thaliana* auxin signaling pathway. Auxin regulates plant development via a system of transcriptional corepressors, the Aux/IAA proteins (referred to as IAAs), which are degraded in the presence of the molecule auxin. Auxin stabilizes the interaction between the degron domain of an IAA and an auxin F-box protein (AFB). The result is the degradation of the IAA via polyubiquitination18. The IAAs exhibit a range of degradation rates and sensitivities to auxin that are determined, in part, by the sequence of their degron domains and in part by the AFB16,19. The degradation dynamics of a large range of auxin degrons with multiple AFBs have been previously studied and thoroughly characterized in yeast16. By using this signaling modality as the basis for our communication system, we avoid using any native yeast (or mammalian) signal transduction machinery associated with adverse phenotypes7. Additionally, the primary components of the pathway, AFBs and IAAs, have been shown to function in several different mammalian cells15, suggesting that our system may be broadly portable.

To maximize modularity, we engineered auxin responsiveness into CRISPR transcription factors (CTFs). CTFs consist of a nuclease null Cas9 protein (dCas9) fused to a transcriptional effector domain. The dCas9 can be programmed to target a locus by coexpressing a small guide RNA (gRNA) that has complementarity to the target locus at a site that is adjacent to an 'NGG' sequence, called the PAM sequence. This strategy, as demonstrated by Farzadfard et al.20 and Qi et al.21, has the benefit of modularity through easily programmable specificity: dCas9 requires only the expression of a new guide RNA for retargeting. In contrast, zinc finger or TAL DNA binding domains require the design of a new protein for each target22,23. These characteristics make CTFs an ideal candidate for signal reception and processing, as they can be targeted to any promoter in the genome that has a suitable PAM site20, can either activate or repress gene expression, and can be layered to form more complex networks23,24. In the present case, CTFs fused to the VP64 strong activator domain were targeted to a promoter upstream of GFP. In addition, these CTFs were fused to Aux/IAA degron domains and co-expressed with AFBs thereby producing auxin-degradable CRISPR transcription factors, or ADCTFs. An ADCTF is thus a modular, coupled sensor-actuator, which should allow cell-to-cell communication to be easily rewired to arbitrary outputs.

Signal production and reception in cell-cell communication is ideally tunable to achieve a broad range of sensitivities and other functions. To implement and tune auxin production

in the sender, we integrated the bacterial iaaH gene from *Agrobacterium tumefaciens* into yeast under the control of a constitutive promoter (GPD). Upon the addition indole-3-acetamide (IAM), sender cells produced a strong enough auxin signal to affect gene regulation via the ADCTFs in co-cultured receiver cells. The concentration of auxin produced can be tuned via the concentration of exogenously added IAM. For increased tunability, we developed a library of ADCTFs, each with a different degron and/or degron location, that displays a range of degradation kinetics and sensitivities to auxin. The sensitivity of the ADCTFs can be further
tuned by the selection of either the AFB2 or TIR1 F-box as the auxin receptor. Thus, components of the ADCTFs, the auxin degron, and the transcriptional effector domain can all be swapped to obtain, respectively, a range of auxin sensitivities, and repression versus activation.
In summary, the combination of sender and receiver modules described here forms the foundation of an orthogonal, modular, and tunable cell-cell communication framework for yeast. We demonstrate each of these aspects of the system below by describing how the senders and receivers behave in isolation, and that they can be combined in co-culture to form a simple communication channel.

Results

Synthetic, scalable, auxin-modulated transcription factors

To link an auxin sensor to diverse transcriptional responses and targets, we designed auxin degradable CRISPR transcription factors (ADCTFs) with three modular domains (Figure 1A). The core component of the ADCTFs is the CRISPR-based transcription factor described by Farzadfard et al18, wherein a deactivated Cas9 protein functions as a programmable DNA binding module. The dCas9 was fused to a transcriptional effector domain, in this case the transcriptional activator VP64, and to an IAA degron. In the presence of an AFB, ADCTFs should be ubiquitinated and degraded when exposed to auxin. We tested the ADCTFs by targeting them to activate the expression of EGFP from a minimal CYC1 promoter and observed deactivation of fluorescence upon the addition of auxin. In the absence of auxin, functional ADCTFs acted identically to controls lacking a gRNA (Figure 1B). When a functional (coexpressed with
gRNA) activator ADCTF was degraded in the presence of auxin, fluorescence dropped to levels at or below the control (no gRNA) levels. Auxin dependent regulation was independent of the promoter being regulated by the ADCTF (Supplementary Figure 1B). The observed effect was also reversible: when auxin was removed from the system, reporter expression returned to its activated state (Supplementary Figure 1A).
One design consideration for building the ADCTFs was the position of the degron within the fusion position. We hypothesized that degron position could alter accessibility to the AFB or otherwise interfere with protein folding thus modulating auxin sensitivity. We explored several possible positions for the degron relative to the other domains (Figure 1C). In all cases, the degron was flanked by flexible linkers composed of five repeats of the amino acid sequence "GS" to limit fusion-associated misfolding. Changing the position of the degron dramatically altered the sensitivity range, defined as the range of

auxin concentrations between which steady-state fluorescence drops from 90% of maximum to 10%
(Figure 1D). Position one is sensitive to the lowest levels of auxin, but also saturates earlier than positions two and three. Placing the degron on either side of dCas9 (positions one and two) resulted in higher auxin sensitivity than position three where the degron was placed at the C-terminal end of the fusion. The percentage drop from maximal activation upon auxin induction was directly correlated to auxin sensitivity, with position one dropping to basal levels at steady state, and positions two and three having progressively smaller effects post induction (Supplementary Fig 2). Altering the position of the degron coarsely tuned the upper and lower bounds of the sensitivity range of the ADCTF. However, since the position one variant was the most sensitive to auxin and had the highest fold change, we chose to fuse degrons in all further ADCTF variants at position one.

Engineered ADCTF variants exhibit a broad range of auxin sensitivities and degradation kinetics

The Aux/IAA family of 29 transcriptional corepressors have been shown to exhibit a large range of degradation rates and sensitivities to auxin in yeast16. This range of responses to the same auxin signal is thought to result in part from the sequence of the different IAA degron domains, and in part from the varying activities different auxin-signaling F-box proteins, each showing different affinities for specific
IAAs. We built a library of ADCTFs using degrons from IAA14, IAA15, and IAA17 and coexpressed them with either of two AFBs (AFB2 or TIR1). These degrons have been previously characterized as encompassing a range of auxin-induced degradation rates. In general, AFB2 promotes faster degradation of IAAs than TIR1. In addition, we included a recently characterized mutant of TIR1, TIR1-DM25, which
has been shown to greatly accelerate auxin-induced TIR1 degradation.
All pairwise combinations of ADCTFs and F-box proteins were tested for their temporal response and
dose response to auxin. Temporal responses, all performed with 30 μM auxin induction, exhibited a range of degradation kinetics that depended on both the choice of ADCTF degron and the F-box protein (Figure
2B). The kinetics, characterized by the time to 50% degradation, can be coarsely tuned by the choice of F-Box protein used, with TIR1-DM being the fastest overall, followed by AFB2 and TIR1. Within this coarse tuning, the choice of degron allowed for smaller changes in kinetics. The ADCTF with the degron from IAA15 (ADC15) seemed to have the overall fastest kinetics. The only exception to this trend was the
interaction between AFB2 and ADC17, which had the fastest degradation rate. All the ADCTFs had
approximately the same percentage change from maximal activation upon auxin induction at steady state. Thus, tuning kinetics by swapping F-box proteins or degrons had a minimal effect on the steady state response to auxin. Most variants dropped to approximately seventy 75% of maximal activation at steady state with a few between ten percent higher or lower than the mean (Supplementary Fig 3D). The
ADCTFs exhibited varied sensitivity to auxin that depended on the combination of the

degron on the ADCTF and the F-box protein. Swapping F-box proteins allowed for more coarse grain tuning of sensitivity range with TIR1 conferring the broadest sensitivity range overall and TIR1-DM conferring the narrowest (Figure 2C). Swapping degrons allows smaller changes, as was observed within the AFB2
variants wherein there is a progressively narrower sensitivity range from ADC14 to ADC17. The dynamics and steady state behavior of the ADCTFs in response to auxin correspond to the behavior of previously characterized IAA proteins, from which the degrons were taken, in yeast14. The only exception being the
degron 17 variant, which had much slower degradation kinetics in the ADCTF context in a TIR1-DM background. This result suggests that the auxin responsive behavior may be predictably tuned by swapping degron and F-box protein variants.

Yeast produce tunable levels of auxin via expression of iaaH from *Agrobacterium tumefaciens*

To generate an auxin producing strain, we integrated half of the IAM pathway from *Agrobacterium tumefaciens* into yeast26. The IAM pathway is a two-step enzymatic process that converts tryptophan to IAM and then into auxin. The first step is via tryptophan-2-monooxygenase (iaaM, not examined here). The second step is catalyzed by indole-3-acetamide hydrolase (iaaH). To test whether yeast could produce auxin from IAM using only the second enzyme (iaaH), we integrated the iaaH gene from the *Agrobacterium tumefaciens* Ti plasmid27 into an auxin reporting yeast strain (Figure 3A) containing a IAA- YFP fusion protein. After adding IAM, reporter degradation rates were measured via time-lapse cytometry (Figure 3B). Upon the addition of IAM, sender strains produced an auxin response comparable to that of native auxin (Figure 3C). In addition, for a given concentration of IAM, the steady state fluorescence values converge to those of auxin (Figure 3D). There was no significant delay between the addition of
IAM and the production of auxin, so the transport and production of auxin from IAM can be assumed to be faster than the reporter's dynamics. We then investigated intercellular auxin production by coculturing the sender strain with an auxin sensor strain that could be distinguished via its mCherry signal (Figure 4A). Rather than a dose response of IAM, increasing fractions of sender cells were cocultured with sensor strains in constant amount of the auxin precursor (300 μM) to test the dependence of auxin production on sender cell concentration (Figure 4B). Greater concentrations of sender cells produced a greater auxin
response in sensor cells, though both the kinetic and steady state behavior suggest that there is less
auxin in the media than in sender cells.(Figure 4C, Figure 4D).

Sender cells produce a tunable auxin response in receiver cells

Sender and receiver cells were cocultured in different ratios to measure the effect of sender cell concentration on auxin signal production. Senders constitutively express iaaH and the receivers expressed an activating ADCTF and a gRNA targeting a minimal CYC1 promoter driving EGFP (Figure
5A). After adding a saturating amount of the IAM and growing the coculture overnight,

we observed a reduction in gene activation in the receiver strain comparable to direct addition of auxin (Figure 5B). Three different receiver strains with a range of responses to auxin were tested with the sender strain. All the receiver strains produced an auxin response and behaviors were consistent to those observed via the direct addition of auxin, suggesting that the sender module is compatible with any ADCTF-based receiver module (Supplementary Figure 4). In addition, a 10% fraction of sender cells is sufficient to a significant change in fluorescence in receiver cell at steady state and a 50% fraction produces a nearly saturating signal (Figure 5C, Figure 5D).

Discussion

Our system is based on a signal transduction modality that is unique to plants and so is orthogonal to native yeast signal transduction pathways, as well as to mammalian cells15. The simplicity of the system will hopefully allow it to be ported to other contexts, such as mammalian cells. The ADCTF library allows the generation of a range of responses to the same auxin signal, and can in principle be connected to any gene of interest, or to another synthetic gene circuit. Additionally, auxin production levels can also be tuned by titrating in different amounts of IAM. It may also be possible to tune the diffusivity of auxin in yeast26, or to harness the sequestration and turnover pathways of auxin found in plants. Our approach of detecting small molecules via F-Box mediated degradation of a transcription factor is potentially scalable as there are other plant hormones such as jasmonate that use a very similar signaling pathway28. Current work involves building on these characteristics to produce more complex multicellular behaviors. For example, feedback systems can be built through regulation of the iaaH gene via the ADCTFs. More generally, our results form the basis platform for implementing distributed decision making, pattern formation, and other complex cell-to-cell communication based multicellular behaviors.

Methods

Strain construction

Building off the work of Farzadfard et al20, the reporter is a yeast-enhanced green fluorescent protein driven by a truncated CYC1 promoter. This reporter was integrated at the URA3 locus in the genome of the W303-1A ADE2 strain of *Saccharomyces cerevisiae* and this reporter strain was used as the parent for all ADCTF strains. All gRNA was driven by an ADH1 promoter driven construct that consists of a gRNA flanked on each side by a hammerhead and an HDV ribozyme, facilitating expression from an RNA polymerase II promoter. All the gRNA constructs were integrated at the HIS3 locus. AFB2, TIR1 and TIR1-DM were integrated, respectively, at the LEU2 loci, and were driven by the GPD promoter. The ADCTFs were constructed by fusing an SV40 nuclear localization tag, a VP64 activation domain, and an auxin degron to a nuclease null

version of the Cas9 protein from *Streptococcus pyogenes*. The auxin degron used for all characterization, unless otherwise mentioned, was the t1 truncation of the degron from IAA17 from *Arabidopsis* that was characterized previously to have the fastest speed of degradation in the presence of AFB2 degradation machinery16. The other degrons used were the domain two regions from IAA14 and IAA15. The ADCTF is driven by a beta-estradiol inducible version of the GAL1 promoter
integrated at the TRP locus in the genome in all strains29. The iaaH gene was amplified via PCR from the Ti plasmid of *Agrobacterium tumefaciens* and cloned via the Gateway™ method into a single-integrating HIS3 plasmid behind the strong TDH3 promoter. The integrating plasmid cassette was produced via digestion of the plasmid by PmeI and integrated into an auxin reporter strain via a standard lithium
acetate transformation method30.

Cytometry

All cytometry measurements were acquired with an Accuri C6 cytometer with attached CSampler apparatus using 488 nm and 640 nm excitation lasers and a 533 nm (FL-1: YFP/GFP) emission filter. Experiments involving time course data were taken during log phase via the following preparation: 16 hours of overnight growth in synthetic complete medium in a 30°C shaker incubator followed by 1:100 dilution into fresh, room-temperature medium. After 5 hours of growth under the same incubation conditions, 100 μL aliquots were read periodically until the completion of the experiment. For experiments involving steady state behavior, cultures were grown overnight, then diluted down in the morning 1:100 in fresh media and grown for 5 hours to log phase. They were then induced and allowed to grow for
between five and twenty four hours depending on the experiment and then read on the cytometer. Cytometry data were analyzed using custom R scripts and the flowCore31 package using the following steps: (1) gating for the yeast population, (2) gating for separate sending / receiving strains via the yellow (GFP) and red (mCherry) channels, and the generation of mean fluorescence values.

References

(1) Wilcox, M., Mitchison, G. J., Smith, R. J., and Road, H. (1973) Pattern formation in the blue- green alga, Anabaena I. Basic mechanisms. *J. Cell Sci. 12*, 707–723.
(2) Medzhitov, R. (2007) Recognition of microorganisms and activation of the immune response.
Nature 449, 819–826.
(3) Regot, S., Macia, J., Conde, N., Furukawa, K., Kjellén, J., Peeters, T., Hohmann, S., de Nadal, E., Posas, F., and Solé, R. (2011) Distributed biological computation with multicellular engineered networks. *Nature 469*, 207–11.
(4) Abelson, H., Weiss, R., Allen, D., Coore, D., Hanson, C., Homsy, G., Knight, T. F., Nagpal, R., Rauch, E., and Sussman, G. J. (2000) Amorphous computing. *Commun. ACM 43*, 74–82.

(5) Barcena Menendez, D., Senthivel, V. R., and Isalan, M. (2014) Sender-receiver systems and applying information theory for quantitative synthetic biology. *Curr. Opin. Biotechnol. 31C*, 101–
107.
(6) Jang, S. S., Oishi, K. T., Egbert, R. G., and Klavins, E. (2012) Specification and simulation of synthetic multicelled behaviors. *ACS Synth. Biol. 1*, 365–374.
(7) You, L., Cox, R. S., Weiss, R., and Arnold, F. H. (2004) Programmed population control by cell-cell communication and regulated killing. *Nature 428*, 868–871.
(8) Danino, T., Mondragón-Palomino, O., Tsimring, L., and Hasty, J. (2010) A synchronized quorum of genetic clocks. *Nature 463*, 326–30.
(9) Chen, M.-T., and Weiss, R. (2005) Artificial cell-cell communication in yeast Saccharomyces cerevisiae using signaling elements from Arabidopsis thaliana. *Nat. Biotechnol. 23*, 1551–5.
(10) Sohka, T., Heins, R. a, Phelan, R. M., Greisler, J. M., Townsend, C. a, and Ostermeier, M. (2009) An externally tunable bacterial band-pass filter. *Proc. Natl. Acad. Sci. U. S. A. 106*,
10135–10140.
(11) Basu, S., Gerchman, Y., Collins, C. H., Arnold, F. H., and Weiss, R. (2005) A synthetic multicellular system for programmed pattern formation. *Nature 434*, 1130–4.
(12) Shou, W., Ram, S., and Vilar, J. M. G. (2007) Synthetic cooperation in engineered yeast populations. *Proc. Natl. Acad. Sci. U. S. A. 104*, 1877–82.
(13) Youk, H., and Lim, W. a. (2014) Secreting and sensing the same molecule allows cells to achieve versatile social behaviors. *Science 343*, 1242782.
(14) Zhang, N.-N., Dudgeon, D. D., Paliwal, S., Levchenko, A., Grote, E., and Cunningham, K. W. (2006) Multiple signaling pathways regulate yeast cell death during the response to mating pheromones. *Mol. Biol. Cell 17*, 3409–3422.
(15) Nishimura, K., Fukagawa, T., Takisawa, H., Kakimoto, T., and Kanemaki, M. (2009) An auxin-based degron system for the rapid depletion of proteins in nonplant cells. *Nat. Methods 6*,
917–22.
(16) Havens, K. a, Guseman, J. M., Jang, S. S., Pierre-Jerome, E., Bolten, N., Klavins, E., and Nemhauser, J. L. (2012) A synthetic approach reveals extensive tunability of auxin signaling. *Plant Physiol. 160*, 135–42.
(17) Pierre-Jerome, E., Jang, S. S., Havens, K. a, Nemhauser, J. L., and Klavins, E. (2014) Recapitulation of the forward nuclear auxin response pathway in yeast. *Proc. Natl. Acad. Sci. U. S. A. 111*, 9407–12.
(18) Gray, W. M., Kepinski, S., Rouse, D., Leyser, O., and Estelle, M. (2001) Auxin regulates
SCF(TIR1)-dependent degradation of AUX/IAA proteins. *Nature 414*, 271–276.
(19) Villalobos, C., a, L. I., Lee, S. C., De Oliveira, C., Ivetac, A., Brandt, W., Armitage, L., Sheard, L. B., Tan, X., Parry, G., Mao, H., Zheng, N., Napier, R. M., Kepinski, S., and Estelle, M. (2012) A combinatorial TIR1/AFB–Aux/IAA co-receptor system for differential sensing of auxin *8*.
(20) Farzadfard, F., Perli, S., and Lu, T. (2013) Tunable and multifunctional eukaryotic transcription factors based on CRISPR/Cas. *ACS Synth. Biol. 2*, 604–13.

(21) Qi, L. S., Larson, M. H., Gilbert, L. A., Doudna, J. A., Weissman, J. S., Arkin, A. P., and Lim, W. A. (2013) Repurposing CRISPR as an RNA-Guided Platform for Sequence-Specific Control of Gene Expression. *Cell 152*, 1173–1183.
(22) Khalil, A. S., Lu, T. K., Bashor, C. J., Ramirez, C. L., Pyenson, N. C., Joung, J. K., and Collins, J. J. (2012) A Synthetic Biology Framework for Programming Eukaryotic Transcription Functions. *Cell 150*, 647–658.
(23) Kiani, S., Beal, J., Ebrahimkhani, M. R., Huh, J., Hall, R. N., Xie, Z., Li, Y., and Weiss, R. (2014) CRISPR transcriptional repression devices and layered circuits in mammalian cells. *Nat. Methods 11*, 723–6.
(24) Nielsen, A. A. K., and Voigt, C. A. (2014) Multi-input CRISPR/Cas genetic circuits that interface host regulatory networks. *Mol. Syst. Biol. 10*, 1–12.
(25) Yu, H., Moss, B. L., Jang, S. S., Prigge, M., Klavins, E., Nemhauser, J. L., and Estelle, M. (2013) Mutations in the TIR1 auxin receptor that increase affinity for auxin/indole-3-acetic acid proteins result in auxin hypersensitivity. *Plant Physiol. 162*, 295–303.
(26) Zhao, Y. (2010) Auxin biosynthesis and its role in plant development. *Annu. Rev. Plant Biol.*
61, 49–64.
(27) P?curar, D. I., Thordal-Christensen, H., P?curar, M. L., Pamfil, D., Botez, M. L., and Bellini, C. (2011) Agrobacterium tumefaciens: From crown gall tumors to genetic transformation. *Physiol. Mol. Plant Pathol. 76*, 76–81.
(28) Turner, J. G., Ellis, C., and Devoto, A. (2002) The jasmonate signal pathway. *Plant Cell 14*
Suppl, S153–S164.
(29) McIsaac, R. S., Silverman, S. J., McClean, M. N., Gibney, P. a, Macinskas, J., Hickman, M. J., Petti, A. a, and Botstein, D. (2011) Fast-acting and nearly gratuitous induction of gene expression and protein depletion in Saccharomyces cerevisiae. *Mol. Biol. Cell 22*, 4447–59.
(30) Gietz, R. D., and Woods, R. A. (2002) Transformation of yeast by lithium acetate/single- stranded carrier DNA/polyethylene glycol method. *Methods Enzymol. 350*, 87–96.
(31) Hahne, F., LeMeur, N., Brinkman, R. R., Ellis, B., Haaland, P., Sarkar, D., Spidlen, J., Strain, E., and Gentleman, R. (2009) flowCore: a Bioconductor package for high throughput flow cytometry. *BMC Bioinformatics 10*, 106.

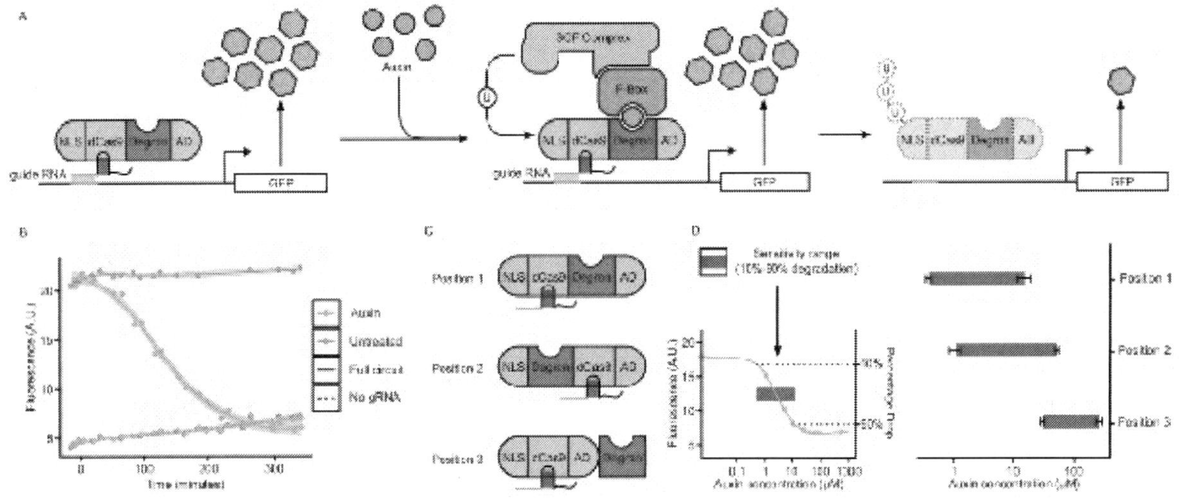

Figure 1. A) The ADCTF design and the molecular mechanism behind its function. An ADCTF is made up of a dCas9 protein fused to an NLS, an activation domain and an auxin sensitive degron. In the presence of auxin, the degron recruits an Auxin Sensing F-box (AFB) protein to form an SCF complex (an E3 ubiquitin ligase). The subsequent ubiquination and degradation of the ADCTF deregulates the gene targeted by the ADCTF. B) Time-lapse cytometry of ADCTF cells with a GFP-producing gRNA target following the addition of auxin or no treatment as well as with and without a guide RNA. The gray ribbon indicates the 95% confidence interval. Following treatment with auxin, the GFP level of the strain expressing gRNA dropped to basal levels (equivalent to a strain with no gRNA). C) Schematic representation of the three fusion proteins tested for the effect of degron position on ADCTF properties.

D) Sensitive range characterization of the three degron position variants at steady state. Horizontal bars indicate the range of auxin concentrations between which mean steady-state fluorescence (measured via cytometry) drops from 90% of maximum to 10%. A larger sensitive range correlated with higher maximum fold changes upon induction (Supplementary Figure 2).

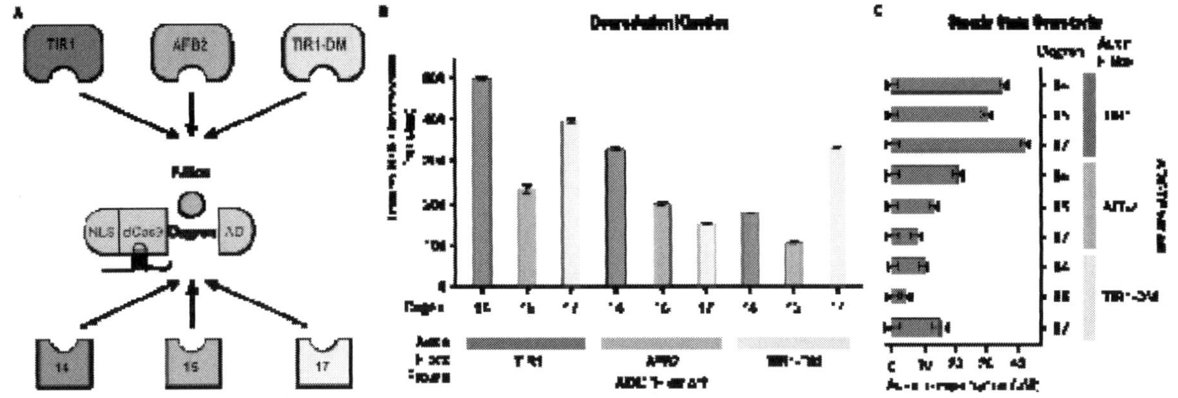

Figure 2. A) The ADCTF (auxin receiver) strain library was generated from all pairwise combinations of

three Auxin Sensing F-box protein variants (AFB2, TIR1, TIR1-DM) and three auxin degron variants (from IAA14, IAA15 and IAA17). B) Receiver strain library degradation kinetics measured via time-lapse cytometry. The kinetics of ADCTF responses to auxin were characterized by the time at which fluorescence dropped to fifty percent of maximum: a smaller time implies a faster response. The ADCTF library displays a wide range of degradation kinetics that were modulated by both the choice of F-box protein and the auxin degron. C) Auxin sensitivity ranges for the ADCTF library. Blue bars represent the auxin sensitivity range at steady state as defined in Figure 1.

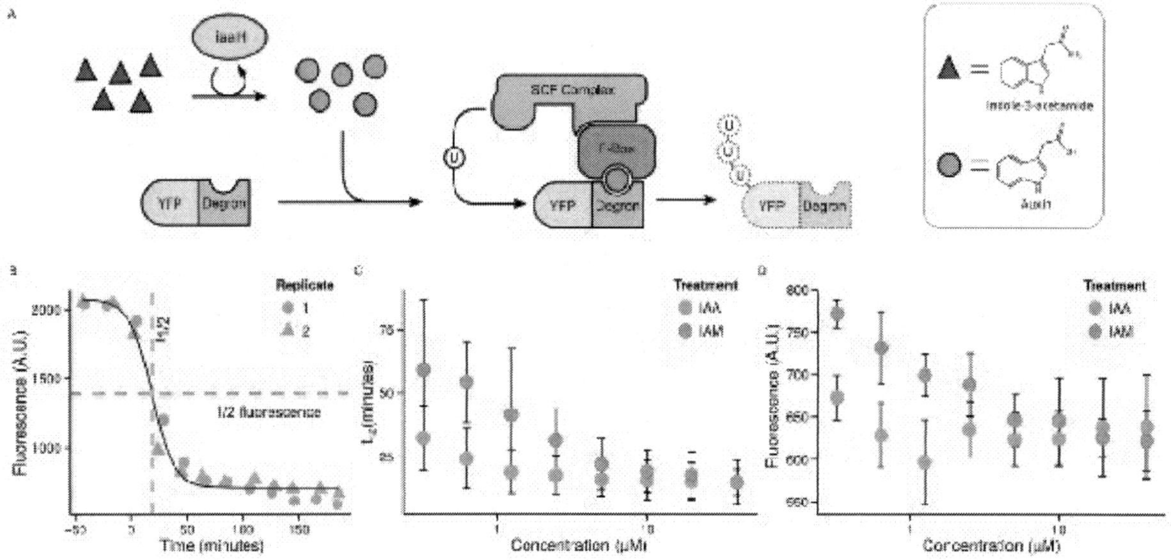

Figure 3. A) Auxin sender strain design. The iaaH enzyme of *Agrobacterium tumefaciens* catalyzes the conversion of indole-3-acetamide (IAM) into auxin, inducing the degradation of proteins fused to an auxin degron. The iaaH enzyme (sender cells) was integrated into an auxin reporting strain (the EYFP-IAA17|AFB2 strain from *Havens et al*16) to test for internal auxin production. B) Kinetic auxin response to IAM addition in sender strains. Following the addition of IAM, the fluorescence of sender cells decreased to basal levels. The time to half-maximal (t1/2) fluorescence was used to measure the rate of reporter degradation. C) Auxin-induced degradation rate in response to varying doses of either IAM or auxin. Sender cells were treated with either auxin or IAM and read at regular intervals producing time courses as in part B. Nonlinear fitting was used to generate t1/2 values. For a given molarity, treatment with IAM produces an auxin-induced degradation similar to, but weaker than, direct treatment with auxin. D) Steady state fluorescence in response to varying doses of either IAM or auxin taken from the same dataset as part C. As the concentration of IAM was increased, a lower steady state fluorescence was produced.

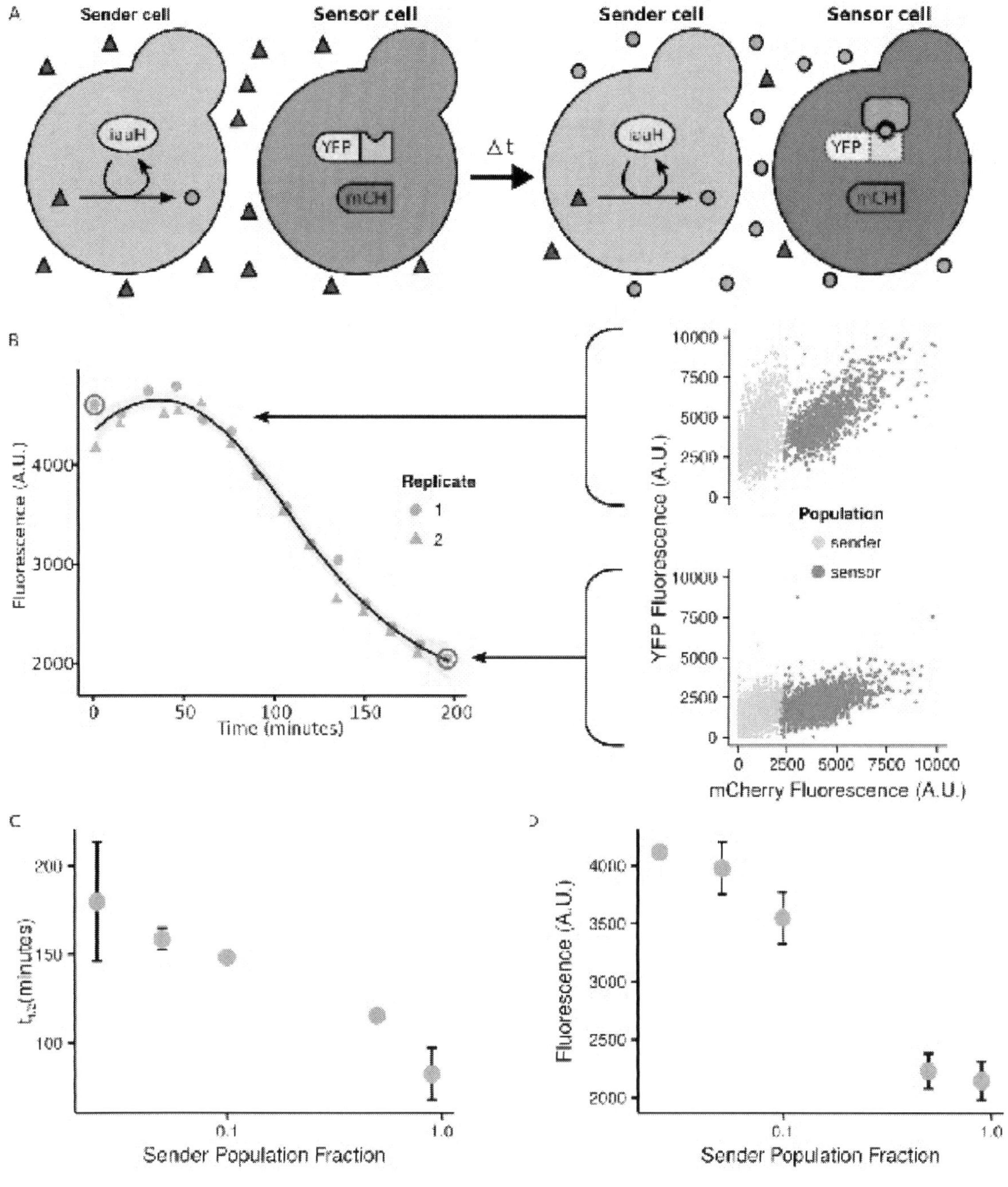

Figure 4. A) Sender-sensor multicellular auxin signaling strains. Sender cells are identical to those in Figure 3 and therefore produce auxin upon the addition of exogenous IAM and sense auxin production via an EYFP-IAA17 reporter. Sensor cells express an EYFP-IAA17 and TIR1 and are distinguished experimentally through the expression of mCherry. In coculture, IAM diffuses into sender cells where it is converted into diffusible auxin that then degrades EYFP- degron proteins in either the sender or sensor cell types. B) Auxin-induced degradation of

EYFP-IAA17 in sensor cells cocultured with sender cells in 300 μM IAM. Data for sensor cells can be separated from sender cells via their mCherry signal. The line represents a LOESS fit

and the light orange ribbon represents a 95% confidence interval of the fit. C) Sender cell fraction dose response. Each fraction had the same volume, so a larger fraction indicates a larger concentration of sender cells in coculture. As the sender cell population increases, the degradation rates decreases. D) Steady state fluorescence in response to varying doses of either IAM or auxin taken from the same dataset as part C. As the concentration of sender cells was increased, a lower steady state fluorescence was produced, flatting out at around a 50:50 split.

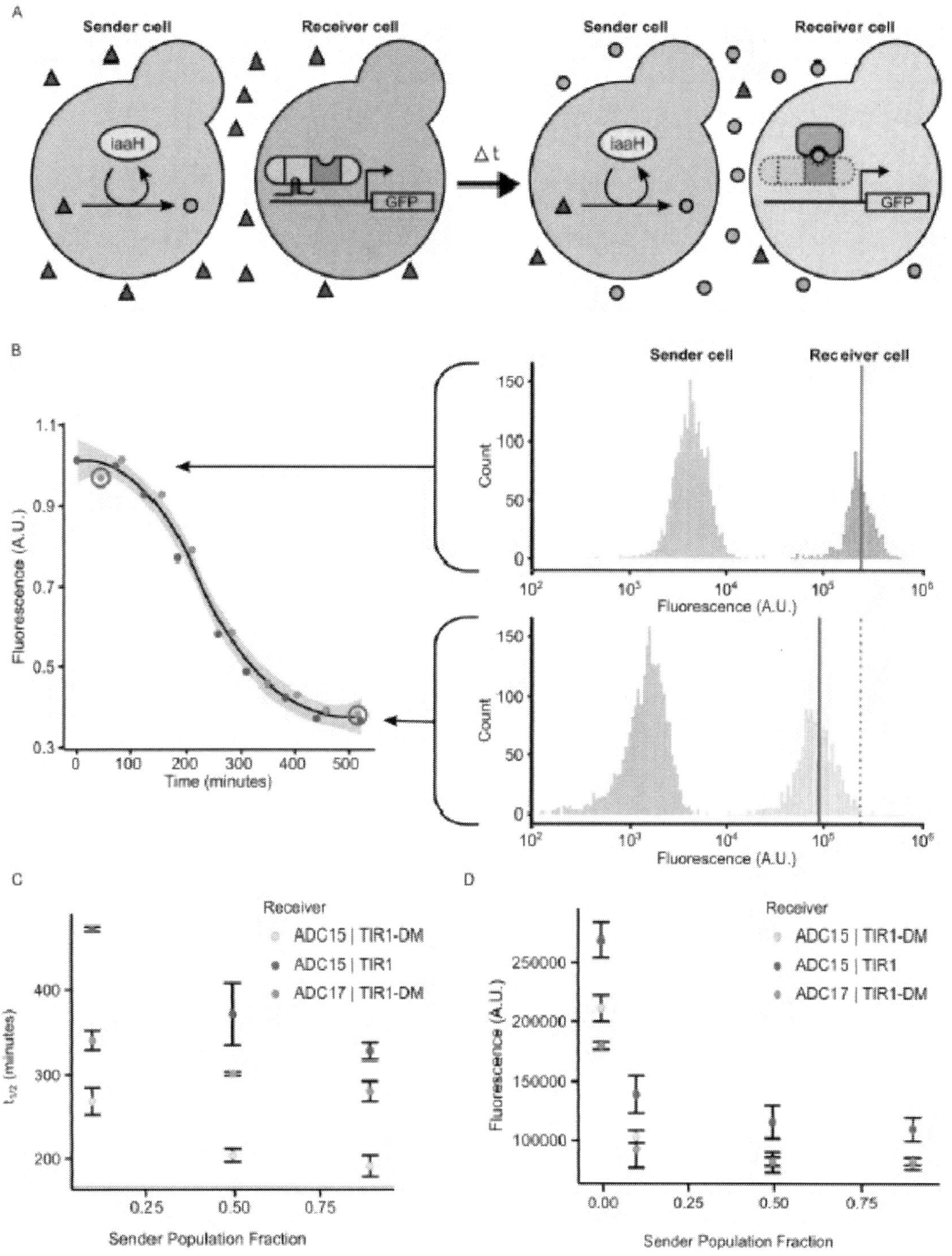

Figure 5. A) Coculture of sender and receiver strains. Sender cells convert IAM into auxin that then diffuses out of sender cells and into receiver cells where it causes the degradation of ADCTFs, producing a drop in fluorescence. B) Time course data for two replicates (shown in blue and purple dots) of a coculture of equal concentrations of sender and receiver cells is plotted on the left. The line represents

a LOESS fit and the gray ribbon represents a 95% confidence interval of the fit. On the right, histograms display distinct populations of sender (gray, left) and receiver (green, right) cells. In the presence of sender cells treated with IAM, receiver cells dropped in fluorescence over time. As in figures

3 and 4, sender cells express also an EYFP-IAA and AFB auxin reporter and therefore also show a decrease in fluorescence. Without IAM, receiver cells did not show a significant decrease in fluorescence. C) Degradation rates (measured as t1/2) in receiver strains in response to sender cell concentration. As
the fraction of sender cells increased, there is a more dramatic auxin effect in receiver cells that saturates at approximately even fractions of send to receive.

Supplementary Information

Supplementary Figure 1.

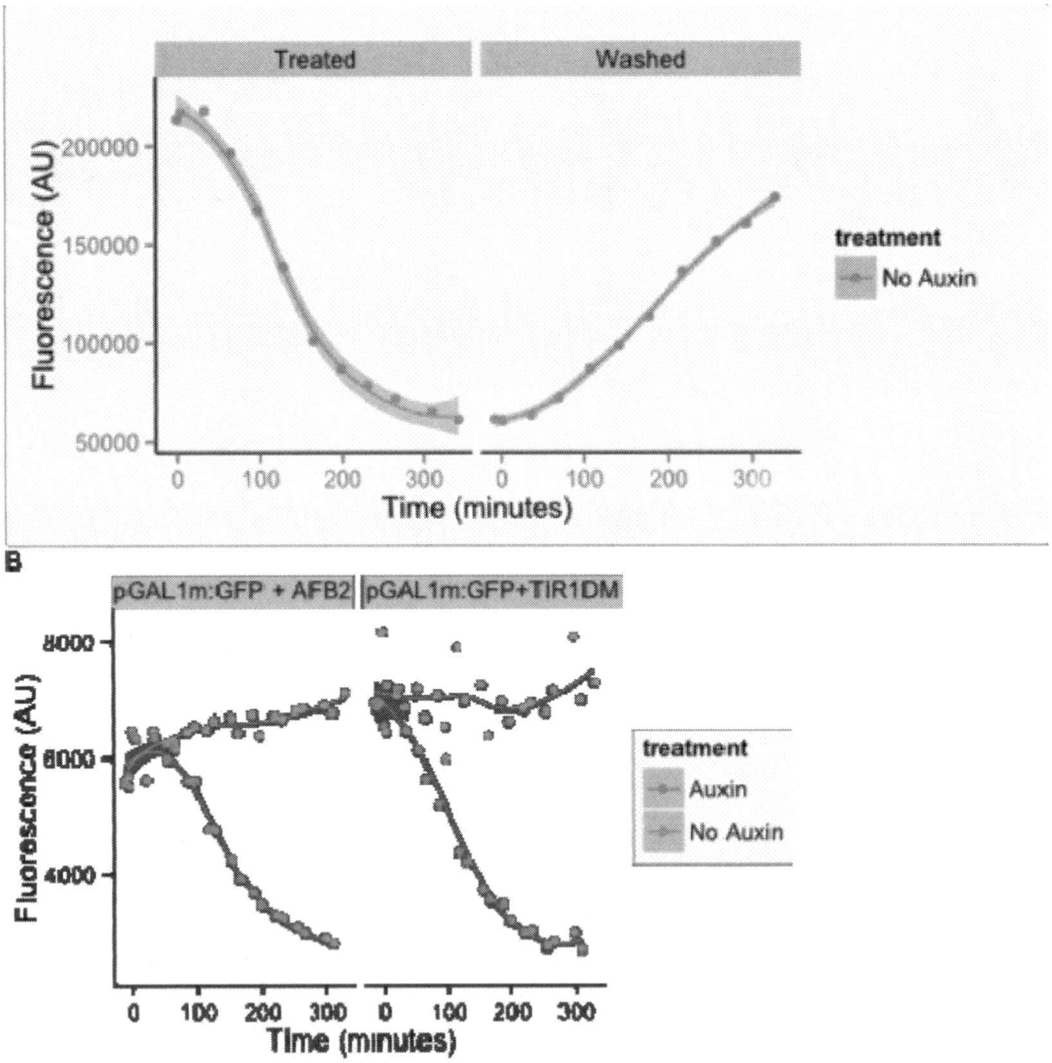

A)Raw data of a time course where receiver cells with ADC17 and AFB2 were treated

with Auxin at time zero and then fluorescence was observed until the sample reached steady state, after which they were washed to remove auxin and recovery was observed over a similar period as induction. The lines are LOESS fits to the data and the gray ribbon represents a 95% confidence interval of the fit.
B) Raw data of two time course replicates of receiver strains with an ADC TF with degron 17 and either
AFB2 or TIR1-DM F boxes. Both these strains have a minimal pGAL promoter driving GFP production
and a guide RNA that targets this promoter. Cultures treated with 30uM auxin show a comparable release of regulation to when pCYC1 was targeted with the ADC TF. The lines are LOESS fits to the data and the gray ribbon represents a 95% confidence interval of the fit.

Supplementary Figure 2.

A

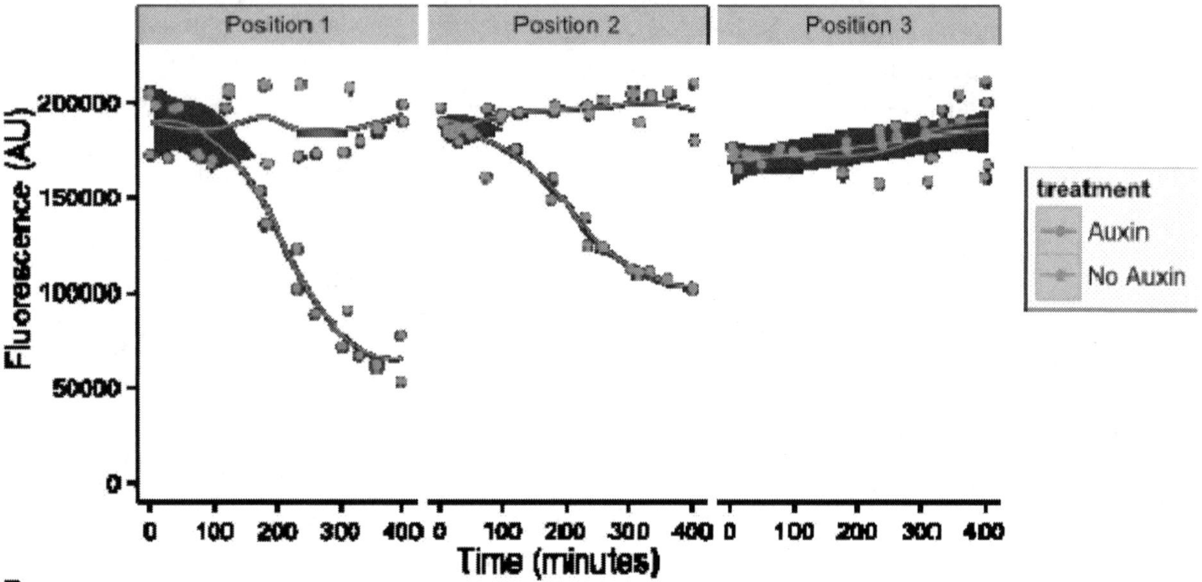

B

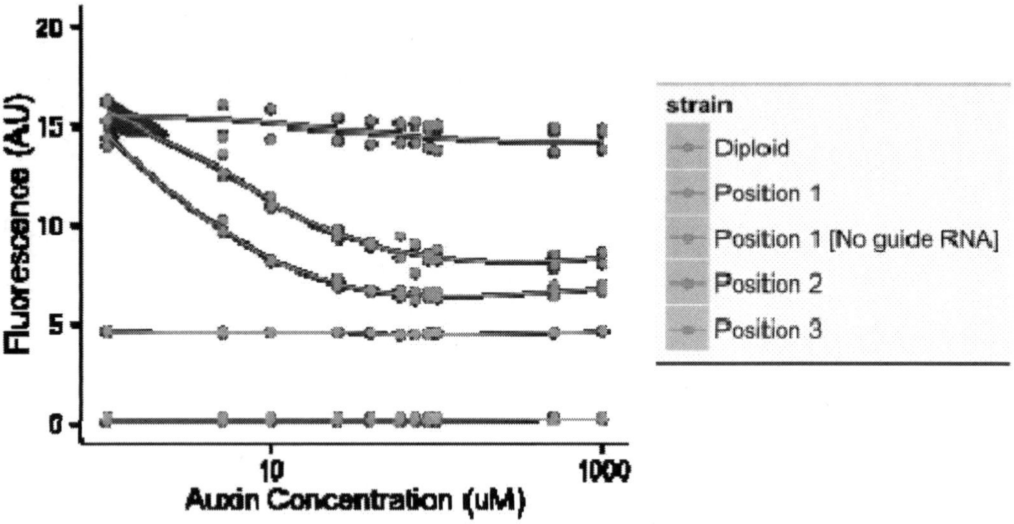

A) Raw data of two time course replicates of the three positional variants of the ADC TFs for cultures that were treated with 10uM auxin, as well as parallel untreated controls. The auxin treated cultures have a consistent drop in fluorescence, with position one having the largest drop at steady state. Position three has no noticeable drop at this auxin concentration. The lines are LOESS fits to the data and the gray ribbon represents a 95% confidence interval of the fit.

B) Raw data of two dose response replicates of the three positional variants of the ADC TFs five
hours after induction with auxin. Position 1 has the highest sensitivity to auxin, and consequently saturates first, followed by position 2 and then position 3. The lines are LOESS fits to the data and the gray ribbon represents a 95% confidence interval of the fit.

Supplementary Figure 3

A

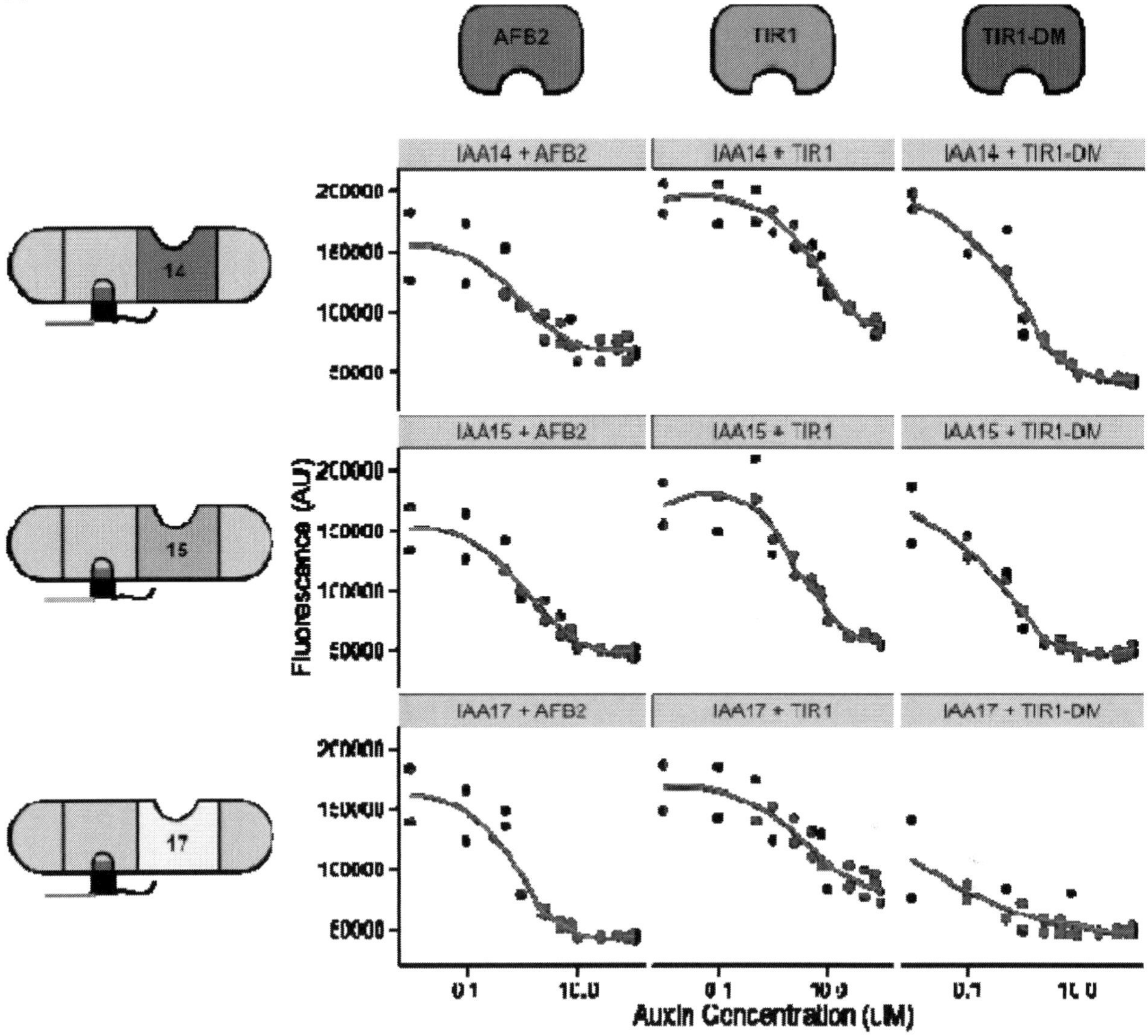

B

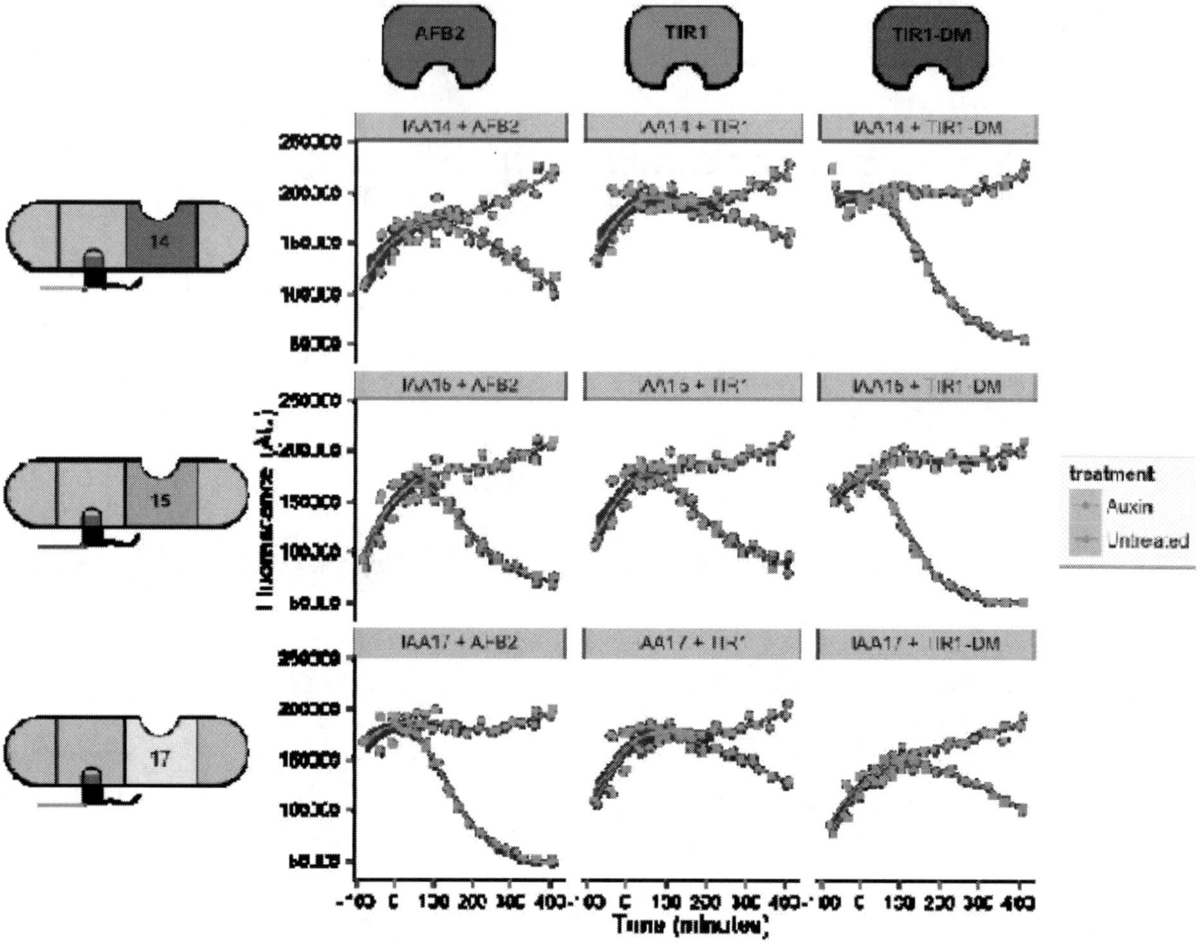
AFB2
TIR1
TIR1-DM
IAA14 + AFB2
IAA14 + TIR1
IAA14 + TIR1-DM
IAA15 + AFB2
IAA15 + TIR1
IAA15 + TIR1-DM
IAA17 + AFB2
IAA17 + TIR1
IAA17 + TIR1-DM
14
15
17
250000
200000
150000
100000
50000
Fluorescence (A.U.)
-100 0 100 200 300 400
Time (minutes)
treatment
Auxin
Untreated

C

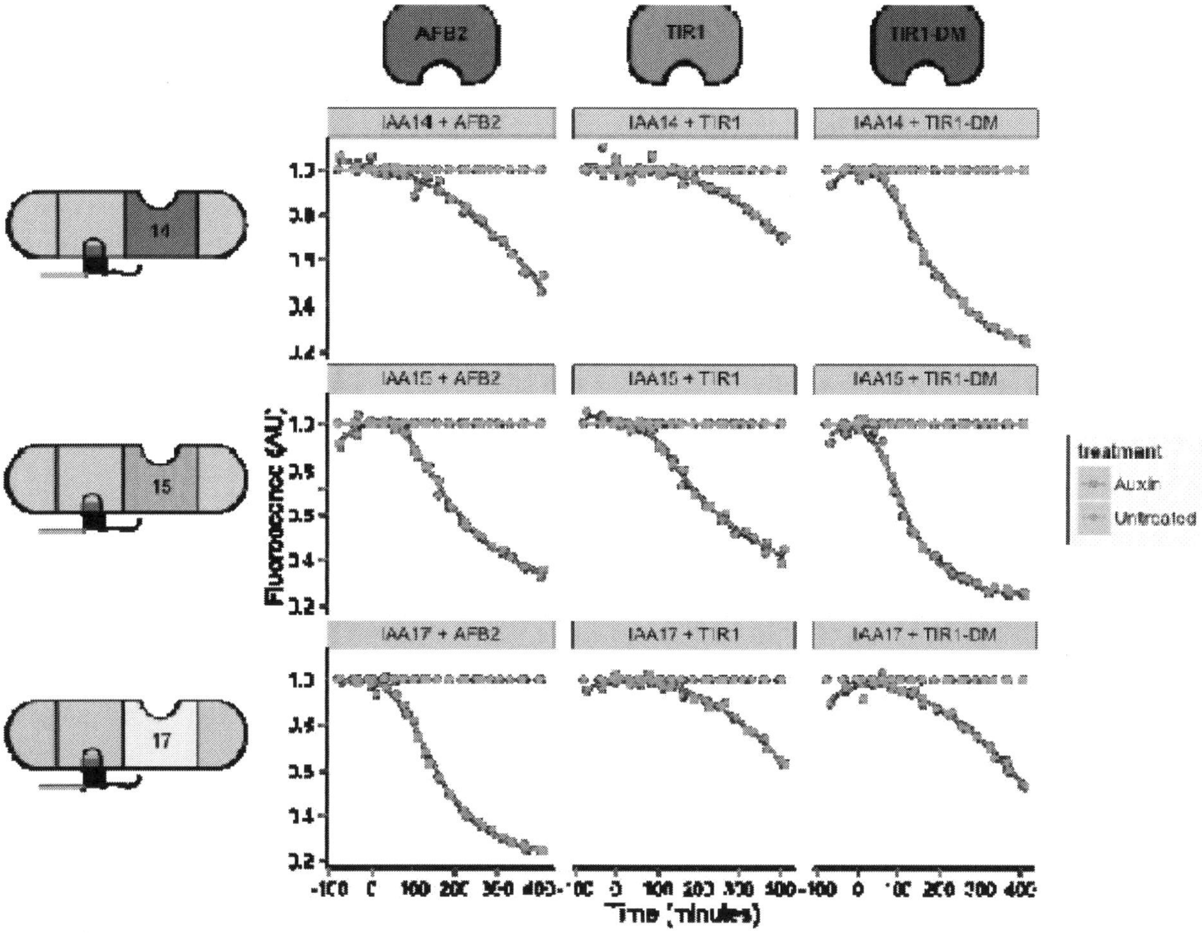

D

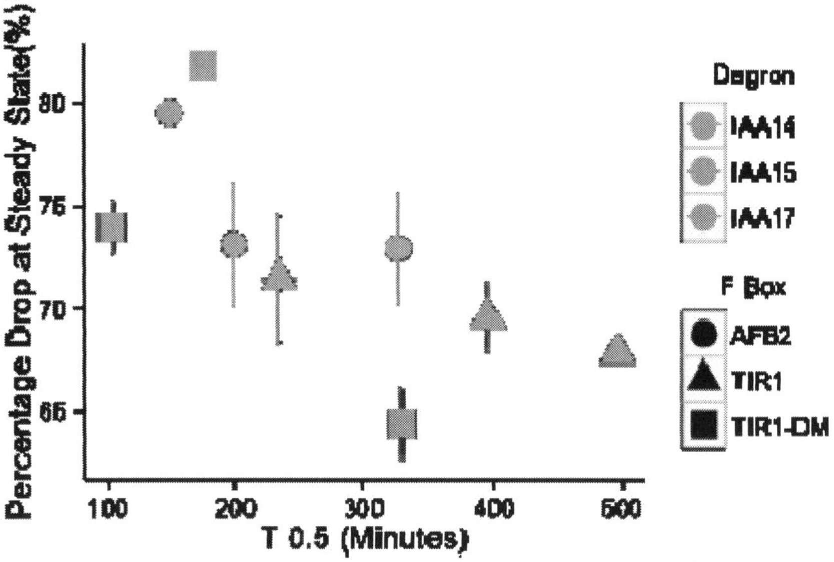

A) Raw data of two dose response replicates of all possible pairwise combination of different degrons on ADC TFs with the three F Boxes AFB2, TIR1 and TIR1-DM twenty four hours after induction with auxin. A range of auxin sensitivities are represented in the library, TIR1 being the most insensitive, and AFB2 and TIR1-DM being much more

sensitive, depending on the degron being used. The lines are LOESS fits to the data and the gray ribbon represents a 95% confidence interval of the fit.
B and C) Raw data and normalized data repectively of two time course replicates of all possible pairwise combination of different degrons on ADC TFs for cultures that were treated with 30uM auxin at time zero. A range of different degradation kinetics are observed with some reaching steady state within 400 minutes such as IAA15 + TIR1-DM, whereas other are still decreasing. The lines are LOESS fits to the data and the gray ribbon represents a 95% confidence interval of the fit.
D) The raw time course data from plot C is summarized here by plotting the T 0.5, time taken for the fluorescence to drop to fifty percent of its maximal drop, versus the percentage drop at steady state, measured twenty four hours after induction. While there appears to be a large range of degradation rates achievable by using different combinations of degrons and ADC-TFs, the steady state percentage changes are all approximately equal, with a variance of 20% between
the highest and the lowest, but with most clustering at approximately 75%.

Supplementary Figure 4

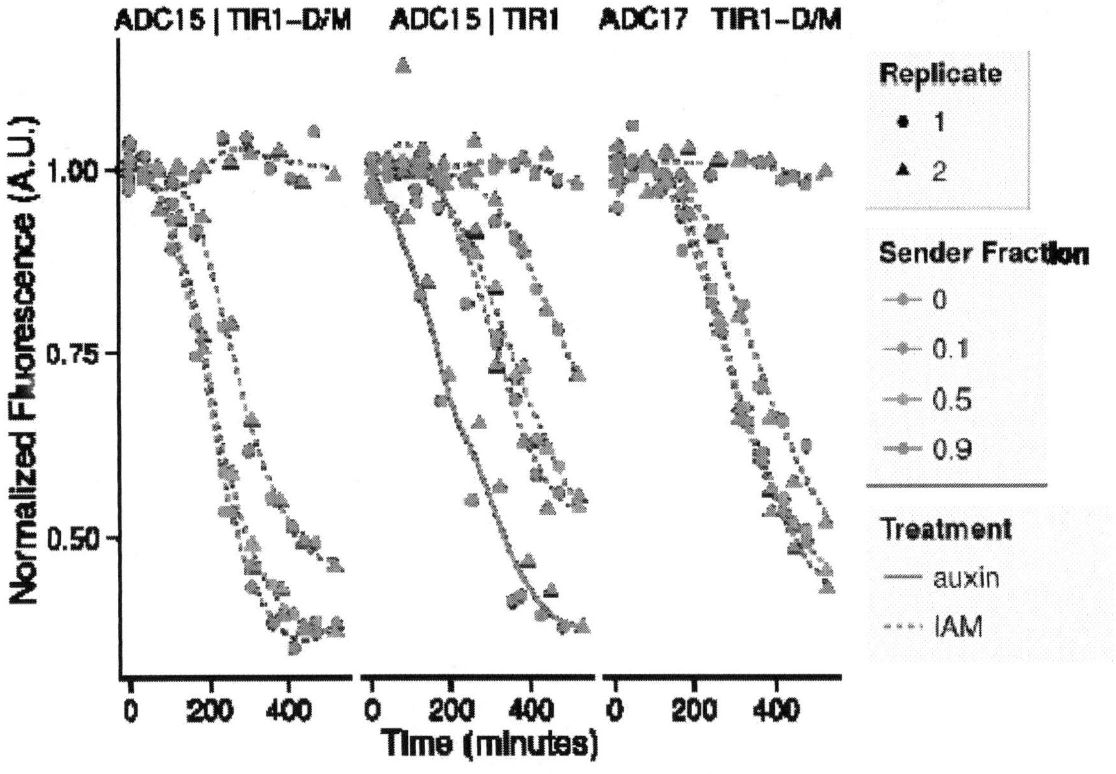

Normalized time course data for two replicates of sender receiver coculture experiments corresponding to Figure 5 in the main text. Sender fractions are indicated by different line and point colors. Lines represent LOESS fits to the combined mean fluorescence data from both replicates. Solid lines represent auxin treatment while dashed represent IAM treatment.

Experimental Setup

Cytometry

All cytometry measurements were acquired with an Accuri C6 cytometer with attached CSampler apparatus using 488 nm and 640 nm excitation lasers and a 533 nm (FL-1: YFP/GFP) emission filter. For experiments involving steady-state ADC behavior (Supplementary figures 1 and 2) cultures were grown overnight in SC media with 1nM beta-estradiol, for ADC expression induction. The next morning they were diluted down 1:100 the next morning and then grown up for 5 hours and then induced with varying auxin concentrations from stock solutions. These cultures were then grown overnight and fluorescence was measured the next day. For the replicates, untreated cultures from the previous days experiments were used as overnights to begin the next days cultures. For time course experiments, cultures were grown overnight in SC media with 1nM beta-estradiol, for ADC expression induction. The next morning
two sets of parallel cultures were set up by diluting down the overnights 1:100. These were then grown up for 5 hours and then induced with 30uM auxin and reads were taken every 30 minutes for approximately seven hours. A final steady state reading was taken the next day. Cytometry data were analyzed using custom R scripts and the flowCore package by first gating for the yeast diploid population and then generating mean fluorescence values for every point. These datasets were then collectively fit using the LOESS fitting function in R and metrics and error were derived from these fits. For the recovery experiments, the same protocol was used except the overnight cultures were grown up in SC with 1nM beta-estradiol and 30uM auxin.

Strains

All receiver strains are mated diploids. They were constructed by mating a Mat-a strain that has a minimal CYC1 promoter driven yeGFP and a guide RNA flanked by two ribozymes driven by the ADH1 promoter, with a Mat alpha strain that has the ADCTF driven by a modified GAL promoter that has a binding site for the Z4 zinc finger, an F Box driven by an ACT1 promoter and the beta-estradiol inducible transcription factor Z4EV that drives the production of the ADCTF.

Strain	Integrations
ADC14 + TIR1	URA3 - pCYC1:yeGFP HIS3 - pADH1:RGR-C3gRNA TRP1 - pGAL(Z4):dCas9-IAA14-VP64 LEU2 - pGPD:TIR1 HO - pACT1:Z4EV

ADC15 + TIR1	URA3 - pCYC1:yeGFP HIS3 - pADH1:RGR-C3gRNA TRP1 - pGAL(Z4):dCas9-IAA15-VP64 LEU2 - pGPD:TIR1 HO - pACT1:Z4EV
ADC17 + TIR1	URA3 - pCYC1:yeGFP HIS3 - pADH1:RGR-C3gRNA TRP1 - pGAL(Z4):dCas9-IAA17-VP64 LEU2 - pGPD:TIR1 HO - pACT1:Z4EV
ADC14 + AFB2	URA3 - pCYC1:yeGFP HIS3 - pADH1:RGR-C3gRNA TRP1 - pGAL(Z4):dCas9-IAA14-VP64 LEU2 - pGPD:AFB2 HO - pACT1:Z4EV
ADC15 + AFB2	URA3 - pCYC1:yeGFP HIS3 - pADH1:RGR-C3gRNA TRP1 - pGAL(Z4):dCas9-IAA15-VP64 LEU2 - pGPD:AFB2 HO - pACT1:Z4EV
ADC17 + AFB2	URA3 - pCYC1:yeGFP HIS3 - pADH1:RGR-C3gRNA TRP1 - pGAL(Z4):dCas9-IAA17-VP64 LEU2 - pGPD:AFB2 HO - pACT1:Z4EV

ADC14 + TIR1-DM	URA3 - pCYC1:yeGFP HIS3 - pADH1:RGR-C3gRNA TRP1 - pGAL(Z4):dCas9-IAA14-VP64 LEU2 - pGPD:TIR1-DM HO - pACT1:Z4EV
ADC15 + TIR1-DM	URA3 - pCYC1:yeGFP HIS3 - pADH1:RGR-C3gRNA TRP1 - pGAL(Z4):dCas9-IAA15-VP64 LEU2 - pGPD:TIR1-DM HO - pACT1:Z4EV
ADC17 + TIR1-DM	URA3 - pCYC1:yeGFP HIS3 - pADH1:RGR-C3gRNA TRP1 - pGAL(Z4):dCas9-IAA17-VP64 LEU2 - pGPD:TIR1-DM HO - pACT1:Z4EV
Sender	HIS3 – pGPD:iaaH TRP1 – pGPD:EYFP-IAA17 LEU2 - pGPD:AFB2
Sensor	HIS3 – pGPD:mCherry-Stop TRP1 – pGPD:YFP-IAA17 LEU2 – pGPD:TIR1

Plasmids

1. pMOD_U_pCYC1:yeGFP
2. pMOD_H_pADH1:RGR-C3gRNA
3. pMOD_T_TRP - pGAL(Z4):dCas9-IAA14-VP64
4. pMOD_T_TRP - pGAL(Z4):dCas9-IAA15-VP64
5. pMOD_T_TRP - pGAL(Z4):dCas9-IAA17-VP64
6. pGP5G-TIR1
7. pGP5G-AFB2
8. pGP5G-TIR1-DM
9. pMODKan-HO-pACT1-Z4EV
10. pGP8G-iaaH

11. pGP4GY-IAA17
12. pGP8G-mCherry-Stop

Visualising Individual Molecular Features with Fluorescence Digital Molecular Imaging (DMI)

Dai, M.,
Jungmann, R.,
Yin, P., Harvard Medical School

Visualising individual molecular features with fluorescence Digital Molecular Imaging (DMI)

Mingjie Dai1,2, Ralf Jungmann1,3, Peng Yin*1,3

1. Wyss Institute, Harvard University, 2. Biophysics Program, Harvard University, 3. Department of Systems Biology, Harvard Medical School

* Corresponding author, email peng_yin@hms.harvard.edu

Abstract

Understanding individual molecular information from a large biomolecular complex, while maintaining its native environment, represents a key challenge in biology. Recent advances in fluorescence super-resolution microscopy have shown images of sub-cellular features and synthetic nanostructures down to ~15 nm in size, but direct optical observation of individual molecular-scale (~5 nm) features within a macromolecular complex (?Digital Molecular Imaging?, or DMI) has yet to be demonstrated. In this paper, we establish three technical requirements for meeting this challenge with localisation based microscopy, and demonstrate our ability to achieve DMI with DNA-PAINT (Point Accumulation for Imaging in Nanoscale Topography), a method which utilises programmable transient oligonucleotide hybridisation, on synthetic DNA nanostructures. In particular, we examined the effect of high photon count, high blinking statistics and appropriate blinking duty cycle on imaging quality, and reported fluorescence imaging of a densely packed triangular lattice pattern with 5 nm point-to-point distance. Using oligonucleotide conjugated small labelling agents, this imaging capability potentially provides a high-resolution, highly quantitative and single-molecule approach for studying structural heterogeneity, modifications and interaction networks towards a systems understanding of cellular biology with molecular-scale resolution.

Project summary

A key challenge for super-resolution fluorescence microscopy is to directly visualise individual molecular targets *in situ*, which are mostly around 5 nm in size (protein subunits, membrane receptors), and are in close proximity to each other. A method that addresses these challenges must:

(1) Provide ultra-high, ?molecular? resolution (~5 nm target-to-target distance)
(2) Visualise individual molecular targets (clear separation close targets)
(3) Operate from within compact native environment (dense cluster of targets)

Despite considerable recent development (PALM, STORM, STED, etc.), the limited imaging resolution of currently demonstrated methods (down to 15 nm) still failed to address this challenge and allow visualisation of closely packed and interacting molecular components.

We describe here a framework which enables stochastic localisation based super-resolution microscopy methods to achieve the above challenge, a scenario which we

termed *Digital Molecular Imaging (DMI)*, and present our result of the first super-resolution image achieving such quality on a synthetic DNA nanostructure template (5 nm point-to-point distance, each individual target is visualised, and within densely labelled grid pattern).
We propose a *framework* for achieving DMI of four technical requirements,

Summary Figure: Principle, requirements and demonstration of super- resolution Digital Molecular Imaging (DMI). (a) Concept of DMI. Top: schematics and point array description of a representative four- component biomolecular complex. Bottom: point array representation and schematics of scenarios (heterogeneity, modification, molecular interaction) that could be better studied with DMI. (b) Three technical requirements for achieving DMI, illustrated in 1D super-resolution. (1) Intensity profile, blue arrow indicates uncertainty of Gaussian fitted centre position. (2) localisation histogram, signal measures the peak-to-valley difference in 2-Gaussian fit, noise measures stochastic fluctuation of peak histogram count. (3) Localisation time trace, orange crosses indicate observed localisations, grey crosses indicate true molecular positions that are shadowed by false localisations. (**c**) DNA-PAINT demonstration of DMI with 5 nm lattice standard. (**i**) Schematic of 5 nm

4x6 triangular grid standard. (**ii**) A representative single-molecule 5 nm grid image with DMI quality. (**iii**) Automatic fitting of image in (ii). Overlaid crosses indicate Gaussian fitted centers (green) and regular grid fitted centers using the green crosses as targets (blue). Root mean square (RMS) deviation between the green and blue crosses is less than 0.7 nm in distance. (iv) More examples of 5 nm grid images. Scale bars: 10 nm.

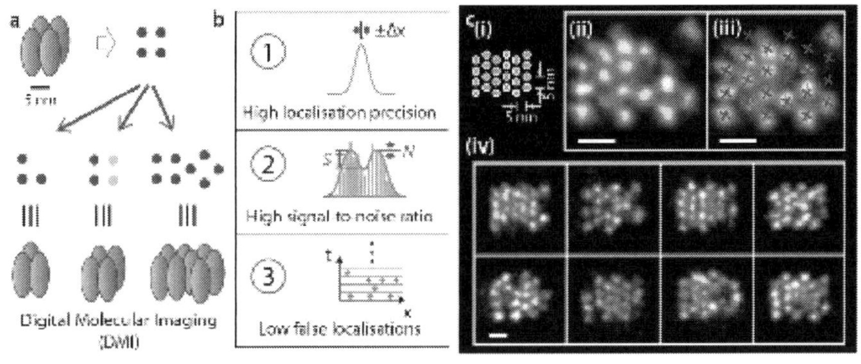

and demonstrate their effects *in silico* on improving imaging quality. We show that well-controlled blinking kinetics supports high localisation precision (down to 1 nm), high image Signal-to-Noise Ratio (SNR), and low fraction of false localisations (<5%). We show *implementation* of the above framework with DNA-PAINT method (Jungmann et al. *Nature Methods*, 2014), by exploiting the flexibility in blinking kinetics control offered by tuneable oligonucleotide hybridisation. We also develop a novel DNA nanostructure based drift correction algorithm that achieved <2 nm residual drift on a commercial microscope setup. We finally demonstrate the DMI imaging quality on a regular triangular pattern with 5 nm point-to-point distance constructed on a synthetic DNA nanostructure.

Understanding individual molecular information from a large biomolecular complex, while maintaining its native environment, represents a key challenge in biology. Recent advances in fluorescence super-resolution microscopy have shown images of sub-cellular features and synthetic nanostructures down to ~15 nm in size, but direct optical observation of individual molecular-scale (~5 nm) features within a macromolecular complex ("Digital Molecular Imaging", or DMI) has yet to be demonstrated. In this paper, we establish three technical requirements for meeting

this challenge with localisation based microscopy, and demonstrate our ability to achieve DMI with DNA-PAINT (Point Accumulation for Imaging in Nanoscale Topography), a method which utilises programmable transient oligonucleotide hybridisation, on synthetic DNA nanostructures. In particular, we examined the effect of high photon count, high blinking statistics and appropriate blinking duty cycle on imaging quality, and reported fluorescence imaging of a densely packed triangular lattice pattern with 5 nm point-to-point distance. Using oligonucleotide conjugated small labelling agents, this imaging capability potentially provides a high-resolution, highly quantitative and single-molecule approach for studying structural heterogeneity, modifications and interaction networks towards a systems understanding of cellular biology with molecular-scale resolution.

Project summary

A key challenge for super-resolution fluorescence microscopy is to directly visualise individual molecular targets *in situ*, which are mostly around 5 nm in size (protein subunits, membrane receptors), and are in close proximity to each other. A method that addresses these challenges must:

(1) Provide ultra-high, "molecular" resolution (~5 nm target-to-target distance)

(2) Visualise individual molecular targets (clear separation close targets)

(3) Operate from within compact native environment (dense cluster of targets)

Despite considerable recent development (PALM, STORM, STED, etc.), the limited imaging resolution of currently demonstrated methods (down to 15 nm) still failed to address this challenge and allow visualisation of closely packed and interacting molecular components.

We describe here a framework which enables stochastic localisation based super-resolution microscopy methods to achieve the above challenge, a scenario which we termed *Digital Molecular Imaging (DMI)*, and present our result of the first super-resolution image achieving such quality on a synthetic DNA nanostructure template (5 nm point-to-point distance, each individual target is visualised, and within densely labelled grid pattern).

We propose a *framework* for achieving DMI of four technical requirements, and demonstrate their effects *in silico* on improving imaging quality. We show that well-controlled blinking kinetics supports high localisation precision (down to 1 nm), high image Signal-to-Noise Ratio (SNR), and low fraction of false localisations (<5%).

We show *implementation* of the above framework with DNA-PAINT method (Jungmann et al. *Nature Methods*, 2014), by exploiting the flexibility in blinking kinetics control offered by tuneable oligonucleotide hybridisation. We also develop a novel DNA nanostructure based drift correction algorithm that achieved <2 nm residual drift on a commercial microscope setup. We finally demonstrate the DMI imaging quality on a regular triangular pattern with 5 nm point-to-point distance constructed on a synthetic DNA nanostructure.

Tunable Protein Degradation for Control of Synthetic and Endogenous Bacterial Systems

Cameron, D. E., Massachusetts Institute of Technology
Collins, J. J., Massachusetts Institute of Technology

Tunable control of protein degradation in bacteria would provide a powerful research tool to develop synthetic gene circuits and probe natural cellular systems. Here we use components of the *Mesoplasma florum* tmRNA system to create a targeted degradation system in *E. coli* that provides control of both the initial steady-state level and the inducible degradation rate of the targeted protein. We demonstrate the system's use in synthetic circuit development and in exogenous control of core bacterial processes including peptidoglycan biosynthesis, cell division and chemotactic motility. We transfer the system to *Lactococcus lactis* to establish its broad functionality in bacteria, and we create a 238-member library of tagged essential proteins in *E. coli* that can serve as both a research tool to study essential gene function and an applied system for antibiotic discovery. The synthetic degradation system is modular, requiring a small peptide tag and a single protease gene, does not require disruption of host systems, and can be transferred to other bacteria with minimal modification.

Design and Implementation of Reprogrammable dCas9 Transcription Factor Wires
Gander, M. W., University of Washington
Voje, W. Jr., University of Washington
Carothers, J.,
Klavins, E.,

Design and Implementation of Reprogrammable dCas9 Transcription Factor Wires

Miles W. Gander*, William Voje Jr.â? , James Carothersâ? , Eric Klavins*

*Department of Electrical Engineering, University of Washington.
â? Department of Chemical Engineering, University of Washington.

Keywords: ***CRISPR, Synthetic Gene Regulation, RNA-guided multiplex gene regulation, RNA***

folding design, Ribozymes

Presently, the size and complexity of engineered synthetic gene networks is limited by the small number well-characterized and robust genetic components. Here, we introduce and demonstrate a framework for creating orthogonal repressing transcription factor wires which exploit the programmability of the CRISPR/Cas9 system in *S. cerevisiae*. Genetic logic circuits such as NOR gates and layered cascades have been implemented in bacteria and eukaryotes using CRISPRi, but are limited by non-cooperative response functions of the guide RNA (gRNA), leading to leak and signal degradation as the number of layers increase [1,2]. To achieve leak free off states, we use a eukaryotic system and a dCas9 fused with a strong chromatin remodeling repression domain (RD). In our system, a transcription factor wire consists of 1) a dCas9-RD fusion protein that when recruited to DNA represses transcription of its target and 2) a guide RNA (gRNA) which when complexed with the dCas9-RD binds a specific DNA sequence. These wires link the output of Pol II gRNA expression cassettes to their cognate gRNA responsive Pol II promoters (pGRRs). Pol II promoters are used as wire nodes, instead of non-coding RNA Pol III promoters, because of their variety and engineerability in *S. cerevisiae* [3]. To consistently express gRNA from Pol II promoters novel computational design methodologies, which consider minimum free energy and kinetic simulations of RNA structure, are used to augment and extend existing ribozyme-flanked gRNA expression methods [4]. Our synthetic pGRRs contain two discrete gRNA target sites and function as NOR gates. Therefore, systems of wires with minimal crosstalk can create any logic function. Building an orthogonal set of gRNAs and pGRRs is challenging due

to the promiscuity of dCas9-gRNA complexes formed with imperfect DNA complements [5,6]. We developed a computational design strategy that generates sets of functional and orthogonal transcription factor wire libraries, enabling the construction of larger and more complex gene networks. To demonstrate the functionality of these wires, we created a library of 20 NOR gates and created a 4- component sequential repression cascade in *S. cerevisiae.* This work advances the
state of the art for cellular logic and provides a framework for creating more complex circuits.

References

[1] Nielsen, Alec AK, and Christopher A. Voigt. "Multi-input CRISPR/Cas genetic circuits that interface host regulatory networks." *Molecular systems biology*10.11 (2014).

[2] Kiani, Samira, et al. "CRISPR transcriptional repression devices and layered circuits in mammalian cells." *Nature methods* (2014).

[3] Blazeck, John, et al. "Controlling promoter strength and regulation in Saccharomyces cerevisiae using synthetic hybrid promoters." *Biotechnology and bioengineering* 109.11 (2012): 2884-2895.

[4] Gao, Yangbin and Yunde Zhao. â??Self-Processing of ribozyme-flanked RNAs into guide RNAs in vitro and in vivo for CRISPR-mediated genome editing.â? *Journal of integrative plant biology* 56.4 (2014): 343-349.

[5] Mali, Prashant, et al. "CAS9 transcriptional activators for target specificity screening and paired nickases for cooperative genome engineering." *Nature biotechnology* (2013).

[6] Doench, John G., et al. "Rational design of highly active sgRNAs for CRISPR-Cas9-mediated gene inactivation." *Nature biotechnology* (2014)

Considering Dynamic Constraints during Strain Optimization

Amin, S.,
Yousofshahi, M., Tufts University
Hassoun, S., Tufts University

Microbial cells have been successfully engineered to produce a large variety of biomolecules useful as biofuels, drugs and drug-precursors, and bioplastics. One challenge in maximizing the production of a target metabolite within a microbial cell is identifying gene modifications in the form of up-regulation, down-regulation, or knockout. While many such interventions are identified in an *ad hoc* manner based on experimental expertise, several computational tools have been developed for strain optimization. Strain optimization is formulated as an optimization problem specified in terms of two variables: flux variables and control (decision) variables that correspond to the presence or absence of regulation for each reaction and in each direction (up/down). Importantly, solutions to strain optimization problems must respect new bounds imposed due to regulation modification. A gene knock out modifies the upper and lower flux bounds to be zero, while up/down regulation defines new upper/lower bounds on reaction fluxes. Changes in bounds due to gene modifications impose new steady-state constraints, referred to as dynamic constraints, on the system as a whole. Another challenge in strain optimization is identifying an optimal fold change (e.g., 0.5, 2x, 5x, or 10x), instead of just identifying the fold change direction (up or down regulation). Prior computational approaches do not consider updating the steady state boundaries nor identify gene fold modification.

We propose a new strain optimization formulation that identifies fold changes required to maximize cellular yield. The fold changes are treated as a random variable with a probability distribution reflecting uncertainty in implementing engineering interventions. Each fold change can be one of three values, 2x, 5x, or 10x, compared to a 1x non-regulated fold change. Simulated Annealing (SA) is used to identify the optimal interventions and their fold changes. Flux Balance Analysis is utilized as a fitness function, evaluating the product yield using updated steady state conditions. We have applied this method to maximize the production of antibody in the Chinese Hamster Ovary (CHO) cell, and pyruvate in *Escherichia coli*. Our results show that SA is capable of identifying several intervention sets with different fold changes values that result in the same yield value of a desired product. Our results also show increased predicted maximum yield value due to utilizing dynamic constraints that update steady-state bounds.

Engineering Bacterial Finite State Machines
Sexton, J. T., Ginkgo Bioworks
Tabor, J. J., Rice University

Engineering Bacterial Finite State Machines

John T. Sexton1, Jeffrey J. Tabor1,2

1. Department of Bioengineering, Rice University, Houston, TX, USA
2. Department of BioSciences, Rice University, Houston, TX, USA

Cells use genetic circuits to make complex decisions such as differentiation, sporulation, or the prioritized utilization of sugars to optimize energy payoffs. The ability to engineer cellular decision making would have widespread applications in basic research and engineering. Finite state machines (FSMs) may serve as a useful tool for engineering complex cellular decisions1. FSMs are a framework whereby decisions can be explicitly specified as a collection of states and a mapping of the transitions among states. In computer engineering, FSMs are implemented using layered networks of digital logic gates (AND, OR, NOT). Digital-like logic gates can be implemented inside of cells using repressors (or activators) to modulate transcription from promoters. Thus, decision-making circuits can be implemented inside of cells as layered networks of promoters and repressors.

The implementation of complex decision-making circuits inside of cells requires many orthogonal promoter-repressor modules. We use the CRISPRi transcriptional repression system2 as a scalable platform for generating a large number of mutually orthogonal repressors in *E. coli*. CRISPRi allows for programmed repression of target promoters via the expression of a protein (dCas9) and a small guide RNA (sgRNA) with ~20 base pairs of homology to the target promoter DNA sequence. To ensure that CRISPRi repressors uniquely target only their intended cognate promoters, we developed a standardized procedure whereby an sgRNA target site (operator) is randomly generated in the spacer region of an otherwise invariant promoter. We demonstrate that this approach allows for the construction of families of promoters that can be strongly repressed by partner sgRNAs with complementar y sequences, while retaining very similar overall transcription rates. We have used this approach to construct 9 highly orthogonal promoters and cognate sgRNAs. We achieve fold repression values ranging from ~25x to several hundred or more, and present a set of 7 promoter-repressor units with minimum orthogonal range of 50x (as defined by Nielsen *et al.*3). We also extended the above design to incorporate a second input sgRNA, resulting in NOR logic.

Working towards more complex logic circuits, we next incorporated one NOT gate and three NOR gates to construct a 2-to-1 multiplexer circuit (MUX) in *E. coli*. The MUX is a signal selection device commonly used in electrical engineering which selects one of

several possible input signals and forwards it as the output signal. In a biological context, the MUX allows us to assess the activities of two different promoters using a single reporter gene (e.g. GFP) based on the presence of an input selector signal such as a chemical inducer.
In addition to the combinatorial circuits (i.e. circuits whose outputs depend only on the current inputs) we have constructed to date, we also plan to use the CRISPRi system to construct sequential digital logic circuits with feedback and memory (e.g. toggle switches and flip-flops). Together, these technologies will be leveraged to implement finite state machines inside of bacteria.

Funding Source: Supported in part by the National Science Foundation Graduate Research Fellowship Program under Grant No. DGE-0940902.

References: [1] K. Oishi, E. Klavins, *ACS Synth. Biol.* **3**, 652-665 (2014). [2] L. S. Qi *et al.*, *Cell* **152**, 1173-1183 (2013).

[3] A. A. Nielsen, T. H. Segall-Shapiro, C. A. Voigt, *Curr. Opin. Chem. Biol.* **17**, 878-892 (2013). [4] A. A. Nielsen, C. A. Voigt, *Mol. Syst. Biol.* **10**, 763 (2014).

[5] B. C. Stanton *et al.*, *Nature Chem. Biol.* **10**, 99-105 (2014).

Approaches to Improving the Potential of Azotobacter Vinelandii As a Biofertilizer

Barney, B. M., University of Minnesota
Plunkett, M., University of Minnesota
Natarajan, V.,
Eberhart, L., University of Minnesota

The industrial Haber-Bosch process responsible for a large portion of ammonia production comes at the expense of fossil fuels. *Azotobacter vinelandii* is a widely studied model diazotrophic (nitrogen fixing) bacterium, and also an obligate aerobe, differentiating it from many other diazotrophs that require environments low in oxygen for the function of the nitrogenase enzyme. This property makes *A. vinelandii* a key target strain for a range of nitrogen biosynthetic processes, including co-culture in mass algal production facilities. Our laboratory has been developing approaches and tools to screen for elevated nitrogen production through rational pathway manipulation and random insertional mutagenesis approaches. Several approaches have resulted in elevated production of specific nitrogen compounds that are sufficient to support the growth of either algae or traditional agricultural crops. Current efforts to expand this approach are studying the potential of developing new symbiotic relationships between *A. vinelandii* and target crops, while further efforts to utilize either plant exudates, residuals or cellulosic by-products would improve the application of biofertilizers further. Studies of transcriptional changes in a high nitrogen production strain are providing a glimpse of how the cell self regulates various processes that enable the flow of significant metabolic flux towards nitrogen fixation, and new tools developed recently should result in further strain enhancements that will improve the application for important nitrogen containing bioproducts. Finally, studies demonstrating the potential to apply new biofertilizer strains to both algae and traditional crops will be presented.

Synthetic nucleic acid nanostructures regulate native triplex and CRISPR/Cas genomic recombination, restriction, and repair potentials

Reza, F.,

CONFERENCE:

AIChE Society for Biological Engineering,
Synthetic Biology: Engineering, Evolution, and Design (SEED) 2015 Conference,
Boston, MA, USA June 10-13, 2015

SUBMISSION HEREIN:

Conference Abstract for Speaker Slot/Platform Talk

SUITABLE CONFERENCE TRACK:

DNA and RNA Based Synthetic Biology

AUTHOR:

Faisal Reza, Ph.D.

CORRESPONDING AUTHOR:

Email: faisal.reza@yale.edu
Synthetic nucleic acid nanostructures regulate native triplex and CRISPR/Cas genomic recombination, restriction, and repair potentials

Faisal Reza, Ph.D.â?

â? Yale University, School of Medicine, Department of Therapeutic Radiology, New Haven, CT, USA
*Address all correspondence to:
Faisal Reza; Yale University, School of Medicine, Department of Therapeutic Radiology, 333 Cedar Street, HRT 316, #208040, New
Haven, CT, 06520-8040, USA; Email: faisal.reza@yale.edu.

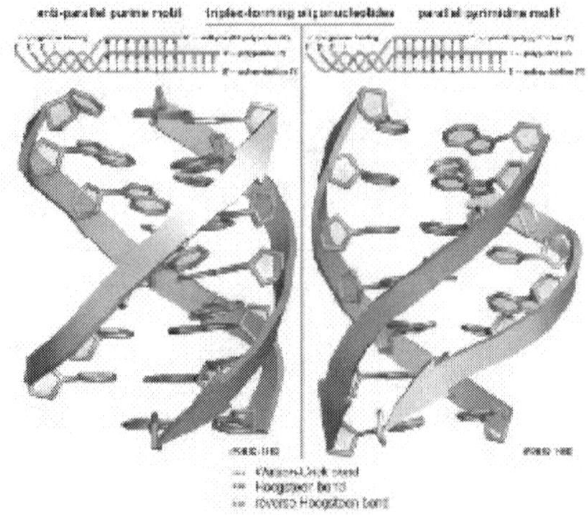

Synthetic oligo- and peptide- nucleic acid nanostructures have been more predictably designed and externally delivered into the intracellular milieu. These deployed technologies interact with, and influence, the cytoplasmic and nuclear molecular machinery in order to regulate potentials involved in genomic targeting and editing. It is demonstrated that synthetic nucleic acid nanostructures composed of various nucleobase and backbone modifications can regulate the genomic recombination rate, the sequence-specific restriction of a locus, and the endogenous repair pathways. The formation of a triplex nanostructure, by exogenously introduced PNA molecules with the duplex chromosomal and episomal DNA, is shown to elevate the cellâ??s targeted recombination potential. Alternatively, the formation of a
RNA-guided CRISPR/Cas nanostructure, again by exogenously introduced RNA molecules with Cas nucleases, is shown to initiate the cellâ??s targeted double-strand restriction potential. Recombinagenic donor DNA molecules co-opt these elevated recombination or initiated restriction potentials to form competing nanostructures that act as homology-dependent templates, sans edits to be introduced, thus potentiating repair. Safety and efficacy of these nanostructures is achieved by leveraging the performance profile of the cell's own endogenous recombination, restriction, and repair machineries in concert with these sequence-specific and localizing-in-tandem molecules. Progenitor cells drugged with designed molecules, and primed with chemical cell modulators, safely and effectively redesign the genome, which are then propagated to cellular progeny. These molecular technologies are developed to remediate the underlying genomic causes of monogenic human diseases, engineer living genetic codes, improve crop characteristics, and defend against outbreaks, and thus has well-positioned technology profiles in the synthetic biology of healthcare, biotechnology, agrotechnology, and national security.

KEY WORDS: triplex, CRISPR/Cas, recombinagenic donor DNA, genome engineering, precision medicine

ACKNOWLEDGEMENTS: This work was supported

by a National Institutes of Health (NIH) National Institute of Diabetes and Digestive and Kidney Diseases (NIDDK) Experimental and Human Pathobiology Postdoctoral Fellowship from NIH grant T32DK007556. The author declared no conflict of interest.

Sbol Visual: Standard Schematic Diagrams for Synthetic Genetic Constructs

Quinn, J., Autodesk Inc.
Cox, R. S. III,
Adler, A., SBOL Visual Working Group
Beal, J., BBN Technologies
Bhatia, S.,
Cai, Y.,
Chen, J., Joint BioEnergy Institute, Lawrence Berkeley National Laboratory
Clancy, K.,
Galdzicki, M.,
Hillson, N. J., Joint BioEnergy Institute
Maheshwari, A.,
Myers, C. J.,
P, U.,
Pocock, M., Newcastle University
Rodriguez, C.,
Soldatova, L.,
Stan, G. B., Imperial College London
Swainston, N.,
Wipat, A.,
Sauro, H.,

Synthetic Biology Open Language (SBOL) Visual is an effort to standardize visual representations of synthetic biological constructs. SBOL Visual 1.0 consists of a set of standardized glyphs for representing designs of synthetic genetic systems as contiguous constructs. Each symbol corresponds to a sequence element in physical DNA, including simple regulatory elements such as promoter or restriction sites. The symbols can be used as stencils to make illustrations on the computer or by hand, for rendering in a web browser, as images for software design, or as a formal symbology for aiding communication and instruction. SBOL Visual symbols are currently used by a number of commercial and academic software tools. SBOL Visual is particularly suited to represent the design of multi-component artificial genetic constructs, and we cite several examples where SBOL Visual has been employed for scientific descriptions of such constructs.

SBOL Visual is a free and open standard with resources made available for the broadest possible use. The symbols are provided for personal, academic, and commercial use in a variety of image formats, including PNG and SVG. SBOL Visual is developed by members of SBOL Developers Group as a visual, human-readable counterpart to the SBOL standard data format for representing the design of genetic constructs to ensure cross-compatibility with other biological standards. We encourage any interested practitioners to download the images, use and modify them freely, suggest changes, and join the SBOL Visual community to participate in developing the standard: http://www.sbolstandard.org/visual

A "Basic" Approach to the DNA Assembly Bottleneck: A New Standard Designed for Flexibility, Reliability and Automation

Storch, M., Imperial College London
Casini, A., Imperial College
Mackrow, B., Imperial College London
Ellis, T., Imperial College London
Baldwin, G. S., Imperial College London

The ability to quickly and reliably assemble DNA constructs is one of the key enabling technologies for synthetic biology. We have developed BASIC (Biopart Assembly Standard for Idempotent Cloning), a new physical standard for DNA parts which exploits the principle of orthogonal linker-based DNA assembly. The standard employs a new robust method for assembly, based on type IIs restriction cleavage and ligation of linker oligonucleotides with single stranded overhangs that determine the assembly order. BASIC achieves efficient, parallel assembly with great accuracy: small (4 parts) constructs are routinely assembled with near-100% accuracy, while larger (7 parts) assemblies achieved 90% accuracy. Additionally the linkers can be used as composable parts for RBS tuning or the creation of fusion proteins. This standard format is compatible with a wide range of DNA parts since it only requires a single forbidden restriction site and does not employ any PCR amplifications, and is also conceptually simple at both the design and experimental levels, being based on an idempotent and single-tier organisation where all parts and composite constructs are maintained in the same format. Finally BASIC has been designed with automation in mind, using a workflow that is easily performed on robotic liquid handling platforms.

Design of Modular Parts for Synthetic Systems Inspired By Anaerobic Fungi

Gilmore, S. P., University of California, Santa Barbara
Haitjema, C. H., University of California, Santa Barbara
Sexton, J. A., University of California, Santa Barbara
Solomon, K. V., University of California, Santa Barbara
Henske, J. K., University of California, Santa Barbara
Kivenson, V.,
Theodorou, M. K., Harper Adams University
Valentine, D., University of California, Santa Barbara
O'Malley, M. A., University of California, Santa Barbara

Anaerobic fungi in the hindgut of large herbivores are among the most robust organisms at degrading crude lignocellulose. They achieve this efficiency through the production of large, multi-enzyme complexes called fungal cellulosomes. The fungi also act synergistically with other microorganisms in the microbiome, such as archaea, bacteria, and protozoa. By elucidating the parts responsible for efficient biomass degradation at both the protein and cellular level, we seek to replicate this efficiency in synthetic systems.

At the protein level, fungal cellulosomes are similar to bacterial cellulosomes in that the protein-protein interactions are mediated through parts termed the dockerin and cohesin. However, many differences exist. The dockerin domains exist in tandem repeats and bear no species specificity like those in the bacterial systems. Furthermore, the exact sequence for the cohesin module has yet to be established. Through analysis of transcriptomic data for three fungal isolates, patterns governing the native placement of dockerin domains on fungal cellulases were characterized. By recombinantly grafting these dockerin domains onto similar enzymes from other organisms, the original activity of the enzymes were retained while allowing for incorporation of these exogenous enzymes into fungal cellulosomes. This was demonstrated for the TmCel5A, TmManB, TmXynA, and TmXynB from *Thermotoga maritima.* These incorporated enzymes demonstrated a greater level of synergy with the native cellulosomes when compared to the catalytic domain without the grafted dockerins. Similarly, the newly grafted dockerin modules were replaced with carbohydrate binding modules, demonstrating that this approach can be extended to other protein binding domains. The eventual goal is to create entirely synthetic cellulosomes, which could be applied to any biocatalytic process.

At the cellular level, the anaerobic fungi have also been shown to interact closely with methane producing archaea (methanogens). The methanogens siphon hydrogen and other metabolites from the fungi, allowing the fungi to more efficiently produce energy by increasing the flux through their hydrogenosomes. This increased energy is hypothesized to increase production of cellulases, accelerating the degradation of lignocellulose in co-culture. To further investigate this mechanism, native fungal/methanogen co-cultures were isolated from herbivore fecal materials. These co-cultures were maintained together and also separated into monocultures, effectively creating parts for synthetic co-cultures. By introducing the methanogens into cultures of other well-characterized anaerobic fungi, stable synthetic co-cultures were established. With this proof of concept, other parts to the consortia can be introduced, such as non-native methanogens capable of funneling other accumulating metabolites like acetate. These stable synthetic consortia should increase the efficiency of conversion of crude biomass into sustainable chemicals.

Seeing the Invisible, and Considering Context: Exploiting Crosstalk to Measure Genetic Construct Activity

Hecht, A., National Institute of Standards and Technology
Munson, M. S., National Institute of Standards and Technology
Endy, D., Stanford University
Salit, M., National Institute of Standards and Technology

â??Seeing the Invisible, and Considering Context: Exploiting Crosstalk to Measure Genetic
Construct Activityâ?
Ariel Hecht, Matthew S. Munson, Drew Endy, and Marc Salit
The ability to quantify the activity of genetic parts that do not produce spectroscopically quantifiable products would support the reusability of parts in a forward engineering context. A potential solution, described here, is a competitive assay performed in an *in vitro* transcription- translation (IVTT) system. In an *in vitro* environment, expression of orthogonal genes are coupled by competition for limited resources, which can be exploited for measuring the activity of â??invisibleâ? genetic constructs. Several new results from previously presented work are highlighted here:
1. The activity of the non-fluorescent construct of interest can be measured by exploiting resource-based coupling in a competitive assay.
2. A linear relationship between the activities of two constructs is predicted only under specific conditions.
3. The relative activities of constructs are context-dependent: assay measurements cannot yet be ported between expression systems.
The above results suggest that, while the competitive assay is feasible, appropriate conditions that define a context must be identified. Future work will focus on identifying assay conditions using a validated model-based framework, potentially leading to a procedure for characterizing the capacity of both *in vitro* and *in vivo* expression contexts.

Improvement of DNA Synthesis
Horlacher, M., Thermo Fisher Scientific GeneArt AG
Kuhn, P.,
Heil, K.,
Wagner, K.,
Bauer, C., Thermo Fisher Scientific
Poehmerer, T.,
Liss, M., Thermo Fisher Scientific
Trefzer, A.,

Improvement of DNA synthesis

M. Horlacher, P. Kuhn, K. Heil, K. Wagner, C. Bauer, T. Poehmerer, M. Liss, A. Trefzer

In the few last years the price for DNA synthesis has decreased rapidly, opening up new markets for commercial gene synthesis suppliers. However, the current technology is reaching a limit with respect to cost reduction, constrained by synthesis scale and error rate.
In our efforts to keep offering our customers top quality synthetic DNA products, we continuously improve our oligonucleotide and gene synthesis workflows. We further develop parts of the process regarding cost reduction, time reduction, higher stability and higher efficiency. This allows us to target the market demand of lower costs for synthetic genes.
We will show data illustrating our efforts to reduce error rate both by improved oligo synthesis as well as error correction of assembled genes. In combination these developments continue to drive overall improvements in DNA synthesis.

Thermal Effects on Cell Death Induced by High-Intensity, Ultrashort Electric Pulses

Jiahui Song
Department of Electrical Engineering and Technology
Wentworth Institute of Technology
Boston, MA 02115, USA
songj@wit.edu
Ravindra Joshi
Department of Electrical & Computer Engineering
Texas Tech University Lubbock, TX 79409, USA ravi.joshi@ttu.edu

Abstractâ??Nanosecond and high intensity electric pulse induced irreversible electroporation has been shown to eliminate melanoma tumor cells with needle electrodes, while thermal effects are typically not considered. Our results of self-consistent analyses of cells show that thermal effects due to high-intensity, ultrashort pulses could facilitate poration over small sections of the membrane. Short-term temperature increases of several degrees Celsius above the physiological temperature could allow a considerable reduction in the pulse duration and amplitude for pore formation. Thermally launched bioresponses are similar in nature to those triggered electrically. Molecular Dynamics simulations indicate an increased probability for pore formation in the plasma membrane with lower pulse amplitude or smaller pulse durations at elevated temperatures. Synergistic effects enhance the benefits at a lower energy cost and better utilize bioelectric effects for therapeutic applications. Therapeutic applications include the treatment of tumors in internal organs by adding thermal effects as a synergistic mechanism. The treatments use antennas to reach internal organs and require the application of even shorter subnanosecond pulses, in order to obtain a reasonably small focal volume.

Intracellular spectral recompositioning of light for improving photosynthetic efficiency in the marine diatom *Phaeodactylum tricornutum*

Weiqi Fu and Kourosh Salehi-Ashtiani*
Division of Science and Math, and Center for Genomics and Systems Biology (CGSB), New York University Abu Dhabi, P.O. Box 129188, Abu Dhabi, UAE

*Corresponding author

Marine diatoms that usually accumulate high amounts of lipids generate as much as 40% of total organic carbon produced each year in the sea. Marine diatoms are expected to be a promising resource for future clean energy supply, as well as for sustainable production of bioactive compounds such as high value nutraceuticals and active pharmaceutical ingredients. The light energy conversion efficiency is a decisive factor in intensive cultivation of diatoms in photobioreactors (PBRs) for determining the economical feasibility and limits of diatom-based cell factories and relevant industrial applications. Maximizing the photosynthetic efficiency may significantly improve biomass productivity and reduce the overall direct and indirect energy costs while imposing high-intensity illumination on algal cultures. We hypothesized and show here that Intracellular Spectral Recompositioning of light (or ISR) can increase the quantum yield of light if the otherwise wasted portion of blue light spectrum is shifted to green which diatoms have evolved to harvest through accessory photosynthetic pigments. We demonstrate that ISR can be employed to largely improve photosynthetic efficiency of the diatom *Phaeodactylum tricornutum* in flat- panel PBRs. We show that an addition of 2 ÂµM of BODIPY fluorescent dye increased the photosynthetic efficiency from 4.9Â±1.0 % to 7.5Â±1.2 % in high-density culture of *P. tricornutum* under combined red and blue light-emitting diode illumination, on the basis of cell growth rate over a 24-hr period. To biogenically implement ISR, green fluorescent proteins (GFPs) as well as other FPs were elected with high quantum yield, photostability and extinction coefficient, and heterologously expressed in *P. tricornutum cells*. Transformants were pooled and then quantitatively screened using fluorescence-activated cell sorting (or FACS) for high expression of fluorescent proteins. Evaluation of transformant *P. tricornutum* cells is in progress and it is expected to optimize both intracellular and intercellular light delivery and re- distribution in flat-panel PBRs through integrating selected fluorescent proteins into the existing light harvesting complexes. The carotenoid metabolism in transformants and cell tolerance to light stress will also be studied for optimizing biomass productivity as well as photosynthetic efficiency. This study may also provide a method to minimize cell damage caused by light stress of photosynthetically active radiation.

RNA Aptamer-Based Biosensors for Multiplexed in Vitro Diagnostics and Circuit Engineering

Burke, C., University of Washington
Carothers, J.,

The need for metabolic sensors spans both *in vitro* and *in vivo* applications, particularly in the design, evolution and optimization of pathways and genetic control systems. In addition, there is great demand for orthogonal sensors to detect and propagate signals independently while maintaining the ability to be coupled for complex diagnostics or pathway design. We are developing RNA aptamer-based sensors that process molecular inputs into programmable nucleic acid outputs, with the goal of generating a robust multiplexed small molecule sensing assay on common laboratory equipment such as a qPCR machine. These sensors are engineered using a multi-state RNA folding design strategy where transient but metastable intermediate states are engineered along the co-transcriptional folding trajectory. Binding of the small molecule input to the aptamer domain directs the folding pathway towards an active state and generates a fluorescent signal through a nucleic acid strand displacement reaction. We are currently expanding this work to include sensors for a broader panel of small molecules, while simultaneously optimizing the nucleic acid circuits for multi-signal detection *in vitro,* with the goal of generating complete biosensor devices for the eventual use in mammalian cell systems.

Design Space Exploration of 1,4-Butanediol (BDO) Biosynthesis Pathway in a Cell-Free Transcription-Translation (TX-TL) System

Wu, Y.,
Culler, S., Genomatica
Khandurina, J.,
Van Dien, S., Genomatica, Inc.
Murray, R. M., California Institute of Technology

Current methods for assembling metabolic pathways require a process of repeated trial and error and have a long design-build-test cycle. Further, it remains a challenge to precisely tune enzyme expression levels for maximizing target metabolite production. Recently it was shown that a cell-free transcriptional-translation system (TX-TL) can be used to rapidly prototype novel complex biocircuits as well as metabolic pathways. TX-TL systems allow protein expression from multiple DNA pieces, opening up the possibility of modulating concentrations of DNA encoding individual pathway enzymes and testing the related effect on metabolite production. In this work, we demonstrate TX-TL as a platform for exploring design space of metabolic pathways using a 1,4-BDO biosynthesis pathway. Using TX-TL, we verified enzyme expression and enzyme activity and identified the conversion of 4-hydroxybutyrate to downstream metabolites as a limiting step of the 1,4-BDO pathway. We further tested combinations of various enzyme expression levels and found increasing downstream enzyme expression levels improved 1,4-BDO production.

A Single Plasmid System to Addict Bacteria to Non-Canonical Amino Acids

Tack, D., University of Texas at Austin
Thyer, R., UT Austin
Ellefson, J., University of Texas at Austin
Ellington, A. D., University of Texas at Austin

A single plasmid system to addict bacteria to non-canonical amino acids.
Genetic incorporation of non-canonical amino acids (NCAAs) allows the addition of new and novel chemistries to the genetic code. This can be accomplished by using an orthogonal aminoacyl tRNA synthetase:tRNA pair derived from an evolutionarily distant organism to reassign the amber (UAG) stop codon. Unfortunately, global UAG suppression imposes a significant fitness cost on the host cell, leading to deletion or inactivation of the orthogonal translation machinery during serial culture experiments. To solve this problem, and ensure maintenance of functional orthogonal translation machinery, we engineered an essential protein, Î²-lactamase, to require incorporation of NCAAs. Following library selections with various 3-substituted tyrosine analogues, clones conferring Î²-lactam resistance were counter-screened to eliminate those active in the presence of common *E. coli* suppressor tRNAs. We show that by iteratively decreasing the number of canonical amino acid solutions, enzymes can be evolved with a strong dependence on NCAA incorporation. An evolved Î²-lactamase has been shown to maintain orthogonal translation machinery in *E. coli* for more than 250 generations, without detectable escape from NCAA dependence. Unlike other approaches that rely on extensive manipulation of the host genome, this system is compact and broadly compatible. To demonstrate the utility of this approach the sub-circuit containing the addicted lactamase and the orthogonal synthetase and tRNA pair were transferred on a broad host range vector to different bacterial species, including *Shigella* and *Salmonella*, which immediately resulted in the non-revertible addiction of these strains to the NCAA without the alteration of their underlying codes via other synthetic biology approaches. The use of our or similar addiction modules should quickly allow the introduction of NCAAs into a much wider variety of bacterial species, including industrially-relevant fermentation and production strains.

Biomolecular Implementation of a Quasi Sliding Mode Controller Using an Ultrasensitive Cell Signalling Pathway

Sawlekar, R., University of Warwick
Montefusco, F., University of Padua
Kulkarni, V., University of Warwick
Bates, D., University of Warwick

A fundamental aim of synthetic biology is to achieve the capability to design and implement robust embedded biomolecular feedback control circuits. An approach to realize this objective is facilitated by using abstract "chemical reaction networks" (CRNs). This approach can impart a good degree of predictive functionality to the wet-lab circuits since these networks can model a large class of chemical and biological processes, and can be well analysed using advanced mathematical techniques. The existing work within this framework concerns the design and implementation of linear time-invariant systems only: an example of such systems is the well-known "proportional+integrator" (PI) controller derived recently by Yordanov et al. In practice, however, the PI controllers must be combined with a nonlinearity to overcome the wind-up effect associated with the integrator action. Here, we extend this approach to allow the implementation of nonlinear controllers using the sliding mode control theory whose strong performance and robustness characteristics have been widely recognised for a number of decades in the more traditional control engineering application domains. We show how a signalling cycle with ultrasensitive response dynamics can provide a biomolecular implementation of a nonlinear quasi sliding mode controller. We implement our controller on a prototype of a biological pathway, specifically a first-order low pass filter and demonstrate that our nonlinear controller outperforms a PI controller by facilitating faster response dynamics without introducing overshoots in the transient response.

Serine Integrase Recombinational Assembly (SIRA) for Rapid Optimization of Metabolic Pathways

Merrick, C. A.,
Colloms, S. D., University of Glasgow
Rosser, S. J., Edinburgh University

A major target in Synthetic Biology is to build microbial cell factories for greener and more economical production of chemicals of therapeutic and industrial importance. Building these cell factories requires genetic engineering of metabolic pathways to produce high yields of compounds for large-scale production. However, engineered metabolic pathways require considerable optimization to limit interference with existing metabolic networks, diversion of cellular resources, and accumulation of toxic intermediates which compromise strain performance. DNA assembly is one of the biggest rate-limiting steps in metabolic pathway optimization. Serine Integrase Recombinational Assembly (SIRA) is a site-specific recombination-based DNA assembly method that uses a one-pot reaction for rapid multi-part assembly[1]. With SIRA technology, genes and regulatory sequences in the form of linear DNA parts integrate into a landing site on target DNA. The assembly process can easily construct gene assemblies using rational or combinatorial approaches, include multiple gene variants at all positions, and facilitate varied gene expression levels by incorporating degenerate ribosome binding sites. Using the carotenoid and violacein biosynthetic pathways as model systems, we have shown assembly and optimization of two fully functional pathways in just two days. Once DNA parts are assembled, SIRA allows targeted addition, deletion, and replacement of genes and DNA elements. This permits much easier optimization of complex genetic systems than was previously possible. This method is currently being applied to enhancing production of therapeutically and industrially important isoprenoids in *E. coli*.

1. Colloms SD, Merrick CA, Olorunniji FJ, Stark WM, Smith MCM, Osbourn A, Keasling KJ & Rosser SJ (2013) Rapid metabolic pathway assembly and modification using serine integrase site-specific recombination. Nucleic Acids Res 42: e23

Logic Gates As sRNA Dependent Transcriptional Attenuators: A Pause for Thought

Harvey, P. J.,

Within the field of synthetic biology, genetic circuits have become progressively more complex, requiring more sophisticated regulation and leading to a drive to generate logic gates within biological systems that are orthogonal, scalable, respond rapidly and perform multilayer logical operations. The reprogrammable nature of RNA folding and binding combined with rapid synthesis and degradation within biological systems provides useful tools to achieve this.

It is possible to dispense with the requirement for proteins to propagate a signal through a logical operation which in the past has acted as a constraint on the size of orthogonal libraries, using instead sRNA dependent transcriptional attenuators. Here we present a new genetic learning algorithm for the *de novo* design and synthesis of the sequences required for sRNA dependent transcriptional attenuators. The Algorithm uses a combination of RNA folding predictions and dimer formation kinetics, aided by the use of transcriptional pausing sites to increase the time window for sRNA-Attenuator interactions.

Sequences generated by the algorithm are currently being tested and work to date will be presented.

Predictable Design in Biological Engineering: Debugging of Synthetic Circuits By in Vivo and in silico Approaches

Pasotti, L.,
Zucca, S.,
Casanova, M.,
Politi, N.,
Massaiu, I.,
Cusella, G.,
Magni, P.,

The bottom-up design process of biological systems is one of the hallmarks of synthetic biology. This approach is adopted in all the fields of engineering and, similarly, it will significantly boost the potential of living systems engineering. However, our ability to predictably engineer genetic systems from sets of quantitatively characterized parts is still limited due to unpredicted interactions among circuit components. The evaluation of the bottom-up approach in a wide number of model systems can elucidate the actual predictability boundaries where the designed systems can behave as intended. The subsequent study of non-functional or unpredictable systems offers the opportunity to determine specific features, parts and conditions that significantly affect predictability. In this work, a set of increasingly complex model systems was considered. All of them were bottom-up designed from sets of pre-characterized parts, but the quantitative characterization of the final systems resulted to be unpredictable, thus requiring a debugging process, herein reported, based on both experimental procedures and mathematical modelling.

The model systems included: 1) a library of novel synthetic repressible devices, 2) transcriptional regulators cascades, 3) a feedback-controlled circuit, 4) enzyme production systems, and 5) circuits controlled by small RNAs. Mathematical models describing the quantitative behaviour of the systems were developed during the design step. Such models were used to investigate the unpredictable experimental output of the systems, measured via population-based, single-cell fluorescence measurements, or enzyme assays. Since the bottom-up design relies on parameter estimates, obtained during biological parts characterization, sensitivity analysis was coupled with Monte Carlo simulations to study the propagation of the uncertainty of parameter values towards the final output of the system. The contribution of biological noise in interconnected circuits was also studied *in silico*, via stochastic models, and *in vivo*, via single-cell measurements, to evaluate its importance in output predictability. The re-characterization of parts in a multi-faceted fashion was finally carried out to deepen the knowledge of parts functioning and identify possible conditions where specific modules do not have predictable behaviour. Crosstalk among parts like transcription factor-promoter pairs, identified via ad-hoc experiments, have also been crucial to elucidate previously unpredicted and non-modelled molecular interactions.

Taken together, the studied systems spanned a wide range of design architectures, parts and strains, and their debugging process enabled to decouple the contributions of context-dependent variability of biological parts, cell-to-cell variability, parameter estimation uncertainty, circuit mutations due to genetic instability, crosstalk, and metabolic burden.

Targets of Opportunity : Hypergraph Analysis of the Retrosynthetic Design Space

Olson, B., Lawrence Berkeley National Laboratory
Dehal, P., Lawrence Berkeley National Laboratory

As biomanufacturing progresses towards the goal of producing a wide array of natural product compounds in vivo, many retrosynthetic design methods have emerged for finding and comparing pathways to a specific target compound. Little attention, however, has been given analyzing the space of all potential natural product pathways accessible through retrosynthesis. Rather, studies focus on a specific target compound, selecting heterologous genes to express in a chassis organism in order to generate a synthetic natural product pathway. This work examines the Retrosynthetic Design Space (RDS) containing all heterologous pathways which could potentially be added to a specific chassis organism. By examining the entire RDS, we directly compare the costs and benefits of pathways across different target compounds rather than only between pathways for the same compound. This analysis reveals which compounds are more easily accessible today and directs the efforts of the synthetic biology community to the most pressing challenges in biomanufacturing.

We present an analysis of the RDS for the chassis organism E. Coli constructed from publicly available databases of biochemical reactions and enzymes. The RDS is represented as a hypergraph with metabolites as nodes and biochemical transformations as hyperedges. By traversing the hypergraph we determine the set of Pareto-optimal pathways which could be potentially added to the chassis organism and which target compounds these pathways would theoretically produce. By analyzing these pathways with graph-theoretic algorithms, we are able to compare targets with inter-pathway metrics, such as centrality and clustering, as well as intra-pathways metrics such as thermodynamics and toxicity. Specifically we: (1) Rank target compounds and compound classes by difficulty using multiple scoring objectives, including, pathway length, pathway thermodynamics, enzyme availability, intermediate toxicity, and RDS centrality. (2) Identify hub compounds in the RDS which are key intermediates for multiple target pathways, including key host metabolites which which serve as precursors. (3) Identify gaps in the RDS for which further investigation could unlock additional natural product targets. From this analysis we identify "targets of opportunity" which are not only easier to reach, but also serve as intermediates for producing more difficult compounds in the future.

A Synthetic Developmental Circuit to Investigate the Effects of Adhesion

Kan, A.,
del Valle, I.,
Federici, F.,
Rudge, T.,
Haseloff, J., University of Cambridge

The design and engineering of multicellular systems is an important step for the future of synthetic biology. A crucial hurdle to the engineering of such systems is the understanding of the self-organisation that shapes the organism, often involving a multitude of intercellular interactions. In order to begin to simplify and disentangle such interactions, we express a primitive synthetic developmental circuit in *E. coli* colonies to investigate cell-cell adhesion. Adhesion is an important developmental mechanism, and here we isolate and observe its effects in a well-defined context. We first characterize a method to segregate bacterial colonies into spatially distinct domains of gene expression. This is achieved by transforming the cells with two similar inducible copy number plasmids in the high copy number regime. The inducer is then removed so plasmid copy number falls whereby plasmid incompatibility ensures that each cell is eventually left with only one of the original two plasmids. The segregating plasmids also contain an inducible *ag43* protein that causes adhesion between cells. By observing and measuring boundaries between domains with differing adhesive properties, we demonstrate the effects of adhesion on the boundary formation process. *CellModeller* software is used to model the cells and their physical interactions, providing insight into the mechanistic details.

Optimization of Cellular Resource Allocation Using Global mRNA Decay

Venturelli, O.,
Arkin, A. P., University of California at Berkeley

The behavior of synthetic circuits is shaped by the availability of cellular resources and the context of the host cell as the heterologous pathway relies on host transcription-translation machinery. Global control mechanisms for host gene expression could be used to optimize cellular resource allocation by decoupling the synthetic pathway and the host cell. Inspired by viral strategies that hijack cellular resources using global mRNA decay, we developed a control system to modulate resource allocation between the host and a synthetic pathway. This mechanism exhibits a fast response and can re-distribute ribosomes. A sequence-specific RNA degrading toxin from *E. coli* was used to trigger global mRNA decay while protecting the target gene(s) of interest. Our approach demonstrates a significant enhancement of protected gene expression and relies on the presence of negative feedback loop and specific induction timing. To optimize and understand the system, genome-wide measurements of RNA and protein levels were used to identify critical target genes for protection and provide key insights into the cellular response to the toxin. These measurements revealed a two-phase dynamic behavior in which transcript abundance is significantly reduced followed by a delayed activation in a set of target genes.

Interfacing Microbial Styrene Production with a Biocompatible Cyclopropanation Reaction

Wallace, S.,
Balskus, E. P.,

Iron(III) phthalocyanine has been identified as a highly active and biocompatible carbene-transfer catalyst capable of cyclopropanating styrene generated by engineered *E. coli* metabolism. The biocompatible reaction is non-toxic to the producing organism and performs alongside cellular styrene output, enabling the near-quantitative production of various phenyl cyclopropanes directly from D-glucose in single vessel fermentations.

Use of Pacific Biosciences Single Molecule Sequencing to Assay DNA Ligase Fidelity

Lohman, G. J. S., New England BioLabs, Inc.
Canton, B.,
Cahoon, D.,
Knight, T. F., Ginkgo Bioworks
Evans, Jr., T. C.,

DNA ligases, key enzymes in molecular biology and synthetic biology, catalyze the joining of single strand breaks in duplex DNA (nicks) as well as the joining of multiple dsDNA fragments. Ligation fidelity has important consequences for the yield, accurate assembly, and maximum number of fragments that can be assembled concurrently in synthetic biology applications such as Golden Gate cloning. In this study, we have applied Pacific Biosciences single-molecule sequencing technology to develop a new method to characterize the fidelity of ligation when joining DNA fragments with short (three or four base) 5'-overhangs. Pools of substrates containing all possible overhang sequences were ligated with T4 DNA ligase, and the ligation products were read directly by sequencing individual molecules. This work has allowed us to determine the overall mismatch ligation frequency, the mismatch frequency for each individual subsequence, and to characterize the particular mismatches that have a high propensity for ligation. This methodology is easily adaptable to other dsDNA substrate structures, and provides a method to screen for ligases or buffer conditions that provide higher fidelity ligation reaction outcomes. The information generated by this study will allow the optimization of DNA fragment assembly protocols, both by the development of high fidelity ligases and by the avoidance of non-complementary overhang pairs prone to misligation.

Reconstructing a Plant Hemicellulose Biosynthetic Pathway in the Heterologous Fungal Host Pichia Pastoris

Levesque-Tremblay, G.,
Schultink Alex, A., UCBerkeley
Mansoori, N.,
Pauly, M.,

The plant cell wall is made of a variety of polysaccharides including the hemicellulose xyloglucan, which is composed of a backbone of β-1,4-linked glucan that is highly substituted with xylose as well as other monosaccharides. Xyloglucan has a wide range of applications including thickening, gelling, and stabilizing agents in the food industry, during weaving of textiles, and as adhesive and binding agents in the pharmaceutical industry.

In the last decade, the enzymes involved in the biosynthesis of the xyloglucan backbone and substitutions were identified. To provide a testing platform for candidate genes involved in xyloglucan biosynthesis and to investigate the activities of these genes independently from endogenous plant systems, we attempt to reconstruct the plant xyloglucan biosynthetic pathway in the heterologous fungal host Pichia pastoris.

Our efforts resulted in the production of a glucan backbone in the yeast Pichia pastoris. Subsequent addition of a cytosolic UDP-glucose dehydrogenase as well as a UDP-glucuronic acid decarboxylase made the synthesis of the nucleotide sugar UDP-xylose possible, which is required as a donor substrate for backbone xylosylation. Finally, the addition of a UDP-xylose transporter to transfer UDP-xylose into the lumen of the golgi to the previously mentioned gene resulted in the production of a xylosylated glucan as determined by HPAEC-PAD, Q-TOF and GC-MS analysis.

Organised Natural Structures Using Synthetic Biology

Pothoulakis, G.,
Ellis, T., Imperial College London

Engineering living organisms that can be utilized in a variety of applications (e.g. valuable compound biosynthesis, applications in materials) is now the main focus of the novel field of Synthetic Biology. The idea is to take advantage of *S.cerevisiae*'s unique growth and adhesion characteristics to create cell patterns such as fractal formations that are interesting and useful. Specifically, we are implementing novel gene networks that control S.cerevisiae's budding behavior to induce pseudohyphal growth in both haploid and diploid strains. Through the characterization and genetic control of the key natural regulators that control pseudohyphal growth we aim to reprogram cells to grow in controllable branched fractal patterns. For a true fractal pattern to occur, cells are needed to differentiate as the colonies grow and therefore novel regulatory mechanisms to control the timing and/or location where genes are switched on and off are being implemented. By controlling cell patterns available surface area is also controlled affecting the availability of growth media, oxygen and other factors to the organism. Ultimately, unique applications in fields such as biomaterials and metabolic engineering are enabled.

Massively Parallel Combinatorial Genetics to Probe and Treat Human Diseases

Lu, T. K., Massachusetts Institute of Technology

Combinatorial gene sets are important in coordinately regulating complex biological phenotypes and human diseases. For example, multiple biological factors are needed to differentiate stem cells into distinct cell types and to reset somatic cells into induced pluripotent stem cells. Targeting multiple pathways via combinatorial drug therapies can yield enhanced efficacy over standard monotherapies. Moreover, even though genome-wide association studies have identified multiple individual loci involved in multifactorial phenotypes, such loci can only account for a minor fraction of disease heritability on their own. Genetic interactions may be significant in explaining this missing heritability but existing strategies for systematically mapping the functions of high-order gene combinations are limited in the order of genetic complexity that can be studied and the throughput and scale that can be achieved.

To overcome these bottlenecks and enable massively parallel characterization of genetic combinations, we created a broadly applicable technology for the rapid and scalable assembly of high-order barcoded combinatorial genetic libraries that can be readily quantified in pooled screens with next-generation sequencing. We have applied this platform, Combinatorial Genetics En Masse (CombiGEM), to bacterial and human systems to discover novel combinatorial gene sets that underlie important diseases.

For example, new therapies are urgently needed to drug-resistant bacterial infections, which constitute a significant growing threat to global health. Thus, we sought to identify novel combinatorial genetic perturbations that were effective against drug-resistant bacteria. Using CombiGEM, we built ~34,000 overexpression constructs containing nearly all pairwise combinations of *Escherichia coli* transcription factors (TFs). By leveraging high-throughput pooled screens together with Illumina sequencing, we discovered diverse genetic combinations that modulated antibiotic-resistance phenotypes in carbapenem-resistant Enterobacteriaceae (CRE). Specifically, we found multiple TF pairs that enhanced antibiotic killing by up to 1,000,000-fold. By delivering these specific genetic combinations into target bacteria via phagemids and co-treating with antibiotics, we significantly increased the killing of highly drug-resistant *E. coli* harboring New Delhi metallo-betalactamase-1 (NDM-1). Furthermore, we assembled libraries of three-wise TF combinations with over four million unique members and validated that these could be quantified via high-throughput sequencing, thus demonstrating the scalability of CombiGEM.

Furthermore, we have established CombiGEM for high-throughput combinatorial screening in human cells. For example, we constructed high-coverage barcoded libraries containing 1,521 two-wise and 51,770 three-wise combinations of 39 human microRNA (miRNA) precursors. We carried out systematic screens with these libraries to discover combinatorial miRNA effectors that sensitized drug-resistant ovarian cancer cells to chemotherapy and/or inhibited cancer cell proliferation. Highly ranked hits from these pooled screens were validated for their ability to suppress drug resistance and block cancer cell growth. Thus, this effort uncovered novel insights into miRNA networks underlying drug-resistance and cell-proliferation phenotypes in cancer. Furthermore, this work identified new miRNA combinations that may be useful as effective anti-cancer therapeutics.

In summary, our work establishes a powerful technology for massively parallel combinatorial genetics using the tools of synthetic biology. We envision that CombiGEM will be useful for the high-throughput profiling of multifactorial genetic combinations that regulate a myriad of phenotypes relevant to biomedical, biotechnology, and basic science applications.

Codon Compression Algorithms for Saturation Mutagenesis

Pines, G., CU Boulder
Garst, A., University of Colorado
Zeitoun, R. I., University of Colorado Boulder
Lynch, S. A., University of Colorado at Boulder
Gill, R. T., University of Colorado Boulder

Saturation mutagenesis is employed in protein engineering and genome-editing efforts to generate libraries that span amino acid design space. Traditionally, this is accomplished by using degenerate/compressed codons such as NNK (N = A/C/G/T, K = G/T), which covers all amino acids and one stop codon. These solutions suffer from two types of redundancy: (a) different codons for the same amino acid lead to bias, and (b) wild type amino acid is included within the library. These redundancies increase library size and downstream screening efforts. Here, we present a dynamic approach to compress codons for any desired list of amino acids, taking into account codon usage. This results in a unique codon collection for every amino acid to be mutated, with the desired redundancy level. Finally, we demonstrate that this approach can be used to design precise oligo libraries amendable to recombineering and CRISPR-based genome editing to obtain a diverse population with high efficiency.

Towards Construction of Programmable Cells to Prevent Infectious Diseases

Shopera, T., Washington University in St. Louis
Hoynes-O'Connor, A. G., Washington University in St. Louis
Immethun, C., Washington University in St. Louis
Henson, W. R., Washington University in St. Louis
Ng, K., Washington University in St. Louis
Lee, Y. J., Washington University in St. Louis
Moon, T. S., Washington University in St. Louis

Signal sensing is critical in bacteria that live in fluctuating environments. However, even a group of isogenic bacteria, subject to identical culture conditions, often exhibit heterogeneity in phenotypes. This multi-stability can be explained by stochastic transitioning among multiple stable states of gene expression which are often the results of positive feedback loops in gene regulatory networks. Although such a "bet-hedging" strategy can increase bacterial fitness in unpredictable environments, such stochastic processes can prevent synthetic biologists from predictably engineering microbial cells that need to face uncontrolled environments. Thus, to fully exploit microbes in synthetic biology applications, complex gene regulatory systems need to be better understood. To this end, we propose to construct "robust" genetic sensors and circuits that endow bacteria with the ability to sense environmental signals (e.g., temperature, light, pH, and oxygen level), perform computational operations, and implement actuations (i.e., generation of toxins to kill parasite eggs).

By building sensors and circuits from the bottom-up, we have gained design principles for construction of programmable cells. Specifically, we have built robust logic gates that contain modular sensors for temperature or oxygen levels that can be more meaningful signals in the environment than chemical inducers that are used in the lab. Furthermore, we have constructed bistable switches that can generate noise-tolerant responses to fluctuating environmental inputs. These switches are based on the sequestering interaction between a transcription factor (an activator) and an anti-activator along with multiple positive feedback loops. Importantly, these switches are functional over a wide range of parameters and endow the bacteria with a long-lasting memory (~weeks). We are assembling these modular sensors, logic gates, and memory devices to construct robust programmable probiotic strains that can reproduce in the intestine where parasite eggs are produced, and come out of the body with the parasite eggs and kill them. We will present progress towards development of such engineered microbes, which can be programmed to kill parasite eggs only when user-defined conditions are met.

Genomically Encoded Analog Memory with Precise in Vivo DNA Writing in Living Cell Populations

Farzadfard, F.,

Cellular memory is crucial to many natural biological processes and sophisticated synthetic biology applications. Existing cellular memories rely on epigenetic switches or recombinases, which are limited in scalability and recording capacity. In this work, we use the DNA of living cell populations as genomic "tape recorders" for the analog and distributed recording of long-term event histories. We describe a platform for generating single-stranded DNA (ssDNA) in vivo in response to arbitrary transcriptional signals. When coexpressed with a recombinase, these intracellularly expressed ssDNAs target specific genomic DNA addresses, resulting in precise mutations that accumulate in cell populations as a function of the magnitude and duration of the inputs. This platform could enable long-term cellular recorders for environmental and biomedical applications, biological state machines, and enhanced genome engineering strategies.

A Worldwide Baseline for Reproducibility in Synthetic Biology

Beal, J., Raytheon BBN Technologies
Haddock, T.,
de Mora, K.,
Lizarazo, M.,
Salit, M., National Institute of Standards and Technology
Munro, S., NIST
Rettberg, R.,

Rapid improvements in our ability to genetically engineer biological organisms offer the potential for revolutionary applications for medicine, manufacturing, agriculture, and the environment A major barrier to transition from principle to practice, however, is the frequent sensitivity of biological systems to small changes in their cellular or environmental context. This makes it difficult to reproduce or build on prior results in the lab, let alone to ensure desirable behavior in a deployed application.

A first step toward addressing these issues is to quantify the degree of variability exhibited by engineered genetic constructs across multiple laboratories. To this end, we present the results of the first large-scale interlab study of reproducibility in synthetic biology, carried out by students at 45 institutions worldwide during the 2014 International Genetically Engineered Machine (iGEM) Competition. Participants were asked to measure fluorescence from three BioBrick constructs constitutively expressing GFP in E. coli. Only around half of the teams were able to produce valid measurements for all three constructs, and nearly all measurements were in non-comparable relative units. Those teams with valid measurements, however, exhibited a high degree of consistency in observed fluorescence, with a mean error of only 1.9-fold. Teams also exhibited a high degree of variation in the details of how measurement protocols were executed.

These results are extremely promising, showing that engineered genetic constructs can exhibit a high degree of consistency in behavior even in the face of variations in context. At the same time, the limits encountered in this study highlight the need for adoption of calibrated measurements producing standardized units and for protocol automation to reduce the impact of "cultural art" in laboratory methods. Finally, this study also demonstrates how critical issues in science and reproducibility can be addressed through "citizen science" by involving undergraduates around the world.

Deletion of Glyoxylate Shunt Pathway Genes Results in a 3-Hydroxybutyrate Overproducing Strain of Ralstonia Eutropha

Brigham, C.,
Budde, C.,
Palmer, J.,
Rha, C.,
Sinskey, A. J., MIT

The glyoxylate shunt pathway was examined in the polyhydroxyalkanoate (PHA) producing bacterium, *Ralstonia eutropha.* The deletion of the *aceB* gene, encoding malate synthase, which catalyzes the second reaction of the glyoxylate shunt, resulted in severe growth defects in cultures containing acetate or fatty acids as the main carbon source. Further deletions of *phaB* genes, encoding acetoacetyl-CoA reductase enzymes, resulted in little or no growth of *R. eutropha* on acetate. Metabolomic studies revealed that the *aceB* deletion strain overproduced pyruvate and 3-hydroxybutyrate, both at 15X higher levels than the wild-type strain, when cultured in acetate as the main carbon source. To further enhance production of these compounds, the PHA biosynthetic pathway was disabled in the *R. eutropha aceB* mutant by an in-frame deletion of the *phaC* (PHA synthase) gene. The resulting strain produced even higher levels of 3-hydroxybutyrate at the onset of nitrogen limitation. This study has provided insight to the construction of a strain of *R. eutropha* that potentially produces value-added compounds when grown on carbon feedstocks containing acetate or fatty acids.

Role of Positive and Negative Feedback in Dynamic Expression of E.coli MarA

Wang, T., The University of Vermont
Dunlop, M., The University of Vermont

Bacteria can stochastically express stress response genes, allowing a subset of a given population to survive sudden changes to the environment. This 'bet-hedging' approach maximizes fitness in variable environments without overtaxing the cells. The multiple antibiotic resistance activator (MarA) in *Escherichia coli* is a transcriptional activator that can trigger more than 40 downstream genes related to stress response. Preliminary data indicates that MarA's expression is dynamic, using a series of stochastic pulses. We hypothesize that the stochastic pulsing behavior of MarA is driven by the interaction between positive and negative feedback loops controlling the *marRAB* operon. However, it is not yet clear how the interaction between the two feedback loops affects the pulsing dynamics. Our research goal is to examine how positive and negative feedback regulating *marRAB* may cause pulsing in MarA expression. To test this, we are constructing systems which contain either only positive or only negative feedback using transversion mutations in the MarR or MarA binding sites in the promoter of the *marRAB* operon. Single-cell microscopy will be employed to measure the expression of MarA expression levels over time. Previous computational results from our group indicate that pulsing in MarA expression may be caused by the interlinked positive and negative feedback loops. However, with only positive or only negative feedback controlling the *marRAB* operon, MarA expression will lose the stochastic pulsing behavior, and instead exhibit continuous high or low levels of MarA expression.

Spatial and kinetic modeling for targeted fusion proteins

Robinson-Mosher, A., Harvard Medical School
Brouwers, N.,
Way, J. C., Harvard University
Silver, P. A., Harvard Medical School

Interferon alpha is a cytokine that has shown efficacy in treating cancer, but its use is limited due to severe side effects. We wish to target the anti-proliferative effects of interferon to cancer cells by following the chimeric activator approach. This consists of making a fusion protein between interferon alpha and an antibody fragment that binds to EGFRvIII, an epitope specific for glioblastoma cells, and also mutating residues in the receptor binding sites of interferon to weaken its affinity. The cell specificity of interferon's activity is then driven primarily by the antibody fragment binding to EGFRvIII. We have built a model to predict the activity of a given antibody fragment-interferon fusion based on the properties of the peptide linker between the two domains and on the strength of the mutation to interferon's binding site. A constrained Brownian dynamics component of the model predicts the effect of the linker properties on cooperative binding kinetics. Those results are used in an ordinary differential equation model of binding state transitions to make quantitative predictions of experimental results. Our antibody fragment-interferon fusion construct achieves a 100-fold increase in specificity on target cells, as measured by shift in EC50 in a proliferation assay, and is well-predicted by our model.

Directed Evolution of CRISPR Endoribonucleases Using Phage-Assisted Continuous Evolution (PACE)

Davis, K. M., Harvard University
Liu, D. R., Howard Hughes Medical Institute

A panel of sequence-specific RNA-cleaving enzymes would enhance our ability to manipulate RNA structure and function and aid in the analysis of coding and non-coding RNAs found in human and other genomes. We have developed a selection system that enables the phage-assisted continuous evolution (PACE) of RNA-cleaving enzymes. Using PACE, we have evolved variants of the CRISPR endoribonuclease Csy4 that demonstrate altered substrate specificity.

Developing Selective Conjugation Reactions for Applications in > Cellular Imaging and Synthetic Biology

Devaraj, N. K.,

Our research group is interested in the development of of in-situ bioconjugation reactions through the pursuit of 2 major research efforts: 1. The study of new bioorthogonal reactions for applications in cellular imaging, and 2. Developing synthetic cells through artificial phospholipid synthesis and modification. For cellular imaging applications, we have focused on advancing bioorthogonal tetrazine ligations with strained dienophiles. We have developed novel synthetic routes to tetrazines, leading to the synthesis of fluorogenic probes that increase over 100-times in fluorescence intensity after ligation. Metabolic imaging applications using tetrazines have been made possible by new dienophiles, such as methyl-cyclopropene mini-tags. Recently, we have introduced tetrazine-mediated transfer (TMT) reactions, which enable DNA/RNA template driven turnover of fluorogenic reactions and the detection of clinically relevant microRNAs. We also have a strong interest in applying covalent coupling reactions to the formation and modification of phospholipid membranes. We have utilized chemoselective reactions, such as copper-catalyzed triazole formation or the native chemical ligation, to drive the de novo synthesis of phospholipid membranes. To interface synthetic membranes proteins, we have recently developed protein reactive membrane anchors utilizing SNAP-tags. Light driven spatiotemporal control over proteoliposome formation is possible using photocaged membrane anchors. Continued development of selective bioconjugation reactions will facilitate the interfacing of synthetic molecules with biological systems for new capabilities.

Bistable Phenotype Switching to Improve Bioprocess Productivity

Venayak, N.,
Jaydeep, R.,
Mahadevan, R., University of Toronto

Metabolic engineering has proven crucial for improving fermentation metrics including yield, titer, and productivity. Historically, there has been a focus on yield improvement using knockout strategies, which do not involve the dynamic redirection of fluxes (by modulating enzyme expression or activity). Since many engineered strains have significant growth defects, these static strategies often lead to strains with low productivities. Due to advances in synthetic biology, dynamic manipulation of fluxes has become more realistic, and provides new avenues to control metabolism and address these issues. These methods have been shown to improve not only productivity, but also the yield and titer of bioprocesses.

To directly overcome the difficulties arising from static engineering strategies, we have developed a bistable genetic controller which allows for phenotype switching in a lactate producing strain. To construct this strain, two competing fermentative pathways, alcohol dehydrogenase (*adhE*) and phosphoacetyltransferase (*pta*), were deleted to improve lactate yield at the expense of growth rate. By heterologously expressing these genes, subject to the bistable controller, we have demonstrated the ability to switch between two distinct phenotypes: a wild phenotype where cells grow quickly and produce mixed products, and a production phenotype with reduced growth rate and improve lactate yield.

The application of this bistable metabolic controller will allow for the implementation of two-stage fermentation for a multitude of products, and has the potential to significantly improve productivity and reduce capital expenditure.

The repeatability of evolutionary trajectories and implications for strain optimization

Miller, A. W.,

Synthetic biologists often liken microbes to chassis in which engineered circuits are placed. From this perspective, the genetic background and various functions of the cell are an unfortunate necessity, and often times a hindrance to the desired product of a given genetic circuit. However, cellular fitness is important for many applications, and much can be gained from improving the background itself. One traditional and unbiased way of doing this is to simply evolve microbes in relevant environments for an extended period of time. We have done this for 96 populations of haploid *S. cerevisiae*, evolved for 300 generations using an array of miniature chemostat bioreactors in glucose, phosphate and sulfate-limited conditions. For this study, we performed inline measurements of fitness every 50 generations by competing the evolving population against a GFP expressing ancestral strain. We found that populations grown in both sulfate and phosphate limited conditions reached increased levels of fitness (avg. 15% increased) after just 50 generations. These populations eventually grew up to 60% more fit than the ancestral strain after 300 hundred generations of growth. However, yeast populations grown in glucose-limited media acquired significantly lower increases in fitness (~10% on avg.) after 300 generations of growth. Thus, a given background will be variably adaptable to different environments – however in some cases (sulfate and phosphate limitation for *S. cerevisiae*) one can obtain strains with greatly increased fitness in relatively little time. We have also performed whole genome sequencing of these 96 evolved populations. Our sequencing results show that (1) nutrient transporter genes *SUL1*, *PHO84,* and *HXT6,7* are frequently amplified or otherwise mutated in sulfate, phosphate, and glucose-limited environments. Frequently recurrent large effect mutations such as nutrient transporter amplifications could be used to guide traditional targeted strain engineering approaches. (2) Hundreds of other mutations are observed in these evolution experiments having varied and unknown effects on fitness. These run the gamut from canonical nutrient sensing/signaling genes to more varied and hard to rationalize mutations across most other gene functions. The large number of mutations and continual increases in fitness throughout 300 generations of evolution in a constant environment suggest that multiple mutations seen in our evolution experiments may be adaptive. The experimental evolution approach can help us bypass our lack of understanding of the additive effect of many mutations and arrive at more fit strains in relatively little time.

Autonomous DNA Nanosystem for Interrogating Molecular Interactions

Mathur, D., Iowa State University
Henderson, E. R., Iowa State University

We describe the use of DNA to create a self-assembling nanosystem for interrogating molecular interactions. The nanosystem contains a rigid supporting frame, a cylindrical central core and a mobile ring that is coaxial with the core. Motion of the ring is influenced by several control elements whose force generating capability is based on the transition of single-stranded DNA to double-stranded DNA. These forces can be directed to act in opposition to adhesive forces between the ring and the frame. We interrogate base stacking adhesion and demonstrate detection of a soluble nucleic acid target using this nanosystem.

An Innovative Cloning Platform for Pathway Engineering Henrique C. DePaoli, Gerald A. Tuskan and Xiaohan Yang BioSciences Division, Oak Ridge National Laboratory, Oak Ridge, TN

Project Goals: **Crassulacean acid metabolism (CAM) is a specialized mode of photosynthesis that features a temporal CO2 pump with nocturnal CO2 uptake, facilitates increased water-use efficiency (WUE), and enables CAM plants to inhabit water-limited environments such as semi-arid deserts or seasonally dry environments. CAM provides an excellent opportunity for engineering both enhanced WUE and photosynthetic performance into bioenergy crops. This project has two main goals: 1) to identify the**

CAM-associated genes using systems biology approaches and 2) to engineer CAM gene modules into C3 species using

synthetic biology approaches. The success of the project will allow biomass production on semi-arid, abandoned, marginal or degraded agricultural lands.

Abstract: To establish synthetic biology capability for transferring CAM gene modules into C3 species, we developed an innovative method for assembling DNA fragments *in vitro*. Several cloning strategies are available to support genetic engineering (DePaoli *et al.*, 2014). However, the current availability of cloning methods imposes several limitations to seamless cloning of large and multi-gene constructs. Our technology, called â??**TNT-cloning**â? , assembles DNA parts in a simple, fast, efficient and flexible manner. Our system combines DNA fragments to be cloned in one single universal library, leaves no undesirable sequences behind, allows â??one-potâ? reaction with up to 3 fragments to be joined at once and automatically maintains the open reading frame (ORF) in-frame between genes of interest. By combining all cloning elements into one single universal library, the method allows a pre-determined assembly without the need of sequence homology or linkers/adaptors, which results in a â??scar-freeâ?? product. Importantly, any fragment regardless the sequence is compatible with our system. In addition to the cloning system, a new buffer, called â??**TNT-buffer**â??, was developed to allow quick and simultaneous processing of DNA fragments, enhancing the efficiency several times compared with current commercial products. The vectors represent a binary platform, making the final gene construct reusable as well as immediately ready for plant transformation. Additionally, the set of plasmids support secondary and tertiary assembling in a loop format, with an exponential reduction of steps for assembling a large number of DNA parts. We demonstrated that the system is wholly functional by cloning, assembling and testing several fragments ranging from 30 bp to 12 kb. Because this technique is compatible with isothermal (Gibson) assembly, virtually any fragment can be used as an element in the library and circularized without the need to carry the binary- backbone, expanding the technologyâ??s use to other systems and making it of special interest for constructing new plasmids/circular molecules. This novel cloning platform will accelerate the creation of multiple gene constructs necessary for CAM engineering and will greatly support the construction and tuning of multi-combinatorial pathways in a wide range of organisms.
Note: The TNT cloning system is protected under the invention disclosure 201403357, DOE S-124,978 and is patent pending. The TNT-buffer formulation is protected under the invention disclosure 201403356, DOE S-124,977 and is patent pending.

References

1. DePaoli, H. C., Borland, A. M., Tuskan, G. A., Cushman, J. C., & Yang, X. (2014). Synthetic biology as it relates to CAM photosynthesis: challenges and opportunities. Journal of Experimental Botany, 65:3381-3393.

This material is based upon work supported by the Department of Energy, Office of Science, Genomic Science Program (under award number DE-SC0008834). Oak Ridge National Laboratory is managed by UT-Battelle, LLC for the US Department of Energy (under contract number DE-AC05-00OR22725).

AUTHOR INDEX

AUTHOR INDEX

AUTHOR INDEX

AUTHOR INDEX

AUTHOR INDEX

AUTHOR INDEX